Reference use only
not for loan

KIRK-OTHMER

ENCYCLOPEDIA OF
CHEMICAL
TECHNOLOGY

FOURTH EDITION

VOLUME **16**

MASS TRANSFER
TO
NEUROREGULATORS

EXECUTIVE EDITOR
Jacqueline I. Kroschwitz

EDITOR
Mary Howe-Grant

KIRK-OTHMER

ENCYCLOPEDIA OF CHEMICAL TECHNOLOGY

FOURTH EDITION

VOLUME **16**

MASS TRANSFER
TO
NEUROREGULATORS

A Wiley-Interscience Publication
JOHN WILEY & SONS

New York • Chichester • Brisbane • Toronto • Singapore

Coventry University

S/0

Library of Congress Cataloging-in-Publication Data

Encyclopedia of chemical technology/executive editor, Jacqueline
  I. Kroschwitz; editor, Mary Howe-Grant.—4th ed.
    p. cm.
  At head of title: Kirk-Othmer.
  "A Wiley-Interscience publication."
  Includes index.
  Contents: v. 16, Mass Transfer to neuroregulators
  ISBN 0-471-52685-1 (v. 16)
  1. Chemistry, Technical—Encyclopedias. I. Kirk, Raymond E.
(Raymond Eller), 1890–1957. II. Othmer, Donald F. (Donald
Frederick), 1904– . III. Kroschwitz, Jacqueline I., 1942– .
IV. Howe-Grant, Mary, 1943– . V. Title: Kirk-Othmer encyclopedia
of chemical technology.
TP9.E685 1992   91-16789
660′.03—dc20

Printed in the United States of America

10 9 8 7 6 5 4 3 2 1

CONTENTS

EDITORIAL STAFF
FOR VOLUME 16

Executive Editor: **Jacqueline I. Kroschwitz**
Editor: **Mary Howe-Grant**
Associate Managing Editor: **Lindy Humphreys**
Copy Editor: **Lawrence Altieri**

CONTRIBUTORS
TO VOLUME 16

Jeffrey A. Ahlgren, *United States Department of Agriculture, Peoria, Illinois,* Microbial polysaccharides

Richard W. Baker, *Membrane Technology & Research, Inc., Menlo Park, California,* Membrane technology

Mark C. Bean, *D. D. Bean & Sons Company, Jaffrey, New Hampshire,* Matches

Oswald R. Bergmann, *E. I. du Pont de Nemours & Co., Inc., Deepwater, New Jersey,* Explosively clad metals (under Metallic coatings)

James A. Brient, *Merichem Company, Houston, Texas,* Naphthenic acids

Clark A. Briggs, *Abbott Laboratories, Abbott Park, Illinois,* Neuroregulators

Charlie R. Brooks, *University of Tennessee, Knoxville,* Case hardening (under Metal surface treatments)

K. Chawla, *New Mexico Institute of Mining and Technology, Socorro,* Metal-matrix composites

Robert Chorvat, *E. I. du Pont de Nemours & Co., Inc., Wilmington, Delaware,* Memory-enhancing drugs

Gregory L. Cote, *United States Department of Agriculture, Peoria, Illinois,* Microbial polysaccharides

Simon Davies, *Davy Process Technology, London, England,* Methanol

Stephen DeVito, *United States Environmental Protection Agency, Washington, D.C.,* Mercury

Douglas V. Doane, *Consultant, Ann Arbor, Michigan,* Molybdenum and molybdenum alloys

John C. Dobson, *Rohm and Haas Company, Spring House, Pennsylvania,* Methacrylic acid and derivatives

Mary Noon Doyle, *Shepherd Chemical Company, Cincinnati, Ohio,* Naphthenic acids

Paul Duby, *Columbia University, New York, New York,* Extractive metallurgy (under Metallurgy)

Alan English, *John Brown E & C, Houston, Texas,* Methanol

Mary G. Enig, *Enig Associates, Inc., Silver Springs, Maryland,* Mineral nutrients

Andrew W. Gross, *Rohm and Haas Company, Spring House, Pennsylvania,* Methacrylic acid and derivatives

Kenneth Hacias, *Parker Amchem, Madison Heights, Michigan,* Pickling (under Metal surface treatments)

Carl W. Hall, *Engineering Information Services, Arlington, Virginia,* Milk and milk products

Cal Hallada, *Climax Molybdenum Company, Ypsilanti, Michigan,* Molybdenum and molybdenum alloys

Robert Hart, *Parker Amchem, Madison Heights, Michigan,* Chemical and electro-chemical conversion treatments (under Metal surface treatments)

Ramesh R. Hemrajani, *Exxon Research and Engineering Company, Florham Park, New Jersey,* Mixing and blending

Brent Hiskey, *University of Arizona, Tucson,* Survey (under Metallurgy)

Mark W. Holladay, *Abbott Laboratories, Abbott Park, Illinois,* Neuroregulators

Joseph P. Hornak, *Rochester Institute of Technology, Rochester, New York,* Medical imaging technology

Laurence A. Jackman, *Teledyne Allvac, Monroe, North Carolina,* Metal treatments

Peter K. Johnson, *APMI International, Princeton, New Jersey,* Powder metallurgy (under Metallurgy)

Curtis R. Kates, *Advanced Aromatics, Baytown, Texas,* Naphthalene derivatives

James F. Kerwin, Jr., *Abbott Laboratories, Abbott Park, Illinois,* Neuroregulators

Charles T. Kresge, *Mobil Research and Development Corporation, Paulsboro, New Jersey,* Molecular sieves

Gerald A. Krulik, *Applied Electroless Concepts, Inc., San Clemente, California,* Survey (under Metallic coatings)

Günter H. Kühl, *Consultant, Cherry Hill, New Jersey,* Molecular sieves

Leonard Lamberson, *Western Michigan University, Kalamazoo, Michigan,* Materials reliability

Patricia M. Lesko, *Rohm and Haas Company, Spring House, Pennsylvania,* Methacrylic polymers

Thomas A. Liederbach, *Electrode Corporation, Chardon, Ohio,* Metal anodes

Nenad V. Mandich, *HBM Engineering Company, Lansing, Illinois,* Survey (under Metallic coatings)

Robert T. Mason, *Koppers Industries, Inc., Pittsburgh, Pennsylvania,* Naphthalene

Walter C. McCrone, *McCrone Research Institute, Chicago, Illinois,* Microscopy

Donald P. Murphy, *Parker Amchem, Madison Heights, Michigan,* Cleaning (under Metal surface treatments)

D. R. Nagaraj, *CYTEC Industries, Stamford, Connecticut,* Minerals recovery and processing

Ronald W. Novak, *Rohm and Haas Company, Spring House, Pennsylvania,* Methacrylic polymers

Milton Nowak, *Troy Chemical Corporation, Newark, New Jersey,* Mercury compounds

John Osepchuk, *Raytheon Company, Lexington, Massachusetts,* Microwave technology

Michael Petschel, *Parker Amchem, Madison Heights, Michigan,* Chemical and electrochemical conversion treatments (under Metal surface treatments)

S. Raharjo, *Colorado State University, Fort Collins,* Meat products

John P. N. Rosazza, *University of Iowa, Iowa City,* Microbial transformations

Jerry Rovner, *John Brown E & C, Houston, Texas,* Methanol

Carlos N. Ruiz, *Teledyne Allvac, Monroe, North Carolina,* Metal treatments

Glenn R. Schmidt, *Colorado State University, Fort Collins,* Meat products

Oldrich K. Sebek, *Consultant, Kalamazoo, Michigan,* Microbial transformations

William Singer, *Troy Chemical Corporation, Newark, New Jersey,* Mercury compounds

Edward I. Stiefel, *Exxon Research and Engineering Company, Annandale, New Jersey,* Molybdenum compounds

James P. Sullivan, *Abbott Laboratories, Abbott Park, Illinois,* Neuroregulators

Mannan Talukder, *Advanced Aromatics, Baytown, Texas,* Naphthalene derivatives

James T. Tanner, *North Carolina State University, Asherville,* Mica

Curt Thies, *Washington University, St. Louis, Missouri,* Microencapsulation

George A. Timmons, *Consultant, Ann Arbor, Michigan,* Molybdenum and molybdenum alloys

Arthur Usmani, *Firestone, Carmel, Indiana,* Medical diagnostic reagents

Peter J. Wessner, *Merichem Company, Houston, Texas,* Naphthenic acids

Jack Westbrook, *Brookline Technologies, Ballston Spa, New York,* Materials standards and specifications

Michael Williams, *Abbott Laboratories, Abbott Park, Illinois,* Neuroregulators

Robert Zaczek, *E. I. du Pont de Nemours & Co., Inc., Wilmington, Delaware,* Memory-enhancing drugs

NOTE ON CHEMICAL ABSTRACTS SERVICE REGISTRY NUMBERS AND NOMENCLATURE

Chemical Abstracts Service (CAS) Registry Numbers are unique numerical identifiers assigned to substances recorded in the CAS Registry System. They appear in brackets in the *Chemical Abstracts* (CA) substance and formula indexes following the names of compounds. A single compound may have synonyms in the chemical literature. A simple compound like phenethylamine can be named β-phenylethylamine or, as in *Chemical Abstracts*, benzeneethanamine. The usefulness of the *Encyclopedia* depends on accessibility through the most common correct name of a substance. Because of this diversity in nomenclature careful attention has been given to the problem in order to assist the reader as much as possible, especially in locating the systematic CA index name by means of the Registry Number. For this purpose, the reader may refer to the CAS Registry Handbook—Number Section which lists in numerical order the Registry Number with the *Chemical Abstracts* index name and the molecular formula; eg, **458-88-8**, Piperidine, 2-propyl-, (S)-, $C_8H_{17}N$; in the *Encyclopedia* this compound would be found under its common name, coniine [*458-88-8*]. Alternatively, this information can be retrieved electronically from CAS Online. In many cases molecular formulas have also been provided in the *Encyclopedia* text to facilitate electronic searching. The Registry Number is a valuable link for the reader in retrieving additional published information on substances and also as a point of access for on-line data bases.

In all cases, the CAS Registry Numbers have been given for title compounds in articles and for all compounds in the index. All specific substances indexed in *Chemical Abstracts* since 1965 are included in the CAS Registry System as are a large number of substances derived from a variety of reference works. The CAS Registry System identifies a substance on the basis of an unambiguous computer-language description of its molecular structure including stereochemical detail. The Registry Number is a machine-checkable number (like a Social Security number) assigned in sequential order to each substance as it enters the registry system. The value of the number lies in the fact that it is a concise and unique means of substance identification, which is independent of, and therefore bridges, many systems of chemical nomenclature. For polymers, one Registry Number may be used for the entire family; eg, polyoxyethylene (20) sorbitan monolaurate has the same number as all of its polyoxyethylene homologues.

Cross-references are inserted in the index for many common names and for some systematic names. Trademark names appear in the index. Names that are incorrect, misleading, or ambiguous are avoided. Formulas are given very frequently in the text to help in identifying compounds. The spelling and form used, even for industrial names, follow American chemical usage, but not always the usage of *Chemical Abstracts* (eg, *coniine* is used instead of *(S)-2-propylpiperidine*, *aniline* instead of *benzenamine*, and *acrylic acid* instead of *2-propenoic acid*).

There are variations in representation of rings in different disciplines. The dye industry does not designate aromaticity or double bonds in rings. All double bonds and aromaticity are shown in the *Encyclopedia* as a matter of course. For example, tetralin has an aromatic ring and a saturated ring and its structure

appears in the *Encyclopedia* with its common name, Registry Number enclosed in brackets, and parenthetical CA index name, ie, tetralin [*119-64-2*] (1,2,3,4-tetrahydronaphthalene). With names and structural formulas, and especially with CAS Registry Numbers, the aim is to help the reader have a concise means of substance identification.

CONVERSION FACTORS, ABBREVIATIONS, AND UNIT SYMBOLS

SI Units (Adopted 1960)

The International System of Units (abbreviated SI), is being implemented throughout the world. This measurement system is a modernized version of the MKSA (meter, kilogram, second, ampere) system, and its details are published and controlled by an international treaty organization (The International Bureau of Weights and Measures) (1).

SI units are divided into three classes:

BASE UNITS

length	meter[†] (m)
mass	kilogram (kg)
time	second (s)
electric current	ampere (A)
thermodynamic temperature[‡]	kelvin (K)
amount of substance	mole (mol)
luminous intensity	candela (cd)

SUPPLEMENTARY UNITS

plane angle	radian (rad)
solid angle	steradian (sr)

[†]The spellings "metre" and "litre" are preferred by ASTM; however, "-er" is used in the *Encyclopedia*.

[‡]Wide use is made of Celsius temperature (t) defined by

$$t = T - T_0$$

where T is the thermodynamic temperature, expressed in kelvin, and $T_0 = 273.15$ K by definition. A temperature interval may be expressed in degrees Celsius as well as in kelvin.

DERIVED UNITS AND OTHER ACCEPTABLE UNITS

These units are formed by combining base units, supplementary units, and other derived units (2–4). Those derived units having special names and symbols are marked with an asterisk in the list below.

Quantity	Unit	Symbol	Acceptable equivalent
*absorbed dose	gray	Gy	J/kg
acceleration	meter per second squared	m/s^2	
*activity (of a radionuclide)	becquerel	Bq	1/s
area	square kilometer	km^2	
	square hectometer	hm^2	ha (hectare)
	square meter	m^2	
concentration (of amount of substance)	mole per cubic meter	mol/m^3	
current density	ampere per square meter	$A//m^2$	
density, mass density	kilogram per cubic meter	kg/m^3	g/L; mg/cm^3
dipole moment (quantity)	coulomb meter	C·m	
*dose equivalent	sievert	Sv	J/kg
*electric capacitance	farad	F	C/V
*electric charge, quantity of electricity	coulomb	C	A·s
electric charge density	coulomb per cubic meter	C/m^3	
*electric conductance	siemens	S	A/V
electric field strength	volt per meter	V/m	
electric flux density	coulomb per square meter	C/m^2	
*electric potential, potential difference, electromotive force	volt	V	W/A
*electric resistance	ohm	Ω	V/A
*energy, work, quantity of heat	megajoule	MJ	
	kilojoule	kJ	
	joule	J	N·m
	electronvolt[†]	$eV^†$	
	kilowatt-hour[†]	$kW·h^†$	
energy density	joule per cubic meter	J/m^3	
*force	kilonewton	kN	
	newton	N	$kg·m/s^2$

[†]This non-SI unit is recognized by the CIPM as having to be retained because of practical importance or use in specialized fields (1).

Quantity	Unit	Symbol	Acceptable equivalent
*frequency	megahertz	MHz	
	hertz	Hz	1/s
heat capacity, entropy	joule per kelvin	J/K	
heat capacity (specific), specific entropy	joule per kilogram kelvin	J/(kg·K)	
heat-transfer coefficient	watt per square meter kelvin	W/(m²·K)	
*illuminance	lux	lx	lm/m²
*inductance	henry	H	Wb/A
linear density	kilogram per meter	kg/m	
luminance	candela per square meter	cd/m²	
*luminous flux	lumen	lm	cd·sr
magnetic field strength	ampere per meter	A/m	
*magnetic flux	weber	Wb	V·s
*magnetic flux density	tesla	T	Wb/m²
molar energy	joule per mole	J/mol	
molar entropy, molar heat capacity	joule per mole kelvin	J/(mol·K)	
moment of force, torque	newton meter	N·m	
momentum	kilogram meter per second	kg·m/s	
permeability	henry per meter	H/m	
permittivity	farad per meter	F/m	
*power, heat flow rate, radiant flux	kilowatt	kW	
	watt	W	J/s
power density, heat flux density, irradiance	watt per square meter	W/m²	
*pressure, stress	megapascal	MPa	
	kilopascal	kPa	
	pascal	Pa	N/m²
sound level	decibel	dB	
specific energy	joule per kilogram	J/kg	
specific volume	cubic meter per kilogram	m³/kg	
surface tension	newton per meter	N/m	
thermal conductivity	watt per meter kelvin	W/(m·K)	
velocity	meter per second	m/s	
	kilometer per hour	km/h	
viscosity, dynamic	pascal second	Pa·s	
	millipascal second	mPa·s	
viscosity, kinematic	square meter per second	m²/s	
	square millimeter per second	mm²/s	

Quantity	Unit	Symbol	Acceptable equivalent
volume	cubic meter	m^3	
	cubic diameter	dm^3	L (liter) (5)
	cubic centimeter	cm^3	mL
wave number	1 per meter	m^{-1}	
	1 per centimeter	cm^{-1}	

In addition, there are 16 prefixes used to indicate order of magnitude, as follows:

Multiplication factor	Prefix	Symbol	Note
10^{18}	exa	E	
10^{15}	peta	P	
10^{12}	tera	T	
10^9	giga	G	
10^6	mega	M	
10^3	kilo	k	
10^2	hecto	h[a]	[a]Although hecto, deka, deci, and centi
10	deka	da[a]	are SI prefixes, their use should be
10^{-1}	deci	d[a]	avoided except for SI unit-multiples
10^{-2}	centi	c[a]	for area and volume and nontech-
10^{-3}	milli	m	nical use of centimeter, as for body
10^{-6}	micro	μ	and clothing measurement.
10^{-9}	nano	n	
10^{-12}	pico	p	
10^{-15}	femto	f	
10^{-18}	atto	a	

For a complete description of SI and its use the reader is referred to ASTM E380 (4) and the article UNITS AND CONVERSION FACTORS which appears in Vol. 24.

A representative list of conversion factors from non-SI to SI units is presented herewith. Factors are given to four significant figures. Exact relationships are followed by a dagger. A more complete list is given in the latest editions of ASTM E380 (4) and ANSI Z210.1 (6).

Conversion Factors to SI Units

To convert from	To	Multiply by
acre	square meter (m^2)	4.047×10^3
angstrom	meter (m)	$1.0 \times 10^{-10\dagger}$
are	square meter (m^2)	$1.0 \times 10^{2\dagger}$

†Exact.

To convert from	To	Multiply by
astronomical unit	meter (m)	1.496×10^{11}
atmosphere, standard	pascal (Pa)	1.013×10^5
bar	pascal (Pa)	$1.0 \times 10^{5\dagger}$
barn	square meter (m²)	$1.0 \times 10^{-28\dagger}$
barrel (42 U.S. liquid gallons)	cubic meter (m³)	0.1590
Bohr magneton (μ_B)	J/T	9.274×10^{-24}
Btu (International Table)	joule (J)	1.055×10^3
Btu (mean)	joule (J)	1.056×10^3
Btu (thermochemical)	joule (J)	1.054×10^3
bushel	cubic meter (m³)	3.524×10^{-2}
calorie (International Table)	joule (J)	4.187
calorie (mean)	joule (J)	4.190
calorie (thermochemical)	joule (J)	$4.184^{\dagger}$
centipoise	pascal second (Pa·s)	$1.0 \times 10^{-3\dagger}$
centistokes	square millimeter per second (mm²/s)	$1.0^{\dagger}$
cfm (cubic foot per minute)	cubic meter per second (m³/s)	4.72×10^{-4}
cubic inch	cubic meter (m³)	1.639×10^{-5}
cubic foot	cubic meter (m³)	2.832×10^{-2}
cubic yard	cubic meter (m³)	0.7646
curie	becquerel (Bq)	$3.70 \times 10^{10\dagger}$
debye	coulomb meter (C·m)	3.336×10^{-30}
degree (angle)	radian (rad)	1.745×10^{-2}
denier (international)	kilogram per meter (kg/m)	1.111×10^{-7}
	tex‡	0.1111
dram (apothecaries')	kilogram (kg)	3.888×10^{-3}
dram (avoirdupois)	kilogram (kg)	1.772×10^{-3}
dram (U.S. fluid)	cubic meter (m³)	3.697×10^{-6}
dyne	newton (N)	$1.0 \times 10^{-5\dagger}$
dyne/cm	newton per meter (N/m)	$1.0 \times 10^{-3\dagger}$
electronvolt	joule (J)	1.602×10^{-19}
erg	joule (J)	$1.0 \times 10^{-7\dagger}$
fathom	meter (m)	1.829
fluid ounce (U.S.)	cubic meter (m³)	2.957×10^{-5}
foot	meter (m)	$0.3048^{\dagger}$
footcandle	lux (lx)	10.76
furlong	meter (m)	2.012×10^{-2}
gal	meter per second squared (m/s²)	$1.0 \times 10^{-2\dagger}$
gallon (U.S. dry)	cubic meter (m³)	4.405×10^{-3}
gallon (U.S. liquid)	cubic meter (m³)	3.785×10^{-3}
gallon per minute (gpm)	cubic meter per second (m³/s)	6.309×10^{-5}
	cubic meter per hour (m³/h)	0.2271

†Exact.
‡See footnote on p. xiii.

To convert from	To	Multiply by
gauss	tesla (T)	1.0×10^{-4}
gilbert	ampere (A)	0.7958
gill (U.S.)	cubic meter (m^3)	1.183×10^{-4}
grade	radian	1.571×10^{-2}
grain	kilogram (kg)	6.480×10^{-5}
gram force per denier	newton per tex (N/tex)	8.826×10^{-2}
hectare	square meter (m^2)	$1.0 \times 10^{4\dagger}$
horsepower (550 ft·lbf/s)	watt (W)	7.457×10^2
horsepower (boiler)	watt (W)	9.810×10^3
horsepower (electric)	watt (W)	$7.46 \times 10^{2\dagger}$
hundredweight (long)	kilogram (kg)	50.80
hundredweight (short)	kilogram (kg)	45.36
inch	meter (m)	$2.54 \times 10^{-2\dagger}$
inch of mercury (32°F)	pascal (Pa)	3.386×10^3
inch of water (39.2°F)	pascal (Pa)	2.491×10^2
kilogram-force	newton (N)	9.807
kilowatt hour	megajoule (MJ)	$3.6^{\dagger}$
kip	newton (N)	4.448×10^3
knot (international)	meter per second (m/S)	0.5144
lambert	candela per square meter (cd/m^3)	3.183×10^3
league (British nautical)	meter (m)	5.559×10^3
league (statute)	meter (m)	4.828×10^3
light year	meter (m)	9.461×10^{15}
liter (for fluids only)	cubic meter (m^3)	$1.0 \times 10^{-3\dagger}$
maxwell	weber (Wb)	$1.0 \times 10^{-8\dagger}$
micron	meter (m)	$1.0 \times 10^{-6\dagger}$
mil	meter (m)	$2.54 \times 10^{-5\dagger}$
mile (statute)	meter (m)	1.609×10^3
mile (U.S. nautical)	meter (m)	$1.852 \times 10^{3\dagger}$
mile per hour	meter per second (m/s)	0.4470
millibar	pascal (Pa)	1.0×10^2
millimeter of mercury (0°C)	pascal (Pa)	$1.333 \times 10^{2\dagger}$
minute (angular)	radian	2.909×10^{-4}
myriagram	kilogram (kg)	10
myriameter	kilometer (km)	10
oersted	ampere per meter (A/m)	79.58
ounce (avoirdupois)	kilogram (kg)	2.835×10^{-2}
ounce (troy)	kilogram (kg)	3.110×10^{-2}
ounce (U.S. fluid)	cubic meter (m^3)	2.957×10^{-5}
ounce-force	newton (N)	0.2780
peck (U.S.)	cubic meter (m^3)	8.810×10^{-3}
pennyweight	kilogram (kg)	1.555×10^{-3}
pint (U.S. dry)	cubic meter (m^3)	5.506×10^{-4}
pint (U.S. liquid)	cubic meter (m^3)	4.732×10^{-4}

[†]Exact.

To convert from	To	Multiply by
poise (absolute viscosity)	pascal second (Pa·s)	$0.10^\dagger$
pound (avoirdupois)	kilogram (kg)	0.4536
pound (troy)	kilogram (kg)	0.3732
poundal	newton (N)	0.1383
pound-force	newton (N)	4.448
pound force per square inch (psi)	pascal (Pa)	6.895×10^3
quart (U.S. dry)	cubic meter (m³)	1.101×10^{-3}
quart (U.S. liquid)	cubic meter (m³)	9.464×10^{-4}
quintal	kilogram (kg)	$1.0 \times 10^{2\dagger}$
rad	gray (Gy)	$1.0 \times 10^{-2\dagger}$
rod	meter (m)	5.029
roentgen	coulomb per kilogram (C/kg)	2.58×10^{-4}
second (angle)	radian (rad)	$4.848 \times 10^{-6\dagger}$
section	square meter (m²)	2.590×10^6
slug	kilogram (kg)	14.59
spherical candle power	lumen (lm)	12.57
square inch	square meter (m²)	6.452×10^{-4}
square foot	square meter (m²)	9.290×10^{-2}
square mile	square meter (m²)	2.590×10^6
square yard	square meter (m²)	0.8361
stere	cubic meter (m³)	$1.0^\dagger$
stokes (kinematic viscosity)	square meter per second (m²/s)	$1.0 \times 10^{-4\dagger}$
tex	kilogram per meter (kg/m)	$1.0 \times 10^{-6\dagger}$
ton (long, 2240 pounds)	kilogram (kg)	1.016×10^3
ton (metric) (tonne)	kilogram (kg)	$1.0 \times 10^{3\dagger}$
ton (short, 2000 pounds)	kilogram (kg)	9.072×10^2
torr	pascal (Pa)	1.333×10^2
unit pole	weber (Wb)	1.257×10^{-7}
yard	meter (m)	$0.9144^\dagger$

†Exact.

Abbreviations and Unit Symbols

Following is a list of common abbreviations and unit symbols used in the *Encyclopedia*. In general they agree with those listed in *American National Standard Abbreviations for Use on Drawings and in Text* (*ANSI Y1.1*) (6) and *American National Standard Letter Symbols for Units in Science and Technology* (*ANSI Y10*) (6). Also included is a list of acronyms for a number of private and government organizations as well as common industrial solvents, polymers, and other chemicals.

Rules for Writing Unit Symbols (4):

1. Unit symbols are printed in upright letters (roman) regardless of the type style used in the surrounding text.
2. Unit symbols are unaltered in the plural.
3. Unit symbols are not followed by a period except when used at the end of a sentence.
4. Letter unit symbols are generally printed lower-case (for example, cd for candela) unless the unit name has been derived from a proper name, in which case the first letter of the symbol is capitalized (W, Pa). Prefixes and unit symbols retain their prescribed form regardless of the surrounding typography.
5. In the complete expression for a quantity, a space should be left between the numerical value and the unit symbol. For example, write 2.37 lm, *not* 2.37lm, and 35 mm, *not* 35mm. When the quantity is used in an adjectival sense, a hyphen is often used, for example, 35-mm film. *Exception:* No space is left between the numerical value and the symbols of degree, minute, and second of plane angle, degree Celsius, and the percent sign.
6. No space is used between the prefix and unit symbol (for example, kg).
7. Symbols, not abbreviations, should be used for units. For example, use "A," not "amp," for ampere.
8. When multiplying unit symbols, use a raised dot:

$$\text{N·m} \quad \text{for} \quad \text{newton meter}$$

In the case of W·h, the dot may be omitted, thus:

$$\text{Wh}$$

An exception to this practice is made for computer printouts, automatic typewriter work, etc, where the raised dot is not possible, and a dot on the line may be used.
9. When dividing unit symbols, use one of the following forms:

$$\text{m/s} \quad or \quad \text{m·s}^{-1} \quad or \quad \frac{\text{m}}{\text{s}}$$

In no case should more than one slash be used in the same expression unless parentheses are inserted to avoid ambiguity. For example, write:

$$\text{J/(mol·K)} \quad or \quad \text{J·mol}^{-1}\text{·K}^{-1} \quad or \quad \text{(J/mol)/K}$$

but *not*

$$\text{J/mol/K}$$

10. Do not mix symbols and unit names in the same expression. Write:

$$\text{joules per kilogram} \quad or \quad \text{J/kg} \quad or \quad \text{J·kg}^{-1}$$

but *not*

$$\text{joules/kilogram} \quad nor \quad \text{joules/kg} \quad nor \quad \text{joules·kg}^{-1}$$

ABBREVIATIONS AND UNITS

A	ampere	AOAC	Association of Official Analytical Chemists
A	anion (eg, HA)		
A	mass number	AOCS	American Oil Chemists' Society
a	atto (prefix for 10^{-18})		
AATCC	American Association of Textile Chemists and Colorists	APHA	American Public Health Association
		API	American Petroleum Institute
ABS	acrylonitrile–butadiene–styrene	aq	aqueous
abs	absolute	Ar	aryl
ac	alternating current, *n.*	*ar-*	aromatic
a-c	alternating current, *adj.*	*as-*	asymmetric(al)
ac-	alicyclic	ASHRAE	American Society of Heating, Refrigerating, and Air Conditioning Engineers
acac	acetylacetonate		
ACGIH	American Conference of Governmental Industrial Hygienists		
		ASM	American Society for Metals
ACS	American Chemical Society	ASME	American Society of Mechanical Engineers
AGA	American Gas Association		
Ah	ampere hour	ASTM	American Society for Testing and Materials
AIChE	American Institute of Chemical Engineers	at no.	atomic number
AIME	American Institute of Mining, Metallurgical, and Petroleum Engineers	at wt	atomic weight
		av(g)	average
		AWS	American Welding Society
		b	bonding orbital
AIP	American Institute of Physics	bbl	barrel
		bcc	body-centered cubic
AISI	American Iron and Steel Institute	BCT	body-centered tetragonal
		Bé	Baumé
alc	alcohol(ic)	BET	Brunauer-Emmett-Teller (adsorption equation)
Alk	alkyl		
alk	alkaline (not alkali)	bid	twice daily
amt	amount	Boc	*t*-butyloxycarbonyl
amu	atomic mass unit	BOD	biochemical (biological) oxygen demand
ANSI	American National Standards Institute		
		bp	boiling point
AO	atomic orbital	Bq	becquerel

C	coulomb	DIN	Deutsche Industrie Normen
°C	degree Celsius		
C-	denoting attachment to carbon	dl-; DL-	racemic
		DMA	dimethylacetamide
c	centi (prefix for 10^{-2})	DMF	dimethylformamide
c	critical	DMG	dimethyl glyoxime
ca	circa (approximately)	DMSO	dimethyl sulfoxide
cd	candela; current density; circular dichroism	DOD	Department of Defense
		DOE	Department of Energy
CFR	Code of Federal Regulations	DOT	Department of Transportation
cgs	centimeter-gram-second	DP	degree of polymerization
CI	Color Index	dp	dew point
cis-	isomer in which substituted groups are on same side of double bond between C atoms	DPH	diamond pyramid hardness
		dstl(d)	distill(ed)
		dta	differential thermal analysis
cl	carload		
cm	centimeter	(E)-	entgegen; opposed
cmil	circular mil	ϵ	dielectric constant (unitless number)
cmpd	compound		
CNS	central nervous system	e	electron
CoA	coenzyme A	ECU	electrochemical unit
COD	chemical oxygen demand	ed.	edited, edition, editor
coml	commercial(ly)	ED	effective dose
cp	chemically pure	EDTA	ethylenediaminetetra-acetic acid
cph	close-packed hexagonal		
CPSC	Consumer Product Safety Commission	emf	electromotive force
		emu	electromagnetic unit
cryst	crystalline	en	ethylene diamine
cub	cubic	eng	engineering
D	debye	EPA	Environmental Protection Agency
D-	denoting configurational relationship		
		epr	electron paramagnetic resonance
d	differential operator		
d	day; deci (prefix for 10^{-1})	eq.	equation
d	density	esca	electron spectroscopy for chemical analysis
d-	dextro-, dextrorotatory		
da	deka (prefix for 10^1)	esp	especially
dB	decibel	esr	electron-spin resonance
dc	direct current, n.	est(d)	estimate(d)
d-c	direct current, adj.	estn	estimation
dec	decompose	esu	electrostatic unit
detd	determined	exp	experiment, experimental
detn	determination	ext(d)	extract(ed)
Di	didymium, a mixture of all lanthanons	F	farad (capacitance)
		F	faraday (96,487 C)
dia	diameter	f	femto (prefix for 10^{-15})
dil	dilute		

FAO	Food and Agriculture Organization (United Nations)	hyd	hydrated, hydrous
		hyg	hygroscopic
		Hz	hertz
fcc	face-centered cubic	^{i} (eg, Pr^i)	iso (eg, isopropyl)
FDA	Food and Drug Administration	i-	inactive (eg, i-methionine)
		IACS	International Annealed Copper Standard
FEA	Federal Energy Administration	ibp	initial boiling point
FHSA	Federal Hazardous Substances Act	IC	integrated circuit
		ICC	Interstate Commerce Commission
fob	free on board		
fp	freezing point	ICT	International Critical Table
FPC	Federal Power Commission	ID	inside diameter; infective dose
FRB	Federal Reserve Board		
frz	freezing	ip	intraperitoneal
G	giga (prefix for 10^9)	IPS	iron pipe size
G	gravitational constant = 6.67×10^{11} N·m^2/kg^2	ir	infrared
		IRLG	Interagency Regulatory Liaison Group
g	gram		
(g)	gas, only as in $H_2O(g)$	ISO	International Organization Standardization
g	gravitational acceleration		
gc	gas chromatography	ITS-90	International Temperature Scale (NIST)
gem-	geminal		
glc	gas–liquid chromatography	IU	International Unit
g-mol wt; gmw	gram-molecular weight	IUPAC	International Union of Pure and Applied Chemistry
GNP	gross national product	IV	iodine value
gpc	gel-permeation chromatography	iv	intravenous
		J	joule
GRAS	Generally Recognized as Safe	K	kelvin
		k	kilo (prefix for 10^3)
grd	ground	kg	kilogram
Gy	gray	L	denoting configurational relationship
H	henry		
h	hour; hecto (prefix for 10^2)	L	liter (for fluids only) (5)
ha	hectare	l-	$levo$-, levorotatory
HB	Brinell hardness number	(l)	liquid, only as in NH_3(l)
Hb	hemoglobin	LC$_{50}$	conc lethal to 50% of the animals tested
hcp	hexagonal close-packed		
hex	hexagonal	LCAO	linear combination of atomic orbitals
HK	Knoop hardness number		
hplc	high performance liquid chromatography	lc	liquid chromatography
		LCD	liquid crystal display
HRC	Rockwell hardness (C scale)	lcl	less than carload lots
		LD$_{50}$	dose lethal to 50% of the animals tested
HV	Vickers hardness number		

LED	light-emitting diode	N-	denoting attachment to
liq	liquid		nitrogen
lm	lumen	n (as n_D^{20})	index of refraction (for
ln	logarithm (natural)		20°C and sodium light)
LNG	liquefied natural gas	^{n} (as Bu^n),	
log	logarithm (common)	n-	normal (straight-chain
LOI	limiting oxygen index		structure)
LPG	liquefied petroleum gas	n	neutron
ltl	less than truckload lots	n	nano (prefix for 10^9)
lx	lux	na	not available
M	mega (prefix for 10^6);	NAS	National Academy of
	metal (as in MA)		Sciences
M	molar; actual mass	NASA	National Aeronautics and
$\overline{M}_w$	weight-average mol wt		Space Administration
$\overline{M}_n$	number-average mol wt	nat	natural
m	meter; milli (prefix for	ndt	nondestructive testing
	10^{-3})	neg	negative
m	molal	NF	*National Formulary*
m-	meta	NIH	National Institutes of
max	maximum		Health
MCA	Chemical Manufacturers'	NIOSH	National Institute of
	Association (was		Occupational Safety and
	Manufacturing Chemists		Health
	Association)	NIST	National Institute of
MEK	methyl ethyl ketone		Standards and
meq	milliequivalent		Technology (formerly
mfd	manufactured		National Bureau of
mfg	manufacturing		Standards)
mfr	manufacturer	nmr	nuclear magnetic
MIBC	methyl isobutyl carbinol		resonance
MIBK	methyl isobutyl ketone	NND	New and Nonofficial Drugs
MIC	minimum inhibiting		(AMA)
	concentration	no.	number
min	minute; minimum	NOI-(BN)	not otherwise indexed (by
mL	milliliter		name)
MLD	minimum lethal dose	NOS	not otherwise specified
MO	molecular orbital	nqr	nuclear quadruple
mo	month		resonance
mol	mole	NRC	Nuclear Regulatory
mol wt	molecular weight		Commission; National
mp	melting point		Research Council
MR	molar refraction	NRI	New Ring Index
ms	mass spectrometry	NSF	National Science
MSDS	material safety data sheet		Foundation
mxt	mixture	NTA	nitrilotriacetic acid
μ	micro (prefix for 10^{-6})	NTP	normal temperature and
N	newton (force)		pressure (25°C and 101.3
N	normal (concentration);		kPa or 1 atm)
	neutron number		

NTSB	National Transportation Safety Board	qv	quod vide (which see)
O-	denoting attachment to oxygen	R	univalent hydrocarbon radical
o-	ortho	*(R)-*	rectus (clockwise configuration)
OD	outside diameter	*r*	precision of data
OPEC	Organization of Petroleum Exporting Countries	rad	radian; radius
o-phen	*o*-phenanthridine	RCRA	Resource Conservation and Recovery Act
OSHA	Occupational Safety and Health Administration	rds	rate-determining step
		ref.	reference
owf	on weight of fiber	rf	radio frequency, *n.*
Ω	ohm	r-f	radio frequency, *adj.*
P	peta (prefix for 10^{15})	rh	relative humidity
p	pico (prefix for 10^{-12})	RI	Ring Index
p-	para	rms	root-mean square
p	proton	rpm	rotations per minute
p.	page	rps	revolutions per second
Pa	pascal (pressure)	RT	room temperature
PEL	personal exposure limit based on an 8-h exposure	RTECS	Registry of Toxic Effects of Chemical Substances
		ˢ (eg, Buˢ);	
pd	potential difference	*sec-*	secondary (eg, secondary butyl)
pH	negative logarithm of the effective hydrogen ion concentration	S	siemens
		(S)-	sinister (counterclockwise configuration)
phr	parts per hundred of resin (rubber)	*S-*	denoting attachment to sulfur
p-i-n	positive-intrinsic-negative		
pmr	proton magnetic resonance	*s-*	symmetric(al)
p-n	positive-negative	s	second
po	per os (oral)	(s)	solid, only as in $H_2O(s)$
POP	polyoxypropylene	SAE	Society of Automotive Engineers
pos	positive		
pp.	pages	SAN	styrene-acrylonitrile
ppb	parts per billion (10^9)	sat(d)	saturate(d)
ppm	parts per million (10^6)	satn	saturation
ppmv	parts per million by volume	SBS	styrene–butadiene–styrene
ppmwt	parts per million by weight	sc	subcutaneous
PPO	poly(phenyl oxide)	SCF	self-consistent field; standard cubic feet
ppt(d)	precipitate(d)		
pptn	precipitation	Sch	Schultz number
Pr (no.)	foreign prototype (number)	sem	scanning electron microscope(y)
pt	point; part		
PVC	poly(vinyl chloride)	SFs	Saybolt Furol seconds
pwd	powder	sl sol	slightly soluble
py	pyridine	sol	soluble

soln	solution	*trans-*	isomer in which substituted groups are on opposite sides of double bond between C atoms
soly	solubility		
sp	specific; species		
sp gr	specific gravity		
sr	steradian		
std	standard	TSCA	Toxic Substances Control Act
STP	standard temperature and pressure (0°C and 101.3 kPa)	TWA	time-weighted average
		Twad	Twaddell
sub	sublime(s)	UL	Underwriters' Laboratory
SUs	Saybolt Universal seconds	USDA	United States Department of Agriculture
syn	synthetic		
^{t} (eg, Bu^t), *t-, tert-*	tertiary (eg, tertiary butyl)	USP	*United States Pharmacopeia*
		uv	ultraviolet
T	tera (prefix for 10^{12}); tesla (magnetic flux density)	V	volt (emf)
		var	variable
t	metric ton (tonne)	*vic-*	vicinal
t	temperature	vol	volume (not volatile)
TAPPI	Technical Association of the Pulp and Paper Industry	vs	versus
		v sol	very soluble
		W	watt
TCC	Tagliabue closed cup	Wb	weber
tex	tex (linear density)	Wh	watt hour
T_g	glass-transition temperature	WHO	World Health Organization (United Nations)
tga	thermogravimetric analysis	wk	week
THF	tetrahydrofuran	yr	year
tlc	thin layer chromatography	(Z)-	zusammen; together; atomic number
TLV	threshold limit value		

Non-SI (Unacceptable and Obsolete) Units		Use
Å	angstrom	nm
at	atmosphere, technical	Pa
atm	atmosphere, standard	Pa
b	barn	cm^2
bar†	bar	Pa
bbl	barrel	m^3
bhp	brake horsepower	W
Btu	British thermal unit	J
bu	bushel	m^3; L
cal	calorie	J
cfm	cubic foot per minute	m^3/s
Ci	curie	Bq
cSt	centistokes	mm^2/s
c/s	cycle per second	Hz

†Do not use bar (10^5 Pa) or millibar (10^2 Pa) because they are not SI units, and are accepted internationally only for a limited time in special fields because of existing usage.

Non-SI (Unacceptable and Obsolete) Units		Use
cu	cubic	exponential form
D	debye	C·m
den	denier	tex
dr	dram	kg
dyn	dyne	N
dyn/cm	dyne per centimeter	mN/m
erg	erg	J
eu	entropy unit	J/K
°F	degree Fahrenheit	°C; K
fc	footcandle	lx
fl	footlambert	lx
fl oz	fluid ounce	m^3; L
ft	foot	m
ft·lbf	foot pound-force	J
gf den	gram-force per denier	N/tex
G	gauss	T
Gal	gal	m/s^2
gal	gallon	m^3; L
Gb	gilbert	A
gpm	gallon per minute	(m^3/s); (m^3/h)
gr	grain	kg
hp	horsepower	W
ihp	indicated horsepower	W
in.	inch	m
in. Hg	inch of mercury	Pa
in. H_2O	inch of water	Pa
in.-lbf	inch pound-force	J
kcal	kilo-calorie	J
kgf	kilogram-force	N
kilo	for kilogram	kg
L	lambert	lx
lb	pound	kg
lbf	pound-force	N
mho	mho	S
mi	mile	m
MM	million	M
mm Hg	millimeter of mercury	Pa
mμ	millimicron	nm
mph	miles per hour	km/h
μ	micron	μm
Oe	oersted	A/m
oz	ounce	kg
ozf	ounce-force	N
η	poise	Pa·s
P	poise	Pa·s
ph	phot	lx
psi	pounds-force per square inch	Pa
psia	pounds-force per square inch absolute	Pa
psig	pounds-force per square inch gage	Pa
qt	quart	m^3; L
°R	degree Rankine	K
rd	rad	Gy
sb	stilb	lx
SCF	standard cubic foot	m^3
sq	square	exponential form
thm	therm	J
yd	yard	m

BIBLIOGRAPHY

1. The International Bureau of Weights and Measures, BIPM (Parc de Saint-Cloud, France) is described in Appendix X2 of Ref. 4. This bureau operates under the exclusive supervision of the International Committee for Weights and Measures (CIPM).
2. *Metric Editorial Guide (ANMC-78-1)*, latest ed., American National Metric Council, 5410 Grosvenor Lane, Bethesda, Md. 20814, 1981.
3. *SI Units and Recommendations for the Use of Their Multiples and of Certain Other Units (ISO 1000-1981)*, American National Standards Institute, 1430 Broadway, New York, 10018, 1981.
4. Based on *ASTM E380-89a (Standard Practice for Use of the International System of Units (SI))*, American Society for Testing and Materials, 1916 Race Street, Philadelphia, Pa. 19103, 1989.
5. *Fed. Reg.*, Dec. 10, 1976 (41 FR 36414).
6. For ANSI address, see Ref. 3.

R. P. LUKENS
ASTM Committee E-43 on SI Practice

Continued

MASS TRANSFER. See SUPPLEMENT.

MATCHES

The word match is of uncertain origin. In common parlance, a match is a short, slender, elongated piece of wood or cardboard, suitably impregnated and tipped to permit, through pyrochemical action between dry solids with a binder, the creation of a small transient flame. The word match also is used for fuse lines which after ignition on one end serve as fire-transfer agents in fireworks and for explosives (qv). Such items belong in the field of pyrotechnics (qv).

The development of the ordinary match followed thousands of years of firemaking by laborious means. It has been perfected in the twentieth century and its formulations have remained basically unchanged. Progress has been achieved in selection of modifying components, mainly in the control of the drying process of the freshly dipped matches, also in the mixing procedures. The mechanical equipment for cutting out the match stems and assembling the match books has become more and more efficient as to precision and speed of production.

The history of the modern match has been well presented (1). White (also called yellow) phosphorus, discovered by Hennig Brand (1669) and described as easily ignited on slight warming or rubbing, was first applied by Robert Boyle (1680) to the ignition of sulfur-tipped wood splints. Between 1780 and 1830, numerous match-like contrivances used this ignition principle. After the discovery of potassium chlorate by Berthollet (ca 1786), its combination with white phosphorus and modifying ingredients led to the manufacture of the friction (strike-anywhere) match which became the most popular means of ignition in the United States until July 1, 1913 when an Internal Revenue tax of two cents per one hundred matches terminated its production. The act followed numerous

1

prohibitions and similar punitive taxation in other countries because of the hazard that the phosphorus constituted to the health of the workers during the days of primitive hand-dipping methods. Vapors of the white phosphorus entering the body, mainly through defective teeth, caused a permanent necrotic destruction of the bones (phossy jaw). Also, these matches were sometimes used in suicide attempts or caused the death of children.

A direct descendant of these matches is the nontoxic modern double-tipped strike-anywhere (SAW), the large "kitchen" match version, or the smaller "penny box" variety. It is based on the invention of two Frenchmen, Henri Sévène and Emile David Cahen, who used the nonpoisonous compound tetraphosphorus trisulfide, P_4S_3, as a phosphorus substitute and acquired a U.S. patent in 1898.

Other early match-like devices were based on the property of various combustible substances mixed with potassium chlorate to ignite when moistened with strong acid. More important was the property of chlorates to form mixtures with combustibles of low ignition point which were ignited by friction (John Walker, 1827). However, such matches containing essentially potassium chlorate, antimony sulfide, and later sulfur (lucifers), rubbed within a fold of glass powder-coated paper, were hard to initiate and unreliable.

The modern safety match owes its qualities to the discovery by Schrötter (1844) of the red, nonpoisonous but easily ignitable variety of phosphorus called red phosphorus. Pasch in Sweden and Böttger in Germany (1845) prepared striking surfaces containing the new material, thus separating the two principal fire-producing components, the chlorate in the match head as the oxidizer and the most sensitive fuel-type material in the striker. This type of match was much improved and made an article of commerce by J. E. Lundstrom in Jonköping, Sweden (1855). However, the United States was quite slow to accept this safety match. The "one-hand" phosphorus match and its successor, the double-tipped SAW match, were easier to handle and more reliable, whereas the early safety matches were often sputtering, hard-striking, and explosive.

The final step in the development of the modern match was the invention of the safety-type cardboard match ascribed to Joshua Pusey (1892), now called the book match. It dominates the American match industry and is gaining in popularity in other countries although it was rather slow in gaining acceptance because it was somewhat more difficult to ignite than the wood-splint match.

Mechanism of Fire Production

The essential chemical reaction takes place on contact of potassium chlorate [3811-04-9] and red phosphorus [7723-14-0] which by itself is one of the most unpredictably hazardous dry reactions in pyrochemistry (see CHLORINE OXYGEN ACID AND SALTS, HClO₃; PHOSPHORUS). This reaction has been the cause of serious injury to chemistry students who mix the two materials without permission, only vaguely aware of their explosive potential. In the match head, and separately in the striker, each of two materials is embedded in a matrix of glue so that, on striking under mild friction, a few particles of both materials come harmlessly in contact and react with formation of well-contained sparks. The modifying materials in the match head function as sensitizers (sulfur or rosin), burning-rate modifiers (potassium dichromate [7778-50-9] or lead thiosulfate [13478-50-7]), and ash-formers (diatomaceous earth, powdered glass, etc); the latter serve to hold the glowing residue safely together by a sintering process. The glue, starch, and paraffin in the stem below the head act as flame-forming fuels and the neutralizers account for the practically indefinite storage stability of well-made matches. In the striker, the glass powder controls proper bite and sensitivity. The binder is insolubilized to prevent staining of clothing caused by rain or perspiration.

The SAW match is similar to the safety match except that it is richer in fuel, and gives a billowing somewhat wind-resistant flame. The phosphorus sulfide [1314-85-8] in the tip provides the ignitability on any solid surface, and a little of the same material in the base bulb adds to wind resistance, but otherwise the base is underbalanced in active materials to prevent self-ignition from rubbing during transportation.

Manufacture

The low price of book matches is mainly the result of high speed, mechanized production methods. Book matches are punched from 1-mm thick, lined chipboard in strips of one hundred splints of ca 3.2 mm width each. In an eight-hour shift, a single machine can produce about 20×10^6 match splints and deliver them half an hour later as completed, strikable matches, ready for cutting and stapling into books. In this half hour, the tips of the punched-out splints are first immersed in molten paraffin wax, without which no persistent flame and fire transfer is possible (see WAXES). Immediately following wax application, the tip composition is affixed by dipping the ends of the strips into a thick but smoothly fluid suspension carried on a cylinder rotating in a relatively small tank at the same speed as the match strips move over it in the clamps of an endless chain. As soon as an evenly rounded match tip has been formed, the matches enter a dryer where the main object is not so much the speedy removal of the water in the match composition as the prior congealing of the match head, which takes place at a temperature of about 24°C and a relative humidity of 45 to 55%. This drying technique is one of the most important parts of the manufacture, and it is responsible for the amazingly uniform quality and ease of ignition of the modern match.

The match-cover board is an approximately 0.4-mm thick, coated or lined chipboard, or sometimes a fancy grade of a variety of decorative and more expensive cardboards, on which a striking strip is printed by a roller coating process from a thin slurry of a composition described below. The cover is more or less elaborately printed with an advertising message. Although the technique of printing varies, it is possible on some presses to apply all the colors, as well as a one-color underprint (inside of cover), on one printing press in one successive operation.

The most common size is a book of 20 or 28 (30) matches, and the 40-match size offers additional advertising area. "Ten-strike" matches are included in military food packages and are sometimes used for advertising purposes.

Wooden matches can be made by a veneering method whereby aspen wood is peeled from a section of a log and cut into splints which have a square cross section of 0.25–0.4 cm thickness, depending on the length of the match. The alternative method consists in cutting round splints from selected blocks of white pine by means of rows of cutting dies each resembling a large darning needle of which the eye is the cutter. The splints in both types of operation are forced into holes in cast-iron plates and are thus transported through the various dipping operations.

A third type of commercial matches popular in South America is the wax vestas with a center of cotton threads or of a rolled and compressed thin and tough paper surrounded by and impregnated with wax; each match is a miniature candle of long (ca 1 min) burning time.

Two processes precede the affixing of the heads for wooden matches. The first one is glow-proofing of the splint by impregnation with ammonium phosphate or a mixture of it with boric acid. In paper matches, the impregnation is conveniently done during the fabrication of the paper. This suppresses continuation of glowing of the carbonized splint after discard and also prevents the burned part with the still-hot tip from falling off and singeing clothing (see also FLAME RETARDANTS). The second impregnation is the soaking up of paraffin wax into the stem for a certain length to assure flame forming and fire transfer to the wood. Head formation is similar to the process described for book matches except that for SAW matches a smaller second tip is affixed to the larger bulb. The rollers in the dipping tank over which the splints travel are grooved, the first roller deeply for the base tip, the second roller shallow-grooved for the SAW tip. The same equipment, simply leaving out the second dipping, can be used for wooden safety matches. A formaldehyde bath, sometimes also employed with paper matches, aids in congealing and subsequent proper drying of the match head. The wooden match industry, still prevalent in Europe, has been described (2).

Nonstandard and Military Matches. Because match manufacture is a series of high speed and highly mechanized operations, any variation that involves dimensional or incisive procedural changes is a significant undertaking which is only warranted if continual high production is to result. Hence, specialties that occasionally appear on the market are actually fireworks items, made laboriously and at relatively high cost by hand-dipping with limited mechanization. Such matches produce a colored flame, give off perfume or fumigating vapors, or furnish a persistent glow or flame for the purpose of burning in a strong draft. In order to do these things effectively, an enlarged elongated bulb is necessary.

An interesting variation of the regular match is the pull match. It is a paper match, considerably thinner and narrower than a regular book match because it needs very little stiffness when being used. The tip part of the match is enclosed in a strip of corrugated paper glued to a flat cardboard (such as a box of cigarettes) and the inside of the corrugated board is covered with striking material. On pulling the match fast enough out of the corrugation, the tip passes and engages the striker and becomes lit.

A curious item is the repeatedly ignitable match. It resembles a tiny pencil, the center part being a safety match composition which is surrounded by a cool-burning chemical mixture whose essential ingredient is nearly always metaldehyde. Notwithstanding the exaggerated claims for its performance, this match, although ignitable a few times in succession, is economically not competitive with the book match and definitely technically inferior.

The principle of the safety match is also used in the pull-wire fuse lighter used to start a fuse train for the ignition of fireworks items or more frequently for blasting work. This is a reversed pull match whereby the striker material is coated on a pull wire, and the match head material is within a small metal cup in a cardboard tube. Pulling the coated wire vigorously out of the device ignites the match mixture in the tube for fire transfer to the tubular fuse train.

Match buttons and strikers are built-in components of certain flares such as the well-known red-burning railroad fusee (3) and of some fire-starting devices invented during World War II to help marooned military personnel to light a fire with a minimum of effort.

During World War II, the Quartermaster Corps of the United States requested development of a SAW match that would withstand at least six hours submersion in water. Although no match is strikable after prolonged exposure to extremely high humidity, it is possible to prevent infiltration of moisture temporarily, and especially attack by liquid water, by coating the match head and part of the stem with nitrocellulose lacquer. This is impractical for the safety match because it makes striking progressively more difficult the thicker the coating becomes. It is, however, possible to protect SAW matches so as to withstand 6 to 10 h of submersion in water and still be readily ignitable. Large numbers of matches protected in this manner were made during World War II by at least two large wood match manufacturers in the United States. Federal specifications for matches include a Type III, Class 2, water-resistance strike-anywhere wooden match for performance after two hours submersion in water (4). For prolonged exposure to high humidity, the only safe protection is heat-sealing in plastic pockets and then canning (see also WATERPROOFING AND WATER REPELLANCY).

Formulations.   Because match manufacturer has become a combination of high speed machinery design and operation, commercial art work and engraving, as well as of pyrotechnical processing, the latter is only a small though highly important part of the match business. Formulations are by no means a closely guarded secret, mainly since they are only a starting point on the way to producing a satisfactory match and striker adapted to the specific conditions of manufacture. The formulations given in (Tables 1 and 2) with some adjustments permit the preparation of workable but not necessarily salable matches (5).

European matches, mostly of brown or black tips, are basically identical with U.S. matches in their formulations, except that they contain in addition red iron oxide or manganese dioxide of pigment grade in the match heads (2). Match materials, testing methods, and related matters have been reviewed (7,8).

Economic Aspects

For 1976 the Consumer Product Safety Commission estimated that the total of all flame producers, ie, book matches (made from cardboard), individual (wooden) stick matches, and lighters, amounted to 645×10^9 of such fire resources or "lights." Paper matches accounted for about 65% or 420×10^9, lighters for 25% or 160×10^9, and wooden matches, the remaining 10% or 65×10^9.

The rising popularity of disposable butane lighters and especially the increasing volumes of very inexpensive imported lighters has resulted in a tremendous decline in the use of matches. By 1992 the total lights market had grown to 695×10^9; however, the relative market share of lighters had grown to approximately 86% or 600×10^9, whereas the share of paper matches had declined to 12% or 83×10^9, and wooden matches to just 2% or 12×10^9.

Table 1. Composition of Commercial Safety Matches and Strikers

Component	%
Match	
animal (hide) glue	9–11
starch	2–3
sulfur	3–5
potassium chlorate	45–55
neutralizer (ZnO, $CaCO_3$)	3
diatomaceous earth	5–6
other siliceous fillers (powdered glass, "fine" silica)	15–32
burning-rate catalyst ($K_2Cr_2O_7$ or PbS_2O_3)	to suit
water-soluble dye	to suit
Strikers	
animal glue or casein[a]	16
red phosphorus	50
calcium carbonate	5
powdered glass	25
carbon black	4

[a]Suitably insolubilized according to Ref. 6, or lightly brushed with diluted formaldehyde after application.

Table 2. SAW (Strike-Anywhere) Match Composition

Component	Tip composition, %	Base composition, %
animal glue	11	12
extender (starch)	4	5
paraffin		2
potassium chlorate	32	37
phosphorus sesquisulfide, P_4S_3	10	3
sulfur		6
rosin	4	6
dammar gum		3
infusorial earth (diatomite)		3
powdered glass and other fillers	33	21.5
potassium dichromate		0.5
zinc oxide	6	1

Some 98% of all these light sources are used for tobacco products and although disposable lighters continue to gain market share from matches, the overall lights market has begun to decline. Annual sales for the U.S. match industry in 1992 were approximately $60 \times \$10^6$.

The cost and selling price for matches increase considerably with higher quality cover paper, elaborateness of printed messages on and inside the cover (and sometimes even on the splint), and size of the order. In any case, the customer receives exactly the same high quality matches and striking strip.

Book matches are an important medium of advertising since they represent a truly utilitarian item which is more often given away than sold. Advertisers seeking national distribution pay for the message on many millions of books without entering otherwise in the sale of the matches (resale advertising match). Large merchandisers such as supermarkets, convenience stores, drug stores, and discount chains frequently purchase matches bearing their own advertising for resale (resale private label).

Individual establishments, including restaurants, hotels, banks, trucking companies, etc, buy the books with their personal message for their own distribution (special reproduction). On the smallest scale, a generalized so-called stock design permits establishments, who may need only a few thousand books a year, to buy and give away matches carrying their message by grouping several small orders in one production run, thereby avoiding a larger outlay for customized artwork and printing plates.

Toxicity and Other Safety Aspects

Because small children may suck on matches, the question of toxicity is often raised and the lingering, vague, though unwarranted idea of phosphorus poisoning may cause concern to laymen and even to physicians. Potassium chlorate is the only active material that can be extracted in more than traces from a match head and only 9 mg are contained in one head. This, even multiplied by the content of a whole book, is far below any toxic amount (19) for even a small child. No poisonous properties whatsoever can be imputed to the striking strip. SAW matches are similarly harmless but, because of their easy flammability, they should be entirely kept out of a household with smaller children. The same warning may apply to all wooden matches.

Safety-match strips arranged for decorative purposes in the form of flower pots are especially undesirable items, since an accidental ignition causes a dangerous flaring of many matches at one time.

Safety matches can be ignited by friction alone only when, with some deftness, they are rubbed on cardboard or glass. Accidental ignition is nearly always due to careless or absent-minded handling. In packaged condition, large numbers of safety matchbooks, if ignited in one spot, flare momentarily and harmlessly, often not igniting the adjoining matchbooks.

Exploding matches (other than for practical jokes) have virtually disappeared with modern manufacturing methods, ie, drying of the heads under controlled conditions. The occasional match that ignites and ejects sparking portions is generally a result of excessive pressure during lighting or of an accidentally cracked match head.

Sometimes a match ignites promptly but only a weak and unsatisfactory flame follows. This is the result of prolonged exposure of the matches to a temperature above 54°C in storage. The defect is caused by gradual dissipation of the paraffin wax throughout the splint and is evidenced by the disappearance of the line of demarcation, which is clearly visible in book matches. Otherwise, all safety matches tolerate exposure to elevated temperatures until about 177°C is reached. A U.S. Government specification requires that safety matches withstand exposure for two hours at 90°C (4). Earlier specifications stipulated a minimum acceptable nonignition temperature of 170°C. SAW matches are much more heat sensitive but still tolerate heating below a self-ignition temperature of 120 to 150°C.

BIBLIOGRAPHY

"Matches" in *ECT* 1st ed., Vol. 8, pp. 819–824, by C. K. Wolfert, The Diamond Match Co., in *ECT* 2nd ed., Vol. 13, pp. 160–166, by H. Ellern, UMC Industries, Inc.; in *ECT* 3rd ed., Vol. 15, pp. 1–8, by H. Ellern, Consultant.

1. M. F. Crass, Jr., *J. Chem. Ed.* **18**(3,6,8–9), (1941).
2. H. Hartig, *Zündwaren*, 2nd revised and extended ed., VEB Fachbuchverlag, Leipzig, Germany, 1971.
3. *Specifications for Red Railroad Fusees or Red Highway Fusees*, Bureau of Explosives, Washington, D.C., revised May 1, 1959.
4. *Fed. Spec. EE-M-101 J*, 1978.
5. H. Ellern, *Military and Civilian Pyrotechnics*, Chemical Publishing Co., Inc., New York, 1968.
6. U.S. Pat. 2,722,484 (Nov. 1, 1955), I. Kowarsky (to UMC Corp.).
7. I. Kowarsky, "Matches," in F. D. Snell and L. S. Ettre, eds., *Encyclopedia of Industrial Chemical Analysis*, Vol. 15, John Wiley & Sons, Inc., New York, 1972.
8. C. A. Finch, in B. Elvers, S. Hawkins, and G. Schulz, eds., *Ullmann's Encyclopedia of Industrial Chemistry*, Vol. A16, VCH Verlagsgesellschaft GmbH, Weinheim, Germany, 1990, pp. 163–169.
9. A. Osol and G. E. Farrar, *The Dispensatory of the United States*, J. B. Lippincott Co., Philadelphia, Pa., 1950.

MARK C. BEAN
D.D. Bean & Sons Company

MATERIAL SAFETY DATA SHEETS.   See   INDUSTRIAL   HYGIENE; PLANT SAFETY; TOXICOLOGY.

MATERIALS RELIABILITY

Reliability is a parameter of design like a system's performance or load ratings and is concerned with the length of failure-free operation. It is difficult to conceptualize reliability as part of the usual design calculations. Further complications are the complexity of organizations needed to produce the large systems of today and the usual time and financial constraints on research and development. Reliability as it relates to products or equipment can be measured in various ways. Since it is a design parameter, it has to be addressed early in the design cycle.

Terminology

Reliability. The reliability of a system is defined as the probability that the system will perform its intended function satisfactorily for a specified interval of time when operating under stated environmental conditions. It has to be realized that supposedly identical products fail at different times, thus reliability can be quantified only as a probability. For any product there is some underlying function that describes this success pattern. Typical reliability functions are shown in Figure 1 for two different products. These products can be compared at the same reliability level R_1 or the reliability levels can be compared for any selected time period, t_2.

In applying the definition of reliability, the concept of adequate performance must be established clearly. Products usually do not fail suddenly, but degrade over time. Gasket leaks on equipment, for example, may start as a slow weep and increase in volume over time. The point at which this undesirable occurrence is called a failure must be clear before reliability can be measured objectively. Changing the failure definition for a product changes its reliability level, although the product itself has not changed.

The reliability level of a product also depends on the operating or environmental conditions, which may produce a variety of failure modes. Reliability can only be assessed relative to a defined environment. Unless these points are established clearly, confusion surrounds any quoted reliability number for a product.

Because of the interrelationship of the system measures, reliability should not be considered by itself since, if taken alone, it does not express the totality of attributes that contribute to system effectiveness. However, in practice, reliability has gained the most acceptance and uniformity of definition. The other concepts described are not always defined uniformly from group to group and are sometimes used interchangeably. Further discussion of these concepts is found in References 1 and 2.

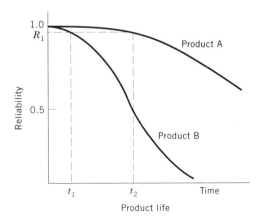

Fig. 1. Product reliability functions.

System Effectiveness. A system is designed to perform some intended function in a prescribed fashion. This overall capability is termed system effectiveness. Figure 2 illustrates the design trade-offs that constitute the components of system effectiveness.

From the standpoint of a military product, system effectiveness is the probability that the system meets successfully an operational demand within a given time when operating under specified conditions. From the standpoint of commercial products, system effectiveness is harder to define, but basically means customer satisfaction. There are several system parameters that are important to the customer. Some of these parameters are defined below.

Maintainability. Maintainability is a characteristic of design, installation, and operation, usually expressed as the probability that a system can be restored to specified operable conditions within a specified interval of time when maintenance is performed in accordance with prescribed procedures. The ease of fault detection, isolation, and repair are all influenced by system design and are principal factors contributing to maintainability. Also contributing is the supply of spare parts, the supporting repair organization, and preventative maintenance practices. Maintainability must be designed into the equipment. Some factors to consider follow.

Accessibility. Accessibility means having sufficient working space around a component to diagnose, troubleshoot, and complete maintenance activities safely and effectively.

Captive Hardware and Quick Attach/Detach. Captive and quick attach/detach hardware provides for rapid and easy replacement of components, panels, brackets, and chassis.

Color Coding. New machinery and equipment must conform to OSHA standards and OEM specifications for color coding. Color coding can also help

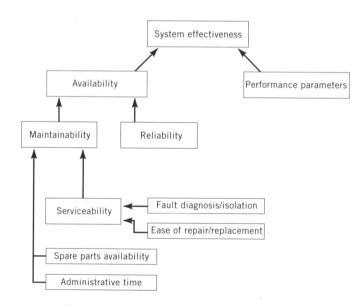

Fig. 2. Components of system effectiveness.

to speed up maintenance procedures. Examples include lubrication information, orientation, timing marks, torque requirements, etc.

Common Tools. Specialty tooling for maintenance repairs should be avoided. Standard tools readily available to the maintenance organization should be used.

Diagnostics. Diagnostic devices indicating the status of equipment should be built into the system to aid maintainability.

Modularity. Modularity requires that designs be divided into physically and functionally distinct units to facilitate removal and replacement. Modularity allows design of components as removable and replaceable units for minimum downtime.

Standardization. Design systems that incorporate component parts that are commercial standard, readily available, and common from system to system contribute to enhanced maintainability and to greatly reduced investment in spare-parts inventories.

Serviceability. Serviceability is defined as the degree of ease (or difficulty) with which a system can be repaired. This measure specifically considers fault detection, isolation, and repair. Repairability considers only the actual repair time, and is defined as the probability that a failed system is restored to operation in a specified interval of active repair time. Access covers, plug-in modules, or other features to allow easy removal and replacement of failed components improve the repairability and serviceability (see also ELECTRICAL CONNECTORS).

Availability. The system attributes of maintainability and reliability must both be considered. The trade-offs are rather complex and difficult to capture with any one measure. However, the term availability has been used to quantify these attributes simultaneously. The availability is sometimes related by inherent availability:

$$A = \frac{MTBF}{MTBF + MTTR} \tag{1}$$

The mean time between failures *MTBF* is used as a measure of system reliability, whereas the mean time to repair *MTTR* is taken as a measure for maintainability. For example, a system with an *MTBF* of 1200 h and a *MTTR* of 25 h would have an availability of 0.98. Furthermore, if only an *MTBF* of 800 h could be achieved, the same availability would be realized if the maintainability could be improved to the point where the *MTTR* was 16 h. Such trade-offs are illustrated in Figure 3, where each curve is at a constant availability.

Design Reliability

Since reliability and the related measures are essentially design parameters, improvements are most easily and economically accomplished early in the design cycle. Useful techniques for design reliability improvement are described below.

Design Review. A design review is a formalized, documented, and systematic audit of a design by senior company personnel. It addresses the complex design trade-offs and assures early design maturity. It should be multiphased and

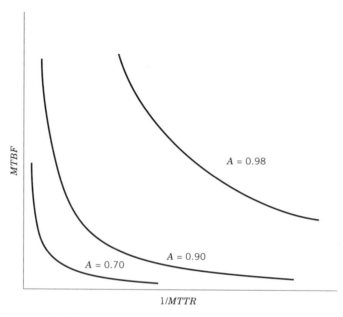

Fig. 3. System availability trade-off curves. $MTBF$ = mean time between failures; $MTTR$ = mean time to repair.

performed at various stages of the product development cycle. The parameters contributing to product availability must be a recognized input to this process.

Definite and known procedures for follow-up must be provided for, with the design group assessing the value of each idea and suggestion presented by the review committee. The actions taken are known to the committee and subject to further review. With such organization, the trade-offs can be acted upon at the appropriate level.

Failure Mode and Effects Analysis.   The system design activity usually emphasizes the attainment of performance objectives in a timely and cost-efficient fashion. The failure mode and effects analysis (FMEA) procedure considers the system from a failure point of view to determine how the product might fail. The terms design failure mode and effects analysis (DFMEA) and failure mode effects and criticality analysis (FMECA) also are used. This FMEA technique is used to identify and eliminate potential failure modes early in the design cycle, and its success is well documented (3,4).

The FMEA begins with the selection of a subsystem or component and then documents all potential failure modes. Their effect is traced up to the system level. A documented worksheet similar to Figure 4 is used on which the following elements are recorded.

Function.   This describes in a concise, short statement the exact function(s) the component/subsystem must perform. A component/subsystem may have more than one function.

Failure Mode.   The failure mode identifies how the component/subsystem can fail to perform each required function. A function may have more than one failure mode.

Failure mode and effects analysis

Part/Subsystem name —————   Prepared by —————

Part/Subsystem number —————   Date —————

Primary design responsibility ————— Revision No. —————

Part name/ function	Failure mode	Cause(s) of failure	Effect(s) of failure		Criticality analysis			RPN	Recommended action(s)
			Local	Global	Occ	Sev	Det		

Fig. 4. FMECA documentation.

Failure Cause. The failure cause is the physical, chemical, electrical, thermal, or other design deficiency which caused the failure. The agent, physical process, or hardware deficiency causing the failure mode must be identified, ie, what caused the failure for each failure mode. There may be more than one cause.

Failure Effect. The failure effect is the local effect on the immediate component/subsystem and the global effect on system performance/operation. In commercial products, the effect on the customer, ie, the global effect, must be addressed.

Criticality Analysis. The criticality assessment provides a figure-of-merit for each failure mode. This figure of merit is based on the likelihood of occurrence of the failure mode (Occ), the criticality (severity) of the failure mode on system performance (Sev), and the detectability of the failure mode by the user prior to occurrence (Det).

The purpose of the criticality rating is to provide guidance as to which failure modes require resolution. However, critical modes of failure resulting in unsafe operation should be given special attention, and design/verification actions should be taken to ensure that they never occur.

The most popular scheme among commercial companies is the assignment of a risk priority number (RPN) based on probability of occurrence, detectability, and severity of a particular failure mode. The factors (Occ, Sev, and Det) are each rated on a 1 to 10 scale and then an RPN is based on the product of the three rating values.

These procedures ensure early design maturity. Performing an FMA on purchased equipment may eliminate maintenance problems and provide a plan for spare-parts inventories.

Life-Cycle Cost. The total cost of ownership of a system during its operational life can be accounted for. The cost of ownership not only includes the initial design and acquisition cost but also cost of personnel training, spare-parts inventories, repair, operations, etc. A complete projection of system costs might point out the wisdom of investing more initially in order to forego high maintenance costs owing to poor reliability and serviceability, as illustrated in Figure 5.

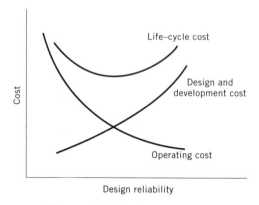

Fig. 5. Life-cycle cost concept.

System Reliability Models

Static reliability models are used in preliminary analyses to determine necessary reliability levels for subsystems and components. A subsystem is a particular low level grouping of components. Some trial and error is usually necessary to obtain reasonable groupings for any particular system. Early identification of potential system weaknesses facilitates corrective action.

A reliability block diagram can be developed for the system from the definition of adequate performance. The block diagram represents the effect of subsystem or component failure on system performance. In this preliminary analysis, each subsystem is assumed to be either a success or failure. A reliability value is assigned to each subsystem where the application and a specified time period are given. The reliability values for each subsystem and the functional block diagram are the basis for the analysis.

Series Systems. The series configuration is the most commonly encountered in practice. In a series system, all subsystems must operate successfully for the system to be successful. The reliability block diagram is given in Figure 6.

Fig. 6. Series block diagram.

The system reliability is

$$R_s = \prod_{i=1}^{n} R_i \tag{2}$$

where R_i is the reliability for the ith subsystem, and R_s is system reliability. It can be seen that

$$R_s \leq \min_i \{R_i\} \tag{3}$$

or the reliability of the system is never greater than the least reliable subsystem. In this analysis it is assumed that subsystems fail independently.

In a series system, if each subsystem had an exponential time to failure given by

$$f(t) = \lambda_i e^{-t\lambda i}, \qquad t \geq 0 \tag{4}$$

where λ_i is the failure rate for the ith subsystem. The system failure rate is

$$\lambda_s = \sum_{i=1}^{n} \lambda_i \tag{5}$$

or if *MTBF*s are used, then

$$\frac{1}{\theta_s} = \sum_{i=1}^{n} (1/\theta_i) \tag{6}$$

where $\theta_s = 1/\lambda_s$. Failure rates are sometimes more convenient to use in high reliability systems and are simply apportioned by equation 5.

Example 1. A gear pump is to be designed for use as an emergency backup system. The pump is driven by a small gasoline engine. Electronic sensing and starting circuitry are provided to automatically start the system during a power failure. Figure 7 gives a possible reliability block diagram for the system. For this application the reliability values are as follows: $R_1 = 0.9999$; $R_2 = 0.95$; $R_3 = 0.90$; $R_4 = 0.999$. This would give an overall system reliability of $R_s = 0.9999 \times 0.95 \times 0.90 \times 0.999 = 0.8541$.

If the information is insufficient to select the R_i values for this application, failure rates can be obtained from available sources (5,6). The failure rates obtained might be as follows: $\lambda_1 = 2.67 \times 10^{-6}/\text{h}$, $\lambda_2 = 591 \times 10^{-6}/\text{h}$, $\lambda_3 = 9.03 \times 10^{-6}/\text{h}$, $\lambda_4 = 4.45 \times 10^{-6}/\text{h}$. Then

$$\lambda_s = 607.15 \times 10^{-6}/\text{h}$$

where λ_s is the sum of λ_1, λ_2, λ_3, and λ_4. For an operating period of 12 h the reliability as calculated from equation 11 is

$$R(12 \text{ h}) = \exp[-12 \times 607.15 \times 10^{-6}] = 0.9927$$

In using these failure rates an exponential distribution for time to failure was assumed. Such an assumption should be made with caution.

Parallel Systems. A parallel (or redundant) system is not considered to be in a failed state unless all subsystems have failed. The system reliability is calculated as

$$R_s = 1 - \prod_{i=1}^{n} (1 - R_i) \tag{7}$$

System reliability is improved by providing alternative means for performing the same task. For example, automobiles were equipped with hand cranks even though they had electric starters. This back-up equipment was provided because at that time starters were unreliable. In contemporary system design, factors such as added cost, weight, and space may prohibit the use of redundant systems.

Systems can have both parallel and series subsystems. Reliability is calculated by successively reducing the system using the basic series or parallel formulas. This is illustrated in Example 2.

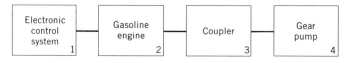

Fig. 7. Parallel block diagram.

Example 2. Figure 8 shows a system block diagram indicating subsystem reliabilities. Applying equation 7 to part A of Figure 8 gives

$$R_a = 1 - (0.20)(0.25)(0.30) = 0.985$$

For part B:

$$R_b = 1 - (0.40)(0.15) = 0.94$$

Then the series equation is applied to give the system reliability

$$R_s = 0.999 \times 0.985 \times 0.99 \times 0.94 = 0.916$$

Some systems cannot be represented by a simple combination of series and parallel subsystems. The systems are more complex in nature and the concept of coherent systems must be used in a more general and powerful treatment (7).

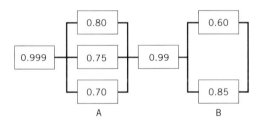

Fig. 8. Parallel and series combinations.

Reliability Measures

The reliability function $R(t)$ is defined as

$$R(t) = P(\mathbf{t} > t) = 1 - F(t) \qquad (8)$$

where $\mathbf{t}$ is the time-to-failure random variable and $F(t)$ is the cumulative distribution. In terms of the probability density function $f(t)$, the reliability function is given by

$$R(t) = \int_t^\infty f(u)\, du \qquad (9)$$

For example, if the time to failure is given as an exponential distribution, then

$$f(t) = \lambda e^{-\lambda t}, \qquad t \geqq 0, \qquad \lambda > 0 \qquad (10)$$

and the reliability function is found as follows:

$$R(t) = \int_t^\infty e^{-\lambda u}\,du = e^{-\lambda t}, \qquad t \geq 0 \qquad (11)$$

Life Expectancy of Devices. The expected or average life of devices is defined as

$$E(\mathbf{t}) = \int_{-\infty}^\infty u f(u)\,du \qquad (12)$$

where $f(t)$ is the probability density function (PDF) for the time-to-failure random variable $\mathbf{t}$. The expected life also can be found from

$$E(\mathbf{t}) = \int_0^\infty R(t)\,dt, \qquad t \geq 0 \qquad (13)$$

The expected life is sometimes used as an indicator of system reliability; however, it can be a false indication and should be used with caution. In most test situations the chance of surviving the expected life is not 50% and depends on the underlying failure pattern. For example, considering the exponential as used in equation 10, the expected life would be

$$E(\mathbf{t}) = \int_0^\infty t e^{-\lambda t}\,dt = 1/\lambda \qquad (14)$$

and the chance of surviving this time can be found from the reliability function

$$R(t = 1/\lambda) = e^{-1} = 0.368 \qquad (15)$$

That is, in this case there is only a 36.8% chance of surviving the mean life. If the distribution were other than exponential, the chance of survival would change. Since the mean life is not associated with constant reliability, the expected life should not be the only indicator of reliability, particularly when comparing products.

Failure Rate and Hazard Function. The failure rate is defined as the rate at which failures occur in a given time interval. Considering the time interval $[t_1, t_2]$, the failure rate is given by

$$\frac{R(t_1) - R(t_2)}{(t_2 - t_1)R(t_1)} \qquad (16)$$

and this is the rate of failure for those surviving at the beginning of the interval. This formula can be used to calculate failure rate from empirical life-test data.

The hazard function is defined as the limit of the failure rate as the interval of time approaches zero. The resulting hazard function $h(t)$ is defined by

$$h(t) = \frac{f(t)}{R(t)} \tag{17}$$

The hazard function can be interpreted as the instantaneous failure rate. The quantity $h(t)\Delta t$ for small Δt represents the probability of failure in the interval Δt, given that the device was surviving at the beginning of the interval.

The failure rate changes over the lifetime of a population of devices. An example of a failure-rate vs product-life curve is shown in Figure 9 where only three basic causes of failure are present. The quality-, stress-, and wearout-related failure rates sum to produce the overall failure rate over product life. The initial decreasing failure rate is termed infant mortality and is due to the early failure of substandard products. Latent material defects, poor assembly methods, and poor quality control can contribute to an initial high failure rate. A short period of in-plant product testing, termed burn-in, is used by manufacturers to eliminate these early failures from the consumer market.

The flat, middle portion of the failure-rate curve represents the design failure rate for the specific product as used by the consumer market. During the useful-life portion, the failure rate is relatively constant. It might be decreased by redesign or restricting usage. Finally, as products age they reach a wearout phase characterized by an increasing failure rate.

In real-life applications, many other failure mechanisms are present and this type of curve is not necessarily obtained. For example, in a multicomponents system the quality related failures do not necessarily all drop out early but might be phased out over a longer period of time.

Hazard function, PDF, and reliability function are related for any theoretical failure distribution. The relationships are

$$f(t) = h(t) \exp\left[-\int_0^t h(u)\, du \right] \tag{18}$$

and

$$R(t) = \exp\left[-\int_0^t h(u)\, du \right] \tag{19}$$

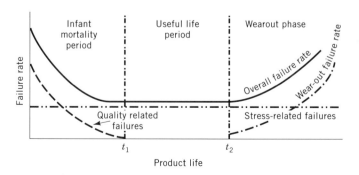

Fig. 9. Failure rate vs product life.

Conditional Failure Probability.   The concept of conditional probability of failure is useful to predict the chances of survival for a device that has been in operation for a period of time and is not in a failed state. Such information is helpful for maintenance planning.

If a device has a reliability function $R(t)$ and has been successfully operating for a period of time T, the conditional reliability function is given by

$$R(t \mid \mathbf{t} > T) = \frac{R(t)}{R(T)}, \qquad t > T \qquad (20)$$

The use of this concept is illustrated in Example 3.

Example 3.   A centrifugal pump moving a corrosive liquid is known to have a time-to-failure that is well approximated by a normal distribution with a mean of 1400 h and a standard deviation of 120 h. A particular pump has been in operation for 1080 h. In order to plan maintenance activities the chances of the pump surviving the next 48 h must be determined.

Applying equation 20 gives

$$R(1128 \text{ h} \mid \mathbf{t} > 1080 \text{ h}) = \frac{R(1128 \text{ h})}{R(1080 \text{ h})}$$

To determine $R(t)$ for the normal distribution, a standard normal variate must be calculated by the following formula:

$$z = \frac{t - \mu}{\sigma} \qquad (21)$$

where μ is the mean time to failure, and σ is the standard deviation. Applying this formula for $t = 1080$ h gives

$$z = (1080 - 1400)/120 = -2.67$$

Then this value of z is used with any readily available normal table to find

$$R(1080 \text{ h}) = 0.99621$$

Similarly

$$R(1128 \text{ h}) = 0.98840$$

which is the unconditional probability of surviving 1128 h. The conditional probability of survival is then

$$R(1128 \text{ h} \mid t > 1080 \text{ h}) = 0.98840/0.99621 = 0.99216$$

In this application, based on the consequences, management has a rule to plan a replacement when the reliability over the next 48 h period drops below 0.99. In this case they would forego scheduling the replacement.

Example 3 illustrated the use of the normal distribution as a model for time-to-failure. The normal distribution has an increasing hazard function which means that the product is experiencing wearout. In applying the normal to a specific situation, the fact must be considered that this model allows values of the random variable that are less than zero whereas obviously a life less than zero is not possible. This problem does not arise from a practical standpoint as long as $\mu/\sigma \geq 4.0$.

Exponential Distribution

The exponential distribution has proved to be a reasonable failure model for electronic equipment (8–13). Since the field of reliability emerged, owing to problems encountered with military electronics during World War II, exponential distribution has had considerable attention and application. However, like any failure model, it has limitations which should be well understood.

Basic Statistical Properties. The PDF for an exponentially distributed random variable **t** is given by

$$f(t, \lambda) = \lambda e^{-\lambda t}, \qquad t \geq 0 \tag{22}$$

where λ is the failure-rate parameter. The quantity $\theta = 1/\lambda$ is the mean or expected life, also expressed as *MTBF*. The PDF is shown in Figure 10.

The reliability function is given by

$$R(t) = e^{-\lambda t}, \qquad t \geq 0 \tag{23}$$

or

$$R(t) = e^{-t/\theta}, \qquad t \geq 0 \tag{24}$$

whereas the hazard function is

$$h(t) = \lambda = \frac{1}{\theta} \tag{25}$$

The hazard function is a constant which means that this model would be applicable during the midlife of the product when the failure rate is relatively stable. It

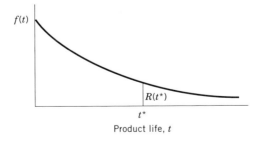

Fig. 10. PDF for the exponential failure model.

would not be applicable during the wearout phase or during the infant mortality (early failure) period.

On complex systems, which are repaired as they fail and placed back in service, the time between system failures can be reasonably well modeled by the exponential distribution (14,15).

Point Estimation.   The estimator for the mean life parameter θ is given by

$$\hat{\theta} = \frac{T}{r} \tag{26}$$

where T is total accumulated test time considering both failed and unfailed (or suspended) items; and r is total number of failures. The reliability function is then estimated by

$$\hat{R}(t) = e^{-t/\hat{\theta}}, \qquad t \geq 0 \tag{27}$$

Example 4.   A particular microprocessor (MPU) is assigned for a fuel-injection system. The failure rate must be estimated, and 100 MPUs are tested. The test is terminated when the fifth failure occurs. Failed items are not replaced. This type of testing, where n is the number placed on test and r is the number of failures specified, is termed a Type II censored life test.

Assuming that the above test produces the following data (failure time in hours), 84.1; 240.1; 251.9; 272.2; 291.9, the *MTBF* is estimated by using equation 26:

$$\hat{\theta} = \frac{84.1 + 240.1 + 251.9 + 272.2 + 291.9 + 95(291.9)}{5} = 5774 \text{ h}$$

From equation 8 it was shown that the chance of surviving the mean life was 36.8% for the exponential distribution. However, this fact must be used with some degree of rationality in applications. For example, in the above situation the longest surviving MPU that was observed survived for 291.9 hours. The failure rate beyond this time is not known. What was observed was only a failure rate of $\hat{\lambda} = 1.732 \times 10^{-4}$ failures per hour over approximately 292 hours of operation. In order to make predictions beyond this time, it must be assumed that the failure rate does not increase because of wearout and aging.

The reliability function in this example could be estimated as

$$\hat{R}(t) = e^{-t \times 1.732 \times 10^{-4}/\text{h}}$$

Since these MPUs are used to control fuel-injection systems, it might be interesting to know the 24,000-km reliability (the warranty period). Assuming an average speed of 80 km/h, 300 h of use are obtained. The reliability would be estimated as

$$\hat{R}(300 \text{ h}) = 0.949$$

or about 5.0% failures can be expected over the warranty period.

Example 5. There are six dynamometers available for engine testing. The test duration is set at 200 h which is assumed to be equivalent to 20,000 km of customer use. Failed engines are removed from testing for analysis and replaced. The objective of the test is to analyze the emission-control system. Failure is defined as the time at which certain emission levels are exceeded.

The testing situation where the duration is specified (ie, time-truncated) is termed Type I censored life testing.

Assuming that this test produces five failures, the *MTBF* would be estimated as

$$\hat{\theta} = \frac{120,000 \text{ km}}{5 \text{ failures}} = 24,000 \text{ km}$$

or the failure rate is

$$\hat{\lambda} = 4.17 \times 10^{-5} \text{ failures/km}$$

Again, these estimates must be used with caution. The system is obviously a mixture of electrical and mechanical components, and it can be assumed that wearout starts well beyond the 20,000 km period. If this is a reasonable assumption based on experience, then reliability predictions can be made over the 20,000-km period. For example, the 6000-km reliability might be estimated as

$$R(6000 \text{ km}) = 0.79$$

However, a 50,000-km reliability estimate might not be reasonable based on this testing scheme.

Confidence-Interval Estimates. Confidence-interval estimates for the expected life or reliability can be obtained easily in the case of the exponential. Here only the limits for failure-censored (Type II) and time-censored (Type I) life testing are given. It is possible to specify a test as either time- or failure-truncated, whichever occurs first. The theory for such tests is explained in References 16 and 17.

Time-Censored Life Tests. In this case the total test time T is specified. From the test, r failures are observed. The $100(1 - \alpha)\%$ two-sided confidence interval for the expected life is

$$\frac{2T}{\chi^2_{\alpha/2,\, 2(r+1)}} \leqq \theta \leqq \frac{2T}{\chi^2_{1-\alpha/2,\, 2r}} \tag{28}$$

The quantities $\chi^2_{\beta,\, \nu}$ are the $(1 - \beta)$ percentiles of a chi square distribution with ν degrees of freedom and are found readily in chi square tables.

Frequently, only a one-sided lower confidence limit is desired. In this case the limit is

$$\frac{2T}{\chi^2_{\alpha,\, 2(r+1)}} \leqq \theta \tag{29}$$

This is a $100(1 - \alpha)\%$ lower confidence limit.

If these limits on the expected life are designated by L and U for the lower and upper, respectively, then the $100(1 - \alpha)\%$ confidence interval on the reliability is

$$e^{-t/L} \leqq R(t) \leqq e^{-t/U} \tag{30}$$

Failure-Censored Life Tests. In this testing situation, the number of failures r is specified with n items initially placed on test $(r \leqq n)$. The test produces failure times $t_1, t_2 \ldots t_r$. The $100(1 - \alpha)\%$ confidence interval for the expected life is calculated by

$$\frac{2T}{\chi^2_{\alpha/2, 2r}} \leqq \theta \leqq \frac{2T}{\chi^2_{1-\alpha/2, 2r}} \tag{31}$$

Here again the quantity $\chi^2_{\beta, \nu}$ is the $(1 - \beta)$ percentile of a chi square distribution with ν degrees of freedom.

If only a $100(1 - \alpha)\%$ lower confidence limit is desired, it can be calculated from

$$\frac{2T}{\chi^2_{\alpha, 2r}} \leqq \theta \tag{32}$$

The confidence limits for the reliability function can be found from equation 30.

The Nonzero Minimum-Life Case. In many situations, no failures are observed during an initial period of time. For example, when testing engine bearings for fatigue life no failures are expected for a long initial period. Some corrosion processes also have this characteristic. In the following it is assumed that the failure pattern can be reasonably well approximated by an exponential distribution.

The PDF for the two-parameter exponential distribution is given by

$$f(t, \theta, \delta) = \frac{1}{\theta} e^{-(t-\delta)/\theta}, \qquad t \geqq \delta \geqq 0, \qquad \theta > 0 \tag{33}$$

The reliability function is

$$R(t) = e^{-(t-\delta)/\theta}, \qquad t \geqq \delta \geqq 0 \tag{34}$$

The expected life is $(\delta + \theta)$. The quantity δ is referred to as the minimum life parameter.

Point Estimation. This is a Type II censored life-testing situation where n items are placed on test and the test is terminated at the time of the rth failure. The life test produces the ordered failure times $t_1, t_2 \ldots t_r$. The estimator for θ is

$$\hat{\theta} = \frac{\sum\limits_{i=2}^{r}(t_i - t_1) + (n - r)(t_r - t_1)}{(r - 1)} \tag{35}$$

and the estimator for δ, the minimum life, is

$$\hat{\delta} = t_1 - \frac{\hat{\theta}}{n} \tag{36}$$

The reliability function is then estimated as

$$\hat{R}(t) = e^{-(t-\hat{\delta})/\hat{\theta}}, \qquad t \geq \hat{\delta} \tag{37}$$

Confidence Limits.   The $100(1 - \alpha)\%$ confidence interval for the parameter θ is

$$\frac{2(r-1)\hat{\theta}}{\chi^2_{\alpha/2, \, 2(r-1)}} \leqq \theta \leqq \frac{2(r-1)\hat{\theta}}{\chi^2_{1-\alpha/2, \, 2(r-1)}} \tag{38}$$

and the $100(1 - \beta)\%$ confidence interval for the minimum life δ is

$$t_1 - \frac{\hat{\theta}}{n} F_{\beta, \, 2, \, 2(r-1)} \leqq \delta \leqq t_1 \tag{39}$$

The quantity $F_{\beta, \, \nu_1, \, \nu_2}$ is the $(1 - \beta)$ percentile of an F-distributed random variable with ν_1, ν_2 degrees of freedom and is readily obtainable from F-tables.

  Example 6.   A return spring used on a butterfly-valve mechanism must have a high reliability. In order to determine the spring reliability, fifty springs are randomly selected and placed on life test. The test is terminated when the tenth spring fails. The data are given in the left column of Table 1. For the right column, equation 35 is applied.

  The estimate of θ is

$$\hat{\theta} = \frac{116.1 + (40)(22.6)}{9} = 113.3 \times 10^3 \text{ cycles}$$

Table 1. Cycles to Failure

$t_i \times 10^3$	$(t_i - t_1) \times 10^3$
61.0	0
64.1	3.1
64.6	3.6
66.2	5.2
73.9	12.9
75.0	14.0
77.4	16.4
79.8	18.8
80.5	19.5
83.6	22.6
	$\Sigma = 116.1$

and the minimum life is estimated from equation 36 as

$$\hat{\delta} = 61.0 - \frac{113.3}{50} = 58.7 \times 10^3 \text{ cycles}$$

Since the minimum life is critical in this application, a confidence limit estimate would be more appropriate, which can be calculated with the help of equation 39. For a 90% confidence limit, the required value of F is

$$F_{0.10, 2, 18} = 2.62$$

and substituting into the confidence interval equation gives

$$\left[61.0 - \frac{113.3}{50}(2.62) \right] \times 10^3 \text{ cycles} \le \delta \le 61.0 \times 10^3 \text{ cycles}$$

or

$$55.1 \times 10^3 \text{ cycles} \le \delta \le 61.0 \times 10^3 \text{ cycles}$$

In order to ensure virtually failure-free operation, a policy of changing this spring at 50,000 cycles of operation might be adopted.

In test planning, the number to be placed on test n and the number of failures r must be determined. The operating characteristic curves in Reference 18 can be used to specify the test, and to control the errors.

The Weibull Distribution

The Weibull distribution is a more versatile failure model than the exponential one. It is a popular model and widely used to estimate product reliability because it can be analyzed graphically with Weibull probability paper. Although the graphical form of analysis is presented here, other procedures are available (19–21).

Basic Statistical Properties. The reliability function for the three-parameter Weibull distribution is given by

$$R(t) = \exp\left[-\left(\frac{t - \delta}{\theta - \delta} \right)^\beta \right], \qquad t \ge \delta \ge 0, \qquad \beta > 0, \qquad \theta > \delta \qquad (40)$$

where δ is minimum life, θ is characteristic life, and β is Weibull slope.

The two-parameter Weibull has a minimum life of zero and the reliability function is

$$R(t) = e^{-(t/\theta)^\beta}, \qquad t \ge 0 \qquad (41)$$

The hazard function for the two-parameter Weibull is

$$h(t) = \frac{\beta}{\theta^\beta} t^{\beta - 1}, \qquad t \ge 0 \qquad (42)$$

This hazard function decreases with $\beta < 1$, increases with $\beta > 1$, and remains constant for $\beta = 1$. The value of β can give some indication of wearout or infant mortality.

The expected life for the two-parameter Weibull distribution is

$$\mu = \theta\Gamma(1 + 1/\beta) \tag{43}$$

where $\Gamma(\cdot)$ is a gamma function and can be found in gamma tables. The variance for the Weibull is

$$\sigma^2 = \theta^2\left[\Gamma\left(1 + \frac{2}{\beta}\right) - \Gamma^2\left(1 + \frac{1}{\beta}\right)\right] \tag{44}$$

The characteristic life parameter θ has a constant reliability associated with it. Evaluating the reliability function at $t = \theta$ gives

$$R(\theta) = e^{-1} = 0.368$$

and this is the same for any parameter value. Thus it is a constant for any Weibull distribution.

Parameter Estimation.   Weibull parameters can be estimated using the usual statistical procedures; however, a computer is needed to solve readily the equations. A computer program based on the maximum likelihood method is presented in Reference 22. Graphical estimation can be made on Weibull paper without the aid of a computer; however, the results cannot be expected to be as accurate and consistent.

The two-parameter cumulative Weibull distribution is

$$F(t) = 1 - e^{-(t/\theta)^\beta} \tag{45}$$

which, after rearranging and taking logarithms twice becomes

$$\ln\left(\ln\frac{1}{1 - F(t)}\right) = \beta \ln t - \beta \ln \theta \tag{46}$$

This would give a straight line plot on rectangular graph paper. Weibull graph paper plots $[F(t), t]$ as a straight line. Figure 11 illustrates a typical Weibull paper.

In using Weibull graph paper, a plotting position $p_j = F(t_j)$ for the jth-ordered observation has to be decided. The mean or median are the principal contenders. The median can be conveniently approximated (23) by

$$p_j = \frac{j - 0.3}{n + 0.4} \tag{47}$$

and the mean is given by

$$p_j = \frac{j}{n + 1} \qquad (48)$$

The failure points (t_j, p_j) are plotted and a straight line is fitted to estimate the Weibull population.

Example 7. In order to illustrate graphical parameter estimation, five failure times are considered: 24,000 km, 39,000 km, 52,000 km, 64,000 km, and 82,000 km. These times-to-failure were obtained by placing five items on test and allowing them to go to failure.

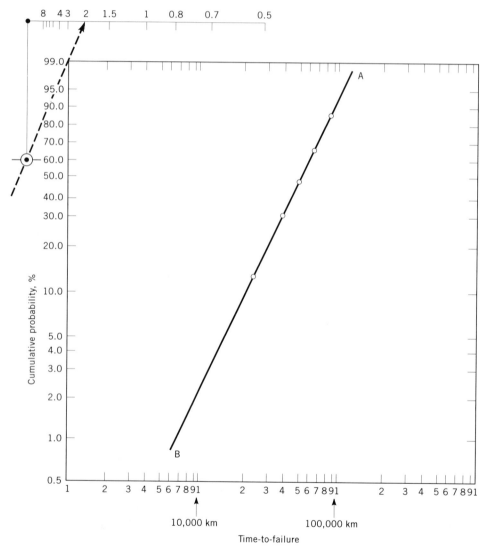

Fig. 11. Weibull probability paper: A, estimate of population; B, estimate of β (line drawn parallel to population line).

The median-rank plotting positions are obtained from equation 47. Tables such as found in Reference 24 can be used also. The data ready for plotting are given in Table 2 and are plotted on the Weibull paper in Figure 11. The Weibull slope parameter is estimated as $\hat{\beta} = 2.0$, and this implies an increasing failure rate. The characteristic life is estimated using the 63% point on the cumulative scale which gives $\hat{\theta} = 61,000$ km. Confidence limits can be also placed about this line; however, special tables are needed (24). The population line can be used to estimate either percent failure at a given time or the time at which a given percentage will fail.

In plotting on Weibull paper, a downward concave plot implies a nonzero minimum life. Values for $\hat{\delta} < t_1$ can be selected by trial and error. When they are subtracted from each t_i, a relatively straight line is produced. This essentially translates the three-parameter Weibull distribution back to a two-parameter distribution.

As can be seen from Figure 11, the graphical method does provide a good visual means for analyzing life data and is easily understood and explained. If used with discretion, graphical analysis can provide a useful means for data analysis.

Table 2. Weibull Paper Plotting Data

Order number, j	Failure times, t_j, km	Cumulative frequency, p_j, %
1	24,000	13
2	39,000	31
3	52,000	50
4	64,000	69
5	82,000	87

Binomial Distribution

To determine in the laboratory if a component survives in use, a test bogey is frequently established based on past experience. The test bogey is correlated with the particular test used to duplicate (or simulate) field conditions. The bogey can be stated in cycles, hours, revolutions, stress reversals, etc. A number of components are placed on test and each component either survives or fails. The reliability for this situation is estimated.

The failure model is the binomial distribution given by

$$p(y) = \binom{n}{y} R^y (1 - R)^{n-y}, \qquad y = 0, 1, 2 \ldots n \tag{49}$$

where R is the product reliability; n, the total number of products placed on test; and y, the number of products surviving the test. Furthermore

$$\binom{n}{y} = \frac{n!}{y! \, (n - y)!}$$

The quantity $p(y)$ is the probability that exactly y out of n components survive the test where the component reliability is R.

Reliability Estimation. Both a point estimate and a confidence interval estimate of product reliability can be obtained.

Point Estimate. The point estimate of the component reliability is given by

$$\hat{R} = \frac{y}{n} \tag{50}$$

Confidence Limit Estimate. An exact $100(1 - \alpha)\%$ lower confidence limit on the reliability is given by

$$R_L = \frac{y}{y + (n - y + 1)F_{\alpha, 2(n-y+1), 2y}} \tag{51}$$

where $F_{\alpha, 2(n-y+1), 2y}$ is easily obtained from tables for values of F.

A convenient approximate limit based on the normal distribution given by

$$R_L = \frac{(y - 1)}{n + z_\alpha \sqrt{\frac{n(n - y + 1)}{(y - 2)}}} \tag{52}$$

where z_α is the upper $(1 - \alpha)$ percentile of the standard normal distribution as is readily obtained from normal tables.

Example 8. There are 40 components placed on an accelerated 80-h life test. A 75% lower confidence limit on the reliability is desired.

To use equation 51, a value of F must be looked up. In this case, $n = 40$ and $y = 37$, and the required value is

$$F_{0.25, 8, 74} = 1.31$$

The lower confidence limit is calculated by

$$R_L = \frac{37}{37 + (4 \times 1.31)} = 0.876$$

or the 75% lower confidence limit on the reliability is

$$0.876 \leq R$$

If the approximate limit given by equation 52 is used, the value for $z_{0.25}$ is 0.67. The limit would be calculated as

$$R_L = \frac{36}{40 + 0.67\sqrt{\frac{40(4)}{35}}} = 0.869$$

As can be seen, the approximation is reasonably close. This approximation is better with large degrees of freedom for the value of F.

Success Testing. Acceptance life tests are sometimes planned with no failures allowed. This gives the smallest sample size necessary to demonstrate a reliability at a given confidence level. The reliability is demonstrated relative to the test employed and the testing period.

For the special case where no failures are allowed ($y = 0$) the $100(1 - \alpha)\%$ lower confidence limit on reliability is given by

$$R_L = \alpha^{1/n} \qquad (53)$$

where α is the level of significance, and n is the sample size. If $C = 1 - \alpha$ is taken as the desired confidence level, then the required sample size to demonstrate a minimum reliability of R is

$$n = \frac{\ln(1 - C)}{\ln R} \qquad (54)$$

For example, if a reliability level of $R = 0.85$ is to be demonstrated at 90% confidence, the required sample size is

$$n = \frac{\ln(0.10)}{\ln(0.85)} = 15$$

where no failures are allowed.

BIBLIOGRAPHY

"Materials Reliability" in *ECT* 3rd ed., Vol. 15, pp. 9–31, by L. Lamberson, Wayne State University.

1. W. H. Von Alven, ed., *Reliability Engineering*, Prentice-Hall, Inc., Englewood Cliffs, N.J., 1964.
2. *AMCP 706-133, Engineering Design Handbook, Maintainability Engineering Theory and Practice*, U.S. Army Materiel Command, Washington, D.C., 1976.
3. *MIL-STD-1629 (SHIPS), Procedures for Performing a Failure Mode and Effects Analysis for Shipboard Equipment*, Department of the Navy, Naval Ship Engineering Center, Hyattsville, Md., 1974.
4. *ARP-926, Design Analysis Procedure for Failure Mode, Effects and Critically Analysis (FMECA)*, Society of Automotive Engineers, Inc., New York, 1967.
5. Reliability Analysis Center, *NPRD-1, Nonelectronic Parts Reliability Data*, Rome Air Development Center, Griffiss AFB, N.Y., 1978.
6. *MIL-STD-217E, Reliability Prediction of Electronic Equipment*, U.S. Superintendent of Documents, Washington, D.C., 1978.
7. R. E. Barlow and F. Proschan, *Statistical Theory of Reliability and Life Testing Probability Models*, Holt, Rinehart and Winston, Inc., New York, 1975.
8. B. Epstein, *Ann. Math. Stat.* **25**, 555 (1954).
9. B. Epstein, *Technometrics* **2**(4), 447 (1960).
10. Ref. 9, p. 435.
11. Ref. 9, p. 403.
12. B. Epstein and M. Sobel, *J. Am. Stat. Assoc.* **48**, 486 (1953).

13. B. Epstein and M. Sobel, *Ann. Math. Stat.* **25**, 373 (1954).
14. R. F. Drenick, *J. Soc. Ind. Appl. Math.* **8**(21), 125 (1960).
15. *Ibid.*, **8**(4), 680 (1960).
16. D. J. Bartholomew, *Technometrics* **5**, 361 (1963).
17. G. Yang and M. Sirvanci, *J. Am. Stat. Assoc.* **72**, 444 (1977).
18. K. H. Schmitz, L. R. Lamberson, and K. C. Kapur, *Technometrics* **21**(4), 539 (1979).
19. L. J. Bain, *Statistical Analysis of Reliability and Life-Testing Models*, Marcel Dekker, Inc., New York, 1978.
20. D. I. Gibbons and L. C. Vance, *A Simulation Study of Estimators for the Parameters and Percentiles in the Two-Parameter Weibull Distribution, General Motors Research Publication No. GMR-3041*, General Motors, Detroit, Mich., 1979.
21. N. Mann, R. Schafer, and N. D. Singpurwalla, *Methods for Statistical Analysis of Reliability and Life Tests*, John Wiley & Sons, Inc., New York, 1974.
22. D. R. Wingo, *IEEE Trans. Reliab.* **R-22**(2), (1973).
23. A. Benard and E. C. Bos-Levenbach, *Statistica* **7**, (1953).
24. K. C. Kapur and L. R. Lamberson, *Reliability in Engineering Design*, John Wiley & Sons, Inc., New York, 1977.
25. E. B. Haugen, *Probabilistic Approaches to Design*, John Wiley & Sons, Inc., New York, 1968.
26. D. Kececioglu and D. Cormier, *Proc. Third Ann. Aerospace Reliab. Maintainab. Conf.*, 546 (1964).
27. D. Kececioglu and E. B. Haugen, *Ann. Assurance Sci.—Seventh Reliab. Maintainab. Conf.*, 520 (1968).
28. C. Mischke, *J. Eng. Ind.* 537 (Aug. 1970).
29. A. H. Bowker and G. J. Lieberman, *Engineering Statistics*, 2nd ed., Prentice-Hall, Inc., Englewood Cliffs, N.J., 1972.

General References

R. E. Barlow and F. Proschan, *Mathematical Theory of Reliability*, John Wiley & Sons, Inc., New York, 1965.

I. Bazovsky, *Reliability Theory and Practice*, Prentice-Hall, Inc., Englewood Cliffs, N.J., 1961.

R. Billinton, *Power System Reliability Evaluation*, Gordon and Breach Science Publishers, New York, 1970.

R. Billinton, R. J. Ringlee, and A. J. Wood, *Power-System Reliability Calculations*, The MIT Press, Cambridge, Mass., 1973.

J. H. Bompas-Smith, in R. H. W. Brook, ed., *Mechanical Survival: The Use of Reliability Data*, McGraw-Hill, New York, 1973.

DARCOM-P-702-4, Reliability Growth Management, U.S. Army Materiel Development and Readiness Command, Alexandria, Va., 1976.

D. K. Lloyd and M. Lipow, *Reliability: Management Methods and Mathematics*, Prentice-Hall, Inc., Englewood Cliffs, N.J., 1962.

M. L. Shooman, *Probabilistic Reliability: An Engineering Approach*, McGraw-Hill, New York, 1968.

U. S. Army Materiel Command, *Engineering Design Handbooks-Development Guide for Reliability, Part 2: Design for Reliability (AMCP 706-196); Part 3: Reliability Prediction (AMCP 706-197); Part 4: Reliability Measurement (AMCP 706-298)*, National Technical Information Service, Springfield, Va., 1976.

H. Ascher and H. Feingold, *Repairable Systems Reliability*, Lecture Notes in Statistics, No. 7, Marcel Dekker, Inc., New York, 1984.

R. Dovich, *Reliability Statistics*, ASQC Quality Press, Milwaukee, Wis., 1990.

D. Kececioglu, *Reliability Engineering Handbook*, Vols. 1 and 2, Prentice-Hall, Inc., Englewood Cliffs, N.J., 1991.

D. J. Klinger, Y. Nakada, and M. Menendez, *AT&T Reliability Manual*, Van Nostrand Reinhold, New York, 1990.

Society of Automotive Engineers, *Reliability and Maintainability Guideline for Manufacturing Machinery and Equipment*, SAE Order No. M-110, Society of Automotive Engineers (SAE), Warrendale, Pa., 1993.

LEONARD LAMBERSON
Western Michigan University

MATERIALS STANDARDS AND SPECIFICATIONS

A standard is a document, definition, or reference artifact intended for general use by as large a body as possible; a specification, which involves similar technical content and similar format, usually is limited in both its intended applicability and its users.

Standards have been a part of technology since building began, both at a scale that exceeded the capabilities of an individual, and for a market other than the immediate family. Standardization minimizes disadvantageous diversity, assures acceptability of products, and facilitates technical communication. There are many attributes of materials that are subject to standardization, eg, composition, physical properties, dimensions, finish, and processing. Implicit to the realization of standards is the availability of test methods and appropriate calibration techniques. Apart from physical or artifactual standards, written or paper standards also must be considered, ie, their generation, promulgation, and interrelationships.

The International Organization for Standardization (ISO) defines a standard as the result of the standardization process: "the process of formulating and applying rules for an orderly approach to a specific activity for the benefit and with the cooperation of all concerned and in particular for the promotion of optimum overall economy taking due account of functional conditions and safety requirements" (1). Standardization involves concepts of units of measurement, terminology and symbolic representation, and attributes of the physical artifact, ie, quality, variety, and interchangeability. A specification, however, is defined as "a document intended primarily for use in procurement which clearly and accurately describes the essential technical requirements for items, materials, or services including the procedures by which it will be determined that the requirements have been met" (2). The ISO defines a specification as "a concise statement of a set of requirements to be satisfied by a product, a material or a process indicating, whenever appropriate, the procedure by means of which it may be determined whether the requirements given are satisfied. Notes—(*1*) A specification may be a standard, a part of a standard, or independent of a standard. (*2*) As far as practicable, it is desired that the requirements are expressed numerically in terms of appropriate units, together with their limits." A specification may also be viewed as the technical aspects of the legal contract between the purchaser of the material, product, or service and the vendor of the same and defines what each may expect of the other.

Standards

Objectives and Types. The objectives of standardization are economy of production by way of economies of scale in output, optimization of varieties in input material, and improved managerial control; assurance of quality; improvement of interchangeability; facilitation of technical communication; enhancement of innovation and technological progress; and promotion of the safety of persons, goods, and the environment. The likely consequences of choosing a material that is not standard, other than in exceptional circumstances, are that the selected "special" would be unusually costly; require an elaborate new specification; be available from few sources; be lacking documentation for many ancillary properties other than that for which it was chosen; be unfamiliar to others, eg, purchasers, vendors, production workers, maintenance personnel, etc; and contribute to the proliferation of stocked varieties and thus exacerbate problems of recycling, mistaken identity, increased purchasing costs, etc.

Physical or artifactual standards are used for comparison, calibration, etc, eg, the national standards of mass, length, and time maintained by the National Institute of Standards and Technology (NIST) or the standard reference materials (SRMs) collected and distributed by NIST. Choice of the standard is determined by the property it is supposed to define, its ease of measurement, its stability with time, and other factors (see FINE CHEMICALS).

Paper or documentary standards are written articulations of the goals, quality levels, dimensions, or other parameter levels that the standards-setting body seeks to establish. *Value standards* are a subset of paper standards and usually relate directly to society and include social, legal, political, and to a lesser extent, economic and technical factors. Such standards usually result from federal, state, or local legislation.

Regulatory standards most frequently derive from value standards but also may arise on an ad hoc or consensus basis. These include industry regulations or codes that are self-imposed; consensus regulatory standards that are produced by voluntary organizations in response to an expressed governmental need, especially where well-defined engineering practices or highly technical issues are involved; and mandatory regulatory standards that are developed entirely by government agencies. Examples of regulatory standards from the materials field include safety regulations, eg, those of the OSHA; clean air and water laws of the U.S. EPA; or rulings related to exposure to radioactive substances. Regulatory standards may be deliberately set in advance of the state of the art in the relevant technology, eg, fuel efficiency of automobiles, in contrast to the other types of standards (see REGULATORY AGENCIES).

Voluntary standards are especially prevalent in the United States and are generated by various consortia of government and industry, producers and consumers, technical societies and trade associations, general interest groups, academia, and individuals. These standards are voluntary in their manner of generation and in that they are intended for voluntary use. Nonetheless, some standards of voluntary origin have been adopted by governmental bodies and are mandatory in certain contexts. Voluntary standards include those which are recommended but which may be subject to some interpretation and those conventions as to units, definitions, etc, that are established by custom.

Product standards may stipulate performance characteristics, dimensions, quality factors, methods of measurement, and tolerances; and safety, health, and environmental protection specifications. These are introduced principally to provide for interchangeability and reduction of variety. The latter procedure is referred to as rationalization of the product offering, ie, designation of sizes, ratings, etc, for the attribute range covered and the steps within the range. The designated steps may follow a modular format or a preferred number sequence.

Public and private standards also may be distinguished. Public standards include those produced by government bodies and those published by other organizations but promoted for general use, eg, the ASTM standards. Private standards are issued by a private company for its own interests and generally are not available to parties other than its vendors, customers, and subcontractors.

Consensus standards are the key to the voluntary standards system because acceptance and use of such standards follow directly from the need for them and from the involvement in their development of all those who share that need. Consensus standards must be produced by a body selected, organized, and conducted in accordance with due process procedures. All parties or stakeholders are involved in the development of the standard and substantial agreement is reached according to the judgment of a properly constituted review board. Other aspects of due process involve proper issuance of notices, record keeping, balloting, and attention to minority opinion.

Generation, Administration, and Implementation. The development of a good standard is a lengthy and involved process, whether for a private organization, a nation, or an international body. The generic aspects of the development of a standard are shown in Figure 1. Once the need for a standard has been determined, information relevant to the subject must be gathered from many sources, eg, libraries and specialists' knowledge, field surveys, and laboratory results. Multidisciplinary and multifunctional teams must digest this information, array and analyze options, achieve an effective compromise in the balanced best interests of all concerned, and participate in the resolution of issues and criticisms arising from the reviews and appeals process. Among its functions, the administrative arm of the pertinent standards organization sets policy, allocates resources, establishes priorities, supervises reviews and appeals procedures, and interacts with organizations external to itself. The affected entities usually comprise a large, diverse, and overlapping group of interests, ie, economic sectors: industry, government, business, construction, chemicals, energy, etc; functions: planning, development, design, production, maintenance, etc; and organizations and groups of individuals: manufacturers, consumers, unions, investors, distributors, etc. No standard can be fully effective in meeting its objectives unless

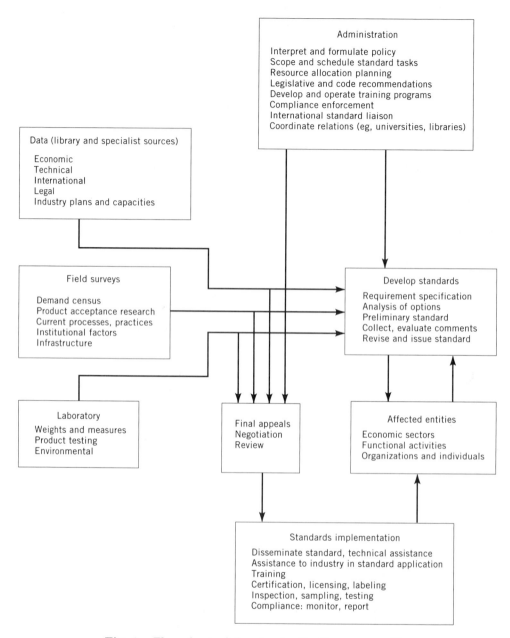

Fig. 1. Flow chart of the standardization process (3).

attention is paid to the implementation function which includes promulgation, education, enforcement of compliance, and technical assistance. Usually in the choice of a standard for a given purpose, the more encompassing the population to which the standard applies, ie, from the private level to the trade association or professional society to the national level or to the international level, the more effective and the less costly the application of the standard. Finally it must be recognized that standardization is a highly dynamic process. It cannot function without continued feedback from all affected parties and it must provide for constant review and adaptation to changing circumstances, improved knowledge, and control.

Standard Reference Materials. An important development in the United States, relative to standardization in the chemical field, is the establishment by NIST of standard reference materials (SRMs), originally called standard samples (4). The objective of this program is to provide materials that may be used to calibrate measurement systems and to provide a central basis for uniformity and accuracy of measurement. SRMs are well-characterized, homogeneous, stable materials or simple artifacts with specific properties that have been measured and certified by NIST. Their use with standardized, well-characterized test methods enables the transfer, accuracy, and establishment of measurement traceability throughout large, multilaboratory measurement networks. More than 1000 materials are included, eg, metals and alloys, ores, cements, phosphors, organics, biological materials, glasses, liquids, gases, radioactive substances, and specialty materials (5). The standards are classified as standards of certified chemical composition, standards of certified physical properties, and engineering-type standards. Although most of these are provided with certified numerical characterizations of the compositions or physical properties for which they were established, some others are included even where provision of numerical data is not feasible or certification is not useful. These latter materials do, however, provide assurance of identity among all samples of the designation and permit standardization of test procedures and referral of physical or chemical data on unknown materials to a known or common basis.

Shifts in the nature of the materials included in the SRM inventory have occurred. In the compositional SRMs, increased attention is paid to trace-organic analysis for environmentally, clinically, or nutritionally important substances; to trace-element analysis in new high technology materials, eg, alloys, plastics, and semiconductors; and for bulk analyses in the field of recycled, nuclear, and fibrous materials (see TRACE AND RESIDUE ANALYSIS). Concern not just for certification of the total concentration of individual elements but for the levels of various chemical states of those species is expected. With regard to the development of physical property SRMs, density standards, dimensional standards at the micrometer and submicrometer level, and materials relative to standardization of optical properties should be among the more active areas. Developments in SRMs for engineering properties should include materials suitable for nondestructive testing (qv), evaluation of durability, standardizing computer and electronic components, and workplace hazard monitoring.

Standard reference materials provide a necessary but insufficient means for achieving accuracy and measurement compatibility on a national or international scale. Good test methods, good laboratory practices, well-qualified personnel, and proper intralaboratory and interlaboratory quality assurance procedures are equally important. A systems approach to measurement compatibility is illustrated in Figure 2. The function of each level is to transfer accuracy to the level below and to help provide traceability to the level above. Thus traversing the hierarchy from bottom to top increases accuracy at the expense of measurement efficiency.

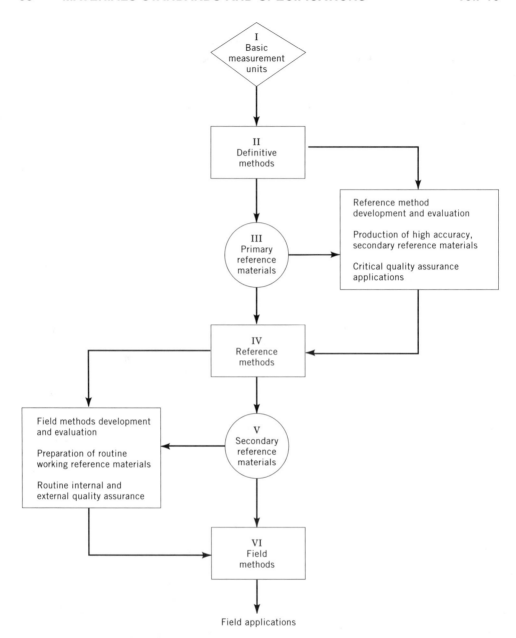

Fig. 2. Measurements standards hierarchy (4).

Analytical standards imply the existence of a reference material and a recommended test method. Analytical standards other than for fine chemicals and for the NIST series of SRMs have been reviewed (6). Another sphere of activity in analytical standards is the geochemical reference standards maintained by the U.S. Geological Survey and by analogous groups in France, Canada, Japan, South Africa, and Germany (7).

Chronological standards are needed for an extremely diverse range of fields, eg, astrophysics, anthropology, archaeology, geology, oceanography, and art. The techniques employed for dating materials include dendrochronology, thermoluminescence, obsidian hydration, varve deposition, paleomagnetic reversal, fission tracks, racemization of amino acids, and a variety of techniques related to the presence or decay of radioactive species, eg, ^{14}C, ^{10}Be, ^{18}O, and various decay products of the U and Th series. Because the time periods of interest range from decades to millions of years and the available materials may be limited, no one technique presents a general solution. Some progress has been made on age standardization and calibration through the efforts of the Sub-Commission on Geochronology of the International Union of Geological Sciences (8). Establishment of a physical bank of chronological standards that are analogous to those standards set by NIST and the U.S. Geological Survey for compositional and physical properties would greatly benefit a broad range of scientific and cultural communities.

Standard Reference Data. In addition to standard reference materials, the materials scientist or engineer frequently requires access to standard reference data. Such information helps to identify an unknown material, describe a structure, calibrate an apparatus, test a theory, or draft a new standard specification. Data are defined as that subset of scientific or technical information that can be represented by numbers, graphs, models, or symbols. The term, standard reference data, implies a data set or collection that has passed some screening and evaluation by a competent body and warrants the body's imprimatur and promotion. Such a data set may be generated expressly for this purpose by especially careful measurements made on a standard reference material or other well-characterized material, eg, the series of standard x-ray diffraction patterns generated by NIST. In some cases, a reference data set may not represent a specific set of real experimental observations but a recommended, consistent set of stated reliability that is synthesized from limited, fragmentary, and conflicting literature data by review, analysis, adjustment, and interpolation (9). A journal specifically devoted to recording techniques for evaluating data and archiving the results of such evaluation is the *Journal of Physical and Chemical Reference Data*, jointly sponsored by NIST, the American Chemical Society, and the American Institute of Physics. The biennial proceedings of the international conferences run by the Committee on Data (CODATA) of the International Council of Scientific Unions (ICSU) also contain many contributions on data evaluation and standard reference data. A convenient reference (10) summarizing and classifying some 1200 data sources relevant to materials has been described (11).

Standards for Nondestructive Evaluation. Nondestructive evaluation (NDE) standards are important in materials engineering for evaluating the structure, properties, and integrity of materials and fabricated products. Such standards apply to test methods, artifactual standards for test calibration, and comparative graphical or pictorial references. These standards may be used as inspection guides, to define terms describing defects, to describe and recommend test methods, for qualification and certification of individuals and laboratories working in the NDE field, and to specify materials and apparatus used in NDE testing. NDE standards have been reviewed with regard to what standards are available, what are satisfactory, what are lacking, and what need improvement (12). Other references include useful compilations of standards and specifications in NDE (13,14) (see NONDESTRUCTIVE EVALUATION).

Traceability. Measurements are traceable to designated standards if scientifically rigorous evidence is produced on a continuing basis to show that the measurement process is producing data for which the total measurement uncertainty is quantified relative to national or other designated standards through an unbroken chain of comparisons (15). The intent of traceability is to assure an accuracy level sufficient for the need of the product or service. Although calibration is an important factor, measurement traceability also requires consideration of measurement uncertainty that arises from random error, ie, variability within the laboratory, and from systematic error of that laboratory relative to the reference standard. Although the ultimate metrological standards are those of mass, length, and time, maintained at NIST and related to those defined by international standardizing bodies, there are literally thousands of derived units, for only a fraction of which primary reference standards are maintained.

Traceability also is used by materials engineers for the identification of the origin of a material. This attribution often is necessary where knowledge of composition, structure, or processing history is inadequate to assure the properties required in service. Thus critical components of industrial equipment may have to be related to the particular heat of the steel that is used in the equipment apart from having to meet the specification. The geographical derivation of certain ores and minerals often is specified where analytical measurements are inadequate, and much recycled material can be used only if traceability to the original form can be established.

Basic Standards for Chemical Technology. There are many numerical values that are standards in chemical technology. A brief review of a few basic and general ones is given herein. Numerical data and definitions quoted are taken from References 16–19 (see UNITS AND CONVERSION FACTORS) and are expressed in the International System of Units (SI). A comprehensive guide for the application of SI has been published by ASTM (20).

Atomic Weight. As of this writing (ca 1994) the definition of atomic weights is based on carbon-12 [7440-44-0], ^{12}C, the most abundant isotope of carbon, which has an atomic weight defined as exactly 12 (21).

Temperature. Temperature is the measurement of the average kinetic energy, resulting from heat agitation, of the molecules of a body. The most widely used scale is the Celsius scale for which the freezing and boiling points of water are used as defining points. The ice point is the temperature at which macroscopic ice crystals are in equilibrium with pure liquid water under air that is saturated with moisture at standard atmospheric pressure (101.325 kPa). One degree on the Celsius scale is 1.0% of the range between the melting and boiling points of water. The unit of thermodynamic temperature is the Kelvin, defined as 1/273.16 of the thermodynamic temperature of the triple point of water. The relation of the Kelvin and Celsius scales is defined by the International Temperature Scale of 1990 (22). The international temperature scale between -259 and $962°C$ is based on a number of defining fixed points, use of a standard platinum resistance thermometer, and the following formula for the resistance, R, as a function of temperature, t, above 0°C,

$$R_t = R_0[1 + At + Bt^2 + Ct^3]$$

where A, B, and C are arbitrary constants (see TEMPERATURE MEASUREMENT).

Pressure. Standard atmospheric pressure is defined to be the force exerted by a column of mercury 760-mm high at 0°C. This corresponds to 0.101325 MPa (14.695 psi). Reference or fixed points for pressure calibration exist and are analogous to the temperature standards cited (23). These points are based on phase changes or resistance jumps in selected materials. For the highest pressures, the most reliable technique is the correlation of the wavelength shift, $\Delta\lambda$, with pressure of the ruby, R_1, fluorescence line and is determined by simultaneous specific volume measurements on cubic metals correlated with isothermal equations of state which are derived from shockwave measurements (24). This calibration extends from 6–100 GPa (0.06–1 Mbar) and may be represented by the following:

$$P = \frac{1904}{5}\left\{ \left[\frac{\lambda_o + \Delta\lambda}{\lambda_o} \right]^5 - 1 \right\}$$

where λ_o is the wavelength measured at 100 kPa (1 bar) and P is in units of GPa (see PRESSURE MEASUREMENT).

Length. One meter is defined as the length of path traveled by light in a vacuum during a time interval of 1/299,792,458 of a second (25).

Mass. The unit of mass is the kilogram and is the mass of a particular cylinder of Pt–Ir alloy which is preserved in France by the International Bureau of Weights and Measures.

Time. The unit of time in the International System of units is the second: "the duration of 9,192,631,770 periods of the radiation corresponding to the transition between the two hyperfine levels of the fundamental state of the atom of cesium-133" (25). This definition is experimentally indistinguishable from the ephemeris-second which is based on the earth's motion.

Standard Cell Potential. A large class of chemical reactions are characterized by the transfer of protons or electrons. Substances losing electrons in a reaction are said to be oxidized, those gaining electrons are said to be reduced. Many such reactions can be carried out in a galvanic cell which forms a natural basis for the concept of the half-cell, ie, the overall cell is conceptually the sum of two half-cells, one corresponding to each electrode. The half-cell potential measures the tendency of one reaction, eg, oxidation, to proceed at its electrode; the other half-cell of the pair measures the corresponding tendency for reduction to proceed at the other electrode. Measurable cell potentials are the sum of the two half-cell potentials. Standard cell potentials refer to the tendency of reactants in their standard state to form products in their standard states. The standard conditions are 1 M concentration for solutions, 101.325 kPa (1 atm) for gases, and for solids, their most stable form at 25°C. Since half-cell potentials cannot be measured directly, numerical values are obtained by assigning the hydrogen gas–hydrogen ion half-reaction the half-cell potential of zero V. Thus, by a series of comparisons referred directly or indirectly to the standard hydrogen electrode, values for the strength of a number of oxidants or reductants can be obtained (26), and standard reduction potentials can be calculated from established values (see BATTERIES; ELECTROCHEMICAL PROCESSING).

Standard cell potentials are meaningful only when these are calibrated against an emf scale. To achieve an absolute value of emf, electrical quantities must be referred to the basic metric system of mechanical units. If the current unit, A, and the resistance unit, Ω, can be defined, then the volt may be defined by Ohm's law as the voltage drop across a resistor of one standard ohm when passing one standard ampere of current. In the ohm measurement, a resistance is compared to the reactance of an inductor or capacitor at a known frequency. This reactance is calculated from the measured dimensions and can be expressed in terms of the meter and second. The ampere determination measures the force between two interacting coils where these carry the test current. The force between the coils is opposed by the force of gravity acting on a known mass; hence, the ampere can be defined in terms of the meter, kilogram, and second. Such a means of establishing a reference voltage is inconvenient for frequent use and reference is made to a previously calibrated standard cell.

Ideally a standard cell is constructed simply and is characterized by a high constancy of emf, a low temperature coefficient of emf, and an emf close to one volt. The Weston cell, which uses a standard cadmium sulfate electrolyte and electrodes of cadmium amalgam and a paste of mercury and mercurous sulfate, essentially meets these conditions. The voltage of the cell is 1.0183 V at 20°C. The a-c Josephson effect, which relates the frequency of a superconducting oscillator to the potential difference between two superconducting components, is used by NIST to maintain the unit of emf. The definition of the volt, however, remains as the Ω/A derivation described (see SUPERCONDUCTING MATERIALS).

Concentration. The basis unit of concentration in chemistry is the mole which is the amount of substance that contains as many entities, eg, atoms, molecules, ions, electrons, protons, etc, as there are atoms in 12 g of ^{12}C, ie, Avogadro's number $N_A = 6.0221367 \times 10^{23}$. Solution concentrations are expressed on either a weight or volume basis. Molality is the concentration of a solution in terms of the number of moles of solute per kilogram of solvent. Molarity is the concentration of a solution in terms of the number of moles of solute per liter of solution.

A particular concentration measure of acidity of aqueous solutions is pH which usually is regarded as the common logarithm of the reciprocal of the hydrogen-ion concentration (see HYDROGEN-ION ACTIVITY). More precisely, the potential difference of the hydrogen electrode in normal acid and in normal alkali solution (-0.828 V at 25°C) is divided into 14 equal parts or pH units; each pH unit is 0.0591 V. Operationally, pH is defined by pH = pH (soln) + E/K, where E is the emf of the cell:

$$H_2 \,|\, \text{solution of unknown pH} \,\|\, \text{saturated KCl} \,\|\, \text{solution of known pH} \,|\, H_2$$

and $K = 2.303\,RT/F$, where R is the gas constant, 8.314 J/(mol·K) (1.987 cal/(mol·K)), T is the temperature in Kelvin, and F is the value of the

Faraday, 9.64853×10^4 C/mol. pH usually is equated to the negative logarithm of the hydrogen-ion activity (qv), although there are differences between these two quantities outside the pH range 4.0–9.2:

$$-\log q_{H^+} m_{H^+} = \text{pH} + 0.014 \,(\text{pH} - 9.2) \text{ for pH} > 9.2$$

$$-\log q_{H^+} m_{H^+} = \text{pH} + 0.009 \,(4.0 - \text{pH}) \text{ for pH} < 4.0$$

Energy. The SI unit of energy is the joule which is the work done when the point of application of a force of one newton is displaced a distance of one meter in the direction of the force. The newton is that force which, when applied to a body having a mass of one kilogram, accelerates that body one meter per second squared.

Specifications

Objectives and Types.   A specification establishes assurance of the fitness of a material, product, process, or service for use. Such fitness usually encompasses safety and efficiency in use as well as technical performance. Material specifications may be classified as to whether they are applied to the material, the process by which it is made, or the performance or use that is expected of it. Product or design specifications are not relevant to materials. Within a company, the specification is the means by which engineering conveys to purchasing what requirements it has for the material to be supplied to manufacturing. It has its greatest utility prior to and at the time of purchase. Yet a properly written and dated specification with accompanying certificates of test, heat, or lot numbers, vendor identification, and other details pertinent to the actual material procurement constitutes an important archival document. Material specification records provide information regarding a proven successful material that can be used in a new product. Such records also are useful in the rebuilding of components and as defense evidence in a liability suit.

Content.   Although formats of materials specifications may vary according to the need, the principal elements are title, statement of scope, requirements, quality assurance provisions, applicable reference documents, and preparations for delivery, notes, and definitions. The scope statement comprises a brief description of the material, possibly its intended area of application, and categorization of the material by type, subclass, and quality grade. Requirements may include chemical composition, physical properties, processing history, dimensions and tolerances, and/or finish. Quality assurance factors are test methods and equipment including precision; accuracy and repeatability; sampling procedures; inspection procedures, ie, acceptance and rejection criteria; and test certification. Reference documents may include citation of well-established specifications, codes and standards, definitions and abbreviations, drawings, tolerance tables, and test methods. Preparations for delivery are the instructions for packing, marking, shipping mode, and unit quantity of material in the shipment. The notes section is intended for explanations, safety precautions, and other details not covered elsewhere. Definitions are specifications of terms in the document which differ from common usage.

Strategy and Implementation.   Great reliance used to be placed on compositional specifications for materials, and improvements in materials control were sought by increasing the number of elements specified and by decreasing the allowable latitude, eg, maximum, minimum, or range, in their concentration. However, the approach is fallible: the purchaser assumes enough knowledge about materials behavior to completely and unerringly associate the needed properties with composition; analyses must be made by purchaser, vendor, or both for each element specified; and the purchaser bears responsibility for materials failures when compositional requirements have been met. Property requirements alone or in combination with a less exacting compositional specification usually is a more effective approach. For example, the engineer may specify a certain class of low alloy steel and call for a particular hardenability but leave the vendor considerable latitude in determining composition to achieve the desired result.

The most effective specification is that which accomplishes the desired result with the fewest requirements. Properties and performance should be emphasized rather than how the objectives are to be achieved. Excessive demonstration of erudition on the part of the writer or failure to recognize the usually considerable processing expertise held by the vendor results in a lengthy and overly detailed document that generally is counterproductive. Redundancy may lead to technical inconsistency. A requirement that cannot be assessed by a prescribed test method or quantitative inspection technique never should be included in the specifications. Wherever possible, tests should be easy to perform and highly correlatable with service performance. Tests that indicate service life are especially useful. Standard test references, eg, ASTM methods, are the most desirable, and those that are needed should be selected carefully and the numbers of such references should be minimized. To eliminate unnecessary review activity by the would-be complier, the description of a standard test should not be paraphrased or condensed unless the original test is referenced.

Effective specification control often can be established other than through requirements placed on the end use material, ie, the specification may bear on the raw materials, the process used to produce the material, or ancillary materials used in its processing. Related but supplementary techniques are approved vendor lists, accredited testing laboratories, and preproduction acceptance tests.

Economic Aspects

A proper assessment of the costs and benefits associated with standardization depends on having suitable baseline data with which to make a comparison. Several surveys have shown typical dollar returns for the investment in standardization in the range of 5:1–8:1 with occasional claims made for a ratio as high as 50:1.

Savings include reduced costs of materials and parts procurement; savings in production and drafting practice; reduction in engineering time, eg, design, testing, quality control, and documentation; and reduction in maintenance, field service, and in-warranty repairs. In most companies a small number of individuals are authorized to write checks on the company's funds, but a large number of people are permitted to specify materials, parts, processes, services, etc, which just as definitely commit company resources. Furthermore, actions involving specification and standard setting frequently lack adequate control and may not be monitored regularly. Thus awareness, appreciation, and involvement of top management in any industrial standardization program are essential to its success.

The DOD estimates conservatively that materials and process specifications represent almost 1% of the total hardware acquisition costs. The operation of a single ASTM committee dealing with engine coolants has been estimated at over $150,000/yr (2). Costs of generation of a single company specification range from a few hundred to several thousand dollars. The total U.S. cost for material and process specifications is greater than 3×10^8. Because these costs can be so large, it is imperative to ensure that monies are not spent unwisely in the specification and standardization field. Although there are justifiable instances for "specials" or documents intended to fill the needs of an individual company or other institution, savings are usually realized by adopting a standard already established by an organization at a higher hierarchical level, ie, a trade association, or national or international standard.

The ideal specification regards only those properties required to assure satisfactory performance in the intended application and properties that are quantitative and measurable in a defined test. Excessively stringent requirements not only involve direct costs for compliance and test verification, but also constitute indirect costs by restricting the sources of the material. Reducing the margin between the specification and the production target increases the risk that an acceptable product is unjustly rejected because the test procedure gives results that vary from laboratory to laboratory. A particularly effective approach is to recognize within a specification or related set of specifications the different levels of quality or reliability required in different applications. Thus the U.S. military recognizes class A, class B, and class S design allowables where, on the A basis $\geqslant$99% of values are above the designated level with a 95% confidence; on the B basis $\geqslant$90% are above with a 95% confidence; and on the S basis, a value is expected which exceeds the specified minimum (27).

From the customer's point of view, there is an optimal level of standardization. Increased standardization lowers costs but restricts choice. Furthermore, if a single minimal performance product standard is rigorously invoked in an industry, competition in a free market ultimately may lead the manufacturer of a superior product to save costs by lowering his product quality to the level of the standard, thus denying other values to the customer. Again, excessive standardization, especially as applied to design or how the product performance is to be achieved, effectively can limit technological innovation.

Legal Aspects

The increasing incidence of class action suits over faulty performance, the trend toward personal accountability and liability, and the increasing role of consumerism have all affected standardization. Improvement in the technical quality of standards, the involvement of all of the possible stakeholders in standards creation, and endorsement by larger standardizing bodies help to minimize the legal exposure of the individual engineer or company. A particular embodiment of these attitudes is the certification label, ie, a symbol or mark on the product indicating that it has been produced according to the standards of a particular organization. The Underwriters Laboratories seal on electrical equipment is a familiar indication that the safety features of the product in question have met the exacting standards of that group. Similarly, the symbol of the International Wool Secretariat on a fabric attests to the fiber content and quality of that material, and the American Petroleum Institute (API) monogram on piping, fittings, chain, motor-oil cans, and other products carries analogous significance.

Antitrust laws sometimes have been invoked in opposition to the collaborative activities of individual companies or private associations, eg, ASTM, in the development of specifications and standards. Although such activities should not constitute restraint of trade, they must be conducted so that the charge can be refuted. Therefore all features of due process proceedings must be observed. Actions aimed at strengthening the voluntary standards system have begun (28). A recommended national standards policy has been generated by an advisory committee that was initiated by, but is independent of, the ANSI (29). The Federal Office of Management and Budget has issued a circular establishing a uniform policy for federal participation and the use of voluntary standards (30). In general, the circular calls for federal agency participation in the development, production, and coordination of voluntary standards and encourages the use, whenever possible, of applicable voluntary standards in federal procurement. As a result, more internal government standards in agencies, such as the Department of Defense and General Services Administration, are being canceled than are being created. Almost 5000 industry standards have been adopted by DOD, a number that is certain to increase.

Education

Seminars, workshops, and short courses sponsored by professional societies and trade associations provide the needed training in materials standards and specifications. Familiarization with sources of information in the field, how to prepare specifications and standards, how to tailor requirements for cost effectiveness, and the cross-referencing and correlation of specifications and standards are covered.

Trends and Outlook

International Standards.   International trade is increasing rapidly in volume, in complexity, and in its significance to individual national economies. Thus the move toward more extensive adoption of international standards as well as cross-referencing of equivalent national specifications is understandable. Historically international trade was comprised principally of raw materials sold by undeveloped countries to industrialized countries in exchange for manufactured products. This is no longer so. The 1990s U.S.-designed car may be equipped with a German engine and French tires, and be built in part from Japanese steel and Dutch plastics. This composite implies a need for materials standardization accepted on an international level.

International standardization began formally in 1904 with the formation of the International Electrotechnical Commission (IEC) and involves the national committees of more than 40 member nations who represent their countries' interests in electrical engineering, electronics, and nuclear energy. In 1947, the International Organization for Standardization (ISO) was formed to review standardization activities in fields other than electrical. ISO is comprised of more than 80 member countries. Both organizations are autonomous but maintain a coordinating committee to answer jurisdictional questions. Both are located in the same building in Geneva, Switzerland. The activities of the IEC and ISO have increased many fold over the years. The United States is represented in ISO by ANSI and in IEC by a U.S. National Committee that is a part of ANSI. Although the two organizations are dominant in drafting documentary standards, the influence and activities of other international organizations are substantial. Among these are IUPAC, NATO, and Comité Européen de Normalisation (CEN), the European Economic Community (EEC), and the Pan American Standards Commission (COPANT). The CEN is supported by the separate national standards authorities, and its influence has been much strengthened following the 1992 adoption of a European Common Market by the EEC. Another organization, the Committee on Data (CODATA) of the International Council of Scientific Unions, should also be mentioned. CODATA concentrates its attention on the evaluation of data and the methodology of compilation, presentation, manipulation, and dissemination of data in all fields of science and technology. Much of its work consists of appraisal of standard data and standards for presentation of data (31).

The problems of existing materials designation systems and the need for schemes to demonstrate equivalencies (or the lack thereof) have been discussed (32–34). A comprehensive international standard for materials designations is highly desirable, but may only be realizable for such new classes of materials as advanced ceramics (qv) (35). Although the international treaty, the General Agreement on Tariffs and Trade (GATT Code), is intended primarily to prevent it, there is also the hazard of the intentional or unintentional use of standards as technical barriers to trade. The implications of the GATT Code for the predominantly voluntary standard programs used in the United States have been reviewed (36).

Increased requirements for quality and reliability in all products, especially for those of high dollar value and in components of highly integrated technological systems, has led to the formation and broad adoption of the ISO 9000 series of international quality standards for products and services. ISO 9000, which has now been adopted by over 50 countries (37), is actually a series of five integrated standards developed during the 1980s to provide uniform, worldwide quality assurance requirements. ISO 9000 is the road map to the series and also defines key terms; ISO 9001 relates to design and servicing; ISO 9002 to production and installation; ISO 9003 to final inspection and testing; and finally ISO 9004 provides guidance on implementing these standards. As of this writing (ca 1994), there is a trend toward use of third-party registrars to certify that ISO 9000 requirements for a quality system have indeed been implemented and documented.

Increasing concern over the environment and safety issues has led to new standards for exposure of organisms to materials, noise, and electromagnetic radiation (see ENVIRONMENTAL IMPACT; INDUSTRIAL HYGIENE; TOXICOLOGY). The decreasing availability of natural resources forces industry to make use of leaner ores and apply materials that are in short supply more frugally (see MINERAL PROCESSING AND RECOVERY). This usage is expected to result in new analytical standards and compositional specifications. The use of specifications in coping with problems of residual and additive elements in both virgin and recycled materials has been reviewed (38) (see RECYCLING).

Computerization.   The computerization of all aspects of industry and commerce, from management to engineering and manufacturing, and from purchasing to sales, has made it vital to standardize the ways materials information is incorporated into machine-readable systems (see COMPUTER-AIDED DESIGN AND MANUFACTURING; COMPUTER-AIDED ENGINEERING). The designation of materials, the recording of properties, as well as auxiliary information, can all be computerized. More than a dozen standards in this area have been developed by ASTM's Committee E49 who have also prepared a guide to the building of materials databases (qv) (39). A particularly important issue is standardization to facilitate the exchange of digital information. Internationally, work toward this goal is carried out under the aegis of ISO-STEP (standard for the exchange of product data). STEP, known formally as ISO 10303, covers all aspects of information needed to describe manufactured products including shape, product configuration, and process description. Materials are covered in Part 45 which treats material structure, properties, and measurement conditions in such a way that the information is fully integratable with other parts of the STEP standard. Implementation for materials awaits the development of particular application protocols (APs), eg, composite part design or polymer testing. The STEP model has been more fully described (39,40). Status is available from the ISO-STEP secretariat at NIST.

Another standardization matter relative to computerization of materials information is that of terminology (41) (see NOMENCLATURE). Full terminological standardization is not expected to be realized until the twenty-first century, but the hazards of lack of such standardization are exacerbated in computerized systems.

Nonlaboratory Environments.   New technology such as that of fusion energy (qv) introduces demands for standards and specifications of increased quality. Extension of temperature and pressure capabilities in the laboratory and factory demand new accepted standards of calibration. Pressure equipment in the GPa range and the tokamak nuclear fusion apparatus having operating temperatures of 10^6 °C dictate the need, but the state of the art of standards in these fields is far behind such values. Microminiaturization of the active components of electronic equipment and the ability to detect material in picogram quantities requires updated standards for purity (see ULTRAPURE MATERIALS), smoothness, and dimensional and compositional measurement and control within tens of nanometers. Standards and specifications work is expected to affect the biomaterials field, eg, regarding laboratory-created microorganisms (see GENETIC ENGINEERING).

Environments deviating significantly from that of the laboratory, a selection of which is presented in Table 1, yet in which all the usual engineering functions must be performed, also pose problems and opportunities for material standards. Sensors (qv) must measure the attributes of these environments,

Table 1. Characteristics of Various Environments[a]

Space	Ocean	Human body	Nuclear reactor	Laboratory
extreme vacuum	high pressure	moist	high	high temperature
radiation	nearly constant	complex and	temperature	high pressure
nonpenetrating	temperature	diverse	neutron flux	high magnetic
penetrating	saline water	electro-	reactive	fields
temperature	silt and colloidal	chemistry of	coolants	plasmas
ascent	suspensions	various	radioactive	
reentry	marine life	body	sources	
ambient	mechanical	fluids	high thermal	
lack of normal	instability	complex	flux	
gravitational	waves	flexural	inaccessibility	
field	tides	behavior		
micrometeorites	currents	multicomponent		
long-term missions	opaque to EM[b]	composite,		
inaccessibility	radiation	highly damped		
	inaccessibility	in the		
		mechanical		
		and electrical		
		sense		
		multielement		
		constitution of		
		body fluids		
		gases		
		wastes		
		nutrients		
		antibodies		
		hormones		
		enzymes		
		reactive to		
		foreign		
		materials		
		inaccessibility		

[a]Ref. 42.
[b]EM = electromagnetic.

construction materials must withstand the exposure regimes, and performance criteria must be specified.

Units. The SI system of units and conversion factors (qv) has been formally adopted worldwide, with the exception of Brunei, Burma, Yemen, and the United States. The participation of the United States in the metrication movement is evident by the passage of the Metric Acts of 1866 and 1975 and the subsequent establishment of the American National Metric Council (private) and the U.S. Metric Board (public) to plan, coordinate, monitor, and encourage the conversion process.

Sources

There are many hundreds of standards-making bodies in the United States. These comprise branches of state and federal government, trade associations, professional and technical societies, consumer groups, and institutions in the safety and insurance fields. The products of their efforts are heterogeneous, reflecting parochial concerns and different ways of standards development. However, by evolution, blending, and accreditation by higher level bodies, many standards originally developed for private purposes eventually become de facto, if not official, national standards. Individuals seeking access to standards and specifications are referred to the directories listed in References 43–45. Selected organizations, principally from these sources, whose work is especially relevant to chemistry and chemical technology are listed below (see INFORMATION RE-TRIEVAL).

Equipment and Instrumentation Standards

Instrument Society of America
400 Stanwix Street
Pittsburgh, Pa. 15222
Standards Library for Measurement and Control, 12th ed., 1994. Instrumentation standards and recommended practices abstracted from those of 19 societies, the U.S. Government, the Canadian Standards Association, and the British Standards Institute. Covers control instruments, including rotameters, annunciators, transducers, thermocouples, flow meters, and pneumatic systems (see FLOW MEASUREMENT).

American Institute of Chemical Engineers
345 East 47th Street
New York, NY 10017
Standard testing procedures for plate distillation (qv) columns, evaporators, solids mixing equipment, mixing equipment, centrifugal pumps (qv), dryers, absorbers, heat exchangers, etc (see EVAPORATION; HEAT-EXCHANGE TECHNOLOGY; MIXING AND BLENDING).

Scientific Apparatus Makers Association
1140 Connecticut Avenue, NW
Washington, D.C. 20036
Standards for analytical instruments, laboratory apparatus, measurement and test instruments, nuclear instruments, optical instruments, process measurement and control, and scientific laboratory furniture and equipment (see ANALYTICAL METHODS).

General Sources

American National Standards Institute (ANSI)
11 West 42nd Street, 13th floor
New York, NY 10036
ANSI, previously the American Standards Association and the United States of America Standards Institute, is the coordinator of the U.S. federal national standards system and acts by assisting participants in the voluntary

system to reach agreement on standards needs and priorities, arranging for competent organizations to undertake standards development work, providing fair and effective procedures for standards development, and resolving conflicts and preventing duplication of effort.

Most of the standards-writing organizations in the United States are members of ANSI and submit the standards that they develop to the Institute for verification of evidence of consensus and approval as American National Standards. There are ca 11,000 ANSI-approved standards, and these cover all types of materials from abrasives (qv) to zirconium as well as virtually every other field and discipline. Presently, ANSI adopts the standard number of the developing organization, eg, ASTM. ANSI also manages and coordinates participation of the U.S. voluntary-standards community in the work of nongovernmental international standards organizations and serves as a clearinghouse and information center for American National Standards and international standards.

The American Society for Testing and Materials (ASTM)
1916 Race Street
Philadelphia, Pa. 19103
The ASTM *Annual Book of ASTM Standards* contains all up-to-date formally approved (ca 9000) ASTM standard specifications, test methods, classifications, definitions, practices, and related materials, eg, proposals. These are arranged in 15 sections plus an index volume as follows.

Section 1.   Iron and Steel Products (7 vols.)
Section 2.   Non-ferrous Metal Products (5 vols.)
Section 3.   Metals Test Methods and Analytical Practices (6 vols.)
Section 4.   Construction (10 vols.)
Section 5.   Petroleum Products, Lubricants, and Fossil Fuels (5 vols.)
Section 6.   Paints, Related Coatings, and Aromatics (4 vols.)
Section 7.   Textiles (2 vols.)
Section 8.   Plastics (4 vols.)
Section 9.   Rubber (2 vols.)
Section 10.   Electrical Insulation and Electronics (5 vols.)
Section 11.   Water and Environmental Technology (4 vols.)
Section 12.   Nuclear, Solar, and Geothermal Energy (2 vols.)
Section 13.   Medical Devices and Services (1 vol.)
Section 14.   General Methods and Instrumentation (3 vols.)
Section 15.   General Products, Chemical Specialties and End-use Products (1 vol.)
Index   (1 vol.)

Defense Logistics Agency
Defense Industrial Supply Center
700 Robbins Avenue
Philadelphia, Pa. 19111
Publishes *Department of Defense Index of Specifications and Standards*, a monthly with annual accumulations; available from Superintendent of Documents, GPO, Washington, D.C. 20402.

General Services Administration
Federal Supply Service
18th and F Streets
Washington, D.C. 20406
Publishes *Index of Federal Specifications and Standards*, 41CFR 101-29.1.

Global Engineering Documentation Services, Inc.
2805 McGaw Avenue, P.O. Box 19539
Irvine, Calif. 92714
An information broker, not an issuer of standards. The world's largest library of government, industry, and technical society specifications and standards, including obsolete documents dating from 1946. Publishes an annual *Directory of Engineering Documentation Sources*.

MTS Systems Corp.
Box 24012
Minneapolis, Minn. 55424
Publishes *Standards Cross-Reference List* that includes standards issued by one agency but adopted and renumbered or redesignated by another agency, compiled and cross-referenced to aid in their location and identification.

National Institute of Standards and Technology (NIST)
Standards Information Service
Gaithersburg, Md. 20899
Maintains a reference collection on standardization, engineering standards, specifications, test methods, recommended practices, and codes obtained from U.S., foreign, and international standards organizations. Publishes various indexes and directories; for example the *Directory of International and Regional Organizations Conducting Standards-Related Activities* (NIST SP 767), and *Standards Activities of Organizations in the United States* (NIST SP 806). Copies are not available from NIST but from NTIS or Global Engineering.

National Standards Association, Inc.
5161 River Road
Bethesda, Md. 20816
Maintains a standards and specifications database for online searching of government and industry standards, specifications, and related documents (see DATABASES).

National Technical Information Service (NTIS)
5285 Port Royal Road
Springfield, Va. 22161
A for-profit organization, spun off from the federal government, that acts as an archiving and distribution agency for public sale of technical information emanating from a wide variety of government agencies and contractually supported technical programs, as well as for some foreign technical reports and other analyses prepared by national and local government agencies, their contractors, or grantees. A bibliographic database of >1.4 million titles is maintained.

Visual Search Microfilm Files (VSMF)
Information Handling Services
15 Inverness Way East
Englewood, Colo. 80150

VSMF carries government specifications, ASTM, AMS, and many other specifications and standards. Copies of these may be obtained on an individual basis or broad categories of this service may be obtained on a subscription basis.

Journals

ANSI Reporter and Standards Action
American National Standards Institute

The monthly *ANSI Reporter* provides news of policy-level actions on standardization taken by ANSI, the international organizations to which it belongs, and the government. *Standards Action*, biweekly, lists for public review and comment standards proposed for ANSI approval. It also reports on final approval actions on standards, newly published American National Standards, and proposed actions on national and international technical work. These two publications replace *The Magazine of Standards* which ANSI, formerly The American Standards Association, discontinued in 1971.

ASTM Standardization News (formerly *Materials Research and Standards* and, earlier, *ASTM Bulletin*)
American Society for Testing and Materials

A monthly bulletin which covers ASTM projects, national and international activities affecting ASTM, reports of new relevant technology, and ASTM letter ballots on proposed standards.

Journal of the American Society of Safety Engineers
American Society of Safety Engineers
850 Busse Hwy.
Park Ridge, Ill. 60068
A monthly that reviews safety standards.

Journal of Research of the National Institute of Standards and Technology
The journal is published in four parts: (*1*) physics and chemistry, (*2*) mathematics and mathematical physics, (*3*) engineering and instrumentation, and (*4*) radio science.

Journal of Physical and Chemical Reference Data
American Chemical Society
1155 16th Street, NW
Washington, D.C. 20036: quarterly

Journal of Testing and Evaluation
American Society for Testing and Materials

A bimonthly in which data derived from the testing and evaluation of materials, products, systems, and services of interest to the practicing engineer are presented. New techniques, new information on existing methods, and new data are emphasized. It aims to provide the basis for new and improved standard methods and to stimulate new ideas in testing.

Metrologia
International Committee of Weights and Measures (CIPM)
Pavillon de Breteuil
Parc de St. Cloud, France
Includes articles on scientific metrology worldwide, improvements in measuring techniques and standards, definitions of units, and the activities of various bodies created by the International Metric Convention.

Standards Engineering
Standards Engineering Society
6700 Penn Avenue South
Minneapolis, Minn. 55423
A bimonthly in which general news and technical articles dealing with all aspects of standards and U.S. and foreign articles on standard materials and calibration and measurement standards are presented.

Materials

Abrasives
Abrasives Engineering Society
1700 Painters Run Road
Pittsburgh, Pa. 15243

Biochemical Compounds
National Research Council
Committee on Biological Chemistry
National Academy of Science
Washington, D.C. 20418
Specifications and Criteria for Biochemical Compounds

Carbides
Cemented Carbide Producers Association
712 Lakewood Center North
Cleveland, Ohio 44107
Standards Developed by Cemented Carbide Producers Association, ie, standard shapes, sizes, grades, and designations and defect classification.

Castings
Investment Casting Institute
8521 Clover Meadow
Dallas, Tex. 75243

American Die Casting Institute
2340 Des Plaines Ave.
Des Plaines, Ill. 60018

American Foundrymen's Society
Gulf and Wolf Roads
Des Plaines, Ill. 60016

Steel Founders Society of America
455 State Street
Des Plaines, Ill. 60016

Cement and Concrete
 Cement Statistical and Technical Association
 Malmo
 Sweden

 American Concrete Institute
 P.O. Box 19150
 Detroit, Mich. 48219

Ceramic Tile
 Methods and Materials Standards Association
 c/o H. B. Fuller Co.
 315 South Hicks Road
 Palatine, Ill. 60067

Chemicals
 Chemical Manufacturers Association
 2501 M Street, NW
 Washington, D.C. 20037
 Manual of Standard and Recommended Practice for chemicals, containers,
tank car unloading, and related procedures.

 Chemical Specialties Manufacturers Association
 1001 Connecticut Avenue, NW
 Washington, D.C. 20036
 Standard Reference Testing Materials for insecticides (see INSECT CONTROL
TECHNOLOGY), cleaning products, sanitizers, brake fluids, corrosion inhibitors
(see CORROSION AND CORROSION CONTROL), antifreezes, polishes, and floor waxes.

Color
 Color Association of the United States
 24 East 39th Street
 New York, NY 10016
 Color standards for fabrics, paints, wallpaper, plastics, floor coverings,
automotive and aeronautical materials, china, chemicals, dyestuffs, cosmetics,
etc.

 American Association of Textile Chemists and Colorists
 P.O. Box 12215
 Research Triangle Park, NC 27709

Inter-Society Color Council
U.S. Army Natick R&D Center
Att: STRNC-ITC
Natick, Mass. 01760

Friction Materials

Friction Materials Standards Institute
E210, Route 4
Paramus, NJ 07652

Leather

Tanners' Council of America
2501 M Street, NW
Washington, D.C. 20037

American Leather Chemists Association
c/o University of Cincinnati
Cincinnati, Ohio 45221
Chemical and physical test methods for leather (qv).

Metals and Alloys

Aluminum Association
900 19th Street, NW
Washington, D.C. 20006

Standards for wrought and cast aluminum and aluminum alloy products, including composition, temper designation, dimensional tolerance, etc.

Society of Automotive Engineers (SAE)
400 Commonwealth Drive
Warrendale, Pa. 15096

SAE Handbook, an annual compilation of more than 500 SAE standards, recommended practices, and information reports on ferrous and nonferrous metals, nonmetallic materials, threads, fasteners, common parts, electrical equipment and lighting for motor vehicles and farm equipment, power-plant components and accessories, passenger cars, trucks, buses, tractor and earth-moving equipment, and marine equipment.

AMS Index, a listing of more than 1000 SAE Aerospace Material Specifications (AMS) on tolerances; quality control and process; nonmetallics; aluminum, magnesium, copper, titanium, and miscellaneous nonferrous alloys; wrought carbon steels; special-purpose ferrous alloys; wrought low alloy steels; corrosion- and heat-resistant steels and alloys; cast-iron and low alloy steels; accessories, fabricated parts, and assemblies; special property materials; refractory and reactive materials.

Copper Development Association
P.O. Box 1840
Greenwich, Conn. 06836

Standards for wrought and cast copper and copper alloy products; a standards handbook is published with tolerances, alloy data, terminology, engineering data, processing characteristics, sources and specifications cross-indexes for six coppers and 87 copper-based alloys that are recognized as standards.

Tin Research Institute
1353 Perry Street
Columbus, Ohio 43201

Zinc Institute
292 Madison Avenue
New York, NY 10017

Lead Industries Association
292 Madison Avenue
New York, NY 10017

American Iron and Steel Institute
1123 15th Street, NW
Washington, D.C. 20005
Standards for steel compositions, steel products, manufacturing tolerances, inspection methods, etc.

Ferroalloys Association
1612 K Street, NW
Washington, D.C. 20006

Metal Powder Industries Federation
P.O. Box 2054
Princeton, NJ 08540

Gold Institute
1001 Connecticut Avenue, NW
Washington, D.C. 20036

Silver Institute
1001 Connecticut Avenue, NW
Washington, D.C. 20036

Paper
Technical Association of the Pulp and Paper Industry
Technology Park/Atlantic
P.O. Box 105113
Atlanta, Ga. 30348
TAPPI Standards and *TAPPI Yearbook* cover all aspects of pulp (qv) and paper (qv) testing and associated standards.

American Paper Institute
260 Madison Avenue
New York, NY 10016
Physical standards, sizes, gauges, definitions of paper and paperboard.

Petroleum Products
American Petroleum Institute
1220 L Street, NW
Washington, D.C. 20005
Fosters development of standards, codes, and safe practices in petroleum industries and publishes the same in its journals and reference publications.

Plastics
Society of the Plastics Industry
355 Lexington Avenue
New York, NY 10017

Refractories
Refractories Institute
Suite 1517
301 Fifth Avenue
Pittsburgh, Pa. 15222

Steam
International Association for Properties of Steam
National Institute of Standards and Technology
Gaithersburg, Md. 20899

Textiles
International Bureau for Standardization of Man-made Fibers
Lautengartenstrasse 12
CH-4010 Basle
Switzerland

Treating and Finishing
Metal Treating Institute
1311 Executive Center, Suite 200
Tallahassee, Fla. 32301

National Association of Metal Finishers
111 E. Wacker Drive
Chicago, Ill. 60601

Welding
American Welding Society
P.O. Box 351040
550 NW Le Jeune Road
Miami, Fla. 33135

Codes, Standards and Specifications, a complete set of codes, standards, and specifications published by the Society and continuously updated. Covers fundamentals, training, inspection and control, and process and industrial applications.

Wood

American Lumber Standards Committee
20010 Century Blvd.
Germantown, Md. 20767

American Wood Preservers Bureau
2772 S. Randolph Street
Arlington, Va. 22206

National Hardwood Lumber Association
332 S. Michigan Avenue
Chicago, Ill. 60604

National Standards, Worldwide.   Most countries have a national standards organization that both leads the standardization activities in that country and acts within its own country as sales agent and information center for the other national standardizing bodies. In the United States, the ANSI performs that function. The organizations for the leading industrial countries of the world are as follows.

Australia: SAA, AS, Standards Association of Australia, 80–86 Arthur Street, North Sidney NSW 2060.

Austria: ON, ONORM, Oesterreichlisches Normungsinstitut, Leopoldsgasse 4, A-1021 Wein 2.

Belgium: IBN, Institut Belge de Normalisation, 29 Avenue de la Braban-conne B-1040 Bruxelles 4.

Brazil: ABNT, NB, EB, Associacao Brasileira de Normas Tecnicas, Caixa Postal 1680, Rio de Janeiro.

Canada: CSA, Standards Council of Canada, 2000 Argentia Road, Suite 2-401, Mississauga, Ontario.

China: China Association for Standardization, PO Box 820, Beijing, People's Republic of China.

Czechoslovakia: Urad pro normalizaci a mereni, Vaclavske namesti 19, 113-47 Praha 1, Czechoslovakia.

Denmark: DS, Dansk Standardiseringsraad, Aurehjvej 12, DK-29000, Hellerup.

Finland: SFS, Suomen Standardisoimisliitto, Box 205 SF-00121 Helsinki 12.

France: AFNOR, NF, Association Francaise de Normalisation, Tour Europe, Cedex 7, 92080 Paris-La Defense.

Germany: DIN, Deutsches Institut fur Normung, 4-10 Burggrafenstrasse, D-1000 Berlin 30.

India: ISI, IS, Indian Standards Institution, Manak Bhavan, 9 Bahadur Shah Zafar Marg, New Delhi 110002.

Iran: Institute of Standards and Industrial Research of Iran, Ministry of Industries and Mines, PO Box 2937, Tehran.

Ireland: IIRS, I.S., Institute for Industrial Research and Standards, Blasnevin House, Ballymun Road, Dublin-9.

Israel: SII, Standards Institution of Israel, 42 University Street, Tel Aviv 69977.

Italy: UNI, Ente Nazionale Italiano de Unificazione, Piazza Armando Diaz 2, 120123 Milano.

Japan: JISC, JIS, Japanese Industrial Standards Committee, Agency of Industrial Science and Technology, Ministry of International Trade and Industry, 1-3-1 Kusumigaseki Chiyoda-Ku, Tokyo 100.

Mexico: DGN, Diraccion General de Normas, Tuxpan No. 2, Mexico 7 DF.

Netherlands: NNI, Nederlands Normalisatie-instituut, Polakweg, 5 Rijswijk (ZH)-2280.

Poland: Polski Komitet Normalizacji Miar i Jakosci, Ul. Elektoraina 2, 00-139 Warszawa.

Romania: Institutul Roman de Standardizare, Casuta Postala 63-87, Bucarest 1.

Russia: GOST, Gosudarstvennyj Komitet Standartov, Leninsky Prospekt 9b, Moskva 11 7049.

Spain: Instituto Nacional de Racionalizacion y Normalizacion, Aurbano 46, Madrid 10.

Sweden: SIS, Standardiseringskommission i Sverige, Tegnergatan 11, Box 3295, Stockholm S 10366.

United Kingdom: BSI, BS, British Standards Institution, 2 Park Street, London W1 A 2BS, England.

Nuclear Standards

American Nuclear Society (ANS)
555 N. Kensington Avenue
La Grange Park, Ill. 60525

American National Standards Institute
11 West 42nd Street, 13th floor
New York, NY 10036

National Institute of Standards and Technology
Index of U.S. Nuclear Standards
W. I. Slattery, ed.
National Bureau of Standards, Special Pub. 483 (1977)
Washington, D.C.

Safety Standards

The American Society of Mechanical Engineers (ASME)
United Engineering Center
345 East 47th Street
New York, NY 10017

The ASME Boiler and Pressure Vessel Code, under the cognizance of the ASME Policy Board, Codes, and Standards, considers the interdependence of design procedures, material selection, fabrication procedures, inspection, and test methods that affect the safety of boilers, pressure vessels, and nuclear-plant components, whose failures could endanger the operators or the public (see NUCLEAR REACTORS). It does not cover other aspects of these topics that affect operation, maintenance, or nonhazardous deterioration.

American Insurance Association (AIA)
85 John Street
New York, NY 10038
Handbook of Industrial Safety Standards, Association of Casualty and Surety Companies, New York, 1962. Compilation of industrial safety requirements based on codes and recommendations of the ANSI, the National Fire Protection Association (now part of AIA), the ASME, and several government agencies.

National Fire Protection Association
470 Atlantic Avenue
Boston, Mass. 02210
National Fire Codes, 1987 ed., issued in 11 volumes. One volume is devoted exclusively to hazardous chemicals, but most other volumes have some coverage of material hazards, use of materials in fire prevention or extinguishing, hazards in chemical processing, etc. More than 200 standards are described.

American Society of Safety Engineers
850 Busse Highway
Park Ridge, Ill. 60068

American Public Health Association
1015 18th Street, NW
Washington, D.C. 20036
Standard Methods for Examination of Water and Wastewater, 16th ed., 1985; *Methods of Air Sampling and Analysis*, 3rd ed., 1988.

National Safety Council
444 North Michigan Avenue
Chicago, Ill. 60611
Industrial safety data sheets on materials and materials handling and safe operation of equipment and processes.

Underwriters Laboratories
333 Pfingsten Road
Northbrook, Ill. 60062
Standards for Safety is a list of more than 200 standards that provide specifications and requirements for construction and performance under test and in actual use of a broad range of electrical apparatus and equipment, including household appliances, fire-extinguishing and fire protection devices

and equipment, and many other nongenerally classifiable items, eg, ladders, sweeping compounds, waste cans, and roof jacks for trailer coaches.

Safety Standards
U.S. Department of Labor
GPO
Washington, D.C. 20402
Industrial safety hazards.

American Conference of Governmental Industrial Hygienists
P.O. Box 1937
Cincinnati, Ohio 45201
Practices, analytical methods, guides to codes and/or regulations, threshold limit values.

Factory Mutual Engineering Corp.
1151 Boston-Providence Turnpike
Norwood, Mass. 02062
Standards for safety equipment, safeguards for flammable liquids, gases, dusts, industrial ovens, dryers, and for protection of buildings from wind and other natural hazards.

Code of Federal Regulations
Title 49, Transportation, Parts 100 to 199
Superintendent of Documents
GPO
Washington, D.C. 20402
Safety regulations related to transportation of hazardous materials and pipeline safety.

Code of Federal Regulations
Title 29, Occupational Safety and Health
Superintendent of Documents
GPO
Washington, D.C. 20402
Safety regulations and standards issued by OSHA.

Code of Federal Regulations
Title 40, Environmental Protection Administration
Superintendent of Documents
GPO
Washington, D.C. 20402
Safety regulations and standards issued by the U.S. EPA.

Code of Federal Regulations
Title 21, Radiological Health
Superintendent of Documents
GPO
Washington, D.C. 20402

Weights and Measures

National Conference on Weights and Measures
c/o National Institute of Standards and Technology
Gaithersburg, Md. 20899

National Conference on Standards Laboratories
c/o National Institute of Standards and Technology
Boulder, Colo. 80303

U.S. Metric Association
Sugarload Star Route
Boulder, Colo. 80302

U.S. Metric Board
1815 N. Lynn Street
Arlington, Va. 22209

American National Metric Council
5410 Grosvenor Lane
Bethesda, Md. 20814

Metrology and Fundamental Constants
A.F. Milone and P. Giacomo, eds.
North Holland Publishing Co.
Amsterdam, the Netherlands, 1980
The proceedings of an international course that was organized to review metrology comprehensively and to illustrate links between metrology and the fundamental constants. Status of research is presented and future work and priorities are outlined.

International Bureau of Weights and Measures
Pavillion de Breteuil
F-92310
Sevres, France

International Standards

The role of the U.S. government in international standardization activities has been examined by a special ASTM task force (46). The addresses of these organizations or their subsidiary standards groups are as follows.

International Bureau of Weights and Measures
Pavillon de Breteuil
F-92310, Sevres, France

International Electrotechnical Commission (IEC)
1 rue de Varembe
1211 Geneve 20
Switzerland

International Organization for Standardization (ISO)
1 rue de Varembe
CH 1211, Geneve 20
Switzerland

North Atlantic Treaty Organization (NATO)
Military Committee
Conference of National Armaments Directors
1110 Brussels, Belgium

European Committee for Standardization (CEN)
rue Brederode 2, Bte 5
1000 Brussels, Belgium

European Economic Community (EEC)
200 rue de la Loi
1049 Brussels, Belgium

COPANT
Avenue Pte. Roque Soenz Pena 501
7 Piso
OF 716, Buenos Aires
Argentina

CODATA
51 Boulevard de Montmorency
75016 Paris, France

International Union for Pure and Applied Chemistry (IUPAC)
Bank Court Way, Cowley Centre
Oxford OX4 3YF
United Kingdom
Among its publications in the standards field are *Manual of Symbols and Terminology for Physico-chemical Quantities and Units*, D. H. Whiffen, ed., Pergamon, New York, 1979, and *Nomenclature of Inorganic Chemistry*, Pergamon, New York, 1977.

Directories and Cross-References

A variety of directories and cross-references to standards and specifications are available worldwide (47–70).

BIBLIOGRAPHY

"Materials Standards and Specifications" in *ECT* 3rd ed., Vol. 15, pp. 32–61, by J. H. Westbrook, General Electric Co.

1. *ISO Standardization Vocabulary*, Geneva, Switzerland, 1977.
2. N. E. Promisel and co-workers, *Materials and Process Specifications and Standards*, NMAB Report 33, Washington, D.C., 1977.
3. D. Lebel and K. Schultz, *Technos*, **4** (Apr./June 1975).
4. G. A. Uriano, *ASTM Standard. News* **7**, 8 (Sept. 1979).
5. *NIST Standard Reference Materials Catalog*, Superintendent of Documents, U.S. Government Printing Office, Washington, D.C.
6. G. W. Latimer, Jr., in C. T. Lynch, ed., *Handbook of Materials Science*, Vol. 1, CRC Press, Boca Raton, Fla., 1974, p. 667.
7. F. J. Flanagan, *Geochim. Cosmochim. Acta* **37**, 1189 (1973).
8. R. H. Steiger and E. Jaeger, *Planet. Sci. Lett.* **36**, 359 (1977).
9. C-Y. Ho and Y. S. Touloukian, *Proceedings of the 5th Biennial International CODATA Conference*, Boulder, Colo., 1977, pp. 615–627.
10. H. Wawrousek, J. H. Westbrook, and W. Grattidge, *Data Sources of Mechanical and Physical Properties of Engineering Materials*, Physik Daten No. 30-1, Fachinformationszentrum-Karlsruhe, Germany, 1989.
11. H. Wawrousek, J. H. Westbrook, and W. Grattidge, in J. G. Kaufman and J. S. Glazer, eds., *Computerization and Networking of Materials Databases*, Vol. 2, ASTM STP 1106, ASTM, Philadelphia, Pa., 1991, p. 142.
12. H. Berger, *Non-destructive Testing Standards—A Review*, ASTM STP 624, ASTM, Philadelphia, Pa., 1977.
13. *Handbook for Standardization of Nondestructive Testing Methods*, Vols. 1 and 2, MIL. HDBK-33, Dept. of Defense, Washington, D.C., 1974; R. E. Englehardt, "Bibliography of Standards, Specifications and Recommended Practices," in *Nondestructive Testing Information Analysis Center Handbook*, Nondestructive Testing Information Analysis Center, Texas Research Institute, Austin, Tex., Mar. 1979, p. 212.
14. D. P. Thompson and D. E. Chimenti, eds., *Review of Progress in Quantitative NDE*, Vol. 8, Plenum Press, New York, 1989; "Nondestructive Testing and Quality Control," *Metals Handbook*, 9th ed., Vol. 17, ASM, Metals Park, Ohio, 1989; *Non-destructive Testing Handbook*, American Society for Non-destructive Testing, Columbus, Ohio, 1986.
15. B. C. Belanger, *ASTM Standard. News*, 8 (Sept. 1979).
16. E. R. Cohen and B. N. Taylor, *J. Res. NBS* **92**, 85 (1987).
17. E. R. Cohen and B. N. Taylor, "The 1986 Adjustment of the Fundamental Physical Constants," *CODATA Bull.* (63) (1986).
18. I. Mills, ed., *Quantities, Units, and Symbols in Physical Chemistry*, Blackwell Scientific Publishing, London, 1988.
19. *ISO Standards Handbook 2: Units of Measurement*, ISO, Geneva, Switzerland, 1982.
20. *Standard Practice for Use of the International System of Units (The Modernized Metric System)*, ASTM E380-93, ASTM, Philadelphia, Pa., 1993.
21. *Pure Appl. Chem.* **64**, 1519 (1992).
22. *Metrologia* **27**, 3 (1990).
23. F. P. Bundy and co-workers, in B. D. Timmerhaus and M. S. Barber, eds., *High Pressure Science and Technology*, Vol. 1, Plenum Press, New York, 1979, pp. 773 and 805.
24. H. K. Mao, P. M. Bell, J. W. Shaver, and D. J. Steinberg, *J. Appl. Phys.* **49**, 3276 (1978).

25. *Proceedings of the General Conference on Weights and Measures, International Bureau of Weights and Measures*, BIPM, Parc de Saint-Cloud, France, 1991.
26. P. Vanýsek, in D. R. Lide, ed., *Handbook of Chemistry and Physics, 1993–1994*, CRC Press, Boca Raton, Fla., pp. 8–21.
27. "Metallic Materials and Elements for Aerospace Vehicle Structures," *Military Handbook 5-F*, Dept. of Defense, Washington, D.C., 1991.
28. *The Voluntary Standards System of the United States of America—An Appraisal by the American Society for Testing and Materials*, ASTM, Philadelphia, Pa., 1975.
29. *ASTM Standard. News* **16**, 8 (May 1978); *Fed. Reg.*, 7 (Dec. 1978).
30. "Federal Participation in Voluntary Standards," OMB Circular A-119, *ASTM Standardization News*, Mar. 1980, p. 21.
31. S. A. Rossmassler and D. G. Watson, eds., *Data Handling for Science and Technology*, North Holland, Amsterdam, the Netherlands, 1980.
32. J. H. Westbrook, in J. S. Glazman and J. R. Rumble, Jr., eds., *Computerization and Networking of Materials Data Bases*, Vol. 1, ASTM STP 1017, ASTM, Philadelphia, Pa., 1989, p. 23.
33. K. W. Reynard, in Ref. 11, p. 57.
34. K. W. Reynard, in T. I. Barry and K. W. Reynard, eds., *Computerization and Networking of Materials Databases*, Vol. 3, ASTM STP 1140, ASTM, Philadelphia, Pa., 1992, p. 413.
35. *ISR/VAMAS Unified Classification Scheme for Advanced Ceramics*, ASTM, Philadelphia, Pa., 1993.
36. D. L. Peyton, *Implementing the GATT Standards Code*, ANSI, New York, 1979.
37. M. Jenkins, *ASTM Standard. News*, 50 (July 1993); *ISO 9000, International Standards for Quality Management*, 5th ed., ISO, Geneva, Switzerland, 1994.
38. J. H. Westbrook, *Phil. Trans. Roy. Soc.* **A295**, 25 (1980).
39. C. H. Newton, ed., *Manual on the Building of Materials Databases*, ASTM MNL 19, ASTM, Philadelphia, Pa., 1993.
40. J. R. Rumble, Jr., in Ref. 34, p. 141.
41. J. H. Westbrook and W. Grattidge, in Ref. 34, p. 15.
42. J. H. Westbrook, in A. B. Bronwell, ed., *Science and Technology in the World of the Future*, John Wiley & Sons, Inc., New York, 1970, p. 329.
43. S. J. Chumas, *Standards Activities of Organizations in the United States*, NBS SP 681, NBS, Gaithersburg, Md., 1984.
44. E. J. Struglia, *Standards and Specifications—Information Sources*, Gale Research, Detroit, Mich., 1973.
45. Technical data, Technical Indexes, Ltd., Bracknall, U.K., updated monthly.
46. *ASTM Standard. News* **8**, 16 (Apr. 1980).
47. *British and Foreign Specifications for Steel Castings* (1980) Steel Casting Research and Trade Association, Sheffield, U.K., 1980.
48. *Department of Defense Index of Specifications and Standards*, Superintendent of Documents, U.S. Government Printing Office, Washington, D.C., published annually.
49. *European Committee for Iron and Steel Standardization (ECISS), Index of Standards*, PSP2, British Steel, plc., Head Office Standards, London, U.K., 1986.
50. J. P. Frick, *Woldman's Engineering Alloys*, 8th ed. ASM International, Materials Park, Ohio, 1994.
51. *German Standards*, Beuth, Germany, English translation of more than 4750 popular DIN Standards plus 1500 DIN-EN, DIN-EC and DIN-ISO Standards.
52. *Handbook of Comparative World Steel Standards*, International Technical Information Institute, Tokyo, Japan, 1990.
53. H. Hucek and M. Wahl, *1990 Handbook of International Alloy Compositions and Designations*, Vol. 1, *Titanium*, MCIC HB-09, Metals and Ceramics Information Center, Battelle Columbus Laboratories, Columbus, Ohio, 1990.

54. W. Hufnagel, ed., *Key to Aluminium* (in German), 4th ed., Aluminium AG, Dusseldorf, Germany, 1991.

55. *Index of Aerospace Materials Specifications*, Society of Automotive Engineers, Warrendale, Pa., 1986.

56. *Iron and Steel Specifications*, 7th ed., British Steel plc., Head Office Standards, London, 1989. Steel grades to BS970 together with French, German, Japanese and Swedish standards.

57. M. Kehler, *Handbook of International Alloy Compositions and Designations*, Vol. 3, *Aluminium*, Aluminium Verlag, Heyden, Germany, 1981.

58. *Metals and Alloys in the Unified Numbering System*, 6th ed., ASTM DS-56E, ASTM, Philadelphia, Pa., 1993.

59. D. L. Potts and J. G. Gensure, eds., *International Metallic Materials Cross-Reference*, 3rd ed., Genium Publishing Co., Schenectady, NY, 1988.

60. K. W. Reynard, ed., *Inventory of Materials Designation Systems*, Versailles Advanced Materials and Standards, TWA-10, National Physical Laboratory, Teddington, U.K., 1992.

61. P. L. Ricci and L. Perry, *Standards: A Resource and Guide for Identification and Acquisition*, Stirz, Minneapolis, Minn., 1991.

62. R. B. Ross, ed., *Metallic Materials Specification Handbook*, 4th ed., Chapman and Hall, London, 1992.

63. H. Schmitz, *Stahl-Eisen-Listen / Steel-Iron Lists*, 7th ed., Verein Deutscher Eisenhuttenleute (VDEh), Dusseldorf, Germany, 1987. 2500 German and 1000 foreign materials covered, also in a computerized database.

64. W. F. Simmons and R. B. Gunia, *Compilation and Index of Trade Names, Specifications, and Producers of Stainless Alloys and Superalloys*, ASTM Publication, DS-45, ASTM, Philadelphia, Pa., 1972.

65. *Standards Cross-Reference List*, 2nd ed., MTS Corp., Minneapolis, Minn., 1977.

66. P. M. Unterweiser, ed., *Worldwide Guide to Equivalent Irons and Steels*, 2nd ed., ASM International, Materials Park, Ohio, 1987.

67. P. M. Unterweiser, *Worldwide Guide to Equivalent Nonferrous Metals and Alloys*, 2nd ed., ASM International, Materials Park, Ohio, 1987.

68. C. W. Wegst, *Key to Steel*, 17th ed., Verlag Stahlschussel, Marbach/Neckar, Germany, 1995.

69. *World Metal Index*, Sheffield City Libraries, Sheffield, U.K. A service covering 70,000 standard grades of metallic materials worldwide.

70. *World Steel Standard Specifications*, 1974, Foreign Technology Division, U.S. Airforce, FTD-NC-23-856-74.

JACK H. WESTBROOK
Brookline Technologies

MEAT PRODUCTS

The new dietary guidelines for Americans and the new food guide pyramid issued by the U.S. Department of Agriculture (USDA) and the Department of Health and Human Services (DHHS) recommend a diet low in fat, saturated fat, and cholesterol. Following the guidelines does not mean omitting animal products from diets. In the United States the per capita consumption of meat, poultry, and fish (boneless trimmed equivalent) combined has steadily increased from 77.7 kg (170.9 lbs) in 1975 to 83.6 kg (184.0 lbs) in 1991 (1). Meat is not only a flavorful product, but it also provides protein and essential minerals and vitamins, especially B vitamins (2). Meat consumption varies with social, economic, political, and geographical differences on a worldwide basis. The meat production from other countries (3) is listed in Table 1 and the consumption of meat and poultry in the 10 leading countries in 1991 (1) is listed in Table 2.

In the United States, red meat production remains steady while poultry production is growing more rapidly (Fig. 1). Beef consumption reached its highest point in 1976 (40.4 kg) and subsequently decreased to 32.8 kg in 1980 and to 28.9 kg in 1991 (Fig. 2). Initially, this decrease was a result of consumers turning to pork as an alternative. After declining steadily through the 1980s, per capita pork consumption turned upward again in 1987. However, the rise was short-lived as consumption decreased again in 1990. The real winners in the shift away from beef have been poultry products. Since 1970, per capita poultry consumption has increased from 15.4 to 25.8 kg in 1991. The shift in consumption patterns is due to the fact that consumers are becoming more health-conscious and some media or popular press articles have labeled red meat as bad for health and longevity.

Income is also an important factor affecting demand for meat. Demand generally increases with higher income, but consumption tends to level off and may even decline at the highest incomes. Increasing incomes also change the types of meat demanded. Lower incomes may lower meat consumption or bring about a switch to lower priced meats. As consumers eat more food away from home, the markets for fast foods such as fried chicken and hamburgers also increase (4).

This article discusses several aspects of processed meat products including (1) health and safety concerns; (2) meat processing ingredients, procedures, and machinery; (3) hazard analysis critical control point; (4) fat reduction in meat products; (5) sous-vide processing; and (6) nutritional labeling.

Health and Safety Concerns

Fat Intake.   Consumers have been warned that a diet high in fat increases the probability of chronic health problems and diseases including coronary heart disease (CHD). It seems the message is getting through as indicated by increasing public awareness on the link between CHD and high fat intake (5). Unfortunately, consumers often equate animal fat with saturated fat. This is misleading because there is no fat that is 100% saturated. Fat always consists of different proportions of saturated and unsaturated fatty acids. Pork fat (lard) and beef fat (tallow) have about 40 and 43% saturated fatty acids, respectively. In fact,

Table 1. Meat Production, 10³ t in Specified Countries in 1990[a]

Country	Beef and veal	Pork[b]	Mutton, lamb, and goat meat	Total production
Argentina	2,650	0[c]	100	2,750
Australia	1,695	305	666	2,666
Austria	212	405	0	617
Belgium-Luxembourg	326	759	7	1,092
Brazil	4,180	1,050	0	5,230
Bulgaria	137	422	77	636
Canada	922	1,140	0	2,062
China	1,250	22,700	1,090	25,040
Colombia	799	118	0	917
Costa Rica	82	0	0	82
Czechoslovakia	498	962	10	1,470
Denmark	200	1,200	2	1,402
Dominican Republic	51	14	0	65
Egypt	408	0	81	489
El Savador	28	0	0	28
Finland	110	178	0	288
France	1,710	1,870	160	3,740
German, unified	2,121	3,811	50	5,982
Greece	80	150	129	359
Guatemala	59	14	0	73
Honduras	23	0	0	23
Hong Kong	0	23	0	23
Hungary	101	931	4	1,036
India	670	0	572	1,242
Ireland	486	157	85	728
Israel	39	0	0	39
Italy	1,180	1,280	83	2,543
Japan	545	1,560	0	2,105
Korea, South	122	440	1	563
Mexico	1,790	792	76	2,658
Netherlands	540	1,672	14	2,226
New Zealand	470	43	499	1,012
Panama	57	0	0	57
Philippines	138	665	0	803
Poland	799	1,814	28	2,641
Portugal	123	218	27	368
Romania	212	620	70	902
Saudi Arabia	28	0	96	124
Singapore	0	76	0	76
South Africa	670	0	212	822
Spain	450	1,738	240	2,428
Sweden	146	298	0	444
Switzerland	160	275	0	435
Taiwan	5	1,000	0	1,005
Turkey	310	0	355	665
Soviet Union[d]	8,700	6,800	1,000	16,500
United Kingdom	997	980	371	2,348
United States	10,464	6,964	165	17,593

Table 1. (*Continued*)

Country	Beef and veal	Pork[b]	Mutton, lamb, and goat meat	Total production
Uruguay	349	0	74	423
Venezuela	345	110	0	455
Yugoslavia	302	771	69	1,142
Total	*47,739*	*64,325*	*6,413*	*118,477*

[a]Ref. 3. Carcass-weight basis, excludes offals, rabbit, and poultry meat.
[b]Includes edible pork fat, but excludes lard and edible greases (except United States).
[c]Less than 1000 metric tons.
[d]Slaughter weight basis, includes fats and offals.

the levels of saturated fatty acids in animal fats are similar to the amounts of saturated fatty acids in many commercial hydrogenated vegetable fats used for shortenings and margarines (3) (see FATS AND FATTY OILS). The misleading designation saturated fat has misinformed the general public; consequently, consumers may eat less meat in order to prevent CHD, cancer, and other illnesses linked to meat in the diet. More recent recommendations suggest that regular consumption of a moderate amount of lean meats is a healthful practice (6–8).

The American Dietetic Association, the American Heart Association, and the National Heart, Lung and Blood Institute recommend 142–198 g (5–7 oz) of lean, trimmed meat daily. It was also pointed out that trimmed meat, especially red meat, provides large amounts of essential nutrients such as iron, zinc, vitamin B_{12}, and balanced protein. The idea that the risk of CHD and cancer can be greatly reduced by avoiding a meat-centered diet have prompted some consumer groups to demand healthy meat products. In response, meat producers began to produce leaner beef with the use of growth hormones, and meat processors developed various types of low fat meat products (9).

Growth Promotants. Livestock can be exposed to many chemicals used to promote growth, improve feed utilization, or enhance meat acceptability. In the late 1960s, the greatest concern to the public was diethylstilbestrol [56-53-1] (DES), a synthetic estrogen used to promote weight gain in cattle. This became a focus of attention when residues of DES were occasionally detected in beef livers. In the 1990s DES is known to be carcinogenic and associated with reproductive disorders in humans when administered in high doses, and its use to promote weight gains in livestock has been banned in the United States (10). Since the early 1980s, bovine somatotropin [66419-50-9] (BST) and porcine somatotropin [9061-23-8] (PST) have been extensively studied. Somatotropin [9002-72-6] is a growth hormone that occurs naturally in animals (see HORMONES, ANTERIOR PITUITARY HORMONES). The safety of beef for human consumption from cattle treated with BST was determined in 1984 by the Food and Drug Administration (FDA). Some of the findings were (1) the protein structure of synthetic BST and that produced by cattle is virtually the same, and (2) BST has no biological effects on humans and is degraded in the digestive process, as are meat proteins (11). However, not everyone accepts the FDA findings. Some groups or individuals have argued that more testing is needed. The use of BST has been approved in the dairy industry, but the use of PST in the pork industry has not been approved by FDA for commercial use in the United States.

Table 2. Per Capita Consumption, kg/yr, of Red Meat and Poultry in 10 Leading Countries from 1985 to 1991[a]

Country	1985–1989	1990	1991
Beef and veal			
Argentina	77.7	69.1	70.0
Uruguay	63.2	58.6	55.9
United States	48.2	44.1	44.1
Australia	40.1	38.6	38.2
Canada	40.5	37.7	36.4
New Zealand	40.5	36.4	35.0
France	30.5	30.0	30.0
Soviet Union	29.5	31.4	30.0
Italy	27.7	26.8	26.8
Switzerland	26.8	25.5	25.5
Pork[b]			
Hungary	84.1	69.5	66.8
Denmark	63.2	67.3	65.5
Czechoslovakia	55.5	57.7	53.2
Austria	51.8	52.7	52.7
Poland	45.0	49.1	52.7
Germany, unified	55.5	54.1	48.6
Belgium-Luxembourg	49.1	46.8	47.7
Spain	38.6	47.3	46.8
Bulgaria	45.5	46.8	44.5
Netherlands	42.7	44.1	41.8
Poultry			
United States	35.5	41.8	43.2
Israel	36.8	36.4	37.3
Hong Kong	29.1	34.1	35.0
Singapore	36.8	34.5	34.1
Canada	26.4	27.7	28.6
Saudi Arabia	27.7	25.9	25.9
Australia	23.6	24.5	25.5
Taiwan	20.0	23.2	23.6
Spain	21.8	23.2	23.2
Hungary	23.2	22.3	22.3
Lamb, mutton, and goat[b]			
New Zealand	38.2	23.2	25.0
Uruguay	7.7	24.5	24.1
Australia	23.2	22.7	23.6
Greece	13.6	14.5	14.5
Ireland	6.8	8.6	9.5
Bulgaria	10.0	8.6	7.7
United Kingdom	6.8	7.3	7.3
Spain	5.9	6.4	6.8
Turkey	6.8	6.4	6.4
South Africa	5.5	5.9	5.9

[a]Ref. 1. Carcass-weight equivalent for red meat; ready-to-cook equivalent for poultry.
[b]U.S. per capita consumption of pork was 30 kg (66 lbs) per person in 1991; lamb and mutton consumption was 0.9 kg (2 lbs) per person.

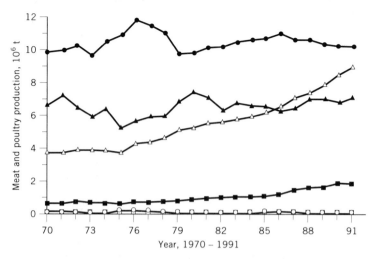

Fig. 1. Red meat and poultry production in the United States (1). Red meat is based on carcass weight, poultry is based on ready-to-cook weight, and 1991 production is preliminary data. (○) Veal, (●) beef, (△) chicken, (▲) pork, (□) lamb, and (■) turkey.

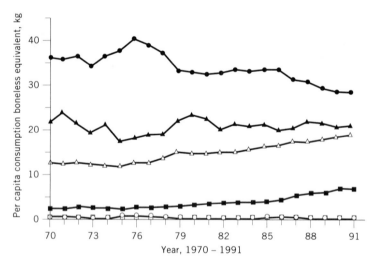

Fig. 2. Per capita consumption of red meat and poultry in the United States (1). Per capita consumption of red meat and poultry in 1991 is preliminary data. (○) Veal, (●) beef, (△) chicken, (▲) pork, (□) lamb, and (■) turkey.

Beta-adrenergic agonists that are known to promote growth such as clenbuterol [37148-27-9], cimaterol [54239-37-1], and L-640,033 improve the growth rate and feed conversion of sheep and poultry. Effects on swine are varied; definitive data on cattle are not yet available (12). β-Estradiol [50-28-2] and zeranol [55331-29-8] are available compounds that occur naturally and are very effective repartitioning agents, enhancing rates of protein and lean tissue production whenever present at effective levels in cattle depositing fat (13). Trenbolone acetate [10161-34-9] is another example of growth promotant, but its precise mechanism of action is unknown (14).

Antibiotics. Antibiotics (qv) may be administered on a one-time basis, for several days, or for longer periods. During the production of meat, the shorter periods of administration are generally for the treatment of a diseased condition; longer use at subtherapeutic dosages is intended to prevent disease, thereby increasing animal productivity while in the feedlot. The industry generally believes that subtherapeutic levels of antibiotics in the feed are essential to prevent economic losses under current husbandry practices (15). However, the use of antibiotics in livestock production has caused serious public concern that the hazardous antibiotic residues in meat are contributing to health problems in humans. Some scientists and consumer groups support the notion that continuous feeding of penicillin, tetracycline [60-54-8], and other antibiotics to livestock for disease prevention may result in development of antibiotic-resistant strains of bacteria and subsequently contribute to human illness. The National Academy of Sciences reported that it has never found data directly implicating subtherapeutic use of antibiotics in feeds as a risk factor in human illness (16). However, the public health implications associated with use of such compounds warrant continuing evaluation and monitoring.

Pathogens. Meat and meat products have a wide variety of microorganisms which could cause product spoilage or illnesses in humans. Occurrence of the microbial contamination varies with the location and the types of processing conditions. Pathogenic and spoilage microorganisms can be transferred to the meat during post-slaughter processing, storage, and handling. During slaughtering, many pathogens that may be present in the intestinal contents of the animals can contaminate the carcass and subsequently the processing tables and other equipment (17). *Salmonella typhimurium* can be transferred from raw poultry skin to other surfaces (18). *Staphylococcus aureus* can be transferred by human contact with the meat during processing. *Staphylococcus aureus* is a microorganism that produces severe gastrointestinal food poisoning through production of several toxins. Other pathogenic bacteria such as *Clostridium botulinum*, *Listeria monocytogenes*, *Escherichia coli*, *Yersinia enterolitica*, and *Bacillus cerreus* have also been found in contaminated meat products. Sufficient application of heat during cooking, however, destroys pathogenic and meat spoilage microorganisms and produces meat products that are commercially stable at ambient or refrigeration temperature. In addition, the heat treatment must be sufficiently severe to not only destroy the contaminating bacteria but also certain bacterial spores or toxins (19).

Trichinosis. Trichinosis is caused by parasitic nematode *Trichinella spiralis* that localizes in the muscles of pigs (see ANTIPARASITIC AGENTS). People become infected by eating undercooked meat, most commonly pork. When ingested, infected meat is digested releasing the larval trichina into the intestine where they rapidly mature into adults, mate, and produce a large number of offspring. The larval offspring leave the intestine, enter the blood stream, and invade the muscles where they migrate extensively before becoming encapsulated within a microscopic cyst. When only a few larvae are ingested, the infection is so light as to go unnoticed. Heavier infections produce symptoms associated with the parasite in the intestines and in the muscles. Diarrhea followed by fever, generalized swelling, muscle pain, and extreme fatigue are characteristic symptoms of trichinosis. The heaviest infections may be fatal, usually because the heart or brain is severely damaged (20). For many years, hotels, restaurants, institutional food suppliers, and consumers cooked pork to 82°C to ensure the destruction of *T. spiralis*. Other methods including freezing (-30°C for at least 16 h), irradiation (19 to 750 krads), and curing (combined with up to 3.5% salt) have also been used for the destruction of *T. spiralis* (21).

Meat Processing Ingredients

Meat. The primary ingredient in processed meats is meat itself. The contents of myofibrillar, sacroplasmic, and stromal proteins within the meat products determine the characteristics of the finished meat product. The ability of various meat ingredients to provide extractable protein for functionality in binding water and fat as well as in the cohesion of meat particles varies tremendously. The structure and composition of muscle varies greatly with the anatomy of individual animal as well as with the species. Certain aspects of the anatomy that are high in collagen provide ingredients that are of little value from the standpoint of protein functionality. If this meat is to be used in processing of comminuted meat products, it is often necessary to combine it with meats that have a lower content of stromal protein.

The sacroplasmic proteins myoglobin and hemoglobin are responsible for much of the color in meat. Species vary tremendously in the amount of sacroplasmic proteins within skeletal muscle with cattle, sheep, pigs, and poultry listed in declining order of sarcoplasmic protein content. Fat is also an important component of meat products. The amount of fat in a portion of meat varies depending on the species, anatomy, and state of nutrition of the animal. The properties of processed meat products are greatly dependent on the properties of the fat included. Certain species, such as sheep, have a relatively higher proportion of saturated fat, whereas other species, such as poultry, have a relatively lower proportion of saturated fat. It is well known that the characteristic flavors of meat from different species are in part determined by their fat composition.

Salt. Salt is a common nonmeat ingredient added to meat products. Meat products may vary in salt content from 1–8%. In addition to enhancing the solubilization of the myofibrillar protein, salt gives flavor and has a preservative effect by retarding bacterial growth. The amount of salt used depends on the finished product characteristics desired by the meat processor. The vast majority of cooked sausage products contain approximately 2–3% salt. Myofibrillar proteins, which significantly affect meat product texture, are soluble only in salt solutions. The effect of ionic strength on meat protein solubilization plays a significant role in the production of sectioned and formed, minced, and finely comminuted meat products.

Water. Water is often added to processed meat products for a variety of reasons. It is an important carrier of various ionic components that are added to processed meat products. The retention of water during further processing of meat is necessary to obtain a product that is juicy and has higher yields. The amount of water added during the preparation of processed meat products depends on the final properties desired. Water may be added to a meat product as a salt brine or as ice during the comminution step of sausage preparation.

Phosphate.   Sodium tripolyphosphate [7758-29-4], sodium pyrophosphate [7722-88-5], disodium phosphate [7558-79-4], and sodium acid pyrophosphate [7758-16-9] alone or in combination are used at varying levels in meat products. Generally, the use of phosphate is restricted to an amount that results in not more than 0.5% in the finished meat product. When used in combination with salt, alkaline phosphates enhance the ability of myofibrillar protein to bind water during heat processing (22). It is thought that the mechanism of action of the alkaline phosphates is to break the bond between myosin and actin within the myofibril. In addition, the alkaline phosphates affect meat hydration by increasing the pH and ionic strength. By raising the pH of meat there is an increase in negative charges on the myofibrillar proteins. The negative charges on adjacent myofilaments repel each other, thereby allowing more space for water to be entrapped within the gel structure.

Nitrite.   Sodium nitrite [7632-00-0] is added to cured meat products to fix their color and flavor, to inhibit *Clostridium botulinum* growth and toxin formation (23,24), and to stabilize lipids against oxidation (25). When nitrite is added to meat for the purpose of curing, less than 50% of that added can be analyzed after the completion of processing (26). Although the processor may add up to 200 ppm of sodium nitrite, many cured meat items contain only a few parts per million when they are consumed. The nitrite has either been lost from the meat to the atmosphere or remains in the meat as a reaction product undetectable by current analytical methods. When nitrite is added to meat products it reacts with the myoglobin and hemoglobin of trapped red blood cells to stabilize meat color to the cured form. Initially, the color changes from the purple-red of myoglobin to the brown of metmyoglobin. Eventually the color is converted to the rather dark red of nitric oxide myoglobin. When heated, this compound is converted to the pigment of nitrosylhemochrome, which is pink.

Extenders.   Extenders are used in the processing of some meat products. The desirable functional properties of extenders are that they must be good binders of water, good binders of fat, commercially sterile, free from objectionable flavors and taste, appropriately colored, and readily available at competitive prices. Extenders are available from both plant and animal sources. Wheat gluten is a good water binder and holds two to three times its own weight of water. If used beyond the 3% level, it tends to give a rubbery texture to sausage. Rusk, a bread-type product which is recrumbled, is an extremely good water absorber prior to cooking, but tends to exude some of this water after the product is held for some time. Soy flour (50% protein), soy protein concentrate (70% protein), and soy protein isolate (90% protein) are usually used as a powder in finely chopped meat products. Sodium caseinate [9005-46-3], a water-soluble form of the dried milk protein, is a good fat binder and its protein level usually exceeds 90%. Blood plasma is also used in some parts of the world in the processing of meat products. Blood plasma must be prepared and included in meat products under highly sanitary conditions (27).

Seasonings.   Seasonings which include spices, herbs, aromatic vegetables, flavor enhancers, and simulated meat flavors, may influence flavor, appearance, or shelf-life of meat products. The most commonly used spices in meat products are peppers (ground black, white, or red), nutmeg, mace, ginger, cardamom, celery, cumin, dill, and mustard. An example of a flavor enhancer is monosodium glutamate [142-47-2]. It brings out and intensifies the species flavor of the meat product. Hydrolyzed vegetable proteins have more flavor of their own which can be described as a meaty or beefy (27).

Curing Accelerators. The main function of curing accelerators is to accelerate color fixing or to preserve color of cured meat products during storage. Curing accelerator agents permitted in meat processing include ascorbic acid [50-81-7], erythorbic acid [89-65-6], fumaric acid [110-17-8], glucono delta lactone [46-80-2], sodium acid pyrophosphate, sodium ascorbate [134-03-2], sodium erythorbate [7378-23-8], citric acid [77-92-9], and sodium citrate [68-04-2]. Each of these agents has different legal limits of use in different cured products. In addition, curing accelerators must be used only in combination with curing agents.

Starter Cultures. A starter culture is required for the production of fermented sausage and it must possess a unique set of physiological characteristics. A starter culture must (1) be tolerant of salt, (2) grow in the presence of at least 100-μg nitrite per gram of meat, (3) grow in the range of 26.7–43.3°C (80–110°F) and preferably with an optimum around 32.2°C (90°F), (4) be homofermentative, (5) not be proteolytic or lypolitic, (6) not produce off-flavor, and (7) be safe and possess no health risk involved upon its application. Several microorganisms used as starter cultures for fermented meat products include *Lactobacillus plantarum*, *Lactobacillus sake*, *Lactobacillus acidophilus*, *Aeromonas* X, *Aeromonas* 19, *Micrococcus aurantiacus* M-53, and *Pediococcus cerevisiae* (28).

Meat Processing Procedures and Machinery

Mechanical Tenderization. Sophisticated advances have been made in improving meat tenderness. Mechanical tenderization involves the application of blades, knives, pins, or needles to meat via mechanical pressure. The increase in tenderness associated with mechanical tenderization is attributed to the partial destruction of connective tissue or the severance of muscle fibers, which leads to reduced resistance to shear force and mastication. Meat from various species is mechanically tenderized by being passed through a reciprocating blade-type machine. Sanitation is extremely important. The mechanical tenderizer tends to distribute the microorganisms that are on the surface throughout the interior of the meat pieces (29).

Cured Meats. The term meat curing means the addition of salt, nitrite and/or nitrate, sugar, and other ingredients for the purpose of preserving and flavoring meat (30). Cured meat products include ham, bacon, frankfurters, bologna, and some sausages. The slowest rate of meat curing is performed by applying the curing ingredients to intact meat pieces in the form of a dry rub. Such methods of curing are still being used for some meat cuts in certain parts of the world. It takes a long time for the cure to penetrate into the internal portion of larger meat cuts. With the increased costs of materials and labor, the amount of meat that is cured in this manner is declining.

Immersion curing is used as an alternative to dry curing. Immersion curing is still commercially used by some small processors. The meat is placed in a brine solution for an appropriate period of time until the brine penetrates the entire portion of meat. It is important not to keep the brine for too long a period of time because the brine strength is thereby reduced and the brine becomes contaminated with meat juices and bacteria.

With injection curing the brine is pumped into the meat with a needle and a pressurized source of liquid. The brine can be injected either through the arterial system in some large cuts such as hams, or it can be stitch pumped into the meat cuts by using a needle that has holes along its length. Both artery and stitch pumping are performed by hand and are relatively slow procedures. Multineedle injectors are most widely used in the industry for brine injecting bone-in or for boneless pieces of meat. The injected meat cuts are subsequently subjected to a mechanical action such as tumbling or massaging. This mechanical action physically disrupts the muscle structure, allowing the brine to interact more effectively with the extractable, salt-soluble myofibrillar proteins and to maintain the extracted proteins in a solubilized state. When the product is heated or cooked, the solubilized proteins form a gel entrapping the liquid more effectively within the product. This effective entrapment of moisture leads to a higher yield and more tender and juicy finished products (27).

Sectioned and Formed Products. The meats that are utilized to produce sectioned and formed products may be entire muscles, very coarsely ground meat, or flaked meat. Large sections may be produced by cutting muscle chunks into sections by hand or using a dicer. Some particles can be produced by using a plate in a meat grinder that has large kidney-shaped holes. Meat particles can be produced by using a flaking machine that is capable of varying the flake size from very fine to coarsely flaked materials. The mechanical energy that must be applied to the various size of meat pieces and other ingredients to extract myofibrillar proteins can be provided by a mixer, tumbler, or massager. Tumbling generally refers to placing meat inside a stainless steel drum that rotates at such a speed that some of the meat is carried to the top of the drum and drops down at least one meter onto the meat at the bottom of the drum. This impact of meat on meat, as well as the friction of one portion abrading another, has several functions: (1) it aids in abrading the myofibrillar proteins from the meat surface, (2) it makes the meat more pliable, and (3) it increases the rate of cure distribution. Massaging is generally a less severe mechanical treatment than tumbling. Massagers come in many sizes and designs. Most models use a bin similar to a standard meat vat which is equipped with a large motor to power a vertical shaft that has arms attached to it. The massager slowly stirs the large chunks of meat to achieve the same results as the tumbler.

Minced Products. Many meat products are produced by grinding or mincing meat to various particle sizes. The products that are included in this class are sausages of the fresh, fermented, dried, and cooked varieties. The meat ingredients can be either ground in a mincer or chopped in a bowl chopper. If particle definition and size are to be maintained, the meats should be cold when either ground or chopped. In order to obtain uniform particle reduction it is necessary to keep the grinder blades and plates in excellent working condition and maintain very sharp knives in the bowl chopper. The presence of connective tissues must be carefully controlled if a high quality product is to result. If the product is to be cooked and the particle-to-particle binding is to be maintained during cooking, and the maximum amount of fat and water retained, it is necessary to mix the meat ingredients along with salt so as to extract myofibrillar proteins. The extracted myofibrillar proteins act to bind the particles together and to trap water and fat during cooking.

Finely Chopped Products. The manufacture of finely comminuted processed meat products is dependent on the formation of a functional protein matrix within the product. The ability of the protein to successfully entrap moisture and fat is affected by many factors. These factors include the water holding capacity of the meat as well as the levels of meat, water, fat, salt, and nonmeat additives in the formulation. A certain level of fat is important in sausage products since it affects tenderness, juiciness, and flavor. The machines used to reduce the meat particle size are selected based on the variety and volume of the operation. Minced sausage production basically ends after comminution by a grinder, bowl chopper, or flaker, whereas the production of finely chopped sausage requires additional particle size reduction with more time in a bowl chopper or passage through an emulsion mill. In the bowl chopper, comminution and mixing are accomplished by revolving the meat in a bowl past a series of knives mounted on a high speed rotating arbor which is in a fixed position so that the knives pass through the meat as the bowl turns. Emulsion mills operate on a principle of one or more rotating knives traveling at an extremely high speed so that the meat mixture is pulled from a chopper and forced through one or more perforated stationary plates. The meat is drawn through tiny pores in these plates, and the mill therefore has the function of reducing the meat and fat particles to a very small size (2.0 mm or less), producing a smooth batter with paste-like consistency. This type of consistency is often desired for the finely chopped sausages and loaves.

Fermented Products. Fermented meat products such as semidried and dried sausages are generally recognized as safe, if critical points during processing are controlled properly. Some of the sausage processors use a small amount of fermented product as the starter for a new batch of product. This can be a dangerous procedure due to the potential growth of food poisoning bacteria such as *Staphylococcus aureus* (31). This method of inoculation requires a very strict condition to assure the absence of not only bacteria associated with a health hazard but also those associated with product failure (proteolytic, greening, and gas-forming microorganisms).

The use of a straight nitrate cure in sausages such as the Lebanon type requires mixed starters including *Micrococcus aurantiacus* M-53 and *Lactobacillus plantarum*. The *Micrococcus aurantiacus* M-53 ensures color formation by reducing the nitrate to nitrite, while the *Lactobacillus plantarum* is responsible for the decrease in pH (32). The fermentation in sausage involves the conversion of either sucrose or glucose to lactic acid by homofermentative lactic acid bacteria. This biological acidulation can reduce the pH value of the meat mixture from approximately 6.0 to 4.8 or 5.0. Attempts to slowly add lactic acid directly to the meat mixture were not successful, because the fermentation conditions cannot be substituted by direct chemical acidulation (28).

Hazard Analysis Critical Control Point

The hazard analysis critical control point (HACCP) concept is a systematic approach to the identification, assessment, prevention, and control of hazards. The system offers a rational approach to the control of microbiological, chemical, environmental, and physical hazards in foods, avoids the many weaknesses inherent in the inspectional quality control approach, and circumvents the shortcomings of reliance on microbiological testing (33,34). The food industry and government regulatory agencies are placing greater emphasis on the HACCP system to provide greater assurance of food safety. In the 1970s and early 1980s, the HACCP approach was adopted by large food companies and began to receive attention from segments of the food industry other than manufacturing. Reports by the International Commission on Microbiological Specifications for Foods (ICMSF) revealed a growing international awareness of the HACCP concept and its usefulness in dealing with food safety (35).

HACCP Principles.   The National Advisory committee on Microbiological Criteria for Foods established seven principles for the HACCP system (36).

Conduct Hazard Analysis and Risk Assessment.   A hazard is any biological, chemical, or physical property that may cause an unacceptable consumer health risk. All of the potential hazards in the food chain are analyzed, from growing and harvesting or slaughtering to manufacturing, distribution, retailing, and consumption of the product.

Determine Critical Control Points.   A critical control point (CCP) is any point in the process where loss of control may result in an unacceptable health risk. A CCP is established for each identified hazard. The emergence of foodborne pathogens has taught food processors the importance of potential product contamination from the processing environment.

Establish Specifications for Each CCP.   It is necessary to include tolerances at each CCP. Examples of specifications or limits include product pH range, the maximum allowable level of bacterial counts, and the time and temperature range for cooking.

Monitor Each CCP.   It is necessary to establish a regular schedule for monitoring of each CCP. The schedule could be, for example, once per shift, hourly, or even continuous. Preferably, a published testing procedure for the monitored parameter should be available.

Establish Corrective Action.   Corrective actions should be clearly defined beforehand, with the responsibility for action assigned to an individual.

Establish a Recordkeeping System.   It has always been important for the food manufacturer to maintain records of ingredients, processes, and product controls so that an effective trace and recall system is available when necessary.

Establish Verification Procedures.   Verification can be performed independently by the manufacturer and the regulatory agency to determine that the HACCP system within the plant is in compliance with the HACCP plan as designed.

Example of an HACCP System.   The HACCP system can be used to ensure production of a safe cooked, sliced turkey breast with gravy, which has been vacuum packaged in a flexible plastic pouch and subjected to a final heat treatment prior to distribution (37). Raw turkey breasts are trimmed, then injected with a solution containing sodium chloride and sodium phosphate. Next, the meat is placed into a tumbler. After tumbling, the meat is stuffed into a casing, placed onto racks, and moved into a cook tank, where it is cooked to an internal temperature of at least 71.1°C (160°F). After cooking, the meat is chilled. Next, it is sliced and placed into a pouch. Rehydrated gravy is then added, and the pouch is vacuum sealed. The product is then pasteurized. Finally, it is chilled, placed into cartons, and moved to storage for subsequent distribution. This process has six CCPs (ie, cooking, chilling, rehydrating, pasteurization, chilling, and storing–distributing–displaying). The process control objectives are to destroy the normal spoilage microflora and pathogens, and to control the potential for toxin produced by *Clostridium botulinum.*

Each CCP can be divided into three components: conditions, monitoring, and verification. Cooking, for example, could include the following. (*1*) *Conditions*: the internal temperature of 71.1°C (160°F) provides a substantial margin of safety for destroying nonspore-forming pathogenic bacteria. The product is relatively large in diameter and requires a long period of time for heating and chilling at temperatures that are lethal to vegetative cells. To assure compliance, it is necessary to have uniform product thickness and heat distribution within the cook tank. (*2*) *Monitoring*: a sensor is used to monitor the temperature of the water. The minimum internal temperature of the product is monitored by a temperature sensor placed at the center of a turkey roll. The temperature can be continuously measured and recorded. Water circulation or agitation to assure uniform heating can be monitored visually. (*3*) *Verification*: temperature sensors should be periodically calibrated for accuracy. Heat distribution should be tested using multiple temperature sensors placed throughout the cook tank. The frequency of verification depends on experience with the equipment and product, and the potential risk presented to consumers.

Fat Reduction in Meat Products

Consumers not only prefer good tasting foods, but they also are concerned with the nutrition, safety, and wholesomeness of the products they consume. The amount of fat, especially saturated fat and cholesterol in meat products, is of concern to a growing number of health-conscious consumers. The introduction of low fat ground beef sandwiches and hamburgers in fast food chains as well as closer trimming of retail beef cuts and leaner ground beef in supermarkets across the United States demonstrates the meat industry's response to consumer desires for lower fat consumption (38). In order to be labeled as low fat, a meat product must contain no more than 10% fat (39). The palatibility of ground beef, however, is directly related to the fat content. The overall acceptability of ground beef products is maximized at a fat content of approximately 20% (40). As the fat content of ground beef decreases, there is a significant decrease in product juiciness and tenderness (41).

Leaner Cuts.   The most obvious method for decreasing fat content in further processed red meat products is to use more trimmed, boneless cuts or leaner raw materials. A notable example has been the production of restructured or sectioned and formed hams or beef top rounds with less than 5% fat content (more than 95% fat free) in which visible surface and seam fat have been removed. Restructured steaks and chops offer processors greater opportunity to control fat content, portion size, and raw material costs but have different sensory characteristics as the fat content increases. Typically, muscles or trimmings from the chuck, round, or pork shoulder can be defatted, decreased in particle size, blended with ingredients, and shaped into the desired form. As a whole, flavor and overall palatability of restructured steaks and chops are not dramatically different over the 10 to 20% fat range (42). Further reductions in fat below 10% in restructured meats and sausages can be formulated by using less caloric dense ingredients such as fat reduced beef or pork, partially defatted chopped beef or pork, and mechanically separated meat or poultry.

Ingredient Additions and Substitutions. Processed meat products have the greatest opportunity for fat reduction for modification because their composition can be altered by reformulation with a fat replacement (see FAT REPLACERS).

Added Water. Frankfurters and bologna are allowed to contain combinations of fat and added water not to exceed 40% with a maximum fat content of 30%. This allows, for example, a 10% fat frankfurter to be produced with 30% added water. Substitution of large amounts of fat with water alone may not give the optimal sensory and textural properties that consumers want (43). To overcome these shortcomings, several binders can be added to improve water and fat-binding properties, cooking yields, texture, and flavor (27).

Protein-Based Substitutes. Several plant and animal-based proteins have been used in processed meat products to increase yields, reduce reformulation costs, enhance specific functional properties, and decrease fat content. Examples of these protein additives are wheat flour, wheat gluten, soy flour, soy protein concentrate, soy protein isolate, textured soy protein, cottonseed flour, oat flour, corn germ meal, nonfat dry milk, caseinates, whey proteins, surimi, blood plasma, and egg proteins. Most of these protein ingredients can be included in cooked sausages with a maximum level allowed up to 3.5% of the formulation, except soy protein isolate and caseinates are restricted to 2% (44).

Carbohydrate-Based Substitutes. Most of the carbohydrates available for use as fat substitutes in processed meats fall into the category of being a gum (hydrocolloid), starch (qv), or cellulose-based derivative (see GUMS). Carrageenan [9000-07-1] is possibly the most widely used binder in current low fat meat products. There are three types of carrageenan: iota-, kappa-, and lambda-carrageenans. Iota- and kappa-carrageenans act as gelling agents. The lambda type is nongelling, and functions as a thickner. Iota-carrageenan has been recommended (45) for use in formulating low fat ground beef due to its ability to retain moisture, especially through a freeze–thaw cycle which is typical for ground beef patties. Oat bran and oat fiber can also be used to improve moisture retention and mouth feel. Modified starches can be used as binders to maintain juiciness and tenderness in low fat meat products. Maltodextrins (dextrose equivalent less than 20) may be used as binders up to 3.5% in finished meat products. Other carbohydrates such as konjac flour, alginate, microcrystalline cellulose, methylcellulose, and carboxymethylcellulose have also been used in low fat meat products (see CELLULOSE ETHERS).

Functional Blends. The term functional blend refers to various ingredient blends formulated to achieve a certain objective such as fat reduction. An example of this blend consists of water, partially hydrogenated canola oil, hydrolyzed beef plasma, tapioca flour, sodium alginate, and salt. This blend is designed to replace animal fat and is typically used at less than 25% of the finished product. Another functional blend is composed of modified food starch, rice flour, salt, emulsifier, and flavor. A recommended formula is 90% meat (with 10% fat), 7% added water, and 3% seasoning blend (38).

Noncaloric Synthetic Fat Substitutes. For new synthetic fat substitutes to succeed in the preparation of low fat meat products, they must be technically superior to existing substitutes and offer greater versatility while mimicking the taste, texture, and function of fat, but without the calories. Although only few synthetic compounds (ie, polydextrose, sucrose polyester, esterified propoxylated glycerols, dialkyl dihexadecymalonate, and trialkoxytricarballate) are available, they may have greater market potential in the future, because they are microbiologically more stable and contribute less calories than the carbohydrate- or protein-based substitutes (44).

New Technologies for Fat Reduction in Meat Products. *Surimi-Like Process.* Surimi is a wet, frozen concentrate of myofibrillar proteins from fish muscle that is usually prepared by freshwater washing of mechanically deboned fish muscle followed by the addition of ingredients to prevent protein denaturation during freezing. This process also has application in converting meat trimmings or mechanically separated meats into highly functional and nutritious ingredients (46). Production of beef surimi from mechanically separated meat removes up to 99.5% of fat and increases the protein content to 133–155% over the starting materials. The beef surimi is a bland-tasting raw material to which flavorings can be added (47).

Naturalean Process. This process claims to separate fat and cholesterol from conventionally deboned, trimmed lean by a process that finely minces the meat tissues in a high speed chopper, followed by the addition of a small amount of acetic acid to decrease the pH and aggregate proteins; then the fat is solidified on a cold surface heat exchanger (48). The lean component then can be removed from the surface of the fat and used for producing patties, sausages, emulsion products, meat fillings, or toppings.

Supercritical Fluid Extraction. Supercritical fluid (SCF) extraction is a process in which elevated pressure and temperature conditions are used to make a substance exceed a critical point. Once above this critical point, the gas (CO_2 is commonly used) exhibits unique solvating properties. The advantages of SCF extraction in foods are that there is no solvent residue in the extracted products, the process can be performed at low temperature, oxygen is excluded, and there is minimal protein degradation (49). One area in which SCF extraction of lipids from meats may be applied is in the production of low fat dried meat ingredients for further processed items. Its application in fresh meat is less successful because the fresh meat contains relatively high levels of moisture (50).

Fat-Reduced Meat Process. Partially defatted chopped beef (PDCB) is typically produced in a batch process, where the desinewed raw material is heated in tanks prior to fat/lean separation. But temperature gradient from vessel surface to center causes variations in product temperature and process time, which results in partially denatured products with reduced binding, flavor, and nutritional properties. In the fat-reduced meat (FRM) process heat exchange is continuous. Water temperature is tightly controlled to a maximum of 43.3°C. The average tempering time is 10 minutes. After tempering, a proprietary separation process is used to separate the lean portion from fat. The defatted material is then frozen into thin sheets at −6.7°C or below within two minutes.

The FRM can be used in hamburger patties, hot dogs, sausages, luncheon meats, and canned meat products (51).

Microwave Cooking Pads. A simple and effective method of reducing fat in meat products involves the use of microwavable heating pads. These pads, made from nonwoven, melt-blown polypropylene materials, absorb fat lost during the cooking process, minimizing its contact with food, and more fat is allowed to cook out (52).

Enzymatic Conversion of Cholesterol. A decrease of cholesterol in meat products in the future may be possible through the conversion of cholesterol [57-88-5] to coprosterol [360-68-9], which is not absorbed readily in the intestine. Cholesterol reductase can be isolated from alfalfa leaves and cucumber leaves (53). Treatment of meat animals might involve an injection of this enzyme immediately prior to slaughter, allowing for the conversion of a portion of the membrane-bound cholesterol into coprostanol.

Sous-Vide Processing

The term sous-vide (pronounced *sue-veed*) means "under vacuum." In sous-vide processing, meats are cooked slowly in sealed, vacuumed, heat-stable pouches or thermoformed trays, so that the natural flavor, aroma, appearance, moisture, and nutrients are retained within the product (54). Such a method is not new, because early civilization used many ingenious ways of cooking foods in a wrapping (eg, leaves) to retain natural flavor and to maximize juiciness. However, what is new about the sous-vide process is the highly controlled packaging/cooking conditions used. Technically, sous-vide is a modified atmosphere packaging (MAP) or controlled atmosphere packaging (CAP) method. What makes it different from the ordinary MAP/CAP methods is the post-packaging pasteurization step. Sous-vide processing is used extensively in Europe and is gaining in popularity as a food processing method in North America.

Sous-vide processing consists of the preparation of top-quality raw ingredients, precooking (if necessary), packaging in heat-stable air-impermeable bags under vacuum to remove all of the air, sealing, and cooking (pasteurization) at a particular temperature for a certain period of time. The pasteurized product is cooled to 4°C within two to three hours of pasteurization, and stored and distributed under refrigerated conditions (55). A MAP/CAP product gradually deteriorates over time beginning with the day it was packaged. For sous-vide products, under good manufacturing conditions, a shelf life of 21 to 30 days is obtainable. The sous-vide product also facilitates the preparation of tasty meals on reheating for 10–15 minutes in boiling water or four to five minutes in a microwave oven (55). However, a significant concern about these minimally processed products is that they are not shelf stable. Therefore, they could be a potential public health risk if subjected to temperature abuse at any stage of production, storage, distribution, and marketing.

Nutritional Labeling

Descriptive terms which convey information about the nutritional value or quality of a food product are useful to consumers when making product choices in the market place. This information, ie, label, is easily seen when displayed on the food container or package. Obviously, space availability on a product's label is very limited. Therefore, the development of concise and informative labeling terms is important for consumers, food processors, and government regulatory agencies. The USDA's Food Safety and Inspection Service (FSIS) regulates the labeling of meat and poultry products, while FDA has responsibility over all other food labeling. The FDA regulations implement the Nutrition Labeling and Education Act of 1990. FSIS relies on its general authority under the Federal Meat Inspection Act (21 USC 601 *et seq.*) and the Poultry Products Inspection Act (21 USC 451 *et seq.*) as the basis for its nutritional labeling proposal. The FSIS strives to ensure that these products are free from adulteration, properly identified, and correctly labeled before leaving a federally inspected establishment or entering the marketplace (56).

Nutritional Labeling Content. As part of its efforts to harmonize labeling requirements with the FDA proposal, the FSIS mandates that nutrition information include the same 15 declarations required by FDA as well as allowing certain optional disclosures. The mandatory disclosures include calories, calories from total fat, total fat to nearest one-half gram, saturated fat to nearest one-half gram, cholesterol in milligrams, total carbohydrates in grams excluding fiber, complex carbohydrates in grams, sugars in grams including sugar alcohols, dietary fiber in grams, protein in grams, sodium in milligrams, vitamin A as a percentage of reference daily intake (RDI), vitamin C as a percentage of RDI, calcium as a percentage of RDI, and iron as a percentage of RDI. If the particular product contains insignificant amounts of eight nutrients, the abbreviated format should include calories, total fat, total carbohydrates, protein, and sodium. The optional disclosures include calories from saturated fat and unsaturated fat, unsaturated fat to nearest 0.5 gram (this is mandatory if fatty acid and/or cholesterol claims are made), polyunsaturated and/or monounsaturated fat to the nearest 0.5 gram, declaration of sugar alcohols in grams, insoluble and soluble fiber, potassium in milligrams, and thiamin, riboflavin, niacin, and other vitamins or minerals (if a claim regarding these nutrients is made) (57).

Service Size. The label presentation should allow the consumers to understand the nutrition contents of individual meat products, compare nutrition contents across product categories, and choose among relevant food alternatives. The establishment of serving sizes has been the most controversial aspect of the nutritional labeling either for the consumers or manufacturers, because there are wide varieties of product sizes on the market, and it is almost impossible to standardize these sizes. In addition, there is also considerable confusion on the definitions of serving and portion (58). The term serving was defined by FDA as a reasonable quantity of food suited for or practicable of consumption as a part of a meal by an adult male engaged in light physical activity, or by an infant or child under age four when the article purports or is represented to be for consumption by an infant or child under age four (21 CFR 101.9 (b) (1)). In contrast, FDA defined the term portion as the amount of food customarily used only as an ingredient in the preparation of a meal component, ie, one-half tablespoon of cooking oil or one-fourth cup of tomato paste.

In order to resolve this problem, USDA's FSIS proposed three options for establishing standardized reference serving sizes: 1 ounce or 100 grams, a single and uniform reference standard serving size using food consumption data, and a reference standard serving size based on dietary recommendations (59). A 1 oz or 100-g serving size would provide the easiest method for conversion and allow consumers to compare between meat and poultry products easily. However, the consumers may not realize that the information has to be converted to be meaningful in terms of the amounts they eat, because 1 oz or 100 g may not be a commonly consumed amount of meat or poultry products. There was virtually no support for the second option in establishing a single uniform serving size based on food consumption data. The third option would provide nutrition information on the recommended portions of foods. However, it would not provide information on what is actually being consumed. Currently, FDA and USDA's FSIS continue to cooperate and the goal is to establish standards that could be used by food

Table 3. Proposed Descriptors for Nutrition Labeling in Meat and Poultry Products[a]

Descriptors	Criteria
	Ingredient-free
sodium-free	less than 5 mg of sodium per serving
salt-free	must meet the definition of sodium-free per serving
fat-free	less than 0.5 g of fat per serving, and no added fat or oil
percent fat-free	may be used only in describing foods that qualify as low fat
cholesterol-free	less than 2 mg of cholesterol per serving and has 2 g or less of saturated fat per serving
	Low content
low sodium	no more than 140 mg of sodium per serving and per 100 g of food
very low sodium	no more than 35 mg of sodium per serving and per 100 g of food
low calorie	no more than 40 calories per serving and per 100 g of food
low fat	no more than 3 g of fat per serving and per 100 g of food
low in saturated fat	no more than 1 g of saturated fat and no more than 15% of the food's calories come from saturated fat
low in cholesterol	no more than 20 mg of cholesterol per serving and per 100 g of food, and no more than 2 g of saturated fat per serving
	Reduced content
reduced calorie	one-third fewer calories than the comparison food
reduced sodium	no more than half of the sodium of a comparison food
reduced fat	no more than half the fat of a comparison food
	to avoid trivial claims, reduction must exceed 3 g of fat per serving
	Other designation
less	25% less of the nutrient than the comparison food
fewer	25% less calories than the comparison food
light or lite	one-third fewer calories than the industry norm and it may only be used when fat is reduced by at least 50%
lean	less than 10.5 g of fat, of which less than 3.5 g is saturated fat, and less than 94.5 mg of cholesterol per 100 g
extra lean	less than 4.9 g of fat, of which less than 1.8 g is saturated fat, and less than 100 mg of cholesterol per 100 g

[a]Ref. 59.

manufacturers to determine label serving sizes and whether a claim such as low sodium meets criteria for the claim (59).

Nutritional Labeling Descriptors.   In order to avoid confusion, descriptive terms must be accompanied by definitions which adequately explain the terms. In the case of nutrition-related claims, analytical sampling offers a means of assuring the accuracy of the stated claims. The USDA's FSIS has proposed a list of descriptors relevant for meat and poultry products (Table 3).

BIBLIOGRAPHY

"Meat and Meat Products" in *ECT* 1st ed., Vol. 8, pp. 825–839, by H. R. Kraybill, American Meat Institute Foundation; in *ECT* 2nd ed., Vol. 13, pp. 167–184, by W. J. Aunan and O. E. Kolari, American Meat Institute Foundation; "Meat Products" in *ECT* 3rd ed., Vol. 15, pp. 62–74, by G. R. Schmidt and R. F. Mawson, Colorado State University.

1. J. J. Putnam and J. E. Allshouse, *Food Consumption, Prices, and Expenditures, 1970–90*, U.S. Department of Agriculture (USDA), Statistical Bulletin No. 840, Washington, D.C., 1992, pp. 32–78.
2. G. M. Briggs and B. S. Schweigert, in A. M. Pearson and T. R. Dutson, eds., *Advances in Meat Research: Meat and Health*, Vol. 6, Elsevier Applied Science, New York, 1990, pp. 2–20.
3. USDA, *Agricultural Statistics 1991*, U.S. Government Printing Office, Washington, D.C., 1991, pp. 288–289.
4. S. D. Shagam and L. Bailey, *Agricultural Outlook*, 12–15 (Apr. 1992).
5. R. B. Shekelle and S. C. Liu, *J. Am. Med. Assoc.* **240**(8), 756–758 (1978).
6. American Dietetic Association, *J. Am. Diet. Assoc.* **86**, 1663 (1986),
7. R. M. Mullis and P. Pirie, *J. Am. Diet. Assoc.* **88**, 191–195 (1988).
8. G. F. Watts and co-workers, *Brit. Med. J.* **296**, 235–237 (1988).
9. D. Putler and E. Frazao, *Food Rev.* **14**(1), 16–20 (1991).
10. Council for Agricultural Science and Technology, (CAST) *Hormonally Active Substances in Foods: A Safety Evaluation*, Task Force Report No. 66, Ames, Iowa, 1977.
11. D. P. Blayney, R. F. Fallert, and S. D. Shagam, *Food Rev.* **14**(4), 6–9 (1991).
12. L. A. Muir, in National Research Council, *Designing Foods: Animal Product Options in the Marketplace*, National Academy Press, Washington, D.C., 1988, pp. 184–193.
13. F. M. Byers, H. R. Cross, and G. T. Schelling, in Ref. 12, pp. 283–291.
14. R. E. Allen, in Ref. 12, pp. 142–162.
15. Committee on the Scientific Basis of the Nation's Meat and Poultry Inspection Program, *Meat and Poultry Inspection*, National Academy Press, Washington, D.C., 1985, pp. 44–47.
16. U.S. Assembly of Life Sciences, *The Effects on Human Health of Subtherapeutic Use of Antimicrobials in Animal Feeds*, National Academy of Sciences, Washington, D.C., 1980.
17. R. C. Benedict, *Reciprocal Meat Conf. Proc.* **41**, 1–6 (1988).
18. M. O. Carson, H. S. Lillard, and M. K. Hamdy, *J. Food Prot.* **50**, 327–329 (1987).
19. A. M. Pearson and J. I. Gray, in Ref. 2, pp. 517–542.
20. P. M. Schantz, *Food Technol.* **37**(3), 83–86 (1983).
21. A. W. Kotula, *Food Technol.* **37**(3), 91–94 (1983).
22. R. Hamm, in J. M. deMan and P. Melnychyn, eds., *Symposium: Phosphates in Food Processing*, AVI Publishing Co., West Port, Conn., 1970, p. 65.
23. J. N. Sofos, F. F. Busta, and C. E. Allen, *Appl. Environ. Microbiol.* **37**, 1103–1109 (1979).

24. J. N. Sofos and co-workers, *J. Food Sci.* **44**, 668–672 (1979).
25. M. P. Zubillaga and G. Maerker, *J. Am. Oil Chem. Soc.* **64**, 757–760 (1987).
26. J. H. Hotchkiss and R. G. Cassens, *Food Technol.* **41**(4), 127–136 (1987).
27. G. R. Schmidt, in H. R. Cross and A. J. Overby, eds., *Meat Science, Milk Science and Technology*, Elsevier Science Publisher B. V., New York, 1988, pp. 83–114.
28. R. H. Deibel, *Proceedings of the Meat Industry Research Conference,* American Meat Institute, Arlington, Va., 1974, pp. 57–60.
29. K. J. Boyd, H. W. Ackerman, and R. F. Plimpton, *J. Food Sci.* **43**, 670–672, 676 (1978).
30. J. Bard and W. E. Townsend, in J. F. Price and B. S. Schweigert, eds., *The Science of Meat and Meat Products*, W. H. Freeman Co., San Francisco, 1973, pp. 452–483.
31. L. E. Barber and R. H. Deibel, *Appl. Microbiol.* **24**, 891–898 (1972).
32. F. P. Niinivaara, *Reciprocal Meat Conf. Proc.* **44**, 59–63 (1991).
33. F. L. Bryan, *Hazard Analysis Critical Control Point Evaluations*, World Health Organization, Geneva, Switzerland, 1992, p. 4.
34. W. H. Sperber, *Food Technol.* **45**(6), 116–118, 120 (1991).
35. B. Simonsen and co-workers, *Intl. J. Food Microbiol.* **4**, 227–247 (1987).
36. National Advisory Committee on Microbiological Criteria for Foods, (NACMCF) *Hazard Analysis and Critical Control Point System*, Food Safety and Inspection Service, USDA, Washington, D.C., 1989.
37. C. E. Adams, *Food Technol.* **45**(4), 148–151 (1991).
38. G. H. Taki, *Food Technol.* **45**(11), 70–74 (1991).
39. D. E. Pzczola, *Food Technol.* **45**(11), 60–66 (1991).
40. W. R. Egbert and co-workers, *Food Technol.* **45**(6), 64–73 (1991).
41. B. W. Berry and K. F. Leddy, *J. Food Sci.* **49**, 870–875 (1984).
42. C. A. Costello, M. P. Penfield, and M. J. Riemann, *J. Food Sci.* **50**, 685–688 (1985).
43. J. R. Claus, M. C. Hunt, and C. L. Kastner, *J. Muscle Foods* **1**, 1–21 (1989).
44. J. T. Keeton, *Reciprocal Meat Conf. Proc.* **44**, 79–90 (1991).
45. D. L. Huffman and co-workers, *Reciprocal Meat Conf. Proc.* **44**, 73–78 (1991).
46. CAST, *Food Fats and Health*, Task Force Report No. 118, Ames, Iowa, 1991.
47. P. Whitehead, *Prepared Foods* **161**(6), 102 (1992).
48. *Food Engineering* **60**(9), 91 (1988).
49. R. R. Chao and co-workers, *J. Food Sci.* **56**, 183–187 (1991).
50. A. D. Clarke, *Reciprocal Meat Conf. Proc.* **44**, 101–106 (1991).
51. *Food Engineering* **61**(2), 53–54 (1991).
52. C. A. Costello, W. C. Morris, and J. R. Barwick, *J. Food Sci.* **55**, 298–300 (1990).
53. U.S. Pat. 4,921,710 (May 1, 1990), D. C. Beitz, J. W. Young, and S. S. Dehal (to Iowa State University Research Foundation, Inc.).
54. *Prepared Foods* **161**(8), 98–100 (1992).
55. J. P. Smith, H. S. Ramaswamy, and B. K. Simpson, *Trends Food Sci. Technol.* **1**, 111–118 (1990).
56. K. F. Leddy, *Reciprocal Meat Conf. Proc.* **41**, 21–24 (1988).
57. P. Olsson and D. Johnson, *Meat and Poultry* **38**(2), 24–36 (1992).
58. D. V. Porter and R. O. Earl, *Nutrition Labeling: Issues and Directions for the 1990s*, National Academy Press, Washington, D.C., 1990, pp. 203–216.
59. USDA Food Safety and Inspection Service, *Fed. Reg.* **56**(229), 60302–60364 (1991).

Glenn R. Schmidt
S. Raharjo
Colorado State University

MECHANICAL TESTING.   See Materials Reliability.

MEDICAL DIAGNOSTIC REAGENTS

Purified enzymes are widely used in medical diagnostic reagents in the measurement of analytes in urine, plasma, serum, or whole blood. Enzymes are very specific catalysts that can be derived from plants and animals, although microbial fermentation (qv) is the most popular production method. Enzymes are used extensively in diagnostics, immunodiagnostics, and biosensors (qv) to measure or amplify signals of many specific metabolites. Purified enzymes are expensive. This is the main reason for the increasing utilization of reusable immobilized enzymes in clinical analyses (see ENZYME APPLICATIONS; IMMUNOASSAY).

The main development in medical diagnostic reagents since the 1960s has been the steady growth of dry (solid-phase) chemistry systems. Dry chemistry systems have made substantial gains over wet clinical analysis in the number of tests performed in hospitals, laboratories, and homes because of ease, reliability, and accuracy.

Wet chemistry methods for analysis of body analytes, eg, blood glucose or cholesterol, require equipment and trained analysts (see AUTOMATED INSTRUMENTATION). In contrast, dry chemistry systems can be used at home. Millions of people with diabetes check their blood glucose levels and are able to obtain results in a matter of a few minutes. An insulin delivery system that can respond to changes in the blood glucose level is not available. Injected insulin does not automatically adjust, and therefore the dose required to mimic the body's response must be adjusted daily or even hourly depending on diet and physical activity. Self-monitoring of blood glucose levels, essential for diabetics, has become possible owing to the advent of dry chemistry systems (1–8) (see INSULIN AND OTHER ANTIDIABETIC DRUGS). By regular and accurate monitoring of her blood glucose level by dry chemistry, an expectant diabetic mother can have a normal pregnancy and give birth to a healthy child. Athletes with diabetes can self-test their blood glucose to avoid significant problems. Dry chemistry systems are useful not only to diabetics, but also to patients having other medical problems. These systems are also used in animal diagnosis, food, fermentation, agriculture, and environmental and industrial monitoring.

The principles and biochemical reactions involved in diagnostic reagents are described herein. Construction of dry chemistry systems and advances are also addressed, as are biosensors.

Enzyme-Catalyzed Reactions in Solution

Measurement Considerations. A prototype enzyme-catalyzed reaction where one substrate (S) produces only one product (P) may be described by

$$E + S \underset{k_{-1}}{\overset{k_1}{\rightleftharpoons}} ES \overset{k_2}{\longrightarrow} E + P$$

where E is enzyme, ES is the enzyme–substrate complex, and k_i represents the reaction rate constants. The reaction can be followed by monitoring the loss of substrate or the formation of product. A graph of the concentration of substrate, or product, vs time gives an exponential curve. In the equilibrium method (end point) used for S determination, data are collected when the concentration of S or P are time-independent. Methods where data are obtained from the early linear part of the curve are known as kinetic methods (see KINETIC MEASURE-MENTS). The reaction rate of the enzyme-catalyzed reaction shown is given by the Michaelis-Menten equation:

$$v = V[S]/(K_m + [S])$$

where v represents the rate of reaction; $V = k_2[E_0]$, the maximum rate of the reaction when $[S]$ is the initial substrate concentration and $[E_0]$ is the initial enzyme concentration; and K_m is the Michaelis constant for the enzyme and this particular substrate.

For measurement of a substrate by a kinetic method, the substrate concentration should be rate-limiting and should not be much higher than the enzyme's K_m. On the other hand, when measuring enzyme activity, the enzyme concentration should be rate-limiting, and consequently high substrate concentrations are used (see CATALYSIS).

Glucose [50-99-7], urea [57-13-6] (qv), and cholesterol [57-88-5] (see STEROIDS) are the substrates most frequently measured, although there are many more substrates or metabolites that are determined in clinical laboratories using enzymes. Co-enzymes such as adenosine triphosphate [56-65-5] (ATP) and nicotinamide adenine dinucleotide [53-84-9] in its oxidized (NAD^+) or reduced (NADH) [58-68-4] form can be considered substrates. Enzymatic analysis is covered in detail elsewhere (9).

Assays using equilibrium (end point) methods are easy to do but the time required to reach the end point must be considered. Substrate(s) to be measured reacts with co-enzyme or co-reactant (C) to produce products (P and Q) in an enzyme-catalyzed reaction. The greater the consumption of S, the more accurate the results. The consumption of S depends on the initial concentration of C relative to S and the equilibrium constant of the reaction. A change in absorbance is usually monitored. Changes in pH and temperature may alter the equilibrium constant but no serious errors are introduced unless the equilibrium constant is small. In order to complete an assay in a reasonable time, for example several minutes, the amount and therefore the cost of the enzyme and co-factor may be relatively high. Sophisticated equipment is not required, however.

Indicators. There are certain compounds that are suitable as indicators for sensitive and specific clinical analysis. Nicotinamide adenine dinucleotide (NAD) occurs in oxidized (NAD^+) and reduced (NADH) forms. Nicotinamide adenine dinucleotide phosphate (NADP) also has two states, $NADP^+$ and NADPH. NADH has a very high uv–vis absorption at 339 nm, extinction coefficient = 6300 $(M \cdot cm)^{-1}$, but NAD^+ does not. Similarly, NADPH absorbs light very strongly whereas $NADP^+$ does not.

An example of the application of these compounds as indicators is in the determination of pyruvate.

$$\text{pyruvate} + \text{NADH} + \text{H}^+ \xrightarrow{\text{lactate dehydrogenase}} \text{L-lactate} + \text{NAD}^+$$

The absorbance change (ΔA) at 340 nm can be used to determine the amount of pyruvate remaining. The lactate dehydrogenase [9001-60-9] catalyzed reaction can also be used in the reverse direction to measure lactate. The reaction takes place in a buffer of pH 9–10 that neutralizes liberated H^+.

Two or more linked enzyme reactions can lead to a change in the concentration of NADH or NADPH that is equivalent to the concentration of the original analyte. The reference glucose measurement using hexokinase [9001-51-8] and glucose-6-phosphate dehydrogenase [9001-40-5] is an example:

$$\text{glucose} + \text{ATP} \xrightarrow{\text{hexokinase}} \text{glucose-6-phosphate} + \text{ADP}$$

$$\text{glucose-6-phosphate} + \text{NADP}^+ \xrightarrow[\text{dehydrogenase}]{\text{glucose-6-phosphate}} \text{6-phosphogluconate} + \text{H}^+ + \text{NADPH}$$

The second enzymatic reaction converts $NADP^+$ to NADPH and H^+, and the appearance of NADPH is measured at 340 nm.

In the enzymatic assays of cholesterol, glucose, and urea, oxygen is used and H_2O_2 is formed. The reaction for uric acid [69-93-2] is

$$\text{uric acid} + \text{O}_2 + 2\,\text{H}_2\text{O} \longrightarrow \text{allantoin} + \text{CO}_2 + \text{H}_2\text{O}_2$$

The H_2O_2 generated reacts with a chromogen in the presence of the enzyme peroxidase [9001-05-02] to produce a color change. A frequently used reaction is the peroxidase-catalyzed coupling of H_2O_2 and 4-aminoantipyrine [83-07-8] and phenol [108-95-2] to produce a quinoneimine dye. Methods using peroxidase are prone to interference by compounds such as ascorbic acid, uric acid, and acetylsalicylic acid.

Measurement of Analytes.   Biochemical reactions used in the measurement of selected analytes are commercially available as prepackaged kits of reagents. Measurement of the reactions given plus many other analytes can be made (10,11).

Cholesterol.   The end point for the cholesterol reaction can be determined by following dye formation. Additionally, the amount of oxygen consumed can be measured amperometrically by an oxygen-sensing electrode (see ELECTROANALYTICAL TECHNIQUES). The H_2O_2 produced by cholesterol oxidase requires phenol to produce dye.

$$\text{cholesterol esters} + \text{H}_2\text{O} \xrightarrow{\text{cholesterol esterase}} \text{cholesterol} + \text{fatty acids}$$

$$\text{cholesterol} + \text{O}_2 \xrightarrow{\text{cholesterol oxidase}} \text{cholest-4-en-3-one} + \text{H}_2\text{O}_2$$

$$\text{H}_2\text{O}_2 + \text{chromogen} + \text{phenol} \xrightarrow{\text{peroxidase}} \text{dye} + \text{H}_2\text{O}$$

A popular alternative to the step utilizing chromogen is to substitute *p*-hydroxybenzenesulfonate for phenol in the reaction with the pyridine nucleotide:

$$\text{H}_2\text{O}_2 + \text{ethanol} \xrightarrow{\text{catalase}} \text{acetaldehyde} + 2\,\text{H}_2\text{O}$$

$$\text{acetaldehyde} + \text{NADP}^+ \xrightarrow{\text{aldehyde dehydrogenase}} \text{acetate} + \text{H}^+ + \text{NADPH}$$

Free cholesterol can also be determined, if cholesterol esterase is omitted.

Citrate. The citrate reaction is followed by monitoring the decrease in the concentration of NADH. Oxaloacetate instantly decarboxylates to pyruvate.

$$\text{citrate} \xrightarrow{\text{citrate lyase}} \text{oxaloacetate} + \text{acetate}$$

$$\text{oxaloacetate} + \text{NADH} + \text{H}^+ \xrightarrow{\text{malate dehydogenase}} \text{L-malate} + \text{NAD}^+$$

$$\text{pyruvate} + \text{NADH} + \text{H}^+ \xrightarrow{\text{lactate dehydrogenase}} \text{L-lactate} + \text{NAD}^+$$

Creatinine. The most widely used creatinine methods are based on reaction between creatinine and picrate ions formed in an alkaline medium.

$$\text{creatinine} + \text{H}_2\text{O} \xrightarrow{\text{creatinase}} \text{creatine}$$

$$\text{creatine} + \text{ATP} \xrightarrow{\text{creatine kinase}} \text{creatine phosphate} + \text{ADP}$$

$$\text{ADP} + \text{phosphoenolpyruvate} \xrightarrow{\text{pyruvate kinase}} \text{ATP} + \text{pyruvate}$$

$$\text{pyruvate} + \text{NADH} + \text{H}^+ \xrightarrow{\text{lactate dehydrogenase}} \text{L-lactate} + \text{NAD}^+$$

The loss of NADH is followed for determination of the enzyme creatine kinase.

Galactose.

$$\text{galactose} \xrightarrow{\text{galactose oxidase}} \text{galactolactone} + \text{H}_2\text{O}_2$$

$$\text{H}_2\text{O}_2 + \text{chromogen} \xrightarrow{\text{peroxidase}} \text{dye} + \text{H}_2\text{O}$$

Glucose.

$$\text{glucose} + \text{O}_2 + \text{H}_2\text{O} \xrightarrow{\text{glucose oxidase}} \text{gluconic acid} + \text{H}_2\text{O}_2$$

$$\text{H}_2\text{O}_2 + \text{chromogen} + \text{phenol} \xrightarrow{\text{peroxidase}} \text{dye} + \text{H}_2\text{O}$$

The reaction can be followed by measurement of dye formation. The rate of oxygen depletion can be measured using an oxygen electrode. Additional reagents can be added to prevent the formation of oxygen from the generated H_2O_2. The reactions are as follows:

$$\text{H}_2\text{O}_2 + \text{ethanol} \xrightarrow{\text{catalase}} \text{acetaldehyde} + 2\,\text{H}_2\text{O}$$

$$\text{H}_2\text{O}_2 + 2\text{H}^+ + 2\,\text{I}^- \xrightarrow{\text{molybdate}} \text{I}_2 + 2\,\text{H}_2\text{O}$$

Lactate. The formation of NADH is followed by measuring the absorbance at 340 nm.

$$\text{L-lactate} + \text{NAD}^+ \xrightarrow{\text{lactate dehydrogenase}} \text{pyruvate} + \text{NADH} + \text{H}^+$$

Triglycerides.   The loss of NADH is followed at 340 nm. This is one of over 20 variations of the method.

$$\text{triglyceride} + 3\ H_2O \xrightarrow{\text{lipase}} \text{glycerol} + 3\ \text{fatty acids}$$

$$\text{glycerol} + \text{ATP} \xrightarrow{\text{glycerol kinase}} \text{glycerol-3-phosphate} + \text{ADP}$$

$$\text{ADP} + \text{phosphoenolpyruvate} \xrightarrow{\text{pyruvate kinase}} \text{pyruvate} + \text{ATP}$$

$$\text{pyruvate} + \text{NADH} + H^+ \xrightarrow{\text{lactate dehydrogenase}} \text{L-lactate} + \text{NAD}^+$$

Blood Urea Nitrogen.

$$\text{urea} + H_2O \xrightarrow{\text{urease}} 2\ NH_3 + CO_2$$

$$NH_3 + \alpha\text{-ketoglutarate} + \text{NADPH} + H^+ \xrightarrow{\text{glutamate dehydrogenase}} \text{NADP}^+ + H_2O + \text{glutamate}$$

Assay of Enzymes.   In body fluids, enzyme levels are measured to help in diagnosis and for monitoring treatment of disease. Some enzymes or isoenzymes are predominant only in a particular tissue. When such tissues are damaged because of a disease, these enzymes or isoenzymes are liberated and there is an increase in the level of the enzyme in the serum. Enzyme levels are determined by the kinetic methods described, ie, the assays are set up so that the enzyme concentration is rate-limiting. The continuous flow analyzers, introduced in the early 1960s, solved the problem of the high workload of clinical laboratories. In this method, reaction velocity is measured rapidly; the change in absorbance may be very small, but within the capability of advanced kinetic analyzers.

Enzymes, measured in clinical laboratories, for which kits are available include γ-glutamyl transferase (GGT), alanine transferase [*9000-86-6*] (ALT), aldolase, α-amylase [*9000-90-2*], aspartate aminotransferase [*9000-97-9*], creatine kinase and its isoenzymes, galactose-1-phosphate uridyl transferase, lipase, malate dehydrogenase [*9001-64-3*], 5′-nucleotidase, phosphohexose isomerase, and pyruvate kinase [*9001-59-6*]. One example is the measurement of aspartate aminotransferase, where the reaction is followed by monitoring the loss of NADH:

$$\text{L-aspartate} + \alpha\text{-ketoglutarate} \xrightarrow{\text{aspartate aminotransferase}} \text{oxaloacetate} + \text{L-glutamate}$$

$$\text{oxaloacetate} + \text{NADH} + H^+ \xrightarrow{\text{malate dehydrogenase}} \text{L-malate} + \text{NAD}^+$$

A second is the measurement of creatine kinase:

$$\text{creatine phosphate} + \text{ADP} \xrightarrow{\text{creatine kinase}} \text{creatine} + \text{ATP}$$

$$\text{ATP} + \text{glucose} \xrightarrow{\text{hexokinase}} \text{glucose-6-phosphate} + \text{ADP}$$

$$\text{glucose-6-phosphate} + \text{NADP}^+ \xrightarrow[\text{dehydrogenase}]{\text{glucose-6-phosphate}} \text{6-phosphogluconate} + \text{NADPH} + H^+$$

This reaction is followed by monitoring NADPH formation.

Immobilized Enzymes in Diagnostic Reagents. The use of immobilized, instead of soluble, enzymes for measurement of analytes has received considerable attention, especially for clinical analyses (12–17). Use of immobilized enzymes offers the advantages of greater accuracy, stability, and convenience. Only a few methods utilizing immobilized enzymes have become commercially available, although these methods may not have achieved full potential in clinical chemistry. Dry chemistry has outpaced all other methods combined in clinical chemistry. Biosensors are gaining momentum, however, and are expected to continue to increase in usage as a result of advances in redox polymers.

Dry Chemistry

Diagnostic medicine is placing demands on technology for newer materials and application techniques, and polymers are finding ever-increasing use in diagnostic medical reagents (7).

Dry or solid-phase chemistry has origins that reach back to the ancient Greeks. Some 2000 years ago, copper sulfate was an important ingredient in tanning and preservation of leather (qv). Dishonest traders were adulterating valuable copper sulfate with iron salts. The first recorded use of a dry chemistry system was described by Pliny. The method for detection of iron involved soaking reeds of papyrus in a plant gall infusion or a solution of gallic acid. The papyrus would turn black in the presence of iron. In 1830, a filter paper impregnated with silver carbonate was used to detect uric acid qualitatively.

Self-testing, eg, measuring one's own self, is likewise not a new concept. Around the turn of the century diabetics were encouraged to monitor their glucose level by testing their urine with Benedict's qualitative test (18). When insulin became available in the early 1920s, self-testing of urine became necessary. In the mid-1940s, a dry tablet was compounded consisting of sodium hydroxide, citric acid, sodium carbonate, and cupric sulfate (19). Adding this tablet to a small urine sample resulted in boiling of the solution and reduction of the blue cupric sulfate to a yellow or orange color if glucose was present. Glucose urine strips, impregnated and based on enzymatic reactions using glucose oxidase, peroxidase, and an indicator were introduced around 1956 (20). Dry chemistry blood glucose test strips coated with a semipermeable membrane to which whole blood could be applied and wiped off were introduced in 1964. As of the mid-1990s, low cost, lightweight plastic-housed reflectance meters having advanced data management capabilities are available and extensively used.

Impregnated dry reagents have coarse texture, high porosity, and uneven large pore size, resulting in nonuniform color development in the reacted strip. In the early 1970s, a coating-film type of dry reagent was developed by applying an enzymatic coating onto a plastic support. This gave the surface a smooth, fine texture, resulting in uniform color development (21). Nonwipe dry chemistry systems, wherein the devices handle the excess blood by absorption or capillary action, appeared in the marketplace in the late 1980s.

Discrete multilayered coatings, developed by the photographic industry (see PHOTOGRAPHY), were adapted in the late 1970s to coat dry reagent chemistry formats for clinical testing (22,23). Each zone of a multilayered coating provides a unique environment for sequential chemical and physical reactions. These devices consist of a spreading layer, separation membrane, reagent zone, and reflective zone, coated onto the base support. The spreading layer wicks the sample and applies it uniformly to the next layer. The separation layer can hold back certain sample components, eg, red blood cells, allowing only the desired metabolites to pass through to the reagent zone that contains all the necessary reagent components. Other layers may be incorporated for filtering, reflecting, or eliminating interfering substances.

Background.   Enzymes are essential in user-friendly diagnostic dry chemistry systems. In a typical glucose-measuring dry reagent, glucose oxidase (GOD) and peroxidase (POD), along with a suitable indicator eg, 3,3′,5,5′-tetramethylbenzidine (TMB), are dissolved or dispersed in a latex or water-soluble polymer. This coating is applied to a lightly pigmented plastic film and dried to a thin film. The coated plastic can be cut to a suitable size, eg, 0.5 cm × 0.5 cm, and mounted on a plastic handle. The user applies a drop of blood on the dry reagent pad, allowing contact for 60 s or less. The blood can be wiped off manually with a swab or by the device itself. The developed color is then read by a meter or visually compared with the predesignated printed color blocks to determine the precise glucose level in the blood.

Dry chemistry test kits are available in thin strips that are usually disposable. They may be either film-coated or impregnated. The most basic diagnostic strip consists of a paper or plastic base, polymeric binder, and reactive chemistry components consisting of enzymes, surfactants, buffers, and indicators. Diagnostic coatings or impregnation must incorporate all reagents necessary for the reaction. The coating can be either single or multilayer in design. A list of analytes, enzymes, drugs, and electrolytes assayed by dry chemistry diagnostic test kits follows (24):

Analytes
glucose
urea
urate
cholesterol (total)
triglycerides
bilirubin (total)
ammonium ions
creatinine
calcium
hemoglobin
HDL cholesterol
magnesium(II)
phosphate (inorganic)
albumin
protein in cerebrospinal fluid

Enzymes
alkaline phosphate (ALP)

lactate dehydrogenase (LDH)
creatine kinase
MB isoenzyme (CK–MB)
lipase
amylase (total)

Drugs
phenobarbitone
phenytoin
theophylline
carbamazepine

Electrolytes
sodium ion
chloride
carbon dioxide

Dry chemistry systems are widely used in physician's offices and hospital laboratories, and by millions of patients in their own homes worldwide. These systems
are used for routine urinalysis, blood chemistry determinations, and immuno-
logical and microbiological testing. The main advantage of this technology is
elimination of the need for reagent preparation and many other manual steps
common to liquid reagent systems. This yields greater consistency and reliability
of test results. Furthermore, dry chemistry systems have longer shelf stability
and hence there is a reduced waste of reagents. Each test unit contains all the
reagents and reactants necessary to perform assays.

Dry chemistry tests are used for the assay of metabolites by concentration
or by activity in a biological matrix. In general, reactive components are present
in amounts in excess of the analyte being determined to make sure that the
reactions go to completion quickly. Other enzymes or reagents are used to drive
the reactions in the desired direction (25). Glucose and cholesterol are the
analytes most commonly measured.

Components of Dry Chemistry Systems. The basic components of typical
dry chemistry systems that utilize reflectance measurement are a base support
material, a reflective layer, and a reagent layer which can be either single-
or multilayered. The base layer serves as a building base for the system and
usually is a thin, rigid thermoplastic film. The reflective layer, usually a white,
pigment-filled plastic film, coating, foam, or paper, reflects whatever light is not
absorbed as a result of the chemistry, to the detector. The reagent layer contains
the integrated reagents for a specific set of reactions. Typical materials include
paper or fiber matrix and coating film as well as various combinations.

A test system having a single-layer coating reagent that effectively excludes
red blood cells (RBC) has an emulsion-based coating containing all the reagents
for a specific chemistry. The emulsion is coated onto a lightly filled thermoplastic
film and dried. For glucose measurement, the coating should contain GOD, POD,
and TMB. It may also contain a buffer for pH adjustment, minor amounts of
ether–alcohol type organic coalescing agent, and traces of a hindered phenol-
type antioxidant to serve as a color signal ranging compound. Whole blood is
applied and allowed to remain in contact for 60 s or less. Excess blood is wiped
off and the developed color is read visually or by meter. Another type of dry
chemistry system consists of a reagent-impregnated paper in between a film
membrane and a reflective support layer.

The schematic of a typical multilayer coating dry test system, in this case a
test for blood urea nitrogen (BUN) is shown in Figure 1. Sample containing BUN
is spread uniformly by the top layer, a spreading and reflective layer. Immediately underneath, the first reagent layer is a porous coating film containing the
enzyme urease [9002-13-15] and buffer (pH 8.0). A semipermeable membrane
coating allows the NH_3 formed in the first reagent layer to permeate to the
second reagent layer excluding OH^-. The second reagent layer is composed of a
porous coating film containing a pH indicator. The indicator color develops when
NH_3 reaches the semipermeable coating film. Typically, such dry reagents are
slides (2.8-cm $\times$ 2.4-cm) having an application area of 0.8 cm^2, and the spreading layer is about 100 μm thick. Touch-and-drain dry chemistry construction
is shown in Figure 2 (6). This construction is useful in evaluation of diagnostic
coating films and may also find acceptance by users.

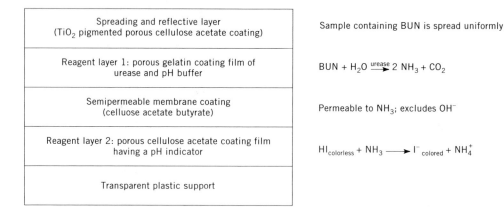

Spreading and reflective layer (TiO_2 pigmented porous cellulose acetate coating)	Sample containing BUN is spread uniformly
Reagent layer 1: porous gelatin coating film of urease and pH buffer	$BUN + H_2O \xrightarrow{urease} 2\ NH_3 + CO_2$
Semipermeable membrane coating (celluose acetate butyrate)	Permeable to NH_3; excludes OH^-
Reagent layer 2: porous cellulose acetate coating film having a pH indicator	$HI_{colorless} + NH_3 \longrightarrow I^-_{colored} + NH_4^+$
Transparent plastic support	

Fig. 1. A multilayer coating dry chemistry test for blood urea nitrogen (BUN) where HI and I^- represent the acid base forms of a pH indicator, respectively (24). See text. Courtesy of the American Chemical Society.

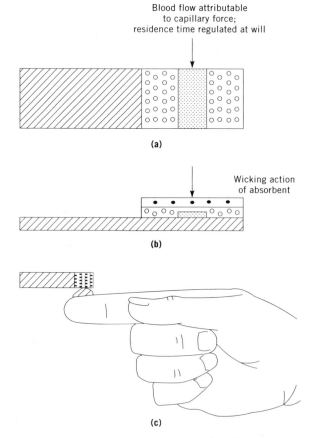

Fig. 2. Touch-and-drain dry chemistry construction: (**a**) dry coated surface; (**b**) cross-section of dry coated surface, adhesive, and cover piece; (**c**) contact with blood drop results in blood filling the cavity. After desired reaction time, blood is drained off by touching end of cavity with absorbent material (6). Courtesy of the American Chemical Society.

Polymers and Coatings.   Advances in polymer chemistry have resulted in many successful medical devices, including diagnostic assays (26). Polymers (qv), which can be manufactured in a wide range of compositions, are used to enhance speed, sensitivity, and versatility of both biosensors and dry chemistry systems to measure vital analytes. Their properties can be regulated by composition variations and modifications. Furthermore, polymers can be configured into simple to complex shapes.

Polymers.   In most dry chemistry systems, polymers account for more than 95% of the strips. Polymer chemistry has been linked to biochemistry in order to develop improved dry chemistries (6,24). The polymer binder incorporates the system's chemistry components in the form of either a coating or impregnation. The reagent matrix must be carefully selected to mitigate or eliminate nonuniformity in the concentrations of reagents resulting from improper mixing, settling, or nonuniform coating thickness. Aqueous-based emulsion polymers and water-soluble polymers (qv) are extensively used. A list of commonly used matrix binders follows (27,28).

Emulsion polymers	Water-soluble polymers
acrylics	poly(vinyl alcohol)
poly(vinyl acetate):	polyvinylpyrrolidinone
homo and copolymers	highly hydroxylated acrylic
styrene acrylics	poly(vinylethylene glycol acrylate)
polyvinyl propionate:	polyacrylamide
homo and copolymers	hydroxyethyl cellulose
ethylene vinyl acetate	other hydrophilic cellulosics
lightly cross-linkable acrylics	various copolymers
polyurethanes	

Polymers must be carefully screened and selected to avoid interfering with the analyte chemistry. The polymer properties, eg, composition, solubility, viscosity, solid content, surfactants (qv), residual initiators, film-forming temperature, and particle size are all important to the dry chemistry system (29). In general, the polymer should be a good film former and have good adhesion to the support substrate. Furthermore, it should have no or minimal tack for handling purposes during manufacturing of the strip. The coated matrix or impregnation must have the desired pore size and porosity to allow penetration of the analyte being measured, as well as the desired gloss, swelling characteristics, and surface energetics. Swelling of the polymer binder owing to the absorption of the liquid sample may or may not be advantageous, depending on the system. Emulsion polymers have a distinct advantage over water-soluble ones because of high molecular weight, superior mechanical properties, and potential for adsorbing enzymes and indicators by micellar forces. Polymeric binders used in multilayered coatings include various emulsion polymers; gelatin (qv); polysaccharides such as agarose; water-soluble polymers such as polyacrylamide, polyvinylpyrrolidinone, poly(vinyl alcohol), and copolymers of vinylpyrrolidinone and acrylamide;

and hydrophilic cellulose derivatives, eg, hydroxyethylcellulose and methylcellulose.

Water-Borne Coatings.   The most important coating film is that of water-borne coatings because enzymes are water soluble. During the late 1980s a tough coating film which would allow whole blood to flow over the film was developed (7). This system is shown in Figure 3. Soak-through technology is no longer utilized. Red blood cells (RBC) must roll over and not stick to the coating film. Investigations involving almost all types of emulsion and water-soluble polymers showed a styrene–acrylate emulsion polymer to be most suitable. A coating containing 51 g styrene–acrylate emulsion in water (50% solids), 10 g linear alkylbenzene sulfonate (15 wt % in water), 0.1 g GOD, 0.23 g POD, 0.74 g TMB, 10.0 g hexanol, and 13.2 g 1-methoxy-2-propanol gave a good dose response for measurement of blood glucose, as seen in Figure 4. This coating film gave a linear dose response, in transmission mode, up to 1500 mg/dL glucose. The coating was slightly tacky and the measurements were found to be highly dependent on the blood residence time.

To mitigate or eliminate tack, ultrafine mica was found to be effective (see MICAS). This made rolling and unrolling of coated plastic for automated strip construction possible. In devices wherein blood residence time is regulated, the residence time-dependent coating is not a problem. Dependence of results on blood residence time creates a serious problem where residence time cannot be regulated by the strip. To make the coating film residence time-independent,

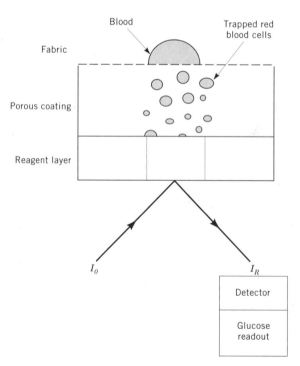

Fig. 3.   Design of topover diagnostic coatings, where I_0 and I_R represent the initial and reflected light intensities, respectively.

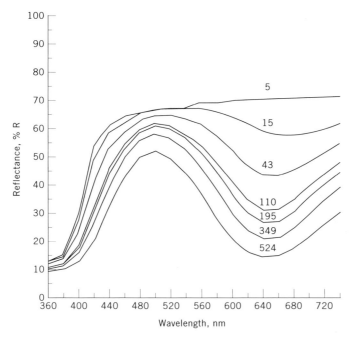

Fig. 4. Dose response for blood glucose measurement using a dry chemistry system having a water-borne, tough coating film. The numbers represent glucose concentrations in mg/dL. A 120-s blood residence time was used.

continuous leaching of color from films during exposure time is essential. Enzymes and indicator can leach rapidly near the blood/coating film interface and develop color that can be drained off. The color formed on the coating film is then independent of the exposure time. It was found that this gave a coating film having minimum time dependency. A low molecular weight polyvinylpyrrolidinone, eg, PVP K-15 from GAF or PVP K-12 from BASF, gave coatings having no residence time dependency. Such a coating consists of 204 g styrene–acrylate emulsion (50% solids), 40 g linear alkylbenzene sulfonate (15% solution in water), 0.41 g GOD (193 unit/mg), 1.02 g POD (162 unit/mg), 2.96 g TMB, 7.4 g PVP K-12, 32 g micromica C-4000, 20 g hexanol, and 6 g Igepal CO-530 (a surfactant). The nonionic surfactant serves as a surface modifier to eliminate RBC retention.

Excellent correlation was found when results at 660 nm and 749 nm were compared using a reference hexokinase glucose method (27). The dose response was excellent up to 300 mg/dL glucose. In general, water-borne coatings do not lend themselves to ranging by antioxidants (qv).

Nonaqueous Coatings. Since the 1970s, dry reagent coatings have been exclusively water-borne because of the belief that enzymes function only in aqueous medium. Nonaqueous enzymatic coatings for dry chemistries have been researched, developed, and refined, however (24,30); red blood cells do not adhere to such coatings. Additionally, quick end points are obtained. These coatings give superior thermostability. Furthermore, these coatings can be easily ranged by antioxidants, whereas water-borne coatings are difficult to range. Nonaqueous hydroxylated acrylic polymers have been synthesized which have good hydrophilicity and hydrogel character. The enzymes GOD and POD are insoluble in organic solvents but become extremely rigid and can be dispersed with ease. Dispersions of less than 1 μm were made using an Attritor mill or a ball mill. To prepare nonaqueous coatings, polymer solution, TMB, mica, surface modifiers, and solvents were added to the enzyme dispersion and slightly mixed on a ball mill. Ranging compound can be post-added. The composition of a typical nonaqueous coating useful for low range blood glucose measurement, gram basis, is 33.29 hydroxyethyl methacrylate–butyl methacrylate–dimethylaminoethyl methacrylate (65:33:2) polymer (40% solid), 2.38 TMB, 1.17 GOD, 2.68 POD, 3.28 sodium dodecyl benzene sulfonate [25155-30-0], 26.53 xylene, 26.53 1-methoxypropanol, and 1.96 cosmetic-grade C-4000 ultrafine mica. Many surfactants and surface modifiers that eliminated RBC retention were investigated. Antioxidants that function as ranging compounds in these nonaqueous systems include 3-amino-9-(aminopropyl)-carbazole dihydrochloride, butylated hydroxy toluene [128-37-0] (BHT), and a combination of BHT–propyl gallate. These ranging compounds are effective in a ranging compound-to-TMB indicator molar ratio of 1:2.5–1:20.

The long-term stability of the nonaqueous coating films under elevated temperature and moderate humidity is reported to be better than aqueous coatings (30). Furthermore, color resolution and sensitivity of reacted nonaqueous coating films are excellent.

Molded Dry Chemistry.   In general, most enzymes are very fragile and sensitive to pH, solvent, and elevated temperatures. The catalytic activity of most enzymes is reduced dramatically as the temperature is increased. Typical properties of diagnostic enzymes are given in Table 1. Common enzymes used in diagnostics, eg, GOD and POD, are almost completely deactivated around 65°C in solid form or in aqueous solution. Thermal analysis work on these biopolymers was reported in 1991 (31). The differential scanning calorimetry (dsc) results indicating glass-transition temperature (T_g), melting temperature (T_m), and decomposition temperature (T_d) are shown in Table 2. Below T_g, the enzymes are in a glassy state and should be thermally stable. Around T_g, onset of the rubbery state begins, and the enzyme becomes prone to thermal instability. When the enzymes melt around T_m, all the tertiary structures are destroyed, thus making the enzyme completely inactive. The presence of ionic salts and other chemicals can considerably influence enzyme stability. The redox center, flavin adenine dinucleotide (FAD), in GOD can conduct electrons and is catalytically relevant. To keep or sustain enzymatic activity, the redox centers must remain intact. The bulk of the enzyme, polymeric in composition, is an insulator, thus altering it does not reduce the enzyme's catalytic activity. An enzymatic compound containing GOD, POD, TMB, a linear alkylbenzene sulfonate, and polyhydyroxyethyl methacrylate (PHEMA) compression molded between 105–150°C has given a response to glucose (32). Molding at 200°C resulted in enzyme deactivation. A mechanism has been proposed where the enzymes are protected by the tight PHEMA coils. It has been suggested that molding of strips using reaction injection molding (RIM) may lead to useful chemistries, including biosensors, in the future.

Application of Diagnostic Technology in Monitoring Diabetes.   Very frequent measurements of blood glucose to manage diabetes are one of the most important applications of diagnostic reagents. It has been estimated that there

Table 1. Properties of Diagnostic Enzymes

Parameter	Cholesterol oxidase (CO)	Cholesterol esterase (CE)	Glucose oxidase (GOD)	Peroxidase (POD)
source	*Streptomyces*	*Pseudomonas*	*Aspergillus*	horseradish
EC	1.1.3.6	1.1.13	1.1.3.4	1.11.1.7
CAS Registry Number	[9028-76-6]	[9026-00-0]	[9001-37-0]	[9001-05-02]
molecular weight	34,000	300,000	153,000	40,000
isoelectric point	5.1 ± 0.1, 5.4 ± 0.1	5.95 ± 0.05	4.2 ± 0.1	
Michaelis constant, M	4.3×10^{-5}	2.3×10^{-5}	3.3×10^{-2}	
inhibitor	Hg^{2+}, Ag^{+}	Hg^{2+}, Ag^{+}	Hg^{2+}, Ag^{+}, Cu^{2+}	CN^{-}, S^{2-}
pH, optimum	6.5–7.0	7.0–9.0	5.0	6.0–6.5
temperature, optimum, °C	45–50	40	30–40	45
pH stability[a]	5.0–10.0	5.0–9.0	$4.0–6.0^{b}$	5.0–10.0

[a]At 25°C for 20 h, unless otherwise indicated.
[b]At 40°C for 1 h.

Table 2. DSC Analysis of Diagnostic Enzymes

Enzyme	Source	T_g, °C	T_m, °C	T_d, °C
cholesterol oxidase	*Nocardia*	50	98	210
cholesterol oxidase	*Steptomyces*	51	102	250
cholesterol esterase	*Pseudomonas*	43	88	162
glucose oxidase	*Aspergillus*	50	105	220
peroxidase	horseradish	50	100	225

are about 15×10^6 diabetics in the United States, although only half that number have been diagnosed with the disease. More than 1.5 million diabetics are treated with injected insulin. The rest are treated with weight loss, diet, and oral antidiabetic drugs, eg, the sulfonylureas Tolbutamide, Tolazamide, Chloropropamide, Glipizide, and Glyburide.

The U.S. market for drugs to control blood glucose totals about 1×10^9, equally divided between insulin and all other antidiabetic drugs (33). Insulin sales are expected to grow by about 10% annually, whereas the antidiabetic drug market as a whole is expected to shrink by about 3%. The blood glucose monitoring market totals about 7.5×10^8 in the United States and is expected to grow at a rate of 10% annually.

Biosensors

Biosensors (qv) and DNA probes are relatively new to the field of diagnostic reagents. Additionally, a near-infrared (nir) monitoring method (see INFRARED TECHNOLOGY AND RAMAN SPECTROSCOPY), a reagentless, noninvasive system, is under investigation. However, prospects for a nir detection method for glucose and other analytes are uncertain.

In the early 1960s, a promising approach to glucose monitoring was developed in the form of an enzyme electrode that used oxidation of glucose by the enzyme GOD (34). This approach has been incorporated into a few clinical analyzers for blood glucose determination. Three detection methods used in the glucose enzyme electrode are shown in Figure 5. Oxygen consumption, measured in the earliest method (Fig. 5a), requires a reference, nonenzymatic electrode to provide an amperometric signal. The second approach (Fig. 5b) detects H_2O_2 but requires an applied potential of about 650 mV and an inside permselective membrane. The third-generation biosensor (Fig. 5c) takes advantage of the fact that the enzymatic reaction happens in two steps. The GOD enzyme is reduced by glucose, and then the reduced enzyme is oxidized by an electron acceptor, ie, a mediator, specifically, a redox polymer. Direct electron transfer between GOD and the electrode occurs extremely slowly; therefore an electron acceptor mediator is required to make the process rapid and effective (35).

Much work has been done on exploration and development of redox polymers that can rapidly and efficiently shuttle electrons. In several instances an enzyme has been attached to the electrode using a long-chain polymer having a dense array of electron relays. The polymer which penetrates and binds the enzyme is also bound to the electrode.

Extensive work has been done on osmium-containing polymers (see METAL-CONTAINING POLYMERS; SUPPLEMENT). Large numbers of such polymers have been made and evaluated (36). The most stable and reproducible redox polymer of this kind is a poly(4-vinyl pyridine) (PVP) to which $Os(bpy)_2Cl_2$, where bpy = 2,2'-bipyridine, has been attached to 1/16th of the pendant pyridine groups. The resultant redox polymer is water insoluble and biologically compatible by partial quaternization of the remaining pyridine groups using 2-bromoethyl

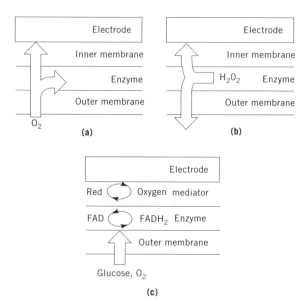

Fig. 5.   Detection methods for glucose enzyme electrode based on (**a**) oxygen, (**b**) hydrogen peroxide, and (**c**) a mediator. See text.

amine. The newly introduced quaternized amine groups can react with a water-soluble epoxy, eg, polyethylene glycol diglycidyl ether, and GOD to produce a cross-linked biosensor coating film. Such coating films produce high current densities and a linear response to glucose up to 600 mg/dL. The synthesis and application of osmium polymers have been refined (37). Osmium monomers that can also shuttle electrons much as the polymer does have also been made (38). A schematic depiction of these polymer–GOD hydrogel films is shown in Figure 6 (38).

Flexible polymer chains have also been used for relays (39,40) to provide communication between GOD's redox centers and the electrode. These ferrocene-modified siloxane polymers are stable and nondiffusing. Biosensors based on these redox polymers gave good response and superior stability. Commercial electrochemical microbiosensors, eg, Exactech (Medisense) and a silicon-based 6+ system (*i*-Stat) have appeared in the marketplace. These newer technologies should certainly impact rapid blood chemistry determinations by the year 2000. A typical important example is that of blood glucose determination using very small (<5-μL) blood volumes obtained by finger-stick. These detection instruments can be designed more compactly than optoelectronic systems. There is

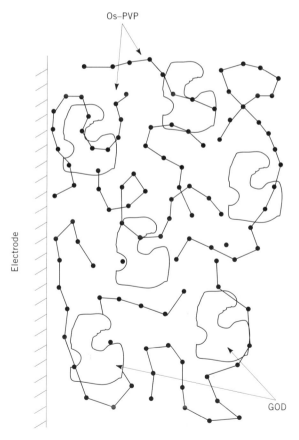

Fig. 6.   Depiction of (-•••-) Os–PVP polymer–GOD hydrogel film on an electrode.

certainly a market for small, disposable electrochemical tests to be used in the emergency room and in surgical and critical care units, as well as at home.

A compound which is a good choice for an artificial electron relay is one which can reach the reduced $FADH_2$ active site, undergo fast electron transfer, and then transport the electrons to the electrodes as rapidly as possible. Electron-transport rate studies have been done for an enzyme electrode for glucose (G) using interdigitated array electrodes (41). The following mechanism for redox reactions in osmium polymer–GOD biosensor films has been proposed.

$$GOD(FAD) + G \underset{k_{-1}}{\overset{k_1}{\rightleftharpoons}} GOD(FAD){\cdot}G \overset{k_2}{\longrightarrow} GOD(FADH_2) + \text{glucolactone}$$

$$GOD(FADH_2) + 2\ Os(III) \overset{k_3}{\longrightarrow} GOD(FAD) + 2\ Os(II) + 2\ H^+$$

$$Os_1(II) + Os_2(III) \overset{k_e}{\longrightarrow} Os_1(III) + Os_2(II)$$

$$Os_2(II) \overset{fast}{\longrightarrow} Os_2(III) + e^-$$

The next generation of amperometric enzyme electrodes may well be based on immobilization techniques that are compatible with microelectronic mass-production processes and are easy to miniaturize (42). Integration of enzymes and mediators simultaneously should improve the electron-transfer pathway from the active site of the enzyme to the electrode.

Functionalized conducting monomers can be deposited on electrode surfaces aiming for covalent attachment or entrapment of sensor components. Electrically conductive polymers (qv), eg, polypyrrole, polyaniline [25233-30-1], and polythiophene [25233-34-5], can be formed at the anode by electrochemical polymerization. For integration of bioselective compounds or redox polymers into conductive polymers, functionalization of conductive polymer films, whether before or after polymerization, is essential. In Figure 7, a schematic representation of an amperometric biosensor where the enzyme is covalently bound to a functionalized conductive polymer, eg, β-amino(polypyrrole) or poly[N-(4-aminophenyl)-2,2$'$-dithienyl]pyrrole, is shown. Entrapment of ferrocene-modified GOD within polypyrrole is shown in Figure 8.

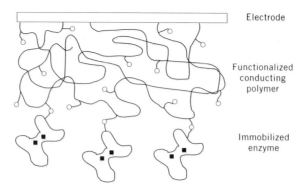

Fig. 7.   Schematic representation of enzyme covalently bound to a functionalized conductive polymer where ($\circ$) represents the functional group on the polymer and ($\blacksquare$) the active site on the enzyme (42). Courtesy of the American Chemical Society.

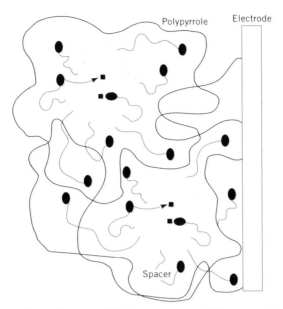

Fig. 8.   Entrapment of mediator-modified enzymes within a conductive polymer film where (●) represents the mediator ferrocene and (■) the active site of the enzyme glucose oxidase (GOD) (42).

There is a pressing need for an implantable glucose sensor for optimal control of blood glucose concentration in diabetics. A biosensor providing continuous readings of blood glucose would be most useful at the onset of hyper- or hypoglycemia, enabling a patient to take corrective measures. Furthermore, incorporating such a biosensor into a closed-loop system having a microprocessor and an insulin infusion pump could provide automatic regulation of the patient's blood glucose. Two novel technologies have been used in the fabrication of a miniature electroenzyme glucose sensor for implantation in the subcutaneous tissues of humans with diabetes (43). An electrodeposition technique has been developed to electrically attract GOD and albumin onto the surface of the working electrode. The resultant enzyme–albumin layer was cross-linked by glutaraldehyde [111-30-8]. A biocompatible polyethylene glycol–polyurethane copolymer has also been developed to serve as the outer membrane of the sensor to provide differential permeability of oxygen relative to glucose, in order to avoid oxygen deficit encountered in physiologic tissues.

BIBLIOGRAPHY

"Medical Diagnostic Reagents" in *ECT* 3rd ed., Vol. 15, pp. 74–92, by H. E. Spiegel, Hoffmann-La Roche Inc.

1. B. Walter, *Anal. Chem.* **55**, 449A (1983).
2. A. H. Free and H. M. Free, *Lab. Med.* **15**, 1595 (1984).
3. T. K. Mayer and N. K. Kubasik, *Lab. Mgmt.*, 43 (April 1986).
4. J. A. Jackson and M. E. Conrad, *Am. Clin. Products Rev.* **6**, 10 (1987).

5. A. F. Azhar, A. D. Burke, J. E. DuBois, and A. M. Usmani, *Polymer Mater. Sci. Eng.* **59**, 1539 (1988).
6. E. Diebold, M. Rapkin, and A. Usmani, *Chem. Tech.* **21**, 462 (1991).
7. M. T. Skarstedt and A. M. Usmani, *Polym. News* **14**, 38 (1989).
8. R. S. Campbell and C. P. Price, *J. Intern. Fed. Clin. Chem.* **3**, 204 (1991).
9. H. U. Bergmeyer, ed., *Methods of Enzymatic Analysis*, 3rd ed., Academic Press, New York, 1983.
10. J. F. Zilva and P. R. Pannall, *Clinical Chemistry in Diagnosis and Treatment*, 3rd ed., Lloyd-Luke, London, 1984.
11. L. G. Whitby, I. W. Percy-Robb, and A. F. Smith, *Lecture Notes on Clinical Chemistry*, Blackwell, London, 1984.
12. P. W. Carr and L. D. Bowers, *Immobilization of Enzymes in Analytical and Clinical Chemistry*, John Wiley & Sons, Inc., New York, 1980.
13. C. G. Guilbault, *Handbook of Enzymatic Methods of Analysis*, Marcel Dekker, Inc., New York, 1976.
14. A. H. Free, *Ann. Clin. Lab. Sci.* **7**, 479 (1987).
15. T. T. Ngo, *Int. J. Biochem.* **11**, 459 (1980).
16. L. D. Bowers, *Trends Anal. Chem.* **1**, 191 (1982).
17. H. A. Mottola, *Anal. Chim. Acta.* **145**, 27 (1983).
18. E. P. Joslin, H. P. Root, P. White, and A. Marble, *The Treatment of Diabetes*, 7th ed., Lea and Febiger, Philadelphia, Pa., 1940.
19. U.S. Pat. 2,387244 (1945), W. A. Compton and J. M. Treneer.
20. A. H. Free, E. C. Adams, and M. L. Kercher, *Clin. Chem.* **3**, 163 (1957).
21. U.S. Pat. 3,630957 (1971), H. G. Rey, P. Rieckmann, H. Wielinger, and W. Rittersdorf (to Boehringer Mannheim).
22. T. L. Shirey, *Clin. Biochem.* **16**, 147 (1983).
23. U.S. Pat. 3,992157 (1976), E. P. Przybylowicz and A. G. Millikan (to Eastman Kodak).
24. A. M. Usmani, in A. M. Usmani and N. Akmal eds., *Diagnostic Biosensor Polymers*, ACS Symposium Series 556, ACS Books, Washington, D.C., 1994, p. 2.
25. N. W. Tietz, *Fundamental of Clinical Chemistry*, W. B. Saunders, Philadelphia, 1976.
26. A. F. Azhar and A. M. Usmani, in A. G. Maadhah ed., *Handbook of Polymer Degradation*, Marcel Dekker, Inc., New York, 1992, p. 575.
27. A. M. Usmani, Biotechnology Symposium, Atlanta, 1992.
28. U.S. Pat. 4,006403 (1978), B. J. Bruschi, (to Eastman Kodak).
29. W. Scheler, *Makromol. Chem. Symp.* **12**, 1 (1987).
30. U.S. Pat. 5,260195 (Nov. 9, 1993), A. F. Azhar and co-workers (to Boehringer Mannheim).
31. J. E. Kennamer and A. M. Usmani, *J. App. Polym. Sci.* **42**, 3073 (1991).
32. J. E. Kennamer, A. D. Burke, and A. M. Usmani, in C. G. Gebelein and C. E. Carraher eds., *Biotechnology and Bioactive Polymers*, Plenum Publishing Corp., New York, 1992.
33. S. C. Stinson, *Chem. Eng. News*, 635, (Sept. 30, 1991).
34. L. C. Clark and C. Lyons, *Ann. N.Y. Acad. Sci.* **102**, 29 (1962).
35. G. Reach and G. S. Wilson, *Anal. Chem.* **64**, 381A (1992).
36. B. A. Gregg and A. Heller, *J. Phys. Chem.* **95**, 5970 (1991).
37. Technical data, A. M. Usmani and D. Deng, Boehringer Mannheim, Indianapolis, Ind., 1992.
38. Technical data, A. M. Usmani, N. A. Surridge and E. R. Diebold, Boehringer Mannheim, Indianapolis, Ind., 1992.
39. L. Boguslavsky, P. Hale, T. Skotheim, H. Karan, H. Lee, and Y. Okamoto, *Polym. Mater. Sci. Eng.* **64**, 322 (1991).
40. H. I. Karan, H. L. Lan, and Y. Okamoto, in Ref. 24, p. 169.
41. N. A. Surridge, E. R. Diebold, J. Chang, and G. W. Neudeck, in Ref. 24, p. 47.

42. W. Schuhmann, in Ref. 24, p. 110.

43. K. W. Johnson and co-workers, Ref. 24, p. 84.

ARTHUR M. USMANI
Bridgestone/Firestone, Inc.

MEDICAL IMAGING TECHNOLOGY

Medical imaging is the application of nonsurgical techniques to produce images of internal organs and tissues. Imaging has become an indispensable tool to the medical community, providing information otherwise obtainable only by invasive exploratory procedures. Imaging technology can be used to determine if a persistent headache is caused by a brain tumor, or if a fetus is developing normally. This technology can sometimes go beyond locating pathology; in some cases it can identify the type of pathology. For example, it is often possible to distinguish between images of benign and malignant tumors, thus eliminating the need for some invasive procedures. Imaging technology has also extended the frontiers of medicine in many areas. Mapping brain functions and fetal surgery *in utero* are two examples (see also NONDESTRUCTIVE EVALUATIONS).

The five principal imaging technologies involve optical, x-ray, ultrasound, radio frequency (rf), or nuclear techniques. Additionally, medical imaging relies heavily on hundreds of ancillary chemical, computer, detector, electronic, film, and magnetic technologies developed in the latter twentieth century. The discussion herein includes basic imaging principles, and endoscopic, x-ray, ultrasound, magnetic resonance, and nuclear imaging as found in hospitals. Endoscopy is a general term used to describe any optical medical imaging technique producing images of the inside of the body by the insertion into the body of an optical imaging device, such as a small video camera, fiber optic viewer, or light pipe. X-ray imaging includes plane film and computerized tomography (CT) imaging techniques, both based on the absorption of x-rays passing through the body. X-ray imaging is used for hard tissues, such as bones, and soft tissues. Ultrasound imaging, based on the reflection of ultrasonic sound waves by the tissues of the body, is primarily used for soft tissue imaging. Magnetic resonance imaging (MRI) is based on the absorption of radio frequency waves by certain nuclei in the body when placed in a magnetic filed. MRI is used to image both soft and hard tissues of the human body, as well as to determine brain function. Nuclear imaging results from measuring γ-rays given off by radioactive compounds, ie, radiopharmaceuticals (qv), introduced into the body. Nuclear imaging is primarily used to image the functionality of an organ which has an affinity for an exogenous radioactive tracer.

Basic Imaging Principles

An image is a matrix of picture elements (pixels) representing the magnitude of the imaged quantity in a given location. Images may be produced by absorption, emission, or reflection of energy by body tissue. Absorption techniques gather information by passing electromagnetic radiation through the body. The spatial variation in intensity of the energy absorbed by the tissues is used to produce an image of the internal organs. Reflection techniques rely on variations in the reflected energy from a tissue or organ. Emission techniques are based on variations in the intensity of energy emitted by the body resulting from the excitation of tissues by an external stimulus, or the introduction of an emissive substance which collects in a specific tissue.

The signal in an image is defined as the intensity of the energy arising from the imaged tissue. The contrast between two tissues in an image is the difference between the signals of the two tissues. The signal-to-noise ratio (SNR) of a tissue in a medical image is the ratio of the signal intensity of that tissue to the noise level in the image. The SNR is not the best indicator of image quality. Rather, the contrast-to-noise ratio (CNR) between adjacent tissues is the factor which determines the utility of an image, provided sufficient signal exists.

Medical images are annotated with the conventional medical nomenclature for the directions of left, right, superior (toward the head), inferior (toward the feet), anterior (toward the front), and posterior (toward the back) of the body. There are three standard planes: one which is perpendicular to the long axis of the body and divides the body into superior and anterior parts is referred to as an axial plane, one which divides the body into left and right halves is called a sagittal plane, and one dividing the anterior from the posterior is referred to as a coronal plane. An oblique plane is one which lies between the three standard planes. For example, an oblique axial–sagittal plane is one which lies between an axial and sagittal plane. Anatomy further away from a reference point is distal and that which is closer is proximal.

Information from an imaging session may be presented as a projection, tomographic, or volume image. Projection images represent the energy coming out of the imaged anatomy. This image is similar to the shadow obtained by placing a hand in front of a light bulb. A single projection image contains no information about the depth of the absorbing tissues. However, anterior-to-posterior and posterior-to-anterior projection images together contain some depth information owing to the point source nature of the source. Tomographic images are images of a thin slice of thickness (Thk) through the body. The thin slice is composed of small-volume elements (voxels), the signal content of which is represented as intensity in the corresponding pixel. Tomographic images are referred to as axial, sagittal, coronal, or oblique. Volume images are three-dimensional (3-D) representations of the organs and tissues. In general, tomographic and volume images require more elaborate imaging hardware than do projection images.

Endoscopy

Endoscopy, the use of optical instruments to image the inside of the human body, was perhaps the first medical imaging modality (1). In the early 1800s small, hollow, rigid tubes were used to examine the larynx and pharynx. These early instruments passed candle light down the tubes to illuminate the organs being viewed. In 1868 Kussmaul saw the inside of the stomach by having the subject, a sword swallower, swallow a 13-mm diameter 47-cm length rigid tube (2). The light source was dim, and the view quickly obscured by the digestive fluids in the stomach. By 1879 the Nitze-Leiter cystoscope was available, but extremely cumbersome for examination of the urinary bladder (1). Minor advances in the 1890s made the endoscope bendable through small angles and improved the light source. The principal breakthrough for endoscopy came in the 1950s with the development of fiber optics (qv). A flexible bundle of fiber optics placed in the body gave images of organs not imagable by rigid pipes. The fiber optics could also be used to provide a light source to view the imaged organ. Moreover, the introduction of flexible fiber optic endoscopes lowered the probability of puncturing an organ which sometimes occurred with rigid light pipes. The next breakthrough in optical imaging was the development in the 1980s of the charge-coupled device (CCD) video camera. A small CCD camera measuring a few millimeters in diameter inserted into the body produces clear images of the internal organs.

Theory and Equipment.   Many diseases of the human body can be identified by visual appearance. Tumors in the upper gastrointestinal (GI) tract, for example, possess a characteristic salmon pink color (3). The presence of such a color can be an indication of disease. Endoscopy is the medical imaging tool used to detect such colors in the inside of hollow internal organs such as the rectum, urethra, urinary bladder, stomach, colon, etc. An endoscope is the instrument used to perform endoscopy. Endoscopic imaging involves the production of a true color picture of the inside of the human body using lenses and either hollow pipes, a fiber optic bundle, or a small CCD camera. All three use a large field-of-view, sometimes referred to as a fish eye, lens to allow a 180° field of view.

The hollow pipe approach uses a small-diameter hollow pipe through which an image from a carefully designed optical system is transmitted. An image is viewed at the end of the pipe external to the body using the appropriate optical eyepiece or video camera. Some devices possess flexible elbows and internal mirrors which allow the pipe to be bent by small angles. Illumination of the object at the end of the pipe is accomplished by sending light down the pipe.

Alternatively, a fiber optic bundle can be used in place of the pipe. In a fiber optic bundle, a matrix of small (50–100 μm) fiber optic strands are arranged such that the ordering of the strands at one end is equivalent to that on the other end. Therefore an image focused on one end with lenses is transmitted to the other end. Light is typically sent down some of the fibers not used for image transmission to provide illumination.

Endoscopes containing a CCD camera replace the fiber optic bundle with a small monochrome CCD chip at the focal point of the fish eye lens (4). The chip contains approximately 512×512 picture elements (pixels). The tissues being imaged are illuminated with light from a few of the fiber optic strands. Color images are produced by alternately illuminating with red, green, and blue light. Data from the CCD chip is therefore a series of red, green, blue, red, green, blue, etc, images which are processed to produce the color video. Endoscopes of this form typically have a camera system connected to an external TV monitor, a fiber optic light source, a tube for rinsing the camera lens with water, and a small tube for insertion of a needle or forceps device for collecting biopsy samples. This combination of implements fits into a flexible 1-cm tube.

Applications. Endoscopy finds applications in a number of investigations of the inside of the human body. It is typically performed using a local anesthetic. Arthroscopy is the examination of a joint using an optical device called an arthroscope which is inserted into a joint through a hole in the skin (5). The arthroscope may be purely optical in construction or contain a miniature electronic camera external to the arthroscope. Cystoscopy is the use of an endoscope to examine the entire urinary tract from the urethra to renal calyx (6). Laparoscopic surgery is surgery using small slender surgical instruments inserted through incisions in the abdomen (7). The surgery is guided with a laparoscope, a slender rigid optical instrument having an external CCD camera inserted into the body. Several surgical techniques can be performed in the peritoneal cavity with only minor incisions for the laparoscope and operating tools: fetoscopy is the visualization of the fetus using a small-diameter needlescope (8); bronchoscopy is the examination of the bronchial tree in the lungs (9); and laryngoscopy is the examination of the larynx (10).

The largest use of endoscopic techniques is in the examination of the gastrointestinal tract. Upper intestinal endoscopy is the examination of the esophagus, stomach, and proximal duodenum. Colonoscopy is the examination of the colon, large intestine, and in some cases the distal parts of the small intestine. Cholangiopancreatography is the examination of the biliary tree and pancreas.

Each of the endoscopic imaging procedures is relatively risk free and painless when performed by competent and well-trained individuals using a local anesthetic. Fetoscopy has the highest risk. There is a 10% increased probability of premature delivery and 10% higher fetal loss rate.

X-Ray Imaging

X-ray medical imaging is the most mature and widely used of the diagnostic imaging modalities (see X-RAY TECHNOLOGY). X-ray imaging began with the discovery of x-rays in 1895 by Wilhelm Röntgen (11). Combining x-ray technology and fluorescent screen technology allowed views of the inside of objects. X-ray imaging had a great and rapid impact on society and Röntgen was awarded a Nobel Prize six years after his discovery. Applications of x-ray imaging to soft tissue came in the 1910s with the development of contrast agents. Compounds of barium were found to absorb x-rays more than the soft tissues of the GI tract, allowing the intestine of a patient who ingested barium to be imaged. The imaging of blood vessels was first demonstrated in 1896 by introducing an x-ray opaque solution into the blood vessels of a cadaveric hand (12). The less invasive form of blood vessel imaging known as angiography was developed in the late 1920s. In the 1970s, driven by safety concerns, higher efficiency intensifier screens and photographic films were developed reducing the required dose of x-ray radiation by as much as 20 times that used previously. In 1972, Hounsfield (13) constructed the first practical (14) computerized tomographic scanner and reconstructed a tomographic image using a mathematical method developed by

Cormack (15) in 1963. They were awarded the Nobel Prize in medicine in 1979. Subsequent developments in scanner technology decreased imaging time from several minutes to a few seconds. Developments in computerized tomography x-ray scanners led to three-dimensional (3-D) imaging instruments.

Theory and Equipment.   The field of x-ray medical imaging can be divided into plane film and CT imaging. Plane film imaging produces projection images of an object placed between a source and a detector which in most cases is a sheet of photographic film (see PHOTOGRAPHY). CT imaging produces tomographic images of a transaxial slice through the body. CT utilizes a source of x-rays and an electronic detector which converts the x-radiation into an electrical signal. Many references on x-ray medical imaging are available (12–28).

X-rays are high energy photons falling into the wavelength, λ, and frequency, ν, ranges of $2.1 \times 10^{-11} < \lambda < 10^{-11}$ m and $1.5 \times 10^{19} < \nu < 2.9 \times 10^{19}$ s^{-1}, respectively. X-rays are produced by high speed electrons directed at a piece of metal. After collision between the electrons and the metal, the metal emits x-rays. The bombarding electrons interact with electrons in the innermost shells of the target metal. The interaction either promotes electrons to higher energy levels or ejects them completely from the metal atom. Relaxation of the electrons back into the vacant inner shell is accompanied by the emission of discrete high energy photons in the x-ray region of the electromagnetic spectrum. The specific wavelength emitted depends on the target metal and the energy of the electron. Other photons are given off when the high velocity electrons are decelerated as they pass through the metal target. This latter form of emission is continuous in nature and is referred to as Bremsstrahlung radiation.

The continuous and discrete emissions from an x-ray tube cover a broad range of frequencies. It is necessary to image using a narrow band of x-rays because the attenuation of x-rays by body tissues is frequency dependent. Narrowing the band of x-rays emanating from a source is accomplished by sending the beam through filters composed of aluminum, copper, or zirconium. This process is referred to as hardening of the beam.

X-ray vacuum tubes contain a resistively heated tungsten cathode (16). A negative potential (cathode) and a positive potential (anode) accelerate electrons from the cathode to a high velocity before collision with the tungsten anode. The geometry of the anode and cathode focus the beam of electrons in an area having approximate dimensions of 1×8 mm. Typical acceleration voltages range from 60 to 120 keV. The efficiency of x-ray generation by this technique is extremely low ($\sim$1%), therefore an enormous amount of heat is generated in the anode. To further complicate matters, the heat is concentrated in a small area of the anode where the electrons are focused. In some tubes this heat is dissipated by water or oil cooling the anode. Other tube designs incorporate a rotating anode which spreads the heat out over a larger mass of metal. The x-ray beam is typically pulsed rather than being a continuous wave (CW). Typical pulse widths are 1/4 to 1/30 of a second depending on the imaged anatomy.

X-rays interact with matter in three ways: photoelectric absorption, Compton scattering, and pair production. Photoelectric absorption is analogous to the absorption of visible light by an electron in an atom or molecule. The difference is that rather than adsorption taking place by an outer shell electron, x-ray absorption dislodges an inner electron from its orbital. Compton scattering involves the interaction of an x-ray with a weakly bound electron. The interaction causes the x-ray to impart some of its energy to the electron in the form of kinetic energy and momentum. Pair production is an interaction between the nucleus of an atom and an x-ray which produces an electron–positron pair. Pair production occurs when the energy of the x-ray photon is $\geq 2\ mc^2 = 1.02$ MeV, where m is the mass of the photon and c the speed of light.

All three interactions occur when x-rays are absorbed by the human body. The first two dominate, however, owing to the lower energy of the x-rays. If N_0 x-ray photons incident on a tissue and N are transmitted, the absorbance, A, is proportional to the thickness, z, and linear attentuation coefficient μ, of the absorbing tissue.

$$A = \ln(N/N_0) = \mu z \tag{1}$$

Some commonly encountered materials arranged in order of decreasing μ are lead > $BaSO_4$ > bone > muscle > blood > liver > fat > air.

Several additional terms related to the absorption of x-radiation require definition: energy of a x-ray photon is properly represented in joules but more conveniently reported in eV; fluence is the sum of the energy in a unit area; intensity or flux is the fluence per unit time; and the exposure is a measure of the number of ions produced in a mass of gas. The unit of exposure in medicine is the Röntgen, R, defined as the quantity of radiation required to produce 2.58×10^{-4} C/kg of air. The absorbed dose for a tissue is a measure of energy dissipated per unit mass. The measure of absorbed dose most prevalent in the medical literature is the rad in the cm–gram–second system of dimensions. The International System of units for absorbed dose is the gray (Gy) which equals 1 J/kg. One rad is defined as the number of ergs absorbed per gram of tissue, or 0.01 J/kg.

Because of their short wavelength, x-rays are not conveniently focused using the principles of refraction, as are longer wavelength, ie, visible, photons. Therefore most imaging schemes utilize point sources of radiation. Images are projection or shadow images of the object placed between the source and the detector. Projection images assume x-ray photons travel a straight line from the source to the detector. Magnification of the imaged object can be achieved by placing the object further away from the screen. A problem associated with projection imaging is scattering of radiation as it interacts with matter. Scattered radiation contributes to intensity on the film in regions not along the original trajectory of the unscattered x-ray photon. A lead grid having a thickness approximately eight times the hole size is typically placed between the imaged object and the detector to minimize the detection of scattered radiation (17,18). Such a grid is referred to as a Bucky diaphragm (Fig. 1).

Detectors.   Two general types of detectors are used in x-ray medical imaging: scintillation and gas ionization. Scintillation detectors are used for both conventional projection and computerized tomographic imaging. Ionization detectors have been used only in CT applications. All detectors used in detection of x-ray radiation must be linear and have a maximum efficiency at the wavelength of the x-ray photon to be detected.

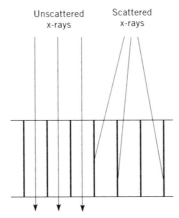

Fig. 1. Magnified view of a part of a lead Bucky diaphragm (grid) used to diminish scattered radiation from striking a detector. The grid is placed between the imaged object and the detector.

Scintillation detectors are substances which fluoresce when struck by x-radiation. Scintillation can, therefore, serve to convert x-ray photons into visible or ultraviolet light. Scintillation materials include thallium-activated crystals of sodium iodide, NaI(Tl), potassium iodide, KI(Tl), or cesium iodide, CsI(Tl); crystals of stilbene (α, β-diphenylethylene) [588-59-0] and anthracene [120-12-7], $C_{14}H_{10}$; bismuth germanium oxide [12233-56-6], $Bi_4Ge_3O_{12}$; barium fluoride [7787-32-8], BaF_2; calcium tungstate [7790-75-2], $CaWO_4$; barium lead sulfate, $BaSO_4$ and $PbSO_4$; zinc sulfide [7789-17-5], ZnS; and cadmium sulfide [1306-23-6], CdS. The application determines the specific material used, as do factors such as cost, durability, x-ray-to-visible light conversion efficiency, scintillation wavelength, transparency at scintillation wavelength, and x-ray attenuation coefficient. The visible or ultraviolet light from the scintillator is detected by photographic film (19,20), electronic camera (21), or photomultiplier tube imaging systems (22) (see PHOTODETECTORS).

The imaging system responsible for the largest number of medical x-ray images utilizes photographic film. Direct exposure of film to x-rays results in an image, however, only 1% of the incident x-rays are absorbed by the emulsion. The use of double-sided emulsions doubles the sensitivity, but this is insufficient to produce high quality images with a minimal dose of x-rays. Scintillators absorb approximately 10 to 70% of the x-rays and convert them to visible light, therefore scintillation materials are placed adjacent to photographic film to increase the efficiency of plane film x-ray imaging techniques. These scintillator plates are called intensifying screens (18,23). The scintillation material and photographic film combination are chosen such that the wavelength of maximum sensitivity of the film matches that emitted by the scintillator. Intensifier screens are typically used on both sides of the film resulting in an overall sensitivity approximately 50 times greater than for the film alone. One drawback associated with intensifying screens is a blurring of the image. Intensifying screens are typically 50–500 μm in thickness. When an x-ray photon strikes the film, visible photons may be emitted in all directions. This emission pattern results in a blurring approximately equal to the thickness of the intensifying screen.

Whereas scintillation screens can be used to directly view x-rays passing through the body, the dose to a patient is too high. Instead, scintillation screens are combined with video cameras or electronic intensifier tubes and video cameras. This detection scheme is used in digital radiography where digital x-ray images are recorded, and fluoroscopy where temporal processes are to be followed.

Scintillators are also used in the detectors of CT scanners. Here an electronic detector, the photomultiplier tube, is used to produce an electrical signal from the visible and ultraviolet light photons. These imaging systems typically need fast scintillators with a high efficiency.

A gas ionization detector consists of a tube filled with a high pressure gas and two electrodes. A tube filled with 2 MPa (20 atm) of xenon is common. The gas in the tube ionizes when x-rays pass through the tube causing a current to flow between a high voltage potential placed across the electrodes. This concept is similar to that used in a Geiger tube detector. Gas ionization detectors are utilized in some CT scanners.

Computerized Tomography. Computerized tomography (CT) imaging is based on obtaining a series of one-dimensional (1-D) projection x-ray images which encompass 180° of projection angles with respect to the imaged object. Each 1-D projection represents the absorption of x-ray radiation along the line from the source to the detector. The 1-D projection images are back-projected using computer programs to produce an image of the internal contents of the original object. The mathematics that describe the signal-generating process in CT imaging are called the Radon and inverse Radon transforms (24). The Radon transform describes the collection of projection functions $P_\theta(t)$ from an object $f(x,y)$. The Radon transform of $f(x,y)$ is

$$P_\theta(t) = \int_{L(\theta,t)} f(x,y)\,ds \qquad (2)$$

where the line $L(\theta,t)$ is defined by Figure 2. The integration at values of θ and t is performed along line $L(\theta,t)$. $P_\theta(t)$ is the projection of function $f(x,y)$ onto line t where θ is the angle between t and the y-axis. The inverse Radon transform describes the back-projection of the Radon transform of an object to obtain an image of the original object. The inverse Radon transform is

$$f(x,y) = \int_0^\pi P_\theta(t)\,d\theta \qquad (3)$$

where the integration is over projection angles between zero and 180°.

The first scanner consisted of an x-ray tube and single-detector combination which moved in unison along the object being imaged (13). In addition to this translational motion, the source/detector assembly was rotated 360° about the imaged object in one-degree steps. At each step the source/detector assembly

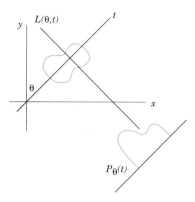

Fig. 2. Description of the functions represented in the mathematical definition of the Radon and inverse Radon transforms. See text.

scanned across the object. The projection images at each step were used to reconstruct a tomographic image of the object. Second generation scanners used a small-angle fan-beam source of x-rays and a detector array. The source and detector array still underwent the translational motion to record the 1-D projection data, but because the array of detectors recorded more information than single-point detectors, fewer rotational steps were required to produce an image. Third generation systems use a wider detector array and an x-ray source having a 30° beam. As a consequence, projection data from the entire object can be recorded at one instant without the need for the translational motion. Additionally, the rotational motion can be performed more quickly and the x-ray source can be pulsed rather than continuous wave. Fourth generation detectors utilize a stationary 360° ring of detectors and a rotating x-ray source having a 30° spread. Experimental scanners are being developed with more sources and detectors which allow dynamic x-ray CT imaging of a beating heart (25). A 3-D image of an object may be obtained by moving either the patient or the source/detector gantry axially with respect to each other. To minimize the x-ray exposure to the patient, two-dimensional (2-D) detector arrays are being used which also eliminate the need for axial motion of the object.

Applications. Applications of x-ray imaging span the entire discipline of medicine. Some of the more common applications are angiography, mammography, and GI, muscular skeletal, neuro, and dental imaging. Muscular skeletal imaging primarily utilizes plane film x-ray imaging. A common use is in gathering the information necessary to set broken bones. Other nonplane film procedures for muscular skeletal imaging are bone density measurements utilizing dual energy CT scans.

Angiography is the study of the blood vessels of the human body. X-ray images possess poor contrast between blood and tissues. As a consequence a contrast agent is typically introduced into the circulatory system to provide the necessary contrast. One of the first angiograms was performed in 1986 using an amputated hand and a mercuric sulfide [1344-48-5] containing contrast agent (12). Fortunately less invasive and safer protocols are available. Angiographic contrast agents generally contain iodine. These agents are usually retained by the body for less than 24 hours. One such contrast agent is N,N'-bis(2,3-dihydroxypropyl)-5-[N-(2,3-dihydroxypropyl)-acetamido]-2,4,6-triiodoisophthalamide [*66108-95-0*], $C_{19}H_{26}I_3N_3O_9$, known as iohexol.

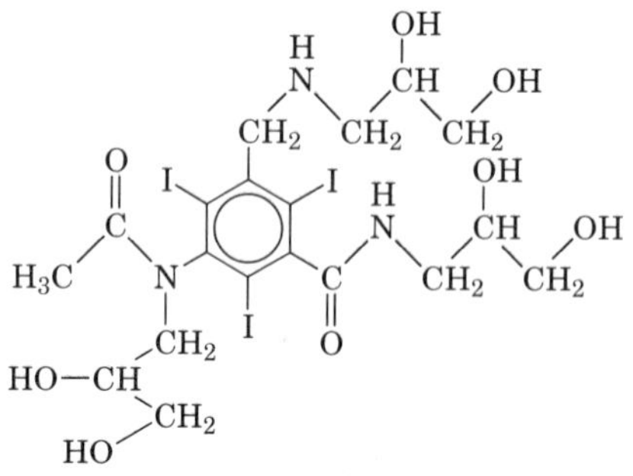

X-ray mammograms are one component of a strategy for lowering a woman's risk of dying from breast cancer. Microcalcification is usually associated with some breast lesions and can be imaged with x-rays. The breast is compressed between two parallel plates and a projection image obtained. Early mammograms were made using industrial x-ray film. As of this writing mammography systems utilize intensifier screens, grids, and high speed film. The x-ray energy is typically 50 keV. One of the shortfalls of mammography is that lesions are not readily detected in dense breasts, that is breasts which are predominantly parenchymal tissue. The overall false negative rate for finding breast cancer by x-ray mammography is 15% for all women, and 40% for women with dense breasts (26,27).

Gastrointestinal x-ray imaging is the imaging of the small and large intestines and the colon. The contrast between the various tissues found in the abdomen is poor. As a consequence, a contrast agent is introduced into the digestive tract which absorbs more x-rays than do the tissues in the abdomen. A common contrast agent is barium sulfate [*7727-43-7*], although iodinated compounds have been used. Owing to the much higher linear attenuation coefficient of the contrast agent, a higher (120 keV) energy x-ray typically is used.

Neuro imaging focuses on the brain and spine and is primarily performed using CT scanning technology. Neuro imaging is often performed with the aid of contrast agents having an affinity for certain types of tissues in the brain. Typical contrast agents are gadolinium-based compounds that possess a higher linear attenuation coefficient than soft tissues.

Dental x-rays provide valuable information on the health of teeth which cannot be obtained by any other medical imaging modality. Dental x-ray procedures use a piece of film placed in the mouth between the tongue and the teeth. A 60 to 70 keV source of x-rays, located outside the mouth, is directed at the film. Metal fillings attenuate x-rays striking the film and therefore appear white in a projection image. Tooth decay appears dark as it attenuates x-rays less than normal tooth enamel.

Safety. X-rays are classified as ionizing radiation. These photons possess sufficient energy to ionize molecules leading to bond breakage and the formation of free radicals. There has been increasing concern about the safety of x-ray medical imaging since the 1970s. The most significant concern centers on radiation dose obtained in x-ray imaging session. In addition to the short-term effects of the radiation, there also is concern about the long-term effects. For example, the abdomen of a premenopausal woman is rarely imaged because of potential damage to the ovaries. Significant advances have been made in decreasing the exposure from x-rays during a plane film imaging session. Radiation exposure during plane film imaging decreased approximately 20-fold from 1960 to 1994. Radiation exposure from CT imaging has, however, increased by a factor of two to three since the first generation CT scanners (28). Table 1 lists the approximate dose obtained from various x-ray imaging procedures to specific body tissues.

An ancillary concern arises from the use of contrast agents, eg, gadolinium complexes during CT scans, barium for GI images, and iodine complexes during angiography. Much research has gone into complexing gadolinium, barium, or iodine to minimize health risks. Despite these efforts, health risks from the use of contrast agents cannot be trivialized. Moreover, most individuals being imaged are already ill and some of the contrast agents are difficult for their bodies to handle. The information gained from a medical imaging procedure must always be balanced against health risks of imaging procedure.

All x-ray equipment must be periodically inspected and the output monitored and calibrated to minimize the chance of accidental overexposure. Another concern involves radiation accumulation by medical personnel operating x-ray equipment. Although the dose to any one patient may be low, the accumulated dose to a clinician performing multiple exams each day over the course of a year is great. Therefore, personnel working with x-ray equipment must take precau-

Table 1. Specific Organ Doses from X-Ray Procedures[a]

Procedure	Organ or tissue dose, μGy[b]				
	Thyroid	Marrow	Breast	Lung	Ovaries
angiogram, carotid	3,000	15,000			
barium enema		3,000			7,900
chest x-ray			140	200	
CT					
brain	960	1,400			
upper abdomen	540	1,700	1,480	6,800	
lower abdomen		2,600	210	1,130	9,500
dental x-ray					
intraoral	10	25			
pantomography	70	50			
mammogram			1,000		
rib x-ray			4,100	3,000	
upper GI tract x-ray		1,200		5,000	

[a] Ref. 21.
[b] To convert μGy to rads, multiply by 1.0×10^{-4}.

tions constantly to minimize and monitor exposure. Lead aprons and film badges are used to minimize exposure and to monitor accumulated dose, respectively.

Ultrasound Imaging

Sound waves having a frequency above 20 kHz, which were audible to a dog but not to humans, were discovered in 1876 (29) and referred to as ultrasound waves (see ULTRASONICS). The piezoelectric effect, discovered in 1880 by Jacques and Pierre Curie (30), and the reverse effect discovered one year later, opened the way for the production and detection of high frequency ultrasound waves (31). After the development of sonar technology in the 1940s, medical ultrasound imaging began in the early 1950s. Soft tissues in the body were imaged in 1952 (32), lesions were shown to be identifiable in 1954 (33), and the precursor of the first clinical ultrasound imager was developed in 1958 (34). Advances in ultrasound hardware and clinical applications occurred throughout the 1960s and 1970s making medical ultrasound imaging an indispensable, inexpensive, and respectable medical imaging modality. More recent advances in ultrasound research have led to the development of Doppler ultrasound imaging of blood flow (35).

      Theory. Ultrasound medical imaging is performed by sending a pulse of ultrasound energy into the body and listening for reflections or echoes. This procedure is similar to that used to estimate the distance across a canyon by measuring the time for sound to travel across the canyon and back. Just as sound waves in air have a characteristic speed or propagation constant, ultrasound waves have a characteristic propagation constant in tissue. The propagation rate, v, in an average soft tissue is 1540 m/s. Values for a few specific tissues can be found in Table 2. Boundaries between tissues having different propagation

Table 2. Ultrasound Properties of Human Tissues[a]

Tissue	v, m/s	Attenuation coefficient, α, at 1 MHz, dB/cm
air	331	
adipose	1450	0.63
water	1480	
humor		
aqueous	1500	0.1
vitreous	1520	0.1
brain	1541	0.85
liver	1541	0.94
spleen	1566	
kidney	1561	1.0
blood	1575	0.18
muscle	1550–1625	1.3–3.3
eye lens	1620	
cartilage	1655	
tendon	1750	
skull bone	4080	20

[a]Refs. 36 and 37.

contants reflect ultrasound, just as the walls of the canyon reflect sound waves. The time, τ, between an ultrasound pulse and a reflection is an indication of the thickness of a tissue. The distance, z, from the source to the boundary is

$$z = v\tau/2 \tag{4}$$

For example, an echo detected 13 μs after the ultrasound pulse represents a boundary 10 cm away from the source.

For ultrasound technology to be useful in medicine, a method is needed for creating an image from the reflected ultrasound energy. When measuring distance between canyon walls, the average distance is determined because of variations in the wall distance. Sending out a focused beam of sound gives a more accurate measurement of distance relative to the focal point of the sound. Similarly, in ultrasound medical imaging the distance between an ultrasound source and a tissue boundary is best determined using a focused beam of ultrasound waves. This 1-D image is not, however, very useful for clinical purposes. A 2-D image is necessary to provide relevant information.

Ultrasound images are typically tomographic images with a slice thickness of 1–2 mm and a field of view of 20–30 cm. A tomographic ultrasound image is generated by sending a series of ultrasound pulses into the portion of anatomy being imaged. Each ultrasound pulse is sent out at a different angle from the source so as to sweep through the anatomy to be imaged in a manner similar to radar sweeping across the sky for airplanes. There are various methods of sweeping the ultrasound beam across the imaged plane.

The source and detector of ultrasound in an ultrasound medical imager is called a transducer. The transducer is a piezoelectric crystal which physically changes its dimensions when a potential is applied across the crystal (38). The application of a force to the piezoelectric crystal which changes its dimensions creates a voltage in the crystal. Application of an oscillating potential to the crystal causes the dimensions of the crystal to oscillate and hence create a sound at the frequency of the oscillation. The application of an oscillating force to the crystal creates an alternating potential in the crystal.

Typical piezoelectric materials are ceramic crystals and copolymers, such as poly(vinylidene fluoride-co-trifluoroethylene), $(-CH_2-Cl_2-)_n-(-CF_2-CFH-)_m$ (39). Ceramic crystals have a higher piezoelectric efficiency. Their high acoustic impedance compared to body tissues necessitates impedance matching layers between the piezoelectric and the tissue. These layers are similar in function to the antireflective coatings on a lens. Polymer piezoelectric materials possess a more favorable impedance relative to body tissues but have poorer performance characteristics. Newer transducer materials are piezoelectric composites containing ceramic crystals embedded in a polymer matrix (see COMPOSITE MATERIALS, POLYMER-MATRIX; PIEZOELECTRICS).

The required sweeping of the ultrasound beam across the imaged plane may be accomplished by one of three methods. The transducer may be physically moved through a series of angles to obtain the image, the transducer may be pointed at an ultrasound mirror that rotates through the desired angles, or a linear array of transducers may be employed (40). Each element in the array of transducers is fired at a different time so as to focus the beam of ultrasound at the desired angle. Although smaller arrays are used, typical arrays are approximately 10 cm in length and may contain as many as 300 elements.

The transducers on most ultrasound imaging systems operate at a frequency between 1 and 20 MHz. The attenuation, A, of ultrasound by tissues is both frequency and tissue dependent. The attenuation coefficient, α, of a tissue is defined by equation 5:

$$A = 10 \log(P/P_0) = \alpha f z_i \tag{5}$$

where P is the power of an ultrasound wave of frequency f and initial power P_0 after traversing a distance z_i in tissue i. Typical values of α for various biological tissues are found in Table 2. The frequency dependence of the ultrasound attenuation allows the ultrasound imager to select the depth to image. Higher frequencies are used to view shorter distances from the transducer. The tissue dependency of the attenuation opens up the possibility of tissue classification using ultrasound intensities.

The resolution in an ultrasound image is, among other things, related to the duration of the ultrasound pulse, ie, the shorter the pulse the better the resolution. Imaging may not be performed when the pulse duration is longer than the time to receive an echo. The shorter the ultrasound pulse the more difficult it is to discern it from noise, and the poorer the SNR of the image. As the pulse duration is decreased, the power of the ultrasound pulse is typically increased to compensate for the poorer SNR.

Another factor affecting the SNR in an ultrasound image is interference between reflected signals from small scatters in the tissues. The ultrasound signals reflected from small reflecting spots in the tissues can constructively and destructively interfere causing a speckle pattern in the image. This speckle pattern manifests itself as a degradation in the SNR of the images.

Contrast in an ultrasound image is related to differences in propagation constants for the tissues. A boundary between two tissues having a large difference in propagation constant reflects large amounts of ultrasound. Ultrasound contrast agents are substances that are introduced into a tissue to change the propagation constant and hence reflect more ultrasound energy (41,42). Typical ultrasound contrast agents are lipid-stabilized microbubbles having a diameter of 1–5 micrometers.

Applications. Ultrasound imaging is used for imaging of soft tissues. Its primary advantages are low cost and safety compared to other medical imaging modalities. Ultrasound imaging finds its greatest applications in obstetrics and gynecology for studying the uterus and a fetus, in cardiology for studying the function of the heart, and for imaging of the abdomen.

Ultrasound energy reflected from a tissue contains both amplitude and frequency information. Generally only the amplitude information is utilized to create an image. The frequency information, typically discarded, contains information on the velocity of the tissues. Doppler ultrasound uses this frequency information. The velocity of blood flowing through the veins and arteries of the body (35) is obtained using the Doppler effect. Blood flowing toward the transducer reflects higher frequency ultrasound than in the incident pulse; blood flowing away from the transducer reflects lower frequency ultrasound waves. This frequency information is typically processed and presented as pseudo-color overlaid on top of a conventional ultrasound image.

Special small ultrasound transducers, often referred to as endoscopic transducers, have been designed which can be inserted into blood vessels to examine blockages in arteries (43). These transducers operate at approximately 20 MHz and have a viewing distance of less than a centimeter. Such devices are capable of producing ultrasound images of the inside of arteries and veins. The quality of the ultrasound image is sufficient to determine the type of blockage.

Safety. High power ultrasound waves can cause local heating and transient cavitation in water (44). Transient cavitation is a process in which microscopic gas bubbles expand and collapse as a consequence of the ultrasound wave. The rapid collapse can be adiabatic causing the energy to be transferred to bond-breaking processes that create free radicals and give rise to the health concern (45,46). Typical ultrasound contrast agents are lipid-stabilized microbubbles having a diameter of 1–5 micrometers. These microbubbles, when exposed to ultrasound, may behave the same way as the ultrasound-generated bubbles and create transient free radicals. The difference in using the contrast media is that the concentration of free radicals is much higher because of the introduced bubbles. The rule of thumb in ultrasound medical imaging is to utilize a power level that is as low as reasonably possible.

Magnetic Resonance Imaging

As of this writing, magnetic resonance imaging (MRI) is the newest medical imaging technique available (47–49). The technique upon which MRI is based, nuclear magnetic resonance (nmr), was developed independently in the 1940s by Bloch and Purcell (see MAGNETIC SPIN RESONANCE). In the 1950s and 1960s nmr was used extensively for chemical and physical analysis. In the 1970s several developments occurred which initiated the development of MRI. The first was the discovery that an nmr property of tissues called the spin-lattice relaxation time could be used to distinguish between a tumor and healthy tissue (50). Another factor was the implementation of x-ray-based computed tomography in a clinical setting, showing that hospitals were willing to purchase expensive equipment vital for a diagnosis. A third development was the demonstration of imaging using nmr and the CT back-projection technique (51). The final development which allowed MRI to become viable was the development of phase and frequency encoding the MRI (52).

The 1980s saw the first demonstration of a whole body imager and the first commercially available magnetic resonance imagers. Additionally, development of rapid imaging sequences pushed the imaging time for a single image from five minutes in 1982 to five seconds in 1986, and to video rates in 1989. Resolution on nmr microscopes was pushed to approximately 10 μm. Magnetic resonance angiography was developed in the late 1980s. The 1990s saw the awarding of the Nobel Prize in chemistry to Richard Ernst for his contributions to pulsed nmr and MRI. The most recent development was the discovery in 1993 that echo planar MRI could be used for functional imaging of the brain.

Theory and Equipment.   MRI is conceptually more difficult to understand than other medical imaging modalities because it is not based on simple optical principles of absorption and reflection of electromagnetic radiation. MRI is based on the principles of nmr, therefore any description of the principles of MRI should be preceded by a good understanding of nmr. This fact makes MRI very interesting to chemists. A brief review of the nmr is presented herein. The less knowledgeable reader is directed to the entry on magnetic spectroscopies (see MAGNETIC SPIN RESONANCE) or one of the many references for a more detailed description (53–55).

Nuclear Magnetic Resonance.   Nuclear magnetic resonance is based on a property of the atomic nucleus called spin which can be thought of as a simple magnetic moment. When a nucleus with spin is placed in an external magnetic field, the magnetic moment can take on one of two possible orientations, one low energy orientation aligned with the field and one high energy orientation opposing the field. A photon with an energy equal to the energy difference between the two orientations or states can cause a transition between the states. The greater the magnetic field the greater the energy difference and hence the frequency of the absorbed photon. The relationship between the applied magnetic field B_0 and the frequency of the absorbed photon ν is linear.

$$\nu = \gamma B_0 \tag{6}$$

The proportionality constant γ is called the gyromagnetic ratio which is a function of the magnitude of the nuclear magnetic moment. Therefore each isotope having a net nuclear spin possesses a unique γ. The γ of some biologically relevant nuclei can be found in Table 3.

The remainder of the nmr description adopts a macroscopic perspective of the spin system in which the B_0 field is applied along the z-axis. Groups of nuclei experiencing the same B_0 are called spin packets. When placed in a B_0 field, spin packets precess about the direction of B_0 just as a spinning top on earth precesses about the direction of the gravitational field. The precessional frequency, also called the Larmor frequency, ω, is equal to $2\pi\nu$. The direction of the precession is clockwise about B_0, and the symbol ω_0 is reserved for spin

Table 3. Gyromagnetic Ratios for Biologically Relevant Nuclei[a]

Nuclei	γ, MHz/T[b]	Natural abundance, %
^{1}H	42.58	99.99
^{13}C	10.71	1.108
^{14}N	3.08	99.63
^{23}Na	11.27	100
^{25}Mg	2.61	10.13
^{31}P	17.25	100
^{35}Cl	4.17	75.53
^{39}K	1.99	93.10
^{43}Ca	2.86	0.13

[a] Ref. 56.
[b] To convert T to Gauss, multiply by 1.0×10^4.

packets experiencing exactly B_0. It is often helpful in nmr and MRI to adopt a rotating frame of reference to describe the motion of magnetization vectors. This frame of reference rotates about the z-axis at ω_0. The axes in the rotating frame of reference are referred to as z, x', and y'.

An nmr sample contains millions of spin packets, each having a slightly different Larmor frequency. The magnetization vectors from all these spin packets form a cone of magnetization around the z-axis. At equilibrium, the net magnetization vector M from all the spins in a sample lies in the center of the cone along the z-axis. Therefore the longitudinal magnetization M_z equals M and the transverse magnetization M_{xy} equals zero at equilibrium. Net magnetization, perturbed from its equilibrium position, wants to return to its equilibrium position. This process is called spin relaxation.

The return of the z component of magnetization to its equilibrium value is called spin-lattice relaxation. The time constant which describes the exponential rate at which M_z returns to its equilibrium value M_{z0} is called the spin-lattice relaxation time, T_1. Spin-lattice relaxation is caused by time-varying magnetic fields at the Larmor frequency. These variations cause transitions between the spin states and hence change M_z. Time-varying fields are caused by the random rotational and translational motions of the molecules in the sample possessing a magnetic moment. The frequency distribution of random motions in a liquid varies with temperature and viscosity. In general, relaxation times tend to get longer as B_0 and ν increase because there are fewer relaxation-causing frequency components present in the random motions of the molecules as ν increases.

At equilibrium, the transverse magnetization M_{xy} equals zero. A net magnetization vector rotated off the z-axis creates transverse magnetization. This transverse magnetization decays exponentially with a time constant called the spin–spin relaxation time T_2. Spin–spin relaxation is caused by fluctuating magnetic fields which perturb the energy levels of the spin states and dephase the transverse magnetization. T_2 is inversely proportional to the number of molecular motions less than and equal to the Larmor frequency. T_2 has two components: a pure T_2 resulting from molecular interactions, and one resulting from spatial inhomogeneities in the B_0 field. The overall T_2^* (referred to as T_2 star) is defined as

$$1/T_2^* = 1/T_2 + 1/T_{2\,\text{inhomogeneous}} \tag{7}$$

In pulsed nmr and MRI, radio-frequency (r-f) energy is put into a spin system by sending rf into a resonant LC circuit, the inductor of which is placed around the sample. The inductor must be oriented with respect to the B_0 magnetic field so that the oscillating r-f field created by the rf flowing through the inductor is perpendicular to B_0. The r-f magnetic field is called the B_1 magnetic field. When the r-f inductor, or coil as it is more often called, is placed around the x-axis, the B_1 field oscillates back and forth along the $\pm x$-axis.

In pulsed nmr spectroscopy, it is the B_1 field which is pulsed. Turning on a B_1 field for a period of time τ causes the net magnetization vector to precess in ever widening circles around the z-axis. Eventually, the vector reaches the xy plane. If B_1 is left on longer, the net magnetization vector reaches the negative

z-axis. In the rotating frame of reference, this vector appears to rotate away from the z-axis. The rotation angle θ, which is measured clockwise about the direction of B_1 in radians, is proportional to γ, B_1, and τ.

$$\theta = 2\pi\gamma B_1\tau \tag{8}$$

Any transverse magnetization M_{xy} precesses about the direction of B_0. An nmr signal is generated from transverse magnetization rotating about the z-axis. This magnetization reduces a current in a coil of wire placed around the x- or y-axis. As long as there is transverse magnetization changing with respect to time, there is an induced current in the coil. For a group of nuclei having one identical chemical shift, the signal is an exponentially decaying sine wave which decays with a time constant T_2^*. It is predominantly the inhomogeneities in B_0 which cause the spin packets to dephase. Net magnetization which has been rotated away from its equilibrium position along the z-axis by exactly 180° does not create transverse magnetization and hence does not give a signal. The time-domain signal from a net magnetization vector in the xy plane is called a free induction decay (FID). This time domain signal must be converted to a frequency domain spectrum to be interpreted for chemical information. The conversion is performed using a Fourier transform. The hardware in most nmr spectrometers and magnetic resonance imagers detects both M_x and M_y simultaneously. This detection scheme is called quadrature detection. These two signals are equivalent to the real and imaginary signals, therefore the input to the Fourier transform is complex. Sampling theory dictates that digitization of the FID at frequency f complex points per second gives a spectrum of frequency width f.

In pulsed Fourier transform nmr spectroscopy, short bursts of r-f energy are applied to a spin system to induce a particular signal from the spins within a sample. A pulse sequence is a description of the types of r-f pulses used and the response of the magnetization to the pulses. The simplest and most widely used pulse sequence for routine nmr spectroscopy is the 90-FID pulse sequence. As the name implies, the pulse sequence is a 90° pulse followed by the acquisition of the FID. The net magnetization vector, which at equilibrium is along the positive z-axis, is rotated by 90° down into the xy plane. The rotation is accomplished by choosing an r-f pulse width and amplitude so that the rotation equation for a 90° pulse is satisfied. At this time the net magnetization begins to precess about the direction of the applied magnetic field B_0. Assuming $T_2 < T_1$, the net magnetization vector begins to dephase as the vectors from the individual spin packets in the sample precess at their own Larmor frequencies. Eventually, M_{xy} equals zero and the net magnetization returns to its equilibrium value along z. The signal which is detected by the spectrometer is the decay of the transverse magnetization as a function of time. The FID or time domain signal is Fourier transformed to yield the frequency domain nmr spectrum.

Magnetic Resonance Imaging. Magnetic resonance imaging (MRI) is a tomographic imaging modality. The basis of MRI, equation 6, states that the resonance frequency of a nucleus is proportional to the magnetic field it is experiencing. If a spatially varying magnetic field is set up across a sample, the nuclei within the sample resonate at a frequency related to their positions. For example, if a 1-D linear magnetic field gradient G_z is set up in the B_0 field along the z-axis, the resonant frequency ν is defined by

$$\nu = \gamma(B_0 + zG_z) \tag{9}$$

The origin of the xyz coordinate system is the point in the magnet where the field is exactly equal to B_0 and spins resonate at ν_0. This point is referred to as the isocenter of the magnet. Equation 9 explains how a simple 1-D imaging experiment can be performed. The sample to be imaged is placed in a magnetic field B_0. A 90° pulse of r-f energy is applied to rotate magnetization into the xy plane. A 1-D linear magnetic field gradient G_z is turned on after the r-f pulse and the FID is immediately recorded. The Fourier transform of the FID yields a frequency spectrum that can be converted to a spatial, z, spectrum as shown in equation 10:

$$z = (\nu-\nu_0)/\gamma G_z \tag{10}$$

This simple concept of a 1-D image can be expanded to a 2-D image employing back-projection technology similar to that used in CT imaging (51). If a series of 1-D images, or projections of the signal in a sample, are recorded for linear 1-D magnetic field gradients applied along several different trajectories in a plane, the spectra can be transformed into a 2-D image using an inverse Radon transform (24) or a back-projection algorithm. This procedure is seldom used. Instead, Fourier-based imaging techniques are used in most MRI (52,57).

A Fourier-based imaging technique collects data from the k-space (2-D Fourier transform of the image) of the imaged object. Figure 3 depicts a timing diagram for a Fourier imaging sequence using simple 90-FID sequence. The timing diagram describes the application of rf and three magnetic field gradients called the slice selection (G_s), phase encoding (G_ϕ), and frequency encoding (G_f) gradients. The first step in the Fourier imaging procedure is slice selection of those spins in the object for which the image is to be generated. Slice selection is accomplished by the application of a magnetic field gradient at the same time the r-f pulse is applied. A r-f pulse, having frequency width $\Delta\nu$ centered at ν, excites when applied in conjunction with a field gradient G_z, spins centered at $z = (\nu-\nu_0)/\gamma G_z$ with a spread of spins at z values $\Delta z = \Delta\nu/\gamma G_z$. Spins experiencing a magnetic field strength not satisfying the resonance condition are not rotated by the r-f pulses, and hence slice selection is accomplished. The image slice thickness is given by Δz. For a clean slice, ie, all spins along the slice thickness are rotated by the prescribed rotations, the frequency content of the pulse must be equal to a rectangular-shaped function. Therefore, the r-f pulse must be shaped as a sinc ($\sin x/x$) function in the time domain.

The next step in the Fourier imaging procedure is to encode some property of the spins as to the location in the selected plane. Spins could easily be encoded as to their x position by applying a gradient G_x after the r-f pulse and during the acquisition of the FID. The difficulty is in encoding the spins with information as to their y location. This is accomplished by encoding the phase of the precessing spin packets with y position (see Fig. 3). Phase encoding is accomplished by

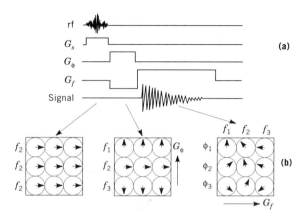

Fig. 3. Timing diagram with corresponding magnetization vector presentation for a 90-FID imaging sequence: (**a**) timing diagram which depicts the time during which the rf, G_s, G_ϕ, and G_f are applied and the signal is acquired; (**b**) vector diagram describing the evolution of nine magnetization vectors during the sequence. See text.

turning on a gradient in the y direction immediately after the slice selection gradient is turned off and before the frequency-encoding gradient is turned on. The spins in the excited plane now precess at a frequency dependent on the y position. After a period of time τ the gradient is turned off and the spins have acquired a phase ϕ equal to

$$\phi = 2\pi\gamma\tau y G_y \qquad (11)$$

Figure 3 describes for nine magnetization vectors the effect of the application of a phase-encoding gradient, G_y, and a frequency-encoding gradient, G_x. The phase-encoding gradient assigns each y position a unique phase. The frequency-encoding gradient assigns each x position a unique frequency. If the phase and frequency of a spin packet could be assessed independently, its position could be assigned in the xy plane. Unfortunately, this cannot be accomplished with a single pulse and signal. The phase-encoding gradient must be varied in amplitude so a 2π radian phase variation between the isocenter and the first resolvable point in the y direction can be achieved, as well as a 256π radian variation from center to edge of the imaged space for 256-pixel resolution in the phase-encoding direction. The result is to traverse, line by line, the k-space of the image. The negative lobe on the frequency-encoding gradient (see Fig. 3), which was not previously described, shifts the center of k-space to the center of the signal acquisition window.

The B_0 field is created by a large-diameter solenoidal-shaped superconducting magnet. The gradient fields are created by room temperature gradient coils located within the bore of the magnet. These coils are driven by high current audio frequency amplifiers. The B_1 field is introduced into the patient by means of a large LC circuit which surrounds the anatomy to be imaged (58). The same or a separate LC circuit is used to detect the signals from the precessing spins in the body.

The field of view (FOV) is dependent on the quadrature sampling rate, R_s, during the application of G_f, and the magnitude of G_f.

$$\text{FOV} = R_s/\gamma G_f \tag{12}$$

The 2-D k-space data set is Fourier transformed, and the magnitude image generated from the real and imaginary outputs of the Fourier transform.

In routine clinical imaging the thickness of a slice is approximately 3 mm and the in-plane resolution approximately 0.8 mm for a 20-cm FOV, 256×256 pixel image. The volume of a voxel is, therefore, approximately two mm^3. That is, taking into account the size of organs and composition of tissues in the body, a voxel is often comprised of more than one substance. As a consequence, the nmr signal from a voxel is a summation of the nmr signals from the substances found in the voxel. Any variation in the signal resulting from the relative amounts of the components found in a voxel is referred to as a partial volume effect.

The most abundant spin-bearing nucleus in the human body is hydrogen. The two most abundant forms of hydrogen are fat and water hydrogens. These hydrogens yield one signal in the image, as chemical shift and spin–spin splitting information is generally not utilized. Occasionally the different chemical shifts for water and fat hydrogens can lead to an artifact in an image called a chemical shift artifact. Hydrogens associated with proteins and the other building blocks of tissues have very short T_2 value and do not contribute directly to the signal. Magnetic resonance images have also been recorded for sodium and phosphorus.

A problematic artifact associated with MRI arises when the imaged subject moves during acquisition of the k-space data. Such motion may result in a discontinuity in the frequency-encoded or phase-encoding direction data of k-space. When Fourier transformed, such a discontinuity causes a blurred band across the image corresponding to the object that moved. Such an artifact in an image is referred to as a motion artifact.

In Figure 3, the slice selection gradient is applied along the z-axis and the phase and frequency encoding gradients along the y and x axes, respectively. In practice the gradients can be applied along any three orthogonal axes, the only restrictions being that the slice selection gradient be perpendicular to the imaged plane.

The most routinely used imaging sequence is the spin-echo. Its popularity is attributable to its ability to produce images which display variations in T_1, T_2, and spin density, ρ, of tissues. This sequence consists of 90° and 180° r-f pulses repeated every TR (repetition time of the sequence) seconds. These pulses are applied in conjunction with the slice selection gradients. The phase-encoding gradient is applied between the 90° and 180° pulses. The frequency-encoding gradient is turned on during the acquisition of the signal. The signal is referred to as an echo because it comes about from the refocusing of the transverse magnetization at an echo time (TE) after the application of the 90° pulse. The signal from a voxel in the body is equal to a summation over all the different types of spins, i, in the voxel.

$$S = \sum_i \rho_i (1 - e^{-TR/T_{1i}}) e^{-TE/T_{2i}} \tag{13}$$

The goal of the MRI scientist is to maximize the contrast-to-noise ratio between tissues. Examination of equation 13 reveals that by varying TR and TE, the clinician has a tremendous amount of flexibility to select the desired contrast between two tissues.

The contrast between any two tissues may be maximized by prudent choice of the imaging parameters. Clinicians have adopted nomenclature for the various types of images produced as a consequence of the choice of imaging parameters. A T_1-weighted image is one in which image contrast displays differences in T_1 of the tissues. A spin-echo sequence produces a T_1-weighted image when TR $\leq T_1$ and TE $< T_2$. A T_2-weighted image is one in which contrast between the tissues is primarily because of differences in T_2 of the tissues. A spin-echo sequence produces a T_2-weighted image when TR $> T_1$ and TE $\geq T_2$. Spin density weighting is, as expected, an image where contrast displays differences in spin density of the tissues. A spin-echo sequence produces a ρ-weighted image when TR $> T_1$ and TE $< T_2$.

The clinician may also change the contrast in an image using a chemical contrast agent (59). A contrast agent is typically a paramagnetic substance that is introduced into the body and has an affinity for certain tissue types. When a contrast agent comes in contact with the tissue, it changes the T_1 and T_2 of the tissue. The magnetic resonance signals from those tissues can, therefore, be altered relative to other tissues with a lesser affinity to the contrast agent. Typical MRI contrast agents contain gadolinium. The gadolinium is chelated with a ligand such as ethylenediaminetetraacetic acid [60-00-4] (EDTA), diethylenetriaminepentaacetic acid [67-43-6] (DTPA), or tetraazacyclododecan-etetraacetic acid [60239-18-1] (DOTA) to lower its toxicity.

Applications.   Magnetic resonance imaging finds its greatest use in neuro imaging (60). MRI has excellent soft tissue specificity and can, therefore, be used to identify many types of lesions in the brain and spinal cord. The utility of MRI in providing structural information about these areas has surpassed that of CT. In addition to the structural information, MRI can also provide functional information (61,62). Previously, functional imaging required the use of positron emission tomography (PET).

Brain functional imaging using MRI is based on the fact that during brain activity the flow of blood to the region of activity increases. This in turn causes a change in the contrast between regions with activity and those without. Because this change occurs in fractions of a second, the imaging of an entire slice through the brain must be performed in a period of time less than this. The technique which allows this is called echo planar imaging. An echo planar imaging sequence allows images to be recorded at video rates, ie, 1/15 to 1/30 second. The faster image acquisition rate is the result of traversing all k-space for an image during a single echo. The frequency-encoding gradient is rapidly switched between a positive and negative value while the phase-encoding gradient is pulsed at each switching of G_f. This procedure permits the rastering through k-space at video rates.

The second largest application of MRI is in muscular skeletal imaging (63) of joints such as the knee, shoulder, hips, and wrist. Torn ligaments and rips in the cartilage between the tibia and femur are readily seen using MRI. This method is preferred by the patient over arthroscopic imaging or surgery where patients are either injected in their joint with approximately 10 cc of air and x-ray imaged or given exploratory surgery. A football player injured in a weekend game can be imaged by MRI the same day and given a diagnosis immediately.

Another imaging procedure in which MRI is challenging traditional x-ray procedures is magnetic resonance angiography (MRA) (64,65). Unlike x-ray-based angiography, MRA does not require the injection of contrast agents into the blood stream. MR angiography images flowing blood as opposed to a contrast agent in a blood vessel. As a consequence MRA can detect locations having poor flow which appear normal on an x-ray angiogram because of the presence of contrast agent in the static blood. Magnetic resonance angiography may be performed by one of two techniques: time-of-flight (64) and phase-contrast (65) angiography. Time-of-flight angiography is performed using a two-step process. First, an image is recorded when the magnetization from spins flowing into the imaged slice is set to zero by a r-f pulse referred to as a saturating pulse. The second step is to subtract this image from a conventional magnetic resonance image without the saturating pulse. The difference is an image of the flowing blood. Phase-contrast angiography utilizes an extra gradient pulse called a bipolar gradient pulse which imparts a phase shift to the precessional motion of flowing spins. Subtracting an image recorded having a positive bipolar gradient pulse from one recorded having a negative pulse produces an image of flowing blood. Both techniques are routinely used to image flowing blood.

Safety. Because of the relatively young age of MRI there is concern regarding its safety (66). Users are trying to err on the side of caution. The principal safety concerns are related to the static magnetic field B_0, changing magnetic fields dB_0/dt, tissue heating from r-f power deposition, and acoustic noise. The United States Food and Drug Administration guidelines on static magnetic field limits B_0 to less than 2 T (2×10^4 G). The greatest concern about the health effects of strong magnetic fields are those effects caused by ferromagnetic objects being pulled into the imager while a patient is inside, or the torques created on a ferromagnetic object which might be in the patient's body. A ferromagnetic pen could be pulled into the magnet with sufficient velocity to puncture the body. A ferromagnetic object, such as a metal sliver or an aneurysm clip, located in the patient's body could become reoriented in the field and damage tissues.

The dB/dt is limited to 6 T/s out of concern that larger values could cause nerve stimulation. The r-f exposure is limited to a specific absorption rate (SAR) of 0.4 W/kg for the whole body, 0.32 W/kg averaged over the head, and less than 8.0 W/kg spatial peak in any one gram of tissue. These numbers are designed to limit the temperature rise to less than 1°C and localized temperature of no greater than 38°C head, 39°C trunk, and 40°C in the extremities.

Magnetic resonance imagers produce a loud knocking sound when the magnetic field gradients are turned on or off. The acoustic noise levels can be high in the bore of the magnet. Patients are usually given ear plugs which can decrease the sound of the knocking by upward of 26 dB.

Nuclear Medicine Imaging

Nuclear medicine imaging involves the use of exogenous radioactive materials to image the body. The first scientific discovery which made the development of this imaging modality possible was radioactivity by Bacquerel in 1896. Bacquerel shared the Nobel Prize in physics with Pierre and Marja Curie in 1903 for their discoveries related to radioactivity. Another discovery which led to the development of nuclear imaging was that of technetium in 1937. Technetium-labeled molecules, known as radiopharmaceuticals, which have specific biological functions, are used. The first imaging procedures were performed in the 1940s. The presence of a brain tumor was detected and normal and abnormal thyroid functions were investigated. Both experiments used [131]I . The Anger scintillation camera, developed in the 1950s, is the primary detector in nuclear medicine. Although the groundwork development of single photon emission computed tomography (SPECT) and positron emission tomography (PET) preceded x-ray-based CT, it was not until the development of CT imaging, and the related reconstruction algorithms, that SPECT and PET were developed to the state of clinical utility.

Theory and Equipment. The basic principle behind nuclear medical imaging is that a radiopharmaceutical can be introduced into the body which emits radiation detectable outside of the body. Radiopharmaceuticals are biologically active and have a short half-life ($T_{1/2}$). The detectable radiation is typically a γ-ray photon. The radiopharmaceutical must be introduced in sufficient concentration to produce detectable signals outside of the body, but not large enough to be lethal. Some radiopharmaceuticals emit γ-rays directly. Other radiopharmaceuticals emit positrons, β^+. Shortly after being emitted, the positron is annihilated when it collides with an electron. Two 511-keV photons are simultaneously produced from the annihilation and possess trajectories 180° apart from each other. The more common radioactive nuclei used in radiopharmaceuticals are listed in Table 4. With the exception of xenon, these nuclei are typically bonded to other atoms or complexed with chelates to form the radiopharmaceutical. The specific structure of the radiopharmaceutical depends on the application.

Table 4. Radioactive Nuclei Used in Radiopharmaceuticals[a]

Nucleus	Radioactive decay product	γ-Ray energy, keV	$T_{1/2}$	Production[b]
[201]Tl	γ	70	73 h	CPB
[133]Xe	γ	81	5.27 d	fission
[131]I	γ	364	8.05 d	fission
[123]I	γ	159	13 h	CPB
[111]In	γ	171, 245	67.9 h	CPB
[99]Tc^m	γ	140.5	6.03 h	[99]Mo decay
[82]Rb	β^+	511	1.2 min	[82]Sr decay
[67]Ga	γ	93, 184, 300	78.3 h	CPB
[18]F	β^+	511	110 min	CPB
[15]O	β^+	511	2 min	CPB
[13]N	β^+	511	10 min	CPB
[11]C	β^+	511	20.5 min	CPB

[a] Refs. 21 and 67.
[b] CPB = charged-particle bombardment.

The radiopharmaceuticals used in nuclear imaging may be produced by either neutron capture, nuclear fission, charged-particle bombardment (CPB) or radioactive decay (21). Neutron capture or activation requires a nuclear reactor, CPB requires an accelerator for production, and radioactive decay production requires a device containing parent nuclei which decay into the desired radiopharmaceutical. Such a device is referred to as a radionuclide generator. Radiopharmaceuticals having a relatively long half-life may be generated off-site or using a radioisotope generated on-site. However, those having shorter half-lives, such as the β^+ emitters, must be generated on-site and thus require a charged-particle accelerator or nuclear reactor.

The radiation emitted by the radiopharmaceutical is most often detected using scintillation detectors of NaI(Tl) (21). The light emitted by a NaI(Tl) scintillator has a wavelength of 410 nm. The intensity of the emitted light is proportional to the energy of the γ-ray photon, therefore a scintillation detector can be used to measure the number of γ-rays and their energy. This violet photon is amplified and converted into an electrical response by a photomultiplier tube. Other γ-ray detectors that have been used in nuclear imaging are multiwire proportional chambers and the lithium-drifted germanium gamma camera.

Gamma radiation may be detected and processed to produce a 2-D planar or a CT emission image. Planar images were first produced using a single detector which moved across the body in a rectilinear motion to produce the planar image of the emitted radiation. Later detectors employed a linear array of 10 or more detectors moved in a linear fashion along the length of the body. More recent designs consist of a flat array of 37 to 91 detectors. The scintillation material is faced with a lead collimator grid, as is done in x-ray imaging. The grid serves as a lens in that it causes the detector to be sensitive to emissions from a specific region of the body. SPECT detectors utilize an array of detectors that move in a manner similar to CT x-ray imaging detectors around the anatomy being imaged.

The scheme used to detect the two 511-keV γ-rays from a β^+ emitter incorporates principles of coincidence detection. The signals from two detectors pointing toward each other along a straight line are processed by circuitry which only produces output when signals are instantaneously detected from each detector. Lead collimators are placed in front of each detector to minimize scattered and random coincidence. Although planar images could be obtained of the positron emissions from a radiopharmaceutical introduced into the body, positron emission images are typically tomographic. A ring of detectors is arranged around the imaged anatomy. The detectors are connected to an elaborate coincidence logic which registers a signal when any two detectors having a line of sight path between them detects a coincidence. Because a region having a high concentration of β^+ emitter emits many β^+ particles and hence γ-ray coincidences in all directions, a tomographic image may be constructed using the information from all the detectors and the Radon transform. The sensitivity of PET is approximately 20 times greater than that of SPECT because of the coincidence-detection scheme.

Applications. Brain and central nervous system imaging are common applications of nuclear imaging. Technium-99^{m}-glucoheptonate [68128-55-2] and ^{99}Tc^{m}-DTPA [80908-06-1] are used to study the integrity of the blood brain barrier. Tumors which disrupt the barrier cause these radiopharmaceuticals to pass into the brain. ^{15}O$_2$ [95682-06-7] and C^{15}O$_2$ [85401-75-8] can be used to study oxygen uptake by regions of the brain. $H_2^{15}O$ [*24286-21-3*] and ^{133}Xe [*14932-42-4*] are used to study brain circulation (68). Because the oxygen metabolism and microcirculation in the brain increases during brain activity, these molecules can be used with PET to determine functionality of the various sections of the brain (69). Cerebral blood volume is measured using $^{99}Tc^m$-red blood cells and SPECT.

Cardiac nuclear imaging using $^{99}Tc^m$-red blood cells can measure the fraction of blood pumped by the heart during each beat. $^{99}Tc^m$-DTPA and sodium o-iodohippurate, $C_9H_7^{123}INNaO_3$, are used to measure renal function of the kidney. The enhanced or diminished uptake of technecium-99m-methylenediphosphonate [*101488-09-9*], $^{99}Tc^m(O_3PCH_2PO_3)$, by bones associated with some skeletal abnormalities allows skeletal nuclear imaging. ^{127}Xe [*13994-19-9*] and $^{81}Kr^m$ [*15678-91-8*] are used for studies of the respiratory system. $Na^{123}I$ and sodium pertechnetate [*13718-28-0*], $Na^{99}Tc^mO_4$, are used for imaging of the thyroid gland. Gallium-67-citrate [*41183-64-6*] is used for tumor imaging. Several other applications of nuclear medicine can be found in the literature (67).

Safety.   The principal concerns regarding nuclear medical imaging are those associated with the radiopharmaceuticals. Much research has gone into the selection of radiopharmaceuticals exhibiting minimal toxicities, rapid elimination from the body, and short half-life. The radioisotope must be nontoxic or capable of being made nontoxic by chelation. These isotopes should have a short half-life so as not to make the body radioactive for a long period of time. Heavy metals, eg, technetium, must be in a form eliminated from the body by the kidneys. The radiopharmaceutical also should not emit harmful amounts of radiation. For example, ^{131}I, a γ- and β-emitter, in small doses has diagnostic utility in imaging the thyroid. In larger doses, it has therapeutic utility in that it can cause either partial or complete ablation of the thyroid owing to the β-emissions. In addition to health concerns specific to the patient, attention must be paid to minimizing accidental exposure or ingestion of radiopharmaceuticals by the clinical personnel suppressing the imaging procedure.

BIBLIOGRAPHY

1. D. D. Gibbs, K. F. R. Schiller and P. R. Salmon, eds., *Modern Techniques in Gastrointestinal Endoscopy*, Year Book Medical Publishers, Chicago, 1976, pp. 1–14.
2. A. Kussmaul, *Berichte naturforsch. Ges. Freiburg i. B.* **5**, 112 (1868).
3. M. O. Blackstone, *Endoscopic Interpretation—Normal and Pathologic Appearances of the Gastrointestinal Tract*, Raven Press, New York, 1984.
4. J. F. Rey, M. Albuisson, M. Greff, J. M. Bidart, and J. M. Monget, *Endoscopy* **20**, 8–10 (1988).

5. D. Drez, Jr., *Clin. Sports Med.* **4**, 275–278 (1985).

6. J. W. Segura, *J. Urol.* **132**, 1079–1084 (1984).

7. M. Ohligisser, Y. Sorokin, and M. Hiefetz, *Obstet. Gynecol. Surv.* **40**, 385–396 (1985).

8. R. J. Benzie, *Clin. Obstet. Gynecol.* **7**, 439–460 (1980).

9. P. D. Phillon and J. V. Collins, *Postgrad. Med. J.* **60**, 213–217 (1984).

10. W. Steiner, *Endoscopy* **1**, 51–59 (1979).

11. W. C. Röntgen, *Erste Mitt. Sitzgsber. physik.—med. Ges. Würzburg* **137** (Dec. 1895); transl. A. Stanton, *Nature (Lond.)* **53**, 274–276 (Jan. 23, 1896).

12. E. Hascheck and O. T. Lindenthal, *Wien. klin. Wochenschr.* **9**, 63 (1896).

13. G. N. Hounsfield, *Br. J. Radiol.* **46**, 1016–1022 (1973).

14. H. H. Barrett, W. G. Hawkins, and M. L. G. Joy, *Radiology* **147**, 172 (1983).

15. A. M. Cormack, *J. Appl. Phys.* **34**, 2722–2727 (1963).

16. W. L. Bloom, J. L. Hollenbach, and J. A. Morgan, *Medical Radiographic Technique*, Charles C. Thomas Publishing, Springfield, Ill., 1965.

17. L. F. Squire and R. A. Novelline, *Fundamentals of Radiology*, Harvard University Press, Cambridge, Mass., 1975.

18. T. R. Eastman, *Radiographic Fundamentals and Technique Guide*, C. V. Mosby, St. Louis, Mo., 1979.

19. H. E. Seemann, in J. M. Sturge, ed., *Niblette's Handbook of Photography and Reprography Materials, Processes and Systems*, Van Nostrand Reinhold, New York, 1977, pp. 550–561.

20. L. Erickson and H. R. Splettstosser, in T. H. James, ed., *The Theory of the Photographic Process*, Macmillan, New York, 1977, pp. 662–671.

21. J. T. Bushberg, J. A. Seibert, E. M. Leidholdt Jr., and J. M. Boone, *The Essential Physics of Medical Imaging*, Williams and Wilkins, Baltimore, Md., 1994.

22. Z. H. Cho, J. P. Jones, and M. Singh, *Fundamentals of Medical Imaging*, John Wiley & Sons, Inc., New York, 1993.

23. M. M. Ter-Pogossian, *The Physical Aspects of Diagnostic Radiology*, Harper & Row, New York, 1967.

24. J. L. C. Sanz, E. B. Hinkle, and A. K. Jain, *Radon and Projection Transform-Based Computer Vision*, Springer-Verlag, Berlin, 1988.

25. K. P. Peschmann and co-workers, *Appl. Opt.* **24**, 4052–4060 (1985).

26. J. N. Wolfe, *AJR* **126**, 1130–1137 (1976).

27. J. N. Wolfe, K. A. Buck, M. Salane, and N. J. Parekh, *Radiology* **165**, 305–311 (1987).

28. J. E. Gray, *Proceedings of the Eighteenth Annual Meeting of the National Council on Radiation Protection and Measurements, Radiation Protection and New Medical Diagnostic Approaches*, National Academy of Sciences, Washington, D.C., 1982, pp. 117–129.

29. S. C. Bushong and B. R. Archer, *Diagnostic Ultrasound, Physics, Biology, and Instrumentation.*, Mosby Year Book, St. Louis, Mo., 1991.

30. P. Curie and J. Curie, *Comptes rendus hebdomadaires des séances de l'Académie des sciences* **91**, 294 (1880).

31. *Ibid.*, **93**, 1137 (1881).

32. D. H. Howry and W. R. Bloss, *J. Lab. Clin. Med.* **40**, 579–592 (1952).

33. J. J. Wild and J. M. Reid, *Cancer Res.* **14**, 277–283 (1954).

34. I. Donald, J. MacVicar, and T. G. Brown, *Lancet* **1**, 1188–1194 (1958).

35. R. W. Gill, *Ultrasound Med. Biol.* **11**, 625–641 (1985).

36. J. C. Bamber and M. Tristam, in S. Webb, ed., *The Physics of Medical Imaging*, IOP Publishing, Philadelphia, Pa., 1988.

37. P. N. T. Wells, *Physical Principles of Ultrasonic Diagnosis*, Academic Press, London, 1969.

38. T. Ikeda, *Fundamentals of Piezoelectricity*, Oxford University Press, New York, 1990.

39. K. Kimura, N. Hashimoto, and H. Ohigashi, *IEEE Trans. Sonics Ultrason.* **32**, 566–573 (1985).
40. O. T. vonRamm and S. W. Smith, *IEEE Trans. Biomed. Eng.* **30**, 438–452 (1983).
41. Y. Nomura, Y. Matsuda, I. Yabuuchi, M. Nishioka, and S. Tarui, *Radiology* **187**, 353–356 (1993).
42. R. H. Simon, S. Y. Ho, S. C. Lange, D. F. Uphoff, and J. S. Darrigo, *Ultrasound Med. Biol.* **19**, 123–125 (1993).
43. W. J. Gussenhoven, N. Bom, and J. Roelandt, eds., *Intravascular Ultrasound*, Kluwer, Dordrecht, the Netherlands, 1991.
44. E. A. Neppiras, *Ultrasonics* **24**, 25–28 (1984).
45. F. W. Kremkau, *Clin. Obstet. Gynaecol.* **10**, 395–405 (1983).
46. K. Makino and M. M. Mossoba, *Radiation Res.* **96**, 416–421 (1983).
47. D. D. Stark and W. G. Bradly, *Magnetic Resonance Imaging*, Mosby, Lanham, Md., 1988.
48. C. L. Partain, R. R. Price, J. A. Patton, M. V. Kulkarni, and A. E. James, *Magnetic Resonance Imaging*, Saunders, Philadelphia, Pa., 1988.
49. J. P. Hornak and L. M. Fletcher, in E. R. Dougherty, ed., *Digital Image Processing Methods*, Marcel Dekker, New York, 1994.
50. R. Damadian, *Science* **171**, 1151 (1971).
51. P. G. Lauterbur, *Nature* **242**, 190–191 (1973).
52. A. Kumar, D. Welti, and R. R. Ernst, *J. Magn. Reson.* **18**, 69–83 (1975); *Naturwissenschaften* **62**, 34 (1975).
53. C. P. Slichter, *Principles of Magnetic Resonance*, Springer-Verlag, Berlin, 1980.
54. T. C. Farrar and E. D. Becker, *Pulse and Fourier Transform NMR*, Academic Press, Inc., New York, 1971.
55. E. Fukushima and S. B. W. Roeder, *Experimental Pulse NMR*, Addison-Wesley, Reading, Mass., 1981.
56. R. C. Weast, ed., *CRC Handbook of Chemistry and Physics*, 53rd ed., CRC Press, Cleveland, Ohio, 1972.
57. S. L. Smith, *Anal. Chem.* **57**, A595–A607 (1985).
58. C. E. Hayes, W. A. Edelstein, and J. F. Schenck, in C. L. Partain, R. R. Price, J. A. Patton, M. V. Kulkarni, and A. E. James, eds., *Magnetic Resonance Imaging*, Saunders, Philadelphia, Pa., 1988.
59. G. M. Bydder, in D. D. Stark and W. G. Bradley, eds., *Magnetic Resonance Imaging*, Mosby, Lanham, Md., 1988.
60. J. H. Bisese, *Cranial MRI: A Teaching File Approach*, McGraw-Hill Book Co., Inc., New York, 1991.
61. S. Ogwa and co-workers, *Proc. Mat. Acad. Sci.* **89**, 5951–5955 (1992).
62. P. Mansfield, *J. Phys. C: Solid State Phys.* **10**, L55 (1977).
63. M. Dunitz, *MRI Atlas of the Musculoskeletal System*, CRC Press, Boca Raton, Fla., 1989.
64. D. G. Nishimura, *Magn. Reson. Med.* **14**, 194–201 (1990).
65. C. L. Dumoulin, S. P. Souza, M. F. Walker, and W. Wagle, *Magn. Reson. Med.* **9**, 139–149 (1989).
66. J. Leigh, *Resonance Newsletter SMRM*, **20**, 9 (1990).
67. R. J. Ott, M. A. Flower, J. W. Babich, and P. K. Marsden, in Ref. 36.
68. M. E. Raichle, *Sci. Am.* **270**, 58–64 (1994).
69. K. F. Hubner, J. Collmann, E. Buonocore, and G. W. Kabalka, *Clinical Positron Emission Tomography*, Mosby Year Book, St. Louis, Mo., 1991.

Joseph P. Hornak
Rochester Institute of Technology

MELAMINE FORMALDEHYDE RESINS. See AMINO RESINS AND PLASTICS.

MELTING and FREEZING TEMPERATURES. See TEMPERATURE MEASUREMENT; THERMAL, GRAVIMETRIC, AND VOLUMETRIC ANALYSIS.

MEMBRANE TECHNOLOGY

Membranes have gained an important place in chemical technology and are being used increasingly in a broad range of applications. The key property that is exploited in every application is the ability of a membrane to control the permeation of a chemical species in contact with it. In packaging applications, the goal is usually to prevent permeation completely. In controlled drug delivery applications, the goal is to moderate the permeation rate of a drug from a reservoir to the body. In separation applications, the goal is to allow one component of a mixture to permeate the membrane freely, while hindering permeation of other components. Since the 1960s, membrane science has grown from a laboratory curiosity to a widely practiced technology in industry and medicine. This growth is likely to continue for some time, particularly in the membrane gas separation and pervaporation separation areas. Membranes will play a critical role in the next generation of biomedical devices, such as the artificial pancreas and liver. The total membrane market grew from $10 million to the $1–2 billion level in the 30 years prior to 1994. Spectacular growth of this magnitude is unlikely to continue, but a doubling in the size of the total industry to the $2–4 billion level during the decade following is likely.

Historical Development

Systematic studies of membrane phenomena can be traced to the eighteenth century philosopher scientists. For example, Abbé Nolet coined the word osmosis to describe permeation of water through a diaphragm in 1748. Through the nineteenth and early twentieth centuries, membranes had no industrial or commercial uses but were used as laboratory tools to develop physical/chemical theories.

For example, the measurements of solution osmotic pressure made with membranes by Traube and Pfeffer were used by van't Hoff in 1887 to develop his limit law, which explains the behavior of ideal dilute solutions. This work led directly to the van't Hoff equation. At about the same time, the concept of a perfectly selective semipermeable membrane was used by Maxwell and others in developing the kinetic theory of gases.

Early investigators experimented with any type of diaphragm available to them, such as bladders of pigs, cattle, or fish, and sausage casings made of animal gut. Later, collodion (nitrocellulose) membranes were preferred, because they could be made reproducibly. In 1907, Bechhold devised a technique to prepare nitrocellulose membranes of graded pore size, which he determined by a bubble test (1). Other workers (2–4) improved on Bechhold's technique, and by the early 1930s microporous collodion membranes were commercially available. During the next 20 years, this early microfiltration membrane technology was expanded to other polymers, notably cellulose acetate. Membranes found their first significant application in the filtration of drinking water samples at the end of World War II. Drinking water supplies serving large communities in Germany and elsewhere in Europe had broken down, and filters to test for water safety were needed urgently. The research effort to develop these filters, sponsored by the U.S. Army, was later exploited by the Millipore Corp., the first and still the largest microfiltration membrane producer.

By 1960, the elements of modern membrane science had been developed, but membranes were used in only a few laboratory and small, specialized industrial applications. No significant membrane industry existed, and total annual sales of membranes for all applications probably did not exceed $10 million in 1990 dollars. Membranes suffered from four problems that prohibited their widespread use as a separation process: they were too unreliable, too slow, too unselective, and too expensive. Partial solutions to each of these problems have been developed since the 1960s, and in the 1990s membrane-based separation processes are commonplace.

The seminal discovery that transformed membrane separation from a laboratory to an industrial process was the development, in the early 1960s, of the Loeb-Sourirajan process for making defect-free, high flux, asymmetric reverse osmosis membranes (5). These membranes consist of an ultrathin, selective surface film on a microporous support, which provides the mechanical strength. The flux of the first Loeb-Sourirajan reverse osmosis membrane was 10 times higher than that of any membrane then available and made reverse osmosis practical. The work of Loeb and Sourirajan, and the timely infusion of large sums of research dollars from the U.S. Department of Interior, Office of Saline Water (OSW), resulted in the commercialization of reverse osmosis (qv) and was a primary factor in the development of ultrafiltration (qv) and microfiltration. The development of electrodialysis was also aided by OSW funding.

The 20-year period from 1960 to 1980 produced a significant change in the status of membrane technology. Building on the original Loeb-Sourirajan membrane technology, other processes, including interfacial polymerization and multilayer composite casting and coating, were developed for making high performance membranes. Using these processes, membranes with selective layers as thin as 0.1 μm or less can be made. Methods of packaging membranes into spiral-wound, hollow-fine fiber, capillary, and plate-and-frame modules were also developed, and advances were made in improving membrane stability. By 1980, microfiltration, ultrafiltration, reverse osmosis, and electrodialysis were all established processes with large plants installed around the world.

The principal development in the 1980s was the emergence of industrial membrane gas-separation processes. The first significant development was the Monsanto Prism membrane for hydrogen separation, developed in the late 1970s (6). Within a few years, Dow was producing systems to separate nitrogen from air, and Cynara and Separex were producing systems to separate carbon dioxide from methane. Gas-separation technology is evolving and expanding rapidly, and further substantial growth will be seen in the 1990s. The final development of the 1980s was the introduction by GFT, a small German engineering company, of the first commercial pervaporation systems for dehydration of alcohol. By 1990, GFT had sold more than 100 plants. Many of these plants are small, but the technology has been demonstrated and a number of other pervaporation applications are at the pilot-plant stage.

Types of Membrane

Although this article is limited to synthetic membranes, excluding all biological structures, the topic is still large enough to include a wide variety of membranes that differ in chemical and physical composition and in the way they operate. In essence, a membrane is a discrete, thin interface that moderates the permeation of chemical species in contact with it. This interface may be molecularly homogeneous, that is, completely uniform in composition and structure, or it may be chemically or physically heterogeneous, for example, containing holes or pores of finite dimensions. A normal filter meets this definition of a membrane, but, by convention, the term membrane is usually limited to structures that permeate dissolved or colloidal species, whereas the term filter is used to designate structures that separate particulate suspensions. The principal types of membrane are shown schematically in Figure 1.

Isotropic Microporous Membranes. A microporous membrane is very similar in its structure and function to a conventional filter. It has a rigid, highly voided structure with randomly distributed, interconnected pores. However, these pores differ from those in a conventional filter by being extremely small, of the order of 0.01–10 μm in diameter. All particles larger than the largest pores are completely rejected by the membrane. Particles smaller than the largest pores, but larger than the smallest pores are partially rejected, according to the pore size distribution of the membrane. Particles much smaller than the smallest pores pass through the membrane. Thus separation of solutes by microporous membranes is mainly a function of molecular size and pore size distribution. In general, only molecules that differ considerably in size can be separated effectively by microporous membranes, for example, in ultrafiltration and microfiltration.

Nonporous Dense Membranes. Nonporous, dense membranes consist of a dense film through which permeants are transported by diffusion under the driving force of a pressure, concentration, or electrical potential gradient. The separation of various components of a solution is related directly to their relative transport rate within the membrane, which is determined by their diffusivity and solubility in the membrane material. An important property of nonporous, dense membranes is that even permeants of similar size may be separated when their concentration in the membrane material (ie, their solubility) differs significantly. Most gas separation, pervaporation, and reverse osmosis membranes use dense membranes to perform the separation. However, these membranes usually have an asymmetric structure to improve the flux.

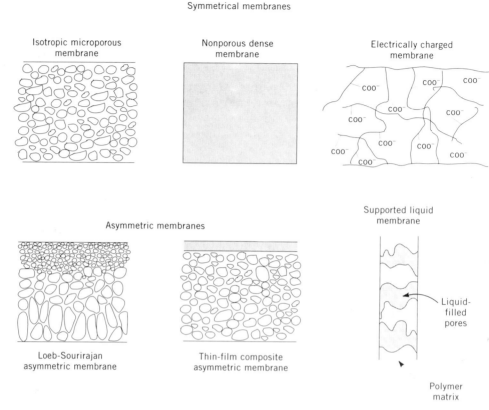

Fig. 1.   Schematic diagrams of the principal types of membrane.

Electrically Charged Membranes.   Electrically charged membranes can be dense or microporous, but are most commonly microporous, with the pore walls carrying fixed positively or negatively charged ions. A membrane with positively charged ions is referred to as an anion-exchange membrane because it binds anions in the surrounding fluid. Similarly, a membrane containing negatively charged ions is called a cation-exchange membrane. Separation with charged membranes is achieved mainly by exclusion of ions of the same charge as the fixed ions of the membrane structure, and to a much lesser extent by the pore size. The separation is affected by the charge and concentration of the ions in solution. For example, monovalent ions are excluded less effectively than divalent ions and, in solutions of high ionic strength, selectivity decreases. Electrically charged membranes are used for processing electrolyte solutions in electrodialysis.

Asymmetric Membranes.   The transport rate of a species through a membrane is inversely proportional to the membrane thickness. High transport rates are desirable in membrane separation processes for economic reasons; therefore, the membrane should be as thin as possible. Conventional film fabrication technology limits manufacture of mechanically strong, defect-free films to about 20 μm thickness. The development of novel membrane fabrication techniques to produce asymmetric membrane structures was one of the breakthroughs of membrane technology during the past 30 years. Asymmetric membranes consist of an extremely thin surface layer supported on a much thicker porous, dense substructure. The surface layer and its substructure may be formed in a single operation or formed separately. The separation properties and permeation rates of the membrane are determined exclusively by the surface layer; the substructure functions as a mechanical support. The advantages of the higher fluxes provided by asymmetric membranes are so great that almost all commercial processes use such membranes.

Ceramic, Metal, and Liquid Membranes. The discussion so far implies that membrane materials are organic polymers and, in fact, the vast majority of membranes used commercially are polymer based. However, interest in membranes formed from less conventional materials has increased. Ceramic membranes, a special class of microporous membranes, are being used in ultrafiltration and microfiltration applications, for which solvent resistance and thermal stability are required. Dense metal membranes, particularly palladium membranes, are being considered for the separation of hydrogen from gas mixtures, and supported or emulsified liquid films are being developed for coupled and facilitated transport processes.

Preparation of Membranes and Membrane Modules

Because membranes applicable to diverse separation problems are often made by the same general techniques, classification by end use application or preparation method is difficult. The first part of this section is, therefore, organized by membrane structure; preparation methods are described for symmetrical membranes, asymmetric membranes, ceramic and metal membranes, and liquid membranes. The production of hollow-fine fiber membranes and membrane modules is then covered. Symmetrical membranes have a uniform structure throughout; such membranes can be either dense films or microporous.

Dense Symmetrical Membranes. These membranes are used on a large scale in packaging applications (see FILMS AND SHEETING; PACKAGING MATERIALS). They are also used widely in the laboratory to characterize membrane separation properties. However, it is difficult to make mechanically strong and defect-free symmetrical membranes thinner than 20 μm, so the flux is low, and these membranes are rarely used in separation processes. For laboratory work, the membranes are prepared by solution casting or by melt pressing.

In solution casting, a casting knife or drawdown bar is used to spread an even film of an appropriate polymer solution across a glass plate. The casting knife consists of a steel blade, resting on two runners, arranged to form a precise gap between the blade and the plate on which the film is cast. A typical hand-casting knife is shown in Figure 2. After the casting has been made, it is left to stand, and the solvent evaporates to leave a uniform polymer film.

The polymer casting solution should be sufficiently viscous to prevent the solution from running over the casting plate, so typical casting solution concentrations are in the range of 15 to 20 wt % polymer. Solvents with high boiling points are inappropriate for solvent casting, because their low volatility demands long evaporation times. During an extended evaporation period, the

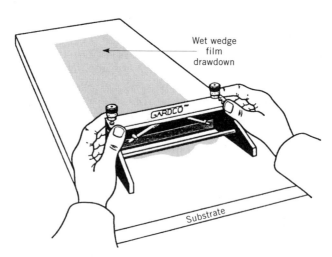

Fig. 2.   A typical hand-casting knife. Courtesy of Paul N. Gardner Co., Inc.

cast film can absorb sufficient atmospheric water to precipitate the polymer, producing a mottled, hazy surface.

Many polymers, including polyethylene, polypropylene, and nylons, do not dissolve in suitable casting solvents. In the laboratory, membranes can be made from such polymers by melt pressing, in which the polymer is sandwiched at high pressure between two heated plates. A pressure of 13.8–34.5 MPa (2000–5000 psi) is applied for 0.5 to 5 minutes, at a plate temperature just above the melting point of the polymer. Melt forming is commonly used to make dense films for packaging applications, either by extrusion as a sheet from a die or as blown film.

Microporous Symmetrical Membranes.   These membranes, used widely in microfiltration, typically contain pores in the range of 0.1–10 μm diameter. As shown in Figure 3, microporous membranes are generally characterized by the average pore diameter, d, the membrane porosity, ϵ (the fraction of the total membrane volume that is porous), and the tortuosity of the membrane, τ (a term reflecting the length of the average pore through the membrane compared to the membrane thickness). The most important types of microporous membrane are those formed by one of the solution–precipitation techniques discussed in the next section under asymmetric membranes; about half of all microporous membranes are made in this way. The remainder is made by various proprietary techniques, the more important of which are discussed in the following.

Irradiation.   Nucleation track membranes were first developed by the Nuclepore Corp. (7). The two-step preparation process is illustrated in Figure 4. A polymer film is first irradiated with charged particles from a nuclear reactor or other radiation source; particles passing through the film break polymer chains and leave behind sensitized or damaged tracks. The film is then passed through an etch solution, which etches the polymer preferentially along the sensitized nucleation tracks, thereby forming pores. The length of time the film is exposed to radiation in the reactor determines the number of pores in the film;

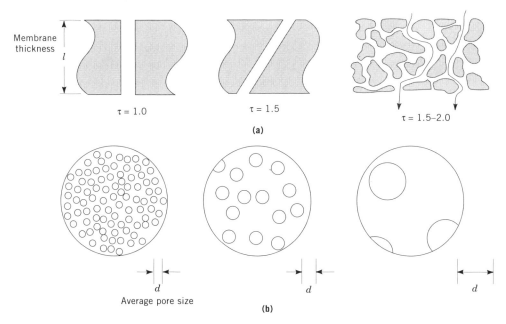

Fig. 3. Microporous membranes are characterized by tortuosity, τ, porosity, ϵ, and their average pore diameter, d. (**a**) Cross-sections of porous membranes containing cylindrical pores. (**b**) Surface views of porous membranes of equal ϵ, but differing pore size.

the etch time determines the pore diameter. Because of the unique preparation techniques used to make nucleation track membranes, the pores are uniform cylinders traversing the membrane almost at right angles. The membrane tortuosity is, therefore, close to 1.0. The membrane porosity is usually relatively low, about 5%, so fluxes are low. However, because these membranes are very close to a perfect screen filter, they are used in analytical techniques that require filtration of all particles above a certain size from a fluid so that the particles can be visualized under a microscope.

Expanded Film. Expanded-film membranes are made from crystalline polymers by an orientation and stretching process. In the first step of the process, a highly oriented film is produced by extruding the polymer at close to its melting point coupled with a very rapid drawdown (9,10). After cooling, the film is stretched a second time, up to 300%, at right angles to the original orientation of the polymer crystallites. This second elongation deforms the crystalline structure of the film and produces slit-like voids 20 to 250 nm wide between crystallites. The process is illustrated in Figure 5. This type of membrane was first developed by Hoechst-Celanese and is sold under the trade name Celgard; a number of companies make similar products. The membranes made by W. L. Gore, sold under the trade name Gore-Tex, are made by this type of process (11).

The original expanded film membranes were sold in rolls as flat sheets. These membranes had relatively poor tear strength along the original direction of orientation and were not widely used as microfiltration membranes. They

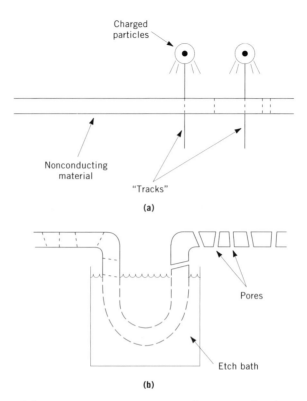

(a)

(b)

Fig. 4. Diagram of the two-step process to manufacture nucleation track membranes. (**a**) Polycarbonate film is exposed to charged particles in a nuclear reactor. (**b**) Tracks left by particles are preferentially etched into uniform cylindrical pores (8). Courtesy of Corning Costar Corp., Nucleopore Division.

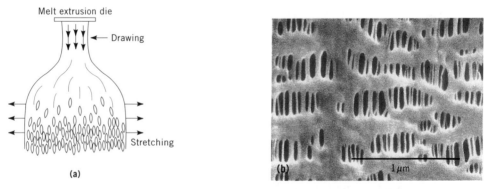

Fig. 5. (**a**) Preparation method and (**b**) scanning electron micrograph of a typical expanded polypropylene film membrane, in this case Celgard. Courtesy of Hoechst-Celanese Corp., Separation Products Division.

142

did, however, find use as porous inert separating barriers in batteries and some medical devices. More recently, the technology has been developed to produce these membranes as hollow fibers, which are used as membrane contactors (12,13).

Template Leaching. Template leaching offers an alternative manufacturing technique for insoluble polymers. A homogeneous film is prepared from a mixture of the membrane matrix material and a leachable component. After the film has been formed, the leachable component is removed with a suitable solvent and a microporous membrane is formed (14,15). The leachable component could be a soluble low molecular weight solid or liquid, or even a polymeric material such as poly(vinyl alcohol) or poly(ethylene glycol). The same general method is used to prepare microporous glass (16). In this case, a two-component glass melt is formed into sheets or small tubes, after which one of the components is leached out by extraction with an alkaline solution.

Asymmetric Membranes. In industrial applications other than microfiltration, symmetrical membranes have been displaced almost completely by asymmetric membranes, which have much higher fluxes. Asymmetric membranes have a thin, permselective layer supported on a more open porous substrate. Hindsight makes it clear that many of the membranes produced in the 1930s and 1940s were asymmetric, although this was not realized at the time. The importance of the asymmetric structure was not recognized until Loeb and Sourirajan prepared the first high flux, asymmetric, reverse osmosis membranes by what is now known as the Loeb-Sourirajan technique (5). This discovery was a critical breakthrough in membrane technology. The reverse osmosis membranes produced were an order of magnitude more permeable than any symmetrical membrane produced previously. More importantly, demonstration of the benefits of the asymmetric structure paved the way for the development of other types of asymmetric membranes. Improvements in asymmetric membrane preparation methods and properties were accelerated by the increasing availability in the late 1960s of scanning electron microscopes, which allowed the effects of changes in the membrane formation process on structure to be assessed easily.

Phase Inversion (Solution Precipitation). Phase inversion, also known as solution precipitation or polymer precipitation, is the most important asymmetric membrane preparation method. In this process, a clear polymer solution is precipitated into two phases: a solid polymer-rich phase that forms the matrix of the membrane, and a liquid polymer-poor phase that forms the membrane pores. If precipitation is rapid, the pore-forming liquid droplets tend to be small and the membranes formed are markedly asymmetric. If precipitation is slow, the pore-forming liquid droplets tend to agglomerate while the casting solution is still fluid, so that the final pores are relatively large and the membrane structure is more symmetrical. Polymer precipitation from a solution can be achieved in several ways, such as cooling, solvent evaporation, precipitation by immersion in water, or imbibition of water from the vapor phase. Each technique was developed independently; only since the 1980s has it become clear that these processes can all be described by the same general approach based on polymer–solvent–nonsolvent phase diagrams. Thus, the Loeb-Sourirajan process, in which precipitation is produced by immersion in water, is a subcategory of the general class of phase-inversion membranes. The theory behind the preparation of membranes by all of these techniques has been reviewed in a number of monographs and review articles (17–20).

Polymer Precipitation by Cooling. The simplest solution–precipitation technique is thermal gelation, in which a film is cast from a hot, one-phase polymer solution. When the cast film cools, the polymer precipitates, and the solution separates into a polymer-matrix phase containing dispersed pores filled with solvent. The precipitation process that forms the membrane can be represented by the phase diagram shown in Figure 6. The pore volume in the final membrane is determined mainly by the initial composition of the cast film, because this determines the ratio of the polymer to liquid phase in the cooled film. However, the spatial distribution and size of the pores is determined largely by the rate of cooling and, hence, precipitation, of the film. In general, rapid cooling produces membranes with small pores (21,22).

Polymer precipitation by cooling to produce microporous membranes was first commercialized on a large scale by Akzo (23). Akzo markets microporous polypropylene and poly(vinylidine fluoride) membranes produced by this technique under the trade name Accurel. Polypropylene membranes are prepared from a solution of polypropylene in N,N-bis(2-hydroxyethyl)tallowamine. The amine and polypropylene form a clear solution at temperatures above 100–150°C. Upon cooling, the solvent and polymer phases separate to form a microporous structure. If the solution is cooled slowly, an open cell structure of the type shown in Figure 7a results. The interconnecting passageways between cells are generally in the micrometer range. If the solution is cooled and precipitated rapidly, a much finer structure is formed, as shown in Figure 7b. The rate of cooling is, therefore, a key parameter determining the final structure of the membrane (21).

A schematic diagram of the polymer precipitation process is shown in Figure 8. The hot polymer solution is cast onto a water-cooled chill roll, which cools the solution, causing the polymer to precipitate. The precipitated film is

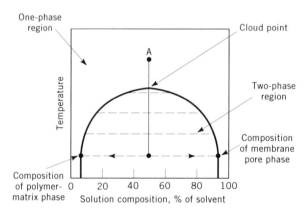

Fig. 6. Phase diagram showing the composition pathway traveled by the casting solution during precipitation by cooling. Point A represents the initial temperature and composition of the casting solution. The cloud point is the point of fast precipitation. In the two-phase region tie lines linking the precipitated polymer phase and the suspended liquid phase are shown.

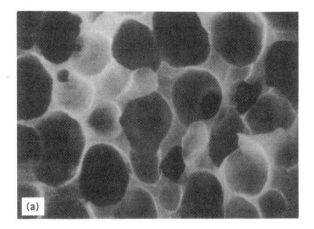

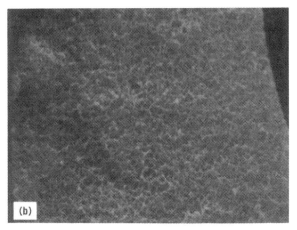

Fig. 7. Polypropylene structures (**a**) Type I Open Cell structure formed at low cooling rates (2400×). (**b**) Type II "Lacy" structure formed at high cooling rates (2000×) (21).

passed through an extraction tank containing methanol, ethanol or 2-propanol to remove the solvent. Finally, the membrane is dried, sent to a laser inspection station, trimmed, and rolled up. The process shown in Figure 8 is used to make flat-sheet membranes. The preparation of hollow-fiber membranes (qv) by the same general technique has also been described.

Polymer Precipitation by Solvent Evaporation. This technique was one of the earliest methods of making microporous membranes (1–4). In the simplest form of the method, a polymer is dissolved in a two-component solvent mixture consisting of a volatile solvent, such as acetone, in which the polymer is readily soluble, and a less volatile nonsolvent, typically water or an alcohol. The polymer solution is cast onto a glass plate. As the volatile solvent evaporates, the casting solution is enriched in the nonvolatile solvent. The polymer precipitates, forming the membrane structure. The process can be continued until the membrane has completely formed, or it can be stopped, and the membrane structure fixed, by

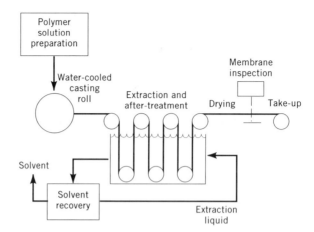

Fig. 8. Equipment to prepare microporous membranes by the polymer precipitation by cooling technique (23).

immersing the cast film into a precipitation bath of water or other nonsolvent. Scanning electron micrographs of some membranes made by this process are shown in Figure 9 (24).

Many factors determine the porosity and pore size of membranes formed by the solvent evaporation method. As Figure 9 shows, if the membrane is immersed in a nonsolvent after a short evaporation time, the resulting membrane will be finely microporous. If the evaporation step is prolonged before fixing the structure by immersion in water, the average pore size will be larger. In general, increasing the nonsolvent content of the casting solution, or decreasing the polymer concentration, increases porosity. It is important that the nonsolvent be completely incompatible with the polymer. If partly compatible nonsolvents are used, the precipitating polymer phase contains sufficient residual solvent to allow it to flow and collapse as the solvent evaporates. The result is a dense rather than microporous film.

Polymer Precipitation by Imbibition of Water Vapor.   Preparation of microporous membranes by simple solvent evaporation alone is not practiced widely. However, a combination of solvent evaporation with precipitation by imbibition of water vapor from a humid atmosphere or by water-vapor imbibition in combination with solvent evaporation are the basis of many commercial phase-inversion processes. The processes often involve proprietary casting formulations that are not normally disclosed by membrane developers. However, during the development of composite membranes at Gulf General Atomic, this type of membrane was prepared and the technology described in some detail in a series of Office of Saline Water Reports (25). These reports remain the best published description of the technique. The type of equipment used is shown in Figure 10. The casting solution typically consists of a blend of cellulose acetate and cellulose nitrate dissolved in a mixture of volatile solvents, such as acetone, and nonvolatile nonsolvents, such as water, ethanol, or ethylene glycol. The polymer solution is cast onto a continuous stainless steel belt. The cast film then passes through a series of environmental chambers; hot, humid air is usually circulated through the first chamber. The film loses the volatile solvent by evaporation and simul-

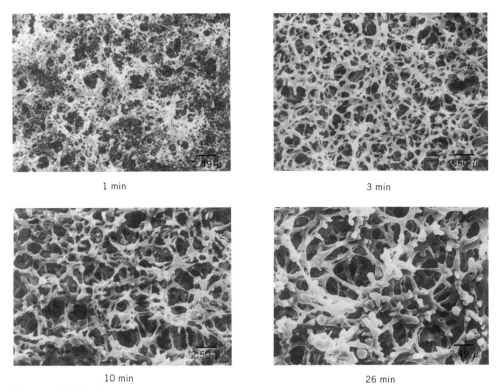

Fig. 9. SEM photographs of cellulose acetate membranes cast from a solution of acetone (volatile solvent) and 2-methyl-2,4-pentanediol (nonvolatile solvent). The evaporation time before the structure is fixed by immersion in water is shown (24).

taneously absorbs water from the atmosphere. The total precipitation process is slow, taking about 10 minutes to complete. The resulting membrane structure is fairly symmetrical. After precipitation, the membrane passes to a second oven, through which hot dry air is circulated to evaporate the remaining solvent and dry the film. The formed membrane is then wound on a take-up roll. Typical casting speeds are of the order of 0.3–0.6 m/min. This type of membrane is widely used in microfiltration applications.

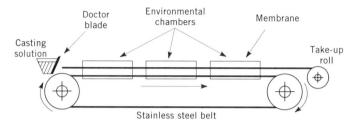

Fig. 10. Schematic of casting machine used to make microporous membranes by water-vapor imbibition. A casting solution is deposited as a thin film on a moving stainless steel belt. The film passes through a series of humid and dry chambers, where the solvent evaporates from the solution, and water vapor is absorbed from the air. This precipitates the polymer, forming a microporous membrane that is taken up on a collection roll (25).

Polymer Precipitation by Immersion in a Nonsolvent Bath. This is the Loeb-Sourirajan process, the single most important membrane preparation technique, and almost all reverse osmosis, ultrafiltration, and many gas separation membranes are produced by this procedure or a derivative of it. A schematic of a casting machine used in the process is shown in Figure 11. A typical membrane casting solution contains approximately 20 wt % of dissolved polymer. This solution is cast onto a moving drum or paper web, and the cast film is precipitated by immersion in a water bath. The water precipitates the top surface of the cast film rapidly, forming an extremely dense, permselective skin. This skin slows down the entry of water into the underlying polymer solution, which precipitates much more slowly, forming a more porous substructure. Depending on the polymer, the casting solution and other parameters, the dense skin varies from 0.1 to 1.0 μm thick. This process was originally developed for reverse osmosis (5). Later the technique was adapted to make membranes for other applications, including ultrafiltration and gas separation (6,20,26).

A great deal of work has been devoted to rationalizing the factors affecting the properties of asymmetric membrane made by the Loeb-Sourirajan technique and, in particular, to understanding those factors that determine the thickness of the membrane skin that performs the separation. The goal is to make this skin as

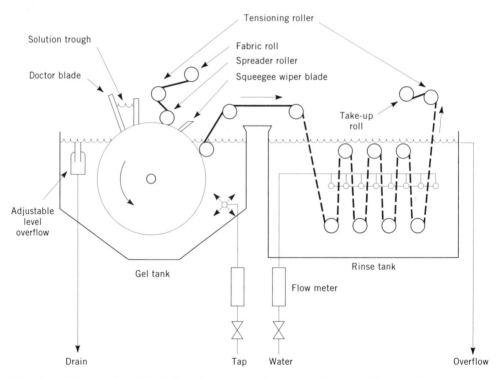

Fig. 11. Schematic of Loeb-Sourirajan membrane casting machine used to prepare reverse osmosis or ultrafiltration membranes. A knife and trough is used to coat the casting solution onto a moving fabric or polyester web which enters the water-filled gel tank. After the membrane has formed, it is washed thoroughly to remove residual solvent before being wound up.

thin as possible, but still defect-free. The skin layer can be dense, as in reverse osmosis or gas separation, or finely microporous with pores in the 10–50 nm diameter range, as in ultrafiltration. In good quality membranes made by this technique, a skin thickness as low as 50–100 nm can be achieved. A scanning electron micrograph of a Loeb-Sourirajan membrane is shown in Figure 12.

The phase-diagram approach has been widely used to rationalize the preparation of these membranes (17–20,26). The ternary phase diagram of the three-component system used in preparing Loeb-Sourirajan membranes is shown in Figure 13. The corners of the triangle represent the three components, polymer, solvent, and precipitant; any point within the triangle represents a mixture of three components. The system consists of two regions: a one-phase region, where all components are miscible, and a two-phase region, where the system separates into a solid (polymer-rich) phase and a liquid (polymer-poor) phase. Although the one-phase region in the phase diagram is thermodynamically continuous, for practical purposes it can conveniently be divided into a liquid and solid gel region. Thus, at low polymer concentrations, the system is a low viscosity liquid, but as the concentration of polymer is increased, the viscosity of the system also increases rapidly, reaching such high values that the system can be regarded as a solid. The transition between liquid and solid regions is, therefore, arbitrary, but can be placed at a polymer concentration of 30–40 wt %. In the two-phase region of the diagram, tie lines link the polymer-rich and polymer-poor phases.

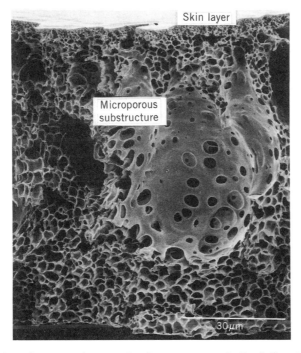

Fig. 12. Scanning electron micrograph of an asymmetric Loeb-Sourirajan membrane. Courtesy of Membrane Technology and Research, Inc.

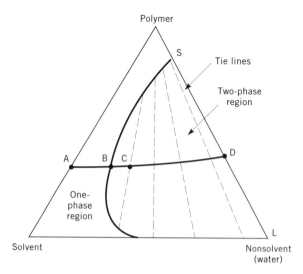

Fig. 13. Phase diagram showing the composition pathway traveled by a casting solution during the preparation of porous membranes by solvent evaporation. A, initial casting solution; B, point of precipitation; and C, point of solidification. See text.

Unlike low molecular weight components, polymer systems in the two-phase region are often slow to separate into different phases and metastable states are common, especially when a polymer solution is rapidly precipitated.

The phase diagram in Figure 13 shows the precipitation pathway of the casting solution during membrane formation. During membrane formation, the system changes from a composition A, which represents the initial casting solution composition, to a composition D, which represents the final membrane composition. At composition D, the two phases are in equilibrium: a solid (polymer-rich) phase, which forms the final membrane structure, represented by point S, and a liquid (polymer-poor) phase, which constitutes the membrane pores filled with precipitant, represented by point L. The position D on the line S–L determines the overall porosity of the membrane. The entire precipitation process is represented by the path A–D, during which the solvent is exchanged by the precipitant. The point B along the path is the concentration at which the first polymer precipitates. As precipitation proceeds, more solvent is lost and precipitant is imbibed by the polymer-rich phase, so the viscosity rises. At some point, the viscosity is high enough for the precipitated polymer to be regarded as a solid. This composition is at C in Figure 13. Once the precipitated polymer solidifies, further bulk movement of the polymer is hindered. The rate and the pathway A–D taken by the polymer solution vary from the surface of the polymer film to the sublayer, affecting the pore size and porosity of the final membrane at that point. The nature of the casting solution and the precipitation conditions are important in determining the kinetics of this precipitation process, and detailed theoretical treatments based on the ternary phase diagram approach have been worked out.

In the Loeb-Sourirajan process formation of minute membrane defects may occur. These defects, caused by gas bubbles, dust particles, and support fabric imperfections, are often difficult to eliminate. These defects may not significantly affect the performance of asymmetric membranes used in liquid separation operations, such as ultrafiltration and reverse osmosis, but can be disastrous in gas separation applications. Membrane defects can be overcome by coating the membrane with a thin layer of relatively permeable material (6,27). If the coating is sufficiently thin, it does not change the properties of the underlying permselective layer, but it does plug membrane defects, preventing simple convective gas flow through defects. This concept has been used to seal defects in polysulfone Loeb-Sourirajan membranes with silicone rubber (6). The form of these membranes is shown in Figure 14. The silicone rubber layer does not function as a selective barrier but rather plugs up defects, thereby reducing nondiffusive gas flow. The flow of gas through the portion of the silicone rubber layer over the pore is very high compared to the flow through the defect-free portion of the membrane. However, because the total area of the membrane subject to defects is very small, the total gas flow through these plugged defects is negligible. When this coating technique is used, the polysulfone skin layer of the Loeb-Sourirajan membrane no longer has to be completely free of defects; the coated membrane can be made with a thinner skin than is possible with an uncoated membrane. The increase in flux brought about by decreasing the thickness of the permselective skin layer more than compensates for the slight reduction in flux due to the silicone rubber sealing layer.

Cellulose acetate Loeb-Sourirajan reverse osmosis membranes were introduced commercially in the 1960s. Since then, many other polymers have been made into asymmetric membranes in attempts to improve membrane properties. In the reverse osmosis area, these attempts have had limited success, the only significant example being Du Pont's polyamide membrane. For gas separation and ultrafiltration, a number of membranes with useful properties have been made. However, the early work on asymmetric membranes has spawned numerous other techniques in which a microporous membrane is used as a support to carry another thin, dense separating layer.

Interfacial Composite Membranes. A method of making asymmetric membranes involving interfacial polymerization was developed in the 1960s. This technique was used to produce reverse osmosis membranes with dramatically improved salt rejections and water fluxes compared to those prepared by the Loeb-Sourirajan process (28). In the interfacial polymerization method, an aqueous solution of a reactive prepolymer, such as polyamine, is first deposited in the pores of a microporous support membrane, typically a polysulfone ultrafiltration

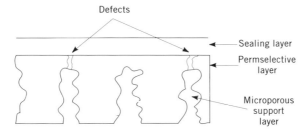

Fig. 14. Schematic of coated gas-separation membranes (6).

membrane. The amine-loaded support is then immersed in a water-immiscible solvent solution containing a reactant, for example, a diacid chloride in hexane. The amine and acid chloride then react at the interface of the two solutions to form a densely cross-linked, extremely thin membrane layer. This preparation method is shown schematically in Figure 15. The first membrane made was based on polyethylenimine cross-linked with toluene-2,4-diisocyanate (28). The process was later refined at FilmTec Corp. (29,30) and at UOP (31) in the United States, and at Nitto (32) in Japan.

Membranes made by interfacial polymerization have a dense, highly cross-linked interfacial polymer layer formed on the surface of the support membrane at the interface of the two solutions. A less cross-linked, more permeable hydrogel layer forms under this surface layer and fills the pores of the support membrane. Because the dense cross-linked polymer layer can only form at the interface, it is extremely thin, on the order of 0.1 μm or less, and the permeation flux is high. Because the polymer is highly cross-linked, its selectivity is also high. The first reverse osmosis membranes made this way were 5–10 times less salt-permeable than the best membranes with comparable water fluxes made by other techniques.

Interfacial polymerization membranes are less applicable to gas separation because of the water swollen hydrogel that fills the pores of the support membrane. In reverse osmosis, this layer is highly water swollen and offers little resistance to water flow, but when the membrane is dried and used in gas separations the gel becomes a rigid glass with very low gas permeability. This glassy polymer fills the membrane pores and, as a result, defect-free interfacial composite membranes usually have low gas fluxes, although their selectivities can be good.

Solution-Cast Composite Membranes.   Another important type of composite membrane is formed by solution casting a thin (0.5–2.0 μm) film on a suitable

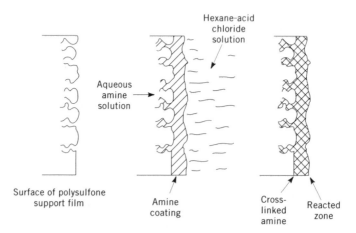

Fig. 15. Schematic of the interfacial polymerization process. The microporous film is first impregnated with an aqueous amine solution. The film is then treated with a multivalent cross-linking agent dissolved in a water-immiscible organic fluid, such as hexane or Freon-113. An extremely thin polymer film forms at the interface of the two solutions.

microporous film. Membranes of this type were first prepared at General Electric (27,33) and at North Star Research (34) using a type of Langmuir trough system (33,34). In this system, a dilute polymer solution in a volatile water-insoluble solvent is spread over the surface of a water-filled trough. The thin polymer film formed on the water surface is then picked up on a microporous support. This technique was developed into a semicontinuous process at General Electric but has not proved reliable enough for large-scale commercial use.

Most solution-cast composite membranes are prepared by a technique pioneered at UOP (35). In this technique, a polymer solution is cast directly onto the microporous support film. The support film must be clean, defect-free, and very finely microporous, to prevent penetration of the coating solution into the pores. If these conditions are met, the support can be coated with a liquid layer 50–100 μm thick, which after evaporation leaves a thin permselective film, 0.5–2 μm thick. This technique was used to form the Monsanto Prism gas separation membranes (6) and at Membrane Technology and Research to form pervaporation and organic vapor–air separation membranes (36,37) (Fig. 16).

Other Asymmetric Membrane Preparation Techniques. A number of other methods of preparing membranes have been reported in the literature and are used on a small scale. Table 1 provides a brief summary of these techniques.

Metal and Ceramic Membranes. Palladium and palladium alloy membranes can be used to separate hydrogen from other gases. Palladium membranes were studied extensively during the 1950s and 1960s, and a commercial plant to separate hydrogen from refinery off-gas was installed by Union Carbide (53). The plant used palladium–silver alloy membranes in the form of 25-μm thick films. The plant was operated for some time, but a number of problems, including long-term membrane stability under the high temperature operating conditions, were encountered, and the plant was later replaced by pressure-swing adsorption systems. Small-scale palladium membrane systems, marketed by Johnson Matthey

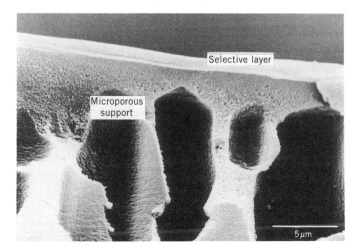

Fig. 16. Scanning electron micrograph of a silicone rubber composite membrane. Courtesy of Membrane Technology and Research, Inc.

Table 1. Less Widely Used Membrane Preparation Techniques

Preparation technique	Membrane characteristics	References
plasma polymerization	monomer is plasma polymerized onto the surface of a support film; resulting chemistry is complex	38–41
reactive surface treatment	an existing membrane is treated with a reactive gas of monomer to form an ultrathin surface layer	42–44
dynamically formed membranes	a colloidal material is added to the feed solution of an ultrafiltration membrane; a gel forms on the membrane surface and enhances the membrane selectivity	45,46
molecular sieve membranes	an ultrafine microporous membrane is formed from a dense, hollow-fiber polymeric membrane by carbonizing or from a glass hollow fiber by chemical leaching; pores in the range 0.5–2 nm are claimed	47–50
microporous metal membranes by electrochemical etching	aluminum metal, for example, is electrochemically etched to form a porous aluminum oxide film; membranes are brittle but uniform with small pore size 0.02–2.0 μm	51,52

and Co., are still used to produce ultrapure hydrogen for specialized applications. These systems use palladium–silver alloy membranes, based on those originally developed (54). Membranes with much thinner effective palladium layers than were used in the Union Carbide installation can now be made. One technique is to form a composite membrane comprising a polymer substrate onto which is coated a thin layer of palladium or palladium alloy (55). The palladium layer can be applied by vacuum methods, such as evaporation or sputtering. Coating thicknesses on the order of 100 nm or less can be achieved.

Ceramic Membranes. A number of companies have developed ceramic membranes for ultrafiltration and microfiltration applications. Ceramic membranes have the advantages of being extremely chemically inert and stable at high temperatures, conditions under which polymer films fail. Ceramic membranes can be made by three processes: sintering, leaching, and sol–gel techniques. Sintering involves taking a colloidal suspension of particles, forming a coagulated thin film, and then heat treating the film to form a continuous, porous structure. The pore sizes of sintered films are relatively large, on the order of 10–100 μm. In the leaching process, a glass sheet or capillary incorporating two intermixed phases is treated with an acid or alkali that dissolves one of the phases. Smaller pores can be obtained by this method, but the uniformity of the structure is difficult to control. The preparation of ceramic membranes by sol–gel techniques is the newest approach, and offers the greatest potential for making finely porous membranes.

Figure 17 summarizes the available sol–gel processes (56). The process on the right of the figure involves the hydrolysis of metal alkoxides in a water–alcohol solution. The hydrolyzed alkoxides are polymerized to form a chemical gel, which is dried and heat treated to form a rigid oxide network held

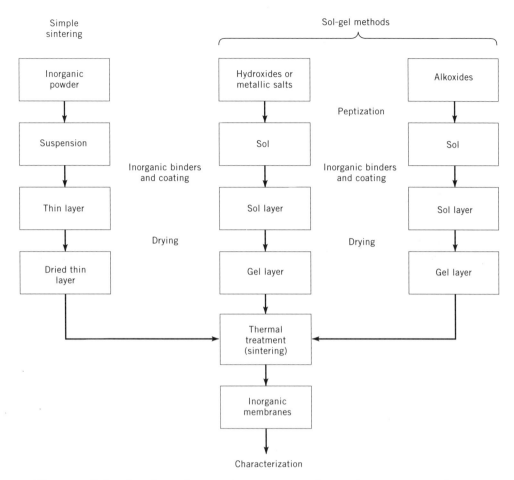

Fig. 17. Sol–gel and simple sintering process used to make ceramic membranes.

together by chemical bonds. This process is difficult to carry out, because the hydrolysis and polymerization must be carefully controlled. If the hydrolysis reaction proceeds too far, precipitation of hydrous metal oxides from the solution starts to occur, causing agglomerations of particulates in the sol.

In the process in the center of Figure 17, complete hydrolysis is allowed to occur. Bases or acids are added to break up the precipitate into small particles. Various reactions based on electrostatic interactions at the surface of the particles take place: the result is a colloidal solution. Organic binders are added to the solution and a physical gel is formed. The gel is then heat treated as before to form the ceramic membrane.

The sol–gel technique has been used mostly to prepare alumina membranes. Figure 18 shows a cross section of a composite alumina membrane made by slip coating successive sols with different particle sizes onto a porous ceramic support. Silica or titanium membranes could also be made by the same principles. Unsupported titanium dioxide membranes with pore sizes of 5 nm or less have been made by the sol–gel process (57).

Fig. 18. Cross-sectional scanning electron micrograph of a three-layered alumina membrane/support (pore sizes 0.2, 0.8, and 12 μm, respectively). Courtesy of U.S. Filter Corp.

Liquid Membranes. A number of reviews summarize the considerable research effort in the 1970s and 1980s on liquid membranes containing carriers to facilitate selective transport of gases or ions (58,59). Although still being explored in a number of laboratories, the more recent development of much more selective conventional polymer membranes has diminished interest in processes using liquid membranes.

Hollow-Fiber Membranes. Most of the techniques described in the foregoing were developed originally to produce flat-sheet membranes, but the majority can be adapted to produce membranes in the form of thin tubes or fibers. Formation of membranes into hollow fibers has a number of advantages, one of the most important of which is the ability to form compact modules with very high surface areas. This advantage is offset, however, by the generally lower fluxes of hollow-fiber membranes compared to flat-sheet membranes made from the same materials. Nonetheless, the development of hollow-fiber membranes at Dow Chemical in 1966 (60), and their later commercialization by Dow, Monsanto, Du Pont, and others represents one of the most significant events in membrane technology (see HOLLOW-FIBER MEMBRANES).

Hollow fibers are usually on the order of 25 μm to 2 mm in diameter. They can be made with a homogeneous dense structure, or preferably with a microporous structure having a dense permselective layer on the outside or inside surface. The dense surface layer can be integral, or separately coated onto a support fiber. The fibers are packed into bundles and potted into tubes to form a membrane module. More than a kilometer of fibers may be required to form a membrane module with a surface area of one square meter. A module can have no breaks or defects, requiring very high reproducibility and stringent quality control standards. Fibers with diameters 25 to 200 μm are usually called hollow-fine fibers. The feed fluid is generally applied to the outside of the fibers and the permeate removed down the bore. Fibers with diameters in the 200 μm to 2 mm range are called capillary fibers. The feed fluid is commonly applied to the inside bore of the fiber, and the permeate is removed from the outer shell.

Hollow-fiber fabrication methods can be divided into two classes (61). The most common is solution spinning, in which a 20–30% polymer solution is extruded and precipitated into a bath of a nonsolvent, generally water. Solution spinning allows fibers with the asymmetric Loeb-Sourirajan structure to be made. An alternative technique is melt spinning, in which a hot polymer melt is extruded from an appropriate die and is then cooled and solidified in air or a quench tank. Melt-spun fibers are usually relatively dense and have lower fluxes than solution-spun fibers, but because the fiber can be stretched after it leaves the die, very fine fibers can be made. Melt spinning can also be used with polymers such as poly(trimethylpentene), which are not soluble in convenient solvents and are difficult to form by wet spinning.

Solution (Wet) Spinning.   In the most widely used solution spinnerette system (60) the spinnerette consists of two concentric capillaries, the outer capillary having a diameter of approximately 400 μm and the central capillary having an outer diameter of approximately 200 μm and an inner diameter of 100 μm. Polymer solution is forced through the outer capillary while air or liquid is forced through the inner one. The rate at which the core fluid is injected into the fibers relative to the flow of polymer solution governs the ultimate wall thickness of the fiber. Figure 19 shows a cross section of this type of spinnerette.

A complete hollow-fiber spinning system is shown in Figure 20. Fibers are formed almost instantaneously as the polymer solution leaves the spinnerette. The amount of evaporation time between the solution exiting the spinnerette and entering the coagulation bath is a critical variable. If water is forced through the inner capillary, an asymmetric hollow fiber is formed with the skin on the inside. If air under pressure, or an inert liquid, is forced through the inner capillary to maintain the hollow core, the skin is formed on the outside of the fiber by immersion into a suitable coagulation bath (62).

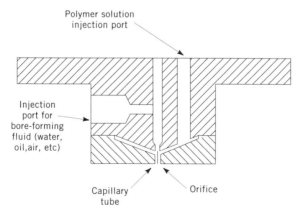

Fig. 19.   Twin-orifice spinnerette design used in solution-spinning of hollow-fiber membranes. Polymer solution is forced through the outer orifice, while bore-forming fluid is forced through the inner capillary.

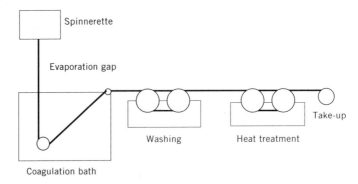

Fig. 20. A hollow-fiber solution-spinning system. The fiber is spun into a coagulation bath, where the polymer spinning solution precipitates forming the fiber. The fiber is then washed, dried, and taken up on a roll.

Wet spinning of this type of hollow fiber is a well-developed technology, especially in the preparation of dialysis membranes for use in artificial kidneys. Systems that spin more than 100 fibers simultaneously on an around-the-clock basis are in operation. Wet-spun fibers are also used widely in ultrafiltration applications, in which the feed solution is forced down the bore of the fiber. Nitto, Asahi, Microgon, and Romicon all produce this type of fiber, generally with diameters of 1–3 mm.

Melt Spinning. In melt spinning, the polymer is extruded through the outer capillary of the spinnerette as a hot melt, the spinnerette assembly being maintained at a temperature between 100 and 300°C. The polymer can be extruded either as a pure melt or as a blended dope containing small amounts of plasticizers and other additives. Melt-spun fibers are usually stretched as they leave the spinnerette, to form very thin fibers. Formation of such small-diameter fibers is a significant advantage of melt spinning over solution spinning. The dense nature of melt-spun fibers leads to lower fluxes than can be obtained with solution-spun fibers, but because of the enormous membrane surface area of these fine hollow fibers, this may not be a problem.

Membrane Modules. A useful membrane process requires the development of a membrane module containing large surface areas of membrane. The development of the technology to produce low cost membrane modules was one of the breakthroughs that led to the commercialization of membrane processes in the 1960s and 1970s. The earliest designs were based on simple filtration technology and consisted of flat sheets of membrane held in a type of filter press: these are called plate-and-frame modules. Systems containing a number of membrane tubes were developed at about the same time. Both of these systems are still used, but because of their relatively high cost they have been largely displaced by two other designs: the spiral-wound module and the hollow-fiber module.

Spiral-Wound Modules. Spiral-wound modules were used originally for artificial kidneys, but were fully developed for reverse osmosis systems. This work, carried out by UOP under sponsorship of the Office of Saline Water (later the Office of Water Research and Technology) resulted in a number of spiral-wound designs (63–65). The design shown in Figure 21 is the simplest and most com-

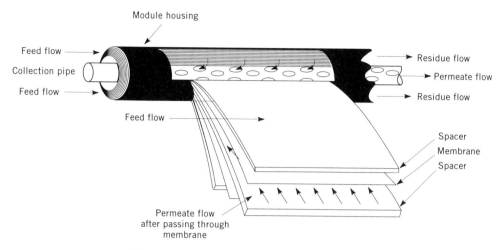

Module housing

Feed flow → ← Residue flow

Collection pipe → Permeate flow

Feed flow → Residue flow

Feed flow →

Spacer
Membrane
Spacer

Permeate flow
after passing through
membrane

Fig. 21. Spiral-wound membrane module.

mon, and consists of a membrane envelope wound around a perforated central
collection tube. The wound module is placed inside a tubular pressure vessel, and
feed gas is circulated axially down the module across the membrane envelope.
A portion of the feed permeates into the membrane envelope, where it spirals
toward the center and exits through the collection tube.

Small laboratory spiral-wound modules consist of a single membrane en-
veloped wrapped around the collection tube. The membrane area of these modules is typically 0.6–1.0 m². Commercial spiral-wound modules are typically
100–150 cm long and have diameters of 10, 15, 20, and 30 cm. These modules
consist of a number of membrane envelopes, each with an area of approximately
2 m², wrapped around the central collection pipe. This type of multileaf design
is illustrated in Figure 22 (64). Such designs are used to minimize the pressure
drop encountered by the permeate fluid traveling toward the central pipe. If a
single membrane envelope were used in these large-diameter modules, the path
taken by the permeate to the central collection pipe would be 5–25 meters depending on the module diameter. This long permeate path would produce a very
large pressure drop, especially with high flux membranes. If multiple, smaller
envelopes are used in a single module, the pressure drop in any one envelope is
reduced to a manageable level.

Hollow-Fiber Modules.   Hollow-fiber membrane modules are formed in two
basic geometries. The first is the shell-side feed design illustrated in Figure 23**a**
and used, for example, by Monsanto in their hydrogen separation systems or by
Du Pont in their reverse osmosis fiber systems. In such a module, a loop or a
closed bundle of fiber is contained in a pressure vessel. The system is pressurized
from the shell side; permeate passes through the fiber wall and exits through the
open fiber ends. This design is easy to make and allows very large membrane
areas to be contained in an economical system. Because the fiber wall must
support a considerable hydrostatic pressure, these fibers are usually made by
melt spinning and usually have a small diameter, on the order of 100 μm ID
and 150–200 μm OD.

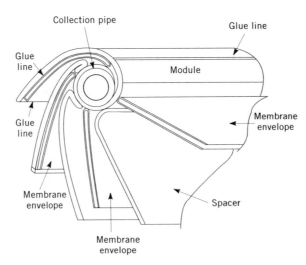

Fig. 22.   Multileaf spiral-wound module, used to avoid excessive pressure drops on the permeate side of the membrane. Large, 30-cm diameter modules may have as many as 30 membrane envelopes, each with a membrane area of about 2 m².

The second type of hollow-fiber module is the bore-side feed type illustrated in Figure 23**b**. The fibers in this type of unit are open at both ends, and the feed fluid is usually circulated through the bore of the fibers. To minimize pressure drops inside the fibers, the fibers often have larger diameters than the very fine fibers used in the shell-side feed system and are generally made by solution spinning. These so-called capillary fibers are used in ultrafiltration, pervaporation, and in some low to medium pressure gas applications. Feed pressures are usually limited to less than 1 MPa (150 psig) in this type of module.

Plate-and-Frame Modules.   Plate-and-frame modules were among the earliest types of membrane system; the design originates from the conventional filter-press. Membrane, feed spacers, and product spacers are layered together between two end plates. A number of plate-and-frame units have been developed for small-scale applications, but these units are expensive compared to the alternatives, and leaks caused by the many gasket seals are a serious problem. Plate-and-frame modules are generally limited to electrodialysis and pervaporation systems and a limited number of highly fouling reverse osmosis and ultrafiltration applications.

Tubular Modules.   Tubular modules are generally limited to ultrafiltration applications, for which the benefit of resistance to membrane fouling because of good fluid hydrodynamics overcomes the problem of their high capital cost. Typically, the tubes consist of a porous paper or fiber glass support with the membrane formed on the inside of the tubes, as shown in Figure 24.

Module Selection.   The choice of the appropriate membrane module for a particular membrane separation balances a number of factors. The principal factors that enter into this decision are listed in Table 2.

Cost, although always important, is difficult to quantify because the actual selling price of membrane modules varies widely, depending on the application. Generally, high pressure modules are more expensive than low pressure or

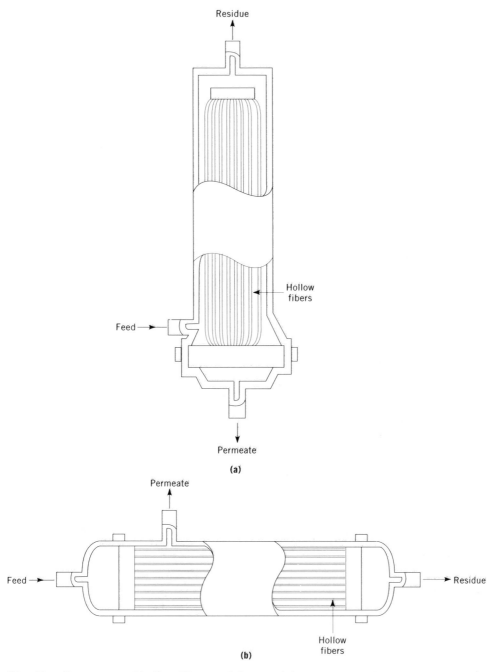

Fig. 23. Two types of hollow-fiber modules used for gas separation, reverse osmosis, and ultrafiltration applications. (**a**) Shell-side feed modules are generally used for high pressure applications up to ~7 MPa (1000 psig). Fouling on the feed side of the membrane can be a problem with this design, and pretreatment of the feed stream to remove particulates is required. (**b**) Bore-side feed modules are generally used for medium pressure feed streams up to ~1 MPa (150 psig), where good flow control to minimize fouling and concentration polarization on the feed side of the membrane is desired.

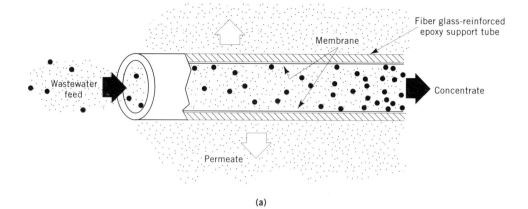

(a)

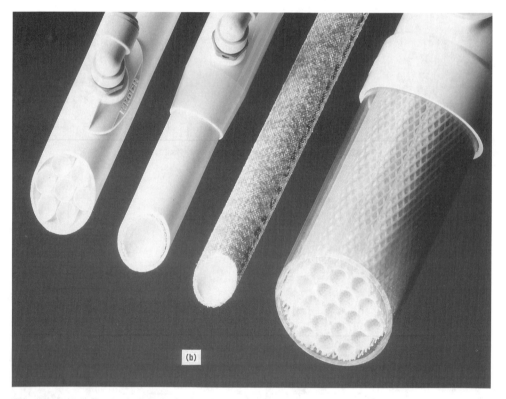

(b)

Fig. 24. (**a**) Typical tubular ultrafiltration module design. In the past, modules in the form of 2–3 cm diameter tubes were common; more recently, 0.5–1.0 cm diameter tubes, nested inside a simple pipe (**b**), have been introduced. Courtesy of Koch Membrane Systems, Inc.

vacuum systems. The selling price also depends on the volume of the application and the pricing structure adopted by the industry. For example, spiral-wound modules for reverse osmosis of brackish water are produced by many manufacturers, resulting in severe competition and low prices, whereas similar modules

Table 2. Characteristics of Module Designs

Property	Hollow-fine fibers	Capillary fibers	Spiral-wound	Plate and frame	Tubular
manufacturing cost, $/m^2	5–20	20–100	30–100	100–200	50–200
resistance to fouling	very poor	good	moderate	good	very good
parasitic pressure drop	high	moderate	moderate	low	low
suitability for high pressure operation	yes	no	yes	can be done with difficulty	can be done with difficulty
limitation to specific types of membrane	yes	yes	no	no	no

for use in gas separation are much more expensive. An estimate of module manufacturing cost is given in Table 2; the selling price is typically two to five times higher.

A second factor determining module selection is resistance to fouling. Membrane fouling is a particularly important problem in liquid separations such as reverse osmosis and ultrafiltration. In gas separation applications, fouling is more easily controlled. Hollow-fine fibers are notoriously prone to fouling and can only be used in reverse osmosis applications if extensive, costly feed-solution pretreatment is used to remove all particulates. These fibers cannot be used in ultrafiltration applications at all.

A third factor is the ease with which various membrane materials can be fabricated into a particular module design. Almost all membranes can be formed into plate-and-frame, spiral, and tubular modules, but many membrane materials cannot be fabricated into hollow-fine fibers or capillary fibers. Finally, the suitability of the module design for high pressure operation and the relative magnitude of pressure drops on the feed and permeate sides of the membrane can sometimes be important considerations.

In reverse osmosis, most modules are of the hollow-fine fiber or spiral-wound design; plate-and-frame and tubular modules are limited to a few applications in which membrane fouling is particularly severe, for example, food applications or processing of heavily contaminated industrial wastewater. Hollow-fiber designs are being displaced by spiral-wound modules, which are inherently more fouling resistant, and require less feed pretreatment. Also, thin-film interfacial composite membranes, the best reverse osmosis membranes available, have not been fabricated in the form of hollow-fine fibers.

For ultrafiltration applications, hollow-fine fibers have never been seriously considered because of their susceptibility to fouling. If the feed solution is extremely fouling, tubular or plate-and-frame systems are still used. Recently, however, spiral-wound modules with improved resistance to fouling have been developed, and these modules are increasingly displacing the more expensive plate-and-frame and tubular systems. Capillary systems are also used in some ultrafiltration applications.

For high pressure gas-separation applications, hollow-fine fibers appear to have a large segment of the market. Hollow-fiber modules are clearly the lowest cost design per unit membrane area, and the poor resistance of hollow-fiber modules to fouling is not a problem in many gas-separation applications. Also, gas-separation membrane materials are often rigid glassy polymers such as polysulfones, polycarbonates, and polyimides, which can be easily formed into hollow-fine fibers. Of the principal companies servicing this area only Separex and W. R. Grace use spiral-wound modules. Both companies use these modules to process natural gas streams, which are relatively dirty, often containing oil mist and condensable components that would foul hollow-fine fiber modules rapidly.

Spiral-wound modules are much more commonly used in low pressure or vacuum gas-separation applications, such as the production of oxygen-enriched air, or the separation of organic vapors from air. In these applications, the feed gas is at close to ambient pressure, and a vacuum is drawn on the permeate side of the membrane. Parasitic pressure drops on the permeate side of the membrane and the difficulty in making high performance hollow-fine fiber membranes from the rubbery polymers used to make these membranes both work against hollow-fine fiber modules for this application.

Pervaporation operates under constraints similar to low pressure gas-separation. Pressure drops on the permeate side of the membrane must be small, and many prevaporation membrane materials are rubbery. For this reason, spiral-wound modules and plate-and-frame systems are both in use. Plate-and-frame systems are competitive in this application despite their high cost, primarily because they can be operated at high temperatures with relatively aggressive feed solutions, for which spiral-wound modules might fail.

Membrane Applications

The principal use of membranes in the chemical processing industry is in various separation processes. These can be classified into technologies that are developed, developing, or to-be-developed, as shown in Table 3. Membranes, or rather films, are also used widely as packaging materials. The use of membranes in various biomedical applications, for example, in controlled-release technology and

Table 3. Membrane Separation Technologies

Uses	Processes	Status
developed technologies	microfiltration ultrafiltration reverse osmosis electrodialysis	well-established unit processes; no significant breakthroughs seem imminent
developing technologies	gas separation pervaporation	a number of plants have been installed; market size and number of applications served is expanding rapidly
to-be-developed technologies	facilitated transport	significant problems remain to be solved before industrial systems will be installed

in artificial organs, such as the artificial kidney, lung, pancreas, etc, are only mentioned briefly here (see CONTROLLED-RELEASE TECHNOLOGY).

The four developed separation processes are microfiltration (MF), ultrafiltration (UF), reverse osmosis (RO), and electrodialysis (ED). All are well established, and the market is served by a number of experienced companies. In microfiltration, ultrafiltration, and reverse osmosis a solution containing dissolved or suspended solids is forced through a membrane filter. The solvent passes through the membrane; the solutes are retained. The three processes differ principally in the size of the particles separated by the membrane. Microfiltration is considered to refer to membranes with pore diameters from 0.1 μm (100 nm) to 10 μm. Microfiltration membranes are used to filter suspended particulates, bacteria, or large colloids from solutions. Ultrafiltration refers to membranes having pore diameters in the range 2–100 nm. Ultrafiltration membranes can be used to filter dissolved macromolecules, such as proteins, from solution. Typical applications of ultrafiltration membranes are concentrating proteins from milk whey, or recovering colloidal paint particles from electrocoating paint rinse waters.

In reverse osmosis membranes, the pores are so small, in the range 0.5–2 nm in diameter, that they are within the range of the thermal motion of the polymer chains. The most widely accepted theory of reverse osmosis transport considers the membrane to have no permanent pores at all. Reverse osmosis membranes are used to separate dissolved microsolutes, such as salt, from water. The principal application of reverse osmosis is the production of drinking water from brackish groundwater or seawater. Figure 25 shows the range of applicability of reverse osmosis, ultrafiltration, microfiltration, and conventional filtration.

The fourth fully developed membrane process is electrodialysis, in which charged membranes are used to separate ions from aqueous solutions under the driving force of an electrical potential difference. The process utilizes an electrodialysis stack, built on the plate-and-frame principle, containing several hundred individual cells formed by a pair of anion- and cation-exchange membranes. The principal current application of electrodialysis is the desalting of brackish groundwater. However, industrial use of the process in the food industry, for example to deionize cheese whey, is growing, as is its use in pollution-control applications.

Of the two developing membrane processes listed in Table 3, gas separation and pervaporation, gas separation is the more developed. At least 20 companies worldwide offer industrial membrane-based gas-separation systems for a variety of applications. In gas separation, a mixed gas feed at an elevated pressure is passed across the surface of a membrane that is selectively permeable to one component of the feed. The membrane separation process produces a permeate enriched in the more permeable species and a residue enriched in the less permeable species. Important well-developed applications are the recovery of hydrogen from refinery and petrochemical purge gases (eg, hydroprocessing unit purge gas); the separation of hydrogen from nitrogen, argon, and methane in ammonia plants; the production of nitrogen from air; the separation of carbon dioxide from methane in natural gas operations; and the separation and recovery of organic vapors from air streams. Gas separation is an area of considerable

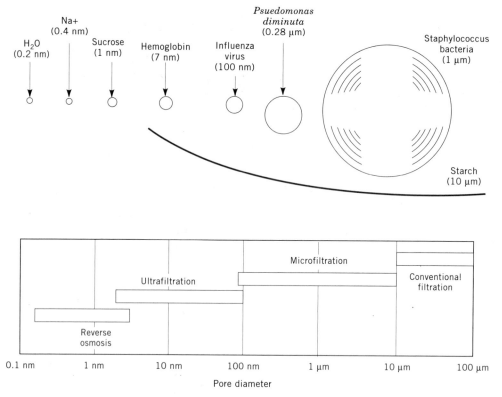

Fig. 25. Reverse osmosis, ultrafiltration, microfiltration, and conventional filtration are related processes differing principally in the average pore diameter of the membrane filter. Reverse osmosis membranes are so dense that discrete pores do not exist; transport occurs via statistically distributed free volume areas. The relative size of different solutes removed by each class of membrane is illustrated in this schematic.

current research interest; the number of applications is expected to increase rapidly over the next few years.

Pervaporation is a relatively new process with elements in common with reverse osmosis and gas separation. In pervaporation, a liquid mixture contacts one side of a membrane, and the permeate is removed as a vapor from the other. Currently, the only industrial application of pervaporation is the dehydration of organic solvents, in particular, the dehydration of 90–95% ethanol solutions, a difficult separation problem because an ethanol–water azeotrope forms at 95% ethanol. However, pervaporation processes are also being developed for the removal of dissolved organics from water and the separation of organic solvent mixtures. These applications are likely to become commercial after the year 2000.

The final membrane process listed in Table 3 is facilitated transport. No commercial plants employing this process are installed or are likely to be installed by 2004. Facilitated transport usually employs liquid membranes containing a complexing or carrier agent. The carrier agent reacts with one permeating component on the feed side of the membrane and then diffuses across the membrane to release the permeant on the product side of the membrane. The carrier agent is then reformed and diffuses back to the feed side of the membrane. Thus the carrier agent acts as a shuttle to selectively transport one component from the feed to the product side of the membrane.

Facilitated transport membranes can be used to separate gases; membrane transport is then driven by a difference in the gas partial pressure across the membrane. Metal ions can also be selectively transported across a membrane driven by a flow of hydrogen or hydroxyl ions in the other direction. This process is sometimes called coupled transport.

Because the facilitated transport process employs a specific reactive carrier species, very high membrane selectivities can be achieved. These selectivities are often far higher than those achieved by other membrane processes. This one fact has maintained interest in facilitated transport since the 1970s, but the problems of the physical instability of the liquid membrane and the chemical instability of the carrier agent are yet to be overcome.

Microfiltration. This process is defined as the separation of particulates between 0.1 and 10 μm by a membrane. Two principal types of membrane filter are used: depth filters and screen filters. Typical particulate retention curves for the two types of filter are shown in Figure 26. Screen filters collect retained particulates on the surface; depth filters collect them within the membrane. Depth filters, therefore, have a much larger surface area available for filtration and usually have a much larger particulate holding capacity before fouling than screen filters. On the other hand, screen filters generally give a cleaner, sharper separation (8,66). Figure 27 compares typical pore sizes of depth and screen filters.

Depth filters are usually preferred for the most common type of microfiltration system, illustrated schematically in Figure 28. In this process design, called "dead-end" or "in-line" filtration, the entire fluid flow is forced through the membrane under pressure. As particulates accumulate on the membrane surface or in its interior, the pressure required to maintain the required flow

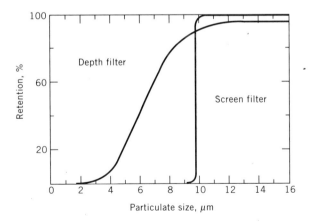

Fig. 26. Screen filters contain pores of a uniform size and retain all particulates greater than the pore diameter at the surface of the membrane. Depth filters contain a distribution of pore sizes. Particulates entering the membrane are trapped at constrictions within the membrane. Both types of filters are rated 10 μm (8). Courtesy of Corning Costar Corp., Nucleopore Division.

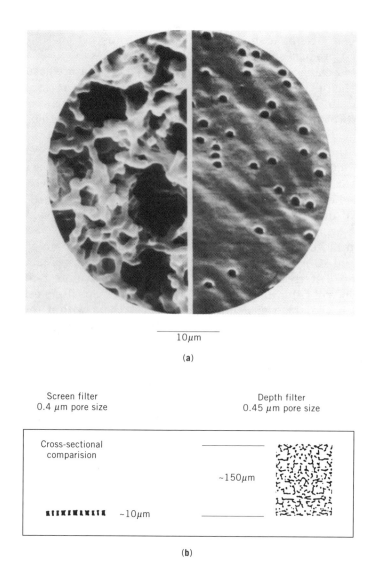

10μm

(a)

Screen filter   Depth filter
0.4 μm pore size   0.45 μm pore size

Cross-sectional
comparision   ————————
                        ~150μm
                    ~10μm   ————————

(b)

Fig. 27. Scanning electron micrograph (**a**) and cross-sectional comparison (**b**) of screen and depth filters both having a nominal particulate cut-off of 0.4 μm. The screen filter (a Nuclepore radiation track membrane) captures particulates at the surface. The phase-inversion cellulosic membrane traps the particulates at constrictions in the interior (8). Courtesy of Corning Costar Corp., Nucleopore Division.

increases until, at some point, the membrane must be replaced. The useful life of the membrane is proportional to the particulate loading of the feed solution. In-line microfiltration of solutions as a final polishing step prior to use is a typical application (66,67).

Increasingly, screen membranes are preferred for the type of cross-flow microfiltration system shown in Figure 28. Cross-flow systems are more complex than the in-line, dead-end filter system because they require a recirculation

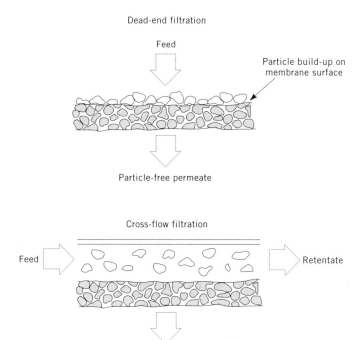

Fig. 28. Schematic representation of dead-end and cross-flow filtration with microfiltration membranes. The equipment used in dead-end filtration is simple, but retained particles plug the membranes rapidly. The equipment required for cross-flow filtration is more complex, but the membrane lifetime is longer.

pump, valves, controls, and so on. However, a screen membrane has a much longer lifetime than a depth membrane and, in principle, can be regenerated by back flushing. Cross-flow filtration is being adopted increasingly for microfiltration of high volume industrial streams containing significant particulate levels (68).

Ultrafiltration. The term ultrafiltration (qv) was coined in the 1920s to describe the collodion membranes available at that time. The process first became widely used in the 1960s when Amicon Corp. adapted the then recently discovered Loeb-Sourirajan asymmetric membrane preparation technique to the production of ultrafiltration membranes (25). These membranes had pore sizes in the range 2–20 nm and found an immediate application in concentrating and desalting protein solutions in the laboratory. Later, Romicon, Abcor, and other companies developed the technology for a wide range of industrial applications. Early and still important applications were the recovery of electrocoat paint from industrial coating operations and the clarification of emulsified oily wastewaters in the metalworking industry. More recent applications are in the food industry for concentration of proteins in cheese production and for juice clarification (69,70). The current ultrafiltration market is in the range $150–250 million/yr.

Ultrafiltration membranes are usually asymmetric membranes made by the Loeb-Sourirajan process. They have a finely porous surface or skin supported on a microporous substrate. The membranes are characterized by their molecular weight cut-off, a loosely defined term generally taken to mean the molecular weight of the globular protein molecule that is rejected 95% by the membrane. A series of typical molecular weight cut-off curves are shown in Figure 29. Globular proteins are usually specified for this test because the rejection of linear polymer molecules of equivalent molecular weight is usually much less. Apparently, linear flexible molecules are able to snake through the membrane pores, whereas rigid globular molecules are retained.

A key factor determining the performance of ultrafiltration membranes is concentration polarization due to macromolecules retained at the membrane surface. In ultrafiltration, both solvent and macromolecules are carried to the membrane surface by the solution permeating the membrane. Because only the solvent and small solutes permeate the membrane, macromolecular solutes accumulate at the membrane surface. The rate at which the rejected macromolecules can diffuse away from the membrane surface into the bulk solution is relatively low. This means that the concentration of macromolecules at the surface can increase to the point that a gel layer of rejected macromolecules forms on the membrane surface, becoming a secondary barrier to flow through the membrane. In most ultrafiltration applications this secondary barrier is the principal resistance to flow through the membrane and dominates the membrane performance.

The phenomenon of concentration polarization, which is observed frequently in membrane separation processes, can be described in mathematical terms, as shown in Figure 30 (71). The usual model, which is well founded in fluid hydrodynamics, assumes the bulk solution to be turbulent, but adjacent to the membrane surface there exists a stagnant laminar boundary layer of thickness (δ) typically 50–200 μm, in which there is no turbulent mixing. The concentration of the macromolecules in the bulk solution concentration is c_b, and the concentration of macromolecules at the membrane surface is c_s. Within the boundary layer, there is a concentration gradient of retained macromolecules.

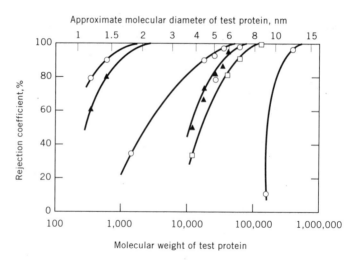

Fig. 29. Rejection of test proteins as a function of molecular weight, in a series of ultrafiltration membranes with different weight cut-offs (69).

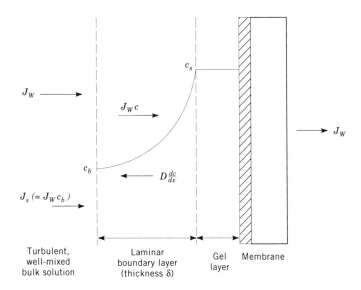

Fig. 30. Illustration of the mathematical model of concentration polarization.

At any point within the boundary layer, the convective flux of the macro-molecule solute to the membrane surface is given by the volume flux, J_W, of the solution multiplied by the concentration of retained solute, c. At steady state, this convective flux within the laminar boundary layer is balanced by the diffusive flux of retained solute in the opposite direction. This balance can be expressed by equation 1:

$$J_W \cdot c = D \cdot \frac{dc}{dx} \qquad (1)$$

where D is the diffusion coefficient of the macromolecule in the boundary layer. Integration of equation 1 over the boundary layer thickness, δ, gives equation 2:

$$\frac{c_s}{c_b} = \left(\frac{J_W \delta}{D} \right) \qquad (2)$$

where c_s is the concentration of retained solute at the membrane surface and c_b is the concentration in the bulk solution.

In ultrafiltration, the flux, J_W, through the membrane is large and the diffusion coefficient, D, is small, so the ratio c_s/c_b can reach a value of 10–100 or more. The concentration of retained solute at the membrane surface, c_s, may then exceed the solubility limit of the solute, c_{gel}, and a precipitated semisolid gel forms on the surface of the membrane. This gel layer is an additional barrier to flow through the membrane.

The formation of a gel layer at the membrane surface provides a limit to the flux through an ultrafiltration membrane, J_{max}, that cannot be exceeded by increasing the applied pressure or by using a more permeable membrane. Some typical results are given in Figure 31, which shows the water flux through a

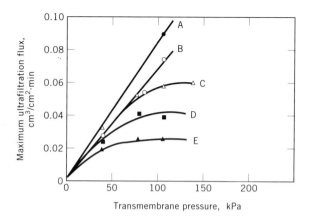

Fig. 31. Effect of pressure on flux in stirred batch-cell ultrafiltration experiments. As the protein concentration in the feed solution increases, the maximum achievable flux decreases. The maximum flux can be adjusted by changing the turbulence in the cell (71). A, 0.9% saline; B, 0.65% protein (1830 rpm); C, 3.9% protein (1830 rpm); D, 6.5% protein (1830 rpm); and E, 6.5% protein (880 rpm). To convert kPa to psi, multiply by 0.145.

membrane used to concentrate blood protein solutions of different concentrations (71). At low pressures, the flux increases with applied pressure, and the concentration of retained solute at the membrane surface increases. However, once the concentration of solute reaches the gel point, the secondary membrane forms and the flux reaches its maximum value. Further increases in pressure increase the thickness of the gel layer, but not the flux. The transmembrane flux is now determined by the rate of diffusion of protein from the membrane surface. However, as shown in Figure 31 for the 6.5% protein concentration, the maximum membrane flux can be increased by increasing the turbulence of the feed solution, which reduces the boundary layer thickness, δ, in equation 2. Techniques to maximize the turbulence of the feed solution to control concentration polarization are important in ultrafiltration.

Figure 31 also shows that the point at which the gel layer forms and the flux reaches a maximum depends on the concentration of the macromolecule in the solution. The more concentrated the solution, the lower the flux at which the gel layer forms. The exact relationship between the maximum flux and macromolecule concentration can be obtained from equation 2, expressing the concentration at the membrane surface, c_s, as c_{gel}, at which point J becomes J_{max} giving equation 3.

$$J_{max} = -\frac{D}{\delta}\left(\ln c_b - \ln c_{gel}\right) \tag{3}$$

Typical plots of the maximum J_{max} as a function of $\ln c_b$ are shown in Figure 32. The flux J_{max} approaches zero as c_b approaches c_{gel}.

Reverse Osmosis. This was the first membrane-based separation process to be commercialized on a significant scale. The breakthrough discovery that made reverse osmosis (qv) possible was the development of the Loeb-Sourirajan

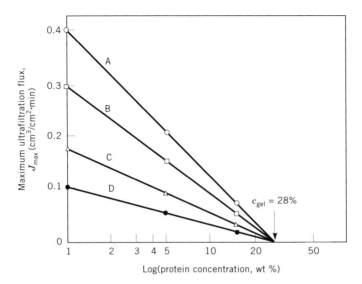

Fig. 32. Maximum flux J_{max} obtained with various protein solutions as a function of protein concentration according to equation 3. Feed flow rates, cm^3/min = A, 3000; B, 2000; C, 1000; and D, 500. The flux decreases exponentially as the protein concentration increases. The extrapolated protein concentration at no flux is the gel point for this type of solution (approx 28%). These results were obtained in a flow-through cell and demonstrate the dependence of flux on fluid flow rate through the cell (69).

asymmetric cellulose acetate membrane. This membrane made desalination by reverse osmosis practical; within a few years commercial plants were installed. The total worldwide market for reverse osmosis membrane modules is about $200 million/yr, split approximately between 25% hollow-fiber and 75% spiral-wound modules. The general trend of the industry is toward spiral-wound modules for this application, and the market share of the hollow-fiber products is gradually falling (72).

The first reverse osmosis modules made from cellulose diacetate had a salt rejection of approximately 97–98%. This was enough to produce potable water (ie, water containing less than 500 ppm salt) from brackish water sources, but was not enough to desalinate seawater efficiently. In the 1970s, interfacial composite membranes with salt rejections greater than 99.5% were developed, making seawater desalination possible (29,30); a number of large plants are in operation worldwide.

The performance characteristics of the membranes available in the 1990s are shown in Figure 33. Hollow-fine fiber membranes made by Dow and Du Pont have relatively low fluxes, but because large membrane areas can be made so economically in hollow-fiber form, these membranes can still compete. The highest flux, highest performance membrane is the cross-linked polyether membrane made by Toray. However, this membrane is unstable to oxidation, and all free oxygen and chlorine must be removed from the feed water to the membrane, a process that is expensive and subject to failure. As a result, most of the reverse osmosis membrane market is divided between various types of thin-film interfacial composite membranes and cellulose diacetate Loeb-Sourirajan membranes.

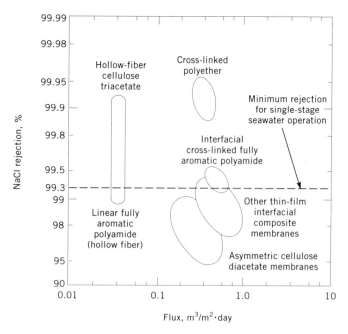

Fig. 33. Performance characteristics of membranes operating on seawater at 5.5 MPa (56 kg/cm^2) and 25°C (72).

Cellulose diacetate membranes still retain a fraction of the market because of their greater chemical and mechanical stability compared to interfacial composites. This advantage is gradually disappearing as improved interfacial composite membranes are developed (73).

The performance of reverse osmosis membranes is generally described by the water and salt fluxes (74,75). The water flux, J_W, is linked to the pressure and concentration gradients across the membrane by equation 4:

$$J_W = A(\Delta p - \Delta \pi) \tag{4}$$

where A is a constant, Δp is the pressure difference across the membrane, and $\Delta \pi$ is the osmotic pressure differential across the membrane. As this equation shows, at low applied pressure when $\Delta p < \Delta \pi$, water flows from the dilute to the concentrated salt-solution side of the membrane by normal osmosis. When $\Delta p = \Delta \pi$, there is no flow. When the applied pressure is higher than the osmotic pressure $\Delta p > \Delta \pi$, water flows from the concentrated to the dilute salt-solution side of the membrane.

The salt flux, J_s, across a reverse osmosis membrane can be described by equation 5 where B is a constant and c_1 and c_2 are the salt concentration differences across the membrane.

$$J_s = B(c_1 - c_2) \tag{5}$$

It follows from these two equations that the water flux is proportional to the applied pressure, but the salt flux is independent of pressure. This means the membrane becomes more selective as the pressure increases. Selectivity can be measured in a number of ways, but conventionally, it is measured as the salt rejection coefficient, R, defined in equation 6.

$$R = \left[1 - \frac{c_2}{c_1}\right] \times 100\% \tag{6}$$

Some data illustrating the effect of pressure on the water and salt fluxes and the salt rejection of a good quality reverse osmosis membrane are shown in Figure 34 (76).

Although the principal application of reverse osmosis membranes is still desalination of brackish water or seawater to provide drinking water, a significant

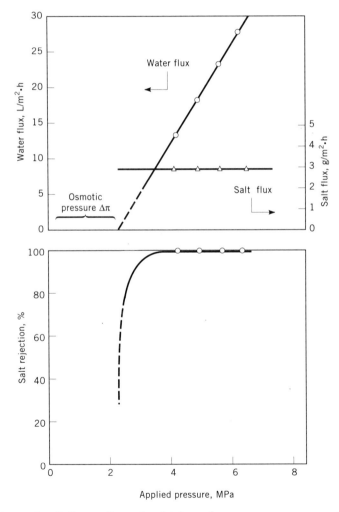

Fig. 34. Water and salt fluxes through a high performance reverse osmosis membrane, when tested with a 3.5% NaCl feed solution. The water flux increases, whereas the salt flux is essentially independent of applied pressure (76). To convert MPa to psig, multiply by 145.

market is production of ultrapure water. Such water is used in steam boilers or in the electronics industry, where huge amounts of extremely pure water with a total salt concentration significantly below 1 ppm are required to wash silicon wafers.

Electrodialysis.   In this electrochemical separation process, a gradient in electrical potential is used to separate ions with charged, ionically selective membranes. A schematic of the simplest type of electrodialysis system is shown in Figure 35 (77–79). The process uses an electrodialysis stack, built on the plate-and-frame principle and containing several hundred cells each formed by a pair of anion- and cation-exchange membranes. Anion-exchange membranes contain fixed, positively charged entities, such as quaternary ammonium groups, fixed to the polymer backbone. These membranes are permeable to negatively charged ions, but positive ions are excluded from permeation by the fixed charges. Similarly, cation-exchange membranes contain fixed, negatively charged groups, such as sulfonic acid groups. Cationic membranes are permeable to positively charged ions, but not to negatively charged ions. The arrangement of the membranes in an electrodialysis stack is such that every second cell becomes depleted of salt, while the adjacent cells become concentrated in salt. The degree of concentration is determined by the rate of flow of solution through the stack.

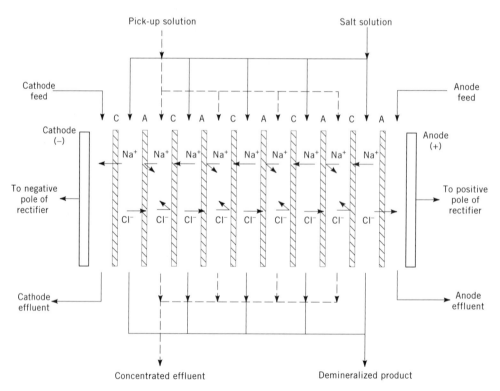

Fig. 35.   Schematic diagram of a plate-and-frame electrodialysis stack. Alternating cation- and anion-permeable membranes are arranged in a stack of up to 100 cell pairs. C, cation-exchange membrane; A, anion-exchange membrane.

Electrodialysis is used widely to desalinate brackish water, but this is by no means its only significant application. In Japan, which has no readily available natural salt brines, electrodialysis is used to concentrate salt from seawater. The process is also used in the food industry to deionize cheese whey, and in a number of pollution-control applications.

In the past, the principal problem inhibiting the use of electrodialysis was slow deterioration of the membranes by chemical degradation and scaling. The introduction in the 1970s of a process called polarity reversal reduced the scaling problem significantly. In this process, the flow of current through the electrodialysis stack is reversed periodically by reversing the polarity of the electrodes. When the polarity of the electrodes is reversed, the concentrated stream becomes the demineralized product stream and the demineralization stream becomes the concentrated stream; automatic valves are used to switch the streams. When the current is reversed, scale deposited on the membranes in the previous cycle is dissolved. Typically, the current of an electrodialysis stack is reversed every 15–20 minutes. One or two minutes production of the system is lost after each reversal, but the reduced scaling and fouling of the membranes more than compensate for this loss in productivity.

One of the most attractive features of electrodialysis is its energy efficiency. The electric current needed to desalinate a solution is directly proportional to the quantity of ions transported through the membranes and is given by equation 7 where z is the electrochemical valence, F is the Faraday constant, Q is the feed solution flow rate, Δc is the difference in concentration between the feed and product solution, and ξ is the current utilization factor. The current utilization factor is always less than 100% because of losses in the stack. Because the membranes are not perfectly semipermeable, some co-ions diffuse across the membrane. Some water is also transferred across the membrane by osmotic flow. Finally, some current flows through the stack manifold and is dissipated as electrical heating. Nonetheless, electrodialysis uses significantly less energy than competitive processes such as evaporation or reverse osmosis, especially for low concentration feed solutions.

$$A = zFQ\,\Delta c/\xi \qquad (7)$$

Gas Separation.   During the 1980s, gas separation using membranes became a commercially important process; the size of this application is still increasing rapidly. In gas separation, one of the components of the feed permeates a permselective membrane at a much higher rate than the others. The driving force is the pressure difference between the pressurized feed gas and the lower pressure permeate.

Both porous and dense membranes can be used as permselective barriers; Figure 36 illustrates the mechanism of gas permeation through both classes. Three types of porous membranes, differing in pore size, are shown. If the pores are relatively large, in the range $0.1–10~\mu m$, gases permeate the membrane by convective flow, and no separation occurs. If the pores are smaller than $0.1~\mu m$, then the pore diameter is the same size or smaller than the mean free path of the gas molecules. Diffusion through such pores is governed by Knudsen diffusion,

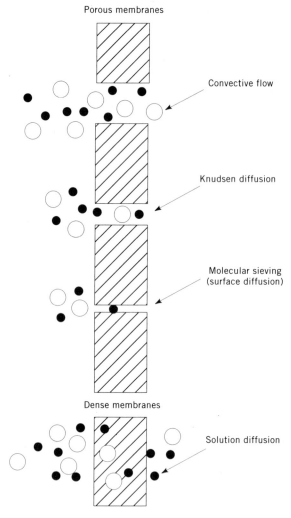

Fig. 36. Mechanisms for permeation of gases through porous and dense gas-separation membranes.

and the transport rate of different gases is inversely proportional to the square root of the molecular weight. This relationship, sometimes called Graham's law of diffusion, was exploited on a massive scale in the separation of $^{235}UF_6$ and $^{238}UF_6$ during the Manhattan Project. Finally, if the membrane pores are very small, of the order 0.5–2 nm, then molecules are separated by molecular sieving. Actual transport mechanisms through this type of membrane are complex and include both diffusion in the gas phase and diffusion of adsorbed species on the surface of the pores (surface diffusion). Nonetheless, ceramic and ultramicroporous glass membranes have been prepared with extraordinarily high separations for similar molecules (47–50).

Although microporous membranes are a topic of research interest, all current commercial gas separations are based on the fourth type of mechanism shown in Figure 36, namely diffusion through dense polymer films. Gas transport through dense polymer membranes is governed by equation 8 where J_i is the flux of component i, $p_{i(o)}$ and $p_{i(l)}$ are the partial pressure of the component i on either side of the membrane, l is the membrane thickness, and P_i is a constant called the membrane permeability, which is a measure of the membrane's ability to permeate gas. The ability of a membrane to separate two gases, i and j, is the ratio of their permeabilities, α_{ij}, called the membrane selectivity (eq. 9).

$$J_i = \frac{P_i(p_{i(o)} - p_{i(l)})}{l} \tag{8}$$

$$\alpha_{ij} = \frac{P_i}{P_j} \tag{9}$$

Permeability P_i, can be expressed as the product of two terms. One, the diffusion coefficient, D_i, reflects the mobility of the individual molecules in the membrane material; the other, the Henry's law sorption coefficient, k_i, reflects the number of molecules dissolved in the membrane material. Thus equation 9 can also be written as equation 10.

$$\alpha_{ij} = \left[\frac{D_i}{D_j} \right]\left[\frac{k_i}{k_j} \right] \tag{10}$$

The ratio D_i/D_j is the ratio of the diffusion coefficients of the two gases and can be viewed as the mobility selectivity, reflecting the different sizes of the two molecules. The ratio k_i/k_j is the ratio of the Henry's law sorption coefficients of the two gases and can be viewed as the sorption or solubility selectivity, reflecting the relative condensabilities of the two gases. If molecule i is larger than j, then the mobility selectivity will always be less than one. The sorption selectivity, however, is normally greater than one, reflecting the higher condensability of large molecules compared to small ones. The balance between the sorption selectivity and the mobility selectivity determines whether a membrane material is selective for large or small molecules in a gas mixture.

In all polymer materials, the diffusion coefficient decreases with increasing molecular size, because large molecules interact with more segments of the polymer chain than do small molecules. Hence, the mobility selectivity always favors the passage of small molecules over large ones. However, the magnitude of the mobility selectivity term is different for glassy and rubbery materials (80), as the data in Figure 37 show. With increasing permeant size, diffusion coefficients in glassy materials decrease much more rapidly than diffusion coefficients in rubbers, in which the polymer chains can rotate freely. For example, the mobility selectivity of natural rubber for nitrogen over pentane is approximately 10. The mobility selectivity of poly(vinyl chloride), a rigid glassy polymer, for nitrogen over pentane is more than 100,000.

The second factor affecting the overall membrane selectivity is the sorption or solubility selectivity. The sorption coefficient of gases and vapors, which is a measure of the energy required for the permeant to be sorbed by the polymer,

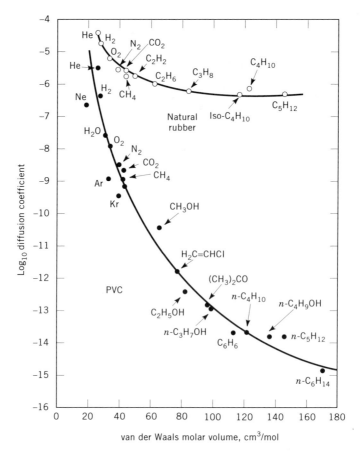

Fig. 37. Diffusion coefficient as a function of molar volume for a variety of permeants in natural rubber and in poly(vinyl chloride) (PVC) (81–83).

increases with increasing condensability of the permeant. This dependence on condensability means that the sorption coefficient also increases with molecular diameter, because large molecules are normally more condensable than smaller ones. The Henry's law sorption coefficient can, therefore, be plotted against boiling point or molar volume; such plots show that sorption selectivity favors the larger, more condensable molecules, such as hydrocarbon vapors, over the permanent gases, such as oxygen and nitrogen (81). The difference between the sorption coefficients of permeants in rubbery and glassy polymers is far less marked than the differences in the diffusion coefficients.

It follows from the discussion above that the balance between the mobility selectivity term and the sorption selectivity term in equation 10 is different for glassy and rubbery polymers. This difference is illustrated by the data in Figure 38. In glassy polymers, the mobility term is usually dominant, permeability falls with increasing permeant size, and small molecules permeate preferentially. When used to separate organic vapors from air, therefore, glassy membranes are air selective. In rubbery polymers, the sorption selectivity term is usually dominant, permeability increases with increasing permeant size, and

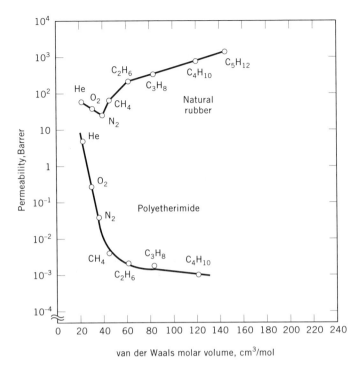

Fig. 38. Permeability as a function of molar volume for a rubbery and glassy polymer, illustrating the different balance between sorption and diffusion in these polymer types. The rubbery membrane is highly permeable; the permeability increases rapidly with increasing permeant size because sorption dominates. The glassy membrane is much less permeable; the permeability decreases with increasing permeant size because diffusion dominates (84). $1 \text{ Barrer} = \frac{0.335 \text{ mmol}}{\text{m·s·TPa}}\left(\frac{10^{-10}\text{cm}^3(\text{STP})\text{·cm}}{\text{cm}^2\text{·s·cm Hg}}\right)$

larger molecules permeate preferentially. The separation properties of polymer membranes for a number of the most important gas-separation applications have been summarized (85).

Both hollow-fiber and spiral-wound modules are used in gas-separation applications. Spiral-wound modules are favored if the gas stream contains oil mist or entrained liquids as in vapor separation from air or natural gas separations.

Table 4 summarizes commercial and precommercial gas separation applications (86,87). The first large-scale commercial application of gas separation was the separation of hydrogen from nitrogen in ammonia purge gas streams. This process, launched in 1980 by Monsanto, was followed by a number of similar applications, such as hydrogen–methane separation in refinery off-gases and hydrogen–carbon monoxide adjustment in oxo-chemical synthetic plants.

Following Monsanto's success, several companies produced membrane systems to treat natural gas streams, particularly the separation of carbon dioxide from methane. The goal is to produce a stream containing less than 2% carbon dioxide to be sent to the national pipeline and a permeate enriched in carbon dioxide to be flared or reinjected into the ground. Cellulose acetate is the most widely used membrane material for this separation, but because its carbon dioxide–methane selectivity is only 15–20, two-stage systems are often required to

Table 4. Gas-Separation Application Areas for Membranes

Gas separation	Applications
O_2/N_2	nitrogen from air, oxygen enrichment of air
H_2/hydrocarbons	refinery hydrogen recovery
H_2/CO	syngas ratio adjustment
H_2/N_2	ammonia purge gas
CO_2/CH_4	acid gas treatment of natural gas
hydrocarbons/air	hydrocarbon recovery, pollution control
H_2O/air	air dehumidification
H_2O/CH_4	natural gas dehydration
H_2S/CH_4	sour gas treatment of natural gas
He/CH_4	helium separation from natural gas
He/N_2	helium recovery

achieve a sufficient separation. The membrane process is generally best suited to relatively small streams, but the economics have slowly improved over the years and more than 100 natural gas treatment plants have been installed.

By far the largest gas separation process in use is the production of nitrogen from air. The first membranes used for this process were based on polysulfone, poly(trimethylpentene), and ethylcellulose. These materials had oxygen–nitrogen selectivities of 4–5, and the economics of the process were marginal. The second-generation materials used in the 1990s have selectivities in the range 7–8 and significantly higher fluxes. With these membranes, the economics of nitrogen production from air are very favorable, especially for small plants producing 0.14–14 m^3/min (5–500 SCFM) of nitrogen. In this range, membranes are the low cost process, and most new nitrogen plants use membrane systems.

A growing application of membrane systems is the removal of condensable organic vapors from air and other streams. Unlike the processes described above, organic vapor separation uses rubbery membranes, which are more permeable to the organic vapor. More than 50 organic vapor recovery plants have been installed. In Europe, most of the plants recover gasoline vapors from air vented during transfer operations; in the United States, most plants recover chlorinated and fluorinated hydrocarbons from refrigeration or chemical processing streams.

Pervaporation. In this separation process a multicomponent liquid stream is passed across a membrane that preferentially permeates one or more of the components (Fig. 39). As the feed liquid flows across the membrane surface, the preferentially permeated component passes through the membrane as a vapor. Transport through the membrane is induced by maintaining a vapor pressure on the permeate side of the membrane that is lower than the vapor pressure of the feed liquid. The pressure difference is achieved by cooling the permeate vapor to below the temperature of the feed stream, causing it to condense. This spontaneously generates a partial vacuum on the permeate side of the membrane. The condensate is then removed as a concentrated permeate fraction; the residue, depleted of the permeating component, exits on the feed side of the membrane. The process can be applied to the removal of dissolved water from organic solvents, to the extraction of organic solvents from water, and to the separation of mixed organic solvents.

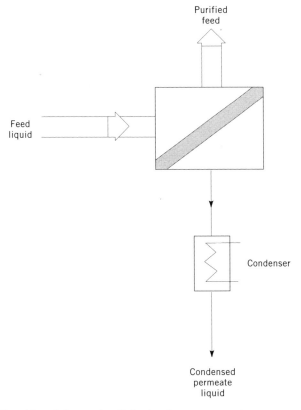

Fig. 39. Schematic of the basic pervaporation process.

The separation, β_{pervap}, achieved by a pervaporation process can be defined in the conventional way as in equation 11 where c'_i and c'_j are the concentrations of components i and j on the feed liquid side and c''_i and c''_j are the concentrations of components i and j on the permeate side of the membrane. Because the permeate is a vapor, c''_i and c''_j can be replaced by p''_i and p''_j, the vapor pressures of components i and j on the permeate side of the membrane. The separation achieved by the membrane can then be expressed by equation 12.

$$\beta_{\text{pervap}} = \frac{c''_i/c''_j}{c'_i/c'_j} \tag{11}$$

$$\beta_{\text{pervap}} = \frac{p''_i/p''_j}{c'_i/c'_j} \tag{12}$$

The most convenient mathematical method of describing pervaporation is to divide the overall separation processes into two steps, as shown in Figure 40. The first is evaporation of the feed liquid to form a (hypothetical) saturated vapor phase on the feed side of the membrane. The second is permeation of this vapor through the membrane to the low pressure permeate side of the

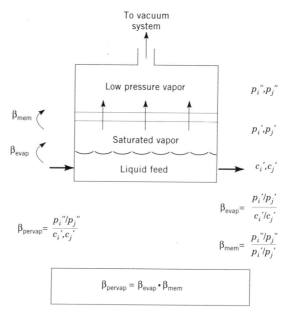

Fig. 40. The separation performed by a pervaporation membrane, β_{pervap}, is the product of the independent processes of evaporative separation, β_{evap}, and membrane permeation separation, β_{mem}.

membrane. Although no evaporation actually takes place on the feed side of the membrane during pervaporation, this approach is mathematically simple and is thermodynamically completely equivalent to the physical process. The evaporation step from the feed liquid to the saturated vapor phase produces a separation, β_{evap}, which can be defined (eq. 13) as the ratio of the concentrations of the components in the feed vapor to their concentrations in the feed liquid where p'_i and p'_j are the partial vapor pressures of the components i and j in equilibrium with the feed solution.

$$\beta_{\text{evap}} = \frac{p'_i/p'_j}{c'_i/c'_j} \tag{13}$$

The second step, permeation of components i and j through the membrane, is related directly to conventional gas permeation. The separation achieved in this step, β_{mem}, can be defined as the ratio of components in the permeate vapor to the ratio of components in the feed vapor (eq. 14).

$$\beta_{\text{mem}} = \frac{p''_i/p''_j}{p'_i/p'_j} \tag{14}$$

From the definitions given in equations 11–14, equation 15 can be written:

$$\beta_{\text{pervap}} = \beta_{\text{evap}} \cdot \beta_{\text{mem}} \tag{15}$$

This equation shows that the separation achieved in pervaporation is proportional to the product of the separation achieved by evaporation of the liquid and the separation achieved by permeation of the components through a membrane. To achieve good separations both terms should be large. It follows that, in general, pervaporation is most suited to the removal of volatile components from relatively involatile components, because β_{evap} will then be large. However, if the membrane is sufficiently selective and β_{mem} is large, nonvolatile components can be made to permeate the membrane preferentially (88).

The selectivity of pervaporation membranes varies considerably and has a critical effect on the overall separation obtained. The range of results that can be obtained for the same solutions and different membranes is illustrated in Figure 41 for the separation of acetone from water using two types of membrane (89). The figure shows the concentration of acetone in the permeate as a function of the concentration in the feed. The two membranes shown have dramatically different properties. The silicone rubber membrane removes acetone selectively, whereas the cross-linked poly(vinyl alcohol) (PVA) membrane removes water selectively. This difference occurs because silicone rubber is hydrophobic and rubbery, thus permeates the acetone preferentially. PVA, on the other hand, is hydrophilic and glassy, thus permeates the small hydrophilic water molecules preferentially.

The acetone-selective, silicone rubber membrane is best used to treat dilute acetone feed streams and concentrate most of the acetone in a small volume of

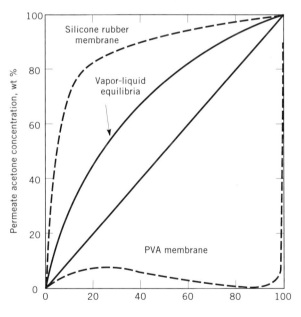

Fig. 41. The pervaporation separation of acetone–water mixtures achieved with a water-selective poly(vinyl alcohol) (PVA) membrane and with an acetone-selective silicone rubber membrane. The PVA membrane is best suited to removing small amounts of water from a concentrated acetone solution, whereas the silicone rubber membrane is best suited to removing small amounts of acetone from a dilute acetone stream (89).

permeate. The water-selective, poly(vinyl alcohol) membrane is best used to treat concentrated acetone feed streams containing only a few percent water. Most of the water is then removed and concentrated in the permeate. Both membranes are more selective than distillation, which relies on the vapor–liquid equilibrium to achieve a separation.

Pervaporation has been commercialized for two applications. The first and most developed is the separation of water from concentrated alcohol solutions. GFT of Neunkirchen, Germany, the leader in this field, installed their first important plant in 1982. More than 100 plants have been installed by GFT for this application (90). The second application is the separation of small amounts of organic solvents from contaminated water (91). In both of these applications, organics are separated from water. This separation is relatively easy, because organic compounds and water, due to their difference in polarity, exhibit distinct membrane permeation properties. The separation is also amenable to membrane pervaporation because the feed solutions are relatively nonaggressive and do not chemically degrade the membrane.

A flow scheme for an integrated distillation–pervaporation plant operating on 5% ethanol feed from a fermentation mash is shown in Figure 42. The distillation column produces an ethanol stream containing 85–90% ethanol, which is fed to the pervaporation system. To maximize the vapor pressure difference across the membrane, the pervaporation module usually operates at a temperature of 80°C with a corresponding feed stream vapor pressure of 405–608 kPa (4–6 atm). Despite these harsh conditions, the membrane lifetime is good and qualified guarantees for up to four years are given.

Figure 42 shows a single-stage pervaporation unit. In practice, at least three pervaporation stages are used in series, with additional heat being supplied to the ethanol feed between each stage. This compensates for pervaporative cooling of the feed and maintains the feed at 80°C. The heat required is obtained by thermally integrating the pervaporation system with the condenser of the

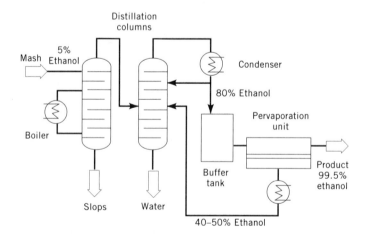

Fig. 42.   Integrated distillation/pervaporation plant for ethanol recovery from fermentors. The distillation columns concentrate the ethanol–water mixture from 5 to 80%. The pervaporation membrane produces a 99.5% ethanol product stream and a 40–50% ethanol stream that is sent back to the distillation column.

final distillation column. Most of the energy used in the process is, therefore, low grade heat. Generally, about 0.5 kg of steam is required for each kilogram of ethanol produced. The energy consumption of the pervaporation process is, therefore, about 559 kJ/L (2000 Btu/gal) of product, less than 20% of the energy used in azeotropic distillation, which is typically in the range 3–3.4 MJ/L (11,000–12,000 Btu/gal). Moreover, pervaporation uses very low grade steam, which is available in most industrial plants at very little cost.

Although most of the installed solvent dehydration systems have been for ethanol dehydration, dehydration of other solvents including 2-propanol, ethylene glycol, acetone, and methylene chloride, has been considered.

No commercial systems have yet been developed for the separation of the more industrially significant organic–organic mixtures. However, technology makes development of pervaporation for these applications possible, and a number of laboratories are actively developing the process. It can only be a matter of time before commercially significant organic–organic separations are attempted using pervaporation. The first pilot-plant results for an organic–organic application, the separation of methanol from methyl *tert*-butyl ether–isobutylene mixtures, was reported by Air Products (92). This is a particularly favorable application and currently available cellulose acetate membranes give good separation. Exxon and Texaco are also working on organic–organic separation, particularly the separation of aromatics from aliphatic hydrocarbons in refinery process streams.

Other Membrane Separation Techniques.　The six membrane separation processes described above represent the bulk of the industrial membrane separation industry. A seventh process, dialysis (qv), is used on a large scale to remove toxic metabolites from blood in patients suffering from kidney failure (93). The first successful artificial kidney was based on cellophane (regenerated cellulose) membranes and was developed in 1945. Many changes have been made since then. In the 1990s, most artificial kidneys are based on hollow-fiber modules having a membrane area of about 1 m^2. Cellulose fibers are still widely used, but are gradually being displaced by fibers made from polycarbonate, polysulfone, and other polymers, which have higher fluxes or are less damaging to the blood. As shown in Figure 43, blood is circulated through the center of the fiber, while isotonic saline, the dialysate, is pumped countercurrently around the outside of the fibers. Urea, creatinine, and other low molecular weight metabolites in the blood diffuse across the fiber wall and are removed with the saline solution. The process is quite slow, usually requiring several hours to remove the required amount of the metabolite from the patient, and must be repeated one to two times per week. Nonetheless, 100,000 patients use these devices on a regular basis.

In terms of membrane area used and dollar value of the membrane produced, artificial kidneys are the single largest application of membranes. Similar hollow-fiber devices are being explored for other medical uses, including an artificial pancreas, in which islets of Langerhans supply insulin to diabetic patients, or an artificial liver, in which adsorbent materials remove bilirubin and other toxins.

One final membrane separation technique, yet to be used on a commercial scale, is carrier facilitated transport. In this process, the membrane used to perform the separation contains a carrier which preferentially reacts with one

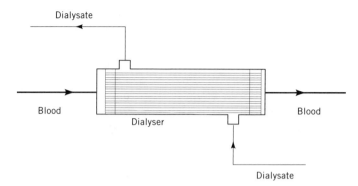

Fig. 43. Schematic of a hollow-fiber artificial kidney dialyser used to remove urea and other toxic metabolites from blood. Several million of these devices are used every year.

of the components to be transported across the membrane. Most of the work on carrier facilitated transport has employed liquids containing a dissolved complexing agent (58,59). Membranes are formed by holding the liquids by capillary action in the pores of a microporous film. The carrier agent reacts with one permeating component on the feed side of the membrane and then diffuses across the membrane to release the permeant on the product side of the membrane. The carrier agent is then reformed and diffuses back to the feed side of the membrane. Thus the carrier agent acts as a selective shuttle to transport one component from the feed to the product side of the membrane. Facilitated transport membranes can be used to separate gases; membrane transport is then driven by a difference in the gas partial pressure across the membrane. Metal ions can also be transported selectively across a membrane, driven by a flow of hydrogen or hydroxyl ions in the other direction; this process is sometimes called coupled transport. Examples of facilitated transport processes for gas and metal ion transport are shown in Figure 44.

Because the facilitated transport process employs a reactive carrier species, very high membrane selectivities can be achieved. These selectivities are often far larger than the selectivities achieved by other membrane processes and they have maintained the interest in facilitated transport. However, no significant commercial applications exist or are likely to exist before the twenty-first century. The principal problems are the physical instability of the membrane and the chemical instability of the carrier agent.

Controlled Drug Delivery.   Membranes have found a significant application in moderating the release of biologically active agents, such as insecticides, fertilizers, and most importantly, drugs. Although the concept of controlled drug release using a rate-controlling membrane to moderate drug delivery can be traced to the 1950s, the founding of the Alza Corp. in the late 1960s gave the entire technology a decisive thrust. The products developed by Alza during the subsequent 25 years stimulated the entire pharmaceutical industry (94,95) (see CONTROLLED RELEASE TECHNOLOGY).

Controlled release can be achieved by a wide range of techniques; a simple but important example is illustrated in Figure 45. In this device, pure drug is contained in a reservoir surrounded by a membrane. With such a system,

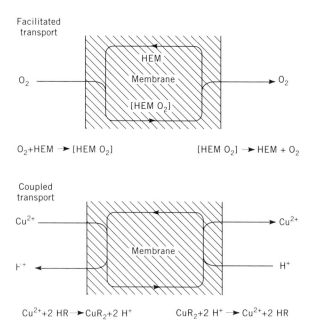

$O_2 + HEM \rightarrow [HEM\ O_2]$　　　　　$[HEM\ O_2] \rightarrow HEM + O_2$

$Cu^{2+} + 2\ HR \rightarrow CuR_2 + 2\ H^+$　　　$CuR_2 + 2\ H^+ \rightarrow Cu^{2+} + 2\ HR$

Fig. 44. Schematic examples of facilitated transport of gases and metal ions. The gas-transport example shows the transport of oxygen across a membrane using hemoglobin (HEM) as the carrier agent. The ion-transport example shows the transport of copper ions across the membrane using a liquid ion-exchange reagent as the carrier agent.

the release of drug is constant as long as a constant concentration of drug is maintained within the device. Such a constant concentration is maintained if the reservoir contains a saturated solution and sufficient excess of solid drug. Systems that operate using this principle are commonly used in transdermal patches to moderate delivery of drugs such as nitroglycerine (for angina), nicotine (for smoking cessation), and estradiol (for hormone replacement therapy) through the skin. A patch of this type is illustrated in Figure 45. Other devices using osmosis or biodegradation as the rate-controlling mechanism are also produced as implants and tablets.

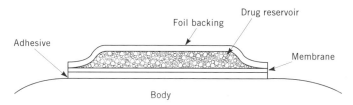

Fig. 45. Schematic of transdermal patch in which the rate of delivery of drug to the body is controlled by a polymer membrane. Such patches are used to deliver many drugs including nitroglycerine, estradiol, nicotine, and scopalamines.

BIBLIOGRAPHY

"Membrane Technology" in *ECT* 3rd ed., Vol. 15, pp. 92–131, by D. R. Paul, University of Texas, and G. Morel, Université de Paris-Nord.

1. H. Bechhold, *Z. Physik Chem.* **60**, 257 (1907).
2. W. J. Elford, *Trans. Far. Soc.* **33**, 1094 (1937).
3. R. Zsigmondy and W. Bachmann, *Z. Anorg. Chem.* **103**, 119 (1918).
4. J. D. Ferry, *Chem. Rev.* **18**, 373 (1936).
5. S. Loeb and S. Sourirajan, "Sea Water Demineralization by Means of an Osmotic Membrane," in *Saline Water Conversion-II, Advances in Chemistry Series Number 28*, American Chemical Society, Washington, D.C., 1963.
6. J. M. S. Henis and M. K. Tripodi, *Sep. Sci. & Tech.* **15**, 1059 (1980).
7. R. L. Fleischer, P. B. Price, and R. M. Walker, *Sci. Am.* **220** (June 30, 1969).
8. H. W. Ballew, *Basics of filtration and Separation*, Nucleopore Corp., Pleasonton, Calif. 1978.
9. H. S. Cierenbaum and co-workers, *Ind. Eng. Chem., Proc. Res. Develop.* **13**(1), 2 (1974).
10. U.S. Pat. 4,585,604 (Apr. 29, 1986), K. Okuyama and H. Mizutani (to Mitsubishi Petrochemical Co. Ltd.).
11. U.S. Pat. 4,187,390 (May 5, 1980), R. W. Gore (to W. L. Gore and Associates Inc.).
12. K. K. Sirkar, in W. S. W. Ho and K. K. Sirkar, eds., *Membrane Handbook*, Chapman & Hall, New York, 1992, pp. 885–912.
13. M. C. Yang and E. L. Cussler, *AIChE J.* **32**, 1910 (1986).
14. U.S. Pat. 4,708,799 (Nov. 24, 1987), K. Gerlach, E. Kessler, and W. Henne.
15. U.S. Pat. 4,708,800 (Nov. 24, 1987), T. Ichikawa and co-workers (to Terumo Kahushiki Kaisha).
16. U.S. Pat. 4,778,499 (Oct. 18, 1988), R. P. Beaver (to PPG Industries, Inc.).
17. H. Strathmann and co-workers, *Desalination* **16**, 179 (1975).
18. H. Strathmann and K. Kock, *Desalination* **21**, 241 (1977).
19. J. G. Wijmans and C. A. Smolders, in P. M. Bungay, H. K. Lonsdale, and M. N. de Pinho, eds., *Synthetic Membranes: Science, Engineering and Applications*, D. Reidel, Dordrecht, the Netherlands, 1986.
20. I. Pinnau and W. J. Koros, *J. Poly. Sci. Phys. Ed.* **31**, 419 (1993).
21. W. C. Hiatt and co-workers, in D. R. Lloyd, ed., *Materials Science of Synthetic Membranes*, ACS Symposium Series 269, American Chemical Society, Washington, D.C., 1985, pp. 229–244.
22. D. R. Lloyd, S. S. Kim, and K. E. Kinzer, *J. Memb. Sci.* **64**, 1 (1991).
23. U.S. Pat. 4,247,498 (Jan. 27, 1981), A. J. Castro (to Akzo Inc.).
24. L. Zeman and T. Fraser, *J. Memb. Sci.* **87**, 267 (1994).
25. R. L. Riley and co-workers, "Development of Ultrathin Membranes," *Office of Saline Water Report No. 386*, PB# 207036, Washington, D.C., Jan. 1969.
26. U.S. Pat. 3,615,024 (Oct. 26, 1971), A. S. Michaels (to Amicon Corp.).
27. U.S. Pat. 3,980,456 (Sept. 14, 1976), W. R. Browall (to General Electric Co.).
28. L. T. Rozelle and co-workers, in S. Sourirajan, ed., *Reverse Osmosis and Synthetic Membranes*, National Research Council of Canada Pub. NRCC 15627, Ottawa, 1977, pp. 249–262.
29. J. E. Cadotte, in D. R. Lloyd, ed., *Materials Science of Synthetic Membranes*, ACS Symposium Series 269, American Chemical Society, Washington, D.C., 1985, pp. 273–294.
30. R. J. Petersen, *J. Memb. Sci.* **83**, 81 (1993).
31. R. L. Riley and co-workers, *Desalination* **19**, 113 (1976).
32. Y. Kamiyama and co-workers, *Desalination* **51**, 79 (1984).
33. W. J. Ward, W. R. Browall, and R. M. Salemme, *J. Memb. Sci.* **1**, 99 (1976).

34. U.S. Pat. 3,551,244 (Dec. 29, 1970), R. H. Forester and P. S. Francis (to North Star Research and Development Institute).
35. U.S. Pat. 4,234,701 (Jan. 1984), R. L. Riley and R. L. Grabowsky (to Universal Oil Products).
36. U.S. Pat. 4,871,378 (Oct. 3, 1989), I. Pinnau (to Membrane Technology and Research, Inc.).
37. U.S. Pat. 4,553,983 (Nov. 19, 1985), R. W. Baker (to Membrane Technology and Research, Inc.).
38. H. Yasuda, *J. Memb. Sci.* **18**, 273 (1984).
39. H. Yasuda, in Ref. 28, pp. 263–294.
40. A. R. Stancell and A. T. Spencer, *J. Appl. Poly. Sci.* **16**, 1505 (1972).
41. M. Kawakami and co-workers, *J. Memb. Sci.* **19**, 249 (1984).
42. U.S. Pat. 4,657,564 (Apr. 14, 1987), M. Langsam; U.S. Pat. 4,759,776 (July 26, 1988), M. Langsam and A. C. Savoca (to Air Products and Chemicals, Inc.).
43. M. Langsam, M. Anand, and E. J. Karwacki, *Gas Sep. & Purif.* **2**, 162–170 (1988).
44. J. M. Mohr and co-workers, *J. Memb. Sci.* **56**, 77–98 (1991).
45. K. A. Kraus, A. J. Shor, and J. S. Johnson, *Desalination* **2**, 243 (1967).
46. J. S. Johnson and co-workers, *Desalination* **5**, 359 (1968).
47. M. B. Rao and S. Sircar, *J. Memb. Sci.* **85**, 253 (1994).
48. K. Keizer and co-workers, *J. Memb. Sci.* **39**, 285 (1988).
49. J. E. Koresh and A. Soffer, *Sep. Sci. Techn.* **18**, 723 (1983).
50. U.S. Pat. 5,288,304 (Feb. 22, 1994), W. J. Koros and C. W. Jones (to The University of Texas System).
51. U.S. Pat. 4,717,455 (Jan. 5, 1988) and U.S. Pat. 4,722,771 (Feb. 2, 1988), M. Textor, M. Werner, and W. Franschitz (to Swiss Aluminum, Ltd.).
52. R. C. Furneaux and co-workers, *Nature* **337**(6203), 147–149 (1989).
53. R. B. McBride and D. L. McKinley, *Chem. Eng. Prog.* **61**, 81 (1965).
54. J. B. Hunter, *Platinum Met. Rev.* **4**, 130 (1960).
55. U.S. Pat. 4,857,080, (Aug. 15, 1989), R. W. Baker and co-workers (to Membrane Technology and Research, Inc.).
56. H. P. Hsieh, in K. K. Sirkar and D. R. Lloyd, eds., *New Membrane Materials and Processes for Separation*, AIChE Symposium Series, Vol. 84, 1988, p. 1.
57. M. A. Anderson, M. S. Gieselmann, and Q. Xu, *J. Memb. Sci.* **39**, 243 (1988).
58. R. W. Baker and I. Blume, in M. C. Porter, ed., *Handbook of Industrial Membrane Technology*, Noyes Publication, Park Ridge, N.J., 1990, pp. 511–588.
59. E. L. Cussler, in D. R. Paul and Y. P. Yampol'skii, eds., *Polymeric Gas Separation Membranes*, CRC Press, Boca Raton, Fla., 1994, pp. 273–300.
60. U.S. Pat. 3,228,876 (1966) and U.S. Pat. 3,228,877 (1966), H. I. Mahon (to Dow Chemical Co.).
61. B. Baum, W. Holley, Jr., and R. A. White, in P. Meares, ed., *Membrane Separation Processes*, Elsevier, Amsterdam, 1976, pp. 187–228.
62. O. M. Ekiner and G. Vassilatos, *J. Memb. Sci.* **53**, 259 (1990).
63. U.S. Pat. 3,367,504 (Feb. 6, 1968), J. C. Westmoreland (to Gulf General Atomic Inc.).
64. U.S. Pat. 3,417,870 (Dec. 24, 1968), D. T. Bray (to Gulf General Atomic Inc.).
65. S. S. Kremen, Ref. 28, pp. 371–386.
66. R. W. Baker and co-workers, *Membrane Separation Systems*, Noyes Data Corp., Park Ridge, N.J., 1991.
67. R. H. Davis, in Ref. 11, pp. 480–505.
68. R. H. Meltzer, *Filtration in the Pharmaceutical Industry*, Marcel Dekker, New York, 1987.
69. M. C. Porter, in P. A. Schweitzer, ed., *Handbook of Separation Techniques for Chemical Engineers*, McGraw Hill Book Co., Inc., New York, 1979, pp. 2.1–2.101.

70. M. Cheryan, *Ultrafiltration Handbook*, Technomic Publishing Co., Lancaster, Pa., 1986.
71. R. W. Baker and H. Strathmann, *J. Appl. Poly. Sci.* **14**, 1197 (1970).
72. R. L. Riley, in R. W. Baker and co-workers, eds., *Membrane Separation Systems-Recent Developments and Future Directions*, Noyes Data Corp., Park Ridge, N.J., 1991, pp. 276–328.
73. B. S. Parekh and co-workers, *Reverse Osmosis Technology*, Marcel Dekker, New York, 1988.
74. U. Merten, in U. Merten, ed., *Desalination by Reverse Osmosis*, MIT Press, Cambridge, Mass., 1966, pp. 15–54.
75. H. K. Lonsdale, U. Merten, and R. L. Riley, *J. Appl. Poly. Sci.* **9**, 1344 (1965).
76. P. Erikson, Film Tech. Corp., private communication, 1994.
77. H. Strathmann, *Sep. and Purif.* **14**, 41 (1985).
78. M. S. Mintz, *Ind. and Eng. Chem.* **55**, 18 (1963).
79. E. Korngold, in G. Belfort, ed., *Synthetic Membrane Processes*, Academic Press, Inc., New York, 1984, pp. 192–220.
80. S. A. Stern, *J. Memb. Sci.* **94**, 1 (1994).
81. G. J. Van Amerongen, *J. Appl. Phys.* **17**, 972 (1946).
82. A. R. Berens and H. B. Hopfenberg, *J. Memb. Sci.* **10**, 283 (1982).
83. R. W. Baker and J. G. Wijmans, in Ref. 59, p. 360.
84. R. W. Baker and J. G. Wijmans, in Ref. 59, p. 362.
85. L. M. Robeson, *J. Memb. Sci.* **62**, 165 (1991).
86. D. R. Paul and Y. P. Yampol'skii, eds., in Ref. 59.
87. R. Spilmann, *Chem. Eng. Prog.* **85**, 41 (1989).
88. J. G. Wijmans and R. W. Baker, *J. Memb. Sci.* **79**, 101 (1993).
89. A. H. Ballweg and co-workers, "Pervaporation Membranes," *Proceedings of the Fifth International Alcohol Fuels Symposium, Auckland, New Zealand, May 13–18, 1992*, John McIndoe, Dunedin, New Zealand, 1982.
90. H. E. A. Brüschke, in R. Bakish, ed., "State of the Art of Pervaporation," in *Proceedings of the Third International Conference on Pervaporation, Nancy, France*, Bakish Materials Corp., Englewood, N.J., 1988.
91. I. Blume, J. G. Wijmans, and R. W. Baker, *J. Memb. Sci.* **49**, 253 (1990).
92. M. S. K. Chen, G. S. Markiewicz, and K. G. Venugopal, *AIChE Symposium Series* **85**(272), 82–88 (1989).
93. T. Daugirdas and T. S. Ing, eds., *Handbook of Dialysis*, Little Brown and Co., Boston, 1988.
94. R. W. Baker, *Controlled Release of Biologically Active Agents*, John Wiley & Sons, Inc., New York, 1987.
95. L. F. Prescott and W. S. Nimmo, eds., *Novel Drug Delivery and Its Application*, John Wiley & Sons, Inc., Chichester, U.K., 1989.

General References

Recent Monographs on Membrane Technology

S. T. Hwang and K. Kammemeyer, *Membranes in Separations*, John Wiley & Sons, Inc., New York, 1975.
E. L. Cussler, *Diffusion*, Cambridge, London, 1984.
G. Belfort, ed., *Synthetic Membrane Processes*, Academic Press, Inc., Orlando, Fla., 1984.
R. E. Kesting, *Synthetic Polymeric Membranes*, 2nd ed., John Wiley & Sons, Inc., New York, 1985.
P. M. Bungay, H. K. Lonsdale, and M. N. de Pinho, eds., *Synthetic Membranes: Science and Engineering Applications*, D. Reidel Publishers, Dordrecht, the Netherlands, 1986.

M. C. Porter, ed., *Handbook of Industrial Membrane Technology*, Noyes Data Corp., Park Ridge, N.J., 1988.

R. Rautenbach and R. Albrecht, *Membrane Processes*, John Wiley & Sons, Inc., Chichester, U.K., 1989.

M. Mulder, *Basic Principles of Membrane Technology*, Kluwer Academic Publishers, Dordrecht, the Netherlands, 1991.

W. S. W. Ho and K. K. Sirkar, eds., *Membrane Handbook*, Chapman & Hall, New York, 1992.

Specialized Monographs

K. S. Spiegler and A. D. K. Laird, eds., *Principles of Desalination*, Academic Press, Inc., New York, 1980.

T. D. Brock, *Membrane Filtration*, Sci. Tech. Inc. Publishing, Madison, Wis., 1983.

M. Cheryan, *Ultrafiltration Handbook*, Technomic Publishing Co., Lancaster, Pa., 1986.

R. W. Baker, *Controlled Release of Biologically Active Agents*, John Wiley & Sons, Inc., New York, 1987.

B. S. Parekh, ed., *Reverse Osmosis Technology*, Marcel Dekker Inc., New York, 1988.

R. Bhave, ed., *Inorganic Membrane Synthesis Characteristics and Applications*, Van Nostrand, New York, 1991.

R. Bakish, ed., *Proceedings of the International Conference of Pervaporation Processes*, Heidelberg, Germany, 1991, and Ottawa, Canada, 1992.

D. R. Paul and Y. P. Yampol'skii, eds., *Polymeric Gas Separation Membranes*, CRC Press, Boca Raton, Fla., 1994.

F. H. Weller, ed., *Electrodialysis (ED) and Electrodialysis Reversal (EDR) Technology*, Ionics, Inc., Watertown, Mass.

RICHARD W. BAKER
Membrane Technology & Research, Inc.

MEMORY-ENHANCING DRUGS

Memory enhancement therapy can be viewed as generally beneficial to many individuals, not only those whose ability to function on a day to day basis has been compromised. For various reasons, however, drugs potentially useful as memory or cognition enhancers are exclusively being developed to treat patients who have been diagnosed with some form of mnemonic or cognitive impairment. Thus, potential memory-enhancing drugs are discussed herein predominantly from the standpoint of treatments that intervene in one or more processes associated with the development of dementia.

Dementia is a condition characterized by impairments in short-term memory, language, visuospatial skills, and alertness resulting from reduced intellectual functioning. Alzheimer's disease (AD) is the most prevalent form of dementia. It is the fourth leading cause of death in the United States, and as of 1990 was estimated to cost $82 billion, annually (1). AD is a neurodegenerative disease of unknown etiology leading to a primary lesion in the cerebral cortex and hippocampus. The functional deficits resulting from the loss of cortical neurons are exacerbated by the loss of several subcortical neuronal systems that project to the cerebral cortex. The systems include the cholinergic, dopaminergic, serotonergic, and noradrenergic. These subcortical afferents play an important role in regulating cortical excitability and resulting cortical function. The degradation and eventual breakdown of the functional connectivity within the cerebral cortex lead to the cognitive impairments seen in AD. Whereas AD is associated with the severest forms of compromised cognitive function and memory, some loss of these functions is generally present in later life even for AD-free individuals and is accepted as a condition known as age-associated memory-impairment (AAMI). In many cases, the causes of AD and AAMI may be similar, such that only the degree of affliction serves as the differentiating factor. Thus efficacious therapies in the severest form of this affliction may also be beneficial in the milder ones.

As of this writing (ca 1994) no drugs are available to address the etiology of neuronal loss and consequent memory impairment. There are, however, a number of drugs used throughout the world that enhance cerebral metabolism or that palliate cognitive dysfunction through modulation of neurotransmitter systems. Whereas there is considerable controversy surrounding the clinical efficacy of these agents, cognition enhancers are sold worldwide and comprise an annual market estimated to be between $1 and $2 billion (2). Widespread usage results largely from availability and the absence of alternative therapy. The hope is that these agents can provide some benefit to patients, no matter how small the probability of efficacy or magnitude of effect.

The compounds used to palliate the mnemonic and cognitive decline associated with dementia include cerebral vasodilators and the so-called nootropic agents. These materials enhance cerebral metabolism. Agents which enhance neurotransmitter function are in most cases cholinergic.

Cerebral Metabolism Enhancers

Whereas the majority of agents being evaluated for treatment of dementia have activities associated with specific neuronal systems, cerebral metabolism enhancers have undefined or varied mechanisms. Hydergine (**1**), vinpocetine [42971-09-5] (**2**), and nimodipine [66085-59-4] (**3**) initially had been thought to exert their activity through cerebral vasodilation (Fig. 1). However, these are used to treat patients with dementia or other age-related symptoms of compromised cognitive function based on other mechanisms and without a clear understanding of the reasons for beneficial actions. The other agent in this group, acetyl-L-carnitine [14992-62-2] (**4**) is thought to exert its beneficial effects by its positive influence on energy metabolism in the mitochondria, as well as on cholinergic activity.

Hydergine (Du Pont), also referred to as ergoloid mesylates (**1**), is a combination of four dehydrogenated derivatives of the ergot alkaloid ergotoxine [8006-25-5] (see ALKALOIDS). Usage is restricted to treatment of patients with

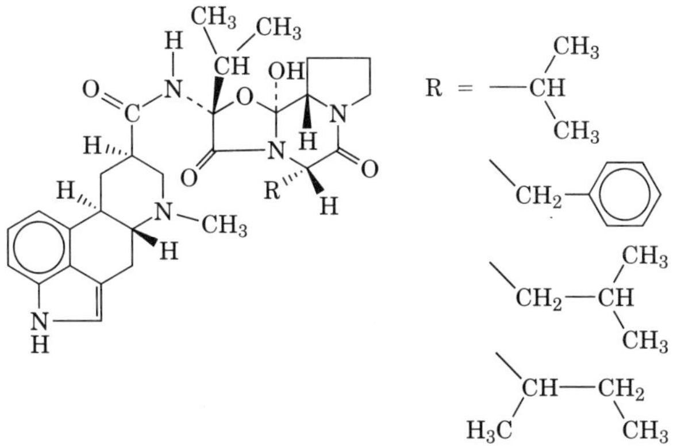

Fig. 1. Structures of cerebral metabolism enhancers.

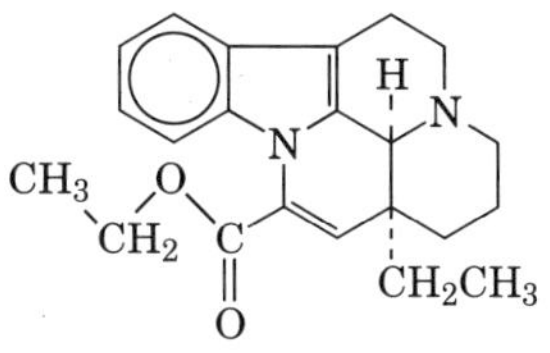

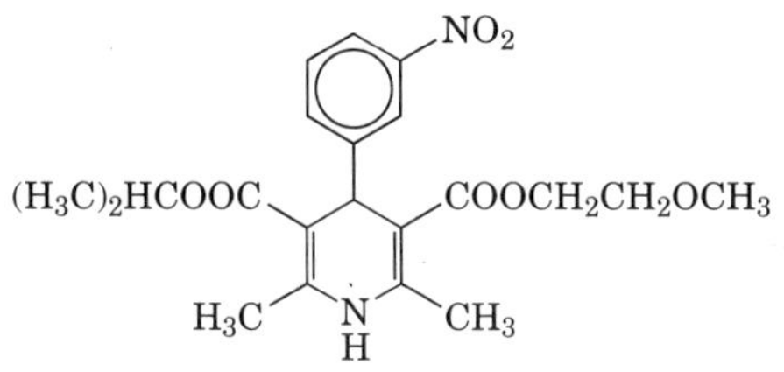

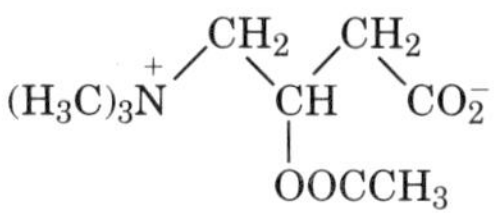

compromised cognitive functions. There is only limited clinical evidence to support its efficacy (3). The effects in patients with possible AD have been modest at best. Moreover, benefits have been associated with behavioral rather than cognitive measures. Beneficial effects appear to be stronger in a subgroup having vascular dementia than in a subgroup having possible AD.

The primary evidence for the cerebral metabolic-enhancing activity of Hydergine is its ability to improve brain oxygen consumption (P_{O_2}) and electrical activity reduced by oligemic hypovolemia, a reduction in blood circulation (4). Hydergine has been reported to increase neuronal noradrenaline release, but the drug itself acts to block these effects at the post-synaptic α_1-adrenoceptors. In addition, Hydergine has partial agonist activity at dopaminergic and serotonergic systems (5). More recently it has also been shown by microdialysis techniques that Hydergine enhances the release of acetylcholine in the hippocampus in a dose-dependent manner (6). This is a response similar to both dopamine D_1 and D_2 receptor antagonists. These latter properties may in part explain the behavioral changes observed in AD patients that have been treated with the drug. Because of the history of poor efficacy in AD, the U.S. Food and Drug Administration (FDA) had been pressured to remove Hydergine from the market.

Vinpocetine (**2**), another drug initially categorized as a cerebral vasodilator, is a member of the vinca alkaloid family of agents (7). However, interest in this compound as a potential drug for learning and memory deficits comes from its ability to act as a neuronal protectant. This compound was evaluated in 15 patients with AD over a one-year period and was ineffective in improving cognitive deficits or slowing the rate of decline (8). However, in studies of patients with chronic vascular senile cerebral dysfunction (9) and organic psychosyndrome (10), vinpocetine showed beneficial results.

The neuroprotective properties of vinpocetine may be related to its anticonvulsant properties (11). It has been suggested that convulsions (*status epilepticus*) cause neuronal loss by excessive intracellular calcium produced by neuronal burst firing (12). This firing is believed to be caused by an excessive stimulation of *N*-methyl-D-aspartic acid (NMDA) [*6384-92-5*] glutamate receptors leading to calcium influx and cell loss. This process may be common to convulsions, cerebral ischemia, and neurodegenerative disorders (13). Compounds such as vinpocetine that have the ability to inhibit ischemia-induced neuronal death (14) may also have a neuronal protectant effect in diseases like AD and AAMI rather than immediately improving the symptoms.

Consistent with the ability of vinpocetine to act as an anticonvulsant is its ability to inhibit cellular reuptake of adenosine (15) which has been described as the brain's endogenous anticonvulsant because of its ability to inhibit calcium influx. Thus the property of vinpocetine to inhibit adenosine reuptake may be responsible for the neuroprotective actions of the drug.

In addition, vinpocetine selectively inhibits a specific calcium, calmodulin-dependent cyclic nucleotide phosphodiesterase (PDE) isozyme (16). As a result of this inhibition, cyclic guanosine 5′-monophosphate (GMP) levels increase. Relaxation of smooth muscle seems to be dependent on the activation of cyclic GMP-dependent protein kinase (17), thus this property may account for the vasodilator activity of vinpocetine. A review of the pharmacology of vinpocetine is available (18).

Nimodipine (**3**), a member of the dihydropyridine series of calcium channel blockers, has been shown to cause cerebral vessel dilation and increase cerebral blood flow in animals and humans (19–21). This drug decreases the severity of neurological deficits and reduces mortality and morbidity of patients with subarachnoid hemorrhage, an indication for which it is marketed in the United States (22). The ability of this agent to reduce the frequency of vasospasm was initially thought to be the basis of its pharmacological action. This has not been demonstrated, however, either angiographically or by noninvasive cerebral blood flow studies. These observations suggest that nimodipine may increase microcirculatory or collateral blood flow to underperfused regions, or provide a direct neuronal protective effect.

The interest in nimodipine for the treatment of individuals with compromised cognitive function is based, in part, on suggestions that blocking neuronal calcium channels may be an effective treatment for memory impairments associated with brain injury as well as age-related memory failure (23). Clinical studies have attempted to demonstrate the benefit of the highly lipophilic, and thus blood brain barrier penetrating nimodipine in a randomized, double-blind, placebo-controlled, multicenter study of 227 AD patients. The drug-treated group was reported to experience a prophylactic benefit across eight measures when contrasted with disease progression seen among placebo recipients (24). Nimodipine also improved clinical symptomatology and cognitive functions in patients having primary degenerative dementia (25). The patients with multiinfarct dementia were less favorably affected. The divergent therapeutic responses of these groups suggest that the protection of neuronal tissue from calcium over-load rather than cerebral vasodilatation may be the reason for the neuroprotec-tive effects of the agent.

The death of cholinergic cell bodies originating in the nucleus basalis of Mynert is a principal neuropathological find in AD. A pharmacological strategy to slow the rate of cholinergic neuronal death should be protective and thus ef-fective in the treatment of AD (26). Because increases in cytosolic free calcium triggers the neuronal death mechanisms, agents that inhibit this rise through calcium channel blockade may prove to retard the progress of this disease. More-over, even in the normal aging process, changes in cellular calcium regulation may be disrupted (27). Because nimodipine has been shown to reduce neuronal degeneration in a variety of toxic conditions, and increase neuronal firing of aged neurons in addition to its cerebrovascular effect, this drug appears to have promise for patients with compromised age-related mental deficits.

Acetyl-L-carnitine (**4**) is marketed in Italy for dementia; as of this writing it is also in Phase III clinical trials in the United States and Europe. In a double-blind, placebo-controlled clinical trial over a one-year period involving 130 patients with clinically diagnosed AD, a slower rate of deterioration in 13 of the 14 outcome measures was observed in the drug-treated group (28). Earlier smaller scale pilot studies in demented patients had also shown some improvement of various behavioral and cognitive functions (29).

Acetyl-L-carnitine is an endogenous substance involved in the uptake of ac-tivated long-chain fatty acids into mitochondria. Studies in rats have also shown that this compound increases acetyl-coenzyme A (acetyl-CoA) and choline acetyl-transferase activities, choline uptake, and acetylcholine release (30), supporting earlier studies that demonstrated the central cholinergic effects of the drugs (31). Thus a beneficial effect of the drug on cognitive function may be associated with its positive influence on energy metabolism in the mitochondria and on cholin-ergic activity. The pharmacology of this agent in the central nervous system (CNS) has been reviewed (32).

More recently, acetyl-L-carnitine has been shown to enhance the response of rat PC12 cells to nerve growth factor (NGF) stimulating the synthesis of NGF re-ceptors (33). This agent may rescue aged neurons by increasing their responsive-ness to neurotrophic factors in the CNS. In rats having impairment of cholinergic activity resulting from transection of the fimbria fornix, 150 mg/(kg·d) acetyl-L-carnitine was found to increase the level of NGF as well as choline acetyl-transferase, an index of cholinergic processes, in the septum and frontal cortex (34). These data are suggestive of a neurotrophic property exerted by the drug on those central cholinergic pathways typically damaged by aging. Agents like acetyl-L-carnitine that mimic the trophism exerted by NGF have been proposed as therapeutic treatments for AD (35).

Nootropics

The term nootropic has been used to describe a class of compounds defined by the ability of its members to facilitate learning (36). The compounds are most effective in animals that have had their cognitive abilities compromised in some way. The molecular mechanism underlying the cognitive-enhancing effects of this class of molecules is unknown, although interaction with the excitatory amino acid network (37–39), muscarinic M-1 receptors (40,41), or enzymes such as prolylendopeptidase (42) have been suggested. Piracetam [*7491-74-9*] (**5**) is the classic representative of the group and many other acetams, such as aniracetam [*72432-10-1*] (**6**), oxiracetam [*62613-82-5*] (**7**), pramiracetam [*68497-12-1*] (**8**), nebracetam (**9**), and nefiracetam (**10**) are undergoing or have undergone clinical evaluation. At least 18 different nootropic agents, including certain cerebral vasodilators, are being evaluated worldwide by various companies. The nootropics are the largest class of compounds being considered for patients with AD or compromised memory and cognition function. Structures are shown in Figure 2.

The mechanism of action of nootropic agents has been proposed to be their ability to facilitate information acquisition, consolidation, and retrieval (36). No one particular effect has been observed with any consistency for these agents, thus whereas a considerable amount of diverse preclinical pharmacological behavioral data has been generated using these compounds, the significance of these results in predicting clinical efficacy has not been established (43,44). Reviews on the biochemical and behavioral effects of nootropics are available (45–47).

Piracetam (**5**) and related analogues facilitate selected aspects of learning and memory as indicated in a variety of animal studies (43,45,48). Human studies, however, have not been as definitive (43,49). There has been indication of efficacy in patients with mild to moderate dementia (50) as well as in AD patients (51), but the results are not compelling. Piracetam's mechanism of action has been related to effects on cholinergic neurotransmission (46), binding to glutamate receptors (52), activation of brain adenylate cyclase (53),

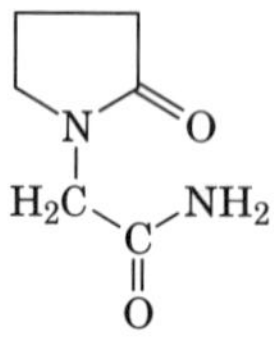

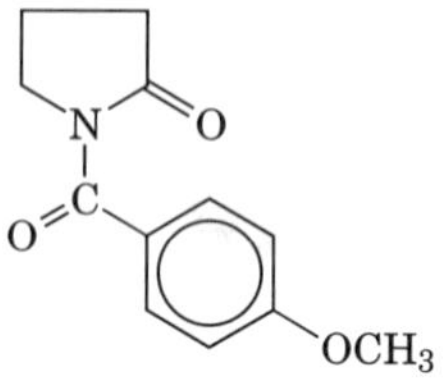

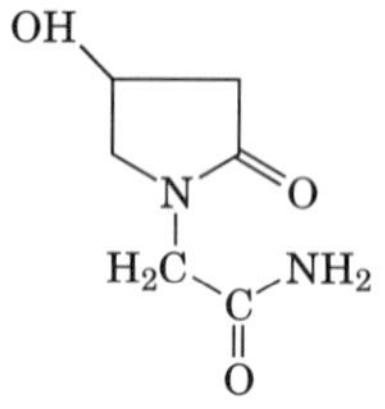

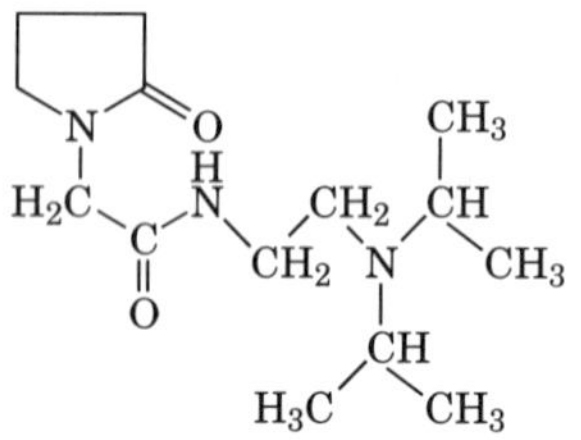

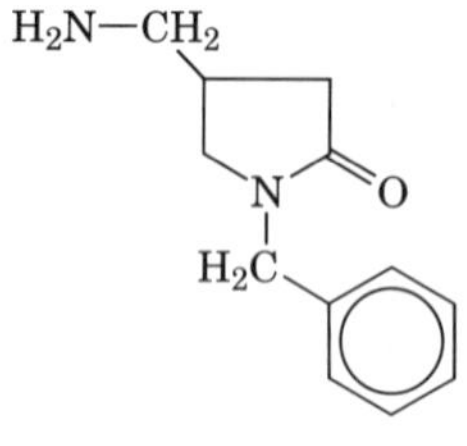

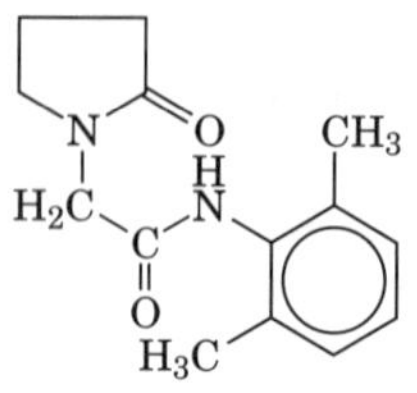

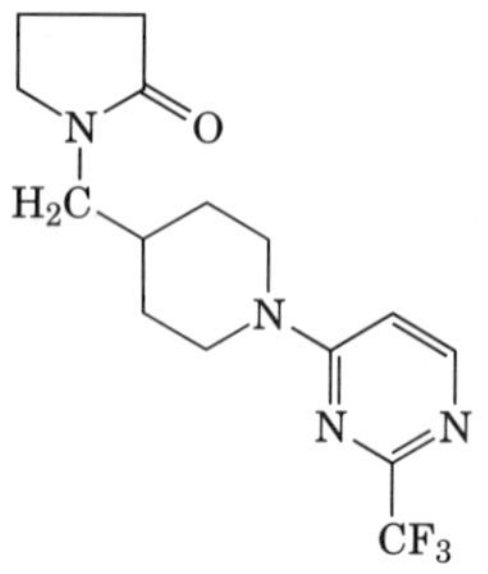

Fig. 2. Structures of nootropic agents.

increases in cerebral glucose utilization (54), and potentiated increase in adenosine $5'$-monophosphate (AMP)-induced calcium influx (37). However, demonstration of these effects often occurs at concentrations or dosages much higher than the serum or brain levels of drug achieved in humans. Thus the lack of definitive cognitive-enhancing action using piracetam may result from the inability to achieve sufficient plasma concentrations of drug to trigger or sustain these types of responses. In spite of this questionable efficacy, piracetam has been marketed in 85 countries beginning in 1973. In addition to use for the symptomatic treatment of AD, it is also indicated for cerebro-vascular injury or insufficiency, ie, specific learning disabilities such as dyslexia, alcoholism, and vertigo. A comprehensive review on the biochemical, pharmacological, and pharmacokinetic properties and clinical effectiveness of piracetam and structurally related nootropics has been published (55).

Oxiracetam (**7**) is the 4-hydroxy derivative of piracetam. This agent was initially launched in Italy in 1987 and although it has not produced convincing results in AD patients (56,57), beneficial effects have been reported in patients with multiinfarct dementia after chronic use (57–59). Various studies suggest that the action of oxiracetam may involve NMDA receptors (60). An indirect mechanism is likely. Like piracetam, oxiracetam increases the density of specific binding sites for dl,α-amino-3-hydroxy-5-methyl-4-isoxazolepropionic acid (AMPA) in synaptic membranes from rat cortex, and does not act on metabotropic glutamate receptors (37). In addition, oxiracetam stimulates choline uptake into isolated hippocampal slices from spontaneously hypertensive rats having sodium chloride-induced cerebrovascular lesions (61). The drug also enhances K^+-induced acetylcholine release from rat hippocampal slices and stimulates choline acetyltransferase (CAT) (62). Each or all of these actions may contribute to the nootropic effect of oxiracetam. However, a review has described the potential therapeutic effects in the context of the influence of endogenous steroid levels, and proposes that these hormones should be considered well in advance of nootropic therapy (47).

Aniracetam (**6**), launched in 1993 in both Japan and Italy for the treatment of cognition disorders, is in Phase II trials in the United States as of this writing. In clinical studies it has been shown to cause some improvement in elderly patients with mild to moderate mental deterioration (63), and in geriatric patients with cerebral insufficiency (64). In a multicenter double-blind placebo-controlled trial involving 109 patients with probable AD, positive effects were observed in 36% of patients after six months of treatment (65), a result repeated in a separate study of 115 patients (66). A review of the biological and pharmacokinetic properties, and clinical results of aniracetam treatment in cognitively impaired individuals is available (49).

Electrophysiological studies indicate that aniracetam prolongs the time course and increases the peak amplitude of the fast excitatory post-synaptic currents (EPSCs) and strongly reduces glutamate receptor desensitization (67,68). Other actions include recruitment of a subset of AMPA-sensitive glutamate receptors which normally do not contribute to synaptic transmission, as suggested for oxiracetam (37). A large number of *in vivo* pharmacological studies have demonstrated that aniracetam also influences cholinergic neurotransmission. In addition, effects on the dopaminergic, adrenergic, and serotonergic systems have also been observed (55). Aniracetam inhibits prolylendopeptidase (PEP), an enzyme associated with the degradation of endogenous proline-containing neuropeptides, that may have beneficial actions on memory and learning (69). The multifacted pharmacological profile of this agent makes it an intriguing prospect, although the therapeutic utility is still undergoing evaluation.

Pramiracetam (**8**), a piracetam derivative having a dialkylaminoalkyl group on the acetamide nitrogen, was launched in Italy in 1993 for the treatment of attention and memory deficits resulting from degenerative or vascular disorders. Whereas the drug was reported to show some benefit in male patients with memory and cognitive problems resulting from head trauma (70), it was without benefit in AD patients (44). More recently, in a multicenter open trial involving 104 elderly patients with cognitive or memory impairment of probable vascular origin, pramiracetam showed better efficacy in patients with moderate compared to those with mild impairment (71).

Unlike aniracetam, pramiracetam does not appear to interact with dopaminergic, serotonergic, or adrenergic neurotransmission (72). The agent inhibits prolylendopeptidase in certain brain areas, but its inhibition constant, K_i, is only 11 μM (69). The absence or weak activity of this compound with various neuronal systems appears to make it less likely to be of significant therapeutic value than other members of this class of agents.

Other nootropic agents in some stage of clinical development include nebracetam (**9**), nefiracetam (**10**), and BMY 21502 (**11**). Nebracetam, an aminomethyl pyrrolidinone derivative, is expected to be approved in Japan in 1994 (73). In clinical studies involving patients having cerebrovascular or senile dementia of the Alzheimer's type, clinical symptoms such as spontaneous or emotional expression were enhanced in up to 71% of cases. Long-term treatment using nebracetam in patients with cerebral infarction also afforded marked improvement in most cases with few side effects (74). A review of this compound has been published (75).

Unlike the other pyrrolidinone nootropic agents, various studies support a cholinergic mechanism of action for nebracetam. In rat brain membranes, nebracetam (**9**) possesses affinity for cholinergic receptors (40). The agent also has direct actions on nicotinic and muscarinic acetylcholine receptors expressed in Xenopus oocytes (76). The impaired working memory and learning acquisition induced in rats by AF64A, a neurotoxic choline analogue, were ameliorated by the drug (77). Nebracetam has also been shown to reverse scopolamine- and AF64A-induced memory impairments in rats (77–79). In addition, studies also support the involvement of limbic and hippocampal noradrenergic mechanisms in the cognition-enhancing effects of the drug (79).

Nefiracetam (**10**) has been reported to show beneficial clinical results apparently arising from effects on neurochemical processes involving GABA and acetylcholine (80–83). A variety of *in vivo* pharmacological studies demonstrating the effect of nefiracetam on various types of chemically or physically induced amnesia have been reported (55,84). Whereas many of these studies are associated with the GABAergic system, the results are difficult to interpret because of the uncertainty about GABA and the memory process (85). Other studies have demonstrated an involvement with acetylcholine neurotransmission, and nefiracetam also causes an increase in choline uptake in rat cortex (86). However, there appears to be some uncertainty regarding significant beneficial effects of nefiracetam on patients with compromised cognitive function compared to other acetams.

Limited clinical data on BMY 21502 (**11**) suggest that some benefit may be provided to patients with dementia (87,88). This expectation is based on the ability of (**11**) to increase the level of arousal and attention in patients with certain types of dementia. BMY 21502 enhances long-term potentiation (LTP) in hippocampal slices (89,90). LTP is believed to be a critical step in memory acquisition, and agents that possess the ability to augment this process *in vivo* are expected to be of benefit in memory enhancement. However, this compound has only demonstrated this property *in vitro*. In addition, because (**11**) does not appear to affect other neurotransmitter systems, as do other nootropics, the potential of BMY 21502 as a memory-enhancing agent is questionable.

There appear to be a number of clinical studies that support the efficacy of various nootropic agents in patients with some form or degree of dementia, but the results are not particularly convincing (50,51,57–59,63,74).

Cholinomimetics

One of the earliest identified and most consistent neurochemical changes observed in AD is the profound loss of neocortical cholinergic innervation (91–94). This loss correlates with the degree of dementia. Experiments in animals have also pointed to the importance of cholinergic function to learning and memory (95–97). These observations have led to what has been called the cholinergic hypothesis of AD (98) which suggests that the cholinergic losses observed in AD lead directly to the observed cognitive and mnemonic deficits.

The wide range of neurochemical alterations documented in AD (99–101) indicates that the cholinergic hypothesis is an oversimplification. Furthermore, studies of animals having excitotoxin lesions of the basal forebrain cholinergic cell group suggest that the cholinergic projection from these cells is not as important to learning and memory as was first thought (102). This projection may, in fact, be more important to attention than to learning and memory (103). However, the role of cholinergic dysfunction in memory impairment and symptoms of dementia is well supported, although this represents only one factor of this disease.

Although controversy exists over the cholinergic involvement in AD dementia, as of 1993 the only AD therapy approved by the U.S. FDA was the cholinesterase inhibitor, tacrine [*321-64-2*], $C_{13}H_{14}N_2$, sold as Cognex (Warner-Lambert).

Several cholinergic strategies, other than cholinesterase inhibition, have been employed with the intention of ameliorating the symptoms of AD. These include precursor loading acetylcholine release enhancement, and direct activation of both muscarinic and nicotinic receptors.

Acetylcholine Precursors. Early efforts to treat dementia using cholinomimetics focused on choline [*62-49-7*] (**12**) supplement therapy (Fig. 3). This therapy, analogous to L-dopa [*59-92-7*] therapy for Parkinson's disease, is based on the hypothesis that increasing the levels of choline in the brain bolsters acetylcholine (ACh) synthesis and thereby reverses deficits in cholinergic function. In

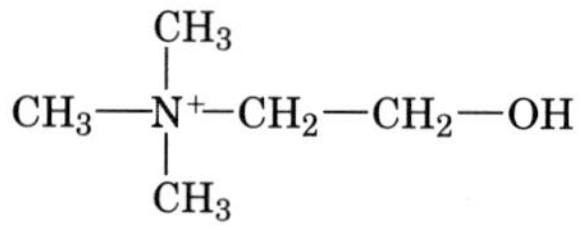

Fig. 3. Structures of acetylcholine precursors.

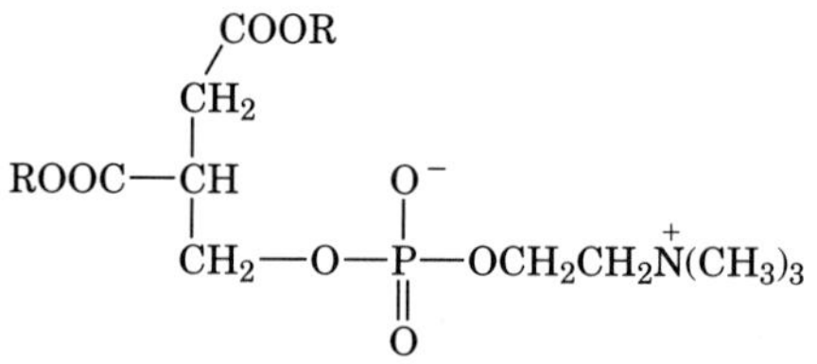

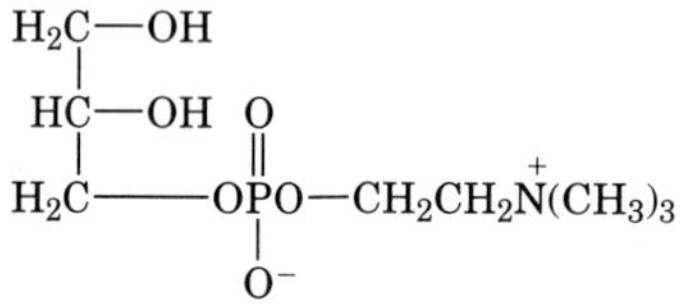

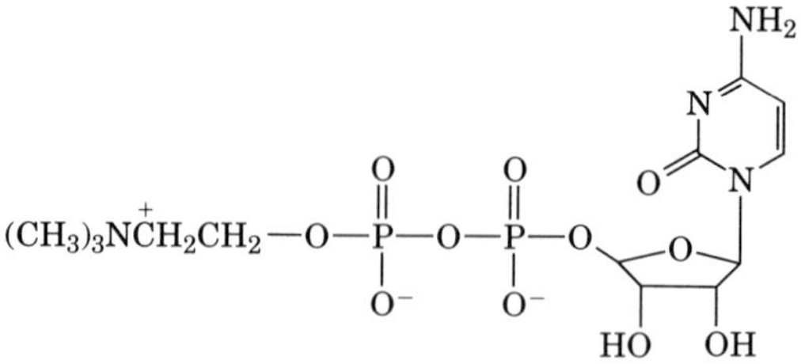

addition, because choline is a precursor of phosphatidylcholine as well as ACh, its supplementation may be neuroprotective in conditions of choline deficit (104).

Precursor loading using choline (qv) or lecithin (qv) (**13**) failed to have a significant effect on AD symptoms (98,105–107). These negative results may, in part, be related to the observation that lecithin does not alter central cholinergic activity in AD (108).

α-Glycerylphosphorylcholine (α-GFC) (**14**) and cytidine-5-diphosphate-choline (CDP-choline) (**15**) are two more recently studied choline-delivering agents. The former has been reported to increase ACh production and release, and to reverse scopolamine-induced behavioral deficits in rats (109) as well as to reverse behavioral deficits in old and excitotoxin-lesioned rats (110). The latter has been shown to be effective in improving behavioral performance in compromised animals (111). α-GFC has been reported to have positive effects in treating patients with multiinfarct dementia (112,113), and CDP-choline has been reported to be effective in treating patients with vascular dementia (112) and AD (114). However, clinical trials assessing the effects of α-GFC and CDP-choline on dementia did not employ double-blind designs.

Acetylcholinesterase Inhibitors. The greatest activity in the area of cholinomimetic treatments for AD has been in the development of agents that retard the degradation of acetylcholine (ACh) through the blockade of acetyl-cholinesterase (AChE) activity. Acetylcholinesterase inhibitors (AChEI) are generally effective in increasing performance in rodent models of learning and memory, especially those in which cholinergic deficits are created (115). The first AChE inhibitor to be tested in dementia, physostigmine [57-47-6] (**16**), was potent, but its efficacy was limited because of a short half-life. This limitation has been addressed in the design of newer generation compounds. Specificity of AChE inhibitors has also been a problem because many AChE inhibitors also are potent inhibitors of plasma butyrylcholinesterase, an activity which might contribute, in part, to several of the side effects associated with this class of molecules. However, the most recent compounds under investigation are relatively specific for brain AChE. Structures of AChEI are shown in Figure 4.

Physostigmine (**16**), an alkaloid, has been the most extensively studied AChE inhibitor. Through its reactive carbamoyl group, (**16**) acylates the catalytic site of AChE, thereby inhibiting the enzyme. This acylation, however, is readily reversible and physostigmine is considered a reversible inhibitor of the enzyme (116). Clinical studies demonstrated that oral physostigmine led to small cognitive improvements in a subpopulation of AD patients (116), but a narrow therapeutic window was observed. Side effects of physostigmine included gastrointestinal disturbances as well as cardiovascular effects. Other shortcomings of physostigmine are a short half-life and variable bioavailability (117).

Heptylphysostigmine (eptastigmine) (**17**) has been shown to be as active as physostigmine in AChE inhibition, but superior to physostigmine in terms of oral bioavailability and half-life (118–120). However, further clinical evaluation of this compound has been halted because of drug-related hematological toxicity.

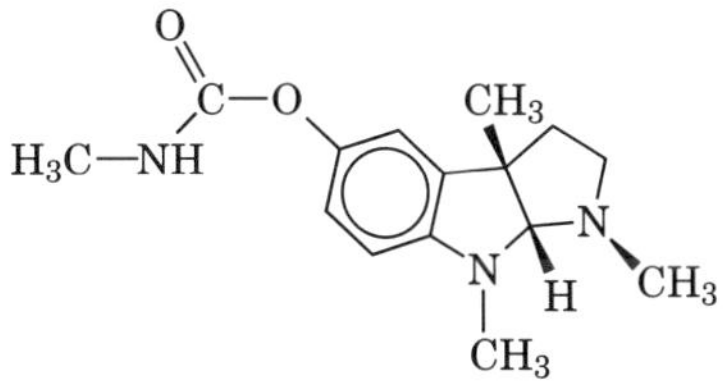

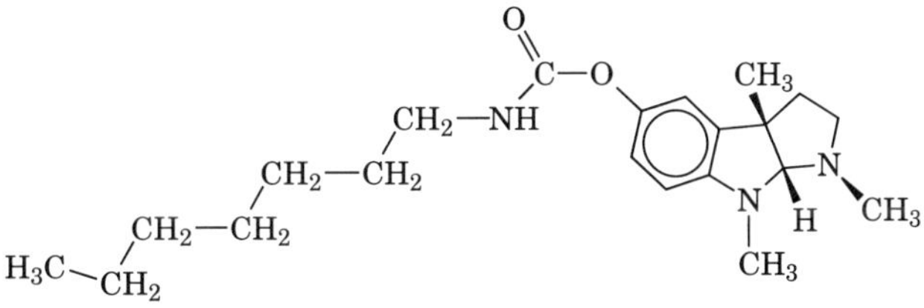

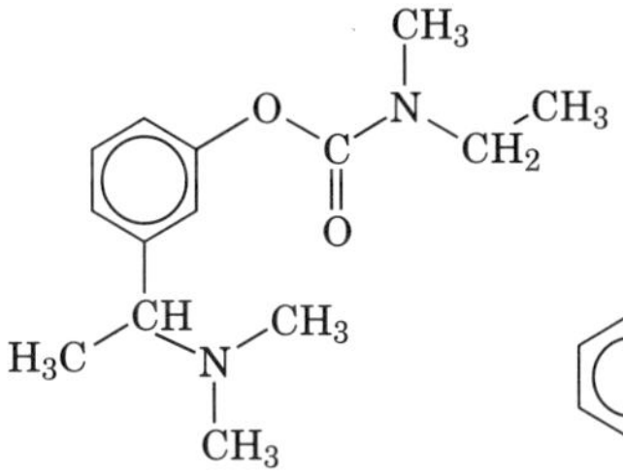

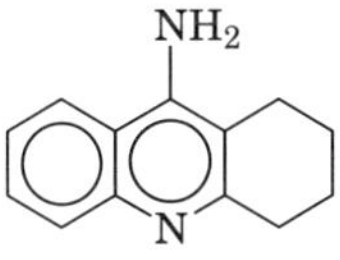

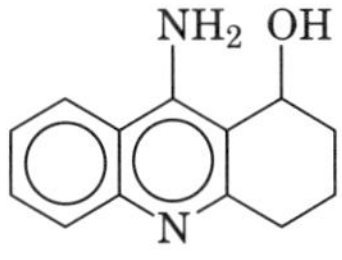

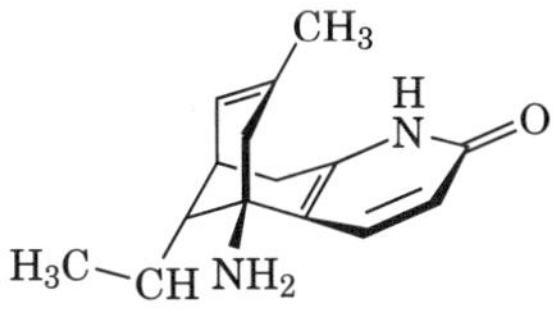

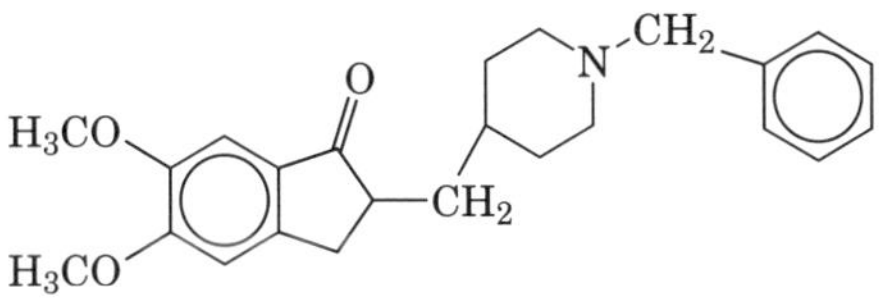

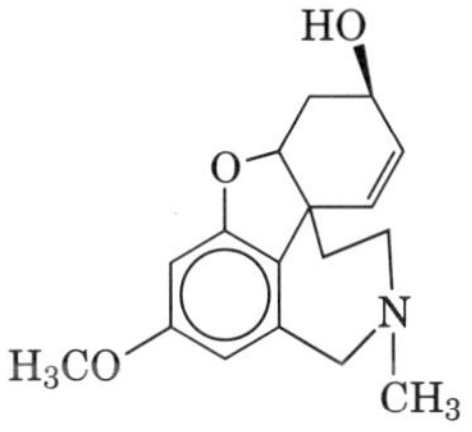

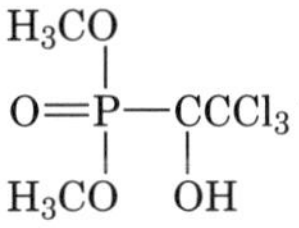

Fig. 4. Structures of acetylcholinesterase inhibitors.

SDZ ENA 713 (**18**) is another long-acting carbamate-containing molecule being investigated for AChE inhibition and AD therapy. The advantage claimed for this compound over physostigmine, heptylphysostigmine, and tacrine (**19**) is the CNS specificity of SDZ ENA 713 relative to the other AChE inhibitors (121). This selectivity may serve to reduce peripheral side effects while maintaining clinical efficacy. Central activity of SDZ ENA 713 has been observed in normal human subjects in the absence of peripheral side effects.

The aminoacridines, tacrine (**19**) and its 1-hydroxy metabolite, velnacrine (**20**), are reversible inhibitors of AChE. Tacrine was synthesized in the 1940s and has been used clinically for the treatment of myasthenia gravis and tardive dyskinesia (115). Placebo-controlled studies have indicated modest efficacy of tacrine to treat AD dementia (122,123) and in 1993 the drug was recommended for approval by the FDA under the trade name Cognex. Tacrine (**19**) has been shown to interact with sites other than AChE, such as potassium channels (124) and muscarinic receptors. However, these interactions are comparatively weak and are not thought to contribute to the biological activity of the drug at therapeutic levels (115).

Serious hepatotoxicity of tacrine has been documented. More recent data suggest, however, that this toxicity can be reduced by carefully monitoring serum alanine aminotransferase levels (125). The side effects of tacrine also include gastrointestinal disturbances and emesis, and alternative AChE therapies are being advanced. Velnacrine (**20**), a metabolite of tacrine, was expected to have reduced hepatotoxicity. However, its limited efficacy and side-effect profile, which includes drug-related hematological changes, caused it to be dropped from further development.

Three structurally unrelated AChE inhibitors being pursued for AD treatment are huperzine A (**21**), E2020 (**22**), and galanthamine [*357-70-0*] (**23**). Huperzine A is an alkaloid extracted from the Chinese herb *Huperzia serrata*. It is an effective treatment for myasthenia gravis (126) and has been suggested as an effective treatment for aged individuals with memory impairment (127). The drug has a long duration of action (128) and a favorable side-effect profile (129). At the present time the compound is being tested in broader clinical trials.

E2020 (**22**) is a relatively specific brain acetylcholinesterase inhibitor. It is over 500 times more selective for AChE than for butyrylcholinesterase (130). In addition, E2020 inhibits brain cholinesterase in a dose-dependent manner without a significant effect on enzyme activity in the intestine or heart. E2020 has an extremely long elimination half-life of about 60 h in young subjects and 104 h in elderly individuals (131). The specificity of this compound may provide a much better safety profile than other AChE inhibitors. The long half-life of the compound may complicate dosing, however.

Galanthamine (**23**) is an alkaloid extracted from the common snowdrop *Galanthus nivalis*. This compound is a long-acting, competitive AChE inhibitor which appears to be somewhat more specific for acetylcholinesterase than plasma butyrylcholinesterase (132). It is well tolerated during long-term treatment (133) and is being evaluated clinically for AD (134).

Metrifonate [*52-68-6*] (**24**) is itself not an AChE inhibitor, but is nonenzymatically converted into an active irreversible inhibitor of the enzyme. The compound is relatively specific for AChE over butyrylcholinesterase (135) and the irreversible nature of its inhibition gives rise to an extended duration of action. Some clinical experience has been gained through its use to treat schistosomiasis (136,137) and it is undergoing clinical evaluation for AD.

Receptor Agonists

Muscarinic Receptor Agonists. Acetylcholine indirect agonists such as the AChE inhibitors and ACh-releasing agents only have value in treating dementia if enough of the cholinergic arbor in the hippocampus and cortex of affected individuals remains functional. As the cholinergic innervation declines, as is the case upon progression of AD, these therapies lose efficacy. However, there is evidence that post-synaptic receptors actually are preserved in AD (138,139). Thus direct muscarinic agonists should remain effective even as the presynaptic cholinergic terminals decline in number.

Initial attempts to treat AD using direct cholinergic agonists were limited by low efficacy and side-effect issues (140–142). Thus trials using RS-86 (**25**), oxotremorine [70-22-4] (**26**), arecoline [63-75-2] (**27**), and pilocarpine [92-32-7] (**28**) to treat AD were equivocal (Fig. 5). However, the identification of multiple subtypes of muscarinic receptors has stimulated a search for subtype specific muscarinic agonists which may limit side effects while increasing efficacy.

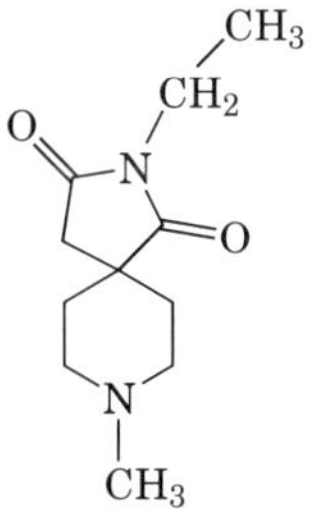

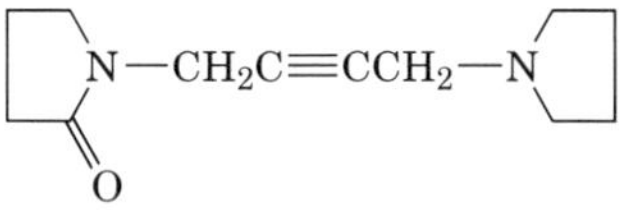

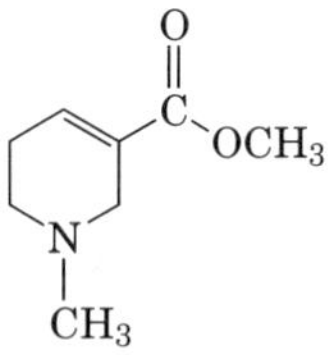

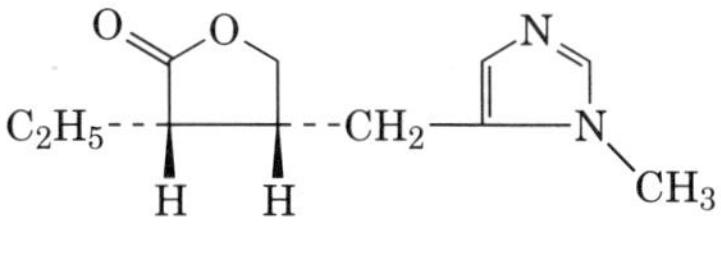

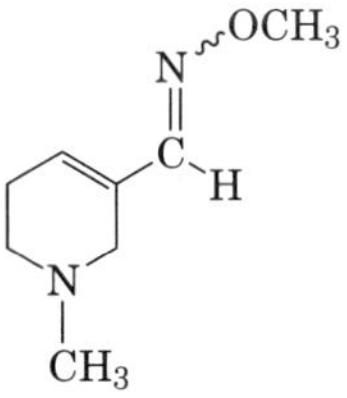

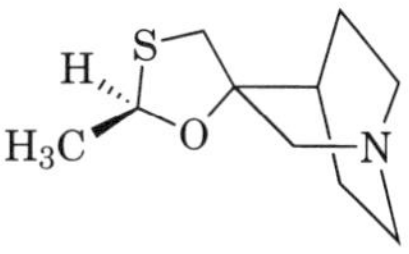

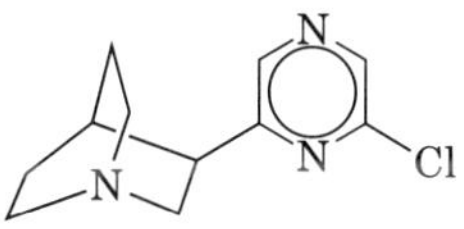

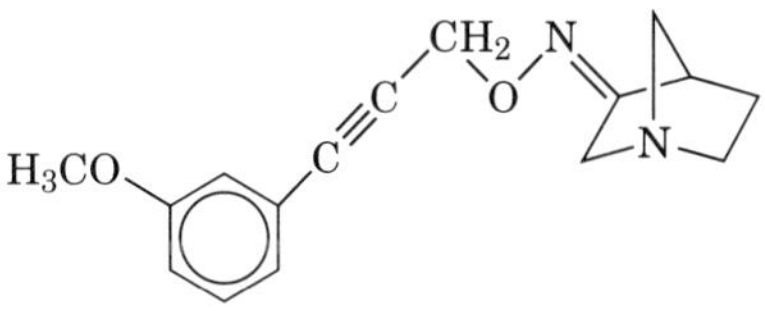

Fig. 5. Structures of muscarinic agonists.

Five distinct muscarinic receptors have been identified (143) designated m1 to m5. The m1 receptor is believed to be important for increasing cerebral cortical tone, and therefore may be an important target for AD therapy (144). The m2 receptor, on the other hand, is thought to be associated with cholinergic side effects such as emesis and bradycardia. It has been determined that many of the first muscarinic agonists evaluated were more potent for m2 receptors. More recently, however, balanced m1/m2 receptor agonists, as well as m1 selective agonists, have been or are being tested.

CI-979 (**29**) is a balanced muscarinic agonist having equal affinities for cloned m1 and m2 receptors (144). However, unlike prototypical muscarinic compounds such as (**25**), (**29**) increases central muscarinic tone, as indicated by behavioral and electroencephalogram (EEG) parameters, at doses lower than those required to produce gastrointestinal effects (144). CI-979 is well tolerated in humans up to a dose of 1 mg. Dose-limiting side effects such as stomach pain and emesis were observed at a dose of 2 mg.

Whereas balanced muscarinic agents having acceptable therapeutic indexes may be of clinical value, more hope is held for subtype specific agents. AF-102b (**30**) and L-689,660 (**31**) appear to be low efficacy muscarinic drugs that display a functional specificity for m1 and m3 receptors (145). These compounds act as antagonists at m2 receptors (145). AF-102b has similar affinity for both m1 and m2 receptors, and its specificity is based on its functional activity at these receptors (144). Unlike AF102b, PD 142505 (**32**) has a threefold higher affinity for m1 over m2 receptors (144). PD 142505 has been shown to enhance performance in a spatial working memory task in mice at doses of 1 and 3.2 mg/kg po. Moreover, the compound does not cause increases in gastrointestinal motility at doses as high as 178 mg/kg in the rat (144).

Nicotinic Receptor Agonists. There has been significant activity in the development of muscarinic cholinergic receptor agonists for dementia. In addition, agents that interact with nicotinic cholinergic receptors may also have therapeutic value. Nicotinic receptors have been reported to be reduced in AD, and pilot clinical data on the use of nicotine [54-11-5] (**33**) in AD have suggested some benefit of the drug (146). However, the gastrointestinal and cardiovascular side effects of nicotine limit its therapeutic value. Thus efforts to discover brain specific nicotinic agonists for AD treatment led to ABT 418 (**34**). This compound was shown to be 3–10 times more potent than nicotine in enhancing performance of laboratory animals in paradigms designed to measure learning and memory. In contrast, ABT 418 was less potent than nicotine in producing emesis (147). ABT 418 is being evaluated in human clinical trials.

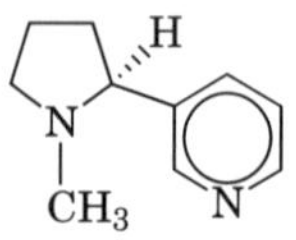

(**33**) 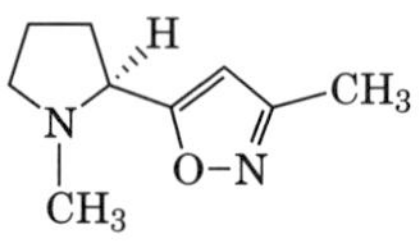   (**34**)

Acetylcholine Release Modulators

An alternative approach to stimulate cholinergic function is to enhance the release of acetylcholine (ACh). Compounds such as the aminopyridines increase the release of neurotransmitters (148). The mechanism by which these compounds modulate the release of acetylcholine is likely the blockade of potassium channels. However, these agents increase both basal (release in the absence of a stimulus) and stimulus-evoked release (148). 4-Aminopyridine [504-24-5] was evaluated in a pilot study for its effects in AD and found to be mildly effective (149).

Unlike the aminopyridines, linopirdine (**35**) (AVIVA) enhances evoked and not basal release of acetylcholine (150). In rats, linopirdine has been shown to enhance the acquisition response and reverse the passive avoidance deficits elicited by hypoxia (151,152). Like 4-aminopyridine, (**35**) enhances the release of several neurotransmitters. Linopirdine has been shown to enhance the K^+-stimulated release of [^{3}H]acetylcholine from neocortical, hippocampal, and striatal slices, as well as the K^+-stimulated release of [^{3}H]dopamine and [^{3}H]serotonin from striatal slices without affecting the basal efflux of these neurotransmitters (150). In contrast, the drug has no effect on the release of [^{3}H]norepinephrine from rat neocortical slices (150). Because the functions of multiple neurotransmitter systems are decreased in dementias like AD, the property of compounds such as linopirdine to enhance the release of several neurotransmitters offers an advantage over AD therapies aimed at stimulating the cholinergic system alone.

(**35**)   (**36**)

Another compound that affects parameters relating to several neurotransmitter systems is HP749 (**36**) which is in clinical trials for the treatment of the dementia associated with Alzheimer's disease. In passive avoidance paradigms, (**36**) was found to be active in reversing scopolamine-induced amnesia in mice, and enhancing retention in normal and nucleus basalis lesioned rats (153). This compound has several effects in *in vitro* neurochemical assays including monoamine reuptake blockade, enhancement of NE release, and inhibition of α_2-adrenergic and muscarinic receptor binding.

Age-related syndromes of cognitive and memory decline ultimately may be treated by agents that slow or stop the progression of dementia, but available therapy as of this writing offers only symptomatic relief. Of the many approaches to palliative treatment for dementia that have been attempted, the greatest effort has been in the areas of cerebral metabolism (cerebral vasodilators and nootropics) and neurotransmission (primarily cholinergic) enhancers. Only the acetylcholinesterase inhibitor tacrine is accepted as having therapeutic effect in a subpopulation of those suffering with AD dementia.

BIBLIOGRAPHY

"Memory-Enhancing Agents and Antiaging Drugs" in *ECT* 3rd ed., Vol. 15, pp. 132–143, by J. S. Bindra, Pfizer, Inc.

1. Robertson, *Inter. J. Health Serv.* **20**, 429 (1990).
2. *Current Drugs: Neurodegenerative Disorders*, NDG4 (1994).
3. L. S. Schneider and J. T. Olin, *Arch. Neurol.* **51**, 787 (1994).
4. N. Meyer-Rouge and co-workers, *Pharmacology* **16**, 45 (1978).
5. R. Markstein, *J. Pharmacol* **16**, 1 (1985).
6. A. Imperato and co-workers, *Neuro Report* **5**, 674 (1994).
7. T. Imamoto, M. Tanabe, N. Shimamoto, K. Kawazoe, and M. Hirata, *Arzneimit-telforschung* **34**, 161 (1984).
8. L. J. Thal, D. P. Salmon, B. Lasler, D. Bower, and M. R. Klanber, *J. Am. Geriatr. Soc.* **37**, 515 (1989).
9. R. Balestreri and R. Fontana, *J. Am. Geriatr. Soc.* **35**, 425 (1987).
10. L. Blaha, H. Erizgkeit, A. Adamczyk, S. Freytag, and R. Schaltenbrand, *Human Psychopharmacol.* **4**, 103 (1989).
11. K. L. Keim and P. C. Hall, *Drug. Dev. Rev.* **11**, 107 (1987).
12. B. S. Meldrum, in I. F. C. Rose, ed., *Metabolic Disorders of the Nervous System*, Pitman, London, 1983, pp. 175–187.
13. W. F. Margos, T. Greenamyre, J. B. Penney, and A. B. Young, *Trends Neurosci.* **10**, 65 (1987).
14. J.-C. Lamar, M. Beaughard, C. Bromont, and H. Poignet, in I. J. Kriegelstein, ed., *Pharmacology of Cerebral Ischemia*, Elsevier, Amsterdam, the Netherlands, 1986, pp. 334–339.
15. B. B. Fredholm, E. Lindgren, L. Lindstrom, and L. Vernet, *Acta. Pharmacol. Toxicol.* **52**, 236 (1983).
16. T. Hagiwara, T. Endo, and H. Hidaka, *Biochem. Pharmacol.* **33**, 453 (1984).
17. S. H. Francis, B. D. Noblett, B. W. Todd, J. N. Wells, and J. D. Corbin, *Mol. Pharmacol.* **34**, 506 (1988).
18. C. D. Nicholson, *Psychopharmacology* **101**, 147 (1990).
19. A. Scriabine, T. Schuurman, and J. Traber, *FASEB J.* **3**, 1799 (1989).
20. M. S. Langley and E. M. Sorkin, *Drugs* **37**, 669 (1989).
21. A. N. Wadworth and D. McTavish, *Drugs Aging* **2**, 262 (1992).
22. F. B. Meyer, *Neurosurg. Clin. N. Am.* **1**, 367 (1990).
23. M. Sandin, S. Jasmin, and T. E. Levere, *Neurobiolog. Aging* **11**, 573 (1990).
24. G. D. Tollefson, *Biol. Psychiatry* **27**, 1133 (1990).
25. P. K. Fischhof, *Methods Find Exp. Clin. Pharmacol.* **15**, 549 (1993).
26. R. J. Branconnier, M. E. Branconnier, T. M. Walshe, C. McCarthy, and P. A. Morse, *Psychopharmacol. Bull.* **28**, 175 (1992).
27. M. C. deJonge and J. Traber, *Clin. Neuropharm.* **16**, 525 (1993).
28. A. Spanoli and co-workers, *Neurology* **41**, 1726 (1991).
29. D. Cucinotta and co-workers, *Drug Dev. Res.* **14**, 213 (1988).
30. A. Imperato, M. T. Ramacci, and L. Angelucci, *Neurosci. Lett.* **107**, 251 (1989).
31. M. Onofri, I. Bodis-Wollner, P. Pola, and M. Calvani, *Drugs Exp. Clin. Res.* **9**, 161 (1983).
32. L. Janiri and E. Tempesta, *Int. J. Clin. Pharmacol. Res.* **3**, 295 (1983).

33. G. Taglialatela and co-workers, *Dev. Brain Res.* **59**, 221 (1991).
34. P. Piovesan, L. Pacifics, G. Taglialatela, M. T. Ramacci, and L. Angelucci, *Brain Res.* **633**, 77 (1994).
35. C. H. Phelps and co-workers, *Neurobiol. Aging* **10**, 205 (1989).
36. C. Giurgea, *Drug. Dev. Res.* **2**, 441 (1982).
37. A. Copani and co-workers, *J. Neurochem* **58**, 1199 (1992).
38. L. J. Vyklicky, D. K. Patneau, and M. L. Mayer, *Neuron.* **7**, 971 (1991).
39. S. Ozawa, M. Iino, and M. Abe, *Neurosci. Res.* **12**, 72 (1991).
40. Y. Kitamura, S. Hayashi, and Y. Nomura, *Jpn. J. Pharmacol.* **52**, 597 (1990).
41. Y. Kitamura, T. Kaneda, and Y. Nomura, *Jpn. J. Pharmacol.* **55**, 177 (1991).
42. T. Yashimoto and co-workers, *J. Pharmacobio-Dyn.* **10**, 730 (1987).
43. M. W. Vernon and E. M. Sorkin, *Drugs Aging* **1**, 17 (1991).
44. J. J. Calus and co-workers, *Neurology* **41**, 570 (1991).
45. E. Gamzu, T. Hoover, S. Gracon, and M. Ninteman, *Drugs Dev. Res.* **18**, 177 (1989).
46. G. Pepeu and G. Spignoli, *Prog. Neuro-Psychopharmacol. Biol. Psych.* **13**, 577 (1989).
47. C. Mondadori, *Behavioral Brain Res.* **59**, 1 (1993).
48. U. Schindler, *Prog. Neurosychopharmacol. Biol. Psych.* **13**, (Suppl.), 99 (1989).
49. X. Rabasseda, N. Mealy, and N. Presti, *Drugs Today* **30**, 9 (1994).
50. W. M. Hermann and K. Stephan, *Alzheimer Dis. Assoc. Disorder* **5** (Suppl. 1), 7 (1991).
51. M. A. Passeri, *Symposium on Piracetam: 5 Years' Progress in Pharmacology and Clinics*, Tenicas Grarficas Formas, Madrid, 1990, p. 75.
52. B. Bering and W. E. Müller, *Arzneim.-Forsch./Drug Res.* **35**, 1350 (1985).
53. V. J. Nicholson and O. L. Wolthuis, *Biochem, Pharmacol.* **25**, 2241 (1976).
54. M. Grau, J. L. Monter, and J. Balasch, *Gen. Pharmacol.* **18**, 205 (1987).
55. A. H. Gouliaev and A. Senning, *Brain Res. Revs.* **19**, 180 (1994).
56. A. Falsaperla, P. A. Monici-Petri, and C. Oliani, *Clin. Ther.* **12**, 376 (1990).
57. L. Parnetti and co-workers, *Neuropsycholpharmacology* **22**, 97 (1989).
58. B. Baumel and co-workers, *Prog. Neuropsychopharmacol. Biol. Psych.* **13**, 673 (1989).
59. G. Maina and co-workers, *Neuropsychobiology* **21**, 141 (1989).
60. M. Marchi, E. Basana, and M. Raiteri, *Eur. J. Pharmacol.* **185**, 247 (1990).
61. C. Nardella and co-workers, *Farmaco* **46**, 1051 (1991).
62. D. Mochizuki, G. Sugiyama, Y. Shinoda, *Nippon Yagurigaku Zasshi* **99**, 27 (1992).
63. V. Canonico and co-workers, *Riv. Neurol.* **61**, 92 (1991).
64. G. Forloni, N. Angeretti, D. Amorsoso, A. Addis, and S. Consolo, *Brain Res.* **530**, 156 (1990).
65. U. Senin and co-workers, *Neurophyschopharmacology* **1**, 511 (1991).
66. L. Parnetti and co-workers, *Dementia* **2**, 262 (1991).
67. C. M. Tang, O. Y. Shi, A. Katchman, and G. Lynch, *Science* **254**, 288 (1991).
68. J. S. Isaacson and R. A. Nicoll, *Proc. Natl. Acad. Sci. U.S.A.* **88**, 10936 (1991).
69. T. Yashimoto and co-workers, *Pharmacobio-Dyn.* **10**, 730 (1987).
70. A. McLean, D. D. Cardenas, D. Burgess, and E. Gamzu, *Brain Inj.* **5**, 375 (1991).
71. P. Scarpazza and co-workers, *Adv. Ther.* **10**, 217 (1993).
72. T. A. Pugsley, Y. H. Shih, L. Coughenour, and S. F. Stewart, *Drug Dev. Res.* **3**, 402 (1983).
73. *Scrip* **20**, 1833 (1993).
74. T. Kinoshita and co-workers, *Jpn. Pharmacol. Ther.* **19**, 179 (1991).
75. *Drugs Future* **18**, 18 (1993).
76. H. Aoshima, R. Shingai, and T. Ban, *Arzneim-Forsch-Drug Res.* **42**, 775 (1992).
77. M. Hashimoto, T. Hashimoto, and T. Kuriyama, *Eur. J. Pharmacol.* **209**, 9 (1991).
78. M. Ohno, T. Yamamoto, I. Kitajima, and S. Ueki, *Jpn. J. Pharmacol* **54**, 53 (1990).
79. K. Iwasaki, Y. Matsumoto, and M. Fujiwara, *Jpn. J. Pharmacol.* **58**, 117 (1992).

80. T. Nabeshima, *Neuropsychopharmacology* **9** (Suppl.), 110 (1993).

81. M. Yoshii, S. Watabe, *Brain Res.* **642**, 123 (1994).

82. S. Watanabe, H. Yamaguchi, and H. Ashida, *Jpn. J. Pharmacol.* **52** (Suppl. 1), 294P (1990).

83. S. Watanabe, H. Yamaguchi, and H. Ashida, *Eur. J. Pharmacol.* **238**, 303 (1993).

84. M. Tanaka, K. Takasuna, and S. Takayama, *Arzneim.-Forsch./Drug Res.* **44**, 193 (1994).

85. M. Sarter, *Trends. Pharm. Sci.* **12**, 456 (1991).

86. S. Watanabe, H. Yamaguchi, and S. Ashida, *Neurosci. Abstr.* **15**, 601 (1989).

87. A. Berardi and co-workers, *Neurosci. Abst.* **17**, 698 (1991).

88. A. Berardi and co-workers, *J. Neurosci. Abstr.* **18**, 1243 (1992).

89. V. K. Gribkoff, L. A. Bauman, and C. P. Vandermaelen, *Soc. Neurosci Abst.* **14**, 207 (1988).

90. V. K. Gribkoff, L. A. Bauman, and C. P. Vandermaelen, *Neuropharmacology* **20**, 1001 (1990).

91. P. Davies and A. J. F. Maloney, *Lancet* **11**, 1403 (1976).

92. E. K. Perry and co-workers, *Br. Med. J.* **2**, 1427 (1978).

93. P. Whitehouse and co-workers, *Science* **215**, 237 (1982).

94. J. T. Coyle, D. Price, and M. DeLong, *Science* **219**, 1184 (1983).

95. S. R. El-Defrawy and co-workers, *Neurobiol Aging* **6**, 325 (1985).

96. M. Watson, T. W. Vickroy, H. C. Fibiger, W. R. Roeske, and H. I. Yamamura, *Brain Res.* **346**, 387 (1985).

97. D. J. Hepler, G. L. Wenk, B. L. Cribbs, D. S. Olton, and J. T. Coyle, *Brain Res.* **346**, 8 (1985).

98. R. T. Bartus, R. L. Dean, B. Beer, and A. S. Lippa, *Science* **217**, 408 (1982).

99. D. L. Price, *Ann Rev Neurosci.* **9**, 489 (1986).

100. R. J. D'Amato and co-workers, *Ann. Neurol.* **22**, 229 (1987).

101. R. G. Struble and co-workers, *J. Neuropathol Exp. Neurol.* **46**, 567 (1987).

102. S. B. Dunnett and H. C. Fibiger, in A. C. Cuello, ed., *Cholinergic Function and Dysfunction*, Vol. 98, Elsevier Science Publishers BV, Amsterdam, the Netherlands, 1993, p. 413.

103. M. L. Voytko and co-workers, *J. Neurosci.* **14**, 167 (1994).

104. R. J. Wurtman, *TINS* **15**, 117 (1992).

105. L. J. Thal, W. Rosen, N. S. Sharpless, and H. Crystal, *Neurobiol. Aging* **2**, 205 (1981).

106. A. Little, R. Levy, K. P. Chuaqui, and D. Hand, *J. Neurol. Neurosurg. Psychiatry* **48**, 736 (1985).

107. A. Heymen and co-workers, *J. Neural Transm. Suppl.* **24**, 279 (1987).

108. N. Pomara and co-workers, *J. Clin. Psychiatry* **44**, 293 (1983).

109. C. M. Lopez and co-workers, *Pharmacol. Biochem. Behav.* **39**, 835 (1991).

110. F. Drago and co-workers, *Pharmacol. Biochem. Behav.* **41**, 445 (1992).

111. F. Drago and co-workers, *Brain Res. Bull.* **31**, 485 (1993).

112. P. R. Di and co-workers, *J. Internat. Med. Res.* **19**, 330 (1991).

113. S. G. Barbagallo, M. Barbagallo, M. Giordano, M. Meli, and R. Panzarasa, *Annal. N.Y. Acad. Sci.* **717**, 253 (1994).

114. J. Caamano, M. J. Gomez, A. Franco, and R. Cacabelos, *Meth. Findings Exp. Clin. Pharmacol.* **16**, 211 (1994).

115. J. C. Jaen and R. E. Davis, *Neurodegenerative Disorders*, 87–101 (1993).

116. L. J. Thal, in R. Becker and E. Giacobini, eds., *Physostigmine in Alzheimer's Disease*, Birkhauser, Boston, Mass., 1991.

117. L. J. Thal, P. A. Fuld, D. M. Masur, and N. S. Sharpless, *Ann. Neurol.* **13**, 491 (1983).

118. L. L. Iversen and co-workers, in R. Becker and E. Giacobini, eds., *Cholinergic Basis for Alzheimer Therapy*, Birkhauser, Boston, Mass., 1991, p. 297.

119. P. DeSarno, M. Pomponi, E. Giacobini, X. C. Tang, and E. Williams, *Neurochem. Res.* **14**, 971 (1989).
120. E. Messamore, U. Warpman, N. Ogane, and E. Giacobini, *Neuropharmacology* **32**, 745 (1993).
121. A. Enz, R. Amstutz, H. Boddeke, G. Gmelin, and J. Malanowski, *Prog. Brain Res.* **98**, 431 (1993).
122. M. Farlow and co-workers, *JAMA* **268**, 2523 (1992).
123. K. L. Davis and co-workers, *New Eng. J. Med.* **327**, 1253 (1992).
124. S. E. Freeman and R. M. Dawson, *Prog. Neurobiol.* **36**, 257 (1991).
125. P. B. Watkins, H. J. Zimmerman, M. J. Knapp, S. I. Gracon, and K. W. Lewis, *JAMA* **271**, 992 (1994).
126. Y. Cheng and co-workers, *New Drugs Clin. Rem.* **5**, 197 (1986).
127. S. Zhang, *New Drugs Clin. Rem.* **5**, 260 (1986).
128. X. C. Tang, P. DeSarno, K. Sugaya, and E. Giacobini, *J. Neurosci. Res.* **24**, 276 (1989).
129. X. F. Yan, W. H. Lu, W. J. Lou, and C. Tang, *Acta Pharamacol. Sin.* **8**, 117 (1987).
130. S. Araki, Y. Yamanishi, T. Kosasa, H. Oguru, and K. Yamatsu, *Jpn. J. Pharmacol.* **49** (Suppl.), (1989).
131. A. Ohnishi and co-workers, *J. Clin. Pharmacol.* **33**, 1086 (1993).
132. T. Thomsen and H. Kewitz, *Life Sci.* **46**, 1553 (1990).
133. T. Thomsen, U. Bickel, J. P. Fischer, and H. Kewitz, *Dementia* **1**, 46 (1990).
134. P. Dal-Bianco and co-workers, *J. Neural. Trans.* **33** (Suppl.), 59 (1991).
135. R. E. Becker, P. Moriearty, and L. Unni, in Ref. 118, p. 263.
136. P. R. Mason, S. A. Tswana, P. Jenks, and C. E. Priddy, *J. Tropical Med. Hygiene* **94**, 180 (1991).
137. A. F. Mgeni and co-workers, *Bull. World Health Org.* **68**, 721 (1990).
138. D. Mash, D. Flynn, and L. Potter, *Science* **228**, 1115 (1985).
139. B. Pearce and L. T. Potter, *Alz. Dis. Assoc. Disord.* **5**, 163 (1991).
140. M. M. Mouradian, E. Mohr, J. A. Williams, and T. N. Chase, *Neurology* **38**, 606 (1988).
141. K. L. Davis and co-workers, *Am. J. Psychiatry* **144**, 468 (1987).
142. S. L. Read and co-workers, *Arch. Neurol.* **47**, 1025 (1990).
143. T. L. Bonner, N. J. Buckley, A. C. Young, and M. R. Brann, *Science* **237**, 527 (1987).
144. R. Davis and co-workers, *Prog. Brain Res.* **98**, 439 (1993).
145. L. L. Iversen, in Ref. 102, pp. 423–426.
146. P. A. Newhouse and co-workers, *Psychopharmacol. Berlin* **95**, 171 (1988).
147. M. W. Decker and co-workers, *J. Pharmacol. Experimen. Therap.* **270**, 319 (1994).
148. R. L. Buyukuysal and R. J. Wurtman, *J. Nuerochem.* **54**, 1302 (1990).
149. H. Wesseling and co-workers, *N. Engl. J. Med.* **310**, 988 (1984).
150. V. J. Nickolson, S. W. Tam, M. J. Meyers, and L. Cook, *Drug Dev. Res.* **19**, 285 (1990).
151. N. V. Cook, L. Steinfels, K. W. Rohrbach, and V. J. DeNoble, *Drug Devel. Res.* **19**, 301 (1990).
152. V. J. DeNoble and co-workers, *Pharmacol. Biochem. Behav.* **36**, 957 (1990).
153. M. Cornfeldt and co-workers, *Soc. Neurosci. Abstr.* **20**, 252.6 (1990).

Robert Zaczek
Robert J. Chorvat
The Du Pont Merck Research Laboratories

MENDELEVIUM. See Actinides and Transactinides.

MERCURY

Mercury [7439-97-6], Hg, atomic number 80, atomic weight 200.59, also called quicksilver, is a heavy, odorless metal belonging to Group 12 (IIB) of the Periodic Table. Elemental mercury is a liquid at room temperature, having a characteristic bright, silvery appearance. The symbol Hg is taken from the Latin word *hydrargyrum*, meaning liquid silver. Unlike the other two Group 12 elements, mercury exhibits two valences, mercurous, Hg^+, and mercuric, Hg^{2+}. Samples of mercury have been found in graves dating to the fifteenth or sixteenth century BC, where cinnabar [19122-79-3], HgS, was used as a pigment for cave and body decoration.

The first recorded mention of mercury metal was by Aristotle in the fourth century BC, at a time when it was used in religious ceremonies. By the time of Pliny, mercury was a familiar substance, and its preparation by roasting cinnabar was well known. The ancient Egyptians, Greeks, and Romans used mercury for cosmetic and medical preparations, and for amalgamation. Although it resembles a metal, mercury's liquid nature usually caused it to be classified among the "waters." It was not until sometime in the period 500–700 AD that mercury was accepted as a metal, and the astrological symbol that had been assigned to tin was given to mercury. The ability of mercury to impart a silvery color to other metals also gave it a special position among the alchemists (1).

In the late fifteenth century AD mercury was successfully used as a treatment for syphilis. In the late sixteenth century the development of the Patio process for the recovery of silver by amalgamation (see SILVER AND SILVER ALLOYS) greatly increased the consumption of mercury. Usage of mercury increased in 1643 when Torricelli invented the barometer, and again in 1720 when Fahrenheit invented the mercury thermometer. Other scientific and medical applications followed. Industrial usage after 1900, particularly in electrical applications, expanded rapidly, offsetting the sharp decline in its use in amalgamation.

Applications of mercury include use in batteries (qv), chlorine and caustic soda manufacture (see ALKALI AND CHLORINE PRODUCTS), pigments (see PIGMENTS, INORGANIC), light switches, electric lighting, thermostats, dental repair (see DENTAL MATERIALS), and preservative formulations for paints (qv) (1–3). As of the end of the twentieth century, however, increased awareness of and concern for mercury toxicity has resulted in both voluntary and regulatory reduction of mercury usage (see also MERCURY COMPOUNDS).

Occurrence

Mercury metal is widely distributed in nature, usually in quite low concentrations. The terrestrial abundance is on the order of 50 parts per billion (ppb), except in mercuriferous belts and anthropogenically contaminated areas. In soils, the average mercury content is about 100 ppb, ranging from 30 to 500 ppb. In rocks, mercury content ranges from 10 to 20,000 ppb. Surface waters, except where special geological conditions prevail or where anthropogenic sources occur, generally contain less than 0.1 ppb total mercury. The average mercury content of sea waters has been found to range from 0.1 to 1.2 ppb (2). The most important mineral of mercury is cinnabar, found in rocks near recent volcanic activity or hot spring areas and in mineral veins or fractures as impregnations (2–4). In addition to cinnabar, corderoite [53237-82-4], $Hg_3S_2Cl_2$; livingstonite [12532-29-5], $HgSb_4S_8$; montroydite [24401-75-0], HgO; terlinguaite [12394-37-5], Hg_2OCl; calomel [10112-91-1], $HgCl$; and metacinnabar [23333-45-1], a black form of cinnabar, are commonly found in mercury deposits (5). The numerous other mercury minerals are rare and not commercially significant.

Mercury ore deposits occur in faulted and fractured rocks, such as limestone, calcareous shales, sandstones, serpentine, chert, andesite, basalt, and rhyolite. Deposits are mostly epithermal in character, ie, minerals were deposited by rising warm solutions at comparatively shallow depths from 1–1000 m (6).

Properties

Mercury has a uniform volume expansion over its entire liquid range (see Table 1). This, in conjunction with high surface tension, ie, inability to wet and cling to glass, makes mercury extremely useful for barometers, manometers, thermometers, and many other measuring devices. Mercury also has a propensity to form alloys (amalgams) with almost all other metals except iron, and at high temperatures, even with iron. Because of low electrical resistivity, mercury is rated as one of the best electrical conductors among the metals. Mercury has a high thermal neutron-capture cross section (360×10^{-28} m^2 (360 barns)), enabling it to absorb neutrons and act as a shield for atomic devices. Its high thermal conductivity also permits it to act as a coolant. The physical and chemical properties of mercury are given in Table 1, and the distribution of the stable isotopes in Table 2. A number of artificial isotopes also are known (see RADIOISOTOPES).

The volume expansion, V_t, of mercury may be calculated over its entire liquid range by

$$V_t = V_0(1 + 0.18182 \times 10^{-3} \, t + 0.0078 \times 10^{-6} \, t^2)$$

where V_0 is the known volume of mercury at a given temperature, t, in °C.

The specific heat varies with temperature. For solid mercury it increases but in liquid mercury it drops such that the specific heat at 210°C is the same as that at −75°C (Table 3). Up to 50°C, specific heat in J/g is given by the formula $0.1418 - 0.0004343(t + 36.7)$ for temperatures t in °C; from 50–150°C, the formula should be modified by an additional corrective factor of $-0.000025(t - 50)$; and between 150 and 250°C, $-0.0000126(t - 150)$. Values for temperatures above 150°C are not as accurate as those for lower temperatures.

The vapor pressure of mercury, P, also behaves irregularly but may be obtained for temperatures from 0–150°C by the following:

$$\log P = (-3212.5/T) + 7.150$$

and from 150°–400°C by

$$\log P = (-3141.33/T) + 7.879 - 0.00019 \, t$$

where P is in kPa (101.3 kPa = 760 mm Hg), T is the absolute temperature in

Table 1. Physical and Chemical Properties of Mercury[a]

Property	Value
accommodation coefficient, at -30 to $60°C$, $(T_3 - T_1)/(T_2 - T_1)$[b]	1.00
angle of contact of glass at $18°C$, deg	128
atomic distance, nm	0.30
melting point, $°C$	-38.87
boiling point, $°C$	356.9
triple point, $°C$	-38.84168
bp rise with pressure, $°C/kPa$[c]	0.5595
compressibility (volume) at $20°C$, MPa^{-1}[d]	39.5×10^{-6}
condensation on glass, $°C$	-130 to -140
conductivity, thermal, $W/(cm^2 \cdot K)$	0.092
critical density, g/cm^3	3.56
critical pressure, MPa[d]	74.2
critical temperature, $°C$	1677
crystal system	rhombohedral
density, g/cm^3	
at melting point	14.43
$-38.8°C$ (solid)	14.193
$0°C$	13.595
$20°C$	13.546
emf, of hot Hg at $100°C$, mV[e]	-0.60
volume expansion coefficient at $20°C$, $°C^{-1}$	182×10^{-6}
latent heat of fusion, J/g[f]	-11.80
hydrogen overvoltage, V	1.06
ionization potentials, eV	
1st electron	10.43
2nd electron	18.75
3rd electron	34.20
magnetic moment, ^{199}Hg, J/T[g]	4.63×10^{-24}
magnetic volume susceptibility	
at $18°C$, cm^3/g	1.885×10^{-6}
potential, V	
contact Hg/Sb	-0.26
contact Hg/Zn	$+0.17$
pressure, internal, MPa[d]	1.317
reflectivity, degree at 550 nm	71.2
refractive index at $20°C$	$1.6-1.9$
temperature coefficient of resistance at $20°C$, $\Omega/°C$	0.9×10^{-3}
resistivity, at $20°C$, $\Omega \cdot cm$	95.8×10^{-6}
soly in water, $\mu g/L$	$20-30$
surface tension temp coefficient, $mN/(m \cdot °C)(=dyn/(cm \cdot °C))$	-0.19
viscosity at $20°C$, $mPa \cdot s(=cP)$	1.55

[a] Ref. 7.
[b] T_1 is the temperature of a gas molecule striking a surface which is at temperature T_2, and T_3 is the temperature of a gas molecule leaving the surface.
[c] To convert kPa to mm Hg, multiply by 7.5.
[d] To convert MPa to atm, divide by 0.101.
[e] Relative to Pt–cold junction, $0°C$.
[f] To convert J to cal, divide by 4.184.
[g] Value is equivalent to $0.4993 \mu_B$.

Table 2. Distribution of Stable Isotopes[a]

Isotope	CAS Registry Number	Abundance, %	Isotope	CAS Registry Number	Abundance, %
^{196}Hg	[14917-67-0]	0.146	^{201}Hg	[15185-19-0]	13.22
^{198}Hg	[13891-21-0]	10.02	^{202}Hg	[14191-86-7]	29.80
^{199}Hg	[14191-87-8]	16.84	^{204}Hg	[15756-14-6]	6.85
^{200}Hg	[15756-10-2]	23.13			

[a]Ref. 7.

Table 3. Thermodynamic Properties of Mercury[a]

Property	Value
entropy, S_{298}, J/mol[b]	76.107
heat of fusion, J/mol[b]	2,297
heat of vaporization, ΔH_v, J/mol[b]	59,149
liquid mercury, 25–357°C, J/mol[b,c]	
heat capacity, Cp	27.66
$H_T - H_{298}$	$-1971 + 6.61\ T$
$F_T - H_{298}$	$-1971 + 661\ T \ln T + 26.08\ T$
gaseous mercury, 357–2727°C, J/mol[b,c]	
heat capacity, Cp	20.79
$H_T - H_{298}$	$13,055 + 4.969\ T$
$F_T - H_{298}$	$13,055 - 4.969\ T \ln T - 8.21\ T$
heat of vaporization at 25°C, J/mol[b]	61.38
latent heat of vaporization, J/g[b]	271.96
specific heat, J/g[b]	
solid	
-75.6°C	0.1335
-40°C	0.1410
-263.3°C	0.0231
liquid	
-36.7°C	0.1418
210°C	0.1335

[a]Ref. 8.
[b]To convert J to cal, divide by 4.184.
[c]T in K.

K, and t is the temperature in °C. The accuracy of these formulas is believed to be within 1%.

Another valuable property of mercury is its relatively high surface tension, 480.3 mN/m(=dyn/cm) at 0°C, as compared to 75.6 mN/m for water. Because of its high surface tension, mercury does not wet glass and exhibits a reverse miniscus in a capillary tube.

Electrode reduction potentials of mercury are given in Table 4.

At ordinary temperatures, mercury is stable and does not react with air, ammonia (qv), carbon dioxide (qv), nitrous oxide, or oxygen (qv). It combines readily with the halogens and sulfur, but is little affected by hydrochloric acid, and is attacked only by concentrated sulfuric acid. Both dilute and concentrated

Table 4. Electrode Reduction Potentials of
Mercury, V

Equation	E^0, volts
$Hg^{2+} + 2\,e^- \rightleftharpoons Hg$	0.851
$Hg_2{}^{2+} + 2\,e^- \rightleftharpoons 2\,Hg$	0.7961
$2\,Hg^{2+} + 2\,e^- \rightleftharpoons Hg_2^{2+}$	0.905

nitric acid dissolve mercury, forming mercurous salts when the mercury is in excess or no heat is used, and mercuric salts when excess acid is present or heat is used. Mercury reacts with hydrogen sulfide in the air and thus should always be covered.

The only metals having good or excellent resistance to corrosion by amalgamation with mercury are vanadium, iron, niobium, molybdenum, cesium, tantalum, and tungsten (8). The diffusion rates of some metals in mercury are given in Table 5.

Table 5. Diffusion of Metals in Mercury[a]

Metal	CAS Registry Number	Diffusion rate, cm²/s × 10⁻⁵	Metal	CAS Registry Number	Diffusion rate, cm²/s × 10⁻⁵
lithium	[7439-93-2]	0.9	silver	[7440-22-4]	1.1
sodium	[7440-23-5]	0.9	gold	[7440-57-5]	0.7
potassium	[7440-09-7]	0.7	zinc	[7440-66-6]	1.57
rubidium	[7440-17-7]	0.5	cadmium	[7440-43-9]	2.07
cesium	[7440-46-2]	0.6	thallium	[7440-28-0]	1.03
calcium	[7440-70-2]	0.6	tin	[7440-31-5]	1.68
strontium	[7440-24-6]	0.5	lead	[7439-92-1]	1.16
barium	[7440-39-3]	0.6	bismuth	[7440-69-9]	0.99
copper	[7440-50-8]	1.06			

[a]Ref. 7.

Production and Shipment

Primary Production. Mercury ore is mined by both surface and underground methods. The latter furnishes about 90% of the world's production. Mercury is recovered also as a by-product in the mining and processing of precious and base metals (see MINERAL RECOVERY AND PROCESSING). Small quantities of mercury have been produced by processing ground under and adjacent to the sites of less efficient ore-burning furnaces used in earlier-day mercury recovery operations. Mercury is also produced by working mine dumps and tailing piles, particularly those accumulated during turn-of-the-century mining operations. The average grade of mercury ore mined from large mines throughout the world in the latter part of the twentieth century has ranged from 4 to 20 kg/t. Total recovery has approached 95%. The average grade of ore has generally declined as a result of the practice of mining the richest parts of ore bodies to realize a higher profit and because prices have generally increased, allowing lower grade ores to be exploited. Since 1990, the only mine production of mercury in the United

States has come as a by-product of gold mining operations (14) (see GOLD AND GOLD COMPOUNDS).

Secondary Production. Smaller quantities of mercury are produced each year from industrial scrap and waste materials, such as discarded dental amalgams, batteries, lamps, switches, measuring devices, control instruments, and wastes and sludges generated in laboratories and electrolyic refining plants (3). Mercury in these materials may be recovered *in situ* or by firms specializing in secondary recovery (see RECYCLING). Traditionally, secondary production has depended on the price of mercury. Secondary production of mercury in the United States has increased steadily in the 1990s, however, as a result of the rising costs of hazardous waste disposal and increased restrictions on mercury disposal.

Shipping. Prime virgin-grade mercury is packaged in wrought iron or steel flasks containing 34.5 kg of the metal. Mercury of greater purity, produced by multiple distillation or other means, may be marketed in flasks but is usually packaged in small glass or plastic containers.

Processing

Primary. Mercury metal is produced from its ores by standard methods throughout the world. The ore is heated in retorts or furnaces to liberate the metal as vapor which is cooled in a condensing system to form mercury metal (9). Retorts are inexpensive installations for batch-treating concentrates and soot. These require only simple firing and condensing equipment. For larger operations, either continuous rotary kilns or multiple-hearth furnaces with mechanical feeding and discharging devices are preferred. Using careful control at properly designed plants, 95% or more of the mercury in the ore can be recovered as commercial-grade (99.9% purity) mercury.

Other recovery methods have been used (10). These include leaching ores and concentrates using sodium sulfide [1313-82-2] and sodium hydroxide [1310-73-2], and subsequently precipitating with aluminum [7429-90-5], or by electrolysis (11). In another process, the mercury in the ore is dissolved by a sodium hypochlorite [7681-52-9] solution, the mercury-laden solution is then passed through activated carbon [7440-44-0] to absorb the mercury, and the activated carbon heated to produce mercury metal. Mercury can be extracted from cinnabar by electrooxidation (12,13).

Secondary. Scrap material, industrial and municipal wastes, and sludges containing mercury are treated in much the same manner as ores to recover mercury. Scrap products are first broken down to liberate metallic mercury or its compounds. Heating in retorts vaporizes the mercury, which upon cooling condenses to high purity mercury metal. Industrial and municipal sludges and wastes may be treated chemically before roasting.

Economic Aspects

Mercury toxicity from environmental pollution and occupational exposure has become an area of concern throughout the world. Since the 1980s much effort has been devoted both in terms of legislation and voluntary action, to limiting the production and use of mercury, and to finding technologies designed to supplant the need for mercury. As a result, overall demand for mercury has decreased throughout the industrialized world. Table 6 gives a breakdown by country of worldwide mercury production.

Spain, which until 1989 was the world's largest producer of mercury, ceased mine production after 1990, although 100 t of mercury were produced in 1991 from stockpiled materials. China, Mexico, and Russia were the largest producers in 1992 (14).

The last operating mercury mine in the United States closed in November of 1990. Since 1990, the only mine production of mercury in the United States has come as a by-product of gold-mining operations. Mercury occurs with most gold ores and must be recovered to avoid release into the environment (14). Secondary production of mercury in the United States has been increasing as a result of the growing cost restrictions on the disposal of mercury-containing wastes (3,14). U.S. imports and exports have also increased since 1989 (14). In the early 1990s over 50% of the exported U.S. mercury went to the Netherlands including 700 t in 1992 (14).

The U.S. Defense Logistics Agency (DLA), which maintains the U.S. National Defense Stockpile (NDS), sold 267 t of stockpiled mercury in 1992 (14). The DLA also sold 103 t of secondary mercury from the Department of Energy (DOE) stocks at Oak Ridge, Tennessee. The DLA accepts bids for prime virgin mercury on a daily basis, and for secondary mercury once a month. Inventories on December 31, 1992 were 4766 t of mercury in the NDS and 121 t of DOE mercury (14). The goal for both is zero.

Production and use of mercury is expected to continue to decrease, owing to mercury toxicity. Many of the traditional uses have either declined or have been eliminated (14).

Table 6. World Mercury Production by Country, 1978–1992, t[a]

Country	1978	1988	1989	1990	1991	1992
Algeria	1055	662	587	637	431	425
People's Republic of China	689	940	1200	1000	950	950
former Czechoslovakia	196	168	131	126	75	70
Mexico	76	345	651	735	720	700
Spain	1020	1716	1380	425	100	
Turkey	173	97	197	60	25	5
former USSR	2068	850	850	800	750	700[b]
United States	833	379	414	562	58	64
former Yugoslavia		70	51	37	30	30[c]
other	142	130	159	141	74	70
Total	*6252*	*5357*	*5620*	*4523*	*3213*	*3014*

[a]Ref. 14.
[b]Sum of mercury production in Russia, Ukraine, and Kyrgyzstan (USSR dissolved Dec. 1991).
[c]Yugoslavia dissolved Apr. 1992. Amount shown in representative of Slovenia.

Grades, Specifications, and Quality Control

The commercial-grade (99.9%) purity marketed as prime virgin-grade mercury has a clean bright appearance and contains less than 1 ppm of dissolved base metals. Prime virgin-grade mercury of lower purity is brought up to specification by filtration, redistillation, or electrolytic processing.

Triple-distilled mercury is of highest purity, commanding premium prices. It is produced from primary and secondary mercury by numerous methods, including mechanical filtering, chemical and air oxidation of impurities, drying (qv), electrolysis, and most commonly multiple distillation.

The purity of mercury can be estimated by its appearance. Because mercury has such a high specific gravity, almost all impurities, including amalgams, are lighter and float on the surface, causing the bright, mirror-like surface to become dull and black. The coating of oxides of base metal is called film or scum and a clean, bright appearance usually signifies an impurity content of <5 ppm (15,16). Other tests measure the residue remaining after the mercury is evaporated (17,18).

Analytical Methods

Ore.   The assay of mercury ores is not simple, owing to the difficulties encountered in obtaining representative ore samples. Crystalline cinnabar is extremely brittle causing it to break loose from adjacent rock and fall into the sample being collected. This uncontrollable salting of the sample can give results as much as several hundred percent over the actual mercury content of the sample.

The two assay methods preferred employ distillation–amalgamation or distillation–titration techniques. In the former, the weighted sample is heated with a flux such as iron [7439-89-6] filings to volatilize the mercury, which is then amalgamated with silver [7440-22-4] or gold [7440-57-5] foil. The mercury content is calculated from the change in foil weight. In the titration method, the sample is heated and the mercury volatilized and collected as metal, which is then dissolved in hot nitric acid. Potassium permanganate [7722-64-7] is added to oxidize the mercury and a peroxide to destroy excess permanganate. After ferric sulfate [10028-22-5], a nitrate indicator, is added to the solution, it is then titrated with standard potassium thiocyanate [333-20-0] solution to a faint pink end point. When a 0.5-g sample is titrated with a 1/400-N solution, each milliliter of titrating solution is equivalent to 0.41 kg/t of mercury.

Commercial Product.   The efficiency of furnace and condensation operations can be checked by spot-testing the condensed mercury for precious and base metal content, the most common contaminants from mining operations. Gold is determined by dissolving a 10-g sample of mercury in equal parts of distilled water and 15-N nitric acid. The residue is evaporated with aqua regia [8007-56-5]. If no precipitate results using sodium hydroxide and hydrogen peroxide [7722-84-1], no gold is present. Silver content is determined by spot-testing using trivalent manganese reagent. Base metals are determined by dissolving the sample in concentrated hydrochloric acid, volatilizing the mercuric chloride [7847-94-7], and analyzing the residue gravimetrically.

Ultrapure (triple distilled) mercury is commonly tested by evaporation or spectrographic analysis. In the former, a composite sample is evaporated and the residue weighed. In spectrographic analysis, a sample is dissolved and evaporated, the residue mixed with graphite [7782-42-5], and the emission spectrum determined with a spectrograph.

Trace Mercury. There are a number and variety of methods and instruments to determine trace quantities of both inorganic and organic mercury in natural or synthetic substances (19) (see also TRACE AND RESIDUE ANALYSIS). Literature describing numerous techniques and trace element analysis of a myriad of mercury-containing substances is available (20). Only the most commonly used methods are mentioned herein.

Atomic Absorption Spectroscopy. Mercury, separated from a measured sample, may be passed as vapor into a closed system between an ultraviolet lamp and a photocell detector or into the light path of an atomic absorption spectrometer. Ground-state atoms in the vapor attenuate the light decreasing the current output of the photocell in an amount proportional to the concentration of the mercury. The light absorption can be measured at 253.7 nm and compared to established calibrated standards (21). A mercury concentration of 0.1 ppb can be measured by atomic absorption.

Neutron Activation Analysis. A measured sample activated by neutron bombardment emits gamma rays that are used to determine the mercury content by proton-spectrum scanning. Mercury concentrations as low as 0.05 ppb have been determined by this method.

X-Ray Methods. In x-ray fluorescence the sample containing mercury is exposed to a high intensity x-ray beam which causes the mercury and other elements in the sample to emit characteristic x-rays. The intensity of the emitted beam is directly proportional to the elemental concentration in the sample (22). Mercury content below 1 ppm can be detected by this method. X-ray diffraction analysis is ordinarily used for the qualitative but not the quantitative determination of mercury.

Colorimetric Method. A finely powdered sample treated with sulfuric acid, hydrobromic acid [10035-10-6], and bromine [7726-95-6] gives a solution that when adjusted to pH 4 may be treated with dithizone [60-10-6] in *n*-hexane [110-54-3] to form mercuric dithizonate [14783-59-6] (20). The resultant amber-colored solution has a color intensity that can be compared against that of standard solutions to determine the mercury concentration of the sample. Concentrations below 0.02 ppm have been measured by this method.

Mercury in the Environment

Releases. Global releases of mercury into the environment may be either natural or anthropogenic. Natural mercury sources are mainly from volcanoes and volatilization from waters, soils, flora, and fauna (3); anthropogenic releases result principally from the production of metals and combustion of coal (qv), municipal refuge, sewage sludge, and wood (qv). Annual global releases of mercury into the atmosphere from natural sources have been estimated to be on the order of 2700–6000 t; annual anthropogenic releases of mercury into the atmosphere have been reported from 630–6800 t (23,24). Hence, natural and anthropogenic releases of mercury into the atmosphere are roughly equivalent.

Global anthropogenic releases of mercury into the biosphere are approximately 11,000 metric tons per year (24). The release of mercury into aquatic ecosystems results from electricity production, manufacturing processes, domestic wastewater and sewage sludge, mining, and fallout of atmospheric mercury (see METALLURGY, EXTRACTIVE; WASTES, INDUSTRIAL). The release of mercury into soils results from mining and smelting activities, coal fly ash and bottom ash, wood wastes, commercial wastes, sewage sludge, urban refuse, fallout of atmospheric mercury, and other sources (3).

Chlor–alkali production is the largest industrial source of mercury release in the United States (see ALKALI AND CHLORINE PRODUCTS). For the 1991 reporting year, chlor–alkali facilities accounted for almost 20% of the facilities that reported releases of mercury to the U.S. Environmental Protection Agency (EPA) for inclusion onto the Toxics Release Inventory (TRI) (25).

Regulations. In order to decrease the amount of anthropogenic release of mercury in the United States, the EPA has limited both use and disposal of mercury. In 1992, the EPA banned land disposal of high mercury content wastes generated from the electrolytic production of chlorine–caustic soda (14), accompanied by a one-year variance owing to a lack of available waste treatment facilities in the United States. A thermal treatment process meeting EPA standards for these wastes was developed by 1993. The use of mercury and mercury compounds as biocides in agricultural products and paints has also been banned by the EPA.

The State of New Jersey has passed a law restricting the sale and disposal of batteries (qv) containing mercury, requiring manufacturers to reduce the mercury content of each battery to 1 ppm by weight by 1995, and to establish a collection program for spent batteries (14). Another New Jersey law bans the sale of products having cadmium, mercury, or other toxic materials in the packaging (14) (see CADMIUM AND CADMIUM ALLOYS; CADMIUM COMPOUNDS; MERCURY COMPOUNDS).

California and Minnesota have placed restrictions on the disposal of fluorescent light tubes, which contain from 40–50 mg of mercury per tube, depending on size. After batteries, fluorescent lamps are the second largest contributor of mercury in solid waste streams in the United States (3,14). A California law classifies the disposal of 25 or more fluorescent lamp tubes as hazardous waste. In Minnesota, all waste lamps generated from commercial sources are considered hazardous waste. Private homes are, however, exempt from the law (14). Other states have proposed similar regulations. Several companies have developed technologies for recovering mercury from spent lamps (14).

A goal of reducing total mercury releases in the United States by 33% between 1988 to 1992, and 50% by 1995 was set by the EPA. The 1992 goal was more than achieved: United States reportable mercury releases were reduced by 39% by 1991 (26). In the United States, discards of mercury in municipal solid waste streams were approximately 643 t in 1989 (3). As a result of increased restrictions on the use and disposal of mercury, by the year 2000 mercury in municipal solid waste streams is expected to be about 160 t (3).

Health and Safety Factors

Exposure. The exposure of humans and animals to mercury from the general environment occurs mainly by inhalation and ingestion of terrestrial and aquatic food chain items. Fish generally rank the highest (10–300 ng/g) in food chain concentrations of mercury. Swordfish and pike may frequently exceed 1 μg/g (27). Most of the mercury in fish is methyl mercury [593-74-8]. Worldwide, the estimated average intake of total dietary mercury is 5–10 μg/d in Europe, Russia, and Canada, 20 μg/d in the United States, and 40–80 μg/d in Japan (27).

In occupational settings mercury exposure results predominantly from direct dermal contact or inhalation of mercury vapors (27). Chlor–alkali plants are one of the principal sources for occupational exposure to mercury (28,29). Mining and refining of mercury contribute to exposure, and the processing of cinnabar can result in high exposures to the skin and lungs, producing poisoning in a relatively short time. Exposure also occurs in the manufacture and use of mercury-containing instruments through breakage, spilling, or carelessness. Dentists and dental assistants can be exposed to significant quantities of mercury during the preparation and use of mercury amalgam (30) (see DENTAL MATERIALS). People who have mercury amalgam dental restorations may be exposed to 1–2 μg/d of mercury as a result of its dissolution in saliva. Exposure to mercury from amalgam restorations, however, is considered unlikely to pose a health risk to the patient (30).

Recommended safety measures which minimize occupational exposure to mercury include the use of efficient respirators, adequate ventilation and air-exhaust systems, employee warning signs and messages, training in accident emergency procedures, immediate and thorough cleanup of spills, airtight storage of mercury-containing wastes, and frequent monitoring of mercury levels in the work area. Additional precautions include coverall-type work clothes, maintenance of floors, work surfaces, and equipment such that crevices that might hold spilled mercury are not present, and adequate washing facilities and provision for showering at the end of the work period. Medical and dental examination at frequent intervals is also recommended.

Toxicity.   The toxic effects of mercury and mercury compounds (qv) are well known, and several detailed discussions on mercury toxicity are available (31–35). The propensity of mercury to accumulate in the food chain has resulted in many outbreaks of mercury toxicity in humans throughout the world. Elemental mercury is not particularly toxic when ingested, owing to very low levels of absorption from the gastrointestinal tract. Inhaled mercury vapor, however, undergoes complete absorption by the lung and is then oxidized to the divalent mercuric cation, Hg^{2+}, by catalase enzymes in erythrocytes (31). Within a few hours the deposition of inhaled mercury vapor resembles that which occurs after ingestion of mercuric salts, but with one important difference. Because mercury vapor crosses membranes much more rapidly than does divalent mercury, a significant amount of the vapor enters the brain before it is oxidized. Therefore, toxicity to the central nervous system is more prominent after exposure to mercury vapor than to divalent mercury (31–35).

Short-term exposure to mercury vapor may produce symptoms within several hours. These symptoms include weakness, chills, metallic taste, nausea, vomiting, diarrhea, labored breathing, cough, and a feeling of tightness in the chest. Pulmonary toxicity may progress to an interstitial inflammation of the lung with severe compromise of respiratory function. Recovery, although usually complete, may be complicated by residual interstitial growth of unnecessary fibrous tissue (31).

Chronic exposure to mercury vapor produces an insidious form of toxicity that is manifested by neurological effects and is referred to as the asthenic vegetative syndrome (31). The syndrome is characterized by tremor, psychological depression, irritability, excessive shyness, insomnia, emotional instability, forgetfulness, confusion, and vasomotor disturbances such as excessive perspiration and uncontrolled blushing (31,32). Common features of chronic mercury vapor intoxication are severe salivation and gingivitis (31). In fact, the triad of increased excitability, tremors, and gingivitis has been recognized historically as the primary manifestations of exposure to mercury vapor when mercuric nitrate [7783-34-8] was used in the fur, felt, and hat industries (31). Renal dysfunction has also been reported to result from long-term industrial exposure to mercury vapor (31–33).

The biochemical basis for the toxicity of mercury and mercury compounds results from its ability to form covalent bonds readily with sulfur. Prior to reaction with sulfur, however, the mercury must be metabolized to the divalent cation. When the sulfur is in the form of a sulfhydryl (—SH) group, divalent mercury replaces the hydrogen atom to form mercaptides, X—Hg—SR and $Hg(SR)_2$, where X is an electronegative radical and R is protein (36). Sulfhydryl compounds are called mercaptans because of their ability to capture mercury. Even in low concentrations divalent mercury is capable of inactivating sulfhydryl enzymes and thus causes interference with cellular metabolism and function (31–34). Mercury also combines with other ligands of physiological importance such as phosphoryl, carboxyl, amide, and amine groups. It is unclear whether these latter interactions contribute to its toxicity (31,36).

The affinity of mercury for thiols provides the basis for treatment of mercury poisoning using chelating agents (qv) such as dimercaprol [59-52-9], for high level exposures or symptomatic patients, or penicillamine [52-67-5], for low level exposures or asymptomatic patients (32). Dimercaprol (a dithiol) reacts more readily with mercury than do monothiols such as cysteine [52-90-4], but dimercaprol does not easily form a ring complex with mercury, because the bond angle of divalent mercury is 180° (36). Detailed discussions on the treatment of mercury poisoning are available (31,32). Removal of the patient from the source of mercury exposure and measurement of the concentration of mercury in the blood should be accomplished as quickly as possible.

Uses

Mercury consumption in the United States is summarized through 1992 in Table 7. Overall worldwide consumption of mercury declined in the 1980s and early 1990s. A detailed discussion of the uses and applications of mercury is available (3).

Electrolytic Preparation of Chlorine and Caustic Soda. The preparation of chlorine [7782-50-5] and caustic soda [1310-73-2] is an important use for mercury metal. Since 1989, chlor–alkali production has been responsible for the largest use for mercury in the United States. In this process, mercury is used as

Table 7. U.S. Mercury Consumption, 1988–1992, t[a,b]

Use	1988	1989	1990	1991	1992
batteries	448	250	106	18	16
chemical and allied products[c]	86	40	33	12	18
chlorine–caustic soda production	354	379	247	184	209
dental preparations	53	39	44	27	37
electrical applications[d]	207	172	103	54	124
general laboratory	26	18	32	16	18
industrial and control instruments	77	87	108	70	52
paints	197	192	22	6	[e]
other	55	32	25	165	148
Total	*1503*	*1212*	*720*	*554*	*621*

[a] Ref. 14.
[b] Use of mercury and mercury compounds for agricultural purposes was banned by the EPA in 1978, and use in paper and pulp is believed to have been discontinued.
[c] Includes uses such as catalysts, pharmaceuticals, and amalgamation.
[d] Includes electric lighting, wiring devices, and switches.
[e] By 1992, use of mercury and mercury compounds in paints was completely banned by the EPA.

a flowing cathode in an electrolytic cell into which a sodium chloride [7647-14-5] solution (brine) is introduced. This brine is then subjected to an electric current, and the aqueous solution of sodium chloride flows between the anode and the mercury, releasing chlorine gas at the anode. The sodium ions form an amalgam with the mercury cathode. Water is added to the amalgam to remove the sodium [7440-23-5], forming hydrogen [1333-74-0] and sodium hydroxide and relatively pure mercury metal, which is recycled into the cell (see ALKALI AND CHLORINE PRODUCTS).

Batteries.   Many batteries intended for household use contain mercury or mercury compounds. In the form of red mercuric oxide [21908-53-2], mercury is the cathode material in the mercury–cadmium, mercury–indium–bismuth, and mercury–zinc batteries. In all other mercury batteries, the mercury is amalgamated with the zinc [7440-66-6] anode to deter corrosion and inhibit hydrogen build-up that can cause cell rupture and fire. Discarded batteries represent a primary source of mercury for release into the environment. This industry has been under intense pressure to reduce the amounts of mercury in batteries. Although battery sales have increased greatly, the battery industry has announced that reduction in mercury content of batteries has been made and further reductions are expected (3). In fact, by 1992, the battery industry had lowered the mercury content of batteries to 0.025 wt % (3). Use of mercury in film pack batteries for instant cameras was reportedly discontinued in 1988 (3).

Electric Lighting.   Mercury is used in several types of electric lamps (light bulbs). Mercury-containing lamps include fluorescent lamps and high intensity discharge (HID) lamps such as mercury vapor, metal halide, and high pressure sodium lamps. Applications include street lighting, industrial and office lighting, floodlighting, photography, underwater lighting, insect lamps, and sun lamps

(3). Fluorescent mercury content varies with both bulb size and wattage. Lamp redesign and improvements in manufacturing process control have reduced mercury content by approximately 25%. The average mercury content in fluorescent and HID lamps as of 1992 was estimated to be 55 and 25 mg, respectively (3).

Electric Light Switches and Thermostats. Mercury electric light switches have been manufactured since the 1960s. Less than one million mercury switches are produced each year, and are in decline (3). Mercury switches may be used wherever lighting is used. The mercury is inside a metal encapsulation. Mercury thermostats have been used for many years. Since the early 1980s, however, mercury thermostats have begun to be replaced by programmable electronic thermostats, which do not contain mercury. The nonmercury, programmable devices are substantially higher in cost.

Dentistry. Mercury is used in dental amalgams for fillings in teeth (see DENTAL MATERIALS). Dental uses have accounted for 2–4% of total U.S. mercury consumption since 1980 and generally 3–6% before that time (3). Dental amalgams used to fill cavities in teeth are approximately 50% mercury by weight. Dental use of mercury can be expected to continue to decrease, in part because of more effective cavity prevention as well as development and increasing use of alternative dental materials such as plastics and ceramics, and increasing awareness of the environmental and health effects of mercury.

Industrial and Control Instruments. Mercury is used in many industrial and medical instruments to measure or control reactions and equipment functions, including thermometers, manometers (flow meters), barometers and other pressure-sensing devices, gauges, valves, seals, and navigational devices (see PRESSURE MEASUREMENTS; PROCESS CONTROL; TEMPERATURE MEASUREMENT). Whereas mercury fever thermometers are being replaced by digital thermometers, most fever thermometers in residential use are expected to continue to contain mercury because of the higher cost of digital thermometers and the relatively infrequent home use of fever thermometers.

Pigments. Mercury has a long history of use in the pigment industry (see PIGMENTS). Mercury sulfide [1344-48-5], in the form of cinnabar ore, has been used as a colorant since antiquity. Most of the mercury in pigments is used in plastics, often in combination with cadmium. There are strong pressures to reduce or eliminate all heavy metals from pigments, and the use of mercury in pigments has been declining steadily. Production of mercury-containing pigments in the United States was discontinued in 1988, but it is likely that some mercury pigments are imported into the United States (3). Production of mercury-containing pigments outside of the United States is also expected to decline (3).

Catalysts. Mercury is or has been used in the catalysis (qv) of various plastics, including polyurethane [26778-67-6], poly(vinyl chloride) [9002-86-2], and poly(vinyl acetate) [9003-20-7]. Most poly(vinyl chloride) and poly(vinyl acetate) is manufactured by processes that do not use mercury (3).

Explosives. Mercury, in the form of organic complexes, eg, mercury fulminate [628-86-4], has had long usage in explosives (see EXPLOSIVES AND PROPELLANTS). In the United States all mercury for use in explosives is diverted to military uses. An explosive based on mercuric 5-nitrotetrazole [60345-95-1] has been developed, but its use is on a small scale and in research and development only (3).

Special Paper Coating. Mercury bromide [10031-18-2] and mercury acetic acid [1600-27-7] are used in the coating of a specialized paper and film (see PAPER; MEDICAL IMAGING TECHNOLOGY). The coating, which also contains silver, is applied to paper that is used when scanning off of a cathode ray tube. A very high resolution is obtained from the process. This type of printing occurs in hospitals and newspaper publishing, and is utilized in microfiche printers. By 1995 mercury is expected to be eliminated entirely from this application (3).

Pharmaceuticals. A variety of mercury compounds have had pharmaceutical applications over the years, eg, mercury-containing diuretics and antiseptics. Whereas some mercury compounds remain available for use as antiseptics such as merbromin [129-16-8], mercuric oxide, and ammoniated mercury [10124-48-8], or as preservatives such as thimerosal [54-64-8] in drugs and cosmetics, most have been supplanted by more effective substances. A detailed discussion of mercury-containing antiseptics is available (37). Many hospitals use mercury metal to serve as weight for keeping nasogastric tubes in place within the stomach.

Discontinued Uses. *Agricultural Products.* Mercury and mercury compounds were at one time used extensively as seed disinfectants (see FUNGICIDES, AGRICULTURAL). In the United States, these uses were greatly restricted by the Federal Insecticide, Fungicide, and Rodenticide Act (FIFRA). In 1978, FIFRA was amended such that the only acceptable uses for mercury under FIFRA are for treatment of outdoor textiles, and to control brown mold on freshly sawn lumber, Dutch elm disease, and snow mold. Although allowable under FIFRA, it is doubtful whether mercury compounds are used for these purposes. No use of mercury or mercury compounds is permitted on food crops in the United States (3). Additional amendments to FIFRA have prohibited the importation of foodstuffs containing residues of banned pesticides. Also, the U.S. government in turn is required to notify foreign countries of health hazards before allowing the export of canceled or restricted pesticides (3).

Paints. For many years mercury-based biocides such as phenylmercuric acetate [62-38-4], 3-(chloromethoxy)propylmercuric acetate, di(phenylmercury) dodecenyl succinate [27236-65-3], and phenylmercuric oleate [104-60-9], were registered as biocides in interior and exterior paints, and in antifouling paints (see PAINT). In 1972 the use of mercury in anitfouling paint formulations was banned. As of July 1990, most registrations for mercury biocides used in interior and exterior paints and coatings (qv) were voluntarily canceled by the registrants (3). In 1991, EPA announced the voluntary cancelation of the remaining mercury biocide registrations (3). No mercury is allowed in the manufacture of paints in the United States.

Other Uses. Other uses of mercury that have been discontinued include coatings for mirrors, manufacture of certain types of glass, treatment of felt, and as a fungicide in paper.

This article has been reviewed by the Office of Pollution Prevention and Toxics, U.S. Environmental Protection Agency, and approved for publication. Approval does not signify that the contents necessarily reflect the views and policies of the Agency, nor does mention of commercial products constitute endorsement or recommendation for use.

BIBLIOGRAPHY

"Mercury" in *ECT* 1st ed., Vol. 8, pp. 808–882, by G. A. Roush, *Mineral Industry*; in *ECT* 2nd ed., Vol. 13, pp. 218–235, by G. T. Engel, U.S. Department of the Interior, Bureau of Mines; in *ECT* 3rd ed., Vol. 15, pp. 143–156, by H. J. Drake, U.S. Bureau of Mines.

1. H. M. Leicester, *The Historical Background of Chemistry*, Dover, Inc., New York, 1979, p. 43.
2. W. P. Iverson and F. E. Brinckman, in R. Mitchell, ed., *Water Pollution Microbiology*, Vol. 2, John Wiley & Sons, Inc., New York, 1978, pp. 201–232.
3. *Characterization of Products Containing Mercury in Municipal Solid Waste in the United States, 1970–2000*, OSW No. EPA530-R-92-013 (NTIS No. PB92-162 569), U.S. Environmental Protection Agency, Washington, D.C., 1992.
4. E. H. Bailey, A. L. Clark, and R. M. Smith, *U.S. Geol. Surv. Prof. Pap.* **820**, 401 (1973).
5. V. A. Cammarota, *U.S. Bur. Mines Bull.* **667**, 669 (1975).
6. E. H. Bailey, A. L. Clark, and R. M. Smith, *U.S. Geol. Surv. Prof. Pap.* **820**, 407 (1973).
7. C. L. Gordon and E. Wickers, *Ann. N.Y. Acad. Sci.* **65**, 382 (1957).
8. C. E. Wicks and R. E. Block, *U.S. Bur. Mines Bull.* **605**, (1962).
9. J. W. Pennington, *U.S. Bur. Mines Inform. Circ.* **7941**, 29 (1959).
10. "Mercury", in *U.S. Bureau of Mines Minerals Yearbook*, Vol. 1, Washington, D.C., 1964–1977.
11. J. N. Butler, *Studies in the Hydrometallurgy of Mercury Sulfide*, Nevada Bureau of Mines, Rept. No. 5.
12. B. J. Scheiner, D. L. Pool, and R. E. Lindstrom, *U.S. Bur. Mines Rept. Invest.* **7660** (1972).
13. E. S. Shedd, B. J. Scheiner, and R. E. Lindstrom, *U.S. Bur. Mines Rept. Invest.* **8083** (1975).
14. *Mercury in 1992, Annual Review*, Mineral Industry Surveys, U.S. Department of the Interior, Bureau of Mines, Washington, D.C., July 27, 1993.
15. E. Wichers, *Chem. Eng. News* **20**, 1111 (1942).
16. *Reagent Chemicals—American Chemical Society Specifications*, 6th ed., American Chemical Society, Washington, D.C., 1980, p. 368.
17. *The United States Pharmacopeia XX (USPXX-NFXV)*, The United States Pharmacopeial Convention, Inc., Rockville, Md., p. 1073.
18. *Am. Dental Assoc. J.*, 409 (Mar. 1932).
19. R. G. Smith, in R. Hartung and B. D. Dinman, eds., *Environmental Mercury Contamination*, Ann Arbor Science Publishers, Inc., Ann Arbor, Mich., 1972, pp. 97–136.
20. F. N. Ward, *U.S. Geol. Surv. Prof. Pap.* **713**, 46 (1970).
21. I. R. Jonasson, J. J. Lynch, and L. J. Trip, *Geol. Surv. Can. Pap.* **73-21**, 2 (1973).
22. H. H. Heady and K. G. Broadhead, *U.S. Bur. Mines Inf. Circ.* **8714r**, 4 (1977).
23. T. C. Hutchinson and K. M. Meema, eds., *Lead, Mercury, Cadmium, and Arsenic in the Environment*, John Wiley & Sons, Inc., New York, 1987, 360, pp.
24. J. O. Nriagu and J. M. Pacyna, *Nature* **333**, 134–139 (1988).
25. *1991 Toxics Release Inventory Public Data Release*, EPA-745-R-93-003, U.S. Environmental Protection Agency, Washington, D.C., 1993.
26. *EPA's 33/50 Program Fourth Progress Update*, EPA-745-R-93-005, U.S. Environmental Protection Agency, Washington, D.C., 1993.
27. A. M. Fan, in L. Fishbein, A. Furst, and M. A. Mehlman, eds., *Advances in Modern Environmental Toxicology*, Vol. XI. *Genotoxic and Carcinogenic Metals: Environmental and Occupational Occurrence and Exposure*, Princeton Scientific Publishing Co., Inc., Princeton, N.J., 1987, pp. 185–210.
28. L. Barregard, B. Hultberg, A. Schultz, G. Sallsten, *Occup. Environ. Health* **61**, 65–69 (1988).

29. A. Cardenas and co-workers, *Brit. J. Indust. Med.* **50**, 17–27 (1993).
30. Y. K. Fung and M. P. Molvar, *Clin. Toxicol.* **30**, 49–61 (1992).
31. C. D. Klaassen, in A. G. Gilman, T. W. Rall, A. S. Nies, and P. Taylor, eds., *Goodman and Gilman's The Pharmacological Basis of Therapeutics*, 8th ed., Pergamon Press, Inc., New York, 1990, pp. 1598–1602.
32. R. A. Goyer, in M. O. Amdur, J. Doull, and C. D. Klaassen, eds., *Casarett and Doull's Toxicology, The Basic Science of Poisons*, 4th ed., Pergamon Press, Inc., New York, 1991, pp. 646–651.
33. T. W. Clarkson, in A. S. Prasad, ed., *Essential and Toxic Trace Elements in Human Health and Disease*, Alan R. Liss, Inc., New York, 1988, pp. 631–643.
34. T. W. Clarkson, *Ann. Rev. Pharmacol. Toxicol.* **32**, 545–571 (1993).
35. V. Foa, *Drug Chem. Toxicol. (Neurotoxicol.)* **3**, 323–343 (1985).
36. S. C. Harvey, in L. S. Goodman and A. Gilman, eds., *The Pharmacological Basis of Therapeutics*, 4th ed., Macmillan Publishing Co., Inc., New York, 1970, pp. 974–977.
37. S. C. Harvey, in A. G. Gilman, L. S. Goodman, and A. Gilman, eds., *Goodman and Gilman's The Pharmacological Basis of Therapeutics*, 6th ed., Macmillan Publishing Co., Inc., New York, 1980, pp. 975–976.

General References

M. J. Ebner, "A Selected Bibliography on Quicksilver 1811–1953", *U.S. Geol. Surv. Bull.*, 1019A (1954).
An Exposure and Risk Assessment for Mercury, EPA-440/4-85-011, U.S. Environmental Protection Agency, Washington, D.C., 1981.
Mercury Health Effects Update, Health Issue Assessment, EPA-600/8-84-019F, U.S. Environmental Protection Agency, Washington, D.C., 1984.
Health Effects Assessment for Mercury, EPA-540/1-86-042, U.S. Environmental Protection Agency, Washington, D.C., 1984.
Ambient Water Quality Criteria for Mercury—1984, EPA-440/5-84-026 (NTIS No. PB85-227452), U.S. Environmental Protection Agency, Washington, D.C., 1985.
Review of National Emission Standards for Mercury, EPA-450/3-84-014b, U.S. Environmental Protection Agency, Washington, D.C., 1987.
Toxicological Profile for Mercury, ATSDR/TP-89/16, Agency for Toxic Substances and Disease Registry, U.S. Public Health Service, Atlanta, Ga., 1989.
For a comprehensive review of mercury in marine seafood, see S. Margolin, *Wld. Rev. Nutr. Diet.* **34**, 182–265 (1980).

STEPHEN C. DeVito
U.S. Environmental Protection Agency

MERCURY COMPOUNDS

Mercury salts exist in two oxidation states: mercurous, Hg^+, and mercuric, Hg^{2+}. The former exist as double salts; for example, mercurous chloride [*10112-91-1*] may be represented as Hg_2Cl_2 in both solution and the solid state, as shown by conductance studies and x-ray analysis. Standard oxidation electrode potentials at 298.15 K (1–3) are as follows:

$$Hg_2^{2+}(aq) + 2\,e^- = 2\,Hg^0 \qquad E_{298}^0 = 0.7960\ V$$

$$Hg^{2+}(aq) + 2\,e^- = Hg^0 \qquad E_{298}^0 = 0.8535\ V$$

$$2\,Hg^{2+}(aq) + 2\,e^- = Hg_2^{2+} \qquad E_{298}^0 = 0.9110\ V$$

Many mercury compounds are labile and easily decomposed by light, heat, and reducing agents. In the presence of organic compounds of weak reducing activity, such as amines (qv), aldehydes (qv), and ketones (qv), compounds of lower oxidation state and mercury metal are often formed. Only a few mercury compounds, eg, mercuric bromide [7789-47-1], mercurous chloride, mercuric sulfide [1344-48-5], and mercurous iodide [15385-57-6], are volatile and capable of purification by sublimation. This innate lack of stability in mercury compounds makes the recovery of mercury from various wastes that accumulate with the production of compounds of economic and commercial importance relatively easy (see RECYCLING).

The toxic nature of mercury and its compounds has caused concern over environmental pollution, and governmental agencies have imposed severe restrictions on release of mercury compounds to waterways and the air (see MERCURY). Methods of precipitation and agglomeration of mercurial wastes from process water have been developed. These methods generally depend on the formation of relatively insoluble compounds such as mercury sulfides, oxides, and thiocarbamates. Metallic mercury is invariably formed as a by-product. The use of coprecipitants, which adsorb mercury on their surfaces facilitating removal, is frequent.

Mercury from these accumulated wastes is generally best recovered by total degradation in stills, where metallic mercury is condensed and collected. The recovery costs are amply compensated by the value of the metal recovered. Moreover, disposal problems are either eliminated or severely diminished.

Concurrent with requirements for low levels of mercurials in discharge water is the problem of their determination. The older methods of wet chemistry are inadequate, and total reliance is placed on instrumental methods. The most popular is atomic absorption spectrophotometry, which relies on the absorption of light by mercury vapor (4). Solutions of mercury compounds not stabilized with an excess of acid tend to hydrolyze to form yellow-to-orange basic hydrates. These frequently absorb onto the walls of containers and may interfere with analytical results when low levels (ppm) of mercury are determined.

The covalent character of mercury compounds and the corresponding ability to complex with various organic compounds explains the unusually wide solubility characteristics. Mercury compounds are soluble in alcohols, ethyl ether, benzene, and other organic solvents. Moreover, small amounts of chemicals such as amines, ammonia (qv), and ammonium acetate can have a profound solubilizing effect (see COORDINATION COMPOUNDS). The solubility of mercury and a wide variety of mercury salts and complexes in water and aqueous electrolyte solutions has been well outlined (5).

Owing to legal restrictions, use of mercury compounds has declined. The most important areas of mercury compound usage as of the mid-1990s are as preservatives and fungicides in coating compositions, as catalysts, as intermediates in the formation of other compounds, and as a component of compositions used as semiconductors (see COATINGS; CATALYSIS; FUNGICIDES, AGRICULTURAL; SEMICONDUCTORS). The use of mercurials as biocides in protective coatings has been prohibited in the United States since 1992. Seed treatment by mercurials is no longer permitted in the United States. As of this writing, however, bis-(2-methoxyethyl mercuric) silicate is used for this purpose in parts of Europe. Some pharmaceutical uses remain, eg, in ophthalmic preparations and antiseptics (see DISINFECTANTS AND ANTISEPTICS).

All mercury compounds should be stored in amber bottles or otherwise protected from light. In manufacture, glass-lined equipment is preferred, although stainless steel may be used. Stainless steel may cause some discoloration at high temperatures if concentrated acetic acid is used.

Mercury Salts

Mercuric Acetate.   Mercuric acetate [1600-27-7], $Hg(C_2H_3O_2)_2$, is a white, water-soluble, crystalline powder, soluble in water and many organic solvents. It is prepared by dissolving mercuric oxide in warm 20% acetic acid. A slight excess of acetic acid is helpful in reducing hydrolysis.

Another method of preparing mercuric acetate is the oxidation of mercury metal using peracetic acid dissolved in acetic acid. Careful control of the temperature is extremely important because the reaction is quite exothermic. A preferred procedure is the addition of approximately half to two-thirds of the required total of peracetic acid solution to a dispersion of mercury metal in acetic acid to obtain the mercurous salt, followed by addition of the remainder of the peracetic acid to form the mercuric salt. The exothermic reaction is carried to completion by heating slowly and cautiously to reflux. This also serves to decompose excess peracid. It is possible and perhaps more economical to use 50% hydrogen peroxide instead of peracetic acid, but the reaction does not go quite as smoothly.

The primary use of mercuric acetate is as a starting material for the manufacture of organic mercury compounds.

Mercuric Carbonate.   Basic mercuric carbonate [76963-38-7], $HgCO_3 \cdot 3HgO$, may be prepared by the addition of sodium carbonate to a solution of mercuric chloride. The brown precipitate, which lacks usefulness, is generally not isolated; rather, the slurry is refluxed, whereupon the carbonate decomposes to red mercuric oxide.

Mercuric Cyanides.   Mercuric cyanide [592-04-1], $Hg(CN)_2$, is a white tetragonal crystalline compound, little used except to a small degree as an antiseptic. It is prepared by reaction of an aqueous slurry of yellow mercuric oxide (the red is less reactive) with excess hydrogen cyanide. The mixture is heated to 95°C, filtered, crystallized, isolated, and dried. Its solubility in water is 10% at 25°C.

Mercuric oxycyanide [1335-31-5], or basic mercuric cyanide, $Hg(CN)_2 \cdot HgO$, is prepared in the same manner as the normal cyanide, except that the mercuric oxide is present in excess. The oxycyanide is white and crystalline but only one-tenth as soluble in water as the normal cyanide. Because this compound is explosive, it normally is supplied as a 1:2 mixture of oxycyanide to cyanide.

Mercuric Fulminate.   Mercuric fulminate [20820-45-5], $Hg(ONC)_2$, is used as a catalyst in the oxynitration of benzene to nitrophenol (see NITRATION). Its most common use is as a detonator for explosives (see EXPLOSIVES AND PROPELLANTS).

Mercury Fluorides.   See FLUORINE COMPOUNDS, INORGANIC–MERCURY.

Mercurous Chloride.   Mercurous chloride [10112-91-1], Hg_2Cl_2, also known as calomel, is a white powder, insoluble in water. It sublimes when heated in an open container, but this probably occurs at least in part as a result of dissociation to mercury metal and mercuric chloride:

$$Hg_2Cl_2 \longrightarrow Hg^0 + HgCl_2$$

Its relatively low toxicity is probably the result of very low aqueous solubility (0.002 g/L) and lack of reactivity with acidic (HCl) digestive fluids.

The compound is generally prepared in one of two ways. The purer grade is made by the direct oxidation of mercury by a quantity of chlorine gas insufficient to produce mercuric chloride, which is always a side product. The chlorine gas is run into a heated silica retort. The mouth of the retort leads to a large chamber made of chlorine-resistant material. Lead is frequently used, but great care must be exercised to exclude moisture; otherwise lead chloride could be formed as well. The chamber should be large, approximately 5.1 m^3 (180 ft^3), because cooling occurs by convection and dissipation of heat by conduction throughout the walls. The mercury burns with a green flame and the product settles to the floor of the chamber from which it is subsequently removed. The correct balance of mercury to chlorine yields about 70–80% mercurous chloride. The remainder is mercuric chloride. No unreacted mercury should remain, as mercury metal gives a grayish tinge to the product and cannot be washed out. The material from the chamber is slurried with water and washed several times by decantation. It is filtered and washed on the filter until it is free of soluble chloride.

The second method of preparation involves precipitation from a cold acidic solution of mercurous nitrate. Mercurous chloride is isolated after washing in a manner similar to the chamber method described. This product, which generally contains small amounts of occluded sodium nitrate, is satisfactory as a technical-grade material. Difficulty may be encountered in having it pass NF or reagent-grade specifications (see FINE CHEMICALS).

For the preparation of mixtures of mercurous and mercuric chlorides used to control turf-fungus diseases, the precipitated product of the second method may be mixed with the required amount of mercuric chloride. Alternatively, the chamber material, if the ratios of mercurous and mercuric chloride are correct, may be used directly.

The mercury contained in the mother liquid and washings of either method is recovered by treatment with sodium hydroxide solution. Yellow mercuric oxide is precipitated and filtered. The filtrate is treated further to remove the last traces of mercury before it is discarded.

Mercuric Chloride. Mercuric chloride [7487-94-7], $HgCl_2$, is also known as corrosive sublimate of mercury or mercury bichloride. It is extremely poisonous, and is particularly dangerous because of high (7 g/L at 25°C) water solubility and high vapor pressure. It sublimes without decomposition at 300°C, and has a vapor pressure of 13 Pa (0.1 mm Hg) at 100°C, and 400 Pa (3 mm Hg) at 150°C. The vapor density is high (9.8 g/cm^3), and therefore mercuric chloride vapor dissipates slowly (5).

In addition to high aqueous solubility (7% at 30°C and 38% at 100°C), $HgCl_2$ is very soluble in methyl alcohol (53% at 36°C), ethyl alcohol (34% at 31°C), and amyl alcohol (ca 10% at 30°C). It also is soluble in acetone, formic acid, the lower acetate esters, and other polar organic solvents.

The preparation of mercuric chloride is identical to the chamber method for mercurous chloride, except that an excess of chlorine is used to ensure complete reaction to the higher oxidation state. Very pure product results from this method. Excess chlorine is absorbed by sodium hydroxide in a tower.

Mercuric chloride is widely used for the preparation of red and yellow mercuric oxide, ammoniated mercury [10124-48-8] USP, mercuric iodide, and as an intermediate in organic synthesis. It has been used as a component of agricultural fungicides. It is used in conjunction with sodium chloride in photography (qv) and in batteries (qv), and has some medicinal uses as an antiseptic.

Until about 1980, mercuric chloride was used extensively as a catalyst for the preparation of vinyl chloride from acetylene (7). Since the early 1980s, vinyl chloride and vinyl acetate have been prepared from ethylene instead of acetylene, and the use of mercuric chloride as a catalyst has practically disappeared.

Mercurous Bromide. Mercurous bromide [15385-58-7], Hg_2Br_2, is a white tetragonal crystalline powder, very similar to the chloride, and prepared in much the same way, ie, by the direct oxidation of mercury by bromine or by precipitation from mercurous nitrate by sodium bromide. It is sensitive to light, less stable than the chloride, and is not of appreciable commercial importance.

Mercuric Bromide. Mercuric bromide [7789-94-7], $HgBr_2$ is a white crystalline powder, considerably less stable than the chloride, and also much less soluble in water (0.6% at 25°C). Therefore, it is prepared easily by precipitation, using mercuric nitrate and sodium bromide solution. Drying of the washed compound is carried out below 75°C. Mercuric bromide has a few medicinal uses.

Mercurous Iodide. Mercurous iodide [7783-30-4], Hg_2I_2, is a bright yellow amorphous powder, extremely insoluble in water and very sensitive to light. It has no commercial importance but may be prepared by precipitation, using mercurous nitrate and potassium iodide. Care must be taken to exclude mercuric nitrate, which may cause the formulation of the water-insoluble mercuric iodide.

Mercuric Iodide. Mercuric iodide [7774-29-0], HgI_2, is a bright red tetragonal powder, only slightly soluble in water. It dissolves in alkalies to form complex salts. Both sodium iodomercurate [7784-03-4], Na_2HgI_4, and potassium iodomercurate [7783-33-7], K_2HgI_4, are known. Mercuric iodide, made by precipitation from a solution of mercuric chloride and potassium iodide, is used in the treatment of skin diseases and as an analytical reagent.

The range of uses of mercuric iodide has increased because of its ability to detect nuclear particles. Various metals such as Pd, Cu, Al, Tri, Sn, Ag, and Ta affect the photoluminescence of HgI_2, which is of importance in the preparation of high quality photodetectors (qv). HgI_2 has also been mentioned as a catalyst in group transfer polymerization of methacrylates or acrylates (8).

Complex Halides. Mercuric halides (except the fluoride) form neutral complex salts with metallic halides. Those made with alkali metal salts frequently are more soluble in water than the mercuric halide itself, and take the form of $MHgX_3$ and M_2HgX_4.

Iodomercurates. Potassium iodomercurate dihydrate, $K_2HgI_4 \cdot 2H_2O$, is a yellow, water-and-alcohol-soluble compound prepared by dissolving one mole of mercuric iodide in a solution of two moles of potassium iodide in distilled water. After filtering off any insoluble particles, the filtrate is evaporated to dryness to isolate the compound. The filtrate, known as Mayer's reagent, has uses as an antiseptic and as a precipitant for alkaloids (qv). In strongly alkaline solution, called Nessler's reagent, it is used for the detection and determination of low levels of ammonia.

Cuprous iodomercurate [13876-85-2], Cu_2HgI_4, is a bright red water-insoluble compound prepared by precipitation from a solution of K_2HgI_4 with cuprous chloride. It is used in temperature-indicating paints because it reversibly changes color to brown at 70°C (see CHROMOGENIC MATERIALS).

Silver iodomercurate [36011-71-9], Ag_2HgI_4, is a bright yellow compound and is prepared similarly to the cuprous salt where silver nitrate is the precipitant. The silver salt, which darkens reversibly at 50°C, is used for the same application as cuprous iodomercurate.

Mercurous Nitrate. Mercurous nitrate [10415-75-5], $Hg_2N_2O_6$ or $Hg_2(NO_3)_2$, is a white monoclinic crystalline compound that is not very soluble in water but hydrolyzes to form a basic, yellow hydrate. This material is, however, soluble in cold, dilute nitric acid, and a solution is used as starting material for other water-insoluble mercurous salts. Mercurous nitrate is difficult to obtain in the pure state directly because some mercuric nitrate formation is almost unavoidable. When mercury is dissolved in hot dilute nitric acid, technical mercurous nitrate crystallizes on cooling. The use of excess mercury is helpful in reducing mercuric content, but an additional separation step is necessary. More concentrated nitric acid solutions should be avoided because these oxidize the mercurous to mercuric salt. Reagent-grade material is obtained by recrystallization from dilute nitric acid in the presence of excess mercury.

Mercuric Nitrate. Mercuric nitrate [10045-94-0], $Hg(NO_3)_2$, is a colorless deliquescent crystalline compound prepared by the exothermic dissolution of mercury in hot, concentrated nitric acid. The reaction is complete when a cloud of mercurous chloride is not formed when the solution is treated with sodium chloride solution. The product crystallizes upon cooling. Mercuric nitrate is used in organic synthesis as the starting material and for the formulation of a great many other mercuric products.

Mercuric Oxide. Mercuric oxide [21908-53-2], HgO, is a red or yellow water-insoluble powder, rhombic in shape when viewed microscopically. The color and shade depend on particle size. The finer particles (<5 μm) appear yellow; the coarser particles (>8 μm) appear redder. The product is soluble in most acids, organic and inorganic, but the yellow form, which has greater surface area, is more reactive and dissolves more readily. Mercuric oxide decomposes at 332°C and has a high (11.1) specific gravity.

Yellow mercuric oxide may be obtained by precipitation from solutions of practically any water-soluble mercuric salt through the addition of alkali. The most economical are mercuric chloride or nitrate. Although yellow HgO has some medicinal value in ointments and other such preparations, the primary use is as a raw material for other mercury compounds, eg, Millon's base [12529-66-7], Hg_2NOH, which is formed by the reaction of aqueous ammonia and yellow mercuric oxide.

Red mercuric oxide generally is prepared in one of two ways: by the heat-induced decomposition of mercuric nitrate or by hot precipitation. Both methods require careful control of reaction conditions. In the calcination method, mercury and an equivalent of hot, concentrated nitric acid react to form mercurous nitrate:

$$6 \; Hg^0 + 8 \; HNO_3 \longrightarrow 3 \; Hg_2(NO_3)_2 + 2 \; NO + 4 \; H_2O$$

After the water and nitrogen oxide are driven off, continued heating drives off vapors of nitric acid, additional water, NO_2, and some mercury–metal vapor:

$$Hg(NO_3)_2 \longrightarrow 2 \; HgO + 2 \; NO_2$$

This secondary reaction starts at about 180°C, but the mass must be heated to 350–400°C to bring the reaction to completion and produce a nitrate-free product. The off-gases are extremely corrosive and poisonous, and considerable attention and expense is required for equipment maintenance and caustic-wash absorption towers. Treatment of the alkaline wash liquor for removal of mercury is required both for economic reasons and to comply with governmental regulations pertaining to mercury in plant effluents.

In the hot precipitation method, sodium carbonate solution is added slowly to a refluxing solution of mercuric chloride, followed by an additional reflux period of 1 to 2 h. The washed precipitate is then dried. A variation allows the substitution of mercuric nitrate for the chloride if substantial quantities of sodium chloride are used. Sodium hydroxide, in the presence of sodium carbonate, is the precipitant.

Red mercuric oxide, identical chemically to the yellow form, is somewhat less reactive and more expensive to produce. An important use is in the Ruben-Mallory dry cell, where it is mixed with graphite to act as a depolarizer (see BATTERIES). The overall cell reaction is as follows:

$$Zn^0 + HgO \longrightarrow ZnO + Hg^0$$

Yellow mercuric oxide is considered less suitable because it is less dense and would not permit adequate packing in the cell casing.

Mercurous Sulfate. Mercurous sulfate [7783-36-0], Hg_2SO_4, is a colorless-to-slightly-yellowish compound, sensitive to light and slightly soluble in water (0.05 g/100 g H_2O). It is more soluble in dilute acids. The compound is prepared by precipitation from acidified mercurous nitrate solution and dilute sulfuric acid. The precipitate is washed with dilute sulfuric acid until nitrate-free. Its most important use is as a component of Clark and Weston types of standard cells.

Mercuric Sulfate. Mercuric sulfate [7783-35-9], $HgSO_4$, is a colorless compound soluble in acidic solutions, but decomposed by water to form the yellow water-insoluble basic sulfate, $HgSO_4 \cdot 2HgO$. Mercuric sulfate is prepared by reaction of a freshly prepared and washed wet filter cake of yellow mercuric oxide with sulfuric acid in glass or glass-lined vessels. The product is used as a catalyst and with sodium chloride as an extractant of gold and silver from roasted pyrites.

Mercuric Sulfide. Mercuric sulfide [1344-48-5], HgS, exists in two stable forms. The black cubic tetrahedral form is obtained when soluble mercuric salts and sulfides are mixed; the red hexagonal form is found in nature as cinnabar (vermilion pigment). Both forms are very insoluble in water (see PIGMENTS, INORGANIC). Red mercuric sulfide is made by heating the black sulfide in a concentrated solution of alkali polysulfide. The exact shade of the pigment varies with concentration, temperature, and time of reaction.

Mercury Telluride. Compounds of mercury with tellurium have gained importance as semiconductors with applications in infrared detection (9) and solar cells (10). The ratio of the components is varied, and other elements such as cadmium, zinc, and indium are added to modify the electronic characteristics.

Organomercury Compounds

Phenylmercurics. Phenylmercuric acetate [62-38-4] (PMA), $HgC_8H_8O_2$, melts at 149°C and is slightly soluble in water, but much more soluble in solutions of ammonium acetate in aqueous ammonia. Such solutions are articles of commerce, and may contain 30% phenylmercuric acetate (PMA-30) or the equivalent of 18% mercury as metal. Phenylmercuric acetate is also soluble in various organic solvents. The compound is prepared by refluxing a mixture of mercuric acetate and acetic acid in a large excess of benzene. This is generally referred to as a mercuration reaction. A large excess of benzene is necessary because more than one hydrogen on the benzene ring can be replaced:

$$C_6H_6 + n\,[Hg(OOCCH_3)_2] \longrightarrow C_6H_{6-n}(HgOOCCH_3)_n + n\,HOOCCH_3$$

where n may be anywhere from 1–4. It is generally desirable to limit the amount of polymercurated benzene formed. The technical grade of phenylmercuric acetate contains about 85% pure compound. The remaining 15% is di- and trimer-curated product. Polymercurated products are less soluble than monomercurated material and are removed by recrystallization. Solvents such as water, acetone, benzene, and benzene ethylene glycol monoethyl ether may be used.

The reaction is complete in about 15 hours, as indicated by the formation of a white precipitate of phenylmercuric sulfide [20333-30-6] when sulfide is added to an ammoniacal solution of the reaction mixture. The product is isolated after distillation of excess benzene and acetic acid.

The ammoniacal solution of phenylmercuric acetate contains polymercurates. The low solubility of polymercurates is an advantage in exterior coatings, where bactericidal and fungicidal activity is unimpaired and these materials are leached out more slowly.

Prior to the 1990s phenylmercuric acetate was the primary bactericide and fungicide in latex and waterborne paints. Because of the increasing concerns of mercury toxicity and the potential for high consumer and occupational exposures to mercury when present in paints, the U.S. Environmental Protection Agency (EPA) induced U.S. manufacturers of PMA and other mercury compounds to withdraw their registrations for use of these substances as biocides in paints (see MERCURY). Mercury compounds are used only for very limited, specific purposes, such as the use of phenylmercuric nitrate [55-68-5] as a bactericide in cosmetic eye preparations (see COSMETICS).

Phenylmercuric acetate is used as the starting material in the preparation of many other phenylmercury compounds, which are generally prepared by double-decomposition reactions using the sodium salts of the desired acid groups in aqueous solution. The lower alkylate esters, such as the propionate and butyrate, are prepared directly in the same manner as the acetate. Another double-decomposition method uses phenylmercuric hydroxide [100-57-2], prepared by reaction of phenylmercuric acetate and hot dilute sodium hydroxide. Other phenylmercury compounds in use are the oleate [104-60-9], dodecenyl succinate, propionate [103-27-5], nitrate, and dimethyldithiocarbamate [32407-99-1]. All of these are toxic. The N,N'-dimethyldithiocarbamate is the least soluble and exhibits the highest tolerated levels in humans. The phenylmercury compounds are less toxic than soluble inorganic mercury, perhaps because the phenylmercury forms the highly insoluble phenylmercuric chloride in the stomach. Phenylmercury compounds serve as catalysts for the manufacture of certain polyurethanes (see URETHANE POLYMERS).

3-Chloro-2-Methoxypropylmercuric Acetate. 3-Chloro-2-methoxypropylmercuric acetate, $ClCH_2C(OCH_3)HCH_2HgOOCCH_3$, is difficult to isolate and generally is sold as an ammoniacal solution containing 10% mercury as metal, in much the same way as phenylmercuric acetate solution is sold. It is prepared by the reaction of allyl chloride, methanol, and mercuric acetate in acetic acid, followed by the addition of ammonia and water. It has many of the same applications as phenylmercuric acetate and was used as a preservative or bactericide for aqueous systems. Because of its superior solubility and compatibility, the compound is not precipitated by anionic dispersants, as is phenylmercuric acetate, and therefore lower levels of this compound often may be used to achieve the same protective biocidal effect.

Alkyl Mercuric Compounds. Alkyl mercuric compounds, RHgX, are no longer manufactured in most of the world because of the long-lasting toxic hazards and destructive effect on the brain and central nervous system of animals, where these tend to accumulate. Until 1970, they were, however, widely used as seed disinfectants. They have some utility in organic synthesis and in the preparation of other organometallics (qv). In general, they are white stable solids of appreciable volatility and are often prepared by a Grignard reaction in ethyl ether:

$$RMgX + HgX_2' \longrightarrow RHgX' + MgXX'$$

Miscellaneous Compounds of Pharmaceutical Interest

Antiseptics. Ammoniated mercury [10124-48-8], $Hg(NH_2)Cl$, is a white odorless powder that is insoluble in water and has a specific gravity of 5.38. It is formed by the reaction of aqueous ammonia (qv) and mercuric chloride.

o-(Chloromercuri)phenol [90-03-9] (mercarbolide) (1) is prepared in the same way as PMA, but from phenol instead of benzene. The phenol group is highly activating and thus the reaction proceeds quickly. The product is precipitated by aqueous sodium chloride and purified from hot water as a white, leafy crystalline compound.

Merbromin [129-16-8], disodium 2,7-dibromo-4-hydroxymercurifluorescein, (2), commonly called mercurochrome, is prepared by refluxing dibromo-fluorescein with mercuric acetate in acetic acid. The precipitate is dissolved in water containing the stoichiometric amount of sodium hydroxide and evaporated.

Merthiolate [54-64-8] (3), sodium ethylmercurithiosalicylate, known also as thimersol, is prepared from a 1:1 ratio of ethylmercuric chloride [107-27-7] and disodium thiosalicylate in ethanol. After removal of the sodium chloride by filtration, the free acid is pecipitated by acidification with dilute sulfuric acid. Purification is achieved by recrystallization from 95% ethanol, and the product, merthiolate, is obtained by neutralization with a stoichiometric amount of sodium hydroxide.

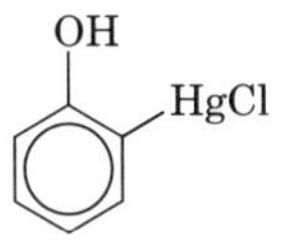

(1) 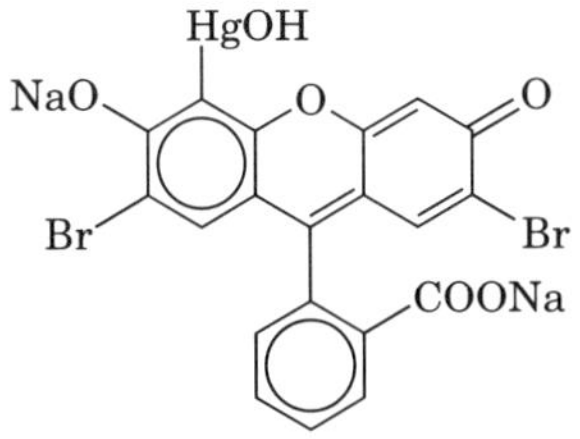   (2) 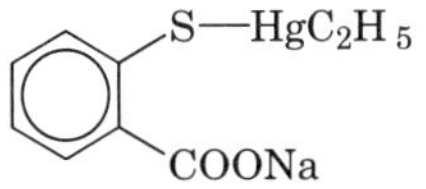   (3)

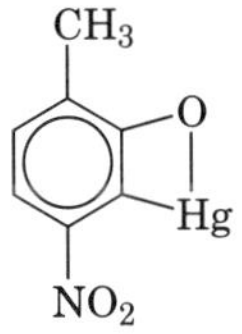

(4) 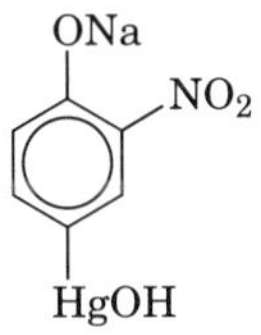   (5)

Nitromersol [133-58-4] (**44**) and mercurophen [52486-78-9] (**5**) are prepared by the same mercuration reaction as phenylmercuric acetate, only 4-nitro-o-cresol and o-nitrophenol are used, respectively, instead of benzene. The second step is reaction with sodium hydroxide to form the anhydride or sodium salt, respectively.

Other organic mercurials used as antiseptics include mercocresol [8063-33-0], acetomeroctol [584-18-9], acetoxymercuri-2-ethylhexylphenosulfonate [1301-13-9], and sodium 2,4-dihydroxy-3,5-dihydroxymercuribenzophenone-2-sulfonate [6060-47-5] (see DISINFECTANTS AND ANTISEPTICS).

Antisyphilitics.   Mercuric salicylate [5970-32-1] (**6**) and mercuric succinimide [584-43-0] (**7**) are simple salts prepared by the reaction in water of mercuric oxide and salicylic acid or succinimide, respectively. Use as antisyphilitics has been substantially eliminated by virtue of the discovery of more potent and effective nonmetallic biocides.

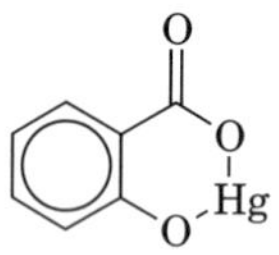

(**6**) 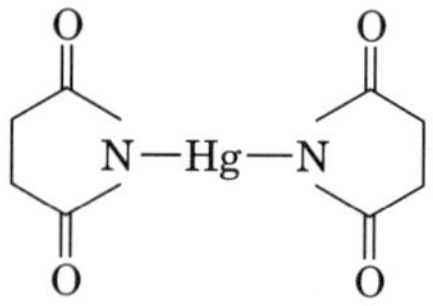   (**7**)

Diuretics.   Chlormeodrin [62-37-3] (methoxy(urea)propylmercuric chloride) (**8**), is prepared in the same sort of reaction used for chloromethoxypropylmercuric acetate. Allyl urea is used instead of allyl chloride, together with methanol and mercuric acetate. The product, after dilution with water and neutralization, is precipitated with sodium chloride:

$$H_2NCNHCH_2CH{=}CH_2 \ + \ CH_3OH \ + \ Hg(OC{+}CH_3)_2 \ \longrightarrow$$

$$H_2NCNHCH_2CHCH_2HgO\,CCH_3 \ + \ CH_3COH \ \xrightarrow{NaCl} \ H_2NCNHCH_2CHCH_2HgCl$$

(**8**)

Other organic mercurials similar in chemical structure to chlormerodrin are meralluride [104-20-5], mercaptomerin [20223-84-1], and mersalyl [486-67-9]. Mercury-based diuretics (qv) are no longer in use.

Health and Safety Factors

Upon discovery of the biomethylation of mercury in 1968 (11,12), the mercurials industry, involving both manufacture and application of mercury and mercury compounds, underwent an extensive change. This change resulted in part from the rapid development of powerful analytical tools capable of detection, identification, and analysis of compounds and elements in the ranges of fractional parts per billion (ppb). Coincidentally, certain chemical moieties were shown to be not only toxic sources, but also potential carcinogens and mutagens (see ANALYTICAL METHODS; TRACE AND RESIDUE ANALYSIS; TOXICOLOGY). Realization that mercury compounds were being accumulated in the environment by biodegradation to methylmercury [16056-34-1], CH_3Hg, led to the development and enforcement of rules and regulations concerning the safe preparation, use, and disposal of mercury compounds.

The toxic effects of mercury and mercury compounds as well as their medicinal properties have been known for many centuries. In the first century AD, Pliny indicated the use of mercuric sulfide (cinnabar or vermilion) in medicine and in cosmetics. This compound was probably known to the Greeks in the time of Aristotle (13).

Galen, a physician whose views outlived him by about a thousand years, died about 200 AD. He believed that mercurials were toxic, and did not use any mercury compound therapeutically. However, as a result of Arabian influence, the therapeutic uses of mercury were slowly recognized by Western Europe. In the thirteenth century mercury ointments were prescribed for treating chronic diseases of the skin. Mercury and its compounds, such as mercurous chloride, mercuric oxide, mercuric chloride, and mercuric sulfide, were used widely from the fifteenth to the nineteenth centuries, and to some extent in the twentieth century. During the first half of the twentieth century, the primary therapeutic uses of mercury included bactericidal preparations, such as mercuric chloride, mercuric oxycyanide, and mercuric oxide; and diuretics, such as aryl HgX (Novasural) and mercurated allyl derivatives (14).

Alkyl mercury compounds were used widely in the United States as seed disinfectants until prohibited in 1970. Subsequently, in 1972, the EPA prohibited the use of all mercury compounds in agriculture (15).

Toxicity. Inorganic mercury compounds, aryl mercury compounds, and alkoxy mercurials are generally considered to be quite similar in their toxicity. Alkyl mercury compounds are considered to be substantially more toxic and hazardous. Mercury and its compounds can be absorbed by ingestion, absorption through the skin, or by inhalation of the vapor. The metal itself, however, rarely produces any harmful effects when ingested (16).

After inorganic mercuric salts are absorbed and dissociated into the body fluids and in the blood, they are distributed between the plasma and erythrocytes. Aryl mercuric compounds and alkoxy mercuric compounds are decomposed to mercuric ions, which behave similarly.

Alkyl mercury compounds in the blood stream are found mainly in the blood cells, and only to a small extent in the plasma. This is probably the result of the greater stability of the alkyl mercuric compounds, as well as their peculiar solubility characteristics. Alkyl mercury compounds affect the central nervous system and accumulate in the brain (17,18). Elimination of alkyl mercury compounds from the body is somewhat slower than that of inorganic mercury compounds and the aryl and alkoxy mercurials. Methylmercury is eliminated from humans at a rate indicating a half-life of 50–60 d (19); inorganic mercurials leave the body according to a half-life pattern of 30–60 d (20). Elimination rates are dependent not only on the nature of the compound but also on the dosage, method of intake, and the rate of intake (21,22).

Environmental Factors.   The control, recovery, and disposal of mercury-bearing waste products are as important to the mercurials industry as the manufacturing process. The difficulties involved in removing mercury from waste-product streams and the problems of recovery or disposal have resulted in a substantial reduction in the number of manufacturers of mercury compounds as well as in the variety of mercury compounds being manufactured. Moreover, the manufacturing process used for a mercury compound may not necessarily be the most efficient or economical. Rather, the choice may depend on the nature of the by-products, the toxic hazard of the process, and the ease of recovery of the mercury from the waste-product stream.

Safety.   The maximum acceptable concentration (MAC) for mercury in all forms except alkyl compounds is 0.05 mg Hg/m^3 air (23). For alkyl mercury compounds the TLV is set at 0.01 mg Hg/m^3 air.

Suitable ventilating equipment, consisting mainly of carbon absorbers which effectively absorb mercury vapor from recirculated air, must be employed to maintain standards below the value permitted in the occupational environment. When the possibility of higher exposures exists, small disposable masks utilizing a mercury vapor absorbent may be employed.

Most inorganic mercury compounds have very low vapor pressures, and generally do not contribute to high mercury vapor readings. Metallic mercury is the most potent and troublesome in this respect. Organic mercurials also contribute to mercury vapor readings, possibly by virtue of the presence of extremely small amounts of metallic mercury present as an impurity.

To safeguard the health of persons working in plants producing mercurials, the following precautions should be observed: maintaining adequate ventilation; use of disposable uniforms so that a contaminated uniform is not a source of absorption through the skin; use of disposable mercury vapor-absorbing masks; careful attention to good housekeeping, eg, avoidance of spills, and prompt and proper cleaning if a spill occurs; ensuring that all containers of mercury and its compounds are kept tightly closed; ensuring that floors are washed on a regular basis with dilute calcium sulfide solution or other suitable reactant; use of floors that are nonporous; requiring that all workers directly involved in the plant operation shower thoroughly each day before leaving; and conducting periodic medical examinations, including analysis of blood and urine for mercury content, of all workers directly involved in production of mercurials, or otherwise exposed to contact with mercury compounds or mercury vapor.

Mercury spills should be cleaned up immediately by use of a special vacuum cleaner. The area should then be washed with a dilute calcium sulfide solution. Small quantities of mercury can be picked up by mixing with copper metal granules or powder, or with zinc granules or powder. To avoid or minimize spills, some plants use steel trays as pallets so that a spill, whether of mercury or a mercury compound, is contained on the steel tray.

Mercury vapor discharge from vents of reactors or storage tanks at normal atmospheric pressure is controlled readily by means of activated carbon. Standard units (208-L (55-gal) drums) of activated carbon equipped with proper inlet and outlet nozzles can be attached to each vent. To minimize the load on the carbon-absorbing device, a small water-cooled condenser is placed between the vent and the absorber.

The control of mercury in the effluent derived from the manufacturing processes used in the preparation of inorganic and organic mercurials is mandated by law in the United States. The concentrations and the total amounts vary with the industry and the location, but generally it is required that the effluent contain not more than 0.01 mg Hg/L. However, individual states and individual publicly owned sewage-treatment plants have set up their own standards.

Mercury Removal From Gases. Removal of mercury from ambient air in the workplace and prevention of discharge of mercury vapor are of the greatest importance in the protection of the health of the worker. Mercury vapor, as metallic mercury or in the form of its various compounds, can be absorbed readily by a number of different media (see AIR POLLUTION CONTROL METHODS). Whereas activated charcoal adsorbs mercury, it is not particularly efficient. Specially treated charcoal containing sulfur compounds is fairly effective but cannot be regenerated after it is saturated (see ADSORPTION, GAS SEPARATION). Because of the problems involved in disposing of mercury-containing wastes such as unregenerated adsorbents, it is preferable to use other systems.

Mercury amalgamates readily with gold and silver, and systems have been developed using these metals distributed on various carriers to remove mercury vapor from an airstream. When the system is saturated, the mercury can be removed easily and recovered by heating the unit and condensing the mercury. Other metals, such as copper and zinc, can also be used.

When the mercury present in the atmosphere is primarily in the form of an organic mercury compound, it may be preferable to utilize an aqueous scrubber. This method is particularly useful for control of emissions from reactors and from dryers. For efficient and economical operation, an aqueous solution of caustic soda, sodium hypochlorite, or sodium sulfide is recirculated through the scrubber until the solution is saturated with the mercury compound.

From Liquids. The chlor-alkali producers employing mercury cathode electrolytic cells for the production of chlorine and caustic soda face the greatest problem in removal of mercury from aqueous effluent streams, and most of the patent literature is concerned with the processes for treatment of mercury-containing brine so produced (see ALKALI AND CHLORINE PRODUCTS). One procedure involves the use of a bed of activated carbon impregnated with silver (24). The alkali-brine solution is passed through a bed of this material which is combined with supporting material, such as nickel turnings or polyethylene shreds. Another process allows the brine solution to contact a strong anion-exchange organic resin of the quaternary ammonium cross-linked type (25) (see ION EXCHANGE).

Soluble sulfides such as sodium sulfide, potassium sulfide, and calcium polysulfides have been used to precipitate mercury salts from alkaline solutions. When this procedure is used, exercise of caution is required to maintain the pH within a given alkaline range so as to prevent evolution of H_2S. Because the solubility of mercuric sulfide in water is 12.5 μg/L at 18°C or 10.7 ppb of mercury, use of this method for removal of mercury is adequate for most purposes. However, the presence of excess alkali, such as sodium hydroxide or sodium sulfide, increases the solubility of mercuric sulfide as shown:

Na$_2$S, g/100 g soln	HgS solubility, g/100 g soln
0.95	0.21
1.50	0.57
2.31	1.45
3.58	2.91
4.37	4.12
6.07	7.27
9.64	15.59

Thus, at a concentration of 0.95 g Na$_2$S/100 g solution, the solubility of mercuric sulfide has increased to 2100 ppm. It is customary to use no greater than a 20% excess of the alkali sulfide. Because the particle size of the precipitated mercuric sulfide is so small, it is helpful to add a ferric compound such as ferric chloride or ferric sulfate to effect flocculation. Sometimes other flocculating agents (qv) may also be added, eg, starch or gum arabic.

Another method of removing mercury compounds from aqueous solution is to treat them with water-soluble reducing agents, thus liberating metallic mercury (26). The use of formaldehyde (qv) at a pH of 10–12 also is recommended.

Problems of removal of mercury from aqueous effluents are more complicated in plants that manufacture a variety of inorganic and organic mercury compounds; it is generally best to separate the effluent streams of inorganic and organic mercurials. When phenylmercuric acetate is precipitated from its solution in acetic acid by addition of water, the filtrate is collected and reused for the next precipitation. This type of recycling is necessary not only for economic reasons but also to minimize recovery operations.

When an aqueous effluent stream containing organomercurials cannot be recycled, it may be treated with chlorine to convert the organomercury to inorganic mercury. The inorganic compounds thus formed are reduced to metallic mercury with sodium borohydride. The mercury metal is drained from the reactor, and the aqueous solution discarded. The process utilizing sodium borohydride is known as the Ventron process (27).

BIBLIOGRAPHY

"Mercury Compounds" in *ECT* 3rd ed., Vol. 15, pp. 157–171, by W. Singer and M. Nowak, Troy Chemical Corp.

1. L. G. Hepler and G. Olofsson, *Chem. Rev.* **75**, 585 (1975).

2. C. E. Vanderzee and J. A. Swanson, *J. Chem. Termodyn.* **6**, 827 (1974).

3. H. L. Clever, S. A. Johnson, and M. Elizabeth Derrick, *J. Phys. Chem. Ref. Data*, **14**(3), 631–679 (1985).

4. Hatch and Ott, *Anal. Chem.* **40**, 2085 (1968).

5. W. F. Linke, *Solubilities: Inorganic and Metal-Organic Compounds*, Vol. 1, D. Van Nostrand Co., Princeton, N.J., 1958, pp. 1179–1253.

6. A. N. Pushnyak, K. Bong Hoang, and M. M. Chobanu, *Zh. Prikl. Khim.* (*Leningrad*) **62**(4), 749–753 (1989).

7. X. J. Bad and co-workers, *Mater. Res. Soc. Symp. Proc.*, 163 (1990).

8. U.S. Pat. 5013358 (May 7, 1991), D. Ball and co-workers; D. C. Harris, R. A. Nissan, and T. H. Kelvin, *J. Inorg. Chem.* **26**, 765–768 (1987).
9. Jpn. Kokai Tokkyo Koho J.P. 01,201,011 [89,201,011] (Aug. 14, 1989), K. Nakamura, T. Vemura, and S. Okamoto.
10. A. Jermelov, Fisheries Research Board of Canada Translation Series No. 1352; *Vatten* **24**(4), 360 (1968).
11. S. Jensen and A. Jernelov, *Nature* **223**, 753 (Aug. 16, 1969).
12. L. Clendening, *Source Book of Medical History*, Dover Publications, New York, 1960; W. Singer and E. A. Underwood, *A Short History of Medicine*, 2nd. ed., Oxford University Press, New York, 1962.
13. H. L. Friedman, *Ann. N.Y. Acad. Sci.* **65**, 461 (1957).
14. *Fed. Res.* **37**(61), 9373 (Mar. 29, 1972).
15. G. F. Nordberg, M. H. Berlin, and C. A. Grant, *Proc. 16th Int. Congr. Occup. Health (Tokyo)*, 234 (Sept. 22–27, 1969).
16. G. F. Phillips, B. E. Dixon, and R. G. Lidzey, *J. Sci. Food Agric.* **10**, 604 (1959).
17. L. Fiberg, *Arch. Indust. Health* **20**, 42 (1959).
18. L. J. Goldwater, *Proceedings of 15th International Congress on Occupational Health*, Paper BIII-17, Vienna, Sept. 19–24 1966.
19. T. Rahola, T. Hattula, A. Kosclainen, and J. K. Miettinene, *Scand. J. Clin. Lab. Forest. Abstr.* **27** (suppl. 116), 77 (1971).
20. *Hygienic Guide Series*, American Industrial Hygiene Association, Westmont, N.J.
21. L. J. Goldwater, M. B. Jacobs, and A. C. Ladd, *Arch. Environ. Health* **5**, 537–541 (1962).
22. Ref. 21, p. 540.
23. U.S. Pat. 3,502,434 (Mar. 24, 1970), J. B. MacMillan (to Canadian Industries Ltd., Canada).
24. U.S. Pat. 3,213,006 (Oct. 19, 1965), G. E. Grain and R. H. Judice (to Diamond Alkali Co.).
25. U.S. Pat. 2,885,282 (May 5, 1959), M. P. Niepert and C. D. Bon (to The Dow Chemical Co.).
26. *Chem. Eng.*, 71 (Feb. 27, 1971).
27. L. J. Goldwater, *Mercury*, York Press, 1972, p. 157.

MILTON NOWAK
WILLIAM SINGER
Troy Chemical Corporation

MESITYLENE.   See POLYMETHYLBENZENES.

MESITYL OXIDE.   See KETONES.

METAL ALKYLS.   See ORGANOMETALLICS.

METAL ANODES

In electrolytic processes, the anode is the positive terminal through which electrons pass from the electrolyte. Anode design and selection of anode materials of construction have traditionally been the result of an optimization of anode cost and operating economics, in addition to being dependent on the requirements of the process. Most materials used in metal anode fabrication are characteristically expensive; use has, however, been justified by enhanced performance and reduced operating cost. An additional consideration that has had increasing influence on selection of the appropriate anode is concern for the environment (see ELECTROCHEMICAL PROCESSING).

Industrial metal anodes can generally be classified in one of two groups. The first group, chlorine-generating anodes, find application primarily in the manufacture of chlorine and caustic, sodium chlorate, and sodium hypochlorite (see ALKALI AND CHLORINE PRODUCTS; CHLORINE OXYGEN ACIDS AND SALTS). The second group, consisting of the oxygen-evolving anodes, do not generate a saleable product directly, but rather facilitate the desired cathodic reaction. Commercial uses include high speed electrogalvanizing of steel (qv), electrowinning of base metals, plating operations, cathodic protection, electrophoretic painting, copper (qv) foil treatment, and, more recently, the primary production of copper foil itself (see ELECTROPLATING; MACHINING METHODS, ELECTROCHEMICAL).

Historical Development

Anodes of zinc, lead, silver, and other metals have found commercial application in a variety of uses since the early 1900s. Then, in 1950, a breakthrough in coated anode technology took place when the development of jet aircraft and the birth of the space industry expanded the demand for and supply of titanium, bringing the price of this metal into a more affordable range (see TITANIUM AND TITANIUM ALLOYS). Whereas titanium [7440-32-6] was valued in air transport and space programs for its strength-to-weight characteristic, other attributes were important for usage as metal anodes in the chlor alkali industry. These are (1) titanium's self-oxidizing valve metal quality which passes current in one direction only; (2) titanium's excellent corrosion-resistance to a wet chlorine environment; and (3) titanium's superior conductivity when compared to graphite.

Because titanium is subject to oxidation under anodic conditions, an electrically resistive layer is formed on its surface. Thus this material is, in itself, not the complete answer to the search for a metal anode suitable for use in chlor-alkali service. The principal participants in the worldwide chlorine industry therefore embarked on programs to develop appropriate coatings (qv) that would make the use of the titanium structure more fully practical. These programs were predicated on the use of platinum-group metals (qv) as the active ingredient of the coating. Patents were issued in the late 1950s for platinized titanium anodes utilizing catalysts on titanium supports (1,2).

During the 1960s, a group of British patents were issued (3–5) wherein technology effective in maintaining the precious metal in the coating in its metallic rather than oxidized state was emphasized. Realization of this objective was achieved by the specification of a reducing atmosphere during thermodeposition, by use of reducing agents in the coating formulation, or by electrolytic deposition. Whereas these metallic coatings (qv) have found application in cathodic protection service and, to some extent, in the manufacture of sodium chlorate, these coatings have not proven commercially effective in chlorine manufacture.

Commercial metal anodes for the chlorine industry came about after the late 1960s when a series of worldwide patents were awarded (6–8). These were based not on the use of the platinum-group metals (qv) themselves, but on coatings comprised of platinum-group metal oxides or a mixture of these oxides with valve metal oxides, such as titanium oxide (see PLATINUM-GROUP METALS, COMPOUNDS; TITANIUM COMPOUNDS). In the case of chlor-alkali production, the platinum-group metal oxides that proved most appropriate for use as coatings on anodes were those of ruthenium and iridium.

Many competitive programs to perfect a metallic anode for chlorine arose. In one, Dow Chemical concentrated on a coating based on cobalt oxide rather than precious metal oxides. This technology was patented (9,10) and developed to the semicommercial state, but the operating characteristics of the cobalt oxide coatings proved inferior to those of the platinum-group metal oxide.

Commercial Uses

Chlorine-Generating Anode.   *Chlorine and Chlorine Oxygen Salts.*   Coated metal anode technology finds its most dramatic industrial application in the chlorine industry. Metal anode technology for chlorine [7782-50-5], Cl_2, manufacture, commercialized as the dimensionally stable anode (DSA) (Electrode Corp.), replaced essentially all graphite anodes worldwide in chlorine plants between 1972 and 1982. Advantages to DSA users relative to the older graphite anode include lower power consumption resulting from low overpotential and dimensional stability; reduced cell renewal activity owing to long anode life; cleaner products of electrolysis; elimination of some environmentally detrimental materials used with the graphite anode; and more consistent and stable cellroom operation.

Success in the chlorine industry led to the incorporation of DSA in sodium chlorate [7775-09-9], $NaClO_3$, manufacture. The unique structural characteristics of the anode allowed for innovative designs in cell hardware, which in turn contributed to the extensive worldwide expansion of the sodium chlorate industry in the 1980s.

Just as straightforward was the incorporation of coated metal anodes into systems for the production of sodium hypochlorite [7681-52-9], NaClO. On-site production units fed with either salt brine or seawater are used in far-reaching applications, ranging from cooling water treatment, drinking water disinfection, wastewater treatment, swimming pool disinfection, and secondary oil recovery, to marine sewage treatment and bleach manufacture (see CHLORINE OXYGEN ACIDS AND SALTS, DICHLORINE MONOXIDE, HYPOCHLOROUS ACID, AND HYPOCHLORITES).

Cathodic Protection Systems.   Metal anodes using either platinum [7440-06-4] metal or precious metal oxide coatings on titanium, niobium [7440-03-1], or tantalum [7440-25-7] substrates are extensively used for impressed current cathodic protection systems. A prime application is the use of platinum-coated titanium anodes for protection of the hulls of marine vessels. The controlled feature of these systems has created an attractive alternative to the use of sacrificial anodes and frequent dry docking for painting and repair (see COATINGS, MARINE). These cathodic protection systems are also used on deep-sea oil-drilling platforms (see PETROLEUM).

Metal anodes using platinum and precious metal oxide coatings are also incorporated into a variety of designs of impressed current protection for pipeline and deep well applications, as well as for protection of condenser water boxes in power generating stations (see PIPELINES; POWER GENERATION).

Electroplating. Platinized titanium-on-niobium anodes are preferred for use in electroplating precious metals. These anodes find wide application in the electronics industry and in the creation of fine jewelry.

Oxygen-Evolving Anode. Research efforts to incorporate the coated metal anode for oxygen-evolving applications such as specialty electrochemical synthesis, electrowinning, impressed current, electrodialysis, and metal recovery found only limited applications for many years.

Two specific uses of oxygen anodes that represent comprehensive and growing market demands have surfaced. In the first, occasioned by the commercialization of the high speed electrogalvanizing process (11), the electrolytic hardware was designed to utilize DSA rather than the traditional dimensionally unstable lead or zinc anodes. This Andritz Ruthner technology has been well accepted by the worldwide electrogalvanizing industry and has been incorporated into production plants in North America, Europe, and South America, creating an instant demand for the coated titanium metal anode. In addition, plants originally designed for the use of lead or zinc anodes have been converting to coated metal anodes. Advantages include electric power savings, consistency of operation, and elimination of many environmental concerns.

The second use involves the application of DSA coating on a titanium mesh for cathodically protecting steel reinforcing in concrete structures, such as highway bridges and parking garages (12). This technology finds value in geographic areas where salt is used to prevent roadway icing and in marine shore locations. In practice, salt solution finds its way through the concrete to the reinforcing bar and promotes oxidation of the steel (qv). The rust thus formed occupies a greater volume than the steel, creating stress and eventually cracking the concrete. This continuing process can lead to virtual destruction of the structure if maintenance is delayed. By embedding the coated titanium mesh between the reinforcing steel and the surface of the concrete and by supplying a potential on the mesh to maintain the reinforcing steel cathodic, this destructive process can be eliminated and structure life significantly extended. This technology is seen as having a significant future of widespread utilization.

Several more traditional materials have found specific though limited commercial application as metal anodes. Examples are lead [7439-92-1] and zinc [7440-66-6] in the electrogalvanizing practice. Lead dioxide [1309-60-0] and manganese dioxide [1313-13-9] anode technologies have also been pursued. Two industrial electrolytic industries, aluminum [7429-90-5] and electric arc steel, still use graphite anodes. Heavy investment has been devoted to research and development to bring the advantages of DSA to these operations, but commercialization has not been achieved.

The commercial status of metal anodes for oxygen-evolving applications may be summarized as follows:

Application	Anode used
electrogalvanizing	DSA, lead, zinc
cathodic protection	platinized titanium, DSA, zinc, magnesium, lead, silver, aluminum
electrolysis of sodium sulfate and other inorganic salts	DSA
electrophoretic painting	stainless steel, DSA
copper foil production	lead, DSA
copper foil treatment	DSA, steel, platinized titanium
zinc electrowinning	DSA, lead
copper electrowinning	lead
aluminum anodizing	DSA, lead
metal finishing/recovery	DSA, lead, platinized titanium
electroplating of precious metals	platinized titanium or niobium

Coating Structure and Morphology

The crystal structure of metal anode coatings has been investigated using x-ray diffraction studies, x-ray fluorescence analysis, and microprobe studies in conjunction with scanning electron micrographs (see MICROSCOPY; X-RAY TECHNOLOGY). However, the role of the titanium metal substrate, or that of the oxides formed when the substrate is anodized or heated in air, is not completely clear for coatings that contain titanium in their solution. Most coatings, even those not containing titanium in the formulation, exhibit traces of titanium dioxide, present either in rutile [1317-80-2] or anatase [1317-70-0] form.

Ruthenium–Titanium Oxides. The x-ray diffraction studies of ruthenium–titanium oxide coatings show that the coating components are present as the metal dioxides, each in the rutile form as well as in solid solution with each other (13). The development of the crystal structure begins to occur at a bake temperature of about 400°C. By following the d_{110} diffraction line for the rutile structure, an increase in crystallinity can be seen as temperatures are increased to the 600–700°C range. Above these temperatures, the d_{110} peak begins to separate into two separate peaks, indicative of phase separation into individual rutile oxides, one rich in ruthenium and one rich in titanium.

It appears that the titanium metal substrate on which the coating is deposited plays an important role in the structure and morphology of the coating. The surface layer of rutile titanium dioxide normally found on oxidized titanium metal apparently acts as a seed to initiate growth of the rutile form of the oxide, rather than the anatase form. Interfacial layers of titanium suboxides, known to be electrically conductive, also act to effect a gradual transition from pure metal to pure rutile oxides. Without this interfacial layer, the large stresses in the titanium crystal structure created in the transition of metal to oxide would occur over a sharp boundary and reduce the adhesion of the coating to the substrate. Thus careful control of the bake temperature and rate of change, ie, the temperature profile, are necessary to ensure optimum adhesion of the oxides to the base metal.

Additional x-ray studies indicate some degree of lattice distortion in coatings prepared from chloride-containing coating solutions. This correlates with an analysis of 3–5% chloride in the coating, which is reduced to near zero if the coating is heated to 800°C.

Coatings produced at bake temperatures of 400–600°C are incompletely crystallized solid solutions of rutile titanium dioxide and ruthenium dioxide [12036-10-1] having lattice defects caused by the presence of chloride. The high degree of crystalline disorder contributes greatly to the electrical conductivity of the coating, the presence of active catalytic sites, and the high surface area of the coating.

Scanning electron micrographs of ruthenium–titanium oxide coatings show a characteristic microcracked surface (13). This cracking occurs early in the coating preparation, as solvent evaporates from the surface to form a gel of unreacted ruthenium and titanium compounds. As the coating is baked at higher temperatures, the cracks increase in size because of volume contraction of the gel. A fully baked anode coating has the appearance shown in Figure 1 and a surface-area factor of 180 to 230 times the geometrical area, as measured by Brunauer-Emmett-Teller (BET) nitrogen adsorption. This large surface area contributes to the low chlorine discharge potential of these coatings, providing a large number of catalytic sites for gas evolution while minimizing concentration polarization.

Iridium Oxide. Iridium dioxide [12030-49-8] coatings, typically used in combination with valve metal oxides, are quite similar in structure to those of ruthenium dioxide coatings. X-ray diffraction shows the rutile crystal structure of the iridium dioxide; scanning electron micrographs show the micro-cracked surface typical of these thermally prepared oxide coatings.

Platinum–Iridium. There are two distinct forms of 70/30-wt % platinum–iridium coatings. The first, prepared as prescribed in British patents

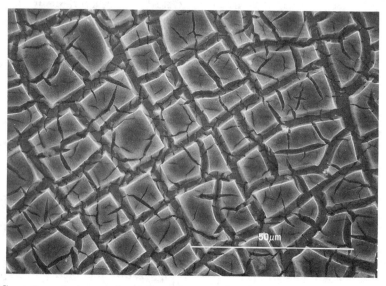

Fig. 1. Scanning electron microscope photograph of DSA ruthenium oxide coating, showing typical cracked surface.

(3–5), consists of platinum and iridium metal. X-ray diffraction shows shifted Pt peaks and no oxide species. The iridium [*7439-88-5*] is thus present in its metallic form, either as a separate phase or as a platinum–iridium intermetallic. The surface morphology of a platinum–iridium metal coating shown in Figure 2 is cracked, but not in the regular networked pattern typical of the DSA oxide materials.

The second form consists of Pt metal but the iridium is present as iridium dioxide. Iridium metal may or may not be present, depending on the baking temperature (14). Titanium dioxide is present in amounts of only a few weight percent. The analysis of these coatings suggests that the platinum metal acts as a binder for the iridium oxide, which in turn acts as the electrocatalyst for chlorine discharge (14). In the case of thermally deposited platinum–iridium metal coatings, these may actually form an intermetallic. Both the electrocatalytic properties and wear rates are expected to differ for these two forms of platinum–iridium-coated anodes.

Spinel Cobalt Oxides. The cobalt mixed oxide [*1308-06-1*], Co_3O_4, containing Co(II) and Co(III) ions, has the spinel structure. The x-ray analysis shows that divalent zinc ions substitute for Co(II) in the tetrahedral sites of the lattice, causing an expansion of the lattice and formation of a normal $Zn_xCo_{3-x}O_4$ spinel structure (15). In these coatings containing zirconium [*7440-67-7*], a separate, partially crystalline phase of zirconium dioxide [*1314-23-4*] occurs. The coating has large microscopic pores, providing a high surface area which appears to be related to the presence of zirconium. Spinel oxides applied without the presence of zirconium are dense and closely packed coatings. Thermal studies indicate that decomposition of zirconyl nitrate [*13826-66-9*], the source of zirconium in the coating, produces NO_2. This gas is released into the cobalt–zinc oxide as it forms, presumably producing the open-pore, spongy morphology of the coating. Surface-area measurements confirm this model. The presence of zirconium

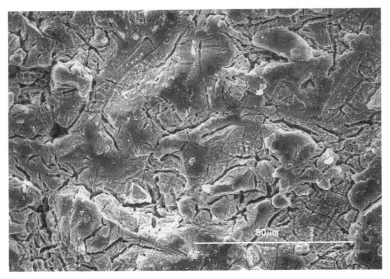

Fig. 2.   Scanning electron microscope photograph of platinum–iridium metal coating on titanium.

increases the surface-area factor from approximately 250 to 1880, determined by BET measurements (15).

Operating Performance of Coatings

The key yardstick of performance for a coated metal anode is the period of time, measured in ampere-hours, that the coating operates before it reaches an unacceptable voltage, as measured by the single electrode potential (SEP). This SEP is characteristically constant over an extended period of operating time, but at some point escalates at increased and varying rates, depending on specific operating conditions. Factors influencing this escalation include accumulated average lifetime, operating current density, electrolyte conditions, exposure to oxygen in the anolyte compartment, and cell design. The magnitude of this escalation dictates the point at which the anode coating is considered to have failed, ie, to have reached the end of its operable life, and requires replacement.

Chlorine Anodes.   In chlorine manufacture, anode operating life is limited by coating wear. The wear rate varies according to cell type. The longest anode coating life has been achieved in diaphragm chlorine cells. When these were first commercialized, it was expected that 5–6 years would represent a fully acceptable anode coating life. In practice, however, lives of 10–15 years and, in some cases, beyond, have been realized. This extended performance has been achieved in spite of the incorporation of voltage-saving techniques that position the anode only a few millimeters from the cathode surface. In some plants, the anode has been allowed to bear directly on the diaphragm itself. Whether this latter arrangement shortens coating operating life has yet to be determined.

Because mercury (qv) chlorine cells operate at higher current densities and because the mercury cell anode can be adjusted during operation to minimize the anode-to-cathode gap, anode coating life in these cells is much shorter. Lifetimes range from less than one year at high current densities and using close gaps, to periods in excess of five years at lower current densities and using greater gaps. In cellrooms that have converted from the traditional mesh- or rod-type structures to the newer blade-type anode, anode coating lifetimes have been extended.

Because of limited commercial experience with anode coatings in membrane cells, commercial lifetimes have yet to be defined. Expected lifetime is 5–8 years. In some cases as of this writing (ca 1995), 10-years performance has already been achieved. Actual lifetime is dictated by the membrane replacement schedule, cell design, the level of oxygen in the chlorine gas, and by the current density at which the anode is operated.

Metal anode coatings commercially used for manufacture of sodium chlorate include not only the ruthenium oxide coatings, but also platinum–iridium coatings. Whereas ruthenium oxide coatings might be preferred for a longer performance life and higher resistance to process upsets, platinum–iridium coatings generally operate at higher efficiencies during the first months of operation. The sodium chlorate industry has experienced only a limited coating replacement requirement. It is generally expected that coating life in chlorate cells approximates 10 years. Poor electrolyte circulation patterns and exposure to impurities in the anolyte have been known to shorten this life dramatically.

Oxygen-Evolving Anodes. In the case of oxygen-evolving anodes employing iridium dioxide-based coatings, the most significant commercial experience has been in the electrogalvanizing industry where coated titanium anodes have supplanted lead and zinc anodes. The coated anodes have achieved operation at current densities as high as 15 kA/m^2, eliminating lead contamination of the product and waste streams (16). Similar anodes are also being utilized in other oxygen-evolving metal-plating systems, such as copper foil production, electrotinning, and electrowinning.

In contrast to the coating wear limitation on anode life experienced in chlorine cells, passivation of the substrate beneath the coating is typically the limiting factor for oxygen-evolving anodes. As a result, technology has been introduced to either maintain or modify the titanium surface (17,18) to increase coating adhesion and significantly improve lifetimes. Anodes having operating lives >65,000 kA·h/m^2 are readily available (19). Care must be taken, however, to avoid both foreign deposits on the anode surface and operation in reverse current modes, if extended lifetimes are to be expected (20).

Structure Design

Each electrolytic application demands a unique approach to anode structure design and fabrication. Factors such as current distribution, gas release, ability to maintain structural tolerances, electrical resistance, and the practicality of recoating must be taken into account. At the outset, chlorine anodes were designed to assume the same configuration as that of a new graphite electrode so that retrofit challenges were minimized. But a variety of newer structure design concepts have been presented to achieve power savings by reduction of the electrolytic gap between the anode and cathode of diaphragm chlorine cells. The most commercially accepted design is that of the expandable anode (Fig. 3) (21).

An expandable anode involves compression of the anode structure using clips during cell assembly so as not to damage the diaphragm already deposited on the cathode (Fig. 3a). When the cathode is in position on the anode base, 3-mm diameter spacers are placed over the cathode and the clips removed from the anode. The spring-actuated anode surfaces then move outward to bear on the spacers, creating a controlled 3-mm gap between anode and cathode (Fig. 3b). This design has also been applied to cells for the production of sodium chlorate (22).

In mercury chlorine cells, anode surfaces were originally fabricated from expanded titanium mesh or from appropriately spaced titanium rods. More recently it was found that cells operate at lower voltages when fitted with anode structures comprised of triangular rods, as embodied in the runner technology, or of vertical blades (23). Various mercury cell anode designs are shown in Figure 4. In some cases anolyte circulation baffles are incorporated into the anode structure to achieve added power savings by maintaining a higher average brine strength in the electrolytic gap.

The dimensionally stable characteristic of the metal anode made the development of the membrane chlorine cell possible. These cells are typically arranged in an electrolyzer assembly which does not allow for anode-to-cathode gap adjustment after assembly. Also, very close tolerances are required. The latitude

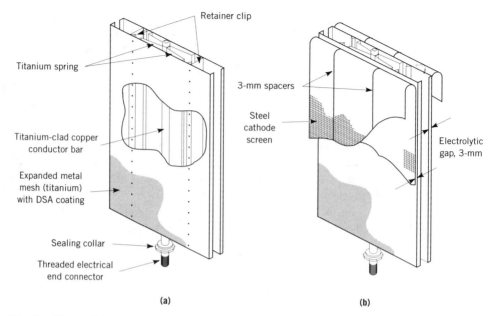

Retainer clip

Titanium spring

3-mm spacers

Steel
cathode
screen

Titanium-clad copper
conductor bar

Electrolytic
gap, 3-mm

Expanded metal
mesh (titanium)
with DSA coating

Sealing collar

Threaded electrical
end connector

(a)   (b)

Fig. 3.   Expandable anode for diaphragm chlorine cells: (**a**) clamped and (**b**) expanded.

that titanium affords the cell designer has made a wide variety of monopolar
and bipolar membrane cell designs possible.

When used in radial cells such as for the production of copper foil and in
some electrogalvanizing operations, the anode must be curved to meet the shape
dictated by the cathodic drum. The anode must also achieve exact tolerances
to assure that a constant anode-to-cathode gap is created. Maintaining the
dimensional stability of a rigid anode structure is nearly impossible when that
unit must endure multiple cycles of high temperature through the coating bake
operation. To remedy this situation, an anode plate substrate is fabricated with
an arc somewhat greater than that of the drum to which it is to form a match.
At this stage, close tolerances are not required, and coating this plate poses no
dimensional stability problems. The anode support structure, on the other hand,
is fabricated using appropriately demanding accuracy so that when the anode
plate is fitted into the support, it can be drawn into position by studs extending
from its rear, inactive surface. It thus assumes a fixed position which attains
the exact arc that has been specified (24). This design is shown in Figure 5.

Manufacturing Technology

Manufacturing techniques for metal-coated anodes have been developed to a high
level of sophistication. Whereas in the early 1970s, anode coatings were applied
by hand and curing was achieved in batch ovens, by the mid-1990s commercial
plants were producing or recoating in excess of 50,000 anodes annually. This
procedure requires continuous coating processes (qv) and the use of electrostatic
spray application, robotics (see ROBOTS, PROCESSING), and strict process control
(qv). The high cost of the coating itself demands high utilization efficiency.

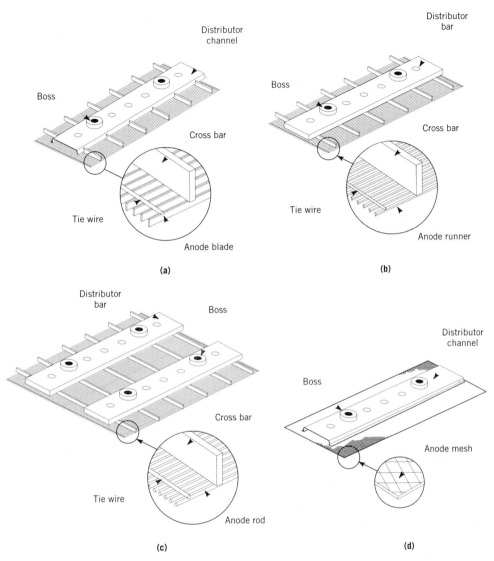

Fig. 4. Configurations of mercury chlorine cell anodes: (**a**) blade type; (**b**) runner technology; (**c**) rod type; and (**d**) mesh type.

Quality requirements include consistent coating distribution, strong adherence, and a good surface appearance.

The coating process, outlined in Figure 6, starts either with the specification of appropriate titanium substrate material for new anodes, or with removal of existing coating in the case of anode recoating. Coating removal is especially critical in the case of anodes intended for electrogalvanizing service. Using an inappropriate method to remove existing coating can distort or even destroy the anode structure. It can also render the substrate surface incapable of achieving an effective bond with the coating. In the next step, the structure is submitted to a sulfuric or hydrochloric acid etch to prepare the surface for receipt of the new

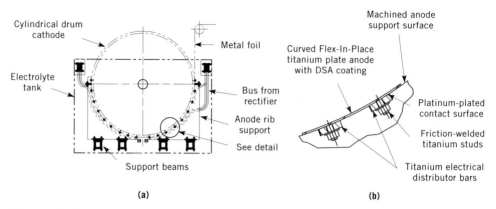

Fig. 5. Radial design for metal anode showing (**a**) cross-sectional view and (**b**) area of detail on drum rim.

coating. The application of an intermediate layer to inhibit passivation of the substrate may also be prescribed. This also is of importance in oxygen-evolving service.

The coating itself is applied in a series of thin layers. A baking step follows each coat. Strict attention is given to the thermal profile of the preheat/bake/post-cool cycle and to the oven atmosphere to assure effective oxidation. Coating formulations vary considerably, based on the desired use of the anode and upon the proprietary technology of the coating supplier. Traditionally, ruthenium oxide coatings and variations of these coatings have been used in the diaphragm and mercury chlorine cell anodes; iridium may be included to achieve longer coating life in membrane cells. Oxygen-evolving applications require iridium oxide coatings. There is also some latitude in selecting whether water- or alcohol-based solvents are to be used.

The series of high temperature baking steps requires that consideration be given to the dimensional stability of the anode structure. If it is not, thermal warping can render a structure useless or, at a minimum, necessitate excessive and expensive anode-straightening techniques. When the anode structure includes titanium-clad copper components for enhanced conductivity, it is also possible to thermally damage the copper–titanium bond, thus introducing a high resistance to the structure. In some cases, the anode surface to be coated, whether mesh, bar, plates, or screen, etc, must be removed and coated separately in order to avoid damage to the structure's integrity.

In order to ensure that an electrolysis plant maintains a high on-stream factor, continuous and ongoing anode recoating programs are utilized based on removing and recoating the anode prior to coating failure. Allowing coatings to suffer catastrophic failure while in operation can result in a difficult situation owing both to the logistics of anode replacement in the operating circuit as well as the amount of time consumed in reprocessing the structure at the recoating facility. Only a limited number of spare anodes are available. Because as many as 30,000 anodes are required at a world-scale chlorine-producing facility, coating replacement is necessarily an ongoing activity, especially when coating life is relatively short, as in mercury chlorine cells and in electrogalvanizing lines. In

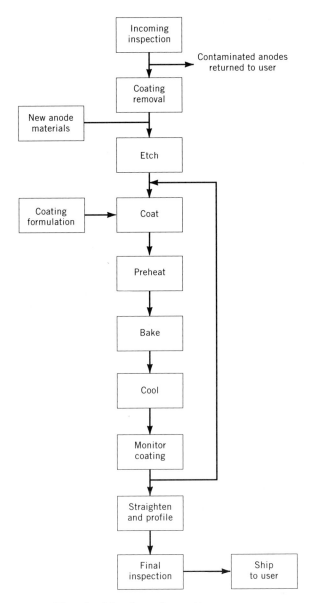

Fig. 6. Metal anode coating process.

cases where coating life extends for a period of 10 years or more, as in diaphragm chlorine cells, coating replacement can involve the use of a periodic recoating program which should be instituted well in advance of any anticipated plant-wide coating failure. For a large chlorine plant, such a program might continue for a period of four to five years.

Environmental and Safety Factors

The environmental challenge at the metal anode fabricating and coating plant originates more from the outside than from within the plant itself. The coating plant's process systems must be, of course, designed and operated to control vent stack emissions and impurities in wastewater streams. Sludges containing precious metal must also be appropriately recycled. But the specific tasks and facilities involved in providing these are matters of straightforward shop design and housekeeping.

The primary environmental concern for the coating plant is actually the residual material on the anode structures being returned for recoating. Therefore the anode user must enact effective cleaning procedures prior to shipment. For example, anodes in chlorine use must be cleaned of all traces of mercury and asbestos (qv). Anodes used in electrogalvanizing or in copper-foil production must similarly be cleaned to remove all traces of process materials. If cleaning at the user's plant is not done effectively, the anode may well be shipped back to the user for appropriate action before it is considered for recoating.

Whereas DSA coating operations demand only normal shop procedures for safety and environmental concerns, the commercial realization of this technology has had a most positive impact in the user's plant environment. The replacement of graphite anodes created a cleaner chlorine product in both the diaphragm and mercury cell plants, eliminating both a purification step and a chlorinated sludge disposal problem. Moreover, the dimensionally stable characteristic and long life of the DSA has facilitated the utilization of polymeric, nonasbestos diaphragms and has led to realization of the membrane chlorine cell. This same stability has eliminated the need for frequent adjustment of mercury cell anodes to compensate for wear. Overall, the DSA has made chlorine manufacture cleaner, more consistent, simpler, and therefore safer.

In electrogalvanizing, copper foil, and other oxygen-evolving applications, the greatest environmental contribution has been the elimination of lead-contaminated waste streams through replacement of the lead anode. In addition, the dimensionally stable characteristic of the metal anode introduces greater consistency and simplification of the process, thus creating a measure of predictability, and a resultant increased level of safety.

Economic Aspects

For the most part, coated metal anodes are made available by long-term lease rather than through outright sale. There has been a single-source supply protected by patents in most geographic areas. In the majority of cases, the supplier is either a privately owned company or a small unit of a large, complex organization. As the cost-saving and environmental value of this technology continues to increase, upward expectation in respect to prices is created. However, the appearance of new anode producers, together with expiration of some early patents, has created a competitive situation in some applications. This competition can be expected in turn to create a counterbalancing pressure to lower prices.

BIBLIOGRAPHY

"Metal Anodes" in *ECT* 3rd ed., Vol. 15, pp. 172–183, by H. S. Holden, Diamond Shamrock Technologies, SA, and J. M. Kolb, Diamond Shamrock Corp.

1. Brit. Pat. 855,107 (Apr. 3, 1958), H. B. Beer (to N. V. Curacaosche Exploitatie Maatschappij Uto).
2. Brit. Pat. 877,901 (Feb. 14, 1958), J. B. Cotton, E. C. Williams, and A. H. Barber (to Imperial Chemical Industries, Ltd.).
3. Brit. Pat. 885,819 (Dec. 28, 1961), C. H. Angell and M. G. Deriaz (to Imperial Chemical Industries Ltd.).
4. Brit. Pat. 984,973 (Mar. 3, 1965), C. H. Angell and M. G. Deriaz (to Imperial Chemical Industries, Ltd.).
5. U.S. Pat. 3,177,131 (Apr. 6, 1965), C. H. Angell, S. Coldfield and M. G. Deriaz (to Imperial Chemical Industries Ltd.).
6. S. Afr. Pat. 66/2667 (May 9, 1966), H. B. Beer (to Chemnor Aktiengesellschaft).
7. U.S. Pat. 3,632,498 (Jan. 4, 1972), H. B. Beer (to Chemnor Aktiengesellschaft).
8. U.S. Pat. 3,711,385 (Jan. 16, 1973), H. B. Beer (to Chemnor Corp.).
9. U.S. Pat. 3,977,958 (Aug. 31, 1976), D. L. Caldwell and R. J. Fuchs, Jr. (to Dow Chemical Co.).
10. U.S. Pat. 4,061,549 (Dec. 6, 1977), M. J. Hazelrigg, Jr., and D. L. Caldwell (to Dow Chemical Co.).
11. U.S. Pat. 4,469,565 (Sept. 4, 1984), J. Hampel (to Andritz Ruthner Industrie AG).
12. U.S. Pat. 4,900,410 (Feb. 13, 1990), J. E. Bennet, G. R. Pohto, and T. A. Mitchell.
13. K. J. O'Leary and T. J. Navin, "Morphology of Dimensionally Stable Anodes," paper presented at the *Chlorine Bicentennial Symposium*, San Francisco, Calif., May 1974.
14. P. S. S. Hayfield and W. R. Jacob, "Platinum–Iridium-Coated Titanium Anodes in Brine Electrolysis," paper presented at *Advances in Chlor-Alkali Technology*, London, 1979.
15. D. L. Caldwell and M. J. Hazelrigg, "Cobalt Spinel-Based Chlorine Anodes," paper presented at *Advances in Chlor-Alkali Technology*, London, 1979.
16. K. L. Hardee, L. K. Mitchell, and E. J. Rudd, *Plat. Surf. Finish.* **76**(4), 68 (1989).
17. U.S. Pat. 5,167,788 (Dec. 1, 1992), K. L. Hardee, L. M. Ernes, R. C. Carlson, and D. E. Thomas.
18. U.S. Pat. 5,262,040 (Nov. 16, 1993), K. L. Hardee and co-workers.
19. K. L. Hardee, L. M. Ernes, C. W. Brown, and T. E. Moore, in *Proceedings of the AESF Continuous Steel Strip Plating Symposium*, May 1993.
20. K. L. Hardee, L. K. Mitchell, and R. C. Carlson, in *Proceedings of the AESF Sixth Continuous Steel Strip Coating Symposium*, May 1990.
21. U.S. Pat. 3,674,676 (July 4, 1972) E. Fogelman.
22. U.S. Pat. 5,100,525 (Mar. 31, 1992) G. R. Pohto, C. P. Tomba, E. L. Cimino, and T. A. Liederbach.
23. U.S. Pat. 4,263,107 (Apr. 21, 1981), A. Pelligri (to Oroneio Denora Impianti Electrochemici SpA).
24. U.S. Pat. 5,017,275 (May 21, 1991), A. J. Niksa (to Eltech Systems Corp.).

THOMAS A. LIEDERBACH
Electrode Corporation

METAL-CONTAINING POLYMERS. See SUPPLEMENT.

METALDEHYDE. See ACETALDEHYDE.

METAL FIBERS. See COMPOSITE MATERIALS; FIBERS, ACRYLIC.

METALLIC COATINGS

SURVEY

Metallic coatings provide an inexpensive way to modify or control the properties of a base material. Coatings (qv) may be functional or decorative, permanent or temporary, sacrificial, or noble. The metallic coating may be continuous, or it may be patterned into discontinuous functional or decorative areas. The base material may be metallic, nonmetallic inorganic, or organic such as plastic, paper, or fiber. The common criterion for the purposes herein is that the metallic coating is functionally bonded to the base material (1–4). Galvanized zinc items, gold clad jewelry, electroplated materials, and semiconductor chips are among the most common types of materials having metallic coatings (see ELECTROPLATING; ELECTRONICS, COATINGS; INTEGRATED CIRCUITS; and SEMICONDUCTORS).

Some types of metallic coating have been used since ancient times, initially for jewelry and decorative applications. One of the earliest techniques, used by the ancient Egyptians, was the use of gold foil. Gold foil, produced by beating gold to thicknesses of <0.1 μm, can be easily bonded to all types of metals, paper (qv), leather (qv), and other materials. Many early artifacts originally thought to have been fabricated by means of a chemical displacement coating of a precious metal, were probably made by selective dissolution of the less noble components of an alloy. Glass mirrors formerly were made by amalgamating a thin tin sheet with mercury, pressing it to a glass sheet, and heating to vaporize the mercury and form a direct metal-to-glass bond (5). Soldering is an ancient metal coating technique used for joining two metals, used as of the 1990s in highest volume in the electronics industry.

Some of the first electroplating applications in the early 1800s involved attempts to coat inexpensive base metals with thin layers of pure gold, or even more cheaply, gold-colored brass. Economic reasons are almost always foremost in the decision of coating selection. Exceptions are where environmental or legislative measures require more expensive, less toxic coatings, as in replacement of cadmium [7440-67-7]. The use of metallic coatings often conserves rare metals, allowing the more abundant materials to serve as substrates. Bulk corrosion-resistant materials such as stainless steel or Monel use large amounts of scarce nickel and chromium. One metal may serve many different functions, depending on the application. Thus gold [7440-57-5] may be the permanent decorative coating on jewelry, a permanent functional coating on electrical connectors (qv) or anodes, or a temporary functional and protective coating for solder joint connections on printed circuit boards in which the thin gold layer protects the underlying nickel from oxidation then dissolves in the molten solder to allow solder bonding to the nickel (see SOLDERS AND BRAZING ALLOYS).

Metallic coatings are most often selected for protective function. Decorative ability is a common secondary function. Most coatings applied by hot dipping, such as galvanizing or aluminizing, use a film of a more chemically reactive metal over a less reactive material such as iron alloy. These are sacrificial coatings because the zinc or aluminum slowly dissolves instead of the underlying steel. The sacrificial metal protects the underlying iron even if the coating is scratched or broken. In this case the slowly dissolving metal acts as a battery, keeping the iron at a cathodic potential and preventing the iron from rusting. In contrast, a protective coating of a more noble metal such as gold, must be absolutely break-free. If a break occurs in this type of coating, the battery effect makes the underlying metal anodic, thus causing accelerated corrosion of the base metal (4,6–9).

Decorative coatings must be sufficiently corrosion resistant to maintain an attractive appearance during the anticipated life of the composite product. Composites of multiple metal layers or mixtures of metallic and nonmetallic coatings are becoming increasingly common as material properties are selectively tailored. Aluminum [7429-90-5] coatings on mirrors, laser disks, and compact disks are given a protective sealer coating of transparent plastic to inhibit the sacrificial oxidation of the shiny aluminum to the dulled aluminum oxide. Thin transparent rhodium [7440-16-6] coatings are used to make tarnish-resistant silver jewelry. Chromium [7440-47-3] coatings combine both decoration and corrosion resistance.

Another important function of metallic coatings is to provide wear resistance. Hard chromium, electroless nickel, composites of nickel and diamond, or diffusion or vapor-phase deposits of silicon carbide [409-21-2], SiC; tungsten carbide [56780-56-4], WC; and boron carbide [12069-32-8], B_4C, are examples. Chemical resistance at high temperatures is provided by alloys of aluminum and platinum [7440-06-4] or other precious metals (10–14).

The most technologically demanding types of coatings are those grown in precisely defined multiple layers. Experimental work is being done on multiple metal layer films using repetitive layers of two metals such as copper [7440-50-8] and nickel [7440-02-0], deposited in very thin and accurately repeated layers. These alloy-modulated layered electrodeposits may have increased tensile strength and electrical resistivity, higher saturated magnetization, or increased elastic modulus (15) (see MAGNETIC MATERIALS). All semiconductor chips start with a base layer of silicon or other semiconductor, then use a combination of bulk and selective deposition steps to give multiple layers of metals and insulators that form the three-dimensional devices. Printed circuit board processing is similar but on a much larger scale, beginning with laminated copper–epoxy sheets and ending with a complex multilayered structure of epoxy–glass, laminated copper, electroless plated and electroplated copper, plated nickel and gold, and molten applied solder.

Plating and galvanizing are the most common methods of applying metallic coatings, but many other processes, such as hot dipping, cementation, thermal spraying, sputtering, and chemical vapor deposition, have been developed (see also COATING PROCESSES). All metallizing techniques are potential sources of pollution, a subject of increasing ecological, economic, legislative, and public health interest. Metallizing has been combined with metal removal and associated functions such as physical or chemical surface modification, by the U.S. EPA as part of the "Metal Finishing" category for pollution control legislative purposes.

There are a plethora of commercially useful methods for applying a metallic coating. Many more techniques have been demonstrated in the laboratory. Each method has different critical parameters, ie, maximum and minimum coating thickness producible, substrate temperature, bulk or imagewise deposition, coating adhesion value, type and cost of coating equipment, labor requirement, scrap rate, rework capability, and safety and waste disposal aspects. For convenience, metallic coating methods may be divided into classes defined by the general way in which the metal is applied, eg, liquid-phase, gas-phase, and vacuum-phase metallizing, as well as metallizing by direct physical or thermal bonding.

The two largest volume processes, in terms of amount of metal used and amount of surface area coated, are the liquid-phase metallizing processes hot dip and coating electroplating (qv). One of the newer developments in hot dip galvanizing is its use on whole fabricated auto bodies giving lower cost resulting from ease of initial fabrication followed by complete zinc coverage, including into hidden and welded areas. This process, especially applicable to complex-shaped parts, had been used only on much smaller assemblies.

Many of the newer coating methods give metallized coatings that are quite distinct from simple single or multiple coatings of metals or alloys. Many of these newer coatings, which cannot be made except by one or a small family of processes, are defined as much by process method as physical and chemical structure. Examples of these coatings include ion-plated surfaces, laser heat-treated surfaces, modifications of existing surfaces to predetermined depths by selective additions of atoms or heat; and Teflon–electroless nickel composite coatings (6,7,12–14,16–19).

Liquid-Phase Metallizing Techniques

Hot Dip Galvanizing.   The largest single type of metallic coating process in terms of amount of metal used is hot dip galvanizing. In 1990 zinc [7440-66-6] production was ca 2 million metric tons (see ZINC AND ZINC ALLOYS). Approximately 40% of all the zinc mined (8×10^5 t) is used for zinc coatings of all sorts. Hot dip galvanizing accounts for over 90% ($>7 \times 10^5$ t) of the coatings usage and the proportion of steel (qv) which is coated with zinc is slowly rising. As of this writing (ca 1995) about 6% or ca 40 million metric tons of world steel production is zinc coated. Half of all galvanized zinc is used on steel coils, one-third is employed for coating of parts after fabrication, and the remainder for wires and tubes. The other uses, in approximate order, are electroplating, zinc-filled paints, zinc spray processes, zinc foils having conductive adhesive, and mechanical zinc plating (1,4,8,9,11,12,18,20,21).

In hot dip galvanizing, zinc is applied to iron (qv) and steel parts by immersing the parts into a bath of molten zinc. Whereas in principle almost any metal could be coated with molten zinc, this coating serves no worthwhile purpose on most metals. The combination of zinc and ferrous materials are almost uniquely suited to each other. Aluminum and cadmium are the only other similar combinations. Zinc provides iron parts with better corrosion protection by developing a coating of zinc and zinc compounds on the base metal surface. Protection is achieved for the following reasons. (1) Zinc has a much lower corrosion rate in comparison to that of iron in the atmosphere as well as in water (see

CORROSION AND CORROSION CONTROL). Zinc develops adherent corrosion products and iron does not. (2) Zinc is anodic to iron, thus corroding in place of iron. (3) The durability and wear resistance of zinc and the basal iron–zinc intermetallic compounds, Fe_3Zn_{10} [12182-98-8], $FeZn_7$ [12023-07-3], and $FeZn_{13}$ [12140-55-5], develop successively on the iron surface. The intermetallic compound formation is about one-tenth of the coating thickness and is essential for complete adherent bonding of zinc and steel. Any residual oxides or dirt inhibit this reaction. In addition, subsequent painting can easily and inexpensively be accomplished to increase protection further, or for decorative reasons. The best results are obtained using steels that contain <0.25% carbon, <0.05% phosphorus, <1.35% manganese, and <0.5% silicon. The primary factor in determining the degree of corrosion protection provided by a zinc coating, is thickness, which depends on the duration of immersion, the speed of withdrawal from the bath, and the temperature of the bath.

The effect of galvanizing on the mechanical properties of steels varies with the alloy and treatment. There is no effect on hot rolled steels, but increased strength resulting from cold working or heat treating is reduced by galvanizing. Impact toughness and fatigue strength are slightly reduced, as well as formability. Cracks can develop in the zinc coat upon bending. Residual stresses are reduced by 50 to 60% in welded structures. Hydrogen embrittlement does not result from hot dip galvanizing, unlike zinc electroplating where embrittlement can be a problem.

The pretreatment of the parts to be hot dip galvanized includes degreasing in a heated alkaline cleaner followed by acid pickling in dilute sulfuric or hydrochloric acid. The steel surface must be clean and completely wettable by the molten zinc. Usually the cleaned steel is immersed in an acid flux bath just prior to immersion in the molten zinc, to ensure instant wetting. Much zinc coated steel is marketed as being coated with raw zinc. The rest of the galvanized steel is given a post-treatment which depends on the application and the environmental conditions. A very thin chromate layer is used for relatively short-term protection, and where weldability is important. A chromate, or preferably a phosphate and a chromate coating, is used when the parts must be painted (see METAL SURFACE TREATMENTS).

The zinc coating is sacrificial, dissolving preferentially before the iron. The most important feature of this process is that holes and breaks in the coating have little effect as long as there is some zinc within a centimeter or so of any exposed iron. The zinc and iron, in the presence of any moisture and electrolyte such as absorbed carbon dioxide or salt, form a battery where the zinc is the anode and steel the cathode. This active passivation lasts as long as there is metallic zinc to corrode within transfer range of electrons donated to the iron (19). Zinc coatings are therefore superior to coatings of gold or other insoluble metals more noble than steel. If steel is coated with gold a different type of battery forms at every pinhole and break in the coating where gold is the cathode. Then the large cathode-to-anode ratio leads to extremely rapid steel corrosion at every pore.

The essential protective film on the zinc surface is that of basic zinc carbonate, which forms in air in the presence of carbon dioxide and moisture (Fig. 1). If wet conditions predominate the normally formed zinc oxide and zinc

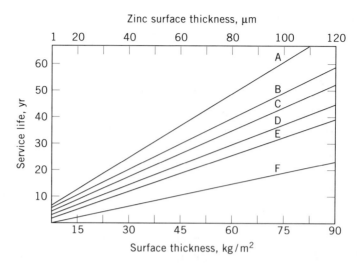

Fig. 1. Relationship between thickness of galvanized zinc surface and service life where service life is defined as the time to 5% rusting of the steel surface and A, use in a rural environment; B, tropical marine; C, temperate marine; D, suburban; E, moderately industrial; and F, heavy industrial. To convert kg/m^2 to oz/ft^2, multiply by 3.278.

hydroxide, called white rust, do not transform into a dense protective layer of adhesive basic zinc carbonate. Rather the continuous growth of porous loosely adherent white rust consumes the zinc then the steel rusts.

Fresh zinc surfaces cannot be painted because of the activity of pure zinc. Diffusion of moisture and oxygen through a paint film results in a reaction with the fresh zinc to form hydroxide ions, which causes the paint to blister and delaminate. Zinc surfaces must be allowed to age before painting to form a protective layer of basic zinc carbonate, or passivation treatments such as phosphating and chromating must be used (see METAL SURFACE TREATMENTS). Painting can also give protection against electrolytic corrosion caused by contact with more noble metals or alloys such as stainless steel, copper, and brass, as long as the nonconductive paint separates the metals.

Dry galvanizing, developed in Europe, and wet galvanizing, more commonly used in North America, are two types of conventional batch galvanizing in wide use. In the dry galvanizing process the material is degreased, acid pickled, immersed in an aqueous flux solution, and dried before immersion into molten zinc. Wet galvanized parts are not prefluxed after cleaning and pickling, but are placed into the molten zinc through a top flux blanket on the kettle. The purpose of the flux blanket is to remove the small amounts of impurities left on the surface after degreasing and pickling. The usual operating temperature of the molten zinc bath is 445–465°C, but higher (550°C) temperatures are used for high silicon steels. Whereas the immersion time controls the alloy layer coating thickness, the immersion speed influences coating uniformity. The rate of withdrawal determines the thickness of the unalloyed zinc layer which is left on the parts. When a low rate of withdrawal to decrease the thickness of the zinc layer is not practical, the parts can be spun or blown using live steam.

Compressed air is normally used on continuous coil galvanizing at speeds up to about 200 m/min. Various feedback controls are used to maintain the desired coating thickness.

The surface of galvanized steel is characteristically rough and irregular. Various surface finishes are also available on continuous sheet galvanized steel, from full spangle, ie, large zinc crystals, to minimum spangle, ie, small crystals, to wire brushed smooth surfaces. The spangle, or zinc crystal size, results from the rapid crystallization of the molten zinc, and from the condition of the bare steel. Minor additions to the molten zinc can lower the melting point of the zinc, and decrease the spangle size. One-side galvanized steel is made by coating one side of the sheet with a sodium silicate solution which is nonwettable in molten zinc. The silicate is then removed with a hydrofluoric acid-based etchant (see FLUORINE COMPOUNDS, INORGANIC). Galvanized steel coatings contain globules of lead (qv), forced to the surface during zinc crystallization. The lead is an impurity in the zinc, usually at <0.5%. Lead does not form an alloy with zinc so it segregates upon cooling. Another process, galvannealing, is used to heat treat the zinc coating on-line to convert the whole coating to an iron–zinc intermetallic compound for better paintability. Numerous commercial grades of galvanized thicknesses are available, from thinner than the standard commercial-grade of 0.275 kg/m^2 (0.19 μm thickness per side) to coatings three times as thick.

Over half of the total tonnage of metallic coated parts is produced by galvanizing of steel strip. The majority of this is produced on continuous high speed coil coating lines, often operating at up to 15 km/h or more. An additional large quantity of electrogalvanized steel coil is produced by electroplating from aqueous zinc sulfate or chloride baths. This thinner but smoother zinc coating is used where a higher quality or painted finish is necessary, and where lesser corrosion resistance is tolerable (Fig. 2). Electroplated steel strip is preferred when used as prefinished, fully painted strip stock sold for final forming into finished parts by the fabricator. The automotive industry consumes a large amount of these materials. Newer developments in electrogalvanized steel are the use of more corrosion-resistant sacrificial zinc alloys such as those containing iron, nickel, and cobalt.

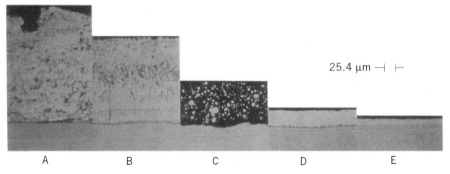

25.4 µm ⊣ ⊢

A   B   C   D   E

Fig. 2. Microstructure of various zinc coatings where A is metallized; B, hot dip galvanized; C, zinc-rich paint; D, galvanized sheet; and E, electroplated.

Hot Dipping Using Other Metals.   *Coating with Aluminum.*   The numerous methods of applying aluminum coatings include continuous and batch hot dipping, pack diffusion, slurry processes, thermal spraying, and cladding, as well as vacuum, chemical, and ion-vapor deposition. Regardless of method, the essential factor for successful coating is proper preparation of the steel surface. The iron oxide scale, as well as adsorbed moisture and gas, must be removed.

Hot dipping of aluminum is normally applied only to steel. Aluminum coatings are harder and more durable than zinc coatings and give some galvanic protection, but must be about twice the thickness of zinc for acceptable performance. The high melting point of aluminum compared to that of zinc can give more softening of cold worked steel when hot dipped (12,14).

The most widely used method for continuous hot dip coating is the Sendzimir process, which consists of oxidizing the cleaned steel surface, reducing the surface oxide in a reducing atmosphere, and immersing the steel into molten aluminum or zinc. In this method no flux is needed, but the strip must be cooled and protected by hydrogen until it enters the molten metal. Other processes may also use a reducing atmosphere but without a preliminary oxidation to form a more uniform oxide. The Lundin process uses aqueous fluxes instead of a reducing atmosphere. Hot dip coatings decrease the tensile strength of cold drawn wires.

In batch hot dip coating, cleaning is achieved by hot alkaline cleaning, water rinsing, abrasive blasting, acid pickling, and drying. Cleaning is followed by fused salt fluxing, dipping into molten aluminum, centrifuging, and air blasting or shaking. Coatings consist of an intermetallic layer of iron−aluminum alloy phases, FeAl and Fe_3Al, and a pure aluminum overlay. The thickness of the intermetallic layer depends mainly on the immersion time, but also on the composition of the steel and the aluminum bath. The total coating thickness is determined by the viscosity of the aluminum bath and the speed of withdrawal. Aluminum hot dipping is difficult to control because the intermetallic layer can grow very quickly, even leading to spalling of the coating. Coating times are generally 2−15 seconds.

The corrosion protection of aluminum depends more on aluminum passivation than on sacrificial corrosion. Thus aluminum coatings are less effective than zinc in inhibiting red rusting around breaks in the coating. Aluminum coatings last appreciably longer than zinc ones and are easier to paint. Aluminum passivation results from formation of layers of various hydrous aluminum oxides and hydroxides giving an adherent protective layer which greatly reduces the corrosion rate of the aluminum. Aluminum can be given a chromate or phosphate plus chromate conversion coating to further decrease the corrosion rate and to give a better surface for painting.

Coating with 55% Aluminum−Zinc Alloy.   A coating with a nominal composition of 55% aluminum, 43.4% zinc, and 1.6% silicon provides the durability and high temperature resistance of aluminum coatings, and the corrosion protection of zinc coatings. This coating is known as Galvalume or Zincalume. Silicon is added to control the growth of the intermetallic boundary layer. The alloy owes its superior corrosion protection to a particular dendritic microstructure which consists of 80 vol % aluminum, 22 wt % zinc-rich interdentritic aluminum−zinc eutectic alloy, and a few silicon-rich particles. The zinc-rich interdendritic material corrodes preferentially; the aluminum-rich dendrites act as a mechanical trap to retain zinc corrosion products. The coating is bonded to steel by a layer of a quaternary Fe–Al–Zn–Si intermetallic compound, which comprises about 10% of the total coating thickness (20 μm). The coating process is similar to hot dip galvanizing, consisting of cleaning, heat treating, coating, and fast cooling. Heat treating achieves stress relief by recrystallization of steel, while forced air cooling is needed to produce the correct microstructure for maximum corrosion resistance.

Galvalume has been shown to have two to six times the life of an equivalent thickness of zinc, including marine atmospheres. For high temperature oxidation resistance up to 700°C, Galvalume is equivalent to pure aluminum.

Hot Dip Tin Coating of Steel and Cast Iron. Hot dipping of tin [7440-31-5] has been largely superseded by electrolytic coating techniques, especially for sheet. However, hot dipping can be the method of choice for complex and shaped parts. Very thin layers of tin are extensively used to passivate steel used for canned goods. Tin is essentially nontoxic, is nearly insoluble in almost all foods, and easily wets and completely covers steel with a pinhole-free coating.

Hot dip tinning applies a thin (3.8–18 μm) coating of molten tin to provide nontoxic protection for food handling, packaging, and dairy equipment. More than 90% of all tin-coated steel is used for food processing (qv). The tin coating facilitates soldering and assists with metal-to-metal bonding during can forming and lid attachment. Low carbon steels give the best results, but medium to high carbon steels work well given extra care in pickling. High alloy steels, especially high chromium stainless steels, are difficult to coat satisfactorily using molten tin. A wide range of compositions of cast irons tin easily, regardless of the total carbon, silicon, manganese, sulfur, and phosphorus analyses. Difficulties may be experienced when there are large quantities of graphite or buried silicon oxides. The actual interfacial coating is a thin intermetallic alloy of $FeSn_2$ [12023-01-7]. This alloy forms by diffusion so is limited to a practical thickness of about 0.5 μm. Pure tin comprises the remainder of the coating.

The metal parts must be cleaned, degreased, pickled, and fluxed prior to tinning. Iron and steel parts must be free of all surface contaminants such as oil, drawing lubricants, and scale. Abrasive blast cleaning is also used on castings to increase the surface area for better adhesion and retention of tin. Fluxing facilitates the reaction between molten tin and iron or steel by promoting the formation of iron–tin intermetallic compounds. These initial intermetallic layers provide a low energy surface which allows the molten tin to spread in an even, smooth film. The fluxes commonly consist of mixtures of zinc, ammonium, and sodium chlorides plus some free hydrochloric acid. Similar fluxes are used in almost all types of steel coating by hot dipping, and even in soldering of various alloys to copper and nickel. These fluxes may be applied from an aqueous dip tank, from a fused layer floating on top of the molten tin, or as a paste with tin powder.

Those coatings which do not require the highest quality finish or which are used prior to soldering or bonding are produced by single-pot tinning. Higher quality thick coatings are achievable by two-pot tinning in which the second bath is covered with a layer of high flash point oil or grease instead of flux. The flux residues from the first tinning are absorbed or displaced by the oil and the retained thin oil coating protects the items during shipment and storage. The final coating from a two-pot system is much more pure because no iron contaminates the surface.

Hot Dip Terne Coating. Terne coatings are lead or lead alloy coatings, usually containing 2–25% tin. Lead alone does not alloy with iron. Other elements must be added to form an adherent interfacial iron alloy. Tin and antimony are common additives. An iron–tin intermetallic layer provides an excellent metallurgical bond which adheres well to lead. Lead acts as a lubricant to facilitate forming and drawing of coated sheet, provides excellent solderability and weldability, and provides good corrosion resistance. Terne coatings cannot be used in any food handling items and are not recommended on general unpainted consumer parts, owing to the lead content. Terne coatings are used for gasoline tanks for cars, trucks, lawnmowers, and boats; oil filter cans, air filter holders; chassis for radios and other electrical equipment; roofing siding, gutters, and downspouts; and electrical hardware. Terne coatings are also commonly made by electroplating.

All parts must be given a thorough cleaning prior to dipping, including a final flux coating. Coating thickness is generally between 5–15 μm. It is controlled by the bath temperature (325–390°C), immersion time, and degree of shaking or centrifugation before coating solidification. Coating thickness is controlled to 2–5 μm for sheet or strip by an air knife system of high pressure jets of air or nitrogen against the molten coating.

Solder Coatings. Soldering, brazing, and welding (qv), are all used for joining metals (see also SOLDERS AND BRAZING ALLOYS). Two metallic objects are joined in a weld by making the interface molten, as by an electric current, friction, electron beams, etc. Brazing is a special type of welding in which various fluxes are used to help clean and bond the two surfaces. Discussions of brazing and welding are available (22).

Solder and other fusible lead and tin-based coatings are used for several applications in printed circuit production. The most commonly applied coating is a 63:37 tin–lead eutectic alloy, but 90:10, and other coatings can be used. Any alloy or metal which melts at a relatively low temperature and forms a bond with the underlying and overlying metals can be used as a solder. Alloys of tin, lead, bismuth, silver, and indium are the most common combinations. There is much interest in finding an economical substitute for lead-containing solders, in order to reduce environmental lead exposure, and many novel alloys have been introduced. A practical tin–lead solder substitute must possess a number of attributes, including reasonable cost, recyclability, good joint strength, heat and cold resistance to recrystallization, corrosion resistance, acceptable liquidus temperature, easy resolderability for rework purposes, and low toxicity. Thus pure tin is not an acceptable solder because it melts at too high a temperature, recrystallizes to form nonadherent crystals at low temperatures (tin pest), and grows conductive short-circuiting whiskers between active circuits on contaminated circuit boards at high humidities (23–27).

Solder is used either as the final coating allowing permanent bonding of electrical connectors to copper or nickel-coated copper parts, as a final protective layer on copper circuits to inhibit corrosion, or as a temporary etch-resistant coating when applied imagewise to the printed circuit board. In the last case solder or tin is usually applied by imagewise electrodeposition on the final circuit paths prior to photoresist removal. The photoresist is then stripped and the exposed copper is etched using ammoniacal copper etchant or sulfuric acid–hydrogen peroxide to make the final circuits. Finally the protective solder layer is chemically stripped from the copper to allow for final processing and inspection. Any future deposits of solder are made by hot dipping or direct soldering, not by electroplating (see RESIST MATERIALS).

Bulk solder coatings are applied by immersion in fused solder. The surface or electrodeposited surfaces are smoothed and leveled by hot oil, hot air, or infrared reflow. There are cases where electroplating is used to give a functional tin–lead solder. The electrodeposited solder is normally heated to the liquidus temperature (reflowed) by one of the means described to give a smooth, pore-free surface. Several different tin–copper alloys form during reflow or hot dipping, and more slowly at room temperature. These alloys are important to give high wettability and bonding strength. Lead does not form an alloy with copper, so the bulk of the solder becomes enriched in lead.

Solder coatings are more difficult to apply to nickel than to other metals. Nickel passivates easily and quickly to form nickel oxides. The nickel oxides inhibit good bonding and need aggressive treatments for removal. Often a thin (<0.1–0.5 μm) gold layer is used as a temporary protective layer on the nickel. Contact with molten solder instantly dissolves the gold, bringing the solder in contact with the clean, oxide-free nickel surface to which the solder bonds. The disadvantage of using gold is that a gold–tin intermetallic layer forms which weakens solder bonds. The cost of gold is so great that molten solder pots are recycled for their gold content when the gold comprises only a few tenths of a percent.

Electroplating. Many developments in electroplating (qv) have been driven by increased coating functionality and economy, others by environmental and legislative compliance. This field has expanded rapidly. Only some of the developments are discussed herein.

The most common electroplated metal is zinc, followed by nickel, copper, and chromium. Many other metals and metaloids can be electroplated, including manganese, iron, cobalt, gallium, germanium, arsenic, selenium, ruthenium, rhodium, palladium, silver, cadmium, indium, tin, lead, bismuth, mercury, antimony, gold, iridium, and platinum. Alloy deposition is common, although the choice of practical alloy coatings is generally somewhat limited to tin–lead, tin–nickel, zinc with cobalt, nickel, or iron, brass (copper–zinc), bronze (copper–tin–zinc), and numerous colored gold alloys. Many other alloys can be electrodeposited, but generally lack commercial utility as a surface coating (3,6,7,15,28,29).

Microcracked and microporous chromiums are a unique way to increase the corrosion resistance of a combination of metals. The top coating is more corrosion resistant than the other metals. A totally pore-free chromium layer may be corrosion resistant, but impossible both to fabricate and to keep in this condition throughout the use of the product. The chromium baths contain additives which deposit this metal in a highly stressed condition, resulting in innumerable tiny pores or cracks in the thin (0.25–1 μm) chromium coating. These openings spread out the cathodic battery current from the passive chromium over a relatively large surface area of exposed metal such as nickel. The currents are either dissipated harmlessly or cause general uniform corrosion instead of pitting corrosion to the base metal.

Semibright and bright nickels are used in combination with chromium baths to increase the effective life of plated parts. The automotive industry has developed the Step test which allows measurement of the electrochemical potential of different nickel coatings. Nickel plating for all uses worldwide annually consumes 81,700 metric tons, about 11–12% of world production of nickel. Of this amount, about 80% is used for decorative plating and 20% for electroforming and engineering plating, such as electroless nickel (see ELECTROLESS PLATING).

Originally decorative chromium coatings consisted of an iron or other metal or plastic base, plated with copper, nickel, and pore-free chromium. Pitting corrosion was common and rapidly produced visible deposits of green copper and red iron corrosion products. As of this writing two separate nickel layers are used, plated from different bath chemistries. The top bright nickel coating is more electrochemically active than the lower layer. The top layer corrodes preferentially instead of the lower, less active semibright nickel layer. Thus the top nickel layer acts as a sacrificial coating whereas the bottom nickel layer continues to provide corrosion resistance.

Environmental Aspects.   Chromium metal seems to be biologically harmless owing to inertness. Trivalent chromium Cr(III), one of the essential mineral nutrients (qv), has not been shown to be harmful. Hexavalent chromium, Cr(VI), found in chromic acid and used in electroplating baths, is very toxic as well as a suspected carcinogen. Two basic types of chromium are plated: hard chromium up to 100 μm or more, and decorative chromium at 0.25–1.0 μm. No replacement coating has been found for thick hard chromium deposits for wear resistance and parts salvage, although electroless nickel can partially substitute for chromium plating baths. Many decorative applications are being converted to Cr(III). Products from the newest Cr(III) baths are essentially equivalent to those from the older decorative Cr(VI) baths, although some cosmetic color differences exist which have prevented complete conversion to the Cr(III) baths.

Cyanide solutions were formerly ubiquitous in electroplating shops. The best cleaners contained sodium cyanide. Zinc, copper, cadmium, gold, silver, and other metals plated easily from cyanide solutions, which were simple to control and tolerant of impurities. The removal of cyanides (qv) from all plating baths has been a general goal since the latter 1980s. As of 1995 cyanide-containing cleaners are rare and used for special purposes. Copper cyanide strike baths are still needed for plating an initial copper layer on iron, zinc, and other active metals which cannot be directly plated in acid copper sulfate plating baths, but several noncyanide copper strike baths have been introduced. Many noncyanide gold baths are known, but often lack the necessary color, hardness (qv), or alloy composition possible using cyanide golds.

The greatest tonnage decrease in cyanide plating has occurred for zinc plating. As of this writing less than one-third of all zinc is plated from cyanide baths. The remainder is plated from alkaline noncyanide, zinc potassium chloride, and zinc ammonium chloride baths. Newer, slower corroding zinc alloys have been developed which plate from alkaline or acidic baths. These include zinc–iron, zinc–nickel, and zinc–cobalt coatings which are being used in the automotive industry.

Chlorinated and fluorinated cleaners have been widely used as degreasers during surface preparation prior to plating. Recognition of the role of many of these solvents as either global warming gases or ozone depletion agents has led to prohibitions in continued use (see AIR POLLUTION; ATMOSPHERIC MODELING). Many new and improved cleaning systems are being developed, such as alcohol-based cleaners, emulsion cleaners, and cleaners based on natural bioproducts such as limonene. During this process a massive reformulation of many aqueous cleaners has also occurred. These changes are more profound than the simple removal of cyanide. The newer formulations are based on less toxic surfactants (qv) and fewer chelating and complexing agents that are difficult to treat as waste. Many cleaners are formulated to displace rather than emulsify oil, and are used with micro- or ultrafiltration (qv) to extend usage life.

Cadmium usage, illegal in most of Europe, is being discouraged elsewhere. The U.S. military has cadmium specifications for electronic, fastener, and marine equipment, which requires only cadmium. Tin is being substituted for tin–lead as a metallic etch resist during printed circuit board production.

Fused Salt Plating. Fused salt electrolysis has been used in mining and metallurgical technology since the 1800s, in order to win alkali and alkaline-earth metals as well as for the production of tantalum and aluminum. Win or winning is a classical term meaning production of metal at an electrode from its soluble salts. Interest in fused salt bath electroplating, ie, electrolytic deposition of metals from salt melts, has grown in the latter 1990s for the following reasons. (1) Certain technically important metals, eg, Nb, Ta, Ti, Zr, W, Mo, or Al, cannot be plated from aqueous electrolytes because of their great affinity for oxygen. (2) Metallic coatings from fused salt baths have properties such as high purity and ductility owing to the undisturbed crystal growth that results from low or no stress, as well as the capability of being deposited as thick films. (3) It is possible to achieve high deposition rates and high throwing power from salt melts because of the high conductivity of the carrier melts and the consequent low electrolytic voltages for deposition. (4) Diffusion layers can be produced whenever the temperature and the deposition rate are adjusted to the diffusion rate of the deposited metal into the basic raw material. Finally, (5) economic usage of expensive metals is feasible. This factor becomes more important as metals rise in price and fused salt technology progresses.

Fused salt electrodeposition of metals is used primarily for providing corrosion resistance, ie, protection from oxidation for the electrochemical activation of certain metals such as titanium. This technique offers a great deal of potential for galvanoplastic electroforming parts from metals and intermetallic compounds which as of this writing can only be formed mechanically with great difficulty, or not at all. This category includes the high temperature resistant metals which are also generally rather brittle, such as molybdenum, Mo, and tungsten, W, the platinum metals iridium, Ir, and rhodium, Rh, and intermetallic compounds such as zirconium diboride [12045-64-6], ZrB_2, and titanium diboride [12045-63-5], TiB_2 (see REFRACTORIES; REFRACTORY COATINGS; TOOL MATERIALS).

The most common fused salt baths are complex mixtures of alkali chlorides, rigorously purified and dried. Fused salt plating must be done under an inert atmosphere. Often argon is used because nitrogen can react with some metals. Inert anodes, eg, Pt-coated titanium or graphite, are used and the plating metal is supplied by additions of an appropriate metal salt.

Diffusion coatings of metals can be applied in the same type of fused salt bath, except without an electric current. The parts are suspended in the molten salt bath containing a metal salt. The metal ions react with the metal surface of the part, are reduced to metal, and diffuse into the bulk metal. One bath for chromizing uses a mixture of 40 mol % each of NaCl and KCl, and 20 mol % $CrCl_2$. Chromium carbides form in the same way as in pack diffusion chromizing. Many other molten baths have been tested or proposed, including baths based on borax, boric acid and other salts, molten calcium, and molten lead. In each case the coating metal has at least 0.1–1 wt % solubility in the molten bath in order to form rapid coatings (30).

Plating from Nonaqueous Solvents.   Metals which are more electrochemically active than water usually cannot be plated from an aqueous solution. There are a few exceptions, such as zinc, where special factors allow plating. Aluminum has been the subject of the most research and application, although almost any metal could be plated from nonaqueous solutions. In general, aqueous solutions are both easier and cheaper to use if there is a choice. The solutions used for aluminum plating have to be oxygen free. These are usually mixtures of an aromatic molecule such as benzene or xylene, and anhydrous aluminum chloride as the metal source. Other chlorides such as ammonium or sodium chlorides are added to give sufficient electrical conductivity. The resultant coatings of aluminum are identical to other plated coatings. Hot dip and vapor or sputtered aluminum are less expensive, but electroplated aluminum may be useful for heat-sensitive or special substrates.

Nonaqueous electrolyte systems typically have a larger voltage window of solvent stability thus there is a greater flexibility in selecting cell operating voltages. Also, salts do not hydrolyze and solute selection is simpler. Nonaqueous electrolytes generally do not react with the substrates that react with water, and can be used when water-reactive metals need to be coated with metals normally deposited from aqueous solutions. The coating of uranium with zinc, tin, or nickel is an example of such an application. Even though aqueous electrolyte systems for electrodepositing these metals exist, the aqueous systems do not provide adequate adhesion.

Organic–solvent-based electrolytes have the advantage that a wide variety of complex ions can exist in solution. Metal–solvent complex formation should not be too strong, however, because although there is sufficient conductivity, deposition may not occur. Examples of generally unsuitable solvents, including those containing acidic hydrogen, are alcohols, ketones, acid anhydrides, amines, and amides.

Disadvantages associated with some organic solvents include toxicity; flammability and explosion hazards; sensitivity to moisture uptake, possibly leading to subsequent undesirable reactions with solutes; low electrical conductivity; relatively high cost; and limited solubility of many solutes. In addition, the electrolyte system can degrade under the influence of an electric field, yielding undesirable materials such as polymers, chars, and products that interfere with deposition of the metal or alloy.

Some inorganic nonaqueous solvents can be used in systems operable at near room temperature, eg, thionyl chloride; others, however, require special handling, eg, liquid ammonia, which must be used below its boiling point of $-33°C$ in a thermally insulated container and in an inert atmosphere.

Brush Plating. A modified electroplating process, done without immersing the part in a plating tank, is called brush plating. Typically, the part is connected to a direct current source as the cathode. A highly concentrated, pH-buffered plating solution is soaked into a saturated pad attached to an insoluble anode. The pad is rubbed over the surface of the part, often while pumping additional solution over the contact area. Metal deposition occurs quickly owing to the concentrated solutions, high current densities, and high agitation at the contact area. Brush plating is useful in repairing nicks in printing press rollers or hydraulic cylinders without disassembling the machines. All platable metals can be applied by this process (3).

Electroless Plating. The metallizing process known as electroless plating (qv) is mainly used for deposition of copper on plastics and for nickel–phosphorus alloy on plastics and metals. A smaller amount of nickel–boron, nickel–copper–phosphorus, palladium–boron, palladium–phosphorus, silver, gold, and gold–boron alloys are deposited. Formaldehyde (qv) is the most common copper reducing agent, giving pure coatings. Sodium hypophosphite is the agent mainly used for electroless nickel. An unusual nickel–phosphorus alloy or solid solution is formed. Unlike electroplating, electroless plating can be used on almost any substrate, metallic or nonmetallic. Often electroless plating is used as the first coating to make glass (qv), ceramic, or plastic conductive, followed by conventional electrolytic plating (10,31–34).

Improvements in this process include electrolytic regeneration for life extension of chromic–sulfuric acid baths used for etching plastics, and alkaline permanganate baths used for etching printed circuit board plastics prior to electroless plating. An initial immersion zinc coating called a zincating bath is used to prepare aluminum for electroless nickel plating. Noncyanide zincates are available. Specially formulated inexpensive electroless nickel strike baths are used to dissolve off the zincate layer and deposit an initial thin nickel layer. The rinsed parts are then placed in the full build electroless nickel bath, which has greater life and better deposit properties owing to elimination of zinc contamination. The inexpensive strike bath is replaced frequently. Newer electroless nickels have improved bath life, tolerance to impurities, and corrosion resistance. Electroless nickel composite coatings are becoming more common. Almost any particulate material can be incorporated into an electroless nickel coating, from fluoropolymers for increased lubricity, to silicon carbide for increased wear resistance. Electroless nickel is used to coat diamonds for better adherence to the bulk material when fabricated into grinding tools.

Many electroless coppers also have extended process lives. Bailout, the process solution that is removed and periodically replaced by liquid replenishment solution, must still be treated. Better waste treatment processes mean that removal of the copper from electroless copper complexes is easier. Methods have been developed to eliminate formaldehyde in wastewater, using hydrogen peroxide (qv) or other chemicals, or by electrochemical methods. Ion exchange (qv) and electrodialysis methods are available for bath life extension and waste

minimization of electroless nickel plating baths (see ELECTROSEPARATIONS, ELEC-
TRODIALYSIS).

Immersion Plating. A simplified aqueous metal deposition process which
does not use electric current, immersion plating, works only when a metal of
higher electromotive force, such as copper, is deposited on a metal of lower
electromotive force, such as iron or aluminum. The coatings are typically very
thin and porous. This process is not often used for applications where durability
is important unless the product is overcoated using a protective transparent film.
One important application is immersion plating of copper from a copper sulfate
bath onto steel wire, to use as a drawing agent. After drawing the steel wire,
the copper is redissolved and reused. Zincate solutions are commonly used to
apply an initial zinc layer on aluminum film prior to electroless nickel plating.
These zinc coatings are dissolved by the electroless nickel plating solution, which
immediately plates a thin nickel layer on the clean, oxide-free aluminum. This
process gives improved adhesion and corrosion resistance of the electroless nickel
deposit. Very thin immersion gold deposits are used for inexpensive jewelry (3).

Miscellaneous Techniques. Lasers (qv) have been used for both electroless and electrolytic plating. A semiconductor illuminated through a plating solution by a laser beam emits electrons which reduce dissolved metal ions to the
metallic form. Whereas electroless plating solutions can also be illuminated, the
heat of the laser beam alone can cause initiation of metal deposition by forming catalytic nuclei of the metal. High intensity light has, however, been used to
inactivate electroless plating catalyst, to allow selective plating.

Selective dissolution has been used from ancient times to give the appearance of a thin plated coating of precious metal. A copper alloy containing some
gold or silver can have part of the copper surface layer dissolved by weak acids,
after which the item can be burnished or buffed to give a dense, rich appearing
coating. More recent uses include pack aluminizing on nickel to form the NiAl
[12003-78-0] intermetallic. The aluminum is then removed with caustic to form
a very high surface area coating of nickel which can be used like Raney nickel
in catalyst applications. The advantage is that foils or screens can be made and
then partially converted to the catalyst.

Mercury layers plated onto the surface of analytical electrodes serve as
liquid metal coatings. These function as analytical sensors (qv) because sodium
and other metals can be electroplated into the amalgam, then deplated and
measured (see ELECTROANALYTICAL TECHNIQUES). This is one of the few ways
that sodium, potassium, calcium, and other active metals can be electroplated
from aqueous solution. In one modification of this technique, a liquid sample can
be purified of trace metals by extended electrolysis in the presence of a mercury
coating (35).

Gas-Phase Metallizing Techniques

Metal or Thermal Spray Coatings. There are several ways to provide a
metal film on a plastic part through the use of atomized metal at atmospheric
pressure. Thermal spray, metal spray, or metallizing are common names for a
group of processes for depositing metallic and nonmetallic coatings. These spe-

cific processes include plasma arc spray, flame spray, laser spray, and electric arc spray, depending on the energy input source. Rods, wires, and powders are used as coating material sources. Typically the coating material is melted, then atomized and forced onto the prepared substrate by a high velocity gas stream. Bonding is most often simply mechanical. The coating particles freeze interlocking on the roughened substrate surface. Some localized diffusion and alloying can also occur. Thermal spray metal and ceramic coatings have diverse properties suitable for numerous applications including corrosion resistance, eg, zinc and aluminum, especially against oxidation or salt water corrosion; high temperature oxidation, eg, nickel, cobalt, and chromium alloys; electrical conductivity, eg, radio frequency interference (RFI) shielding by zinc or tin on nonconductors; electrical resistance, eg, insulating layers in induction heating coils and high temperature strain gauges; wear resistance, eg, chromium–nickel–boron alloys and carbide-containing coatings; catalytic surfaces; nuclear moderators; and dimensional buildup for salvaging worn metal parts (17,36–39).

A great advantage of zinc arc spray is that it can be applied to almost any plastic. The most common alternative technique, electroless plating, is normally only useful on plastics such as acrylonitrile–butadiene–styrene (ABS), polyphenylene oxide, epoxy, polycarbonate, and ABS–polycarbonate alloys, which can be chemically etched to improve the adhesion of deposited metals. Thermoset plastics rarely present a problem when spray metallization is used as long as the application temperature is below the heat-distortion temperature of the plastic. Excessive metal temperature can cause surface degradation, giving poor adhesion. Thermoplastic materials are more difficult to coat. The surface temperature should rise as little as possible to prevent warpage. Even sensitive thermoplastics such as foamed polystyrene can be acceptably coated with molten zinc using sufficient care. The adhesion of the zinc coating to the plastic surface is relatively poor. The plastic parts are sandblasted to provide a more adherent surface or are given a special primer coat that promotes adhesion.

Zinc arc spray, also suitable for prototypes and small lots of materials, is less suited for very small parts and parts having blind holes or complex interior surfaces, or where warpage is a problem. Zinc flame spray is normally applied to a thickness of 0.05–0.10 mm which can lead to appreciable weight gains on a large complex part such as a high speed printer or computer cabinet. Large numbers of parts and small parts are often best metallized by other processes because zinc arc spray is a slow, manual, serial coating process. The coating itself is brittle and can flake off when rubbed or bent. For RFI shielding, zinc coating is normally applied only to the part interior.

Application of an adhesion-promoting paint before metal spraying improves the coating. Color-coded paints, which indicate compatibility with specific plastics, can be applied at 20 times the rate of grit blasting, typically at 0.025-mm dry film thickness. The main test and control method is cross-hatch adhesion. Among the most common plastics coated with such paints are polycarbonate, poly(phenylene ether), polystyrene, ABS, poly(vinyl chloride), polyethylene, polyester, and polyetherimide.

Health and Safety. Zinc arc spraying or flame spray equipment is hardly more hazardous than a welding torch, and only safety goggles and gloves are required. Safety aspects emphasize reduction of noise and vapor inhalation. The gas flow through the jet is nearly at supersonic speed, and because of the high noise level, ear protection is required. The operator normally uses a positive-pressure fresh-air breathing helmet or a full-body protective covering having an integral air supply to eliminate exposure to toxic zinc fumes. Spraying must be done in a high velocity hood, ie, minimum 92 m/min. Metal dust is easily filtered out in a wet or dry collector to eliminate pollution.

Thermal spray processes can be used to give coatings of chromium carbide or nickel chromium for erosion resistance, copper nickel indium for fretting resistance, tungsten carbide cobalt for wear and abrasion resistance, and even aluminum silicon polyester mixtures for abradability.

Thermal or Flame Spray Process.   The earliest experiments in metal spray used molten metal fed to a spray apparatus, where it was dispersed by a high speed air jet into tiny droplets and simultaneously blown onto the surface of the part to be covered. The metal solidified on contact. Modern processes use a more convenient source than premelted metal. Spray heads using a flame or an electrical arc to melt metal wires or powders directly are much more convenient. These are the only types used on a large scale in the United States.

Flame spray utilizes combustible gases such as propane, acetylene, and oxygen–hydrogen mixtures as the heat source to melt the coating material. The spray head contains orifices for the flammable gases in addition to the compressed gas used to blow the molten metal on the surface of the part. The sprayed materials can be used in either rod, wire, or powder form. Flame sprayed coatings exhibit lower bond strength, higher porosity, a narrower working range, and higher heat transmittal to the substrate than coatings obtained by other spray methods. However, this process also has low capital investment, high deposition rates, and high efficiency of deposition, plus relatively low costs of maintenance.

Surface cleanliness of the substrate is important for all thermal spray processes. Degreasing, which formerly often relied on freons or chlorinated hydrocarbons, is done with less environmentally harmful solvents and aqueous cleaners. Surface roughening, commonly used to increase adhesion, is done by rough threading and grit blasting using alumina, sand, crushed steel, or silicon carbide. Thorough cleaning to remove particulates follows grit blasting. The as-sprayed parts are usually rough and porous, requiring some type of finishing treatment which may consist of low viscosity epoxy coating, painting, machining, lapping, or polishing.

A modification of thermal spray processes is the flame spray and fuse in which the coating material is fusible and self-fluxing, requiring post-spray heat treatment. Fusible self-fluxing coatings are typically nickel or cobalt-based alloys which use boron, phosphorus, or silicon, singly or in combination, as melting point depressants and fluxing agents. Spraying is followed by fusing of the surface using flame or torch, induction, or vacuum, inert or hydrogen furnaces. Fusing is usually done between 1010 and 1175°C. These coatings can have very good wear resistance. The hardness can be as high as 65 HR_c (Rockwell hardness) and the wear resistance can be further improved by incorporating tungsten or chromium carbide particles into the alloy powders. These coatings are fully dense, have close-packed crystal lattice, and exhibit good metallurgical bonds. There is a limitation on the substrates to those which can tolerate the high fusing temperatures. The temperature of the part being coated is controlled by the application rate.

Another modification is the detonation gun process, in which mixtures of oxygen and acetylene are exploded in the combustion chamber. Metal powers are metered into the chamber. The shock wave of the supersonic explosion (2770 m/s) propels the powders to a speed of up to 770 m/s. The 3000°C combustion gases also heat the particles. Bond strength is exceptionally high with low porosities. Most coating materials are oxides or carbides (qv), plus bonding materials of Co and NiCr. Disadvantages of this process include line of sight coating, the danger of the high speed particles, and the explosions, which require soundproof enclosures and remote operation. The detonation gun process is one of the most expensive of the spray coating processes, but coating life can outlast conventionally sprayed coatings by up to eight times.

Flame spray metallizing is widely used for the protection of metal against corrosion, especially for *in situ* protection of structural members. The principal metal used for spraying of plastics is zinc. Aluminum and copper are also used. If the distance from the part is too great, the zinc solidifies before it touches the part and adhesion is extremely poor. If the molten zinc oxidizes, conductivity and adhesion are poor. If the distance is too short, the zinc is too hot and the plastic warps or degrades. These coatings are not as dense as electrically deposited coatings because of numerous pores, oxide inclusions, and discontinuities where particles have incompletely coalesced.

Arc Spray Process. The least expensive among the thermal spray processes is the electric arc spray process which uses metal wires as spray materials. A plasma, an excited gas, consisting of equal amounts of free electrons and positive ions (see also PLASMA TECHNOLOGY) is formed by ionizing a primary gas, usually argon or nitrogen, using the electrical discharge from a high frequency arc starter. Two electrically isolated zinc or other metal wires connected to a high voltage transformer are slowly fed into a central chamber where an arc is formed at their intersection. A secondary gas such as nitrogen, helium, or hydrogen can be added to produce higher temperatures at lower power level. Once initiated, plasma can conduct currents as high as 2000 A d-c, at potentials of 30–80 V. A high velocity stream of compressed air simultaneously pushes the vaporized zinc out of the chamber and onto the surface of a part. A typical plasma arc spray gun is shown in Figure 3.

Process variables that must be controlled include the power level, pressure, and flow of the arc gases, and the rate of flow of powder and carrier gas. The spray gun position and gun to substrate distance are usually preset. Substrate temperature can be controlled by preheating and by limiting temperature increase during spraying by periodic interruptions of the spray.

Modified processes represent improvements in capability and quality in respect to the conventional process. By applying a secondary current through both the plasma and substrate, the transferred plasma arc process enables heating and melting of the substrate surface. This promotes improved metallurgical bonding, higher coating density, higher deposition rates, and higher coating thicknesses per pass. In another modification, inert atmosphere chamber spraying confines hazardous materials and restricts the formation of oxides which occur in open air spraying. Inert atmosphere spraying in a low pressure

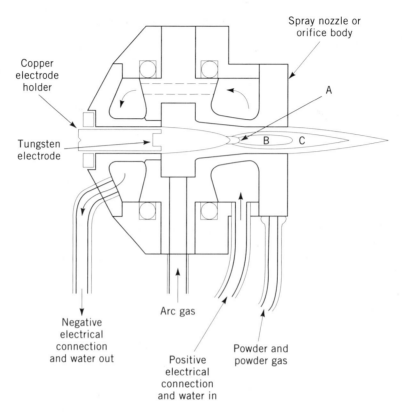

Fig. 3. A plasma arc spray gun where approximate plasma temperatures are A, 7,800°C; B, 10,000–13,000°C; and C, 7,800–10,000°C.

chamber has additional advantages including increase of the bond strength because higher substrate temperatures allow the coatings to diffuse into the substrate minimizing environmental problems such as dust and noise, created by
the very high speed of the atomized coating materials. If an enclosed atmosphere is used, provision must be made for cooling and gas expansion. An alloy
can be deposited by using alloy wires or two dissimilar wires, the only limitation being that ductile electrically conductive wires can be used. Among the
advantages of this process are high bond strength, no need for an external heat
source, low substrate heating, low electrical power requirements, and no need
for expensive combustible gas. The operator should wear a protective suit with
an independent breathing air supply.

Nickel–aluminum has been used for parts repair. Complex alloys of
chromium, aluminum, yttrium, and another metal can be applied for oxidation and corrosion resistance. Before spraying, the part must be cleaned of oil
and dirt. Standard aqueous or solvent cleaners are sufficient. Molded surfaces
must be coated with a special paint to promote metal adhesion or blasted with
fine aluminum oxide grit of 250–177 μm. Use of iron grit may lead to staining of the surface. Grit blasting is difficult to automate, and manual blasting
may give quality control problems because there is little change in appearance

after the operation. Arc spray coating must follow the grit blasting as soon as possible or blasting must be repeated.

Zinc arc spraying is an inexpensive process in terms of equipment and raw materials. Only 55–110 g/m² is required for a standard 0.05–0.10 mm Zn thickness. It is more labor intensive, however. Grit blasting is a slow process, at a rate of 4.5 m²/h. Application of an adhesive paint layer is much quicker, 24 m²/h, although the painted part must be baked or allowed to air dry. Arc sprayed zinc is applied at a rate of 9–36 m²/h to maintain the plastic temperature below 65°C. The actual price of the product depends on part complexity, number of parts, and part size. A typical price in 1994 was in the range of $10–32/m².

One large market for flame spray coatings on plastic is for radio frequency interference (RFI) shielding owing to concern expressed first in 1979 by the FCC for computerized equipment. Because of increasing pollution by or transmission of electromagnetic energy through the environment, insertion of a conductive shield between a part and the environment is often the cheapest and simplest way to eliminate RFI. Zinc arc spraying is an excellent method because of its high conductivity at normally applied thickness. Environmental pollution is minimal and only vaporized zinc particles need to be removed from the air.

Plasma Spray. The plasma spray is similar to arc spray coating, but the coatings of plasma spray are not limited to materials which can be fabricated as conductive wires. A controlled, low voltage arc is struck between a cooled tungsten cathode and a cooled copper anode. The plasma arc reaches temperatures up to 8000°C. The plasma consists of equal amounts of free electrons and positive ions. The plasma arc spray process produces the highest temperatures and particle velocities among the thermal spray processes. Argon, nitrogen, hydrogen, or other gases are fed through the arc, then past a powder feeder. Almost any powder, metallic or nonmetallic, can be used. The gas is so hot that tungsten, molybdenum, WC, and ceramics can be deposited. The porosity of these deposits may be high, but special techniques are available to increase particle velocity from 100–300 m/s up over 1200 m/s, giving denser deposits.

Carburizing and Nitriding. Several commonly used metallurgical surface treatments are applied by gas-phase reactions in a reducing atmosphere for carburizing, and in a nitrogen atmosphere for nitriding. These treatments are used to increase the surface hardness of ferrous alloys by diffusion of carbon and nitrogen at high temperature. As for all processes, good cleaning is necessary prior to treatment. Selective treatment can be done by electroplating cyanide copper onto the areas of the parts which are not to be hardened. The copper is an excellent barrier to carbon and nitrogen, and is easily removed after hardening (13,14) (see METAL SURFACE TREATMENTS, CASE HARDENING).

Pack Diffusion. Pack diffusion or cementation processes are similar to pack carburizing, and are used to coat iron, nickel, cobalt, and copper with chromium, boron, zinc (Sheradizing), aluminum, silicon, titanium, molybdenum, and other metals. It is possible to obtain a surface layer which contains 60% aluminum, but usual limits are 25% on iron-based alloys and 12% on nickel- or cobalt-based alloys. A pure aluminum overlay is never formed in this method, owing to the high temperature of the substrate, on which the deposited aluminum immediately alloys with the substrate. Aluminum is supplied in the pack diffusion process from pure aluminum powder or a ferroalloy powder, aluminum oxide, and an aluminum halide. Cleaned parts and the packing material react in a closed container at 820–1200°C, depending on the base metal, in a reducing atmosphere. The reaction deposits a high concentration of aluminum metal which subsequently diffuses deeply into the substrate. The concentration of aluminum is 50–60% immediately after coating, dropping to 12–25% after the diffusion cycle (18,30).

Sheradizing is an old process which consists of heating cleaned iron parts in a mixture of zinc powder and zinc oxide diluent in a rotating furnace at 350–375°C for 3–12 hours. This diffused coating is $FeZn_7$ with some $FeZn_3$ at the higher temperature. Like most other diffusion coatings, the coating is microcracked but uniform providing excellent sacrificial corrosion protection of the steel. Sheradizing was formerly used widely on nuts, bolts, and washers, but has been largely replaced by electroplating and mechanical plating.

Chromizing and Related Diffusion Processes. Chromizing is similar to aluminizing. A thin corrosion and wear-resistant coating is applied to low cost steels such as mild steel, or to a nickel-based alloy. The pack contains chromium powder, alumina diluent, and an active transfer agent which gasifies at operating temperature, such as ammonium chloride. The pack is heated for about 24 hours at 950–1100°C to give a 150–200-μm thick chromized layer. If there is >0.3 wt % carbon in the steel, a chromium carbide coating forms on the surface. Chromizing is complicated because the Cr can diffuse rapidly in the grain boundaries, can form complex carbide phases, and can decarburize steel, changing its physical properties.

There is also a two-step process of chromizing followed by aluminizing. Above 900°C the chromizing begins to rediffuse and the protective oxide changes to Al_3O_3 from Cr_2O_3. Aluminum oxide is less volatile than chromium oxide and better for high temperature oxidation resistance above 1000°C.

In the related boronizing process, a thin boron alloy is produced for extreme hardness, wear, and corrosion resistance. The powder can be boron, boron carbide, or FeB. A chloride or fluoride salt is used as an activator, along with an inert diluent such as alumina or silicon carbide. The parts are heated to 800–900°C for 6–24 hours. Steel usually needs to be heat treated for strength after boronizing. FeB and Fe_2B form; the former is harder but more brittle. Most of the boronized layer is Fe_2B because of its lower solubility in iron. Plain steel achieves a boronized coating of up to 150 μm in 6 h, but alloy steels containing chromium or other refractory elements achieve only a 10-μm coating. Boronized coatings are much harder than nitrided or carburized coatings, and are unaffected by heating to 1000°C. The Vickers hardness of boronized coatings can be 1600–2000, compared to 600–900 for nitrided and 700–800 for carburized coatings (see HARDNESS).

Siliconizing is yet another process used especially for coating of the refractory metals Ti, Nb, Ta, Cr, Mo, and W (see REFRACTORIES). These metals form silicides which have a surface oxidation protection layer of SiO_2. Siliconizing is especially effective on molybdenum against air oxidation up to 1700°C.

Metals. Aircraft and space vehicles, turbine generators, and other such applications require high strength at high temperature along with excellent oxidation resistance. Superalloys, ie, complex nickel and cobalt-based alloys, and refractory metals, eg, niobium, tungsten, molybdenum, tantalum, and their alloys, are used for applications at temperatures above 1000°C. In many cases the coatings must be resistant both to oxidation and to hot corrosion by sulfidation from sulfur-bearing gases. Testing must be done to assure compatibility between the coatings and the substrate and to avoid undesirable solid-state reactions and interdiffusion, which can produce voids, cracks, and weak layers. Refractory metal coatings must be sufficiently ductile for the anticipated service and environmental pressures and stresses. All coatings must resist thermal cycling and mechanical forces without cracking (see also REFRACTORY COATINGS) (22,40).

Two types of coatings have been used for superalloys: diffusion coatings, in which a layer of nickel, cobalt, platinum, or palladium aluminide, ie, NiAl, CoAl, PtAl, or PdAl, is formed on the surface by diffusion; and overlay coatings, in which a complex coating material such as nickel–cobalt–chromium–aluminum–yttrium, NiCoCrAlY, is applied to the surface. Pack cementation is the most widely used process for applying diffusion coatings to superalloys, but brushing, dipping into, or spraying a prepared mixture of the coating elements followed by high temperature heating is also used. Physical and vapor deposition, plasma spraying, and sputtering are often used for applying the overlay coatings. Pack cementation, fluidized-bed deposition, and spray or dip-and-sinter processes are used for the application of silicide and aluminide diffusion coatings to refractory metals (40,41).

Miscellaneous Methods. Powdered metals such as aluminum, chromium, nickel, and copper, along with various alloys, can be applied to parts by electrostatic deposition. The metal strip containing the attached powdered metal must be further processed by cold rolling and sintering to compact and bond the metal powder.

The laser spray process uses a high power carbon dioxide laser focused onto the surface of the part to be metallized. A carrier gas such as helium blows metal particles into the path of the laser and onto the part. The laser melted particles may fuse to the surface, or may be incorporated into an alloy in a molten surface up to 1-mm thick. The laser can be used for selective alloying of the surface, for production of amorphous coatings, or for laser hardening.

Vacuum-Phase Metallizing Techniques

Vacuum-phase metallizing techniques all depend on the use of a vacuum as part of the metallizing process (see VACUUM TECHNOLOGY). The tonnage of metal deposited by these techniques is insignificant compared to hot dip galvanizing or plating, and the total surface area metallized is much less than that done by plating. However, the economic added value of vacuum metallizing probably exceeds either plating or galvanizing because of the extremely rapid growth and deep market penetration of semiconductor devices. In 1994, the growth rate for all of these devices was 10–15% and projections call for at least a 10% compounded annual growth rate into the twenty-first century. The state of the art in vacuum metallizing is in semiconductors (qv).

The largest value added industry which uses metallizing is by far the semiconductor industry. In terms of total metals usage, this industry is almost insignificant, but the effective value of the metals used is enormous. Almost all metals used in the semiconductor industry are small-volume, superpure, and extremely costly grades (see ULTRAPURE MATERIALS). One 200-mm silicon wafer, when fully utilized to fabricate a technologically demanding multilevel chip, such as the Pentium, may be worth up to $250,000. The total amount of metallic and nonmetallic elements used for metallizing the wafer is ca 0.25 grams.

Semiconductor manufacturing facilities are large and expensive. A modern plant doing 0.5-μm device manufacturing on 200-mm silicon wafers may cost >$1 billion to install. Semiconductor manufacturing uses a variety of vacuum deposition processes resulting in the semiconductor chip. Multiple layers of metal give a three-dimensional circuit in microscopic patterns. The next generation of semiconductor plants are expected to use 300–400-mm wafers at 0.35-μm device size and further shrinkage is in development (16,42–45) (see INTEGRATED CIRCUITS).

Thermal Evaporation. Thermal evaporation is done in a high vacuum to minimize chemical side reactions of the evaporated active metal. The vacuum is necessary because almost any metal reacts with oxygen, water vapor, or nitrogen to form brittle nonmetallic inclusions. Some metals even trap inert gases as the metal is deposited. The metal to be deposited is simply heated in a vacuum to a high temperature to increase its vapor pressure. The vapor deposits on any cold object in the chamber. The rate of deposition depends on the heat input but is normally faster than sputtering. The evaporated molecules travel in straight lines owing to the low pressure, so only line of sight coverage is achieved. This method is difficult to use on temperature-sensitive substrates such as plastics, except when formation of very thin films of metals is required. Adhesion can be marginal unless the object can be heated to about 400°C to remove residual moisture.

Thermal evaporation is inexpensive and efficient. It is used for low cost items such as aluminized plastic sheet and decorative Christmas tinsel, second surface coating of transparent plastic auto parts, and metallization of glass and ceramics (qv). Aluminum is the predominant metal used. Color effects such as gold or brass are achieved by applying a dyed translucent organic protective coating over the aluminum. Variants of this process use different methods to impart the thermal energy. These include electron beam vaporization and laser evaporation.

Sputtering or Glow Discharge. Sputtering can be done using both conductive and nonconductive items. A low pressure atmosphere of argon is used in the deposition vessel. The depositing metal becomes the cathode with a high potential; the metal item is the anode. Alternatively, specially designed anodes allow deposition on nonconductive plastics. Argon becomes ionized and is accelerated into the cathode at a high speed. The energy is transferred to the target metal atoms, which are sputtered off and deposit on the part to be coated. This method gives excellent coating adhesion and more consistent coatings than simple vacuum deposition. It can also be applied to a wider range of materials, including complex metal alloys and oxides. The equipment and operating costs are higher than vacuum deposition, and deposition rates are lower.

Reactive sputtering is a variation in which two or more deposition sources are used. This process can be used to give compounds such as silicon carbide on the surface. Titanium nitride [25583-20-4], TiN, can be applied to aluminum to increase wear resistance, especially sliding resistance. In a related process, a magnetron discharge is used to give higher deposition rates and less radiation damage. This is preferred for thermally sensitive substrates.

Chemical Vapor Deposition. This process is distinct from simple thermal evaporation because chemical vapor deposition (CVD) depends on a chemical reaction at the surface of the part and in that way is analogous to electroless plating. This process uses a gas of one or more chemical species, which react at a heated substrate to form an appropriate film. The film can be metallic, nonmetallic, single element, or compound. The reactions occur at 500–1500°C on the part to be coated, so temperature-sensitive materials cannot be used. Another variant is plasma-assisted chemical vapor deposition, which can use a lower processing temperature of near 300°C. The film-forming reactions can be disproportionation, reduction, displacement, or chemical reaction with formation of a compound. Among the coatings available are copper and aluminum conductors, diamond and tantalum oxide dielectrics, lead zirconium titanate piezoelectrics (qv), and bismuth strontium calcium cuprate superconductors (see also ELECTRONICS, COATINGS; THIN FILMS, FILM FORMATION TECHNIQUES). Some of the largest applications are for coating TiC and TiN on cutting tools (2,46).

Ion Implantation. Also known as ion plating, ion implantation (qv) is a high vacuum process for modifying the surface properties of any material. The equipment consists of an ion source, where an electrical discharge in a low pressure gas ionizes the chemical to be implanted. The ionized atoms are fed into a low pressure ion accelerator tube where they receive sufficient energy to penetrate the target item. The energy imparted to the ionized atoms can be closely controlled over a range of about 100–>3000 keV or 5–200 keV for semiconductor manufacturing, allowing penetration to a controlled and reproducible depth. Typical implantation of metal ions is to a depth of several tens of nanometers. This is a surface modification technique rather than a surface coating technique. Semiconductors are commonly modified using this technique, because ion implantation can be done using total or imagewise scanning of the beam over the surface. Nitrogen, chromium, and phosphorus can be used to harden metals and increase the corrosion resistance. Metastable alloys can be produced that are difficult to make by other processes. For example, niobium can be implanted into 316 stainless steel to improve its corrosion resistance (47–49).

Laser Hardening and Modification. Lasers are used to surface harden ductile steels and improve the toughness to a depth of 0.35 mm or more. The very rapid melting and freezing acts as an extremely quick quench process. Lasers can also be used to bond solid or powder coatings to a surface. The surface layer is melted and mixes with the substrate extremely quickly. Typical coatings are nickel or titanium carbide on iron, and nickel, cobalt, manganese, and titanium carbide [12070-08-5], TiC, on aluminum. Use of lasers with other specialized coating methods is common. Thermally sprayed coatings of Stellite, Co–Cr–WC, on stainless steel, 1 Cr–18 Ni–9 Ti, eliminate porosity, improve chemical homogeneity, increase the coating strength, and develop a better metallurgical bond.

Miscellaneous. Electron beams can be used to decompose a gas such as silver chloride and simultaneously deposit silver metal. An older technique is the thermal decomposition of volatile and extremely toxic gases such as nickel carbonyl [13463-39-3], $Ni(CO)_4$, to form dense deposits or dendritic coatings by modification of coating parameters.

In contrast to the older techniques, a newer method is to use a scanning tunneling electron microscope to deposit metal coatings in microscopic images as small as 0.001 μm. The ultimate surface metallization techniques allow deposition of metals atom by atom in controlled three-dimensional arrays.

Metallizing by Direct Physical or Thermal Bonding

Direct bonding techniques are among the oldest types of metallizing, and the most versatile. Applications for extremely thin films of gold foil range from ancient funerary decorations and medieval manuscripts to coatings of the roofs of many U.S. statehouses. Many methods depend on heat or pressure and an adhesive layer to glue the coating to the substrate. Methods for metallizing on a metallic surface often depend on removal or displacement of a preexisting surface oxide layer. Many metals form intermetallic alloys or self-diffuse into one another even at room temperature, but the surfaces in contact must be clean and oxide free.

Lamination.   The most commonly used process for application of metals to nonmetals is that of lamination (see LAMINATED AND REINFORCED PLASTICS). One large-volume process involves the production of continuous lengths of copper foil by electrodeposition on a rotating polished cylindrical stainless steel mandrel. This gives one very smooth surface, and one surface which is microroughened for attachment to the substrate. The copper foil is the subject of numerous specialized treatments, some of which involve applying tin, bronze, or chromate treatments to one or both sides of the foil. A more recent development is the selective micronodularization of the exposed surface of the foil prior to removal from the cylinder. These foils are laminated under high pressure to epoxy–glass, polyimide–glass, or phenolic sheets to serve as the raw material for printed circuit boards, which can have as few as one layer, or greater than 40 layers of copper and insulator. Copper is also laminated to ceramics for high temperature printed circuit boards.

Die cut metal laminates are also applied to plastics for decorative effects, or for radio frequency interference shielding. These laminates are formed by vacuum deposition, or by attachment of foil to a plastic layer for easier handling. Multiple layer composites of plastic and aluminum are used for specialized packaging, especially for moisture or gas barriers.

Zinc foil coated with a conductive, pressure-sensitive adhesive is used for repair of other zinc coatings or for imparting corrosion resistance at field sites. The 0.08-mm zinc tape or sheet has a 0.025-mm conductive adhesive. The laminate is cut to size and pressed tightly to activate the adhesive. Conductive tape can be wrapped around pipe, especially around welds or connections. The corrosion resistance of this material is intermediate between galvanized or thermally sprayed coatings and zinc-filled paints (21,50).

Mechanical Plating.   Impact or peen plating is a mechanical process whereby the metal powder is compacted and welded to parts by mechanical energy. This process is limited to relatively small parts of no more than about one kilogram. The parts are placed in a specially designed barrel along with water, metal powder sized at 3–10 μm, glass impact beads of several sizes, and various promoter chemicals. The metal particles are plastically deformed by the impact of the glass beads, bonding to the continuously cleaned metal parts. Ductile metals and alloys can be applied by this process. Sequential layers of different metals can be prepared, and mixed co-deposits of metals difficult to prepare by other methods such as cadmium–zinc deposits can easily be made. Among the metals useful with this process are zinc, cadmium, lead, aluminum, silver, indium, gold, tin, copper, and brass. Rate of increase in coating thickness is almost independent of the plating time, but is governed by the amount of metal powder added. The coatings can be semibright but do not have the full brightness of electroplated coatings. This process is especially suitable for high strength steels because there is no hydrogen embrittlement. The coating corrosion resistance can be as good as or better than that from hot dip galvanizing or electroplating (3).

Slurry Coatings.   Many types of slurry coatings are used, including metal-filled paints and metal–glass frit compositions. Zinc-filled paints contain more metal powder by weight than organic binder so are electrically conductive. These are used as inexpensive sacrificial coatings in place of hot dip galvanizing. Aluminum, copper, stainless steel, and brass-filled paints are also common. Many of these paints are used in RFI shielding applications on plastic computer casings to reduce electrical interference (21,51).

Metal powder–glass powder–binder mixtures are used to apply conductive (or resistive) coatings to ceramics or metals, especially for printed circuits and electronics parts on ceramic substrates, such as multichip modules. Multiple layers of aluminum nitride [24304-00-5], AlN, or alumina ceramic are fused with copper sheet and other metals in powdered form. The mixtures are applied as a paste, paint, or slurry, then fired to fuse the metal and glass to the surface while burning off the binder. Copper, palladium, gold, silver, and many alloys are commonly used.

There are several types of slurry or powdered paint methods for coating high temperature processing equipment in the chemical and petroleum industry, and for turbine blades. These last are used for aircraft and power generation (qv) and are exposed to high temperature corrosive gases. The process consists of cleaning, sand blasting, and acid pickling parts, then applying a slurry by dipping or spraying. The parts are dried at 120°C and fired to 930–1200°C. The coating and blasting slurries are made of an aluminum source, eg, pure aluminum powder or 5–12% silicon–aluminum alloy; a vehicle, water or an organic solvent; a clay or gum binder; and an inert material such as alumina or a ceramic oxide. The thickness of the coating depends mostly on the substrate and the firing temperature, with a lesser effect from the time at temperature. For chromium stainless steels the thickness of the diffusion layer increases with the temperature. Nickel and cobalt alloys decrease the diffusion rate of aluminum in alloys owing to the stability of the intermetallic compounds NiAl, Ni_3Al, CoAl, and Co_3Al and the low solid solubility of aluminum in nickel and cobalt. This limits the attainable thickness of the diffusion layer in nickel and cobalt alloy steels.

Specialized alloys are used for high temperature applications on turbine blades, furnace parts, thermocouples, etc. These coatings can be as simple as iron–silicon–chromium or as exotic as chromium–aluminum–hafnium (36, 41,52).

Roll Bonding or Strip Roll Welding. Roll bonding, also known as strip roll welding, was originally applied for fabrication of bimetallic strips for thermostats. It later found application in jewelry, cladding copper coins with cupronickel alloy, forming electrical contacts, conductive springs, cookware, electromagnetic shielding, underground cable wrappings, etc. An especially important application is in production of roll bonded heavy plate and sheet for process vessels in chemical plants, and in shipbuilding. The latter uses a 90:10 copper–nickel alloy cladding for resistance to fouling by marine organisms.

Clad strips are produced by continuous rolling, during which the bond is achieved by pressure welding. The process is relatively simple, but the requirements for surface preparation are strict. Wire brushing is the most commonly used method for surface preparation which is done just prior to rolling. Welding is usually accomplished in a single rolling pass. Subsequent heat treatment may be applied to the strips in order to improve mechanical properties and/or corrosion resistance. Almost all combinations of ductile metals and alloys can be clad by roll welding. However, oxides of some metals make the process difficult. Certain combinations of metals may suffer from formation of intermetallic compounds. The most common combinations of cladding by strip roll welding are copper to steel, nickel, and gold, aluminum to aluminum alloys, tin to copper and nickel, and gold to nickel.

Skiving is a variant in which the base metal surface oxides are mechanically removed followed immediately by pressure rolling of a precious metal or alloy strip. This is commonly used for inlays for electrical contacts and for jewelry fabrication. The common inlay materials include gold, silver, copper, brass, and solder. No heat is needed, and the coating is applied only to designated areas so there is little waste (3,50).

Miscellaneous Processes. Metal strip for cladding can be produced by cold pressing metal powder into a low density green strip, followed by sintering to compact the powder. Alloy powders can be made into strip, along with specialized strip with one powder bonded to a different powder on the opposite side.

Explosive bonding uses the energy of an explosion to drive two metals together (see METALLIC COATINGS, EXPLOSIVELY CLAD). The energy dissipated at the juncture of the metals dissociates or disperses oxides and provides a clean metallurgical joint. This process is economically attractive because no expensive furnaces, presses, and other capital equipment is needed. This process is especially suitable for metals that are reactive with oxygen and/or nitrogen, such as titanium, tantalum, stainless steels, and aluminum. It can be done with metals having a wide difference in melting points, such as aluminum (660°C) and tantalum (2996°C). The metals to be bonded must not be brittle because these materials fracture during the explosion. As little as 5% tensile elongation can be sufficient.

Environmental Concerns

Each type of metallic coating process has some sort of hazard, whether it is thermal energy, the reactivity of molten salt or metal baths, particulates in the air from spray processes, poisonous gases from pack cementation and diffusion, or electrical hazards associated with arc spray or ion implantation. Vacuum or inert gas operations can produce flammable dusts or powders when opened for cleaning. Most of the hazards are confined to the operator and immediate environs in the operating plant. OSHA is the primary regulator of these hazards in the United States, although many local and state agencies, especially fire departments, also regulate coatings plants. Adequate training, documentation, and protective equipment are the minimum requirements. Many companies use worker–management teams, suggestion boxes, consultant surveys, supplier training sessions, and other methods to reduce risk of injuries (see HAZARD ANALYSIS AND RISK ASSESSMENT). The principal regulatory burden falls on wastes and discharges which leave the plant (3,53,54).

The U.S. EPA regards metallizers such as platers, surface finishers, and printed circuit board producers as among the most important point source polluters for metals. Whereas there has been some directed growth to inherently nonwater polluting processes such as vacuum metallizing, as of this writing most metal coatings are still applied by traditional processes. Much production of electroplated items has, however, shifted from the United States to less environmentally stringent countries.

The surviving U.S. plants have embraced all types of waste treatment processes (see WASTES, HAZARDOUS WASTE TREATMENT; WASTES, INDUSTRIAL). The most desired pollution prevention processes are those which reduce the total amount of waste discharged (see WASTE REDUCTION). Treatment and disposal are less strongly emphasized options. Zero wastewater discharge facilities and water recycling processes are becoming more common (55,56).

Some metals used as metallic coatings are considered nontoxic, such as aluminum, magnesium, iron, tin, indium, molybdenum, tungsten, titanium, tantalum, niobium, bismuth, and the precious metals such as gold, platinum, rhodium, and palladium. However, some of the most important pollutants are metallic contaminants of these metals. Metals that can be bioconcentrated to harmful levels, especially in predators at the top of the food chain, such as mercury, cadmium, and lead are especially problematic. Other metals such as silver, copper, nickel, zinc, and chromium in the hexavalent oxidation state are highly toxic to aquatic life (37,57–60).

Discharge limits vary between localities and among plants. Table 1 shows federal EPA maximum discharge limits for a number of metals for a new metal-finishing installation. These limits are for large captive manufacturers who discharge to a publicly owned treatment works (POTW). Limits vary according to the age and size of a plant and type of coating operation.

The waste discharge categories (Table 1) are regulated under the National Pollutant Discharger Elimination System (NPDES), a U.S. EPA program. Local municipalities are free to set more restrictive discharge limits. Typically mercury and arsenic are the most highly regulated, often at limits of 50 ppb for mercury and 150 ppb for arsenic, followed by lead, silver, and cadmium, which have discharge limits of <1 ppm (<1 mg/L). Hexavalent chromium, free and complexed cyanide, and copper are also highly regulated (3,62).

Newer federal limits on metals content of sewage sludge combined with laws on fuller treatment of the sewage sludge and its allowed disposal methods have affected limits. The metallic content of sludges from municipal waste treatment facilities is becoming of great concern. High levels of cadmium and other metals

Table 1. EPA Pretreatment Standards for Aqueous Discharge[a,b]

Material, mg/L	Existing source, PSES[c]		New source, PSNS[c]	
	1 Day	30 Days	1 Day	30 Days
cadmium	0.69	0.26	0.11	0.07
chromium, total	2.77	1.71	2.77	1.71
copper	3.38	2.07	3.38	2.07
lead	0.69	0.43	0.69	0.43
nickel	3.98	2.38	3.98	2.38
silver	0.43	0.24	0.43	0.24
zinc	2.61	1.48	2.61	1.48
cyanide				
total	1.2	0.65	1.2	0.65
treatable	0.86	0.32	0.86	0.32
total toxic organics	2.13		2.13	

[a]Captive manufacturers performing metal finishing, including electroplating, discharging to POTWs (61).
[b]pH equals 6–10.
[c]Maximum value. PSES = pretreatment standards for existing sources; PSNS = pretreatment standards for new sources.

may bioaccumulate when spread over cultivated fields. The U.S. EPA has put strict limits on the metals contents of such sludges, stating how much sludge may be added to fields, pastures, and forests. Zinc is being recognized as one of the most common nonpoint source metal pollutants. The great majority of this zinc comes from weathering of galvanized steel roofs, fences, and other items exposed to the atmosphere, although significant amounts may also come from zinc oxide pigments used in paints and tires. Copper is another common nonpoint source pollutant. Most comes from plumbing fixtures and piping.

Iron is commonly found as the toxic and difficult to treat ferrocyanide complex in plating waste streams, thus ancillary solution used to prepare surfaces or to complete the processes of metallizing surfaces can also be pollutants. Cleaners are significant sources of alkalinity, complexing agents, phosphates, grease and oils, particulates, surfactants, dissolved metals, suspended solids, chemical and biological oxygen demand, and total dissolved solids. Pickling solutions contribute strong acids and dissolved metals. Sandblasting, buffing, grinding, and polishing operations may produce a metal-laden hazardous solid waste. Passivating solutions contain chromates, dissolved metals, and acids. Fluxes often contain zinc and acids, plus dissolved or particulate lead, silver, and tin. Gaseous exhaust from flame spraying, sputtering, ion implantation, and other processes may be toxic, hazardous, flammable, or corrosive. Drosses from hot dipping, zinc or solder, may be a recyclable waste, but these are still hazardous. Even neutral salts can be of regulatory concern as more municipalities practice some type of sewage water recycle and reuse, especially for irrigation or groundwater replenishment (see also GROUNDWATER MONITORING).

Air pollution (qv) is recognized as a significant problem for coating facilities. Chromium emissions are tightly regulated to very low levels, based on the amount of electricity used. Hexavalent chromium is a known carcinogen. Trivalent and metallic chromium are of no special health concern. Many decorative chromium plating facilities have converted to the use of trivalent chromium plating baths, however there is no substitute for hexavalent chromium used for hard chrome plating applications.

Lead, a highly toxic and accumulative material, is also of concern especially with regard to children (53) (see LEAD COMPOUNDS, INDUSTRIAL TOXICOLOGY). Lead, both from fume exposure during coating or melting, and from accidental ingestion from handling the metal or its salts, is problematic. Cadmium is another highly toxic and bioaccumulative metal that has been eliminated from use in many European countries. Its main continuing U.S. use is in military applications, aircraft electronics, and marine applications where it outperforms zinc coatings. Nickel can cause allergic reactions in many people, because the metal easily oxidizes to soluble salts.

The semiconductor industry uses significant quantities of metals combined in highly reactive and volatile compounds as sources for metallizing silicon wafers. These metals include organometallic or volatile inorganic compounds of aluminum, copper, lead, gallium, platinum, barium, strontium, titanium, tungsten, and the semimetals silicon, antimony, and arsenic. The waste gases from ion implantation, sputtering, evaporation, and other vacuum processes must be appropriately treated. Small chemical reactors are available to decompose some gases as these are discharged. The remaining metallic waste is then recycled or the appropriate legal disposal made.

Buffing, polishing, sandblasting, and grinding operations generate both airborne and solid wastes. Personnel exposed to these processes must use adequate protective equipment. Many states such as California require that solid wastes from these processes be tested and certified as to the toxic metals content to allow for appropriate disposal. Trivalent chromium oxides are commonly used in polishing compounds. Care must be taken to keep such wastes separate from oxidizing materials to avoid formation of carcinogenic hexavalent chromium.

Vapor degreasing by means of chlorinated hydrocarbons, freons, and other inert compounds used to be a universal practice. Many of these compounds are prohibited by the Montreal protocol, or are being taxed at progressively higher rates. These gases are often ozone (qv) depletion agents or greenhouse gases which contribute to global warming. New and improved aqueous cleaning processes have been developed as replacements, many based on emulsion chemistries of citrus derivatives or terpenes. Alcohol-based processes can be used for critical processes, including electronic assembly cleaning (3,63).

Flame and other sprayed coatings generate large amounts of potentially toxic fumes. Zinc spraying can expose personnel to fumes, but the workers develop a tolerance to this exposure which disappears quickly once exposure stops. Excess zinc is rapidly excreted from the body. Silicon-containing materials, especially when used for abrasive blasting, have the potential to cause silicosis of the lungs after long exposure.

Both materials and process substitutions have been used to try to limit wastes while providing favorable performance and costs. One problem is that most of the vacuum or controlled atmosphere deposition methods have much higher equipment and operating costs than do electroplating or hot dipping. Moreover, electroplating gives better coverage on complex-shaped parts, and usually is much faster. The focus of most electroplating processes has been to eliminate the use of the most toxic compounds such as cyanides. Cleaners in general have undergone extensive transformations, with emphasis on formulations having smaller amounts of chelating agents (qv) to improve ease of waste treatment. Other cleaners are designed with surfactants (qv) which displace oil for removal by filters or overflow, rather than surfactants for emulsification of oil. Longer cleaner life and a higher oil content waste for energy recovery has resulted.

Electroplating processes in general have moved away from cyanide-based deposition baths. The majority of electroplated zinc is done from zinc sulfate, zinc potassium chloride, zinc ammonium chloride, or alkaline noncyanide zinc processes. Newer baths based on nontoxic chelating agents are replacing copper cyanide strike baths. Noncyanide zincate immersion baths are used for treating aluminum prior to electroless nickel plating. The electroplated metals are also being changed. Zinc–nickel alloy has been suggested as a substitute for toxic cadmium. Plated tin coatings are being used in place of tin–lead in the printed circuit industry, as temporary metallic etch resists (see RESISTS MATERIALS).

Many types of waste treatment and waste minimization processes are in common use in the metallization industry. Air scrubbers are commonly required, even for general acid fume removal from plating shops. Additionally, many newer technologies have been adapted for use in metallizing operations. These include air stripping, antimisting agents, biological destruction, carbon absorption, countercurrent rinsing, crystallization (qv), distillation (qv), Donnan dialysis, electrodialysis, electrowinning, evaporation (qv), filtration (qv), flotation (qv), flocculation, hydrolysis, incineration (see INCINERATORS), ion exchange (qv), metallic replacement, neutralization, oxidation, pH adjustment (see HYDROGEN–ION ACTIVITY), photolysis, precipitation, process modification, reduction, reverse osmosis (qv), salt splitting, sedimentation (qv), solidification, and spray rinsing.

Whereas many of these technologies are not really new, they have never had the regulatory and economic justification for their use in metallizing. Each of these general methods has many variants. Some may be directed to waste treatment, some to recycle, and some to reclaim. An example is filtration, used to prevent release to air of zinc particles from flame spraying, microfiltration of cleaners to extend life, in combination with chemical precipitation to remove metal particles from wastewater, and many other uses.

BIBLIOGRAPHY

"Metallic Coatings" in *ECT* 1st ed., Vol. 8, pp. 898–922, by W. W. Bradley, Bell Telephone Laboratories, Inc.; in *ECT* 2nd ed., Vol. 13, pp. 249–284, by W. B. Harding, The Bendix Corp.; "Survey" under "Metallic Coatings" in *ECT* 3rd ed., Vol. 15, pp. 241–274, by R. C. Krutenat, Exxon Research and Engineering Co.

1. E. A. Brandes, ed., *Smithells Metals Reference Book*, 6th ed., Butterworths, New York, 1983.
2. K. G. Budinski, *Surface Engineering for Wear Resistance*, Prentice Hall, Englewood Cliffs, N.J., 1988.
3. *Metal Finishing Guidebook and Directory*, Vol. 92, No. 1A, Elsevier Publishing, New York, 1944; collected vols., Publications, Inc., Hackensack, N.J., updated yearly.

4. P. Schweitzer, *Corrosion Resistance Tables: Metals, Non-Metals, Coatings, Mortars, Plastics, Elastomers and Linings, and Fabrics*, 3rd ed., Marcel Dekker, New York, 1992.

5. *The Industrial Revolution*, in C. Singer and co-eds., *History of Technology*, Vol. 4, Oxford University Press, New York, 1958.

6. J. W. Dini, *Electrodeposition: The Materials Science of Coatings and Substrates*, Noyes Data Corp., Park Ridge, N.J., 1993.

7. D. R. Gabe, *Principles of Metal Surface Treatment and Protection*, 2nd ed., Pergamon Press, Inc., Elmsford, N.Y., 1978.

8. "Properties and Selection of Nonferrous Alloys and Pure Metals" in *Metals Handbook*, 10th ed., American Society for Metals, Metals Park, Ohio, 1990.

9. H. H. Uhlig, *Corrosion Handbook*, John Wiley & Sons, Inc., New York, 1948.

10. *The Engineering Properties of Electroless Nickel Deposits*, International Nickel Co., New York, 1977.

11. G. Mann, *Fabricator* (Mar. 1992).

12. K. N. Strafford, P. K. Datta, and G. G. Googan, eds., *Coatings and Surface Treatment for Corrosion and Wear Resistance*, Halstead Press, New York, 1984.

13. *Surface Modification Technologies II: Proceedings of the 2nd International Conference on Surface Modification Techniques, Minerals, Metals, and Materials Society*, Warrendale, Mich., 1989.

14. R. B. Waterhouse and A. Niku-Lari, *Metal Treatments Against Wear, Corrosion, Fretting, and Fatigue*, Vol. 6, Pergamon Press, New York, 1988.

15. D. R. Gabe, *Electrochimica Acta* **39**, 1115–1121 (1994).

16. R. F. Bunshak, *Handbook of Deposition Technologies for Films and Coatings*, 2nd ed., Noyes Data Corp., Park Ridge, N.J., 1994.

17. G. E. McGuire, D. C. McIntyre, and S. Hoffman, eds., *18th International Conference on Metallurgical Coatings and Thin Films, Apr. 22–26, 1991, San Diego, Calif.*, Elsevier Sequoia, Lausanne, Switzerland, 1991.

18. V. Sedlacek, *Metallic Surfaces, Films, and Coatings*, Elsevier, New York, 1992.

19. *Surface Engineering*, Vol. 5, American Society of Metals, Metals Park, Ohio, 1994.

20. *International Conference on Zinc and Zinc Alloy Coated Steel Sheet, Sept. 5–7, 1989, Tokyo*, Iron and Steel Institute of Japan, 1989.

21. F. C. Porter, *Zinc Handbook: Properties, Processing, and Use in Design*, International Lead Zinc Research Organization, Marcel Dekker, New York, 1991.

22. "Advanced Joining of Aerospace Metallic Materials," *61st Meeting of Structures and Materials Panel of AGARD, Oberammergau, Germany, Sept. 11–13, 1985*, AGARD #398, Nato Advisory Group for Aerospace Research and Development, Neuilly-sur-Seine, France.

23. R. J. Klein-Wassink, *Soldering in Electronics*, 2nd ed., Electrochemical Publications Ltd., Ayr, Scotland, 1989.

24. H. H. Manko, *Solders and Soldering*, 3rd ed., McGraw-Hill Book Publishing Co., Inc., New York, 1992.

25. M. C. Pecht, ed., *Soldering Processes and Equipment*, John Wiley & Sons, Inc., New York, 1992.

26. A. C. Tan, *Tin and Solder Plating in the Electronic and Semiconductor Industry*, Chapman & Hall, New York, 1993.

27. F. G. Yost, R. M. Hosking, and D. R. Frean, eds., *The Mechanics of Solder Alloy Wetting and Spreading*, Van Nostrand Reinhold, New York, 1993.

28. L. J. Durney, ed., *Electroplating Engineering Handbook*, 4th ed., Van Nostrand Reinhold Co., Inc., New York, 1984.

29. W. H. Safranek, *The Properties of Electrodeposited Metals and Alloys*, 2nd ed., American Electroplaters and Surface Finishers Society, Orlando, Fla., 1986.

30. S. Lyakhovich, ed., *Multicomponent Diffusion Coatings*, Oxonian Press Pvt., New Delhi, India, 1987.
31. American Society of Electroplated Plastics, *Standards and Guidelines for Electroplated Plastics*, 3rd ed., Prentice-Hall, Inc., Englewood Cliffs, N.J., 1984.
32. G. G. Gawrilov, *Chemical (Electroless) Nickel-Plating*, Portcullis Press, Redhill, U.K., 1979.
33. J. Hajdu and G. Mallory, *Electroless Plating: Fundamental and Applications*, American Plating and Surface Finishing Society, Orlando, Fla., 1990.
34. R. Suchentrunch, *Metallization of Plastics*, Finishing Publications Ltd., Herts, U.K., 1993.
35. M. Fleischmann, S. Pons, D. R. Rolison, and P. P. Schmidt, *Ultramicroelectrodes*, Datatech Systems, Inc., Morgantown, N.C., 1987.
36. *Fifth National Conference on Thermal Spray, June 7–11, 1993, Anaheim, Calif.*, American Society for Metals, Metals Park, Ohio, 1993.
37. "Lead–Zinc '90: Proceedings of a World Symposium on Metallurgy and Environmental Control," *119th TMS Annual Meeting, Feb, 18–21, 1990, Warrington, Pa.*, TMS Lead, Zinc, and Tin Committee.
38. R. Manory and A. Grill, *Protective Coatings of Metal Surfaces by Cold Plasma Treatments*, NASA Technical Memorandum 87152, National Technical Information Service, Springfield, Va., 1985.
39. H. S. Savage and R. G. McCormack, *Arc Sprayed Coatings for Electromagnetic Pulse Protection*, USA-CERL Technical Report M-89/15, U.S. Army Corps of Engineers, Washington, D.C., 1989.
40. "Advanced Materials and Coatings for Combustion Turbines," *ASM Congress Week, Oct. 17–21, 1993*, American Society for Metals, Metals Park, Ohio, 1994.
41. N. Birks and G. H. Meir, *Introduction to High Temperature Oxidation of Metals*, E. Arnold, London, 1983.
42. *Advanced Metallization for ULSI Applications, Oct. 20–22, 1992 Tempe, Ariz.*, Materials Research Society, Pittsburgh, Pa., 1993.
43. S. P. Murarka, *Metallization: Theory and Practice for VLSI and ULSI*, Butterworth-Heineman, Boston, Mass., 1993.
44. Technical data, SEMATECH, Austin, Tex.
45. S. R. Wilson, C. J. Tracy, and J. Freeman, eds., *Handbook of Multilevel Metallizations for Integrated Circuits*, Noyes Data Corp., Park Ridge, N.J., 1993.
46. H. K. Pulker, ed., *Wear and Corrosion Resistant Coatings by CVD and PVD*, Halstead Press, New York, 1989.
47. R. Iscoff, *Semiconductor Int.* **17**, 65–72 (Oct. 1994).
48. F. F. Lomarov, *Ion Beam Modification of Metals*, Gordon and Breach Science Publishers, Philadelphia, Pa., 1992.
49. P. D. Prewett and G. L. R. Mair, *Focused Ion Beams from Liquid Metals Ion Sources*, John Wiley & Sons, Inc., New York, 1991.
50. J. A. Mock, *Introduction to Prefinished Metals*, Technomic Publishing Co., Lancaster, Pa., 1983.
51. C. W. Hoppesch, *A Method for Predicting Service Life of Zinc Rich Primers on Carbon Steel*, NASA technical memorandum #93101, Kennedy Space Center, Fla., 1986.
52. *Life Prediction for High Temperature Gas Turbine Materials, Aug, 27–30, 1985, Syracuse, N.Y.*, Syracuse University, 1986.
53. *Air Quality Criteria for Lead: Supplement to the 1986 Addendum*, U.S. EPA, Environmental Criteria and Assessment Office, Washington, D.C., 1990.
54. *Technical Support Document to Proposed Airborne Toxic Control Measure for Emissions of Toxic Metals from Non-Ferrous Metal Melting*, State of California Air Resources Board, Stationary Source Division, Sacramento, Calif., 1992.

55. C. S. Brooks, *Metal Recovery from Industrial Waste*, Lewis Publishers, Chelsea, Mich., 1991.
56. *Hazardous Waste Reduction Checklist and Assessment Manual for the Metal Finishing Industry*, California Department of Health Services, Alternative Technology Division, Toxic Substances Control Program, Sacramento, Calif., 1990.
57. B. L. Carson, H. V. Ellis, and J. L. McCann, *Toxicology and Biological Monitoring of Metals in Humans*, Lewis Publishers, Chelsea, Mich., 1986.
58. T. W. Clarkson, ed., *Biological Monitoring of Toxic Metals*, Plenum Press, New York, 1988.
59. H. K. Dillon and M. H. Ho, *Biological Monitoring of Exposure to Chemicals: Metals*, John Wiley & Sons, Inc., New York, 1991.
60. E. C. Foulkes, ed., *Biological Effects of Heavy Metal*, Vols. 1–2, CRC Press, Boca Raton, Fla., 1990.
61. *Code of Federal Regulations*, Vol. 40, part 433, Washington, D.C.
62. *Eleventh AESF/EPA Conference on Environmental Control for the Surface Finishing Industry, Feb. 5–7, 1990*, American Electroplaters and Surface Finishers Society, Orlando, Fla., 1990; *Environmental Aspects of the Metal Finishing Industry*, United Nations Environmental Program, Technical Report Service No. 1, Industry and Environmental Office, Paris, 1989.
63. L. M. Brown, J. Springer, and M. Bower, *Chemical Substitution for 1,1,1-Trichloroethane and Methanol in an Industrial Cleaning Operation*, U.S. EPA, Cincinnati, Ohio; National Technical Information Service, Springfield, Va., 1992.

General References

Plating and Surface Finishing, Semiconductor International, and *Metal Finishing* provide some of the best information on innovative metallizing methods.

GERALD A. KRULIK
Applied Electroless Concepts, Inc.

NENAD V. MANDICH
HBM Engineering Company

EXPLOSIVELY CLAD METALS

After World War II explosives began to be used in specialized metalworking operations, particularly for metal forming. Explosives provided an inexpensive source of fast-release energy and greatly reduced the need for expensive capital equipment (1,2). Research on explosively bonding metals began during the same period (3–7) (see also EXPLOSIVES AND PROPELLANTS).

Explosive cladding, or explosion bonding and explosion welding, is a method wherein the controlled energy of a detonating explosive is used to create a metallurgical bond between two or more similar or dissimilar metals. No intermediate filler metal, eg, a brazing compound or soldering alloy, is needed to promote bonding and no external heat is applied. Diffusion does not occur during bonding.

In 1962, the first method for welding (qv) metals in spots along a linear path by explosive detonation was patented (8). This method is not, however, used industrially. In 1963, a theory that explained how and why cladding occurs was published (9). Research efforts resulted in process patents which standardized industrial explosion cladding. Several of the patents describe the use of variables involved in parallel cladding which is the most popular form of explosion cladding (10–13). Several excellent reviews on metal cladding have been published (14–16).

The explosive cladding process provides several advantages over other metal-bonding processes:

(*1*) A metallurgical, high quality bond can be formed between similar metals and between dissimilar metals that are incompatible for fusion or diffusion joining. Brittle, intermetallic compounds, which form in an undesirable continuous layer at the interface during bonding by conventional methods, are minimized, isolated, and surrounded by ductile metal in explosion cladding. Examples of these systems are titanium–steel, tantalum–steel, aluminum–steel, titanium–aluminum, and copper–aluminum. Immiscible metal combinations, eg, tantalum–copper, also can be clad.

(*2*) Explosive cladding can be achieved over areas that are limited only by the size of the available cladding plate and by the magnitude of the explosion that can be tolerated. Areas as small as 1.3 cm^2 (17) and as large as 27.9 m^2 (18) have been bonded.

(*3*) Metals having tenacious surface films that make roll bonding difficult, eg, stainless steel/Cr–Mo steels, can be explosion clad.

(*4*) Metals having widely differing melting points, eg, aluminum (660°C) and tantalum (2996°C), can be clad.

(*5*) Metals having widely different properties, eg, copper or maraging steel, can be bonded readily.

(*6*) Large clad-to-backer ratio limits can be achieved by explosion cladding. Stainless steel-clad components as thin as 0.025 mm and as thick as 3.2 cm have been explosion clad.

(*7*) The thickness of the stationary or backing plate in explosion cladding is essentially unlimited. Backers >0.5-m thick and weighing 50 t have been clad commercially.

(*8*) High quality, wrought metals are clad without altering chemical composition.

(*9*) Different types of backers can be clad; clads can be bonded to forged members, as well as to rolled plate.

(*10*) Clads can be bonded to rolled plate that is strand-cast, annealed, normalized, or quench-tempered.

(*11*) Multilayered composite sheets and plates can be bonded in a single explosion, and cladding of both sides of a backing metal can be achieved simultaneously. When two sides are clad, the two prime or clad metals need not be of the same thickness nor of the same metal or alloy.

(*12*) Nonplanar metal objects can be clad, eg, the inside of a cylindrical nozzle can be clad with a corrosion-resistant liner.

(*13*) The majority of explosion-clad metals are less expensive than the solid metals that could be used instead of the clad systems.

Limitations of the explosive bonding process are as follows.

(*1*) There are both inherent hazards in storing and handling explosives and undesirable noise and blast effects from the explosion.

(*2*) Obtaining explosives with the proper energy, form, and detonation velocity is difficult.

(*3*) Metals to be explosively bonded must be somewhat ductile and resistant to impact. Brittle metals and metal alloys fracture during bonding. Alloys having as little as 5% tensile elongation in a 5.1-cm gauge length, and backing steels having as little as 13.6 J (10 ft·lbf) Charpy V-notch impact resistance can, however, be bonded.

(*4*) For metal systems in which one or more of the metals to be explosively clad has a high initial yield strength or a high strain-hardening rate, a high quality bonded interface may be difficult to achieve. Metal alloys of high (>690 MPa (10^5 psi)) yield strength, are difficult to bond. This problem increases when there is a large density difference between the metals. Such combinations often are improved by using a thin interlayer between the metals.

(*5*) Geometries suited to explosive bonding promote straight-line egression of the high velocity jet emanating from between the metals during bonding, eg, for the bonding of flat and cylindrical surfaces.

(*6*) Thin backers must be supported, thus adding to manufacturing cost.

(*7*) The preparation and assembly of clads is not amenable to automated production techniques, and each assembly requires considerable labor.

Theory and Principles

To obtain a metallurgical bond between two metals, the atoms of each metal must be brought sufficiently close so that their normal forces of interatomic attraction produce a bond. The surfaces of metals and alloys must not be covered with films of oxides, nitrides, or adsorbed gases. When such films are present, metal surfaces do not bond satisfactorily (see METAL SURFACE TREATMENTS).

Explosive bonding is a cold pressure-welding process in which the contaminant surface films are plastically jetted from the parent metals as a result of the high pressure collision of the two metals. A jet is formed between the metal plates, if the collision angle and the collision velocity are in the range required for bonding. The contaminant surface films that are detrimental to the establishment of a metallurgical bond are swept away in the jet. The metal plates, which are cleaned of any surface films by the jet action, are joined at an internal point by the high pressure that is obtained near the collision point.

Parallel and Angle Cladding. The arrangements shown in Figures 1 and 2 illustrate the operating principles of explosion cladding. Angle cladding (Fig. 1) is limited to cladding for relatively small pieces (19,20). Clad plates having large areas cannot be made using this arrangement because the collision of long plates at high stand-offs, ie, the distance between the plates, on long runs is so violent that metal cracking, spalling, and fracture occur. The arrangement shown in Figure 2 is by far the simplest and most widely used (10).

Jetting. A layer of explosive is placed in contact with one surface of the prime metal plate which is maintained at a constant distance from and parallel to the backer plate, as shown in Figure 2**a**. The explosive is detonated and, as

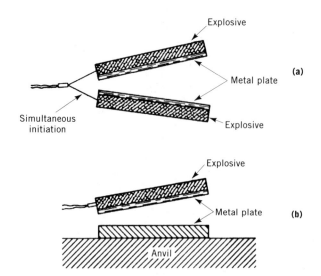

Fig. 1. Angle arrangements to produce explosion clads, where (**a**) represents symmetric angle cladding and (**b**), angle cladding.

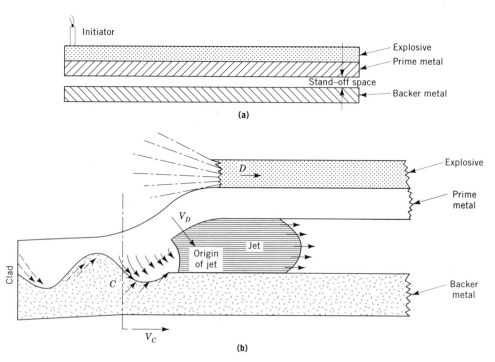

Fig. 2. Parallel arrangement for (**a**) explosion cladding and subsequent collision between the prime and backer metals that leads to (**b**) jetting and formation of the wavy bond zone, where V_D is the detonation velocity and V_C, the collosion velocity.

the detonation front moves across the plate, the prime metal is deflected and accelerated to plate velocity, V_P; thus an angle is established between the two plates. The ensuing collision region progresses across the plate at a velocity equal to the detonation velocity, D. When the collision velocity, V_C, and the angle are controlled within certain limits, high pressure gradients ahead of the collision region in each plate cause the metal surfaces to flow hydrodynamically as a spray of metal from the apex of the angled collision. Jetting is the flow process and expulsion of the metal surface (9). Photographic evidence of jetting during an explosion-bonding experiment is given in Figure 3 (21). The jet, which moves in the direction of detonation, is observed between the deflected prime metal and the backer metal.

Typically, jet formation is a function of plate collision angle, collision-point velocity, cladding-plate velocity, pressure at the collision point, and the physical and mechanical properties of the plates being bonded. For jetting and subsequent cladding to occur, the collision velocity has to be substantially below the sonic velocity of the cladding plates, usually ca 4000–5000 m/s (22). There also is a minimum collision angle below which no jetting occurs regardless of the collision velocity. In the parallel-plate arrangement shown in Figure 2, this angle is determined by the stand-off. In angle cladding (Fig. 1), the preset angle determines the stand-off and the attendant collision angle.

Nature of the Bond.   Extensive metallurgical testing has determined that the best clad properties are obtained when the bond zone is wavy. It is therefore preferable that commercial, explosively bonded metals exhibit a wavy bond-zone interface. The amplitude and wavelength of the bond zone wave structure varies as a function of explosive properties and the stand-off distance, as shown in Table 1 (12). Moreover, the amplitude of the waves is proportional to the square of the collision angle. This latter finding is consistent with the fluid flow analogy theory of bond-zone wave formation (23), where bond-zone wave formation is treated as fluid flowing around an obstacle (see FLOW MEASUREMENT; FLUID MECHANICS). When the fluid velocity is low, the fluid flows smoothly around the obstacle, but above a certain fluid velocity, the flow pattern becomes turbulent, as illustrated in Figure 4. In explosion bonding, the obstacle is the point of highest pressure in the collision region. Because the pressures in this region are many times higher than the dynamic yield strength of the metals, the metals flow plastically, as evidenced by the microstructure of the metals at the bond zone. Electron microprobe analysis across such plastically deformed areas shows that no diffusion occurs because there is extremely rapid self-quenching of the metals (22).

Under optimum conditions, the metal flow around the collision point is unstable and it oscillates, thereby generating a wavy interface. Typical explosion-bonded interfaces between nickel plates made at different collision velocities are illustrated in Figure 4 (23). A typical explosion-bonded interface between titanium and steel is shown in Figure 5. Small pockets of solidified melt form under the curl of the waves; some of the kinetic energy of the driven plate is locally converted into heat as the system comes to rest. These discrete regions are completely encapsulated by the ductile prime and base metals. The direct metal-to-metal bonding between the isolated pockets provides the ductility necessary to support stresses during routine fabrication.

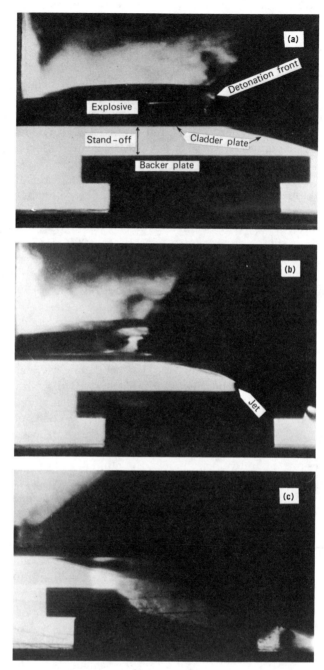

Fig. 3. (**a–c**) Cladding of aluminum to aluminum showing jet formation (21).

The quality of bonding is related directly to the size and distribution of solidified melt pockets along the interface, especially for dissimilar metal systems that form intermetallic compounds. The pockets of solidified melt are brittle and contain localized defects which do not affect the composite properties.

Table 1. Measured Explosion-Cladding Parameters and Bond-Zone Characteristics[a,b]

Collision velocity, m/s	Parallel stand-off, mm	Flyerplate Velocity, m/s	Angle, deg	Type[c]	Wave-length, μm	Ampli-tude, μm	Equivalent melt thick-ness, μm
				Grade A nickel			
1650	1.14	215	7.4	straight and wavy	112	10	<1
2000	1.14	250	7.0		103	11	<1
2500	1.14	270	6.6		236	39	1.6
3600	1.14	410	6.5		254	38	5.1
2000	2.16	310	8.7		318	41	<1
2500	2.16	337	8.2		425	76	3.9
3600	2.16	510	8.25		590	96	9.8
1650	3.96	325	11.2		520	52	<1
2000	3.96	372	10.5		567	88	<1
2500	3.96	407	9.7		671	121	6.0
3600	3.96	625	9.95		739	146	28.8
2000	6.35	420	11.8		790	132	<1
2500	6.35	462	10.8		895	171	9.0
3600	6.35	700	11.2		965	162	24.0
1650	10.54	425	14.8		1018	169	<1
2000	10.54	460	13.0		623	97	<1
3600	10.54	775	12.5		1333	284	59.2
1650	17.78	465	16.5	straight			[d]
				Grade 1 titanium			
2000	1.14	330	9.9		103	8	<1
2500	1.14	400	9.7		254	19	1.2
3600	1.14	580	9.3	MLW	250	23	14.0
2000	2.16	420	12.2	MLW	215	17	<1
2500	2.16	465	11.0	MLW	482	47	3.0
3600	2.16	710	11.3	MLW	468	59	11.6
2000	3.96	495	14.2	MLW	373	31	<1
2500	3.96	520	12.1	MLW	768	89	3.1
3600	3.96	845	13.4	MLW	868	122	9.2
2000	6.35	530	15.2	MLW	610	53	<1
2500	6.35	565	13.0	MLW	1009	130	8.2
3600	6.35	945	15.0	MLW	1228	189	18.5
2000	10.54	560	15.5	MLW	1013	96	<1
2500	10.54	600	14.0	MLW	1300	167	3.6
3600	10.54	1040	16.5	MLW	1360	230	21.8

[a] Ref. 12.
[b] For cladding 3.2-mm metal to 12.7-mm AISI 1008 Carbon Steel.
[c] Wavy unless otherwise noted. MLW = melted layer and waves.
[d] Not detectable.

Explosion-bonding parameters for dissimilar metal systems normally are chosen to minimize the pockets of melt associated with the interface.

When cladding conditions are such that the metallic jet is trapped between the prime metal and the backer, the energy of the jet causes surface melting

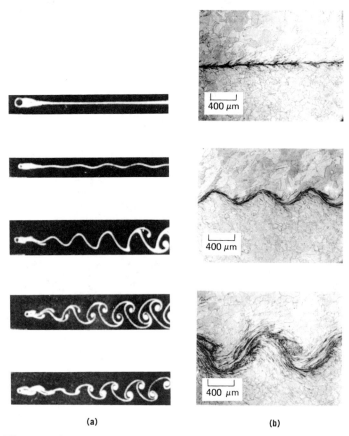

(a)　　　　　　　　　　　　(b)

Fig. 4. (**a**) Photographs of fluid flow behind cylinders at increasing flow velocities top to bottom. (**b**) Photomicrographs of nickel–nickel bond zones made at increasing collision velocities; top, ~1600 m/s; middle, ~1900 m/s; bottom, ~2500 m/s (23).

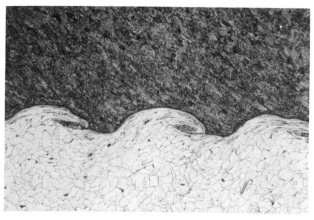

Fig. 5. Photomicrograph of titanium, top, to carbon steel, bottom, explosion clad (100x).

between the colliding plates. In this type of clad, alloying through melting is responsible for the metallurgical bond. As shown in Figure 6, solidification defects can occur and, for this reason, this type of bond is not desirable.

The industrially useful combinations of explosively clad metals that are available in commercial sizes are listed in Figure 7. The list does not include tri-clads or possible combinations not yet explored. The combinations that explosion cladding can provide are virtually limitless (24).

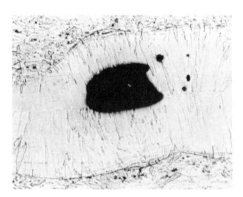

Fig. 6. Solidification defects in the copper–copper explosion-clad evidence the occurrence of melting at the interface (100x) (9).

Processing

Explosives. The pressure, P, generated by the detonating explosive that propels the prime plate is directly proportional to its density, ρ, and the square of the detonation velocity, V_D^2 (25):

$$P = \frac{1}{4}\rho V_D^2$$

The detonation velocity is controlled by adjusting the packing density or the amount of added inert material (26).

The types of explosives that have been used include both high (4500–7600 m/s) and low to medium (1500–4500 m/s) velocity materials (24,26).

High velocity	Low–medium velocity
trinitrotoluene (TNT)	ammonium nitrate
cyclotrimethylenetrinitramine (RDX)	ammonium nitrate prills
pentaerythritol tetranitrate (PETN)	sensitized with fuel oil
composition B	ammonium perchlorate
composition C_4	amatol
plasticized PETN-	amatol and sodatol diluted
based rolled sheet	with rock salt to 30–35%
and extruded cord	dynamites
primacord	nitroguanidine
	diluted PETN

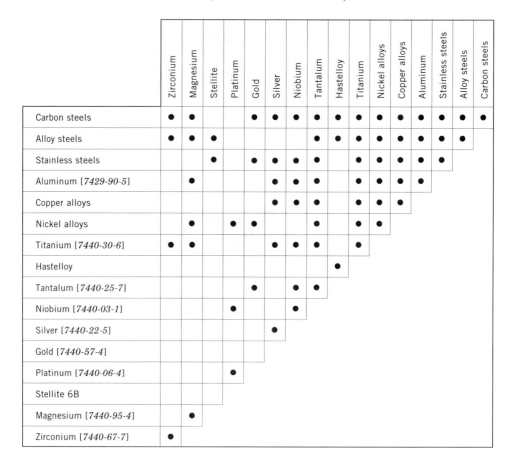

	Zirconium	Magnesium	Stellite	Platinum	Gold	Silver	Niobium	Tantalum	Hastelloy	Titanium	Nickel alloys	Copper alloys	Aluminum	Stainless steels	Alloy steels	Carbon steels
Carbon steels	•	•			•	•	•	•	•	•	•	•	•	•	•	•
Alloy steels	•	•	•					•	•	•	•	•	•	•	•	
Stainless steels			•		•	•	•	•		•	•	•	•	•		
Aluminum [7429-90-5]		•				•	•	•		•	•	•	•			
Copper alloys						•	•	•		•	•	•				
Nickel alloys		•		•	•			•		•	•					
Titanium [7440-30-6]	•	•			•	•	•			•						
Hastelloy									•							
Tantalum [7440-25-7]					•		•	•								
Niobium [7440-03-1]				•			•									
Silver [7440-22-5]					•											
Gold [7440-57-4]																
Platinum [7440-06-4]			•													
Stellite 6B																
Magnesium [7440-95-4]		•														
Zirconium [7440-67-7]	•															

Fig. 7. Commercially available explosion-clad metal combinations.

In commercial practice, powdered explosives on an ammonium nitrate basis are used in most cases. Typical detonation velocities are between 1800 and 3500 m/s depending on the metal system to be bonded. The lower detonation velocity range is preferred for many metal systems in order to minimize the quantity of solidified melt associated with the bond-zone waves (12). In addition, subsonic detonation velocity explosives are required for the parallel cladding technique in order to avoid attached shock waves in the collision region, which preclude formation of a good bond.

Metal Preparation. Preparation of the metal surfaces to be bonded usually is required because most metals contain surface imperfections or contaminants that undesirably affect bond properties. The cladding faces usually are surface ground, using an abrasive machine, and then are degreased with a solvent to ensure consistent bond strength (26). In general, a surface finish that is ≥ 3.8 μm deep is needed to produce consistent, high quality bonds.

Fabrication techniques must take into account the metallurgical properties of the metals to be joined and the possibility of undesirable diffusion at the interface during hot forming, heat treating, and welding. Compatible alloys, ie, those that do not form intermetallic compounds upon alloying, eg, nickel and nickel alloys (qv), copper and copper alloys (qv), and stainless steel alloys clad to steel, may be treated by the traditional techniques developed for clads produced by other processes. On the other hand, incompatible combinations, eg, titanium, zirconium, or aluminum to steel, require special techniques designed to limit the production at the interface of undesirable intermetallics which would jeopardize bond ductility.

Assembly, Stand-Off. The air gap present in parallel explosion cladding can be maintained by metallic supports that are tack-welded to the prime and backer plates or by metallic inserts that are placed between the prime and backer (26–28). The inserts usually are made of a metal that is compatible with one of the cladding metals. If the prime metal is so thin that it sags when supported by its edges, other materials, eg, rigid foam, can be placed between the edges to provide additional support; the rigid foam is consumed by the hot egressing jet during bonding (26–29) (see FOAMED PLASTICS). A moderating layer or buffer, eg, polyethylene sheet, water, rubber, paints, and pressure-sensitive tapes, may be placed between the explosive and prime metal surface to attenuate the explosive pressure or to protect the metal surface from explosion effects (24).

Facilities. The preset, assembled composite is placed on an anvil of appropriate thickness to minimize distortion of the clad product. For thick composites, a bed of sand usually is a satisfactory anvil. Thin composites may require a support made of steel, wood, or other appropriate materials. The problems of noise, air blast, and air pollution (qv) are inherent in explosion cladding, and clad-composite size is restricted by these problems (see NOISE POLLUTION AND ABATEMENT METHODS; SUPPLEMENT). Thus the cladding facilities should be in areas that are remote from population centers. Using barricades and burying the explosives and components under water or sand lessens both noise and air pollution (24). An attractive method for making small-area clads using light explosive loads employs a low vacuum, noiseless chamber (24). Underground missile silos and mines also have been used as cladding chambers (see INSULATION, ACOUSTIC).

Analytical and Test Methods

When the explosion-bonding process distorts the composite so that its flatness does not meet standard flatness specifications, it is reflattened on a press or roller leveler (ASME SA20). However, press-flattened plates sometimes contain localized irregularities which do not exceed the specified limits but which, generally, do not occur in roll-flattened products.

Nondestructive Testing. Nondestructive inspection of an explosion-welded composite is almost totally restricted to ultrasonic and visual inspection. Radiographic inspection is applicable only to special types of composites consisting of two metals having a significant mismatch in density and a large wave pattern in the bond interface (see NONDESTRUCTIVE EVALUATION).

Ultrasonics. The most widely used nondestructive test method for explosion-welded composites is ultrasonic inspection. Pulse-echo procedures (ASTM A435) are applicable for inspection of explosion-welded composites used in pressure applications (see ULTRASONICS).

The acceptable amount of nonbond depends on the application. In clad plates for heat exchangers, >98% bond usually is required (see HEAT-EXCHANGE TECHNOLOGY). Other applications may require only 95% of the total area to be bonded. Configurations of a nonbond sometimes are specified, eg, in heat exchangers where a nonbond area may not be >19.4 cm^2 or 7.6 cm long. The number of areas of nonbond generally is specified. Ultrasonic testing can be used on seam welds, tubular transition joints, clad pipe and tubing, and in structural and special applications.

Radiographic. Radiography is an excellent nondestructive test (NDT) method for evaluating the bond of Al–steel electrical and Al–Al–steel structural transition joints. It provides the capability of precisely and accurately defining all nonbond and flat-bond areas of the Al–steel interface, regardless of size or location (see SURFACE AND INTERFACE ANALYSIS).

The clad plate is x-rayed perpendicular from the steel side and the film contacts the aluminum. Radiography reveals the wavy interface of explosion-welded, aluminum-clad steel as uniformly spaced, light and dark lines with a frequency of one to three lines per centimeter. The waves characterize a strong and ductile transition joint and represent the acceptable condition. The clad is interpreted to be nonbonded when the x-ray shows complete loss of the wavy interface (see X-RAY TECHNOLOGY).

Destructive Testing.   Destructive testing is used to determine the strength of the weld and the effect of the explosion-welding process on the parent metals. Standard testing techniques can be utilized on many composites; however, nonstandard or specially designed tests often are required to provide meaningful data for specific applications.

Pressure-Vessel Standards.   Explosion-clad plates for pressure vessels are tested according to the applicable ASME Boiler and Pressure Vessel Code Specifications. Unfired pressure vessels using clads are covered by ASTM A263, A264, and A265; these include tensile, bend, and shear tests (see TANKS AND PRESSURE VESSELS).

Tensile tests of a composite plate having a thickness of <3.8 cm require testing of the joined base metal and clad. Strengthening does occur during cladding and tensile strengths generally are greater than for the original materials. Some typical shear-strength values obtained for explosion-clad composites covered by ASTM A263, A264, A265, which specify 138 MPa (20,000 psi) minimum, and B432, which specifies 83 MPa (12,000 psi) minimum, are listed in Table 2 (see HIGH PRESSURE TECHNOLOGY).

Chisel.   Chisel testing is a quick, qualitative technique that is widely used to determine the soundness of explosion-welded metal interfaces. A chisel is driven into and along the weld interface, and the ability of the interface to resist the separating force of the chisel provides an excellent qualitative measure of weld ductility and strength.

Ram Tensile.   A ram tensile test has been developed to evaluate the bond-zone tensile strength of explosion-bonded composites. The specimen is designed to subject the bonded interface to a pure tensile load. The cross-section area of the specimen is the area of the annulus between the outer and inner diameters of the specimen. The specimen typically has a very short tensile gauge length and is constructed so as to cause failure at the bonded interface. The ultimate

Table 2. Shear Strengths of Explosively Clad Metals

Cladding metal on carbon steel backers	Shear strength,[a] MPa[b]
stainless steels	448
nickel and nickel alloys	379
Hastelloy alloys	391
zirconium	269
titanium[c]	241
cupronickel	251
copper	152
aluminum (1100-Ah4)	96

[a]See ASTM A263, A264, A265, and B432.
[b]To convert MPa to psi, multiply by 145.
[c]Stress relief annealed at 621°C.

tensile strength and relative ductility of the explosion-bonded interface can be obtained by this technique.

Mechanical Fatigue.   Some mechanical fatigue tests have been conducted on explosion-clad composites where the plane of maximum tensile stress is placed near the bond zone (30).

Thermal Fatigue and Stability.   Explosion-welded plates have performed satisfactorily in several types of thermal tests (18). In thermal fatigue tests, samples from bonded plate are alternately heated to 454–538°C at the surface and are quenched in cold water to less than 38°C. The three-minute cycles consist of 168 s of heating and 12 s of cooling. Weld-shear tests are performed on samples before and after thermal cycling. Stainless steel clads have survived 2000 such thermal cycles without significant loss in strength (18). Similarly welded and tested Grade 1 titanium–carbon steel samples performed in a similarly satisfactory fashion.

Metallographic.   The interface is inspected on a plane parallel to the detonation front and normal to the surface. A well-formed wave pattern without porosity generally is indicative of a good bond. The amplitude of the wave pattern for a good weld can vary from small to large without a large influence on the strength, and small pockets of melt can exist without being detrimental to the quality of the bond. However, a continuous layer of melted material indicates that welding parameters were incorrect and should be adjusted. A line-type interface with few waves indicates that the collision velocity of the plate was not great enough and/or that the collision angle was too high for jetting to occur. A well-defined wave pattern in which the crest of the wave is bent over to form a large melt pocket with a void in the swirl is indicative of a poor bond. In this case, the plate velocity is too high as is the collision angle.

In some materials, eg, titanium and martensitic steels, shear bands are adjacent to the weld interface if the cladding variables are excessive. This is the result of thermal adiabatic shear developed from excessive overshooting energy, and a heat treatment is required to eliminate the hardened-band effect. When the cladding variables and the system energy are optimum, thermal shear bands are minimized or eliminated and heat treatment after cladding is not required. Several types of metal composites require heat treatment after cladding to relieve stress, but intermetallic compounds can form as a result of the treatment. A metallographic examination indicates if the heat treatment of the explosion-bonded composite has resulted in the formation of intermetallic compounds.

Hardness, Impact Strength. Microhardness profiles on sections from explosion-bonded materials show the effect of strain hardening on the metals in the composite (see HARDNESS). Figure 8 illustrates the effect of cladding a strain-hardening austenitic stainless steel to a carbon steel. The austenitic stainless steel is hardened adjacent to the weld interface by explosion welding, whereas the carbon steel is not hardened to a great extent. Similarly, aluminum does not strain harden significantly.

Impact strengths also can be reduced by the presence of the hardened zone at the interface. A low temperature stress-relief anneal decreases the hardness and restores impact strength (30). Alloys that are sensitive to low temperature heat treatments also show differences in hardness traverses that are related to the explosion-welding parameters, as illustrated in Figure 9 (16). Low welding impact velocities do not develop as much adiabatic heating as higher impact velocities. The effect of the adiabatic heating is to anneal and further age the alloys. Hardness traverses indicate the degree of hardening during welding and what, if any, subsequent heat treatment is required after explosion bonding. Explosion-bonding parameters also can be adjusted to prevent softening at the interface, as shown in Figure 9.

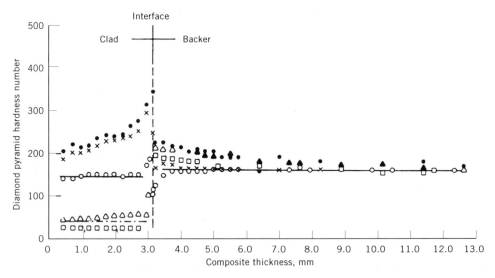

Fig. 8. Microhardness profile across interfaces of two types of explosion clads that show widely divergent response resulting from the inherent cold-work hardening characteristics where (——) represents the 3.2-mm type 304L stainless/28.6-mm, A 516-70 control (before cladding); (•) = clad + flat; (×) = clad + stress relief annealed at 621°C + flat; (○) = clad + normalize at 954°C. (—•—) represents 3.2-mm 1100-H14 aluminum/25.4-mm, A 516-70 control (before cladding); (△) = clad + flat; (□) = clad + flat + stress relief annealed at 593°C (31).

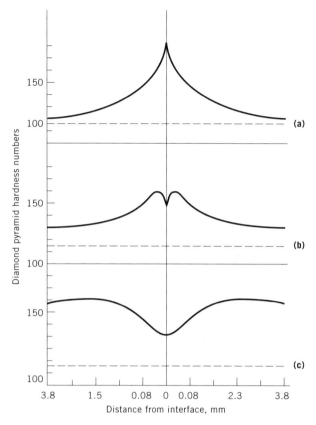

Fig. 9. Microhardness profiles across interface of explosion-clad age-hardenable aluminum alloy 2014-T3 where the initial hardness is shown as (–––) (**a**) low, (**b**) medium, and (**c**) high impact velocity (16).

Safety Aspects

All explosive materials should be handled and used following approved safety procedures in compliance with applicable federal, state, and local laws, regulations, and ordinances. The Bureau of Alcohol, Tobacco, and Firearms (BATF), the Hazardous Materials Regulation Board (HMRB) of the Department of Transportation (DOT), the Occupational Safety and Health Agency (OSHA), and the Environmental Protection Agency (EPA) in Washington, D.C., have federal jurisdiction on the sale, transport, storage, and use of explosives. Many states and local counties have special explosive requirements. The Institute of Makers of Explosives (IME) in New York provides educational publications to promote the safe handling, storage, and use of explosives. The National Fire Protective Association (NFPA) in Boston, Mass., similarly provides recommendations for safe explosives manufacture, storage, handling, and use.

Uses

Cladding and backing metals are purchased in the appropriately heat-treated condition because corrosion resistance is retained through bonding. It is customary to supply the composites in the as-bonded condition because hardening usually does not affect the engineering properties. Occasionally, a post-bonding heat treatment is used to achieve properties required for specific combinations.

Vessel heads can be made from explosion-bonded clads, either by conventional cold- or by hot-forming techniques. The latter involves thermal exposure and is equivalent in effect to a heat treatment. The backing metal properties, bond continuity, and bond strength are guaranteed to the same specifications as the composite from which the head is formed. Applications such as chemical-process vessels and transition joints represent approximately 90% of the industrial use of explosion cladding.

Chemical-Process Vessels. Explosion-bonded products are used in the manufacture of process equipment for the chemical, petrochemical, and petroleum industries where the corrosion resistance of an expensive metal is combined with the strength and economy of another metal. Applications include explosion cladding of titanium tubesheet to Monel, hot fabrication of an explosion clad to form an elbow for pipes in nuclear power plants, and explosion cladding titanium and steel for use in a vessel intended for terephthalic acid manufacture.

Precautions must be taken when welding incompatibly clad systems, eg, hot forming of titanium-clad steel plates must be conducted at 788°C or less. The preferred technique for butt welding involves a batten-strap technique using a silver, copper, or steel inlay (Fig. 10). Precautions must be taken to avoid iron contamination of the weld either from the backer steel or from outside sources. Stress relieving is achieved at normal steel stress-relieving temperatures, and special welding techniques must be used in joining tantalum–copper–steel clads (32,33).

Conversion-Rolling Billets. Much clad plate and strip have been made by hot and cold rolling of explosion-bonded slabs and billets. Explosion bonding is economically attractive for conversion rolling because the capital investment for plating and welding equipment needed for conventional bonding methods is avoided. Highly alloyed stainless steels and some copper alloys, which are difficult to clad by roll bonding, are used for plates made by converting explosion-bonded slabs and billets. Conventional hot-rolling and heat-treatment practices are used when stainless steels, nickel, and copper alloys are converted. Hot rolling of explosion-bonded titanium, however, must be performed below ca 843°C to avoid diffusion and the attendant formation of undesirable intermetallic

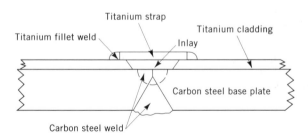

Fig. 10. Double-v inlay, batten-strap technique for fusion welding of an explosion-clad plate containing titanium and zirconium.

compounds at the bond interface. Hot-rolling titanium also requires a stiff rolling mill because of the large separation forces required for reduction.

Perhaps the most extensive application for conversion-rolled, explosion-bonded clads was for U.S. coinage in the 1960s (34) when over 15,900 metric tons of explosion-clad strip that was supplied to the U.S. Mint helped alleviate the national silver coin shortage. The triclad composites consist of 70–30 cupronickel/Cu/70–30 cupronickel.

Transition Joints. Use of explosion-clad transition joints avoids the limitations involved in joining two incompatible materials by bolting or riveting. Many transition joints can be cut from a single large-area flat-plate clad and delivered to limit the temperature at the bond interface so as to avoid undesirable diffusion. Conventional welding practices may be used for both similar metal welds.

Electrical. Aluminum, copper, and steel are the most common metals used in high current–low voltage conductor systems. Use of these metals in dissimilar metal systems often maximizes the effects of the special properties of each material. However, junctions between these incompatible metals must be electrically efficient to minimize power losses. Mechanical connections involving aluminum offer high resistance because of the presence of the self-healing oxide skin on the aluminum member. Because this oxide layer is removed by the jet, the interface of an explosion clad essentially offers no resistance to the current. Thus welded transition joints, which are cut from thick composite plates of aluminum–carbon steel, permit highly efficient electrical conduction between dissimilar metal conductors. Sections can be added by conventional welding. This concept is routinely employed by the primary aluminum reduction industry in anode-rod fabrication. The connection is free of the aging effects that are characteristic of mechanical connections and requires no maintenance. The mechanical properties of the explosion weld, ie, shear, tensile, and impact strength, exceed those of the parent-type 1100 aluminum alloy.

Usually, copper surfaces are mated when joints must be periodically disconnected because copper offers low resistance and good wear. Junctions between copper and aluminum bus bars are improved by using a copper–aluminum transition joint that is welded to the aluminum member. Deterioration of aluminum shunt connections by arcing is eliminated when a transition joint is welded to both the primary bar and the shunting bar.

The same intermetallic compounds that prevent conventional welding between aluminum and copper or steel can be developed in an explosion clad by heat treatment at elevated temperature. Diffusion can be avoided if the long-term service temperature is kept below 260°C for aluminum–steel and 177°C for copper–aluminum combinations. Under short-term conditions, as during welding, peak temperatures of 316 and 232°C, respectively, are permissible. Bond ductility is maintained, although there is a reduction in bond strength as the aluminum is annealed. Bond strength, however, never falls below that of the parent aluminum; therefore nominal handbook values for type 1100 alloy aluminum may be used in design considerations. The bond is unaffected by thermal cycling within the recommended temperature range.

Marine. In the presence of an electrolyte, eg, seawater, aluminum and steel form a galvanic cell and corrosion takes place at the interface. Because the aluminum superstructure is bolted to the steel bulkhead in a lap joint, crevice corrosion is masked and may remain unnoticed until replacement is required. By using transition-joint strips cut from explosion-welded clads, the corrosion problem can be eliminated. Because the transition is metallurgically bonded, there is no crevice in which the electrolyte can act and galvanic action cannot take place. Steel corrosion is confined to external surfaces where it can be detected easily and corrected by simple wire brushing and painting.

Explosion-welded construction has equivalent or better properties than the more complicated riveted systems. Peripheral benefits include weight savings and perfect electrical grounding. In addition to lower initial installation costs, the welded system requires little or no maintenance and, therefore minimizes life-cycle costs. Applications of structural transition joints include aluminum superstructures that are welded to decks of naval vessels and commercial ships as illustrated in Figure 11.

Tubular. Explosion welding is a practical method for providing the means to join dissimilar metal pipes, eg, aluminum, titanium, or zirconium, to steel or stainless steel, using standard welding equipment and techniques. The process provides a strong metallurgical bond which assures that the transition joints provide maintenance-free service throughout years of thermal and pressure/vacuum cycling. Explosion-welded tubular transition joints are being used in many diverse applications in aerospace, nuclear, and cryogenic industries. These oper-

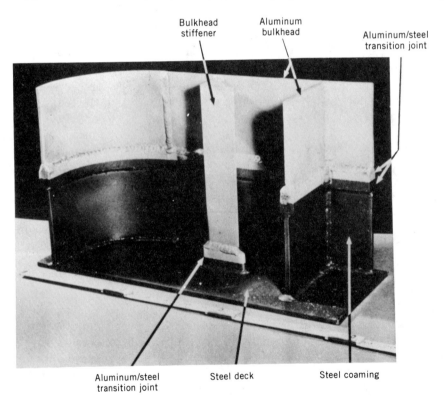

Fig. 11. Sample showing typical aluminum superstructure and deck connection made possible by use of explosion-clad aluminum–carbon steel transition joint.

ate reliably through the full range of temperatures, pressures, and stresses that normally are encountered in piping systems. Tubular transition joints in various configurations can be cut and machined from explosion-welded plate, or made by joining tubes by overlap cladding. Standard welding practices are used to make the final joints.

Nonplanar Specialty Products.   The inside walls of hollow forgings that are used for connections to heavy walled pressure vessels have been metallurgically bonded with stainless steel. These bonded forgings, or nozzles, range from 50 to 610 mm in inner diameter and are up to 1 m long. Large-clad cylinders and internally clad, heavy-walled tubes have been extruded using conventional equipment. Other welding applications have been demonstrated, including those shown in Figure 12.

Tube Welding and Plugging.   Explosion-bonding principles are used to bond tubes and tube plugs to tube sheets. The commercial process resembles the cladding of internal surfaces of thick-walled cylinders or pressure vessel nozzles, as shown in Figure 13; angle cladding is used (35). Countersink machining at the tube entrance provides the angled surface of 10–20° at a depth of 1.3–1.6 cm. The exploding detonator propels the tube or tube plug against the face of the tube-sheet to form the proper collision angle which in turn provides the required jetting and attendant metallurgical bond. Tubes may be welded individually or in groups. Metal combinations that are welded commercially include carbon steel–carbon steel, titanium–stainless steel, and 90–10 cupronickel–carbon steel.

Refractory Metals and Alloys.   Special modifications of the explosion-bonding process have been developed to successfully produce clads of refractory metals on steel and other backer metals. Among the refractory metals that have received particular attention for specialized chemical process equipment are tantalum, molybdenum, and molybdenum–rhenium alloys. In the case of tantalum, thin interlayers of copper are usually clad simultaneously with the tantalum

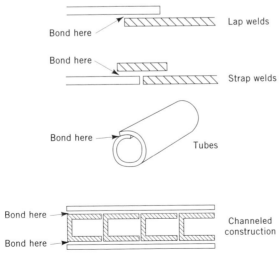

Fig. 12.   Explosion-clad welding applications (7).

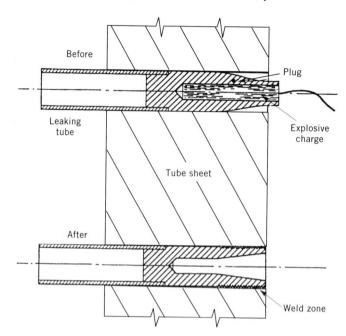

Fig. 13.   Tube-to-tubesheet plugging (35).

sheet onto steel or stainless steel backers. This facilitates production of reliable welds when several explosion-clad plates are joined along their edges to form a larger clad plate for fabrication into large commercial chemical reactors. In the case of molybdenum, and certain molybdenum alloys that have a high brittle-to-ductile transition temperature, it is necessary to use specialized heating techniques during the explosion cladding operation to produce well-bonded clads having good ductility for subsequent forming operations (36–41).

Production and Markets

Explosion-bonded metals are produced by several manufacturers in the United States, Europe, and Japan. The chemical industry is the principal consumer of explosion-bonded metals which are used in the construction of clad reaction vessels and heat-exchanger tube sheets for corrosion-resistant service. The primary market segments for explosion-bonded metals are for corrosion-resistant pressure vessels, tube sheets for heat exchangers, electrical transition joints, and structural transition joints. Total world markets for explosion-clad metals are estimated to fluctuate between 30×10^6 to 60×10^6 annually.

BIBLIOGRAPHY

"Metallic Coatings (Explosively Clad)" in *ECT* 3rd ed., Vol. 15, pp. 275–296, by A. Pocalyko, E. I. du Pont de Nemours & Co., Inc.

1. J. Pearson, *J. Met.* **12**, 673 (1960).
2. R. S. Rinehart and J. Pearson, *Explosive Working of Metals*, MacMillan, New York, 1963.
3. J. J. Douglass, *New England Regional Conference of AIME*, Boston, Mass., May 26, 1960.
4. *Ryan Reporter*, Vol. 21, No. 3, Ryan Aeronautical Co., San Diego, Calif., 1960, pp. 6–8.
5. C. P. Williams, *J. Met.*, 33 (1960).
6. "High Energy Rate Forming" in *Product Engineering and American Machinist/Metalworking Manufacturing*, McGraw Hill, New York, 1961 and 1962.
7. A. H. Holtzman and C. G. Rudershansen, *Sheet Met. Ind.* **39**, 401 (1961).
8. U.S. Pat. 3,024,526 (Mar. 13, 1962), V. Philipchuk and F. Le Roy Bois (to Atlantic Research Corp.).
9. G. R. Cowan and A. H. Holtzman, *J. Appl. Phys.* **34**(Pt. 1), 928 (1962).
10. U.S. Pat. 3,137,937 (June 23, 1964), G. R. Cowan, J. J. Douglass, and A. H. Holtzman (to E. I. du Pont de Nemours & Co., Inc.).
11. U.S. Pat. 3,233,312 (Feb. 8, 1966), G. R. Cowan and A. H. Holtzman (to E. I. du Pont de Nemours & Co., Inc.).
12. U.S. Pat. 3,397,444 (Aug. 20, 1968), O. R. Bergmann, G. R. Cowan, and A. H. Holtzman (to E. I. du Pont de Nemours & Co., Inc.).
13. U.S. Pat. 3,493,353 (Feb. 3, 1970), O. R. Bergmann, G. R. Cowan, and A. H. Holtzman (to E. I. du Pont de Nemours & Co., Inc.).
14. B. Crossland and A. S. Bahrani, *Proceedings of the First International Conference on Center High Energy Forming*, University of Denver, Denver, Colo., 1967.
15. A. A. Ezra, *Principles and Practices of Explosives Metal Working*, Industrial Newspapers, Ltd., London, 1973.
16. S. H. Carpenter and R. H. Wittman, *Ann. Rev. Mater. Sci.* **5**, 177 (1975).
17. J. L. Edwards, B. H. Cranston, and G. Krauss, *Metallic Effects at High Strain Rates*, Plenum Press, New York, 1973.
18. A. Pocalyko, *Mater. Prot.* **4**(6), 10 (1965).
19. U.S. Pat. 3,264,731 (Aug. 9, 1966), B. Chudzik (to E. I. du Pont de Nemours & Co., Inc.).
20. U.S. Pat. 3,263,324 (Aug. 2, 1966), A. A. Popoff (to E. I. du Pont de Nemours & Co., Inc.).
21. O. R. Bergmann, G. R. Cowan, and A. H. Holtzman, *Trans. Met. Soc. AIME* **236**, 646 (1966).
22. A. H. Holtzman and G. R. Cowan, *Weld. Res. Counc. Bull. No. 104*, Engineering Foundation, New York, Apr. 1965.
23. G. R. Cowan, O. R. Bergmann, and A. H. Holtzman, *Met. Trans.* **2**, 3145 (1971).
24. V. D. Linse, R. H. Wittman, and R. J. Carlson, *Defense Metals Information Center, Memo 225*, Columbus, Ohio, Sept. 1967.
25. M. A. Cook, *The Science of High Explosives*, Reinhold Publishing Corp., New York, 1966, p. 274.
26. A. A. Popoff, *Mech. Eng.* **100**(5), 28 (1978).
27. U.S. Pat. 3,140,539 (July 14, 1964), A. H. Holtzman (to E. I. du Pont de Nemours & Co., Inc.).
28. U.S. Pat. 3,205,574 (Sept. 14, 1965), H. M. Brennecke (to E. I. du Pont de Nemours & Co., Inc.).
29. U.S. Pat. 3,360,848 (Jan. 2, 1968), J. J. Saia (to E. I. du Pont de Nemours & Co., Inc.).
30. J. L. DeMaris and A. Pocalyko, *American Society of Tool and Manufacturing Engineers, Paper AD66-113*, Dearborn, Mich., 1966.
31. A. Pocalyko and C. P. Williams, *Weld. J.* **43**, 854 (1964).
32. U.S. Pat. 3,464,802 (Sept. 2, 1969), J. J. Meyer (to Nooter Corp.).

33. U.S. Pat. 4,073,427 (Feb. 14, 1978), H. G. Keifert and E. R. Jenstrom (to Fansteel, Inc.).
34. J. M. Stone, paper presented at *Select Conference on Explosive Welding*, Hove, U.K., Sept. 1968, pp. 29–34.
35. R. Hardwick, *Weld. J.* **54**(4), 238 (1975).
36. U.S. Pat. 5,226,579 (July 13, 1993), O. R. Bergmann, V. M. Felix, W. J. Simmons, and R. H. Tietzen (to E. I. Du Pont de Nemours & Co., Inc.).
37. U.S. Pat. 4,264,029 (Apr. 28, 1981), R. Henne and R. Pruemmer (to Deutsche Forschungs und Versuchsanstalt Fuer Luft und Raumfahrt).
38. Ger. Pat. 1,934,104 (Jan. 21, 1971), A. A. Deribas and V. M. Kudinov (to Institute of Hydrodynamics, Novosibirsk, Russia).
39. R. Koecher, *Chem. Ing. Tech.* **55**(10), 752–762 (1983).
40. U. Gramberg, E. M. Horn, and K. O. Cavalar, "Explosionsplattierte Bleche in Anlagen, Erfahrungen aus der Chemietechnik," VDI Meeting, *Explosionsplattieren-ein modernes Verfahren zur Herstellung von Hochleistungsvebundsystemen*, Duesseldorf, Germany, Dec. 1, 1983, pp. 29–35.
41. U.S. Pat. 5,323,955 (June 28, 1994) O. R. Bergmann, V. M. Felix, W. J. Simmons, and R. H. Tietjen (to E. I. du Pont de Nemours & Co.).

General References

The Joining of Dissimilar Metals, DMIC Report S-16, Battelle Memorial Institute, Columbus, Ohio, Jan. 1968.

S. H. Carpenter, *Nat. Tech. Info. Ser. Rept. No. AMMRC CTR74-69*, Dec. 1978.

C. Birkhoff, D. P. MacDougall, E. M. Pugh, and G. Taylor, *J. Appl. Phys.* **19**, 563 (1948).

L. Zernow, I. Lieberman, and W. L. Kincheloe, *American Society of Tool and Manufacturing Engineers, Technical Paper SP60-141*, Dearborn, Mich., 1961.

R. H. Wittman, *Metallurgical Effects at High Strain Rates*, Plenum Press, New York, 1973.

R. H. Wittman, *American Society of Tool and Manufacturing Engineers, Technical Paper AD-67-177*, Dearborn, Mich.

G. Bechtold, I. Michael, and R. Prummer, *Gold Bull. Chamber Mines S. Afr.* **10**(2), 34 (1977).

B. H. Cranston, D. A. Machusak, and M. E. Skinkle, *West Electr. Eng.*, 26 (Oct. 1978).

A. A. Popoff, *American Society of Manufacturing Engineers, Technical Paper AD77-236*, Dearborn, Mich., 1977.

T. Z. Blazynski, *International Conference on Welding and Fabrication of Non-Ferrous Metals, Eastbourne, May 2 and May 3, 1972, Cambridge, U.K.*, The Welding Institute, 1972.

J. K. Kowalick and D. R. Hay, *Second International Conference of the Center For High Energy Forming*, Estes Park, Colo., June 23–27, 1969.

J. Ramesam, S. R. Sahay, P. C. Angelo, and R. V. Tamhankar, *Weld Res. Suppl.*, 23s (1972).

L. F. Trueb, *Trans. Met. Soc. AIME* **2**, 147 (1971).

Metals Handbook, 9th ed., Vol. 6, American Society for Metals, Metals Park, Ohio, 1983, pp. 705–718.

ASM Handbook, Vol. 6, ASM International, Metals Park, Ohio, pp. 160–164, 303–305, 896–900.

OSWALD R. BERGMANN
E. I. du Pont de Nemours & Co., Inc.

METALLIC SOAPS.   See DRIERS AND METALLIC SOAPS; PAINTS.

METALLOCENES.   See ORGANOMETALLICS.

METALLURGY

SURVEY

The early foundations of metallurgy can be traced to the late Stone Age (ca 8000 BC) when copper (qv) was first used as a substitute for stone. The Metallic Age, when copper was melted and casted into tools, utensils, and weapons, followed. The Mesopotamians, the metallurgists of the ancient world, made exquisitely formed gold and bronze objects that have been dated to around 3000 BC. The Industrial Revolution of the eighteenth and nineteenth centuries was driven to a large degree by metallurgy and the use of basic materials such as iron (qv), steel (qv), and copper. Certainly, the large-scale generation and delivery of electrical energy would not have been possible without copper (see POWER GENERATION). Going into the twenty-first century, numerous advanced materials utilized in contemporary society depend on specialty metals often requiring novel processing techniques as well as a detailed knowledge of chemistry and atomic structure (see ABLATIVE MATERIALS; GLASSY METALS; HIGH TEMPERATURE ALLOYS; SHAPE-MEMORY ALLOYS). Clearly, metallurgy is both an ancient art and a modern science.

The first authoritative cataloging of metallurgical knowledge, by Georgius Agricola in *De Re Metallica* (1556) (1), reflected an early appreciation for metallurgy as encompassing many disciplines. Metal production was known to be achieved by many diverse processes. Early sources define metallurgy as the process of extracting metal from ores. For many metals, the primary source materials as of the 1990s are still crude metalliferous ores. For some metals however, recycled materials contribute significantly to total metal production. For example, in the United States the recycling (qv) rate of all-aluminum used beverage cans is over 50%. For an energy-intensive metal such as aluminum, this represents a substantial energy saving. Recycled aluminum requires only 5% of the energy needed to make aluminum from bauxite ore (see ALUMINUM AND ALUMINUM ALLOYS; RECYCLING, NONFERROUS METALS).

Metallurgy includes not only the treatment of crude ore and scrap, but also the processing of intermediates, ie, concentrates, and wastes, such as, slags, tailings, etc, for contained metal values. The various areas and subdisciplines comprising metallurgy may be summarized as follows:

Extractive metallurgy			
Mineral processing	Chemical metallurgy	Process metallurgy	Physical metallurgy
comminution	hydrometallurgy	alloying	structure–property
classification	pyrometallurgy	casting	relationships
flotation	electrometallurgy	deformation	failure analysis
		processes	corrosion
	corrosion	heat treatment	
		powder metallurgy	
		nuclear metallurgy	

Extractive metallurgy, the initial phase of winning metals from a given raw material, involves both physical and chemical processes. Those steps consisting of physical operations are termed mineral processing or mineral beneficiation, sometimes referred to as ore dressing. Physical operations are often required to liberate and concentrate metal values contained in an ore so that subsequent chemical processing can be performed at higher efficiencies (see also MINERAL RECOVERY AND PROCESSING). Processes involving some form of chemical change, required to free a given metal from associated impurities, embody the field of chemical metallurgy. Chemical metallurgy in turn is conveniently divided into pyrometallurgy, hydrometallurgy (see SUPPLEMENT), and electrometallurgy.

Process metallurgy, the next phase in the metals cycle, is the procedure or sequence of procedures whereby metals are worked or shaped. There are five main methods of working and shaping metals including casting, hot and cold working, machining, electroforming, and powder metallurgy. Powder metallurgy is especially important in working with high melting point metals. Examples are tungsten, mp = 3410°C, rhenium, mp = 3180°C, tantalum, mp = 2996°C, and platinum, mp = 1774°C. Powder processing is vital to the manufacture of products from refractory metals and alloys (see REFACTORIES).

Metallurgy also embraces the scientific study of the structure, properties, and behavior of metals and metal alloys. This branch of metallurgy is referred to as physical metallurgy. The two areas that commonly characterize physical metallurgy are structure–property relationships and failure analysis.

Both powder and nuclear metallurgy have attained a certain status and prominence and, in certain respects, represent main areas within the field of metallurgy (see METALLURGY, POWDER METALLURGY). Nuclear metallurgy involves all the main disciplines in metallurgy as related to the production of fuels, fabrication of fuels, and the fabrication of cladding materials and moderators (see NUCLEAR REACTORS). Corrosion is a discipline which bridges aspects of both chemical and physical metallurgy (see CORROSION AND CORROSION CONTROL).

Definitions

The field of metallurgy has a unique and frequently very specialized vocabulary. Understanding this language helps to clarify certain concepts and processing steps. A complete dictionary of mining, mineral, and related terms has been compiled (2). The definitions and explanations of key terms follow.

Concentrate. An action to intensify in strength or purity by the removal of valueless or unneeded constituents, ie separation of ore or metal from its containing rock or earth. The concentration of ores always proceeds by steps or stages. Liberation of mineral values is often the initial step. Concentrate also means a product of concentration, ie, enriched ore after removal of waste in a beneficiation mill.

Electrometallurgy. A term covering the various electrical processes for the working of metals, eg, electrodeposition, electrorefining and electrowinning, and operations in electric furnaces.

Flotation. The method of mineral separation in which a froth created in water by a variety of reagents floats some finely crushed minerals, whereas other minerals sink.

Gangue. Undesired minerals associated with ore, mostly nonmetallic. Gangue represents the portion of ore rejected as tailings in a separating process. It is usually valueless, but may have some secondary commercial use.

Hydrometallurgy. The treatment of ores, concentrates, and other metal-bearing materials by wet processes, usually involving the solution of some component, and its subsequent recovery from solution.

Leaching. Extracting a soluble metallic compound from an ore by selectively dissolving it in a suitable solvent. The solvent is usually recovered by precipitation of the metal or by other methods.

Mineral. An inorganic substance occurring in nature, though not necessarily of inorganic origin, which has (1) a definite chemical composition, or more commonly, a characteristic range of chemical composition, and (2) distinctive physical properties or molecular structure. In a broad sense, mineral should embrace both inorganic and organic substances, eg, fuel minerals such as coal (qv), oil (see PETROLEUM), and natural gas (see GAS, NATURAL). Ore is often classified as containing ore minerals, valuable constituents, and gangue minerals, ie, waste.

Mineral Dressing. Physical and chemical concentration of raw ore into a product from which a metal can be recovered at a profit.

Ore. A mineral or aggregate of minerals from which a valuable constituent, especially a metal, can be profitably extracted.

Pyrometallurgy. Metallurgy involved in winning and refining metals where heat is used, as in roasting and smelting. Pyrometallurgy is the oldest extractive process and is probably the most important.

Roast. The heating of solids, frequently to promote a reaction with a gaseous constituent in the furnace atmosphere.

Smelt/Smelting. Any metallurgical operation in which metal is separated by fusion from those impurities with which it may be chemically combined or physically mixed, such as in ores.

Ore

The metal content of an ore is typically called the ore grade and is usually expressed as weight percent for most metals. For precious metals, however, grade is usually expressed in g/t (oz/short ton). Because the definition of ore is established by economic considerations, there is no upper limit to grade, ie, the richer, the better. There is frequently a lower limit or cutoff grade, however, based on process efficiency and economics. Table 1 shows the average grade of various metalliferous ores that can be processed economically. Also shown is an estimate of the world total reserve base for each metal. For many metals, ore grade depletion has been a serious problem. This is illustrated for copper by the decline in average copper yield for U.S. copper ores during the 1900s (Fig. 1). The ability of the copper industry to remain competitive while faced with this problem has been a challenge. Technical developments in leaching, and improvements in solution concentration, purification, and metal reduction (solvent extraction and electrowinning), have turned this problem into an opportunity for additional metal production. Leaching of large tonnages of low grade material accounts for about 35% of the U.S. primary copper production.

Modern heap leaching practice has also made possible the treatment of extremely lean (ca 1 g/t) gold ores, which had been considered uneconomic as late as the early 1980s. In addition, changing technology and process innovations have contributed to the extraction of nickel from low grade lateritic ores and to the potential recovery of aluminum from nonbauxitic aluminum resources.

Some metals are produced primarily as a by-product from the mining and refining of other metals. For example, approximately one-half of the world's cobalt supply is produced as a by-product from copper operations in Zaire and

Table 1. Grade of Ore for Economic Processing and Estimated World Reserves

Ore	Grade, wt %	World reserves, t
aluminum	27–29	
chromium	27–34	6,778,000,000[a]
cobalt	1–11	8,340,000
copper	0.5–2	574,000,000
gold	0.0001–0.001	48,600
iron	30–60	230,000,000,000
lead	5–10	120,000,000
manganese	45–55	3,540,000,000
molybdenum	0.6–1.8	13,000,000
nickel	1.5–3	109,000,000
platinum	0.001	100,000[b]
silver	0.04–0.08	420,000
tin	1–5	6,050,000
titanium	2.5–25	
uranium	0.1–0.9	
vanadium	1.6–4.5	16,330,000
zinc	10–30	295,000,000

[a] As chromite [53293-42-8].
[b] Reserves of the platinum group metals (qv), ie Pt, Pd, Rh, Ru, Ir, and Os, together equal 1×10^8 t.

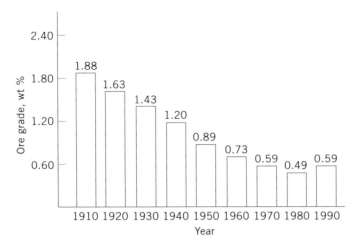

Fig. 1. Average yield of copper ores mined in the United States.

Zambia. Also, significant quantities of cobalt are produced as a by-product of nickel. Molybdenum is found in small (0.005–0.15% Mo) quantities in porphyry copper ores, and for many years high grade molybdenite [1309-56-4], MoS_2, concentrates have been produced as part of copper-milling operations. Other important by-products associated with copper ores are nickel, cobalt, silver, gold, platinum, lead, and zinc.

Product Specifications

Impurities in crude metal can occur as other metals or nonmetals, either dissolved or in some occluded form. Normally, impurities are detrimental, making the metal less useful and less valuable. Sometimes, as in the case of copper, extremely small impurity concentrations, eg, arsenic, can impart a harmful effect on a given physical property, eg, electrical conductivity. On the other hand, impurities may have commercial value. For example, gold, silver, platinum, and palladium, associated with copper, each has value. In the latter situation, the purity of the metal is usually improved by some refining technique, thereby achieving some value-added and by-product credit.

Refined metals, as traded on the open market, vary considerably in composition. However, there are strict specifications for certain impurity elements for a number of metals. The degree of purity of common industrial metals obtained from modern metallurgical practices is shown in Table 2 (3).

Economic Aspects

Metals have influenced and have served as the unit of monetary exchange throughout many parts of the world for millennia. Metal price has a very significant influence on production patterns. Price drives metal exploration and development of new resources, and is the main factor in determining materials substitution. High metal prices stimulate exploration activities and new mine development, which in turn increase supply. High metal prices also encourage the

Table 2. Analyses of Refined Metals,[a] wt %

Metal	Copper[b]	Nickel[c]	Lead[d]	Zinc[e]	Silver
Cu	99.99	0.02	0.0010	0.002	0.01
Ni	0.001	99.8	0.0002		
Pb	0.0005	0.005	99.97	0.003	0.001
Zn	0.0001	0.005	0.0005	99.990	
Ag	0.0025		0.0025		99.99
Sb	0.0004	0.005	0.0005		
As	0.0005	0.005	0.0005		
Fe	0.001	0.02	0.001	0.003	0.001
O	0.0005				
P	0.0003	0.005			
Se	0.0003				0.0005
S	0.0015	0.01			
Sn	0.0002	0.005	0.0005	0.001	

[a]Values are minimum for the primary metal and maximum for impurity concentrations.
[b]Oxygen-free electrolytic copper containing 0.0001 wt % Bi and Cd and 0.00005 wt % Mn.
[c]Refined nickel primarily produced from ore or matte or similar material.
[d]Pure lead for lead–acid battery application.
[e]Special high grade containing only 0.003 wt % Cd and 0.002 wt % Al.

search for substitutes, decreasing demand. The prices in constant 1987 dollars of selected metals are plotted in Figure 2 (4). Metal price fluctuations and cycles are controlled by numerous factors. For example, the price of aluminum (Fig. 2a) indicates supply shortages and a relatively high price during World War I, and a general price decrease following this period. This decreasing trend reflects, in part, improved metallurgical efficiencies and process innovations. The decrease in copper, zinc, and nickel prices (Fig. 2b) after 1920 indicates the successful and widespread application of sulfide ore flotation and advances in electrolytic processing. Swings in gold and silver price (Fig. 2c) since the early 1970s indicate free-market forces and global political and economic effects following suspension of governmental price controls.

There is a general relationship between metal price and terrestrial concentration. Metals present at relatively high concentrations, in the earth's crust, such as iron and aluminum, are the least expensive; rare metals such as gold and platinum are the most valuable. This situation has existed for gold and silver valuation for centuries. The amount of silver in the earth's crust is approximately 20 times that of gold, and the historical price ratio for gold and silver varied between 10 and 16 for over 3000 years. Since 1970 that price ratio has been strongly affected by market forces and investor speculation.

Metal demand has an important influence on price. Both lead and gallium occur in the earth at about 0.0015 wt %. The demand for gallium (1990 U.S. consumption was 10,000 kg) is limited to optoelectronic devices and high performance microelectronics. There appears to be no need to expand supply, which would reduce price. On the other hand, reported 1990 consumption of lead in the United States was 1.25×10^6 t. Lead (qv) production is carried out on a large scale by relatively simple and efficient processes.

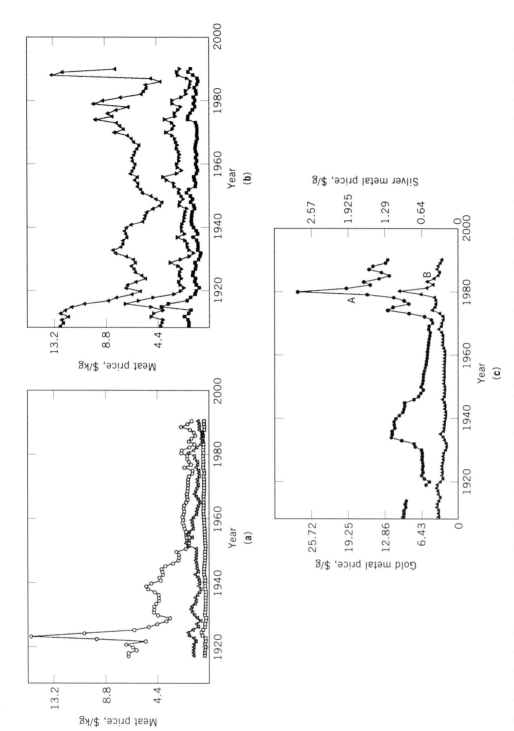

Fig. 2. Metal price in constant 1987 dollars since 1909 for (**a**) (□) iron (hot-rolled steel bars), (○) aluminum, and (△) lead; (**b**) (▼) copper, (■) zinc, and (▲) nickel; and (**c**) A, gold and B, silver (4). To convert $/g to $/troy oz, multiply by 31.10.

319

Approximately three-quarters of the elements in the Periodic Table are metals. The winning, refining, and fabrication of these metals for commercial use together represent the complex and diverse field of metallurgy. Metallurgy has played a vital role in society for thousands of years, yet it continues to advance and to have increasing importance in many areas of science and technology.

BIBLIOGRAPHY

1. G. Agricola, *De Re Metallica* (Basel, 1556), H. C. Hoover and L. L. Hoover, trans., Dover Publications, Inc., New York, 1950.
2. P. W. Thrush, *A Dictionary of Mining, Mineral, and Related Terms*, U.S. Bureau of Mines, Washington, D.C., 1968.
3. *ASTM Standards*, Vol 2.01 and Vol 2.04, American Society for Testing and Materials, Philadelphia, Pa., 1993.
4. *Metal Prices in the United States through 1991*, U.S. Bureau of Mines, Washington, D.C., 1991.

BRENT HISKEY
University of Arizona

EXTRACTIVE METALLURGY

Extractive metallurgy deals with the extraction of metals from naturally occurring compounds and the subsequent refinement to a purity suitable for commercial use. These operations, known as winning and refining of metals, follow mining and beneficiation of the ore. They precede the fabricating processes. Selection and design of the extractive processes depend on the raw materials available, and conditions for the refining steps are related to the ultimate use of the metal. Thus mining, extraction, and fabrication are closely interrelated (see also MINERALS RECOVERY AND PROCESSING).

Production of a metal is usually achieved by a sequence of chemical processes represented as a flow sheet. A limited number of unit processes are commonly used in extractive metallurgy. The combination of these steps and the precise conditions of operations vary significantly from metal to metal, and even for the same metal these steps vary with the type of ore or raw material. The technology of extraction processes was developed in an empirical way, and technical innovations often preceded scientific understanding of the processes.

The scientific basis of extractive metallurgy is inorganic physical chemistry, mainly chemical thermodynamics and kinetics (see THERMODYNAMIC PROPERTIES). Metallurgical engineering relies on basic chemical engineering science, material and energy balances, and heat and mass transport. Metallurgical systems, however, are often complex. Scale-up from the bench to the commercial plant is more difficult than for other chemical processes.

Extractive metallurgy is usually divided into three principal areas: (*1*) pyrometallurgy which consists of high temperatures processes to carry out smelting and refining reactions; (*2*) hydrometallurgy which is characterized by the use of aqueous solutions and inorganic solvents to achieve the desired reactions (see SUPPLEMENT); and (3) electrometallurgy in which electrical energy is used to extract and refine metals by electrolytic processes. Electrometallurgy may be carried out either at high temperature or in aqueous solution. An integrated metallurgical flow sheet may include pyrometallurgical, hydrometallurgical, and/or electrometallurgical steps.

History

Of the seven metals known and widely used in antiquity, copper, gold, silver, tin, lead, iron, and mercury, only the first three occur naturally in the metallic state. The first extractive metallurgy process was probably used in the mountains of western Asia, in the areas that are now Iraq and Iran. There is tangible evidence of the smelting of copper (qv) by the Summerians and Egyptians as early as ca 4000 BC. Whereas discovery of the reduction of copper ore by fire was certainly accidental, the observation of the relationship between the blue azurite [1319-45-5], $Cu(OH)_2 \cdot 2CuCO_3$, or the green malachite [1319-53-5], $Cu(OH)_2 \cdot CuCO_3$, and the red copper metal was a truly remarkable scientific discovery. Ancient peoples also knew how to treat copper sulfide ores.

Gold was used extensively in antiquity and the operation of gold mines and mills by the Egyptians is well documented (see GOLD AND GOLD COMPOUNDS). Mercury (qv) also was known to the ancients. Preparation from cinnabar [19122-79-3], HgS, and use in the amalgamation process for the recovery of gold were discovered in prehistory. Silver was found associated with gold, and the gold–silver alloy called electrum was commonly used by the ancients (see SILVER AND SILVER ALLOYS). Silver also was obtained from argentiferrous lead ores, and the cupellation process for the separation of precious metals from base metals was another early metallurgical achievement.

The production of metallic lead (qv) from galena, PbS, is a simple metallurgical operation, and both this metal and its compounds were used in antiquity. Tin, used mainly in bronze, was probably obtained by mixing the tin ore cassiterite [1317-45-9], SnO_2, and copper ore. Some iron (qv) of meteoric origin was found by the ancients, but use of wrought iron was also widespread. Iron was obtained easily by reduction of iron ore by carbon, followed by hammering of the sponge iron, and forging.

During the Middle Ages an active mining industry developed in central Europe, but it produced few technological breakthroughs. Several books on mining and metallurgy were written. The extensive treatise by Agricola (1) illustrates the state of the art of the industry in the mid-sixteenth century.

Further progress was made in the eighteenth and early nineteenth centuries. Many metals were discovered upon the development of experimental chemistry. The modern metallurgical industry was born with the invention of steelmaking in 1856 (see STEEL). Industrial processes for making zinc (see ZINC AND ZINC ALLOYS), aluminum (see ALUMINUM AND ALUMINUM ALLOYS), and copper followed before the end of the nineteenth century. These processes made possible the industrial revolution and the development of an industrial society relying heavily on the use of metals.

Sources of Metals and Ore Preparation

All metals come originally from natural deposits present in the earth's crust. These ore deposits result from a geological concentration process, and consist mainly of metallic oxides and sulfides from which metals can be extracted. Seawater and brines are another natural source of metals, eg, magnesium (see CHEMICALS FROM BRINE; MAGNESIUM AND MAGNESIUM ALLOYS; OCEAN RAW MATERIALS). Metal extracted from a natural source is called primary metal.

Scrap is also an important raw material for the metallurgical industry. The metal produced by a recycling (qv) process is called secondary metal.

The concentration of most metals in the earth's crust is very low, and even for abundant elements such as aluminum and iron, extraction from common rock is not economically feasible. An ore is a metallic deposit from which the metal can be economically extracted. The amount of valuable metal in the ore is the tenor, or ore grade, usually given as the wt % of metal or oxide. For precious metals, the tenor is given in grams per metric ton or troy ounces per avoirdupois short ton (2000 pounds). The tenor and the type of metallic compounds are the main characteristics of an ore. The economic feasibility of ore processing, however, depends also on the nature, location, and size of the deposit; the availability and cost of a suitable extraction process; and the market price of the metal.

Valuable metal is present in the ore in a form called the ore mineral, an inorganic substance of given composition. The ore mineral may be native metal or a compound of the metal. Most ores contain several ore minerals. The minerals without economic value are called the gangue minerals. Because designation as ore minerals or gangue minerals depends on the economics of processing the ore, the same mineral may be classified in either group. In particular, silica, alumina, and iron oxides are usually gangue minerals except when the concentration is high enough to warrant extraction.

The direct treatment of an ore by a chemical extraction process is usually not economically feasible. Instead the first step after mining is the ore preparation, also called ore dressing, milling, mineral processing, or beneficiation. The purpose of ore dressing is to separate the ore minerals from the gangue minerals by physical methods. Milling yields a concentrate containing most of the valuable metals to be recovered and a tailing containing most of the gangue minerals to be discarded.

The first step in ore processing is the liberation of finely disseminated minerals. This is accomplished by comminution, or crushing and grinding the ore (see SIZE REDUCTION). It is followed by screening and/or classification (see SIZE CLASSIFICATION). The main type of mineral beneficiation is concentration, achieved by making use of different physical properties of the mineral particles: magnetic and electrostatic separation (see SEPARATIONS, MAGNETIC), gravity concentration, and flotation (qv).

Table 1 gives the average metal content of the earth's crust, ore deposits, and concentrates. With the exceptions of the recovery of magnesium from seawater and alkali metals from brines, and the solution mining and dump or heap leaching of some copper, gold, and uranium (see URANIUM AND URANIUM COMPOUNDS), most ores are processed through mills. Concentrates are the raw materials for the extraction of primary metals.

Table 1. Metal Content of Igneous Rocks, Ore Deposits, and Concentrates, wt %

Metal	Igneous rocks	Ore deposits	Concentrates
aluminum	8.13	27–29	
iron	5.01	30–60	55
magnesium	2.09		
titanium	0.63	2.5–25	
manganese	0.10	45–55	
nickel	0.020	1.5–3	10
vanadium	0.017	1.6–4.5	
copper	0.010	0.5–5	30
uranium	0.008	0.1–0.9	
tungsten	0.005		50
zinc	0.004	10–30	50
lead	0.002	5–10	70
cobalt	0.001	1–11	
beryllium	0.001		3–4
molybdenum	2×10^{-4}	0.6–1.8	50
tin	4×10^{-4}	1.5	40–75
mercury	5×10^{-5}	0.1–0.5	
silver	2×10^{-6}	0.04–0.08	
platinum	5×10^{-7}	0.001	
gold	10^{-7}	0.001	

The successive stages of mining, mineral beneficiation, smelting, and refining usually require large energy inputs in the form of fuels, electrical energy, and energy-intensive reagents and supplies. Some metals are energy intensive because their ores are low grade; other metals are energy intensive because the reduction process is difficult. Among the commonly used metals, steel is the least energy intensive because of high grade ore and ease of reduction. Titanium is the most energy intensive because of low grade ore and complex processing chemistry. Although recovering metals from common rocks is technologically feasible, the energy requirements are prohibitive. Production of secondary metals from metal scrap, on the other hand, requires much less energy than the production of primary metals. Secondary metal production conserves metallic resources and reduces the net cost of waste disposal.

Pyrometallurgy

The essential operations of an extractive metallurgy flow sheet are the decomposition of a metallic compound to yield the metal followed by the physical separation of the reduced metal from the residue. This is usually achieved by a simple reduction or by controlled oxidation of the nonmetal and simultaneous reduction of the metal. This may be accomplished by the matte smelting and converting processes.

In a simple pyrometallurgical reduction, the reducing agent, R, combines with the nonmetal, X, in the metallic compound, MX, according to a substitution reaction of the following type:

$$MX + R \longrightarrow M^0 + RX \tag{1}$$

In order to achieve a spontaneous process, the Gibbs energy change for equation 1 must be negative. This condition is obtained not merely by the choice of the appropriate reducing agent, but also by adjusting the chemical activities of the reactants and products and by selecting the optimum temperature and pressure for a suitable thermodynamic driving force and acceptable reaction rates.

Often the starting raw material to be reduced is an oxide. The standard Gibbs energies of formation, under a pressure of 100 kPa (1 bar) of a few oxides of metals along with those of water, carbon monoxide, and carbon dioxide are represented in Figure 1. A metal oxide is reduced by another metal, or C, H_2, or CO, if the Gibbs energy of formation at the given temperature is less negative than that of the oxide of the reductor. Similar diagrams are available for sulfides and halides (2,3). These diagrams provide a simple way of estimating the value of standard Gibbs energy changes for reactions such as that shown in equation 1.

The selection of a particular type of reduction depends on technical feasibility and the economics of the process as well as on physicochemical considerations. In particular, the reducing agent should be inexpensive relative to the value of the metal to be reduced. The product of the reaction, RX, should be easily separated from the metal, easily contained, and safely recycled or disposed of. Furthermore, the physical conditions for the reaction should be such that a suitable reactor can be designed and operated economically.

Reduction processes are characterized either by the reducing agent selected or by the physical state of the metallic product. The separation of reaction products determines the choice and design of the furnace. Reduction processes are classified according to the physical state of the reduced metal.

Usually, the ore or concentrate cannot be reduced to the metal in a single operation. An additional preparation process is needed to modify the physical or chemical properties of the raw material prior to its reduction. Furthermore, most pyrometallurgical reductions do not yield a pure metal and an additional step, refining, is needed to achieve the chemical purity that is specified for the commercial use of the metal.

The preparation, reduction, and refining operations are very much interdependent, and for a given metal must be considered as parts of a single flow sheet. To illustrate the principles of extractive metallurgy, however, it is convenient to discuss the various operations separately.

Preparatory Processes. Several pyrometallurgical operations are used to change the chemical and physical properties of the ore or concentrate in order to make it more suitable for the main extraction process. The chemistry of these preparatory processes, which involve mainly gas–solid reactions, is relatively simple.

Drying and Calcination. The simplest pyrometallurgical operation is the evaporation of free water and the decomposition of hydrates and carbonates. A typical reaction is the decomposition of pure limestone [1317-65-3], $CaCO_3$, to calcium oxide [1305-78-8] and carbon dioxide:

$$CaCO_3 \text{ (s)} \longrightarrow CaO \text{ (s)} + CO_2 \text{ (g)} \tag{2}$$

This reaction is strongly endothermic, and the equilibrium pressure of CO_2, equal to the equilibrium constant of the reaction, increases exponentially with temperature (see LIME AND LIMESTONE).

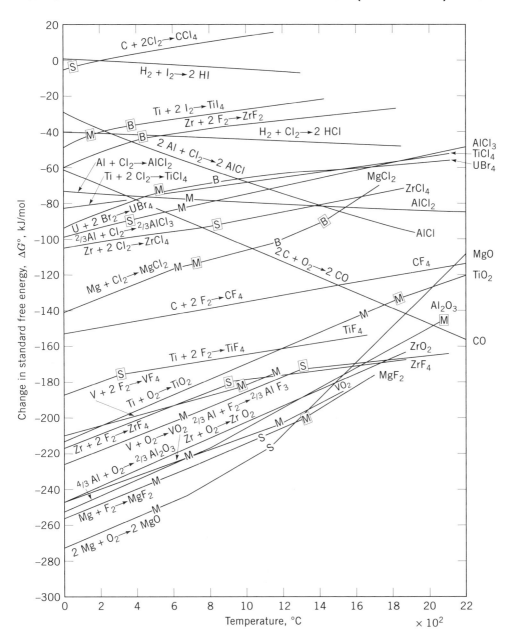

Fig. 1. Standard Gibbs energy of formation vs temperature where changes in state are denoted as M, B, and S for melting, boiling, and sublimation points, respectively, for elements, and $\boxed{M}$, $\boxed{B}$, and $\boxed{S}$ for oxides or halides. To convert kJ to kcal, divide by 4.184.

Lime [1305-78-8] is an important raw material for the metallurgical industry. It is used primarily as a flux in smelting and converting, but it is also a neutralizing agent for hydrometallurgical processes. The calcination of magnesite, $MgCO_3$, yields magnesia, MgO, which is an essential raw material for furnace refractories. The calcination of dolomite, $CaCO_3 \cdot MgCO_3$, also yields a calcine used as a feed for the preparation of magnesium metal.

Examples of similar processes are the decomposition of precipitated aluminum trihydroxide to alumina, which is the feed for the electrolytic production of aluminum metal, and the drying of wet sulfide concentrates in preparation for flash roasting (see ALUMINUM AND ALUMINUM ALLOYS).

Roasting of Sulfides. Most nonferrous metals occur in nature mainly as sulfides. These cannot be easily reduced directly to the metal. Burning metallic sulfides in air transforms them into oxides or sulfates which are more easily reduced. The sulfur is released as sulfur dioxide, as shown by the following typical reaction for a divalent metal, M:

$$MS \text{ (s)} + 3/2 \text{ } O_2 \text{ (g)} \rightleftharpoons MO \text{ (s)} + SO_2 \text{ (g)} \tag{3}$$

This reaction is strongly exothermic and proceeds spontaneously from left to right for most common metallic sulfides under normal roasting conditions, ie, in air, because $P_{SO_2} + P_{O_2} = {\sim}20$ kPa (0.2 atm) at temperatures ranging from 650 to 1000°C. The physical chemistry of the roasting process is more complex than indicated by equation 3 alone. Sulfur trioxide is also formed,

$$SO_2 \text{ (g)} + 1/2 \text{ } O_2 \text{ (g)} \rightleftharpoons SO_3 \text{ (g)} \tag{4}$$

and the oxide–sulfate equilibrium must be considered:

$$MO \text{ (s)} + SO_3 \text{ (g)} \rightleftharpoons MSO_4 \text{ (s)} \tag{5}$$

Very little SO_3 is produced at 1000°C, but the equilibrium of equation 4 shifts to the right at lower temperatures favoring the formation of sulfates according to equation 5. This behavior is illustrated by the schematic diagram in Figure 2.

Most roasting is carried out to obtain an oxide for reduction by carbon or carbon monoxide (qv), or for leaching in sulfuric acid solution followed by electrowinning. This roasting to completion, which eliminates all the sulfur and produces the metal oxide, is called dead roast or sweet roast. Incomplete roasting is used to remove excess sulfur in preparation of copper and nickel sulfides for the matte smelting process. A sulfating roast yields water-soluble sulfates in preparation for leaching.

The sulfur dioxide produced by the process is usually converted to sulfuric acid, or sometimes liquified, and the design of modern roasting facilities takes into account the need for an efficient and environmentally clean operation of the acid plant (see SULFURIC ACID AND SULFUR TRIOXIDE).

Chlorination. In some instances, the extraction of a pure metal is more easily achieved from the chloride than from the oxide. Oxide ores and concentrates react at high temperature with chlorine gas to produce volatile chlorides of the metal. This reaction can be used for common nonferrous metals, but it is particularly useful for refractory metals like titanium (see TITANIUM AND TITANIUM ALLOYS) and zirconium (see ZIRCONIUM AND ZIRCONIUM COMPOUNDS), and for reactive metals like aluminum.

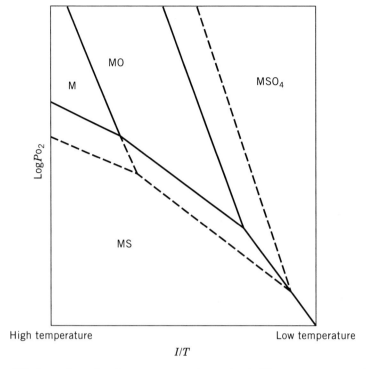

Fig. 2. Equilibria and predominance areas at constant SO_2 pressure as a function of temperature for the system M–S–O where (—) represents $P_{SO_2} = 101$ kPa (1 atm), and (---), $P_{SO_2} = 10$ kPa (0.1 atm) (2).

The chlorination of titanium requires a reducing agent such as carbon,

$$TiO_2 \text{ (s)} + C^0 \text{ (s)} + 2\,Cl_2 \text{ (g)} \longrightarrow TiCl_4 \text{ (g)} + CO_2 \text{ (g)} \tag{6}$$

This reaction is slightly exothermic, but additional heat is required to maintain the operating temperature at 800–900°C.

Sintering and Pelletizing. The beneficiation of ores by physical techniques requires the liberation of metal-bearing minerals. This is usually achieved by reducing the raw material to a very fine size. The high throughput reduction processes, such as the iron blast furnace, however, cannot handle a finely divided feed. Sintering and pelletizing are techniques for agglomerating finely divided solid particles into a coarse material suitable for charging a blast furnace.

Sintering consists of heating a mixture of fine materials to an elevated temperature without complete fusion. Surface diffusion and some incipient fusion cause the solid particles in contact with one another to adhere and form larger aggregates. In the processing of hematite, Fe_2O_3, or magnetite, Fe_3O_4, fines to prepare a suitable feed for the iron blast furnace, the oxide ore is mixed with 5–8% of coal (qv) or coke to supply the fuel requirement. Often limestone is added to produce the self-fluxing sinter, and it is calcined during the sintering process. In the processing of sulfide ores, such as the treatment of lead sulfide concentrates, the contained sulfur acts as the fuel after ignition of the charge, and a single process achieves both roasting and sintering of the feed.

Pelletizing was developed for treating low grade iron ores. In particular, taconite ores consist of finely divided magnetite or hematite mixed with the gangue. The agglomeration of this material after fine comminution and separation is accomplished by balling in a rotating device such as a disk, drum, or cone followed by heating to a final temperature of ~1300°C; this process is called induration. The induration achieves not only the agglomeration of the fines, but also the vaporization of free or combined water, the calcination of limestone, and the oxidation of magnetite to hematite. The pelletizing process has many advantages over sintering, and it is applied to high grade ores. The use of computers has made possible the automatic control of operations in modern pelletizing plants (see PROCESS CONTROL). The induration process can also be carried out under reducing conditions to yield partially reduced pellets for the blast furnace or fully reduced pellets for direct charging to steelmaking furnaces.

Equipment. Drying and calcination are usually carried out in various types of kilns such as rotary kilns, shaft furnaces, and rotary hearths. The induration step in a pelletizing plant requires similar equipment, but sometimes the charge passes through successive stages, ie, traveling grates for drying, a rotary hearth for preheating, and a shaft furnace for the chemical reactions and sintering (see also DRYING; FURNACES).

The Dwight-Lloyd continuous sintering machine is commonly used. It consists of a series of grates mounted as an endless traveling belt. The feed is spread as a bed (10–20-cm deep), and the top is ignited as it passes through a fuel-fired ignition box. Air passes through the bed into a suction box and fan below the grate, and the combustion zone progresses through the bed as material moves to the discharge end of the belt. Downdraft machines are used for sintering iron ores, and updraft machines are preferred for sinter-roasting of lead and zinc concentrates.

The older roasting furnace is the multiple-hearth roaster consisting of 8–12 hearths enclosed in a cylindrical shell. The ore or concentrate fed at the top drops from hearth to hearth by the moving action of a rotating rabble attached to a central shaft. The sulfide particles are roasted by contact with the rising gases and are discharged at the bottom as calcine. In a suspension roaster, the pulverized concentrate is dried on the top hearths, drops through a combustion chamber, and is collected at the bottom. If the roasting process occurs while a very fine concentrate is mixed with the gas stream in the hot chamber, it is called flash roasting.

Fluid-bed roasters are well suited for gas–solid reactions. Usage in metallurgical operations started in 1950 and developed rapidly thereafter. Fluid-bed roasters have no moving parts in the hot zone, and the temperature and gas composition can be easily controlled. Air is pumped upward through a perforated hearth at a rate sufficient to expand the bed of sulfide particles, and the gas–solid reaction proceeds very fast. Circulating fluidized beds, elutriation reactors, and cyclones are commonly used for treating fine feed materials (see FLUIDIZATION).

Reduction to Liquid Metal. Reduction to liquid metal is the most common metal reduction process. It is preferred for metals of moderate melting point and low vapor pressure. Because most metallic compounds are fairly insoluble in molten metals, the separation of the liquified metal from a solid residue or from another liquid phase of different density is usually complete and relatively simple. Because the product is in condensed form, the throughput per unit volume of reactor is high, and the number and size of the units is minimized. The common furnaces for production of liquid metals are the blast furnace, the reverberatory furnace, the converter, the flash smelting furnace, and the electric-arc furnace (see FURNACES, ELECTRIC).

The Iron Blast Furnace. The reduction of iron oxides by carbon in the iron (qv) blast furnace is the most important of all extractive processes, and the cornerstone of all industrial economies. Better understanding of the reactions taking place within the furnace has made possible a more efficient operation through better preparation of the burden, higher blast temperature, and sometimes increased pressure. Furnace capacity has doubled since the 1800s, whereas coke consumption has been reduced by about half. The ratio of coke to iron produced on a per weight basis is ca 0.5 to 1.

The iron blast furnace (Fig. 3) is a circular shaft furnace ca 30-m high having a maximum internal diameter of ca 9 m. It is a steel shell lined with refractory bricks (see REFRACTORIES). The furnace is charged at the top and preheated air is introduced through the tuyeres above the hearth. At the tuyere level, the coke burns in the air. This strongly exothermic reaction (eq. 16) is the main source of heat for the process. Equation 17, the Boudouard reaction, supplies the reducing gas, carbon monoxide. This reaction is endothermic, and the value of its equilibrium constant determines the composition of the gas phase in the furnace. This furnace is a countercurrent reactor in which the iron oxides descending through the upper part of the stack are heated and reduced in stages as shown (Fig. 3). In the lower part of the stack, where the temperature is >1000°C, the Boudouard reaction (eq. 17) is shifted to the right and proceeds simultaneously with equation 11, yielding

$$\text{FeO (s)} + \text{C}^0 \text{ (s)} \longrightarrow \text{Fe}^0 \text{ (s)} + \text{CO (g)} \tag{13}$$

referred to as the direct reduction of wustite, FeO. The fluxes consisting mainly of limestone, or sometimes dolomite, are heated and calcined on the way down the shaft. In the lower part of the furnace, called the bosh, these melt with the gangue material to form the slag. The iron blast furnace slag consists mainly of lime, alumina, and silica and has smaller amounts of magnesia, manganese oxide, iron oxide, and calcium sulfide. The lime is added as a flux to produce a slag having a melting point of 1300–1400°C and a relatively low viscosity. The reduced iron is also molten at the tuyere level. It drops as globules through the slag and settles at the bottom of the furnace.

The dissolution of carbon in molten iron in the lower part of the furnace, leads to the reduction of manganese oxide (eq. 15) and some silica (eq. 14), both in the slag, whereby the subsequent dissolution of these metals occurs in the molten iron.

$$\text{C}^0 \text{ (s)} \longrightarrow \text{C}^0_{\text{Fe}} \text{ (l)} \tag{18}$$

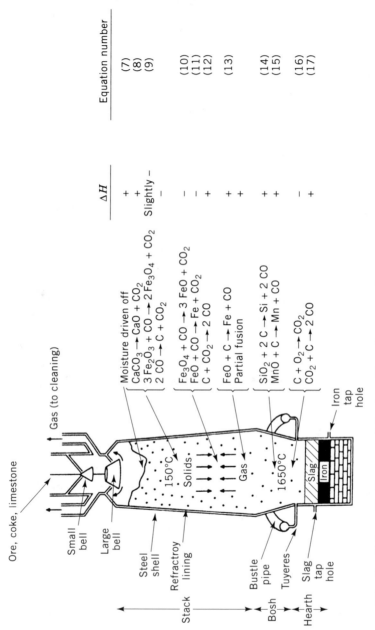

Fig. 3. Schematic diagram of the iron blast furnace indicating some of the chemical reactions (3).

	Reaction	ΔH	Equation number
	Moisture driven off	+	(7)
	$CaCO_3 \rightarrow CaO + CO_2$	+	(8)
	$3\,Fe_2O_3 + CO \rightarrow 2\,Fe_3O_4 + CO_2$	Slightly −	(9)
	$2\,CO \rightarrow C + CO_2$	−	
	$Fe_3O_4 + CO \rightarrow 3\,FeO + CO_2$	−	(10)
	$FeO + CO \rightarrow Fe + CO_2$	−	(11)
	$C + CO_2 \rightarrow 2\,CO$	+	(12)
	$FeO + C \rightarrow Fe + CO$	+	(13)
	Partial fusion	+	
	$SiO_2 + 2\,C \rightarrow Si + 2\,CO$	+	(14)
	$MnO + C \rightarrow Mn + CO$	+	(15)
	$C + O_2 \rightarrow CO_2$	−	(16)
	$CO_2 + C \rightarrow 2\,CO$	+	(17)

Ore, coke, limestone

Gas (to cleaning)

Small bell
Large bell
Steel shell
Refractroy lining
Bustle pipe
Tuyeres
Slag tap hole
Iron tap hole

Stack
Bosh
Hearth

150°C
Solids
Gas
1650°C
Slag
Iron

The molten slag and the molten iron, called hot metal or pig iron, are tapped from the hearth of the blast furnace. A modern blast furnace yields 5000–9000 t/d of iron. The compositions of the pig iron and the slag are determined by the furnace temperature, the composition of the ore, and the added flux. Pig iron always contains 3.5–4.5 wt % carbon, variable amounts of silicon, manganese, sulfur, and phosphorus.

The main technique for regulating the operating temperature of the blast furnace is the preheating of the blast. Progress in regulation has been made with the addition of natural gas, pulverized coal, or oil to the blast, oxygen enrichment, and addition of steam. Preheating and prereduction of the ore is also considered.

Reduction Smelting of Nonferrous Metals. The traditional method of lead (qv) smelting is reduction of an oxide in a blast furnace similar in principle to that used for iron, but very different in design and operation. The lead furnace consists of a shaft of rectangular cross section, 9-m high, 7-m long, and tapering in width from ca 2.5 m at the top to ca 1.5 m at the bottom. The feed, added at the top, consists mainly of roasted and sintered lead concentrates, coke, and fluxes to form a slag with the gangue material. Air is blown through tuyeres near the bottom, and coke burns according to equations 16 and 17 (Fig. 3). The reduction proceeds by contact between the hot reducing gas and the lump charge.

$$PbO \ (s) + CO \ (g) \longrightarrow Pb^0 \ (l) + CO_2 \ (g) \tag{19}$$

The lead is molten (mp = 327°C) and trickles down to the crucible below the tuyeres. The slag melts when it reaches the high temperature zone of the tuyeres and floats on top of the lead. The slag consists mainly of SiO_2, CaO, FeO, with lesser amounts of Al_2O_3, MgO, ZnO, and other oxides. Most lead ores contain zinc because zinc-plant residues are often charged into lead blast furnaces. The amount of zinc in the slag is generally sufficient to justify recovery by treatment in a separate fuming furnace. Molten lead, ie, lead bullion, is alloyed with other easily reducible metals, mainly copper, tin, antimony, arsenic, gold, and silver present in the charge. The capacity of furnaces ranges from 100 to 400 t/d.

Many nonferrous metals can be extracted by reduction smelting, eg, copper, tin, nickel, cobalt, silver, antimony, and bismuth. Blast furnaces are sometimes used for the smelting of copper or tin, but flash and reverberatory furnaces are more common for metals other than lead.

The reverberatory furnace consists of a shallow, usually rectangular, hearth, side-walls, end-walls, and a roof or arch. The charge rests on the hearth and the furnace is fired using gas, oil, or pulverized coal which burns in the space between the charge and the roof. Heat transfer is achieved mainly by radiation directly from the flame and by reflection from the arched refractory roof. There is little or no reaction between the charge and the waste gases which leave the furnace at very high temperature. The feed often consists of fine-particle concentrates, and the reducing agent is usually coke. Both are added through ports on the side or the top of the furnace. Slag and metals are removed through tap holes, and the smelting can be carried out either continuously or as a batch process. Reverberatory furnaces are made in many sizes. Small furnaces hold only about one metric ton of metal; large ones such as those employed in copper smelting hold several hundred tons.

Electric heat provided by a resistance or by an electric arc can be substituted for the burning of a fuel. Electric furnaces can be designed in a variety of shapes and are more versatile than fuel-heated furnaces. The furnace atmosphere can be controlled independently of the chemistry of the combustion (see FURNACES, ELECTRIC).

Matte Smelting and Converting.   Rich oxide ores of copper or roasted sulfide concentrates can be treated by reduction smelting in a blast or reverberatory furnace to yield an impure copper (qv) known as black copper. Except for a few plants treating mainly copper scrap, this process is obsolete. Most primary copper is produced by matte smelting, an operation yielding a molten sulfide of copper and iron, called matte, which is further oxidized in the converting step to yield metallic copper. A similar process is used for the extraction of nickel (see NICKEL AND NICKEL ALLOYS).

Copper is easily reduced from its compounds. The most important copper mineral is chalcopyrite [1308-56-1], $CuFeS_2$, and the separation of copper from iron is one of the objectives of the smelting operation. The sulfur in the ore has value as a fuel as well as a reducing agent. The traditional matte smelting is carried out in a reverberatory furnace in which a solid charge consisting of concentrates, calcines, converter slags, and fluxes is heated to 1150–1250°C. This melting produces two separate immiscible liquid phases, ie, a slag floating on top of a copper-rich matte. Matte consists mainly of Cu_2S and FeS and ranges from 30 to 60 wt % Cu. The smelting slag is a mixture of iron oxides and silica having smaller amounts of alumina and lime. This slag is usually discarded. It should contain as little copper as possible.

The electric furnace is an alternative to the reverberatory furnace in environmentally sensitive areas where electricity costs are not too high. The electric furnace is versatile, produces small volumes of effluent gases, and the SO_2 concentration can be easily controlled. Operating costs, however, are high.

Since the mid-1960s most matte smelting capacity has been provided by flash smelting furnaces. In flash smelting, dry concentrates are blown with oxygen, hot air, or a mixture of both, into a hearth-type furnace. Inside the furnace, the sulfide particles react rapidly with the gases, causing a large evolution of heat and controlled partial oxidation of the concentrates. The main reactions, which are exothermic, can be summarized by the following:

$$2\ CuFeS_2\ (s) + 5/2\ O_2\ (g) \longrightarrow Cu_2S{\cdot}FeS\ (l) + FeO\ (l) + 2\ SO_2\ (g) \tag{20}$$

$$FeS\ (s) + 3/2\ O_2\ (s) \longrightarrow FeO\ (l) + SO_2\ (g) \tag{21}$$

$$2\ FeO\ (l) + 2\ SiO_2\ (s) \longrightarrow 2\ FeO{\cdot}SiO_2\ (l, slag) \tag{22}$$

The reacting particles melt rapidly, and the droplets fall to the slag layer. The sulfide drops settle through it to form the matte phase. Any oxidized copper is reduced to the matte by the following reaction:

$$Cu_2O\ (l, slag) + FeS\ (l) \longrightarrow FeO\ (l, slag) + Cu_2S\ (l) \tag{23}$$

Flash smelting is efficient because the fuel value of the sulfur and iron in the charge is fully used, and the productivity (8–12 t/d of charge processed per square meter of hearth) is higher than that of the reverberatory or electric furnace.

In addition to copper, iron, and sulfur, the matte produced by smelting contains oxygen and various impurities, eg, As, Bi, Ni, Pb, Sb, Zn, Au, and Ag, depending on the composition of the concentrate. The converting operation removes the iron and the sulfur along with some of the other impurities. The molten matte is charged to a converter where it is oxidized by blowing air. This is universally done in a Pierce-Smith converter, or a similar reactor consisting of a cylindrical steel shell lined with magnesite refractory bricks. The shell is placed horizontally and rotates to achieve the different positions for charging, blowing through submerged tuyeres, and pouring. Typical inside dimensions are 4 m dia and 9 m in length for an output of 100–200 t/d.

The converting operation takes place in two successive stages. In the slag-forming stage, the main reaction is the oxidation of FeS to FeO and Fe_2O_3 which combine with a silica flux to form a silicate slag by reactions similar to those of equations 21 and 22. The matte is added in several batches, each addition followed by oxidation and discharge of the slag. In the copper making stage, the remaining sulfur is oxidized to sulfur dioxide.

$$Cu_2S \ (l) + O_2 \ (g) \longrightarrow 2 \ Cu^0 \ (l) + SO_2 \ (g) \tag{24}$$

Some copper oxide is formed which then reacts with the remaining sulfide to form metallic copper:

$$2 \ Cu_2O \ (l) + Cu_2S \ (l) \longrightarrow 6 \ Cu^0 \ (l) + SO_2 \ (g) \tag{25}$$

The equilibrium constants for these reactions are such that copper is not appreciably oxidized by oxygen until most sulfur has been removed. This makes possible the production of blister copper, 98.6–99.5% Cu that is low in both sulfur (0.02–0.1%) and oxygen (0.5–0.8%). The converter slag, however, contains a significant amount of copper and must be recycled to the smelting stage.

All the operations in the winning of copper from sulfide ores are controlled oxidations using air or oxygen. An important effort has been made to carry out the operation as a continuous single-step process. The Noranda submerged tuyere process, carried out in a long cylindrical vessel, produced copper for several years. In order to improve impurity elimination and to increase capacity, it is used to produce a high grade copper matte (70% Cu). The Outokumpu flash smelting (see COPPER) is the only single-furnace direct-to-blister process. Blister copper can be obtained by precise control of the input rate of oxygen relative to that of the concentrate. The advantage is a single source of SO_2 emission and potentially lower costs. The significant drawback is the high copper content of the slag (15–35% of the copper input) which has to be reduced in an electric slag cleaning furnace. The Mitsubishi process uses three furnaces interconnected by a continuous flow of matte and slag, ie, a smelting furnace, an electric slag cleaning furnace, and a converting furnace. Its advantages are a high productivity achieved by O_2-enrichment and an efficient recovery and use of high strength SO_2 streams.

Two processes, developed for the direct processing of lead sulfide concentrates to metallic lead (qv), have reached commercial scale. The Kivcet process combines flash smelting features and carbon reduction. The QSL process is a bath-smelting reactor having an oxidation zone and a reduction zone. Both processes use industrial oxygen. The chemistry can be shown as follows:

$$PbS\ (s) + 2\ C^0\ (s) + 5/2\ O_2 \longrightarrow Pb^0\ (l) + SO_2\ (g) + CO\ (g) + CO_2\ (g) \qquad (26)$$

Reduction to Gaseous Metal.   Volatile metals can be reduced and easily and completely separated from the residue before being condensed to a liquid or a solid product in a container physically separated from the reduction reactor. Reduction to gaseous metal is possible for zinc, mercury, cadmium, and the alkali and alkaline-earth metals, but industrial practice is significant only for zinc, mercury, magnesium, and calcium.

Zinc is produced by reduction of zinc oxide, usually a calcine obtained by roasting zinc sulfide concentrates. Carbon is used in the absence of air at 1200–1300°C, well above the boiling point of the metal (906°C).

$$ZnO\ (s) + C^0\ (s) \longrightarrow Zn^0\ (g) + CO\ (g) \qquad (27)$$

The original process, patented in 1810 and known as the Belgian retort process, was the main process for winning zinc for more than a century. This process, now obsolete, is historically significant. The reaction was carried out in small retorts, each producing daily batches of 25–50 kg of zinc. The gases produced (eq. 27) exited through a condenser, the liquid zinc was collected, and carbon monoxide vented and burned in air. The retorts and condensers, made of fireclay, were placed horizontally in a fuel-fired furnace. This process was labor intensive and had a poor metallurgical efficiency.

The development in the 1950s of a zinc–lead blast furnace (Imperial smelting process) is the most recent breakthrough in zinc smelting as of this writing. Both lead bullion and zinc metal are recovered by roasting and reducing mixed lead and zinc concentrates. The main feature of the process is the rapid cooling of zinc vapor from 1100 to 550°C in a spray of molten lead in a specially designed condenser. This prevents reoxidation of the zinc vapor by carbon dioxide in the furnace gases.

A vacuum-retort process (Pidgeon process) was used during World War II for the production of magnesium and calcium. Silicon, in the form of ferrosilicon, was used as the reducing agent instead of carbon to avoid the problem of cooling magnesium vapor in the presence of carbon dioxide:

$$2\ CaO \cdot MgO\ (s) + Fe_x Si(s) \longrightarrow 2\ Mg^0\ (g) + (CaO)_2 \cdot SiO_2\ (s) + x\ Fe^0\ (s) \qquad (28)$$

At 1200°C, the equilibrium pressure of magnesium vapor for this reaction was well below atmospheric pressure, and a high vacuum had to be maintained in the condenser. The reduction was carried out in small retorts made from Cr–Ni–Fe alloys and placed horizontally in a furnace. A newer process using a vertical electric-arc furnace (Magnetherm process) has been developed, but most magnesium is produced by electrolysis (see MAGNESIUM AND MAGNESIUM ALLOYS).

Reduction to Solid Metal. Metals having very high melting points cannot be reduced in the liquid state. Because the separation of a solid metallic product from a residue is usually difficult, the raw material must be purified before reduction. Tungsten and molybdenum, for instance, are prepared by reduction of a purified oxide (WO_3, MoO_3) or a salt, eg, $(NH_4)_2MO_4$, using hydrogen. A reaction such as

$$WO_3 \text{ (s)} + 3 \text{ } H_2 \text{ (g)} \longrightarrow 3 \text{ } H_2O \text{ (g)} + W^0 \text{ (s)} \tag{29}$$

is used. The metallic product consists of fine particles fabricated into various shapes by the techniques of powder metallurgy (see METALLURGY; MOLYBDENUM AND MOLYBDENUM ALLOYS; POWDER METALLURGY; TUNGSTEN AND TUNGSTEN ALLOYS).

Very reactive metals, eg, titanium or zirconium, which in the liquid state react with all the refractory materials available to contain them, also require reduction to solid metal. Titanium is produced by metallothermic reduction of its chloride using liquid magnesium at 750°C (Kroll process).

$$TiCl_4 \text{ (g)} + 2 \text{ } Mg_0 \text{ (l)} \longrightarrow Ti^0 \text{ (s)} + 2 \text{ } MgCl_2 \text{ (l)} \tag{30}$$

A steel reaction vessel is partly filled with solid magnesium, sealed, and flushed with helium or argon. The reactor is placed in a furnace and heated to melt the magnesium. Pure liquid titanium tetrachloride is slowly fed to the vessel, where it vaporizes and is reduced by molten magnesium. The reaction (eq. 30) is exothermic and proceeds to completion. Some of the liquid magnesium chloride is drained, and the vessel is cooled to room temperature. The reaction mass, interlocking crystals of metallic titanium, magnesium chloride, and some magnesium metal, is removed by boring. The magnesium chloride and the magnesium are separated from the titanium by vacuum distillation or by leaching with dilute acid. About 700–1400 kg of titanium is produced in one batch. This process is summarized in Figure 4 (see TITANIUM AND TITANIUM ALLOYS).

Less reactive metals can also be produced by reduction below their melting point. Several processes, referred to as direct reduction, have been developed for the reduction of iron oxides to solid iron. The reducing agents used commercially are hydrogen, carbon monoxide, natural gas, carbon, and carbonaceous fuels. The reaction can be carried out in a rotary kiln, a shaft furnace, or in the case of gaseous reduction, in a fluidized-bed reactor. These processes require a relatively pure iron ore. The metallic product is impure metal (90–95 wt % Fe), usually charged to a steelmaking furnace. The advantage of the direct reduction process is that it can be operated on a small scale and the capital investment is much lower than that of a blast-furnace plant (see IRON BY DIRECT REDUCTION).

Refining Processes. All the reduction processes yield an impure metal containing some of the minor elements present in the concentrate, eg, cadmium in zinc, or some elements introduced during the smelting process, eg, carbon in pig iron. These impurities must be removed from the crude metal in order

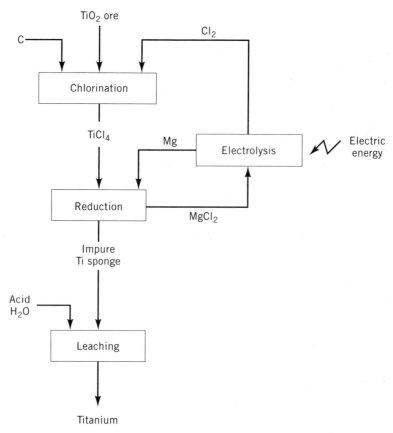

Fig. 4. Flow sheet for the processing of titanium ore by chlorination followed by reduction with magnesium.

to meet specifications for use. Refining operations may be classified according to the kind of phases involved in the process, ie, separation of a vapor from a liquid or solid, separation of a solid from a liquid, or transfer between two liquid phases. In addition, they may be characterized by whether or not they involve oxidation–reduction reactions.

   Volatilization.   In this simplest separation process, the impurity or the base metal is removed as a gas. Lead containing small amounts of zinc is refined by batch vacuum distillation of the zinc. Most of the zinc produced by smelting processes contains lead and cadmium. Crude zinc is refined by a two-step fractional distillation. In the first column, zinc and cadmium are volatilized from the lead residue, and in the second column cadmium is removed from the zinc (see ZINC AND ZINC ALLOYS).

   Impurities can be removed by formation of a gaseous compound, as in the fire-refining of copper (qv). Sulfur is removed from the molten metal by oxidation with air and evolution of sulfur dioxide. Oxygen is then removed by reduction with C, CO, H_2, or CH_4 in the form of natural gas, reformed natural gas, or wood. Final impurity contents of ca 0.001 wt % sulfur and 0.05–0.2 wt % oxygen are usually achieved.

The iodide or van Arkel-de Boer process is a volatilization process involving transfer of an involatile metal as its volatile compound. It is used for the purification of titanium. The reaction of iodine gas with impure titanium metal at 175°C yields gaseous titanium iodide and leaves the impurities in the solid residue.

$$Ti^0 \text{ (s)} + 2 \, I_2 \text{ (g)} \rightleftharpoons TiI_4 \text{ (g)} \tag{31}$$

The equilibrium is reversed at high temperature. The iodide is decomposed by passing the vapor over an electrically heated wire (1300–1400°C), yielding purified solid titanium and iodine gas which is recycled. The iodide process also applies to the purification of zirconium, hafnium, and silicon.

The vapometallurgical refining of nickel is based on the reaction of the metal with carbon monoxide to form gaseous nickel tetracarbonyl [*13463-39-3*], $Ni(CO)_4$, at 50°C. Decomposition occurs at ca 230°C.

Precipitation. In the simplest case, the solubility of an impurity in the liquid metal changes with temperature. Thus the impurity may precipitate as a solid phase upon cooling. For instance, the removal of iron from tin and of copper from lead are achieved by precipitation. When the solid is lighter than the liquid, it floats as a dross on the surface of the melt where it is easily removed by scraping. The process is called drossing.

Precipitation can also occur upon chemical reaction between the impurity and a precipitating agent to form a compound insoluble in the molten metal. The refining of crude lead is an example of this process. Most copper is removed as a copper dross upon cooling of the molten metal, but the removal of the residual copper is achieved by adding sulfur to precipitate copper sulfide. The precious metals are separated by adding zinc to liquid lead to form solid intermetallic compounds of zinc with gold and silver (Parkes process). The precious metals can then be recovered by further treatment (see LEAD).

Slag Refining. Unwanted constituents can be removed by transfer into a slag phase. Slag refining is also used for operations in which the liquid metal is maintained in contact with a slag or a molten salt. This second immiscible liquid is usually more oxidizing than the metallic phase and selective oxidation of the impurities renders them soluble in the slag or molten salt. Impurities that are less easily oxidized remain in the liquid metal.

Slag refining is effective for the purification of the less reactive nonferrous metals, eg, gold, silver, copper, lead, and others. In the recovery of precious metals from anode slimes, one of the steps is a fusion, usually in a small reverberatory furnace, where impurities are removed by oxidation with air and dissolution into successive slags: an oxide slag, a sodium carbonate slag, and a sodium nitrate slag. The product of this operation is a silver–gold alloy, ca 90% Ag and 9% Au, called doré metal, which is further refined by electrolysis. In the refining of lead, one of the steps is the removal of arsenic, antimony, and tin in a molten mixture of sodium hydroxide and sodium nitrate (Harris process). Slag refining of copper is effective in removing impurities less noble than copper, but has been replaced by electrorefining for the production of primary copper.

Steelmaking. Steelmaking is the most economically important slag refining process (see STEEL). Pig iron contains up to 4% carbon, 1% manganese, 1%

silicon, and variable amounts of phosphorus and sulfur. The removal of these impurities is based on preferential oxidation and control of the slag metal equilibrium. Modern steelmaking started with the invention of the Bessemer process in 1856. Blowing air through the bottom of a converter containing pig iron oxidizes silicon first, then manganese, and finally carbon. The reactions are exothermic, and the heat released is sufficient to maintain the final product, a low carbon steel, molten at ~1600°C. The original Bessemer process uses an acid slag high in silica, in which silicon oxide and manganese oxide are soluble. Carbon escapes as carbon monoxide. Phosphorus is also oxidized, but it does not transfer into an acid slag. The Thomas or Basic Bessemer process uses a basic slag high in lime which removes phosphorus and sulfur by the following reactions:

$$2\,P_{Fe}^0\,(l) + 3\,CaO\,(l, slag) + 5\,FeO\,(l, slag) \longrightarrow P_2O_5 \cdot 3CaO\,(l, slag) + 5\,Fe_0\,(l) \qquad (32)$$

$$S_{Fe}^0\,(l) + CaO\,(l, slag) \longrightarrow CaS\,(l, slag) + O_{Fe}^0\,(l) \qquad (33)$$

Steel produced by the converter process contains dissolved nitrogen and oxygen. The steel is usually deoxidized by small additions of elements like aluminum, manganese, and silicon. Vacuum degassing is used to remove residual oxygen, nitrogen, and hydrogen. The Siemens-Martin or open-hearth process, developed ca 1880, used a reverberatory furnace which is charged with solid steel scrap, limestone, iron ore, and molten pig iron. As of the 1990s, electric-arc furnaces are used. These provide the higher temperatures and flexibility of operation required for the production of alloy steels.

Shortly after World War II, the top-blown converter was invented in Austria (L-D process), and oxygen lancing was applied to the open-hearth process. Top-blown oxygen steelmaking, common practice in the 1990s, reduces the time for processing of high carbon iron into steel from hours to minutes. The process is autogeneous. Larger (250–300 t capacity) converters can handle 25–35% scrap. Rotating converters like the Kaldo and Rotor processes have also been developed (4). The invention of the Savard/Lee gas-shielded tuyere made possible the injection of oxygen through tuyeres in the bottom of a steel converter. The bottom injection of argon is another improvement in the production of stainless steel by the argon–oxygen decarburization process.

Hydrometallurgy

The treatment of ores by dissolution in aqueous solutions is a fairly simple operation which imitates natural leaching processes. The development of hydrometallurgy started when aqueous physical chemistry became better understood in the late nineteenth and early twentieth centuries. Hydrometallurgical processes are preferred when the pyrometallurgical route is impossible or impractical. If the metal to be extracted is more reactive than the impurities to be removed, eg, aluminum, or if the grade of the ore is very low and cannot be upgraded by physical beneficiation, eg, gold and uranium, hydrometallurgy becomes important.

In some respects, hydrometallurgy can be described as wet analytical chemistry carried out on a large scale. Many different flow sheets can be designed with various types of unit operations and most metals can be extracted from a

complex ore and recovered at the desired level of purity. A viable hydrometallurgical process, however, must achieve that goal at an economically acceptable cost.

The continuing decrease in grade, the increasingly complex nature of available ores, and the need for high purity metallic products has favored the development of hydrometallurgy which has found applications for advanced materials,
including ceramics (qv), composites (see COMPOSITE MATERIALS), and nanostruc-
ture materials. Concern for air pollution (qv) caused by pyrometallurgical plants
and the cost of preventive devices are incentives to consider hydrometallurgy in
cases where water pollution can be controlled.

The three main steps are leaching or dissolution of the metal in a suitable
aqueous solvent, purification or removal of impurities and/or concentration of
the solution, and recovery or precipitation of the reduced metal or its compound
from solution. In exceptional cases such as the treatment of seawater or brines
for metal values, the first step has been taken by nature (see CHEMICALS FROM
BRINES; OCEAN RAW MATERIALS). Most hydrometallurgical processes, however,
involve several steps for the purification and/or concentration of the solution
before the recovery of a pure product becomes feasible. In addition, steps are
required for the physical separation of the solid phases from the liquid, ie,
washing, clarification, thickening, filtering, drying (qv), evaporation (qv), etc.
Last but not least, the solvent is usually too valuable to be discarded, and must
be regenerated and recycled (see SOLVENT RECOVERY). Hydrometallurgical plants
operate most efficiently in a closed circuit which limits water pollution (see
SUPPLEMENT).

Leaching Chemistry.   The purpose of the leaching operation is to dissolve
the desired mineral and separate it from the gangue material. The reaction
should be selective and fast, the solvent inexpensive or easily regenerated. Several leaching agents are commonly used.

Water.   Because most metallic chlorides and sulfates are fairly soluble in
water, water can be used to leach calcines from chloridizing or sulfating roasts.

Acid Solutions.   Dilute sulfuric acid is the most important solvent for oxide
ore and for dead roasted sulfide concentrates. For instance, the leaching of zinc
oxide, described by the following equation, can be written as follows:

$$ZnO \text{ (s)} + 2\,H^+ \text{ (aq)} \longrightarrow Zn^{2+} \text{ (aq)} + H_2O \text{ (l)} \tag{34}$$

Other acids, eg, hydrochloric or nitric acid, are more seldom used because of
higher costs and corrosion problems.

Alkaline Solutions.   The most important example of alkaline leach is the
digestion of hydrated alumina from bauxite by a sodium hydroxide solution at
160–170°C, ie, the Bayer process (see ALUMINUM AND ALUMINUM ALLOYS).

$$Al_2O_3 \cdot 3H_2O \text{ (s)} + 2\,OH^- \text{ (aq)} \longrightarrow 2\,AlO_2^- \text{ (aq)} + 4\,H_2O \text{ (l)} \tag{35}$$

Complex-Forming Solutions.   The solubility of a metal can be enhanced
by complexation using a suitable ligand. The dissolution of copper oxide in
ammoniacal solutions is an example:

$$CuO \text{ (s)} + 4\,NH_4^+ \text{ (aq)} + 2\,OH^- \text{ (aq)} \longrightarrow Cu(NH_3)_4^{2+} \text{(aq)} + 3\,H_2O \text{ (l)} \tag{36}$$

The alkalinity of the solution prevents the attack of a carbonate gangue which would be soluble in an acid medium.

Oxidizing Solutions. In many leaching processes the mineral must be oxidized, as for instance, in the leaching of copper sulfides by ferric sulfate or ferric chloride solutions.

$$Cu_2S \text{ (s)} + 2 \, Fe^{3+} \text{ (aq)} \longrightarrow CuS \text{ (s)} + Cu^{2+} \text{ (aq)} + 2 \, Fe^{2+} \text{ (aq)} \qquad (37)$$

$$CuS \text{ (s)} + 2 \, Fe^{3+} \text{ (aq)} \longrightarrow S \text{ (s)} + Cu^{2+} \text{ (aq)} + 2 \, Fe^{2+} \text{ (aq)} \qquad (38)$$

Bacterial leaching is another example of oxidizing dissolution whereby specific bacteria either directly attack the sulfide mineral or indirectly enhance the regeneration of the oxidant.

Oxidizing and Complex-Forming Solutions. The leaching of gold or silver can be achieved only by oxidation of the metal by air followed by formation of a stable cyanide complex.

$$2 \, Au_0 \text{ (s)} + 4 \, CN^- \text{ (aq)} + 1/2 \, O_2 \text{ (g)} + H_2O \text{ (l)} \longrightarrow 2 \, [Au(CN)_2]^- \text{ (aq)} + 2 \, OH^- \text{ (aq)} \qquad (39)$$

Another important example is the leaching of nickel sulfide under ammonia and oxygen pressure to form hexaammine nickel (Sherritt-Gordon process).

$$NiS \text{ (s)} + 2 \, O_2 \text{ (g)} + 6 \, NH_3 \text{ (g)} \longrightarrow [Ni(NH_3)_6]^{2+} \text{ (aq)} + SO_4^{2-} \text{ (aq)} \qquad (40)$$

Leaching Techniques. Leaching techniques vary depending on the type of reaction and the characteristics of the ore. Dissolution occurs at the interface between the solid and the solution. The rate depends on the rate of transport of the leaching agent from the solution to the solid and on the rate of chemical reaction. These rates are enhanced by increasing the surface area of the solid, usually by fine grinding of the ore or concentrate. For some low grade material, however, little preparation is justified beyond fracture and rubblization of the rock.

In Situ Leaching. Copper and uranium ores are sometimes leached in place by circulating acidified mine water through the underground deposit. This process is known as solution mining.

Heap and Dump Leaching. Heap leaching is practiced mainly for oxide copper ores, and for gold and silver. The crushed ore is placed in large heaps and sprayed with the leaching solution. Mine wastes containing copper values as sulfides are also leached in a similar way known as dump leaching. *In situ* and dump leaching of copper sulfide values rely on the presence of iron in natural waters. The source of this iron and the acidity of the solution results from the aqueous oxidation of pyrite. The iron is oxidized by contact with the atmosphere, and the ferric ions act as oxidant for the dissolution of sulfides by reactions similar to equations 37 and 38. Bacterial activity in the bed has been found to enhance reaction rates. This type of leaching, however, is a slow process carried out over periods of several years.

Percolation Leaching. Ground material coarse enough to permit circulation of a solution through a bed of particles can be leached by percolation of the solvent through the material placed in a tank or vat. The process usually takes several days.

Agitation Leaching. Very fine ore products, slimes, are treated by agitation of a pulp, which is a suspension of the ore particles in the solution. The process is carried out in a tank with mechanical or air agitation. In a leaching process for which oxygen is needed, eg, gold leaching, air agitation is preferred. Residence time is several hours.

Pressure Leaching. Leaching at elevated (90–250°C) temperature is carried out in autoclaves under pressure. In some instances the purpose is to modify equilibrium conditions and to increase the solubility of the ore, eg, digestion of bauxite. In cases requiring a gaseous reagent, eg, oxygen or ammonia, the higher pressure increases the solubility of the gas and enhances the reaction rates. The increased temperature also increases reaction rates, and residence times are one or two hours. This process has been applied successfully to oxidation of sulfidic refractory gold ores prior to cyanidation.

Liquid–Solid Separation. The separation of the solution containing the dissolved metal, ie, the pregnant solution, from the leach residue is the final leaching step. Most leaching operations are carried out in several tanks in series to permit countercurrent flow. The pregnant solution leaves at one end after contacting fresh ore, and the depleted solids exit at the other end after contacting the fresh solvent. Additional steps such as washing the residue, clarification of the solution, and filtering are required.

Recovery of Metal from Solution. The same chemical principles govern the recovery of an element or compound from solution, whether it is the main metal to be processed or an impurity to be removed.

Electrowinning. When it is possible, electrolytic deposition is the most efficient way of recovering a valuable metal from solution. It is quite selective and usually yields a pure product which can be marketed directly as cathodes, or after casting into commercial shapes. It is, however, the most expensive method.

Precipitation. The precipitation of aluminum trihydroxide in the recovery step of the Bayer process is achieved either by lowering the temperature or by diluting the pregnant liquor and reducing its pH. Both methods reverse the direction of equation 35, but seeding with previously precipitated crystals is required in order to initiate nucleation.

The removal of copper from the pregnant nickel solution in the Sherritt-Gordon process is an example of purification by precipitation of a fairly insoluble compound. First, in the copper boil step, ammonia is driven off by heating the solution, and some copper sulfide precipitates. The residual copper is removed by adding hydrogen sulfide for the chemical precipitation of more copper sulfide.

$$Cu^{2+} \text{ (aq)} + H_2S \text{ (g)} \longrightarrow CuS \text{ (s)} + 2\,H^+ \text{ (aq)} \tag{41}$$

Cementation. A metal can be removed from solution by displacing it with a more active metal. This simple, inexpensive method has been commonly used to recover copper from dilute (1–3 kg/m^3) solution using shredded iron and de-tinned iron cans as reducing agent.

$$Cu^{2+} \text{ (aq)} + Fe^0 \text{ (s)} \longrightarrow Cu^0 \text{ (s)} + Fe^{2+} \text{ (aq)} \tag{42}$$

A similar reaction leads to the precipitation of gold or silver from cyanide solution using zinc powder.

$$2 \text{ Au(CN)}_2^- \text{ (aq)} + \text{Zn}^0 \text{ (s)} \longrightarrow 2 \text{ Au}^0 \text{ (s)} + \text{Zn(CN)}_4^{2-} \text{ (aq)} \tag{43}$$

Cementation is also an efficient way of purifying a pregnant solution by removing impurities that are more noble than the metal being processed. An example is the cementation of copper, cadmium, cobalt, and nickel from zinc solutions prior to electrowinning.

The cementation of gold and the purification of the zinc electrolyte are usually carried out in cylindrical vessels using mechanical agitation. The cementation of copper is carried out in long narrow tanks called launders, in rotating drums, or in an inverted cone precipitator (see COPPER).

Gas Reduction. The use of a gaseous reducing agent is attractive because the metal is produced as a powder that can easily be separated from the solution. Carbon dioxide, sulfur dioxide, and hydrogen can be used to precipitate copper, nickel, and cobalt, but only hydrogen reduction is applied on an industrial scale. In the Sherritt-Gordon process, the excess ammonia is removed during the purification to achieve a 2:1 ratio of NH_3:Ni in solution. Nickel powder is then precipitated by

$$\text{Ni(NH}_3)_2^{2+} \text{ (aq)} + \text{H}_2 \text{ (g)} \longrightarrow \text{Ni}^0 \text{ (s)} + 2 \text{ NH}_4^+ \text{ (aq)} \tag{44}$$

The reaction proceeds in an autoclave at 200°C under ca 3 MPa (30 atm) of hydrogen pressure.

Ion Exchange. Metallic ions can be removed from an aqueous solution by ion exchange (qv) at the surface of an organic resin. Both anionic and cationic exchangers are available. Anionic ion-exchange resins are used for concentration and purification of the dilute pregnant solutions obtained by leaching uranium ores with sulfuric acid. The ion-exchange reaction is represented by

$$\text{UO}_2(\text{SO}_4)_3^{4-} \text{ (aq)} + 4 \text{ RX (s)} \longrightarrow \text{R}_4\text{UO}_2(\text{SO}_4)_3 \text{ (s)} + 4 \text{ X}^- \text{ (aq)} \tag{45}$$

where X^- is an anion, usually NO_3^- or Cl^-, and R represents the organic resin. After the resin is loaded, it is eluted with a strong solution of the X^- ion. This reverses the equilibrium conditions for equation 45, transfers the complex uranyl ion back to an aqueous phase, and regenerates the resin. The technology for this process, borrowed from the water-treatment industry, uses mainly packed columns (see WATER). The resin-in-pulp process is typical of uranium hydrometallurgy. The leaching pulp and ion-exchange resin are agitated together and both reactions proceed in the same reactor. The residue and the loaded resin are then separated.

Solvent Extraction. Liquid–liquid extraction, well known in the chemical industry, was first used in extractive metallurgy for the processing of uranium.

When a dilute solution of uranium is contacted with an extractant such as di(2-ethylhexyl) phosphoric acid (D2EHPA) or R_2HPO_4, dissolved in kerosene, the uranyl ion is transferred to the organic phase.

$$2\ R_2HPO_4\ (\text{org}) + UO_2^{2+}\ (\text{aq}) \longrightarrow UO_2(R_2PO_4)_2\ (\text{org}) + 2\ H^+\ (\text{aq}) \tag{46}$$

The pregnant organic solvent is stripped by agitation with a strong carbonate solution which removes uranium as the stable $UO_2(CO_3)_3^{4-}$ aqueous complex.

The development of selective extractants for copper has made extraction from dilute solutions ($1-5\ kg/m^3$) economically feasible. Transfer of the copper by stripping to a more concentrated sulfuric acid solution, ie, $30-40\ kg/m^3$ for Cu^{2+} and $150-170\ kg/m^3$ for H_2SO_4, from which the copper is recovered by electrowinning. The simplified reaction,

$$Cu^{2+}\ (\text{aq}) + 2\ HR\ (\text{org}) \longrightarrow CuR_2\ (\text{org}) + 2\ H^+\ (\text{aq}) \tag{47}$$

is carried out in two or three extraction mixer-settlers followed by one or two stripping mixer-settlers. The organic extractants are hydroxyphenyl oximes, typically $5-10$ vol % of salicylaldoximes and ketoximes in purified kerosene (4).

In addition, solvent extraction is applied to the processing of other metals for the nuclear industry and to the reprocessing of spent fuels (see NUCLEAR REACTORS). It is commercially used for the cobalt–nickel separation prior to electrowinning in chloride electrolyte. Both extraction columns and mixer-settlers are in use.

Hydrometallurgical Flow Sheets. The various hydrometallurgy operations can be combined in many ways to design processes appropriate for specific metals.

Aluminum. All primary aluminum as of 1995 is produced by molten salt electrolysis, which requires a feed of high purity alumina to the reduction cell. The Bayer process is a chemical purification of the bauxite ore by selective leaching of aluminum according to equation 35. Other oxide constituents of the ore, namely silica, iron oxide, and titanium oxide remain in the residue, known as red mud. No solution purification is required and pure aluminum hydroxide is obtained by precipitation after reversing reaction 35 through a change in temperature or hydroxide concentration; the precipitate is calcined to yield pure alumina.

Uranium. The hydrometallurgical treatment of uranium ores is a concentration and purification process. Typical ore grade is $0.1-0.5\%$ U_3O_8, and pregnant solutions contain ca $1\ kg/m^3$ of U_3O_8. The dissolution requires the presence of an oxidant, either oxygen or a ferric salt.

$$U_3O_8\ (\text{s}) + 6\ H^+\ (\text{aq}) + 1/2\ O_2\ (\text{g}) \longrightarrow 3\ UO_2^{2+}\ (\text{aq}) + 3\ H_2O\ (\text{l}) \tag{48}$$

The solvent, a solution of either sulfuric acid or sodium carbonate, forms the stable complex uranyl ions $UO_2(SO_4)_2^{2-}$, $UO_2(SO_4)_3^{4-}$, and $UO_2(CO_3)_3^{4-}$. The pregnant solution is concentrated and purified by ion exchange or solvent extraction, yielding a stripping solution of ca 50 kg/m^3 U$_3$O$_8$. Uranium is then precipitated chemically. Pure U$_3$O$_8$ is obtained by calcination (see URANIUM AND URANIUM COMPOUNDS).

Gold. Gold is produced exclusively by hydrometallurgy. The traditional steps are leaching in alkaline cyanide solution (eq. 39) and cementation on zinc powder (eq. 43). In modern plants, the dissolved gold in the pregnant solution is adsorbed on carbon from which it can be transferred into a more concentrated solution suitable for electrowinning. The need to process refractory ores, so-called because these contain organic carbon or metal sulfides which make direct leaching less efficient, has led to the development of preparatory steps including aqueous chlorine oxidation of carbon, bacterial leaching, oxidizing autoclave leach, and roasting of sulfides.

Nickel. Most nickel is produced by smelting sulfide ores. Several hydromet-allurgical processes have been developed. Among them the Sherritt-Gordon process stands out as the first successful commercial application of pressure hydrometallurgy to a complex feed. The raw material is a pentlandite, NiS·FeS, concentrate. The flow sheet in Figure 5 illustrates how several hydrometallurgi-cal operations are integrated into a complex process. Nickel is leached selectively in an ammoniacal solution according to equation 40, iron remains in the residue as Fe$_2$O$_3$. Copper is removed from solution by the purification steps mentioned earlier with equation 41. Cobalt remains in solution and nickel is recovered by hydrogen reduction following equation 44. Similar flow sheets are quite appro-priate for the treatment of complex raw materials containing several different metal values.

Electrometallurgy

The use of electricity for winning and refining metals could not have been developed without an understanding of the basic principles of electrochemistry and the availability of cheap industrial electric power. Electrometallurgy is a more powerful tool than the other extractive processes. It supplies energy to the system in a way that enables a reaction to proceed against its chemical affinity and in the direction of its electrochemical affinity. Chemically reactive metals are more easily recovered by electrometallurgy than by chemical reduction, and can be obtained in high purity. Electrometallurgy, however, has practical limitations. First, is expense. Second, the electrochemical process occurs at an electrolyte–metal interface so that the output of an electrochemical reactor is directly proportional to the area of that interface. Improving the energy efficiency and increasing the electrode area per unit volume of reactor remain a challenge.

The theoretical energy requirement for an electrochemical process is given by

$$nFE = -\Delta G \qquad (49)$$

where E is the reversible electromotive force (emf), ΔG is the Gibbs free energy change, n is the number of electrons involved in the process, and F is Faraday's constant = 96,485 C/eq. For an electrolytic process, ΔG is positive and E is neg-

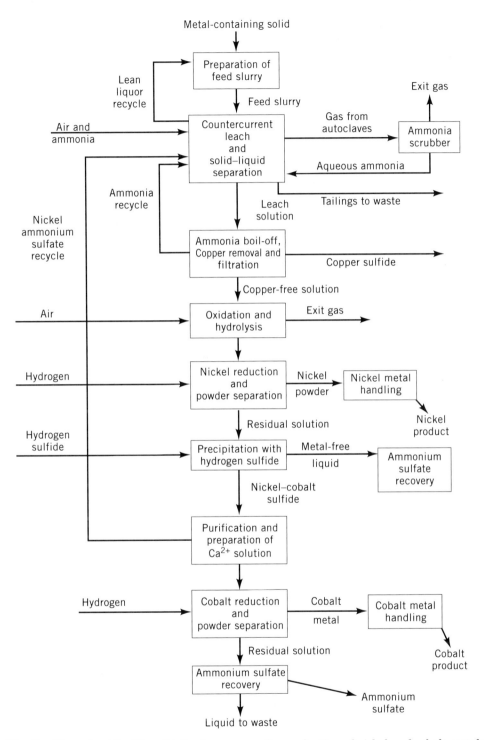

Fig. 5. Flow sheet for Sherritt-Gordon process for production of nickel and cobalt metals from sulfide ore.

ative, ie, energy must be supplied to the system. The actual voltage to be applied to the cell, V, is larger than the reversible emf because of the irreversibility of the process, namely electrochemical overpotentials, η, and ohmic losses owing to the resistance of the circuit, R:

$$V = -E + \sum_i \eta_i + RI \tag{50}$$

where I is the current. The voltage efficiency, ϵ_V, is defined as the ratio between the arithmetic value of the emf and the applied voltage:

$$\epsilon_V = \frac{-E}{V} \tag{51}$$

The theoretical amount of metal produced by electrolysis is directly proportional to the amount of electricity according to Faraday's law. Because of losses by chemical or electrochemical processes, the actual amount is less. It is characterized by the current efficiency, ϵ_I, defined by the following:

$$\epsilon_I = \frac{nF}{It} \tag{52}$$

where t is the time required to produce one mole of metal. Combining equations 51 and 52 leads to the relationship between ΔG and the energy actually used, W, and the definition of the energy efficiency, ϵ_w:

$$W = VIt = \frac{\Delta G}{\epsilon_I \epsilon_V} = \frac{\Delta G}{\epsilon_w} \tag{53}$$

Usually the specific energy consumption, w, is reported in kWh/kg of product:

$$w = \frac{VnF}{\epsilon_I M} \tag{54}$$

where M is the atomic weight of the metal in kg.

Electrowinning from Aqueous Solutions.   Electrowinning is the recovery of a metal by electrochemical reduction of one of its compounds dissolved in a suitable electrolyte. Various types of solutions can be used, but sulfuric acid and sulfate solutions are preferred because these are less corrosive than others and the reagents are fairly cheap. From an electrochemical viewpoint, the high mobility of the hydrogen ion leads to high conductivity and low ohmic losses, and the sulfate ion is electrochemically inert under normal conditions.

The generalized flow sheet of an aqueous electrowinning process consists of at least three main steps. (1) The metal is put into solution by leaching of a calcine, ie, a roasted concentrate of sulfide ore, or by direct leaching of low grade ores containing oxidized minerals or weathered sulfides. (2) The pregnant solution is purified to remove metallic impurities more noble than the metal to be electrowon, and impurities that could reduce the current efficiency. Sometimes, the pregnant solution is treated by solvent extraction to produce a more concentrated electrolyte. (3) The purified solution is fed to the electrolysis tanks where the metal is plated on a cathode and oxygen is evolved at an inert anode, usually made of lead or one of its alloys. The sulfuric acid is regenerated by the anodic process and is recycled to leaching.

Zinc. The electrowinning of zinc on a commercial scale started in 1915. Most newer facilities are electrolytic plants. The success of the process results from the ability to handle complex ores and to produce, after purification of the electrolyte, high purity zinc cathodes at an acceptable cost. Over the years, there have been only minor changes in the chemistry of the process to improve zinc recovery and solution purification. Improvements have been made in the areas of process instrumentation and control, automation, and prevention of water pollution.

Zinc sulfide ores are concentrated by flotation, roasted, and then leached in sulfuric acid. The leaching involves at least two stages. In a neutral leach, an excess of calcine maintains a pH of ca 5. Using aeration and sometimes addition of manganese dioxide, iron is removed by precipitation of ferric hydroxide. The solid residue undergoes further dissolution of the zinc in one or several stages of cold or hot acid leach, and the final residue goes to a lead smelter. Improvements have resulted in the precipitation of iron as goethite, $FeOOH$, or jarosite, $NaFe_3(SO_4)_2(OH)_6$. Impurities such as arsenic, antimony, and germanium coprecipitate with iron. Copper, cadmium, cobalt, and nickel are removed by cementation on zinc powder. A simplified flow sheet of the process is shown in Figure 6.

The standard electrode potential for zinc reduction (-0.763 V) is much more cathodic than the potential for hydrogen evolution, and the two reactions proceed simultaneously, thereby reducing the electrochemical yield of zinc. Current efficiencies slightly above 90% are achieved in modern plants by careful purification of the electrolyte to bring the concentration of the most harmful impurities, eg, germanium, arsenic, and antimony, down to ca 0.01 mg/L. Addition of organic surfactants (qv) like glue, improves the quality of the deposit and the current efficiency.

The zinc electrolyte contains ca 60 kg/m^3 zinc as sulfate and ca 100 kg/m^3 free sulfuric acid. It is electrolyzed between electrodes suspended vertically in lead or plastic-lined, eg, poly(vinyl chloride), concrete tanks. The insoluble anodes are made of lead with small amounts of silver. The anodic reaction produces oxygen and regenerates sulfuric acid which is recycled to the acid leach. The cathodes are aluminum sheets, from which the zinc deposits are stripped every 24–48 hours. The effective cathode area varies from about 1.5 m^2 in traditional plants to 3.2 m^2 for the newer jumbo cathodes. The current density is usually 300–600 A/m^2. The electrodes are in parallel in each cell and the cells are connected in series. The average voltage across a cell is ca 3.5 V. The reversible emf calculated for the cell reaction,

$$Zn^{2+} \text{ (aq)} + H_2O \text{ (l)} \longrightarrow Zn^0 \text{ (s)} + 1/2 \, O_2 \text{ (g)} + 2 \, H^+ \text{ (aq)} \qquad (55)$$

is ca 2 V; the remaining 1.5 V is accounted for by the anodic overvoltage (about 0.9 V), the cathodic overvoltage, and the ohmic drop in the electrolyte and at

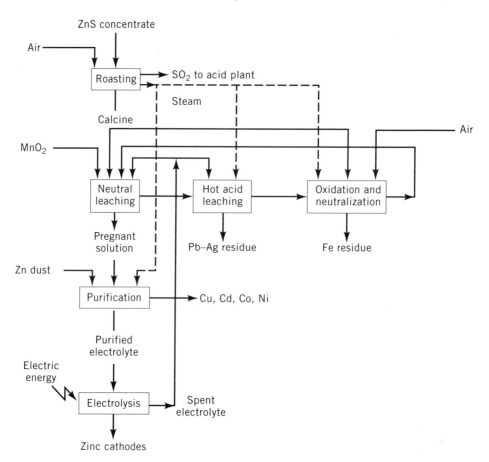

Fig. 6. Roast-Leach electrowinning process for the production of zinc metal from a sulfide concentrate.

the electrode contacts. The voltage efficiency is ca 57% and the energy efficiency is usually slightly above 50%. The energy consumption is 3.3 kWh/kg zinc. The relatively high energy cost of the process is acceptable in view of the special high grade zinc produced (99.995% zinc).

Copper. Copper is economically extracted by smelting of a chalcopyrite concentrate. A copper electrowinning process was developed commercially in 1912 for the treatment of lean ores. It is also suitable for treatment of copper oxide or sulfate obtained by roasting of the concentrate. Most electrowon copper comes from the direct leaching of low grade oxidized copper ores. It accounted for approximately one-tenth of the world output of copper-from-ore in 1994 (4). Dilute copper solutions from dump leaching or solution mining are concentrated by solvent extraction to a level suitable for electrolysis. In the United States in 1994, ca 35% of the primary copper was produced by electrowinning. The electrowinning of copper is carried out in sulfate solutions containing 25–35 kg/m^3 Cu^{2+} and ca 160 kg/m^3 free H$_2$SO$_4$ at current densities equal to ca 200 A/m^2.

The anodes are made of lead alloyed with antimony, and copper is deposited on copper starting sheets.

Other Nonferrous Metals. Cadmium, and in much smaller quantities thallium and indium, are by-products of zinc and are recovered by electrolysis in sulfate solutions. When associated with copper, cobalt is also electrowon by a similar process. About 10% of the primary nickel output is produced by electrolysis in chloride and sulfate solutions. Some cobalt is electrowon from chloride electrolytes. Metallic chromium and manganese, as distinct from the ferroalloys, are electrowon from ammonium sulfate solutions. These metals are more difficult to electroplate because of the more negative electrode potential and complex chemistry. The required close control of the process variables is achieved by separating the anolyte and catholyte by a diaphragm. Gallium is produced in the liquid state or as an amalgam by electrolysis of a caustic solution.

Electrolysis of Molten Salts. Metals more active than zinc and manganese cannot be recovered by electrodeposition from aqueous solutions. Most of them, however, can be electrowon from a molten electrolyte. A compound of the metal to be electrodeposited is dissolved in a mixture of salts of more active metals in order to achieve a low melting point, a suitable viscosity and density, and a high conductivity. Solid electrodeposits obtained from molten salts are mostly dendritic or powdery, and are difficult to separate from the melt. Reduction of the metal as a separate liquid phase having a density different from that of the electrolyte is preferable. The recombination of reaction products is the most common cause of faradaic inefficiency. Because recombination is often related to the solubility of the reduced metal in the molten electrolyte, the cells must be designed to protect the reduced metal from contact with the anodic reaction products and with air. The cell electrolyte cannot be recycled through leaching and purification as in aqueous electrowinning, and the feed must be carefully purified and dehydrated.

Aluminum. The electrowinning of aluminum, developed in 1886 by Hall in the United States and by Héroult in France, is the single most important electrometallurgical process and remains the only commercial process for the production of aluminum. The purified alumina, obtained by the Bayer process, is dissolved in an electrolyte consisting mainly of cryolite, $AlF_3 \cdot 3NaF$, and other fluorides in smaller amounts, eg, CaF_2 and LiF. The electrolysis cells, called pots or furnaces, are made of steel lined with insulating material and carbon bricks that make electrical contact to the molten aluminum cathode. The fused salt electrolyte floats on top of the molten metal and carbon anodes dip into the bath from above, leaving a spacing of ca 5 cm to the cathode. Two types of anodes, prebaked blocks and self-baking Soderberg anodes, are used, but the prebaked anodes are preferred in modern plants.

The electrolytic decomposition of alumina yields oxygen which reacts with the carbon anode for an overall cell reaction:

$$Al_2O_3 \text{ (melt)} + 3/2 \ C^0 \text{ (s)} \longrightarrow 2 \ Al^0 \text{ (l)} + 3/2 \ CO_2 \text{ (g)} \tag{56}$$

The theoretical value of the reversible electromotive force for this reaction is 1.5 V, but the cell operates under a total voltage from ca 4 to 4.5 V. The difference is accounted for by the overvoltage at the electrodes, including the anode effect, and the ohmic drop through the electrolyte and the electrode contacts. The heat generated by irreversible electrode processes and by ohmic losses keeps the operating temperature of the cell at the desired level of ca 960°C. Current densities range from 6 to 12 kA/m^2, and the total current of modern pots is ca 100–300 kA. Alumina is periodically added to the melt to maintain its concentration between 2 and 5%. The current efficiency of modern plants is ca 90%; the main losses are due to the reoxidation of aluminum by carbon dioxide, yielding alumina and carbon monoxide. The energy consumption is ca 14 kWh/kg of aluminum.

The chemistry of the process has not changed significantly since its invention. Improvements in the energy efficiency and operation of modern plants have been achieved mainly by improved cell design and better control of the electrolysis through the use of computers to monitor the pot conditions. The electrolysis of an all-chloride mixture in a cell of a new bipolar design has been tested successfully on a small commercial scale with an energy consumption below that of the Hall-Héroult cell, but problems encountered in the production of the aluminum chloride feed from bauxite have prevented further development as of this writing (see ALUMINUM AND ALUMINUM ALLOYS).

Magnesium.   The electrowinning of magnesium, another well-established process, accounts for ca 80% of the metal output. The electrolyte is a mixture of chlorides of potassium, sodium, calcium, and magnesium. Because the liquid metal has a lower density than the electrolyte, it floats to the surface. The cells are designed so that the reduced metal does not come into contact with air or with the chlorine gas produced at the anode. The temperature of the cells is kept between 700 and 750°C by external heating. The cell voltage ranges from 5 to 8 V, and the energy consumption is ca 17–22 kWh/h of magnesium (see MAGNESIUM AND MAGNESIUM ALLOYS).

Other Metals.   All the sodium metal produced comes from electrolysis of sodium chloride melts in Downs cells. The cell consists of a cylindrical steel cathode separated from the graphite anode by a perforated steel diaphragm. Lithium is also produced by electrolysis of the chloride in a process similar to that used for sodium. The other alkali and alkaline-earth metals can be electrowon from molten chlorides, but thermochemical reduction is preferred commercially. The rare earths can also be electrowon but only the mixture known as mischmetal is prepared in tonnage quantity by electrochemical means. In addition, beryllium and boron are produced by electrolysis on a commercial scale in the order of a few hundred t/yr. Processes have been developed for electrowinning titanium, tantalum, and niobium from molten salts. These metals, however, are obtained as a powdery deposit which is not easily separated from the electrolyte so that further purification is required.

Electrorefining.   Electrolytic refining is a purification process in which an impure metal anode is dissolved electrochemically in a solution of a salt of the metal to be refined, and then recovered as a pure cathodic deposit. Electrorefining is a more efficient purification process than other chemical methods because of its selectivity. In particular, for metals such as copper, silver, gold, and lead, which exhibit little irreversibility, the operating electrode potential is close to the reversible potential, and a sharp separation can be accomplished, both at the anode where more noble metals do not dissolve and at the cathode where more active metals do not deposit.

Because the same electrochemical reaction proceeds in opposite directions at the anode and the cathode, the overall chemical changes is small change in the activity of the metal at the two electrodes. Therefore, the reversible emf is practically zero. There is some polarization of the electrodes during electrolysis, but the main component of the voltage across the operating cell is the ohmic drop through the electrolyte and at the electrode contacts. Normal cell voltages are ca 0.2 V. The power consumption is correspondingly very small, and electrorefining is much less sensitive to the cost of electric power than other electrometallurgical processes. When a diaphragm is used to separate the anodic and cathodic solutions, the cell voltage increases up to ca 1.2 V, and the power consumption rises accordingly.

Copper. The first electrolytic copper refinery was started in 1871, making electrorefining the oldest commercial electrometallurgical process (see COPPER). Most copper is electrorefined. The refining process ensures that the metal meets the specifications of the electrical industry which are stringent because minor quantities of some impurities lower the electrical conductivity of copper markedly. Furthermore, silver and gold, common constituents of copper ores, follow along through all the pyrometallurgical steps. Removal by the electrolytic refining is an important economic asset of the process.

Impure copper is cast in the shape of anodes ~0.9 by 1.0 m and 3.5–4.5-cm thick, weighing 300–400 kg. These anodes are cast with lugs that support them from the walls of the cell or from bus bars and make electrical contact. The starting cathodes used to be pure copper sheets, made by electrodeposition on smooth starting blanks of copper or titanium. The copper starting sheets have been replaced by permanent stainless steel cathodes. The electrode spacing from anode center to anode center is ca 10 cm. The current density is usually 200–250 A/m^2. The anodes are consumed and replaced at regular intervals of 20–28 days, and two successive cathodes are produced during the same period. The solution contains ca 45 kg/m^3 of copper as copper sulfate and ca 200 kg/m^3 of sulfuric acid. The temperature is maintained at 55°C to lower the resistance of the electrolyte which is circulated through the cell. The lead or plastic-lined tanks of wood or concrete used hold ~30–35 anodes and one more cathode than anode.

Metals less noble than copper, such as iron, nickel, and lead, dissolve from the anode. The lead precipitates as lead sulfate in the slimes. Other impurities such as arsenic, antimony, and bismuth remain partly as insoluble compounds in the slimes and partly as soluble complexes in the electrolyte. Precious metals, such as gold and silver, remain as metals in the anode slimes. The bulk of the slimes consist of particles of copper falling from the anode, and insoluble sulfides, selenides, or tellurides. These slimes are processed further for the recovery of the various constituents. Metals less noble than copper do not deposit but accumulate in solution. This requires periodic purification of the electrolyte to remove nickel sulfate, arsenic, and other impurities.

Nickel. Most nickel is also refined by electrolysis. Both copper and nickel dissolve at the potential required for anodic dissolution. To prevent plating of the dissolved copper at the cathode, a diaphragm cell is used, and the anolyte is circulated through a purification circuit before entering the cathodic compartment (see NICKEL AND NICKEL ALLOYS).

Other Metals.   Although most cobalt is refined by chemical methods, some is electrorefined. Lead and tin are fire refined, but a better removal of impurities is achieved by electrorefining. Very high purity lead is produced by an electrochemical process using a fluosilicate electrolyte. A sulfate bath is used for purifying tin. Silver is produced mainly by electrorefining in a nitrate electrolyte, and gold is refined by chemical methods or by electrolysis in a chloride bath.

The electrorefining of many metals can be carried out using molten salt electrolytes, but these processes are usually expensive and have found little commercial use in spite of possible technical advantages. The only application on an industrial scale is the electrorefining of aluminum by the three-layer process. The density of the molten salt electrolyte is adjusted so that a pure molten aluminum cathode floats on the electrolyte, which in turn floats on the impure anode consisting of a molten copper–aluminum alloy. The process is used to manufacture high purity aluminum.

BIBLIOGRAPHY

"Extractive Metallurgy" in *ECT* 3rd ed., Vol. 9, pp. 739–767, by P. Duby, Columbia University.

1. G. Agricola, *De Re Metallica*, Basil, 1556, transl. H. C. Hoover and L. L. Hoover, Dover Publications, Inc., New York, 1950.

2. T. Rosenqvist, *Principles of Extractive Metallurgy*, 2nd ed., McGraw-Hill Book Co., Inc., New York, 1983.

3. J. W. Evans and L. C. DeJonghe, *The Production of Inorganic Materials*, Macmillan Publishing Co., New York, 1991.

4. A. K. Biswas and W. G. Davenport, *Extractive Metallurgy of Copper*, 3rd ed., Pergamon Press Ltd., Oxford, U.K., 1994.

5. J. R. Boldt, Jr. and P. Queneau, *The Winning of Nickel*, D. Van Nostrand, Princeton, N.J., 1967.

General References

References 2–5.

C. B. Alcock, *Principles of Pyrometallurgy*, Academic Press, London, 1976.

C. Bodsworth, *The Extraction and Refining of Metals*, CRC Press, Inc., Boca Raton, Fl., 1994.

A. R. Burkin, *The Chemistry of Hydrometallurgical Processes*, D. Van Nostrand, Princeton, N.J., 1966.

P. L. Claessens and G. B. Harris, *Electrometallurgical Plant Practice*, Pergamon Press, Inc., New York, 1990.

W. H. Dennis, *Extractive Metallurgy*, Philosophical Library, New York, 1965.

J. D. Gilchrist, *Extractive Metallurgy*, 3rd ed., Pergamon Press Ltd., Oxford, U.K., 1989.

R. I. L. Guthrie, *Engineering in Process Metallurgy*, Oxford University Press, New York, 1989.

F. Habashi, *Principles of Extractive Metallurgy*, Gordon and Breach, New York, Vol. 1, "General Principles," 1969; Vol. 2, "Hydrometallurgy," 1970; Vol. 3, "Pyrometallurgy," 1986.

C. R. Hayward, *An Outline of Metallurgical Practice*, Van Nostrand Reinhold Co., New York, 1952.

E. Jackson, *Hydrometallurgical Extraction and Reclamation*, Ellis Horwood Ltd., Chichester, U.K., 1986.

A. Kuhn, ed., *Industrial Electrochemical Processes*, Elsevier Scientific Publishing Co., Inc., Amsterdam, the Netherlands, 1971.

W. C. Cooper, in *McGraw-Hill Encyclopedia of Science and Technology*, McGraw-Hill, Inc., New York, 1992, Vol. 14, pp. 552–558.

H. H. Kellogg, in M. B. Bever, ed., *Encyclopedia of Materials Science and Engineering*, Vol. 2, Pergamon Press Ltd., Oxford, U.K., 1986, pp. 1528–1535.

C. L. Mantell, *Electrochemical Engineering*, 4th ed., McGraw-Hill Book Co., Inc., New York, 1960.

J. J. Moore, *Chemical Metallurgy*, 2nd ed., Butterworths & Co., Ltd., London, 1990.

J. Newton, *Extractive Metallurgy*, John Wiley & Sons, Inc., New York, 1959.

R. H. Parker, *An Introduction to Chemical Metallurgy*, 2nd ed., Pergamon Press Ltd., Oxford, U.K., 1978.

R. D. Pehlke, *Unit Processes of Extractive Metallurgy*, Elsevier Scientific Publishing Co., Inc., New York, 1973.

J. T. Tien and J. F. Elliott, eds., *Metallurgical Treatises*, The Metallurgical Society of AIME, Warrendale, Pa., 1981.

G. D. Van Arsdale, *Hydrometallurgy of Base Metals*, McGraw-Hill Book Co., Inc., New York, 1953.

PAUL DUBY
Columbia University

POWDER METALLURGY

Powder metallurgy (P/M) is both an ancient art and a modern technology on the cutting edge of new materials and manufacturing processes. Powder metallurgy, a worldwide industry, uses metal powders as the raw material from which to manufacture all kinds of products, but chiefly precision metal shapes and parts. These parts perform in many consumer and industrial products such as automobile engines and transmissions, aircraft engines, aerospace hardware, washing machines, power tools, riding lawn mowers, computer disk drives, and surgical implements.

Metal powders are consolidated into shapes through a number of different production processes (Fig. 1). The most widely used powder metallurgy process covers three basic steps for producing conventional density parts: mixing, compacting, and sintering. Elemental or prealloyed metal powders are first mixed with lubricants or other alloy additions to produce a homogeneous mixture of ingredients (see LUBRICATION AND LUBRICANTS). A controlled amount of mixed powder is automatically gravity fed into a precision die and compacted, usually at room temperature, at pressures as low as 138 MPa (20,000 psi) or as high as 827 MPa (120,000 psi), depending on the density requirements of the part. Compacting pressures normally range from 414 to 690 MPa (60,000–100,000 psi).

Compacting a loose powder produces a green compact. Using conventional pressing techniques this compact has the size and shape of the finished part

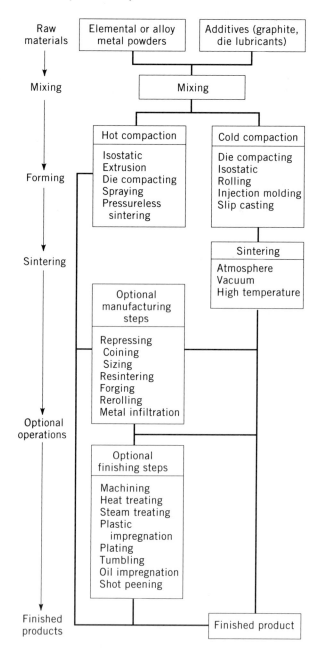

Fig. 1. The P/M process.

when ejected from the die, and sufficient strength for in-process handling and transport to a sintering furnace. Typical compacting techniques use rigid dies set into special mechanical or hydraulic presses.

Tool sets are made of either hardened steel (qv) and/or carbides (qv), and consist of a die body or mold, an upper and lower punch, and in some cases one or more core rods. The pressing cycle for producing a simple part is shown in

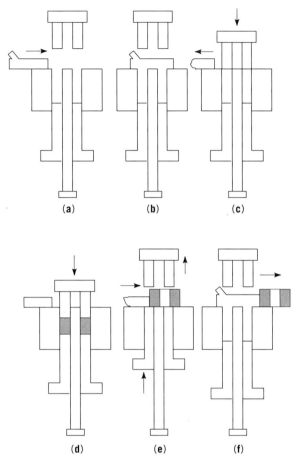

Fig. 2. Pressing cycle for a single level part: (**a**) cycle start; (**b**) charging (filling) die with powder; (**c**) compaction begins; (**d**) compaction completed; (**e**) ejection of part; and (**f**) recharging die.

Figure 2. More specialized compacting and alternative forming methods, such as isostatic pressing, extrusion, injection molding, and slip casting, are sometimes used.

In sintering, the green compact is placed on a wide-mesh belt and slowly moves through a controlled atmosphere furnace (Fig. 3). The parts are heated to below the melting point of the base metal, held at the sintering temperature, and cooled. Basically a solid-state process, sintering transforms mechanical bonds, ie, contact points, between the powder particles in the compact into metallurgical bonds which provide the primary functional properties of the part.

History

Powder metallurgy dates to prehistoric times when artisans learned to form a solid structure by hammering gold or iron (qv) particles into metallic objects (1,2). This occurred long before the advent of furnaces that could even approach the melting point of the metal. Egyptian implements were made from metal particles

Fig. 3. Sintering of the green compact in a controlled atmosphere furnace.

as early as 3000 BC. Pure iron oxide was heated in a charcoal fire intensified by air blasts from a bellows. The iron was reduced to a spongy metallic form. A more or less solid metallic structure was produced by hammering this porous metal while it was still hot. Final shapes were obtained by forging techniques. The direct reduction of iron oxide without fusion dates to 300 AD when the famous Delhi Pillar, weighing almost 5.5 t, was produced (see IRON BY DIRECT REDUCTION).

The Incas and their predecessors used powder techniques to form articles from grains of platinum, gold, and silver (3). Until the end of the eighteenth century all platinum was fabricated from granules by variations of the process used by the Incas (see PLATINUM-GROUP METALS). Woolaston worked on producing platinum metal products from sponge powder without liquid-phase sintering in the early nineteenth century. Methods of powder preparation and compacting were investigated and a horizontal toggle press was developed. In 1830 Osann was recognized for work with copper powder, making shapes by pressing and sintering. The process was also used to make medals out of silver and lead.

The search for a durable filament for the incandescent electric light bulb opened up development work in osmium, tantalum, and tungsten powders. These materials were mixed with a binder, extruded into wire, and sintered. In the early 1900s it was discovered that tungsten could be worked in a certain temperature range yet keep its ductility at room temperature (see TUNGSTEN AND TUNGSTEN ALLOYS). Sintered tungsten ingots were annealed, swaged, and drawn into very thin and ductile lamp filaments. Next came cemented tungsten carbide, a mixture of tungsten carbide and cobalt powders used to manufacture cutting, forming, and mining tools and wear parts (see TOOL MATERIALS).

In the 1920s automotive engineers, recognizing the interesting characteristics of powder metallurgy products, designed the oil-impregnated self-lubricated P/M bearing. Conventional P/M bearings can absorb from 10 to 30 vol % of additive-free nonautomotive engine oils. Impregnation takes place by soaking the part in heated oil, or by vacuum techniques. When friction heats the part, the oil expands and flows to the bearing surface. On cooling, the oil returns into the pores of the metal by capillary action (see BEARING MATERIALS). World War II brought about further commercial applications of powder metallurgy, especially for iron, steel, and copper base P/M parts and products, such as sintered iron bearings, paraffin-impregnated iron driving bands for military ammunition, magnets, bullet cores and fuse-body parts, and magnetic cores (4).

By the 1950s P/M parts were used in postage meters and home appliances. In the 1970s P/M superalloys for aerospace applications were used, followed by steel P/M forgings. The 1980s opened the way for rapid solidification processing, P/M tool steels, and metal injection molded parts.

U.S. annual production has grown from several thousand tons in the late 1940s to several hundred thousand tons in the 1990s. P/M parts, applications, and materials have expanded from simple shapes into complicated components used in high performance environments. The automotive field has been the principal driving force behind the rapid growth of P/M. Each passenger car or light truck contains more than 13.6 kg of P/M parts. Vehicular engines, transmissions, steering, brakes, suspension, seats, and windshield washers contain a myriad of P/M parts such as valves, levers, and gears.

In the 1980s and 1990s P/M forgings, P/M tool steels, rolled-compacted strip, dispersion-strengthened copper (qv), high strength aluminum alloys, metal-matrix composites (MMC) (qv), and materials such as aluminides (see GLASSY METALS) have all opened up markets for powder metallurgy (5).

The hot forging of P/M products, known as powder forging (P/F), is a recognized technology used to form parts for critical applications, and it is expected to continue to grow. As of this writing, automotive manufacturers are designing P/F connecting rods for operation in new engines. Metal injection molding (MIM) holds great promise for producing complex shapes in large quantities. Spray forming, a single-step gas atomization and deposition process, produces near-net shape products. In this process droplets of molten metal are collected and solidified onto a substrate. Potential applications include tool steel end mills, superalloy tubes, and aerospace turbine disks (6,7).

The market for lighter weight P/M materials such as aluminum and titanium aluminides is expected to grow, especially for uses in automobiles. P/M processing of titanium aluminides results in more consistent product quality than the conventional casting process, and offers novel alloy/microstructure possibilities and improved ductility. Processing trends include use of high (1200–1350°C) temperature sintering to improve mechanical properties of steel and stainless steel parts.

Advantages

The P/M process is cost effective in producing simple or complex parts at, or very close to, final dimensions at hourly production rates that range from a few hundred to several thousand parts. As a result, only minor or no machining is required. P/M parts also may be sized for closer dimensional control and/or coined for both higher density and strength (8). Both ferrous and nonferrous parts can be oil impregnated to function as self-lubricating bearings.

Most P/M parts weigh <2.27 kg (5 lbs), although parts weighing as much as 15.89 kg can be fabricated in conventional P/M equipment. Many early P/M parts such as bushings and bearings were very simple shapes, in contrast to the complex contours and multiple levels often produced economically in the 1990s. The P/M process is not shape-sensitive and normally does not require draft. Parts such as cams, gears, sprockets, and levers are economically produced.

Powder metallurgy is basically a chipless metalworking process that typically uses more than 97% of the starting raw material in the finished part (see METAL TREATMENTS). Thus P/M is an energy as well as a materials conserving process. The P/M process eliminates or minimizes machining and scrap losses; maintains close dimensional tolerances; permits use of a wide variety of alloy systems; produces good surface finishes; provides materials which may be heat-treated for increased strength or increased wear resistance; provides controlled porosity for self-lubrication or filtration; facilitates manufacture of complex or unique shapes which would be impractical or impossible with other metalworking processes; is suited to moderate-to-high volume component production requirements; offers long-term performance reliability in critical applications; and is cost effective (9).

Powder metallurgy competes with such conventional metal-forming processes as casting, stamping, screw-machining, forging, and permanent mold casting. Although most metals are more expensive in powdered form than in other forms, the powders themselves are less expensive to fabricate into finished precision metal products. Metal powders are precisely engineered materials that meet a wide range of performance requirements. These powders are available in numerous types and grades designed for the P/M process.

Powder Characteristics

Individual Particles. *Size.* The precise determination of particle size, usually referred to as the particle diameter, can actually be made only for spherical particles. For any other particle shape, a precise determination is practically impossible and particle size represents an approximation only, based on an agreement between producer and consumer with respect to the testing methods (see SIZE MEASUREMENT OF PARTICLES).

Particles smaller than 44 μm (-325 mesh) are called fines; 44 μm is the finest sieve used on a large-volume basis (U.S. Standard). Size determination of fines is described elsewhere (10,11).

Shape. Metal powder particles are produced in a variety of shapes, as shown in Figure 4. The desired shape usually depends to a large extent on the method of fabrication. Shape can be expressed as a deviation from a sphere

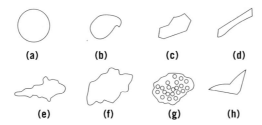

Fig. 4. Shapes of metal powder particles: (**a**) spherical; (**b**) rounded; (**c**) angular; (**d**) acicular; (**e**) dendritic; (**f**) irregular; (**g**) porous; and (**h**) fragmented.

of identical volume, or as the ratio between length, width, and thickness of a particle, as well as in terms of some shape factors.

Density. The density of a metal powder particle is not necessarily identical to the density of the material from which it is produced because of the particles' internal porosity. Therefore, it is difficult to determine the actual particle size and specific BET surface area.

Surface. Any reaction between two powder particles starts on the surface. The amount of surface area compared to the volume of the particle is, therefore, an important factor in powder technology. The particle–surface configuration, whether it is smooth or contains sharp angles, is another. The particle surface area depends strongly on the method of production, as shown in Table 1. The method of production usually determines the particle shape.

Microstructure. Powder particles may consist of a single crystal or many crystal grains of various sizes. The microstructure, ie, the crystal grain size, shape, and orientation, depend also on the method of powder fabrication. However, in many cases a correlation exists between particle size and grain size. In metal powders made by atomization, the smaller atomized particles solidify faster than the larger ones, and therefore in small-sized atomized powder particles there is usually less grain growth, producing smaller grains rather than coarse ones. The grain structure affects the activity of the powder particle, and to a certain extent the amount of material transported by grain-boundary diffusion.

A faster cooling rate increases dendrite nuclei formation, resulting in smaller dendrites. Small dendrites produce a microstructure that is easier to

Table 1. Particle Shapes and Surface Areas of Fabricated Powder Particles[a]

Process	Particle shape	Approximate surface area[b]
carbonyl process	uniform spherical	πd^2
atomization	round irregular spheroids	$1.5-2\pi d^2$
reduction of oxides	irregular spongy	$7-12\pi d^2$
electrolytic process	dendritic	
mechanical comminution		
crushing	angular	$3-4\pi d^2$
ball milling	flakes or leaves	varies over wide range

[a]Ref. 12.
[b]d = diameter.

homogenize during sintering. The finer the constituents, the more uniform the properties of the powder.

Surface Oxide Layer.   The possibility of reducing surface oxide layers in a reducing gas atmosphere (Cu, Fe, Ni, W, Mo, etc) depends on the type of metal. For a given thickness of the oxide layer, the amount of oxide in a powder changes with the particle size (Table 2). The thickness of the oxide layer on an individual particle depends on the conditions under which oxidation occurs. For metals where the oxide layers can be reduced during sintering, the type and amount of oxide crystals converted to metal crystals greatly affect the activity of the particle surface. This is because of the increased mobility of atoms during the conversion of the crystal structure from the oxide to the metallic state. In many cases, an optimum amount of oxide determines the sinterability of the respective powders and their physical properties after sintering.

Particle Activity.   Particle activity determines the type and rate of the reaction of a powder particle with its environment.

The total activity of a powder particle consists of the activity of the bulk and that of the surface. Both depend on the type and number of defects in the crystal structure. These defects are usually present in large numbers on the surface. Surface activity increases with increasing ratio of surface area-to-volume. Small particles, therefore, usually show greater activity than larger ones. The shape and surface configuration of the particle also influence the activity. Particles having sharp-angled corners are more active than particles having rounded smooth surfaces.

Because grain boundaries contribute to the activity of a particle, particles having small crystal grains are more active than particles consisting of larger grains. With respect to diffusion, the activity of particles consisting of single grains is much lower than that of multicrystal particles. The activity of a powder particle determines the rate of material transport by bulk and surface diffusion, the rate of adsorption (qv) and absorption (qv), and other reactions with the environment.

Powder Mass.   A mass of powder consists of a large quantity of particles. The characteristics of some iron powders are shown in Table 3. The most important properties of a good molding-grade powder are flow rate, particle size, and size distribution, apparent density, green strength, compressibility, and dimensional stability during sintering. A powder must flow well in order to fill all parts of the die cavity evenly and move through the automatic equipment (see POWDERS HANDLING). The particle size and size distribution must maximize the compact density. There is a close relationship between particle size distribution and such factors as powder flow, apparent density, and compressibility. Because the amount of powder needed for each compact is charged to a cavity of constant

Table 2. Particle Thickness and Oxide Content of Aluminum Flake Powder

Particle thickness, μm	Oxide content, vol %	Particle thickness, μm	Oxide content, vol %
0.1	30	100	0.03
1.0	3	1000	0.003
10	0.3		

Table 3. Characteristics of Iron Powders[a]

Fabrication method	Average particle size, μm	Particle size distribution,[b,c] %				Specific surface area,[b] cm^2/g	Apparent density,[b] g/cm^3
		>100 μm (150)	>74–<100 μm (200 to −150)	>44–<74 μm (325 to −200)	<44 μm (−325)		
reduction	68	28.5	15.5	54.5	1.5	516	3.03
	51		6.5	81.5	12.0	945	2.19
	6	3.5	2.0	13.5	81.0	5160	0.97
from iron carbonyl	7	2.5	0.1	1.0	96.4	3460	3.40
atomization	75	28.3	20.4	29.5	21.8	355	2.90
	83	30.7	21.5	26.7	21.1	320	2.95
	89	35.4	25.5	25.3	13.8	284	3.00

[a]Refs. 4–12.
[b]Values given are typical of iron powder samples.
[c]Numbers in parentheses represent mesh sizes.

volume, the apparent density becomes extremely important. Although changes in the depth of a die cavity can be made with ease, it is most desirable that the powder has uniform apparent density batch-to-batch and hour-to-hour.

High compressibility is another desired property of metal powders. Compressibility is the density to which a powder may be pressed at any given pressure. As the compressibility of a powder increases, the pressure needed to obtain any given density decreases. Lower pressures result in lower tool and machine wear. Under the same compacting conditions, higher compressibility powders compact to a higher green strength. This is desirable because the compact must have enough strength to be transported either mechanically or by hand. Green strength of a pressed compact, like density, is generally considered a property of the powder.

Average Particle Size.   Average particle size refers to a statistical diameter, the value of which depends to a certain extent on the method of determination. The average particle size can be calculated from the particle-size distribution (see SIZE MEASUREMENT OF PARTICLES).

Particle Size Distribution.   For many P/M processes, the average particle size is not necessarily a decisive factor, whereas the distribution of the particles of various sizes in the powder mass is. The distribution curve can be irregular, show a rather regular distribution with one maximum, have more than one maximum, or be perfectly uniform.

Specific Surface.   The total surface area of 1 g of powder measured in cm^2/g is called its specific surface. The specific surface area is an excellent indicator for the conditions under which a reaction is initiated and also for the rate of the reaction. It correlates in general with the average particle size. The great difference in surface area between 6-μm reduced iron powder and 7-μm carbonyl iron powder (Table 3) cannot be explained in terms of particle size, but mainly by the difference between the very irregular-shaped reduced and the spherical carbonyl iron powders.

Determination of the specific surface area can be made by a variety of adsorption measurements or by air-permeability determinations. It is customary to calculate average particle size from the values of specific surface by making assumptions regarding particle size distribution and particle shape, ie, assume it is spherical.

Apparent Density.   This term refers to the weight of a unit volume of loose powder, usually expressed in g/cm^3 (13). The apparent density of a powder depends on the friction conditions between the powder particles, which are a function of the relative surface area of the particles and the surface conditions. It depends, furthermore, on the packing arrangement of the particles, which depends on the particle size, but mainly on particle size distribution and the shape of the particles.

The characteristics of a powder that determine its apparent density are rather complex, but some general statements with respect to powder variables and their effect on the density of the loose powder can be made. (1) The smaller the particles, the greater the specific surface area of the powder. This increases the friction between the particles and lowers the apparent density but enhances the rate of sintering. (2) Powders having very irregular-shaped particles are usually characterized by a lower apparent density than more regular or spherical ones. This is shown in Table 4 for three different types of copper powders having identical particle size distribution but different particle shape. These data illustrate the decisive influence of particle shape on apparent density. (*3*) In any mixture of coarse and fine powder particles, an optimum mixture results in maximum apparent density. This optimum mixture is reached when the fine particles fill the voids between the coarse particles.

Tap Density.   Tapping a mass of loose powder, or more specifically, the application of vibration to the powder mass, separates the powder particles intermittently, and thus overcomes friction. This short-time lowering of friction results in an improved powder packing between particles and in a higher apparent density of the powder mass. Tap density is always higher than apparent density. The amount of increase from apparent to tap density depends mainly on particle size and shape (see Table 4).

Flow.   The free flow of a powder through an orifice depends on the orifice which is standardized for the testing of the powder (*14*). Flow, therefore, depends not only on friction between powder particles, but also on friction between the particles and the wall of the orifice. Flow is usually expressed by the time necessary for a specific amount of powder (usually 50 g) to flow through the orifice.

Inasmuch as friction conditions determine the flow characteristics of a powder, coarser powder particles of spherical shape flow fastest and powder particles of identical diameter but irregular shape flow more slowly. Finer particles may start to flow, but stop after a short time. Tapping is needed in order to start the flow again. Very fine powders (<ca 20 μm) do not flow at all. Addition of fine powder particles to coarser ones may increase the apparent density, but usually decreases the flow quality. Metal powders having a thin oxide film may flow well. When the oxide film is removed and the friction between the particles therefore increases, these powders may flow poorly.

The free flow of a powder is necessary for automatically filled compacting dies. Powders having low flow rates need vibratory filling in order to overcome friction. Powders that do not flow at all can be used only for manual filling of the die cavity.

Table 4. Effect of Particle Shape on Apparent and Tap Density[a]

Particle shape	Density, g/cm^3		Increase, %
	Apparent	Tap[b]	
spherical	4.5	5.3	18
irregular	2.3	3.14	36
flake	0.4	0.7	75

[a]Ref. 4.
[b]Density after tapping the sample.

Manufacture

The manufacture of metal in powder form is a complex and highly engineered operation. It is dominated by the variables of the powder, namely those that are closely connected with an individual powder particle, those that refer to the mass of particles which form the powder, and those that refer to the voids in the particles themselves. In a mass of loosely piled powder, $\geq 60\%$ of the volume consists of voids. The primary methods for the manufacture of metal powders are atomization, the reduction of metal oxides, and electrolytic deposition (15,16). Typical metal powder particle shapes are shown in Figure 5.

In atomization, a stream of molten metal is struck with air or water jets. The particles formed are collected, sieved, and annealed. This is the most common commercial method in use for all powders. Reduction of iron oxides or other compounds in solid or gaseous media gives sponge iron or hydrogen-reduced mill scale. Decomposition of liquid or gaseous metal carbonyls (qv) (iron or nickel) yields a fine powder (see NICKEL AND NICKEL ALLOYS). Electrolytic deposition from molten salts or solutions either gives powder directly, or an adherent mass that has to be mechanically comminuted.

Metals can be precipitated from the liquid or gas phase. For example, nickel ammonium carbonate gives nickel powder when subjected to hydrogen in an autoclave. Copper, cobalt, molybdenum, and titanium powders can also be formed by precipitation.

In spinning or rotating electrode metal powder formation, molten metal droplets are centrifuged from a spinning electrode in a closed chamber. In high

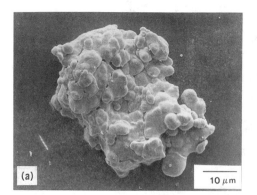
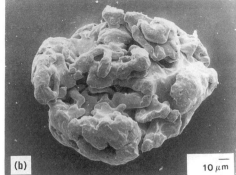

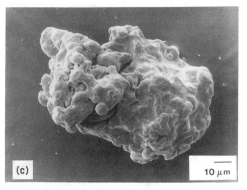

Fig. 5. Metal powder particle shapes: (**a**) atomized copper; (**b**) sponge iron; and (**c**) atomized iron.

energy impaction, brittle coarse shapes are impinged against a tungsten carbide target at high velocity and ambient or lower temperatures.

Mechanical comminution may be used to form metal powders. Relatively coarse particles are produced by machining, whereas ball mills, impact mills, gyratory crushers, and eddy mills give fine powders of brittle materials.

Condensation of metal vapors followed by deposition on cooler surfaces yields metal powders as does decomposition of metal hydrides. Vacuum treatment of metal hydrides gives powders of fine particle size. Reaction of a metal halide and molten magnesium, known as the Kroll process, is used for titanium and zirconium. This results in a sponge-like product.

Using rapid solidification technology (RST), molten metal is quench cast at a cooling rate up to $10^6 °C/s$ as a continuous ribbon. This ribbon is subsequently pulverized to an amorphous powder. RST powders include aluminum alloys, nickel-based superalloys, and nanoscale powders. RST conditions can also exist in powder atomization.

Other methods of metal powder manufacture are also employed for specific metals. Selective corrosion of carbide-rich grain boundaries in stainless steel, a process called intergranular corrosion, also yields a powder.

Processing

Consolidation. Metal powders are consolidated by heat or by pressure followed by heat, or by heating during the application of pressure (17). Consolidation produces a coherent mass of definitive size and shape for further working, heat treating, or use as is.

During pressure application, the powder particles are first rearranged or packed. Then elastic and plastic deformation takes place, and finally, the particles are cold-worked. Many attempts have been made to develop a mathematical relationship between compacting pressure and density and strength of the compact and strength of the sintered compact (18–21). Such a relationship has not been found to be of great practical significance. Particles bond and form a coherent mass under pressure because of liquid surface cementation; interatomic forces such as surface adhesion, cold welding, and surface tension; and mechanical interlocking of particles (22).

Plasticity of the metal crystals plays a dominant role in interparticle bonding. This is a characteristic of each metal. It is affected to a large extent by the condition of the individual powder particles. Gold, silver, lead, and iron are highly ductile (plastic) metals, whereas chromium and tungsten deform with great difficulty. Plasticity also depends on nature and history of the powder, impurities present (especially on the surface), and friction conditions between the particles. Powders that exhibit a high degree of plastic deformation form many areas of metal contact and therefore many interparticle bonds, whereas powders composed of harder particles form relatively few such bonds.

Probably the most important powder property governing the formation of atomic bonds is the surface condition of the particles, especially with respect to the presence of oxide films. If heavy oxide layers are present, they must be penetrated by projections on the particles. This results in only local rather than widespread bonding. A ductile metal such as iron which has a heavy oxide layer may not form as strong or as many bonds as a less ductile metal.

Mechanical Interlocking.   The fact that irregularly shaped powder particles form denser and stronger compacts under pressure than regularly shaped powder particles leads to the theory of interlocking particles. Interlocking is probably the principal strengthening mechanism in the compact prior to sintering and is probably also responsible for providing increased surface contact for cold welding. This mechanism plays an important role in the strengthening of compacts made from metals such as tungsten and chromium, which normally are not plastic at room temperature.

The characteristics of a pressed compact are influenced by the characteristics of the powder: rate and manner of pressure application, maximum pressure applied and for what period of time, shape of die cavity, temperature during compaction, additives such as lubricants and alloy agents, and die material and surface condition. The effect of various compaction variables on the pressed compact are shown in Figure 6.

Consolidation Techniques.   *Unidirectional Compaction.*   Uniaxial pressing of metal powders in a die of specific dimensions and configuration is the most frequently used technique for the consolidation of powders in the manufacture of P/M products. Powder flows automatically into a die cavity; the bottom punch

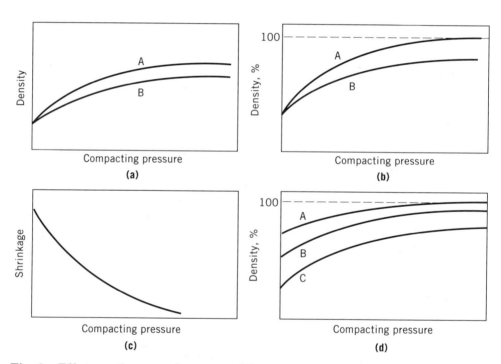

Fig. 6.   Effects on the pressed compact of (**a**) speed, where A is low and B, high speed compacting; (**b**) powders, where A is soft and B, hard powders; (**c**) dimensional change after sintering; and (**d**) sintering temperatures, where A is high, B, medium, and C room temperature.

or punches act as its bottom. An upper punch seals the top and pressure can be applied parallel to the direction of powder flow into the cavity by forcing the punches together. In single action either the bottom or the top punch is held rigid, whereas the other corresponding punch moves to press the powder. In double action both punches move. Double-acting punches produce a more uniformly dense compact.

After pressure application, the top punch is removed and the compact is ejected from the cavity by the bottom punch. The cavity is then refilled and is ready for another charge. This cycle is repeated automatically at a rate that varies with the part and size and the complexity and flowability of the powder. Pressing equipment producing relatively small, simple parts can operate at up to 200 parts/min. Rotary presses with multiple die sets are even faster. Table 5 gives the ranges of pressures used for various materials during die compaction.

Compacting tools must be properly designed, constructed, and fitted to the press. These may be made of heat-treated steel or cemented carbide, depending on the economics and number of parts to be produced. Carbide tools are more expensive; however, they can be used much longer than steel tools.

Isostatic Pressing. Isostatic pressing generally is used to produce P/M parts to near-net sizes and shapes of varied complexity (8,23). Unlike conventional press compaction or molding, isostatic pressing is performed in a pressurized fluid such as oil, gas, or water. As shown in Figure 7, a flexible membrane or hermetic container surrounds the powder mass and provides a pressure differential between its contents and the pressurizing medium. Among the benefits of isostatic pressing are capability to produce complex shapes; minimized expensive powder input; applicability to difficult-to-compact materials; and uniform density and properties of products.

Cold Isostatic Pressing. Cold or room temperature compaction, known as cold isostatic pressing (CIP), is carried out in liquid systems at pressures commonly reaching 414 MPa (60,000 psi). Metal powder can be packed into

Table 5. Pressure Requirements and Compression Ratios[a] for Powder Products

Type of compact	Pressure, MPa[b]	Compression ratio
brass parts	413–690	2.4–2.6:1
bronze bearings	207–276	2.5–2.7:1
carbon products	138–165	3.0:1
carbides	138–414	2.0–3.0:1
iron bearings	207–345	2.2:1
iron parts		
low density	345–483	2.0–2.4:1
medium density	483–552	2.1–2.5:1
high density	483–828	2.4–2.8:1
iron powder cores	138–690	1.5–3.5:1
tungsten	69–138	2.5:1
tantalum	69–138	2.5:1

[a]Compression ratio is the dimensional relationship between the loose and compacted powder at a given compacting pressure.
[b]To convert MPa to psi, multiply by 145.

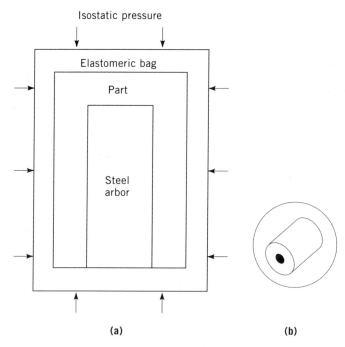

Fig. 7.   (**a**) Isostatic compacting and (**b**) the shape of the completed part.

complex shaped rubber or elastomeric molds before compacting. Free of die frictional forces, the powder compact reaches a higher and more uniform density at its external surfaces. Powders having spherical or rounded particles are not cold compacted because of the inability to form a strong green body. Sintering can be performed by any of the conventional methods.

   Hot Isostatic Pressing.   Hot isostatic pressing (HIP) is performed in a gaseous (inert argon or helium) atmosphere contained within the pressure vessel. Usually the gaseous atmosphere as well as the part to be pressed are heated by a furnace within the vessel (8,24,25). Common pressure levels extend upward to 103 MPa (15,000 psi) and combined temperatures to about 1250°C. Processing volumes reach to diameters of 1.2 m by 2.7 m long. Higher pressures and temperatures may be used but at the expense of reduced work volumes. The powder to be hot compacted is hermetically sealed, usually within an iron or glass container or envelope which deforms plastically at elevated temperatures. Prior to sealing, the container is evacuated to eliminate residual gases in the powder mass.

   In contrast to the cold isostatic pressing process, the hot process can readily employ powders having spherical or noninterlocking particles. The powder is simply poured in and vibration packed into a container of desired shape. The powder mass is then simultaneously compacted and bonded during the treatment.

   Densification to what is practically full density is achievable for most materials. The resulting mechanical properties are equivalent to wrought parts in similar condition. In some materials, the properties of the HIP product are superior because of reduced anisotropy.

Depending on the material being compacted, removal of the HIP container may be by chemical leaching, machining, or other mechanical methods. As for cold isostatic pressing, mandrels can form internal contours. Parting agents are generally used to prevent bonding. Differential thermal contraction eases mandrel removal for simple, regular shapes. Mandrels with locking-in features are removable by methods used for container removal (26).

Powder Rolling. Powder conveyed either horizontally or vertically through a set of steel rolls is compacted in the roll gap and emerges as a porous sheet (27). This sheet is then sintered, rerolled (warm or cold, depending on the material), annealed, and rolled again in a finishing operation. The rolling mills and furnaces are arranged for continuous production. Green strength of the sheet, as it emerges from the roll gap, is the limiting factor for both the thickness and width of the sheet. Roll size, roll gap, roll speed, and rate of powder feed have to be controlled. Compaction occurs essentially in only one direction because very little pressure is transmitted laterally.

Assuming a cost-effective starting material, powder rolling eliminates much of the equipment needed for the usual melting, casting, and rolling to produce thin sheet. Sheet can be rolled closely to finished size with a minimum loss of material. Most scrap generated by this process can be reclaimed as powder. This technique also permits the production of clad materials by using a metal strip as the carrier for the powder through the roll gap. Roll compacting of nickel powder produces strips from which blanks are made for coins. In the United States, large quantities of nickel and nickel–iron–cobalt powders are rolled into strip for glass-sealing alloys and various electronic applications. Multiple-ply strips may also be rolled directly by metallurgical sintering of the layered components in the roll gap. This process is used commercially to make tri-ply strip box composite bearings.

P/M Forging. Even after conventional repressing of a P/M component, it is still difficult to increase density above 95%. However, full density in a P/M part improves its properties. Hot isostatic pressing in autoclaves works well, especially for titanium and superalloy components, but the capital equipment is expensive and production rates are slow.

For ordinary materials and higher production rates, P/M forging can be used (26,28). After parts are compacted and sintered to medium density, they are reheated, lubricated, and fed into a hot-forming or P/M-forging press. The part is formed by one stroke of the press in a closed precision die. A typical hot-forming press setup includes die sets, automatic die cooling and lubrication, transfer mechanism, an induction heating unit for preforms, and controls.

P/M forged parts with dimensions up to 120 cm^2 and weighing up to 4.5 kg are produced on hot-forming presses in the 500–2500 t range. Usually the process is used for parts weighing 0.35–4.5 kg. Properties are generally equal or superior to those of conventional forgings. P/M forgings usually have a minimum density of 99% of theoretical and depending on alloy composition, can exhibit tensile strengths >1.38 GPa (200,000 psi) after heat treatment. Oil quenching and tempering at 230°C, for example, produces tensile strengths of ca 1.5 GPa (215,000 psi) (ultimate) and ca 1.25 GPa (185,000 psi) (yield) with 7%

elongation and RC 43 hardness (qv). Heat treating at 650°C reduces strength and hardness values by 50% but triples elongation. The cups and cones for tapered roller bearings are forged from grade 4600 powder (2 wt % nickel, 1/2 wt % molybdenum alloy), which corresponds to AISI 4600 grade steel. These perform at least equal to and up to eight times longer than the same part produced from wrought steel.

The automotive industry is the principal user of P/M forgings, primarily for transmission and differential components, but also for engine parts. Other markets are in power tools and farm machinery. Cost effectiveness is generally the reason for substituting P/M forgings for conventionally forged, cast, or machined parts.

High Energy Rate Compaction. Metal powders can be rapidly compacted in rigid dies by the application of high pressures at high speed (29). Special presses employ a pressure upsetting system for rapid movement of the compacting tools. Explosives are also used. Single- and double-acting presses are capable of generating pressures on the order of several GPa (10^6 psi) for a time. The rams in presses employing explosives move at the rate of more than 1000 m/s, causing compaction to occur in microseconds. Shock fronts from explosives generate pressures close to 14 GPa (2×10^6 psi).

Higher energy rate-forming techniques have been used mainly for laboratory studies or to produce compacts with special properties, but these techniques are not of commercial interest.

Slip Casting. Slip casting of metal powders into useful articles is an interesting process but has only limited industrial application (30,31). It is sometimes used to produce large, very complicated parts from refractory metals (see REFRACTORIES).

Slip casting of metal powders closely follows ceramic slip casting techniques (see CERAMICS). Slip, which is a viscous liquid containing finely divided metal particles in a stable suspension, is poured into a plaster-of-Paris mold of the shape desired. As the liquid is absorbed by the mold, the metal particles are carried to the wall and deposited there. This occurs equally in all directions and equally for metal particles of all sizes which gives a uniformly thick layer of powder deposited at the mold wall.

Vibratory Consolidation. Powders are vibrated in a mold or other container in which they will be sintered, or in a metal container that will be used for extrusion or other metalworking process (31). Vibratory consolidation produces packings of UO_2 particles up to 95% of theoretical density.

Hot Pressing. Hot pressing may be used either to consolidate a powder that has poor compactability at room temperature, or to combine compaction and sintering in one operation. The technique is essentially the same as described for unidirectional die compacting. The powder is heated by either heating the entire die assembly in a furnace or by induction heating. In most instances, a protective atmosphere must be supplied.

Hot pressing produces compacts that have superior properties, mainly because of higher density and finer grain size. Closer dimensional tolerances than can be obtained with pressing at room temperature are also possible. Hot pressing is used only where the higher cost can be justified. It has been useful in producing reactive materials. One use is the combination of P/M and composites to produce hot-pressed parts that are fiber reinforced. However, the technique is mainly employed in the laboratory.

Extrusion, Swaging, or Rolling. Metal powders first formed into ingots by isostatic pressing, or metal powders encased in a suitable container may be subjected to any number of operations such as extrusion, swaging, or rolling. In the absence of a container, a protective atmosphere is essential. This type of consolidation is usually performed at elevated temperature. For contained powder, the case becomes a sheath during working, which is subsequently removed by either machining or chemical methods. Canned extrusion is used commercially for tool steels, superalloys, and beryllium.

Metal Injection Molding. Metal injection molding (MIM) offers a manufacturing capability for producing complex parts in large quantities (8,26). The process utilizes fine metal powders, typically <20 μm, which are intimately mixed with various thermoplastics, waxes (qv), and other ingredients. In contrast to conventional powder metallurgy, these polymeric binder materials may comprise as much as 40 vol % of the mixture. The resulting feedstock is granulated and can then be fed into a conventional injection molding machine. Multiple cavity tooling is commonly employed. Most but not all of the binder is then removed from the molded (green) components by thermal or solvent processing, or a combination of both. The exact method chosen is dependent on the binders used for molding and the cross sections of the part. Following binder removal, parts are sintered in either an atmosphere or vacuum, during which the remaining traces of binder are removed. Sintering temperatures are usually in excess of 1260°C. Final relative densities are generally 95–98% with interconnected porosity <1%.

Parts weighing up to 100 g are commonly produced. Cross sections are generally under 6.35 mm. Parts are not restricted to this combination of mass/cross section, however, and larger parts are in production. Processing is normally to a tolerance of ±0.3%, although specific dimensions may be held as close as ±0.1%. Common secondary operations, such as plating, coloring, or heat treating, are often employed. Because interconnected porosity is so low, parts do not have to be resin impregnated for plating and close control over case depth is possible in carburizing.

The injection molding process eliminates the restriction of straight-sided components required when parts are ejected from a die, and offers opportunities for external undercuts and threads. A wide variety of alloys can be processed, including alloy steels and stainless steels. Material properties of injection molded parts are available (32).

Process details may be summarized: powder sizes are fine (usually <20 μm); low (generally <69 MPa (10,000 psi)) injection pressure; low (ca 149°C) molding temperature; shrinkage (molded part to finished size) typically 20%; final part densities are usually 95–98%+ of maximum pore-free density; and ductility is exceptionally high, elongation values are ≥30%.

Compacting Lubricants. The surface area of most moldable metal powders is in the range of 500–700 cm^2/g. Finer powders can have a surface area as high as 1500 cm^2/g (33). A very large number of individual particles is involved. For example, 1 cm^3 uniformly filled with 2-μm spherical particles having a surface area of 1200 cm^2/g contains ca 1.2 × 10^9 particles. Because of this large surface area, a considerable amount of friction has to be overcome during powder consolidation. To one degree or another, friction is present in all consolidation methods.

Dry lubricants are usually added to the powder in order to decrease the friction effects. The more common lubricants include zinc stearate [557-05-1], lithium stearate [4485-12-5], calcium stearate [1592-23-0], stearic acid [57-11-4], paraffin, graphite, and molybdenum disulfide [1317-33-5]. Lubricants are generally added to the powder in a dry state in amounts of 0.25–1.0 wt % of the metal powder. Some lubricants are added by drying and screening a slurry of powder and lubricant. In some instances, lubricants are applied in liquid form to the die wall.

Lubricants protect die and punch surfaces from wear and burn-out of the compact during sintering without objectionable effects or residues. They must have small particle size, and overcome the main share of friction generated between tool surfaces and powder particles during compaction and ejection. They must mix easily with the powder, and must not excessively impede powder flow (see LUBRICATION AND LUBRICANTS).

Sintering. Basically a solid-state process, sintering transforms compacted mechanical bonds between the powder particle into metallurgical bonds (23,34–38).

During the sintering treatment, which usually occurs below the melting point of the metal powder except for liquid-phase sintering of some powder mixtures, material transport takes place in the solid state. This results in some changes of the properties of the compacted powder, as shown in Figure 8. With increasing temperature and time, the strength of the powder mass increases, electrical resistivity and porosity decrease, and density, therefore, increases. The grain structure also undergoes some changes, and recrystallization and grain growth occur. In order to avoid oxidation of the metal powder mass during the high temperature treatment, either a neutral or a reducing atmosphere is provided. The movement or transport of material during sintering is caused by surface diffusion, volume or lattice diffusion, grain-boundary diffusion, evaporation and condensation, and plastic or viscous flow (37,39). Probably several

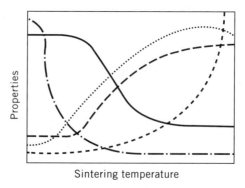

Fig. 8. Effect on sintering on (——) porosity, (— — —) density, (— •—) electrical resistivity, (· · ·) strength, and (– – –) grain size.

of these mechanisms act simultaneously, depending on the type of powder, its particle size and shape, and especially the temperature.

Diffusion is based mainly on the diffusion of vacancies; grain boundaries may act as sinks for these vacancies. This vacancy movement and annihilation cause the porosity of the powder compact to decrease during sintering.

Pressed powder (green) compacts are characterized by a porosity, or total pore volume, of approximately 10–40%. The number and size of the pores can be correlated with the size and shape of the powder particles from which the compact has been prepared, and the pressure applied during compacting. During sintering, the porosity undergoes a number of changes: with increasing sintering temperature or time, the total porosity decreases, the pores that originally are irregular or angular in shape become spherical; the average pore size becomes larger; the total number of pores decreases; the smaller pores disappear first; and the number of larger pores increases slightly.

The decrease of porosity during sintering results in shrinkage and, accordingly, in the densification of the powder compact. Density and densification rate during sintering are strongly affected by the particle size, the pressure applied during compacting (Fig. 9), and the sintering temperature and time. They depend further on the type of metal from which the powder has been prepared. The average sintering temperatures for various types of metals depend on their melting points (Table 6).

Liquid-Phase Sintering. Sintering in the liquid state refers to the sintering of a powder mixture of two or more components, of which at least one has a melting temperature lower than the others. The sintering temperature is then selected in such a manner that a liquid phase is formed in which the solid powder particles of the other components rearrange. A high density powder compact is the result.

The properties of the sintered product depend to a large extent on the surrounding atmosphere. There are reducing atmospheres (hydrogen, dissociated ammonia, carbon monoxide, nitrogen plus hydrocarbons, or hydrocarbons) and neutral atmospheres (vacuum, argon, helium, or nitrogen). Metal powders susceptible to surface oxidation must be sintered in a reducing atmosphere in order

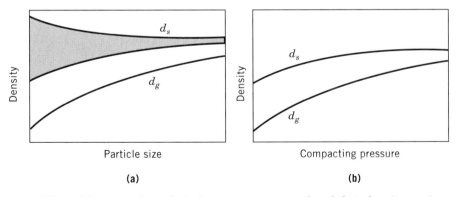

Fig. 9. Effect of density, where d_g is density as compacted and d_s is density as sintered, for (**a**) particle size and (**b**) compacting pressure.

Table 6. Material Sintering Temperature and Time[a,b]

Material	Temp, °C	Time, min
bronze	760–871	10–20
copper	843–899	10–30
brass	843–899	10–30
iron, iron–graphite, etc	1010–1149	15–30
nickel	1010–1149	30–45
stainless steel	1288–1316	20–30
Alnico magnets	1204–1302	121–150
90% tungsten–6% nickel–4% copper	1343–1593	10–120
tungsten carbide	1426–1482	20–30
molybdenum	2054	120[c]
tungsten	2343	480[c]
tantalum	2400[c]	480[c]

[a]Ref. 23.
[b]In high heat chamber.
[c]Value given is approximate.

to remove the oxide films. The selection of the atmosphere depends on the type of material to be sintered, whether a reaction between the metal and the atmosphere is desired, and the type of reaction desired. Cost is also a factor.

Both batch and continuous furnaces may be employed. The maximum temperature that can be reached in a sintering furnace depends on the furnace and the heating methods (Table 7) (23).

Infiltration. In infiltration, the pores of a sintered solid are filled with a liquid metal or alloy (24). The most common application is the infiltration of a porous sintered steel matrix with copper, to give a copper-infiltrated P/M part. The aims of infiltration are to obtain higher strength and a pore-free structure for plating, machining, and sealing of pressure-tight applications. The

Table 7. Heating Methods and Element Material Temperature Limits[a]

Method	Temperature, °C
gas	1150
resistance wire heating	
Nichrome A	1175
Kanthal A	1400
molybdenum	1550
tungsten	2500
silicon carbide	1538
carbon short-circuit tube[b]	3300
induction heating	3000

[a]Temperature limit depends on furnace atmosphere.
[b]Current goes directly through the furnace tube and no other heating elements are necessary.

liquid and the solid must not react to form a solid compound or alloy having a specific volume as great as or greater than their combined preinfiltration specific volumes. The porosity in the porous matrix should be interconnected. Ideally, the matrix material should be insoluble in the liquid infiltrant. Simple liquid-phase sintering does not give pore-free structures; infiltration does, provided the pores are interconnected.

The process is used for ferrous P/M structural parts that have densities of at least 7.4 g/cm^3 and mechanical properties superior than those of parts that have been only compacted and sintered. Depending on the application, the porous matrix may be infiltrated only partially or almost completely. Copper-base alloy infiltrants have been developed to minimize erosion of the iron matrix.

In one method, the matrix component has already been sintered and the green compact of infiltrant powders is positioned on or underneath it. The assembly is then heated to the infiltration temperature. Alternatively, sintering of the matrix and infiltration may be combined into one operation by positioning the green compact of the infiltrant on the green compact of the iron or steel matrix. This operation is called sintrating. By controlling the rate of heating, the matrix is adequately sintered by the time the melting point of the infiltrant is reached. If only part of the matrix is to be infiltrated, eg, the teeth of an infiltrated gear, the matrix may be positioned in a graphite container in which space is provided adjacent to the gear teeth to be preferentially infiltrated. The space is then filled with the appropriate amount of infiltrant in powder form. Tungsten for electrical contacts is often infiltrated in this manner with copper or silver.

Post-Sintering Treatments. The sintering process concludes the powder metallurgy processes of production and consolidation. However, some P/M parts may require a number of further operations.

Working Treatments. Restricted plastic deformation takes place entirely within the confines of a closed die cavity. A sintered part may be placed in the die cavity and pressure applied to the part. This pressure generally is of the same magnitude as the original compaction pressure. This second application of pressure can be categorized as follows.

Sizing. The desired size is obtained by a final pressing of a sintered compact. During sintering, the compact may have expanded, shrunk, or changed dimensions slightly, which is corrected by the sizing operation.

Repressing. Pressure is applied to a previously pressed and sintered compact, usually for the purpose of improving a physical property, such as tensile strength, hardness, or density.

Coining. The sintered compact is pressed to obtain a definite surface configuration which changes the shape of the article. In some instances, the sintered piece is used as a blank with most of the surface configuration produced by coining such as in striking coins or medallions.

In general, sizing, repressing, and coining are performed at room temperature. With elevated temperatures, a protective atmosphere must be provided.

Unrestricted plastic deformation includes all of the metalworking procedures generally applied to cast metals. P/M ingots made by die pressing and sintering, or isostatic pressing and sintering, may be forged, swaged, drawn, rolled, or extruded. These processes may also be applied to a mass of loose or loosely sintered powder. In most instances, the P/M ingot being worked has lower density and lower workability than cast ingots; temperatures are generally higher. In many cases a neutral atmosphere is provided by enclosing the operation in a protective gas or in a sheath and providing the proper atmosphere within the sheath.

Heat Treatments.   If the inherent porosity is taken into consideration, heat treatments performed on P/M parts do not differ substantially from the same treatments performed on cast or wrought metal. Results are also similar.

The most common heating operation is resintering, usually under conditions similar to those of the first sintering operation, but at higher temperature. Resintering relieves stress or removes the effects of cold work imparted during coining or repressing. Resintering is also undertaken for further densification. Cold-working, as in repressing or coining, reduces porosity and ruptures oxide films within the compact, which creates new metal-contact areas. A resumption of sintering enables diffusion to proceed with fewer obstructions at a lower temperature. In general, it increases ductility and density with some loss in strength; however, hardness, tensile strength, and ductility are usually much improved.

Sintered metals may be softened or hardened by a number of procedures common in treating cast or wrought metals. Pure copper or silver may be annealed by heating above the recrystallization temperature. Certain copper-base alloys are precipitation hardened, such as copper–beryllium, copper–chromium, and copper–zirconium by solution-treatment quenching and coining (see COPPER ALLOYS). Precipitation hardening is practiced on aluminum alloys that contain up to 5 wt % copper and 1.5 wt % silicon, or 0.5 wt % magnesium and 0.5 wt % manganese.

P/M steels can be heat treated in the same manner as cast or wrought steels. They may be austenitized, quenched, and tempered. Surface hardening includes pack or gas carburization or nitriding, ie, heating in a nitrogen-containing atmosphere. Because of the greater amount of exposed surface area in the form of porosity, a protective atmosphere is needed (see METAL SURFACE TREATMENTS).

Finishing.   Finishing treatments include shaping operations such as machining, broaching, sizing, burnishing, grinding, straightening, deburring, and abrading, as well as surface treatments, such as steam oxidizing, coloring, plating, impregnating, dipping, or spraying.

Steam treatment imparts increased corrosion resistance for ferrous P/M parts. The parts are heated to 400–600°C and then exposed to superheated steam. After cooling, the parts are usually oil dipped to further increase corrosion and wear resistance, and to enhance appearance (see CORROSION AND CORROSION CONTROL). Heat treated parts are seldom steam treated because annealing reduces hardness and tensile strength.

The blue-black iron oxide formed in this process fills some of the interconnecting porosity and much of the surface. Hence the density is increased, resulting in higher compressive strength. Furthermore, the oxide coating increases hardness (qv) and wear resistance.

Bluing.   Ferrous P/M parts can be colored by several methods. To enhance corrosion resistance, the parts are blued, ie, blackened, by heating in a furnace to the bluing temperature and then cooled. Oil dipping deepens the color and improves corrosion resistance. A dry-to-touch oil may leave a dry film on the parts.

Ferrous P/M parts can also be blackened chemically, using one of several commercial liquid salt baths. If the parts are below a density of 7.3 g/cm^3, entrapment of salt is avoided by impregnating the parts before blackening with a resin that does not break down in the bath. Nickel- or copper-bearing parts tend to adversely affect most blackening baths. Furthermore these materials seriously affect color. As for furnace blackening, an oil dip improves appearance and corrosion resistance. However, the process causes a slight change in size and makes the parts more brittle and more difficult to machine.

Plating. All types of plating in general use, including copper, nickel, chromium, cadmium, and zinc, can be applied on P/M parts. High (7.2 g/cm^3) density and infiltrated parts can be plated by the same methods as wrought parts. To avoid entrapment of plating solutions in the pores, lower density parts should be sealed with resin. Before resin impregnation, the oil must be removed from all pores and surfaces. Electroless nickel plating can also be used, and peen (mechanical) plating is applicable to nonimpregnated ferrous parts with a density of 6.6–7.2 g/cm^3 (see ELECTROLESS PLATING; ELECTROPLATING).

Economic Aspects

Powder metallurgy is a worldwide industry. The P/M parts and products industry in North America has an estimated annual sales of about $3 billion. It is comprised of approximately 150 companies that make conventional P/M parts and products from iron and copper-base powders plus about 50 other companies that make specialty P/M products. Specialty products include superalloys, tool steels, porous products, friction materials, strip for electronic applications, high strength permanent magnets, magnetic powder cores and ferrites (qv), tungsten carbide cutting tools and wear parts, tungsten alloy parts, and metal injection molded parts.

The United States has the largest metal powder producing and consuming industry of any country. The value of U.S. metal powder shipments, including paste and flake, was $1.243 billion in 1992. Approximately 307,000 t of iron and steel powder and 21,000 t of copper and copper-base powders were shipped in North America in 1994. Estimated annual world metal powder production exceeds 950,000 t.

About 88% of iron powder production is used in the manufacture of P/M parts and friction materials. Detailed statistical data are available (16). About 86% of the copper and copper-base powder produced is used in the production of self-lubricating bearings and P/M parts.

In the production of automotive parts, powder metallurgy offers strong competition to fine-blanking, stamping, screw machining, gray iron, and die casting. Automotive applications of P/M parts are expected to continue to dominate the industry through the end of twentieth century. Auto makers are specifying more P/M parts in engines, transmissions, and subassemblies. P/M also offers weight reduction in parts, a characteristic that is becoming the single most important way to boost auto fuel economy.

Health and Safety

Metal powders possess an immensely high ratio of specific surface area to volume. This characteristic contributes to several potentially hazardous properties such as pyrophoricity, explosiveness, and toxicity (40–42). The problems associated with the fine particles can be minimized or eliminated with proper handling and good housekeeping procedures, such as storage in appropriate containers, processing in sparkproof equipment, avoiding exposure to open flames, and minimizing airborne particulates (see POWDERS HANDLING). Investigations of various commercial metal powders with respect to ignition temperature, minimum explosive concentration, minimum ignition temperature for dust clouds, maximum pressure, and rate of pressure rise are available (40,43). The prevention of dust explosions is covered in codes issued by the National Fire Protection Association (42).

The toxicity of a metal in powder form may vary from that of the massive metals in that fine particles can be ingested or inhaled more readily (41). The metal powder producing or consuming industries must conform to OSHA requirements. The limits of airborne particulates are set by NIOSH.

There are no fumes or effluents generated in the processing of powders and the requirements of state and federal environmental protection agencies are met without difficulty. However, the production of powders has been subjected to the same concerns as most other metal refining and smelting operations.

Safety in the P/M industry workplace is also a concern regarding the operation of compacting presses. Guarding devices are required by OSHA to prevent injuries. Those devices applying specifically to metal powder compacting presses are described in a standard issued by the Metal Powder Industries Federation.

Applications

Structural Parts.   P/M parts can be made from a range of materials including iron, steels, low and high brass, bronze, nickel and nickel-base alloys, copper, aluminum, titanium, and various alloys including refractory metals (Fig. 10). P/M parts can be made smaller than a ball point on a pen or as large as bearing weighing more than 50 kg. Because parts are formed in precision dies under high pressure at room temperature, reproducibility is a great advantage.

The tensile properties of sintered (nonheat treated) conventional P/M parts are generally similar to cast metals of the same composition. Fatigue strength is about 38% of tensile values, whereas in wrought metals, such as carbon and alloy steels, it is about 50%. Toughness, as a measure of impact strength determined by Izod or Charpy methods, is density dependent. The higher the density, the higher the toughness.

Most P/M parts are manufactured on a custom engineered basis and orders are filled according to customer specifications. In addition, most manufacturers offer secondary operations such as plating, machining, heat treating, steam treating, plastic impregnation, sizing or coining, or grinding. Physical properties can match those of machined, cast, or even forged parts. Properties range from low density, porous, self-lubricating bearings, to high density structural parts

Fig. 10. Structural P/M parts made from iron- and copper-based powders. Other P/M parts can be made from a range of materials. See text.

with tensile strengths exceeding 1241 N/mm^2 (180,000 psi). P/M parts can be made hard and dense in one section and porous for self-lubrication in another. Although the P/M process is most amenable for the production of large quantities of parts in order to amortize tooling costs, short runs in quantities of below 10,000 pieces are possible.

Most properties of the P/M part are closely related to the final density of the part expressed in g/cm^3. Normally, density of mechanical and structural parts is reported on a dry unimpregnated basis; density of bearings is reported on a fully oil impregnated basis. Density may be calculated by any of several means. The commonly used method is given (44).

Density may also be expressed as relative density, defined as the ratio of the P/M part density to that of its pore-free equivalent. Generally, P/M parts <75% or relative density are considered low density, those >90% are high density, and those in between these two ranges are medium density. Structural and mechanical parts usually have relative densities ranging from 80 to >95%. Forgings and HIP products often exceed 99%. Many self-lubricating bearings have relative densities on the order of 75%; filter parts usually have relative densities of 50%. As for wrought and cast metals, chemical composition of P/M parts strongly influences the mechanical properties. In P/M parts, however, density, particle size, extent of sintering, pore size, shape, and distribution also play a role.

Processing. Metal powders for structural parts are blended or mixed with each other and with a lubricating agent, then fed into automatic molding presses where the mixture is shaped into a compact. The compact is heated at a specific temperature under a protective gas atmosphere, generally 1120°C for iron and steel, 1260°C for stainless steel, and 982°C for copper and copper alloys (qv). In some instances, a separate presintering treatment (burn-off) at a lower temperature is applied in order to volatilize the pressing lubricant and to sinter the compact to a specific degree for subsequent repressing. In general, the part is completed after a single sintering. Occasionally, coining or various heat treatments are necessary.

Changes in configuration may occur during sintering. Dimensional changes are generally a function of the powder; shape changes are a function of density inequalities. Configuration changes are controlled either by a combination of powder properties or post-sintering treatments such as sizing or machining; the latter, however, increase costs.

Porous Materials. In porous materials, the void space that determines the porosity is controlled as to amount, type, and degree of interconnection (25,45–47). Porous parts include self-lubricating bearings, bushings, certain types of P/M parts, and metallic filters. Their manufacture represents an important aspect of the P/M industry. The main applications for porous metals are in filters for separating combinations of solids, liquids, and gases; surge dampeners and flame arrestors for use with gases and liquids; metering devices and distribution manifolds; and storage reservoirs for liquids. These last include self-lubricating bearings. In addition, porous metals are used for the diffusion of air for aeration (qv) of liquids, and the physical separation of immiscible liquids such as gasoline and water, pressure gauge equalizers in instrumentation, flow control, flame arrestors, and sound deadeners.

Porous parts and bearings are made by both the press and sinter techniques, whereas filters are made by loose powder sintering. The metals most commonly used for P/M porous products are bronze, stainless steel (type 316), nickel-base alloys (Monel, Inconel, nickel), titanium, and aluminum.

Processing.   Porous metal products are made by compacting and sintering, isostatic compacting, gravity sintering, sheet forming, and metal spraying.

Gravity (loose powder sintering) is employed to make porous metal parts from powders that bond easily by diffusion, eg, bronze. No outside pressure is applied to shape the part. The appropriate material, graded for size, is poured into a steel or graphite mold that is heated to the sintering temperature of the metal; at this point bonding takes place. The part tolerances are necessarily liberal, although the inside diameters tend to be predictable because the material usually shrinks to the core during sintering. Design tolerances of ±2% for the outside diameter and ±3% for the part length are typical.

In the sheet-forming process, stainless steel, bronze, nickel-base alloys, or titanium powders are mixed with a thermosetting plastic and presintered to polymerize the plastic. Sintering takes place in wide, shallow trays. The specified porosity is achieved by selecting the proper particle size of the powder. Sheet is available in a variety of thicknesses between 16 × 30 mm and as much as 60 × 150 cm. A sheet can be sheared, rolled, and welded into different configurations.

Porous metal structures can also be created by spraying molten metal onto a base. Porosity is controlled by spraying conditions or by an additive that may be removed later.

Porous P/M products can be sinter bonded to solid metals. They can also be welded, brazed, or soldered. Filling the voids with flux or molten metal has to be avoided. P/M porous products can be machined, but blocking of the porous passages has to be avoided. Press fitting and epoxy bonding are commonly used.

Self-Lubricating Parts. Self-lubrication depends on the presence of oil within the pores of the bearing or bushing. A built-in oil reservoir provides a protective oil film that separates the bearing from the shaft, for example in a motor, and prevents metal-to-metal contact. During operation, the rotating shaft draws the oil in the bearing to the surface through capillary action. When not under load, most of the oil is drawn back into the bearing, leaving a layer of oil between the two metal parts.

Most bearing materials (qv) are made from bronze powders. Compositions may be 5–12 wt % tin and up to 6 wt % graphite. Standard compositions are 90.5 wt % Cu–7 wt % Sn–2.5 wt % graphite; 89 wt % Cu–7 wt % Sn–4 wt % graphite; 93 wt % Cu–7 wt % Sn; 96 wt % Cu–4 wt % graphite; and 100 wt % Cu. Compositions containing copper with up to 8 wt % Pb or up to 40 wt % Zn are also available.

Self-lubricating bearings are also made from iron usually containing 36 wt % Cu and 4 wt % Sn. Iron–lead alloys contain 2–6 wt % Pb and 2–4 wt % graphite. Iron-base materials offer increased hardness and strength; in addition, their coefficient of thermal expansion is close to that of the steel shaft. However, iron-base materials are generally not rated as high as copper-base materials for self-lubricating bearing materials.

Processing. Simple self-lubricating parts are produced on an automatic, rapid, high volume basis. The powder mixtures usually contain a few percent of organic lubricants. They are molded unidirectionally in a closed die cavity at pressures of 276–552 MPa (40,000–80,000 psi). During sintering, the lubricant volatilizes which facilitates some control over interconnection of the voids. Porosity is increased by volatile pore-forming agents that are added during sintering.

The sintering of bronze bearings is an example of liquid-phase sintering in which the liquid is present only during part of the sintering time. Alloying causes the liquid to solidify. Bronze bearings are sometimes presintered at 370–480°C in order to control the degassing lubricants. During this treatment, the bronze alloy is formed by the diffusion of liquid tin into the copper, and it solidifies in the furnace. When the temperature is raised, the tin-rich phase liquefies and is absorbed by the copper-rich matrix.

After sintering, P/M bearings are usually sized for dimensional accuracy on the same type of equipment used for molding. This produces a smooth surface and involves forcing the part back into a cavity of the dimensions desired. Lubricating oil is impregnated into the part either before or after sizing. The resulting part is an oil-impregnated bearing containing approximately 20 vol % oil.

A broad range of ferrous and nonferrous compositions is utilized in self-lubricating finished P/M parts. Among these materials are iron, brass, low carbon and low nickel steel, copper, nickel–silver, and stainless steel. Processing is essentially the same as for conventional bearings. P/M parts differ from conventional bearings only in shape and in the capability to withstand shear and compressive stresses considerably greater than those that bearings must withstand.

An important application of self-lubricating P/M parts is the oil-pump gear used in automotive engines. This part is made from a steel of eutectic composition containing 0.3 wt % carbon and 2 wt % copper; it is hardened after sintering. The gear teeth are resistant to wear and plastic deformation. The oil flowing through the interconnected pores lubricates the gear and increases corrosion resistance. High strength nonporous sections can be combined with other sections having lower density and self-lubricating characteristics such as ratchet-and-cam combinations.

Filters.   Compared to screens, metallic filters offer a tortuous path through which the fluid must flow and thus provide depth filtration (qv). Compared to glass (qv) or ceramic filters, P/M filters are strong and particles of the filtered material cannot break away in service and enter the filtrate. The P/M filters are ductile and do not fracture under mechanical or thermal shocks. They can be fitted into housings by rolling, machining, press-fitting, or welding. In vibrating applications, the limited fatigue resistance of P/M filters must be taken into consideration (25). A wide range of design possibilities, permitting filtration at 0.5–200 μm and flow rates from a trickle to high volumes, is possible.

Porosity Measurement.   Porosity is difficult to define precisely because the total interconnected porosity reflects not only a range of hole diameters but also passages of varying lengths and tortuosity. Maximum pore size is determined by the bubble-point test. It measures the pressure at which a bubble forms at the part surface when submerged in alcohol. The micrometer rating is roughly related to the ability of the filter to retain or remove particles. The lower the rating, the finer the powder used to make the part and the higher the pressure drop. For filter design, the maximum size particles in micrometers that is acceptable in the filtrant is specified. The minimum pressure drop at that micrometer rating should be attained.

P/M Tool Steels.   In conventionally produced high alloy tool steels (slowly cooled cast ingots), carbide tends to segregate (48). Segregated clusters of carbide persist even after hot working, and cause undesirable effects on tool fabrication and tool performance. P/M tool steels, on the other hand, provide very fine and uniform carbides in the compact, the final bar stock, and the tools. Several tool steel suppliers consolidate gas-atomized tool steel powder by HIP to intermediate shapes, which are then hot-worked to final mill shapes. Water-atomized tool steel powder is also available (see also TOOL MATERIALS).

Small complex tool steel parts are being made by conventional compaction and sintering in vacuum to near theoretical density. Applications include spade drills, knife blades, slotting cutters, insert blades for gear cutters, reamer blades, and cutting tool inserts.

Friction Materials.   Sintered friction materials are classified as metal–nonmetal combinations (49,50). These are best manufactured by the P/M process. Clutch plates, brake bands, brake blocks, and packing compositions are examples of friction materials (see BRAKE LININGS AND CLUTCH FACINGS).

A sintered friction material is composed of a metal matrix, generally mainly copper, to which a number of other metals such as tin, zinc, lead, and iron are added. Important constituents include graphite and friction-producing components such as silica, emery, or asbestos.

Copper, with its high heat conductivity, resists frictional heat during service and is readily moldable. It is generally used as a base metal, at 60–75 wt %, whereas tin or zinc powders are present at 5–10 wt %. Tin and zinc are soluble in the copper, and strengthen the matrix through the formation of a solid solution during sintering.

Iron or other higher melting metals insoluble in copper are added in amounts of 5–10 wt %. Harder particles, such as iron embedded in the soft copper base, increase the coefficient of friction and also exert a scouring action on the surface. Lead, which is also insoluble in copper, is added in amounts of 5–15 wt %. Dispersed in the matrix, it acts as a lubricant during molding and as a lubricating film during operation of the friction material if the surface temperature is such that the lead liquefies. In addition, lead enhances the smooth engagement of sliding surfaces, which prevents erratic brake or clutch action. Graphite, 5–10 wt %, has effects similar to lead. Silica, emery, or other similar friction-producing materials are added at 2–7 wt %. Minute quantities affect the coefficient of friction. A typical composition of friction–P/M material, in wt %, is Cu, 62; Pb, 12; Fe, 8; graphite, 7; and silica, 4.

Friction materials have to be able to wear the surface away in a uniform manner in order to provide new friction-producing conditions at the surface.

Processing. Friction materials are manufactured by cold pressing a mixture of the ingredients at relatively low (138–276 MPa (20,000–40,000 psi)) pressures. Large parts are sintered in bell-type furnaces, whereas small parts are sintered in conventional furnaces. The pressed compact is bonded to a backing plate during sintering. A number of compact steel plate assemblies are stacked on top of each other and pressure is applied to the stacked assemblies during sintering. Post-sintering treatments such as bending to shape, drilling holes, or machining to dimensions are usually necessary.

Electrical and Electronic Applications. *Contact Materials.* Electrical contact materials are produced by either slicing rod made from metal powder, infiltrating a porous refractory skeleton, or compaction and sintering of powders (see ELECTRICAL CONNECTORS) (51–53).

Tungsten contacts cut from rod are resistant to deformation under a large number of cycles at relatively high forces, have high hardness and strength, the ability to switch high currents without detrimental arcing or welding, and vaporize very little in an arc (if one should form). These contacts meet the requirements of many automotive, aviation, and appliance applications; they are, however, limited in use because an insulating oxide film forms during switching.

Copper and silver combined with refractory metals, such as tungsten, tungsten carbide, and molybdenum, are the principal materials for electrical contacts. A mixture of the powders is pressed and sintered, or a previously pressed and sintered refractory matrix is infiltrated with molten copper or silver in a separate heating operation. The composition is controlled by the porosity of the refractory matrix. Copper–tungsten contacts are used primarily in power-circuit breakers and transformer-tap charges. They are confined to an oil bath because of the rapid oxidation of copper in air. Copper–tungsten carbide compositions are used where greater mechanical wear resistance is necessary.

Tungsten–silver contacts are made similarly, but can be operated in air because of the greater stability of silver. The three standard compositions of this class include tungsten–silver, tungsten carbide–silver, and molybdenum–silver.

A third group includes silver–nickel, silver–cadmium oxide, and silver–graphite combinations. These materials are characterized by low contact resistance, some resistance to arc erosion, and excellent nonsticking characteristics. They can be considered intermediate in overall properties between silver alloys and silver or copper–refractory compositions. Silver–cadmium oxide compositions, the most popular of this class, have wide application in aircraft relays, motor controllers, and line starters and controls.

Figure 11 illustrates the superior conductivity of P/M silver–nickel or silver–cadmium oxide contacts when compared with contacts made by standard melting techniques and formed from solid-solution alloys.

So-called sliding contacts transfer current in motors and generators. Sliding contacts are often called brushes. High electrical conductivity, wear resistance, lubricating qualities, and some arc-erosion resistance are the characteristics required by high current applications. Compositions are usually of copper and 5–7% graphite, silver and graphite, or bronze and graphite.

Magnets. Permanent magnets known as Alnico magnets are made by pressing and sintering powder mixtures (51). These materials are alloys of aluminum, nickel, and iron with additives such as cobalt, copper, and titanium. Sintered Alnico as compared to cast Alnico offers greater mechanical strength and closer tolerances without costly finishing operations. Other P/M magnetic materials include barium and strontium hard ferrites (qv) and iron–neodymium–boron, Fe–Nd–B.

Soft magnetic parts include iron pole pieces for small d-c motors or generators, armatures for generators, and sintered and rolled iron–nickel alloys for radio transformers, measuring instruments, and similar applications. The highest (358.1 T·kA/m (45 MG·Oe)) energy product magnets, made from Fe–Nd–B, are used in cars, appliances, and electric motors (see MAGNETIC MATERIALS).

Rare-Earth Magnets. The combination of rare earths and cobalt exhibits magnetic flux 4–19 times greater than conventional Alnico magnets (54). Because it is necessary to press the metal powders in a magnetic field to align the fine particles, ≤10 micrometers in size, powder metallurgy is the only practical way to obtain such superalloys. The rare-earth–cobalt alloy powders are compacted in a mechanical press and sintered in argon or nitrogen at 1100–1200°C (see COBALT AND COBALT ALLOYS). Samarium, mischmetal, and praseodymium are mostly used (see LANTHANIDES).

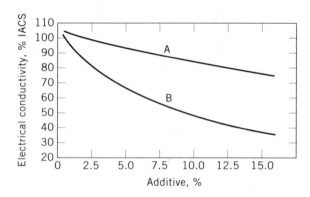

Fig. 11.   Conductivity of A, silver semirefractory compositions and B, solid-solution alloys of the same metals. Copper gives 100% IACS (International Annealed Copper Standard).

Cores. Iron powder cores are manufactured for a-c self-inductance coils for high frequency applications in telephone, radio, and television systems. In the production of radio and television cores, iron powder is coated with an electrically insulating material, compacted in conventional P/M presses, ejected, and baked to fuse the coated particles together. Such a core can afford a large change of inductance by a simple movement in one direction in or out of a wire-wound coil. Fine iron powders of the electrolytic or carbonyl types are employed. These cores exhibit minimum eddy-current and hysteresis losses and the magnetic permeability returns to its original value after application of large magnetizing forces. Cores are also made from molybdenum–permalloy materials.

Electrolytic Capacitors. Tantalum, because of its high melting point of 2850°C, is produced as a metal powder. As such, it is molded, sintered, and worked to wire and foil, and used to build certain types of tantalum capacitors (51). Other capacitors are made by compacting and sintering the tantalum powder.

Batteries. In the nickel–cadmium rechargeable storage battery, the nickel powder forms a porous sheet around a woven wire grid (see BATTERIES). The sheet, with up to 90% porosity, is used as a reservoir for nickel and cadmium compounds and as an electron take-off. It offers a large surface area per unit weight or volume. The porous sheet is made either by sifting nickel powder into a ceramic mold of the correct size and containing the woven screen, or by applying a slurry to a thin perforated nickel-plated iron sheet. The composite is then sintered at a temperature that effects strengthening but not densification.

Dry-battery mercury anodes are pressed compacts of zinc–mercury amalgam. They were first developed and produced during World War II for walkie-talkie communication systems. Practically all hearing aids employ this type of battery in the 1990s.

Incandescent Lamps, Electronic Tubes, and Resistance Elements. Articles fashioned in any form from molybdenum and tungsten usually fall within the bounds of powder metallurgy. These metals normally are first produced as a powder. Both molybdenum and tungsten are used as targets in x-ray tubes, for structural shapes such as lead and grid wires in electron tubes, and as resistance elements in furnaces.

Iron–nickel and iron–nickel–cobalt alloys are prepared in wire and sheet form by either rolling an isostatic-pressed ingot or by directly rolling and sintering the metal powders. These materials have numerous applications in the electronic field as glass-sealing alloys because their coefficient of thermal expansion is close to that of certain glasses. Moreover, their coefficients can be tailored to the application by varying the composition, an easy accomplishment for the powder metallurgist.

Refractory Materials. Extremely high melting metals and those that are more resistant to deformation when hot are considered refractory metals (55). This generally refers to melting above the melting point of iron (1536°C). Tungsten, rhenium, tantalum, molybdenum, and niobium belong to this group and also osmium, boron, titanium, thorium, zirconium, and vanadium. Refractory materials also include alloys of these metals (see HIGH TEMPERATURE ALLOYS; REFRACTORIES).

Refractory metals are associated with powder metallurgy because these metals are not easily melted. Therefore in smelting the ores, the metal is recovered in powder form rather than melted. Refractory metals are used mainly to produce filament wire for incandescent lamps.

Processing. Tungsten powder of approximately 5 μm average particle size produced from the hydrogen reduction of tungsten oxides, is compacted into ingots. The bar is extremely weak and must be heated in hydrogen in a standard muffle furnace in order to strengthen it before sintering in a bell-shaped furnace under a hydrogen atmosphere. Heating is by resistance and is carried on until the sintered bar is dense enough to be heated and worked in air. The bar is then worked first at 1550–1650°C and then at gradually decreasing temperatures. Metallurgically, this is cold working because recrystallization does not take place. The material is swaged and drawn with intermediate annealings to wire filaments. During working, the sintered bar increases in tensile strength and ductility and becomes 100% dense.

Metalworking, such as swaging, drawing, rolling, etc, may also be performed on slabs or ingots of other metals prepared by any of the consolidation and sintering techniques described.

Refractory Metal Alloys. Alloys of refractory or nonrefractory metals are made by direct mixing of the elemental metal powders or by the incorporation of compounds of the solute in the processing step for subsequent reduction to the metal (56). In most instances, the metals are alloyed by solid-state diffusion during sintering. Tungsten and molybdenum form a continuous series of solid solutions which offer considerably greater workability compared to pure molybdenum and pure tungsten.

Tungsten with the addition of as much as 5% thoria is used for thermionic emission cathode wires and as filaments for vibration-resistant incandescent lamps. Tungsten–rhenium alloys are employed as heating elements and thermocouples. Tantalum and niobium form continuous solid solutions with tungsten. Iron and nickel are used as alloy agents for specialized applications.

Heavy Metal. Heavy-metal composition is widely used for containers for shielding radioactive materials, counter-balances in airplane controls, and as cores in armor-piercing projectiles. The tungsten–nickel–iron alloy nominally containing 93 wt % W–5 wt % Ni–2 wt % Fe is considered to be a heavy-metal composition, because its sintered density may exceed 17 g/cm^3. It has sufficient ductility to be worked and can be sintered to full density at 1400°C in contrast to pure tungsten. This alloy is utilized when a high mass having good impact strength, ductility, and tensile strength is needed. Pure tungsten has more mass but is brittle.

Cemented Carbides. Cemented carbides contain mostly tungsten carbide and lesser amounts of other hard-metal components, embedded in a matrix of cobalt (see CARBIDES, CEMENTED CARBIDES).

The principal application of cemented carbides is as cutting tools for metals. They are also used in armor-piercing projectile cores and tips, and for carbide shot. Cemented carbides are used for parts requiring corrosion resistance, including burnishing tools and dies, pump valves, sandblast nozzles, gauges, rock and masonry drills, and guides of many types. Carbide compositions containing only tungsten carbide [*12070-12-1*], and cobalt are unsuitable for machining highly resistant materials. However, tungsten carbide can be replaced with solid solutions of tungsten, molybdenum, and titanium carbides to machine such materials. Various rare-earth carbides have been added to tungsten carbide for special-purpose cutting tools. Tungsten-free compositions have also been developed as well as compositions containing diamonds.

Cemented carbides exhibit extreme hardness and toughness which are retained at the elevated temperatures that may occur between tool and work during cutting.

Processing. Tungsten carbide is made by heating a mixture of lampblack with tungsten powder in such proportions that a compound with a combined carbon of 6.25 wt % is obtained. The ratio of free-to-combined carbon is of extreme importance. Tantalum and titanium carbides are made by heating a mixture of carbon with the metal oxide. Multicarbide powders, such as Mo_2C-WC, $TaC-NbC$, and $TiC-TaC-WC$, are made by a variety of methods, the most important of which is carburization of powder mixtures.

These carbides are mixed with fine cobalt powder in a wet ball-milling operation. Water, benzene, or other organic materials may be used. Milling produces an intimate mixture of cobalt attached to the carbide particles. This is followed by mixing with a lubricant or other binder, such as paraffin, camphor, or stearic acid, in a separate operation. Dried and screened, the mass results in a lubricated, flowable powder that can be cold-molded in conventional presses and tooling. A low temperature, presinter treatment is carried out in order to strengthen the part through sintering of the binder metal and to evaporate the pressing lubricant. The presintered part is then formed by machining into the desired shape and dimensions. This is followed by high temperature sintering between 1350 and 1550°C for 1–2 h. Sintering is characterized by solution of the carbides in the liquid phase, and precipitation at areas of low energy. Recrystallized grains of carbide form and are embedded in a matrix of a solution of carbides and matrix metal. Linear dimensional shrinkage of 21–25% occurs.

Cermets. High temperature applications in space and nuclear technology created a need for materials to fill the gap between cobalt- and nickel-base superalloys and refractory metals such as tungsten and molybdenum. Pure ceramics are strong at elevated temperatures but lack sufficient ductility to be worked at room temperature, whereas metal alloys, although ductile at room temperature, are not strong enough at the higher temperatures. Mixtures of ceramics and metals have both the high temperature strength of ceramics, and sufficient ductility and thermal conductivity contributed by the metal to provide resistance from thermal shock at high temperatures and workability at room temperature. These compositions are termed ceramals and cermets (see CERAMICS; HIGH TEMPERATURE ALLOYS). Actually, the cemented carbides can be considered cermets.

The cermet class of materials contains a large number of compositions (57). Most cermets are carbide-based, eg, WC and titanium carbide [*12070-08-5*], TiC. Cemented tungsten carbides are widely used for cutting tools and car parts.

Dispersions of flake aluminum powders having surface oxide up to 14 wt % Al_2O_3 have been pressed, sintered, and worked to a material known as sintered aluminum powder (SAP). This product exhibits high strength at elevated temperatures. Nickel containing small additions of thoria, known as TD-nickel, is also a high temperature cermet.

Thoria-dispersed nickel products are obtained by precipitating basic nickel compounds, whereby thoria particles of ca 100 nm are coated with layers of nickel to the extent that the product has a 2% thoria dispersion.

Dispersion-strengthened copper is made by dispersing a thoria or alumina phase through copper powder. The resulting P/M product retains its strength at elevated temperatures. It is used, for example, as the conductor or lead wire that supports the hot filament inside incandescent lamps.

Mechanical alloying is another method of producing dispersion-strengthened metals. In this process, the powdered constituents of the alloy are treated in an attrition mill. A finely distributed layer of the dispersed phase is distributed on particles of the base metal. Subsequent pressing and sintering strengthens the dispersion (25).

Metal-Matrix Composites. A metal-matrix composite (MMC) is comprised of a metal alloy, less than 50% by volume that is reinforced by one or more constituents with a significantly higher elastic modulus. Reinforcement materials include carbides, oxides, graphite, borides, intermetallics or even polymeric products. These materials can be used in the form of whiskers, continuous or discontinuous fibers, or particles. Matrices can be made from metal alloys of Mg, Al, Ti, Cu, Ni or Fe. In addition, intermetallic compounds such as titanium and nickel aluminides, Ti_3Al and Ni_3Al, respectively, are also used as a matrix material (58,59). P/M MMC can be formed by a variety of full-density hot consolidation processes, including hot pressing, hot isostatic pressing, extrusion, or forging.

MMCs can offer improved strength and elastic modulus especially at elevated temperatures when compared with metal alloys. For example, titanium–silicon carbide composites have a high strength-to-weight ratio, good elevated temperature creep-rupture strength, and a higher modulus than conventional titanium alloys. Potential applications include gas turbine engine components, and airframe structures in advanced aircraft and auto engine valves and turbocharger rotors (see METAL-MATRIX COMPOSITES).

Space Applications. The growth of powder metallurgy in space technology has arisen from the difficulty of handling many materials in conventional fusion-metallurgy techniques, the need for controlled porosity, and the requirement of many special and unique properties (60,61). Powder metallurgy is applied in low density components with emphasis on porous tungsten for W–Ag structures, beryllium compounds, titanium and dispersion-hardening systems.

Plates of beryllium metal were used as heat-shield shingles on the cylindrical section of the Mercury spacecraft in some of the first suborbital flights (see ABLATIVE MATERIALS). A fairing for the Agena-B space-vehicle interstage structure used beryllium machined from a pressed compact 180-cm in dia and 70-cm high. Beryllium has also been used in the nose cap of the ascent shroud in several of the Ranger-Mariner space probe series and in air- and spaceborne mirrors (62). Beryllium has a favorable stiffness-to-density ratio and good thermal properties which make it acceptable as a shielding material against micrometeorites.

Special tungsten powders have been developed for space applications. Spherical tungsten particles are used to form porous tungsten bodies in the ionizing surface in ion-propulsion engines. Silver-infiltrated tungsten parts have a density before infiltration of 70–80% of theoretical; 90% of the porosity is infiltrated with silver. These parts are used where the ablative cooling provided by the silver during solid–vapor change of state and transpiration cooling through the resulting porosity are desirable, as in reentry vehicles.

Space technology has always demanded materials that can operate at temperatures between those of superalloys and refractory metals and that have high temperature strength during operation and room temperature ductility for fabrication. The development of dispersion-strengthened and oxide alloy systems has solved part of this problem.

Space technology also utilizes sintered magnetic materials such as Alnico, sintered bronze bushings, electrical contacts and tantalum capacitors, and a variety of P/M porous filters. Aluminum and magnesium powders are blended with solid fuels for enhancing rocket-propulsion systems. The space shuttle Columbia is launched using 90,720 kg of aluminum powder plus ammonium perchlorate and oxidizers in each of its two booster rockets. P/M titanium housings are used in guided missiles.

High vacuum is one of the main characteristics of space. Bearings having liquid lubricants would lose the lubricant through evaporation. Bearings produced by powder metallurgy techniques having embedded MoS_2 give good service.

Nuclear Applications. Powder metallurgy is used in the fabrication of fuel elements as well as control, shielding, moderator, and other components of nuclear-power reactors (63) (see NUCLEAR REACTORS). The materials for fuel, moderator, and control parts of a reactor are thermodynamically unstable if heated to melting temperatures. These same materials are stable under P/M process conditions. It is possible, for example, to incorporate uranium or ceramic compounds in a metallic matrix, or to produce parts that are similar in the size and shape desired without effecting drastic changes in either the structure or surface conditions. Only little post-sintering treatment is necessary.

In nuclear technology, the P/M process is applied to beryllium, zirconium, uranium, and thorium. Uranium and thorium are used as fuel materials, beryllium for moderating purposes, and zirconium as construction material. These metals are used in elemental form, as alloys, or in metal–ceramic combinations. The metal powders must be handled somewhat differently from ordinary metal powders. Beryllium powder is extremely toxic, zirconium is highly pyrophoric, and uranium and thorium powders are both toxic and pyrophoric. The vapors eminating from burning uranium are also extremely toxic. These powders are generally handled in closed containers, called dry boxes, usually filled with a protective atmosphere such as argon or helium. Beryllium powders have extremely poor compactability at room temperature and are therefore generally processed by sintering in vacuum or HIP.

Both zirconium hydride and zirconium metal powders compact to fairly high densities at conventional pressures. During sintering the zirconium hydride decomposes and at the temperature of decomposition, zirconium particles start to bond. Sintered zirconium is ductile and can be worked without difficulty. Pure zirconium is seldom used in reactor engineering, but the powder is used in conjunction with uranium powder to form uranium–zirconium alloys by solid-state diffusion. These alloys are important in reactor design because they change less under irradiation and are more resistant to corrosion.

BIBLIOGRAPHY

"Powder Metallurgy" in *ECT* 1st ed., Vol. 11, pp. 43–64, by W. Leszynski, American Electro Metal Corp.; in *ECT* 2nd ed., Vol. 16, pp. 401–435, by A. R. Poster, Metals Sintering Corp. and H. H. Hausner, Consulting Engineering; in *ECT* 3rd ed., Vol. 19, pp. 28–62, by K. H. Roll, Metal Powder Industries Federation, American Powder Metallurgy Institute.

1. W. D. Jones, *Principles of Powder Metallurgy*, Edward Arnold & Co., London, 1937.
2. G. G. Goetzel, *Treatise on Powder Metallurgy*, Vol. 1, Interscience Publishers, Inc., New York, 1949.
3. R. L. Sands and C. R. Shakespeare, *Powder Metallurgy*, George Newnes, Ltd., London, 1966.
4. *Government Research Report*, Vol. 9, Powder Metallurgy, Ministry of Supply, Department of Scientific and Industrial Research, London, 1951.
5. R. M. German, "Emerging P/M Technologies in the USA," *Int. J. Powder Metall.* **28**(1) (1992).
6. R. D. Payne, M. A. Matteson, and A. L. Moran, "Application of Neural Networks in Spray Forming Technology," *Int. J. Powder Metall.* **29**(4) (1993).
7. S. Annavarapu and R. D. Doherty, "Evolution of Microstructure in Spray Casting," in Ref. 6.
8. *Powder Metallurgy Design Solutions*, Metal Powder Industries Federation, Princeton, N.J., 1993.
9. *Powder Metallurgy Design Manual*, Metal Powder Industries Federation, Princeton, N.J., 1989.
10. R. R. Irani and C. F. Callis, *Particle Size: Measurement, Interpretation and Application*, John Wiley & Sons, Inc., New York, 1963.
11. H. E. Rose, *The Measurement of Particle Size in Very Fine Powders*, Chemical Publishing Co., Inc., New York, 1954.
12. H. H. Hausner, "Basic Characteristics of Metal Powders" in A. R. Poster, ed., *Handbook of Metal Powders*, Reinhold Publishing Corp., New York, 1966.
13. *Determination of Apparent Density of Free-Flowing Metal Powders Using the Hall Apparatus* MPIF Standard No. 04, and *Determination of Apparent Density of Non-Free Flowing Metal Powders Using the Carney Apparatus*, No. 4, Metal Powder Industries Federation, Princeton, N.J., 1992.
14. *Determination of Flow Rate of Free-Flowing Metal Powders Using the Hall Apparatus*, MPIF Standard No. 3, Metal Powder Industries Federation, Princeton, N.J., 1980.
15. J. S. Hirschhorn, *Introduction to Powder Metallurgy*, American Powder Metallurgy Institute, Princeton, N.J., 1976.
16. F. V. Lenel, *Powder Metallurgy—Principles and Applications*, Metal Powder Industries Federation, Princeton, N.J., 1980.
17. H. H. Hausner, *Planseeber. Pulvermet* **12**, 172 (Dec. 1964).
18. D. Train and C. J. Lewis, *Trans. Inst. Chem. Eng. (London)* **40**, 235 (1962).
19. M. J. Donachie, Jr. and M. F. Burr, *J. Metal.* **15**, 849 (Nov. 1963).
20. R. W. Heckel, *Trans. AIME* **221**, 671 (1961).
21. R. W. Heckel, *Trans. AIME* **222**, 1073 (1962).

22. C. G. Goetzel, *Treatise on Powder Metallurgy*, Vol. 1 (1949), 2 (1950), 3 (1952), and 4 (1963), Interscience Publishers, Inc., New York; Vol. 1, pp. 259–312.
23. S. Bradbury, *Powder Metallurgy Equipment Manual*, Powder Metallurgy Equipment Association, Metal Powder Industries Federation, Princeton, N.J., 1986.
24. Ref. 16, Chaps. 12–13, 15, and 21.
25. R. R. Irving, *Iron Age* (July 28, 1980).
26. *Prod. Eng.*, 45 (Aug. 1979).
27. S. Storchheim, *Met. Prog.* **69**, 120 (Sept. 1956).
28. *Product Bulletin #182*, Hoeganaes Corp., Riverton, N.J., 1976.
29. P. D. Peckner, *Mater. Des. Eng.* **51**, 89 (July 1960).
30. H. H. Hausner and A. R. Poster, in W. Leszynski, ed., *Powder Metallurgy*, Interscience Publishers, Inc., New York, 1961, p. 461.
31. H. H. Hausner, "Compacting and Sintering of Metal Powder Without the Application of Pressure," *International Symposium Agglomeration*, Apr. 12–14, 1961.
32. *Materials Standards for Metal Injection Molded Parts*, Metal Powder Industries Federation, Princeton, N.J.
33. I. Ljengberg and P. G. Arbstedt, *Proc. Ann. Meeting Metal Powder Assoc.* **12**, 78 (1956); H. H. Hausner and I. Sheinhartz, *Proc. Ann. Meeting Metal Powder Assoc.* **10**, 6 (1954).
34. H. H. Hausner, ed., "Modern Developments in Powder Metallurgy," *Proceedings of the International Powder Metallurgy Conference of New York*, 1966, Plenum Press, New York, 1966.
35. F. V. Lenel and G. S. Ansell, in Ref. 34, Vol. 1, pp. 281–296.
36. M. H. Tikkanen and S. Ylasaari, in Ref. 34, Vol. 1, pp. 297–309.
37. F. Thummler and W. Thomma, *Met. Rev.* (115).
38. *Definitions of Terms Used in Powder Metallurgy*, MPIF Standard 09, Metal Powder Industries Federation, Princeton, N.J., 1992.
39. H. H. Hausner "Grain Growth During Sintering" in *Special Report No. 58 of the Iron and Steel Institute*, London, 1954, pp. 102–112.
40. H. H. Hausner, *Powder Metall.* **1**(2), (Apr. 1965).
41. I. T. Brakhnova, *Environmental Hazards of Metals*, Consultant's Bureau, New York, 1975.
42. *Codes 651 and 61a*, National Fire Protection Association, Quincy, Mass., 1993, 1989.
43. *Explosibility of Metal Powders*, R. I. 6516R1, U.S. Bureau of Mines, Washington, D.C., 1965.
44. MPIF Standard 42, *Determination of Density Compacted or Sintered Metal Powder Projects*, Metal Powder Industries Federation, Princeton, N.J., 1991.
45. MPIF Standard No. 35, *Materials Standards for P/M Structural Parts*, Metal Powder Industries Federation, Princeton, N.J., 1994–1995; Ref. 22, Vol. 2, pp. 503–542.
46. *Products of Powder Metallurgy*, Engineering Manual E-64, Amplex Division, Chrysler Corp., Detroit, Mich.
47. *Porous Metals Guidebook*, Metal Powder Industries Federation, Princeton, N.J., 1980.
48. B. J. Collins and C. P. Schneider, in Ref. 34, Vol. 3, pp. 160–165.
49. *Met. Prog.* **100** (Jan. 1980);
50. Ref. 22, Vol. 2, pp. 543–558.
51. A. S. Doty, *Proc. Ann. Meeting Metal Powder Assoc.* **12**, 46 (1956).
52. G. A. Meyer, *Met. Prog.* **86**, 95 (June 1965).
53. *Ibid.* **87**, 92 (July 1965).
54. *PM Technology Newsletter*, American Powder Metallurgy Institute, Princeton, N.J., Oct./Nov. 1978.
55. Ref. 45, pp. 3–73.
56. *What are Refractory Metals?* Refractory Metals Association, Metal Powder Industries Federation, Princeton, N.J., 1980, pp. 4–5.

57. *Metals Handbook*, 9th ed., Vol. 7, ASM International, Metals Park, Ohio, 1984.

58. R. M. German and R. G. Iacocca, "Powder Metallurgy Processing and Applications for Intermetallics," *Advances in Powder Metallurgy & Particulate Materials*, Vol. 6, Metal Powder Industries Federation, Princeton, N.J., 1993.

59. J. Moll, C. F. Yolton, and B. J. McTiernan, "P/M Processing of Titanium Aluminides," *Int. J. Powder Metall.* **26**(2), (1990).

60. C. G. Goetzel and J. B. Rittenhouse, *J. Met.* **17**, 876 (Aug. 1965).

61. H. H. Hausner, *J. Met.* **16**, 894 (Nov. 1964).

62. F. H. Froes, "Synthesis of Metallic Materials For Demanding Applications Using Powder Metallurgy Techniques," *P/M in Aerospace and Defense Technologies*, Metal Powder Industries Federation, Princeton, N.J., 1991.

63. H. H. Hausner, *Proc. Ann. Meeting Metal Powder Assoc.* **12**, 27 (1956).

General References

A. Lawley, *Atomization: The Production of Metal Powder*, Metal Powder Industries Federation, Princeton, N.J., 1992.

F. V. Lenel, *Powder Metallurgy—Principles and Applications*, Metal Powder Industries Federation, Princeton, N.J., 1980.

R. M. German, *Powder Metallurgy Science*, 2nd ed., Metal Powder Industries Federation, Princeton, N.J., 1994.

Powder Metallurgy Design Manual, Metal Powder Industries Federation, Princeton, N.J., 1995.

International Journal of Powder Metallurgy (quarterly), APMI International (formerly American Powder Metallurgy Institute), Princeton, N.J.

PM Technology Newsletter (monthly), APMI International (formerly American Powder Metallurgy Institute), Princeton, N.J.

PETER K. JOHNSON
Metal Powder Industries Federation
APMI International

METAL-MATRIX COMPOSITES

A composite material (1) is a material consisting of two or more physically and/or chemically distinct, suitably arranged or distributed phases, generally having characteristics different from those of any components in isolation. Usually one component acts as a matrix in which the reinforcing phase is distributed. When the continuous phase or matrix is a metal, the composite is a metal-matrix composite (MMC). The reinforcement can be in the form of particles, whiskers, short fibers, or continuous fibers (see COMPOSITE MATERIALS).

Types of Metal-Matrix Composites

There are three kinds of metal-matrix composites distinguished by type of reinforcement: particle-reinforced MMCs, short fiber- or whisker-reinforced MMCs, and continuous fiber- or sheet-reinforced MMCs. Table 1 provides examples of some important reinforcements used in metal-matrix composites as well as their aspect (length/diameter) ratios and diameters.

Particle or discontinuously reinforced MMCs have become important because they are inexpensive compared to continuous fiber-reinforced composites and they have relatively isotropic properties compared to the fiber-reinforced composites. Figures 1a and b show typical microstructures of continuous alumina fiber/Mg and silicon carbide particle/Al composites, respectively.

Table 1. Typical Reinforcements Used in Metal-Matrix Composites

Type	Aspect ratio	Diameter, μm	Examples
particle	~1–4	1–25	SiC, Al$_2$O$_3$, BN, B$_4$C
short fiber or whisker	~10–1000	0.1–25	SiC, Al$_2$O$_3$, Al$_2$O$_3$ + SiO$_2$, C
continuous fiber	>1000	3–150	SiC, Al$_2$O$_3$, C, B, W

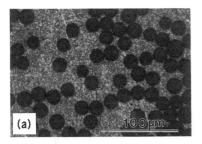 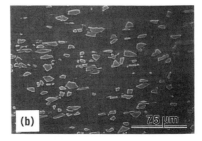

Fig. 1. Typical microstructures of some metal-matrix composites: (**a**) continuous alumina fiber/Mg and (**b**) silicon carbide particle/Al composites.

Processing

There are several important fabrication processes for metal-matrix composites.

Liquid-State Processes. *Casting* or *liquid infiltration* involves infiltration of a fiber bundle by liquid metal. It is not easy to make MMCs by simple liquid-phase infiltration, mainly because of difficulties with wetting ceramic reinforcement by the molten metal. When the infiltration of a fiber preform occurs readily, reactions between the fiber and the molten metal can significantly degrade fiber properties. Fiber coatings (qv) applied prior to infiltration, which improve wetting and control reactions, have been developed and are producing some encouraging results. In this case, however, the disadvantage is that the fiber coatings must not be exposed to air prior to infiltration because surface oxidation alters the positive effects of coating (2). One liquid infiltration process involving particulate reinforcement, called the Duralcan process, has become quite successful (Fig. 2). Ceramic particles and ingot-grade aluminum are mixed

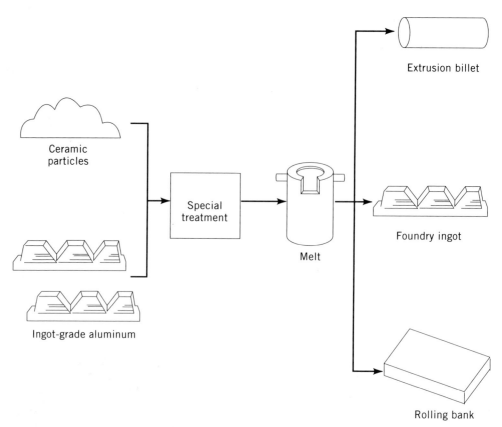

Fig. 2.   Casting or liquid infiltration process.

and melted. The ceramic particles are given a proprietary treatment. The melt
is stirred just above the liquidus temperature (600–700°C). The melt is then
converted into one of the following four forms: extrusion blank, foundry ingot,
rolling bloom, or rolling ingot. The Duralcan process of making particulate composites by a liquid metal casting route involves the use of 8–12 μm particles.
For particles that are too small (2–3 μm), the result is a very large interface
region and thus a very viscous melt. In foundry-grade MMCs, high Si aluminum
alloys (eg, A356) are used, whereas in wrought MMC, Al–Mg alloys type (eg,
6061) are used. Alumina particles are typically used in foundry alloys; silicon
carbide particles are used in wrought aluminum alloys.

Alternatively, tows of fibers can be passed through a liquid metal bath,
where the individual fibers are wet by the molten metal, wiped of excess metal,
and a composite wire is produced. A bundle of such wires can be consolidated
by extrusion to make a composite. Another pressureless liquid metal infiltration
process of making MMCs is the Primex process (Lanxide), which can be used with
certain reactive metal alloys such as Al–Mg to infiltrate ceramic preforms. For
an Al–Mg alloy, the process takes place between 750–1000°C in a nitrogen-rich
atmosphere (2). Typical infiltration rates are less than 25 cm/h.

Squeeze casting or *pressure infiltration* involves forcing the liquid metal into a fibrous preform (Fig. 3). Pressure is applied until the solidification is complete. By forcing the molten metal through small pores of fibrous preform, this method obviates the requirement of good wettability of the reinforcement by the molten metal. Composites fabricated with this method have minimal reaction between the reinforcement and molten metal, and are free from common casting defects such as porosity and shrinkage cavities. Infiltration of a fibrous preform by means of a pressurized inert gas is another variant of the liquid metal infiltration technique. The process is conducted in the controlled environment of a pressure vessel and rather high fiber volume fractions; complex shaped structures are obtainable (3). Although commonly aluminum matrix composites are made by this technique, alumina fiber-reinforced intermetallic matrix composites, eg, TiAl, Ni_3Al, and Fe_3Al matrix materials, have been prepared by pressure casting (4). The technique involves melting of the matrix alloy in a crucible in vacuum; the fibrous preform is heated separately. The molten matrix material (at ~100°C above T_m) is poured onto the fibers and argon gas is introduced simultaneously. Argon gas pressure forces the melt to infiltrate the preform. The melt generally contains additives to aid wetting of the fibers.

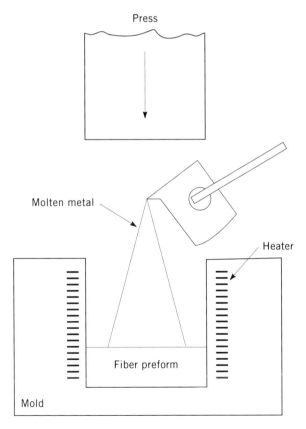

Fig. 3.   Squeeze casting or pressure infiltration process.

Solid-State Processes.   *Diffusion bonding* is a common solid-state welding technique for joining similar or dissimilar metals. Interdiffusion of atoms from clean metal surfaces in contact at an elevated temperature leads to welding. The principal advantages of this technique are the ability to process a wide variety of matrix metals and control of fiber orientation and volume fraction. Among the disadvantages are processing times of several hours, expensive high processing temperatures and pressures, and only objects of limited size can be produced. There are many variants of the basic diffusion bonding process, however, all of them involve a simultaneous application of pressure and high temperature. Matrix alloy foil and fiber arrays (composite wire) or monolayer laminae are stacked in a predetermined order (Fig. 4). Vacuum hot pressing is an important step in the diffusion bonding processes for metal-matrix composites. Hot isostatic pressing (HIP), instead of uniaxial pressing, can also be used. In HIP, gas pressure against a can consolidates the composite piece contained inside the can. With HIP it is relatively easy to apply high pressures at elevated temperatures over variable geometries.

Deformation processing of metal–metal composites involves mechanical processing (swaging, extrusion, drawing, or rolling) of a ductile two-phase material. The two phases co-deform, causing the minor phase to elongate and become fibrous in nature within the matrix. These materials are sometimes referred to as *in situ* composites. The properties of a deformation processed composite depend largely on the characteristics of the starting material which is usually a billet of two-phase alloy that has been prepared by casting or powder metallurgy methods (see METALLURGY, POWDER). Roll bonding is a common technique used

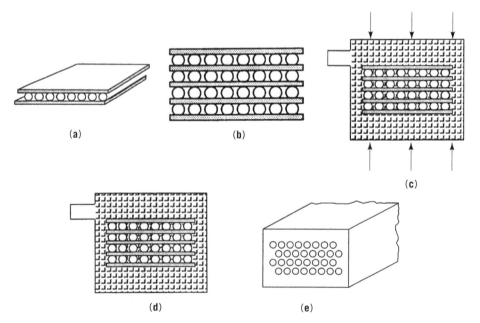

Fig. 4.   Diffusion bonding process: (**a**) apply metal foil and cut to shape, (**b**) lay up desired plies, (**c**) vacuum encapsulate and heat to fabrication temperature, (**d**) apply pressure and hold for consolidation cycle, and (**e**) cool, remove, and clean part.

to produce a laminated composite consisting of different metals in sheet form (5). Such composites are called sheet laminated metal-matrix composites. Roll bonding and hot pressing have also been used to make laminates of Al sheets and discontinuously reinforced MMCs (6,7). Figure 5 shows the roll bonding process of making a laminated MMC. Other examples of deformation processed metal-matrix composites are the niobium-based conventional filamentary superconductors and the silver-based high T_C superconductors. There are two main types of the conventional niobium-based superconductors: Nb–Ti/Cu and Nb_3Sn/Cu. Niobium–titanium (~50–50) form a ductile system. Rods of Nb–Ti are inserted in holes drilled in a block of copper, evacuated, sealed, and subjected to a series of drawing operations interspersed with appropriate annealing treatments to give the final composite superconducting wire. This process is fairly straightforward. In the case of Nb_3Sn/Cu, a process called the bronze route is used to make this composite. Nb_3Sn, an A-15-type brittle intermetallic, cannot be processed like Nb–Ti. Instead, the process starts with a bronze (Cu–13% Sn) matrix, pure niobium rods are inserted in holes drilled in bronze, evacuated, sealed, and subjected to wire drawing operations as in the case of Nb–Ti/Cu. The critical step is the heat treatment (~700°C) that drives out the tin from the bronze matrix to combine with niobium to form stoichiometric, superconducting Nb_3Sn, leaving behind copper matrix. A process that is increasingly gaining importance is called the oxide-powder-in-tube (OPIT) method (8). In this process, the oxide powder of appropriate composition (stoichiometry, phase content, impurities, etc) is packed inside a metal tube (generally silver), sealed, and degassed. Commonly, swaging and drawing are used for making wires and rolling is used for tapes. Heat treatments, intermediate and/or subsequent to deformation, are given to form the correct phase, promote grain interconnectivity and crystallographic alignment of the oxide, and obtain proper oxygenation (8).

Powder processing methods involving cold pressing and sintering, or hot pressing can be used to fabricate MMCs, primarily particle- or whisker-reinforced MMCs (9). The matrix and the reinforcement powders are blended to produce a homogeneous distribution. Most of the information on this critical step is considered proprietary. The blending stage is followed by cold pressing to produce what is called a green body, which is about 80% dense and can be easily handled. The cold pressed green body is canned in a sealed container and degassed to remove any absorbed moisture from the particle surfaces. The final

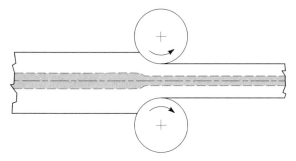

Fig. 5. Roll bonding process of making a laminated MMC; a metallurgical bond is produced.

step is hot pressing, uniaxial or isostatic, to produce a fully dense composite. The hot pressing temperature can be either below or above the matrix alloy solidus.

Deposition techniques for metal-matrix composite fabrication involve coating individual fibers in a tow with the matrix material needed to form the composite followed by diffusion bonding to form a consolidated composite plate or structural shape. The main disadvantage of using deposition techniques is that they are time consuming. However, there are several advantages. (*1*) The degree of interfacial bonding is easily controllable; interfacial diffusion barriers and compliant coatings can be formed on the fiber prior to matrix deposition or graded interfaces can be formed. (*2*) Filament-wound thin monolayer tapes can be produced that are easier to handle and mold into structural shapes than other precursor forms; unidirectional or angle-plied composites can be easily fabricated in this way.

Several deposition techniques are available: immersion plating, electroplating, spray deposition, chemical vapor deposition (CVD), and physical vapor deposition (PVD) (see THIN FILMS). Dipping or immersion plating is similar to infiltration casting except that fiber tows are continuously passed through baths of molten metal, slurry, sol, or organometallic precursors. Electroplating (qv) produces a coating from a solution containing the ion of the desired material in the presence of an electric current. Fibers are wound on a mandrel, which serves as the cathode, and placed into the plating bath with an anode of the desired matrix material. The advantage of this method is that the temperatures involved are moderate and no damage is done to the fibers. Problems with electroplating involve void formation between fibers and between fiber layers, adhesion of the deposit to the fibers may be poor, and there are limited numbers of alloy matrices available for this processing. A spray deposition operation typically consists of winding fibers onto a foil-coated drum and spraying molten metal onto them to form a monotape. The source of molten metal may be powder or wire feedstock which is melted in a flame, arc, or plasma torch. The advantages of spray deposition are easy control of fiber alignment and rapid solidification of the molten matrix. In a CVD process, a vaporized component decomposes or reacts with another vaporized chemical on the substrate to form a coating on that substrate. The processing is generally carried out at elevated temperatures.

***In Situ* Processes.**   In these techniques, the reinforcement phase is formed *in situ*. The composite material is produced in one step from an appropriate starting alloy, thus avoiding the difficulties inherent in combining the separate components as done in a typical composite processing. Controlled unidirectional solidification of a eutectic alloy is a classic example of *in situ* processing. Unidirectional solidification of a eutectic alloy can result in one phase being distributed in the form of fibers, and ribbon in the other. Distribution fineness of the reinforcement phase can be controlled by simply controlling the solidification rate. The solidification rate in practice, however, is limited to a range of 1–5 cm/h because of the need to maintain a stable growth front which requires a high temperature gradient.

The XD process (Martin Marietta) is another *in situ* process which uses an exothermic reaction between two components to produce a third component. Sometimes such processing techniques are referred to as the self-propagating high temperature synthesis (SHS) process. Specifically, the XD process is a pro-

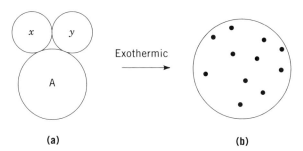

(a)   (b)

Fig. 6. The XD process where in (**a**) x and y are hardening elements and A is the solvent. The result of this process is (**b**), submicrometer-sized xy particles distributed in matrix A.

prietary process developed to produce ceramic particle-reinforced metallic alloy (Fig. 6). Generally, a master alloy containing a rather high volume fraction of reinforcement is produced by the reaction synthesis. This is mixed and remelted with the base alloy to produce a desirable amount of particle reinforcement. Typical reinforcements are SiC, TiB_2, etc, in an aluminum, nickel, or an intermetallic matrix (10). An example of the microstructure of an XD composite (TiB_2/Al) is shown in Figure 7.

    Spray-Forming of Particulate MMCs. Another process for making particle-reinforced MMCs involves the use of spray techniques that have been used for some time to produce monolithic alloys (11). One particular example of this, a co-spray process, uses a spray gun to atomize a molten aluminum alloy matrix, into which heated (for drying) silicon carbide particles are injected (Fig. 8). An optimum particle size is required for an efficient transfer, eg, whiskers are too fine to be transferred. The preform produced in this way is generally quite porous. The co-sprayed metal-matrix composite is subjected to scalping, consolidation, and secondary finishing processes, thus making it a wrought material. The process is totally computer controlled and is quite fast, but it should be noted that it is essentially a liquid metallurgy process. The formation of deleterious reaction products is avoided because the time of flight

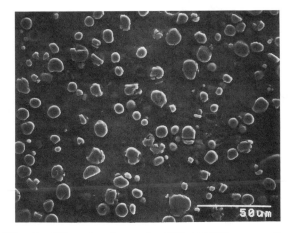

Fig. 7. Microstructure of a TiB_2/Al XD composite.

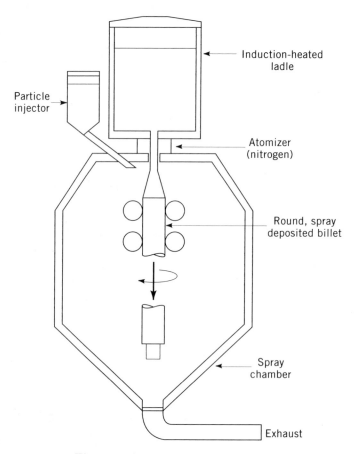

Fig. 8. The spray-forming process.

is extremely short. Silicon carbide particles of an aspect ratio (length/diameter) between 3–4 and volume fractions up to 20% have been incorporated into aluminum alloys. An advantage of the process is the flexibility it affords in making different types of composites, eg, *in situ* laminates can be made using two sprayers or by selective reinforcement. This process is quite expensive, however, mainly because of the capital equipment.

Interfaces in Metal-Matrix Composites

The interface region in a composite is important in determining the ultimate properties of the composite. At the interface a discontinuity occurs in one or more material parameters such as elastic moduli, thermodynamic parameters such as chemical potential, and the coefficient of thermal expansion. The importance of the interface region in composites stems from two main reasons: the interface occupies a large area in composites, and in general, the reinforcement and the matrix form a system that is not in thermodynamic equilibrium.

In crystallographic terms, ceramic–metal interfaces in composites generally are incoherent and high energy interfaces. Accordingly, they can act as efficient vacancy sinks, and provide rapid diffusion paths, segregation sites, sites of heterogeneous precipitation, as well as sites for precipitate-free zones. Among the possible exceptions to this are the eutectic composites (1) and the newer XD-type particulate composites (12).

Some bonding must exist between the ceramic reinforcement and the metal matrix for load transfer from matrix to fiber to occur. Two main categories of bonding are mechanical and chemical. Mechanical keying effects between two surfaces can lead to bonding. This has been confirmed experimentally for tungsten filaments in an aluminum matrix (13). Mechanical gripping effects have also been observed at Al_2O_3/Al interfaces (14) and the results are shown (Fig. 9) in the form of crack density in alumina as a function of strain in the alumina–aluminum composite for different degrees of interface roughness. The

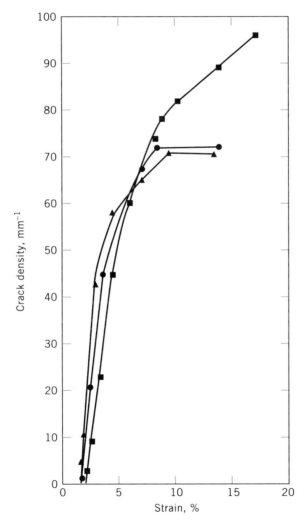

Fig. 9. Number of cracks per mm vs percent strain in an alumina–aluminum composite for different degrees of roughness where (●) represents the polished substrate; (■), steep sided pits, $10^6/cm^2$; and (▲), gently sloping pits, $10^3/cm^2$.

crack density continues to increase for large strain values in the case of a rough interface. Most metal-matrix composite systems, however, are nonequilibrium systems in the thermodynamic sense; that is, there exists a chemical potential gradient across the fiber–matrix interface. This means that given favorable kinetic conditions, which in practice means a high enough temperature or long enough time, diffusion and/or chemical reactions occur between the components. Figure 10 shows a transmission electron microscopy (tem) micrograph of the reaction zone between alumina fiber and magnesium matrix. The interface layer(s) formed because of such a reaction generally have characteristics different from those of either one of the components. However, at times some controlled amount of reaction at the interface (Fig. 10) may be desirable for obtaining strong bonding between the fiber and the matrix; too thick an interaction zone adversely affects the composite properties.

Ceramic–metal interfaces are generally formed at high temperatures. Diffusion and chemical reaction kinetics are faster at elevated temperatures. Knowledge of the chemical reaction products and, if possible, their properties are needed. It is therefore imperative to understand the thermodynamics and kinetics of reactions such that processing can be controlled and optimum properties obtained.

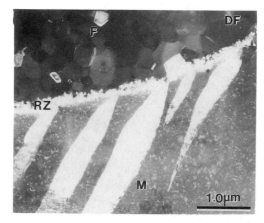

Fig. 10. A dark field (DF) transmission electron micrograph showing interface in a continuous fiber (F) α-Al$_2$O$_3$ (F)/Mg alloy (ZE41A) matrix (M) within the reaction zone (RZ).

Properties

Modulus. Unidirectionally reinforced continuous fiber-reinforced metal-matrix composites show a linear increase in the longitudinal Young's modulus as a function of the fiber volume fraction. Figure 11 shows an example of modulus increase as a function of fiber volume fraction for an alumina fiber-reinforced aluminum–lithium alloy matrix (15). The increase in the longitudinal Young's modulus is in agreement with the rule of mixtures value, whereas the modulus increase in a direction transverse to the fibers is very low. Particle reinforcement also results in an increase in the modulus of the composite; the increase, however,

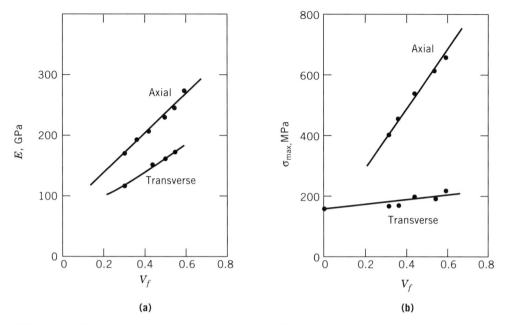

Fig. 11. Modulus increase as a function of fiber volume fraction (V_f) for alumina fiber-reinforced aluminum–lithium alloy matrix for (**a**) E (elastic modulus), and (**b**) σ_{max}. To convert MPa to psi, multiply by 145.

is much less than that predicted by the rule of mixtures. This is understandable inasmuch as the rule of mixtures is valid only for continuous fiber reinforcement. Figure 12 shows increase in Young's modulus in an aluminum composite with a SiC reinforcement volume fraction for different forms of reinforcement, eg, continuous fiber, whisker, or particle (16). There is a loss of reinforcement efficiency

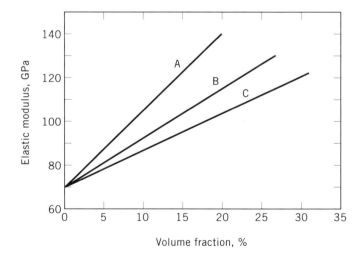

Fig. 12. Young's modulus increase in an aluminum composite with SiC reinforcement volume fraction for different forms of reinforcement: A, continuous fiber; B, whisker; and C, particle (16). To convert GPa to psi, multiply by 145,000.

in going from continuous fiber to particle. Metal-matrix particulate composites such as SiC particle-reinforced aluminum can offer a 50–100% increase in modulus over that of unreinforced aluminum, ie, the modulus equivalent to that of titanium but density about 33% less. Also, unlike the fiber-reinforced composites, the stiffness enhancement in particulate composites is reasonably isotropic.

Strength.   Prediction of MMC strength is more complicated than the prediction of modulus. Consider an aligned fiber-reinforced metal-matrix composite under a load P_c in the direction of the fibers. This load is distributed between the fiber and the matrix:

$$P_c = P_m V_m + P_f V_f$$

where P_m and P_f are loads on the matrix and fiber, respectively. This equation can be converted to the following rule of mixtures relationship under conditions of isostrain, ie, the strain in the fiber, matrix, and composite is the same:

$$\sigma_c = \sigma_f V_f + \sigma_m V_m$$

where σ is the stress, V is the volume fraction, and c, f, and m denote composite, fiber, and matrix, respectively. This rule of mixtures equation says that the strength of the composite is a volume weighted average of the strengths of the fiber and matrix. This is correct, however, the *in situ* values of σ_f and σ_m must also be included. If the fiber remains essentially elastic up to the point of fracture, the fiber strength in the composite, σ_f, is the same as that determined in an isolated test of the fiber. This is particularly true of ceramic fibers, however, the same cannot be said for the metallic matrix. The matrix strength in the composite (*in situ* strength) is not the same as that determined from a test of an unreinforced matrix sample in isolation because the metal matrix can suffer several microstructural alterations during processing and consequently changes in its mechanical properties. In view of the fact that, in general, ceramic reinforcements have a coefficient of thermal expansion greater than that of most metallic matrices, thermal stresses are generated in both the components, the fiber and the matrix. The reinforcement also acts as a constraint on the matrix flow, thus increasing its flow stress.

A series of events can take place in response to the thermal stresses: (*1*) plastic deformation of the ductile metal matrix (slip, twinning, cavitation, grain boundary sliding, and/or migration); (*2*) cracking and failure of the brittle fiber; (*3*) an adverse reaction at the interface; and (*4*) failure of the fiber–matrix interface (17–20).

An important example of an MMC *in situ* composite is one made by directional solidification of a eutectic alloy. The strength, σ, of such an *in situ* metal-matrix composite is given by a relationship similar to the Hall-Petch relationship used for grain boundary strengthening of metals:

$$\sigma = \sigma_0 + k\lambda^{-1/2}$$

where σ_0 is a friction stress term, k is a material constant, and λ is the spacing between rods or lamellae (21). The interfiber spacing, λ, can be varied rather easily by controlling the solidification rate, R, because $\lambda^2 R$ = constant.

Toughness. Toughness can be regarded as a measure of energy absorbed in the process of fracture or more specifically as the resistance to crack propagation, K_{Ic}. The toughness of MMCs depends on the matrix alloy composition and microstructure; the reinforcement type, size, and orientation; and processing insofar as it affects microstructural variables, eg, distribution of reinforcement, porosity, segregation, etc.

For a given V_f, the larger the diameter of the fiber, the tougher the composite. This is because the larger the fiber diameter for a given fiber volume fraction, the larger the amount of tough, metallic matrix in the interfiber region that can undergo plastic deformation and thus contribute to the toughness. Table 2 shows this effect of fiber diameter on the impact fracture energy for a boron fiber–aluminum composite system (22).

Unidirectional fiber reinforcement can lead to easy crack initiation and propagation compared to the unreinforced alloy matrix. Braiding of fibers can make the crack propagation toughness increase tremendously because of extensive matrix deformation, fiber bundle debonding, and pullout (23). Table 3 summarizes this effect of fiber braiding (23). The fracture energy is the maximum for the three-dimensionally arranged alumina fiber–aluminum composites.

The general range of K_{Ic} values for particle-reinforced aluminum-type MMCs is between 15–30 MPa·m$^{1/2}$, whereas short fiber- or whisker-reinforced MMCs have $\approx$5–10 MPa·m$^{1/2}$. Table 4 gives fracture toughness data for silicon carbide whisker (SiC$_w$) and silicon carbide particle (SiC$_p$) reinforced aluminum composites (24).

Thermal Stresses and Properties. In general, ceramic reinforcements (fibers, whiskers, or particles) have a coefficient of thermal expansion greater than that of most metallic matrices. This means that when the composite is

Table 2. Impact Fracture Energy of B/Al 1100[a]

Fiber diameter, μm	Energy, kJ/m^2[b]
100	90
140	150
200	200–300

[a]Ref. 22.
[b]To convert kJ/m^2 to ft·lbf/in.^{2}, divide by 2.10.

Table 3. Effect of Fiber Braiding on Fracture Energy of Alumina Fiber/Al–Li Composite[a]

Material	Initiation energy, J[b]	Propagation energy, J[b]	Total fracture energy, J[b]
Al–2.5 Li alloy	173	145	218
0.34 V_f alumina/Al–Li unidirectional	62	79	141
0.36 V_f alumina/Al–Li 3-D braided	68	196	264

[a]Ref. 23.
[b]To convert J to ft·lb, divide by 1.356.

Table 4. Toughness of SiC Particle- and Whisker-Reinforced 6061 Al[a]

Material	K_{Ic}, MPa·m$^{1/2}$
20 v/o SiC$_w$/6061-T6 Al	7.1
25 v/o SiC$_p$/6061-T6 Al	15.8
6061-T6Al	37.0

[a]Ref. 24.

subjected to a temperature change, thermal stresses are generated in both components.

Work with a single-crystal copper matrix containing large diameter tungsten fibers has shown the importance of thermal stresses in MMCs (17). A dislocation etch-pitting technique was employed to delineate dislocations in a single-crystal copper matrix and it was shown that near the fiber the dislocation density was much higher in the matrix than the dislocation density far away from the fiber. The situation in the as-cast composite can be depicted where a primary plane section of the composite is shown having a hard zone (high dislocation density) around each fiber, and a soft zone (low dislocation density) away from the fiber (Fig. 13). The enhanced dislocation density in the copper matrix near the fiber arises because of the plastic deformation in response to the thermal stresses generated by the thermal mismatch between the fiber and the matrix. The intensity of the gradient in dislocation density depends on the interfiber spacing. The dislocation density gradient decreases with a decrease in the interfiber spacing. The existence of a plastically deformed zone containing high dislocation density in the metallic matrix in the vicinity of the reinforcement has been confirmed by transmission electron microscopy, both in fibrous and particulate metal-matrix composites (25–28).

Thermal expansion mismatch between the reinforcement and the matrix is an important consideration. Thermal mismatch is something that is difficult to avoid in any composite, however, the overall thermal expansion characteristics of a composite can be controlled by controlling the proportion of reinforcement

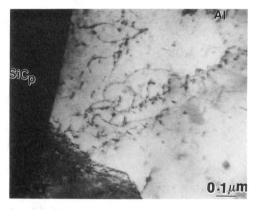

Fig. 13. Transmission electron micrograph (tem) showing dislocations in aluminum in the region near a silicon carbide particle, SiC$_p$.

and matrix and the distribution of the reinforcement in the matrix. Many models have been proposed to predict the coefficients of thermal expansion of composites, determine these coefficients experimentally, and analyze the general thermal expansion characteristics of metal-matrix composites (29–33).

Aging. Frequently the metal-matrix alloy used in a MMC has precipitation hardening characteristics, ie, such an alloy can be hardened by suitable heat treatment called aging treatment. It has been shown that the microstructure of the metallic matrix is modified by the presence of ceramic reinforcement and consequently the standard aging treatment for, eg, an unreinforced aluminum alloy, is not valid (34). The particle- or whisker-type reinforcements such as SiC, B_4C, Al_2O_3, etc, are unaffected by the aging process. These reinforcements, however, can affect the precipitation behavior of the matrix quite significantly. In particular, a higher dislocation density in the matrix metal or alloy than that in the unreinforced metal or alloy is produced. The higher dislocation density in the matrix has its origin in the thermal mismatch, $\Delta\alpha$, between the reinforcement and the metallic matrix. This thermal mismatch can be quite large, for example, in the case of SiC and aluminum it has a high value of 21×10^{-6}/K. The precipitate hardenable aluminum alloy matrix has a much more important role to play in these composites than a nonhardenable matrix material. A considerable strength increment results due to age hardening treatments. The high dislocation density also affects the precipitation kinetics in a precipitation hardenable matrix such as 2XXX series aluminum alloy. Indeed, faster precipitation kinetics have been observed in the matrix than in the bulk unreinforced alloy. This has important practical implications, ie, one may not use the standard heat treatments commonly available in various handbooks, eg, for monolithic aluminum alloys for an alloy used as a matrix in an MMC. Such accelerated aging effects occur in alloy systems where dislocation-stimulated precipitation is important.

Fatigue. This is the phenomenon of mechanical property degradation leading to failure of a material or a component under cyclic loading. Many high volume applications of composite materials involve cycling loading situations, eg, automobile components. Application of conventional approaches, such as the stress vs cycle (S–N) curves or the application of linear elastic fracture mechanics (LEFM) to fatigue of composites, is not straightforward. The main reasons for this are the inherent heterogeneity and anisotropic nature of the composites which result in damage mechanisms in composites being very different from those encountered in conventional, homogeneous, or monolithic material. Novel approaches, such as that epitomized by the measurement of stiffness reduction as a function of cycles, are being used to analyze the fatigue behavior of MMCs. Because many applications of composite MMCs involve temperature changes, it is important that thermal fatigue characteristics of composites be evaluated in addition to their mechanical fatigue characteristics.

Fatigue Crack Initiation. Particle or short fibers can provide easy crack initiation sites. The detailed behavior can vary depending on the volume fraction of the reinforcement, shape, size, and most importantly on the reinforcement/matrix bond strength. Frequently in aluminum matrix composites, especially those made by the casting route, there are particles other than SiC present, such as $CuAl_2$, $(Fe, Mn)_3SiAl_{12}$, and $Cu_2MgSi_6Al_5$. Particle pushing ahead of the solidification front results in SiC particles and the so-called constituent particles decorating the cell boundaries in the aluminum alloy matrix. The slow cooling rate in the conventional castings process leads to rather coarse constituent particles and SiC segregating to the interdendritic region. For example, fatigue crack initiation has been observed at the poles of SiC whiskers in 2124 aluminum and at the unbonded silicon carbide whiskers as well as non-SiC intermetallics (35). Not unexpectedly, a reduction in the clustering of SiC particles and the number and size of the intermetallics leads to an increased fatigue life.

S–N Curves.   A rule-of-thumb approach in fatigue of monolithic metals is to increase their monotonic strength which concomitantly results in an increase in cyclic strength. This rule assumes that the ratio of fatigue strength to tensile strength is about constant. It is generally true that the maximum efficiency in terms of stiffness and strength gains in fiber-reinforced composites occurs when the fibers are continuous, uniaxially aligned, and the properties are measured parallel to the fiber direction. Going off-angle, the role of the matrix becomes more important. Because of the highly anisotropic nature of the fiber-reinforced composites in general, the fatigue strength of off-axis MMCs (just like that of any other kind of off-axis fibrous composite), decreases with increasing angle between the fiber axis and the stress axis. In a study of the fatigue behavior of tungsten fiber-reinforced aluminum–4% copper alloy under tension compression cycling it was found that increasing the fiber volume fraction from 0 to 24% resulted in increased fatigue resistance (36). This was a direct result of increased monotonic strength of the composite as a function of the fiber volume fraction. One significant drawback of studies on fatigue behavior of a material using this S–N approach is that no distinction can be made between the crack initiation phase and the crack propagation phase.

In terms of general S–N curve behavior, a composite such as a silicon carbide particle-reinforced aluminum alloy shows an improved fatigue behavior compared to the reinforced alloy (37–40). Such an improvement in stress controlled cyclic loading or high cyclic fatigue is attributed to the higher stiffness of the composite. However, the fatigue behavior of the composite, evaluated in terms of strain amplitude vs cycles or low cycle fatigue, was found to be inferior to that of the unreinforced alloy (41). This was attributed to the general lower ductility of the composite compared to the unreinforced alloy.

Fatigue Crack Propagation Tests.   Tests on notched or precracked samples, generally conducted in an electrohydraulic closed-loop testing machine on notched samples, provide results in the form of crack growth per cycle, $\log(da/dN)$ vs cyclic stress intensity factor $\log \Delta K$. Crack growth rate, da/dN, is related to the cyclic stress intensity factor range, ΔK, according to the power law relationship first formulated in 1963, where A and m depend on the material and test conditions (42).

$$da/dN = A \, \Delta K^m$$

The applied cyclic stress intensity range is given by

$$\Delta K = Y \Delta \sigma a^{1/2}$$

where Y is a geometric factor, $\Delta\sigma$ is the cyclic stress range, and a is the crack length. The principal concern in this kind of test is to make sure that there is only one dominant crack that is propagating. This is particularly a problem in fibrous composites. In the case of a particulate composite, frequently a dominant crack can be made to propagate and the above relationships can be used. Figure 14 shows some reinforcement and crack tip interactions in a particulate composite. Whether a particle is cracked or debonded from the matrix depends on the particle–matrix interface strength. A schematic representation of crack growth vs cyclic intensity factor for particulate MMCs is shown in Figure 15. Higher threshold values, ΔK_{th}, are observed for composites than for monolithic materials. The slope of da/dN vs ΔK in region II for the composites is generally comparable to the reinforced alloys. It appears that choosing the optimum particle size and volume fractions together with a clean matrix alloy results in a composite with improved fatigue characteristics.

Fatigue crack propagation studies have been done on aligned eutectic or *in situ* composites as well. Because many of these *in situ* composites are meant for high temperature applications in turbines, their fatigue behavior has been studied at temperatures ranging from room temperature to 1100°C. The general consensus is that the mechanical behavior of *in situ* composites, static and cyclic strengths, is superior to that of the conventional cast superalloys (21).

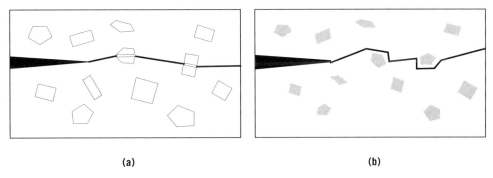

(a)   (b)

Fig. 14. Reinforcement and crack tip interactions in a particulate composite: (**a**) coarse particles in a strong particle–matrix interface, and (**b**) fine particles in a weak particle–matrix interface.

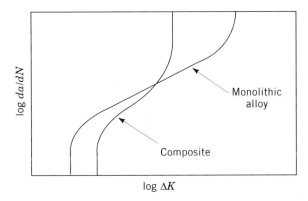

Fig. 15. Crack growth vs cyclic intensity factor for particulate MMCs.

Laminated MMCs.   There are three types of laminated MMCs: (*1*) metallic matrix-containing fibers oriented at different angles in different layers, similar to that of polymeric laminates; (*2*) two or more different metallic sheets bonded to each other; and (*3*) laminated metal/discontinuously reinforced MMC.

In the case of the fibrous laminate not much work has been done, but it has been observed that a significant loss of stiffness in boron–aluminum laminate occurs when cycled in tension–tension (43,44). Also, in a manner similar to that in the laminated PMCs, the ply stacking sequence affects the fatigue behavior. For example, 90° surface plies in a 90°/0° sequence develop damage more rapidly than 0° plies. In the case of laminates made out of metallic sheets, eg, stainless steel and aluminum, further enhanced resistance against fatigue crack propagation than either one of the components in isolation has been observed (45).

Stiffness Loss in Fatigue.   The complexities in composites lead to the presence of many modes of damage, such as matrix cracking, fiber fracture, delamination, debonding, void growth, multidirectional cracking, etc. In the case of the isotropic material, a single crack propagates in a direction perpendicular to the cyclic (mode I) loading axis (Fig. 16). In the fiber-reinforced composite, on the other hand, a variety of subcritical damage mechanisms (Fig. 17) lead to a

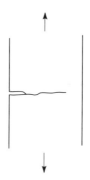

Fig. 16.   Self-similar crack propagation in an isotropic material. The crack propagates in a direction perpendicular to the cyclic loading axis (mode I loading).

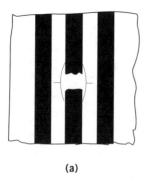  

(a)   (b)   (c)

Fig. 17.   Variety of subcritical damage mechanisms in fiber-reinforced composites, that lead to a highly diffuse damage zone. (**a**) Fiber cracking, (**b**) matrix cracking, and (**c**) interface debonding.

highly diffuse damage zone, and these multiple fracture modes appear rather early in the fatigue life of composites. The various types of subcritical damage result in a reduction of the load carrying capacity of the laminate composite, which in turn manifests itself as a reduction of laminate stiffness and strength (46–50). The stiffness changes in the laminated composites to the accumulated damage under fatigue have been experimentally related. The change in stiffness values is a good indicator of the damage in composites, thus stiffness loss in continuous fiber-reinforced MMCs can be a useful parameter for detecting fatigue damage initiation and growth. This progressive loss of stiffness during fatigue of composites is a characteristic that is very different from the behavior of metals in fatigue. Similar loss of stiffness has been observed on subjecting a fiber-reinforced composite to thermal fatigue (51).

Creep.   The phenomenon of creep refers to time-dependent deformation. In practice, at least for most metals and ceramics, the creep behavior becomes important at high temperatures and thus sets a limit on the maximum application temperature. In general, this limit increases with the melting point of a material. An approximate limit can be estimated to lie at about half of the Kelvin melting temperature. The basic governing equation of steady-state creep can be written as follows:

$$\epsilon = A(\sigma/G)^n \exp(-\Delta Q/kT)$$

where ϵ is the steady-state creep strain rate, σ is the applied strain, n is an exponent, G is the shear modulus, ΔQ is the activation energy for creep, k is the Boltzmann's constant, and T is temperature in kelvin. In general, the creep resistance of metal is improved by the incorporation of ceramic reinforcements. The steady-state creep rate as a function of applied stress for silver matrix and tungsten fiber–silver matrix composites at 600°C is an example (Fig. 18) (52).

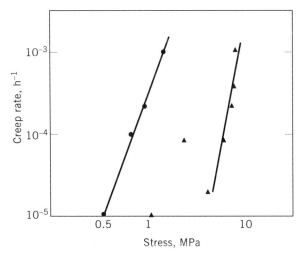

Fig. 18.   Steady-state creep rate as a function of applied stress for silver matrix (●) and tungsten fiber–silver matrix composites (▲) at 600°C. To convert MPa to psi, multiply by 145.

The modeling of creep behavior of MMCs is complicated because in the temperature regime where the metal matrix may be creeping, the ceramic reinforcement is likely to be deforming elastically.

Applications

In aerospace applications, low density coupled with other desirable features, such as tailored thermal expansion and conductivity, high stiffness and strength, etc, are the main drivers. Performance rather than cost is an important item. Inasmuch as continuous fiber-reinforced MMCs deliver superior performance to particle-reinforced composites, the former are frequently used in aerospace applications. In nonaerospace applications, cost and performance are important, ie, an optimum combination of these items is required. It is thus understandable that particle-reinforced MMCs are increasingly finding applications in nonaerospace applications.

Reduction in the weight of a component is a significant driving force for any application in the aerospace field. For example, in the Hubble telescope, a pitch-based continuous carbon fiber-reinforced aluminum was used for waveguide booms because of its light weight; high elastic modulus, E; and low coefficient of thermal expansion, α. Other aerospace applications of MMCs involve replacement of light but toxic beryllium. For example, in the United States Trident missile, beryllium has been replaced by a SiC_p/Al composite.

An important application of MMCs in the automotive area is in diesel piston crowns (53). This application involves incorporation of short fibers of alumina or alumina–silica in the crown of the piston. The conventional diesel engine piston has an Al–Si casting alloy with a crown made of a nickel cast iron. The replacement of the nickel cast iron by aluminum matrix composite results in a lighter, more abrasion resistant, and cheaper product. Another application in the automotive sector involves the use of carbon fiber and alumina particles in an aluminum matrix for use as cylinder liners in the Prelude model of Honda Motor Co.

Particulate metal-matrix composites, especially with light metal-matrix composites such as aluminum and magnesium, also find applications in automotive and sporting goods. In this regard, the price per kg becomes the driving force for application. An excellent example involves the use of Duralcan particulate MMCs to make mountain bicycles. The company Specialized Bicycle in the Unites States sells these bicycles with the frame made from extruded tubes of 6061 aluminum containing about 10% alumina particles. The primary advantage is the gain in stiffness. An important item in this regard has to do with recycling and reclamation, particularly in aluminum matrix composites, because recycling has been extremely successful in the aluminum industry. The two terms, recycling and reclamation, are used advisedly for composites. Recycling (qv) refers to reuse of a worn product as a composite, and involves a reclamation of the individual components, ie, aluminum and ceramic particles separately (16). Cast MMCs can be remelted and reused as a composite. Although some of the issues such as particle/molten metal reaction are the same in remelting, it should be appreciated that scrap sorting and melt cleanliness and degassing can be rather complex problems in MMCs. Special fluxing and degassing techniques have been developed at Duralcan. Powder processed MMCs as well as the reclamation route involve remelting. According to one report (16), by using combined argon and salt fluxing, silicon carbide particles can be removed from the melt and 85–90% of the aluminum reclaimed.

An important potential commercial application of particle-reinforced aluminum composite is in the making of automotive driveshafts. In driveshaft design, the speed at which it becomes dynamically unstable needs to be considered. In terms of geometric parameters, a shorter shaft length and larger diameter gives a higher critical speed, N_c; in terms of material parameters, the higher the specific stiffness (E/ρ), the higher the N_c. Changes in the driveshaft geometry, increase in length or a reduction in the diameter, can be made while maintaining a constant critical speed. A decrease in the driveshaft diameter can be important because of under-chassis space limitations.

Copper-based composites having Nb, Ta, or Cr as the second phase in a discontinuous form are of interest for certain applications requiring high thermal conductivity and high strength, eg, in high heat flux applications in rocket engine thrust chambers. Carbon fiber-reinforced copper has applications in the aerospace industry as a very high thermal conductivity material. The problem with carbon fibers and metallic matrices is that the surface energy considerations preclude wetting of the carbon fibers by the molten metals. The surface energy of carbon fiber is about 100 mJ/m^2 (=dyn/cm), whereas for most molten metals the surface energy is 10 times higher. Wetting occurs when the surface energy of the substrate (fiber) is higher than that of the molten metal. Thus the wetting of carbon fibers by molten copper is not expected.

Filamentary Superconductors.   Niobium–titanium superconductors are used in magnetic resonance imaging (MRI) techniques for medical diagnostics. The superconducting solenoid made from Nb–Ti/Cu composite wire is immersed in a liquid helium cryogenic Dewar flask. Nb–Ti/Cu superconducting composites are also used in various high energy physics applications, such as particle accelerators. Nb_3Sn/Cu superconducting composites are used for magnetic fields greater than 12 T (1.2×10^5 G). Such high fields are encountered in a thermonuclear fusion reactor, representing a sizeable fraction of the capital cost of such a fusion power plant.

Electronic-Grade MMCs.   Metal-matrix composites can be tailored to have optimal thermal and physical properties to meet requirements of electronic packaging systems, eg, cores, substrates, carriers, and housings. A controlled thermal expansion space truss, ie, one having a high precision dimensional tolerance in space environment, was developed from a carbon fiber (pitch-based)/Al composite. Continuous boron fiber-reinforced aluminum composites made by diffusion bonding have been used as heat sinks in chip carrier multilayer boards.

Unidirectionally aligned, pitch-based carbon fibers in an aluminum matrix can have high thermal conductivity along the fiber direction. The conductivity transverse to the fibers is about two-thirds that of aluminum. Such a C/Al composite is useful in heat-transfer applications where weight reduction is an important consideration, eg, in high density, high speed integrated circuit packages for computers and in base plates for electronic equipment. Another possible use of this composite is to dissipate heat from the leading edges of wings in high speed airplanes.

BIBLIOGRAPHY

"Laminated and Reinforced Metals" in *ECT* 3rd ed., Vol. 13, pp. 941–967, by C. C. Chamis, National Aeronautics and Space Administration.

1. K. K. Chawla, *Composite Materials: Science & Engineering*, Springer-Verlag, New York, 1987.
2. C. R. Kennedy, in P. Vincenzine, ed., *Proceedings of the 7th CIMTEC-World Ceramics Congress*, Elsevier, New York, 1991, p. 691.
3. A. J. Cook and P. S. Warner, *Mater. Sci. Eng.* **A144**, 189 (1991).
4. S. Nourbakhsh, F. L. Liang, and H. Margolin, *Met. Trans. A* **21A**, 213 (1990).
5. K. K. Chawla and L. B. Godefroid, *Proceedings of the 6th International Conference on Fracture*, Pergamon Press, Oxford, U.K., 1984, p. 2873.
6. W. H. Hunt, T. M. Osman, and J. J. Lewandowski, *J. Miner. Metal Mater. Soc.*, 30–35 (Mar. 1991).
7. M. Manoharan, L. Ellis, and J. J. Lewandowski, *Scripta Met. et. Mater.* **24**, 1515–1521 (1990).
8. K. H. Sandhage, G. N. Riley, Jr., and W. L. Carter, *J. Miner. Metal. Mater. Soc.* (Mar. 1991).
9. W. H. Hunt, in *Processing and Fabrication of Advanced Materials*, The Minerals and Metal Materials Society, Warrendale, Pa., 1994, pp. 663–683.
10. D. Lewis, *Metal Matrix Composites: Processing and Interfaces*, Academic Press, Inc., San Diego, Calif., 1991, p. 121.
11. E. J. Lavernia, E. Gutierrez, and J. Baram, *Mater. Sci. Eng.* **132A**, 119–133 (1991).
12. R. Mitra, W. A. Chiou, M. E. Fine, and J. R. Weertman, *J. Mater. Res.* **8**, 2300 (1993).
13. R. G. Hill, R. P. Nelson, and C. L. Hellerich, *Proceedings of the 16th Refractory Working Group Meeting*, Seattle, Wash., Oct. 1969.
14. K. K. Chawla and M. Metzger, *Advances in Research on Strength and Fracture of Materials*, Vol. 3, Pergamon Press, New York, 1978, p. 1039.
15. A. R. Champion, W. H. Krueger, H. S. Hartman, and A. K. Dhingra, *Proceedings of the 2nd International Conference on Composite Materials (ICCM/2)*, TMS-AIME, New York, 1978, p. 883.
16. D. J. Lloyd, *Int. Mater. Rev.* **39**, 1 (1994).
17. K. K. Chawla and M. Metzger, *J. Mater. Sci.* **7**, 34 (1972).
18. K. K. Chawla, *Metallography* **6**, 155 (1973).
19. K. K. Chawla, *Philos. Mag.* **28**, 401 (1973).
20. K. K. Chawla, *Fibre Sci. Tech.* **8**, 49 (1975).
21. N. S. Stoloff, *Advances in Composite Materials*, Applied Science Publishers, London, p. 247.
22. D. L. McDanels and R. A. Signorelli, *NASA TN D*, 8204 (1976).
23. A. P. Majidi and T. W. Chou, *Proc. ICCM VI* **2**, 422–430 (1987).
24. D. F. Hasson and C. R. Crowe, *Strength of Metals and Alloys*, Pergamon Press, Oxford, U.K., 1985, pp. 1515–1520.
25. R. J. Arsenault and R. M. Fisher, *Scripta Met.* **17**, 67–71 (1983).
26. T. Christman and S. Suresh, *Mater. Sci. Eng.* **102A**, 211–216 (1988).
27. H. J. Rack, *The Sixth International Conference on Composite Materials*, Elsevier Applied Science Publishers, New York, 1987, p. 382.
28. K. K. Chawla, A. H. Esmaeili, A. K. Datye, and A. K. Vasudevan, *Scripta Met. Mater.* **25**, 1315–1319 (1991).
29. E. H. Kerner, *Proc. Phys. Soc.* **69**, 808 (1956).
30. P. S. Turner, *J. Res. NBS* **37**, 239 (1946).
31. R. A. Schapery, *J. Comp. Mater.* **2**, 311 (1969).

32. R. U. Vaidya and K. K. Chawla, in K. Upadhya, ed., *Developments in Metal and Ceramic Matrix Composites*, The Minerals, Metals and Materials Society, Warrendale, Pa., 1991, pp. 253–272.
33. Z. R. Xu, K. K. Chawla, R. Mitra, and M. E. Fine, *Scripta Met. Mater.* **31**, 1277 (1994).
34. S. Suresh and K. K. Chawla, in S. Suresh, A. Needleman, and A. Mortensen, eds., *Metal Matrix Composites*, Butterworth-Heinemann, Boston, Mass., 1993.
35. D. R. Williams and M. E. Fine, *Proceedings of the 5th International Conference on Composite Materials (ICCM/V)*, TMS-AIME, Warrendale, Pa., 1985, p. 639.
36. M. A. McGuire and B. Harris, *J. Phys. D: Appl. Phys.* **7**, 1788 (1974).
37. K. K. Chawla in *Metal Matrix Composites: Properties*, Academic Press, Inc., San Diego, Calif., 1991.
38. J. K. Shang, W. Yu, and R. O. Ritchie, *Mater. Sci. Eng.* **A102**, 181–192 (1988).
39. D. L. Davidson, *Eng. Fract. Mech.* **33**, 965–977 (1989).
40. S. Kumai and J. F. Knott, *Mater. Sci. Eng.* **A146**, 317 (1991).
41. J. J. Bonnen, C. P. You, J. E. Allison, and J. W. Jones, in *Proceedings of the International Conference on Fatigue*, Pergamon Press, New York, 1990, pp. 887–892.
42. P. C. Paris and F. Erdogan, *J. Basic Eng. Trans. ASME* **85**, 528 (1963).
43. W. S. Johnson, *Damage in Composite Materials*, ASTM STP 775, American Society for Testing and Materials, Philadelphia, Pa., 1982, p. 83.
44. W. S. Johnson, in W. S. Johnson, ed., *Metal Matrix Composites: Testing, Analysis, and Failure Models*, ASTM STP 1032, ASTM, Philadelphia, Pa., 1989, p. 194.
45. K. K. Chawla and P. K. Liaw, *J. Mater. Sci.* **14**, 2143 (1979).
46. H. T. Hahn and R. Y. Kim, *J. Comp. Mater.* **10**, 156 (1976).
47. A. T. Highsmith and K. L. Reifsnider, *Damage in Composite Materials*, ASTM STP 775, ASTM, Philadelphia, Pa., 1982, p. 103.
48. R. Talreja, *Fatigue of Composite Materials*, Technical University of Denmark, Lyngby, 1985.
49. T. K. O'Brien and K. L. Reifsnider, *J. Comp. Mater.* **15**, 55 (1981).
50. S. L. Ogin, P. A. Smith, and P. W. R. Beaumont, *Composites Sci. Tech.* **22**, 23 (1985).
51. Z. R. Xu, A. Neuman, K. K. Chawla, A. Wolfenden, and G. M. Liggett, *High Performance Composites: Commnalty of Phenomena*, The Minerals, Metals, and Materials Society, Warrendale, Pa., 1994, p. 485.
52. A. Kelly and W. R. Tyson, *J. Mech. Phys. Solids*, **14**, 177–186 (1966).
53. T. Donomoto, N. Miura, K. Funatani, and N. Miyake, *SAE Technical Paper*, No. 83052, Detroit, Mich., 1983.

General References

K. K. Chawla, *Composite Materials: Science & Engineering*, Springer-Verlag, New York, 1987.
M. Taya and R. J. Arsenault, *Metal Matrix Composites*, Pergamon Press, Oxford, U.K., 1990.
S. Suresh, A. Needleman, and A. Mortensen, eds., *Metal Matrix Composites*, Butterworth-Heinemann, Boston, Mass., 1993.
T. W. Clyne and P. J. Withers, *An Introduction to Metal Matrix Composites*, Cambridge University Press, Cambridge, U.K., 1993.

K. K. CHAWLA
New Mexico Institute of Mining and Technology

METAL PLATING. See ELECTROPLATING; METALLIC COATINGS.

METAL SURFACE TREATMENTS

CASE HARDENING

The performance of many metallic components can be markedly improved by developing a surface region which is harder than that of the underlying region. Processes to achieve this are called case hardening. The case is the surface region which is hardened, and the region under the case is called the core (Fig. 1a). The original surface region is an intrinsic part of the case, unlike treatments such as electroplating (qv) where material is deposited on the original surface. Although case hardening is at times restricted to processes which change the chemistry of the surface region, eg, carburizing, herein this term is used in a broader sense to represent changing the surface region either physically, such as by cold working or surface heating and quenching, or by altering the chemistry of the surface region by addition of elements, such as by carburizing ion implantation (qv).

There are many characteristics of hard cases that make their development desirable. One is wear resistance. Usually, the process is designed to develop high compressive residual stresses in the surface which counteract tensile stresses induced by the loading condition during use of the component (1) (Fig. 1b).

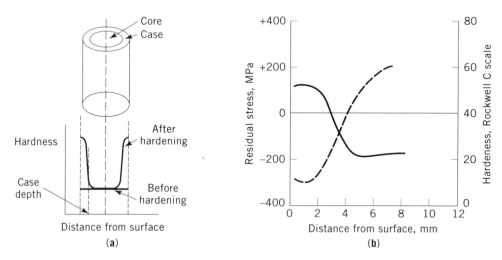

Fig. 1. (a) Schematic diagram showing definition of case hardening and of case depth where (—) is the diameter; (b) residual stress across the radius of a case hardened steel, showing the high compressive residual stress (– – –) at the surface induced by induction heating, as well as the microhardness (—) of the surface (1). To convert MPa to psi, multiply by 145.

Machine components are commonly subjected to loads, and hence stresses, which vary over time. The response of materials to such loading is usually examined by a fatigue test. The cylinder, loaded elastically to a level below that for plastic deformation, is rotated. Thus the axial stress at all locations on the surface alternates between a maximum tensile value and a maximum compressive value. The cylinder is rotated until fracture occurs, or until a large number of cycles is attained, eg, 10^6. The test is then repeated at a different maximum stress level. The results are presented as a plot of maximum stress, σ_{max}, versus number of cycles to fracture. For many steels, there is a maximum stress level below which fracture does not occur called the fatigue strength or fatigue limit. The fatigue fracture begins by formation of a crack at the surface during the tensile stress part of the loading cycle. Failure can be reduced by the presence of compressive residual stresses, and the development of these stresses is an important purpose of case hardening (Fig. 1b). The relationship between the residual stress and fatigue strength is given in Table 1 for three types of steel (qv), the most common metallic material that is case hardened, and the material emphasized herein. The three main characteristics to control in case hardening are case depth, case microstructure, and case hardness and strength. The methods and steels chosen for case hardening are closely related to these factors.

Table 1. Improvement in Material Life as a Result of Case Hardening[a]

Steel	Surface hardness, HRC[b]	Method of hardening	Number of shafts tested	Cycles to failure $\times 10^6$
4140	36–42	through-hardened	20	>0.4
4320	40–46	carburized to 1.0–1.3 mm	6	ca 0.8
1137	42–48	induction hardened to 3.0-mm min effective depth and 40 HRC[b]	5	>1.1

[a]Ref. 2.
[b]HRC = Rockwell C hardness (see HARDNESS).

Conventional Hardening of Steels

The Iron–Carbon Phase Diagram.   The hardening of steels begins with the formation of the high temperature phase austenite, denoted γ, which is iron with a fcc crystal structure. Austenite can dissolve up to about 2 wt % carbon in the interstices between the iron (qv) atoms, where small atoms like carbon locate. Also larger atoms such as manganese and chromium can replace iron atoms on lattice sites. The presence of these alloying elements incorporated into the crystal structure of austenite produces a solid solution. Austenite is stable only above about 723°C (Fig. 2). At lower temperatures, iron has bcc crystal structure, designated α. In this arrangement of iron atoms, the size of the interstices is much smaller than in the austenite, so that carbon solubility is much less, being a maximum of about 0.02 wt % C at 723°C. Iron combines with carbon to form a high melting temperature compound iron carbide [12011-67-5], sometimes called cementite, which has 6.67 wt % C. In heat-treating steels,

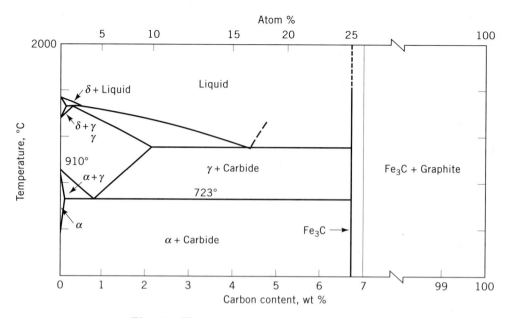

Fig. 2. The iron–carbon phase diagram (3).

the precursor structure is austenite. Austenite, ferrite, and Fe$_3$C phases are the ones of interest herein.

Decomposition of Austenite. In heat-treating steels, the initial step is usually to heat the steel into the austenite region (>723°C) and then control the cooling process to produce the desired structure. The phase diagram (Fig. 2) shows that austenite decomposes into the two phases α and Fe$_3$C upon cooling to below 723°C (3). The size and distribution of these phases, ie, the microstructure, depend on how the austenite is cooled. This has a critical bearing on the hardness because Fe$_3$C is very hard, and the smaller the carbide particles and the finer their distribution, the harder the steel.

When a steel is cooled sufficiently rapidly from the austenite region to a low (eg, 25°C) temperature, the austenite decomposes into a nonequilibrium phase not shown on the phase diagram. This phase, called martensite, is body-centered tetragonal. It is the hardest form of steel, and its formation is critical in hardening. To form martensite, the austenite must be cooled sufficiently rapidly to prevent the austenite from first decomposing to the softer structure of a mixture of ferrite and carbide. Martensite begins to form upon reaching a temperature called the martensite start, M_s, and is completed at a lower temperature, the martensite finish, M_f. These temperatures depend on the carbon and alloy content of the particular steel.

When a component at an austenitizing temperature is placed in a quenchant, eg, water or oil, the surface cools faster than the center. The formation of martensite is more favored for the surface. A main function of alloying elements, eg, Ni, Cr, and Mo, in steels is to retard the rate of decomposition of austenite to the relatively soft products. Whereas use of less expensive plain carbon steels is preferred, alloy steels may be required for deep hardening.

Tempering. Although martensite is hard and strong, it is also brittle, and therefore must be heat treated to improve toughness and ductility. This is achieved by reheating the quenched steel to a temperature in the ferrite $+Fe_3C$ region of the phase diagram, eg, $<723°C$ (Fig. 2). During this treatment, the single-phase, metastable martensite decomposes into a structure of fine particles of Fe_3C in ferrite. This heat treatment is called tempering, and the structure is called tempered martensite. The longer the tempering time and the higher the tempering temperature, the coarser the structure. This reduces the hardness and strength but increases ductility and toughness. Thus the desired mechanical properties are obtained by proper choice of the tempering temperature and time.

Case Hardening by Heat Treatment

No Chemical Change. By proper choice of the quenching process and the steel, a heat-treated component can be produced that is hard in the surface region, ie, all martensite, and softer in the center portion. However, for many applications the desired properties are obtained if the hardened region (case) is relatively shallow, eg, 0.2 cm. It is difficult to achieve this by conventional methods. Instead, the component is heat treated to achieve desired core properties, eg, low hardness but good ductility. Then only the surface region is heated to the austenite range, which is subsequently cooled rapidly to form martensite. This is case hardening without any chemical change in the surface region.

To prevent the core from attaining a high temperature, the surface is heated using a high energy flux for only a relatively short time. Before the core begins to heat, the energy source is removed and the surface is rapidly cooled, eg, by a water spray. The surface region also cools rapidly by conduction of heat into the cold core. This case hardening heat treatment takes only a short time to heat to the austenite region, and then to cool to room temperature. The short time is an essential characteristic of this type of case hardening.

Because the time at high temperature is much less, austenite is produced, which is chemically inhomogeneous especially with undissolved carbides, and has a fine grain crystal size. The formation of the hard martensite requires more rapid cooling than for conventional hardening. Thus case hardening by heat treatment intrinsically requires that the surface region to be hardened be relatively thin and cooled rapidly.

Case Hardening by Flame Heating. In case hardening by heating with a flame, gas burners of appropriate geometry and configuration are used to heat selected localized or general regions of the surface of the steel component. The case depth is controlled by the burner and flame characteristics, distance from the burner face to the surface of the component, and heating time. Figure 3**a** shows a typical heating curve (14). Note that several seconds are required for the surface region to heat to the temperature range of austenite stability (see Fig. 2). When the desired temperature is reached, the flame is removed and the surface is sprayed with water, cooling the case to room temperature in a few seconds. This is sufficiently rapid to produce some martensite in the case for even plain carbon and low alloy steels. Figure 3**b** shows a typical hardness distribution curve. The flame hardening process can be readily automated.

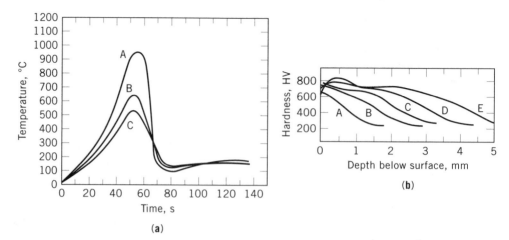

Fig. 3. Flame hardening: (**a**) temperature–time heating curves of a 25 × 50 × 100 mm specimen at a feed of 75 mm/min and burner distance of 8 mm showing temperatures of A, surface; B, 2 mm below surface; and C, 10 mm below surface; (**b**) hardness–depth curves for a 0.50% C steel 25 × 75 × 100 mm specimen at a feed of 50 mm/min, temperatures in °C measured 10 mm below the surface, and burner distances in mm, respectively, of A, 530 and 50; B, 540 and 12; C, 545 and 10; D, 550 and 8; and E, 565 and 6. Flame heating followed by water spray quenching. HV = Vickers hardness. Courtesy of Butterworths (4).

Case Hardening by Induction Heating. Electric current flow can be induced by placing an electrically conducting material in the vicinity of an alternating magnetic field. If the frequency is sufficiently high (eg 10^4–10^6 Hz), this current is concentrated in the surface layers. The flow of current causes resistive or Joule heating, so high frequency (eg, 10^4–10^6 Hz) magnetic fields can be used to selectively heat the surface layer of steel components. The coupling of induction heating to heat the surface region to the austenitizing temperature with subsequent rapid cooling of this region produces a hard surface region. This method is called induction hardening. Following surface hardening, the component can be tempered by induction heating, in some cases using the same equipment.

A wide range of shapes can be induction hardened and a system for quenching can be incorporated into the coil assembly. The process can be automated to harden components in an essentially continuous manner. The case depth is controlled by the coil design, frequency, power, and time of application.

Typical hardness distribution curves obtained by induction hardening are available (5). Case depths of ca 0.05–0.5 cm are commonly obtained.

Case Hardening by Laser Heating. The process of heating surfaces using lasers (qv) is not fundamentally different from that of flame and induction heating. The energy density is much higher, however, producing a very high heating rate. To prevent heating to too high, ie, above melting temperature, the heating time is very short. Usually temperature–time curves are not measured, but calculated (6). The surface region that is heated is considerably thinner than in flame and induction heating. Thus the underlying core remains at room temperature and serves as a heat sink for cooling. Because the surface is intimately connected to the core, the surface is self-quenched, and the cooling rate controlled only by the thermal diffusivity of the steel. Cooling times are of the order of only a few seconds. In some cases, the surface must be coated to increase the amount of light absorbed.

Figure 4 shows a typical hardness distribution (7). The case depth is considerably less than that for flame and induction hardening. The case has a high compressive residual stress, which improves the fatigue properties (8).

Case Hardening by Electron Beam Heating. Surfaces can be heated effectively using electron beams which can be focused by magnetic and electric lenses to control the spot size, and hence the energy density at the surface. Also, the beam can be moved in a controlled fashion to cover areas of the surface to be hardened. The process must be conducted in vacuum because of excessive scattering of electrons by air, thus cooling of the heated surface region is restricted to conduction into the cold core. This limits the case hardened region to a relatively thin layer.

Chemical Change.   *Gas Carburizing.*   In case hardening by carburizing, carbon atoms are deposited by chemical reaction on the surface of the steel while at high temperature in the austenite region. The atoms then move by diffusion into the steel, but are continuously replaced at the surface by more carbon atoms. The rate of deposition exceeds that of diffusion, so that a carbon gradient builds up with time, as depicted in Figure 5. The steel is then quenched to convert this surface region to martensite, forming a hard surface with a favorable compressive residual stress.

If the gas has the correct composition, the carbon content at the surface increases to the saturation value, ie, the solubility limit of carbon in austenite (Fig. 2), which is a function of temperature. Continued addition of carbon to the surface increases the carbon content curve. The surface content is maintained at this saturation value (9) (Fig. 5). The gas carburizing process is controlled by three factors: (*1*) the thermodynamics of the gas reactions which determine the equilibrium carbon content at the surface; (*2*) the kinetics of the chemical

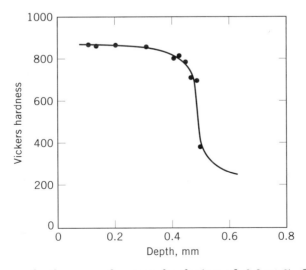

Fig. 4. Hardness–depth curve for case hardening of 0.6 wt % C steel by laser heating (7).

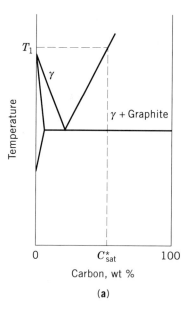

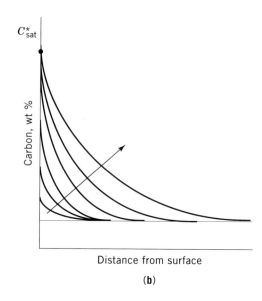

Fig. 5. Schematic illustration of the increase in carbon content with carburizing time. The maximum carbon content at the surface is given by the saturation value, C^*_{sat}. (**a**) Phase diagram; (**b**) plot of carbon content under the surface where the arrow represents increasing time at temperature T_1 (9).

reactions which deposit the carbon; and (*3*) the diffusion of carbon into the austenite.

Chemical Thermodynamics. Consider carburizing using a mixture of CO and CO_2. The chemical reaction can be represented by

$$CO_2 + C^\gamma \longrightarrow 2\,CO$$

where C^γ is the carbon in solution in austenite. The equilibrium concentration of the gaseous species, ie, vol % CO and vol % CO_2, or the partial pressure, p_i, of each species, and the carbon content of the austenite in wt % C are related through an equilibrium constant, K.

$$K = p_{CO}^2/(p_{CO_2} \cdot C^\gamma)$$

The temperature dependence of K is well known experimentally. At a given temperature, if the composition of the gas is fixed, then equilibrium carbon content is fixed. This composition is sometimes referred to as the carbon potential.

The relationship between the gas composition, austenite carbon content, and temperature is known. For example, a gas in equilibrium with austenite containing 0.8% C must have the partial pressure ratio $p_{CO}^2/(p_{CO_2}) = 20$. If the gas is comprised of only CO and CO_2, then the sum of the partial pressures must equal the total pressure of the system. These thermodynamic relations can be presented in several graphical forms (10,11).

In commercial carburizing, the gas mixture contains not only CO and CO_2, but also CH_4, H_2, and H_2O, as well as the inert N_2. Reactions characterized by their equilibrium constants such as

$$CO_2 + H_2 \longrightarrow CO + H_2O$$

occur. For example, at 926°C, where K for this latter equation equals 1.43,

$$1.43 = (X_{H_2O})(X_{CO})/(X_{H_2})(X_{CO_2})$$

where X_i represents mol % of the various species. This equation and the equilibrium equation for CO, CO_2, and the carbon content of the steel, must be simultaneously satisfied. Thus the carbon content of the austenite can be controlled by controlling the water vapor content of the gas, which is usually monitored and reported as the dew point.

Many commercial gases are generated by burning hydrocarbons (qv) eg, natural gas or propanes, in air (see GAS, NATURAL; LIQUIFIED PETROLEUM GAS). The combustion process, especially the amount of air used, determines the gas composition. For a given fuel-to-air ratio, the gas composition can be used to determine the water vapor content required to achieve a desired equilibrium carbon content of the austenite (see COMBUSTION TECHNOLOGY).

The equilibrium carbon content of austenite also depends on the alloy content, eg, Cr or Ni of the steel. A given gas composition equilibrates with a carbon content of the austenite which is different for a plain carbon steel than for an alloy steel.

Diffusion of Carbon.   When carbon atoms are deposited on the surface of the austenite, these atoms locate in the interstices between the iron atoms. As a result of natural vibrations the carbon atoms rapidly move from one site to another, statistically moving away from the surface. Carbon atoms continue to be deposited on the surface, so that a carbon gradient builds up, as shown schematically in Figure 5. When the carbon content of the surface attains the equilibrium value, this value is maintained at the surface if the kinetics of the gas reactions are sufficient to produce carbon atoms at least as fast as the atoms diffuse away from the surface into the interior of the sample.

The rate of diffusion of the carbon atoms is given by Fick's laws of diffusion. In one dimension,

$$\text{number of carbon atoms/(time·area)} = -D[dC/dx]$$

where C is concentration of carbon, x is distance, and D is the diffusion coefficient, which has temperature dependence

$$D = D_0 \exp[-Q/RT]$$

where D_0 is a constant, Q is the activation energy dependent on the carbon and alloy content, R is the ideal gas constant, and T is temperature in Kelvin. Thus the rate of diffusion of carbon is exponentially related to temperature.

Relations exist between the temperature, T, and the average time, t, it takes a carbon atom to move a given distance, x. One approximation is

$$x \simeq (D \cdot t)^{1/2}$$

Similar expressions relate the diffusion distance to the case depth, d, for example, $d \approx (660\ t^{1/2})\ \exp(-8287/T)$, where d is in mm, t in hours, and T in degrees Kelvin. This equation is based on the surface being saturated, ie, at the solubility limit, at each temperature. At 927°C, carburizing for four hours gives a case depth of about 1.3 mm. A higher temperature allows the use of a shorter time, but the risk of an unacceptable increase in the austenite grain size must be considered. Using much lower temperatures may require inordinately long times.

The maximum surface carbon content is usually set by the gas composition via the equilibrium constant. If the gas reaction kinetics deposit carbon at a rate which cannot be equaled by the diffusion of carbon into the steel, then the surface value may be less than the possible equilibrium value.

In some cases, the carbon profile may not provide the necessary hardness or other properties. For example, if the carbon content is too high, quenching to room temperature may not produce all martensite at the surface because the high carbon content places the martensite finish temperature, M_f, below room temperature. This results in the presence of soft retained austenite, and a low surface hardness. Conversion to martensite by subzero cooling to below the M_f temperature can increase the hardness (Fig. 6) (12).

The high carbon content at the surface can be lowered by reheating the steel to the austenite region in a noncarburizing atmosphere. This allows the carbon

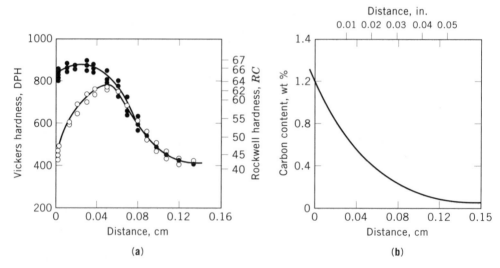

Fig. 6. (a) The effect of subzero cooling on the hardness gradient in a carburized and quenched 3312 steel where (○) is oil quenched from 925 to 20°C and (●) is cooled to −195°C. The initial quench to 20°C does not convert all of the austenite to martensite because the high carbon content in the surface region lowers the M_f temperature below 20°C. Subsequent cooling to −195°C converts most of the retained austenite to martensite, raising the hardness. (b) The carbon gradient for this system (12).

to diffuse toward the center, lowering the carbon content and raising the M_f above room temperature. Then upon quenching all martensite is formed. This second heat treatment, often referred to as a diffusion treatment, can also be achieved at the end of the carburizing time by changing the gas atmosphere to minimize carburizing. This procedure is sometimes called boost diffusion carburizing.

An important advancement in carburizing has been the development of diffusion models to calculate the carbon gradient as a function of time as the gas composition and temperature change (13). Such models can be coupled with computer control of the gas composition and temperature to produce desired carbon profiles.

Selective Carburizing. In most components, it is desirable to carburize only parts of the surface. To prevent other regions from carburizing, they must be protected. For holes, simple plugs of copper may be used. In some cases, copper plating can be applied, but diffusion into the steel must be considered, and the copper may have to be machined off later. Coatings (qv), which can be applied as a paste and then removed after heat treatment, are also available and include copper plating, ceramic coatings, and copper and tin pastes.

Gas Nitriding. The nitrogen atom, similar in size to carbon, is quite soluble in the high temperature austenite phase of iron. Thus nitrogen can be used to form a hard iron–nitrogen martensite. Nitrogen can also be added simultaneously with carbon to austenite to form hard martensite. However, the nitriding temperature is usually well below that of the formation of austenite ($<723°C$) (Fig. 2), and because martensite forms only from austenite, hardening by nitrogen addition occurs by another mechanism. The solubility of nitrogen is very low in ferrite, as is that of carbon (Fig. 2). Sufficient nitrogen cannot, therefore, be added to ferrite to impart significant solid solution hardening. Instead, the nitrogen added to the steel surface reacts with iron or specific alloying elements to form a fine dispersion of very hard nitrogen compounds, ie, nitrides (qv) and the nitrides produce a very hard case.

A problem in nitriding is that a layer of nitrides may form on the surface which does not have desirable properties. For example, if the layer is too thick, it is brittle and may spall off. This is called the white layer or compound layer, and may have to be machined or ground off after nitriding.

Molecular nitrogen, N_2, is stable and relatively inert. It does not decompose to atomic nitrogen to form a nitrogen case. Instead, a gaseous compound containing nitrogen must be used. The common carrier of nitrogen in gas nitriding is ammonia (qv) (4).

Plasma or Ion Nitriding. Nitrogen can be produced for deposition on the steel surface by ionization of gaseous nitrogen to form an ionic plasma. The components to be nitrided are electrically insulated and become the cathode (negative) of a d-c power supply. A metal plate serves as the anode. When a potential of ~ 1000 V is applied to the components at sufficiently low pressure of nitrogen, the gas becomes positively ionized, forming a plasma. These ions are accelerated toward and strike the surface of the components to be nitrided. The plasma gives off visible radiation, the color of which depends on the type of gas being ionized. This process is sometimes called glow discharge nitriding (14) (see PLASMA TECHNOLOGY).

The nitrogen ions have sufficient energy to cause the parts to heat. Heaters can be used if necessary to achieve the correct temperature. The nitrogen ions are neutralized at the surface, forming atomic nitrogen, which then diffuses into the surface and reacts with iron and alloying elements to form the hard nitrides. A distinct advantage of this method is that the plasma generally deposits nitrogen uniformly on the surface, so that less uneven deposition occurs than is sometimes found with nitriding by flowing hydrogen–ammonia gas.

Nitriding Temperature and Time.   Because nitriding is carried out in the ferrite and carbide region of the phase diagram (see Fig. 2) at lower temperatures than in carburizing, the diffusion rate of nitrogen is relatively low and hence the time to attain a suitable case is longer than in carburizing. Figure 7**a** shows typical nitrogen content–depth profiles for nitriding. The hardness profiles in Figure 7**b** shows that the case hardness is very high (3,15). For example, in Figure 6**a** the hardness at the surface of the carburized steel, after cooling to −196°C, was about 900 DPH (Vickers); the maximum hardness for nitriding was about 1100 Vickers. To obtain the case of about 0.12 cm by carburizing (Fig. 6) takes about two hours; to get a case depth of about 0.05 cm by the nitriding process for the data in Figure 7**b** requires about 15 hours.

An advantage of nitriding is that the core properties can be set by prior heat treatment, such as by quenching and tempering, then the surface nitrided. During nitriding the steel is not heated back into the austenite region, thus no phase changes occur in the core during nitriding or following cooling to room temperature. However, the core properties can be affected by coarsening during nitriding. If the core properties are set by quenching and tempering, then the tempering temperature must be about 50°C above the nitriding temperature to prevent any significant changes in core properties from occurring during nitriding.

Properties of Case Nitrided Steels.   Nitrided components possess outstanding fatigue properties. In one example, the fatigue strength of crankshafts made from quenched and tempered steel is about 280 MPa (40,000 psi), whereas the same steel after nitriding has a fatigue strength of ca 790 MPa (115,000 psi). These case nitrided crankshafts also have about the same fatigue strength as those that are induction hardened (16).

Nitriding can impart significant wear resistance to steel surfaces, as illustrated in Figure 8. The resistance to abrasion of an uncase hardened steel compared to that of the same steel nitrided, and the steel having a carburized case, is shown (3,17). Improvement in weight loss is related directly to the hardness of the case.

Case Hardening by Surface Deformation.   When a metallic material is plastically deformed at sufficiently low temperature, eg, room temperature for most metals and alloys, it becomes harder. Thus one method to produce a hard case on a metallic component is to plastically deform the surface region. This can be accomplished by a number of methods, such as by forcing a hardened rounded point onto the surface as it is moved. A common method is to impinge upon the surface fine hard particles such as hardened steel spheres (shot) at high velocity. This process is called shot peening.

The surface may gain a very high (eg, ~1000 Vickers) hardness from this process. Surface deformation also produces a desired high compressive residual

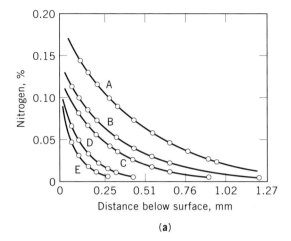

(a)

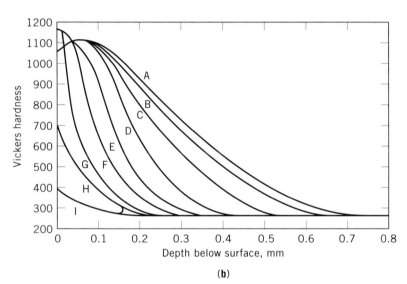

(b)

Fig. 7. (**a**) Nitrogen concentration profiles in a 1015 steel nitrided at 566°C using an aerated bath process where A is nitrided 10 h; B, 3 h; C, 90 min; D, 30 min; and E, 10 min (15), and (**b**) hardness profiles in a steel gas nitrided at 510°C where A is nitrided 90 h; B, 60 h; C, 30 h; D, 15 h; E, 10 h; F, 5 h; G, 3 h; H, 1 h; and I, 0.5 h. Courtesy of Butterworths (4).

stress. Figure 9 illustrates the improvement in fatigue properties of a carburized surface that has been peened (18).

Case Hardening by Ion Implantation

One method of changing the chemistry of the surface of a component without heat treatment or chemical reaction is ion implantation (qv). Not only can elements such as carbon and nitrogen be added, but larger elements such as titanium can be also. A gas of the implant species is produced by vaporization, if

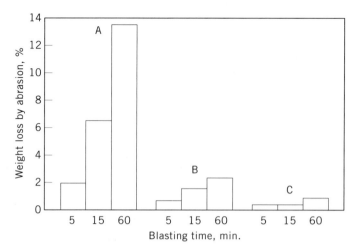

Fig. 8. Illustration of the effect of nitriding on the wear resistance of a steel blasted with steel grit: A, 300 HV steel; B, 750 HV steel case hardened by carburizing; and C, 1100 HV steel nitrided at 500°C for 60 h (17). HV = Vickers' hardness.

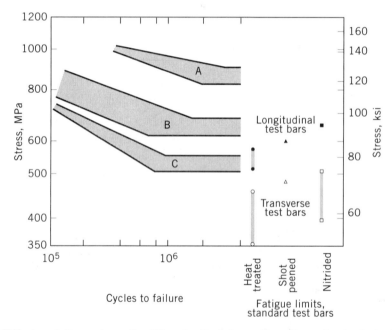

Fig. 9. Effect on fatigue strength of the plastic deformation of a carburized steel surface by shot peening (B) as compared to nitriding (A) and heat treating (C) (18).

necessary, and ionized in some manner, such as by heating or forming a plasma. The ions are removed from the region of formation by a negative electrostatic field, then they enter an accelerator. The ions exit the accelerator with a velocity sufficient to penetrate the surface of the target. The ion beam is focused and its direction controlled by a magnetic or electrostatic lens system. Hence the location of implantation on the surface can be controlled.

The ions not only are implanted in the surface, but cause considerable lattice damage displacing the host atoms. An amorphous layer may be formed and the structure is not an equilibrium one. Thus the solubility of the implanted ions may greatly exceed the solubility limit. All of these effects combine to produce a hard case.

The complexity of the apparatus needed for ion implantation makes this method of case hardening of limited application. Further, the case depth is considerably lower than that produced by carburizing or nitriding. The depth of implantation of nitrogen in a steel is about 0.00006 cm (19), ie, so thin that it is difficult to measure the hardness profile by conventional microhardness measurements.

The improvement in wear resistance from ion implantation is shown in Figure 10 (20). However, the thin case cannot sustain very heavy loads. Hence this application for improved wear resistance is limited to special situations, eg, low loads.

Ion implantation is being used to form a thin hard case on materials other than steels. Titanium alloys have been successfully implanted with nitrogen. The process has been applied to ceramics to modify the surface region.

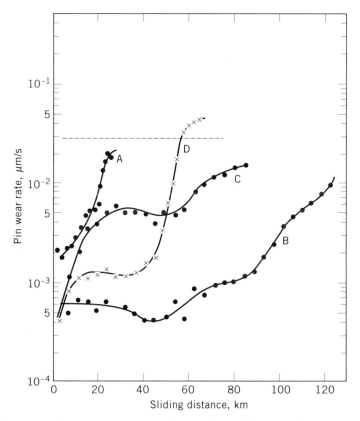

Fig. 10. Wear rates for iron, as measured in a pin-on-disk machine (– – –) unimplanted and implanted with (•) nitrogen at rates in atoms/cm^2 of A, 7×10^{15}; B, 1.7×10^{17}; C, 1.2×10^{18}; and with (×) neon D, at 1.7×10^{17} atoms/cm^2 (20).

Other Methods of Case Hardening

There are a multitude of methods available for case hardening, and the technical details of implementation may be complicated.

Pack Carburizing.   In pack carburizing, the components to be case hardened are placed in the center of a box filled with carbon particles, usually charcoal, and air. Upon heating to the carburizing temperature, the carbon reacts with the air to form CO, which in turn decomposes on the steel surface to deposit carbon. This is the same chemical reaction as described for gaseous carburizing. The CO_2 released in turn is reduced by the carbon in the charcoal, producing more CO for carburizing. To enhance the rate of the chemical reaction, energizers such as barium carbonate, $BaCO_3$, are added to the charcoal. These compounds decompose at relatively low temperature to release CO_2 which then is reduced by the carbon in the charcoal to form CO for carburizing. There is no direct control of the gas composition in pack carburizing, so the surface carbon content is determined by the carburizing temperature and time.

An advantage of pack carburizing is that the charcoal pack supports the work load which minimizes distortion. However, the components must be removed from the pack for quenching, or the mass cooled to room temperature and then the components reaustenitized and quenched.

Drip Carburizing.   The gases CO and CH_4, needed for decomposition to provide carbon at the surface of a component to be carburized, can, in principle, be produced by the decomposition of many carbon-containing organic chemicals. One method is to insert liquid methanol into the hot furnace containing the components, where CH_3OH decomposes to produce the carburizing gas. This process can be used with air where the reaction of the oxygen in the air and the vaporized methanol form the carburizing gas. The reaction rate is controlled by the rate of addition of the liquid methanol, usually by the release of drops, hence drop carburizing.

Vacuum Carburizing.   In vacuum carburizing, part of the carburizing cycle involves heating the carburized component in a vacuum to allow diffusion of carbon to occur without significantly affecting the surface. The steel is heated to the austenitizing (carburizing) temperature under partial vacuum, then a carburizing gas is introduced to the steel and retained until the desired carbon gradient is obtained. The system is then evacuated and the carbon allowed to diffuse inward under the protective vacuum until the gradient is reduced to the desired profile. The steel components are quenched in oil or air under vacuum.

Plasma Carburizing.   Plasma carburizing generates carbon atoms at the surface by ionization of a carbon-containing gas, eg, methane. The process is similar to that described for ion nitriding. Because the process is carried out in partial vacuum, there is less chance of oxidation.

Liquid Carburizing.   Case hardening by carburizing can be achieved by the use of carbon-containing molten salts. Thermal gradients are minimized with the use of molten salts. The most common chemicals in liquid carburizing are cyanide salt baths, in which case the process is referred to as cyaniding. Because nitrogen is also formed from the decomposition of these salts, the case may contain nitrogen, which is favored by the use of high concentrations of the cyanide salts. A critical disadvantage in the use of cyanide salts is that they are extremely poisonous.

Carbonitriding.   A modification of gaseous carburizing where ammonia is added to provide nitrogen is called carbonitriding. The effect of adding both carbon and nitrogen is similar to that of cyaniding. This terminology refers to the addition of carbon and nitrogen to austenite, which is then quenched to form a case of hard martensite.

Liquid Nitriding.   As in gas nitriding, the process is carried out below the austenite region, and hardening is associated with the formation of hard nitrides in the ferrite. Liquid cyanide salts are used with others to provide the source of nitrogen.

Ferritic Nitrocarburizing.   This process is similar to carbonitriding, except that it is carried out in the temperature range of the stability of ferrite and carbide ($<723°C$). Therefore hardening is not by martensite formation, but because of the formation of very hard carbonitrides.

Plasma Nitrocarburizing.   This process is similar to plasma nitriding and plasma carburizing, except that a gas is used which when ionized releases both carbon and nitrogen at the surface of the component. Mixtures of hydrogen, nitrogen, and a carbonaceous gas are used. The process is carried out in the temperature range of stability of ferrite and carbide, and case hardening occurs as a result of the formation of hard carbonitride particles and a hard nitride layer on the surface.

Austenitic Nitrocarburizing.   This is similar to ferritic nitrocarburizing except that the temperature may extend into the austenite range. The case usually consists of hard carbonitride particles, and quenching to achieve hardening is not required.

Boriding or Boronizing.   Boron, also a relatively small atom, can be added to the surface of steels by methods similar to that for carburizing and nitriding. For example, the process can be carried out in a pack of boron-containing salts, by use of a plasma and by gaseous environment. The process can be carried out in the temperature range of the stability of ferrite and carbide ($<723°C$), or in the higher temperature austenite range. In either case the hardening is associated with the formation of hard, eg, $1400-5000$ Vickers hardness, boron compounds.

BIBLIOGRAPHY

"Metal Cementation Process" under "Metal Surface Treatments" in *ECT* 1st ed., Vol. 9, pp. 23–31, by E. S. Greiner, Bell Telephone Laboratories, Inc.; "Case Hardening" under "Metal Surface Treatments" in *ECT* 2nd ed., Vol. 13, pp. 304–315, by A. J. Schwarzkopf, National Aeronautics and Space Administration; in *ECT* 3rd ed., Vol. 15, pp. 313–324, by L. E. Alban, Fairfield Manufacturing Co., Inc.

1. L. J. Vande Walle, ed., *Residual Stress for Designers and Metallurgists*, American Society for Metals, Metals Park, Ohio, 1981.
2. T. H. Spenser and co-workers, *Induction Hardening and Tempering*, American Society for Metals, Metals Park, Ohio, 1964.
3. *Metals Handbook*, 8th ed., Vol. 8, American Society for Metals, Metals Park, Ohio, 1973.

4. K. E. Thelning, *Steel and Its Heat Treatment*, Butterworths, Boston, Mass., 1975.
5. D. L. Martin and F. E. Wiley, *Trans. ASM* **34**, 351 (1945).
6. W. Amende, in H. Koebner, ed., *Industrial Applications of Lasers*, John Wiley & Sons, Inc., New York, 1984.
7. M. F. Ashby and K. E. Easterling, *Acta Metallurgica* **32**, 1935 (1984).
8. H. B. Singh, S. M. Copley, and M. Bass, *Met. Trans.* **12A**, 138 (1981).
9. C. R. Brooks, *Principles of the Surface Treatment of Steels*, Technomic Publishing Co., Lancaster, Pa., 1992.
10. J. B. Austin and M. J. Dary, *Controlled Atmospheres*, American Society for Metals, Metals Park, Ohio, 1942.
11. *ASM Handbook*, Vol. 4, ASM International, Materials Park, Ohio, 1991.
12. H. Scott and J. L. Fisher, in Ref. 10.
13. J. I. Goldstein and A. E. Moren, *Met. Trans.* **9A**, 1515 (1978).
14. T. Spalvins, ed., *Ion Nitriding*, ASM International, Materials Park, Ohio, 1987, p. 1.
15. Ref. 3, Vol. 2, 1964.
16. R. F. Kern and M. E. Suess, *Steel Selection*, John Wiley & Sons, Inc., New York, 1979.
17. R. Lambert, *Revue de Metallurgie*, **52**, 553 (1955).
18. Y. Nakamura, Y. Matsushima, T. Hasegawa, and Y. Nakatami, *Metal Behaviour and Surface Engineering*, IITT-International, Gournay-sur-Marne, France, 1989, p. 187.
19. G. Dearnaley and co-workers, *Proceedings of the Conference on Ion Plating and Allied Techniques*, CEP Consultants, Edinburgh, Scotland, 1979.
20. P. D. Goode, A. T. Peacock, and J. Asher, *Nucl. Inst. Meth.* **209–210**, 925 (1983).

General References

ASM Handbook, Vol. 4, ASM International, Materials Park, Ohio, 1991; excellent and detailed articles of most case hardening methods.
C. R. Brooks, *Principles of the Surface Treatment of Steels*, Technomic Publishing Co., Lancaster, Pa., 1992.
G. Krauss, *Steels: Heat Treatment and Processing Principles*, ASM International, Materials Park, Ohio, 1990.
W. L. Grube and S. Verhoff, in R. Kossowsky, ed., *Surface Modification Engineering*, Vol. II, CRC Press, Boca Raton, Fla., 1989, p. 107.
H. E. Boyer, ed., *Case Hardening of Steel*, ASM International, Metals Park, Ohio, 1987.
S. L. Semiatin, D. E. Sutuz, and I. L. Harry, *Induction Heat Treatment of Steel*, American Society for Metals, Metals Park, Ohio, 1986.
Carburizing and Carbonitriding, American Society for Metals, Metals Park, Ohio, 1977.

CHARLIE R. BROOKS
University of Tennessee

CLEANING

Cleaning, the removal of unwanted matter, is the beginning of the treatment cycle for metal. The unwanted matter may be carbon smut, welding flux, ink, oxidation products, oil, fingerprints, or other material. Cleaners may be classified as solvent-based or aqueous. Within the aqueous class there are many subclasses, the most important of which are the alkaline cleaners. There are also a variety of ways to apply cleaners. As of the mid-1990s, solvent-based cleaner usage is declining.

Alkaline Cleaners

Alkaline cleaners are the most commonly used for metal surfaces. These are typically composed of a blend of alkaline salt builders, such as sodium phosphates, sodium silicates, sodium hydroxide, or sodium carbonate. In addition, they almost always contain detergents, ie, surfactants (qv), and, optionally, wetting agents, coupling agents, chelating agents (qv), and solubilizers. Alkaline cleaners are usually applied in the range of 38–93°C.

The composition of the builders in an alkaline cleaner is dependent on the metal substrate from which the soil is to be removed. For steel (qv) or stainless steel aggressive, ie, high pH, alkaline salts such as sodium or potassium hydroxide can be used as the main alkaline builder. For aluminum, zinc, brass, or tin plate, less aggressive (lower pH) builders such as sodium or potassium silicates, mono- and diphosphates, borates, and bicarbonates are used.

The mechanisms by which an alkaline cleaner removes the soil are saponification, emulsification, and dispersion. These mechanisms can operate independently or in combination. Saponification occurs when alkaline salts react with fatty components of the soil, forming a soluble soap compound.

Emulsification involves the joining of two mutually insoluble materials, such as petroleum oil and water. The surfactant, which usually has a hydrophilic or water-soluble end and a hydrophobic or oil-soluble end, holds the oil and water together in much the same manner that a fastener holds two pieces of material. Often, the emulsion which forms is unstable, subsequently breaking up and releasing the oil from the water. Break-up is actually preferred, because the oil then floats to the surface, whereas the surfactant is free to emulsify more oil.

Dispersion is the process of wetting the surface of the metal, thereby penetrating the oil film. Surfactants can reduce the surface tension and interfacial tension of the cleaning solution at the metal–liquid interface. As the cleaner undercuts and penetrates the oil, the cleaner breaks the oil into small droplets which then float to the surface.

Applications

Cleaners are typically applied either by immersion or by spray. Immersion or soak cleaning involves simply immersing a part or panel into a tank containing the cleaner and letting the metal sit for a period of 1–10 min. A simple variation of immersion cleaning is to recirculate the cleaner so that fresh cleaner is continuously passing by the object to be cleaned. Other variations include electrocleaning, ultrasonic cleaning, and barrel cleaning. Variations of spray cleaning include steam cleaning and power washing. Additionally, solvent cleaners can be applied by vapor degreasing, and both solvent and aqueous cleaners can be applied by mechanical methods such as hand wiping or hand brushing the cleaner onto the soiled area. Because of high evaporation rates and toxic fumes, solvent cleaners are not usually applied by spray.

Immersion Cleaning. The simplest method for using an alkaline cleaner is by immersion. A part is placed on a hook or rack and immersed in the cleaner solution so that all of the part is below the liquid level. A typical concentration, temperature, and process time for an immersion cleaner would be ca 75 g/L at 77°C for 5 min. In addition to being the simplest method, immersion is also among the least expensive in terms of equipment. Only a vessel to contain the cleaning solution and a means of heating the solution are needed.

Electrocleaning. Electrocleaning is a specialized variation of immersion cleaning. Electrocleaners are very similar in composition to immersion cleaners, except that the surfactant levels are usually lower and less foaming. The cleaning action of the cleaner is assisted through the use of direct-current electricity. Electrodes are placed on two sides on the inside of the tank, and in most cases they carry a negative charge (making them the cathode). The part to be cleaned carries a positive charge (anode). The oxygen evolving at the part acts as a mechanical scrubber assisting in removal of the soil. The concentration and temperature are usually a little higher than for a straight immersion cleaner: 75–120 g/L at 77–99°C. The d-c current is supplied by a large rectifier, and the current density on the part varies from $27–160$ mA/cm^2 ($25–150$ A/ft^2). Electrocleaning is usually preceded by an immersion or a spray cleaner to remove the bulk of the soil. The part is typically left in the electrocleaner for 1–3 min to produce an exceptionally clean surface.

Ultrasonic Cleaning. Ultrasonic cleaning is another variation of immersion cleaning. Waves of bubbles are generated in the cleaning solution through the use of sound waves. The sound waves are generated by electrically powered devices called transducers, which vibrate rapidly causing the sound waves and bubbles. This is called cavitation (see ULTRASONICS). The ultrasonic cleaning action literally vibrates or shakes dirt loose from narrow cracks and crevices. This is aided by the cleaning action of the detergents in solution.

The transducers are typically mounted on an outside wall of the cleaning tank, but may also be mounted on the inside of the tank below the solution level in a sealed container. Alkaline cleaning solutions are typically at the same concentration and temperature as for a normal immersion cleaner, but the time required to clean may be less because of the ultrasonic effect. Like electrocleaning, ultrasonic cleaning produces an extremely clean surface. The main drawback is the relatively high cost.

Barrel Cleaning. Barrel cleaning is a minor variation of immersion cleaning in which parts are placed in a six- or eight-sided barrel the sides of which are perforated, allowing the cleaning solution to enter the barrel. The barrel is immersed in the cleaner and rotated around the long axis by an electric motor. The parts inside tumble and rub against each other, thus aiding in the removal of the soil. Electrodes can also be inserted and hooked up to a rectifier to add electrocleaning assistance. The restriction of barrel cleaning is that it is generally used only for small parts such as fasteners, cabinet handles, and screwdriver shafts.

Spray Cleaning. The other principal method of application of cleaners is by spray. In spray cleaning, the cleaning solution is pumped out of a holding tank into a series of pipes (risers) which have other pipes called headers coming out from the risers. These headers have holes drilled into them and nozzles screwed into the holes. The pipes are configured in such a way that a part going by on a conveyor would be sprayed with cleaner from every conceivable angle. This configuration of pipes is housed in a steel hood which keeps the spray mist confined. The hood is generally over the holding tank so that the excess cleaner drains back into the holding tank. The hood has a narrow slit at the top to permit conveyor parts racks to pass through and at least one exhaust stack to remove the steam. Spray pressures for commercial spray cleaner systems are usually in the range of 70–275 kPa (10–40 psi), concentrations are typically 4–30 g/L, and temperatures vary between 21–88°C. The main difference in composition from immersion cleaners is that spray cleaners contain low foaming surfactants.

Steam Cleaning. In steam cleaning, a machine generates high pressure steam and injects a cleaning solution at the nozzle where the steam is exiting. Cleaner concentration is varied by changing the concentration of the cleaner in the reservoir or the rate at which the cleaner feeds into the exiting steam.

Power Washing. Power washing is similar to steam cleaning, except that the cleaner is injected into a high pressure (6.9–35 MPa (1000–5000 psi)) water blast in power washing. The water can be cold, warm, or hot, because the power washer is fed from a water tap.

Solvent Cleaning. Solvent cleaning employs the natural solubilizing properties of various nonaqueous solvents or blends of solvents. Either 100% solvents or aqueous emulsions of solvents can be used. Application is typically by immersion, hand application, or via a vapor degreaser machine, where the solvents are continuously vaporized and condensed. Solvent cleaners are rarely applied by spray because of excessive loss by evaporation and the subsequent environmental considerations. Solvent cleaning also requires a lot of ventilation and an adequate exhaust system. Whereas solvent cleaning is effective, its cost is usually higher than other cleaning methods.

Economic Aspects

In 1990 the value of the metal cleaner industry was 390×10^6 (1). An estimated figure for 1994 was on the order of 450×10^6. Usage of solvent-based cleaners is decreasing; usage of aqueous-based alkaline, acid, or neutral cleaners is increasing. There are over 60 suppliers of metal cleaners in the United States. One of the leading suppliers of aqueous-based cleaners is the Parker Amchem Division of Henkel Corp. Aqueous cleaners, aqueous slurry cleaners, and dry blend cleaners are available worldwide in acid, alkaline, and neutral media. Common cleaners found have the trademarks Parco, P3, Alumiprep, Ridoline, Metalprep, and Prep-N-Cote, to name just a few.

Most cleaners are available for <$2.20/kg either as a dry blend or as a liquid. Liquid cleaners are usually less expensive than the dry blend type. A trend toward liquid cleaners is evident as of this writing (ca 1994) because of convenience features such as automatic additions of the cleaner by chemical feed pump. Safety features such as minimized heat generation upon blending with water to make the desired concentration are also important.

BIBLIOGRAPHY

"Solvent Cleaning, Alkali Cleaning, and Pickling and Etching" in *ECT* 1st ed., under "Metal Surface Treatment," Vol. 9, pp. 1–9, by August Mendizza, Bell Telephone Laboratories, Inc.; "Cleaning, Pickling, and Related Processes" in *ECT* 2nd ed., under "Metal

Surface Treatments," Vol. 13, pp. 284–292, by S. Spring, Oxford Chemical Division, Consolidated Foods Corp.; in *ECT* 3rd ed., Vol. 15, pp. 296–303, by G. L. Schneberger, General Motors Institute.

1. *Chem. Week*, 28 (Aug. 11, 1993).

General References

S. Spring, *Industrial Cleaning*, Prism Press, Melbourne, Australia, 1974.
D. Murphy, *Metals Handbook*, 9th ed., Vol. 5, ASM, Metals Park, Ohio, 1982, pp. 22–25.
W. P. Innes, *Metal Finishing 62nd Guidebook for '94*, Elsevier Science Publishing Co., Inc., New York, 1994, pp. 113–132.
D. Murphy and G. Tupper, *Users' Guide to Powder Coating*, Association for Finishing Processes of the Society of Manufacturing Engineers, Dearborn, Mich., 1985, pp. 33–58.
Davidson and Milwidsky, *Synthetic Detergents*, John Wiley & Sons, Inc., New York, 1967.
A. Pollack and P. Westphall, *An Introduction to Metal Degreasing and Cleaning*, R. Draper Ltd., Teddington, U.K., 1963.
L. I. Osipow, *Surface Chemistry—Theory and Industrial Applications*, R. E. Krieger Publishing Co., Huntington, N.Y., 1977.
A. W. Adamson, *Physical Chemistry of Surfaces*, 5th ed., John Wiley & Sons, Inc., New York, 1990.
M. J. Rosen, Surfactants and Interfacial Phenomena, John Wiley & Sons, Inc., New York, 1978.

Donald P. Murphy
Henkel Corporation

CHEMICAL AND ELECTROCHEMICAL CONVERSION TREATMENTS

Phosphating

Reactive metal surfaces can be chemically treated and covered with inert, amorphous, or crystalline coatings which grow on the base metal. Phosphates represent the most important area of the conversion coatings. These coatings (qv) are applied as preparation for painting, temporary corrosion protection, lubricant carrier in cold forming, friction improver for stamping and drawing, and as insulation on electrical steels (1–3) (see also METALLIC COATINGS; METAL TREATMENTS).

Phosphating processes began with the work of Thomas Watts Coslett. The original patent covered the use of phosphoric acid to which iron filings had been added (4). It took from 2 to 2.5 hours exposure in boiling solution to form a coating on iron and steel articles. Early improvements consisted of controlling the quantity of free phosphoric acid, thus inhibiting attack on the metal, and of accelerating the reaction by using an electric current. Concentrates from which phosphate baths could be prepared by dilution were in use by 1909. The use of zinc in the phosphate bath was patented in 1909 and the use of manganese in 1911 (5). The use of manganese dihydrogen phosphate gave rise to the process which became known as Parkerizing. Reduction in processing time from one hour to 10 minutes came through the addition of copper to the bath. This was the basis for the Bonderizing process. Addition of an oxidizing agent such as nitrate increased the reaction rate by preventing the adsorption of hydrogen on the metal surface. Paint-base phosphate coatings could be applied in two to five minutes. In 1934, this time constraint was shortened even further when phosphate solutions were sprayed onto the metal surface. Processing times as short as 60 seconds became possible.

A crucial development for zinc phosphate coatings came in 1943 when it was found that more uniform and finer crystals would develop if the surface was first treated with a titanium-containing solution of disodium phosphate (6). This method of crystal modification is a prime reason for the excellent paint (qv) adhesion seen on painted metal articles.

Modern phosphating practice involves the treatment of reactive metals with acidic phosphate-containing solutions. This produces a coating which consists mainly of phosphate compounds. Chemically, phosphating processes can be separated into two types. In processes of the first type, the metal ions of the phosphate layer derive almost totally from the substrate. These layers, known as noncoating or iron phosphates, are based on sodium and ammonium dihydrogen phosphate. Processes of the second type, on the other hand, provide metal ions for coating either partially or totally in the phosphate bath. These are the zinc phosphate processes which may contain zinc alone or modifying ions such as nickel, manganese, calcium, as well as several others.

The Iron Phosphating Process.   Dissolution and oxidation of iron(II) to iron(III) and coating development result in the formation of an amorphous coating which contains iron phosphate, $FePO_4$, as the principal coating constituent (7). In addition, some iron oxide, Fe_2O_3, which forms from the rearrangement of ferric hydroxide, and some tertiary iron phosphate called vivianite, $Fe_3(PO_4)_2$, are present. The oxidative condition needed for coating formation is provided by accelerators such as chlorate, nitrate, permanganate, and air entrapped during spraying.

Operation and Control.   The product concentrates can be supplied either as powders or liquids. Powdered iron phosphates based on dihydrogen phosphates of sodium or ammonium form a popular line of clean-and-coat materials used in three-and four-stage lines. These formulations may also include surfactants (qv), sequestrants, solvents, inhibitors, and accelerators. High temperature iron phosphates are applied at 65.6–76.7°C for 40–60 seconds. The systems operate in the pH range of 4.5–5.5. Total acidity ranges from 6–16 points, ie, 6–16 mL of 0.1 N NaOH for a 10 mL sample of the bath determined by titrating to an end point of pH 8.2, which is faint pink with a phenolphthalein indicator. Low temperature iron phosphate processes operate at 40.6–46.1°C, at a total acidity range of 9–18 points. The optimum pH range lies between 3.7 and 4.4, having measurable free acidity, ie, titration with 0.1 N NaOH to an end point of pH 3.8. These formulations are used at concentrations similar to the high temperature variations. However, these formulations must have increased acidity and increased acceleration to compensate for the lower operating temperature. The formation of Na_2HPO_4 as a by-product of the reaction causes the pH to rise, thus acid replenishment is necessary to keep the bath in the operating range.

The application of this type of conversion coating can be by spray or immersion and is easily tailored to the needs of the user (see COATING PROCESSES). The number of stages may vary from two, ie, clean and phosphate then rinse, to as many as five, ie, clean, rinse, phosphate, rinse, and post-treat. The performance requirements dictate the number of operation stages as well as the need for a post-treatment.

Coating Characterization.   As indicated, the coatings are amorphous and can therefore be characterized only by coating weight. Coating weights normally range from 0.1 to 0.8 g/m^2, but special formulations may yield coating weights of 1.2 g/m^2. For coating weight determination, coatings may be stripped from steel surfaces with a solution of chromic acid, whereas aluminum or zinc surfaces are stripped with a solution of ammonium dichromate.

Product Utilization.   Iron phosphating has been the process of choice for applications where cost considerations override maximum performance needs. Advantages are low chemical cost, low equipment cost, good paint bonding, easy control, and minimal sludge. The disadvantages, thin coatings and poorer corrosion resistance than for zinc phosphating, limit the application of iron phosphating to particular industries. Nevertheless, iron phosphates are the most widely applied conversion coatings when surface preparation for paint application is needed. Most recently developments have been aimed at formulating baths which allow coating deposition on a mixture of metals using the same coating bath. The addition of stronger etchants and the use of different accelerator systems brings these systems closer to zinc phosphates in versatility and performance.

The Zinc Phosphating Process.   The zinc phosphating reaction involves acid attack on the substrate metal at microanodes and deposition of phosphate crystals at microcathodes (8). Liberation of hydrogen and the formation of phosphate sludge also occur. The equation for the dissolution of iron together with precipitation of dissolved iron as sludge in a nitrite accelerated system is as follows:

$$2 \, Fe_0 + 2 \, H_2PO_4^- + 2 \, H^+ + 3 \, NO_2 \longrightarrow 2 \, FePO_4 + 3 \, H_2O + 3 \, NO$$

The NO_2 is generated when nitrite oxidizes the hydrogen produced by the acid attack on the metal surface to form water. The total reaction uses up hydrogen ions and reduces acidity. The only source of these ions at the interface is the zinc dihydrogen phosphate in solution. The result is that zinc phosphate crystallizes at the surface–solution interface to form a tightly adhering crystal layer according to the following equation:

$$3 \, Zn^{2+} + 2 \, H_2PO_4^- \longrightarrow Zn_3(PO_4)_2 + 4 \, H^+$$

Combining these equations gives a total coating reaction of

$$4 \, Fe_0 + 3 \, Zn^{2+} + 6 \, H_2PO_4^- + 6 \, NO_2 \longrightarrow 4 \, FePO_4 + Zn_3(PO_4)_2 + 6 \, H_2O + 6 \, NO$$

Hydrogen is prevented from forming a passivating layer on the surface by an oxidant additive which also oxidizes ferrous iron to ferric iron. Ferric phosphate then precipitates as sludge away from the metal surface. Depending on bath parameters, tertiary iron phosphate may also deposit and ferrous iron can be incorporated into the crystal lattice. When other metals are included in the bath, these are also incorporated at distinct levels to generate species that can be written as $Zn_2Me(PO_4)_2$, where Me can represent Ni, Mn, Ca, Mg, or Fe.

Operation and Control. The zinc phosphating process sequence consists of cleaning, rinsing, surface conditioning, phosphating, and final rinsing. Alkaline cleaning is used to remove soils and oils that have accumulated during storage and forming of the parts (9) (see METAL SURFACE TREATMENTS, CLEANING). Rinses between the active stages remove residual chemicals. For zinc phosphates, an effective conditioning rinse before phosphating provides the development of small, tight crystals necessary for improved paint adhesion. Conditioners, dispersions of colloidal titanium phosphate, are formulated to function consistently with equal results in hard and soft water. Finally, improved corrosion performance is most often obtained using a post-treatment or final rinse step. Some of these rinses contain chromium in mixtures of both the +3 and +6 oxidation states. Others contain chromium in the +3 oxidation state alone. Chromium-free post-treatments which match the performance of the chromium-containing post-treatments have been developed. These derive effectiveness from organic polymers which can chelate to the metal surface and to the coating (10).

The bath components for a nitrite–nitrate accelerated bath basic to this conversion coating process are (1) zinc metal or zinc oxide dissolved in acid; (2) phosphate ions added as phosphoric acid; (3) addition of an oxidant such as sodium nitrite; and (4) addition of nitric acid. Other oxidants such as peroxide, chlorate, chlorate in combination with nitrate, or an organic nitro compound may also be used.

Conventional practice is to dilute the liquid, which consists of 40–50% solids, zinc phosphate concentrate, having a density ~1.3 times that of water, to a 2–6% concentration with water. Soda ash or caustic soda is added to obtain the correct free acid value and the bath is then heated to the operating temperature. This is followed by addition of $NaNO_2$ to achieve about a 0.02% concentration. The level of free acid is critical. The free acid must be lowered to the point where zinc phosphate starts to precipitate. This is called the point of incipient precipitation (11). If the bath is neutralized too far, valuable zinc dihydrogen phosphate is lost as sludge. As the reaction proceeds and the crystalline coating is formed on the surface, the zinc dihydrogen phosphate in the bath must be replenished. Practical replenishers operate at phosphate to zinc mole ratios between 1.5 and 2.5:1 because an excess of phosphate is required to precipitate dissolved iron as the phosphate. In modified zinc phosphates, the other metals are also restored by the replenisher according to their individual incorporation in the coating.

Application of the zinc phosphate coating can be by spray or immersion. When the coating is applied by spray, the solution is quickly replaced and the coating formation takes place very rapidly. Replacement at the surface of depleted bath solution during immersion application is diffusion controlled. The rate at which the coating develops is somewhat slower. This difference in the rate of coating development is reflected in the composition of the coatings deposited. Those from the immersion application contain a large amount of iron when applied to steel and have better performance qualities than those from spray application (12). Although control parameters for the two application methods do not vary significantly, there is a difference in the replenisher formulations. This is because of a higher drag-out from the spray bath than from the immersion bath.

Coating Characterization.   Phosphate coatings cannot be fully characterized by any simple methods. Characteristics such as appearance, coating weight, P-ratio (13), porosity (14,15), coating composition (16), and crystal size combine to define a system's performance. However, measurement of any of these factors individually does not provide a sufficient indication of the coating's efficacy in meeting its intended purpose.

Modified Zinc Phosphates.   Coatings on steel have been identified as hopeite, $Zn_3(PO_4)_2 \cdot 4H_2O$, phosphophyllite, $Zn_2Fe(PO_4)_2 \cdot 4H_2O$, and iron hureaulite, $Fe_5H_2(PO_4)_4 \cdot 4H_2O$. Incorporation is also achieved when nickel, manganese, or calcium are added to the phosphating bath. Similar structures have been identified for these metals (17,18). Modification of the hopeite structure is reflected in an increased resistance to alkalinity and a higher dehydration temperature (19). Both of these properties contribute to improved corrosion and adhesion properties of painted steel (qv), zinc, and zinc alloy surfaces (see ZINC AND ZINC ALLOYS).

Product Utilization.   Zinc phosphate coatings form the basis for paint adhesion in a variety of industries. These are used when long-term quality is of concern in applications such as for automotive parts and vehicles, coil-coated products, and appliances.

Testing of Painted Products.   The enhancement of paint adhesion is one of the principal functions of conversion coating (20–22). A group of tests based on product deformation is used to test the painted product. The appliance and coil-coating industries use the mandrel bend, the cross-hatch adhesion test, and the direct and reverse impact tests. Adhesion after a water soak is judged using a cross-hatch test performed on the exposed surface.

Several accelerated corrosion tests are also employed to evaluate the effectiveness of the phosphate coating in the performance of painted products (see CORROSION AND CORROSION CONTROL). These tests are designed to duplicate the service environment as well as the corrosion mechanism for the painted article. In the appliance industry the tests used most often include the salt spray test, detergent immersion, water immersion, and humidity resistance. The coil coating industry uses the salt spray test, exposure in the Q-uv cabinet, ie, a test regimen involving uv light and condensing humidity at a temperature of 60°C, and humidity exposure. Testing in the automotive industry is slightly more complicated because each of the car companies has its own cyclic corrosion test. Among the cyclic tests in use are the Ford APGE Test, the GM 9511 Cyclic Corrosion Test, the GM 9540-P Cyclic Corrosion Test, and the Chrysler Chipping Test. Water immersion, humidity exposure and, to a very limited extent, the salt spray test are also being used. In addition, all industries use outdoor exposure tests for long-term evaluation of corrosion performance. However, test results from both the accelerated and in-service tests function as indicators of relative performance. These are not predictive of application lifetimes.

Other Phosphate Coatings.   Phosphate coatings are also used as surface treatments for wire drawing, tube drawing, and cold extrusion. Zinc phosphates of heavy ($10–35$ g/m^2) coating weights serve as the carrier for drawing compounds and lubricants. It is standard practice to use a manganese phosphate coating as an oil retaining medium on bearings or a sliding surface to eliminate scuffing, galling, and pickup, and to facilitate running in, ie, wear of mating surfaces during initial hours of operation (see BEARING MATERIALS). These coatings are usually applied by immersion at temperatures reaching 93°C at times between 5 and 30 minutes.

Chromating

Chromating or chromatizing has been widely practiced in the metals industry since the late 1940s to improve corrosion resistance and performance of subsequently applied organic finishes (23). Commonly, chromates are used to treat wrought alloys, cast alloys, and coatings comprising aluminum, zinc, magnesium, or cadmium. The aerospace, transportation, architecture, appliance, marine, and electronics industries, among others, utilize chromate coatings extensively (see also COATINGS, MARINE; ELECTRONICS, COATINGS).

Chromate conversion coatings are thin, noncrystalline, adherent surface layers of low solubility phosphorus and/or chromium compounds produced by the reaction of suitable reagents with the metal surface (2,3). The two classes of chromate coatings are chromium phosphates (green chromates) and chromium chromates (gold chromates).

Chromium Phosphate. Chromium phosphate treatment baths are strongly acidic and comprise sources of hexavalent chromium, phosphate, and fluoride ions. Conversion coating on aluminum precedes by the following reactions (24):

Oxide removal	$Al_2O_3 + 6\ HF \longrightarrow 2\ AlF_3 + 3\ H_2O$
Redox reaction	$Cr^{6+} + 3\ e^- \longrightarrow Cr^{3+}$
	$Al^0 \longrightarrow Al^{3+} + 3\ e^-$
Chromate reduction	$8\ H^+ + 2\ HCrO_4^- + 2\ Al^0 \longrightarrow 2\ Al^{3+} + Cr_2O_3 \cdot 3H_2O + 2\ H_2O$
Phosphate reaction	$Cr_2O_3 \cdot 3H_2O + 2\ H_3PO_4 + 2\ H_2O \longrightarrow 2\ CrPO_4 \cdot 4H_2O$

Chromium phosphate coatings can be deposited with very low $(0.05-0.15\ g/m^2)$ weights to give colorless coatings for applications such as a paint base or very heavy $(2.0-5.0\ g/m^2)$ weights for decorative applications.

Operation and Control. Control of a chromium phosphate conversion coating bath requires monitoring chromium and aluminum concentrations, active fluoride level, and temperature. Coating weight is very sensitive to active, ie, uncomplexed, fluoride. An innovative electrochemical method using a silicon electrode (25) is employed for measuring active fluoride. A special precaution in chromium phosphate bath operation is the formation of whitish, powdery coatings resulting from accumulation of dissolved aluminum. Commercially available proprietary chromium phosphating products typically incorporate sodium fluoride and potassium fluoride. These fluoride additions precipitate aluminum from solution as elpasolite, NaK_2AlF_6.

Typical processing steps in chromium phosphate application by either spray or immersion contact are alkaline cleaning (etching or nonetching), water rinse, chromium phosphate treatment, water rinse, and finally a deionized water rinse because acidified water is often used for paint-base applications. Usually desmutting or deoxidizing is not required for chromium phosphating after alkaline cleaning because of the strongly acidic nature of the treatment bath.

Product Utilization.   The principal application for chromium phosphate coatings is as a paint base for painted aluminum extrusions and aluminum beverage can stock. In these applications, extremely demanding performance criteria are met by the chromium phosphate conversion coatings. As an example, the Architectural Aluminum Manufacturer's Association Voluntary Specification 605.2-92 requires humidity and salt spray testing for 3000 hours and allows only minimal incidence of paint failure after testing (26).

Chromium Chromate.   Chromium chromate treatment baths are acidic and made up from sources of hexavalent chromium and complex fluoride, fluorosilicate, fluorozirconate, fluorotitanate, and silicofluorides. Optional additional components added to accelerate coating rate are free fluoride, ferricyanide, and other metal salts such as barium nitrate. Conversion coating on aluminum precedes by the following reactions (2,3,17):

Oxide removal

$$Al_2O_3 + 6\ HF \longrightarrow 2\ AlF_3 + 3\ H_2O$$

Redox reactions

$$HCrO_4^- + 7\ H^+ + 3\ e^- \longrightarrow Cr^{3+} + 4\ H_2O$$

$$Al^0 \longrightarrow Al^{3+} + 3\ e^-$$

$$8\ H^+ + 2\ HCrO_4^- + 2\ Al^0 \longrightarrow 2\ Al^{3+} + Cr_2O_3 \cdot 3H_2O + 2\ H_2O$$

The coating composition is a combination of hydrated chromium and aluminum oxides and hydroxides, eg, $Cr_2O_3 \cdot xH_2O$, $x = 1, 2$.

Chromium chromate coatings that have very low $(0.05\ g/m^2)$ coating weights may be deposited. These give colorless coatings. Moderately heavy $(0.8\ g/m^2)$ gold-colored coatings for maximum corrosion protection may also be deposited. Because of the hydrated nature of chromium chromate coatings, the corrosion protective properties are sensitive to elevated temperatures and prolonged heating.

Operation and Control.   Control of chromium chromate conversion coating baths is accomplished by controlling chromium concentration and pH. The quality of the conversion coating is sensitive to aluminum accumulations in the coating bath as well as to rinse water purity. Sulfate contamination is a particular problem.

Typical processing steps in chromium chromate application by spray or immersion contact are alkaline cleaning (etching or nonetching), water rinse, acid deoxidize/desmut, water rinse, chromium chromate conversion coat, and a final water rinse. The performance of chromium chromate conversion coatings is strongly affected by the presence of oxides and alkali-insoluble alloying elements (smut) on workpiece surfaces. For most applications, an acidic deoxidizing–desmutting operation is required before the conversion coating stage.

Product Utilization.   Applications for chromium chromate coating are extensive. Aircraft components, aluminum castings, zinc castings, magnesium castings, galvanized steel passivation, and aluminum sheet passivation are example applications for chromium chromates. In aerospace applications, the performance criteria for chromium chromate conversion coatings are extremely demanding. The Mil-C-5541E Specification includes performance criteria for salt spray resistance of 168 hours for unpainted conversion coated aluminum alloys without significant signs of corrosion. At the same time, the conversion coating must comply with a low electrical resistivity criteria to perform well in resistance spot-welding operations and electrical and electronic uses (27).

Anodizing

Whereas many metals can be anodized, aluminum is by far the most widely anodized metal. The anodizing process is comprised of several pre- and post-treatment steps. The anodizing step consists of placing the part to be anodized in a tank where a controlled direct current charge can be applied for a predetermined length of time. At the anode, ie, the part, the aluminum is oxidized to aluminum oxide. The hydrogen ion migrates to the cathode, where it is reduced, forming gas bubbles that are given off to the atmosphere. As this process continues, a uniform porous oxide film is formed on the part (28). The resulting material having the anodic film can be used for structural purposes, such as aircraft parts, or for decorative purposes, such as windows or picture frames.

Operation and Control.   The amount of current applied to the part determines the speed of which the anodic film is formed. Generally, a current density of 12.9–25.8 mA/cm^2 (12–24 A/ft^2) is applied to produce a coating thickness of 10–20 μm (0.4–0.8 mils) per 20 minutes. Most coatings range in thickness from 5–50 μm (0.2–2.0 mils).

The temperature of the anodizing solution also has an effect on the anodic film structure. Thus chillers are used to maintain a consistent temperature in the bath while the part is being anodized, and air agitation is continuously applied to ensure a uniform temperature. The result is a consistent anodic film density of uniform pore size.

Several different acids can be used for anodizing. Phosphoric acid is employed primarily to produce a very porous film. During the anodizing step, the phosphoric acid attacks the growing anodic film, etching a portion of it away. The resulting film can be used as a paint base or adhesive bonding preparation having excellent adhesion properties. Chromic acid is used when high corrosion and abrasion resistance is needed. Most aircraft parts receive this type of anodic coating because any residual acid does not corrode the metal. The anodic film has a distinctive dull green color when chromic acid is employed. Oxalic acid is rarely used but produces a very hard, wear-resistant coating. The anodic film has a grayish yellow color. Some movable aluminum engine parts receive this type of anodizing. Sulfophthalic acid, otherwise known as duranodic or integral coloring, is used to produce a decorative anodic film ranging in color from bronze to black. This process was used extensively from 1950 to 1970 for the color of window and building fascia parts. This process consumes large amounts of electricity, however, and has become too expensive to use. Sulfuric acid is the most widely used acid for anodizing. It produces a hard, clear anodic film which can be subsequently colored using inorganic or organic coloring solutions. Sulfuric acid anodizing is used for both wear-resistant coatings and decorative purposes.

Typical processing steps in anodizing are precleaning, alkaline etching, desmutting, anodizing, two-step coloring or dyeing, and sealing. Water rinses are generally interspersed between stages. The precleaning step is critical in anodizing because it removes the shop soils and cutting lubricants before etching. Precleaning solutions are nonetching and nonsilicated cleaners, and usually contain borates to inhibit the etch. If the part to be anodized is not completely cleaned, the subsequent etching step leaves undesirable patterns on the aluminum.

In the alkaline etch step, the aluminum is given the desired appearance by either etching the surface or chemically polishing it. The alkaline etch gives the aluminum a matte finish which has a satin appearance. The desmutting step is used to remove any of the alloying elements left on the surface after the etching process. Many alloying elements, such as copper, iron, magnesium, and silicon, are not soluble in alkaline solutions. Thus these are left behind on the aluminum surface as a smut.

In the anodizing stage electrolytic reactions produce a uniform aluminum oxide layer across the aluminum surface. This anodic film is transparent and porous. The underlying matte or bright surface can be seen. After anodizing, the aluminum part can be colored or sealed.

The two-step coloring method, sometimes called electrolytic coloring, is the most popular coloring method for architectural purposes. The process involves placing the unsealed anodized parts in a bath constructed similarly to the anodizing tank. The two-step bath chemistry contains a metal salt dissolved in an acid. When alternating current is applied to the part, the metal ions are attracted to the anodic film and precipitate in the anodic pore. As the precipitation continues, the anodized part gradually increases in color until a black color is achieved showing complete saturation of the anodic pores. This coloring process can be stopped at any time during the coloring process with varying shades of color.

The use of dyes for coloring is becoming more popular because of the almost infinite range of colors that can be produced. Moreover, dyes do not need to be electrically deposited. Anodized parts are simply placed in a heated dye solution until the pores become saturated with the pigment (see DYES AND DYE INTERMEDIATES; PIGMENTS).

As a final step, anodized parts must be sealed to ensure corrosion resistance of the anodic coating. Sealing involves plugging the anodic pores completely so contaminants cannot reach the base metal. A variety of sealing methods are used by anodizers (see SEALANTS).

Environmental Issues

There are three aspects of conversion coating application which impact the environment. First, high operating temperatures for cleaners and phosphating baths make these processes energy intensive. Efforts aimed at formulating systems from cleaning to post-treatment, which can be applied at temperatures ranging from 38–49°C, have had some success for iron and zinc phosphates.

Secondly, because conversion coatings rely on including such metals as zinc, nickel, and chromium, the possibility of discharge of these metals is likely. Some manufacturing locations already fall under legislation prohibiting the use of these metals. Substitutes for Ni-containing phosphates have been developed, but zinc is still part of these sytems. Zinc phosphates and zinc phosphates modified with nickel are coming under scrutiny as contributors to heavy-metal contamination of water supplies. In some locations, the use of nickel is already prohibited and iron phosphates are the alternative choice. Chromium has been eliminated from post-treatment solutions and chromium-containing coatings on aluminum and zinc are being replaced with nonchromium coatings for some applications without any sacrifice of performance.

Lastly, sludge generation is an expected by-product of the conversion coating reaction. Some types of sludge are not considered hazardous waste. In many locations these can be disposed of in open landfill sites. This is expected to change and the disposal of all sludges is expected to be problematic (29). General restrictions on the disposal of sludge generated when aluminum is treated in a conversion coating system have been in place since the early 1980s. No economically feasible way to recycle conversion-coating sludge is available, although this problem occupies research efforts of companies worldwide.

BIBLIOGRAPHY

"Chemical and Electrochemical Conversion Treatments" under "Metal Surface Treatments" in *ECT* 1st ed., Vol. 9, pp. 9–23, by A. Mendizza, Bell Telephone Laboratories, Inc.; in *ECT* 2nd ed., Vol. 13, pp. 292–303, by L. F. Schiffman, Amchem Products, Inc.; in *ECT* 3rd ed., Vol. 15, pp. 304–312, by G. L. Schneberger, General Motors Institute.

1. W. Rausch, *Die Phosphatierung von Metallen*, Eugene G. Lenz, Verlag, Germany, 1974.
2. G. Lorin, *Phosphating of Metals*, Finishing Publications Ltd., Middlesex, U.K., 1974.
3. D. B. Freeman, *Phosphating and Metal Pretreatment*, Woodhead Faulkner, Cambridge, U.K., 1986.
4. British Pat. 8,667 (1906), T. W. Coslett.
5. U.S. Pat. 1,069,903 (Aug. 12, 1913), R. G. Richards.
6. U.S. Pat. 2,310,239 (Feb. 9, 1943), G. W. Jernstedt.
7. K. Woods and S. Spring, *Metal Finish.* (Sept. 1980).
8. *Ibid.*, (Mar. 1979).
9. "Preparing Steel for Organic Coatings", *Product Finishing Directory*, Gardiner Publications, Inc., Cincinnati, Ohio, 1993.
10. M. Petschel, SAE Paper #892558, *National SAE Meeting*, Society of Automotive Engineers, Warrendale, Pa., 1989.
11. A. T. El-Mallah and M. H. Abbas, *Metal Finish.* (Apr. 1987).
12. S. Maeda, T. Asai, and H. Okada, *Corrosion Eng.* **31** (1982).
13. M. O. W. Richardson and D. B. Freeman, *Tran. IMF* **64**(1), 16 (1986).
14. G. D. Cheever, *J. Paint Technol.* **41**, 259 (1969).
15. D. W. Hardesty, R. F. Reising, and Y. M. Wang, ASM Paper #8812-002, ASM, Metals Park, Ohio, 1988.
16. G. D. Kent and M. Petschel, *Product Finish.* (Sept. 1988).
17. W. A. Roland and K.-H. Gottwald, *Metalloberflaeche* **42**(6) (1987).
18. U. Selvadurai, *Ueber den Einbau von Mn in Hopeit und Phosphophyllit*, Ph.D. dissertation, University of Cologne, Germany, 1987.
19. J. P. Servais, B. Schmitz, and V. Leroy, *Corrosion* **88** (1988).
20. *ASTM Test Methods 1994*, ASTM, Philadelphia, Pa.

21. *SAE Report J1563* (rev. Oct. 1993), Warrendale, Pa.

22. *Technical Manual 1994*, National Coil Coaters Association, Philadelphia, Pa.

23. U.S. Pat. 2,438,877 (Mar. 30, 1948), (to American Chemical Paint Co.).

24. D. B. Freeman and A. M. Triggle, *Trans. Inst. Met. Fin.* **87**, 56–64 (1960).

25. U.S. Pat. 3,329,587 (July 4, 1967), L. Steinbrecher.

26. *AAMA Publication No. AAMA 605.2-92*, AAMA, Palatine, Ill., 1992.

27. *Military Specification Mil-C-5541E*, Defense Quality and Standardization Office, Falls Church, Va., Nov. 30, 1990.

28. F. A. Lowenheim, *Electroplating Fundamentals of Surface Finishing*, McGraw-Hill Book Publishing Co., Inc., New York, 1978.

29. J. Braithwaite, *Environ. Waste Manag. Mag.* **4** (July–Aug. 1992).

MICHAEL PETSCHEL
ROBERT HART
Parker Amchem

PICKLING

Pickling is a term used to describe metal-cleaning operations designed to remove oxides from metal surfaces. These oxide films may be the result of in-process operations such as heat treating, hot rolling, forging, chemical passivation and etching, or simply environmental corrosion. Among the terms commonly used to describe these oxide layers are scale, rust, smut, white corrosion, and black or blue oxide. Although in some cases the oxide films may be removed using alkaline solutions of various compositions, pickling solutions are predominantly acidic, and most often very strongly acidic.

A number of pickling formulations are available. The proper choice is dictated by the chemistry of the base metal as well as of the oxide film itself. Whereas no single pickle formula is generally effective on all metal alloys, sulfuric acid is probably the most versatile of all the acids. The mechanism by which the acid solutions attack and remove the oxide film varies with the metal, metal oxides to be removed, and the acid used. In some cases oxides are removed by the acid, such as sulfuric acid, penetrating film imperfections such as cracks in the oxide layer, then attacking the base metal. Hydrogen gas is formed at the oxide–metal interface. The pressures thus generated blow off the oxides. Many acids, eg, hydrochloric acid, dissolve the base metal as well as the oxide layer. Pickle inhibitors are therefore available for most pickling acids. These inhibitors minimize or prevent the acid from attacking the base metal, yet allow effective removal of the oxides.

The speed of the pickle reaction is also dependent on the concentration and temperature of the pickle, the degree of agitation of either the metal part or the pickle solution, the alloy being pickled, and the acid used. Pickling solutions may be applied by either spray or immersion techniques. However, because of the noxious fumes emitted, there must be adequate ventilation. Sometimes, particularly when spraying techniques are used, an enclosure to contain the fumes and mist is employed.

Metals and Alloys

Carbon and Low Alloy Steels. Sulfuric acid and hydrochloric acid are the predominant pickling acids for ferrous alloys and some stainless steels having less than 10% chrome and nickel (see STEEL). However, organic acids, such as citric acid and tartaric acid, are seeing more use because these latter are nontoxic and less hazardous. Sulfuric acid is generally used at 10–15 vol % of 66°Bé acid at 65–83°C; hydrochloric is generally used at 20–40 vol % of 20°Bé acid at ambient up to 55°C. Pickling inhibitors are often used with these acids to minimize base metal loss, severe etching, and pitting without significantly affecting the rate of oxide removal. Metal loss can be reduced by 97–99% by using an inhibitor formulated specifically for the pickling acid.

High Chrome–Nickel Stainless Steels. Probably the most common descaling or pickling solution is a combination of about 10% by volume nitric acid with 1–4% hydrofluoric acid at about 49°C. For heavy oxide films, this mixed acid pickle may be preceded by a 20% by volume sulfuric acid pickle. Because scale is generated upon annealing, a molten oxidizing salt bath is used to condition the scale and minimize the concentration and temperature required for the mixed acid treatment. Hydrochloric acid is rarely used for stainless steels, particularly for descaling or deoxidizing finished parts, because of its propensity to cause localized pitting. Sulfuric acid fortified with chloride salts or organic acids may be used successfully for removal of thin oxides on certain alloys.

Aluminum Alloys. Oxide removal on aluminum is generally accomplished using solutions of nitric (20–25% vol %) or sulfuric acid (5–10 vol %), containing various amounts of fluoride ions to control the etch and speed of oxide removal. Chromic acid has also been used as an etch retardant in sulfuric acid, but for ecological and health reasons use of chromic acid is becoming rare. Nonetching oxide removers, or bright dip solutions, consisting of fluoride compounds such as hydrofluoric acid or hydrofluorosilicic acid, formulated with organic compounds such as tragacanth gum or certain surfactants, are also effective.

Copper and Copper-Containing Alloys. Either sulfuric or hydrochloric acid may be used effectively to remove the oxide film on copper (qv) or copper-containing alloys. Mixtures of chromic and sulfuric acids not only remove oxides, but also brighten the metal surface. However, health and safety issues related to chromium(VI) make chromic acid less than desirable.

Zinc and Zinc Alloys. Zinc metal is highly reactive in acid solutions such as sulfuric, hydrochloric, and nitric dissolving rapidly at acid concentrations normally used to pickle steel and aluminum. Dilute (1–4%) solutions of these acids can be used with caution to remove zinc oxides. Sulfamic acid at concentrations of 2–6%, in conjunction with the proper proprietary inhibitor, can be effective in removing zinc oxides and corrosion by-products without attacking the zinc metal.

Magnesium Alloys. Heavy oxides are usually removed from magnesium alloys by using a concentrated (ca 20–25%) hydrofluoric acid solution. Because of its highly toxic and corrosive nature, extreme care must be taken when handling or using hydrofluoric acid solutions. Chromic acid at about 350 g/L containing small amounts (0.1–0.5 g/L) sulfate and chloride ions effectively removes surface contaminants and oxides without significantly attacking the base metal.

Alkaline Deoxidizers

In certain applications, and particularly when hydrogen embrittlement caused by acid etching must be prevented, highly caustic alkaline solutions together with complexing agents, eg, gluconates, citric acid, and EDTA, can be used to derust or remove light scale on steel alloys. These solutions are normally operated at concentrations of 120–360 g/L and at temperatures above 93°C. Even under these conditions, alkaline deoxidizers are significantly slower than acids in removing the surface contaminants.

Economic Aspects

The widespread use of chemical etchants to remove oxides and certain other foreign materials from metal surfaces stems primarily from the very low cost of the pickle. Also important is the high efficiency and versatility in cleaning simple configurations such as flat sheet or wire, as well as workpieces of intricate, complex design where mechanical cleaning would be cumbersome and in many cases incomplete. Sulfuric and hydrochloric acids sell at about $0.10–0.13/L, and caustic soda at about $0.53/kg. The cost of pickling or chemical cleaning is thus only a fraction of a dollar for 90 m^2 (1000 ft^2), even when factoring ancillary costs such as energy, vendor margins, and disposal into the process. As an example, in a wire mill processing scaled steel wire at a cost of approximately $4/m^2 ($350/ft^2), the cost of cleaning off the oxide using 10% inhibited sulfuric acid is only about $0.0004/m^2 ($0.04/ft^2). This includes the cost of metal lost via dissolution by the acid; chemical cost, including a proprietary inhibitor at $2.40/L; and waste treatment. Although the cost of pickling or oxide removal of such metals as aluminum, stainless steel, and copper alloys may be more expensive per square unit of metal treated, the cost of cleaning relative to the cost of the metal is similar.

Health and Safety

Acids such as sulfuric, hydrochloric, nitric, and especially hydrofluoric as well as strong alkalies such as caustic soda and caustic potash are extremely corrosive to animal and vegetable tissue. Extreme caution must be taken to prevent skin contact, inhalation, or ingestion. Violent reactions may occur when dissolving or diluting many of these chemicals with water.

Proprietary additives used with or formulated into acids and alkalies to impart desirable characteristics such as inhibition, improved wetting, and chelating may also contain toxic or carcinogenic chemicals. Producers of proprietary metal finishing chemicals are researching ways to replace and eliminate hazardous chemicals traditionally used in the pickling industry. As of this writing, however, new technology has not fully replaced the old. The Material Safety Data Sheet (MSDS) for each chemical or proprietary blend used in a process must be thoroughly read and understood before the process is put into practice.

BIBLIOGRAPHY

"Solvent Cleaning, Alkali Cleaning, and Pickling and Etching" in *ECT* 1st ed., under "Metal Surface Treatment," Vol. 9, pp. 1–9, by A. Mendizza, Bell Telephone Laboratories, Inc.; "Cleaning, Pickling, and Related Processes" in *ECT* 2nd ed., under "Metal Surface Treatments," Vol. 13, pp. 284–292, by S. Spring, Oxford Chemical Division, Consolidated Foods Corp.; in *ECT* 3rd ed., Vol. 15, pp. 296–303, by G. L. Schneberger, General Motors Institute.

General References

Metals Handbook, 9th ed., Vol. 5, ASM International, Materials Park, Ohio.
K. Hacias, *Wire J. Int.*, 86–89 (Jan. 1995).
M. Murphy, ed., *Metals Finishing, Guidebook and Directory*, Vol. 90, Metal Finishing, Hackensack, N.J., 1992.

KENNETH HACIAS
Henkel Corporation

METAL TREATMENTS

Operations performed on metals previously consolidated by processes such as melting and casting are referred to as metal treatments. Most of these treatments are mechanical and/or thermal. Mechanical treatments involve changes in shape by forming or machining. Metal treatments such as joining and coating of metals are not discussed herein (see METALLIC COATINGS; WELDING).

Mechanical Forming

Forming processes and techniques that are available for a particular alloy depend on its workability, which is the ability to be plastically deformed.

Workability Testing. Workability tests measure the amount of deformation that can be tolerated without fracture, or the development of an instability such as buckling or necking. Buckling or wrinkling is shown at the flange of the cup of Figure 1 where the metal was too thin or was insufficiently supported. Local thinning or necking occurred in the walls of the cup. In most workability tests a specimen is deformed to failure at a constant load rate or strain rate in tension, compression, torsion, shear, or bending. The most common technique is a tensile test at a constant strain rate where the load and elongation are measured continuously. Stress and strain are calculated from load and elongation and are presented in the form of an engineering stress–engineering strain diagram (Fig. 2). Yielding represents the transition from elastic deformation, where atomic bonds are being stretched, to plastic or nonrecoverable deformation, where atomic slip is occurring. The yield stress, which specifies this transition, is

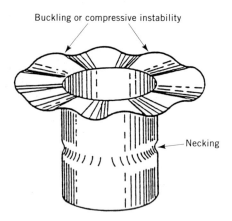

Fig. 1. Examples of instabilities in a deep-drawn cup.

defined usually as the stress (load per area) that produces a small permanent strain, usually 0.002 (0.2% offset yield stress). Following yielding, the stress required for further strain increases, but unlike conditions in the elastic region, stress and strain are not linearly related. Increasing stress with increasing strain in the plastic region is termed strain hardening. The decrease in stress following the maximum load is a result of necking or localized deformation. Engineering stress is decreasing in Figure 2 only because the area in the necked region is rapidly decreasing. The true stress or actual load per unit area continues to

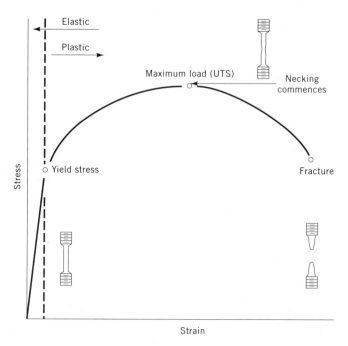

Fig. 2. Schematic stress–strain diagram, where UTS = ultimate tensile stress and (———) represents the demarcation between elastic and plastic behavior. See text.

increase with strain all the way to fracture. The onset of necking is defined by the ultimate tensile stress (UTS) which is the maximum load divided by the initial area. Ductility is measured by % elongation (El) and % reduction in area (RA) which are defined as

$$\% \text{ El} = \left(\frac{L - L_0}{L_0}\right) \times 100$$

where L_0 = initial length; L = final length;

$$\% \text{ RA} = \left(\frac{A_0 - A}{A_0}\right) \times 100$$

where A_0 = initial area; A = final area.

Both % El and % RA are frequently used as a measure of workability. Workability information also is obtained from parameters such as strain hardening, yield strength, ultimate tensile strength, area under the stress–strain diagram, and strain-rate sensitivity.

Tensile testing frequently is used to assess mechanical properties other than workability. However, the strain rate is usually much faster when workability is being measured in order better to simulate forming processes. Standard testing is done at about 10^{-3} s^{-1}, compared to strain rates up to 10^2 s^{-1} for workability testing. An indication of strain-rate sensitivity is given in Figure 3 for a commercial titanium alloy (1) (see TITANIUM AND TITANIUM ALLOYS). Stress–strain diagrams are not unique to tensile testing. These diagrams are also generated by other testing modes such as compression, torsion, shear, and bending.

Temperature strongly influences stress–strain behavior (Fig. 3). Therefore, evaluating hot workability entails testing over a range of temperatures. Hot-torsion data are presented for two nickel-base superalloys, Nimonic 90 and Nimonic 115, and for a 0.48% C steel over a range of temperatures in Figure 4 (2) (see NICKEL AND NICKEL ALLOYS; STEEL). Strength is indicated by torque, and ductility is measured by the number of revolutions to failure. It can be seen that the hot ductility of the nickel-base alloys, particularly N-115, is significantly less than that of steel. Furthermore, the required stresses are substantially greater.

When determining the temperature range for hot working, it is usually not sufficient to merely heat directly to various temperatures. Instead, it is also necessary to acquire cooling data on specimens that are tested after cooling from a temperature corresponding to the furnace temperature in a hot-working operation. Both heating and cooling tensile data are shown in Figure 5 for a Nimonic 115 ingot. The ductility on testing is lower after cooling from 1105 and 1135°C compared to that determined on heating. These tensile tests were performed on a Gleeble machine, where the specimen is resistance-heated.

Plastic Deformation. When plastic deformation occurs, crystallographic planes slip past each other. Slip is facilitated by the unique atomic structure of metals, which consists of an electron cloud surrounding positive nuclei. This structure permits shifting of atomic position without separation of atomic planes

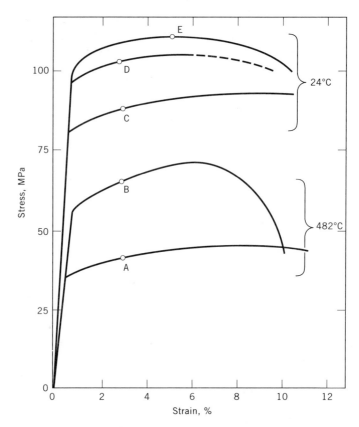

Fig. 3. Effect of temperature and strain rate on stress–strain diagram of Ti–5% Al–2.5% Sn where A–E correspond to the strain rates 1.6×10^{-4}, 5×10^{-1}, 5×10^{-6}, 1.6×10^{-4}, and 5×10 s^{-1}, respectively. The dashed line represents an extrapolation of D (1).

and resultant fracture. The stress required to slip an atomic plane past an adjacent plane is extremely high if the entire plane moves at the same time. Therefore, the plane moves locally, which gives rise to line defects called dislocations. These dislocations explain strain hardening and many other phenomena.

Dislocation may be either edge or screw type. In slipping under the application of the shear stress shown in Figure 6, two distinctively different intermediate stages are possible. The stage shown in Figure **6b** represents an edge dislocation, whereas in Figure **6c** a screw dislocation is shown. An edge dislocation contains an extra half-plane of atoms at the slip plane as shown in Figure **6e**, the front view of Figure **6b**. The term screw dislocation is derived from the fact that following this line defect from above and below the slip plane generates a screw pattern as shown in Figure **6f**. A given dislocation line can be pure edge, pure screw, or any combination of edge and screw components.

An analogy to slip dislocation is the movement of a caterpillar where a hump started at one end moves toward the other end until the entire caterpillar moves forward. Another analogy is the displacement of a rug by forming a hump at one end and moving it toward the other end. Strain hardening occurs because

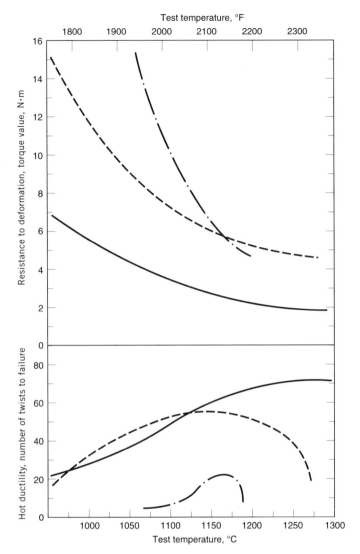

Fig. 4. Torsion properties versus temperature for the nickel alloys (–·–) Nimonic 115 and (–––) Nimonic 90 and for (—) 0.48% C steel (2). To convert N·m to ft·lbf, divide by 1.35.

the dislocation density increases from about 10^7 dislocations/cm^2 to as high as 10^{13}/cm^2. This makes dislocation motion more difficult because dislocations interact with each other and become entangled. Slip tends to occur on more closely packed planes in close-packed directions.

   Hot Working. Plastic deformation at temperatures sufficiently high that strain hardening does not result is termed hot working. The temperature range for successful hot working depends on composition and other factors such as grain size, previous cold working, reduction, and strain rate. Typical hot-working ranges are presented in Figure 7 for various alloys.

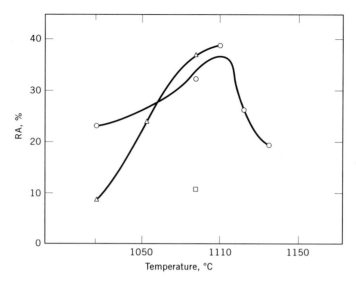

Fig. 5. Hot workability of cast Nimonic 115 as determined by tensile testing using a Gleeble machine: (○), heating; (□), cooling 1135°C; (△), cooling 1105°C. RA = reduction in area.

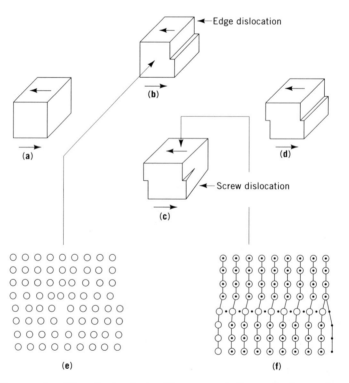

Fig. 6. Continuum (**a–d**) and atomic (**e–f**) representations of edge and screw dislocations. For the screw dislocations, open circles and dots correspond to atoms above and below the slip plane, respectively.

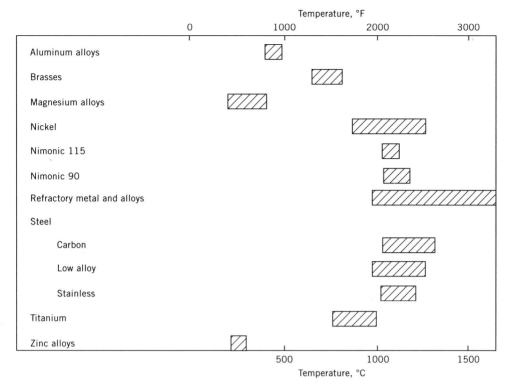

Fig. 7. Hot-working ranges of various metals and alloys.

The lack of strain-hardening results from sufficient thermal energy for recrystallization, which refers to the formation of new grains. Because the new grains have relatively low dislocation densities, strength is not increased. The driving force for recrystallization is the large strain energy associated with the high dislocation density generated during deformation. Recrystallization for hot rolling is shown schematically in Figure 8. The recrystallized grain size is small, but growth occurs with time at a temperature dependent on the alloy.

Hot working permits forming of relatively brittle materials that cannot readily be cold worked. Other advantages are grain refinement, reduction of segregation, healing of defects, such as porosity, and dispersion of inclusions.

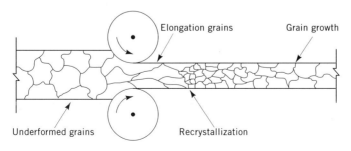

Fig. 8. Recrystallization and grain growth during hot rolling.

Disadvantages are the formation of oxide surface scales and the requirement of heating facilities.

Cold Working.   Cold working involves plastic deformation well below the recrystallization temperature. Required stresses for cold working are greater than for hot working and the amount of strain without heat treatment is limited. Advantages are close dimension control, good surface finish, and increased low temperature strength because of strain hardening. Grain refinement can be achieved by annealing, which entails heating after cold working to temperatures at which recrystallization occurs. The effect of cold working on tensile properties and grain structure and subsequent annealing are shown schematically in Figure 9.

Primary Forming Processes.   Primary forming operations are usually hot-working operations directed toward converting cast ingots into wrought blooms, billets, bars, or slabs (Fig. 10). In primary working operations the large grains typical of cast structures are refined, porosity is reduced, segregation is reduced, inclusions are more favorably distributed, and a shape desirable for subsequent operations is produced. Figure 11 illustrates the principal operations used for ingot breakdown, ie, forging, extruding, and rolling. Extrusion differs from forging and rolling in that more deformation occurs in one pass. Forging and rolling include many passes and some reheating. In addition, intermediate

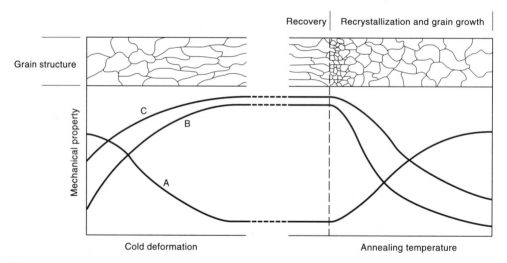

Fig. 9. Variation of tensile properties and grain structure with cold working and annealing: A, elongation; B, yield stress; and C, ultimate tensile stress.

Fig. 10. Distinctions between various intermediate wrought shapes.

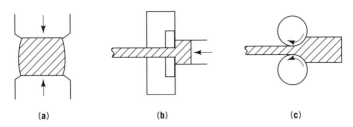

Fig. 11. Primary hot-working processes for ingot breakdown: (**a**) forging; (**b**) extruding; and (**c**) rolling.

conditioning is sometimes necessary. This makes extrusion attractive but it is an expensive operation.

Secondary Forming Processes.   The objective of secondary forming processes, either cold- or hot-working, is to form a shape. Such processes include rolling, open- and closed-die forging, upset forging, extruding, roll forging, ring rolling, deep drawing, spinning, bending, stretching, stamping, drawing, and high velocity forming.

Sheet and semifinished products such as round, rectangular, and shaped bars are produced by rolling. Flat, V-shaped, and swaging dies are used for open-die forging. In closed-die forging, the metal is forced to flow into die cavities to form the impressions of dies attached to the anvil and ram. Forging is performed on both hammers and presses. Hammers have a strain rate of $1-100$ s^{-1} as compared to $0.05-5$ s^{-1} for presses. The energy or stress required for deformation is greater for hammers because of the faster strain rate. (Stress dependance on strain rate is illustrated in Fig. 3). Die contact time and, therefore, die chilling are shorter for hammers. However, the faster strain rate can more readily cause an excessively increased temperature locally, resulting in localized grain growth and possibly incipient melting. The propensity for cracking caused by brittle second-phase particles is less for the slower strain rates of presses.

In a typical process used to manufacture a forged turbine blade, the first three operations involve gathering material for subsequent closed-die forging (Fig. 12). Gathering is accomplished by one extrusion and two upsetting operations. Initial forging results in a perform that is of the correct volume to produce the finished forging in the final operation. A lubricant must be used during closed-die forging to minimize sticking to the die and to promote metal flow. Forging operations must be designed to provide adequate metal flow to prevent critical grain growth.

Extrusion for gathering and producing shapes can entail significantly more than direct forward extrusion. An example of backward extrusion is given in Figure 13**a**.

Roll forging differs from rolling, producing a short length of varying cross section, as opposed to the long, uniform cross sections produced by rolling. In ring rolling, ring-shaped forgings are produced. A seamless ring is produced by reducing the cross section and increasing the circumference of a heated, doughnut-shaped blank between two rotating rolls.

Deep drawing, spinning, bending, stretching, and stamping are cold-working processes applicable to the forming of shapes from sheet and strip

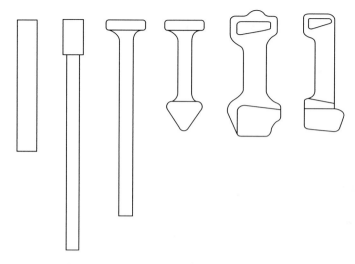

Fig. 12. Typical steps in forging a turbine blade.

(see Fig. 13**b**–**f**). Deep drawing uses a shaped punch to force sheet metal into a die or through a die opening. The drawing metal must have excellent ductility and several draws may be required, with annealing between draws. In spinning, a tool is pressed against one side of a sheet-metal die which is rotated at high speed. The tool gradually makes the disk conform to the shape of the forging instrument that it is forced against. Spinning is used in place of deep drawing if production does not justify the high cost of deep-drawing punches and dies. Stamping is used for cutting blanks from sheet and strip, and usually precedes deep drawing and spinning. Embossing and coining are also stamping operations. In embossing, the impressions of punch and die match each other, whereas in coining they do not, as shown in Figure 13**g** and **h**. Therefore, embossing results in more bending and less flow of metal than coining.

Drawing is a method of reducing the diameter of wire, rod, and tubing. It is similar to extrusion, except that the metal is pulled through the die instead of pushed through it, as shown in Figure 13**i**.

Tubes may be seamed or seamless. Seamed tubes are produced by bending plate, sheet, or strip into the appropriate shape and welding longitudinally. Seamless tubes are manufactured by opening the center of an ingot or billet and working the resulting shell or by working a cast hollow ingot. This may involve extrusion and subsequent drawing. Another method for producing a seamless tube is rotary piercing where metal is rolled over a mandrel, as shown in Figure 13**j**.

High velocity forming has become successful for alloys having poor workability. Examples of such processes are explosive forming, electromagnetic forming, and electrohydraulic forming. In explosive forming, an explosive charge is detonated in a water tank containing the workpiece and die. Shock waves from the explosion propagate throughout the liquid and impact the workpiece with sufficient energy to force it into the die (see METALLIC COATINGS, EXPLOSIVELY CLAD METALS).

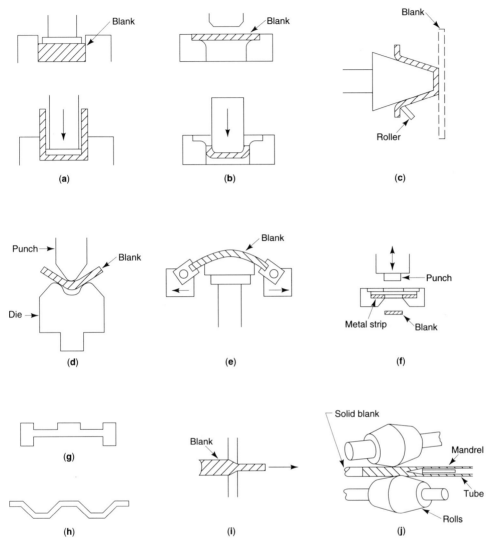

Fig. 13. Various forming operations: (**a**) backward extrusion; (**b**) deep drawing; (**c**) spinning; (**d**) bending; (**e**) stretching; (**f**) stamping; (**g**) coining; (**h**) embossing; (**i**) drawing; (**j**) rotary piercing.

Thermal Treatments

Annealing. In annealing, a cold-worked material is heated to soften it and improve its ductility. The three stages of annealing are recovery, recrystallization, and grain growth (see Fig. 9). Recovery occurs at relatively low temperature and may result in some softening caused mainly by the arrangement of dislocations into a more favorable distribution. Recrystallization is the formation of new grains with a relatively low dislocation density and little internal strain, which replaces strained grains with high dislocation densities. At increasing temperature, the newly formed grains exhibit grain growth. Prolonged exposure at a given temperature also tends to promote grain growth.

Precipitation Hardening.   In precipitation hardening, also called age hardening, fine particles are precipitated from a supersaturated solid solution. These particles impede the movement of dislocations, thereby making the alloy stronger and less ductile. In order for an alloy to exhibit precipitation hardening, it must exhibit partial solid solubility and decreasing solid solubility with decreasing temperatures.

An example of the many alloy systems satisfying these requirements is the aluminum–copper system. The diagram in Figure 14 shows a portion of the equilibrium phase diagram for the binary Al–Cu system, includirg the phases existing under equilibrium conditions at various temperatures as Cu is added to Al. At about 500–600°C, an alloy containing 4.5% Cu consists only of alpha, a solid solution of Cu in Al. Below 500°C, the phase θ (CuAl$_2$) exists in addition to alpha. The objective of precipitation hardening is to distribute the second phase as fine particles which are effective in blocking dislocation motion.

Precipitation hardening consists of solutioning, quenching, and aging. Solutioning entails heating above the solvus temperature in order to form a homogeneous solid solution.

Rapidly quenching to room temperature retains a maximum amount of alloying element (Cu) in solid solution. The cooling rate required varies considerably with different alloys. For some alloys, air cooling is sufficiently rapid, whereas other alloys require water-quenching. After cooling, the alloy is in a relatively soft metastable condition referred to as the solution-treated condition.

In aging, the alloy is heated below the solvus to permit precipitation of fine particles of a second phase θ (CuAl$_2$). The solvus represents the boundary on a phase diagram between the solid-solution region and a region consisting of a second phase in addition to the solid solution.

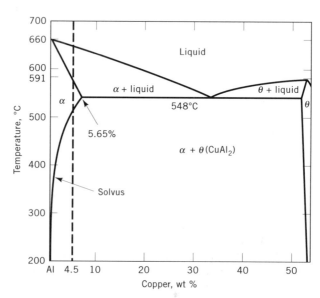

Fig. 14.   Aluminum-rich portion of aluminum–copper-phase diagram.

The precipitation-hardening process and resulting structures are shown in Figure 15. Particles formed initially during aging tend to fit into the lattice of the matrix solid solution, which distorts the lattice at the particle–matrix interface. This accommodation to the matrix phase is termed coherency, and contributes significantly to dislocation blockage, and, therefore, strengthening.

An optimum combination of temperatures and time generates particle size and spacing for the best combination of properties. If aging temperatures are too high, and times too long, particles coalesce and lose coherency, resulting in decreased strength, as shown in Figure 16. Table 1 gives the effect of precipitation hardening on a 95.5% Al–4.5% Cu alloy.

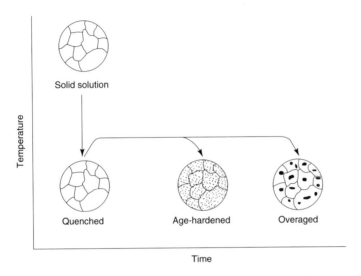

Fig. 15.   Microstructures during precipitation hardening.

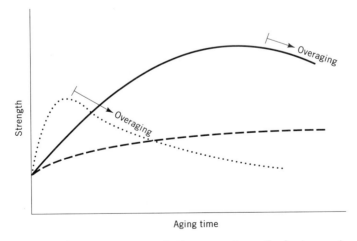

Fig. 16.   Influence of temperature and time on strength during aging where (—) corresponds to the optimum temperature, and (---) and (· · ·) below and above optimum, respectively.

Table 1. Effect of Heat Treatment on Tensile Properties of Al–4.5% Cu[a]

Treatment	Yield strength, MPa[b]	Ultimate tensile strength, MPa[b]	Elongation, %
solution-treated	103	241	30
age-hardened	331	414	20
overaged	69	117	20

[a]Heat treatment consisted of a solution treatment at 515°C, a water quench, followed by aging at 155°C for 10 h.
[b]To convert MPa to psi, multiply by 145.

An example of a precipitation-hardened microstructure is presented in Figure 17 for Udimet 700, which is a precipitation-strengthened nickel-base superalloy. The precipitated phase is $Ni_3(Ti,Al)$, called gamma prime, and is the primary strengthening phase in many commercial superalloys (see HIGH TEMPERATURE ALLOYS). Both coarse and fine gamma prime are present owing to high and low aging temperatures.

Heat Treatment of Steel.   Steels are alloys having up to about 2% carbon in iron plus other alloying elements. The vast application of steels is mainly owing to their ability to be heat treated to produce a wide spectrum of properties. This

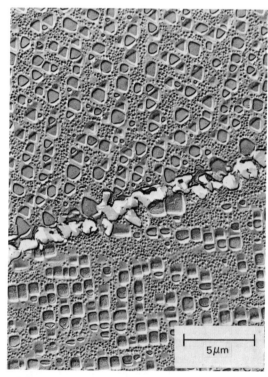

Fig. 17.   Structure of U-700 after precipitation hardening temperature of 1168°C/ 4 h + 1079°C/4 h + 843°C/24 h + 760°C/16 h with air cooling from each temperature. A grain boundary with precipitated carbides is passing through the center of the electron micrograph. Matrix precipitates are γ'-$Ni_3(TiAl)$.

occurs because of a crystallographic or allotropic transformation which takes place upon quenching. This transformation and its role in heat treatment can be explained by the crystal structure of iron and by the appropriate phase diagram for steels (see STEEL).

Iron (qv) exists in three allotropic modifications, each of which is stable over a certain range of temperatures. When pure iron freezes at 1538°C, the body-centered cubic (bcc) δ-modification forms, and is stable to 1394°C. Between 1394 and 912°C, the face-centered cubic (fcc) γ-modification exists. At 912°C, bcc α-iron forms and prevails at all lower temperatures. These various allotropic forms of iron have different capacities for dissolving carbon. γ-Iron can contain up to 2% carbon, whereas α-iron can contain a maximum of only about 0.02% C. This difference in solubility of carbon in iron is responsible for the unique heat-treating capabilities of steel. The solid solutions of carbon and other elements in γ-iron and α-iron are called austenite and ferrite, respectively.

The phase diagram for the iron-rich side of the iron−carbon system is shown in Figure 18. This diagram does not truly represent equilibrium because cementite [12169-32-3], Fe_3C, is a metastable phase. The stable phase for carbon is graphite, but the decomposition of the metastable Fe_3C to graphite and iron is so sluggish that Fe_3C must be treated as the stable phase for most practical purposes. At a composition of 0.77% C and a temperature of 727°C, a eutectoid reaction (Solid 1 → Solid 2 + Solid 3) occurs. The product of this reaction is pearlite, which has a lamellar structure consisting of alternate plates of ferrite and cementite. As the cooling rate increases, the interlamellar spacing decreases and the pearlite becomes finer, as indicated in Figure 19. Pearlite is not a phase in the thermodynamic sense but rather a constituent consisting of two phases, ie, ferrite and cementite.

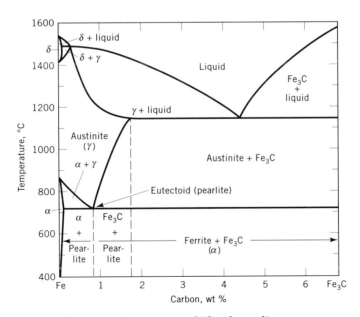

Fig. 18. Iron−iron carbide phase diagram.

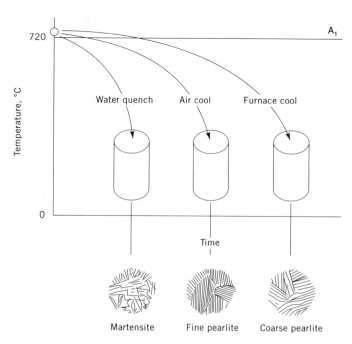

Fig. 19. Effect of cooling rate on structure of a eutectoid steel. A_1 = austenitizing temperature; $>A_1$ = austenite phase.

If the quenching rate is so rapid that pearlite does not form because diffusion cannot occur, another phase, termed martensite, forms, as shown in Figure 19. Martensite is a supersaturated solid solution of carbon in α-iron. It has a body-centered tetragonal crystal structure. Carbon retained in solution distorts the lattice in one edge direction. The strains generated by the carbon in solution impede the movement of dislocations, which results in tremendous strengthening. As shown in Table 2, the strength can increase by almost a factor of 3 by quenching rapidly to form martensite, as compared to cooling slowly to form coarse pearlite. Because martensitic structures have such low ductilities, these are usually tempered following quenching. Tempering entails heating above 100°C to precipitate Fe_3C.

Other common heat-treatment processes are annealing and normalizing. Annealing is usually applied to produce softening and involves heating and cooling. Normalizing is a process in which a steel is heated into the austenite region and then air-cooled. The objective is to obliterate the effects of any

Table 2. Properties of Steel Structures for a Eutectoid Steel

Structure	Yield strength, MPa[a]	Ultimate tensile strength, MPa[a]	Elongation, %
coarse pearlite	372	621	24
fine pearlite	524	1010	20
martensite		1724	low

[a]To convert MPa to psi, multiply by 145.

previous heat treatment or cold working and to ensure a homogeneous austenite on reheating for hardening or full annealing.

Structures that form as a function of temperature and time on cooling for a steel of a given composition are usually represented graphically by continuous-cooling and isothermal-transformation diagrams. Another constituent that sometimes forms at temperatures below that for pearlite is bainite, which consists of ferrite and Fe_3C, but in a less well-defined arrangement than pearlite. There is not sufficient temperature and time for carbon atoms to diffuse long distances, and a rather poorly defined acicular or feathery structure results.

Homogenization. When alloys solidify, substantial segregation occurs. Therefore, ingots are sometimes given a high temperature heat treatment to generate a more homogeneous structure by diffusion in the solid state. Also, homogenization may eliminate undesirable phases that are present in the segregated cast structure.

Thermomechanical Processing. It is possible to develop desirable structures and, therefore, properties by uniquely combining thermal treatments and forming operations. An example of such thermomechanical processing is Minigrain processing (3) to produce fine grains in alloys such as Inconel 718 and Incoloy 901. These alloys are representative of iron–nickel base alloys that precipitate a second phase in addition to that primarily responsible for precipitation strengthening. Figure 20 shows an example of Minigrain processing for Incoloy 901. The primary strengthening phase for Incoloy 901 is a γ' (fcc Ni_3Ti), whereas the phase used for grain-size control is η (hexagonal Ni_3Ti). A conditioning heat treatment is applied which precipitates η in a needle-like registered pattern. This heat treatment is on the order of 8 h at 899°C. Working is carried out at about 954°C, which is below the η-solvus temperature. Final deformation occurs below the recrystallization temperature. A fine-grained structure is generated

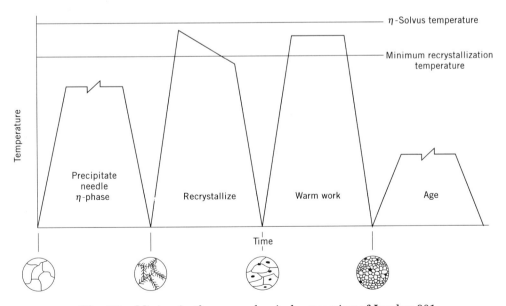

Fig. 20. Minigrain thermomechanical processing of Incoloy 901.

by subsequent recrystallization below the solvus. The needlelike η-phase has become spherical and restricts grain growth. Aging follows standard procedure to precipitate γ'. The resulting fine-grained structure has unique properties, such as excellent low cycle fatigue properties.

Machining

The term machinability is used to indicate the ease or difficulty with which a material can be machined to the size, shape, and desired surface finish. Machining parameters that affect machinability include feed, speed, depth of cut, cutting fluid, cutting-tool material, and cutting-tool geometry (4) (see TOOL MATERIALS). The relative ease of machining materials also depends on the particular machining operation and corresponding material removal characteristics. Machinability index and machinability rating are used as qualitative measures of machinability under specified conditions. Machinability ratings have been based on one or more of the following criteria: tool life, cutting speed, and power consumption.

Tool life tests are used to evaluate the effects of changes in tool materials, cutting variables, processing history, workpiece composition, or workpiece microstructure. Tool life, T, and cutting speed, V_c, can be related by the following equation:

$$V_c T^n = C_t$$

where n and C_t are empirical constants that reflect the cutting conditions under which the tests were made and the machinability of the material, respectively. It is more practical to measure the machinability, C_t, as the cutting speed necessary to cause tool failure within a specified period, than as tool life, T, at a specified speed. To determine C_t, tool life is measured for each of several cutting speeds using standardized cutting conditions and tool shape. These data yield values of n and C_t and the cutting speed that corresponds exactly to a specified tool life can be calculated.

Power consumption during metal cutting can be estimated from forces acting on a tool during the cutting. This can be measured on a dynamometer. In cutting operations the power consumption, P, at the spindle, in watts, is approximately equal to

$$P = V_c \cdot F_c$$

where V_c is the cutting speed in units of m/s and F_c is the force, in newtons, parallel to the cutting direction.

The unit power consumption, P_u, or power required to cut a given quantity of a particular material, can be determined by

$$P_u = P/Q$$

in cutting operations, where Q is the material removal rate, which can be expressed as

$$Q \approx d \cdot f \cdot V_c$$

where d is depth of cut and f is feed rate. Therefore, P_u can be expressed as follows:

$$P_u = F_c/(d \cdot f)$$

where P_u is in units of J/cm^3. The factors that affect the unit power consumption are principally the inherent machinability of the material and friction. The choice of cutting-tool shape, material, and application of coolant have comparatively little effect except for altering the amount of power expended in friction.

Metallurgical factors that affect machinability include chemical composition, previous processing treatments, and physical characteristics. The influence of chemical composition is evident in steels. Carbon and carbide-forming elements such as chromium, tungsten, molybdenum, and vanadium tend to decrease machinability by increasing hardness (qv). Elements which dissolve in ferrite, such as nickel, also usually reduce machinability because of the increased toughness and hardness imparted to the steel. Elements which form hard abrasive particles, such as alumina or silica, are also detrimental to machinability. Elements which form soft inclusions have a beneficial effect. Examples for steels are sulfur, lead, phosphorus, calcium, selenium, and tellurium. In resulfurized steels, sulfur along with manganese is added for the sole purpose of decreasing machining costs through higher machining speeds, improved tool life, or eliminating secondary operations. The manganese sulfide, MnS, inclusions that are formed improve the machining by producing a broken chip compared to a continuous chip. The improved chip formation is a result of the MnS particles that create an interface allowing the chip to separate from the workpiece. In addition, the sulfides prevent the chips from sticking to the tool and undermining the cutting edge.

Processing treatments affecting machinability include hot working, cold working, and heat treatment. The finishing temperatures during hot working influence grain size. For steels, larger grain sizes are preferred for most machining operations. Machining characteristics of most steels and many soft nonferrous metals can be greatly improved by cold working. The lower ductility, caused by cold working, promotes clean shearing and chip breakage. The machining characteristics of most metals are markedly affected by heat treatment. For example, high strength precipitation-strengthened alloys such as nickel-base superalloys are usually machined in the annealed condition prior to final aging.

Physical characteristics of metals have a significant impact on machinability. These include microstructural features such as grain size, mechanical properties such as tensile properties, and physical properties such as thermal conductivity.

Surface Treatments

In some metal-forming operations such as rod and wire drawing, various surface treatments are applied to the workpiece. These include descaling, cleaning the application of lubricant carriers, and the use of lubricants (see LUBRICATION AND LUBRICANTS). Descaling can be mechanical or chemical (pickling). Lubricant carriers are applied by dipping the workpiece in hot solution or slurries such as lime or phosphate coatings. Details of surface treatments are available (5) (see also METAL SURFACE TREATMENTS).

Shot Peening.   A surface treatment that can be used to enhance the surface performance of parts is shot peening. Shot peening is a cold-working process in which the surface of a part is bombarded with small spherical media called shot. Each piece of shot striking the surface of the material imparts a small indentation. To create the indentation, the material surface must yield in tension. Material below the surface tends to restore the surface to its original shape, producing a hemisphere of cold-worked material highly stressed in compression. Overlapping indentations develop an even layer of metal in residual compressive stress. The maximum compressive residual stress produced at or under the surface by shot peening is at least as great as half the yield strength of the material being peened. In addition, the surface hardness of many materials increases as a result of cold working.

Cracks neither initiate nor propagate in a compressively stressed zone. Because nearly all fatigue and stress corrosion failures originate at the surface, compressive residual stresses induced by shot peening provide considerable increases in the resistance to fatigue failures, corrosion fatigue, stress-corrosion cracking, hydrogen-assisted cracking, fretting, galling, and erosion caused by cavitation. Cold working associated with shot peening also provides benefits such as work hardening, intergranular corrosion resistance, surface texturing, closing of porosity, and testing the bond of coatings. Shot peening is often used in the post-machining treatment of superalloy disks in jet engines.

Toxicity in Metal Treatments

Workers in the metals treatment industry are exposed to fumes, dusts, and mists containing metals and metal compounds, as well as to various chemicals from sources such as grinding wheels and lubricants. Exposure can be by inhalation, ingestion, or skin contact. Historically, metal toxicology was concerned with overt effects such as abdominal colic from lead toxicity. Because of the occupational health and safety standards of the 1990s such effects are rare. Subtle, chronic, or long-term effects of metals treatment exposure are under study. An index to safety precautions for various metal treatment processes is available (6). As additional information is gained, standards are adjusted.

Regulation of occupational toxicity is under the supervision of OSHA (7). Industrial employers must provide a written Hazard Communication Program (see HAZARD ANALYSIS AND RISK ASSESSMENT). Elements of this program must include labels and other forms of warning, material safety and data sheets (MSDS), training, a list of hazardous chemicals, a list of hazards of nonroutine tasks, and methods to inform on-site contractors of hazards. Standards are specified as threshold limit values (TLVs) and permissible exposure limits (PELs). The former refer to airborne concentration of substances and represent conditions to which nearly all workers may be repeatedly exposed day after day without known adverse effects. The latter are acceptable airborne concentrations of contaminants adopted by OSHA as standards under U.S. federal laws (29 CFR 1910.1000). Both TLVs and PELs can be specified as time-weighted average (TWA), short-term exposure limit (STEL), or ceiling (C) values. The TWA is the average concentration for a normal 8-h workday. A 15-min TWA exposure which should not be exceeded at any time during the workday is defined as STEL; C represents a concentration that should not be exceeded during any part of the working exposure. Standards vary significantly for the various metals. For example, the TLV–TWA for beryllium is 0.002 mg/m^3, the value for aluminum is 10 mg/m^3 (8).

Developments and Outlook

The principal challenges facing the metals industry as of this writing (ca 1994) are the continuing need for improvements in efficiency in order to meet global competition and environmental requirements; the cost and availability of raw materials; and development of alloys and processing methods to meet the demands of new technologies. The drive for continuous improvement in efficiency has led to development of such processes as strip rolling and strip casting, where sheet and strip are produced directly from the melt, and all hot-rolling steps are eliminated. Operations have also been eliminated by employing continuous casting. About 80% of U.S. crude steel (qv) production comes from continuous casting.

The metals industry relies heavily on raw materials imports. Many strategic elements, such as chromium, nickel, manganese, cobalt, niobium, and tungsten, are imported. For example, no significant chromium deposits exist in the United States. Chromium is used in stainless and specialty steels and superalloys. Efforts to alleviate raw material shortages include stockpiling, extensive recycling (qv), improved melting and processing methods, alloy development and substitution, and use of less expensive raw materials. The widespread use of duplex melting, which utilizes methods such as argon–oxygen decarburization (AOD), has helped greatly in these efforts.

Advanced materials and processes directed toward enhancing the performance and efficiency of aircraft turbines are examples of developments to meet technology demands. These include powder metallurgy techniques, directionally solidified eutectics and single crystals, and composites such as tungsten-reinforced superalloys (see METALLURGY, POWDER; METAL-MATRIX COMPOSITES).

Powder Metallurgy of Superalloys. In addition to providing improved yields and efficiencies, powder metallurgy allows higher alloying additions in wrought products because of reduced segregation, and therefore superplastic structures are readily achieved. This is a significant factor in enabling superalloys which are heavily alloyed to provide superior mechanical properties and oxidation and corrosion resistance for rotating parts exposed to high operating temperatures in aircraft turbine engines.

Superalloy powders are produced by argon atomization, the rotating electrode process, and soluble gas-vacuum atomization (9) (see COATING PROCESSES, POWDER TECHNOLOGY). Consolidation of the powder is accomplished by vacuum hot pressing, forging, extruding, or hot isostatic pressing. In vacuum hot pressing, loose powder is compacted in a cylindrical die with end plugs. Because of the long times involved and the small relative movement between particles, this process is seldom used. In forging, extruding, and hot isostatic pressing, the loose powder is put in a can, usually stainless steel, which is evacuated and sealed. Forging and extrusion are similar to conventional processing. In hot isostatic pressing (HIP) the can is placed in a resistance furnace located inside a water-cooled pressure vessel. Pressures are usually around 103 MPa (15,000 psi), but may be as high as 200 MPa (29,000 psi), and temperatures are up to 1260°C. The unique advantage of this process is its ability to produce near-net shapes or complex parts. Novel canning techniques to achieve these shapes are also available.

Figure 21 compares conventional processing for a superalloy disk to the various processing routes available by powder metallurgy. Figure 21**b** shows the preferred method for powder metallurgy turbine engine disks. This method ensures adequate material consolidation and distribution of micrometer-sized

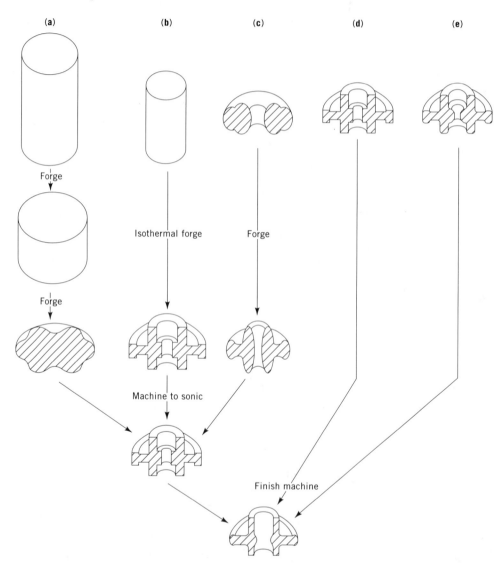

Fig. 21.   Processing of superalloys: (**a**) conventional forging; (**b**) isothermal forging using a powder metallurgy billet; (**c**) forging preform using a powder metallurgy preform; (**d**) hot isotactic processing using a powder metallurgy sonic shape; and (**e**) hot isotactic processing of a near-net powder metallurgy shape.

oxide particles. However, for some applications alternative routes are preferred because high quality parts can be produced using less material and fewer processing steps. The isothermal forging process (Fig. 21**b**) utilizes superplasticity, which refers to the ability of some alloys to exhibit extensive ductility. Elongations >2000% have been seen in tensile testing. The main prerequisite for superplasticity is an extremely fine and stable grain size, which is readily produced by powder metallurgy technicians. The dies are heated to the temperature of the workpiece. Dies are usually made of molybdenum in order to withstand elevated temperatures and, therefore, must be protected from oxidation by forging under a vacuum or inert gas atmosphere. Compared with conventional forging, strain rates are very low. Isothermal forging has an advantage over conventional forging in that it permits the formation of complex shapes in one operation.

A further development in superalloy powder metallurgy is oxide dispersion-strengthened (ODS) material, which contains a finely dispersed oxide that is stable at elevated temperatures. The dispersed oxide provides strength at elevated temperatures where precipitated phases such as γ' are dissolved. Thoria is the dispersed phase in TD nickel and TD nichrome, but emphasis is on dispersion of yttria in alloys such as MA 754 (a Ni–Cr alloy). Oxide dispersion-strengthened materials are produced by mechanical mixing of the oxide and appropriate metal powders in an attritor. Consolidation of these alloys is challenging because it involves thermomechanical processing to achieve large recrystallized grains that are elongated and of a specific crystallographic texture.

Another interesting development which utilizes powder-making techniques and further improves yields is spray forming (Fig. 22) (10). In this process a stream of liquid metal is extracted from a melting operation and passed through an atomizer. The resulting liquid droplets are then sprayed directly onto a collector, and flattened into a layer which builds up a preform. This technique was first explored using aluminum- and iron-based alloys and has some level of

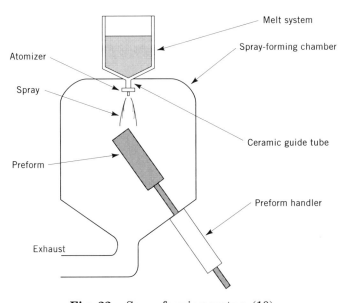

Fig. 22. Spray-forming system (10).

commercial production for stainless steel tubes, rolling mill rolls, and aluminum compressor rotors.

Among the challenges of applying this technology to aircraft-quality metal are the need for reducing porosity, removing ceramic inclusions, and lowering cost, respectively. Studies of atomizing utilizing argon or nitrogen gases have shown that nitrogen atomization lowers porosity to acceptable levels. However, the resulting increases in the nitrogen content of the metal may have detrimental effects. To address ceramic inclusion removal and cost concerns, it has been proposed that the liquid stream be extracted from an electroslag remelting (ESR) furnace through a nonceramic nozzle, as shown in Figure 23 (11,12).

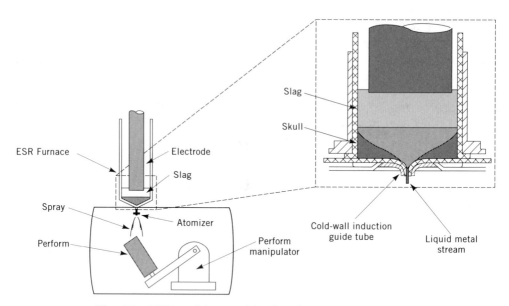

Fig. 23. ESR melting combined with spray forging (11).

BIBLIOGRAPHY

"Metal Treatments" in *ECT* 2nd ed., Vol. 13, pp. 315–331, by H. Herman, University of Pennsylvania, and J. G. Byrne, University of Utah; in *ECT* 3rd ed., Vol. 15, pp. 325–345, by L. A. Jackman, Special Metals Corp.

1. D. K. Allen, *Metallurgy Theory and Practice*, American Technical Society, Chicago, Ill., 1969, p. 503.
2. W. Betteridge and J. Heslop, *The Nimonic Alloys and Other Nickel-Base High-Temperature Alloys*, Crane, Russak & Co., Inc., New York, 1974, p. 130.
3. E. E. Brown, R. C. Boettner, and D. L. Ruckle, *Superalloys—Processing, Proceedings of the Second International Conference*, TMS, Warrendale, Ohio, 1972, pp. L-1–L-12.
4. F. W. Boulger, "Machinability of Steels," *Metals Handbook*, 10th ed., Vol. 1, 1990, p. 591.
5. K. Lang, ed., *Handbook of Metal Forming*, McGraw-Hill Book Co., Inc., New York, 1985, pp. 26.18–26.23.

6. Comprehensive Index for *Metals Handbook*, 10th ed., ASM International, Materials Park, Ohio, 1994.
7. *Occupational Safety and Health Standards for General Industry* (29 CFR Part 1910), Commerce Clearing House, Inc., Washington, D.C., 1991.
8. R. A. Goyer, "Toxicity of Metals", *Metals Handbook*, 10th ed., Vol. 2, ASM International, Materials Park, Ohio, 1990.
9. C. T. Sims and W. C. Hagel, *The Superalloys II*, John Wiley & Sons, Inc., New York, 1987.
10. M. G. Benz and co-workers, *Superalloys 718, 625, 706 and Various Derivatives*, TMS, Warrendale, Ohio, 1994, p. 99.
11. *Ibid.*, p. 106.
12. U.S. Pat. 5,160,532 (Nov. 3, 1992), M. G. Benz and T. F. Sawyer (to General Electric Corp.).

General References

J. Datski, *Material Properties and Manufacturing Processes*, John Wiley & Sons, Inc., New York, 1966.
L. E. Doyle and co-workers, *Manufacturing Processes and Materials for Engineers*, Prentice-Hall, Inc., Englewood Cliffs, N.J., 1985.
M. M. Eisenstadt, *Introduction to Mechanical Properties of Materials*, The Macmillan Co., New York, 1971.
A. G. Guy, *Introduction to Materials Science*, McGraw-Hill Book Co., Inc., New York, 1972.
R. W. Hanks, *Materials Engineering Science*, Harcourt, Brace & World, Inc., New York, 1970.
C. A. Keyser, *Materials Science in Engineering*, Charles E. Merrill Co., Columbus, Ohio, 1968.
F. A. McClintock and A. S. Argon, *Mechanical Behavior of Materials*, Addison-Wesley Publishing Co., Inc., Reading, Pa., 1966.
W. J. Patton, *Materials in Industry*, 3rd ed., Prentice-Hall, Inc., Englewood Cliffs, N.J., 1986.
R. E. Reed-Hill, *Physical Metallurgy Principles*, D. Van Nostrand Co., Inc., Princeton, N.J., 1964.
L. H. Van Vlack, *Materials Science for Engineers*, Addison-Wesley Publishing Co., Reading, Pa., 1970.
Shot Peening Applications, 7th ed., Metal Improvement Co., Inc., Paramus, N.J.

LAURENCE A. JACKMAN
CARLOS N. RUIZ
Teledyne Allvac

METAL VAPOR SYNTHESIS. See LIGHT GENERATION; THIN FILMS.

METANILIC ACID. See AMINES, AROMATIC–ANILINE AND ITS DERIVATIVES.

METHACROLEIN. See ACROLEIN AND DERIVATIVES.

METHACRYLIC ACID AND DERIVATIVES

Methacrylic acid (MAA) was first prepared in 1865 by the hydrolysis of ethyl methacrylate, which was in turn obtained by dehydrating ethyl α-hydroxyisobutyrate (1). The polymerizability of methacrylic acid was first noted in 1880 when a white powder was obtained in a distillation of methacrylic acid (2). The initial recognition of the potential importance of acrylic and methacrylic acids and their derivatives and the foundation for their commercial exploitation can be traced to the doctoral thesis of Otto Röhm at the University of Tübingen in 1901 (3), who described the preparation of colorless, clear, rubbery films which he speculated could be of commercial utility. Although Röhm was granted patent coverage for applications of acrylic polymers in 1914 (4), commercial processes for methacrylate monomer manufacture were not developed until the early 1930s (5,6). The acetone cyanohydrin process for the synthesis of methyl methacrylate via the formation of methacrylamide sulfate, which was patented in 1934 (7), still forms the basis for the bulk of the methyl methacrylate (MMA) currently produced.

Physical Properties

Selected physical properties of various methacrylate esters, amides, and derivatives are given in Tables 1–4. Tables 3 and 4 describe more commercially available methacrylic acid derivatives. Azeotrope data for MMA are shown in Table 5 (8). The solubility of MMA in water at 25°C is 1.5%. Water solubility of longer alkyl methacrylates ranges from slight to insoluble. Some functionalized esters such as 2-dimethylaminoethyl methacrylate are miscible and/or hydrolyze. The solubility of 2-hydroxypropyl methacrylate in water at 25°C is 13%. Vapor–liquid equilibrium (VLE) data have been published on methanol, methyl methacrylate, and methacrylic acid pairs (9), as have solubility data for this ternary system (10). VLE data are also available for methyl methacrylate, methacrylic acid, methyl α-hydroxyisobutyrate, methanol, and water, which are the critical components obtained in the commercially important acetone cyanohydrin route to methyl methacrylate (11).

Reactions

Methacrylic acid and its ester derivatives are α,β-unsaturated carbonyl compounds and exhibit the reactivity typical of this class of compounds, ie, Michael and Michael-type conjugate addition reactions and a variety of cycloaddition and related reactions. Although less reactive than the corresponding acrylates as the result of the electron-donating effect and the steric hindrance of the α-methyl group, methacrylates readily undergo a wide variety of reactions and are valuable intermediates in many synthetic procedures.

Nucleophilic Addition Reactions.   Many nucleophiles, including amines, mercaptans, and alcohols, undergo 1,4-conjugate addition to the double bond of methacrylates (12–14).

Table 1. Properties of Methacrylic Esters, $CH_2{=}C(CH_3)COOR$

R substituent	CAS Registry Number	Bp, °C (at kPa[a])	Refractive index, n_D^T	Sp gr[b]
Alkyl methacrylates				
methyl	[80-62-6]	100 (101)	1.4120^{25}	0.939_4
ethyl	[97-63-2]	119 (101)	1.4116^{25}	0.909_4
propyl	[2210-28-8]	141 (101)	1.4183^{20}	$0.902_4{}^c$
isopropyl	[4655-34-9]	125 (101)	1.4334^{25}	$0.885_4{}^c$
n-butyl	[97-88-1]	162–163 (101)	1.4220^{25}	0.889_4
isobutyl	[97-86-9]	155 (101)	1.4197^{20}	0.882_4
s-butyl	[2998-18-7]	72–73 (6.7)	1.4195^{25}	
tert-butyl	[585-07-9]	52 (4.7)	1.4120^{25}	
n-hexyl	[142-09-6]	204–210 (101)	1.4310^{20}	0.894_4
n-octyl	[2157-01-9]	114 (1.9)	1.4373^{20}	
isooctyl	[28675-80-1]	68 (0.087)	1.4386^{20}	
2-ethylhexyl	[688-84-6]	47 (0.013)	1.4380^{20}	
n-decyl	[3179-47-3]	99–100 (0.17)	1.4425	
dodecyl	[142-90-5]		1.444	0.868^d
tetradecyl	[2549-53-3]	147–154 (0.093)	1.4480^{20}	
octadecyl	[32360-05-7]		1.4502^{25}	
Unsaturated alkyl methacrylates				
vinyl	[4245-37-8]	63 (16.4)		
allyl	[96-05-9]	32 (1.3)	1.4328^{25}	
oleyl	[13533-08-9]	165–170 (0.0027)	1.4607^{20}	
2-propynyl	[13861-22-8]	47–49 (1.5)	1.4483^{20}	
Cycloalkyl methacrylates				
cyclohexyl	[101-43-9]	44 (0.040)	1.4583^{20}	
1-methylcyclohexyl	[76392-14-8]	94–98 (1.3)	1.4588^{25}	
3-vinylcyclohexyl	[76392-15-9]	63–70 (0.013)	1.4692^{25}	
3,3,5-trimethylcyclohexyl	[75673-26-6]	51–52 (0.013)	1.4548^{20}	
bornyl	[4647-84-1]	68–70 (0.040)	1.4739^{25}	
isobornyl	[7534-94-3]	112–117 (0.33)	1.4748^{25}	0.980_4
cyclopenta-2,4-dienyl	[76741-96-3]	115 (0.093)	1.4990^{25}	
dicyclopentenyl	[51178-59-7]	137 (1.7)		
dicyclopentenyloxyethyl	[68586-19-6]	350–360		
Aryl methacrylates				
phenyl	[2177-70-0]	58–61 (0.13)	1.5184^{20}	
benzyl	[2495-37-6]	119–121 (0.1–0.2)	1.5095^{25}	
nonylphenyl	[76391-98-5]	120–127 (0.0040)	1.5020^{25}	
2-phenoxyethyl	[10595-06-9]	130–132 (1.0)		
Hydroxyalkyl methacrylates				
2-hydroxyethyl	[868-77-9]	87 (0.67)	1.4505^{25}	1.064_4
2-hydroxypropyl	[923-26-2]	87 (0.67)	1.4456^{25}	1.028
3-hydroxypropyl	[2761-09-3]	67–69 (0.013)	1.4496^{25}	
2-hydroxybutyl	[13159-51-8]	77–79 (0.26)		
4-hydroxybutyl	[997-46-6]	80–85 (0.013)		
5-hydroxypentyl	[61016-96-4]	85–90 (0.004)		
6-hydroxyhexyl	[13092-57-4]			
3,4-dihydroxybutyl	[62180-57-8]	110–111 (0.033)		
2,3-dihydroxypropyl	[5919-74-4]			

Table 1. (*Continued*)

R substituent	CAS Registry Number	Bp, °C (at kPa[a])	Refractive index, n_D^T	Sp gr[b]
Methacrylates of ether alcohols				
methoxymethyl	[20363-82-0]	54 (2.0)	1.4233[20]	
ethoxymethyl	[76392-16-0]	87–88 (7.3)	1.4216[20]	
allyloxymethyl	[49978-33-8]	80–82 (2.7)	1.4422[20]	
2-ethoxyethoxymethyl	[76392-17-1]	113–114 (2.3)	1.4302[20]	
benzyloxymethyl	[76392-18-2]	136–137 (0.67)	1.5067[20]	
cyclohexyloxymethyl	[76392-19-3]	91–92 (0.27)	1.4599[20]	
1-ethoxyethyl	[51920-52-6]	64–65 (2.7)	1.4182[25]	
2-ethoxyethyl	[2370-63-0]	85 (2.5)		
2-butoxyethyl	[13532-94-0]	104 (2.0)	1.4304[25]	
1-methyl-(2-vinyloxy)ethyl	[76392-20-6]	50 (0.13)	1.4400[20]	
methoxymethoxyethyl	[76392-21-7]	68–70 (0.27)	1.4310[25]	
methoxyethoxyethyl	[45103-58-0]	67–75 (0.13)	1.4397[20]	
vinyloxyethoxyethyl	[76392-22-8]	80–82 (0.13)	1.4515[20]	
1-butoxypropyl	[76392-23-9]	51–53 (0.13)	1.4309[20]	
1-ethoxybutyl	[76392-24-0]	85–88 (3.1)	1.4223[20]	
tetrahydrofurfuryl	[2455-24-5]	59–62 (0.080)	1.4552[20]	
furfuryl	[3454-28-2]	62–63 (0.40)	1.4770[25]	
Heterocyclic methacrylates				
glycidyl	[106-91-2]	75 (1.3)	1.4482[25]	1.073
2,3-epoxybutyl	[68212-07-7]	45–50 (0.033)	1.4422[25]	
3,4-epoxybutyl	[55750-22-6]	55–56 (0.11)	1.4472[25]	1.038
2,3-epoxycyclohexyl	[76392-25-1]	70–73 (0.040)	1.4671[25]	
10,11-epoxyundecyl	[23679-96-1]	115–119 (0.0027)	1.4553[25]	0.949
2-(1-aziridinyl)ethyl	[6498-81-3]	190		
thioglycidyl	[3139-91-1]	59 (0.18)		
3-thietanyl	[21806-33-7]	81–83 (1.3)		
Aminoalkyl methacrylates				
2-dimethylaminoethyl	[2867-47-2]	97.5 (5.3)	1.4396[20]	0.933
2-diethylaminoethyl	[105-16-8]	49 (0.040)	1.4442[20]	
2-*tert*-butylaminoethyl		97 (1.6)	1.4400[25]	0.914
2-*tert*-octylaminoethyl	[14206-24-7]	138–139 (1.6–1.7)	1.4345[25]	0.933
2-dibutylaminoethyl	[2397-75-3]	110 (0.13)	1.4474[20]	
3-diethylaminopropyl	[17577-32-1]	105 (0.20)	1.4770[20]	
3-dimethylaminopropyl	[20602-77-1]	88 (2.13)	1.4418[20]	
4-dimethylaminobutyl	[60238-41-2]	72 (0.26)	1.4417	
7-amino-3,4-dimethyloctyl	[76392-26-2]	115–120 (1.10)	1.4570[25]	0.922
N-methylformamidoethyl	[25264-39-5]	121–123 (0.16)	1.4693[25]	
2-ureidoethyl	[4206-97-7]	74–76 (0.40)		
2-piperidinylethyl	[19416-48-9]	78 (0.04)	1.4682	
3-morpholinylpropyl	[20602-96-4]	100 (0.13)	1.4704	
Glycol dimethacrylates				
methylene	[4245-38-9]	54–57 (0.024)	1.4520[20]	
ethylene glycol	[97-90-5]	96–98 (0.53)	1.4520[25]	
1,2-propanediol	[7559-82-2]	68–72 (0.13)	1.4450	
1,3-butanediol	[1189-08-8]	78–79 (0.053)	1.4523[20]	
1,4-butanediol	[2082-81-7]	88 (0.027)	1.4872[20]	

Table 1. (*Continued*)

R substituent	CAS Registry Number	Bp, °C (at kPa^a)	Refractive index, n_D^T	Sp gr^b
	Glycol dimethacrylates			
2,5-dimethyl-1,6-hexanediol	[76392-00-2]	125–127 (0.13)	1.4567^{20}	
1,10-decanediol	[6701-13-9]	170–178 (0.27)	1.4577^{25}	
diethylene glycol	[2358-84-1]	120–125 (0.27)	1.4550^{25}	
triethylene glycol	[109-16-0]	155 (0.13)	1.4604^{20}	
tetraethylene glycol	[109-17-1]	>200 (0.13)		
neopentyl glycol	[1985-51-9]	112 (0.21)		
1,6-hexanediol	[6606-59-3]	110 (2.6)		
1,12-dodecandiol	[72829-09-5]			
4,4′-isopropylidenediphenol	[3253-39-2]	70 (mp)		
	Polyfunctional methacrylates			
trimethylolpropane trimeth- acrylate	[3290-92-4]	155 (0.13)	1.471	1.06
tris(2-methacryloxyethyl)- amine	[13884-43-0]	155–165 (0.0067)	1.4768^{25}	
pentaerythritol tetrameth- acrylate	[3253-41-6]			
	Carbonyl-containing methacrylates			
R = carboxymethyl	[76391-99-6]	108–111 (0.033)		
R = 2-carboxyethyl	[13318-10-0]	104–106 (0.013)	1.4546^{25}	
R = acetonyl	[44901-95-3]	40 (0.067)	1.4437^{20}	
R = oxazolidinylethyl	[46235-93-2]	83–87 (0.067)	1.4688^{25}	
N-(2-methacryloyloxyethyl)- 2-pyrrolidinone	[946-25-8]	115–128 (0.067)	1.4872^{20}	
N-(3-methacryloyloxyethyl)- 2-pyrrolidinone	[76747-97-4]	127 (0.073)	1.4860^{20}	
N-(methacryloyloxy)- formamide	[76392-27-3]	127 (0.13)	1.4782^{20}	
	Other nitrogen-containing methacrylates			
2-methacryloyloxyethyl- methylcyanamide	[76392-28-4]			
methacryloyloxyethyl- trimethylammonium chloride	[5039-78-1]			
N-(methacryloyloxyethyl)- diisobutylketimine		93–97 (0.027)	1.4543^{25}	
cyanomethyl methacrylate	[7726-87-6]	79–80 (1.3)	1.4381^{25}	
2-cyanoethyl methacrylate	[4513-53-5]	72 (0.067)	1.4459^{20}	
	Methacrylates of halogenated alcohols			
chloromethyl	[27550-73-8]	54–56 (3.1)	1.4434^{25}	
1,3-dichloro-2-propyl	[44978-88-3]	58–60 (0.027)	1.4670^{25}	
4-bromophenyl	[36889-09-5]	80–85 (0.013)		
2-bromoethyl	[4513-56-8]	65 (0.67)	1.4750^{20}	
2,3-dibromopropyl	[3066-70-4]	70–76 (0.0040)	1.5132^{25}	
2-iodoethyl	[35531-61-4]	112–119 (4.0)		
1,1-dihydroperfluoroethyl	[352-87-4]	100 (13)		
1*H*,1*H*,5*H*-octafluoropentyl	[355-93-1]	179		

Table 1. (*Continued*)

R substituent	CAS Registry Number	Bp, °C (at kPa^a)	Refractive index, n_D^T	Sp gr^b
Methacrylates of halogenated alcohols				
1*H*,1*H*,7*H*-dodecafluoro-heptyl	[2261-93-1]	107 (3.0)		
hexafluoroisopropyl	[3063-94-3]	50 (18.6)		
Sulfur-containing methacrylates				
methyl thiolmethacrylate	[52496-39-6]	57 (4.01)		
butyl thiolmethacrylate	[54667-21-9]	57 (0.27)		1.4828^{20}
ethylsulfonylethyl methacrylate	[25289-10-5]	120–132 (0.020)		
ethylsulfinylethyl methacrylate	[3007-24-7]	116–119 (0.013)		1.4902^{25}
thiocyanatomethyl methacrylate	[76392-29-5]	59 (0.020)		1.4899^{25}
methylsulfinylmethyl methacrylate	[76392-30-8]	110–112 (0.04–0.06)		1.4963^{25}
4-thiocyanatobutyl methacrylate	[76392-31-9]	102 (0.015)		1.4861^{25}
bis(methacryloyloxyethyl)-sulfide	[35411-32-6]	115–125 (0.067)		1.4894^{25}
2-dodecylthioethyl methacrylate	[14216-26-3]	155–160 (0.020)		1.4731^{25}
Phosphorus-, boron-, and silicon-containing methacrylates				
2-(ethylenephosphito)propyl methacrylate	[76392-32-0]	73–80 (0.0027)		1.4635^{25}
diethyl methacryloyl-phosphonate	[76392-33-1]	125–165 (0.027)		1.4668^{25}
dimethylphosphinomethyl methacrylate	[41392-09-1]	80–85 (0.0053)		1.4347^{25}
dimethylphosphonoethyl methacrylate	[22432-83-3]	78–84 (0.0093)		1.4426^{25}
dipropyl methacryloyl phosphate	[76392-34-2]	75–76 (1.6)		1.4411^{20}
diethyl methacryloyl phosphite	[3729-12-2]	64 (0.19)		1.4438^{25}
2-methacryloyloxyethyl diethyl phosphite	[817-44-7]	83 (0.13)		1.4483^{20}
diethylphosphatoethyl methacrylate	[814-35-7]	115–124 (0.067)		1.4340^{25}
2-(dimethylphosphato)-propyl methacrylate	[76392-35-3]	85–90 (0.0027)		1.4359^{25}
2-(dibutylphosphono)-ethyl methacrylate	[3729-11-1]	130–140 (0.0027)		1.4390^{25}
2,3-butylene methacryloyl-oxyethyl borate	[76392-36-4]	88–94 (0.0027)		1.4451^{25}
methyldiethoxymethacryl-oyloxyethoxysilane	[76392-37-5]	75–82 (0.0027)		1.4216^{25}

^{a}To convert kPa to mm Hg, multiply by 7.50. ^{b}At 25°C, unless otherwise noted. ^{c}At 20°C.
^{d}Lauryl methacrylate, a mixture.

Table 2. Properties of Amides of Methacrylic Acid

Compound	CAS Registry Number	Bp, °C (at kPa[a])	Mp, °C	Refractive index, n_D^T
N-methylmethacryl-amide	[3887-02-3]	88 (0.47)		1.4740[20]
N-ethylmethacryl-amide	[7370-88-9]	79–82 (0.13)		
N-isopropylmethacryl-amide	[13749-61-6]	112 (15.3)	90–91	
N-butylmethacryl-amide	[28384-61-4]	113 (0.8)		
N-tert-butylmethacryl-amide	[6554-73-0]		57.5	
N-cyclohexylmeth-acrylamide	[2918-67-4]		111	
N-phenylmethacryl-amide	[1611-83-2]		84–85	
N-1-naphthylmeth-acrylamide	[22447-06-9]		114–115	
N-anisylmethacryl-amide	[7274-71-7]		89–90	
N-anilinylmethacryl-amide	[1611-83-2]		87	
N-morpholinylmeth-acrylamide	[5117-13-5]		125	
N-ethoxymethylmeth-acrylamide	[3644-09-5]	100 (0.67)		
N-butoxymethylmeth-acrylamide	[5153-77-5]	110–112 (0.27)		
N-hydroxymethylmeth-acrylamide	[923-02-4]		53.5–54	
N-(2-hydroxyethyl)-methacrylamide	[5238-56-2]	147–157 (0.15)		1.5002[25]
1-methacryloylamido-2-methyl-2-propanol	[74987-95-4]	100–104 (0.013)	74–76	
4-methacryloylamido-4-methyl-2-pentanol	[23878-87-1]	114–119 (0.053)		1.4732[25]
N-(methoxymethyl)-methacrylamide	[3644-12-0]	78–82 (0.03–0.05)		1.4707[25]
N-(dimethylamino-ethyl)methacryl-amide	[13081-44-2]	85–92 (0.1–0.4)		1.4744[20]
N-(3-dimethylamino-propyl)methacryl-amide	[5205-93-6]	92 (0.0053)		1.4789[20]
N-acetylmethacryl-amide	[44810-87-9]	76 (0.16)		1.4835[25]
N-methacryloylmale-amic acid	[76392-01-3]		192–193	
methacryloylamido-acetonitrile	[65993-30-8]	114–118 (0.053)	32–34	
N-(2-cyanoethyl)meth-acrylamide	[24854-94-2]		46–48	

Table 2. (Continued)

Compound	CAS Registry Number	Bp, °C (at kPa^a)	Mp, °C	Refractive index, n_D^T
1-methacryloylurea	[20602-83-9]		138	
N-phenyl-N-phenyl-ethylmethacryl-amide	[76392-02-4]		63–64	
N-(3-dibutylamino-propyl)methacryl-amide	[76392-03-5]	125 (0.017)		1.4731^{20}
N,N-dimethylmeth-acrylamide	[6976-91-6]	68 (0.80)		1.4600^{20}
N,N-diethylmethacryl-amide	[5441-99-6]	71–72 (0.33)		
N-(2-cyanoethyl)-N-methylmethacryl-amide	[76392-04-6]	113–116 (0.15)		1.4755^{25}
N,N-bis(2-diethyl-aminoethyl)meth-acrylamide	[76392-05-7]	122 (0.053)		1.4702^{20}
N-methyl-N-phenyl-methacrylamide	[2918-73-2]	88–96 (0.27)	50	
N,N'-methylenebis-methacrylamide	[2359-15-1]		163–164	
N,N'-ethylenebismeth-acrylamide	[6117-25-5]		170 (dec)	
N-(diethylphosphono)-methacrylamide	[76392-06-8]	105–110 (0.0067)		1.4412^{25}
N-tert-butyl-N-(diethyl-phosphono)meth-acrylamide	[76392-07-9]	80–85 (0.020)		1.4296^{25}

^{a}To convert kPa to mm Hg, multiply by 7.50.

Nucleophilic addition

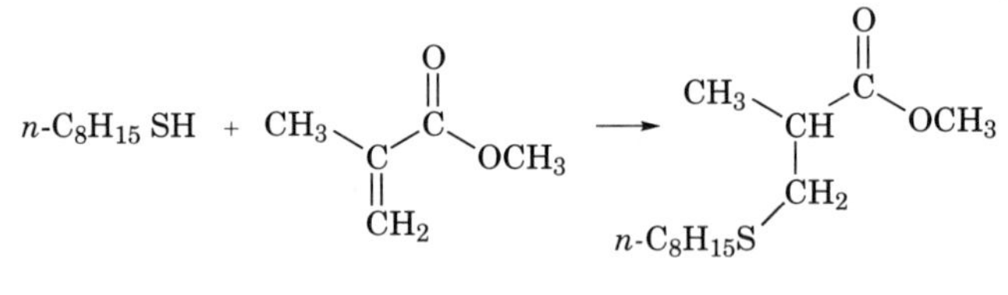

Multifunctional nucleophiles may undergo 1,4-conjugate addition to the double bond, followed by reaction at the carbonyl moiety to give heterocyclic products (15–18).

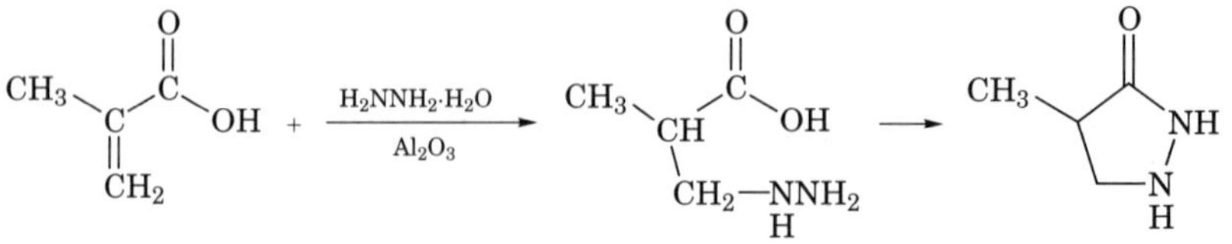

Table 3. Selected Properties of Methacrylates

Compound	Molecular weight	Mp, °C	Viscosity, mPa(=cP)	Flash point, °C	Autoignition temperature, °C
methyl meth-acrylate[a]	100.11	−48	0.53	9[b]	435
ethyl meth-acrylate[a]	114.14	−17	0.92	16[c]	393
butyl meth-acrylate[a]	142.19	−50	0.92	49[d]	294
lauryl meth-acrylate[e]	262	−22		110	277
2-dimethyl-aminoethyl methacrylate[f]	157.2	−30	1.1	75	
2-hydroxyethyl methacrylate[g]	130.14	−12		66	
2-hydroxypropyl methacrylate[g]	144.17	−89	7.1	98	
glycidyl meth-acrylate	142.1	< −60	5	76	

[a] Heat of polymerization = 57.5 kJ/mol (13.7 kcal/mol).
[b] Lower explosion limit (LEL) = 2.1%; upper explosion limit (UEL) = 12.5%.
[c] LEL = 1.8%; UEL to saturation.
[d] LEL = 2.0%; UEL = 8%.
[e] Made from a mixture of higher alcohols, predominantly C-12.
[f] pK_a = 8.4.
[g] Heat of polymerization = ~ 50 kJ/mol (12 kcal/mol).

1,4-Conjugate addition to a β-halogen substituted methacrylate results in an addition–elimination reaction which regenerates the α,β-unsaturated moiety (19–21).

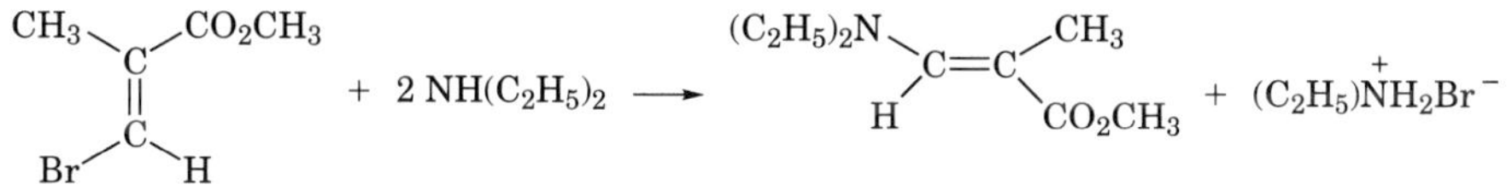

Cycloaddition Reactions. Methacrylates have been widely used as dienophiles in Diels-Alder reactions (22–24).

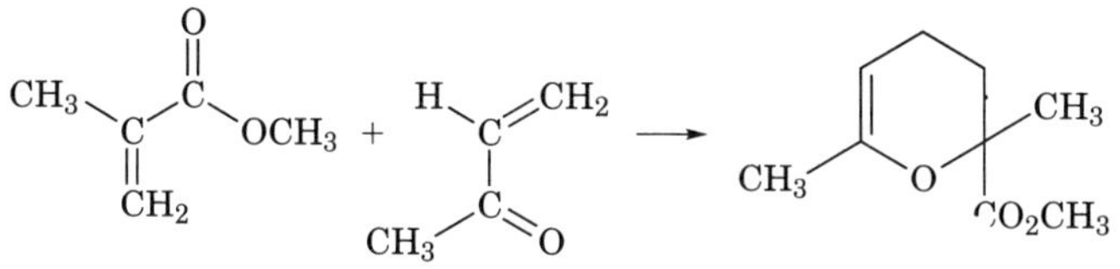

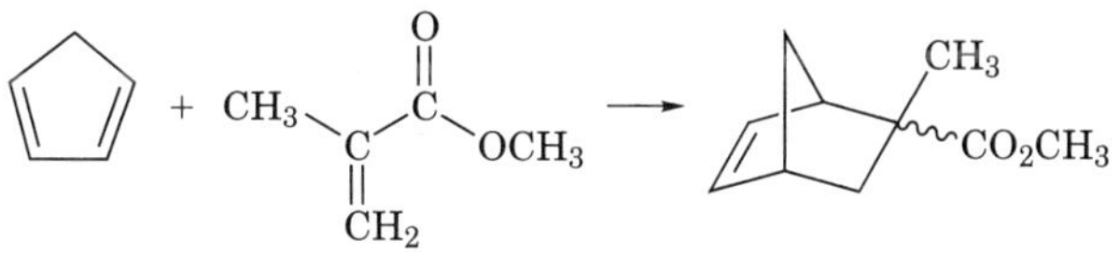

Table 4. Properties of Methacrylic Acid and Nonester Derivatives

Compound	CAS Registry Number	Molecular weight	Melting point, °C	Boiling point, °C[a]	Refractive index	Specific gravity	Viscosity, mPa(=cP)	Solubility in water at 25°C	Flash point, °C
methacrylic acid[b]	[79-41-4]	86.09	14	162	1.4288	1.015	1.3	miscible	67[c]
methacrylic anyhdride	[760-93-0]	154.17		87[d]	1.453	1.035	3	reacts	84
methacrolein	[78-85-3]	70.09	−81	69	1.416	0.847		miscible	−15
methacryloni-trile	[126-98-7]	67.09	−35.8	90.92	1.4	0.8	0.39	2.6%	12
methacryloyl chloride	[920-46-7]	104.54		95.96	1.442	1.07		reacts	2

[a] At 101.3 kPa = 1 atm, unless otherwise indicated.
[b] pK_a = 4.66; heat of polymerization = 66.1 kJ/mol (15.8 kcal/mol).
[c] Autoignition temperature = 400°C.
[d] At 1.7 kPa = 13 mm Hg.

Table 5. Azeotropic Mixtures with Methyl Methacrylate[a]

Component	Pressure, kPa[b]	Bp, °C	MMA, %
water	101	83	86
	27	49	88.4
methanol	101	64.2	15.5
ethanol	27	34.5	18

[a]Ref. 8.
[b]To convert kPa to mm Hg, multiply by 7.50.

1,3-Dipolar cycloaddition reactions with azides, imines, and nitrile oxides afford synthetic routes to nitrogen-containing heterocycles (25–30).

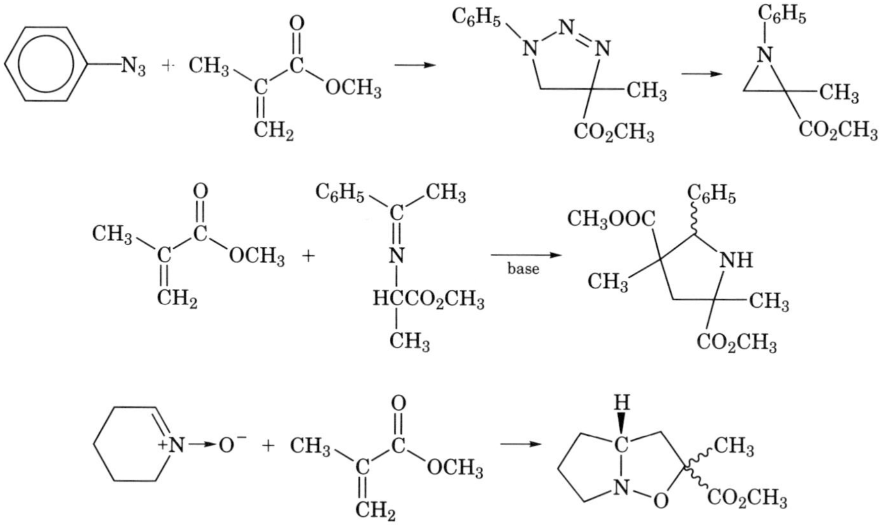

Epoxidation of the double bond as well as cyclopropanation reactions with diazo compounds and metallocarbenes are also well documented (31–34).

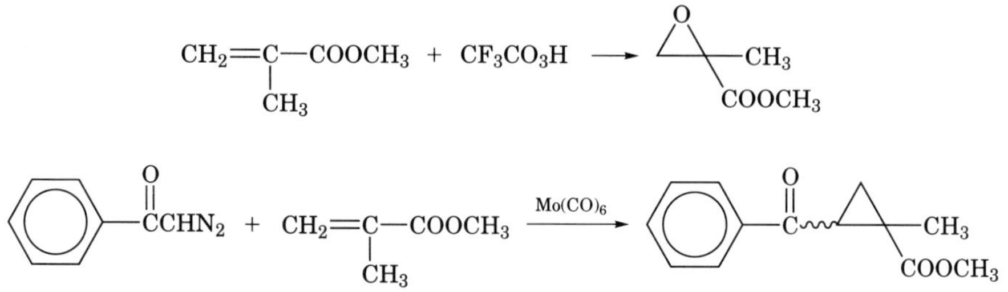

Methacrylates also undergo 2 + 2 photocyclization reactions with a variety of substrates (35,36).

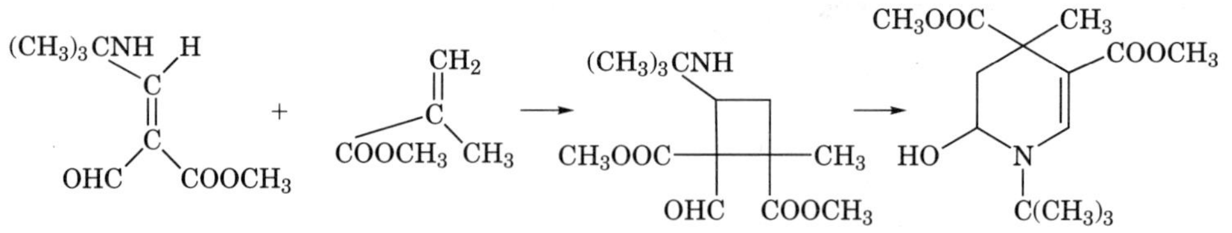

Methacrylates have also found use in diastereoselective -ene reactions. Although not a cycloaddition reaction, this reaction is mechanistically related to the Diels-Alder reaction (37).

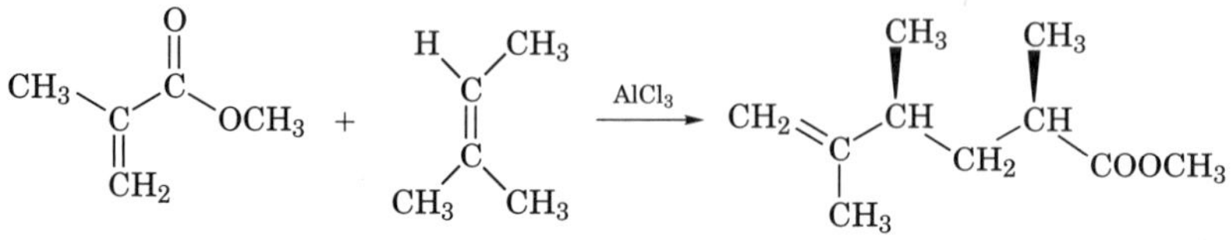

Polymerization

The vast majority of commercial applications of methacrylic acid and its esters stem from their facile free-radical polymerizability (see INITIATORS, FREE-RADICAL; POLYMERIZATION MECHANISMS AND PROCESSES). Solution, suspension, emulsion, and bulk polymerizations have been used to advantage. Although of much less commercial importance, anionic polymerizations of methacrylates have also been extensively studied. Strictly anhydrous reaction conditions at low temperatures are required to yield high molecular weight polymers in anionic polymerization. Side reactions of the propagating anion at the ester carbonyl are difficult to avoid and lead to polymer branching and inactivation (38–44).

Polymerization of methacrylates is also possible via what is known as group-transfer polymerization. Although only limited commercial use has been made of this technique, it does provide a route to block copolymers that is not available from ordinary free-radical polymerizations. In a prototypical group-transfer polymerization the fluoride–ion-catalyzed reaction of a methacrylate (or acrylate) in the presence of a silyl ketene acetal gives a high molecular weight polymer (45–50).

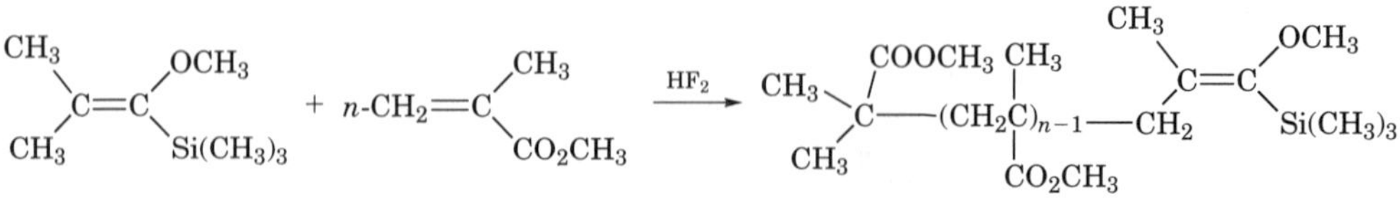

Higher Alkyl and Functional Methacrylates

Most large-scale industrial methacrylate processes are designed to produce methyl methacrylate or methacrylic acid. In some instances, simple alkyl alcohols, eg, ethanol, butanol, and isobutyl alcohol, may be substituted for methanol to yield the higher alkyl methacrylates. In practice, these higher alkyl methacrylates are usually prepared from methacrylic acid by direct esterification or transesterification of methyl methacrylate with the desired alcohol.

Direct esterification of methacrylic acid with alcohols requires the use of strong acid catalysts, such as sulfuric acid, sulfonic acids, or phosphoric acid. Heterogeneous resin catalysts containing strong acid functionality have been used in order to facilitate catalyst removal from the product and to reduce process waste, by eliminating the necessity of a neutralization and separation step (51–53). In a typical direct esterification, a mixture of water and methacrylic acid is taken overhead to shift the equilibrium toward the desired ester. The use of an appropriate solvent to remove water azeotropically allows operation at more moderate temperatures. Direct esterification reactions with alcohols of lower boiling point than water often require special processing techniques. Vaporphase reactions (54) or the use of drying agents such as molecular sieves have been used to give acceptable conversion. Secondary alcohols react more slowly than primary alcohols, and acid-catalyzed esterification of tertiary alcohols often fails as a result of elimination reactions leading to the olefin. Interestingly, the acid-catalyzed reaction of olefins with methacrylic acid may be used to prepare branched alkyl esters (55,56).

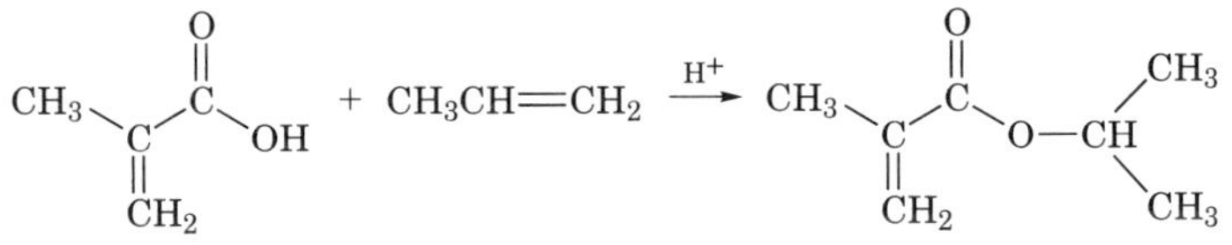

Transesterification of methyl methacrylate with the appropriate alcohol is often the preferred method of preparing higher alkyl and functional methacrylates. The reaction is driven to completion by the use of excess methyl methacrylate and by removal of the methyl methacrylate–methanol azeotrope. A variety of catalysts have been used, including acids and bases and transition-metal compounds such as dialkyltin oxides (57), titanium(IV) alkoxides (58), and zirconium acetoacetate (59). The use of the transition-metal catalysts allows reaction under nearly neutral conditions and is therefore more tolerant of sensitive functionality in the ester alcohol moiety. In addition, transition-metal catalysts often exhibit higher selectivities than acidic catalysts, particularly with respect to by-product ether formation.

Methacrylic acid, methyl methacrylate, and the simple alkyl methacrylates are commonly used building blocks of commercial polymers. Polymers containing methacrylate backbones are more rigid, ie, have a higher glass-transition temperature than the corresponding acrylate polymers. Copolymerization of the appropriate acrylates and methacrylates allows an extensive range of polymer properties to be obtained. Further product differentiation may be achieved by the addition of small amounts (1–10%) of monomers containing specific reactive functional groups that modify or enhance polymer properties or serve to allow subsequent reactions of the polymer, such as cross-linking or grafting. The preparation and use of functional monomers have been reviewed (60).

Functional Monomers.   Hydroxy functional methacrylates are accessible by the reaction of methacrylic acid and ethylene oxide or propylene oxide in the presence of chromium (61), iron (62), or ion-exchange catalysts (63).

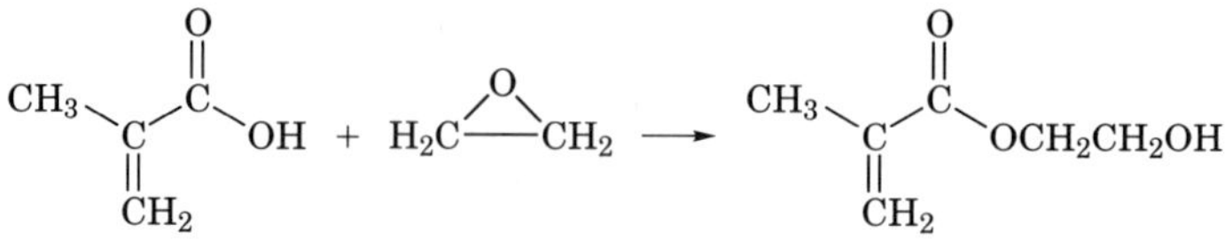

Hydroxy functional methacrylates are used in automotive coatings, dental resins, contact lenses (qv) and a variety of other applications (64).

Reactive groups can be introduced into the polymer backbone by the choice of an appropriate functional monomer. Commercially available examples of such monomers are as follows:

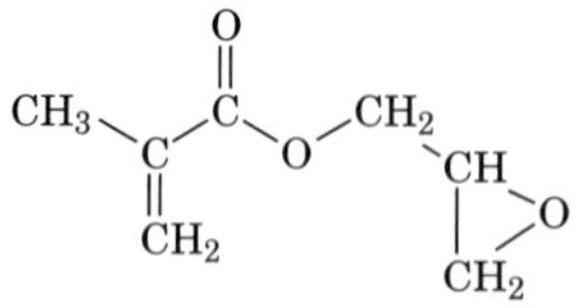

glycidyl methacrylate

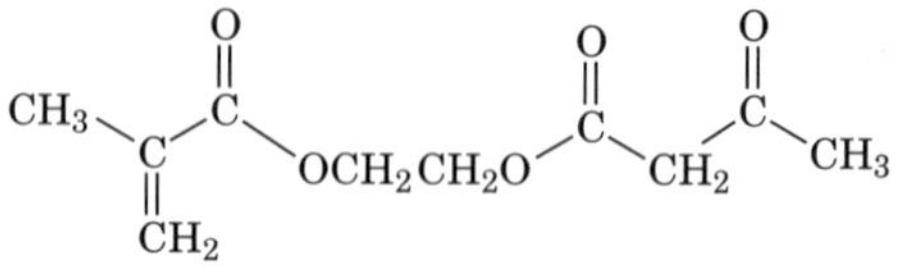

acetoacetoxyethyl methacrylate

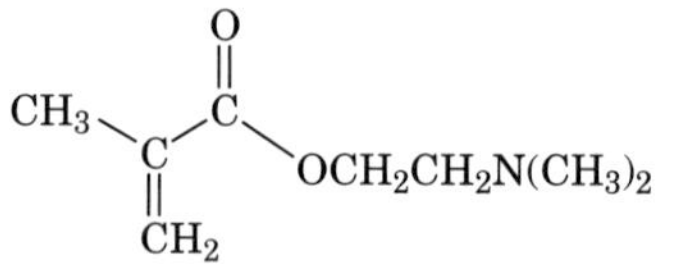

dimethylaminoethyl methacrylate

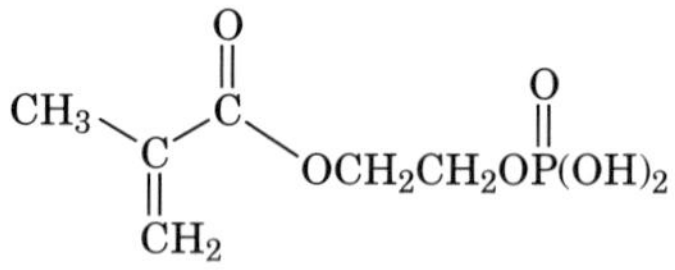

phosphoethyl methacrylate

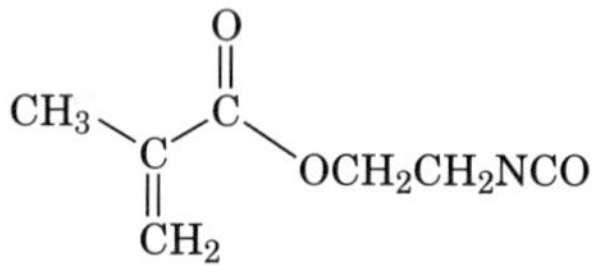

isocyanatoethyl methacrylate

Shown are glycidyl methacrylate to introduce epoxide functionality, aceto-acetoxyethyl methacrylate to introduce active methylene groups, dimethyl-aminoethyl methacrylate to introduce amine functionality, phosphoethyl methacrylate for strong acid functionality, and isocyanatoethyl methacrylate to introduce isocyanate functionality, which may then react with a wide variety of nucleophiles.

Organometallic methacrylate monomers containing tin, silicon, germanium, lead, and titanium have become available in at least laboratory research quantities (65).

Manufacture and Processing

The basic feedstock for the manufacture of methyl methacrylate and methacrylic acid is, ultimately, natural gas or crude oil. It is convenient to categorize the various manufacturing routes in terms of the specific hydrocarbon raw material used. Propylene (C-3) routes require the addition of one carbon atom and are shown schematically in Figure 1. Ethylene (C-2)-based routes require the

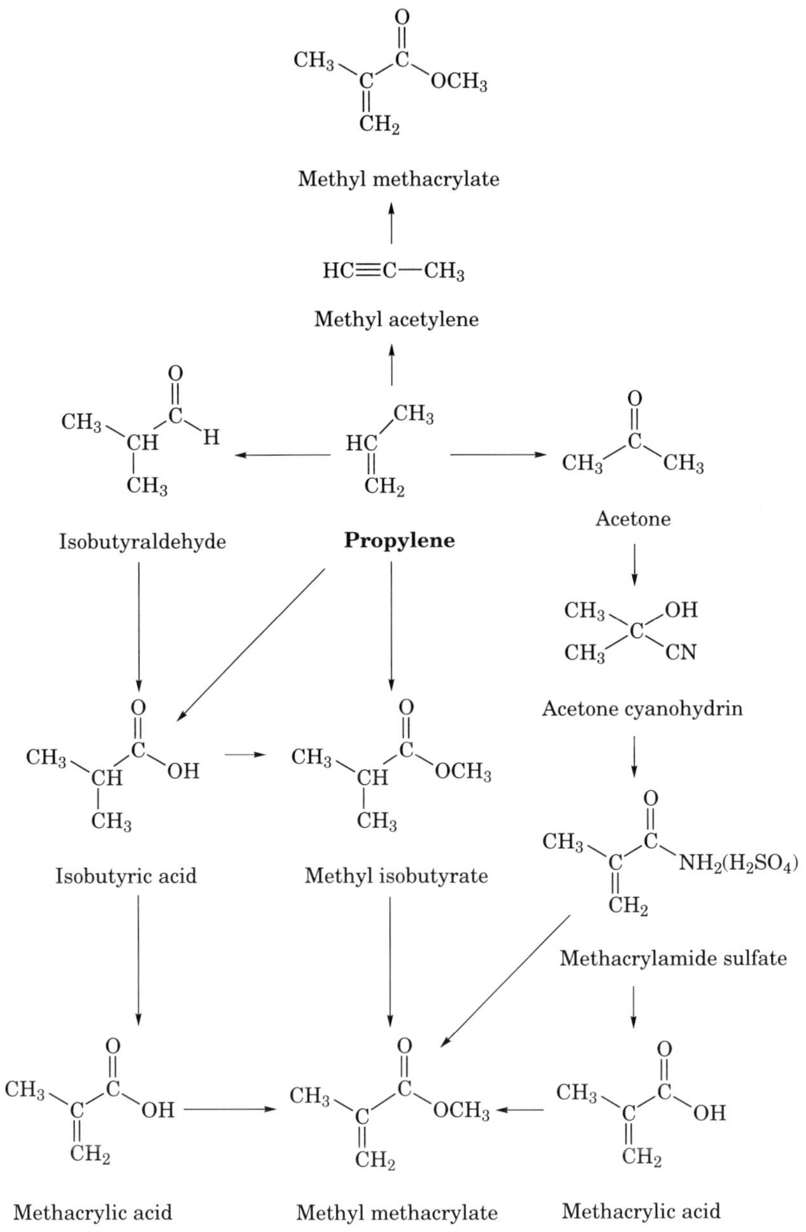

Fig. 1.   C-3 routes to MMA/MAA.

addition of two carbon atoms to create the four-carbon methacrylate backbone. Figure 2 shows the key intermediates and transformations for C-2-based routes. Routes proceeding via the oxidation of isobutylene or *tert*-butyl alcohol (C-4) are shown in Figure 3. The commercial viability of a process is determined by the aggregate of raw material cost and utilization (process yield), operating costs (energy), waste disposal costs, environmental impact, and plant capital investment.

Until 1982, almost all methyl methacrylate produced worldwide was derived from the acetone cyanohydrin (C-3) process. In 1982, Nippon Shokubai Kagaku Kogyo Co. introduced an isobutylene-based (C-4) process, which was quickly followed by Mitsubishi Rayon Co. in 1983 (66). Japan Methacrylic Monomer Co., a joint venture of Nippon Shokubai and Sumitomo Chemical Co., introduced a C-4-based plant in 1984 (67). Isobutylene processes are less economically attractive in the United States where isobutylene finds use in the synthesis of methyl *tert*-butyl ether, a pollution-reducing gasoline additive. BASF began operation of an ethylene-based (C-2) plant in Ludwigshafen, Germany, in 1990, but favorable economics appear to be limited to conditions unique to that site.

Considerable research is currently directed toward development of novel technologies that may present economic advantages with respect to the conventional acetone cyanohydrin (ACH) route. Mitsubishi Gas Chemical Co. has developed and patented a modified acetone cyanohydrin-based route that does not use sulfuric acid and therefore presents the opportunity for reduced waste costs.

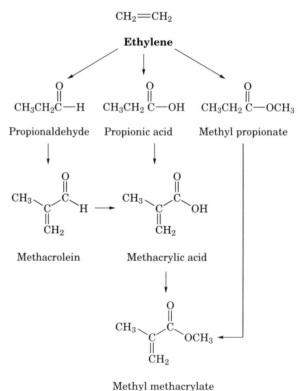

Fig. 2.   C-2 routes to MMA/MAA.

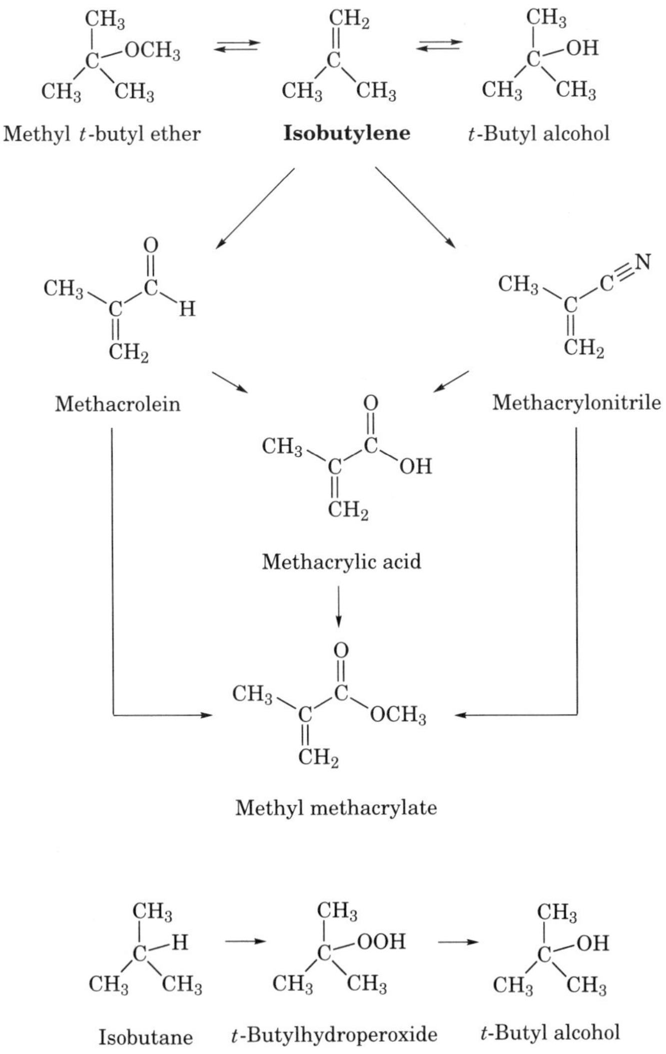

Fig. 3. C-4 routes to MMA/MAA.

A novel C-3 route based on the palladium-catalyzed carbonylation of methylacetylene has been developed by Shell Oil Co. There have been significant improvements in catalysts and resulting yields for key transformations in many routes since the 1980s. It is likely that the commercial dominance of the ACH process will erode over time as new processes become incrementally more attractive and as the primary producers attempt to diversify their raw material feedstocks.

C-3 ROUTES

MMA from Acetone Cyanohydrin (ACH). The process for conversion of acetone cyanohydrin–H_2SO_4 to methyl methacrylate through methacrylamide sulfate [29194-31-8] (68) has been practiced commercially since 1937 and is based on technology patented by ICI in 1934 (7). Acetone cyanohydrin [75-86-5] is prepared via base-catalyzed reaction of acetone and hydrogen cyanide. Acetone and hydrogen cyanide are obtained as by-products from the commercial production of phenol and acrylonitrile, respectively. Hydrogen cyanide is also manufactured directly by catalytic ammoxidation of methane. Sulfuric acid is used in excess (1.4–1.8 mol/mol ACH) and serves as both reactant and solvent in the reaction with acetone cyanohydrin to form methacrylamide sulfate through an α-sulfatoamide intermediate.

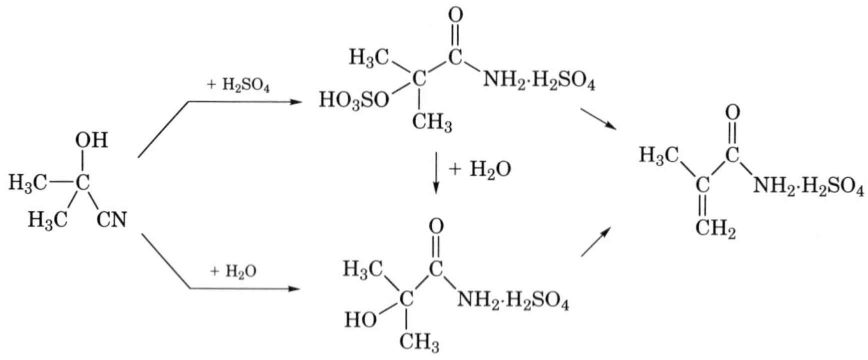

Inhibitors are introduced at specific points in the process to prevent polymerization. Kinetics and mechanistic work studying the elimination of H_2SO_4 from the α-sulfatoamide intermediate has appeared (69). Anhydrous conditions are used to minimize formation of α-hydroxyisobutyramide sulfate, since higher temperatures are required for conversion of α-hydroxyisobutyramide sulfate to methacrylamide sulfate. If insufficient sulfuric acid is used, the reaction mixture becomes difficult to pump and heat transfer is poor, resulting in decomposition of acetone cyanohydrin to by-product acetone sulfates, formamide sulfate, and carbon monoxide. Typically, the initial reaction is carried out at 80–100°C followed by brief thermal cracking at 120–160°C to convert any by-product α-hydroxyisobutyramide sulfate to methacrylamide sulfate. Total reaction time for this stage of the process is ca 1 h.

In the next stage, sulfuric acid serves as catalyst in a combined hydrolysis–esterification of methacrylamide sulfate to a mixture of methyl methacrylate and methacrylic acid. Conversion of methacrylamide sulfate to methyl methacrylate can be carried out using a variety of procedures for the recovery of crude methyl methacrylate and for separation of methanol and methacrylic acid for recycling. One procedure involves treating the methacrylamide sulfate stream with aqueous methanol at pressures of ca 710 kPa (7 atm) in a continuous reactor or reactors operating at 100–150°C. Residence times are under 1 h. Some of the by-products of this stage are dimethyl ether, methyl α-hydroxyisobutyrate, α-hydroxyisobutyric acid, methyl β-methoxyisobutyrate, and methyl formate. The reactor effluent phase separates under pressure, and the lower layer is steam-stripped to recover methacrylic acid for recycling to the hydrolysis–esterification stage. The waste ammonium acid sulfate can be treated with ammonia for conversion to fertilizer, or burned to regenerate sulfuric acid. The upper layer passes through a distillation column for removal of low boiling materials, eg, dimethyl ether and acetone. The bottoms of this column are treated with aqueous ammonia to recover methacrylic acid and methanol for recycling to the hydrolysis–esterification stage. Crude methyl methacrylate is dehydrated and purified in downstream distillation columns to provide product methyl methacrylate in >99% purity. A schematic of the overall process is given in Figure 4. The overall yield based on acetone cyanohydrin is approximately 90%. Most of the world supply of MMA is still produced by this process.

Mitsubishi Gas Chemical Company Process. The commercial MMA manufacturing process based on sulfuric acid and acetone cyanohydrin suffers from the large quantities of ammonium sulfate produced. Because ammonium sulfate has only low value as fertilizer, regeneration of sulfuric acid from ammonium sulfate [7783-20-2] is required. Despite the drawbacks of using sulfuric acid, this technology is still the most widely practiced of the current MMA manufacturing processes. Mitsubishi Gas Chemical (MGC) has developed a novel ACH process that uses no sulfuric acid, with an overall yield of MMA from acetone of ca 84% (70). As a result, no acid sludge is produced and corrosion problems are minimized. The process also recovers and recycles hydrogen cyanide (HCN).

In the MGC process ACH is hydrolyzed to α-hydroxyisobutyramide in the liquid phase using a fixed bed of a modified MnO_2 catalyst (71). The reaction is carried out in acetone (to minimize ACH decomposition) at temperatures near 60°C. Conversion of ACH exceeds 99% with selectivity to α-hydroxyisobutyramide of greater than 90%.

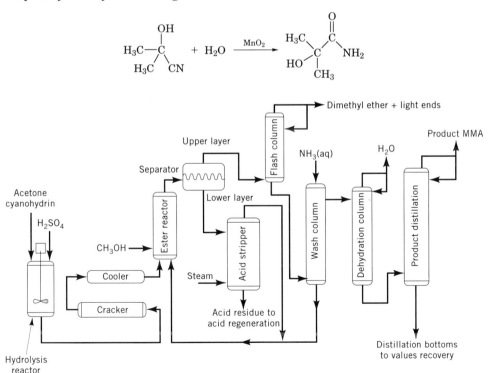

Fig. 4. MMA from acetone cyanohydrin via methacrylamide sulfate.

Arguably the key step in the MGC process is the conversion of α-hydroxy-isobutyramide to methyl α-hydroxyisobutyrate using methyl formate as the methylating agent. Methyl formate is made commercially by MGC via vapor-phase dehydrogenation of methanol (72).

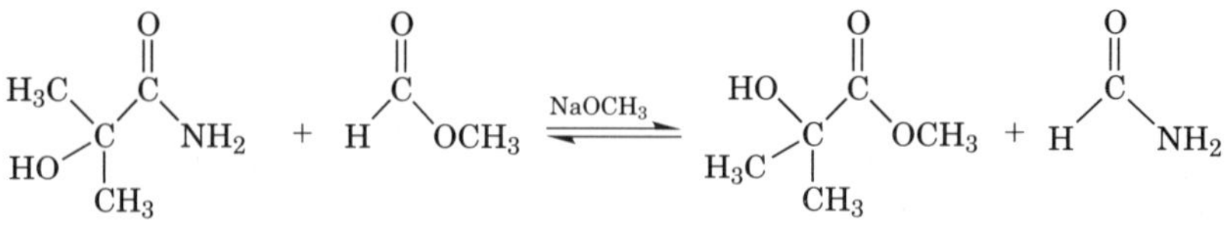

$$
\underset{\substack{\text{HO} \\ | \\ \text{CH}_3}}{\overset{\text{H}_3\text{C}}{\text{C}}}\!\!\!\overset{\text{O}}{\underset{}{\overset{\|}{\text{C}}}}\!\!\text{NH}_2 \quad + \quad \text{H}\overset{\text{O}}{\underset{}{\overset{\|}{\text{C}}}}\text{OCH}_3 \quad \underset{}{\overset{\text{NaOCH}_3}{\rightleftharpoons}} \quad \underset{\substack{\text{H}_3\text{C} \\ | \\ \text{CH}_3}}{\overset{\text{HO}}{\text{C}}}\!\!\!\overset{\text{O}}{\underset{}{\overset{\|}{\text{C}}}}\!\!\text{OCH}_3 \quad + \quad \text{H}\overset{\text{O}}{\underset{}{\overset{\|}{\text{C}}}}\text{NH}_2
$$

The reaction is run for several hours at temperatures typically below 100°C under a pressure of carbon monoxide to minimize formamide decomposition (73). Conversions of α-hydroxyisobutyramide are near 65% with selectivities to methyl α-hydroxyisobutyrate and formamide in excess of 99%. It is this step that is responsible for the elimination of the acid sludge stream characteristic of the conventional H_2SO_4–ACH processes. Because methyl formate, and not methanol, is used as the methylating agent, formamide is the co-product instead of ammonium sulfate. Formamide can be dehydrated to recover HCN for recycle to ACH generation.

The methyl α-hydroxyisobutyrate produced is dehydrated to MMA and water in two stages. First, the methyl α-hydroxyisobutyrate is vaporized and passed over a modified zeolite catalyst at ca 240°C. A second reactor containing phosphoric acid is operated at ca 150°C to promote esterification of any methacrylic acid (MAA) formed in the first reactor (74,75). Methanol is co-fed to improve selectivity in each stage. Conversions of methyl α-hydroxyisobutyrate are greater than 99%, with selectivities to MMA near 96%. The reactor effluent is extracted with water to remove methanol and yield crude MMA. This process has not yet been used on a commercial scale.

MMA from Propyne. Advances in catalytic carbonylation technology by Shell researchers have led to the development of a single-step process for producing MMA from propyne [74-99-7] (methyl acetylene), carbon monoxide, and methanol (76–82).

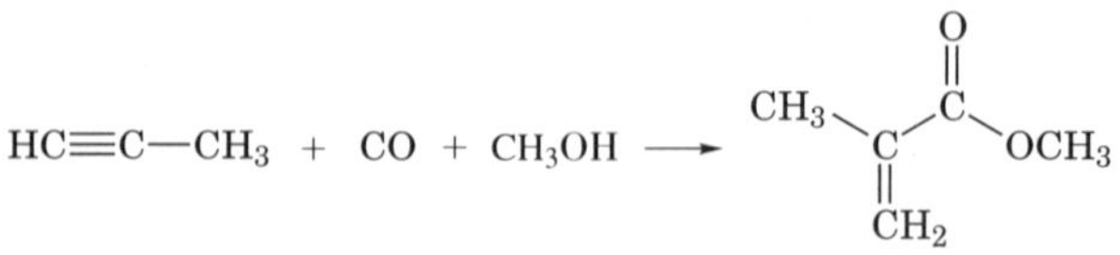

$$
\text{HC}\!\!\equiv\!\!\text{C}\!\!-\!\!\text{CH}_3 \quad + \quad \text{CO} \quad + \quad \text{CH}_3\text{OH} \quad \longrightarrow \quad \underset{\substack{\| \\ \text{CH}_2}}{\overset{\text{CH}_3}{\text{C}}}\!\!\!\overset{\text{O}}{\underset{}{\overset{\|}{\text{C}}}}\!\!\text{OCH}_3
$$

Novel palladium catalysts show marked improvements in both yields and selectivities, compared to nickel carbonyl catalysts utilized in earlier commercial carbonylation processes (83,84). The palladium catalysts are also expected to be less hazardous.

A key feature of the palladium catalyst is a phosphine ligand environment, preferably a phosphine ligand containing an imino nitrogen atom such as 2-(diphenylphosphino)pyridine (85–88). A strong acid possessing a relatively noncoordinating anion such as an alkylsulfonate, eg, methane sulfonic acid, is also present. Second-generation catalyst systems include the addition of a tertiary amine to prolong catalyst life by providing a higher resistance to propadiene contamination of the propyne feed (83).

The carbonylation process is operated at mild temperatures (45–110°C) and elevated pressures (2–6 MPa = 20–60 atm), and can be carried out in MMA or in an inert solvent such as N-methylpyrrolidinone. Selectivities are claimed in excess of 99%, thereby requiring minimal purification to obtain high quality product MMA. The principal by-product is methyl crotonate.

An apparent limitation of this technology is the relative unavailability of plant-scale quantities of propyne. Propyne and propadiene are present in C-3 streams from ethylene plants, but the quantities appear insufficient for MMA plants in excess of 45,000 t/yr. MMA production facilities with capacities in excess of 45,000 t/yr would likely have to rely on propyne supplied from several ethylene plants. Of course, advances in technology or changes in cracker operations may lead to intentionally higher percentages of propyne in the C-3 streams, thus easing the question of an adequate supply of propyne for MMA production. Recovery of propyne from the C-3 streams requires extractive distillation in a polar organic solvent. Propyne and propadiene can be concentrated in the bottoms stream, while propane and propylene are removed overhead (13). Propadiene is a catalyst poison, but can be maintained at low levels (<0.4%) via isomerization to propyne over potassium carbonate (89). This process has not yet been used at commercial scale.

Propylene-Based Routes. The strong acid-catalyzed carbonylation of propylene [115-07-1] to isobutyric acid (Koch reaction) followed by oxidative dehydration to methacrylic acid has been extensively studied since the 1960s. The principal side reaction in the Koch reaction is the formation of oligomers of propylene. Increasing yields of methacrylic acid in the oxydehydration step is the current focus of research. Isobutyric acid may also be obtained via the oxidation of isobutyraldehyde, which is available from the hydroformylation of propylene. The n-butyraldehyde isomer that is formed in the hydroformylation must be separated.

The preparation of isobutyric acid in ca 90% yield via the liquid-phase reaction of propylene, carbon monoxide, and water in the presence of excess sulfuric acid and a metal phthalocyanine co-catalyst has been reported (90). Early work utilizing aqueous hydrofluoric acid, which gave high yields (91,92), never came to commercial fruition, presumably owing to the highly corrosive nature of aqueous hydrofluoric acid. Use of anhydrous hydrofluoride as the catalyst for the reaction of propylene and carbon monoxide affords 90% yields of isobutyryl fluoride (93,94). Hydrolysis of isobutyryl fluoride to isobutyric acid is carried out with less than the stoichiometric amount of water in order to avoid the creation of a corrosive environment (95), but the necessity of handling large volumes of hydrogen fluoride still dictates against commercialization of this route.

The oxidative dehydration of isobutyric acid [79-31-2] to methacrylic acid is most often carried out over iron–phosphorus or molybdenum–phosphorus based catalysts similar to those used in the oxidation of methacrolein to methacrylic acid. Conversions in excess of 95% and selectivity to methacrylic acid of 75–85% have been attained, resulting in single-pass yields of nearly 80%. The use of cesium-, copper-, and vanadium-doped catalysts are reported to be beneficial (96), as is the use of cesium in conjunction with quinoline (97). Generally the iron–phosphorus catalysts require temperatures in the vicinity of 400°C, in contrast to the molybdenum-based catalysts that exhibit comparable reactivity at 300°C (98).

The distillative separation of methacrylic acid from unreacted isobutyric acid is problematic because the boiling points of isobutyric acid (155°C) and methacrylic acid (162°C) are quite close to one another. Alternatively, the crude reaction mixture can be esterified, and the resultant methyl methacrylate (101°C) and methyl isobutyrate (92°C) can be separated.

Although not commercialized, both Elf Atochem and Röhm GmbH have published on development of hydrogen fluoride-catalyzed processes. Norsolor, since acquired by Elf Aquitaine, had been granted an exclusive European license for the propylene–hydrogen fluoride technology of Ashland Oil (99). Röhm has patented a process for the production of isobutyric acid in 98% yield via the isomerization of isopropyl formate in the presence of carbon monoxide and hydrofluoric acid (100).

ETHYLENE-BASED (C-2) ROUTES

MMA and MAA can be produced from ethylene [74-85-1] as a feedstock via propanol, propionic acid, or methyl propionate as intermediates. Propanal may be prepared by hydroformylation of ethylene over cobalt or rhodium catalysts. The propanal then reacts in the liquid phase with formaldehyde in the presence of a secondary amine and, optionally, a carboxylic acid. The reaction presumably proceeds via a Mannich base intermediate which is cracked to yield methacrolein (101,102). Alternatively, a gas-phase, crossed aldol reaction with formaldehyde catalyzed by molecular sieves (qv) may be used to form methacrolein (103). The methacrolein is then oxidized to methacrylic acid as discussed in C-4 routes to MMA. BASF operates a plant at Ludwigshafen with a capacity of 36,000 t based on propanal, which is available at that site.

Propionic acid is accessible through the liquid-phase carbonylation of ethylene over a nickel carbonyl catalyst (104), or via ethylene and formic acid over an iridium catalyst (105). Condensation of propionic acid with formaldehyde over a supported cesium catalyst gives MAA directly with conversions of 30–40% and selectivities of 80–90% (106,107). Catalyst lifetime can be extended by adding low levels (several ppm) of cesium to the feed stream (108).

The reaction of methyl propionate and formaldehyde in the gas phase proceeds with reasonable selectivity to MMA and MAA (ca 90%), but with conversions of only 30%. A variety of catalysts such as V–Sb on silica–alumina (109), P–Zr, Al, boron oxide (110), and supported Fe–P (111) have been used. Methylal (dimethoxymethane) or methanol itself may be used in place of formaldehyde and often result in improved yields. Methyl propionate may be prepared in excellent yield by the reaction of ethylene and carbon monoxide in methanol over a ruthenium acetylacetonate catalyst or by utilizing a palladium–phosphine ligand catalyst (112,113).

Only with propanal are very high conversions (99%) and selectivity (>98%) to MMA and MAA possible at this time. Although nearly 95% selective, the highest reported conversions with propionic acid or methyl propionate are only 30–40%. This results in large recycle streams and added production costs. The propanal route suffers from the added expense of the additional step required to oxidize methacrolein to methacrylic acid.

ISOBUTYLENE-BASED (C-4) ROUTES

Isobutylene [115-11-7] or tert-butyl alcohol can be converted to methacrylic acid in a two-stage, gas-phase oxidation process via methacrolein as an intermediate. The alcohol and isobutylene may be used interchangeably in the processes since tert-butyl alcohol [75-65-0] readily dehydrates to yield isobutylene under the reaction conditions in the initial oxidation. Variations of this process have been commercialized by Mitsubishi Rayon and by a joint venture of Sumitomo and Nippon Shokubai. Nippon Kayaku, Mitsui Toatsu, and others have also been active in isobutylene oxidation research.

The first-stage catalysts for the oxidation to methacrolein are based on complex mixed metal oxides of molybdenum, bismuth, and iron, often with the addition of cobalt, nickel, antimony, tungsten, and an alkali metal. Process optimization continues to be in the form of incremental improvements in catalyst yield and lifetime. Typically, a dilute stream, 5–10% of isobutylene (tert-butyl alcohol) in steam (10%) and air, is passed over the catalyst at 300–420°C. Conversion is often nearly quantitative, with selectivities to methacrolein ranging from 85% to better than 95% (114–118). Often there is accompanying selectivity to methacrylic acid of an additional 2–5%. A patent by Mitsui Toatsu Chemicals reports selectivity to methacrolein of better than 97% at conversions of 98.7% for a yield of methacrolein of nearly 96% (119).

The oxidation of methacrolein to methacrylic acid is most often performed over a phosphomolybdic acid-based catalyst, usually with copper, vanadium, and a heavy alkali metal added. Arsenic and antimony are other common dopants. Conversions of methacrolein range from 85–95%, with selectivities to methacrylic acid of 85–95%. Although numerous catalyst improvements have been reported since the 1980s (120–123), the highest claimed yield of methacrylic acid (86%) is still that described in a 1981 patent to Air Products (124).

In typical processes, the gaseous effluent from the second-stage oxidation is cooled and fed to an absorber to isolate the MAA as a 20–40% aqueous solution. The MAA may then be concentrated by extraction into a suitable organic solvent such as butyl acetate, toluene, or dibutyl ketone. Azeotropic dehydration and solvent recovery, followed by fractional distillation, is used to obtain the pure product. Water, solvent, and low boiling by-products are removed in a first-stage column. The column bottoms are then fed to a second column where MAA is taken overhead. Esterification to MMA or other esters is readily achieved using acid catalysis.

Several variations of the above process are practiced. In the Sumitomo-Nippon Shokubai process, the effluent from the first-stage reactor containing methacrolein and methacrylic acid is fed directly to the second-stage oxidation without isolation or purification (125,126). In this process, overall yields are maximized by optimizing selectivity to methacrolein plus methacrylic acid in the first stage. Conversion of isobutylene or *tert*-butyl alcohol must be high because no recycling of material is possible. In another variation, Asahi Chemical has reported the oxidative esterification of methacrolein directly to MMA in 80% yield without isolation of the intermediate MAA (127,128).

With the advent of the use of methyl *tert*-butyl ether (MTBE) as an oxygenated gasoline additive, there has been considerable emphasis on the synthesis of isobutylene or *tert*-butyl alcohol, which can be used as feedstocks for MTBE production. Isobutylene is a primary by-product in the C-4 stream from ethylene plants. After butadiene has been extracted, the isobutylene may be separated as such from the linear saturated and unsaturated butenes, or the separation may be facilitated by conversion of the isobutylene to *tert*-butyl alcohol. *n*-Butane may be readily isomerized to isobutane, which can be dehydrogenated to isobutylene. *tert*-Butyl alcohol is also available as a by-product in the manufacture of styrene or in the manufacture of propylene oxide. Isobutane is oxidized to produce *tert*-butyl hydroperoxide, which, upon reaction with propylene, yields propylene oxide and *tert*-butyl alcohol (129). The use of isobutylene or *tert*-butyl alcohol as a feedstock for MTBE production creates an economic disincentive for their use as a raw material for MAA and MMA production, because MTBE is the product with the higher value. However, reports have suggested potential environmental problems with MTBE as a fuel additive (130).

METHACRYLONITRILE PROCESS

MAA and MMA may also be prepared via the ammoxidation of isobutylene to give methacrylonitrile as the key intermediate. A mixture of isobutylene, ammonia, and air are passed over a complex mixed metal oxide catalyst at elevated temperatures to give a 70–80% yield of methacrylonitrile. Suitable catalysts often include mixtures of molybdenum, bismuth, iron, and antimony, in addition to a noble metal (131–133). The methacrylonitrile formed may then be hydrolyzed to methacrylamide by treatment with one equivalent of sulfuric acid. The methacrylamide can be esterified to MMA or hydrolyzed to MAA under conditions similar to those employed in the ACH process. The relatively modest yields obtainable in the ammoxidation reaction and the generation of a considerable acid waste stream combine to make this process economically less desirable than the ACH or C-4 oxidation to methacrolein processes.

Economic Aspects

Domestic production of methyl methacrylate from 1971 to 1992 has shown a 5.7% annualized rate of growth:

Year	Production, 10^3 t
1971	210
1977	338
1981	500
1991	641

Worldwide production capacity is shown in Table 6. Economic conditions in the late 1980s and early 1990s led to global overcapacity of methyl methacrylate, which caused many plants to be operated at less than optimum levels.

Historical list prices for bulk quantities of selected methacrylates are given in Table 7 (134). The historical price trends reflect the combined effects of improved manufacturing capability and the market price of crude oil, the basic raw material to which these materials are ultimately tied.

Table 6. World Production Capacity of MMA

Producer	Capacity, 10^3 t	Process[a]
North America		
Röhm and Haas	357	ACH
ICI (formerly Du Pont)	227	ACH
CyRo	98	ACH
Total	*682*	
Europe		
ICI	205	ACH
Rohm	159	ACH
Atochem	135	ACH
DeGussa	65	ACH
BASF	36	C-2[b]
Repsol Quimica	36	ACH
other	30	ACH
Total	*666*	
Japan		
Mitsubishi Rayon	185	ACH/C-4
Japan Methacrylic Monomer (Nippon Shokubai/Sumitomo jv)	80	C-4
Kuraray (formerly Kyowa Gas Chemical)	52	ACH
Kyodo Monomer	40	C-4
other	32	ACH
Total	*389*	
Asia / Latin America		
Koahsiung Monomer Co. (Taiwan)	80	ACH
Fenoquimica (Mexico)	17	ACH
Quimica Metacril (Brazil)	10	ACH
Total	*107*	
1992 Worldwide capacity	*1,889[c]*	

[a]ACH = acetone cyanohydrin; C-4 = *tert*-butyl alcohol–isobutylene; C-2 = ethylene.
[b]Via propionaldehyde.
[c]Includes ~45,000 t from other countries not specifically mentioned.

Table 7. Prices for Methacrylates, $/kg[a]

Year	MMA	EMA	n-BMA	i-BMA	Lauryl	MAA
1955	0.81	1.14	1.21		1.43	
1960	0.64	1.10	1.21		1.43	0.88
1965	0.46	0.95	0.70	0.84	1.43	0.66
1970	0.42	0.95	0.64	0.84	1.43	0.66
1975	0.66	1.25	1.05	1.01	2.31	0.88
1980	1.10	1.94	1.41	1.52	2.42	1.43
1985	1.18	2.33	1.94	1.91	3.78	1.72
1990	1.56	2.71	1.95	2.20	4.48	2.18
1993	1.56	3.08	2.11	2.35	4.14	2.28

[a]Ref. 134.

Uses

Methacrylic acid and methacrylic esters are used in a wide variety of polymers with a broad spectrum of applications. Poly(methacrylic acid) or its neutralized salts are used as additives for detergent builders and rheology modifiers. Methacrylic esters, the most important of which is methyl methacrylate, yield hard, tough polymers in contrast to the softer acrylates. Copolymerization of methacrylic esters with acrylates allows the preparation of hard but flexible polymers for use in paints, polishes, and many other coatings. Methacrylate polymers are prized for their clarity, colorability, color compatibility, weatherability, and ultraviolet light stability, which allows them to be used for both indoor and outdoor applications. The principal end uses of methacrylates are in acrylic sheet and molding resins which find commercial application in signs, displays, glazing compounds, lighting fixtures, building panels, automotive components, plumbing fixtures, and appliances. They are also used as impact modifiers in poly(vinyl chloride) (PVC) siding, film, sheet, and plastic bottle manufacture (135,136). The principal methacrylate market segments are acrylic sheet, 28%; molding products, 26%; coatings, 20%; impact modifiers, 11%; and other, 15%. The relative sizes of these market segments has remained almost unchanged since the late 1960s.

Storage and Handling

Polymerizations of methacrylic acid and derivatives are very energetic (MAA, 66.1 kJ/mol; MMA, 57.5 kJ/mol = 13.7 kcal/mol). The potential for the rapid evolution of heat and generation of pressure presents an explosion hazard if the materials are stored in closed or poorly vented containers. Methacrylic acid polymer is insoluble in the monomer, which may result in the plugging of transfer lines and vent systems. Polymers of the lower alkyl esters are often soluble in the parent monomer and may be detected by an increase in solution viscosity. Alternatively, dilution with a nonsolvent for the polymer such as methanol results in the formation of haze and can be used as a diagnostic tool for determining presence of polymer.

To prolong usable shelf-life, commercially available methacrylic monomers are inhibited with the methyl ether of hydroquinone [95-71-6] (MEHQ). Other commonly used inhibitors are alkylphenols and hydroquinone. Once inhibited, methacrylic acid or its esters may be handled as flammable materials. Methacrylic acid is typically inhibited with MEHQ at levels in the range of 100–250 ppm. At the higher inhibitor level, methacrylic acid is stable for over two years at 38°C. Stability decreases rapidly with increasing storage temperature and recommended storage temperatures are 18–40°C and preferably 20–25°C. Some deterioration in purity is to be expected upon storage. The lower alkyl esters of methacrylic acid are more stable than methacrylic acid itself, and 25 ppm or less of MEHQ is often sufficient to achieve the desired shelf life. Aminomethacrylates and polyfunctional methacrylates are less stable and require higher levels of inhibitors; they possess more limited storage life.

To avoid photoinitiation of polymerization, all methacrylates should be stored with minimal exposure to light. Because dissolved oxygen is required for phenolic inhibitors such as MEHQ to function efficiently as inhibitors (137), methacrylate monomers should never be handled or stored under an oxygen-free atmosphere. Methacrylic acid vapor in air at room temperature does not form a flammable mixture, so air or gas mixtures containing from 5–21% oxygen may be used. Because oxygen is consumed as part of the MEHQ radical trapping mechanism, periodic venting to replenish oxygen is necessary.

Most unwanted polymerization events of methacrylic monomers occur because of overheating, leading to inhibitor depletion or oxygen depletion, and in turn to inhibitor inactivation. Care should be taken to avoid stagnant areas in transfer lines or pump heads where polymerization may begin and which may then act to seed polymerization of the bulk material. Because methacrylic acid polymer is insoluble in the monomer the formation of haze in the monomer is the earliest indicator of instability. The onset of uncontrolled polymerization with concomitant evolution of heat may ensue a short time after the first appearance of haze. Contamination of the monomer, often via corrosion of the storage vessel, which may then introduce metal ions, can act to initiate polymerization. The preferred material of construction for equipment that comes into contact with methacrylic acid is 316 L stainless steel. The use of 316 L stainless is recommended even for the storage of the less corrosive methacrylate esters in order to avoid the formation of rust (iron contamination) that can promote radical generation from the peroxidic contaminants that form upon storage.

The relatively high freezing point of methacrylic acid (15°C) presents special problems. Because the inhibitor tends to partition into the liquid phase upon freezing, thawing of the material tends to create localized pools of uninhibited methacrylic acid which are extremely susceptible to polymerization. Care should be taken to limit thawing temperatures to less than 40°C and to ensure good mixing of the thawed material. No material should be withdrawn from a partially frozen drum, as it may contain little inhibitor or, conversely, it may contain the bulk of the inhibitor necessary for stabilizing the remainder of the drum.

For most polymer applications the removal of the inhibitors from the monomer is unnecessary. Should it be required, the phenolic inhibitors can be removed by an alkaline wash or by treatment with a suitable ion-exchange resin. Uninhibited MMA is sufficiently stable to be shipped under carefully controlled temperature and time restrictions. Uninhibited monomers should be monitored carefully and used promptly.

Health and Safety Factors

The toxicity of methacrylate monomers has been studied extensively and new data are constantly being generated. Based on lethality, acute toxicity testing by the oral, dermal, and inhalation routes shows the methacrylic esters are only slightly toxic after single exposures to rats and rabbits (Table 8). Methacrylates are slightly to severely irritating to skin and eyes and are considered potential skin sensitizers. That is, after first exposure, sensitive individuals may experience severe allergic skin reactions (rashes and swelling) upon subsequent exposures even to small amounts of the same or different monomers. Repeated exposure to the methacrylates results in little systemic toxicity. Nasal tissue is the primary target organ in inhalation studies, with irritation and inflammatory changes resulting at this site of contact. Repeated exposure of animals to near-lethal concentrations results in liver and kidney damage (138,139).

In general, toxicity of the methacrylate esters decreases with increasing molecular weight due to the diminished rate at which higher molecular weight materials are absorbed into the body. Nevertheless, methyl methacrylate serves as a toxicological representative of other methacrylates. The chronic toxicity of methyl methacrylate is well documented. Several lifetime exposure studies by a variety of routes (oral capsule, drinking water, inhalation) in rodents have

Table 8. Acute Toxicity of Methacrylate and Related Monomers

Compound	Acute oral LD_{50}, g/kg (rat)	Inhalation toxicity, LC_{50}	Dermal LD_{50}, g/kg (rabbit)	Remarks[a]
methyl methacrylate	>5	7094 ppm/4 h	>5	allergic response observed in humans (JIHTAB 23,343,41)
ethyl methacrylate	13.3	2700–3200 ppm/8 h	9.1	allergic response observed (JIHTAB 23,343,41)
butyl methacrylate	>5	~30 ppm/4 h	>5	allergic response observed (AIHAAP 30,470,69)
lauryl methacrylate	>5	[b]	>3	slight skin irritant/ mild eye irritant
methacrylic acid	1.06	>1300 ppm	<2	corrosive to skin and eyes (GISAAA 41(4),6,76)
glycidyl methacrylate	0.6		0.47 mg/kg	(GISAAA 50(2), 67,85)
hydroxypropyl methacrylate	>5		>5	allergic response observed
hydroxyethyl methacrylate	>5	[b]	>3	(GISAAA 54(9), 75,89)
dimethylamino- ethylmethacry- late	1.5–3	0.6 ppm/4 h	>3	possible skin sensitizer (85GMAT –, 55,82)

[a] From Material Safety Data Sheets from commercial suppliers in parentheses.
[b] Saturated air for 1 h; not lethal.

shown it to be noncarcinogenic. Methyl methacrylate also is nonteratogenic (did not produce birth defects) after inhalation exposures. Headaches, vomiting, and drowsiness are symptomatic of overexposure to vapors. All methacrylates are hydrolyzed rapidly to methacrylic acid in the process of conversion to CO_2 before excretion. Methacrylic acid is corrosive. Mucous membranes of the eyes, nose, and throat are particularly sensitive, and prolonged exposure can lead to nasal or eye tissue damage. In the workplace, the pungent odor and irritant nature of the methacrylate monomers serves as a warning property and tends to keep exposures low (140).

The methacrylates are slightly to essentially nontoxic to fish and other aquatic species. Hydrolysis data suggest rapid breakdown at alkaline conditions, and studies show that MMA is ultimately biodegradable in sewage sludge samples. Based on this information, the methacrylates are not considered to be a significant environmental hazard.

Monomers containing reactive functional groups in the ester moiety exhibit the toxicological profile of the functional group and should be considered on an individual basis. Consideration of methods of monomer manufacture may be appropriate, as by-products, even at trace levels, may affect the observed biological response.

Appropriate protective clothing and equipment should be worn to minimize exposure to methacrylate liquids and vapors. Chemically resistant clothes and gloves and splash-proof safety goggles are recommended. The working area should be adequately ventilated to limit vapors. Should chemical exposure occur, contaminated clothing should be removed and the affected area washed with copious amounts of water. Medical attention should be sought if symptoms appear. Further information about methyl methacrylate and other methacrylates is available (141).

BIBLIOGRAPHY

"Methacrylic Compounds" in *ECT* 2nd ed., Vol. 13, pp. 331–363, by F. J. Glavis and J. F. Woodman, Rohm and Haas Co.; "Methacrylic Acid and Derivatives" in *ECT* 3rd ed., Vol. 15, pp. 346–376, by J. W. Nemec and L. S. Kirch, Rohm and Haas Co.

1. E. Franklin and B. Duppa, *Ann. Chem.* **136**, 12 (1865).
2. F. Engelhorn, *Ann. Chem.* **200**, 68 (1880).
3. H. von Pechman and O. Röhm, *Chem. Ber.* **34**, 427 (1901).
4. U.S. Pat. 1,121,134 (Dec. 15, 1914), O. Röhm.
5. Brit. Pat. 264,143 (Jan. 11, 1926), H. Matheson and K. Blaikie (to Canadian Electro Products Co. Ltd.).
6. U.S. Pat. 2,100,993 (Nov. 30, 1937), H. A. Bruson (to Rohm and Haas Co.).
7. Brit. Pat. 405,699 (Feb. 12, 1934), J. W. C. Crawford (to Imperial Chemical Industries Ltd.).
8. L. S. Luskin, in F. D. Snell and C. L. Hilton, eds., *Encyclopedia of Industrial Chemical Analysis*, Vol. 4, John Wiley & Sons, Inc., New York, 1967, p. 81.
9. S. Zheng, M. Xia, X. Ying, and Y. Hu, *Huadong, Huagong Xueyuan Xuebao* **12**, 449 (1986); A. Frolov, M. Loginova, A. Saprykina, and A. Kondakova, *Zh. Fiz. Khim.*, **36**, 2282 (1962).
10. A. Frolov, M. Loginova, and B. Ustavshchekov, *Zh. Obshch. Khim.* **36**, 180 (1966).

11. S. Danov, T. Obmelyukima, G. Chubarov, and A. Balashov, *Zh. Prikl. Khim.* (*Leningrad*) **63**, 596 (1990).
12. Jpn. Kokai Tokkyo Koho, 60,156,659 (Aug. 16, 1985), S. Torii, T. Iguchi, and M. Kubota (to Osaka Yuki Kagaku Kogyo Co. Ltd.).
13. U.S. Pat. 4,332,967 (June 1, 1982), N. Thompson, D. Redmore, B. Alink, and B. Outlaw (to Petrolite Corp.).
14. J. Kopecky, J. Smejkal, I. Linhart, V. Hanus, and P. Pech, *Bull. Soc. Chim. Belg.* **95**, 1123 (1986).
15. S. Perri, S. Slater, S. Toske, and J. White, *J. Org. Chem.* **55**, 6037 (1990).
16. T. Rabini and G. Vita, *J. Org. Chem.* **30**, 2486 (1965).
17. H. Stetter and K. Findeinsen, *Chem. Ber.* **98**, 3228 (1965).
18. G. Ponticello, M. Freedman, C. Habecker, M. Holloway, J. Amato, R. Conn, and J. Baldwin, *J. Org. Chem.* **53**, 9 (1988).
19. F. Texier and J. Bourgois, *Bull Soc. Chim. France*, 3–4 (Pt. 2), 487 (1976).
20. I. Knunyants, E. Rokhlin, U. Utebaev, and E. Mysov, *Izu. Akad. Nauk SSSR Ser. Khim.* **1**, 137 (1976).
21. P. Srinivas and L. Burka, *Drug Metab. Disposition* **16**, 449 (1988).
22. U.S. Pat. 4,994,268 (Feb. 19, 1991), H. Weiser, E. Dixon, H. Cerezke, and A. MacKenzie (to University of Calgary).
23. K. Furuta, Y. Miwa, K. Iwanaga, and H. Yamamoto, *J. Amer. Chem. Soc.*, **110**, 6254 (1988).
24. K. Shishido, K. Hiroya, K. Fukumoto, and T. Kametani., *Tetrahedron Lett.*, 1167 (1986).
25. G. Szeimies and R. Huisgen, *Chem. Ber.* **89**, 491 (1966).
26. I. Juranic, S. Husinec, V. Savic, and A. Porter, *Collect. Czech. Chem. Comm.* **56**, 411 (1991).
27. J. Yaozhong, Z. Changyou, W. Shengde, C. Daimo, M. Youan, and L. Guilan, *Tetrahedron* **44**, 5343 (1988).
28. J. Yaozhong, Z. Changyou, and W. Shengde, *Synth. Comm.* **17**, 33 (1987).
29. S. A. Ali, J. Khan, and M. Wazier, *Tetrahedron* **44**, 5911 (1988).
30. U.S. Pat. 4,785,117 (Nov. 15, 1988), V. Georgiev and G. Mullen (to Pennwalt Corp.).
31. W. D. Emmons and A. Pagano, *J. Amer. Chem. Soc.* **77**, 89 (1955).
32. W. Ho, G. Tutweiler, S. Cottrell, D. Morgans, O. Tarhan, and R. Mohrbacher, *J. Med. Chem.* **29**, 2184 (1986).
33. M. Doyle, R. Dorow, and W. Tamblyn, *J. Org. Chem.* **47**, 4059 (1982).
34. J. Herndon and S. Turner, *J. Org. Chem.* **56**, 286 (1991).
35. L.-F. Tietze, A. Bergmann, and K. Bruggemann, *Synthesis*, 190 (1986).
36. T. Nishio, *J. Chem. Soc. Perkin Trans. I*, 1225 (1987).
37. J. Duncia, P. Lansbury, T. Miller, and B. Snider, *J. Amer. Chem. Soc.* **104**, 1930 (1982).
38. D. Glusker, R. Galluccio, and R. Evans, *J. Amer. Chem. Soc.* **86**, 187 (1964).
39. D. Glusker, E. Stiles, and B. Yoncoskie, *J. Polym. Sci.* **49**, 297 (1961).
40. D. Glusker, I. Lysloff, and E. Stiles, *J. Polym. Sci.* **49**, 315 (1961).
41. D. Wiles and S. Bywater, *Polymer* **3**, 175 (1962).
42. B. Cottam, D. Wiles, and S. Bywater, *Can. J. Chem.* **41**, 1905 (1963).
43. G. L'Abbe and G. Smets, *J. Polym. Sci., Part A-1* **5**, 1359 (1967).
44. M. Van Beylen, S. Bywater, G. Smets, M. Szwarc, and D. Worsfold, *Adv. Polym. Sci.* **86**, 87 (1988).
45. O. Webster, W. Hertler, D. Sogah, W. Farnham, and T. RajanBabu, *J. Amer. Chem. Soc.* **105**, 5706 (1983).
46. W. Hertler, D. Sogah, O. Webster, and B. Trost, *Macromolecules* **17**, 1415 (1984).
47. D. Sogah, W. Hertler, O. Webster, and G. Cohen, *Macromolecules* **20**, 1473 (1987).

48. W. Hertler, T. RajanBabu, D. Ovenall, G. Reddy, and D. Sogah, *J. Amer. Chem. Soc.* **110**, 5481 (1988).
49. H.-D. Speikamp and F. Bandermann, *Makromol. Chem.* **189**, 437 (1988).
50. U. Schmalbrock, H. Sitz, and F. Bandermann, *Makromol. Chem.* **190**, 2713 (1989).
51. M. Hino and K. Arata, *Appl. Catal.* **18**, 401 (1985).
52. U.S. Pat. 4,833,267 (May 23, 1989), S. Nakashima, H. Sogabe, and A. Okuba (to Nippon Shokubai Kagaku Kogyo Co.).
53. C. Blandy, D. Gervais, J.-L. Pellegatta, B. Gilot, and R. Giraud, *J. Mol. Catal.* **64**, L1–6 (1991).
54. Ger. Offen. 2,555,901 (June 24, 1976), T. Onoda and M. Ohtake (to Mitsubishi Chemical Industries Co., Ltd.).
55. U.S. Pat. 3,087,962 (Apr. 30, 1963), N. Bortnick (to Rohm and Haas Co.).
56. S. Mektiev and Yu. Saferov, *Azerb. Khim. Zh.* **5–6**, 33 (1973).
57. Can. Pat. 829,337 (Dec. 18, 1969), R. Stanley, D. Brook, and V. Prenna (to British Titan Products Co. Ltd.).
58. M. Marsi, *Inorg. Chem.* **27**, 3062 (1988); B. Weidmann and D. Seebach, *Angew. Chem. Int. Ed. Engl.* **22**, 31 (1983); Jpn. Kokai Tokkyo Koho 01,258,642 (Oct. 16, 1989), M. Kurokawa (to Mitsubishi Gas Chemical Co.).
59. U.S. Pat. 4,202,990 (May 13, 1980), M. Fumiki, S. Teshima, and T. Yokoyama (to Mitsubishi Rayon Co., Ltd.); U.S. Pat. 5,242,877 (Sept. 7, 1993), J. C. Dobson, C. McDade, M. G. Mirabelli, and J. J. Venter (to Rohm and Haas Co.).
60. R. H. Yocum and E. B. Nyquist, eds., *Functional Monomers*, Marcel Dekker Inc., New York, 1973.
61. U.S. Pat. 4,404,395 (Sept. 13, 1983), K. H. Markiewitz (to ICI Americas Inc.).
62. U.S. Pat. 4,365,081 (Dec. 21, 1982), N. Shimizu, H. Yoshida, H. Daigo, S. Matumoto, and H. Uchino (to Nippon Shokubai Kagaku Kogyo Co. Ltd.).
63. U.S. Pat. 4,970,333 (Nov. 13, 1990) J. Rabon, W. Pike, and C. Boriack (to Dow Chemical Co.).
64. *Hydroxy Alkyl Acrylates/Hydroxy Alkyl Methacrylates*, ROCRYL 400 Series, Bulletin 77S5, Rohm and Haas Co., Philadelphia, Pa., Mar. 1992.
65. *Metal–Organic Monomers and Polymers*, Geleste Inc. Product Catalogue, Tullytown, Pa., 1993.
66. T. Nakamura and T. Kita, *CEER Chem. Econ. Eng. Rev.* **15**, 23 (1983).
67. Nippon Shokubai Kagaku Kogyo Co. and Sumitomo Chemical Co., *Hydrocarbon Process.* **64**, 143 (1985).
68. M. Salkind, E. H. Riddle, and R. W. Keefer, *Ind. Eng. Chem.* **51**, 1232 (1959).
69. C. D. Hall, C. J. Leeding, S. Jones, S. Case-Green, I. Sanderson, and M. J. van Hoorn, *Chem. Soc. Perkin Trans.* **2**, 417 (1991).
70. Eur. Pat. 407,811 A2 (Jan. 16, 1991), H. Higuchi and K. Kida (to Mitsubishi Gas Chemical Company).
71. U.S. Pat. 4,987,256 (Jan. 22, 1991), S. Ebata, H. Hirayama, H. Higuchi, T. Kondo, and T. Kida (to Mitsubishi Gas Chemical Company).
72. U.S. Pat. 4,319,037 (Mar. 9, 1982), M. Yoneoka and co-workers (to Mitsubishi Gas Chemical Company).
73. U.S. Pat. 4,983,757 (Jan. 8, 1991), J. Ishikawa, H. Higucki, S. Ebata, and K. Kida (to Mitsubishi Gas Chemical Company).
74. Eur. Pat. 429,800 A2 (June 5, 1991), S. Naito, T. Kouzai, and R. Ikeda (to Mitsubishi Gas Chemical Company).
75. Eur. Pat. 446,446 A2 (Sept. 18, 1991), Y. Shima, T. Abe, H. Higuchi, and K. Kida (to Mitsubishi Gas Chemical Company).
76. E. Drent, P. Arnoldy, and P. H. M. Budzelaar, *J. Organometal. Chem.* **455**, 247–253 (1993).

77. U.S. Pat. 5,158,921 (Oct. 27, 1992), E. Drent, P. H. M. Budzelaar, and W. Jager (to Shell Oil Company).
78. U.S. Pat. 5,099,062 (Mar. 24, 1992), E. Drent, P. H. M. Budzelaar, and W. Jager (to Shell Oil Company).
79. Eur. Pat. 279,477 (Aug. 24, 1988), L. Petrus (to Shell Internationale Research Maatschappij BV).
80. Eur. Pat. 271,144 (June 15, 1988), E. Drent (to Shell Internationale Research Maatschappij BV).
81. U.S. Pat. 4,739,109 (Apr. 19, 1988), E. Drent (to Shell Oil Company).
82. Eur. Pat. 186,228 (July 2, 1986), E. Drent (to Shell Internationale Research Maatschappij BV).
83. U.S. Pat. 3,812,175 (May 21, 1974), J. Happel and co-workers (to National Lead).
84. R. Ugo, in W. Keim, ed., *Catalysis in C_1 Chemistry*, D. Reidel Co., Dordrecht, the Netherlands, 1983, p. 135.
85. U.S. Pat. 5,179,225 (Jan. 12, 1993), E. Drent, P. H. M. Budzelaar, W. Jager, and J. Stapersma (to Shell Oil Company).
86. U.S. Pat. 5,166,116 (Nov. 24, 1992), E. Drent, P. H. M. Budzelaar, W. Jager, and J. Stapersma (to Shell Oil Company).
87. U.S. Pat. 5,028,576 (July 2, 1991), E. Drent and P. H. M. Budzelaar (to Shell Oil Company).
88. U.S. Pat. 5,103,043 (Apr. 7, 1992), E. Drent and P. H. M. Budzelaar (to Shell Oil Company).
89. U.S. Pat. 5,081,286 (Jan. 14, 1992), M. J. Doyle (to Shell Oil Company).
90. U.S. Pat. 4,749,810 (June 7, 1988), S. E. Peterson, S. A. Pesa, and T. A. Haase (to Standard Oil, Ohio).
91. Y. Takezaki and co-workers, *Bull. Jpn. Pet. Inst.* **8**, 31 (1966).
92. U.S. Pat. 2,975,199 (Mar. 14, 1961), B. S. Friedman and S. M. Cotton (to Sinclair Refining Co.).
93. U.S. Pat. 4,499,029 (Feb. 12, 1985), C. Scaccia and J. R. Overley (to Ashland Oil).
94. U.S. Pat. 4,832,878 (May 23, 1989), C. Scaccia (to Ashland Oil).
95. Brit. Pat. 2,134,113 (Aug. 8, 1984), J. E. Corn Jr., D. Grote, and R. Pascoe (to Ashland Oil).
96. J. M. Millet and co-workers, *Appl. Catal.* **76**(2), 209 (1991); U.S. Pat. 4,370,490 (Jan. 25, 1983), W. S. Gruber, and G. Schroeder (to Röhm GmbH).
97. Jpn. Kokai 60-209258 (Oct. 21, 1985), H. Tsuneki and co-workers (to Nippon Shokubai).
98. U.S. Pat. 4,410,729 (Oct. 18, 1983), C. Daniel (to Ashland Oil).
99. *Chem. Week*, 10–11 (Oct. 23, 1985).
100. U.S. Pat. 4,647,696 (Mar. 3, 1987), S. Besecke, G. Schröeder, H.-J. Siegert, and W. Gaenzler (to Röhm GmbH).
101. U.S. Pat. 4,408,079 (Oct. 4, 1983), F. Merger and H.-J. Foerster (to BASF).
102. U.S. Pat. 4,496,770 (Jan. 29, 1985), G. Duembgen, G. Fouquet, R. Krabetz, E. Lucas, F. Merger, and F. Nees (to BASF).
103. U.S. Pat. 4,433,174 (Feb. 21, 1984), G. P. Hagen (to Standard Oil Co., Indiana).
104. U.S. Pat. 3,835,185 (Sept. 10, 1974), H. Hohenschutz, D. Franz, H. Buelow, and G. Dinkhauser (to BASF).
105. Eur. Pat. Appl. 92,350 (Oct. 26, 1983), R. L. Pruett and P. L. Burk (to Exxon Research and Eng. Co.).
106. U.S. Pat. 4,801,571 (Jan. 31, 1989), R. A. Montag and S. T. McKenna (to Amoco).
107. U.S. Pat. 4,599,144 (July 8, 1986), M. O. Baleiko and E. F. Rader (to Standard Oil Co., Indiana).
108. U.S. Pat. 4,943,659 (July 24, 1990), G. P. Hagen (to Amoco).

109. U.S. Pat. 4,743,706 (May 10, 1988), A. T. Guttmann and R. K. Grasselli (to Standard Oil Co., Ohio).
110. U.S. Pat. 4,739,111 (Apr. 19, 1988), J.-Y. Ryu (to Exxon Research and Engineering Co.).
111. U.S. Pat. 4,324,908 (Apr. 13, 1982), R. K. Grasselli and A. T. Guttmann (to Standard Oil Co., Ohio).
112. U.S. Pat. 4,945,179 (July 31, 1990), E. Drent (to Shell Oil Co.).
113. U.S. Pat. 5,028,576 (July 2, 1991), E. Drent and P. H. M. Budzelaar (to Shell Oil Co.).
114. U.S. Pat. 4,354,044 (Oct. 12, 1982), A. Aoshima, R. Mitsui, and N. Hitoshi (to Asahi Chemical Industry Co.).
115. U.S. Pat. 5,166,119 (Nov. 24, 1992), M. Oh-Kita and Y. Tanaguchi (to Mitsubishi Rayon Co.).
116. Eur. Pat. 450,596 (Oct. 9, 1991), T. Kawajiri, H. Hironaka, S. Uchida, and Y. Aoki (to Nippon Shokubai).
117. Eur. Pat. 58,046 (Aug. 18, 1982), K. Ohdan, K. Suzuki, and T. Yamao (to Ube Industries Ltd.).
118. Eur. Pat. 361,372 (Apr. 4, 1990), W. G. Etzkorn and G. G. Harkreader (to Union Carbide).
119. U.S. Pat. 5,138,100 (Aug. 11, 1992), I. Matsuura (to Mitsui Toatsu Chemicals).
120. U.S. Pat. 4,925,823 (May 15, 1990), R. Krabetz, G. Duembgen, F. Nees, F. Merger, and G. Fouquet (to BASF).
121. U.S. Pat. 5,126,307 (June 30, 1992), S. Yamamoto and M. Oh-Kita (to Mitsubishi Rayon Co.).
122. U.S. Pat. 4,469,810 (Sept. 4, 1984), M. Kato, M. Kamogawa, T. Nakano, and J. Furuse (to Mitsubishi Rayon Co.).
123. U.S. Pat. 4,596,784 (June 24, 1986), W. J. Kennelly and L. S. Kirch (to Rohm and Haas Co.).
124. U.S. Pat. 4,272,408 (June 9, 1981), C. Daniel (to Air Products).
125. N. Shimizi, *Petrotech* **6**, 778 (1983).
126. Nippon Shokubai Kagaku Kogyo Co. and Sumitomo Chemical Co., *Hydrocarbon Process.* **64**, 143 (1985).
127. *Eur. Chem. News* **57**, 1500 (Dec. 9, 1991).
128. Ger. Offen. 2,848,369 (May 23, 1979), N. Tamura, Y. Fukuoka, S. Yamamatsu, Y. Suzuki, R. Mitsui, and T. Ibuki (to Asahi Chemical Industry Co.).
129. *Hydrocarbon Process.* **57**, 108 (1978).
130. *Chem. Eng. News*, 28 (Apr. 12, 1993).
131. U.S. Pat. 4,532,083 (July 30, 1985), D. Suresh, R. Grasselli, J. Brazdil, and F. Ratka (to Standard Oil Co. of Ohio).
132. U.S. Pat. 4,473,506 (Sept. 25, 1984), J. Burrington, J. Brazdil, R. Grasselli (to Standard Oil Co. of Ohio).
133. H. Seeboth, J. Freiberg, P. Jiru, M. Krivenek, and H. Seefluth, *Z. Phys. Chem. (Leipzig)* **264**, 1105 (1983).
134. *Chemical Marketing Reporter*, mid-year issues (1955–1993).
135. *Acrylic Monomers*, Bulletin 84C10, Rohm and Haas Co., Philadelphia, Pa., 1991.
136. *Methacrylic/Acrylic Monomers*, Bulletin 84C2, Rohm and Haas Co., Philadelphia, Pa., 1986.
137. A. Nicolson, *Plant/Oper. Progr.* **10**, 171 (1991).
138. C. W. Chung and A. L. Giles, *J. Invest. Demat.* **68**, 187 (1977).
139. *Storage and Handling of Acrylic and Methacrylic Esters and Acids*, Bulletin 84C7, Rohm and Haas Co., Philadelphia, Pa., Dec. 1987.
140. J. M. Smith and co-workers, "Methyl Methacrylate: Subchronic, Chronic and Oncogenic Inhalation Safety Evaluation Studies," *Abstracts of the Eighteenth Annual Meeting of the Society of Toxicology*, New Orleans, La., 1979.

141. G. D. Clayton and F. E. Clayton, eds., *Patty's Industrial Hygiene and Toxicology*, 4th ed., Wiley-Interscience, New York, 1991.

ANDREW W. GROSS
JOHN C. DOBSON
Rohm and Haas Company

METHACRYLIC POLYMERS

The nature of the R group in methacrylic acid ester monomers having the generic formula $CH_2=C(CH_3)COOR$ generally determines the properties of the corresponding polymers. Methacrylates differ from acrylates in that the α-hydrogen of the acrylate is replaced by a methyl group (see ACRYLIC ESTER POLYMERS). This methyl group imparts stability, hardness, and stiffness to methacrylic polymers. The methacrylate monomers are extremely versatile building blocks. They are moderate-to-high boiling liquids that readily polymerize or copolymerize with a variety of other monomers. All of the methacrylates copolymerize with each other and with the acrylate monomers to form polymers having a wide range of hardness; thus polymers that are designed to fit specific application requirements can be tailored readily. The properties of the copolymers can be varied to form extremely tacky adhesives (qv), rubbers, tough plastics, and hard powders. Although higher in cost than many other common monomers, the methacrylates' unique stability characteristics, ease of use, efficiency, and the associated high quality products often compensate for their expense.

Development of the methacrylate esters has followed that of the acrylates. One of the first applications was as laminating resin for safety glass. Acrylates were copolymerized with ethyl methacrylate which led to the commercial production of ethyl methacrylate in 1933. During the development of safety glass interlayering, methyl methacrylate alone was also polymerized in a mold formed by two sheets of glass. Following polymerization, the two glass plates were separated from the polymer revealing a thin sheet of poly(methyl methacrylate) [9011-14-7] (PMMA) (see LAMINATED MATERIALS, GLASS). The commercial value of these sheets was readily apparent and methyl methacrylate soon became the most important member of the acrylate–methacrylate series. Research on cast sheets of poly(methyl methacrylate) was carried out in 1930s in the United States by Rohm and Haas Co. and E. I. du Pont de Nemours & Co., Inc., in Germany by Rohm and Haas A-G, and in England by Imperial Chemicals Industries, Ltd. By 1936, methyl methacrylate [80-62-6] was used to produce an organic glass by a cast polymerization process. The first significant market for methyl methacrylate sheet was in windows and canopies for military aircraft where its lighter weight, high transparency, shatter resistance, and ease of molding offered superior performance. The demand from the plastics industry for a methacrylate molding powder to match the performance, clarity, and strength of the methacrylate sheet resulted in the introduction of molding powders for both compression and injection molding in 1938. During the 1930s, the demand for methyl methacrylate also increased because of its use in adjusting the hardness of acrylate copolymers produced by emulsion and solution polymerization (1,2).

The uniqueness of methyl methacrylate as a plastic component accounts for its industrial use in this capacity, and it far exceeds the combined volume of all of the other methacrylates. In addition to plastics, the various methacrylate polymers also find application in sizable markets as diverse as lubricating oil additives, surface coatings (qv), impregnates, adhesives (qv), binders, sealers (see SEALANTS), and floor polishes. It is impossible to segregate the total methacrylate polymer market because many of the polymers produced are copolymers with acrylates and other monomers. The total 1991 production capacity of methyl methacrylate in the United States was estimated at 585,000 t/yr. The worldwide production in 1991 was estimated at about 1,785,000 t/yr (3).

Physical Properties

The nature of the alkyl group from the esterifying alcohol, the molecular weight, and the tacticity determine the physical and chemical properties of methacrylate ester polymers.

Glass-Transition Temperature. The physical properties of amorphous methacrylic polymers evidence a principal change in the glass-transition region. Chemical reactivity, mechanical and dielectric relaxation, viscous flow, load bearing capacity, hardness (qv), tack, heat capacity, refractive index, thermal expansivity, creep, and diffusion differ markedly below and above the transition region. During the transition, there is no significant absorption of latent heat, but for most polymers there is a change in specific volume, coefficient of expansion, compressibility, specific heat, and refractive index at the glass transition.

In methacrylic ester polymers, the glass-transition temperature, T_g, is influenced primarily by the nature of the alcohol group as can be seen in Table 1. Below the T_g, the polymers are hard, brittle, and glass-like; above the T_g, they are relatively soft, flexible, and rubbery. At even higher temperatures, depending on molecular weight, they flow and are tacky. Table 1 also contains typical values for the density, solubility parameter, and refractive index for various methacrylic homopolymers.

Glass-transition temperatures are commonly determined by differential scanning calorimetry or dynamic mechanical analysis. Many reported values have been measured by dilatometric methods; however, methods based on the torsional pendulum, strain gauge, and refractivity also give results which are in good agreement. Vicat temperature and brittle point yield only approximate transition temperature values but are useful because of the simplicity of measurement. The reported T_g values for a large number of polymers may be found in References 5, 6, 12, and 13.

Table 1. Physical Properties of Methacrylate Homopolymers

Polymer	CAS Registry Number	T_g, °C[a]	Density[b] at 20°C, g/cm³	Solubility parameter[b,c] (J/cm³)^1/2	Refractive index, n_D^{20}[d]
poly(methyl methacrylate)	[9011-14-7]	105	1.190	18.6	1.490
poly(ethyl methacrylate)	[9003-42-3]	65	1.119	18.3	1.485
poly(n-propyl methacrylate)	[25609-74-9]	35	1.085	18.0	1.484
poly(isopropyl methacrylate)	[26655-94-7]	81	1.033		1.552
poly(n-butyl methacrylate)	[9003-63-8]	20	1.055	17.8	1.483
poly(sec-butyl methacrylate)	[29356-88-5]	60	1.052		1.480
poly(isobutyl methacrylate)	[9011-15-8]	53	1.045	16.8	1.477
poly(t-butyl methacrylate)	[25213-39-2]	107	1.022	17.0	1.4638
poly(n-hexyl methacrylate)	[25087-17-6]	-5	1.007^25	17.6	1.4813
poly(2-ethylbutyl methacrylate)	[25087-19-8]	11	1.040		
poly(n-octyl methacrylate)	[25087-18-7]	-20	0.971^25	17.2	
poly(2-ethylhexyl methacrylate)	[25719-51-1]	-10			
poly(n-decyl methacrylate)	[29320-53-4]	-60			
poly(lauryl methacrylate)	[25719-52-2]	-65	0.929	16.8	1.474
poly(tetradecyl methacrylate)	[30525-99-6]	-72			1.47463
poly(hexadecyl methacrylate)	[25986-80-5]				1.47503
poly(octadecyl methacrylate)	[25639-21-8]	-100		16.0	
poly(stearyl methacrylate)	[9086-85-5]			16.0	
poly(cyclohexyl methacrylate)	[25768-50-7]	104	1.100	16.8	1.50645
poly(isobornyl methacrylate)	[28854-39-9]	170(110)	1.06	16.6	1.5000
poly(phenyl methacrylate)	[25189-01-9]	110	1.21		1.571
poly(benzyl methacrylate)	[25085-83-1]	54	1.179	20.3	1.5680
poly(ethylthioethyl methacrylate)	[27273-87-0]	-20			1.5300
poly(3,3,5-trimethylcyclohexyl methacrylate)	[75673-26-6]	79			1.485

[a] Refs. 4–12.
[b] Refs. 5, 12, and 13.
[c] To convert (J/cm³)^1/2 to (cal/cm³)^1/2, divide by 2.05.
[d] Refs. 5, 10, 12, and 13.

The stereoregularity of the backbone of the polymer chain also influences the T_g, as illustrated in Table 2. The syndiotactic and isotactic forms are of scientific interest but have found few commercial uses.

The T_gs of methacrylic polymers may be regulated by the copolymerization of two or more monomers as illustrated in Figure 1. The approximate T_g value for the copolymer can be calculated from the weight fraction of each monomer type and the T_g (in K) of each homopolymer (15). Acrylates with low transition temperatures are frequently used as permanent plasticizers (qv) for methacrylates. Unlike plasticizer additives, once polymerized into the polymer chain, the acrylate cannot migrate, volatilize, or be extracted from the polymer.

Measurement of modulus over an extensive temperature range offers more information than T_g alone (16). Typical modulus–temperature curves are shown in Figure 1. Assuming that the reference temperature is the transition temperature of the copolymer, then curve A of Figure 1 is that of a softer polymer and curve B is that of a harder polymer. Cross-linking of the polymer elevates and extends the rubbery plateau; little effect on T_g is noted until extensive cross-linking has been introduced. In practice, cross-linking of methacrylic polymers is used to decrease thermoplasticity and solubility and to increase resilience.

Molecular Weight. Typically, the mechanical properties of polymers increase as the molecular weight increases. However, beyond some critical molecular weight, often about 100,000 to 200,000 for amorphous polymers, the increase in property values is slight and levels off asymptotically. As an example, the glass-transition temperature of a polymer usually follows the relationship:

$$T_g = T_{g_i} - k/M_n$$

T_{g_i} is the glass-transition temperature at infinite molecular weight and M_n is the number average molecular weight. The value of k for poly(methyl methacrylate) is about 2×10^5; the value for acrylate polymers is approximately the same (9). A detailed discussion on the effect of molecular weight on the properties of a polymer may be found in Reference 17.

Mechanical Properties Related to Polymer Structure. Methacrylates are harder polymers of higher tensile strength and lower elongation than their acrylate counterparts because substitution of the methyl group for the α-hydrogen

Table 2. Glass-Transition Temperatures of Atactic, Syndiotactic, and Isotactic Polymethacrylate Esters, °C[a]

Methacrylate	Atactic	Syndiotactic	Isotactic
methyl	105	105	38
ethyl	65	66	12
n-propyl	35		
isopropyl	84	85	27
n-butyl	20		-24
isobutyl	53		8
sec-butyl	60		
t-butyl	118	114	7
cyclohexyl	104		110

[a]Refs. 13 and 14.

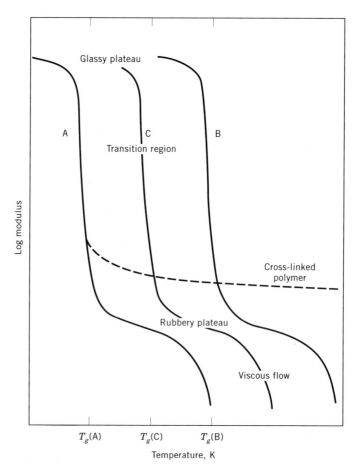

Fig. 1. Modulus–temperature curve of amorphous and cross-linked methacrylic polymers where A is a softer polymer, B, a harder polymer, and C, a 1:1 copolymer of A and B.

on the main chain restricts the freedom of rotation and motion of the polymer backbone. This is demonstrated in Table 3.

At room temperature, the first member of the linear aliphatic methacrylate series, poly(methyl methacrylate), is a hard, fairly rigid material which can

Table 3. Comparison of Mechanical Properties of Polyacrylate and Methyl Methacrylate[a]

	Tensile strength, MPa[b]		Elongation at break, %	
Ester	Polymethacrylate	Polyacrylate	Polymethacrylate	Polyacrylate
methyl	62	6.9	4	750
ethyl	34	0.2	7	1800
butyl	6.9	0.02	230	2000

[a]Refs. 18 and 19.
[b]To convert MPa to psi, multiply by 145.

be sawed, carved, or worked on a lathe. When heated above its T_g, poly(methyl methacrylate) is a tough, pliable, extensible material that is easily bent or formed into complex shapes, and can be molded or extruded. It is primarily the length, flexibility, bulk, and degree of crystallinity of the side chain that determines the T_g of conventional polymethacrylates of a given molecular weight. As the length of the side chain increases, the transition temperature of the polymer decreases. For example, poly(n-hexyl methacrylate) ($T_g = -5°C$) is rubber-like, and is an order of magnitude softer and more extensible than poly(methyl methacrylate). As the ester side-chain length is further increased, the T_g continues to decrease through poly(lauryl methacrylate) ($T_g = -65°C$); further increase in side-chain length allows crystallization of the side chains to occur, which restricts chain motion and results in a more brittle polymer. For example, poly(octadecyl methacrylate) has a brittle temperature of 36°C (5).

Increasing the bulkiness of the alkyl group from the esterifying alcohol in the ester also restricts the motion of backbone polymer chains past each other, as evidenced by an increase in the T_g within a series of isomers. In Table 1, note the increase in T_g of poly(isopropyl methacrylate) over the n-propyl ester and similar trends within the butyl series. The member of the butyl series with the bulkiest alcohol chain, poly(t-butyl methacrylate), has a T_g (107°C) almost identical to that of poly(methyl methacrylate) ($T_g = 105°C$), whereas the butyl isomer with the most flexible alcohol chain, poly(n-butyl methacrylate), has a T_g of 20°C. Further increase in the rigidity and bulk of the side chain increases the T_g. An example is poly(isobornyl methacrylate) which has a T_g of 170°C.

Optical Properties. Poly(methyl methacrylate) transmits light in the range of 360–1000 nm almost perfectly (92% compared to the theoretical 92.3%). The wavelength of visible light falls approximately between 400 and 700 nm. At a thickness of 2.54 cm or less, poly(methyl methacrylate) absorbs virtually no visible light (20). Beyond 2800 nm, essentially all infrared radiation is absorbed (21,22). Commercial grades of poly(methyl methacrylate) often contain uv radiation absorbers that block light in the 290–350 nm range. The absorber thus screens the user from sunburn and, in addition, protects the polymer against long-term degradation from light. Poly(methyl methacrylate)'s transparency to x-rays and radiation has been found to be about the same as that of human flesh or water. Sheets of poly(methyl methacrylate) are opaque to alpha particles, and for thicknesses above 6.35 mm (0.250 in.), the polymer is essentially opaque to beta radiation; poly(methyl methacrylate) is used as a transparent neutron stopper (22). Most formulations of colorless sheet have high transmittance to standard broadcast and television waves as well as to most radar bands (22).

Poly(methyl methacrylate) can serve as a conduit for light (23). When a ray of light moving within the polymer encounters an air interface at the critical angle (42.2° off the normal) or greater, it is not transmitted; rather it is totally reflected back through the plastic at an equal and opposite angle and trapped within the plastic until it emerges at the opposite end. In a clear rectangular piece, light entering one surface may leave only through a parallel and opposite surface. Light may be piped around corners by proper design. Image transmission around bends and corners may be accomplished by reflecting periscopically from facets (23). The ability to mold poly(methyl methacrylate) precisely allows complex optical designs to be manufactured easily and inexpensively. Many items, such as magnifiers, reducers, camera lenses, prisms, and especially complex reflex lenses widely used in automotive taillights, are made from poly(methyl methacrylate).

Electrical Properties. Poly(methyl methacrylate) has specific electrical properties that make it unique (Table 4). The surface resistivity of poly(methyl methacrylate) is higher than that of most plastic materials. Weathering and moisture affect poly(methyl methacrylate) only to a minor degree. High resistance and nontracking characteristics have resulted in its use in high voltage applications, and its excellent weather resistance has promoted the use of poly(methyl methacrylates) for outdoor electrical applications (22).

Table 4. Electrical Properties of 6.35-mm Thick Poly(methyl methacrylate) Sheet[a]

Property	Typical values	ASTM method
dielectric strength	$>16.9-20.9$	D149
short-term test, $V/\mu m$[b]		
dielectric constant, V/mm[c]		D150
at 60 Hz	142–154	
1,000 Hz	130–134	
power factor, V/cm[d]		D150
at 60 Hz	20–24	
1,000 Hz	16–20	
1,000,000 Hz	8–12	
loss factor, V/cm[d]		D150
at 60 Hz	75–87	
1,000 Hz	51–59	
1,000,000 Hz	24–31	
arc resistance	no tracking	D495
volume resistivity, $\Omega \cdot cm$	$1 \times 10^{14}-6 \times 10^{17}$	D257
surface resistivity, $\Omega/square$	$1 \times 10^{17}-2 \times 10^{18}$	D257

[a]Ref. 22.
[b]To convert $V/\mu m$ to V/mil, multiply by 25.4.
[c]To convert V/mm to V/mil, divide by 39.4.
[d]To convert V/cm to V/mil, divide by 394.

Chemical Properties

Methacrylate polymers have a greater resistance to both acidic and alkaline hydrolysis than do acrylate polymers; both are far more stable than poly(vinyl acetate) and vinyl acetate copolymers. There is a marked difference in the chemical reactivity among the noncrystallizable and crystallizable forms of poly(methyl methacrylate) relative to alkaline and acidic hydrolysis. Conventional (ie, free-radical), bulk-polymerized, and syndiotactic polymers hydrolyze relatively slowly compared with the isotactic type (24). Polymer configuration is unchanged by hydrolysis. Complete hydrolysis of nearly pure syndiotactic poly(methyl methacrylate) in sulfuric acid and reesterification with diazomethane yields a polymer with an nmr spectrum identical to that of the original polymers. There is a high proportion of syndiotactic configuration present in conventional poly(methyl methacrylate) which contributes to its high degree of chemical inertness. The chemical resistance of poly(methyl methacry-late) may be summarized as in the following paragraph (20).

PMMA is not affected by most inorganic solutions, mineral oils, animal oils, low concentrations of alcohols; paraffins, olefins, amines, alkyl monohalides; and aliphatic hydrocarbons and higher esters, ie, >10 carbon atoms. However, PMMA is attacked by lower esters, eg, ethyl acetate, isopropyl acetate; aromatic hydrocarbons, eg, benzene, toluene, xylene; phenols, eg, cresol, carbolic acid; aryl halides, eg, chlorobenzene, bromobenzene; aliphatic acids, eg, butyric acid, acetic acid; alkyl polyhalides, eg, ethylene dichloride, methylene chloride; high concentrations of alcohols, eg, methanol, ethanol; 2-propanol; and high concentrations of alkalies and oxidizing agents.

The chemical resistance and excellent light stability of poly(methyl methacrylate) compared to two other transparent plastics is illustrated in Table 5 (25). Methacrylates readily depolymerize with high conversion, ie, 95%, at >300°C (1,26). Methyl methacrylate monomer can be obtained in high yield from mixed polymer materials, ie, scrap.

Table 5. Relative Outdoor Stability of Poly(methyl methacrylate)[a]

	Light transmittance		Haze	
Material	Initial, %	After exposure,[b] %	Initial, %	After exposure,[b] %
poly(methyl methacrylate)	92	92	1	2
polycarbonate	85	82	3	19
cellulose acetate butyrate	89	68	3	70

[a]Ref. 25.
[b]Three-yr outdoor.

Methacrylate Monomers

Properties. Properties of many methacrylic esters are given in Table 1 of the article METHACRYLIC ACID AND DERIVATIVES. Some other properties of commercially important methacrylate monomers are given in Table 6 (27–30). Thermodynamic properties of common methacrylate esters and functional monomers are enumerated in Table 7 (29).

Manufacture of Monomers. The various commercial processes for the manufacture of methyl methacrylate have been reviewed (31) (see METHACRYLIC ACID AND DERIVATIVES).

Handling, Health, and Safety of Methacrylic Monomers. Good ventilation to reduce exposure to vapors, splashproof goggles to avoid eye contact, and protective clothing to avoid skin contact are required for the safe handling of methacrylic monomers. A more extensive discussion of safety factors should be consulted before handling these monomers (28).

Methacrylate monomers are shipped in bulk quantities, tank cars, or tank trucks. Mild steel is the usual material chosen for bulk storage facilities for these monomers, although stainless steel (Types 304 and 316) is also recommended for the esters and is a necessity for the acids. Moisture must be excluded to avoid rusting and contamination of the monomers. Copper (qv) or copper alloys

Table 6. Physical Properties of Commercially Available Methacrylate Monomers $CH_2 = C(CH_3)COOR$[a]

Compound	Mol wt	Mp, °C	Bp, °C	Refractive index, n_D^{25}	Density, d_5^{25} g/cm³	Flash point, °C[b] COC	TOC	Typical inhibitor, ppm[c]
methacrylic acid	86.09	14	159–163[d]	1.4288	1.015	77		100 MEHQ
methyl methacrylate	100.11	−48	100–101[d]	1.4120	0.939		13	10 MEHQ
ethyl methacrylate	114.14		118–119[d]	1.4116	0.909	35	21	15 MEHQ
n-butyl methacrylate	142.19		163.5–170.5[d]	1.4220	0.889	66		10 MEHQ
isobutyl methacrylate	142.19		155[d]	1.4172	0.882		49	10 MEHQ
isodecyl methacrylate	226		120[e]	1.4410	0.878			10 HQ + MEHQ
lauryl methacrylate	262	−22	272–343[d]	1.444	0.868	121		100 HQ
stearyl methacrylate	332	15	310–370[d]	1.4502	0.864	>149		100 HQ
2-hydroxyethyl methacrylate	130.14	−12	95[g]	1.4505	1.064	108		1200 MEHQ
2-hydroxypropyl methacrylate	144.17	<−70	96[g]	1.4456	1.027	121		1200 MEHQ
2-dimethylaminoethyl	157.20	ca −30	68.5[g]	1.4376	0.933		74	200 MEHQ
2-t-butylaminoethyl methacrylate	185.25	<−70	93[g]	1.4400	0.914	11		1000 MEHQ
glycidyl methacrylate	142.1		75[g]	1.4482	0.073		84	25 MEHQ
ethylene glycol dimethacrylate	198.2		96–98[h]	1.4520	1.048	113		60 MEHQ
1,3-butylene dimethacrylate	226		110[e]	1.4502	1.011	124		200 MEHQ
trimethylolpropane trimethacrylate	338	−14	155[i]	1.471	1.06	>149		90 HQ

[a]Refs. 27–30.
[b]COC = Cleveland open cup; TOC = Tagliabue open cup.
[c]MEHQ = monomethylether of hydroquinone; HQ = hydroquinone.
[d]At 101 kPa (1 atm).
[e]At 0.4 kPa[f].
[f]To convert kPa to mm Hg, multiply by 7.5.
[g]At 1.3 kPa[f].
[h]At 0.53 kPa[f].
[i]At 0.13 kPa[f].

Table 7. Thermodynamic Properties of Methacrylates[a]

Methacrylates	Heat of vaporization, kJ/g[b]	Heat capacity, J/(g·K)[b]	Heat of polymerization, kJ/mol[b]
methacrylic acid		2.1–2.3	56.5
methyl methacrylate	0.36	1.9	57.7
ethyl methacrylate	0.35	1.9	57.7
n-butyl methacrylate		1.9	56.5
2-hydroxyethyl methacrylate			49.8
2-hydroxypropyl methacrylate			50.6
2-dimethylaminoethyl methacrylate	0.31		

[a]Ref. 29.
[b]To convert J to cal, divide by 4.184.

(qv) must not be allowed contact with acrylic monomers intended for use in polymerization because copper is an inhibitor (28).

The toxicity of methacrylate monomers has been studied extensively (28,32,33). Based on lethality, acute toxicity tested by oral, dermal, and inhalation routes shows the methacrylic esters to be practically nontoxic after single exposures to rats and rabbits. Methacrylates are slightly to severely irritating to skin and eyes and are considered potential skin sensitizers. Repeated exposure results in little systemic toxicity. Repeated exposure of animals to near-lethal concentrations resulted in liver and kidney damage. In general, the toxicity of the methacrylate esters decreases with increasing molecular weight. Methyl methacrylate serves as a toxicological model for all methacrylates (see METHACRYLIC ACID AND DERIVATIVES).

Manufacture and Processing

RADICAL POLYMERIZATION

Free-radical polymerization processes are used to produce virtually all commercial methacrylic polymers. Usually free-radical initiators (qv) such as azo compounds or peroxides are used to initiate the polymerizations. Photochemical and radiation-initiated polymerizations are also well known. At a constant temperature, the initial rate of the bulk or solution radical polymerization of methacrylic monomers is first-order with respect to monomer concentration, and one-half order with respect to the initiator concentration. Rate data for polymerization of several common methacrylic monomers initiated with 2,2′-azobisisobutyronitrile [78-67-1] (AIBN) have been determined and are shown in Table 8.

Methacrylate polymerizations are accompanied by the liberation of a considerable amount of heat and a substantial decrease in volume. Both of these factors strongly influence most manufacturing processes. Excess heat must be dissipated to avoid uncontrolled exothermic polymerizations. Volume changes are

Table 8. Polymerization Data for Methacrylic Ester Monomers[a]

Methacrylates	$k sp,$[b] 44.1°C	Heat of polymerization, kJ/mol[c]	Shrinkage, vol %
methyl	27[d]	57.7	21.0
ethyl	25[e]	57.7	18.2
butyl	41[d]	59.4	14.9

[a] Ref. 34.
[b] The units of $k sp$ are $L^{1/2}/mol^{1/2} \cdot h^{-1}$. Initial rate of polymerization is calculated from $k sp$ and the concentration of AIBN using the following equation: initial rate of polymerization in %/h = $k sp$/AIBN.
[c] To convert kJ to kcal, divide by 4.184.
[d] In bulk.
[e] 2.5 M solution in benzene.

particularly important in sheet-casting processes where the mold must compensate for the decreased volume. In general, the percent shrinkage decreases as the size of the alcohol substituent increases; on a molar basis, the shrinkage is relatively constant (35).

The free-radical polymerization of methacrylic monomers follows a classical chain mechanism in which the chain-propagation step entails the head-to-tail growth of the polymeric free radical by attack on the double bond of the monomer. Chain termination can occur by either combination or disproportionation, depending on the conditions of the process (36).

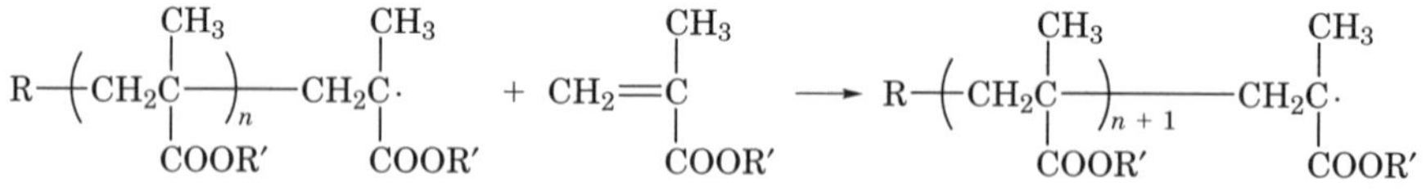

Some details of the chain-initiation step have been elucidated. With an oxygen radical–initiator such as the t-butoxyl radical, both double bond addition and hydrogen abstraction are observed. Hydrogen abstraction is observed at both the α-methyl and the ester alkyl group of methyl methacrylate. The relative proportions are shown (37,38).

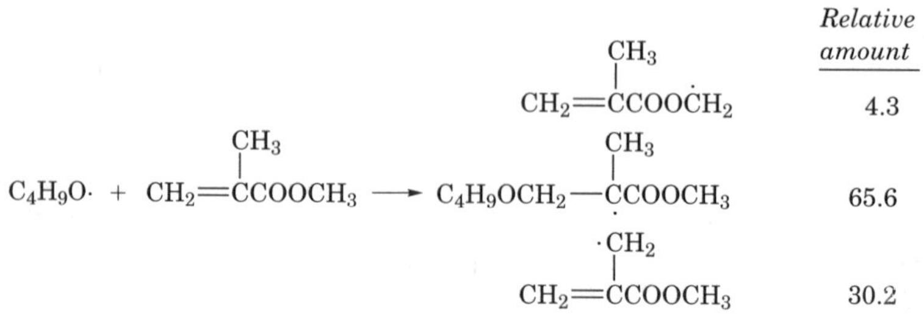

Methacrylate polmerizations are markedly inhibited by oxygen; therefore considerable care is taken to exclude air during the polymerization stages of manufacturing. This inhibitory effect has been shown to be caused by copolymerization of oxygen with monomer forming an alternating copolymer (39,40).

$$
R-CH_2\overset{\overset{\displaystyle CH_3}{|}}{\underset{\underset{\displaystyle COOR'}{|}}{C}}\cdot\ +\ O_2\ \xrightarrow{\text{fast}}\ R\!\left(\!CH_2\overset{\overset{\displaystyle CH_3}{|}}{\underset{\underset{\displaystyle COOR'}{|}}{C}}OO\!\right)_{\!n}\!\!-CH_2\overset{\overset{\displaystyle CH_3}{|}}{\underset{\underset{\displaystyle COOR'}{|}}{C}}OO\cdot
$$

In the presence of any substantial amount of oxygen, this reaction is extremely rapid, but the terminal peroxy radical formed reacts slowly with the monomer and has a relatively rapid termination rate.

$$
R\!\left(\!CH_2\overset{\overset{\displaystyle CH_3}{|}}{\underset{\underset{\displaystyle COOR'}{|}}{C}}OO\!\right)_{\!n}\!\!-CH_2\overset{\overset{\displaystyle CH_3}{|}}{\underset{\underset{\displaystyle COOR'}{|}}{C}}OO\cdot\ +\ CH_2\!\!=\!\!\overset{\overset{\displaystyle CH_3}{|}}{\underset{\underset{\displaystyle COOR'}{|}}{C}}\ \xrightarrow{\text{slow}}\ R\!\left(\!CH_2\overset{\overset{\displaystyle CH_3}{|}}{\underset{\underset{\displaystyle COOR'}{|}}{C}}OO\!\right)_{\!n+1}\!\!-CH_2\overset{\overset{\displaystyle CH_3}{|}}{\underset{\underset{\displaystyle COOR'}{|}}{C}}\cdot
$$

The overall effect, aside from the change in the polymer composition, is a decrease in the rate of monomer reaction, the kinetic chain length, and the polymer molecular weight (41).

A substantial fraction of commercially prepared methacrylic polymers are copolymers. Monomeric acrylic or methacrylic esters are often copolymerized with one another and possibly several other monomers. Copolymerization greatly increases the range of available polymer properties. The all-acrylic polymers tend to be soft and tacky; the all-methacrylic polymers tend to be hard and brittle. By judicious adjustment of the amount of each type of monomer, polymers can be prepared at essentially any desired hardness or flexibility. Small amounts of specially functionalized monomers are often copolymerized with methacrylic monomers to modify or improve the properties of the polymer directly or by providing sites for further reactions. Table 9 lists some of the more common functional monomers used for the preparation of methacrylic copolymers.

Bulk Polymerization. This is the method of choice for the manufacture of poly(methyl methacrylate) sheets, rods, and tubes, and molding and extrusion compounds. In methyl methacrylate bulk polymerization, an autoacceleration is observed beginning at 20–50% conversion. At this point, there is also a corresponding increase in the molecular weight of the polymer formed. This acceleration, which continues up to high conversion, is known as the Trommsdorff effect, and is attributed to the increase in viscosity of the mixture to such an extent that the diffusion rate, and therefore the termination reaction of the growing radicals, is reduced. This reduced termination rate ultimately results in a polymerization rate that is limited only by the diffusion rate of the monomer. Detailed kinetic data on the bulk polymerization of methyl methacrylate can be found in Reference 42.

Three bulk polymerization processes are commercially important for the production of methacrylate polymers: batch cell casting, continuous casting, and

Table 9. Common Functional Monomers for Copolymerization with Acrylic and Methacrylic Esters

Functionality	Monomer	Structure
carboxyl	methacrylic acid	$CH_2{=}C{-}COOH$ $\quad\quad\mid$ $\quad\quad CH_3$
	acrylic acid	$CH_2{=}CHCOOH$
	itaconic acid	CH_2COOH $\mid$ $CH_2{=}C{-}COOH$
amino	2-t-butylaminoethyl methacrylate	CH_3 $\mid$ $CH_2{=}C{-}COO(CH_2)_2NHC(CH_3)_3$
	2-dimethylaminoethyl methacrylate	CH_3 $\mid$ $CH_2{=}C{-}COO(CH_2)_2N(CH_3)_2$
hydroxyl	2-hydroxyethyl methacrylate	CH_3 $\mid$ $CH_2{=}C{-}COOCH_2CH_2OH$
	2-hydroxyethyl acrylate	$CH_2{=}CHCOOCH_2CH_2OH$
N-hydroxymethyl	N-hydroxymethyl acrylamide	$CH_2{=}CHCONHCH_2OH$
	N-hydroxymethyl methacrylamide	CH_3 $\mid$ $CH_2{=}C{-}CONHCH_2OH$
oxirane	glycidyl methacrylate	$CH_3 \quad\quad\quad O$ $\mid \quad\quad\quad\quad /\backslash$ $CH_2{=}C{-}COOCH_2CH{-}CH_2$
multifunctional	1,4-butylene dimethacrylate	CH_3 $\mid$ $\left(CH_2{=}C{-}COOCH_2CH_2\right)_{\!\overline{2}}$

continuous bulk polymerization. Approximately half the worldwide production of bulk polymerized methacrylates is in the form of molding and extrusion compounds, a quarter is in the form of cell cast sheets, and a quarter is in the form of continuous cast sheets.

Sheet. Sheets are produced in widths as narrow as a meter and lengths up to several hundred meters, and in thicknesses ranging from 0.16 to 15 cm. Poly(methyl methacrylate) (PMMA) sheets are mainly produced by three methods: batch cell, continuous, and extruded sheet. The batch cell method is inherently simple and easily adapted for manufacturing a wide variety of grades, colors, and sizes. It is used when the highest surface quality, maximum light transmission, or maximum transparency is required. Continuous casting is more economical and is used to produce relatively thin sheets or highly filled products such as synthetic marble-like materials. Sheets are also produced by extruding poly(methyl methacrylate) molding compounds. For gauges below 0.3 cm and for certain specialty applications, extruded sheets are economically preferred.

In the batch cell method, each sheet is cast in a mold assembled from two sheets of highly polished plate glass separated by a flexible spacer. The glass plates are held against the spacer with a clamping force such as spring clamps.

The tension on the clamping force is such that shrinkage during polymerization is accommodated by movement of the faces of the glass closer together. The mold is filled from one open corner with exact amounts of monomer (or monomer–polymer syrup), initiator, and other additives, then closed and heated to effect cure. An exactly programmed temperature cycle is used in order to achieve a reasonable cure time without initiating an uncontrolled exothermic reaction that would result in monomer vaporization and bubble formation. The heat of polymerization is dissipated with the aid of high velocity air ovens or water baths. Typical cure times range from 10–12 h for thin gauge sheets to several days for thicker sheets. After curing, the molds are cooled and the glass plates separated from the plastic sheet. The glass plates may be cleaned and reused. The plastic sheet may be used as is or annealed by heating to 140–150°C for several hours to reduce strains and achieve maximum dimensional stability. As a protective measure, a masking paper or a plastic film is often applied to the surface of the finished sheets.

Many of the problems encountered in preparing poly(methyl methacrylate) sheets can be reduced by casting with a monomer–polymer syrup instead of straight monomer. The use of a monomer–polymer syrup reduces shrinkage, reduces the amount of heat to be dissipated, shortens the induction period, and lowers the initiation temperature. The casting syrup typically contains about 20% polymer and is prepared by either partially polymerizing monomer in a reactor under controlled conditions or by dissolving poly(methyl methacrylate) powder in the monomer. More detailed descriptions of the batch cell sheet casting process can be found in References 43–45. Numerous process improvements are also described (46,47). Rods and tubes of poly(methyl methacrylate) are also produced by cast molding processes (see PLASTICS PROCESSING).

Probably all U.S. companies known to be producing poly(methyl methacrylate) sheet by continuous casting are using essentially the same process (48). A monomer–polymer syrup is introduced between two parallel stainless steel belts that travel through a curing and an annealing zone (Fig. 2). Within the curing zone, the belts are inclined to the horizontal so that the hydrostatic pressure of the syrup maintains the desired spacing between the belts. Spring-mounted rollers on the top side of the upper belt maintain external pressure. The edges between the belts are sealed by a flexible gasket that travels at the same speed as

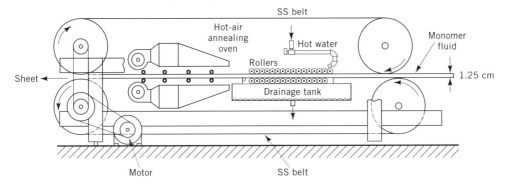

Fig. 2. Continuous process for manufacturing poly(methyl methacrylate) plastic sheet.

the belts. Heating in the curing zone is accomplished by a hot water spray, and in the annealing zone by electrical heaters or hot air. The belts travel at about 1 m/min, which results in a residence time of about 45 min in the curing zone at ca 70°C and about 10 min in the annealing zone at ca 110°C. The resulting sheet is covered with a masking paper before being cut into the desired lengths.

One of the most useful properties of poly(methyl methacrylate) plastic sheet is its formability. Because it is thermoplastic, it becomes soft and pliable when heated, and can be formed into almost any desired shape. As the material cools, it stiffens and retains the shape to which it has been formed. For exact details on forming methods, manufacturers' technical bulletins should be consulted (49,50). These technical bulletins also contain detailed instructions on a wide variety of fabrication methods, such as sawing, drilling, cementing, cleaning, painting, buffing, and polishing.

Molding Powder.   Poly(methyl methacrylate) molding and extrusion compounds are available in numerous grades and sizes. They are used by the plastics industry primarily in extrusion or injection molding processes. The most common form of molding powder is 0.3-cm pellets, although fine beads and granulated powders are also sold. The pellets are prepared by extruding either melted poly(methyl methacrylate) prepared by a separate bulk polymerization, or a monomer–polymer syrup. A devolatilizing extruder can be used to remove any unreacted monomer. The poly(methyl methacrylate) is extruded as a small-diameter rod, which is then chopped into short segments. Details on molding procedures are available from the manufacturers' technical bulletins (51).

Synthetic Marble.   Synthetic marble-like resin products are prepared by casting or molding a highly filled monomer mixture or monomer–polymer syrup. When only one smooth surface is required, a continuous casting process using only one endless stainless steel belt can be used (52,53). Typically on the order of 60 wt % inorganic filler is used. The inorganic fillers, such as aluminum hydroxide, calcium carbonate, etc, are selected on the basis of cost, and such properties as the translucence, chemical and water resistance, and ease of subsequent fabrication (54,55).

Solution Polymerization.   The solution polymerization of methacrylic monomers to form solution polymers or copolymers is an important commercial process for the preparation of polymers for use as coatings, adhesives, impregnates, and laminates. Typically the polymerization is done batchwise by adding monomer to an organic solvent in the presence of a soluble peroxide or azo initiator. This method is suitable for the preparation of polymers with molecular weights of about 2,000–200,000. Polymers with higher molecular weight are not only difficult to prepare by this method, but also have viscosities too high for easy handling. For a review of the quantitative aspects of the solution polymerization of methyl methacrylate, see References 56 and 57.

In general, the polymethacrylate esters of the lower alcohols are soluble in aromatic hydrocarbons, esters, ketones, and chlorohydrocarbons. They are insoluble, or only slightly soluble, in aliphatic hydrocarbons and alcohols. The polymethacrylate esters of the higher alcohols ($>C_4$) are soluble in aliphatic hydrocarbons. Cost, toxicity, flammability, volatility, and chain-transfer activity are the primary considerations in the selection of a suitable solvent. Chain transfer to solvent is an important factor in controlling the molecular weight of polymers prepared by this method. The chain-transfer constants for poly(methyl methacrylate) in various common solvents (C_s) and for various chain-transfer agents are listed in Table 10.

The type of initiator utilized for a solution polymerization depends on several factors, including the solubility of the initiator, the rate of decomposition of the initiator, and the intended use of the polymeric product. The amount of initiator used may vary from a few hundredths to several percent of the monomer weight. As the amount of initiator is decreased, the molecular weight of the polymer is increased as a result of initiating fewer polymer chains per unit weight of monomer, and thus the initiator concentration is often used to control molecular weight. Organic peroxides, hydroperoxides, and azo compounds are the initiators of choice for the preparations of most methacrylic solution polymers and copolymers.

The molecular weight of a polymer can be controlled through the use of a chain-transfer agent, as well as by initiator concentration and type, monomer concentration, and solvent type and temperature (61). Chlorinated aliphatic compounds and thiols are particularly effective chain-transfer agents used for regulating the molecular weight of methacrylic polymers (see Table 10).

Solution polymerizations of methacrylic esters are usually conducted in large stainless steel, nickel, or glass-lined cylindrical kettles designed to withstand at least 446 kPa (65 psi). An anchor-type agitator is suitable for solutions up to 1.0 Pa·s (1000 cP) viscosity; a slow-speed ribbon agitator with close clearance to the walls is required for solutions of higher viscosity. In large kettles, turbine agitators with swept-back blades are often used. Typically production kettles are fitted with a jacket for temperature control, a reflux condenser, inlets for addition of the reaction ingredients, sightglasses, a thermometer, and a rupture disk. A bottom valve is used for discharging the finished product to receiving tanks or drums.

Table 10. Chain-Transfer Constants for Methyl Methacrylate

Type	$C_s \times 10^5$ at 80°C	C_s at 60°C
Solvents[a]		
benzene	0.8	
toluene	5.3	
chlorobenzene	2.1	
2-propanol	19.1	
isobutyl alcohol	2.3	
3-pentanone	17.3	
chloroform	11.3	
carbon tetrachloride	24.2	
Chain-transfer agent[b]		
carbon tetrabromide		0.27
butanethiol		0.66
t-butyl mercaptan		0.18
thiophenol		2.7
ethyl mercaptoacetate		0.63

[a] Refs. 58 and 59.
[b] Ref. 60.

Because polymerizations are accompanied by the liberation of considerable heat, the chances of a violent or runaway reaction must be avoided. This is most easily done by gradual addition of the reactants to the kettle. Usually the monomers are added from weighing or measuring tanks situated close to the kettle. The rate of addition of monomer is adjusted to permit removal of heat. A supply of inhibitor is kept on hand to stop the polymerization if the cooling becomes inadequate.

Initiators, usually from 0.02 to 2.0 wt % of the monomer, are dissolved in the reaction solvents and fed as a separate stream to the kettle. Because oxygen is an inhibitor of methacrylic polymerizations, its presence is undesirable. When the polymerization is carried out below reflux temperatures, oxygen concentration is lowered by a sparge with carbon dioxide or nitrogen, and a blanket of the inert gas is then maintained over the polymerization mixture. The time required to complete a batch is usually 8–24 h (62).

A typical process for the preparation of an acrylic solution terpolymer of composition 27.5% 2-ethylhexyl acrylate–41.3% methyl methacrylate–31.2% hydroxyethyl methacrylate begins with the following charges:

Reactor charge	*Parts, wt %*
xylene	28.4
ethoxyethanol	14.1
Monomer charge	
2-ethylhexyl methacrylate	15.5
methyl methacrylate	23.3
hydroxyethyl methacrylate	17.6
Initiator charge	
t-butyl perbenzoate	1.1

The reactor charge is heated to 140°C under a nitrogen atmosphere and the monomer charge and initiator charge are added uniformly over three hours while maintaining 140 ± 2°C. After the additions are complete, this temperature is maintained for two more hours, then the product is cooled and packaged. A clear, viscous solution of about 58% polymer is obtained (63).

Methacrylic resins are shipped in drum, tank truck, and tank car quantities. Storage tanks, usually of 20–40 m³ capacity, are constructed of stainless steel, although tanks of mild steel can be used if provisions are made to keep them free of water. Moisture in mild steel tanks causes rusting and discoloration of the polymer solution. Typically storage tanks are equipped with a bottom outlet and are dished or sloped to permit complete drainage. Most are also equipped with an access hole so that they can be entered for cleaning. Because the viscosity of most acrylic solution polymers varies with temperature, temperature must be controlled by means of tank location, insulation, and heating or cooling devices. Steel pipe with steel or stainless steel valves is used for the resin transfer lines (64).

Emulsion Polymerization.   The principal markets for aqueous dispersion polymers made by emulsion polymerization of methacrylic esters are the paint (qv), paper (qv), textile, floor polish, and leather (qv) industries where they are used principally as coatings or binders. Copolymers of methyl methacrylate with either ethyl acrylate or butyl acrylate are most common.

Most of the lower alkyl methacrylates readily polymerize in water in the presence of a surfactant and a water-soluble initiator. The final product is an opaque, gray, or milky-white disperson of high molecular weight polymer at a concentration of 30–60 wt % in water. The particle size of methacrylic–acrylic copolymer dispersions ranges from ca 0.1 to 1.0 μm. These emulsion polymerizations are usually rapid and give high molecular weight polymers at high concentration and low viscosity. Difficulties in agitation, heat transfer, and material transfer, which are often encountered in the handling of viscous polymer solutions, are greatly decreased with aqueous dispersions. In addition, the safety hazards and the expense of flammable solvents are eliminated.

The surfactants (qv) used in the emulsion polymerization of acrylic or methacrylic monomers are classified as anionic, cationic, or nonionic. Anionic surfactants, such as alkyl sulfates and alkylarene sulfonates and phosphates, or nonionic surfactants, such as alkyl or aryl polyoxyethylenes, are most common. Mixed anionic nonionic surfactant systems are also widely utilized.

Water-soluble peroxide salts, such as ammonium or sodium persulfate, are the usual initiators. The initiating species is the sulfate radical anion generated from either the thermal or redox cleavage of the persulfate anion. The thermal dissociation of the persulfate anion, which is a first-order process at constant temperature, can be greatly accelerated by the addition of certain reducing agents and small amounts of polyvalent metal salts, or both. By using redox initiator systems, rapid polymerizations are possible at much lower temperatures (25–60°C) than are practical with thermally initiated systems (75–90°C).

Methacrylic emulsion polymerizations are usually conducted by batch processes in jacketed stainless steel or glass-lined kettles designed to withstand an internal pressure of at least 446 kPa (65 psi). Agitators are constructed from the same materials as the reactor. Versatility in controlling agitation is provided by use of variable speed drive. A baffle is sometimes used in the kettle to improve mixing; however, excessive shear must be avoided to control formation of coagulum. The temperature of the reactants is controlled by circulating steam and cold water through the jacket. A schematic diagram of a typical plant installation is given in Figure 3. A feed line for emulsified monomer enters through the top of the kettle along with feed lines for adding aqueous solutions of initiators and activators or both. Additional equipment includes a temperature recorder, manometer, sightglass, and emergency stack equipped with a rupture disk.

Monomer emulsions are prepared in separate stainless steel preemulsification tanks, which are usually equipped with an agitator, manometer level gauge, cooling coils, temperature recorder, rupture disk, flame arrester, and various nozzles for charging the ingredients. Monomer emulsions may be charged in one shot to the reactor or, more commonly, fed continuously throughout the polymerization.

A simple stainless steel drumming tank is used to receive the polymerized emulsion and hold it until it is packaged into drums or tank cars. This tank also

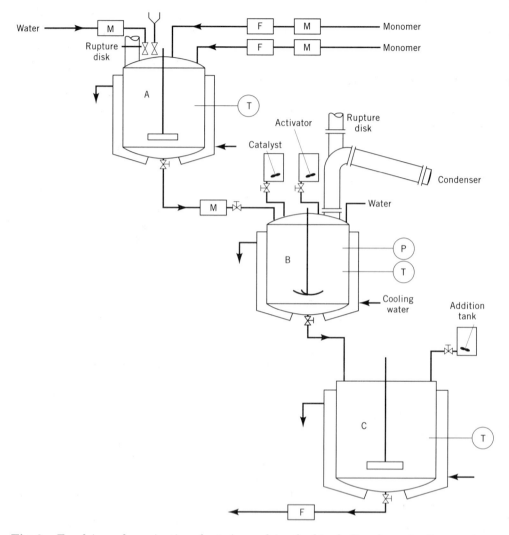

Fig. 3. Emulsion polymerization plant: A, emulsion feed tank; B, polymerization reactor; C, drumming tank; F, filter; M, meter; P, pressure gauge; and T, temperature indication.

is employed to adjust the solids content and pH, for the addition of preservatives, stabilizers, and thickeners, and for blending operations. A paddle-type low speed agitator is used for mixing. A cooling jacket on the drumming tanks permits the finished dispersion to be discharged hot from the reactor, thus increasing the productivity of the kettle. The cooled finished dispersion is passed through a coarse filter prior to packaging.

Numerous recipes have been published which describe the preparation of methacrylate homopolymer and copolymer dispersions (65,66). A typical process for the preparation of a 50% methyl methacrylate, 49% butyl acrylate, and 1% methacrylic acid terpolymer as an approximately 45% dispersion in water begins with the preparation of the monomer emulsion charge.

Monomer emulsion charge	Parts
deionized water	13.65
sodium lauryl sulfate	0.11
methyl methacrylate	22.50
butyl acrylate	22.05
methacrylic acid	0.45

The listed ingredients are added in given order while maintaining good agitation. The reactor charge (deionized water, 30.90 parts; sodium lauryl sulfate, 0.11 parts) is heated with good agitation under a nitrogen atmosphere to 85°C, then the initiator charge (ammonium persulfate, 0.23 parts) is added to the reactor and the monomer emulsion feed is begun. The monomer emulsion is fed uniformly over 2.5 h while maintaining 85°C. After the addition is complete, the temperature is raised to 95°C to complete the conversion of monomer. The product is then cooled to room temperature, filtered, and packaged.

Methacrylic dispersion polymers are shipped in bulk or in drums. Tank trucks and tank cars used for bulk shipment are constructed of stainless or resin-coated steel and are insulated to prevent freezing. Filament-wound glass fiber-reinforced polyester tanks are recommended for storage because of their relatively low cost, ease of installation, and chemical resistance. Usually storage tanks are located in an enclosed and heated environment to prevent freezing during cold weather. Dispersion polymers are subject to the various instability problems common to all colloidal systems, such as sedimentation, skinning (surface film), gritting (solid with the dispersion), gumming (deposits on walls), and sponging (formation of an aerogel). Undesirable changes may be caused by time, drift in pH, evaporation, high or low temperature, shear and turbulence, and foaming (see FOAMS). Oxidative degradation is not usually encountered with methacrylic dispersion polymers, but bacterial attack is common and is avoided by pH adjustment, addition of bactericidal agents, and careful housekeeping (62).

Suspension Polymerization. This method yields polymethacrylates in the form of tiny beads, which are primarily used as molding powders and ion-exchange resins. Most suspension polymers prepared as molding powders are poly(methyl methacrylate); copolymers containing up to 20% acrylate for reduced brittleness and improved processibility are also common. Suspension polymers of poly(methyl methacrylate) copolymerized with an amino or acid functional monomer, and with a di- or trivinyl monomer for cross-linking, are useful as ion-exchange resins.

In a suspension polymerization, monomer is suspended in water as 0.1–5 mm droplets, stabilized by protective colloids or suspending agents. Polymerization is initiated by a monomer-soluble initiator and takes place within the monomer droplets. The water serves as both the dispersion medium and a heat-transfer agent. Particle size is controlled primarily by the rate of agitation and the concentration and type of suspending aids. The polymer is obtained as small beads of about 0.1–5 mm in diameter, which are isolated by filtration or centrifugation.

Suitable protective colloids for the preparation of acrylic suspension polymers include cellulose derivatives, polyacrylate salts, starch (qv), poly(vinyl alcohol), gelatin (qv), talc (qv), clay, and clay derivatives (62,67). These materials are added to prevent the monomer droplets from coalescing during polymerization. Thickeners such as glycerol (qv), glycols, polyglycols, and inorganic salts are also often added to improve the quality of acrylic suspension polymers (62). Other constituents may be added to assist in the formation of uniform beads or to influence the use properties of the polymers through plasticization or cross-linking. These include lubricants, such as lauryl or cetyl alcohol, and stearic acid and cross-linking monomers, such as di- or trivinylbenzene, diallylesters of dibasic acids, and glycol dimethacrylates.

Initiators of suspension polymerization are organic peroxides or azo compounds that are soluble in the monomer phase, but insoluble in the water phase. The amount of initiator influences both the polymerization rate and the molecular weight of the product (62).

Because the polymerization occurs totally within the monomer droplets without any substantial transfer of materials between individual droplets or between the droplets and the aqueous phase, the course of the polymerization is expected to be similar to bulk polymerization. Accounts of the quantitative aspects of the suspension polymerization of methyl methacrylate generally support this model (62,68).

The equipment for methacrylate suspension processes, in general, is similar to an emulsion plant (see Fig. 3). The reactor is a jacketed glass-lined or stainless steel kettle capable of withstanding at least 446 kPa (65 psig) and fitted with similar safety and control devices. The agitator is a propeller or paddle-type rotated by a powerful, adjustable drive capable of maintaining uniform and continuous circulation throughout the polymerization. The rate of agitation is often adjusted to afford the desired characteristics and particle size of the product. The accessories for charging the reaction components are also similar to those of an emulsion process.

In a typical process, water, suspending agents, other minor components, the monomer mixture, and the initiator are charged to the kettle in the stated order. The well-agitated mixture is heated under a nitrogen atmosphere to the desired polymerization temperature. The heat of polymerization often causes a considerable rise in temperature and pressure; the reaction is usually completed rapidly. The slurry is cooled and dropped either to a filter box for filtration through a stainless steel screen, or to a centrifuge with a stainless steel basket and stainless steel screen. The beads are washed thoroughly, then dried on aluminum trays in a circulating air oven at 80–120°C or in a stainless steel rotary vacuum drier (62) (see DRYING).

A typical process for the preparation of a poly(methyl methacrylate) suspension polymer involves charging a mixture of 24.64 parts of methyl methacrylate and 0.25 parts of benzoyl peroxide to a rapidly stirred, 30°C solution of 0.42 parts of disodium phosphate, 0.02 parts of monosodium phosphate, and 0.74 parts of Cyanomer A-370 (polyacrylamide resin) in 73.93 parts of distilled water. The reaction mixture is heated under nitrogen to 75°C and is maintained at this temperature for three hours. After being cooled to room temperature, the polymer beads are isolated by filtration, washed, and dried (69).

Nonaqueous Dispersion Polymerization. Nonaqueous dispersion polymers are prepared by polymerizing a methacrylic monomer dissolved in an organic solvent to form an insoluble polymer in the presence of an amphipathic graft or block copolymer. This graft or block copolymer, commonly called a stabilizer, lends colloidal stability to the insoluble polymer. Particle sizes in the range of 0.1–1.0 μm were typical in earlier studies (70), however particles up to 15 μm have been reported (71).

Acrylic and methacrylic nonaqueous dispersions (NADs) are primarily utilized by the coatings industry to avoid certain difficulties associated with aqueous dispersion (emulsion) polymers. Water as a suspension medium has numerous practical advantages, but also some inherent difficulties: a high heat of evaporation, a low boiling point, and an evaporation rate that depends on the prevailing humidity. Nonaqueous dispersions alleviate these problems, but introduce others such as flammability, increased cost, odor, and toxicity.

There are numerous examples of acrylic NADs, but they are of limited commercial importance. Some typical examples of acrylic NADs are listed in Table 11, and References 70 and 71 offer excellent overviews.

Graft Polymerization. Graft copolymers are prepared by attaching one polymer as a branch to the chain of another polymer of different composition. This is usually accomplished by generating radical sites on the first polymer onto which monomer of the second polymer is grafted. The grafting may be accomplished in bulk, solution, or dispersion systems. The presence of distinct, but chemically bonded segments of two polymers often confers interesting and useful properties. Commercially, the most important methacrylate graft copolymers are the MABS and MBS polymers.

The MABS copolymers are prepared by dissolving or dispersing polybutadiene rubber in a methyl methacrylate–acrylonitrile–styrene monomer mixture. The graft copolymerization is accomplished by either a bulk or suspension process. The final polymer is a two-phase system in which the continuous phase is a terpolymer of methyl methacrylate, acrylonitrile, and styrene grafted onto the dispersed polybutadiene phase. These polymers are used in applications requiring a tough, transparent, highly impact-resistant, thermally formable material. Except for their transparency, the MABS polymers are similar to the more common opaque ABS plastics. Transparency is achieved by matching the

Table 11. Examples of Methacrylic Nonaqueous Dispersion Polymers

Polymer	Diluent	Dispersant	Reference
poly(methyl methacrylate)	aliphatic hydrocarbon	hydroxystearic acid–acrylic graft copolymer	70
poly(methyl methacrylate)	aliphatic hydrocarbon	drying oil-modified polyester	72
poly(methyl methacrylate)	hexane	isobutylene-co-isoprene graft copolymer	73
poly(methyl methacrylate)	methanol	ethylene glycol–methyl methacrylate graft copolymer	74
poly(ocytyl methacrylate)	methanol	ethylene glycol–octyl methacrylate graft copolymer	70
poly(methacrylic acid)	chloroform-ethanol	methacrylate functional polyester	70

refractive indexes of the two phases; this is the primary function of the methyl methacrylate.

MBS polymers are prepared by grafting methyl methacrylate and styrene onto a styrene–butadiene rubber in an emulsion process. The product is a two-phase polymer useful as an impact modifier for rigid poly(vinyl chloride).

A review covers the preparation and properties of both MABS and MBS polymers (75). Literature is available on the grafting of methacrylates onto a wide variety of other substrates (76,77). Typical examples include the grafting of methyl methacrylate onto rubbers by a variety of methods: chemical (78,79), photochemical (80), radiation (80,81), and mastication (82). Methyl methacrylate has been grafted onto such substrates as cellulose (83), poly(vinyl alcohol) (84), polyester fibers (85), polyethylene (86), poly(styrene) (87), poly(vinyl chloride) (88), and other alkyl methacrylates (89).

IONIC POLYMERIZATION

Methacrylate monomers do not generally polymerize by a cationic mechanism. In fact, methacrylate functionality is often utilized as a passive pendent group for cationically polymerizable monomers. Methacrylate monomers also have been used as solvents or cosolvents for cationic polymerizations (90,91).

The anionic polymerization of methacrylic monomers to stereoregular or block copolymers is well known. These polymerizations are conducted in organic solvents, primarily using organometallic compounds as initiators. This technology is of minor commercial significance in the 1990s, but is of interest for the preparation of polymers of narrow molecular weight distribution and controlled molecular architecture. Stereoregular forms of numerous polymethacrylates have been prepared and characterized. These include poly(methyl methacrylate) (92,93), and poly(n-butyl methacrylate) (94). Carefully controlled reaction conditions are usually required to obtain polymers with some measurable degree of crystallinity. In nonpolar solvents, the anionic polymerization of methacrylates generally yields isotactic polymer, whereas in polar solvents syndiotactic polymerization is favored. The physical and chemical properties of the various forms are often quite different. Methacrylate block copolymers formed by the living polymer method are also described in the literature (95). A general review covers these and other aspects of the anionic polymerization of acrylates and methacrylates (96); the polymerization kinetics of methacrylate anionic polymerizations have been described (97); progress in anionic polymerization has also been reviewed (98).

Initiation of these anionic polymerizations is considered to take place by a Michael reaction:

$$R_3C^-M \; + \; CH_2\!\!=\!\!\overset{\displaystyle CH_3}{\underset{\displaystyle |}{C}}\!\!-\!\!COOR \;\longrightarrow\; R_3C\!-\!CH_2\!-\!\overset{\displaystyle CH_3}{\underset{\displaystyle |}{C}}\!\!-\!\!COOR$$

Propagation occurs by head-to-tail addition of monomer:

$$R_3CCH_2\underline{C}-COOR + CH_2{=}\overset{\overset{\displaystyle CH_3}{|}}{C}-COOR \longrightarrow R_3CCH_2\overset{\overset{\displaystyle CH_3}{|}}{C}-CH_2-\overset{\overset{\displaystyle CH_3}{|}}{\underset{\underset{\displaystyle COOR}{|}}{\underset{M^+}{C}}}-COOR$$

Unlike free-radical polymerizations, copolymerizations between acrylates and methacrylates are not observed in anionic polymerizations; however, good copolymerizations within each class are reported (99).

The anionic polymerization of methacrylates using a silyl ketene acetal initiator has been termed group-transfer polymerization (GTP). First reported by Du Pont researchers in 1983 (100), group-transfer polymerization allows the control of methacrylate molecular structure typical of living polymers, but can be conveniently run at room temperature and above. The use of GTP to prepare block polymers, comb-graft polymers, loop polymers, star polymers, and functional polymers has been reported (100,101).

Economic Aspects

The worldwide production capacity of methyl methacrylate in 1991 was estimated to be about 1,785,000 t/yr. Principal producers are Western Europe, Japan, and the United States, which accounted for about 33% of this capacity. The important U.S. producers are as follows (3):

Producer	1991 Capacity, t
Röhm and Haas Co.	585,000
ICI including former Du Pont	353,000
CYRO Industries	89,000

In the United States the total volume of methacrylic acid and the higher alkyl methacrylates is roughly 10% of the volume of methyl methacrylate produced. The significant higher methacrylates produced are n-butyl, 2-ethylhexyl, isobutyl, lauryl, and stearyl. The first three are used primarily in coatings (qv), the latter two in oil and fuel additives. The U.S. market distribution for the use of methyl methacrylate is estimated at forming cast sheet (24%), molding powder (21%), coatings (18%), exports (11%), impact modifiers (10%), emulsion polymers (8%), mineral sheets (3%), higher methacrylates (2%), and other uses. Overall growth predictions are for a decline in demand from the historic 5% to 3–4%/yr through 1995 (3). The 1993 standard price for methyl methacrylate was about $1.35/kg. Methacrylate polymers are generally listed at two to three times the monomer cost.

Analytical Test Methods and Specifications

Plastic Sheet. Poly(methyl methacrylate) plastic sheet is manufactured in a wide variety of types, including clear and colored transparent, clear and colored translucent, and colored semiopaque. Various surface textures are also produced. Additionally, grades with improved weatherability (added uv absorbers), mar resistance, crazing resistance, impact resistance, and flame resistance are available. Selected physical properties of poly(methyl methacrylate) sheet are listed in Table 12 (102).

Table 12. Typical Properties of Commercial Poly(methyl methacrylate) Sheet[a]

Property	Value	ASTM method
specific gravity	1.19	D792 66
refractive index	1.49	D542 50 (1965)
tensile strength		D638 64T
maximum, MPa[b]	72.4	
rupture, MPa[b]	72.4	
elongation, max, %	4.9	
elongation, rupture, %	4.9	
modulus of elasticity, MPa[b]	3103	
flexural strength		D790 66
maximum, MPa[b]	110.3	
rupture, MPa[b]	110.3	
deflection, max, cm	1.52	
deflection, rupture, cm	1.52	
modulus of elasticity, MPa[b]	3103	
compressive strength		D695 68T
maximum, MPa[b]	124.1	
modulus of elasticity, MPa[b]	3103	
compressive deformation under load[c]		D621 64
14 MPa[b] at 50°C, 24 h, %	0.2	
28 MPa[b] at 50°C, 24 h, %	0.5[e]	
shear strength, MPa[b]	62.1	D732 46 (1961)
impact strength		
Charpy unnotched, J/cm^{2}[d]	2.94	D256 56 (1961)
Rockwell hardness	M-93[e]	
hot forming temperature, °C	144–182	
heat distortion temperature		D648 56 (1961)
2°C/min, 1.8 MPa[b], °C	96[e]	
2°C/min, 0.46 MPa[b], °C	107[e]	
maximum recommended continuous service temperature, °C	82–94	
coefficient of thermal expansion, cm/cm/°C $\times$ 10^{-5}		
−40°C	5.0	
−18°C	5.6	
5°C	6.5	
27°C	7.6	
38°C	8.3	
coefficient of thermal conductivity, kW/(m·K)	0.00344	
specific heat, 25°C, kJ/(kg·°C)[f]	0.452	

[a] Ref. 102.
[b] To convert MPa to psi, multiply by 145.
[c] Conditioned 48 h at 50°C.
[d] To convert J/cm^{2} to lbf/in., divide by 0.0175.
[e] Values change with thickness; the reported value is for 0.635 cm.
[f] To convert kJ to kcal, divide by 4.184.

Solution Polymers. Methacrylic solution polymers are usually characterized by their composition, solids content, viscosity, molecular weight, glass-transition temperature, and solvent type. The compositions of methacrylic polymers are most readily determined by physicochemical methods such as spectroscopy, pyrolytic gas–liquid chromatography, and refractive index measurements. The solids content is determined by dilution followed by solvent evaporation to constant weight. Solution viscosities are most conveniently determined with a Brookfield viscometer. Methods for estimating molecular weights by intrinsic viscosity are available (103).

Emulsion Polymers. Methacrylate dispersion polymers are classified as anionic, cationic, or nonionic, depending on the type of surfactants and functional monomers used in synthesis. Typical characterizations include composition, solids content, viscosity, pH, particle-size distribution, glass-transition temperature, and minimum film-forming temperature. Polymer compositions are most readily determined by methods such as spectroscopy, pyrolytic gas–liquid chromatography, and refractive index measurements. Typically, the solids content of dispersions are determined by gravimetric procedures, viscosities with a Brookfield viscometer, and glass-transition temperatures by differential scanning calorimetry. Minimum film-forming temperatures are determined by casting a film on a variable temperature bar and observing the minimum temperature at which a continuous film is obtained. Traditionally, particle-size measurements are performed by light scattering or electron microscope techniques. Instrumentation based on photon correlation spectroscopy is well established (104).

Additionally, mechanical (primarily shear), freeze–thaw, and thermal stability; the tendency to form sediment on long-term standing; and compatibility with other dispersions, salts, surfactants, and pigments of acrylic dispersions are often evaluated. Details on the determination of the properties of emulsion polymers are available (60).

Suspension Polymers. Methacrylate suspension polymers are characterized by their composition and particle-size distribution. Screen analysis is the most common method for determining particle size. Melt-flow characteristics under various conditions of heat and pressure are important for polymers intended for extrusion or injection molding applications. Suspension polymers prepared as ion-exchange resins are characterized by their ion-exchange capacity, density (apparent and wet), solvent swelling, moisture holding capacity, porosity, and salt-splitting characteristics (105).

Health and Safety Factors

In general, methacrylate polymers are considered nontoxic. In fact, various methacrylate polymers are used in food packaging (qv) and handling, in dentures and dental fillings (see DENTAL MATERIALS), and as medicine dispensers and contact lenses. However, care must be exercised because additives or residual monomers present in various types of polymers can display toxicity. For example, some acrylic latex dispersions can be mild skin or eye irritants. This toxicity is usually ascribed to surfactants in the latex and not to the polymer itself.

During manufacture, considerable care is exercised to reduce the potential for violent polymerizations, and to reduce exposure to flammable and potentially toxic monomers and solvents. Recent environmental legislation governing air quality has resulted in completely closed kettle processes for most polymerizations.

Dust explosions ignited by static discharge are a recognized hazard encountered in the handling of poly(methyl methacrylate) powders or in the fabrication of poly(methyl methacrylate) plastic sheet. Methacrylic molding powders and plastic sheet are treated as flammable materials. Most local building codes allow the use of poly(methyl methacrylate) in residential and commercial buildings, with certain restrictions designed to reduce its flammability hazard. Methacrylic solution polymers are treated as flammable mixtures; latex polymers are nonflammable.

Uses

The principal U.S. market for methacrylate resins is for glazing and skylights. Other significant markets include consumer products, transportation signs and lighting fixtures, plumbing (spas, tubs, showers, sinks, etc), and panels and siding (106,107).

Glazing. The uses for polymethacrylates as a glazing, lighting, or decorative material are varied and depend on the polymer's unique combination of light transmittance, light weight, dimensional stability, and formability, and its weather, chemical, impact, and bullet resistance; eg, polymethacrylates are used in aircraft fuselage windows (108), bank teller windows, police cars, panels around hockey rinks, storm doors, bath and shower enclosures, sliding partitions, and showcases. In architecture, the polymethacrylates are used for domes over stadia, pools, tennis courts, glazing, archways between buildings, curved windows, and geodesic domes. Because of their impact resistance, polymethacrylates are used in the maintenance field for the glazing of windows in schools, housing authority buildings, and factories. Colored and clear polymethacrylate sheets are used in decorative applications, eg, window mosaics, side glazing, color coordinated structures, and for solar control in sunscreens (109,110). Sheet and molding pellets have been used for lighting, as a sunscreen, and in signs (111–115).

Medicine. The polymethacrylates have been used for many years in the manufacture of dentures, teeth, denture bases, and filling materials (116,117) (see DENTAL MATERIALS). In the orthodontics market, methacrylates have found acceptance as sealants, or pit and fissure resin sealants which are painted over teeth and act as a barrier to tooth decay. The dimensional behavior of curing bone-cement masses has been reported (118), as has the characterization of the microstructure of a cold-cured acrylic resin (119). Polymethacrylates are used to prepare both soft and hard contact lenses (120,121). Hydrogels based on 2-hydroxyethyl methacrylate are used in soft contact lenses and other biomedical applications (122,123) (see CONTACT LENSES).

Optics. Good optical properties and low thermal resistance make poly(methyl methacrylate) polymers well suited for use as plastic optical fibers. The manufacturing methods and optical properties of the fibers have been reviewed (124) (see FIBER OPTICS). Methods for the preparation of Fresnel lenses and a Fresnel lens film have been reported (125,126). Compositions and methods for the industrial production of cast plastic eyeglass lenses are available (127).

Oil Additives. Long-chain polymethacrylates are used as additives to improve the performance of internal combustion engine lubricating oils and hydraulic fluids (128) (see HYDRAULIC FLUIDS; LUBRICATION AND LUBRICANTS). Long-chain polymethacrylates add little viscosity to oil when it is cold, but increase the viscosity of the oil as temperature is increased. The proper balance of composition and molecular weight of the polymer allows oils of controlled and constant properties to be formulated (129). In addition to improving the viscosity index, sludge dispersancy and antioxidant qualities can be built into the polymer (130) (see ANTIOXIDANTS; ANTIOZONANTS). Polyethylene and polypropylene grafts onto linear, long-chain polymethacrylates, eg, N,N-dialkylaminoalkyl methacrylate copolymers, are claimed to function as a multipurpose lube-oil additive having viscosity index improving properties, pour point depressing properties, and detergent dispersing properties (131).

Other. A significant, high value use for methacrylate resins is for the preparation of synthetic marble sanitary fixtures and thermoformed bathtubs (132,133). The use of opaque and clear methacrylate sheet for the construction of recreational vehicles has been reported (134). Polymethacrylates also are used as electrical insulators. Thermoplastic methacrylate resins are used in lacquer coatings for plastics, in printing inks (qv), as heat seal lacquers for packaging, as screen printing media for decorating procelain, in road marking paints, and to protect structures and buildings from weathering and acid rain (135). Other copolymer uses are discussed in the *Encyclopedia* article ACRYLIC ESTER POLYMERS.

BIBLIOGRAPHY

"Methacrylic Polymers" in *ECT* 3rd ed., Vol. 15, pp. 377–398, by B. B. Kine and R. W. Novak, Rohm and Haas Co.

1. E. H. Riddle, *Monomeric Acrylic Esters*, Reinhold Publishing Corp., New York, 1954.
2. M. Salkind, E. H. Riddle, and R. W. Keefer, *Ind. Eng. Chem.* **51**, 1232, 132B (1959).
3. *Chem. Mkt. Rptr.* **239**(3), 42 (1991).
4. J. W. C. Crawford, *J. Soc. Chem. Ind. London* **68**, 201 (July 1949).
5. D. W. VanKrevelen, *Properties of Polymers*, Elsevier Publishing Co., Amsterdam, the Netherlands, 1976.
6. H. B. Burrell, *Off. Dig. Fed. Soc. Paint Technol* **34**, 131 (Feb. 1962).
7. S. Krause, J. J. Gormley, N. Roman, J. A. Shetter, and W. H. Watanabe, *J. Polym. Sci. Part A* **3**, 3573 (1965).
8. W. A. Lee and G. J. Knight, *Br. Polym. J.* **2**, 73 (1970).
9. R. H. Wiley and G. M. Braver, *J. Polym. Sci.* **3**, 647 (1948).
10. Ref. 9, p. 455.
11. D. H. Klein, *J. Paint Technol.* **42**, 335 (1970).
12. O. G. Lewis, *Physical Constants of Linear Homopolymers*, Springer-Verlag, New York, 1968.
13. J. Brandrup and E. H. Immergut, *Polymer Handbook*, 3rd ed., Wiley-Interscience, New York, 1989.

14. J. A. Shetter, *Polym. Lett.* **1**, 209 (1963).
15. T. G. Fox, Jr. *Bull. Am. Phys. Soc.* **1**, 123 (1956).
16. L. E. Nielsen, *Mechanical Properties of Polymers and Composites*, Vol. 1, Marcel Dekker, New York, 1974.
17. J. R. Martin, J. F. Johnson, and A. R. Cooper, *J. Macromol. Sci. Rev. Macromol. Chem.* **8**, 57 (1972).
18. W. H. Brendley, Jr., *Paint Varn. Prod.* **63**, 23 (July 1973).
19. A. S. Craemer, *Kunststoffe* **30**, 337 (1940).
20. *Plexiglas Molding Pellets, PL-926a*, Rohm and Haas Co., Philadelphia, Pa.
21. *Plexiglas Design and Fabrication Data, Plexiglas Cast Sheet for Lighting, PL-927a*, Rohm and Haas Co., Philadelphia, Pa.
22. *Plexiglas Design and Fabrication Data, PL-53i*, Rohm and Haas Co., Philadelphia, Pa.
23. *Optics, PL-897B*, Rohm and Haas Co., Philadelphia, Pa.
24. F. J. Glavis, *J. Polym. Sci.* **36**, 547 (1959).
25. *Plexiglas Acrylic Plastic Molding Powder, PL-866*, Rohm and Haas Co., Philadelphia, Pa.
26. R. Simha, "Degradation of Polymers," in *Polymerization and Polycondensation Processes, No. 34, Advances in Chemistry Series*, American Chemical Society, Washington, D.C., 1962, p. 157.
27. *Acrylic and Methacrylic Monomers—Typical Properties and Specifications, CM-16*, Rohm and Haas Co., Philadelphia, Pa.
28. *Storage and Handling of Acrylic and Methacrylic Esters and Acids, Bulletin 84C*, Rohm and Haas Co., Philadelphia, Pa.
29. L. S. Luskin, in E. C. Leonard, ed., *High Polymers, Vinyl and Diene Monomers*, Vol. 24, Wiley-Interscience, New York, 1970, part 1.
30. L. S. Luskin, in F. D. Snell and C. L. Hilton, eds., *Encyclopedia of Industrial Chemical Analysis*, Vol. 4, Wiley-Interscience, New York, 1967, p. 181.
31. R. V. Porcelli and B. Juran, *Hydrocarbon Proc. Int. Ed.* **65**(3), 37–43 (1986).
32. F. E. Clayton and G. D. Clayton, eds., *Patty's Industrial Hygiene and Toxicology*, 4th ed., Wiley-Interscience, New York, 1991.
33. Technical data, The Methacrylate Producers Assoc., Washington, D.C.
34. *Preparation, Properties and Uses of Acrylic Polymers, CM-19*, Rohm and Haas Co., Philadelphia, Pa.
35. T. G. Fox, Jr. and R. Loshock, *J. Am. Chem. Soc.* **75**, 3544 (1953).
36. D. Pramanick and co-workers, *Hungarian J. Ind. Chem.* **12**, 1 (1984).
37. P. G. Griffiths, E. Rizzardo, and D. H. Solomon, *J. Macromol. Sci. Chem.* **17**, 45 (1982).
38. P. G. Griffiths, E. Rizzardo, and D. H. Solomon, *Tetrahedron Lett.* **23**, 1309 (1982).
39. G. V. Schulz and G. Henrici, *Makromol. Chem.* **18/19**, 473 (1956).
40. F. R. Mayo and A. A. Miller, *J. Am. Chem. Soc.* **80**, 2493 (1956).
41. M. M. Mogilevich, *Russ. Chem. Rev.* **48**, 199 (1979).
42. J. Shen and co-workers, *Makromol. Chem.* **192**, 2669 (1991).
43. L. S. Luskin, J. A. Sawyer, and E. H. Riddle, in W. M. Smith, ed., *Polymer and Manufacturing and Processing*, Reinhold Publishing Corp., New York, 1964.
44. E. H. Riddle and P. A. Horrigan, in P. H. Groggins, ed., *Unit Processes in Organic Synthesis*, 5th ed., McGraw-Hill, Inc., New York, 1958.
45. J. O. Beattie, *Mod. Plast.* **33**, 109 (1956).
46. U.S. Pat. 3,113,114 (Dec. 3, 1963), R. A. Maginn (to E. I. du Pont de Nemours & Co., Inc.).
47. U.S. Pat. 3,382,209 (May 7, 1968), W. G. Deschert (to American Cyanamid Co.).
48. U.S. Pat. 3,376,371 (Apr. 2, 1968), C. J. Opel (to Swedlow, Inc.).

49. *Forming Plexiglas Sheet, PL-4k*, Rohm and Haas Co., Philadelphia, Pa.
50. *Plexiglas Acrylic Sheet, PL-80M*, Rohm and Haas Co., Philadelphia, Pa.
51. *Plexiglas Molding Manual, PL-710*, Rohm and Haas Co., Philadelphia, Pa.
52. *Mod. Plast.* **47**, 6 (June 1970).
53. U.S. Pat. 3,706,825 (Dec. 19, 1972), N. L. Hall (to Du Pont).
54. *Plast. Ind. News* **32**, 11 (Nov. 1986).
55. U.S. Pat. 3,847,865 (Nov. 1984), R. B. Duggins (to Du Pont).
56. E. L. Madruga and co-workers, *J. Appl. Polym. Sci.* **41**, 1133 (1990).
57. I. Czajlik and T. Foldes-Berezanich, *Eur. Polym. J.* **17**, 131 (1981).
58. L. Maduga, *An. Quim.* **65**, 993 (1969).
59. E. P. Bonsall, *Trans. Faraday Soc.* **49**, 686 (1953).
60. *Emulsion Polymerization of Acrylic Monomers, CM-104*, Rohm and Haas Co., Philadelphia, Pa.
61. "Chain Transfer" in J. I. Kroschwitz, ed., *Encyclopedia of Polymer Science and Engineering*, 2nd ed., Vol. 3, John Wiley & Sons, Inc., New York, 1985, pp. 288–290.
62. *The Manufacture of Acrylic Polymers, CM-107*, Rohm and Haas Co., Philadelphia, Pa.
63. Brit. Pat. 2,097,409A (Apr. 6, 1982), P. Bitler and co-workers (to ICI Co.).
64. *Bulk Storage and Handling of Acryloid Coating Resins, C-186*, Rohm and Haas Co., Philadelphia, Pa.
65. U.S. Pat. 3,458,466 (July 29, 1969), W. J. Lee (to The Dow Chemical Co.).
66. U.S. Pat. 3,344,100 (Sept. 26, 1967), F. J. Donat and co-workers (to B. F. Goodrich Co.).
67. E. Giannetti and co-workers, *J. Poly. Sci. Poly. Chem.* **24**, 2517 (1986).
68. G. S. Whitby and co-workers, *J. Polym. Sci.* **16**, 549 (1955).
69. W. Cooper and co-workers, *J. Poly. Sci.* **34**, 651 (1959).
70. K. E. J. Barrett, ed., *Dispersion Polymerizations in Organic Media*, John Wiley & Sons, Inc., New York, 1975.
71. M. A. Winnik and co-workers, *Makromol. Chem., Macromol. Symp.* **10–11**, 483 (1987).
72. Brit. Pat. 1,002,493 (1965), C. J. Schmidle (to Rohm and Haas Co.).
73. Brit. Pat. 934,038 (1963), C. J. Schmidle and G. L. Brown (to Rohm and Haas Co.).
74. U.S. Pat. 3,514,500 (May 26, 1970), D. W. J. Osmond and co-workers (to ICI Co.).
75. T. O. Purcell, Jr., in N. M. Bikales, ed., *Encyclopedia of Polymer Science and Technology*, Suppl. 1, Wiley-Interscience, New York, 1976, pp. 319–325.
76. R. J. Ceresa, *Block and Graft Copolymers*, Vol. 1, Butterworth, Inc., Washington, D.C., 1962.
77. H. A. J. Battaerd and G. W. Tregear, *Polymer Reviews, Graft Copolymers*, Vol. 16, Wiley-Interscience, New York, 1967.
78. W. Kobryner, *J. Polym. Sci.* **34**, 381 (1959).
79. P. W. Allen and co-workers, *J. Polym. Sci.* **36**, 55 (1959).
80. W. Cooper and co-workers, *J. Appl. Polym. Sci.* **1**, 329 (1959).
81. W. Cooper and co-workers, *J. Poly. Sci.* **34**, 651 (1959).
82. D. J. Angier and co-workers, *J. Polym. Sci.* **20**, 235 (1956).
83. T. Toda, *J. Polym. Sci.* **58**, 411 (1962).
84. U.S. Pat. 3,030,319 (Apr. 17, 1962), K. Kaizirman and G. Mino (to American Cyanamid Co.).
85. S. Lenka, *J. Polym. Sci., Polym. Lett. Ed.* **21**, 281 (1983).
86. A. Chapiro, *J. Polym. Sci.* **29**, 321 (1958).
87. Brit. Pat. 788,175 (Dec. 3, 1957), R. G. Norrish (to Distillers Co., Ltd.).
88. S. P. Rao and co-workers, *J. Polym. Sci. Part A-1* **5**, 2681 (1967).
89. R. K. Graham and co-workers, *J. Polym. Sci.* **38**, 417 (1959).

90. T. Nishikubo, T. Ichijyo, and T. Takoha, *J. Appl. Polym. Sci.* **20**, 1133 (1976).
91. N. G. Gaylord and co-workers, *J. Polym. Sci. Polym., Chem. Ed.* **13**, 467 (1975).
92. J. Trekoval and D. Lim, *J. Polym. Sci. Part C* **4**, 333 (1964).
93. D. Braun and co-workers, *Makromol. Chem.* **51**, 15 (1962).
94. T. Tsuruta and co-workers, *J. Macromol. Chem.* **1**, 31 (1966).
95. T. E. Long and co-workers, *J. Poly. Sci. Polym. Chem.* **27**, 4001 (1989).
96. K. Hatada and co-workers, *Prog. Polym. Sci.* **13**, 189 (1988).
97. A. H. E. Mueller, *American Chemical Society Symposium Series No. 166*, ACS, Washington, D.C., 1981, p. 441.
98. M. Van Beylen and co-workers, *Adv. Polym. Sci.* **86**, 87 (1988).
99. H. Yuki and co-workers, in O. Vogl and J. Furukawan, eds., *Ionic Polymerization*, Marcel Dekker, Inc., New York, 1976.
100. O. W. Webster, *Makromol. Chem., Macromol. Symp.* **70/71**, 75 (1993).
101. W. J. Brittain, *Rubb. Chem. Technol.* **65**(3), 580 (1992).
102. *Plexiglas Acrylic Sheet, PL-783a*, Rohm and Haas Co., Philadelphia, Pa.
103. P. W. Allen, *Technique of Polymer Characterization*, Butterworths Scientific Publications, London, 1959.
104. H. G. Barth and S.-T. Sun, *Anal. Chem.* **65**, 55R (1993).
105. R. Kunin, *Elements of Ion Exchange*, R. E. Krieger Publishing Co., Huntington, N.Y., 1971.
106. A. M. Dave, *Pop. Plast.* **31**(4), 27–30 (1986).
107. R. R. Jobanputra and R. C. Mathur, *Pop. Plast.* **33**(6), 31–32 (1988).
108. R. Hermann, *Mater. Des.* **9**(6), 339 (1988).
109. H. J. Gambino, Jr., *Security You Can See Through, PL-1228*, Rohm and Haas Co., Philadelphia, Pa.
110. *Mod. Plast.* (5), 52 (May 1975).
111. *Plexiglas Molding Pellets, PL-926a*, Rohm and Haas Co., Philadelphia, Pa.
112. P. W. Allen and co-workers, *J. Polym. Sci.* **36**, 55 (1959).
113. *Sun Screen Innovations with Plexiglas, PL-935*, Rohm and Haas Co., Philadelphia, Pa.
114. *Transparent Plexiglas Solar Control Series*, Rohm and Haas Co., Philadelphia, Pa.
115. *Plexiglas DR Sign Manual, PL-1097e*, Rohm and Haas Co., Philadelphia, Pa.
116. W. D. Cook and co-workers, *Biomaterials* **6**(6), 362 (1985).
117. J. M. Antonucci, *ACS Polymer Science and Technology Series* **34**, 277 (1986).
118. J. R. DeWijn, F. C. M. Driessenj, and T. J. J. H. Sloff, *J. Biomed. Mater. Res.* **9**(4), 99 (1975).
119. R. P. Kusy and D. T. Turner, *J. Dent. Res.* **53**, 948 (1974).
120. W. Timmer, *Chem. Tech.* **9**, 1975 (Mar. 1979).
121. B. J. Tighe, *Br. Polym. J.* **18**(1), 8–13 (1986).
122. J. P. Montheard and co-workers, *J. Macromol. Sci., Rev. Macromol. Chem. Phys.* **C32**(1), 1 (1992).
123. E. J. Mack and co-workers, in N. A. Peppas, ed., *Hydrogels in Medicine & Pharmacy*, Vol. 2, CRC, Boca Raton, Fla., 1987, p. 65.
124. C. Emslie, *J. Mater. Sci.* **23**(7), 2281 (1988).
125. I. Kaetsu, K. Yoshida, and O. Okubo, *J. Appl. Polym. Sci.* **24**, 1515 (1979).
126. H. Okubo, K. Yoshida, and I. Kaetsu, *Int. J. Appl. Radiat. Isot.* **30**, 209 (1979).
127. U.S. Pat. 4,146,696 (Mar. 17, 1979), H. M. Bond, D. L. Torgersen, and C. E. Ring (to Buckee-Mears Co.).
128. R. J. Kopko and R. L. Stambaugh, *SAE Paper 750693, Fuels and Lubricants Meetings*, Houston, Tex., June 3–5, 1975, Society of Automotive Engineers, Inc., Warrendale, Pa.
129. U.S. Pat. 2,091,627 (Aug. 31, 1937), H. A. Bruson (to Rohm and Haas Co.).

Table 1. Mica Group Minerals

Mineral	CAS Registry Number	Molecular formula
muscovite	[1318-94-1]	$K_2Al_4(Al_2Si_6O_{20})(OH)_4$
phlogopite	[12257-58-0]	$K_2Mg_6(Al_2Si_6O_{20})(OH,F)_4$
biotite	[1302-27-8]	$K_2(Mg,Fe)_6(Al_2Si_6O_{20})(OH)_4$
lepidolite	[1317-64-2]	$K_2Li_3Al_3(Al_2Si_6O_{20})(OH,F)_4$
rocoelite	[12271-44-2]	$K_2V_4(Al_2Si_6O_{20})(OH)_4$
fuchsite	[12198-09-3]	$K_2Cr_4(Al_2Si_6O_{20})(OH)_4$
paragonite	[12026-53-8]	$Na_2Al_4(Al_2Si_6O_{20})(OH)_4$

Mineralogy

Mica has been classified into three groups: (1) the mica group proper, (2) the clintonite or brittle micas group, and (3) the chlorite group (3). Supplementary to these are the vermiculites, which are hydrated compounds that result from the alteration of any one of the micas, but usually biotite. All minerals in these groups belong to the monoclinic crystal system, and all show plane angles of 60 and 120° on the basal section. The crystals usually form in hexagonal or rhombohedral-shaped scales, prisms, or plates.

The basic structural unit of mica is a layer composed of two silicon tetrahedral sheets with a central octahedral sheet (4). The tips of the un-joined tetrahedrons in each silica sheet point toward the center of the unit and are shared with the octahedral sheet in a single layer with suitable replacement of OH by O. Varying amounts of other minor elements, including calcium, magnesium, iron, chromium, and lithium, can also be present. The general formula describing the composition of micas is $X_2Y_{4-6}Z_8O_{20}(OH,F)_4$, where X is mainly K, Na, or Ca; Y is mainly Al, Mg, or Fe; and Z is mainly Si or Al (5). In many of the well-crystallized micas one-fourth of the silicons are replaced by aluminum ions (4). The unit layers extend indefinitely in the a and b dimension or are stacked in the c dimension. Potassium ions occur between unit layers fitting into the surface perforations. Adjacent layers are stacked in such a way that the potassium ion is equidistant from 12 oxygens, six in each layer.

Muscovite is dioctahedral, having a theoretical composition of 11.8% K_2O, 45.2% SiO_2, 38.5% Al_2O_3, and 4.5% H_2O (4). Two-thirds of the possible octahedral positions are filled, and the octahedral sheet is populated by aluminum only. Biotite-type micas are trioctahedral where the octahedral positions are completely filled mostly by Fe^{2+}, Mg^{2+}, and/or Fe^{3+}. Examples are biotite, $(OH)_4K_2(Si_6Al_2)(Mg,Fe)_6O_{20}$, having varying proportions of iron and magnesium, and phlogopite, $(OH)_4K_2(Si_6Al_2)Mg_6O_{20}$. It has been found that all micas have six possible polymorphic variations (6). The variations occur as a result of the number of silica–alumina–silica units per unit cell, and the manner in which the unit cells are stacked. Unit cells composed of 1, 2, 3, 6, and 24 silica–alumina–silica units are known, with stackings yielding monoclinic, rhombohedral, or triclinic forms.

Muscovite mica formed as a primary mineral in pegmatites and granodiorite differs in physical properties compared to muscovite mica formed by secondary alteration (mica schist) (Table 2). The main differences are in flexibility

Table 2. Physical Properties of Natural and Synthetic Micas[a,b]

Property	Natural			Synthetic fluorophlogopite
	Muscovite[c]	Phlogopite	Biotite	
density, g/cm^3	2.77–2.88	2.76–2.90	2.70–3.30	2.8
hardness[d]				
Mohs'	2.0–3.2	2.5–3.0	2.5–3.0	3.4
Shore	80–150	70–100		
optical axial				
angle, 2 V[e,f], deg	38–47	0–10	0–25	14.6 ± 0.5
refractive index, n_D				
α	1.552–1.570	1.541–1.632	1.541–1.579	1.522
β	1.582–1.607	1.598–1.606	1.574–1.638	1.549
γ	1.593–1.611	1.598–1.606	1.511–1.638	1.549
decomposition, °C	940–980	890–960	1100	1200
specific heat, 25°C, J/g[g]	0.049–0.05	0.049–0.05		0.046
thermal conductivity[h], W/(m·h·K)	0.6910	0.6910		
expansion coefficient °C^{-1} × 10^{-6}				
perpendicular				
at 20–100°C	15–25	1–1000		
100–300°C	15–25	200–20000		
300–600°C	16–36	10–3000		
parallel				
at 0–200°C	8.9	13–14.5		10–11.5
200–500°C	10–12	13–14.5		
strength, MPa[i]	225–296	255–296		310–358
water, wt %[j]	4.5	3.2	~5–6	0
melting point, °C	decomposes	decomposes	decomposes	1387 ± 3
dielectric				
constant	6.5–9.0[k]	5–6		6.5
strength,[k,l] V/μm	235–118	165–83		235–157
power factor, 25°C, %[m]				
1 MHz	0.01–0.02	0.3		0.02
resistivity, Ω·cm	10^{12}–10^{15}	10^{10}–10^{13}	10^{10}–10^{13}	10^{12}–10^{15}
modulus of elasticity, GPa[i]	172	172		
linear coefficient of expansion at 20–600°C, μm/°C	0.58–0.79	0.79–0.97		

[a]Ref. 1,5,7–9. [b]All micas have a vitreous luster; phlogopite luster can range from vitreous to submetallic. [c]Volume resistivity = 2 × 10^{13} to 1 × 10^{17}. [d]Mohs' hardness values may vary; Shore hardness number is derived from rebound height of standard steel ball when dropped on material from standard height. [e]Interior acute angle between optic axes of biaxial mineral. [f]Orientation of optical plane to plane of symmetry is parallel; muscovite is perpendicular. [g]To convert J to cal, divide by 4.184. [h]Perpendicular to cleavage. [i]Tensile strength; compression strength for muscovite and phlogopite is 221 MPa. To convert MPa to psi, multiply by 145. [j]Chemically combined. [k]Not specified in good quality mica. [l]Of material from 25–75 μm thick at 21°C. [m]Value for muscovite at 60 Hz is 0.08–0.09.

and ability to be delaminated. Primary muscovite is not as brittle and delaminates much easier than muscovite formed as a secondary mineral. Mineralogical properties of the principal natural micas are shown in Table 3. The make-up of muscovite, phlogopite, and biotite are as follows:

Oxides	Muscovite, wt %	Phlogopite, wt %	Biotite, wt %
SiO_2	46.5	40.0	37.0
Al_2O_3	34.0	17.0	18.0
K_2O	10.0	10.0	9.0
Na_2O	0.8	0.5	1.0
MgO	0.5	26.0	8.0
CaO	0.3		
Fe_2O_3	2.5	0.2	21.0
FeO	1.0	2.8	2.0
minor elements		0.5	1.0
H_2O	4.5	3.0	3.0

Table 3. Mineralogical Properties of Micas[a,b]

Properties	Muscovite	Biotite	Lepidolite	Phlogopite
specific gravity	2.76–3.0	2.7–3.1	2.8–3.3	2.8–2.9
luster	vitreous–pearly	splendent, sometimes submetallic	pearly	pearly
crystal symmetry	rhombic or hexagonal	pseudo-rhombohedral	hexagonal	hexagonal
colors	gray, brown, pale green, violet, yellow, dark olive-green, and ruby	green, black, and yellow	rose red, violet-gray, lilac, yellowish, grayish white, and white	yellow-white, gray to green, pearly, brown, black

[a]Refs. 10 and 11.
[b]Optical signs of micas are negative, crystal system is monoclinic, and the streak is colorless.

Mica Deposits

Sheet Mica. Pockets of mica crystals ranging in size from a few square centimeters to several square meters are found in pegmatite sills and dikes or granodiorite (alaskite) ore bodies. In order to be used industrially, manufacturers must be able to cut a 6 cm^2 pattern in the mica. "Books" of mica, ranging from 12.9 to 645 cm^2 or more, are cut from the crystals. The books can be punched into various shapes and split into thicknesses varying from 0.0031 to 0.010 cm (12). The highest quality micas may be used in aerospace computers, and those of lower quality find use as insulators in electrical appliances.

The pegmatic and granodiorite ore bodies are found as intrusions in surrounding country rock, including granites, gneiss, and schist. They range in color from pink to almost white, depending on which type of feldspar predominates, eg, microcline, orthoclase, or plagioclase. The granodiorite ore bodies are found as irregularly shaped masses and are white in color. The color of the mica varies from ruby to green to almost colorless or clear. Although the granodiorites are considered a coarse textured granite rock, the grain size of the individual minerals is not as coarse as the minerals associated with pegmatites.

Pegmatites range in size from 4-cm^2 wide stringers to dikes and sills several hundred meters wide and over 300 m long; depths vary from a few centimeters to hundreds of meters. Granodiorites are usually larger and less well defined. A general mineralogical composition of a granodiorite is plagioclase feldspar, 40%; microcline feldspar (the primary feldspar mineral), 20%; quartz, 30%; and mica, 10–20%. Garnet and biotite are found in small amounts. Mica crystals occur in size from a few centimeters to several meters. Other natural minerals found in pegmatites are tourmaline, beryl, rare earths, and uranium minerals.

Phlogopite mica is found in areas of metamorphosed sedimentary rocks into which pegmatite-rich granite rocks have intruded. Only a few deposits exist that contain books of phlogopite mica large enough to be mined economically. Phlogopite ore bodies may be classified into vein, pocket, and contact deposits. Vein deposits are narrow and enclosed in fine-to-medium grained pyroxenite. Contact deposits are the primary source of phlogopite sheet mica.

Sheet mica occurs in pockets within pegmatite. When exploring for sheet mica, test pits are sunk to determine the presence of pockets and quality of the mica. The size, shape, and attitude of the pegmatite is determined by stripping and trenching. These procedures are costly and problematic, but the quality can be determined to a sufficient degree (5).

Scrap and Flake Mica.   Flake mica is found either in hard rock pegmatites and granodiorites or the weathered reminants of these ore bodies. The flakes of mica range in size from "thumbnail" (1.5–2.0 cm) to −44 μm. Scrap mica generally refers to waste trimmings resulting from the preparation of sheet mica for punching and machining into electrical parts. Reconstituted mica is a paper produced by forming a mat of thin, well-delaminated flakes of scrap mica. This mat is usually impregnated with an organic binder, but is available unimpregnated (5,12).

The hard rock deposits are mined mainly for feldspar with mica and quartz being accessory minerals. These deposits are extensive, often covering hundreds of square meters and are recognized by the light-colored, granite-like appearance with shiny mica flakes being a prominent feature. The mica content of these deposits ranges from approximately 6–10 wt %.

The soft weathered granodiorite and pegmatites can vary in color from white to pink, depending on iron content and type of feldspar present. The mica content of these deposits ranges from 6–15% and varies in particle size from tiny (<44 μm) specks to thumbnail size. Large books of mica that weigh several hundred kilograms have been found in these deposits.

Mica schist deposits range in consistency from loosely consolidated soil to fairly hard ore and vary in color from brown to almost white. These ore bodies are noted for their micaceous appearance, which results from a 40–90% mica content. Mica sizes vary from <44 μm to ~1.5 cm.

Weathered and hard rock pegmatites, granodiorites, schist, and gneisses are evaluated for mica content by first core drilling and then extracting available mica by conventional vanning, magnetic, and flotation techniques. Drill cores are normally 3.8–7.6 cm dia and 15–31 m in length. Auger drilling may also be used in soft material. However, it is difficult to obtain a representative sample of a particular section when using this procedure because of the probability of intermixing foreign material with vein material.

Mining

Flake Mica. Flake mica is mined from weathered and hard rock pegmatites, granodiorite, and schist and gneiss by conventional open-pit methods. In soft, residual material, dozers, shovels, scrapers, and front-end loaders are used to mine the ore. Often kaolin, quartz, and feldspar are recovered along with the mica (see also CLAYS; SILICON COMPOUNDS).

Hard rock mining of these ore bodies requires drilling and blasting with ammonium nitrate and dynamite. After blasting, the ore is reduced in size with a drop ball and then loaded on trucks for transportation to the processing plant. Mica, quartz, and feldspar concentrates are separated, recovered, and sold from the hard rock ore.

Sheet Mica. Sheet mica is mined by both underground and open-pit mining procedures. Underground mining is accomplished by driving a shaft, formed with tungsten carbide-tipped air drills, hoists, and explosives (see EXPLOSIVES AND PROPELLANTS; TOOL MATERIALS), through the pegmatite at a suitable angle to the strike and dip. Cross-cuts and raises are developed to follow promising exposures of mica. Small charges of 40–60% strength dynamite are placed around a pocket of mica to shake the mica loose from the host rock without fracturing the books. After blasting, the mica is placed in boxes or bags for transporting to the trimming shed where it is graded, split, and cut to various specified sizes for sale.

Sheet mica is no longer mined in the United States because of high cost, small market, and high capital risk. Most sheet mica is mined in India.

Beneficiation Processes

Flake or Scrap Mica. In the early to mid-1900s, flake or scrap mica was mainly processed by a jigging procedure which consists of hydraulically washing a pile of bulldozed ore across a series of roll crushers and Trommel screens gaped at different size openings (Fig. 1). The ore first passes across a Trommel screen of heavy wire construction with an opening of 2.5–5.0 cm. Large flakes of mica, feldspar, and quartz are discharged from this screen as a mica concentrate. Undersized particles are discarded as waste throughout the procedure. Material that passes through the screen, which still contains considerable mica, flows by gravity across a second Trommel screen having a 0.3-cm opening. This screen concentrates underflow from the first screen by removing most of the water. The

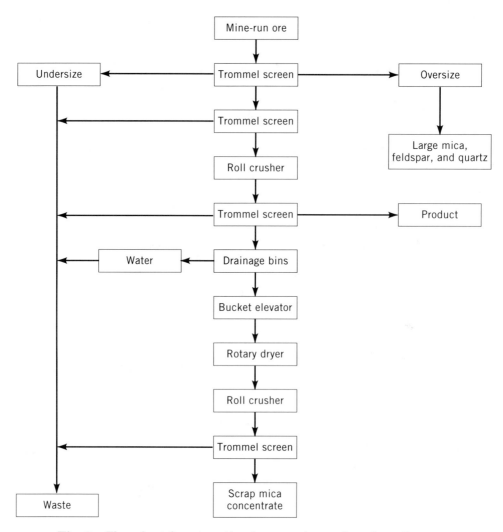

Fig. 1. Flow sheet for conventional scrap mica washer plant. See text.

+0.3-cm screen fraction then flows through a roll crusher and across another Trommel screen having an 0.3-cm opening. The concentrate discharged from this screen is considered product. The minus (undersize) screen fraction continues to additional Trommel screens and roll crushers (normally four or five) with smaller size openings. The final mica concentrates flow together into a central drainage bin of either metal or wooden construction having drainage holes to allow water to escape.

Mica concentrate is transported by either a front-end bucket loader or screw conveyor to a rotary dryer. After drying, the mica may be crushed again across a role crusher, then screened before being transported by bucket elevator or screw conveyor to metal storage bins. The dried mica concentrate is either sold without further processing or hammer milled, screened on various sieves with various size openings depending on the product being made, and bagged in 110-kg (50-lb) multiwall paper bags for shipment. Other types of jigging operations are similar.

The grade of mica produced by jigging is very poor, usually about 75% concentrate, and recovery of available mica low (50%). Specifications on mica have become more stringent, therefore a more efficient processing method has been devised that provides higher quality mica, as well as more efficient recovery. Mica companies have begun to use rod mills, Humphrey spirals, and froth flotation (qv) to concentrate the ore. Although there is an increase in efficiency, the additional processing increases the cost. To offset the higher cost, by-products are extracted, processed, and sold (see also MINERAL RECOVERY AND PROCESSING).

Because of improved mica processing operations, low cost earthen waste impoundment ponds have been built to store solid waste and thereby provide for a relatively cheap means of meeting new federal and state environmental laws.

There are several methods of preparing ore for beneficiation after it arrives at the plant site (Fig. 2). (1) The ore is transferred to rod mills, 1.5 m dia × 2.5–3 m in length, hydraulically from a bin at 40% solids. The rod mill crushes the ore to approximately 0.833 mm (−20 mesh). (2) The ore is slurried after the initial crushing step (jaw and roll crusher) to about 20–25% solids and pumped to either a hydrosizer or cyclones for desliming, removing 0.147 mm (−100 mesh) material. The roughly 1.168–0.833 mm (14–20 mesh) × 0.147 mm (100 mesh) deslimed feed is then either pumped directly or fed from a stockpile to Humphrey spirals where 1.168–0.833 × 0.47 mm (14–20 × 35 mesh) mica is removed. The tailings from the spirals are dewatered on screw classifiers, then transferred to a rod mill. (3) The Humphrey spirals are bypassed and the entire feed fed from the crushed stockpile. (4) The ore is dumped directly into a bin from which a rod mill is fed directly with a screw feeder. Water is added to make ~40% solids slurry feed for the mill.

Sodium silicate (41°Bé, 1:3.22 ratio $Na_2O:SiO_2$) is added in the milling operation to disperse the slime, mostly kaolin. Dispersion also aids the grinding process. The rod mill serves to grind the ore to 0.833 mm (−20 mesh) or to the point where mica, quartz, feldspar, and iron minerals are liberated. Cyclones, or rake, hydraulic, or other types of classifiers, are used after grinding to produce coarse and fine mica fractions that are treated separately.

Coarse mica is separated from coarse quartz on spirals. The quartz tailings from the spiraling process are often upgraded to a commercially saleable product by scrubbing, desliming, and froth flotation. In some cases the quartz may lend itself to being processed for high purity quartz, used in making crucibles in which quartz crystals are grown for computer chips (see ELECTRONIC MATERIALS). Other quartz grades may be used in the manufacture of glass (qv) or porcelain products (see ENAMELS, PORCELAIN OR VITREOUS).

The coarse mica concentrate is either respiraled to produce a product with a grade suitable for use by wet grinding mica producers, or floated to further upgrade it for specialty high grade dry ground mica products.

The fine mica fraction is deslimed over 0.875–0.147-mm (80–100-mesh) Trommel screens or hydrocylcones, or is separated with hydrosizers. The deslimed pulp (≤0.589 mm (−28 mesh)) of mica, feldspar, and quartz is then fed to a froth flotation circuit where these materials are separated from each other either by floating in an acid circuit with rosin amine and sulfuric acid

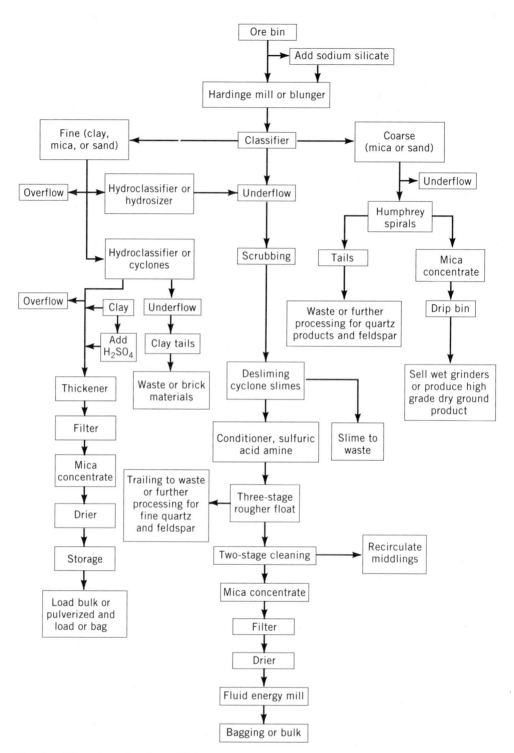

Fig. 2. Flow sheet for the acid circuit processing and recovery of mica from weathered granodiorite ore. An alkaline–cationic circuit may be used by inserting a second conditioner containing lignin sulfonate, adjusting the pH to 8.0, and adding NaOH and DRL (distilled tall oil) fatty acid to the first conditioner.

(2.5–4.0 pH), or an alkaline circuit (7.5–9.0 pH) with tall oil amine, goulac, rosin amine acetate, and caustic soda (see Fig. 2).

The floated mica concentrate is dewatered. After dewatering (qv), the mica is either dried in a fluid-bed rotary drier, flash dried in a fluid energy mill, or sold "drip-dry" to other mica grinders. The dry mica is then ground and screened to a size gradation dictated by the customer.

The slime, consisting of kaolin, fine quartz, and feldspar, is sometimes used as is after being dewatered. This material may be used in the manufacture of light-colored brick or may be further processed to produce a high grade ceramic kaolin used in the manufacture of dinnerware, electrical porcelain, or sanitary-ware (see CERAMICS). Flocs of kaolin may be sold in bulk from the drier or pulverized and sold in a powdered form.

The remaining tailings left over from the clay fractionation step is either flocculated with alum, high molecular weight polymers, or a weak (pH 3.0) solution of sulfuric acid, and stored in settling ponds as waste, or may be filtered and sold to the brick industry as a coating material. It also may be dried and sold as a filler in plastics and textured paint (qv).

Flake mica is also produced as a by-product from processing feldspar ore (hard granodiorite). This process consists of mining hard rock granodiorite, crushing the large pieces of ore developed from the blasting operation, and producing a flotation feed of ≤0.589 mm (−28 mesh). The mica flotation step is followed by a feldspar flotation step. The quartz by-product resulting from the feldspar float is further processed with additional flotation steps to produce either a glass-grade (flat or container glass) product or a feedstock for the production of high purity quartz. Magnetic separation, leaching with a combination of hydrofluoric acid and hydrochloric acid, and chlorination are additional steps necessary to produce the high purity quartz grade.

Flake mica is also produced from mica schist which normally contains from 30–60% recrystallized muscovite mica along with quartz and iron minerals. The quartz is usually not suitable for glass sand or high purity material, however.

Sheet Mica. The preparation of sheet mica for feedstock for various punching and machining operations involves cobbing mica blocks or books to remove dirt, rock, and defective mica, trimming and splitting into sizes and thicknesses suitable for punching and milling to desired shapes, and grading the finished mica sheets according to size and quality. Classifications such as hardness (qv) and visual qualities may be found in Reference 13. The waste mica resulting from cobbing and trimming (scrap mica) is often mixed with flake mica for processing by dry or wet ground procedures.

The grade determines whether the mica can be used in high technology electronic instruments, eg, computer-aided tomography (CAT) scan (see MEDICAL IMAGING TECHNOLOGY), or in low technology devices, eg, a toaster. Many types of insulators, as well as the base for electronic circuits, are formed from the high quality sheets of mica by a punch pressing operation.

Mica splittings are processed from lower quality blocks and from sheets too thin for blocks and unsatisfactory for producing film. The splittings are packed for sale in three ways: book form, which are laminae split to the desired thickness from the same book of mica, then dusted with mica dust, and restacked in book form; pan packed, in which splitting layers are placed evenly in a pan, and each layer separated by a thin sheet of paper then pressed together; and loose packed, in which splittings are sized with screens then padded loosely in a wooden box for shipment.

The splittings may be used by mica processors to reconstitute large mica sheets, produced by overlapping irregularly shaped pieces to the desired thickness (built-up mica). The sheets are then bound with organic and inorganic binders and heat treated under pressure to complete the binding process. The sheets may also be bound to fiber glass and asbestos (qv) by a similar process. After forming, the built-up mica sheets are cut and punched to the desired shape and design. Built-up mica and reconstituted mica are used for products that do not require the special properties only obtained from high priced sheet mica.

Procedures for Production

The general pieces of equipment used in grinding flake mica or mica concentrate into saleable mica products are hammer mills of various types, fluid energy mills, Chaser or Muller mills for wet grinding, and Raymond or Williams high side roller mills. Another method is being developed, called a Duncan mill (J. M. Huber, Inc.), that is similar in many respects to an attrition mill. All of these mills are used in conjunction with sieves, and all but some types of hammer mills incorporate air classifiers as a part of the circuit.

Ground Mica. This constitutes by far the largest commercial use for mica. It is largely produced from the beneficiation of weathered and unweathered pegmatites, granodiorite, and metamorphic schists, although some higher grades are produced from trimmings of sheet mica or Type A (low quality) mica blocks.

Wet ground mica products account for approximately 15% of the total mica market. These products are produced by grinding either scrap or flake mica or both in a Muller or Chaser mill which consists of a 4.57-m diameter tub containing two 1.22-m diameter steel spiked wheels or rollers. A 1.13 t charge of mica is placed in the tub along with enough water to represent 25–35% of the batch. The necessary water content for good delamination and grinding varies for micas from different deposits. After charging the mill, rollers rotate at 30 rpm through the mica pulp for six to eight hours to cause the mica flakes to slide across each other, and thereby cause delamination by friction while producing a grinding action that reduces particle size.

When grinding is complete, the mica pulp is washed from the Mullers into settling tanks. After allowing the mica to settle for approximately 24 hours, the settled portion is pumped from the settling tanks to a holding tank; the unsettled portion is decanted into additional settling tanks. The settlings are pumped to a leaf-type filter press (700 kPa (100 psig)) or Bird centrifuge where partial dewatering takes place. The filtered cakes (62% solids) are dried in a stream tube at ~100°C, then passed through a hammer or attrition mill to reduce the pieces to powder. The powdered mica is screened on sieves containing 0.088–0.074 mm (170–200 mesh) stainless steel screen cloth. Oversized mica from the sieves reports back to the raw feed and is reground.

The Muller mill process has not changed significantly in almost a century (7,14). Although many newer processes have been developed and patented, none has been able to match the quality (slip and sheen) of mica produced by the

Muller process. The Duncan mill was developed to process quality wet ground mica. After drying and screening, the wet ground mica products are packaged in 27.72-kg bags and palletized in 1.13 t lots for shipment.

Exact sizing of mica products coupled with surface treatment procedures have led to a greater use for wet ground mica in plastic compositions, particularly automobile bodies. These quality products demand a high dollar value.

Dry Ground Mica. Dry ground mica concentrate is processed into usable products by several different grinding methods. Relatively coarse particle sizes (1.651–0.147 mm (10–100 mesh)) are used in oil-well drilling muds, some types of welding rod coatings, asphalt (qv), roofing shingles, and some other types of fillers (qv). These products are ground on a hammer mill in closed circuit with a sieve. Roofing micas produced from mica schist are often ground in a Raymond or Williams high side roll mill in closed circuit with an air classifier and a sieve. The finer particle-size micas ≤0.147–0.044 mm (−100 to −325 mesh), used mainly in textured paints and joint cement compounds, are ground on several types of fluid energy mills, but generally a mill of the Majacs type. The finest dry ground mica product is ground with superheated steam (Micronized, KMG Minerals).

In the fluid energy mill wet mica concentrate (8–35% moisture) is fed to the mill with a screw conveyor attached to a feed hopper. The mica is then conveyed to an air classifier operating at 70–300 rpm, depending on desired particle size, with high pressure superheated air. The high pressure air is furnished by a compressor operating at ca 690 kPa (100 psig). It is superheated in a furnace at 371–427°C. The superheated air is introduced into the mill through opposing ventures, called guns, which direct oversize mica particles, rejected by the air classifier, to impinge upon themselves at high velocity, causing a grinding action. The ground mica is continuously removed from the grinding chamber in the high pressure air stream which circulates in a closed circuit through the mill system. Hot air is also continuously pumped from the furnace through a strip of air fan to the air classifier. This air flash dries the mica as it enters the classifier.

The fines are classified out of the feed and pass through the air classifier into a baghouse, from which they are fed onto sieves, containing 165 μm–0.124 mm (105–120 mesh) stainless steel wire cloth. Oversize mica from the sieves returns to the mill's grinding chamber for further grinding.

Mica passing through the sieves is conveyed to storage bins from which it is bagged in 22.2-kg units and palletized in 1.13 t lots for shipment. Tables 4 and 5 show properties of ground mica products. For all forms of ground mica the index of refraction is 1.58 wt %, Mohs' hardness is 2.5, oil absorption (Brit. Stand. 3483) is 60.75%, water solubility (Brit. Stand. 1765) is <0.3%, the phericity factor is 0.01, and the softening point in °C is 1538. For Micronized and wet ground micas the brightness (green filter), pH, and apparent density in kg/m^3 are 75, 5.2, and 160–224, respectively; for dry ground mica, 66–75, 6.2, and 192–561, respectively (1).

Testing of Mica

There are several conventional tests required by consumers of ground mica.

Screening. A 100-g sample of mica is usually used for this test, plus a rack of six Tyler sieves and a pan. The stack of sieves containing the sample is

Table 4. Particle Size Distribution of Ground Mica Products[a]

Sieve analysis, mesh size	Sizes, μm	Water ground mica		Micronized mica	Oil-well		Dry ground mica			
		88 μm (170 mesh)	44 μm (325 mesh)		Fine	Coarse	Joint cement	High K_2O	No. 160	Special[b] plastics
6	3360					0–10				
16	1190					25–55				
20	850				trace					
28	600									
32	500				10–30					
35	475							trace	trace	
60	250				10–50		trace	5–0		1.0
80	180									
100	150	0.0	0.0	0.00	10–70	25–65	1.5 max	10.0	10.0 max	30.0
140	106	0.5–1.0			10–30[c]	10–20[c]				50.0
200	74	5.0 max	1.0 min					10.0 max	65.0	
270	53			0.00			10.0 min			
325	44	7.0–14.0	3.0–10.0							80.0
−325	≤44	75.0–85.0	90% min	99.0 min			50.0 min[e]	40–75	75.0 max[d]	15.0
bulk density, g/cm³		0.2 max	0.2 max	0.2 max	0.4–0.6	0.18–0.26	0.2–0.2	0.29–0.45		5.0

[a]Information furnished by consumers; percent by weight retained.
[b]Micronized mica and wet ground both used in certain types of plastics.
[c]At 0.147 mm (100 mesh).
[d]Retained on 325 mesh.
[e]Usually run 70–75%.

Table 5. Chemical Analysis of Ground Muscovite Mica[a]

Analysis	Dry ground, wt %	Micronized,[b] wt %	Wet ground, wt %
SiO_2	45.57	46.27	48.65
Al_2O_3	36.10	35.24	32.04
K_2O	9.87	9.87	9.87
$Fe_2O_3^3$	2.48	2.48	3.68
Na_2O	0.62	0.60	0.28
TiO_2	0.20	0.20	0.20
CaO	0.21	0.21	0.21
MgO	0.15	0.16	0.16
H_2O	0.10	0.20	0.20
P_2O_5	0.03	0.02	0.018
S	0.01	0.01	0.010
C	0.44	0.44	0.44
LOI[c]	4.30	4.30	4.30
Total	*100.08*	*100.00*	*100.05*

[a]Ref. 1.
[b]Micronized is a trade name of KMG Minerals (formerly English Mica Co.).
[c]LOI = loss on ignition.

rotated, and after screening, the mica remaining on each screen is weighed and the percentage retained is calculated. A combination of wet and dry screening may also be used to determine particle size distribution of fine mica (≤ 0.147 mm (-100 mesh)).

Bulk Density. Bulk density of fine mica is determined with a Scott-Schaeffer-White volumeter using a 2.54-cm cube. A 7.6-cm diameter cylinder $\times$ 12.7-cm long is used to determine the bulk density of coarse mica, eg, oil-well-grade mica.

True Specific Gravity. This is determined with an air comparison or conventional glass pycnometer using distilled water as the displacement liquid.

Chemical Analysis. Chemical analysis is determined with conventional atomic absorption methods. Reliable wet techniques are sometimes employed (see Table 5).

Moisture. Moisture is determined by weighing a 50-g sample, then drying at 110°C. The sample is reweighed after drying. The difference between the two weights divided by the wet weight $\times$ 100 equals the percent moisture.

Free Silica. Free silica down to 1% can be determined with x-ray diffraction techniques (xrd).

Refraction Index. Refraction index is determined with a petrographic microscope by submersing a sample in emersion oils of known refractive index.

Oil Absorption. The Gardner rub-out (ASTM D281) method is used for this evaluation. A 5-g sample of mica is placed on a smooth glass plate, and linseed oil is added by drops. The mica is constantly mixed while the oil is being added. The end point is reached when the mica becomes saturated with oil. The amount of oil per 45.4 kg of mica is calculated as follows:

$$(mL \times \text{specific gravity linseed oil } (0.908))/5 \text{ g} =$$
$$\text{kg oil}/45.4 \text{ kg mica}$$

Brightness. The brightness of mica is determined with a Photovoltmeter (Photovolt Co.) or other suitable reflectance meter using a green 550-μm filter. The mica sample is prepared by pressing it into a smooth layer on a smooth glass surface.

Grit Content. The grit content is determined for fine mica by thoroughly mixing a 50-g sample with 1 L water. After mixing, the mixture is allowed to stand for 15 s, after which one-half of the contents is carefully decanted. The volume is again increased to 1 L, thoroughly mixed, and the slurry allowed to stand for 15 s before one-half of the contents is decanted. This procedure is repeated until the slurry becomes transparent, after which the slurry is allowed to stand for 10 s before being decanted. This is repeated until the water is perfectly clear, at which point the water is carefully decanted as much as possible without removing grit. The grit is washed, dried at 110°C, weighed, and reported as percent grit. In some cases, the grit is screened on a 0.210-mm (70-mesh) sieve and the percentages of $\pm$0.210 mm ($\pm$70 mesh) grit are reported as separate percentages. Another method of determining grit content of fine mica is with a Frantz isodynamic magnetic separator which provides a magnetic field to remove magnetic mica from nonmagnetic grit.

The grit content of coarse mica is determined by the vanning method. A 25-g sample of dry mica is placed on a vanning plaque or a 21.6 $\times$ 28-cm piece of rough cardboard. The cardboard is cupped and the mica rolled so that sand becomes separated from mica. The sand is removed and the percentage of grit is calculated by dividing the weight of the grit by the weight of the sample $\times$ 100.

Aspect Ratio. The aspect ratio of mica is determined with electromicroscopic image analysis techniques.

By-Products of Mica

The main by-products of mica processing plants are kaolin, quartz, and feldspar. Some plants produce all of these products for sale.

Glass-grade silica can be produced from most mica operations with additional beneficiation of the quartz. Ceramic-grade kaolin can be produced from some mica flotation plants by selective mining and additional processing of the clay slime removed prior to mica flotation (see CLAYS).

Mica Market and Consumption

Sheet Mica. Good quality sheet mica is widely used for many industrial applications, particularly in the electrical and electronic industries, because of its high dielectric strength, uniform dielectric constant, low power loss (high power factor), high electrical resistivity, and low temperature coefficient (Table 6). Mica

Table 6. U.S. Mica Imports for Consumption, t × 10³ [a]

Year	Split block Quantity, t	Value, $	Splittings Quantity, t	Value, $	Other <0.55/kg Quantity, t	Value, $	>0.55/kg Quantity, t	Value, $
1989	158	235	1379	1355	2996	714	79	436
1990	381	364	1204	1419	3829	864	30	268
1991	127	155	1244	1249	2734	608	51	204
1992	273	291	1601	1104	2860	626	180	616
1993	184	337	2625	1754	3656	739	147	433

[a]Refs. 8, 15–17.

also resists temperatures of 600–900°C, and can be easily machined into strong parts of different sizes and shapes (1).

Some parts fabricated from sheet include transistor mica, suitable for transistor mounting washers; interlayer insulator mica, used to insulate transformer coils; resistance and potentiometer cards, suitable for winding noninductive resistance cards and in potentiometers; vacuum tube mica, generally replaced by transistors; natural mica bushings and tubes; target and mosaic mica, used in the telecasting industry and in computers; and guided missile micas. High quality natural mica is used for various other special applications, ie, special optical filters, diaphragms for oxygen breathing equipment, washer dials for navigator compasses, microwave windows, and quarterwave plates of optical instruments, pyrometers, neon lasers, and thermal regulators.

Built-Up Mica. When the primary property needed for a particular application is insulation, built-up mica made by binding layered mica splittings together serves as a substitute for the more expensive sheet mica. The principal uses for built-up mica are segment plate, molding plate, flexible plate, heater plate, and tape (7).

Segment plate, used as insulation between copper commutator segments on direct-current universal motors and generators, accounts for the primary use for built-up mica. Phlogopite built-up mica is preferred for these segments because it wears at the same rate as the copper segments.

Some types of built-up mica are bound to special paper (qv), silk (qv), linen, muslin, glass cloth, or plastic. These products are very flexible and are produced in continuous wide sheets. These sheets are either shipped in rolls or cut into ribbons, tapes, or other desired shapes (Table 7).

Wet Ground Mica. Wet ground mica is used because of its unique properties, ie, luster, slip and sheen, and high aspect ratio (1,13).

Wallpaper and Coated Paper. The shiny particles of mica give a silky or pearly luster when applied to paper. The effect is pleasing to the eye, and in some cases simulates fabric.

Nacreous Pigments. Mica is used as a substrate for coatings (qv) of various metal oxides to obtain a pearlescent effect. Mica coated in this fashion is used as filler and as a coloring agent in certain types of plastics.

Rubber. A thin coating of mica acts as a mold-release compound in the priming of rubber goods such as tires. It prevents the migration of sulfur from

Table 7. Built-Up Mica[a] Sold or Used in the United States by Product[b,c]

Product	1991 Quantity, t	Value, $	1992 Quantity, t	Value, $	1993 Quantity, t	Value, $
flexible (cold)	105	685	163	794	104	551
molding plate	305	1708	211	1516	210	1718
segment plate	281	1893	262	1729	213	1382
other[c]	148	1415	216	2036	124	1074
Total[d]	839	6022	852	6075	651	4725

[a]Consists of alternative layers of binder and irregularly arranged and partly overlapped splittings.
[b]Refs. 8, 15–17; quantity and $ value × 10^3.
[c]Includes mica used for heater plate and tape.
[d]Approximate because of independent rounding.

the tire to the air bag when the tire is being vulcanized (see TIRE CORDS). Mica is also dusted on rubber inner tubes to prevent sticking.

Outdoor House Paint.   Mica acts as a reinforcing pigment to reduce checking and cracking while controlling chalking in outside latex, oleoresinous, alkyd, and alkyl-modified latex exterior paints (see PAINT). It also reduces penetration into porous surfaces, provides excellent adhesion, reduces running and sagging, and improves weatherability and brushability.

Aluminum Paints.   Mica is substituted for up to 25% of the aluminum in this type of paint (qv) as an economic measure. Mica is inert which tends to protect the more reactive aluminum from corrosive atmospheres, thus helping the paint to maintain its luster.

Sealers.   Mica is used in all types of sealers for porous surfaces, such as wallboard masonry, and concrete blocks, to reduce penetration and improve holdout (see SEALANTS). It permits a thicker film to be applied and at the same time reduces sagging. Cracking is reduced by the reinforcing action of the flakes, and gaps and holes in rough masonry are bridged by the mica flakes.

Plastics.   Mica is used as a filler in plastics to improve its electrical and thermal resistance and its insulating qualities. Although dry ground mica is also used as a filler in plastics, wet ground enjoys the higher value market and is used because of its extrusion properties, and because it replaces the higher priced fiber glass in many applications (see FILLERS; PLASTICS PROCESSING).

Dry Ground Mica.   *Hammer Mill Mica.*   Dry ground mica produced by hammer milling and screening is used in oil-well drilling, coatings for roofing shingles, roofing felt, and for some types of welding rod flux (see BUILDING MATERIALS, SURVEY).

Certain sizes of hammer milled and screened micas are used as a flux coating on the metal wire of welding rods. Fine particle-size, fluid-energy ground mica is also used for this purpose. High heat resistance, excellent mechanical and thermal strength, low moisture absorption, high surface leakage resistance, unusual chemical inertness, and nonhydroscopic properties are some of the distinctive qualities which make mica an indispensable ingredient for use in welding (qv) electrodes. The fluxing property is attributed to the high K_2O content.

Fluid Energy Produced Mica. The largest use for fine, dry ground mica is in the manufacture of wallboard joint cements. Ground mica that is essentially ≤0.147 mm (−100 mesh) (Table 4) and ~70% passing a 0.044 mm (325 mesh) Tyler sieve is used in the joint compound mixture as a filler and extender. These compounds are used to fill joints between panels of gypsum plaster board (see CALCIUM COMPOUNDS). In this particular use, mica contributes to making a nonabsorbing smooth surface that reduces shrinkage and eliminates cracks. It is also used in the finished coating on ceilings and to prepare thermal insulation and acoustical qualities of ceiling tile and prefabricated concrete.

Fine particle-size dry ground mica is also used as an extender and filler in certain texture and traffic paints. Mica particles are stronger than iron and not brittle like other inerts. It is an antifriction, antifouling, antisettling, anticorrosive, antitarnish, and antisiege agent. It is a superior reinforcing pigment that acts as a sealer over porous surfaces and reduces penetration and flushing (see SEALANTS); moreover, it improves the moisture resistance of protective coatings and adhesion to all types of surfaces.

Micronized Mica. Micronized mica is a trade name (KMG Minerals, formerly English Mica Co.) for a very fine particle-size dry ground product, usually ground with superheated steam in a special fluid energy mill and used as a replacement for wet ground mica in certain types of paints. Micronized mica, preferably calcined, is also used in cosmetic applications, ie, nail varnishes, lipsticks, eyeshadows, and barrier cream, because is has the advantages of high ultraviolet light stability, excellent lubricity, skin adhesion, and compressibility (see COSMETICS). Some of these micas are coated with oxides like titanium and iron (Table 8).

Table 8. Ground Mica Sold or Used in the United States[a,b]

	1991		1992		1993	
Use	Quantity, t	Value, $	Quantity, t	Value, $	Quantity, t	Value, $
joint cement	39	6,173	43	6,819	49	7,549
paint	15	4,428	16	5,227	16	5,416
plastics	1	560	4	1,347	4	1,647
well-drilling mud	4	484	2	281	4	560
other[c]	15	5,642	19	8,082	19	11,814
Total[d]	*75*	*17,289*	*84*	*21,755*	*92*	*26,986*

[a]Refs. 8, 15–17; quantity and $ value × 10^3.
[b]Domestic and some imported scrap. Low quality sercite is not included.
[c]Includes mica used for molded electrical insulation, roofing, rubber, textile and decoration coatings, welding rods, and miscellaneous.
[d]Approximate because of independent rounding.

Economic Aspects

In 1990, North Carolina produced 60% of the total scrap mica; the remainder was produced in Connecticut, Georgia, New Mexico, Pennsylvania, South Carolina, and South Dakota. In 1991, the five largest producers produced 67% of the nation's total output (Table 9) (15).

The United States imports most of its manufactured sheet mica (muscovite) and paper-quality scrap mica from India, with the remainder coming from Canada, China, Japan, Norway, and others (Table 10). Consumption of mica splittings increased 90% to 2625 t in 1993 from 1989. The splittings were fabricated into various built-up mica products. Total production of built-up mica decreased 34% from 1989. Molding plates and segment plates were the primary end products, accounting respectively for 32 and 33% of the total. The combined value of all mica imports increased 42% to $21.2 million in 1993, while the combined value of all mica exports increased 28% to $12.2 million. Sheet mica prices ranged from approximately $4.40 to $1100/kg depending on the grade, while splittings were bought for $1.10–$3.30/kg. Wet ground mica selling prices ranged from $440 to $800/t; dry ground mica sold for $110 to $220/t fob the plant site. The average $/t of U.S. produced wet and dry ground micas in 1993 are as follows:

Type	Price, $
joint cement	155
paint	348
plastics	396
well-drilling mud	209
other	616

Table 9. U.S. Producers of Mica, 1991[a]

Company	Location	Capacity, t/yr	Product	Remarks
Spartan Minerals Corp.	Pacolet, S.C.		dry	sold into the joint compound market
Mineral Mining Corp.	Kershaw, S.C.	22,675	dry	product is sericite for paint industry
Franklin Industrial Minerals (MICA Division)	Taos Country, N.M.	18,140	dry	many improvements in dry and wet processing; new flotation plant
Pacer Corp.	Custer, S.D.		flake	70% muscovite mica for oil-well use
Concord Mica		1,633	wet	for paint and lining rubber molds
Franklin Mineral Products	Hartwell, Ga.		wet	

[a] Ref. 16.

Table 10. U.S. Mica Trade Data[a]

	Scrap and flake mica				Sheet mica			
	Powder		Waste		Unworked		Worked	
Year	Quantity, t	Value, $	Quantity, t	Value, $	Quantity, t	Value, $	Quantity, t	Value, $
				Exports				
1989	3,628	1,634	1,224	555	60	156	415	7,227
1990	4,319	2,050	580	646	148	272	612	7,568
1991	3,420	1,717	874	331	205	309	411	7,454
1992	3,954	2,054	475	204	170	307	436	7,180
1993	4,614	2,604	335	99	292	511	617	9,019
			Imports for consumption					
1989	8,902	4,971	4,185	1,256	1,616	2,054	1,229	6,711
1990	9,142	5,133	4,034	987	1,615	2,051	1,085	7,431
1991	9,725	5,219	3,630	996	1,422	1,608	918	6,835
1992	11,568	7,479	3,786	974	2,054	2,011	1,407	9,011
1993	13,098	8,070	4,765	1,307	2,956	2,524	1,352	9,338

[a]Ref. 16; quantity and $ value $\times 10^3$.

Transportation

Mica is shipped by truck (20.4 t max) or railroad (27.2 t max). Bulk truck-loads and rail shipments, such as palletizing, bagging, and shrink- or stretch-wrapping, are made to avoid extra costs. Mica is usually palletized and shrink- or stretch-wrapped for shipment.

Environmental Regulations

Mica mining is subjected to local, state, and federal laws. The Mining, Safety and Health Administration (MSHA) regularly monitors mica mining operations for safety violations.

Both state air and water environmental departments together with the U.S. EPA regulate and oversee air and water quality associated with mica mining operations. Most states have land management departments that regulate dam safety, erosion, sedimentation, and reclamation. The mica mines must control erosion and sedimentation and restore the mined out areas. This is accomplished either by backfilling or contouring and seeding operations, or in cases where this is impractical or undesirable, lakes for water-related recreation may be built. The Corps of Engineers have jurisdiction over laws governing wetlands.

Health and Safety Factors

Health regulations are supervised by county and state health departments. There are no known health problems caused by the mica crystal, however, most industrial mica products contain some free silica particles that can cause silicosis and some states require employees who work in mica plants to receive an annual x-ray.

Future Applications

The future for mica is in the speciality plastic market, eg, as a molecular barrier in plastic containers and in plastic automobile parts.

Research and testing results have shown that mica can be used success-fully in air-conditioner fan blades, dashboard panels, head lamp assemblies, fan shrouds, and floor panels. Mica can also be substituted for more expensive glass flakes to strengthen lightweight plastic seat backs, load floors, grill panels, igni-tion system parts, and air-conditioning and heater valve housings. Both Ameri-can and Japanese automakers are incorporating ground mica in place of asbestos (qv) in acoustic compounds that change vibrations and eliminate road and en-gine noise. Mica is also an environmentally accepted replacement for asbestos in brake linings (see BRAKE LININGS AND CLUTCH FACINGS).

BIBLIOGRAPHY

"Mica" in *ECT* 1st ed., Vol. 9, pp. 68–75, by P. M. Tyler, Consultant; "Mica, Synthetic" in *ECT* 1st ed., Suppl. Vol., pp. 480–487, by A. Van Valkenburg, National Bureau of

Standards; "Micas, Natural and Synthetic" in *ECT* 2nd ed., Vol. 13, pp. 398–424, by H. R. Shell, U.S. Bureau of Mines; in *ECT* 3rd ed., Vol. 15, pp. 416–439, by A. B. Zlobik, U.S. Bureau of Mines.

1. M. L. Rajgarhia, *Ground Mica*, Mica Manufacturing Co., Private Ltd., Calcutta, India, 1987, p. 30; *British Standards*, British Standards Institute, London.
2. S. W. Mudd, *Industrial Minerals and Rocks*, 2nd ed., AIME, New York, 1949, pp. 551–566.
3. S. W. Bailey, *Reviews in Mineralogy*, Vol. 13, Mineralogical Society of America, Washington, D.C., 1984, pp. 1–12.
4. R. W. Grim, *Clay Mineralogy*, McGraw-Hill Book Co., Inc., New York, 1968, 596 pp.
5. G. P. Chapman, in S. J. Lefond, ed., *Industrial Minerals and Rocks*, 5th ed., Vol. 2, AIME, New York, 1983, pp. 915–929.
6. S. B. Hendricks and M. Jefferson, *Am. Mineral.* **24**, 729–771 (1939).
7. A. B. Zlobik, *Mica*, preprint from bulletin 671, U.S. Bureau of Mines, Washington, D.C., 1980, 16 pp.
8. L. L. Davis, *Minerals Yearbook*, U.S. Bureau of Mines, Washington, D.C., 1991–1993, p. 4, 5, 7–9.
9. *Mica Fabricators*, The Tar Heel Mica Co., Plumtree, N.C., 1942, 3 pp.
10. E. J. Dana, *A Text Book of Mineralogy*, 3rd ed., John Wiley & Sons, Inc., New York, 1945, pp. 659–665.
11. C. W. Chesterman and co-workers, *The Audubon Society Field Guide to North American Rocks and Minerals*, Alfred A. Knopf, New York, 1989, pp. 531–536.
12. L. L. Davis, *Mineral Facts and Problems*, preprint from bulletin 675, U.S. Bureau of Mines, Washington, D.C., 1985, 12 pp.
13. *Quality Classification of Muscovite Block Mica*, ASTM D351, ASTM, Philadelphia, Pa., 1961, p. 1046.
14. J. B. Preston, "Mica," *Pigment Handbook*, John Wiley & Sons, Inc., New York, 1971, 30 pp.
15. Ref. 8, 1991, p. 9.
16. *Ibid.*, 1992, p. 4.
17. *Ibid.*, 1993, p. 8.

General References

National Mica for Industry, The Tar Heel Mica Co., Buffalino, N.C., 1990, 3 pp.
S. D. Broadhurst and L. J. Hash, *The Scrap Mica Resources of North Carolina*, Bulletin No. 66, North Carolina Department of Conservation and Development, 1953, p. 6.
G. M. Clarke, *Ind. Mineral.*, 27–28 (June 1983).

JAMES T. TANNER
North Carolina State University

MICELLULAR FLOODING. See PETROLEUM, ENHANCED OIL RECOVERY.

MICROBIAL
POLYSACCHARIDES

Produced by virtually all microbes, carbohydrate polymers serve as intracellular energy stores and cell wall components, among other roles (see CARBOHYDRATES). They are perhaps most apparent when present as extracellular capsules, sheaths, or slime secretions. Extracellular polysaccharides may confer a survival advantage to microbes by protecting against desiccation, acting as buffers against environmental changes, preventing invasion by bacteriophages, and by helping cells adhere to surfaces and to one another. Technological exploitation of microbial polysaccharides has been limited mostly to those produced extracellularly in substantial quantities. Applications of these exopolysaccharides can be based on the unique chemical functionalities present, or on the bulk physical properties of the biopolymer. For example, vaccines against meningitis caused by *Haemophilus influenzae* are made from the capsular polysaccharide of the bacterium, and rely on its immunochemical properties. By contrast, the usefulness of gellan results from its ability to form clear, firm gels at low concentrations. The physical properties of polymers, including polysaccharides, result from their chemical structures. This article deals with the chemical structures of some microbial exopolysaccharides having actual or potential commercial applications, and describes how the chemical structures confer certain useful properties on those materials (see also BIOPOLYMERS).

Microbial polysaccharides may be categorized into groups based on the types of monomer units present. Two of the most important types of microbial polysaccharides are neutral homopolysaccharides and anionically charged heteropolysaccharides. Other groups also exist, such as charged homopolysaccharides, but are of limited occurrence and not commercially significant as of this writing. Homopolysaccharides consist of a single type of monosaccharide, some examples being pullulan and dextran, both of which are polymers of D-glucose [50-99-7]. Some important anionic heteropolysaccharides are xanthan, which contains both neutral sugar and uronic acid residues within its structure, and alginic acid, which consists of two different types of uronic acid. Structures can be complicated by the presence of noncarbohydrate groups attached to the carbohydrate chains. Such groups include *O*-acetyl esters, and 1-carboxyethylidene (or pyruvate) ketals.

Polysaccharides can be classified by structure, biological origin, or mode of biosynthesis. Most polysaccharides are synthesized by an elaborate sequence of steps catalyzed by cytosolic and membrane-bound enzymes. The usual precursor is a nucleotide phosphate–sugar glycosidic ester. This activated sugar is transferred to a membrane-bound lipid phosphate, and the repeating units are assembled in a blockwise sequence, ie, each repeating unit is added as a single group (1–4). Once assembled, the polysaccharide is released. Another simpler mechanism is used for the synthesis of a few types of neutral homopolysaccharides. In these instances, monosaccharide units from sucrose or a similar glycosyl donor are incorporated directly into the polysaccharide; the glycosidic linkage of the donor contains sufficient stored energy to allow the reaction to proceed. No lipid–phosphate intermediates exist, and usually only a single enzyme is required. The best known examples of this latter mechanism are synthesis of dextran, catalyzed by dextransucrase, and synthesis of levan, catalyzed by levansucrase.

For some applications, microbial polysaccharides have supplemented or replaced those derived from plants or algae; in other instances, microbial polysaccharides have been developed for specific applications that cannot be met by other polysaccharides. Further information is available (5–24).

Alginates

The term alginate refers to the salt forms of alginic acid [9005-32-7], a copolymer of D-mannopyranosyluronic acid [56687-62-8] (**1**) and L-gulopyranosyluronic acid [56688-68-7] (**2**) residues.

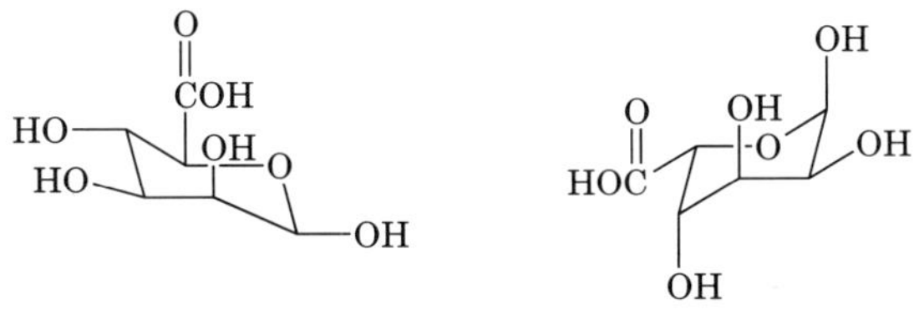

(**1**)   (**2**)

For many years, alginates were derived solely from marine algae, hence the origin of the term algin. However, in 1964 it was reported that a strain of *Pseudomonas* isolated from the sputum of a cystic fibrosis patient produced an extracellular capsular polysaccharide essentially identical to algal alginate (25). Later it was demonstrated that the common nitrogen-fixing soil bacterium *Azotobacter vinelandii* also produces extracellular capsules of alginate (26). The phenomenon is known to be common among strains of *A. vinelandii*, as well as *A. chroococcum* (27,28) and *A. beijerinckii* (29). Many other bacteria have been found to produce extracellular alginate, most notably strains of the opportunistic human pathogen *Pseudomonas aeruginosa* (30), the nonpathogenic *P. mendocina* (31), and numerous plant pathogenic species of *Pseudomonas* (32). Most of these organisms also produce other extracellular polysaccharides, and the relative yields of the different polymers often depend on strain, growth conditions, carbon source, and other such factors. Much has been made of the possibility of using bacterial fermentations to produce alginates on a large scale, but marine algae remain the only significant commercial source. However, a great deal of the research done on bacterial alginate has led to a better understanding of alginate structure and biosynthesis.

Alginic acid is a copolymer of β-D-mannopyranosyluronic acid (**1**) and α-L-gulopyranosyluronic acid (**2**) units linked $1 \rightarrow 4$, with no detectable branching. Bacterial alginates contain *O*-acetyl residues linked to the sugar units. The sequences and proportions of the constituent units can vary, and much research has focused on understanding the factors determining these structural variations. Particularly important is the arrangement of the D-mannuronate and L-guluronate residues within the alginate chains. Sequences of poly-β-D-mannuronate tend to be soluble and nongelling. The introduction of blocks of α-L-guluronate sequences into the chain changes the conformation, so that the polymer is able to complex with divalent cations to form rigid gels. The so-called egg-box model has been used to describe the complexes between polyuronates and cations that result in gel formation (33). To obtain an alginate with good gelling properties it is necessary to grow the bacteria under conditions that favor the biosynthesis of a product with the proper ratio and arrangement of mannuronate and guluronate sequences, which implies a need for understanding the biosynthetic mechanism. It is generally accepted that bacterial alginate, like the algal product, is first assembled as a polymer of poly-β-D-mannuronate, and that a key enzyme, polymannuronic acid C-5 epimerase, then epimerizes some of the D-mannuronate residues to L-guluronate residues (34–36). The extracellular epimerase from *A. vinelandii* has been purified and characterized (37), and is active on alginates from algae as well as bacteria. The enzyme is calcium dependent, and the concentration of calcium can determine the degree and pattern of epimerization. This knowledge has been used to select growth conditions influencing the constituent ratio in alginate produced by *A. vinelandii* fermentations (38,39). O-Acetyl groups in bacterial alginate are also believed to affect structure–property relationships. The C-5 epimerase is incapable of acting on mannuronate residues bearing O-acetyl groups (40,41); acetylation therefore prevents blocks of poly-β-D-mannuronate residues from becoming converted to L-guluronate. Regulation of O-acetylation, which apparently occurs intracellularly, is not fully understood. The chain length, or molecular weight, of the alginate molecule also plays a role in determining its physical properties. Bacterial alginate chains often begin to be broken down at some point during their biosynthesis and subsequent accumulation. This breakdown is usually attributed to the presence of alginate lyase, an enzyme that cleaves the polyuronate chains by an elimination, rather than hydrolysis type of reaction (42). Evidence seems to suggest that these enzymes can arise from infecting bacteriophages within the host organism (43,44), although more recent evidence indicates that the presence of endogenous alginases may be common among strains of *Azotobacter* (45). The depolymerization of alginate causes a decrease in its viscosity and gelling ability.

   Production and Utilization. Uses proposed for bacterial alginate have been the same as those for the algal-derived material, and are based on its gelling, viscosifying, film-forming, and suspension-stabilizing properties. The topic has been reviewed extensively (21,46–49). Much of the potential for bacterially produced alginate has not yet been realized, owing mainly to the difficulty in competing with algal-derived products. Most research has been aimed at developing more efficient, higher yielding fermentations (38,39,50,51) and at strain improvement (52) (see FERMENTATION). The main difficulties in competing with algal alginates are of two types. On the commercial side, there are the usual problems associated with bringing a new process to market, ie, building facilities, finding inexpensive and reliable feedstocks (qv), etc. On the technical side, there are still some unsolved problems. *Azotobacter* possess extremely high rates of respiration. This translates into a need for energy-intensive rates of aeration in fermentations. It also means that much of the carbon feedstock is diverted to metabolic products other than alginate (50). There is also the problem of product degradation by endogenous alginases. One example of the microbial product having unique advantages utilizes cultures of *A. vinelandii* in an *in situ* process to coat ceramic particles for enhanced aqueous dispersion (53) (see CERAMICS). Another suggested application of *A. vinelandii* is the use of bacterially derived polymannuronate C-5 epimerase to improve the gelling qualities of alginates from traditional sources (48).

Bacterial Cellulose

Although cellulose [9004-34-6] is usually thought of as a plant-derived polysaccharide, there does exist one well-known example of cellulose (qv) production by a bacterium. In 1886, a bacterial isolate from what was referred to as the vinegar plant was described. This bacterium, *Acetobacter xylinum*, produced a tough, membranous pellicle in liquid cultures. Using the best methods then available, it was concluded that the pellicular material was cellulose (54), which was later confirmed by chemical methods and x-ray diffraction (55). *A. xylinum* has become a useful model in the study of cellulose biosynthesis (56,57). The extracellular pellicle is composed primarily of microfibrils of cellulose, a $\beta(1 \rightarrow 4)$-linked D-glucan. These microfibrils, which consist of parallel chains of polysaccharide molecules, form ribbon-like arrays, which in turn make up the pellicle itself (56,58). Values for the average molecular weight vary, but are typically in the range of 350,000 to 975,000, corresponding to average chain lengths of approximately 2000 to 6000 glucose residues. The polymer chains are synthesized by a membrane-associated enzyme complex. The precursor is the nucleotide phosphate–sugar ester uridine disphosphate–glucose (UDP–glucose), and synthesis seems to proceed through a lipid-linked intermediate (56), although there is still some disagreement as to the exact mechanism of biosynthesis. Part of the problem with studying cellulose biosynthesis in *A. xylinum* is that the bacteria also produce other polysaccharides (59), some of which also contain β-linked D-glucose units. These include $\beta(1 \rightarrow 2)$-linked D-glucans (60) and heteropolysaccharides such as acetan (61,62). Acetan appears to be structurally related to xanthan as well as to cellulose in that it consists of a $\beta(1 \rightarrow 4)$-linked D-glucan backbone with side-chain units containing D-glucose, D-mannose, and D-glucuronic acid, as well as terminal L-rhamnose residues (62). Because it is possible that these polysaccharides are all produced by similar mechanisms, the difficulty in separating the enzymes and intermediates involved in the biosynthesis of each is considerable.

Production and Utilization. Although bacterial cellulose has been known since the late 1800s, there had been little commercial interest for many years, owing to the abundance and low cost of plant-derived cellulose. However, the ability of bacterial cellulose to form tough, uniform membranes has suggested applications, for example, in ultrafiltration membranes. Bacterial cellulose is used in speaker diaphragms for personal stereo headphones (57), which are reported to have excellent acoustical properties (63). Other applications have been proposed for bacterial cellulose, including nonwoven fabrics, coatings (qv), and suspending agents. In addition, it is claimed that the sheared material possesses good thickening properties (64). Research into improved methods for the production of bacterial cellulose continues (65). Processes have been developed which are said to yield cellulose at a cost of approximately $11–26/kg ($5–$12/lb) (57).

Dextrans

Dextran [9004-54-0] is a term that has traditionally been applied to any extracellular bacterial α-D-glucan synthesized from sucrose [57-50-1] in which $\alpha(1\rightarrow6)$ linkages predominate. Dextrans have been more strictly defined as D-glucans containing chains of D-glucopyranosyl residues consecutively $\alpha(1\rightarrow6)$-linked, with various degrees of branching through $\alpha(1\rightarrow2)$, $\alpha(1\rightarrow3)$, or $\alpha(1\rightarrow4)$ linkages (66). Dextran has been known for many years, and was recognized as a polymer of dextrose (D-glucose) by the latter half of the nineteenth century, hence the origin of the name. Dextran research has been reviewed (66–73), and a bibliography has been published (74). A number of lactic acid bacteria produce dextrans (75), the most notable being *Leuconostoc mesenteroides* and certain *Streptococcus* species. Much of the early interest in dextrans arose as a result of its occurrence in sugar (qv) refineries. Infection of sugar cane, and to a lesser extent, sugar beets, by *L. mesenteroides* during harvesting and processing leads to the production of dextran. Large amounts can give rise to sticky, gummy solutions that foul processing equipment, whereas lesser concentrations inhibit proper crystallization of sugar (71). Food preparations that contain sucrose can also become contaminated by growths of dextran-producing bacteria, leading to sliminess, gumminess, or "ropy" solutions.

Interest in dextrans became intense after 1945, when it was shown that solutions of the polymer were suitable for use as artificial blood plasma substitutes (76,77) (see BLOOD, ARTIFICIAL). In the late 1940s an investigation into the production and properties of dextran was undertaken. This work led to the characterization of dextrans produced by nearly 100 strains of bacteria (78), and to several patents for methods of producing dextran and dextran fractions (79–87). There is great structural diversity among dextrans from different bacterial sources, and some strains of bacteria produce more than one type of dextran (78,88). Among those strains of bacteria that produce two or more types of dextran are *L. mesenteroides* strains (Northern Regional Research Laboratory (NRRL)) B-742, B-1299, B-1355, B-1498, and B-1501, and *Streptobacterium dextranicum* B-1254. Dextrans produced by these bacteria may be separated into two or more fractions by precipitation with varying concentrations of ethanol, or other water-miscible solvents, in water. Designation of the fractions is based on their differing solubilities in water–alcohol mixtures. Those precipitated with lower alcohol concentrations are referred to as *L*-fractions (less soluble); those precipitated at higher alcohol concentrations are referred to as *S*-fractions (more soluble). The most important factor in determining the properties of the various types of dextrans is the percentage of non-$\alpha(1\rightarrow6)$ linkages (78,89); the nature and distribution of the non-$\alpha(1\rightarrow6)$ linkages have been the subject of many studies (90–103).

Much of our understanding of the structure of dextrans is a result of studies carried out during the 1970s and 1980s using gc/mass spectrometry of methylated dextran derivatives (104–107), ^{1}H and ^{13}C-nmr spectrometry (108–117), and Fourier-transform infrared spectroscopy (ftir) (118). According to one proposal (116), dextrans may be classified into three types: class 1 dextrans, which contain a main chain of $\alpha(1\rightarrow6)$-linked D-glucopyranosyl units, with branching through carbon positions 2, 3, or 4 of the D-glucopyranosyl ring; class 2 dextrans, which contain 3-mono-O-substituted D-glucopyranosyl units in nonconsecutive positions, as well as 6-mono- and 3,6-di-O-substituted residues; and class 3 dextrans, which contain consecutive 3-mono-O-substituted D-glucopyranosyl units in addition to 6-mono- and 3,6-di-O-substituted residues. The vast majority of polysaccharides that have been broadly defined as dextrans fall into class 1, and in most of these the branching is through position 3. Only three glucans from the NRRL strains are considered class 2 dextrans, and these are the fraction S-glucans from strains B-1355, B-1498, and B-1501 (112). These three glucans contain alternating sequences of $\alpha(1\rightarrow3)$ and $\alpha(1\rightarrow6)$-linked D-glucopyranosyl residues. Because they do not contain significant linear, consecutive sequences of $\alpha(1\rightarrow6)$-linked D-glucopyranosyl residues, these glucans do not fit the strict definition of dextrans as $(1\rightarrow6)$-α-D-glucans. For this reason, the name alternan has been proposed for this class of polysaccharides (119). Alternan from *L. mesenteroides* strain NRRL B-1355 has been studied in more detail than the others (92,120–122), and continues to be of interest because of its unique immunochemical (123) and physical (124) properties. *L. mesenteroides* strains NRRL B-523 and B-1149, as well as many strains of *Streptococcus*, produce glucans that fall into the category of class 3 dextrans. These dextrans contain linear sequences of $\alpha(1\rightarrow3)$-linked D-glucopyranosyl units, and tend to be insoluble or only slightly soluble in water. The distinction between these α-D-glucans and true dextrans has been noted by most workers in the field, and the name mutan has been proposed for polysaccharides consisting mainly of $\alpha(1\rightarrow3)$-linked D-glucose chains (125). Streptococcal α-D-glucans play an important role in the adhesion of the bacteria to oral surfaces and the subsequent formation of carious lesions (cavities) in the teeth (126), and may also be an important factor in bacterial endocarditis (127). Because of this, much chemical and microbiological research has been done on streptococcal α-D-glucans (75,126,128).

Class 1, or true dextrans, are a structurally diverse group. The percentage of non-$\alpha(1\rightarrow6)$ branch linkages ranges from as low as 3% in *L. dextranicum* NRRL B-1146 (78) to as high as 39% in some preparations of *L. mesenteroides* NRRL B-742 S-fraction dextran (107). Studies done on this latter strain between 1937 and 1986 have resulted in findings that seem to differ significantly (78,89,91,93,105–107,111,116,129–134). Some of these differences can be explained by the fact that strain B-742 makes two distinct dextrans, the proportions of which can vary, depending on fermentation (qv) conditions. Fraction L contains approximately 15% $\alpha(1\rightarrow4)$ branch linkages, whereas fraction S contains a much higher percentage of branch linkages, primarily $\alpha(1\rightarrow3)$ (107). The percentage of $\alpha(1\rightarrow3)$ linkages can vary, depending on the conditions of biosynthesis.

L. mesenteroides strain NRRL B-512F produces a water-soluble dextran with 95% $\alpha(1\rightarrow6)$ main-chain linkages and 5% $\alpha(1\rightarrow3)$ branch linkages (78). This strain was subcultured from NRRL strain B-512, isolated in 1943. Strain NRRL B-512F is the strain used for commercial dextran production in the United States and most other countries. Nearly all of the studies done on the industrial production and utilization of dextran have used this strain.

Biosynthesis.   Unlike most microbial exopolysaccharides, dextrans are enzymatically synthesized directly from sucrose. The enzymes, classified as glycosyltransferases, are known generically as dextransucrase (Enzyme Commission (EC) 2.4.1.5), or α-1,6-glucan:D-fructose 2-glucosyltransferase (see ENZYME APPLICATIONS). Dextransucrase is quite specific in its requirement for sucrose as the glucosyl donor substrate. The only other known glucosyl donors that give rise to dextran are α-D-glucopyranosyl fluoride (135), lactulosucrose (136), and p-nitrophenyl-α-D-glucopyranoside (137). The latter compound yields dextran at less than 1% of the rate for sucrose. Dextransucrase usually occurs as a soluble enzyme in the cell-free culture fluid of bacteria grown in liquid media (138). The streptococcal enzymes are constitutive (126,128), whereas *L. mesenteroides* requires the presence of sucrose to induce enzyme production. Many methods have been reported for the isolation and purification of dextransucrases. The enzymes from *L. mesenteroides* B-1299 have been investigated, and several different isozymic forms isolated and partially characterized (139–142). Interest in alternan from *L. mesenteroides* B-1355 has led to the isolation and partial purification of the enzyme responsible for alternan synthesis (119,143,144); this enzyme, named alternansucrase, is distinct from the dextransucrase also elaborated by this organism (119).

Because of its commercial importance, *L. mesenteroides* B-512F dextransucrase has been studied most extensively. Several methods have been reported for its purification (145–150), and high yields of electrophoretically pure dextransucrase have been obtained from a mutant of B-512F that produces enhanced levels of enzyme activity (151). The purified enzyme had a molecular weight of 158,000, and was reportedly derived from a larger native protein of 177,000 molecular weight by a specific proteolytic cleavage (150,151). The pH optimum of this enzyme is in the range of 5.0–5.5, and its Michaelis constant, K_{m}, for sucrose is approximately 12–16 mM (150).

In addition to polymerizing glucosyl units to form dextran, dextransucrases also catalyze the transfer of α-D-glucopyranosyl units to a wide variety of hydroxyl-bearing acceptor molecules (152–154). These transfers result in the formation of di- and oligosaccharides when sugars are the acceptors, and various glycosides when the acceptor is a noncarbohydrate. A number of sugars and sugar derivatives have been tested for their ability to act as acceptors in reactions catalyzed by B-512F dextransucrase (155–158). These studies have shown that maltose [69-79-4], the most effective acceptor sugar, gives rise to large amounts of the trisaccharide panose and diverts glucosyl transfer away from dextran synthesis. The kinetics of this reaction have been described (159). The acceptor reactions catalyzed by alternansucrase (160) and streptococcal glucansucrases (161–164) have also been investigated. In each case, it was shown that the acceptor substrate and product specificity depend on the enzyme source. One important acceptor reaction to note involves fructose [57-48-7]. Because fructose is released from sucrose during the glucosyl transfer reaction, it is always present during the enzymatic synthesis of dextran. Reactions of fructose catalyzed by B-512F dextransucrase give rise to one major and one minor product. The main product is 5-O-α-D-glucopyranosyl-D-fructose, known by the trivial name leucrose (165,166), and the minor product is 6-O-α-D-glucopyranosyl-D-fructose (167), known as isomaltulose or palatinose.

The molecular mechanism by which dextransucrase synthesizes dextran has been a matter of considerable interest. It was long assumed that dextran synthesis occurred by the transfer of glucosyl residues from sucrose to the 6-hydroxyl group at the nonreducing ends of growing dextran chains (168). An alternative mechanism was proposed in which α-D-glucopyranosyl residues were inserted between the enzyme and the reducing end of the growing dextran chain (169). Experimental evidence has been provided for this type of mechanism (170–174). The mechanism by which acceptor reactions occur can be explained within this context (175), and there is evidence for separate sucrose and acceptor-binding sites on the enzyme (176). It is believed that acceptor reactions play an important role in the formation of branch linkages by dextransucrase (177–180). According to the insertion–mechanism model, branches are formed when glucosyl units or dextranosyl chains are transferred to secondary hydroxyl positions on the dextran chains. Differences in acceptor binding among dextransucrases from various sources account for the differences in branching among the dextrans.

Although sucrose is generally the only natural substrate that bacteria can use to synthesize dextran, there is one notable exception to this rule. *Acetobacter capsulatum* and certain related bacteria are known to produce dextrans from amylodextrins (maltodextrins) [9050-36-6], via an enzyme referred to as dextrandextrinase (EC 2.4.1.2: dextrin, α-1,6-glucan 6-glucosyltransferase). This enzyme synthesizes a typical $\alpha(1 \rightarrow 6)$-linked dextran from $\alpha(1 \rightarrow 4)$-linked D-glucans (181,182). Some glucans, such as native starch, glycogen, and unhydrolyzed amylose or amylopectin, are very poor substrates, whereas starch hydrolyzates such as corn syrup or isolated amylodextrins are readily converted to dextran. The conversion efficiencies are not nearly so high as for dextransucrase. Typically, the reaction reaches equilibrium at a 1:1 mixture of amylodextrin and dextran (183). Several patents have been issued covering dextran production by dextran–dextrinase (184–188), and studies have been done on the properties (189,190) and mechanism (191) of the enzyme. This enzyme has thus far found no commercial applications, probably because of the low conversion efficiencies. The finding that reduced maltodextrins serve as more efficient glucosyl donors (190) suggests that this drawback may not be insurmountable.

Derivatives. Derivatives of dextran may be classified into two categories: those in which the dextran chain has been covalently modified, and those in which the structure is unchanged, but the molecular weight has been lowered, either by alteration of biosynthetic conditions or by depolymerization of high molecular weight dextran. Under the usual conditions of biosynthesis, dextran quickly reaches molecular weights of 10^6 or greater. In some applications, for example blood plasma extenders, dextran of a lower size is required because of its osmotic and rheological properties, and because dextrans of higher molecular weight are not cleared from the bloodstream as quickly. The recommended molecular weight range for blood plasma extenders is 75,000 $\pm$ 25,000. These low molecular weight fractions are thus often referred to as clinical dextrans. Several approaches can be used to obtain dextrans of the desired molecular weight range. The method used most often involves hydrolysis with dilute acid under mild conditions, followed by separation of the variously sized fractions by graded precipitation with ethanol (qv). Drawbacks include the formation of unwanted side-products, eg, colored material; oligosaccharides arising from reversion reactions; and significant amounts of glucose. Nevertheless, the method is relatively inexpensive, and the ethanol precipitation step can remove nearly all of these contaminants. Other routes for obtaining clinical-sized dextrans have been developed, and may be more suitable for some applications. Biosynthesis in the presence of acceptors gives rise to dextran of lowered molecular weight (192); a systematic study of the biosynthetic parameters affecting molecular weight has been done (193). It has been found that at higher sucrose concentrations, overall dextran yield is diminished, and the average molecular weight lower. The addition of such exogenous acceptors as maltose, glucose, or preformed clinical dextrans has been found to be particularly useful for the synthesis of clinical dextrans. Conditions for producing clinical dextrans of suitable molecular weights have been established (194), and several methods have been patented (79,81–83). The clinical dextrans produced by hydrolysis and synthesis in the presence of acceptors are similar in terms of their structures and viscosity behavior (195). Besides acid hydrolysis, there are other ways to degrade dextran to lower molecular weight fractions. Preparations of endodextranases (EC 3.2.1.11), enzymes that split dextran chains in an *endo*-hydrolytic fashion, have been described (196). Several types of Penicillium mold produce large quantities of extracellular endodextranase, and a method for the production of clinical dextrans using endodextranase to hydrolyze dextran to a suitable molecular weight has been patented (86). Co-fermentation of *L. mesenteroides* and a dextranase-producing organism has also been used to synthesize clinical-sized dextran (197). Other methods have been used in the laboratory to produce clinical-sized dextrans, including depolymerization by ultraviolet radiation, ultrasonic treatment (198,199), dry heating (85), and synthesis by a streptococcal species known to produce dextrans of lower molecular weight (87). None of these methods has been used commercially. In the 1990s research has focused on the use of acceptor reactions for the synthesis of clinical dextrans (200–202). Although clinical dextrans are still used in some applications, artificial polymers such as poly(vinylpyrrolidone) are replacing dextran as blood plasma extenders, because dextran tends to elicit an immune response in sensitive individuals.

Many covalently modified derivatives of dextran have been described. Of these, the most important are dextran sulfate [9042-14-2] and cross-linked dextran. The uses and properties of these and other dextran derivatives have been reviewed (66,69,71–73,203).

Production and Utilization.   Dextran sulfate displays anticoagulant properties, and has been investigated as a substitute for heparin (73) (see BLOOD, COAGULANTS AND ANTICOAGULANTS). It has been shown that dextran sulfate can inhibit HIV binding to human T-lymphocytes (204), and is being studied for its potential in the treatment of AIDS and other viruses. Dextran, chemically cross-linked with epichlorohydrin (Sephadex), is useful in gel-filtration chromatography (qv). Derivatives such as diethylaminoethyl–Sephadex and carboxymethyl–Sephadex are used in ion-exchange chromatography.

Cross-linked dextran known as dextranomer (Debrisan), which is similar to Sephadex, has been used in treating wounds. Fluids and small molecules are absorbed into the gel particles, and proteins and cellular material are excluded (205). Complexes of colloidal iron with dextran (206), known as iron–dextran [9004-66-4], are used in treating iron deficiency anemia. This use is limited mainly to animals, especially pigs, because iron–dextran has been listed as a suspected carcinogen. The ability of dextran to form stable complexes with metals is one of its more useful properties. One of the largest markets for dextran is in the manufacture of photographic and x-ray films, where it is used to stabilize silver halide emulsions. Another large market is in aluminum manufacturing, where dextran solutions are sometimes used in the recovery of aluminum from bauxite ores. Dextran has been used as a binder in tobacco products, and its use in shaving creams and other cosmetics (qv) has also been suggested. The U.S. FDA status of dextran as a food additive is not clear. Although GRAS approval of low molecular weight dextran as a direct food additive was dropped in 1977 (207), *L. mesenteroides* is approved for use in fermented foods (see FOOD ADDITIVES). Because many foods of plant origin contain sucrose, it is virtually certain that any foods containing *L. mesenteroides* also contain dextran. In fact, some patents describing food applications of *L. mesenteroides* B-523 and similar proprietary strains rely on the production of insoluble gelling dextran for key properties as food ingredients (209–211). Other food uses have been proposed for unusual dextrans. Alternan and low molecular weight alternan fractions from *L. mesenteroides* B-1355 are being studied as bulking agents for reduced calorie foods (124), and dextran oligomers containing $\alpha(1 \rightarrow 2)$-linkages have been proposed for similar applications (211).

Dextran is produced commercially by fermentation of sucrose with *L. mesenteroides* B-512F (212). Typical media consist of sodium or potassium phosphate, sucrose or a sucrose source such as molasses, and a nitrogen and nutrient source such as yeast extract or corn steep liquor. The initial pH is adjusted to ~7, but unless the pH is controlled, it may fall to below 5 by the time growth is complete, due to lactic acid formation. Sucrose concentrations up to 10% have been used (213), but anything over 2% gives such high viscosities in the final stages of synthesis that 2% is usually the preferred concentration. If a clean product is required, cells must first be removed, typically by centrifugation. The polysaccharide is recovered by precipitation with a suitable solvent, such as methanol, ethanol, or 2-propanol, and dried to a white powder. The only U.S. manufacturer as of this writing is Pharmachem Corp. The price is about $11–13/kg ($5–6/lb), and U.S. sales are on the order of approximately $3 million per year. Another main supplier is Pharmacia–LKB (Sweden), which manufactures clinical-sized dextrans, Sephadex, and other derivatives, as well as industrial-grade dextran.

Although dextran is manufactured using traditional fermentation (qv) methods, there are advantages of using cell-free enzyme preparations to synthesize dextran (212). These advantages include better control over the synthetic process and greater ease of purification of the end product. The profit margin for dextran producers is not large, however, and it may be difficult to justify changes in processes requiring large capital investments. A systematic study of this subject has shown that existing plants could be adapted to separate enzyme and dextran production (214). Scale-up of the enzymatic process can give high dextran yields in shorter reaction times (215) (see ENZYME APPLICATIONS, INDUSTRIAL). Researchers have looked to the use of immobilized enzymes for improved synthesis of dextran. Immobilized dextransucrase is especially well suited for the production of low molecular weight, low viscosity oligosaccharides and clinical-sized dextrans (216). Problems are encountered, however, when larger, high viscosity dextrans are synthesized. The problem is particularly severe if the enzyme is immobilized using a porous gel; the pores become filled with entrapped dextran, and synthesis of new polysaccharide slows down and eventually comes to a halt. Research into new supports and methods of immobilization may overcome these problems.

Emulsan and Liposan

Microorganisms that degrade hydrocarbons and utilize them as carbon sources usually possess some way of rendering the hydrophobic hydrocarbons water soluble, generally by secreting some type of surfactant or emulsifying agent (see SURFACTANTS). Two of the best known compounds in this category are emulsan [80450-55-1] and liposan. Emulsan, first isolated from cultures of the oil-degrading bacterium *Acinetobacter calcoaceticus* strain RAG-1 (217), consists mainly of a heteropolysaccharide with amino sugars substituted by *O*-esterification with long-chain acyl groups (218). This combination of hydrophilic carbohydrate and hydrophobic fatty acid functionalities gives the emulsan molecule an amphiphilic character. The polymeric nature of the emulsan molecules causes the resultant emulsions (qv) to be very stable and not prone to phase separation. Emulsan can stabilize oil-in-water emulsions at concentrations as low as 1:1000 emulsan:oil (219). Many different strains of *Acinetobacter* produce emulsan-type compounds (220). Emulsan from strain RAG-1 (American Type Culture Collection (ATCC) strain 31012) has been shown to contain 2-amino-2-deoxy-D-galacturonic acid, 2,4-diamino-2,4,6-trideoxy-D-glucose, and 2-amino-2-deoxy-D-galactose (galactosamine). The amino groups are at least partially acetylated, and 3-hydroxybutyrate ester groups have been found. The fatty acyl chains are predominantly 2- and 3-hydroxydodecanoate, with one fatty acyl chain per trisaccharide repeating unit. The average molecular weight is approximately one million, corresponding to chain lengths of 3000 to 3500 monosaccharide units (219). In contrast, *A. calcoaceticus* strain BD4 secretes an emulsan composed of L-rhamnose, D-mannose, D-glucose, and D-glucuronic acid residues. Amino sugars and fatty acyl groups are conspicuously absent (21). It has been noted, however, that the purified polysaccharide from this strain possesses no emulsifying ability (221). It has been shown that both the yield and structure of emulsan can vary, depending on the carbon source and growth conditions (222).

Another microbial polysaccharide-based emulsifier is liposan, produced by the yeast *Candida lipolytica* when grown on hydrocarbons (223). Liposan is apparently induced by certain water-immiscible hydrocarbons. It is composed of approximately 83% polysaccharide and 17% protein (224). The polysaccharide portion consists of D-glucose, D-galactose, 2-amino-2-deoxy-D-galactose, and D-galacturonic acid. The presence of fatty acyl groups has not been demonstrated; the protein portion may confer some hydrophobic properties on the complex.

Applications.   Proposed uses of emulsan have included scouring crude oil from tankers and other containers, cleaning oil-handling equipment, dispersing oil-soluble pigments (see DISPERSANTS), enhanced oil recovery, and other emulsion-stabilizing applications. Field tests in the 1980s demonstrated its usefulness in forming easily pumped emulsions for the transport of crude oil (219). It was also found that oil-in-water emulsions of No. 6 fuel oil with emulsan could be used directly in combustion processes (219). Proposed consumer uses include cosmetics (qv), cleaning compounds, and as a food ingredient emulsifier. Emulsan has been marketed by Petroferm (United States).

Gellan

Gellan [*71010-52-1*] was successfully introduced as a replacement for agar in the early 1980s. Synthesized by the bacterium *Pseudomonas elodea* (*Sphingomonas elodea*, *Auromonas elodea*) (225–227), this anionic heteropolysaccharide consists of D-glucopyranosyl (Glc$p$), L-rhamnopyranosyl (Rha$p$), and D-glucopyranosyluronic acid (Glc$p$A) residues linked in repeating units:

$$[\longrightarrow 3)\text{-}\beta\text{-}\text{D-Glc}p\text{-}(1\longrightarrow 4)\text{-}\beta\text{-}\text{D-Glc}p\text{A-}(1\longrightarrow 4)\text{-}\beta\text{-}\text{D-Glc}p\text{-}(1\longrightarrow 4)\text{-}\alpha\text{-}\text{L-Rha}p\text{-}(1\longrightarrow]_n$$

The polysaccharide also contains O-acetyl groups linked to C-6 of approximately half of the 3-linked D-glucopyranosyl units (228), and L-glyceric acid groups esterified to position 2 of the same residues (229). Native gellan forms soft, elastic gels, whereas gels formed by deacylated gellan are more rigid and brittle (225). Gels are formed only in the presence of cations, and the presence of a chelating agent such as EDTA can prevent gel formation (see CHELATING AGENTS). Divalent cations such as calcium or magnesium give far stronger gels than monovalent cations like sodium or potassium. Gellan gels are typically thermoreversible, with 1 wt %/vol gels melting at temperatures just below 100°C and gelling at lower temperatures, usually in the range of 35–50°C (230). The gel properties are strongly dependent on the degree of acyl substitution, type and concentration of cation present, and polysaccharide concentration (225,230). Gellan was developed as a substitute for agar, and many of its properties are similar. The most notable differences are the dependence on cations for gel formation, and the fact that lower concentrations of gellan are required to give gels comparable to agar. For example, 1.5 wt %/vol agar is a typical concentration used in microbiological plate media. Similar gels can be made using gellan at concentrations of 0.5–0.8% in the presence of magnesium salts (225). Calcium–gellan gels are especially stable, and unlike agar gels show no syneresis (shrinkage and loss of water). As with agar, heating is required to achieve dissolution of the dried material. If gellan is dissolved in distilled ion-free water, no gel forms on cooling, but an increase in viscosity is observed (230). Fibers of gellan are formed by extruding heated solutions into a bath containing an aqueous solution of magnesium salts (231).

A left-handed double-helical structure has been proposed for gellan in the crystalline state, based on x-ray diffraction studies (227). The presence of acetyl groups presumably disrupts interchain aggregation, since these groups are postulated to be on the outside of the helices. The role played by acetyl and glyceryl ester groups and their influence on the double-helical structure has been studied using computer models (232).

Production and Utilization. Unlike many newly discovered microbial polysaccharides, gellan has become a commercial success in a relatively short period. Discovered in the late 1970s, gellan was patented in 1982 (233,234). It was first marketed as a replacement for agar in microbiological applications under the trade name GELRITE (Kelco Division, Monsanto (formerly with Merck

& Co.)). The advantages of gellan over agar include higher purity, better clarity, and the ability to obtain strong gels at lower polysaccharide concentrations. Gellan is especially useful in marine microbiology, where agar-degrading microbes are often encountered. Gellan has also found biotechnological applications in plant tissue culture and in cell immobilization (14,230). Another possible use is in air-freshener gels (16). In 1990 the U.S. FDA approved gellan for use as a food additive in icings, frostings, jams, jellies, and fillings, where it can be used as a replacement for agar and carrageenans (235,236). Use of gellan as a general food additive for stabilizing, thickening, and gelling was approved by the FDA in 1992 (237). Additional applications include structured meat products, pet foods, candies, cheeses, yogurt, dressings, sauces, and ice cream and other frozen desserts (13,238) (see FOOD ADDITIVES; MEAT PRODUCTS; MILK AND MILK PRODUCTS).

Gellan is made by fermentation in a medium containing glucose, salts, and a nitrogen source. The bacteria are grown for 2–3 days at 28–30°C, and the polysaccharide is recovered by precipitation with 2-propanol or ethanol (239). Aeration of the culture medium becomes difficult as the gellan concentration builds up. This is one of the primary difficulties in achieving high yields of the product. Gelling of the culture medium also causes problems in product recovery; one way of overcoming this is to dilute the final cultures with water. A method for post-fermentation modification of gellan to give a product containing fewer (<1%) acetyl groups and approximately 3–12% glyceryl groups has been patented (240). As for many commercially important microbial polysaccharides, attention has focused on cloning genes for gellan biosynthesis (241) (see BIOTECHNOLOGY; GENETIC ENGINEERING).

Welan

Welan is produced by an *Alcaligenes* species (ATCC-31555) by aerobic fermentation, and marketed under the trade name BIOZAN (Merck and Co., Inc.); early reports also referred to it as S-130 (229). The polymer is structurally similar to gellan, sharing the same backbone sequence. It has an additional side group of an α-L-rhamnopyranosyl or an α-L-mannopyranosyl (Man$p$) unit linked $(1 \rightarrow 3)$ to a β-D-glucopyranosyl unit in the backbone of the polymer:

$$[\rightarrow 3)\text{-}\beta\text{-D-Glc}p\text{-}(1 \rightarrow 4)\text{-}\beta\text{-D-Glc}p\text{A-}(1 \rightarrow 4)\text{-}\beta\text{-D-Glc}p\text{-}(1 \rightarrow 4)\text{-}\alpha\text{-L-Rha}p\text{-}(1 \rightarrow]_n$$

$$3$$
$$\uparrow$$
$$1$$
$$\alpha\text{-L-Rha}p \text{ or } \alpha\text{-L-Man}p$$

The presence of the L-form of mannose is unusual. The side-chain substitution is randomly distributed (242); approximately two-thirds of the side chains are rhamnose. The repeat unit may also contain an O-acyl group, but the distribution of these units has not been completely determined. The polymer is moderately soluble in water but is insoluble in isopropanol solutions, which are used to obtain the polymer from the culture medium. A method for producing a rapidly hydrating form of welan is available (243).

Solutions of welan are very viscous and pseudoplastic, ie, shear results in a dramatic reduction in viscosity that immediately returns when shearing is stopped, even at low polymer concentrations (230). They maintain viscosity at elevated temperatures better than xanthan gum; at 135°C the viscosity half-life of a 0.4% xanthan gum solution is essentially zero, whereas a welan gum solution has a viscosity half-life of 900 minutes (230). The addition of salt to welan solutions slightly reduces viscosity, but not significantly. It has excellent stability and rheological properties in seawater, brine, or 3% KCl solutions (see RHEOLOGICAL MEASUREMENTS).

Applications. The high heat tolerance and good salt compatibility of welan gum indicate its potential for use as an additive in several aspects of oil and natural gas recovery. Welan also has suspension properties superior to xanthan gum, which is desirable in oil-field drilling operations and hydraulic fracturing projects. It is compatible with ethylene glycol, and a welan–ethylene glycol composition that forms a viscous material useful in the formulation of insulating materials has been described (244).

Levan

Levan [9013-95-0] has been the subject of numerous studies, most of which have focused on structure and biosynthesis (245,246). The term levulan first appeared in 1881 (247), and was used to describe a gum consisting of fructose (levulose) units that had been formed by microbial action on molasses. It was named by analogy with the dextrose-containing gum known as dextran. By 1931, when the first detailed structural characterization was published (248), the name had been shortened to levan. Levan is a homopolysaccharide consisting of $\beta(2\rightarrow6)$-linked D-fructofuranoside units, with branch chains linked to the main chain via $\beta(2\rightarrow1)$ linkages. The β-linked fructose units impart a strongly levoratatory effect on polarized light. Levan is synthesized by a number of bacteria, some of the better known examples being *Erwinia herbicola* (formerly *Aerobacter levanicum*) (249), *Bacillus amyloliquefaciens* (250), *Bacillus subtilis* (251–253), *Bacillus polymyxa* (254), *Zymomonas mobilis* (255), *Actinomyces viscosus* (256), *Gluconobacter oxydans* (257), a *Serratia* species (258), and various species of *Arthrobacter* (259), *Corynebacterium* (260,261), *Streptococcus* (262–264), and *Pseudomonas* (251,265,266). It is synthesized from sucrose by the extracellular enzyme levansucrase (β-2,6-fructan:D-glucose 6-fructosyltransferase, EC 2.4.1.10), which incorporates the fructosyl moiety of sucrose into levan and releases the glucose portion (249,252). In this respect, levan is analogous to the dextran family of polysaccharides; however, levansucrase can also utilize the sucrose-containing trisaccharide raffinose [512-69-6] as a substrate for levan synthesis, releasing the disaccharide melibiose as a side-product. This reaction is often used to differentiate between dextransucrase and levansucrase in unknown microbial preparations.

Levans are water-soluble, nongelling, and generally of lower viscosity than most other gums and polysaccharides, despite their often high molecular weights. The average molecular weight can vary considerably, and depends on the conditions of biosynthesis (267), usually falling in the range between 10^6 and

10^7. The reasons for the variation can be best explained in terms of the mechanism of enzymatic biosynthesis (268,269). Not only does the enzyme levansucrase transfer fructosyl units from sucrose to growing levan chains, it can also transfer fructosyl units from sucrose to a wide variety of carbohydrate acceptors, as well as to water. Moreover, it is also capable of transferring fructosyl residues from the ends of levan chains to water, albeit at a much slower rate. Therefore, the molecular sizes of levan chains depend on the presence or absence of other carbohydrates in the synthetic reaction mixtures and on the length of time the enzyme acts on the levan after all of the sucrose is consumed. The source and purity of the enzyme, reaction temperature, and other factors may also play a role (246).

Levan's chemical structure also varies, especially in the degree of branching. The levan produced by *E. herbicola* was shown by methylation analysis to contain 16–18% $\beta(2 \rightarrow 1)$ branch linkages, whereas levan from a *Corynebacterium* species analyzed in the same manner contained only 3–6% branch linkages. The levan produced by *L. mesenteroides* strain B-512 as a minor fraction of the total extracellular polysaccharide contained between 10 and 22% branching (270). A strain of *Streptococcus salivarius* produces a levan with 11–14% $\beta(2 \rightarrow 1)$ branch linkages (271); the branches have been shown to be at least 4–6 residues in length. A *Bacillus polymyxa* levan analyzed by ^{13}C-nmr spectroscopy and methylation was found to contain 12–13% branching (272). The levan from a species of *Serratia* has been reported to contain a much lower than average degree of branching, as determined by enzymatic and spectroscopic methods (258), but methylation data for this levan are not available. Branching in *Bacillus subtilis* levan has been shown to be dependent on the conditions of biosynthesis (273). It has been reported that levan sometimes bears a nonreducing sucrose group at what would otherwise be considered its reducing end. In the case of levan synthesized from raffinose, the presence of a raffinose end group has been reported. These are presumed to arise from the transfer of fructosyl units (or a levan chain) to sucrose or raffinose. Often such terminal residues are not detected; this may simply be a reflection of the difficulty in detecting a single residue within a large polymer, or it may be due to the fact that water and other molecules can also act as acceptors (246,269), resulting in any one of these acceptors being present at the reducing terminus.

Production and Utilization.   Although many uses have been proposed for levan (246), it is not being manufactured or used for any applications as of this writing. Low inherent viscosity and sensitivity to hydrolysis render it unsuitable as a thickening or viscosifying agent, and many of the applications for which it has been proposed (274) are better served by the more stable and readily available dextran. However, some specialty application for which its unique properties are well-suited may be found. One proposal was to use levan as a blood plasma extender, in a manner similar to clinical dextran (275). The antigenicity of early levan preparations, and the widespread acceptance of dextrans, prevented levan from being more seriously considered for this purpose. A variety of food uses have been proposed for levan (246,276). These include use as fillers (qv), bulking agents, and as a substitute for gum arabic, as well as a source of fructooligosaccharides analogous to those known commercially as Neosugar (Meiji Seika Kaisha, Ltd.).

Two approaches can be used to synthesize levan: classical fermentation (277) and direct enzymatic synthesis (278,279). Both require an organism that produces high levels of levansucrase. A number of researchers have directed their efforts at improving the methods for manufacturing the polysaccharide. A new strain of *B. polymyxa* capable of producing elevated levels of levan in the culture broth (254), conditions for production of a high molecular weight levan by *Z. mobilis* (280), and culture conditions for enhanced secretion of the constitutive extracellular enzyme from *E. herbicola* has been reported (281). In order to produce levan by cell-free enzymatic synthesis, levansucrases from *Bacillus* spp. have been immobilized by a variety of methods, including binding to activated silica (282) and a ceramic matrix (283), and by adsorption on hydroxyapatite (284). Enzymes immobilized by the latter method were found to have increased levels of polymerase activity, and gave good yields of a high molecular weight product.

Pullulan

Pullulan [*9057-02-7*], first described in detail in 1959, is a water-soluble extra-cellular α-D-glucan elaborated by the fungus *Aureobasidium pullulans* (formerly *Pullularia pullulans*) (285). It is a linear polymer of maltotriose units linked from the reducing end of one trisaccharidic unit to the nonreducing end of the next trisaccharidic unit by $\alpha(1 \rightarrow 6)$ linkages (286):

$$[\rightarrow 6)\alpha\text{-D-Glc}p(1 \rightarrow 4)\alpha\text{-D-Glc}p(1 \rightarrow 4)\alpha\text{-D-Glc}p(1 \rightarrow]_n$$

Pullulan also contains some maltotetraose units within the chains, to approximately 5–7% (287,288). Occasional reports of other types of linkages and units in pullulan preparations (71) may be attributable to contamination by other polysaccharides also known to be produced by *Aureobasidium pullulans* (289), although enzymatic studies have suggested that pullulan itself may contain a small percentage of atypical units or branch linkages (290). The molecular weight of pullulan varies from one preparation to the next, with values as low as 10,000 and as high as 25 million having been reported. It is an essentially linear polymer and the viscosity of aqueous solutions is often used as a reliable measure of pullulan molecular weight (291). The relationship between viscosity and molecular size indicates pullulan chains are flexible and coil-like in aqueous solutions (292). Molecular modeling suggests that much of this flexibility arises from the $\alpha(1 \rightarrow 6)$ linkages, which interrupt the $\alpha(1 \rightarrow 4)$ sequences and add an extra degree of rotational freedom (293).

Pullulan is generally produced in liquid fermentations, and its accumulation causes a marked increase in the viscosity of the medium. However, as cultures age, the viscosity usually drops off. This lowering of viscosity results from enzymatic cleavage of the maltotetraosyl regions by an endogenous amylase-like enzyme (287,291). This phenomenon is a key factor in determining the physical properties of pullulan produced by fermentation. The fungus is capable of producing pullulan from a number of carbohydrate substrates (294,295), but the ones generally used are glucose, sucrose, or starch hydrolyzates. Biosynthesis is accomplished via a lipid-linked intermediate (296), with UDP–glucose serving as the initial glucosyl donor. Experimental evidence suggests that UDP–glucose is incorporated into lipid-linked panosyl or isopanosyl intermediates, which are then assembled into pullulan (297).

   Production and Utilization.   There are many patents and publications covering applications for production of pullulan (71,298–301). Hayashibara Biochemical Laboratories, Inc. (Japan) manufactures pullulan for a number of applications; the market price in 1990 was approximately $11.50/kg (21). Pullulan can be used in foods as a viscosifier and low calorie partial replacement for starch, and as a binder. Its unique film-forming and oxygen-barrier properties (299) make it especially useful in protective and adhesive edible coatings (qv) (301). Other applications have been suggested in degradable films and fibers, paper coatings and binders, cosmetics, pharmaceutical tablet coatings, and even in soluble contact lenses (qv) which contain slow-release bioactive medicines (301). Pullulan fractions of narrow molecular weight ranges are available for use as standards in gel-permeation chromatography.

   Much of the research on pullulan is focused on improving methods for its production. A technique has been described for synthesis of pullulan using immobilized cells of *A. pullulans* (302). Other studies have led to improved fermentation conditions (303), bioreactor systems (304), and methods for recovering pullulan from fermentation broths (305,306). It is generally acknowledged that there is a need for strains of the fungus capable of producing high yields of pullulan that has not been degraded by endogenous amylolytic enzymes, and that is free of contamination by the melanin pigments produced by most strains of *Aureobasidium*. Some progress has been made in this area with newly isolated color variants (307,308) and mutant forms of the fungus (309,310).

Scleroglucan

Scleroglucans [*39464-87-4*] are neutral, branched homopolysaccharides composed of glucose residues (311,312). They are produced by fungi of the genus *Sclerotium*, which are plant parasites in the Basidiomycete family. The main chain of scleroglucan consists of β-D-glucopyranosyl residues linked $(1\rightarrow3)$ with every third sugar bearing a single D-glycopyranyl residue linked $\beta(1\rightarrow6)$ (313). The polysaccharide is insoluble in 2-propanol, which is used to isolate and concentrate the material.

   Scleroglucan exists in a triple helical conformation that is highly stable (314). The D-glucopyranosyl side groups project to the outside of the helix (312) and prevent the aggregation of helices, which would result in insolubility, as in the case of curdlan (*vide infra*). The transition from helix to coil occurs at temperatures above 90°C or by increasing the pH to 12 (315,316). Solutions cooled below 8°C form weak gels (317). The molecular weight of the polymer depends on the time the culture is harvested; scleroglucan from a 10-hour culture has an average molecular weight of 1.2×10^6, whereas polymer from a 20-hour culture has an average value of 2.6×10^6 (318). Solutions of scleroglucan show pseudoplastic flow behavior during shear; the viscosity is relatively temperature insensitive and remains virtually constant from 10 to 90°C (312).

Production and Utilization. The most important potential use for scleroglucan is as a mobility control agent for enhanced oil recovery. Many of its rheological properties, such as the production of highly viscous solutions at low concentrations, and excellent long-term stability at elevated temperature and salt concentrations, are similar to those of xanthan gum. In a study of 140 synthetic and native biopolymers, scleroglucan was found to be the most thermostable in a synthetic North Sea brine at 90°C for extended time periods (319). A strategy to improve the dispersibility and filterability of scleroglucan preparations utilized a precipitant to form a coagulum, followed by the addition of a surfactant prior to drying and grinding; the dried mixture was reportedly easier to dissolve than dried native material (320). A problem with native scleroglucan is that the polymer readily adsorbs to rock, which can lead to pore plugging and reduced flow (312).

Scleroglucan, like other β-glucans, can stimulate an immune response and repress the development of some forms of cancer when administered intravenously (312,321). A scleroglucan derived from *Sclerotium glucanicum* stimulated *in vivo* murine macrophage phagocytic activity by 66%, and enhanced murine bone marrow proliferation by up to 300% (321) (see IMMUNOTHERAPEUTIC AGENTS). The polysaccharide is produced commercially by fermentation in a medium containing glucose, corn steep liquor, nitrate, and mineral salts, which is cultured at 28–30°C for approximately 60 hours (312). Two fermentation products are available: native scleroglucan, containing some residual fungal mycelia, and a refined-grade product that has been sheared and filtered to yield mycelium-free scleroglucan.

Curdlan

Curdlan [*54724-00-4*] is a neutral $\beta(1\rightarrow3)$-linked D-glucan produced by several bacteria, primarily *Alcaligenes faecalis* var. *myxogenes*, as well as by *Agrobacterium radiobacter*, *Rhizobium meliloti*, and *R. trifolii* (322). Prior to 1968, when a mutant strain of *A. faecalis* var. *myxogenes* that produced only curdlan was isolated (323), the *A. faecalis* strain used by most investigators produced both curdlan and the acidic polysaccharide succinoglycan. The *Agrobacterium* and *Rhizobium* species that produced curdlan also produced succinoglycan (322). The isolation of the mutant producing only curdlan simplified preparation of the polymer for characterization of its physical properties.

Curdlan is insoluble in water, but does absorb some water to become swollen. It can be dissolved in alkaline solutions, formic acid (see FORMIC ACID AND DERIVATIVES), and certain organic solvents such as dimethyl sulfoxide (322). Suspensions of curdlan heated above 54°C form a firm gel. A closely related polysaccharide, pachyman, which has a $\beta(1\rightarrow3)$-D-glucan main chain and a small number of $\beta(1\rightarrow6)$-linked D-glucose branches, does not form gels when heated, even at higher concentrations. The lack of side groups in curdlan apparently allows the polymer to form aggregates that are insoluble and to form gels at moderately warm temperatures. Curdlan forms a low set gel when heated to 55°C and allowed to cool (324); high set gels obtained by heating curdlan suspensions at higher temperatures have melting temperatures of 140–160°C (325). According to conformational studies, the solid and low set gelled polymer is a single-stranded helix (326); whereas the high set gelled polymer is a triple-stranded helix (322,327,328) with strands from several helices intertwining to form three-dimensional networks.

Applications.   Several food uses have been proposed for curdlan including jellies, jams, noodles, and tofu (322). Its gelling properties make it useful for the preparation of instant puddings and multiple layer puddings. It has been suggested that it may be useful as a stabilizing agent in frozen desserts such as ice creams (322). Curdlan can be added to bind water, add stability, and improve the body and gloss of food products.

Succinoglycan

Succinoglycan is an acidic extracellular polysaccharide produced by several bacteria, including *Alcaligenes faecalis*, *Agrobacterium radiobacter*, *Rhizobium meliloti*, and *R. trifolii* (322). The polymer consists of a repeating octasaccharide, as shown for succinoglycan from *Alcaligenes faecalis* var. *myxogenes* (329).

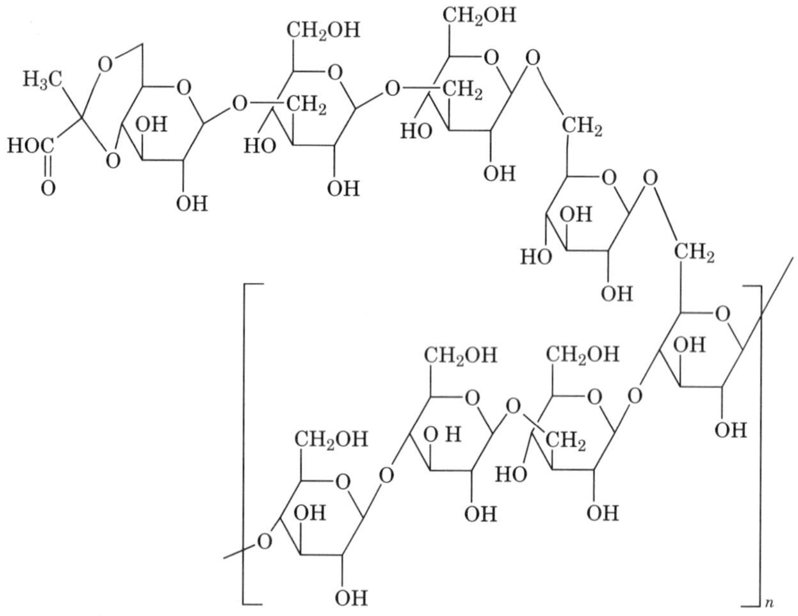

The name succinoglycan was derived from the presence of succinic acid covalently bound to the side chain of the polymer from *A. faecalis* (329,330). In addition, the terminal D-glucosyl residue of the side chain contains pyruvic acid linked as a ketal to the glucopyranose ring (322). Structural determination of the polysaccharide produced by *R. meliloti* and several strains of *Agrobacterium* has shown that the ratio of succinate and pyruvate groups varies slightly between species, and that the succinoglycan produced by some strains also has some covalently bound acetate. Structural characterization by fast-atom bombardment mass spectrometry (qv) of succinoglycan oligosaccharide from *R. meliloti* has determined the location of succinyl and acetyl modifications (331).

Succinoglycan can form viscous solutions if in the free acid or calcium salt form, but produces low viscosity solutions if in the sodium salt form (332). The viscosity of solutions is quite temperature dependent; increasing the temperature leads to a reduction in viscosity, which approaches zero at approximately 60°C (322). The viscosity is relatively stable from pH 3–10 and is compatible with a number of inorganic salts other than sodium. The production of succinoglycan and its potential use in foods and industrial processes as a thickening agent has been described (322).

Xanthan Gum

Xanthan gum [*11138-66-2*] is an anionic heteropolysaccharide produced by several species of bacteria in the genus *Xanthomonas*; *X. campestris* NRRL B-1459 produces the biopolymer with the most desirable physical properties and is used for commercial production of xanthan gum (see GUMS). This strain was identified in the 1950s as part of a program to develop microbial polysaccharides derived from fermentations utilizing corn sugar (333,334). The primary structure of xanthan has been determined by chemical degradation and methylation analysis (335,336); it is composed of repeating units consisting of a main chain of D-glucopyranosyl residues with trisaccharide side chains made up of D-mannopyranosyl and D-glucopyranosyluronic acid residues.

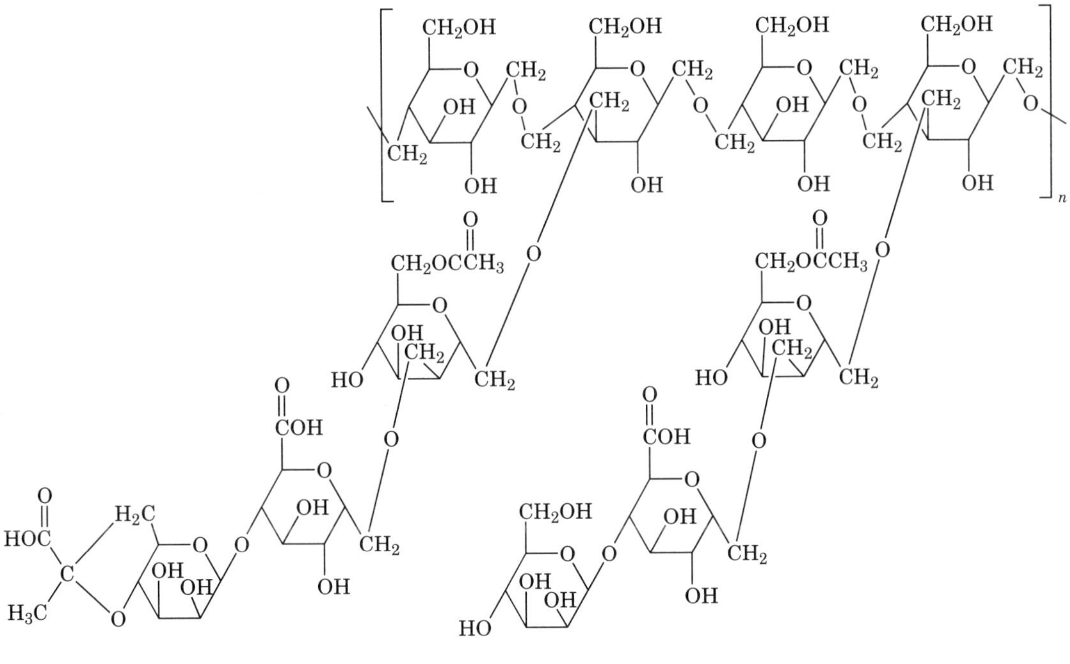

Approximately half of the terminal D-mannosyl residues have pyruvate present as 4,6-O-(1-carboxyethylidene) substituents. The pyruvate content has been shown to vary with culture conditions (337,338) and strains of X. campestris (339). The internal D-mannosyl residue is acetylated at the O-6 position.

Xanthan gum has several desirable physical properties that explain the wide application range developed for this polysaccharide (6,24). The viscosity of xanthan gum solutions is highly pseudoplastic. Relatively low concentrations of the biopolymer produce highly viscous solutions that maintain viscosity over wide ranges of temperature and pH. Mono- and divalent cations enhance the stability of solution viscosity to exposure to elevated temperatures (340). The addition of salt to solutions of xanthan results in essentially stable solution viscosity from pH 1.5 to 13; salt also increases the yield value, ie, suspending power, of xanthan gum solutions (230). There is a surprisingly small effect on viscosity over a relatively large range of salt concentrations even though xanthan gum is a polyelectrolyte. At a polysaccharide concentration of approximately 0.35%, there is essentially no change in solution viscosity between 0.01 and 1% KCl. At concentrations of polysaccharide lower than 0.35%, the presence of salt slightly lowers the viscosity, whereas at concentrations higher than 0.35% gum, the presence of salt results in a slight increase in viscosity (230).

Using a variety of measurement techniques, a relatively wide range of molecular weights have been reported for xanthan gum preparations ($\sim$2–50 $\times$ 10^6). Two preparations of native xanthan gum have molecular weights of 13 and 50 $\times$ 10^6, as determined by light scattering measurements (341). By measuring the contour length of the molecule from electron micrographs, an estimate of 20 $\times$ 10^6 has been made (342). Sedimentation studies have yielded an average value of 7.6 $\times$ 10^6 (343) and a range of values of 4–12 $\times$ 10^6 (344). Using size-exclusion chromatography, average molecular weights of 1.8 and 2.4 $\times$ 10^6 have been measured (345). Low angle laser light scattering has given molecular weight estimations in the range of 4.1–12.2 $\times$ 10^6 (346).

Structure and Conformation.   The conformation of the biopolymer in solution has been a subject of debate. Several analytical methods (qv) have shown that the polymer goes through a phase transition at elevated temperatures, which depends on the concentration of salt. Increasing the temperature at low salt concentrations brings about a conformational shift from an ordered state to a disordered, random coil conformation (334). The change is associated with a decrease in solution viscosity and can also be characterized by changes in optical rotation and circular dicroism (347–349), nuclear magnetic resonance (348), and electron microscopy (342,350,351). The transition temperature increases with increasing sodium or calcium ion concentration, as well as with the acetate content of the polysaccharide, and decreases with increasing pyruvate content. The transition temperature is apparently independent of polymer concentration; this has been interpreted as evidence that the polymer is a single-stranded helix (352), and this conclusion has been supported by spectroscopic analysis of the polymer in solution (353–355). The crystalline structure of xanthan gum has been examined by x-ray diffraction (356) and reported to be a single-stranded helix with a diameter of 2 nm, but the possibility that the molecule could form a double-stranded helix has not been ruled out. Evidence that the native structure of xanthan is a double-stranded helix has been provided by light-scattering experiments (357). As of this writing, it appears that xanthan generally has a double-helical conformation but can be treated in a fashion, eg, prolonged dialysis at low ionic strength, which causes it to assume a single-helical conformation (349).

The structure of xanthan has also been investigated by electron microscopy. A series of micrographs of native and denatured xanthan preparations show stiff, relatively straight rod-shaped structures, ranging from 2 to 10 μm long and 4 nm in diameter (342). Xanthan gum prepared using several methods shows sections of single- and double-stranded structures within a single polymer strand (351); apparently the polymer becomes partly unwound under the particular conditions used to prepare the sample for analysis. The method of preparation and treatment of the sample is important in determining what structural features will be present in the polymer, and treatments exist to change the polymer from one conformation to another. The effect of salt on single- and double-stranded structures has been studied (358), and a model for salt-induced extension and dissociation of native xanthan gum involving a two-step mechanism proposed. Lowering the salt concentration results in an extended double-stranded polymer; at even lower salt concentrations, dissociation of the double helix into single strands occurs (359).

Studies utilizing enzymes capable of hydrolyzing the cellulosic backbone chain indicate that the polymer is incompletely substituted with trisaccharide side chains on alternating glucose residues. An excess of glucose in a high molecular weight fragment obtained by hydrolysis of native xanthan gum using a heat-stable xanthanese mixture (360), suggests that this hydrolytic fragment contains cellulosic regions lacking in side chains. When disordered xanthan gum was hydrolyzed using a fungal cellulase preparation the product had more glucose than the amount expected if the backbone chain were fully substituted (361).

Production and Utilization.   The nutritional requirements of X. *campestris* have been studied in order to optimize the production of xanthan gum. Fermentations for the industrial production of xanthan gum are done at 28°C, and utilize glucose concentrations from 1–5% (362). Higher glucose concentrations do not result in higher levels of gum biosynthesis. Saccharides such as sucrose, starch, and maltodextrins can also be used for gum production. The use of a completely defined media for gum production has been described (230). It has also been shown that some organic acids including pyruvic, succinic, and α-ketoglutaric acids increase the production of xanthan gum. It is necessary to maintain a neutral pH during fermentation in order to obtain maximal yields; during polymer biosynthesis the medium becomes acidic, but can be neutralized by the addition of a suitable base.

A typical commercial production process starts with an innoculum of X. *campestris*, prepared in a yeast extract containing broth, which is added to the production medium consisting of 3% D-glucose, 0.5% potassium phosphate buffer, 0.4% dried distillers solubles, and 0.01% magnesium sulfate (230). The stirred, aerated culture is held at 28°C and after 96 hours approximately 50% of the glucose is converted into polymer (230). The medium is highly viscous, and water is added to reduce the viscosity to improve the removal of cells. The fermentation broth is heated to nearly 100°C to pasteurize the liquid, and after cooling the polysaccharide is precipitated with ethanol or 2-propanol. The process of heating also causes the strands of polysaccharide to separate, and slow cooling allows the strands to reanneal between helices and form networks, improving the viscosity. The recovery of polymer is enhanced if potassium chloride is added to the mixture. The alcohol is removed and the material dried and packaged. Food-grade xanthan is tested during production and packaging to ensure that the material is not contaminated by microorganisms. It has been estimated that in 1989 the yearly production of xanthan gum for food markets was $18-22 \times 10^6$ kg ($8-10 \times 10^6$ lbs), with a value of $\$44-60 \times 10^6$ (230).

The unusual rheological properties of xanthan gum have led to its use in a wide variety of food and industrial applications. Uses of xanthan gum in food products have been reviewed (230). It is frequently the thickener of choice because of its ability to maintain solution viscosity (for emulsion stabilization) in salts and acids, such as are found in salad dressings. It has been used extensively in no oil or low oil salad dressing formulations, and has been shown to provide long-term emulsion stability. U.S. FDA Standards of Identity also permit its use in sauces, puddings, bakery and pie fillings, and dry mixes for beverages. Xanthan gum can be mixed with carrageenan and galactomannans such as locust bean gum to improve gelling and stabilization of frozen dairy products and desserts. The synergistic interactions of xanthan gum and the galactomannans can be used in situations where fast gelling time is desirable or to reduce costs of additives (230).

Large quantities of xanthan gum are used by the oil and natural gas industry in several aspects of hydrocarbon production (230,349) (see GAS, NATURAL; HYDROCARBONS; PETROLEUM). The high viscosity achievable at low concentration and the high suspending power efficiently remove bit cuttings while reducing friction substantially throughout the drill string. Xanthan is compatible with many additives frequently used to prepare drilling fluids, and drilling fluids can be made up of fresh, brackish, or salt water and still maintain viscosity. It can effectively thicken hydraulic fracture fluids that are employed to improve porosity in subterranean formations, where viscosity is required to suspend a propping additive such as sand which is then pumped underground into newly formed fissures. The excellent heat stability and salt compatibility frequently result in its selection over lower cost viscosifiers. Xanthan is also added to solutions pumped underground that are used to displace oil toward a collection well. It improves the sweeping efficiency of these flooding fluids enough to make them cost effective.

Other industrial uses for xanthan gum include thickening textile and carpet printing pastes, suspending pigments in ceramic glazes to improve glaze dispersion, and ink and clay coating formulations in the printing and paper (qv) industries, respectively (230). Agrochemical producers blend herbicides (qv) and insecticides (qv) with xanthan in order to improve application to plants.

Other Microbial Polysaccharides

There are a number of other polysaccharides from fungi and bacteria which have actual or proposed practical applications. These include polysaccharides whose usefulness derives from particular structural features which are important not because of their viscosity or gelling ability, but because they elicit a particular biological response. The *Haemophilus influenzae* vaccine, composed in large part of the capsular polysaccharide of this organism, has prevented many deaths from meningitis since its introduction in the early 1980s. An earlier polysaccharide-based vaccine against *Streptococcus pneumoniae* was successful in preventing many types of bacterial pneumonia, but fell into disuse with the introduction of newer antibiotics (qv). The increased incidence of antibiotic-resistant bacterial infections has resulted in renewed interest in such vaccines, and the use of pneumonia vaccines is on the rise. Other polysaccharide-based vaccines have also been developed, and their chemistry and biology reviewed (363,364).

Some polysaccharides elicit a more general type of immune response. These immunomodulator polysaccharides often fall into two categories: sulfated polysaccharides and β-D-glucans. The biological activity of dextran sulfate has been discussed; other sulfated polysaccharides also show anticoagulant activity as well as immunomodulator activity. Some examples include curdlan sulfate, which exhibits anti-HIV activity (365), and sulfated yeast glucan, which enhances immune resistance against bacterial, viral, and fungal infections, and also shows antitumor activity (366). The β-D-glucans, are not only microbial in origin, but can also be found in plants, fungi, and macroalgae. Many of the studies regarding these polysaccharides have been reviewed (367). One compound of this type is Betafectin, derived from yeast cell wall β-D-glucan (368).

Many researchers continue to search for microbes that produce better polysaccharides, and higher yields (369–372). Many polysaccharides with exceptional qualities have been discovered and investigated, but few have been successfully commercialized (372). Because many of the proposed applications of new polysaccharides are as food ingredients, a primary obstacle to successful introduction of products is the need to obtain regulatory approval. This long and expensive process is a strong disincentive to introduce a product unless it has clear and overwhelming advantages over currently used materials. Some of these advantages might be a significantly lower price or functional properties not yet available. There appears to be a need for inexpensive carbohydrate-based bulking agents for use in artificially sweetened foods, and for a replacement for gum arabic [9000-01-5]. Gum arabic, a plant polysaccharide, is imported to the United States from Africa, and the price and availability are reported to fluctuate unpredictably. A microbially produced replacement could bring down the cost and create a more stable supply.

BIBLIOGRAPHY

"Microbial Polysaccharides" in *ECT* 3rd ed. Vol. 15, pp. 439–458, by M. E. Slodki, Northern Regional Research Center, U.S. Dept. of Agriculture.

1. J. F. Robyt, *Trends Biochem. Sci.* **4**, 47–49 (1979).
2. R. W. Stoddart, *The Biosynthesis of Polysaccharides*, MacMillan, New York, 1984.
3. I. W. Sutherland, *Adv. Microbial. Physiol.* **23**, 79–150 (1982).

4. I. W. Sutherland, in A. Mitsui and C. C. Black, eds., *CRC Handbook of Biosolar Resources*, Vol. I, Part 1, *Basic Principles*, CRC Press, Boca Raton, Fla., 1982, pp. 363–371.

5. T. H. Evans and H. Hibbert, *Adv. Carbohydr. Chem.* **2**, 203–233 (1946).

6. P. A. Sandford and A. Laskin, eds., *Extracellular Microbial Polysaccharides*, American Chemical Society, Washington, D.C., 1977.

7. M. E. Slodki and M. C. Cadmus, *Adv. Appl. Microbiol.* **23**, 19–54 (1978).

8. P. A. Sandford, *Adv. Carbohydr. Chem. Biochem.* **36**, 266–313 (1979).

9. G. G. Geesey, *ASM News* **48**, 9–14 (1982).

10. M. E. Bushell, ed., *Progress in Industrial Microbiology*, Vol. 18, *Microbial Polysaccharides*, Elsevier, Amsterdam, the Netherlands, 1983.

11. P. A. J. Gorin and E. Barreto-Bergter, in G. O. Aspinall, ed., *The Polysaccharides*, Vol. 2, Academic Press, Inc., Orlando, Fla., 1983, pp. 365–409.

12. L. Kenne and B. Lindberg, in Ref. 11, pp. 287–363.

13. J. K. Baird, P. A. Sandford, and I. W. Cottrell, *BioTechnology* **1**, 778–783 (1983).

14. G. T. Colegrove, *Ind. Eng. Chem. Prod. Res. Dev.* **22**, 456–460 (1983).

15. R. H. Marchessault, *Chemtech* **14**, 542–552 (1984).

16. P. A. Sandford, I. W. Cottrell, and D. J. Pettitt, *Pure Appl. Chem.* **56**, 879–892 (1984).

17. I. W. Sutherland, *Ann. Rev. Microbiol.* **39**, 243–270 (1985).

18. M. Yalpani, *Progress in Biotechnology 3, Industrial Polysaccharides, Genetic Engineering, Structure/Property Relations and Applications*, Elsevier, Amsterdam, the Netherlands, 1987.

19. C. Whitfield, *Can. J. Microbiol.* **34**, 415–420 (1988).

20. I. C. M. Dea, *Pure Appl. Chem.* **61**, 1315–1322 (1989).

21. I. W. Sutherland, *Biotechnology of Microbial Exopolysaccharides*, Cambridge University Press, Cambridge, U.K., 1990.

22. J. D. Linton, S. G. Ash, and L. Huybrechts, in D. Byrom., ed., *Biomaterials, Novel Materials from Biological Sources*, Stockton Press, New York, 1991, pp. 215–261.

23. V. J. Morris, *Agrofood Industry Hi-Tech* **3**, 3–8 (1992).

24. R. L. Whistler and J. N. BeMiller, eds., *Industrial Gums, Polysaccharides and their Derivatives*, Academic Press, Inc., San Diego, Calif., 1993.

25. A. Linker and R. S. Jones, *Nature* **204**, 187–188 (1964).

26. P. A. J. Gorin and J. F. T. Spencer, *Can. J. Chem.* **44**, 993–998 (1966).

27. G. L. Cote and L. H. Krull, *Carbohydr. Res.* **181**, 143–152 (1988).

28. M. G. de la Vega, F. J. Cejudo, and A. Paneque, *Appl. Biochem. Biotechnol.* **30**, 273–284 (1991).

29. L. M. Likhosherstov and co-workers, *Carbohydr. Res.* **222**, 233–238 (1991).

30. L. R. Evans and A. Linker, *J. Bacteriol.* **116**, 915–924 (1973).

31. A. J. Hacking, I. W. F. Taylor, T. R. Jarman, and J. R. W. Govan, *J. Gen. Microbiol.* **129**, 3473–3480 (1983).

32. W. F. Fett, S. F. Osman, M. L. Fishman, and T. S. Siebles, *Appl. Environ. Microbiol.* **52**, 466–473 (1986).

33. G. T. Grant, E. R. Morris, D. A. Rees, P. J. C. Smith, and D. Thom, *FEBS Lett.* **32**, 195–198 (1973).

34. B. Larsen and A. Haug, *Carbohydr. Res.* **17**, 287–296 (1971).

35. A. Haug and B. Larsen, *Carbohydr. Res.* **17**, 297–308 (1971).

36. D. F. Pindar and C. Bucke, *Biochem. J.* **152**, 617–622 (1975).

37. G. Skjak-Braek and B. Larsen, *Carbohydr. Res.* **139**, 273–283 (1985).

38. G. Annison and I. Couperwhite, *Appl. Microbiol. Biotechnol.* **25**, 55–61 (1986).

39. H. Obika, J. Sakakibara, and Y. Kobayashi, *Biosci. Biotechnol. Biochem.* **57**, 332–333 (1993).

40. I. W. Davidson, I. W. Sutherland, and C. J. Lawson, *J. Gen. Microbiol.* **98**, 603–606 (1977).

41. G. Skjak-Braek, B. Larsen, and H. Grasdalen, *Carbohydr. Res.* **145**, 169–174 (1985).
42. J. Preiss and G. Ashwell, *J. Biol. Chem.* **237**, 309–316 (1962).
43. L. Pike, R. D. Humphrey, and O. Wyss, *Life Sci.* **15**, 1657–1663 (1974).
44. I. W. Davidson, C. J. Lawson, and I. W. Sutherland, *J. Gen. Microbiol.* **98**, 223–229 (1977).
45. L. Kennedy, K. Mcdowell, and I. W. Sutherland, *J. Gen. Microbiol.* **138**, 2465–2471 (1992).
46. K. Clare, in Ref. 24, pp. 105–143.
47. G. Skjak-Braek, *Biochem. Soc. Trans.* **20**, 27–33 (1992).
48. I. W. Sutherland, in D. Byrom., ed., *Biomaterials, Novel Materials from Biological Sources*, Stockton Press, New York, 1991, pp. 307–331.
49. P. Gacesa, *Carbohydr. Polymers* **8**, 161–182 (1988).
50. L. Deavin, T. R. Jarman, C. J. Lawson, R. C. Righelato, and S. Slocombe, in Ref. 6, pp. 14–26.
51. W.-P. Chen, J.-Y. Chen, S.-C. Chang, and C.-L. Su, *Appl. Environ. Microbiol.* **49**, 543–546 (1985).
52. J. A. M. Fyfe and J. R. W. Govan, in Ref. 10, pp. 45–83.
53. T. Ren, N. B. Pellerin, G. L. Graff, I. A. Aksay, and J. T. Staley, *Appl. Environ. Microbiol.* **58**, 3130–3135 (1992).
54. A. J. Brown, *J. Chem. Soc.* **49**, 432–439 (1886).
55. H. Hibbert and J. Barsha, *Can. J. Res.* **5**, 580–591 (1931).
56. D. P. Delmer, *Adv. Carbohydr. Chem. Biochem.* **41**, 105–153 (1983).
57. D. Byrom, in Ref. 48, pp. 265–283.
58. C. H. Haigler, R. M. Brown, and M. Benziman, *Science* **210**, 903–905 (1980).
59. R. A. Savidge and J. R. Colvin, *Can. J. Microbiol.* **31**, 1019–1025 (1985).
60. A. Amemura, T. Hashimoto, K. Koizumi, and T. Utamura, *J. Gen. Microbiol.* **131**, 301–307 (1985).
61. R. O. Couso, L. Ielpi, and M. A. Dankert, *J. Gen. Microbiol.* **133**, 2123–2135 (1987).
62. P.-E. Jansson, J. Lindberg, K. M. S. Wimalasiri, and M. A. Dankert, *Carbohydr. Res.* **245**, 303–310 (1993).
63. T. J. Canby, *National Geographic* **184**, 36–61 (1993).
64. D. C. Johnson, R. S. Stephens, and J. A. Westland, *Abstr. Am. Chem. Soc. Nat. Meeting 199th*: Abstract No. CARB 13 (1990).
65. S. Masaoka, T. Ohe, and N. Sakota, *J. Ferment. Bioeng.* **75**, 18–22 (1993).
66. J. F. Robyt, in J. J. Kroschwitz, ed., *Encyclopedia of Polymer Science and Engineering*, 2nd ed., Vol. 4, John Wiley & Sons, Inc., New York, 1986, pp. 752–767.
67. T. H. Evans and H. Hibbert, *Adv. Carbohydr. Chem.* **2**, 203–233 (1946).
68. W. B. Neely, *Adv. Carbohydr. Chem. Biochem.* **15**, 341–369 (1960).
69. A. Jeanes, in N. M. Bikales, ed., *Encyclopedia of Polymer Science and Technology*, Vol. 8, John Wiley & Sons, Inc., New York, 1968, pp. 693–711.
70. R. L. Sidebotham, *Adv. Carbohydr. Chem. Biochem.* **30**, 371–444 (1974).
71. A. Jeanes, in Ref. 6, pp. 284–298.
72. R. M. Alsop, in Ref. 10, pp. 1–44.
73. A. N. DeBelder, in Ref. 24, pp. 399–425.
74. A. Jeanes, *Dextran Bibliography*, Misc. Publ. No. 1355, U.S. Dept. Agriculture, Washington, D.C., 1978.
75. J. Cerning, *FEMS Microbiol. Rev.* **87**, 113–130 (1990).
76. A. Gronwall and B. Ingelman, *Nature* **155**, 45 (1945).
77. U.S. Pat. 2,437,518 (Mar. 9, 1948), A. J. T. Gronwall and B. G. Ingelman (to Aktiebolaget Pharmacia).
78. A. Jeanes and co-workers, *J. Am. Chem. Soc.* **76**, 5041–5052 (1954).
79. U.S. Pat. 2,660,551 (Nov. 24, 1953), H. J. Koepsell, H. M. Tsuchiya, and N. N. Hellman (to USDA).

80. U.S. Pat. 2,686,147 (Sept. 27, 1955), H. M. Tsuchiya and H. J. Koepsell (to USDA).
81. U.S. Pat. 2,726,985 (Dec. 13, 1955), N. N. Hellman, H. M. Tsuchiya, and F. R. Senti (to USDA).
82. U.S. Pat. 2,726,190 (Dec. 6, 1955), H. J. Koepsell, N. N. Hellman, and H. M. Tsuchiya (to USDA).
83. U.S. Pat. 2,724,679 (Nov. 22, 1955), H. M. Tsuchiya, N. N. Hellman, and H. J. Koepsell (to USDA).
84. U.S. Pat. 2,712,007 (June 28, 1955), I. A. Wolff, R. L. Mellies, and C. E. Rist (to USDA).
85. U.S. Pat. 2,719,147 (Aug. 10, 1954), I. A. Wolff, P. R. Watson, and C. E. Rist (to USDA).
86. U.S. Pat. 2,776,925 (Jan. 8, 1957), J. Corman and H. M. Tsuchiya (to USDA).
87. U.S. Pat. 2,906,669 (Sept. 29, 1959), E. J. Hehre, H. M. Tsuchiya, N. N. Hellman, and F. R. Senti (to USDA).
88. C. A. Wilham, B. H. Alexander, and A. Jeanes, *Arch. Biochem. Biophys.* **59**, 61–75 (1955).
89. R. Lohmar, *J. Am. Chem. Soc.* **74**, 4974 (1952).
90. N. W. Taylor, H. F. Zobel, N. N. Hellman, and F. R. Senti, *J. Phys. Chem.* **63**, 599–603 (1959).
91. W. M. Pasika and L. H. Cragg, *Can. J. Chem.* **41**, 293–299 (1963).
92. I. J. Goldstein and W. J. Whelan, *J. Chem. Soc.*, 170–175 (1962).
93. H. Suzuki and E. J. Hehre, *Arch. Biochem. Biophys.* **104**, 305–313 (1964).
94. R. W. Bailey, D. H. Hutson, and H. Weigel, *Biochem. J.* **80**, 514–519 (1961).
95. E. J. Bourne, D. H. Hutson, and H. Weigel, *Biochem. J.* **85**, 158–163 (1962).
96. E. J. Bourne, D. H. Hutson, and H. Weigel, *Biochem. J.* **86**, 555–562 (1963).
97. D. Abbott, E. J. Bourne, and H. Weigel, *J. Chem. Soc.* (*C*), 827–831 (1966).
98. D. Abbott and H. Weigel, *J. Chem. Soc.* (*C*), 821–827 (1966).
99. D. Abbott and H. Weigel, *J. Chem. Soc.* (*C*), 816–820 (1966).
100. E. J. Bourne, R. L. Sidebotham, and H. Weigel, *Carbohydr. Res.* **22**, 13–22 (1972).
101. E. J. Bourne, R. L. Sidebotham, and H. Weigel, *Carbohydr. Res.* **34**, 279–288 (1974).
102. O. Larm, B. Lindberg, and S. Svensson, *Carbohydr. Res.* **20**, 39–48 (1971).
103. M. T. Covacevich and G. N. Richards, *Carbohydr. Res.* **54**, 311–315 (1977).
104. F. R. Seymour, M. E. Slodki, R. D. Plattner, and A. Jeanes, *Carbohydr. Res.* **53**, 153–166 (1977).
105. F. R. Seymour, E. C. M. Chen, and S. H. Bishop, *Carbohydr. Res.* **68**, 113–121 (1979).
106. A. Jeanes and F. R. Seymour, *Carbohydr. Res.* **74**, 31–40 (1979).
107. M. E. Slodki, R. E. England, R. D. Plattner, and W. E. Dick, *Carbohydr. Res.* **156**, 199–206 (1986).
108. F. R. Seymour, R. D. Knapp, and S. H. Bishop, *Carbohydr. Res.* **51**, 179–194 (1976).
109. F. R. Seymour, R. D. Knapp, and S. H. Bishop, *Carbohydr. Res.* **72**, 229–234 (1979).
110. F. R. Seymour, R. D. Knapp, and S. H. Bishop, *Carbohydr. Res.* **74**, 77–92 (1979).
111. F. R. Seymour, R. D. Knapp, S. H. Bishop, and A. Jeanes, *Carbohydr. Res.* **68**, 123–140 (1979).
112. F. R. Seymour, R. D. Knapp, E. C. M. Chen, S. H. Bishop, and A. Jeanes, *Carbohydr. Res.* **74**, 41–62 (1979).
113. F. R. Seymour, R. D. Knapp, E. C. M. Chen, A. Jeanes, and S. H. Bishop, *Carbohydr. Res.* **71**, 231–250 (1979).
114. F. R. Seymour, R. D. Knapp, E. C. M. Chen, A. Jeanes, and S. H. Bishop, *Carbohydr. Res.* **75**, 275–294 (1979).
115. F. R. Seymour and R. D. Knapp, *Carbohydr. Res.* **81**, 67–103 (1980).
116. *Ibid.*, pp. 105–129.
117. F. R. Seymour, R. D. Knapp, and B. L. Lamberts, *Carbohydr. Res.* **84**, 187–195 (1980).

118. F. R. Seymour and R. L. Julian, *Carbohydr. Res.* **74**, 63–75 (1979).
119. G. L. Cote and J. F. Robyt, *Carbohydr. Res.* **101**, 57–74 (1982).
120. A. Misaki, M. Torii, T. Sawai, and I. Goldstein, *Carbohydr. Res.* **84**, 273–285 (1980).
121. M. Torii, S. Tanaka, and T. Sawai, *Microbiol. Immunol.* **25**, 969–973 (1981).
122. T. Ogawa and T. Kaburagi, *Carbohydr. Res.* **110**, c12–c15 (1982).
123. A. Jeanes, *Molec. Immunol.* **23**, 999–1028 (1986).
124. G. L. Cote, *Carbohydr. Polym.* **19**, 249–252 (1992).
125. B. Guggenheim and E. Newbrun, *Helv. Odont. Acta* **13**, 84–97 (1969).
126. S. Hamada and H. D. Slade, *Microbiol. Rev.* **44**, 331–384 (1980).
127. C. L. Munro and F. L. Macrina, *Molec. Microbiol.* **8**, 133–142 (1993).
128. T. J. Montville, C. L. Cooney, and A. J. Sinskey, *Adv. Appl. Microbiol.* **24**, 55–84 (1978).
129. F. L. Fowler, I. K. Buckland, F. E. Brauns, and H. Hibbert, *Can. J. Res.* **15B**, 486–497 (1937).
130. I. Levi, W. L. Hawkins, and H. Hibbert, *J. Am. Chem. Soc.* **64**, 1959–1962 (1942).
131. A. Jeanes and C. A. Wilham, *J. Am. Chem. Soc.* **72**, 2655–2657 (1950).
132. A. Jeanes and C. A. Wilham, *J. Am. Chem. Soc.* **74**, 5339–5341 (1952).
133. J. W. Sloan, B. H. Alexander, R. L. Lohmar, I. A. Wolff, and C. E. Rist, *J. Am. Chem. Soc.* **76**, 4429–4434 (1954).
134. G. L. Cote and J. F. Robyt, *Carbohydr. Res.* **119**, 141–156 (1983).
135. D. S. Genghof and E. J. Hehre, *Proc. Soc. Exper. Biol. Med.* **140**, 1298–1301 (1972).
136. E. J. Hehre and H. Suzuki, *Arch. Biochem. Biophys.* **113**, 675–683 (1966).
137. T. P. Binder and J. F. Robyt, *Carbohydr. Res.* **124**, 287–299 (1983).
138. E. J. Hehre, *Science* **93**, 237–238 (1941).
139. M. Kobayashi and K. Matsuda, *Biochim. Biophys. Acta* **370**, 441–449 (1974).
140. M. Kobayashi and K. Matsuda, *Biochim. Biophys. Acta* **397**, 69–79 (1975).
141. M. Kobayashi and K. Matsuda, *Agric. Biol. Chem.* **39**, 2087–2088 (1975).
142. M. Kobayashi and K. Matsuda, *J. Biochem.* **79**, 1301–1308 (1976).
143. A. Lopez-Munguia and co-workers, *Ann. N.Y. Acad. Sci.* **613**, 717–722 (1991).
144. A. López-Munguía and co-workers, *Enz. Microb. Technol.* **15**, 77–85 (1993).
145. J. F. Robyt and T. F. Walseth, *Carbohydr. Res.* **68**, 95–111 (1979).
146. M. Kobayashi and K. Matsuda, *Biochim. Biophys. Acta* **614**, 46–62 (1980).
147. M. Kobayashi and K. Matsuda, *J. Biochem.* **100**, 615–621 (1986).
148. M. Kobayashi, K. Mihara, and K. Matsuda, *Agric. Biol. Chem.* **50**, 551–556 (1986).
149. F. Paul, D. Auriol, E. Oriol, and P. Monsan, *Ann. NY Acad. Sci.* **7**, 267–270 (1984).
150. A. W. Miller, S. Eklund, and J. F. Robyt, *Carbohydr. Res.* **147**, 119–133 (1986).
151. D. Fu and J. F. Robyt, *Preparative Biochem.* **20**, 93–106 (1990).
152. H. J. Koepsell and co-workers, *J. Biol. Chem.* **200**, 793–801 (1953).
153. W. B. Neely, *Arch. Biochem. Biophys.* **79**, 154–161 (1959).
154. F. Yamauchi and Y. Ohwada, *Agric. Biol. Chem.* **33**, 1295–1300 (1969).
155. J. F. Robyt and S. H. Eklund, *Carbohydr. Res.* **121**, 279–286 (1983).
156. D. Fu, M. E. Slodki, and J. F. Robyt, *Arch. Biochem. Biophys.* **276**, 460–465 (1990).
157. D. Fu and J. F. Robyt, *Arch. Biochem. Biophys.* **283**, 379–387 (1990).
158. D. Su and J. F. Robyt, *Carbohydr. Res.* **248**, 339–348 (1993).
159. K. D. Reh, H. J. Jordening, and K. Buchholz, *Ann. N.Y. Acad. Sci.* **613**, 723–729 (1991).
160. G. L. Cote and J. F. Robyt, *Carbohydr. Res.* **111**, 127–142 (1982).
161. R. M. Mayer, M. M. Matthews, C. L. Futerman, V. K. Parnaik, and S. M. Jung, *Arch. Biochem. Biophys.* **208**, 278–287 (1981).
162. M. K. Bhattacharjee and R. M. Mayer, *Bioorg. Chem.* **19**, 445–455 (1991).
163. D. Fu and J. F. Robyt, *Carbohydr. Res.* **217**, 201–211 (1991).
164. M. K. Bhattacharjee and R. M. Mayer, *Carbohydr. Res.* **242**, 191–201 (1993).

165. F. H. Stodola, H. J. Koepsell, and E. S. Sharpe, *J. Am. Chem. Soc.* **74**, 3202–3203 (1952).
166. F. H. Stodola, E. S. Sharpe, H. J. Koepsell, *J. Am. Chem. Soc.* **78**, 2514–2518 (1956).
167. E. S. Sharpe, F. H. Stodola, and H. J. Koepsell, *J. Org. Chem.* **25**, 1062–1063 (1960).
168. E. J. Hehre, *J. Polym. Sci. (C)* **23**, 239–244 (1968).
169. K. H. Ebert and G. Schenk, *Adv. Enzymol.* **30**, 179–221 (1968).
170. J. F. Robyt, B. K. Kimble, and T. F. Walseth, *Arch. Biochem. Biophys.* **165**, 634–640 (1974).
171. J. F. Robyt, in J. J. Marshall, ed., *Mechanisms of Saccharide Polymerization and Depolymerization*, Academic Press, New York, 1980, pp. 43–54.
172. J. F. Robyt and P. J. Martin, *Carbohydr. Res.* **113**, 301–315 (1983).
173. S. L. Ditson and R. M. Mayer, *Carbohydr. Res.* **126**, 170–175 (1984).
174. R. M. Mayer, *Meth. Enzymol.* **138**, 649–661 (1987).
175. J. F. Robyt and T. F. Walseth, *Carbohydr. Res.* **61**, 433–445 (1978).
176. A. Tanriseven and J. F. Robyt, *Carbohydr. Res.* **225**, 321–329 (1992).
177. K. H. Ebert and M. Brosche, *Biopolymers* **5**, 423–430 (1967).
178. J. F. Robyt and H. Taniguchi, *Arch. Biochem. Biophys.* **174**, 129–135 (1976).
179. G. L. Cote and J. F. Robyt, *Carbohydr. Res.* **119**, 141–156 (1983).
180. G. L. Cote and J. F. Robyt, *Carbohydr. Res.* **127**, 95–107 (1984).
181. E. J. Hehre, *Adv. Enzymol.* **11**, 297–337 (1951).
182. S. A. Barker, E. J. Bourne, G. T. Bruce, and M. Stacey, *J. Chem. Soc.*, 4414–4416 (1958).
183. E. J. Hehre and D. M. Hamilton, *J. Biol. Chem.* **192**, 161–174 (1951).
184. U.S. Pat. 2,689,816 (Sept. 21, 1954), E. R. Kooi (to Corn Products Refining Co.).
185. U.S. Pat. 2,810,677 (Oct. 22, 1957), J. S. Gilkison and E. R. Kooi (to Corn Products Refining Co.).
186. U.S. Pat. 2,801,205 (July 30, 1957), E. R. Kooi (to Corn Products Refining Co.).
187. U.S. Pat. 2,833,695 (May 6, 1958), E. R. Kooi (to Corn Products Refining Co.).
188. U.S. Pat. 2,801,204 (July 30, 1957), E. R. Kooi (to Corn Products Refining Co.).
189. K. Yamamoto, K. Yoshikawa, S. Kitahata, and S. Okada, *Biosci. Biotechnol. Biochem.* **56**, 169–173 (1992).
190. K. Yamamoto, K. Yoshikawa, and S. Okada, *Biosci. Biotechnol. Biochem.* **57**, 136–137 (1993).
191. K. Yamamoto, K. Yoshikawa, and S. Okada, *Biosci. Biotechnol. Biochem.* **57**, 47–50 (1993).
192. H. J. Koepsell and co-workers, *J. Biol. Chem.* **200**, 793–801 (1953).
193. H. M. Tsuchiya and co-workers, *J. Am. Chem. Soc.* **77**, 2412–2419 (1955).
194. N. N. Hellman and co-workers, *Ind. Eng. Chem.* **47**, 1593–1598 (1955).
195. V. Gasoiolli, L. Choplin, F. Paul, and P. Monsan, *J. Biotechnol.* **19**, 193–202 (1991).
196. H. M. Tsuchiya, A. Jeanes, H. M. Bricker, and C. A. Wilham, *J. Bacteriol.* **64**, 513–519 (1952).
197. U.S. Pat. 5,229,277 (July 20, 1993), D. F. Day and D. Kim (to Louisiana State University).
198. P. R. Watson and I. A. Wolff, *J. Am. Chem. Soc.* **77**, 196 (1955).
199. S. C. Szu, G. Zon, R. Schneerson, and J. B. Robbins, *Carbohydr. Res.* **152**, 7–20 (1986).
200. F. Paul, E. Oriol, D. Auriol, and P. Monsan, *Carbohydr. Res.* **149**, 433–441 (1986).
201. D. Prat, L. A. Valdivia, P. Monsan, F. Paul, and C. A. Lopez-Munguia, *Biotechnol. Lett.* **9**, 1–6 (1987).
202. M. Remaud, F. Paul, P. Monsan, A. Heyraud, and M. Rinaudo, *J. Carbohydr. Chem.* **10**, 861–876 (1991).
203. K. Gekko, in D. A. Brant, ed., *Solution Properties of Polysaccharides*, ACS Symposium Series 150, American Chemical Society, Washington, D.C., 1981, pp. 415–438.

204. H. Mitsuya and co-workers, *Science* **240**, 646–649 (1988).
205. J. Lonngren, *Pure Appl. Chem.* **61**, 1313–1314 (1992).
206. U.S. Pat. 2,820,740 (Jan. 21, 1958), E. London and G. D. Twiggs (to Benger Laboratories, Ltd.).
207. *Fed. Reg.* **42**(223), 59518–59521 (Nov. 18, 1977).
208. U.S. Pat. 4,399,160 (Aug. 16, 1983), R. D. Schwartz and E. A. Bodie (to Stauffer Chemical Co.).
209. U.S. Pat. 4,877,634 (Oct. 31, 1989), M. J. Pucci and B. S. Kunka (to Microlife Technics).
210. U.S. Pat. 4,933,191 (June 12, 1990), M. J. Pucci and B. S. Kunka (to Microlife Technics).
211. U.S. Pat. 5,141,858 (Aug. 25, 1992), F. B. Paul, A. L. M. Canales, M. M. Remaud, V. P. Pelenc, and P. F. Monsan (to BioEurope).
212. A. Jeanes, *Meth. Carbohydr. Chem.* **5**, 118–132 (1965).
213. J. P. Martinez-Espindola and C. A. Lopez-Munguia, *Biotechnol. Lett.* **7**, 483–486 (1985).
214. R. S. Landon and C. Webb, *Process Biochem.* **25**, 19–23 (1990).
215. N. J. Ajongwen and P. E. Barker, *J. Chem. Tech. Biotechnol.* **56**, 113–118 (1993).
216. P. Monsan, F. Paul, D. Auriol, and A. Lopez, *Meth. Enzymol.* **136**, 239–254 (1987).
217. E. Rosenberg, A. Zuckerberg, C. Rubinovitz, and D. L. Gutnick, *Appl. Environ. Microbiol.* **37**, 402–408 (1979).
218. A. Zuckerberg, A. Diver, Z. Peeri, D. L. Gutnick, and E. Rosenberg, *Appl. Environ. Microbiol.* **37**, 414–420 (1979).
219. D. L. Gutnick, *Biopolymers* **26**, s223–s240 (1987).
220. N. Sar and E. Rosenberg, *Curr. Microbiol.* **9**, 309–314 (1983).
221. N. Kaplan and E. Rosenberg, *Appl. Environ. Microbiol.* **44**, 1335–1341 (1982).
222. B. A. Bryan, R. J. Linhardt, and L. Daniels, *Appl. Environ. Microbiol.* **51**, 1304–1308 (1986).
223. M. C. Cirigliano and G. M. Carman, *Appl. Environ. Microbiol.* **48**, 747–750 (1984).
224. M. C. Cirigliano and G. M. Carman, *Appl. Environ. Microbiol.* **50**, 846–850 (1985).
225. R. Moorhouse, G. T. Colegrove, P. A. Sandford, J. K. Baird, and K. S. Kang, in Ref. 203, pp. 111–124.
226. K. S. Kang, G. T. Veeder, P. J. Mirrasoul, T. Kaneko, and I. W. Cottrell, *Appl. Environ. Microbiol.* **43**, 1086–1091 (1982).
227. R. Moorhouse, in Ref. 18, pp. 187–206.
228. P.-E. Jansson, B. Lindberg, and P. A. Sandford, *Carbohydr. Res.* **124**, 135–139 (1983).
229. M.-S. Kuo, A. J. Mort, and A. Dell, *Carbohydr. Res.* **156**, 173–187 (1986).
230. K. S. Kang and D. J. Pettitt, in Ref. 24, pp. 341–397.
231. U.S. Pat. 5,230,853 (July 27, 1993), G. T. Colegrove and T. A. Lindroth (to Kelco Div. of Merck, Inc.).
232. R. Chandrasekaran and V. G. Thailambal, *Carbohydr. Polym.* **12**, 431–442 (1990).
233. U.S. Pat. 4,326,052 (Apr. 20, 1982), K.S. Kang, G. T. Colegrove, and G. T. Veeder (to Kelco Div. of Merck, Inc.).
234. U.S. Pat. 4,326,053 (Apr. 20, 1982), K. S. Kang and G. T. Veeder (to Kelco Div. of Merck, Inc.).
235. *Fed. Reg.* **55**, 39,613 (1990).
236. J. D. Dziezak, *Food Technol.* **44**, 88–90 (1990).
237. *Fed. Reg.* **57**, 55444–55445 (1992).
238. K. Hannigan, *Food Engin.* **55**, 52–53 (1983).
239. U.S. Pat. 4,377,636 (Mar. 22, 1983), K. S. Kang and G. T. Veeder (to Kelco Div. of Merck, Inc.).
240. U.S. Pat. 5,190,927 (Mar. 2, 1993), H. C. Chang and J. M. Kobzeff (to Kelco Div. Merck, Inc.).

241. G. A. Monteiro, A. M. Fialho, S. J. Ripley, and I. Sacorreia, *J. Appl. Bacteriol.* **72**, 423–428 (1992).
242. P.-E. Jansson and G. Widmalm, *Carbohyd. Res.* **256**, 327–330 (1994).
243. U.S. Pat. 5,175,277 (Mar. 20, 1991), W. G. Rakitsky and D. D. Richey (to Kelco Div. of Merck, Inc.).
244. U.S. Pat. 5,290,768 (Jan. 18, 1991), A. M. Ramsay, G. Trimble, J. M. Seheult and M. S. O'Brien (to Kelco Div. of Merck, Inc.).
245. M. Suzuki and N. J. Chatterton, eds., *Science and Technology of Fructans*, CRC Press, Boca Raton, Fla., 1993.
246. Y. W. Han, *Adv. Appl. Microbiol.* **35**, 171–194 (1990).
247. E. O. von Lippmann, *Chem. Ber.* **14**, 1509–1512 (1881).
248. H. Hibbert, R. S. Tipson, and F. Brauns, *Can. J. Research* **4**, 221–239 (1931).
249. S. Hestrin, S. Avineri-Shapiro, and M. Aschner, *Biochem. J.* **37**, 450–456 (1943).
250. P. Mantsala and M. Puntala, *FEMS Microbiol. Lett.* **13**, 395–399 (1982).
251. D. J. Bell and R. Dedonder, *J. Chem. Soc.* 2866–2870 (1954).
252. F. C. Harrison, H. L. A. Tarr, and H. Hibbert, *Can. J. Res.* **3**, 449–463 (1930).
253. M. Aschner, S. Avineri-Schapiro, and S. Hestrin, *Nature* **149**, 527 (1942).
254. Y. W. Han, *J. Ind. Microbiol.* **4**, 447–452 (1989).
255. E. A. Dawes, D. W. Ribbons, and D. A. Rees, *Biochem. J.* **98**, 804–812 (1966).
256. M. J. Pabst, *Infect. Immun.* **15**, 518–526 (1977).
257. V. I. Elisashvili, *Mikrobiol.* **50**, 69–73 (1981).
258. I. Kojima, T. Saito, M. Iizuka, N. Minamiura, and S. Ono, *J. Ferment. Bioeng.* **75**, 9–12 (1993).
259. E. A. Bodie, R. D. Schwartz, and A. Catena, *Appl. Environ. Microbiol.* **50**, 629–633 (1985).
260. G. Avigad and D. S. Feingold, *Arch. Biochem. Biophys.* **70**, 178–184 (1957).
261. F. Dias and J. V. Bhat, *Antonie van Leeuwenhoek* **28**, 63–72 (1962).
262. C. F. Niven, K. L. Smiley, and J. M. Sherman, *J. Bacteriol.* **41**, 479–484 (1941).
263. E. J. Hehre, *Proc. Soc. Exper. Biol. Med.* **58**, 219–221 (1945).
264. R. J. Gibbons and M. Nygaard, *Arch. Oral Biol.* **13**, 1249–1262 (1968).
265. E. A. Cooper and J. F. Preston, *Biochem. J.* **29**, 2267–2277 (1935).
266. A. Fuchs, *Nature* **178**, 921 (1956).
267. S. S. Stivala, J. E. Zweig, and J. Ehrlich, in Ref. 203, ACS Symposium Series 45, pp. 101–110.
268. T. Tanaka, S. Oi, and T. Yamamoto, *J. Biochem.* **85**, 287–293 (1979).
269. G. L. Cote and J. A. Ahlgren, in Ref. 242, pp. 141–168.
270. B. Lindberg, J. Lonngren, and J. L. Thompson, *Acta Chem. Scand.* **27**, 1819–1821 (1973).
271. R. A. Hancock, K. Marshall, and H. Weigel, *Carbohydr. Res.* **49**, 351–360 (1976).
272. Y. W. Han and M. A. Clarke, *J. Agric. Food Chem.* **38**, 393–396 (1990).
273. T. Tanaka, S. Oi, and T. Yamamoto, *J. Biochem.* **87**, 297–303 (1980).
274. G. Avigad, in Ref. 69, pp. 711–718.
275. U.S. Pat. 3,033,758 (May 8, 1962), W. Kaufmann and K. Bauer (to SchenLabs Pharmaceuticals, Inc.).
276. A. Fuchs, *Biochem. Soc. Trans.* **19**, 555–560 (1991).
277. G. Avigad, *Meth. Carbohydr. Chem.* **5**, 161–165 (1965).
278. E. J. Hehre, *Meth. Enzymol.* **1**, 178–192 (1955).
279. R. Dedonder, *Meth. Enzymol.* **8**, 500–505 (1966).
280. U.S. Pat. 4,863,719 (Sept. 5, 1989), T. D. Mays and E. L. Dally.
281. G. L. Cote, *Biotechnol. Lett.* **10**, 879–882 (1988).
282. P. Perlot and P. Monsan, *Ann. NY Acad. Sci.* **434**, 468–471 (1984).
283. M. Iizuka, H. Yamaguchi, S. Ono, and N. Minamiura, *Biosci. Biotechnol. Biochem.* **57**, 322–324 (1993).

284. R. Chambert and M.-F. Petit-Glatron, *Carbohydr. Res.* **244**, 129–136 (1993).
285. H. Bender, J. Lehmann, and K. Wallenfels, *Biochim. Biophys. Acta* **36**, 309–316 (1959).
286. K. Wallenfels, G. Keilich, G. Bechtler, and D. Freudenberger, *Biochem. Zeitschrift* **341**, 433–450 (1965).
287. B. J. Catley and W. J. Whelan, *Arch. Biochem. Biophys.* **143**, 138–142 (1971).
288. R. Taguchi, Y. Kikuchi, Y. Sakano, and T. Kobayashi, *Agric. Biol. Chem.* **37**, 1583–1588 (1973).
289. R. W. Silman, W. L. Bryan, and T. D. Leathers, *FEMS Microbiol. Lett.* **71**, 65–70 (1990).
290. B. J. Catley, A. Ramsay, and C. Servis, *Carbohydr. Res.* **153**, 79–86 (1986).
291. B. J. Catley, *FEBS Lett.* **10**, 190–193 (1970).
292. T. Kato, T. Okamoto, T. Tokuya, and A. Takahashi, *Biopolymers* **21**, 1623–1633 (1982).
293. D. A. Brant and B. A. Burton, in Ref. 267, pp. 81–99.
294. B. J. Catley, *FEBS Lett.* **20**, 174–176 (1972).
295. B. J. Catley, in R. C. W. Berkeley and co-workers, eds., *Microbial Polysaccharides and Polysaccharases*, Academic Press, London, 1979, pp. 69–84.
296. R. Taguchi, Y. Sakano, Y. Kikuchi, M. Sakuma, and T. Kobayashi, *Agr. Biol. Chem.* **37**, 1635–1641 (1973).
297. B. J. Catley and W. McDowell, *Carbohydr. Res.* **103**, 65–75 (1982).
298. M. S. Deshpande, V. B. Rale, and J. M. Lynch, *Enzyme Microb. Technol.* **14**, 514–527 (1992).
299. S. Yuen, *Process Biochem.* **9**, 7–9,24 (1974).
300. A. LeDuy, L. Choplin, J. E. Zajic, and J. H. T. Luong, in Ref. 66, Vol. 13, 1988, pp. 650–660.
301. Y. Tsujisaka and M. Mitsuhashi, in Ref. 24, pp. 446–460.
302. A. Mulchandani, J. H. T. Luong, and A. LeDuy, *Biotechnol. Bioeng.* **33**, 306–312 (1989).
303. R. Schuster, E. Wenzig, and A. Mersmann, *Appl. Microbiol. Biotechnol.* **39**, 155–158 (1993).
304. P. A. Gibbs and R. J. Seviour, *Biotechnol. Lett.* **14**, 491–494 (1992).
305. H. Yamasaki, M.-S. Lee, T. Tanaka, and K. Nakanishi, *Appl. Microbiol. Biotechnol.* **39**, 26–30 (1993).
306. *Ibid.*, pp 21–25.
307. T. D. Leathers, G. W. Nofsinger, C. P. Kurtzman, and R. J. Bothast, *J. Ind. Microbiol.* **3**, 231–239 (1988).
308. T. J. Pollock, L. Thorne, and R. W. Armentrout, *Appl. Environ. Microbiol.* **58**, 877–883 (1992).
309. L. Tarabasz-Szymanska and E. Galas, *Enzyme Microb. Technol.* **15**, 317–320 (1993).
310. J. Johnson, Jr. and co-workers, *Chem. Ind. (London)*, 820–822 (May 18, 1963).
311. U.S. Pat. 3,301,848 (Jan. 31, 1967), F. E. Halleck (to Pillsbury Co.).
312. G. Brigand, in Ref. 24, pp. 461–474.
313. M. Rinaudo and M. Vincendo, *Carbohydr. Polym.* **2**, 135–144 (1982).
314. T. L. Bluhm, Y. Deslandes, R. H. Marchessault, S. Perez, and M. Rinaudo, *Carbohydr. Res.* **100**, 117–130 (1982).
315. S. Bo, M. Milas, and M. Rinaudo, *Int. J. Biol. Macromol.* **9**, 153–157 (1987).
316. T. Yanaki and T. Norisuye, *Polymer J.* **15**, 389–396 (1983).
317. C. Biver and co-workers, *Polym. Mater. Sci. Eng.* **55**, 582 (1986).
318. D. Lecacheux, Y. Mustiere, R. Panaras, and G. Brigand, *Carbohydr. Polym.* **6**, 477–492 (1986).
319. P. Davison and E. Mentzter, *Soc. Petr. Eng. J.* **22**, 353–362 (1982).

320. U.S. Pat. 5,224,988 (Apr. 17, 1992), R. Pirri, Y. Huet, and A. Donche (to Society National Elf Aquitaine).
321. H. A. Pretus and co-workers, *J. Pharm. Exp. Therapeut.* **257**, 500–510 (1991).
322. T. Harada, M. Terasaki, and A. Harada, in Ref. 24, pp. 427–445.
323. T. Harada, M. Masada, K. Fujimori, and I. Maeda, *Agric. Biol. Chem.* **30**, 196–198 (1966).
324. I. Maeda, H. Saito, M. Masuda, A. Misaki, and T. Harada, *Agric. Biol. Chem.* **31**, 1184–1188 (1967).
325. T. Kuge, N. Suetsuga, and N. Nishiyama, *Agric. Biol. Chem.* **41**, 1315–1316 (1977).
326. H. Saito, Y. Yashioka, Y. Yokio, and Y. Yamada, *Biopolymers* **29**, 1689–1698 (1990).
327. Y. Deslandes, R. H. Marchesault, and A. Sarko, *Macromol.* **13**, 1466–1471 (1980).
328. C. T. Chuah, A. Sarko, Y. Deslandes, and R. H. Marchessault, *Macromol.* **16**, 1375–1382 (1983).
329. M. Hisamatsu, J. Abe, A. Amemura, and T. Harada, *Agric. Biol. Chem.* **44**, 1049–1055 (1980).
330. T. Harada, *Arch. Biochem. Biophys.* **112**, 65–69 (1965).
331. B. B. Reinhold and co-workers, *J. Bacteriol.* **176**, 1997–2002 (1994).
332. T. Harada and T. Yoshimura, *Agric. Biol. Chem.* **29**, 1027–1032 (1965).
333. S. P. Ragovin, R. G. Anderson, and M. C. Cadmus, *J. Biochem. Microbiol. Technol. Eng.* **3**, 51–63 (1961).
334. A. Jeanes, J. E. Pittsley, and F. R. Senti, *J. Appl. Polym. Sci.* **5**, 519–526 (1961).
335. P.-E. Jansson, L. Kenne, and B. Lindberg, *Carbohydr. Res.* **45**, 275–282 (1975).
336. L. D. Melton, L. Mindt, D. A. Rees, and G. R. Sanderson, *Carbohydr. Res.* **46**, 245–257 (1975).
337. R. A. Moraine and P. Rogovin, *Biotechnol. Bioeng.* **15**, 225–237 (1973).
338. I. W. Davidson, *FEMS Microbiol. Lett.* **3**, 347 (1978).
339. M. C. Cadmus and co-workers, *Can. J. Microbiol.* **22**, 942–948 (1976).
340. G. Holzwarth, *Biochemistry* **15**, 4333–4339 (1976).
341. F. R. Dintzis, G. E. Babcock, and R. Tobin, *Carbohydr. Res.* **13**, 257–267 (1970).
342. G. Holzwarth and E. B. Prestridge, *Science* **197**, 757–759 (1977).
343. P. J. Whitcombe and C. W. Macosko, *J. Rheol.* **22**, 493–505 (1978).
344. G. Holzwarth, *Dev. Ind. Microbiol.* **26**, 271–280 (1985).
345. J. Lecourtier and G. Chauveteau, *Macromolecules* **17**, 1340–1343 (1984).
346. E. A. Lange, in G. A. Stahl and D. N. Shulz, eds., *Water Soluble Polymers for Petroleum Recovery*, Plenum Publishing Corp., New York, 1988.
347. D. A. Rees, *Biochem. J.* **126**, 257–273 (1972).
348. E. R. Morris, in Ref. 6, pp. 81–90.
349. K. S. Sorbie, *Polymer-Improved Oil Recovery*, CRC Press, Inc., Boca Raton, Fla., 1991, p. 16.
350. B. T. Stokke, A. Elgsaeter, and O. Smidsrod, *Polym. Mater. Sci. Eng.* **55**, 583–587 (1986).
351. B. T. Stokke, O. Smidsrod, A. B. L. Martinsen, and A. Elgsaeter, in Ref. 346.
352. E. R. Morris, D. A. Rees, G. Young, M. D. Walkinshaw, and A. Darke, *J. Mol. Biol.* **110**, 1–16 (1977).
353. M. Milas and M. Rinaudo, *Carbohydr. Res.* **76**, 189–196 (1979).
354. I. T. Norton, D. M. Goodall, E. R. Morris, and D. A. Rees, *J. Chem. Soc. Chem. Commun.*, 545 (1980).
355. S. A. Frangou, E. R. Morris, D. A. Rees, R. K. Richardson, and S. B. Ross-Murphy, *J. Polym. Sci. Polym. Lett.* **20**, 531–538 (1982).
356. R. Moorhouse, M. D. Walkinshaw, and S. Arnott, in Ref. 6, pp. 90–102.
357. T. Sato, T. Norisuye, and H. Fujita, *Polymer J.* **16**, 341–350 (1984).
358. M. Milas and M. Rinaudo, *Polym. Bull.* **12**, 507–514 (1984).

359. J. Lecourtier, G. Chauveteau, and G. Muller, *Int. J. Biol. Macromol.* **8**, 306–310 (1986).
360. M. C. Cadmus, M. E. Slodki, and J. J. Nicholson, *J. Ind. Microbiol.* **4**, 127–133 (1989).
361. I. W. Sutherland, *Carbohydr. Res.* **131**, 93–104 (1984).
362. P. Rogovin, W. Albrecht, and V. Sohns, *Biotechnol. Bioeng.* **7**, 161–169 (1965).
363. C. T. Bishop and H. J. Jennings, in Ref. 11, Vol. 1, 1982, pp. 291–330.
364. H. J. Jennings, *Adv. Carbohydr. Chem. Biochem.* **41**, 155–208 (1983).
365. Z. Osawa and co-workers, *Carbohydr. Polym.* **21**, 283–288 (1993).
366. D. L. Williams and co-workers, *Immunopharmacology* **22**, 139–156 (1991).
367. T. Trnovec and M. Hrmova, *Biopharm. Drug Dispos.* **14**, 187–198 (1993).
368. J. Alper, *BioTechnology* **11**, 1093 (1993).
369. K. S. Kang, G. T. Veeder, and D. D. Richey, in Ref. 6, pp. 211–219.
370. K. S. Kang and W. H. McNeely, in Ref. 6, pp. 220–230.
371. P. A. Sandford and J. Baird, in Ref. 11, pp. 411–490.
372. J. K. Baird and D. J. Pettitt, in I. Goldberg and R. Williams, eds., *Biotechnology and Food Ingredients*, Van Nostrand Reinhold, New York, 1991, pp. 223–263.

GREGORY L. COTE
JEFFREY A. AHLGREN
U.S. Department of Agriculture

MICROBIAL TRANSFORMATIONS

Microorganisms are of considerable economic importance in the manufacture of antibiotics (qv), alkaloids (qv), vitamins (qv), amino acids (qv), industrial solvents (see SOLVENTS, INDUSTRIAL), organic acids, nucleosides, nucleotides, fermented beverages (see BEER; WINE), and fermented foods (see also FERMENTATION). Microbes also catalyze simple and chemically well-defined reactions involving a variety of other compounds. Some reactions can be carried out more economically by microbial means than by strictly chemical manipulation. Microorganisms have therefore been used in processes that yield a number of important products, eg, L-ascorbic acid, steroid hormones, 6-aminopenicillanic acid, various L-amino acids, L-ephedrine, D-fructose, vinegar (qv), and malt (see BEVERAGE SPIRITS, DISTILLED; MALTS AND MALTING).

Experimental procedures have been described in which the desired reactions have been carried out either by whole microbial cells or by enzymes (1–3). These involve carbohydrates (qv) (4,5); steroids (qv), sterols, and bile acids (6–11); nonsteroid cyclic compounds (12); alicyclic and alkane hydroxylations (13–16); alkaloids (7,17,18); various pharmaceuticals (qv) (19–21), including antibiotics (19–24); and miscellaneous natural products (25–27). Reviews of the microbial oxidation of aliphatic and aromatic hydrocarbons (qv) (28), monoterpenes (29,30), pesticides (qv) (31,32), lignin (qv) (33,34), flavors and fragrances

(35), and other organic molecules (8,12,36,37) have been published (see ENZYME APPLICATIONS, INDUSTRIAL; ENZYMES IN ORGANIC SYNTHESIS; FLAVORS AND SPICES).

Microorganisms and their enzymes have been used to functionalize non-activated carbon atoms, to introduce centers of chirality into optically inactive substrates, and to carry out optical resolutions of racemic mixtures (1,2,37–42). Their utility results from the ability of the microbes to elaborate both constitutive and inducible enzymes that possess broad substrate specificities and also remarkable regio- and stereospecificities.

Reactions

Oxidation.   The ability of microorganisms to catalyze reactions of organic compounds was demonstrated by the conversion of ethanol to acetic acid in the presence of bacteria (43) and the oxidation of polyhydric alcohols to their corresponding keto-sugars by *Acetobacter xylinum* (44). The latter studies led to the formulation of a rule which states that secondary hydroxyl groups of polyols are oxidized to ketones (qv) only when flanked by a primary alcohol group and a secondary alcohol group which is cis to the oxidizable group (45). The oxidation of D-sorbitol (**1**) by *Acetobacter suboxydans* became an important step in the Reichstein-Grussner synthesis of L-ascorbic acid [50-81-7]. As shown in Figure 1 (**1**), obtained by catalytic hydrogenation of glucose (**2**), is converted by regiospecific bacterial (*A. suboxydans*) oxidation to L-sorbose [87-79-6] (**3**) (46). This keto-sugar in turn serves as a substrate for the chemical synthesis of L-ascorbic acid via intermediates which include diacetone-L-sorbose and 2-keto-L-gulonic acid [526-98-7] (**4**). The latter compound can be also prepared by sequential regiospecific oxidation of (**2**) to 2,5-diketo-D-gluconic acid [2595-33-7] (**5**), followed by stereospecific reduction of (**5**) to (**4**) (47). This process was further improved by cloning the gene of the *Corynebacterium*, which encodes the enzyme catalyzing the reduction of (**5**) to (**4**), and expressing it in *Erwinia citreus*, thus permitting a direct conversion of (**2**) to (**4**) in one fermentation step (48) (see BIOTECHNOLOGY; GENETIC ENGINEERING).

Both mono- and polynuclear aromatic hydrocarbons can be oxidized by different microorganisms. Thus, *p*-cymene is converted to cumic acid and *p*-xylene to *p*-toluic acid (49). A high (98%) yield process has been developed in Japan for the production of salicylic acid [69-72-7] from naphthalene using *Pseudomonas aeruginosa* (50). Microorganisms are used to construct chiral synthons, synthetic chemical intermediates, from various substituted benzenes. The technology of accumulating *cis*-dihydrodiols (**6**) produced by the action of dioxygenases on benzenoid substrates is available (51). Mutant *Pseudomonas putida* strains accumulate the chiral diols that permit the highly enantiomerically controlled syntheses of agents such as conduritols (polyols), pinitols, and others (2,52).

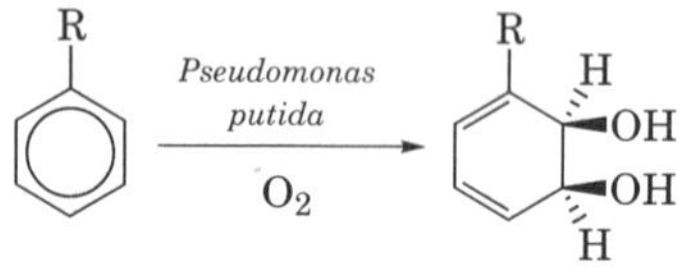

(1)

(**6**)

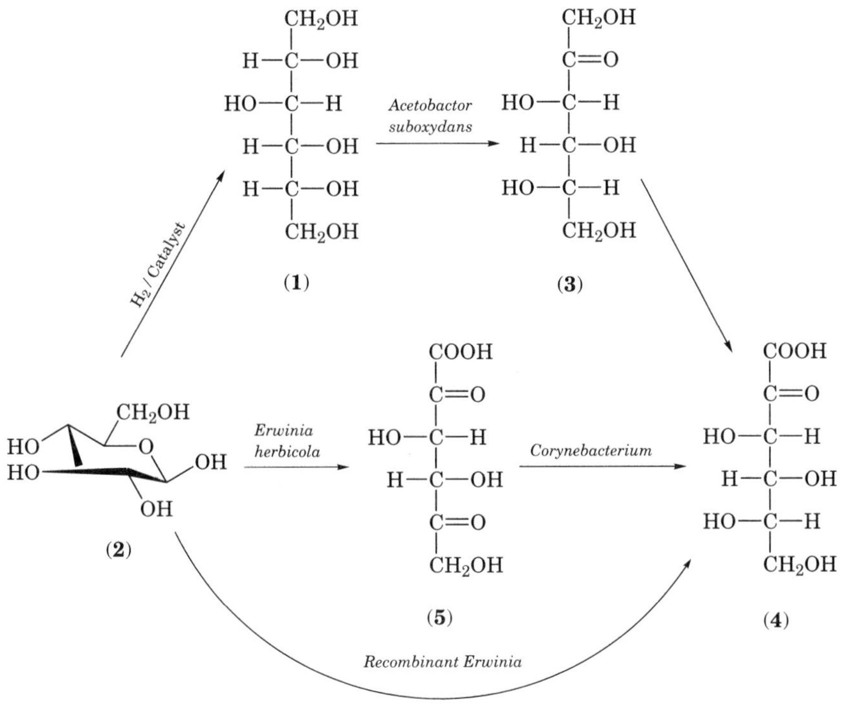

Fig. 1. Conversion of glucose (**2**) to 2-keto-L-gulonic acid (**4**).

Hydroxylation, dehydrogenation, and β-oxidation reactions occur commonly with microorganisms. When cortisone [53-06-05] and hydrocortisone [50-23-7] were identified in 1949 as potent antiinflammatory agents and no adequate synthesis existed to meet the sharply increased demand for these compounds, hydroxylation of steroid intermediates at C-11 became crucial for large-scale production. The problem of introducing functionality at that site was solved upon discovery that progesterone [57-83-0] (**7**) is oxidized to 11-α-hydroxyprogesterone [80-75-1] (**8**) by *Rhizopus arrhizus* and also *Aspergillus niger* (53).

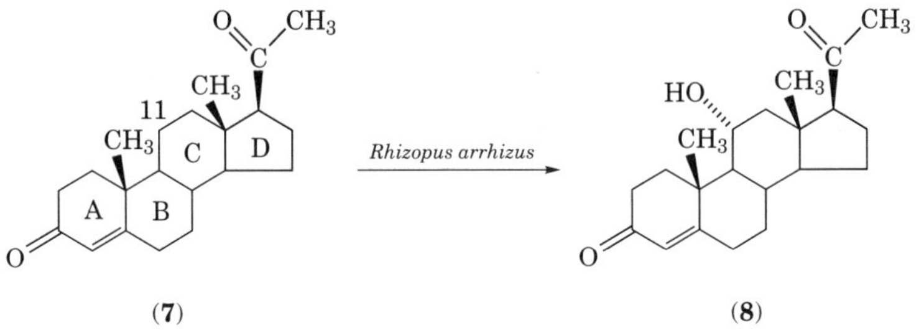

The subsequent observation of the increased efficiency of the reaction in equation 2 when *R. nigricans* was used, led to massive programs intended to use microorganisms to modify other sites of the steroid molecule. The goals were to develop efficient syntheses of steroid hormones and to find new derivatives having more specific physiological activities than the parent compounds. Stereo-specific microbial hydroxylations at practically all available carbon atoms of the steroid molecule were found. Other microbial reactions, eg, dehydrogenation, epoxidation, reduction, side-chain cleavage, aromatization of ring A, lactoniza-tion of ring D, esterification, hydrolysis, and isomerization were also reported (6–11). Of these, 1-dehydrogenation has assumed industrial importance in the synthesis of prednisone [53-03-2], prednisolone [50-24-8] (**9**), and their deriva-tives, all of which are more potent and have fewer side effects than the parent hormones (qv) (see STEROIDS).

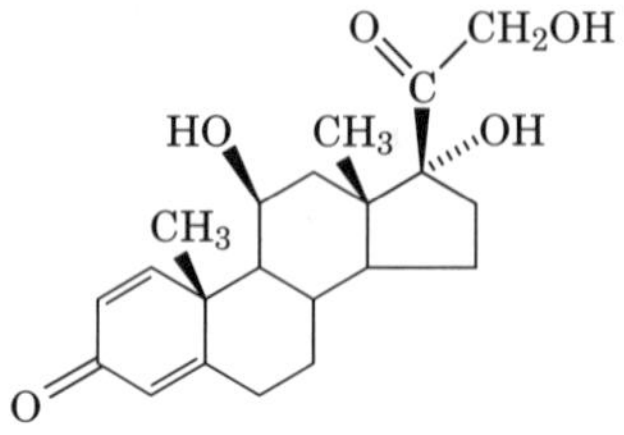

(**9**)

Bacterial removal of sterol side chains is carried out by a stepwise β-oxidation, whereas the degradation of the perhydrocyclopentanophenanthrene nucleus is prevented by metabolic inhibitors (54), chemical modification of the nucleus (55), or the use of bacterial mutants (11,56). β-Sitosterol [83-46-5] (**10**), a plant sterol, has been used as a raw material for the preparation of 4-androstene-3,17-dione [63-05-8] (**13**) and related compounds using selected mutants of the β-sitosterol-degrading bacteria (57) (Fig. 2).

Microbial Baeyer-Villiger reactions are well known in the steroid field and have been applied for synthetic uses (41,58). The D-rings of various steroids were converted into lactones using species of *Penicillium* and *Aspergillus* as shown for the preparation of testololactone [968-93-4] (**12**) (Fig. 2) (59). Many other cyclic ketones are converted to their respective lactones by other organisms. The microbiological reaction displays advantageous features of enantio- and regioselectivity.

Reduction.   Stereospecific microbial reductions of ketonic substrates are often dependent on the size and nature of substituents flanking the ketone func-tional group to be reduced. The reduction of racemic decalone and hexahydroin-danone derivatives and of related di- and tricyclic ketones by the fungus *Curvu-laria falcata* is highly stereospecific giving alcohols of S-absolute configuration when ketones are flanked by large and small groups (60). Models of these reac-tions may help to predict the stereochemistry of optically active alcohols obtain-able from ketone substrates. Oxidoreductases having useful properties continue to be identified, such as the alcohol dehydrogenase from *Pseudomonas* sp. that gives (R)-alcohols (61). Microorganisms can also be used to implement selective reductions of β-diketones which are important in steroid syntheses (62). Reduc-tion of 4-androstene-3,17-dione (**13**) to 17β-hydroxy-4-androsten-3-one (testos-

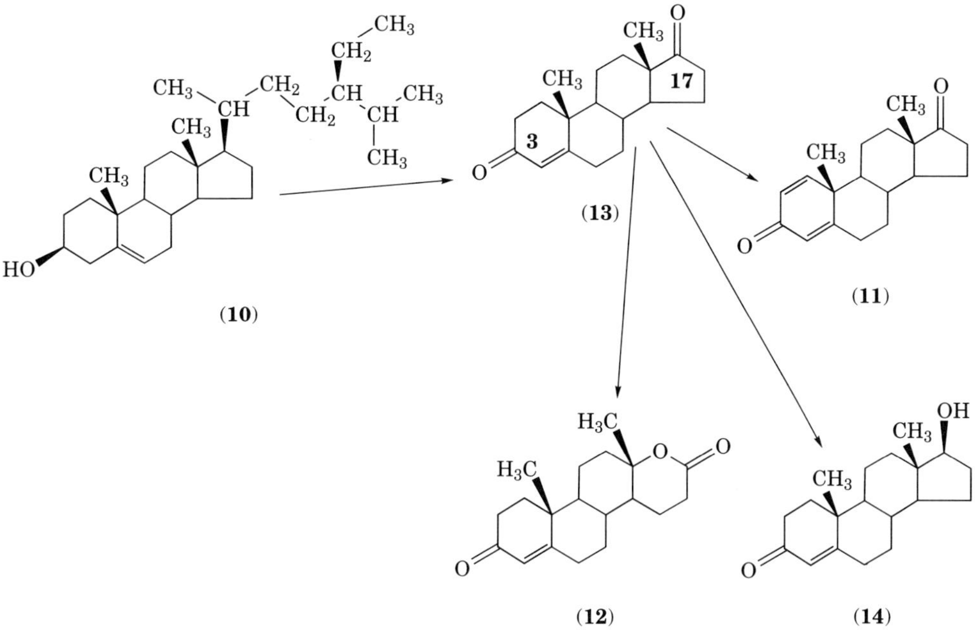

Fig. 2. Oxidation and reduction reactions using microbial transformation for steroid synthesis.

terone) [58-22-0] (14) (Fig. 2) by yeasts (qv) is one of the earliest observed conversions of steroids (63). Oxidoreductases of varying enantio- and regiospecificities are among the most widely used enzymes in synthetic organic chemistry (1,2,39,42). Specific examples of experimental applications to the reduction of cyclic and acyclic β-keto esters, aliphatic and aromatic ketones, as well as double bonds have been summarized (2). The microbial reduction of many simple and substituted aromatic carboxylic acids has been described (64). This unusual reaction is carried out using whole cells of microorganisms such as *Nocardia* sp. and *Aspergillus*. It is useful because of the water solubilities of the substrate and the *in situ* regeneration of reducing co-factors by the living catalysts.

Hydrolysis. Hydrolysis is one of the most widely used microbiological reactions (1,2,12,39,42). Hydrolytic enzymes are generally stable and require no co-factors for catalysis. Hydrolysis of many esters, glycosides, epoxides, lactones, β-lactams, nitriles, and amides has been described. Acetylated steroids have been hydrolyzed with varying degrees of selectivity by numerous organisms. (−)-14-Acetoxycodeine has been converted to (−)-14-hydroxycodeine [467-14-1] (65) and atropine [51-55-8] to tropine [120-29-6]. The sugar moiety has been removed from cardiac glycosides and saponins (19). L-Amino acids have been produced from their optically inactive forms. Tartaric acid [87-69-4] (15) is produced from glucose by *A. suboxydans* (66) and is prepared from maleic anhydride [108-31-6] (16) by hydrolysis of the *cis*-epoxysuccinic acid (17) intermediate in 99.8% yield (67).

$$(3)$$

(16)   **(17)**   **(15)**

Penicillins are inactivated by many bacteria through hydrolysis of the β-lactam ring (see ANTIBIOTICS, β-LACTAMS). The hydrolysis of the amide bond of penicillins to give 6-aminopenicillanic acid [551-16-6] (6-APA) (**18**) is economically valuable because this acid is the principal intermediate in the chemical or enzymatic preparation of semisynthetic penicillins. Although (**18**) was originally produced by direct fermentation in the absence of the side-chain precursors, the yields were low and the presence of penicillins which are also formed complicated its extraction and purification (68) (see FERMENTATION). 6-APA is made on large scale by selective enzymatic hydrolysis of penicillin G [61-33-6] (**19**) (benzylpenicillin) or penicillin V [87-08-1] (**20**) (phenylmethoxypenicillin), which are produced in high yields by direct fermentation.

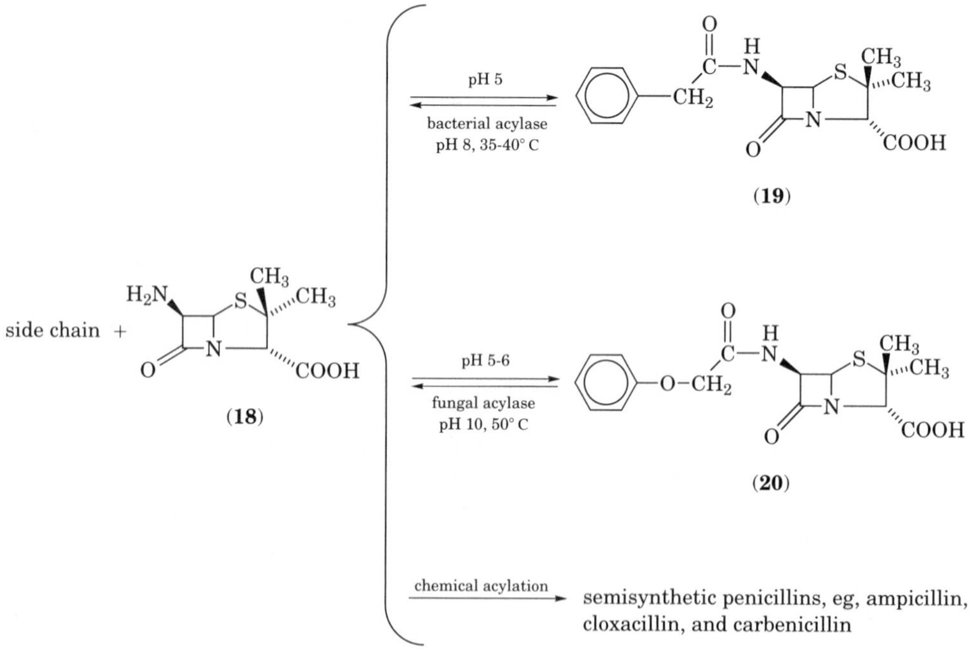

$$(4)$$

Penicillin acylases, which carry out this selective hydrolysis, are generally of two types: those which are produced by bacteria and primarily hydrolyze penicillin G, and those which are elaborated by actinomycetes, fungi, and yeasts and which preferentially hydrolyze penicillin V. The reaction can be reversed at an acidic pH in the presence of suitable acyl donors. Mutants have been selected by systematic culture development to hydrolyze high concentrations of penicillin G. The reaction pH, temperature, and ratio of the cells to the substrate must be carefully regulated to force the equilibrium toward complete deacylation and to minimize undesirable β-lactam ring degradation. 6-APA has been prepared by conventional batch processing and by using immobilized cells, spores, and immobilized enzymes (69–71) (see CATALYSTS, SUPPORTED).

In a similar way, several cephalosporins have been hydrolyzed to 7-amino-deacetoxycephalosporanic acid (72), and nocardicin C to 6-aminonocardicinic acid (73). Penicillin G amidase from *Escherichia coli* has been used in an efficient resolution of a racemic cis intermediate required for a preparation of the synthon required for synthesis of the antibiotic Loracarbef (74). The racemic intermediate (**21**) underwent selective acylation to yield the cis derivative (**22**) in 44% yield; the product displayed a 97% enantiomeric excess (ee).

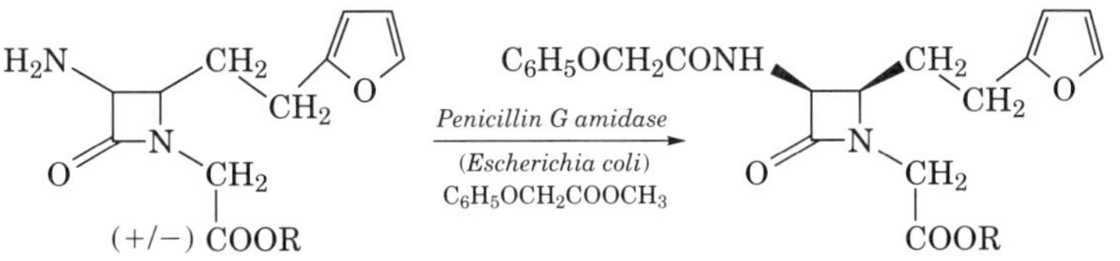

$$(5)$$

(**21**)   (**22**)

Enzymatic hydrolysis is also used for the preparation of L-amino acids. Racemic D- and L-amino acids and their acyl-derivatives obtained chemically can be resolved enzymatically to yield their natural L-forms. Aminoacylases such as that from *Aspergillus oryzae* specifically hydrolyze L-enantiomers of acyl-DL-amino acids. The resulting L-amino acid can be separated readily from the unchanged acyl-D form which is racemized and subjected to further hydrolysis. Several L-amino acids, eg, methionine [63-68-3], phenylalanine [63-91-2], tryptophan [73-22-3], and valine [72-18-4] have been manufactured by this process in Japan and production costs have been reduced by 40% through the application of immobilized cell technology (75). Cyclohexane chloride, which is a by-product in nylon manufacture, is chemically converted to DL-amino-ε-caprolactam [105-60-2] (**23**) which is resolved and/or racemized to (**24**) as shown. Quantitative hydrolysis of the L-isomer (**24**) gives L-lysine (**25**) (76).

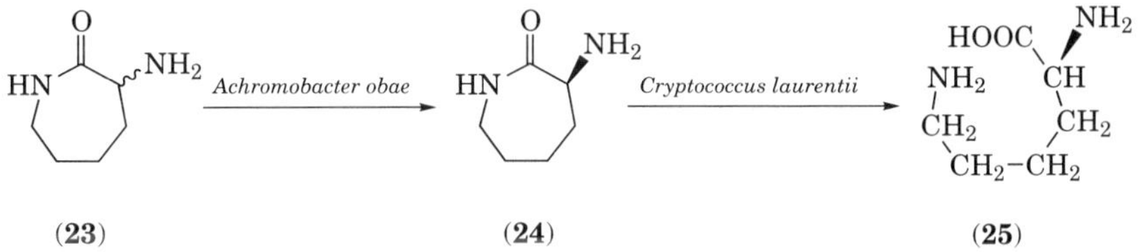

$$(6)$$

(**23**)   (**24**)   (**25**)

Significant advances have been made in the development of nitrile hydrolyzing enzymes. Nitrilases from microorganisms such as *Rhodococcus*

rhodochrous and *Pseudomonas chlororaphis* catalyze complete hydrolysis of the
nitrile (**26**) to the corresponding acid (**27**). This reaction may also be carried out
by the stepwise hydrolysis of (**26**) by a nitrile hydratase to form amide (**28**) and
subsequent hydrolysis of (**28**) using an amidase to yield the corresponding acid
(**27**). These enzymes can be differentially induced, and features of the nitrilase
gene of *R. rhodococcus* have been described (77). Since 1991, nitrile hydratase
has been used in an industrial-scale production (3×10^4 t/yr) of acrylamide
from acrylonitrile.

In this thiamine pyrophosphate-mediated process, benzaldehyde (**29**), added to
fermenting yeast, reacts with acetaldehyde (qv) (**30**), generated from glucose by
the biocatalyst, to yield (*R*)-1-phenyl-1-hydroxy-2-propanone (**31**). The enzymatically induced chiral center of (**31**) helps in the asymmetric reductive (chemical)
condensation with methylamine to yield (1*R*,2*S*)-ephedrine [*299-42-3*] (**32**). Substituted benzaldehyde derivatives react in the same manner (80).

Aldolases catalyze the asymmetric condensation of intermediates common
in sugar metabolism, such as phosphoenolpyruvic acid, with suitable aldehyde
acceptors. Numerous aldolases derived from plants or animals (Class I aldolases)
or from bacteria (Class II) have been examined for applications (81). Efforts to
extend the applications of these enzymes to the synthesis of unusual sugars
have been described (2,81).

Novel aromatic carboxylation reactions have been observed in the anaerobic
transformation of phenols to benzoates (82). A mixed anaerobic microbial con-
sortium apparently transforms phenol (**33**) through an intermediate to benzoic
acid (**34**) via dehydroxylation. This reaction has not yet been widely exploited
for its obvious synthetic value.

Carbon–Carbon Bond Formation.   Asymmetric microbial acyloin condensation was discovered in 1921 (78) and utilized in 1934 in the stereospecific
synthesis of (**32**) (79).

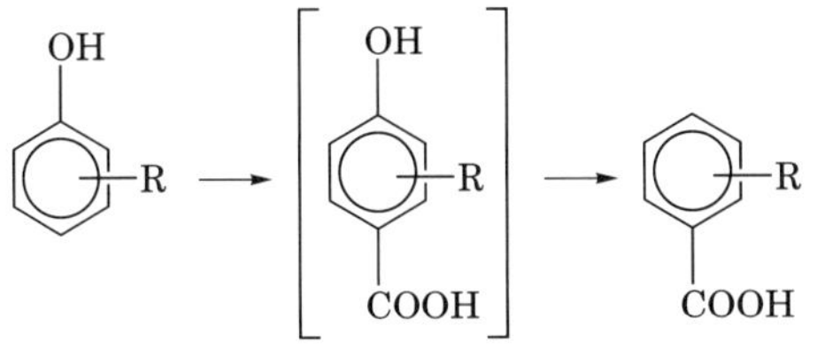

(9)

(33)   (34)

Amination and Hydration. Optically active products which correspond to the naturally occurring L-isomers have been obtained by the asymmetric addition of ammonia or water to fumaric acid [*110-17-8*] (**35**). Thus aspartase-producing bacteria have been used in the manufacture of L-aspartic acid [*56-84-8*] (**36**) (83). Bacteria having high fumarase activity hydrate (**35**) to L-malic acid [*6915-15-7*] (**37**) (84) in 70% yield, especially when pretreated with detergents, eg, bile extracts, in order to suppress the formation of succinic acid. Both reactions have been carried out by cells that are immobilized in polyacrylamide gels whereby the half-lives of the enzymes increase from 11–120 and 53 d, respectively. L-Citrulline [*372-75-8*] (**38**) is manufactured from L-arginine [*74-79-3*] (**39**) by the action of a deaminase (84).

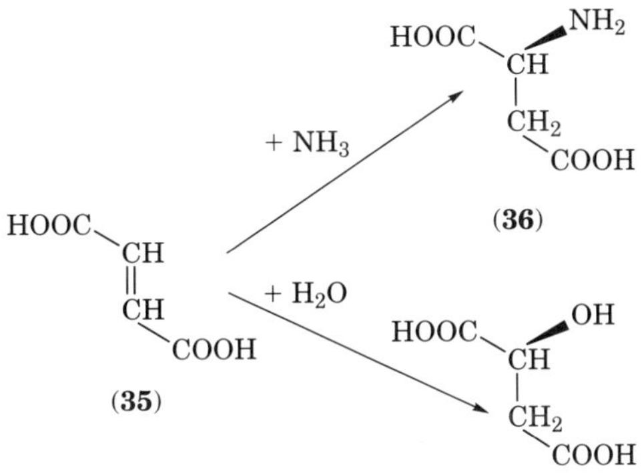

(10)

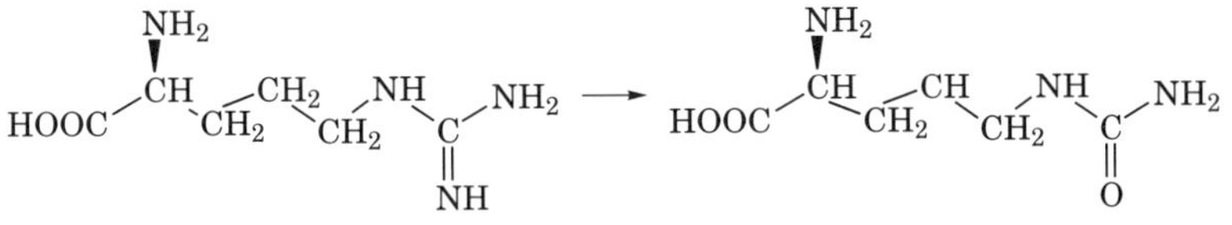

(11)

Other L-amino acids have been obtained using bacterial enzymes. Thus, by the addition of NH_4^+ and pyruvate to the reaction mixtures, L-tyrosine [*60-18-4*] has been produced from phenol, L-DOPA [*59-92-7*] from catechol, L-tryptophan from indole, and 5-hydroxy-L-tryptophan [*145224-90-4*] from 5-hydroxyindole. By hydration of DL-mixtures, L-isomers of a few other amino acids have been generated in 95–100% yields (49). Amination is also involved in the production of 5′-guanosine monophosphate (5′-GMP) [85-32-5], a compound used as a food seasoning, by means of two mutants of *Brevibacterium ammoniagenes*. The first mutant ferments glucose to xanthosine-5′-monophosphate (5′-XMP), which is converted to 5′-GMP by the second mutant, either by sequential operation or by mixed cultures of both mutants (85).

   Deamination, Transamination.   Two kinds of deamination that have been observed are hydrolytic, eg, the conversion of L-tyrosine to 4-hydroxyphenyllactic acid in 90% yield (86), and oxidative (12,87,88), eg, isoguanine to xanthine and formycin A to formycin B. Transaminases have been developed as biocatalysts for the synthetic production of chiral amines and the resolution of racemic amines (89). The reaction possibilities are illustrated for the stereospecific synthesis of (S)-α-phenylethylamine [98-84-0] (ee of 99%) (40) from (41) by an (S)-aminotransferase or by the resolution of the racemic amine (42) by an (R)-aminotransferase.

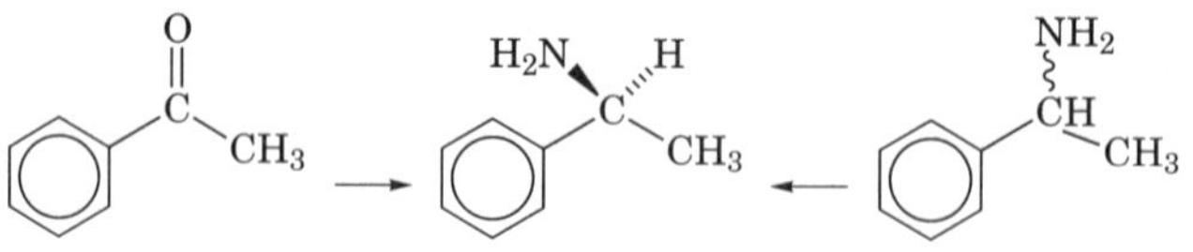

| (41) | (40) | (42) | (12) |

   Dehydration.   Dehydration of hydroxy fatty acids is quite common. Other compounds undergo the same reaction, eg, elymoclavine [548-43-6] to agroclavine [548-42-5], chanoclavine [2390-99-0], and other compounds; and *cis*-terpin hydrate [2541-01-63] to α-terpineol [98-55-5] (19).

   N- and O-Demethylation.   Microbial N- and O-demethylation reactions occur with high regiospecifities for multifunctional natural products. The biocatalytic approach is preferable to chemical methodologies, which require drastic conditions and are nonspecific (17,25). For example, O-demethylation of 10,11-dimethoxyaporphine [18605-42-0] (43) by *Cunninghamella blakesleeana* proceeds with high regioselectivity in 100% yield to isoapocodeine [36507-69-4] (44) (90).

| (43) | (44) | (13) |

Vindoline [2182-14-1] (45), a monomeric *Vinca* alkaloid intermediate important in the synthesis of antineoplastic alkaloids, is selectively converted in good yield to O-desmethylvindoline [68687-22-9] (46) by cultures of *Sepedonium chrysosper-*

mum (17,25), whereas *Streptomyces albogriseolus* removes only the *N*-methyl group to give (**47**) (91) (see CHEMOTHERAPEUTICS, ANTICANCER).

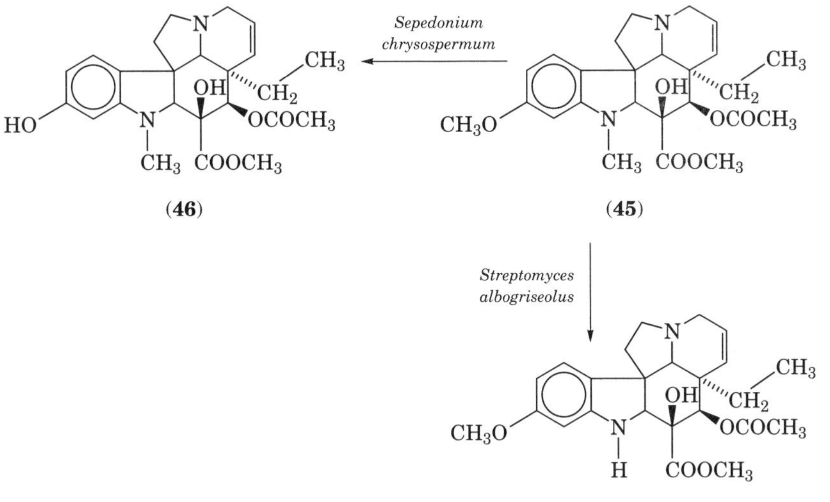

(14)

Decarboxylation.   Decarboxylation of linear and aromatic carboxylic acids and of amino acids is common and of practical interest. L-Lysine [56-87-1] (**48**) can be synthesized by stereospecific decarboxylation of *meso*- (but not DL-) $\alpha\alpha'$-diaminopimelic acid [2577-62-0] (**49**). The reaction is catalyzed by *Bacillus sphaericus* and proceeds in quantitative yields (92).

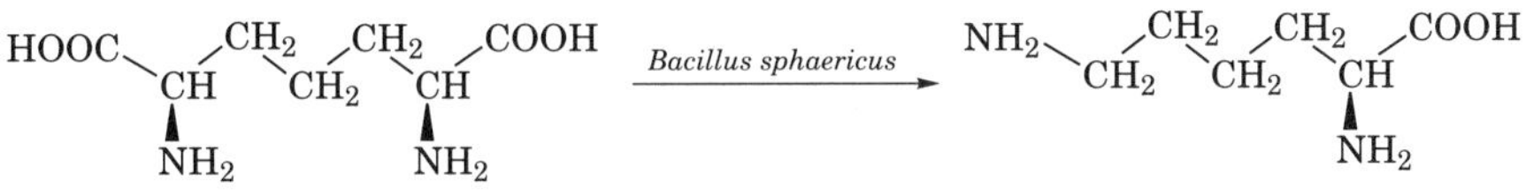

(15)

After recovery of L-lysine, the residual DL-(**49**) is epimerized to a mixture of the DL and meso isomers, and the latter is subjected to the same decarboxylation step. This reaction is a part of a microbial process in which glucose is fermented by a lysine auxotroph of *E. coli* to *meso*-(**49**) which accumulates in the medium. *Meso*-(**49**) is quantitatively decarboxylated to L-lysine by cell suspensions of *Aerobacter aerogenes* (93). However, L-lysine and some other L-amino acids are manufactured much more economically in thousands of tons per year in Japan by simplified fermentations directly from glucose, ethanol, acetic acid, glycerol, or *n*-paraffin, by means of selected auxotrophic, regulatory, and analogue-resistant bacterial mutants (94,95).

N-Acetylation, O-Phosphorylation, and O-Adenylylation.   *N*-Acetylation, *O*-phosphorylation, and *O*-adenylylation provide mechanisms by which therapeutically valuable aminocyclitol antibiotics, eg, kanamycin [8063-07-8], gentamicin [1403-66-3], sisomicin [32385-11-8], streptomycin [57-92-1], neomycin, or spectinomycin are rendered either partially or completely inactive. Thus, eg, kanamycin B [*4696-78-8*] (**50**) can be inactivated by modification at several sites, as shown. The elucidation of these mechanisms has allowed chemical modification of the sites at which the inactivation occurs. Several such bioactive analogues, eg, dibekacin and amikacin have been prepared and are not subject to the inactivation; hence, they inhibit those organisms against which the parent antibiotics are ineffective (96) (see ANTIBACTERIAL AGENTS, SYNTHETIC).

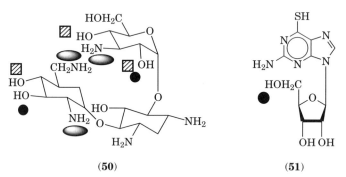

(**50**)   (**51**)

Fig. 3. Sites of possible modification for kanamycin B (**50**) and 6-thioguanosine (**51**) where (◠) represents *N*-acetylation, (●) *O*-phosphorylation, and (▨) *O*-adenylylation.

Acetylation of hydroxyl groups and esterification of carboxyl groups have been observed in a limited number of cases but, in general, have no preparative advantage over chemical methods. By comparison, phosphorylation has been useful in the preparation of modified purine and pyrimidine mononucleotides from their corresponding nucleosides, eg, 6-thioguanosine [*85-31-4*] (**51**) (97).

Transglycosylation. Enzymatic transglycosylations allow preparation of various oligosaccharides that cannot be made readily by strictly chemical means. The reaction can be expressed as

$$R—O—R' + R''OH \rightleftharpoons R—O—R'' + R'OH$$

donor   acceptor   product   by-product

It was first used in 1944 to prepare sucrose from glucose 1-phosphate and fructose using sucrose phosphorylase from *Pseudomonas saccharophila* (98). Corresponding disaccharides are formed when fructose is replaced in the same system by other sugars, eg, xylose, arabinose, or sorbose. The energy required for these reactions is derived from the high energy phosphate bond. The glycosidic linkage has sufficient energy for sucrose to be used as a glucosyl donor. It reacts with sorbose to yield glucosylsorbose when incubated with the *Pseudomonas saccharophila* enzyme (99). Tri- and tetrasaccharides are also formed from sucrose, maltose, lactose, and cellobiose by various bacterial, yeast, and fungal enzymes (12) (see SUGAR).

Isomerization. Isomerization of the double bonds of steroids is well known. Isomerizations of 5-androstene and 5-pregnene have been investigated but the products are not of practical value (100). On the other hand, isomerization is of considerable importance in the manufacture of high fructose syrup, which is used as a food sweetener. A process has been developed whereby inexpensive carbohydrates are hydrolyzed to glucose which is isomerized to fructose in 50% solutions by a *Streptomyces* strain. Upon purification and evaporation, syrup is obtained which contains 75 wt % solids consisting of 55% fructose and 45% glucose. The enzyme has been immobilized and is used in high volume continuous processes in the United States and Japan (94) (see SWEETENERS; SYRUPS).

Methodology

The selection of organisms that carry out biotransformation reactions is of paramount importance. However, because only a few systematic examinations of the action of microorganisms on specific classes of organic compounds have been made, there is no assured way to select the organism that performs the desired reaction with the substrate of interest. In many cases the suitable microorganisms have been uncovered by screening randomly selected cultures isolated from various natural sources, eg, soil, decomposing organic material, or spontaneous fermentations. Others have been found by the enrichment method, which utilizes the substrate in question as the only source of carbon and nitrogen for growth and energy (1,3).

To conduct biotransformation reactions, the selected organism usually is grown in flasks with or without aeration and shaking until a sufficient amount of cell biomass has been generated (see AERATION; CELL CULTURE TECHNOLOGY). The substrate chemical to be transformed is then added to the cells, the incubation is continued, and the progress of the transformation monitored by suitable chromatographic, spectroscopic, or biological methods. When the maximum transformation has been obtained, the reaction is terminated and the product isolated and identified.

There are many variations of this procedure. The substrate may be dissolved in water or in a solvent that is relatively nontoxic to the organisms, eg, ethanol, acetone, dimethylformamide, or dimethylsulfoxide; or it may be finely dispersed in water. When enzyme induction is required, some of the substrate may be added during early growth. When the desired transformation is carried out by constitutive enzymes, the substrate may be added in the late logarithmic or early stationary phase of growth, or to washed, ie, nonproliferating, buffered cell suspensions. A number of useful reactions have been carried out by microorganisms, spores, or isolated enzymes that have been immobilized by ionic binding to a water-insoluble ion exchanger, cross-linking with bifunctional reagents, or entrapping into a polymer matrix (see ION EXCHANGE). Such immobilization permits reuse of the cells and results in a more economical process.

A common means of increasing the rate or extent of biotransformation reactions is to expose the microorganisms to a mutagenic agent (such as x-rays, ultraviolet irradiation, nitrosoguanidine, 5-bromouracil, acridine, half-mustards, or novobiocin) which acts by causing DNA base-pair deletions, inversions, transitions, transversions, or deletion of extrachromosomal DNA elements (plasmids). Mutant strains are subsequently screened for enhanced expression of the desired enzyme. However, these types of mutations are characteristically random and multiple and often result in impaired growth rates and stability of the biocatalyst. By genetic engineering, new strains have been prepared for a variety of transformations either by expression of plasmid-borne genes, by site-specific insertions, or by mutations of the chromosome.

BIBLIOGRAPHY

"Microbial Transformations" in *ECT* 3rd ed., Vol. 15, pp. 459–470, by O. K. Sebek, The Upjohn Co.

1. J. B. Jones, C. J. Sih, and D. Perlman, eds., *Applications of Biochemical Systems in Organic Chemistry*, John Wiley & Sons, Inc., New York, 1976, part 1.
2. H. G. Davies, R. H. Green, D. R. Kelly, and S. M. Roberts, *Biotransformations in Preparative Organic Chemistry*, Academic Press Inc., San Diego, Calif., 1989; K. Faber, *Biotransformations in Organic Chemistry*, Springer-Verlag, New York, 1992; S. M. Roberts, K. Wiggins, and G. Casy, *Preparative Biotransformations: Whole Cells and Isolated Enzymes in Organic Synthesis*, John Wiley & Sons, Inc., New York, 1993; A. R. Battersby, *Enzymes in Organic Synthesis*, Ciba Foundation Symposium 111, Pitman, London, 1985.
3. C. T. Goodhue, J. P. Rosazza, and G. P. Peruzzotti, in A. L. Demain and N. A. Solomon, eds., *ASM Manual of Industrial Microbiology and Biotechnology*, American Society for Microbiology, Washington, D.C., 1985, pp. 97–121.
4. D. G. Drueckhammer and co-workers, *Synthesis*, 499 (1991); G. C. Look, C. H. Fotsch, and C. H. Wong, *Accounts Chem. Res.* **26**, 182 (1993).
5. J. F. T. Spencer and P. A. J. Gorin, *Progr. Ind. Microbiol.* **7**, 178 (1965).
6. W. Charney and H. L. Herzog, *Microbial Transformations of Steroids*, Academic Press, Inc., New York, 1967.
7. H. Iizuka and A. Naito, *Microbial Transformation of Steroids and Alkaloids*, University Park Press, State College, Pa., 1967.
8. L. L. Wallen, F. H. Stodola, and R. W. Jackson, *Type Reactions in Fermentation Chemistry*, ARS-71-13, U.S. Dept. of Agriculture, U.S. Government Printing Office, Washington, D.C., 1959.
9. L. L. Smith, *Spec. Period Rep. Chem. Soc. London* **4**, 394 (1974).
10. K. Kieslich and O. K. Sebek, *Ann. Rep. Ferment. Proc.* **3**, 275 (1979).
11. W. J. Marsheck, *Progr. Ind. Microbiol.* **10**, 49 (1971).
12. K. Kieslich, *Microbial Transformations of Non-Steroid Cyclic Compounds*, G. Thieme, Publishers, Stuttgart, Germany, 1976.
13. O. K. Sebek and K. Kieslich, *Ann. Rep. Ferment. Proc.* **1**, 263 (1977).
14. A. J. Markovetz, *Crit. Rev. Microbiol.* **1**, 225 (1971).
15. P. Jurtshuk and G. E. Cardini, *Crit. Rev. Microbiol.* **1**, 239 (1971).
16. B. J. Abbott and W. E. Gledhill, *Adv. Appl. Microbiol.* **14**, 249 (1971).
17. J. P. N. Rosazza and M. W. Duffel, in R. H. F. Manske and R. G. A. Rodrigo, eds., *The Alkaloids*, Academic Press, Inc., New York, 1986, pp. 324–400.
18. H. L. Holland, in R. G. A. Rodrigo, ed., *The Alkaloids*, Vol. XVIII, Academic Press, Inc., New York, 1981, Chapt. 5, pp. 324–400.
19. R. Beukers, A. F. Marx, and M. H. J. Zuidweg, *Drug Des.* **3**, 1 (1972).
20. H. G. W. Leuenberger, *Pure Appl. Chem.* **62**, 753 (1990).

21. K. Kieslich, *Bull. Soc. Chim. France*, II-9–II-17 (1980); A. L. Margolin, *Enz. Microb. Technol.* **15**, 266 (1993).

22. O. K. Sebek, *Lloydia* **37**, 115 (1974); *Acta Microbiol. Acad. Sci. Hung.* **22**, 381 (1975).

23. M. Shibata and M. Uyeda, *Ann. Rep. Ferment. Proc.* **2**, 267 (1978).

24. K. Kieslich, *Biotransformations*, Verlag Chemie, Deerfield Beach, Fla., 1984.

25. F. S. Sariaslani and J. P. N. Rosazza, *Enz. Microbial Technol.* **6**, 242 (1984); J. P. Rosazza, in J. Cassady and J. D. Douros, eds., *Anticancer Agents Based on Natural Product Models*, Academic Press, Inc., New York, 1981, pp. 437–464.

26. C. H. Tamm, *Angew. Chem., Int. Ed.* **1**, 178 (1962).

27. K. Kieslich, *FEMS Symposium* **13**, 311 (1982).

28. D. T. Gibson, ed., *Microbial Degradation of Organic Compounds*, Marcel Dekker Inc., New York, 1984.

29. I. C. Gunsalus and V. P. Marshall, *Crit. Rev. Microbiol.* **1**, 291 (1971).

30. P. W. Trudgill, *Biodegradation* **1**, 93 (1990).

31. M. M. Haggblom, *FEMS Microbiol. Rev.* **103**, 29 (1992).

32. D. J. Cork and J. P. Krueger, *Adv. Appl. Microbiol.* **36**, 1 (1991).

33. T. K. Kirk and R. L. Farrell, *Ann. Rev. Microbiol.* **41**, 465 (1987).

34. I. D. Reid, *Enz. Microb. Technol.* **11**, 786 (1989).

35. P. S. J. Cheetham, in V. Moses and R. Cape, eds., *Biotechnology, the Science and the Business*, Gordon and Breach, London, 1991, pp. 481–506.

36. G. S. Fonken and R. A. Johnson, *Chemical Oxidations with Microorganisms*, Marcel Dekker, Inc., New York, 1972.

37. R. A. Johnson, *Oxidation in Organic Chemistry*, Academic Press, Inc., New York, 1978, part C, p. 131.

38. P. Dunnill, A. Wiseman, and N. Blakebrough, *Enzymic and Nonenzymatic Catalysis*, Society of Chemical Industry, Ellis Horwood, Ltd., London, 1980.

39. R. Porter and S. Clark, *Enzymes in Organic Synthesis*, CIBA Foundation Symposium III, Pitman Publishers, London, 1985.

40. C. H. Wong and G. M. Whitesides, *Enzymes in Synthetic Organic Chemistry*, Pergamon Press, Tarrytown, N.Y., 1992.

41. H. L. Holland, *Organic Synthesis with Oxidative Enzymes*, VCH Publishers, New York, 1992.

42. C. J. Sih and E. Abushanab, *Ann. Rep. Med. Chem.* **12**, 298 (1977); M. A. Findeis and G. M. Whitesides, *Ann. Rep. Med. Chem.* **19**, 263 (1984); J. B. Jones, *Tetrahedron* **42**, 3351 (1986); F. S. Sariaslani, *Crit. Rev. Biotechnol.* **9**, 171 (1989).

43. L. Pasteur, *Mémoire sur la Fermentation Acétique, Études sur la Vinaigre*, Nasson and Sons, Paris, 1868.

44. G. Bertrand, *Compt. Rend.* **122**, 900 (1896); G. Bertrand, *Ann. Chim. Phys.* **3**, 181 (1904).

45. R. M. Hann, E. B. Tilden, and C. S. Hydson, *J. Am. Chem. Soc.* **60**, 1201 (1938); B. Magasanik, R. E. Franzl, and E. Chargaff, *J. Am. Chem. Soc.* **74**, 2618 (1952).

46. P. A. Wells, J. J. Tubbs, L. B. Lockwood, and E. T. Roe, *Ind. Eng. Chem.* **29**, 1385 (1937); **31**, 1518 (1939).

47. T. Sonoyama and co-workers, *Appl. Environ. Microbiol.* **43**, 1064 (1982).

48. J. F. Grindley, M. A. Payton, H. van de Pol, and K. G. Hardy, *Appl. Environ. Microbiol.* **54**, 1770 (1988); S. Anderson and co-workers, *Science* **230**, 144 (1985); G. Boudrant, *Enz. Microb. Technol.* **12**, 322 (1990).

49. Y. Hirose and H. Okada, in H. J. Peppler and D. Perlman, eds., *Microbial Technology*, Academic Press, Inc., New York, 1979, p. 211.

50. A. Kitai and A. Ozaki, *J. Ferment. Technol.* **47**, 527 (1969).

51. D. T. Gibson, M. Hensley, H. Yoshioka, and T. J. Mabry, *Biochemistry* **9**, 1626 (1970); D. T. Gibson, G. J. Zylstra, and S. Chauhan, in S. Silver, ed., *Pseudomonas:*

Biotransformations, Pathogenesis, and Evolving Biotechnology, American Society for Microbiology, Washington, D.C., 121, 1990.

52. T. Hudlicky and co-workers, *J. Chem. Soc., Perkin Trans I*, 2907 (1991); S. V. Ley and F. Sternfeld, *Tetrahedron* **45**, 3463 (1989).

53. D. H. Peterson and H. C. Murray, *J. Amer. Chem. Soc.* **74**, 1871 (1952); H. C. Murray and D. H. Peterson, U.S. Pat. 2,602,769 (July 8, 1952); J. Fried and co-workers, *J. Amer. Chem. Soc.* **74**, 3962 (1952).

54. J. M. Whitmarsh, *Biochem. J.* **90**, 23P (1964).

55. C. J. Sih, H. H. Tai, Y. Y. Tsong, S. S. Lee, and R. G. Coombe, *Biochemistry* **7**, 808 (1968).

56. W. J. Marsheck, S. Kraychy, and R. D. Muir, *Appl. Microbiol.* **23**, 72 (1972).

57. O. K. Sebek and D. Perlman, in H. J. Peppler and D. Perlman, eds., *Microbial Technology* (2nd ed.), 1979, p. 483.

58. V. Alphand and R. Furstoss, *J. Org. Chem.* **57**, 1306 (1992).

59. J. Fried, R. W. Thoma, and A. Klingsberg, *J. Amer. Chem. Soc.* **75**, 5764 (1953); and D. H. Peterson and co-workers, *J. Amer. Chem. Soc.* **75**, 5768 (1953).

60. W. Acklin and V. Prelog, *Helv. Chim. Acta* **48**, 1725 (1965).

61. C. W. Bradshaw, H. Fu, G.-J. Shen, and C. H. Wong, *J. Org. Chem.* **57**, 1526 (1992).

62. H. Kosmol and co-workers, *Liebigs Ann. Chem.* **701**, 198 (1967).

63. L. Mamoli and A. Vercellone, *Ber. Deutsch. Chem. Ges.* **70**, 470 (1937).

64. H.-A. Arfmann and W.-R. Abraham, *Z. Naturforsch. B* **48c**, 52 (1993).

65. M. Yamada, K. Iizuka, S. Okuda, T. Asai, and K. Tsuda, *Chem. Pharm. Bull.* **11**, 206 (1963).

66. U.S. Pat. 2,314,831 (Mar. 23, 1943), J. Kamlet (to Miles Laboratories, Inc.).

67. U.S. Pat. 3,957,579 (May 18, 1976), E. Sato and A. Yanai (to Toray Industries, Inc.).

68. T. R. Carrington, *Proc. Roy. Soc. London Ser. B.* **179**, 321 (1971).

69. B. J. Abbott, *Ann. Rep. Ferment. Proc.* **2**, 91 (1978).

70. I. Chibata, ed., *Immobilized Cells*, Kodansha, Ltd., Tokyo, and John Wiley & Sons, Inc., New York, 1978, pp. 182, 184.

71. L. B. Wingard, Jr., E. Katchalski-Katzir, and L. Goldstein, eds., *Applied Biochemistry and Bioengineering*, Vol. 2, Academic Press, Inc., New York, 1979.

72. M. Shimizu, T. Masuike, H. Fujita, K. Kimura, R. Okachi, and T. Nara, *Agri. Biol. Chem.* **39**, 1225 (1975).

73. T. Komori, K. Junugita, K. Nakahara, H. Aoki, and H. Imanaka, *Agric. Biol. Chem.* **42**, 1439 (1978).

74. M. Zmijewski, Jr., J. Briggs, J. Thompson, and M. Wright, *Tetrahedron Lett.* **32**, 1621 (1991).

75. I. Chibata, T. Tosa, T. Saot, T. Mori, and T. Matsuo, *Ferm. Technol. Today*, 383 (1972).

76. U.S. Pats. 3,770,585 (Nov. 6, 1973); 3,796,632 (Mar. 12, 1974), T. Fukumura (to Toray Industries, Inc.); T. Fukumura, *Agric. Biol. Chem.* **41**, 1321–1325 and 1327–1330 (1977).

77. M. Kobayashi, H. Komeda, N. Yanaka, T. Nagasawa, and H. Yamada, *J. Biol. Chem.* **267**, 20746 (1992); M. Kobayashi, T. Nagasawa, and H. Yamada, *TIBTECH* (Trends in Biotechnology) **10**, 402 (1992).

78. C. Neuberg and J. Hirsch, *Biochem. Z.* **115**, 282 (1921).

79. U.S. Pat. 1,956,950 (May 1, 1934), G. Hildebrandt and W. Klavehn (to E. Bilhuber, Inc.).

80. U.S. Pat. 3,338,796 (Aug. 29, 1967), J. W. Rothrock, (to Merck and Co., Inc.).

81. C.-H. Wong and G. M. Whitesides, *Enzymes is Synthetic Organic Chemistry*, Pergamon Publishing, Tarrytown, N.Y., 1994, pp. 195–251.

82. B. R. S. Genthner, G. T. Townsend, and P. J. Chapman, *FEMS Microbiol. Lett.* **78**, 265 (1991).

83. I. Chibata, T. Tosa, and T. Sato, *Appl. Microbiol.* **27**, 878 (1974).
84. I. Chibata and T. Tosa, *Adv. Appl. Microbiol.* **22**, 1 (1977).
85. A. Furuya, R. Okachi, K. Takayama, and S. Abe, *Biotechnol. Bioeng.* **19**, 1101 (1977).
86. F. Ehrlich and K. A. Jacobsen, *Ber. Deut. Chem. Gesell.* **44**, 888 (1911).
87. S. Friedman and J. S. Gots, *Arch. Biochem.* **32**, 227 (1951).
88. Y. Sawa and co-workers, *J. Antibiot.* **21**, 334 (1968).
89. U.S. Pat. 4,950,606 (Aug. 21, 1990), D. I. Stirling, A. Zeitlin, and G. W. Matcham (to Celgene Corp.); D. I. Stirling and G. W. Matcham, Proc. Chiral '90, Manchester, England, 1990, pp. 111.
90. J. P. Rosazza, A. W. Stocklinski, M. A. Gustafson, and J. Adrian, *J. Med. Chem.* **18**, 791 (1975).
91. N. Neuss, D. S. Fukuda, D. R. Brannon, and L. L. Huckstep, *Helv. Chim. Acta* **57**, 1891 (1974).
92. B. S. Gorton, J. N. Coker, H. P. Browder, and C. W. DeFiebre, *Ind. Eng. Chem. Prod. Res. Develop.* **2**, 308 (1963).
93. U.S. Pat. 2,771,396 (Nov. 20, 1956), L. E. Casida, Jr. (to Chas. Pfizer & Co., Inc.).
94. K. Arima, *Dev. Ind. Microbiol.* **18**, 79 (1977).
95. K. Yamada, *Biotechnol. Bioeng.* **19**, 1563 (1977).
96. J. Davies and D. I. Smith, *Ann. Rev. Microbiol.* **32**, 469 (1978).
97. Fr. Pat. 1,388,758 (Feb. 12, 1965), K. Mitsugi (to Ajinomoto Co., Inc.).
98. W. Z. Hassid, M. Doudoroff, and H. A. Barker, *J. Amer. Chem. Soc.* **66**, 1416 (1944).
99. M. Doudoroff, H. A. Barker, and W. Z. Hassid, *J. Biol. Chem.* **168**, 725, 733 (1947).
100. P. Talalay and V. S. Wang, *Biochem. Biophys. Acta* **18**, 300 (1955).

OLDRICH K. SEBEK
The Upjohn Company

JOHN P. N. ROSAZZA
University of Iowa

MICROCHEMISTRY. See ANALYTICAL METHODS.

MICROELECTRONICS. See SEMICONDUCTORS; X-RAY TECHNOLOGY; NANOTECHNOLOGY (SUPPLEMENT).

MICROEMULSIONS. See SUPPLEMENT.

MICROENCAPSULATION

Microencapsulation is the coating of small solid particles, liquid droplets, or gas bubbles with a thin film of coating or shell material. In this article, the term microcapsule is used to describe particles with diameters between 1 and 1000 μm. Particles smaller than 1 μm are called nanoparticles; particles greater than 1000 μm can be called microgranules or macrocapsules.

Many terms have been used to describe the contents of a microcapsule: active agent, actives, core material, fill, internal phase (IP), nucleus, and payload. Many terms have also been used to describe the material from which the capsule is formed: carrier, coating, membrane, shell, or wall. In this article the material being encapsulated is called the core material; the material from which the capsule is formed is called the shell material.

Table 1 lists representative examples of capsule shell materials used to produce commercial microcapsules along with preferred applications. The gelatin–gum arabic complex coacervate treated with glutaraldehyde is specified as nonedible for the intended application, ie, carbonless copy paper, but it has been approved for limited consumption as a shell material for the encapsulation of selected food flavors. Shell material costs vary greatly. The cheapest acceptable shell materials capable of providing desired performance are favored,

Table 1. Shell Materials Used to Produce Commercially Significant Microcapsules

Shell material	Regulatory status	Chemical class	Encapsulation process	Applications
gum arabic	edible	polysaccharide	spray drying	food flavors
gelatin	edible	protein	spray drying	vitamins
gelatin–gum arabic[a]	nonedible[b]	protein–polysaccharide complex	complex coacervation	carbonless paper
ethylcellulose	edible	cellulose ether	Wurster process or polymer–polymer incompatibility	oral pharmaceuticals
polyurea or polyamide	nonedible	cross-linked polymer	interfacial polymerization	agrochemicals and carbonless paper
aminoplasts	nonedible	cross-linked polymer	*in situ* polymerization	carbonless paper, fragrances, and adhesives
maltodextrins	edible	low molecular weight carbohydrate	spray drying and desolvation	food flavors
hydrogenated vegetable oils	edible	glycerides	fluidized bed	assorted food ingredients

[a]Treated with glutaraldehyde.
[b]For intended application, ie, carbonless paper.

however, defining the optimal shell material for a given application is not an easy task.

Microcapsules can have a wide range of geometries and structures. Figure 1 illustrates three possible capsule structures. Parameters used to characterize microcapsules include particle size, size distribution, geometry, actives content, storage stability, and core material release rate.

Research and development activity throughout the world dedicated to advancing microcapsulation technology has produced a steadily increasing number of commercially successful products that utilize microcapsules. Most are comparatively small volume products. Only two high volume microcapsule-based products exist currently: carbonless copy paper and a long-acting herbicide (alachlor) formulation. The number of processes used to produce microcapsules continues to expand. Hot melt and aqueous-based coating processes are preferred, because environmental and safety regulations greatly increase the cost of processes that utilize toxic and flammable solvents. There have been a number of reviews on encapsulation technology and applications (1–6).

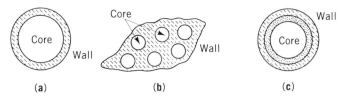

Fig. 1. Schematic diagrams of several possible capsule structures: (**a**) continuous core/shell microcapsule in which a single continuous shell surrounds a continuous region of core material; (**b**) multinuclear microcapsule in which a number of small domains of core material are distributed uniformly throughout a matrix of shell material; and (**c**) continuous core capsule with two different shells.

Encapsulation Process

Classification of the many different encapsulation processes is useful. Previous schemes employing the categories chemical or physical are unsatisfactory because many so-called chemical processes involve exclusively physical phenomena, whereas so-called physical processes can utilize chemical phenomena. An alternative approach is to classify all encapsulation processes as either Type A or Type B processes. Type A processes are defined as those in which capsule formation occurs entirely in a liquid-filled stirred tank or tubular reactor. Emulsion and dispersion stability play a key role in determining the success of such processes. Type B processes are processes in which capsule formation occurs because a coating is sprayed or deposited in some manner onto the surface of a liquid or solid core material dispersed in a gas phase or vacuum. This category also includes processes in which liquid droplets containing core material are sprayed into a gas phase and subsequently solidified to produce microcapsules. Emulsion and dispersion stabilization can play a key role in the success of Type B processes also.

Many Type A and Type B processes are similar. For example, solvent evaporation is a key step in most spray dry encapsulation protocols (Type B) and protocols involving solvent evaporation from an emulsion (Type A). The difference in these protocols is that evaporation in the former case occurs directly from a liquid to a gas phase, whereas in the latter case evaporation involves transfer of a volatile liquid from a dispersed phase to a continuous liquid phase from which it is subsequently evaporated. Another example is encapsulation by gelation. In Type A gelation processes, the droplets that are gelled and become microcapsules are formed by dispersion in a liquid phase and are gelled in this phase. In Type B gelation processes, droplets formed by atomization or extrusion into a gas phase are subsequently gelled either in the gas phase or a liquid gelling bath.

Most Type A processes might be classified as chemical processes, whereas most Type B processes are classified as mechanical processes. Representative examples of both types of processes follow. Type B processes tend to be promoted by organizations that sell and service equipment for producing microcapsules. Most Type A processes are not promoted by equipment manufacturers, but are developed and used by organizations that produce microcapsules.

Type A processes	Type B processes
complex coacervation	spray drying
polymer–polymer incompatibility	fluidized bed
interfacial polymerization at liquid–liquid and solid–liquid interfaces	interfacial polymerization at solid–gas or liquid–gas interfaces
in situ polymerization	centrifugal extrusion
solvent evaporation or in-liquid drying	extrusion or spraying into a desolvation bath
submerged nozzle extrusion	rotational suspension separation (spinning disk)

TYPE A PROCESSES

Complex Coacervation. This process occurs in aqueous media and is used primarily to encapsulate water-immiscible liquids or water-insoluble solids (7). In the complex coacervation of gelatin with gum arabic (Fig. 2), a water-insoluble core material is dispersed to a desired drop size in a warm gelatin solution. After gum arabic and water are added to this emulsion, pH of the aqueous phase is typically adjusted to pH 4.0–4.5. This causes a liquid complex coacervate of gelatin, gum arabic, and water to form. When the coacervate adsorbs on the surface of the core material, a liquid complex coacervate film surrounds the dispersed core material thereby forming embryo microcapsules. The system is cooled, often below 10°C, in order to gel the liquid coacervate shell. Glutaraldehyde is added and allowed to chemically cross-link the capsule shell. After treatment with glutaraldehyde, the capsules are either coated onto a substrate or dried to a free-flow powder.

Any pair of oppositely charged polyelectrolytes capable of forming a liquid complex coacervate can be used to form microcapsules by complex coacervation.

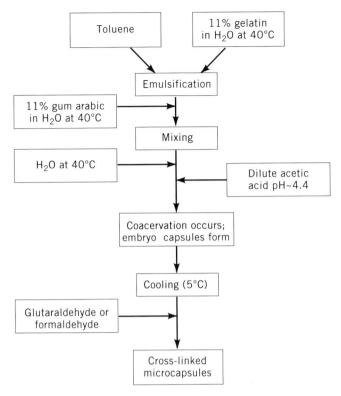

Fig. 2. Flow diagram of a typical encapsulation process based on the complex coacervation of gelatin with gum arabic.

Gelatin normally is the positively charged polyion, because it is readily available and forms suitable complex coacervates with a wide range of polyanions. Polyanions typically used include gum arabic, polyphosphate, poly(acrylic acid), and alginate.

Because glutaraldehyde lightly cross-links gelatin-based complex coacervates, the shell of such capsules immediately after manufacture is a highly swollen hydrogel and difficult to dry without agglomeration. In order to reduce capsule shell hydrophilicity and improve ease of capsule drying, the capsules can be treated at low (2–4) pH with urea and formaldehyde. These materials polymerize within water-swollen complex coacervate capsule shells thereby greatly increasing the degree of chemical cross-linking and enhancing ease of drying to a free-flow powder.

A wide variety of capsules loaded with water-immiscible or water-insoluble materials have been prepared by complex coacervation. Capsule size typically ranges from 20–1000 μm, but capsules outside this range can be prepared. Core contents usually are 80–95 wt %. Complex coacervation processes are adversely affected by active agents that have finite water solubility, are surface-active, or are unstable at pH values of 4.0–5.0. The shell of dry complex coacervate capsules is sensitive to variations in atmospheric moisture content and becomes plasticized at elevated humidities.

Polymer–Polymer Incompatibility. Two chemically different polymers dissolved in a common solvent usually are incompatible. That is, they spontaneously separate into two liquid phases with each phase containing predominately one polymer. When a core material insoluble in the solvent is dispersed in such systems, it is spontaneously coated by a thin film of the liquid phase that contains the polymer designed to be the capsule shell material; this polymer must be preferentially adsorbed by the core material which is not difficult to arrange. Microcapsules are harvested by desolvating this coating either by chemical cross-linking or addition of a nonsolvent (Fig. 3) (8). In the latter case, the embryonic capsule as slurry is often added to a very large excess of nonsolvent in order to minimize capsule agglomeration during isolation.

Polymer–polymer incompatibility encapsulation processes can be carried out in aqueous or nonaqueous media, but thus far have primarily been carried out in organic media. Core materials encapsulated tend to be polar solids with a finite degree of water solubility. Ethylcellulose historically has been the shell material used. Biodegradable shell materials such as poly(D,L-lactide) and lactide–glycolide copolymers have received much attention. In these latter cases, the object has been to produce biodegradable capsules that carry proteins or polypeptides. Such capsules tend to be below 100 μm in diameter and are for oral or parenteral administration (9).

Interfacial Polymerization. Many types of polymerization reactions can be made to occur at interfaces or produce polymers that concentrate at interfaces thereby producing microcapsules. Accordingly, this approach to encapsulation has steadily developed into a versatile family of encapsulation processes.

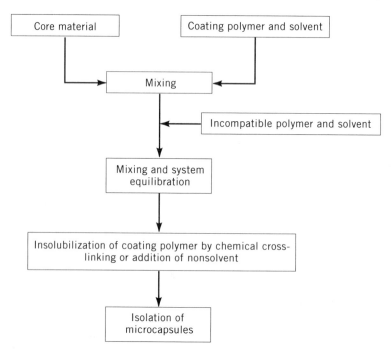

Fig. 3. Flow diagram of typical encapsulation process based on polymer–polymer incompatibility.

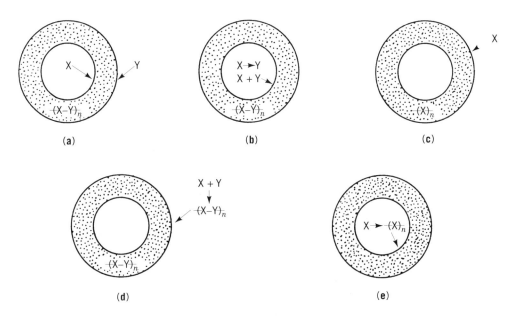

Fig. 4. Schematic diagrams that illustrate the different types of interfacial polymerization reactions used to form microcapsules. Reactants: X, Y; polymerization product: $+X-Y+_{\overline{n}}$ or $+X+_{\overline{n}}$. See text for descriptions of cases (**a**)–(**e**).

Figure 4 schematically illustrates five types of encapsulation processes that utilize these types of reactions.

Figure 4**a** represents interfacial polymerization encapsulation processes in which shell formation occurs at the core material–continuous phase interface due to reactants in each phase diffusing and rapidly reacting there to produce a capsule shell (10,11). The continuous phase normally contains a dispersing agent in order to facilitate formation of the dispersion. The dispersed core phase encapsulated can be water, or a water-immiscible solvent. The reactant(s) and coreactant(s) in such processes generally are various multifunctional acid chlorides, isocyanates, amines, and alcohols. For water-immiscible core materials, a multifunctional acid chloride, isocyanate or a combination of these reactants, is dissolved in the core and a multifunctional amine(s) or alcohol(s) is dissolved in the aqueous phase used to disperse the core material. For water or water-miscible core materials, the multifunctional amine(s) or alcohol(s) is dissolved in the core and a multifunctional acid chloride(s) or isocyanate(s) is dissolved in the continuous phase. Both cases have been used to produce capsules.

Figure 5 illustrates the type of encapsulation process shown in Figure 4**a** when the core material is a water-immiscible liquid. Reactant X, a multifunctional acid chloride, isocyanate, or combination of these reactants, is dissolved in the core material. The resulting mixture is emulsified in an aqueous phase that contains an emulsifier such as partially hydrolyzed poly(vinyl alcohol) or a lignosulfonate. Reactant Y, a multifunctional amine or combination of amines such as ethylenediamine, hexamethylenediamine, or triethylenetetramine, is added to the aqueous phase thereby initiating interfacial polymerization and formation of a capsule shell. If reactant X is an acid chloride, base is added to the aqueous phase in order to act as an acid scavenger.

A key feature of encapsulation processes (Figs. 4a and 5) is that the reagents for the interfacial polymerization reaction responsible for shell formation are present in two mutually immiscible liquids. They must diffuse to the interface in order to react. Once reaction is initiated, the capsule shell that forms becomes a barrier to diffusion and ultimately begins to limit the rate of the interfacial polymerization reaction. This, in turn, influences morphology and uniformity of thickness of the capsule shell. Kinetic analyses of the process have been published (12). A drawback to the technology for some applications is that aggressive or highly reactive molecules must be dissolved in the core material in order to produce microcapsules. Such molecules can react with sensitive core materials.

Figure 4b represents the case where a reactant dissolved in the dispersed phase reacts with the continuous phase to produce a co-reactant. The co-reactant and any remaining unreacted original reactant left in the dispersed phase then proceed to react with each other at the dispersed phase side of the interface and produce a capsule shell. Capsule shell formation occurs entirely because of reaction of reactants present in the droplets of dispersed phase. No reactant is added to the aqueous phase. As in the case of the process described by Figure 4a, a reactive species must be dissolved in the core material in order to produce a capsule shell.

A specific example of the process represented by Figure 4b occurs when a multifunctional isocyanate is dissolved in a liquid, water-immiscible core material and the mixture produced is dispersed in an aqueous phase that contains a dispersing agent. The aqueous phase reacts with some of the isocyanate groups to produce primary amine functionalities. These amino groups react with unreacted isocyanate groups to produce a polyurea capsule shell (13).

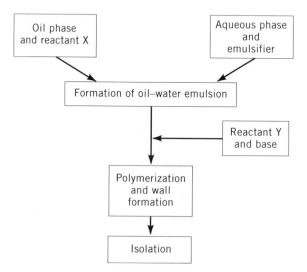

Fig. 5. Flow diagram of typical interfacial polymerization encapsulation process in which reactants X and Y are dissolved in separate mutually immiscible phases (see Fig. 4a).

Figure 4**c** illustrates interfacial polymerization encapsulation processes in which the reactant(s) that polymerize to form the capsule shell is transported exclusively from the continuous phase of the system to the dispersed phase–continuous phase interface where polymerization occurs and a capsule shell is produced. This type of encapsulation process has been carried out at liquid–liquid and solid–liquid interfaces. An example of the liquid–liquid case is the spontaneous polymerization reaction of cyanoacrylate monomers at the water–solvent interface formed by dispersing water in a continuous solvent phase (14). The poly(alkyl cyanoacrylate) produced by this spontaneous reaction encapsulates the dispersed water droplets. An example of the solid–liquid process is where a core material is dispersed in aqueous media that contains a water-immiscible surfactant along with a controlled amount of surfactant. A water-immiscible monomer that polymerizes by free-radical polymerization is added to the system and free-radical polymerization localized at the core material–aqueous phase interface is initiated thereby generating a capsule shell (15).

Figure 4**c** also describes the spontaneous polymerization of *para*-xylylene diradicals on the surface of solid particles dispersed in a gas phase that contains this reactive monomer (16) (see XYLYLENE POLYMERS). The poly(*p*-xylylene) polymer produced forms a continuous capsule shell that is highly impermeable to transport of many penetrants including water. This is an expensive encapsulation process, but it has produced capsules with impressive barrier properties. This process is a Type B encapsulation process, but is included here for the sake of completeness.

Figure 4**d** represents *in situ* encapsulation processes (17,18), an example of which is presented in more detail in Figure 6 (18). The first step is to disperse a

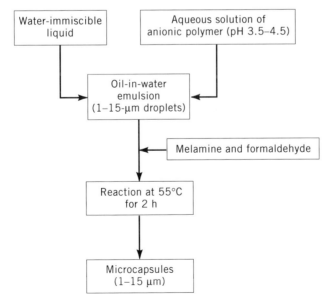

Fig. 6. Flow diagram of microencapsulation process that utilizes acid-catalyzed *in situ* polymerization of melamine or urea with formaldehyde to form a capsule shell (see Fig. 4**d**).

water-immiscible liquid or solid core material in an aqueous phase that contains urea, melamine, water-soluble urea–formaldehyde condensate, or water-soluble urea–melamine condensate. In many cases, the aqueous phase also contains a system modifier that enhances deposition of the aminoplast capsule shell (18). This is an anionic polymer or copolymer (Fig. 6). Shell formation occurs once formaldehyde is added and the aqueous phase acidified, eg, pH 2–4.5. The system is heated for several hours at 40–60°C.

A unique feature of *in situ* encapsulation technology is that polymerization occurs in the aqueous phase thereby producing a condensation product that deposits on the surface of the dispersed core material where polymerization continues. This ultimately produces a water-insoluble, highly cross-linked polymer capsule shell. The polymerization chemistry occurs entirely on the aqueous phase side of the interface, so reactive agents do not have to be dissolved in the core material. The process has been commercialized and produces a range of commercial capsules.

Figure 4**e** represents interfacial polymerization processes in which a solution of water-immiscible vinyl monomer(s), vinyl monomer initiator, and core material is dispersed in an aqueous phase that contains a dispersing agent. Polymerization is then initiated, eg, by heating. The polymer produced within the droplets of core material is designed to precipitate or deposit at the core material–aqueous phase interface thereby producing a microcapsule shell. Although this technology was disclosed many years ago, it has not received much use, presumably because the capsules produced do not have acceptable barrier properties.

Solvent Evaporation. This encapsulation technology involves removing a volatile solvent from either an oil-in-water, oil-in-oil, or water-in-oil-in-water emulsion (19,20). In most cases, the shell material is dissolved in a volatile solvent such as methylene chloride or ethyl acetate. The active agent to be encapsulated is either dissolved, dispersed, or emulsified into this solution. Water-soluble core materials like hormonal polypeptides are dissolved in water that contains a thickening agent before dispersion in the volatile solvent phase that contains the shell material. This dispersed aqueous phase is gelled thermally to entrap the polypeptide in the dispersed aqueous phase before solvent evaporation occurs (21).

Once the active agent is dispersed in the volatile solvent phase, the dispersion produced typically is emulsified into an aqueous phase that contains a dispersing agent. Solvent used to dissolve the shell material is subsequently removed from this emulsion at atmospheric or reduced pressure. The microparticles produced can range in size from below 1 μm to over several hundred micrometers. The technique is used often to form drug-loaded microparticles from biodegradable polymer shell materials (19,20). Significantly, solvent removal from the dispersed phase during a solvent evaporation process occurs by partitioning or extraction of volatile solvent into the continuous phase of the emulsion rather than evaporation. Solvent is removed from the continuous phase by evaporation. Because of this, a range of mass-transfer events can occur during the solvent evaporation process. These often have a negative effect on microcapsule formation, such as causing loss of core material to the continuous phase.

Centrifugal Force and Submerged Nozzle Processes. A variety of processes have used centrifugal force or submerged nozzles to produce microcapsules. For example, a device in which a perforated cup was immersed in an oil bath and spun at a fixed rate has been described (22). An emulsion of core and shell material fed into the spinning cup was thrown in droplet form into the oil outside the spinning cup by centrifugal force. Such droplets were gelled by cooling to thereby yield gel beads loaded with an oil that could subsequently be dried.

Capsules have also been produced by extruding in droplet form a liquid core and liquid shell formulation through a so-called two-fluid nozzle into a moving stream of carrier fluid. Such nozzles are called submerged nozzles. The liquid shell formulation is gelled, usually by cooling, to yield gel beads that can be dried (23). To date, Type A centrifugal force and submerged nozzle encapsulation processes have been used primarily for the encapsulation of vitamin and flavor oils. The capsules tend to be relatively large and are formed from natural shell materials that are capable of being gelled thermally.

TYPE B PROCESSES

Spray Drying. Spray-dry encapsulation processes (Fig. 7) consist of spraying an intimate mixture of core and shell material into a heated chamber where rapid desolvation occurs to thereby produce microcapsules (24,25). The first step in such processes is to form a concentrated solution of the carrier or shell material in the solvent from which spray drying is to be done. Any water- or solvent-soluble film-forming shell material can, in principle, be used. Water-soluble polymers such as gum arabic, modified starch, and hydrolyzed gelatin are used most often. Solutions of these shell materials at 50 wt % solids have sufficiently low viscosities that they still can be atomized without difficulty. It is not unusual to blend gum arabic and modified starch with maltodextrins, sucrose, or sorbitol.

The second step is to disperse the core material being encapsulated in the solution of shell material. The core material usually is a hydrophobic or

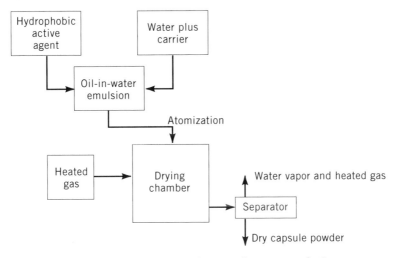

Fig. 7. Flow diagram of a typical spray-dry encapsulation process.

water-immiscible oil, although solid powders have been encapsulated. A suitable emulsifier is used to aid formation of the dispersion or emulsion. In the case of oil core materials, the oil phase is typically reduced to a drop size of 1–3 μm. Once a suitable dispersion or emulsion has been prepared, it is sprayed into a heated chamber. The small droplets produced have a high surface area and are rapidly converted by desolvation in the chamber to a fine powder. Residence time in the spray-drying chamber is 30 s or less. Inlet and outlet air temperatures are important process parameters as is relative humidity of the inlet air stream.

Capsules produced by spray drying typically have a diameter of 10–40 μm and larger. They may be individual particles or aggregates. In either case, capsule geometry is irregular. Core material loadings of commercial spray-dried capsules typically are 20–30 wt %, although loadings as high as 60 wt % have been reported (24). The core material is dispersed as small droplets or particles throughout spray-dried capsules. Hydrocolloid carriers such as gum arabic and modified starch produce spray-dried capsules that are dispersible in cold water, but warm or hot water is needed to dissolve them completely. Maltodextrins and hydrolyzed gelatins give capsules soluble in cold and hot water.

Spray-dry encapsulation is a low cost process capable of producing a range of microcapsules in good yield. It is often used to produce commercial capsules loaded with fragrance or flavor oils. Although this encapsulation technology has steadily improved over the years, the capsules produced may contain a small amount of free or unencapsulated active agent which, if a flavor oil, oxidizes on storage thereby producing an objectionable odor. Another problem is loss of volatiles during the atomization and drying steps. Low boiling and polar active agents such as ethyl acetate and acetaldehyde can be difficult to encapsulate by this technology. Their retention is favored by maximizing initial concentration of carrier material.

A number of research workers have prepared drug-loaded biodegradable capsules by spray drying from nonaqueous media. Spray chilling, cooling, and congealing are essentially variations of conventional spray drying. In these cases, chilled air is used to solidify molten capsule shell material formulations rather than volatilize a solvent as is done in the case of spray drying (26). Various fats, waxes, fatty alcohols, or fatty acids are used as shell materials. In such encapsulation procedures, the active agent is dispersed in a molten shell material with the aid of an emulsifier if necessary (see EMULSIFIERS). This dispersion is atomized through heated nozzles into a cooling chamber analogous to that shown in Figure 7. The shell material is solidified by cooling and solid particles are isolated. If the chamber is at room temperature, the coating material has a melting point between 45 and 122°C. If the chamber is cooled, material melting at 32–43°C can be used. Particles produced by this method have water-insoluble shells. The influence of processing temperature on shell material polymorphism, a phenomenon characteristic of many fats, has been discussed (27).

Fluidized-Bed Coating. Fluidized-bed encapsulation technology involves spraying shell material in solution or hot melt form onto solid particles suspended in a stream of heated gas, usually air (28,29). Two types of fluidized-bed units, top and bottom spray, are used. In top-spray units, hot melt shell materials such as fats and waxes are sprayed onto the top of a fluidized bed of solid particles. The coated particles are subsequently cooled producing capsules with a solid shell. This technology is used to prepare a variety of encapsulated water-soluble food ingredients. Top-spray units generally are not recommended for applying solutions of shell material because the spray droplets move counter-current to the gas stream that suspends the fluidized bed. This favors solvent evaporation and deposition of a poor coating of shell material.

In bottom-spray or Wurster units (Fig. 8), the coating material is sprayed as a solution into the bottom of a column of fluidized particles. The freshly coated particles are carried away from the nozzle by the airstream and up into the coating chamber where the coating solidifies due to evaporation of solvent. At the top of the column or spout, the particles settle. They ultimately fall back to the bottom of the chamber where they are guided once again by the airstream past the spray nozzle and up into the coating chamber. The cycle is repeated until a desired capsule shell thickness has been reached. Coating uniformity and final coated particle size are strongly influenced by the nozzle(s) used to apply the coating formulation. This technology is routinely used to encapsulate solids, especially pharmaceuticals (qv). It can coat a wide variety of particles, including irregularly shaped particles. The technology typically has been used to produce capsules larger than 100–150 μm, but can produce coated particles smaller than 100 μm. An important feature of the Wurster process is that it can be used to apply a wide range of coating materials including hydrocolloids, solvent-soluble polymers, sugars, and pseudolatices. The last are aqueous dispersions of small polymer particles which fuse together during desolvation to form a continuous capsule shell.

Centrifugal Force Processes. Several Type B encapsulation processes utilize centrifugal force to form microcapsules. In one process, two mutually immiscible liquids, the core and shell material, are pumped through a two-fluid nozzle that is spun rapidly (30). The two-fluid column produced breaks up spontaneously into a series of droplets. Each droplet contains a core region surrounded by a continuous film of fluidized shell material. The shell is solidified

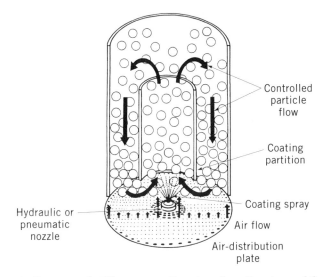

Fig. 8. Schematic diagram of a Wurster coating chamber. Courtesy of the Coating Place.

either by cooling or by immersion in a gelling bath to produce microcapsules that can be harvested. In the former case, the shell material is typically a molten wax or fat that has relatively low melt viscosity and solidifies by cooling as it falls away from the nozzle. It may contain a small amount of polymer added in order to toughen the final capsule shell. The core material is typically water or an aqueous solution. In the latter case, the fluidized shell material typically is an aqueous sodium alginate solution. This solution is gelled ionically by immersion in an aqueous $CaCl_2$ solution to form a gel bead which can subsequently be dried and isolated. The core material is a water-immiscible liquid. This type of encapsulation technology tends to favor production of larger microcapsules. The lower particle size limit is approximately 250 μm; the upper particle size limit is several mm.

A second Type B encapsulation process that uses centrifugal force is called rotational suspension separation (31,32). It utilizes a rapidly spinning disk to produce particles of core material surrounded by shell material. The fluidized shell material formulation, often a wax, fat, or solution of a polymer in these materials, is mixed with particles of core material. The resulting dispersion is fed onto a rotating disk placed at the top of a chamber. Individual particles of liquid-coated droplets are flung off the edge of this disk into the chamber producing solid particles of core material enclosed in a thin film of liquid shell material. The shell formulation solidifies as the embryonic capsules fall to the bottom of the chamber producing solid microcapsules. Droplets of pure coating material are also flung off the edge of the disk, and fall in a circumferential area or zone below the disk at a closer distance to the disk than the capsules.

A number of hot melt shell formulations have been applied in this manner to produce a range of capsules. It is claimed that the process is capable of rapidly producing capsules smaller than 100 μm in large amounts at low cost. Candidate core materials must act like solids on the rotating disk and should have a spherical geometry. Candidate coating formulations must solidify rapidly on cooling and not have high viscosities. For these reasons, preferred shell formulations typically are waxes and fats or polymer solutions in these materials.

Desolvation or Extractive Drying.   This encapsulation technology (Fig. 9) has been used to produce water-soluble capsules loaded with a range of flavor

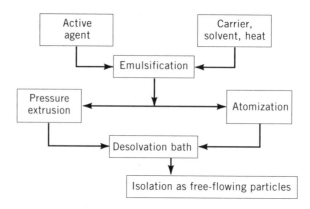

Fig. 9.   Desolvation process.

compounds (see FLAVORS AND SPICES) (33,34). Water-soluble shell materials used are maltodextrins, sugars, and gums (qv). The shell material is dissolved in as little water as possible and the core material is dispersed in this solution. Heat, up to 124°C, is used to reduce the viscosity of the shell formulation so that dispersion can be achieved readily. The dispersion of core material and shell formulation is either extruded or atomized into a desolvation solvent. Preferred desolvation solvents are water-miscible alcohols such as 2-propanol or polyglycols. This step solidifies the particles. Because the core materials used in this process are miscible with the desolvation solvents used, capsules with no free core material are produced. Such capsules have excellent storage stability properties. The problem with this approach to encapsulation has been that it uses solvents other than water which must be recycled. It also produces capsules with relatively low core loadings. Core loadings as high as 15% can be achieved, but forming stable emulsions in the concentrated hot aqueous core solutions used poses serious problems (34). Although the technology has been commercialized, it appears to have lost ground to spray drying in the 1990s.

Other Processes. A steady stream of encapsulation technologies continues to appear in the patent literature. Some are simply modifications or improvements of established technologies, whereas others are novel new technologies, eg, biliquid column (35), electrostatic encapsulation (36), metal vapor deposition (37), powder bed (1), ethylene polymerization (38), very low temperature casting (39), and supercritical fluid extraction (40). Pan coating, a well-established means of producing coated particles, has not been discussed because it produces particles larger than 1000 μm.

Applications

Microcapsules are used in a number of pharmaceutical, graphic arts, food, agrochemical, cosmetic, and adhesive products. Other specialty products also exist, thus the concept of microencapsulation has been accepted by a wide range of industries. In order to illustrate how microcapsules are used commercially, it is appropriate to describe a number of commercial microcapsule-based products and the role that microcapsules play in these products.

Graphic Arts. Carbonless copy paper with global annual sales well in excess of $5 billion is by far the largest single commercial application of microcapsules (2,4). This product consumes thousands of tons of capsules annually. Figure 10, a schematic diagram of a three-part business form, illustrates the concept of carbonless copy paper. The bottom surface of the top sheet of this form (sheet A) and the second sheet (sheet B) is coated with a layer of microcapsules that have a diameter of 3–6 μm. The coating includes inert spacer particles that are larger than the microcapsules. They are often starch particles added to protect the microcapsules from premature rupture. The capsules, filled with a colorless solution of 2–6% leuco dye dissolved in a high boiling organic solvent, rupture under pressures encountered in normal handwriting or impact printing. The dye solution released is transferred from the bottom surface of sheets A and B to the top surface of sheets B and C, respectively, where it reacts to form an image. The reactive coating on the top surface of sheets B and C originally was attapulgite clay (see CLAYS). The use of a phenolic resin, salicylic acid

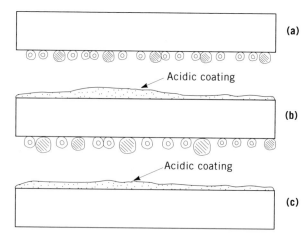

Fig. 10. Cross section of a three-part business form prepared from carbonless copy paper where ⊚ are microcapsules and ⬚ are starch: (**a**), CB sheet; (**b**), CFB sheet; and (**c**), CF sheet.

attapulgite clay, or acid-treated montmorillonite is favored. Any material that contains sufficient acidic sites to generate a color reaction and form a light-stable image could be used. Sheet A in Figure 10 is called a coated back (CB) sheet; sheet B, a coated front back (CFB) sheet; and sheet C, a coated front (CF) sheet. The capsules and reactive coating can be coated on the same paper surface. In this case, the product is called self-contained carbonless copy paper.

Success of all carbonless paper products depends on the microcapsules, leuco dyes, and reactive coating. A number of leuco dyes are available. Originally, a mixture of crystal violet lactone [1552-42-7] (CVL) and N-benzoylleucomethylene blue (N-BLMB) was used. Upon contact with the reactive coating, CVL reacts rapidly to develop an unstable blue color. The N-BLMB reacts more slowly, but forms a more light-stable image. By combining both dyes in the microcapsules, a light-stable image that forms rapidly is produced. This image initially is blue but changes to blue-green on storage. The mechanism of these color-forming reactions has been described (2). A goal of carbonless copy paper producers is to develop a leuco dye that produces a permanent black image like india ink. This has prompted a continuous effort to synthesize leuco dyes other than CVL and N-BLMB. Fluoran and phthalide leuco dyes have resulted from this effort; metal chelate systems have also been developed.

Capsules for carbonless copy paper must be small enough to give a sharp image on the reactive coating of the transfer sheet and large enough to be broken by pressures encountered in normal writing and impact printing. The upper size limit is approximately 20 μm, since capsules larger than this produce a fuzzy image. The lower size limit is approximately 1–2 μm. Capsules smaller than this tend to fall into the crevices of the paper where they are protected from breakage. Capsules that meet these size requirements and are able to retain leuco dye solutions can be produced by complex coacervation, interfacial polymerization, or *in situ* polymerization. Solvents for leuco dyes used in carbonless copy paper are high boiling (>200°C) organic liquids. Solvents mentioned in the patent literature include benzylated ethylbenzene, benzyl butyl phthlate, isopropylbiphenyl, and diisopropylnaphthalene. The choice of solvent is determined by cost, availability, toxicity, ability to dissolve the dye system, produce the desired color intensity on the transfer sheet, and form acceptable microcapsules.

In the 1980s, Mead (Dayton, Ohio) used carbonless paper leuco dye color chemistry to develop an interesting full color imaging system called the Cycolor process (41), but it is no longer commercially available. Many imaging applications that utilize microcapsules have been disclosed in the patent literature. Specific examples include light- and heat-activated papers which contain an encapsulated diazo dye (42), electrophotographic toners (43), thermochromic dyes, and photochromic dyes (see CHROMOGENIC MATERIALS). Microencapsulated inks (qv) for ink-jet printing have also been described. A number of these concepts have been commercialized.

Pharmaceuticals. The concept of microencapsulation has intrigued the pharmaceutical industry for many years, because it offers the possibility of providing a number of important new oral and parenteral dosage forms. Microcapsules in oral dosage forms could conceptually taste-mask bitter pharmaceuticals, provide extended release *in vivo*, provide enteric release, improve the stability of incompatible drug mixtures, provide resistance to oxidation, reduce volatility, and distribute a drug in many small carrier particles so that effects of the drug on the sensitive walls of the stomach are minimized. Microencapsulated parenteral formulations could provide prolonged delivery of drugs with short half-lives *in vivo* and perhaps even achieve targeted drug delivery. For these reasons, microencapsulation has received much attention by pharmaceutical scientists (44). Several microcapsule-based oral pharmaceutical formulations which offer some of these features are available. All have an ethylcellulose shell and are produced by a polymer–polymer phase-separation process carried out in hot cyclohexane (8). The core material is a solid that has finite solubility in water. Encapsulated potassium chloride, KCl, has been used extensively because the KCl dispersed in many small ethylcellulose capsules minimizes high localized concentrations of KCl in the stomach that can irritate the lining of the stomach and induce bleeding. Aspirin encapsulated in ethylcellulose for arthritic patients is an example of using microencapsulation to extend time of release of a drug in the gastrointestinal tract (see GASTROINTESTINAL AGENTS). In this case, the encapsulated aspirin formulation provides overnight relief. Finally, encapsulated oral acetaminophen formulations for children are used to provide taste-masking.

The number of microencapsulated commercial oral formulations available and the volume of these formulations sold annually is comparatively small. This may reflect the difficulty of developing new drug formulations and bringing them successfully to market or the fact that existing microencapsulation techniques have had difficulty economically producing microcapsules that meet the strict performance requirements of the pharmaceutical industry. One application that is a particularly active area of development is microcapsules or microspheres for oral delivery of vaccines (45,46).

Several parenteral microencapsulated products have been commercialized; the core materials are polypeptides with hormonal activity. Poly(lactide– glycolide) copolymers are the shell materials used. The capsules are produced by solvent evaporation, polymer–polymer phase separation, or spray-dry encapsulation processes. They release their core material over a 30 day period *in vivo*, although not at a constant rate.

Injected formulation of leuproleine and triptoreline [*57773-63-4*] are used to treat metastasized cancer of the prostrate, whereas an encapsulated formulation of bromocryptin [*25614-03-3*] is used to inhibit milk production in women after pregnancy. The performance of encapsulated hormonal polypeptides is well documented (47). Injectable biodegradable microcapsules loaded with fertility control agents have been under development for a number of years and have been carried to various stages of clinical development (48) (see CONTRACEPTIVES).

The development of injectable microcapsules for delivery of chemotherapy agents remains another active area of research. The ultimate goal is to achieve targeted delivery of chemotherapy agents to specific sites in the body, ideally by injection of drug-loaded microcapsules that would seek out and destroy diseased cells. Intra-arterial infusion chemotherapy is a direct approach to targeted delivery. The clinical applications of microspheres and microcapsules in embolization and chemotherapy have been assessed (49) (see CHEMOTHERAPEUTICS, ANTICANCER).

Biomedical and Biotechnology.   The use of microcapsules for a variety of biomedical and biological applications has been promoted for many years (50,51). Several biomedical microcapsule applications are in clinical use or have approached clinical use. One application is the use of air-filled human albumin microcapsules as ultrasound contrast agents. Such microcapsules, called microbubbles, are formed by ultrasonicating 5% albumin solutions to produce $4-10$-μm diameter air-filled capsules (52). When injected the capsules act as a useful transpulmonary echo contrast agent (53) that has been approved for use in humans by the U.S. FDA.

Another biomedical application of microcapsules is the encapsulation of live mammalian cells for transplantation into humans. The purpose of encapsulation is to protect the transplanted cells or organisms from rejection by the host. The capsule shell must prevent entrance of harmful agents into the capsule, allow free transport of nutrients necessary for cell functioning into the capsule, and allow desirable cellular products to freely escape from the capsule. This type of encapsulation has been carried out with a number of different types of live cells, but studies with encapsulated pancreatic islets or islets of Langerhans are most common. The alginate–poly(L-lysine) encapsulation process originally developed in 1981 (54) catalyzed much of the cell encapsulation work carried out since. A discussion of the obstacles to the application of microencapsulation in islet transplantation reviewed much of the more recent work done in this area (55). Animal cell encapsulation has also been researched (56).

The microencapsulation or immobilization of a variety of live cells and organisms other than mammalian cells has also received much attention. A wide range of applications has been explored including the use of gel beads to shorten the handling time for preparing champagne (57,58). Capsules prepared for these applications often are larger than 1000 μm and may not have a well-defined capsule shell, nevertheless they are designed to provide many of the capabilities traditionally associated with microcapsules.

Other biomedical and biological applications of microcapsules continue to be developed. For example, the encapsulation of enzymes continues to attract interest even though loss of enzyme activity due to harshness of the encapsulation protocols used has been a persistent problem (59). The use of microcapsules in antibody hormone immunoassays has been reviewed (60). The encapsulation of hemoglobin as a red blood substitute has received much attention because of AIDS and blood transfusions (61).

Food Ingredients.   A number of food ingredients or additives have been encapsulated and are available commercially. Solid ingredients encapsulated are typically water-soluble and are encapsulated with a hydrophobic or hydrophilic coating material usually applied by the Wurster process. Preferred hydrophobic coating materials are partially hydrogenated vegetable oils of varying melting points, monoglycerides, and diglycerides. Hydrogenated vegetable oils used include cottonseed, soybean, and palm. Hydrophilic coating materials tend to be maltodextrins and occasionally gum arabic. Both types of coating materials are well-accepted food-grade products (see FOOD ADDITIVES).

The function of a number of commercially available encapsulated solid food ingredients has been discussed (62). Acidulants like citric and lactic acid encapsulated in partially hydrogenated vegetable oil are used in meat processing where they provide direct acidification and shorten processing time. Sodium acid pyrophosphate encapsulated in hydrogenated vegetable oil is used in frozen cake batters in order to aid mixing and reduce gas release during batter make-up. In both types of applications, release of core material occurs during a heating cycle that melts the shell formulation and releases the core material.

Acidulants like citric, lactic, and fumaric acids encapsulated in a water-soluble maltodextrin shell formulation are used in dry mix beverages and desserts as well as prepared pre-mixes for the baking and dairy industries. The maltodextrin coating is designed to minimize hygroscopicity, reduce dusting, and minimize reactions with incompatible ingredients. It dissolves in the presence of liquid water to rapidly release the contents of the capsules during a mixing cycle. Ferrous sulfate and vitamin C (ascorbic acid) encapsulated in hydrogenated vegetable oil or maltodextrin are used to fortify a nutritional product of some sort. The capsules provide taste-masking, possibly a degree of prolonged release, stabilization of the core material against oxidation, and minimization of reaction with other ingredients in the final product.

Calcium proprionate [4075-81-4], and sodium bicarbonate [144-55-8] encapsulated in hydrogenated vegetable oil are used in chemically leavened products. Release typically occurs during the baking cycle due to melting of the hydrogenated shell material. Sodium bicarbonate encapsulated with maltodextrin is used in dry mix baking and other chemically leavened products (see BAKERY PRODUCTS AND LEAVENING AGENTS). In this case, release occurs during the mixing step. Sodium chloride [7647-14-5] encapsulated in a hydrogenated vegetable shell is used in various meat products, yeast-containing mixes, and assorted types of dough. The capsules are designed to minimize inhibition of yeast activity, rancidity, and excessive salt binding during product storage. Encapsulated sweeteners such as sugar (qv) and aspartame are also available (62).

Liquid food ingredients encapsulated are typically oil-soluble flavors, spices (see FLAVORS AND SPICES), and vitamins (qv). Even food oils and fats are encapsulated (63). These core materials normally are encapsulated with a water-soluble shell material applied by spray drying from water, but fat shell formulations are used occasionally. Preferred water-soluble shell materials are gum arabic, modified starch, or blends of these polymers with maltodextrins. Vitamins are encapsulated with zero bloom strength gelatin by spray drying.

A range of spray-dried flavor-filled capsules primarily with water-soluble shell formulations are used in various dry beverage mixes and other dry food products. Flavors containing ethyl acetate and other low boiling point components pose problems for successful spray-dry encapsulation (64). Such components are either lost during the initial emulsification process or during the actual dewatering step as a result of azeotrope formation. Another problem with spray-dry encapsulation is the formation of free surface oil. The rapid desolvation that occurs in the drying chamber can produce blow holes in the capsules essentially leaving a small amount of flavor oil or free or surface oil which oxidizes on storage and detrimentally affects product quality. An alternative approach to the encapsulation of food flavors is desolvation by pressure extrusion. The rod-like particles produced in this manner typically have a relatively low loading (8–10 wt %) although higher loadings (15%) have been achieved. The advantage of this technology is that it yields essentially defect-free particles with superb shelf-life storage stability. A third approach to the encapsulation of food flavors, spray chilling, should be particularly well suited to the encapsulation of aqueous flavor compositions (64). Release occurs during a product heating cycle, eg, baking.

Microencapsulation has much hidden potential for the food industry which promises to be tapped in the future (62). An interesting discussion of the problems that have been encountered while attempting to develop microcapsule formulations for commercial use in food products has been presented (65) and a review provides a number of references to food encapsulation studies (66).

Agrochemicals. The microencapsulation of pesticides (qv) and herbicides (qv) has been an active area of development that has produced several commercial products. The function of the microcapsules is to prolong activity while reducing mammalian toxicity, volatilization losses, phytotoxicity, environmental degradation, and movement in the soil (67,68). Ideally, encapsulation would also reduce the amount of agrochemical needed. Penncap-M was the first of several encapsulated pesticide formulations commercialized by Atochem North America (Pennwalt) (69). It is sold as an emulsifiable concentrate of encapsulated methyl parathion. The 30–50-μm capsules are prepared by interfacial polymerization. Encapsulation reduces mammalian toxicity of methyl parathion and prolongs its activity. When Penncap-M was first used, a number of massive bee kills were attributed to it. Adjustments to the application protocol and formulation appear to have resolved this problem. The status of microencapsulated insecticides in relation to beekeeping in the United States has been discussed (70) (see INSECT CONTROL TECHNOLOGY).

Encapsulated diazinnon sold as Knox-Out 2FM is a commercial encapsulated pesticide formulation said to have reduced dermal and oral toxicity as well as prolonged effectiveness. The capsules, prepared by interfacial polymerization, are claimed to be highly effective against cockroaches with no objectionable odor and low insect repellence. The capsules are believed to function as a contact poison when insects walk on it and as a stomach poison when insects preen capsules stuck to their legs and ingest them (71).

Encapsulated fonofos, a soil insecticide, was developed to coat seeds before they were planted (72). Encapsulation reduces oral toxicity 100-fold and dermal toxicity 10-fold while extending activity of the fonofos. Other encapsulated pesticides available include permethrin and parathion (69). Significantly, all commercial encapsulated pesticides are prepared by interfacial polymerization.

A significant trend in current agrochemical development work is the production of agrochemicals that are markedly less toxic and more potent than past pesticides. Agents of low mammalian toxicity, such as insect pheromones, growth regulators, and pyrethins, are receiving much attention. Microencapsulation is being used to enhance the effectiveness of such compounds. For example, methoprene [40596-69-8], a mosquito growth regulator, is sold as an encapsulated formulation that provides release over a 5–7 d period in the field. A microencapsulated pyrethin formulation is used to control crawling insects such as cockroaches. The capsules are sold as an aqueous-based suspension that provides protection for at least 30 days. Field tests (73) indicate effectiveness of the encapsulated pyrethin compares favorably with several nonencapsulated pesticides although the capsule formulation leaves a visible residue. Encapsulated fenitrothion [122-14-5] is also available commercially (74). The capsules are 20 μm in diameter and have a polyurethane shell formed by interfacial polymerization.

The encapsulation of herbicides has received much attention. Encapsulated alachlor is a high volume herbicide product generally sold as a liquid formulation, although a dry granule version is also available. The capsules, produced by interfacial polymerization (11), are reported to be spherical with a diameter of 2–15 μm (75). Two thiocarbamate herbicides, EPTC and vernolate [1929-77-7], were encapsulated by interfacial polymerization because they are volatile compounds. When applied in unencapsulated form, they must be incorporated in the soil within two hours in order to provide effective weed control. When applied as a microencapsulated formulation, the rate of volatilization is lower and soil incorporation can be delayed 24 hours (76).

A number of studies have been made to demonstrate the feasibility of using starch to entrap agrochemicals such as herbicides. In one protocol, a starch, water, herbicide mixture is extruded by a twin-screw extruder to form an extrudate that can be ground to produce herbicide-loaded granules (77). Encapsulation is being considered as a means of improving the performance of a variety of biological pest control agents. In such cases, the tendency is to entrap the active agent in relatively large alginate gel beads. Various workers have explored the use of microencapsulation as a means of preparing formulations for feeding fish at various stages of development. Encapsulated products of this nature have been developed and the subject has been discussed (78).

Consumer and Industrial Products. A number of encapsulated consumer and industrial products are enjoying considerable success. Advertising inserts that utilize encapsulated perfumes and flavors contain a coating of scent-filled capsules which break and release scent when the insert is torn open. This product is widely used as a marketing tool, primarily for new perfumes. Various paper products coated with scent-filled capsules that break when the coated area is rubbed have been used for many years. Children's crayons loaded with encapsulated scents are appearing on the market. The capsules break during the drawing process thereby releasing a scent characteristic of the drawn object.

Microcapsules are used in several film coatings other than carbonless paper. Encapsulated liquid crystal formulations coated on polyester film are used to produce a variety of display products including thermometers. Polyester film coated with capsules loaded with leuco dyes analogous to those used in carbonless copy paper is used as a means of measuring line and force pressures (79). Encapsulated deodorants that release their core contents as a function of moisture developed because of sweating represent another commercial application. Microcapsules are incorporated in several cosmetic creams, powders, and cleansing products (80).

A majority of the fasteners used in automobiles in the United States are coated with microcapsules loaded with an adhesive. When the fastener is installed, a fraction of the capsules in the coating rupture releasing the adhesive payload. The adhesive essentially glues the fasteners in place preventing them from becoming loose and causing rattles. The capsules are designed so that only a fraction of them break each time a fastener is taken off and put back. The on/off cycle can be repeated three or four times.

Encapsulated ammonium polyphosphate [10124-31-9] incorporated in plastics acts as a fire-retardant. An example is Hostaflam AP sold by Hoechst AG (Frankfurt am Main, Germany). In Japan, microcapsules loaded with naramycin are incorporated into poly(vinyl chloride) (PVC). The capsule-loaded PVC is used to coat a range of electrical cables. Rodents bite and gnaw on cables coated with PVC free of capsules ultimately causing power interruptions, but when rodents bite into cables coated with capsule-loaded PVC coatings, they break the capsules in the coating thereby releasing the naramycin [66-81-9], a taste-repellent to the rodents.

The oil industry uses microencapsulated oil-field chemicals. For example, microencapsulated breaker is delivered into a subterranean formation where it breaks the fracturing liquid used to stimulate the recovery of fluids such as crude oil or natural gas. Examples of breakers encapsulated include oxidizers, enzymes, and various mineral or organic acids.

New Developments.   Creation of microencapsulation technology and applications of microcapsules are goals that many research and development groups are pursuing globally. Technical deficiencies of current capsules, unanticipated problems in marketing, and poor product design are problems that consistently must be addressed. Nevertheless, the growing number of workers in this field provides good reason to suggest that these problems will increasingly be resolved.

BIBLIOGRAPHY

"Microencapsulation" in *ECT* 2nd ed., Vol. 13, pp. 436–456, by J. A. Herbig, The National Cash Register Co.; in *ECT* 3rd ed., Vol. 15, pp. −493, by R. E. Sparks, Washington University.

1. A. Kondo, *Microcapsule Processing and Technology*, Marcel Dekker, Inc., New York, 1979.
2. W. Sliwka, *Angew. Chem. Internat. Edit.* **14**, 539 (1975).
3. P. B. Deasy, *Microencapsulation and Related Drug Processes*, Marcel Dekker, Inc., New York, 1984.

4. C. A. Finch, *Ullmann's Encyclopedia of Industrial Chemistry*, Vol. A16, 5th ed., VCH Publishers, New York, 1990, p. 575.

5. J. A. Bakan, in L. Lachman, H. A. Lieberman, and J. L. Kanig, eds., *The Theory and Practice of Industrial Pharmacy*, 3rd ed., Marcel Dekker, New York, 1986.

6. C. Thies, *How-to-Make Microcapsules: Lecture and Lab Manual*, Thies Technology, St. Louis, Mo., 1994.

7. U.S. Pat. 2,800,457 (July 23, 1957), B. K. Green and L. Schleicher (to the National Cash Register Co.).

8. U.S. Pat. 3,341,416 (Sept. 12, 1967), J. L. Anderson, G. L. Gardner, and N. H. Yoshida (to the National Cash Register Co.).

9. J. M. Ruiz, B. Tissier, and J. P. Benoit, *Int. J. Pharm.* **49**, 69 (1989).

10. U.S. Pat. 3,577,515 (May 4, 1971), J. E. Vandegaer (to Pennwalt Corp.).

11. U.S. Pat. 4,280,833 (July 28, 1981), G. B. Beestman and J. M. Deming (to Monsanto Co.).

12. S. K. Yadav, A. K. Suresh, and K. C. Khilar, *AIChE J.* **36**, 431 (1990).

13. U.S. Pat. 4,285,720 (Aug. 25, 1981), H. B. Scher (to Stauffer Chemical Co.).

14. D. A. Wood, T. L. Whatley, and A. T. Florence, *Int. J. Pharm.* **8**, 35 (1981).

15. M. Bakhshaee, R. A. Pethrick, H. Rashid, and D. C. Sherrington, *Polym. Comm.* **26**, 185 (1985).

16. W. D. Jayne, in J. E. Vandegaer, ed., *Microencapsulation: Processes and Applications*, Plenum Press, New York, 1973, p. 103.

17. U.S. Pat. 3,516,941 (June 23, 1970), G. W. Matson (to 3M Corp.).

18. U.S. Pat. 4,087,376 (May 2, 1978), P. L. Foris, R. W. Brown, and P. S. Phillips (to the National Cash Register Co.).

19. C. Thies, in M. Donbrow, ed., *Microcapsules and Nanoparticles in Medicine and Pharmacy*, CRC Press, Boca Raton, Fla., 1992, p. 47.

20. D. R. Cowsar, T. R. Tice, R. M. Gilley, and J. P. English, *Methods Enzymol.* **112**, 101 (1985).

21. Y. Ogawa, M. Yamamoto, H. Okada, T. Yashiki, and T. Shimamoto, *Chem. Pharm. Bull. (Tokyo)*, **36**, 1095 (1988).

22. U.S. Pat. 2,299,929 (Oct. 27, 1942), J. A. Reynolds (to Atlantic Vitamin Corp.).

23. U.S. Pat. 3,389,194 (June 1968), G. R. Somerville (to Southwest Research Institute).

24. J. Brenner, *Perfum. Flav.* **8**, 40 (1983).

25. G. A. Reineccius, in S. J. Risch and G. R. Reineccius, eds., *Flavor Encapsulation*, ACS Symposium Series 370, American Chemical Society, Washington, D.C., 1988, p. 55.

26. A. H. Taylor, *Food Flav. Ingred. Proc. Pckg.*, p. 48 (Sept. 1983).

27. R. Lamb, *Food Flav. Ingred. Proc. Pckg.* **9**, 39 (1987).

28. K. Lehmann, in Ref. 19, p. 73.

29. D. M. Jones, in Ref. 25, p. 158.

30. J. T. Goodwin and G. R. Sommerville, *Chemtech* **4**, 623 (1974).

31. U.S. Pat. 4,675,140 (June 1987), R. E. Sparks and N. S. Mason (to Washington University Technology Associates).

32. R. E. Sparks, I. C. Jacobs, and N. S. Mason, in M. A. El-Nokaly, D. M. Piatt, and B. C. Charpentier, eds., *ACS Symposium Series 520*, American Chemical Society, Washington, D.C., 1993, p. 145.

33. U.S. Pat. 3,704,137 (Nov. 28, 1972), E. E. Beck.

34. S. J. Risch, in Ref. 25, p. 103.

35. R. H. Sudekum, in J. R. Nixon, ed., *Microencapsulation*, Marcel Dekker, Inc., New York, 1976, p. 119.

36. G. Langer and G. Kamate, *J. Colloid Interface Sci.* **29**, 450 (1969).

37. U.S. Pat. (June 1, 1976), M. Morishita and co-workers (to Toyo Jozo Co.).

38. D. F. Herman, U. Krause, and J. J. Brancato, *J. Polym. Sci. Part C* **11**, 75 (1965).

39. U.S. Pat. 5,019,400 (May 28, 1991), W. R. Gombotz, M. S. Healy, and L. R. Brown (to Enzytech).
40. U.S. Pat. 5,043,280 (1991), B. W. Muller and W. Fischer (to Scharma Pharmaceutical).
41. US. Pat. 4,483,912 (Nov. 20, 1984), F. W. Sanders (to the Mead Corp.).
42. U.S. Pat. 4,842,979 (June 27, 1989), S. Ishige, T. Usami, H. Kamikawa, and T. Tanaka (to Fuji Photo Film Co.).
43. U.S. Rat. 4,727,011 (Feb. 23, 1988), H. K. Mahabadi, R. Patel, S. W. Webb, and J. D. Wright (to Xerox).
44. C. Thies, *CRC Crit. Rev. Biomed. Eng.* **8**, 335 (1983).
45. J. H. Eldridge and co-workers, *J. Controlled Release* **11**, 205 (1990).
46. C. A. Gilligan and A. Li Wan Po, *Int. J. Pharm.* **75**, 1, 1991.
47. H. Okada and co-workers, *Pharm. Res.* **8**, 787 (1991).
48. H. Ziyan and X. Rubqiu, *Int. J. Gynaecol. Obstet.* **22**, 319 (1984).
49. P. M. J. Flandroy, C. Grandfils, and R. J. Jerome, in A. Rolland, ed., *Pharmaceutical Particulate Carriers*, 1993, p. 321.
50. T. M. S. Chang, *Artificial Kidney, Artificial Liver, and Artificial Cells*, Plenum Publishing Corp., New York, 1978.
51. T. M. S. Chang, *Biomat., Art. Cells Immob. Biotech.* **21**, 291 (1993).
52. M. W. Keller, W. Glasheen, K. Teja, A. Gear, S. Kaul, *J. Am. Coll. Cardiol.* **12**, 1039 (1988).
53. B. Geny and co-workers, *J. Am. Coll. Cardiol.* **22**, 1193 (1993).
54. F. Lim and R. D. Moss, *J. Pharm. Sci.* **70**, 351 (1981).
55. P. De Vos, G. H. J. Wolters, W. M. Fritschy, and R. Van Schilfgaarde, *Int. J. Artif. Organs* **16**, 205 (1993).
56. M. F. A. Goosen, ed., *Fundamentals of Animal Cell Encapsulation and Immobilization*, CRC Press, Boca Raton, Fla., 1993.
57. A. H. Prevost, R. Cachon, J.-F. Cavin, and C. Divies, *Biofutur.*, 42 (Mar. 1994).
58. D. Poncelet and M. Lievremont, *Biomat., Art. Cells Immob. Biotech.* **21** (1993).
59. C. Thies, in K. L. Kadam, ed., *Granulation Technology for Bioproducts*, CRC Press, Boca Raton, Fla., 1991, p. 179.
60. A. M. Wallace, *Drug Targeting Delivery* **1**, 177 (1992).
61. T. M. S. Chang, ed., *Blood Substitutes and Oxygen Carriers*, Marcel Dekker, New York, 1992.
62. J. D. Dziezak, *Food Technol.*, 136 (Apr. 1988).
63. *Food Technol.*, 188 (Apr. 1988).
64. W. M. McKernan, *Flavour Ind.*, 596 (Dec. 1972).
65. S. Hagenbart, *Food Prod. Des.*, 28 (Apr. 1993).
66. R. Arshady, *J. Microencaps.* **10**, 413 (1993).
67. R. C. Koestler, G. Janes, and J. A. Miller, in A. F. Kydonieus, ed., *Treatise of Controlled Drug Delivery*, Marcel Dekker, New York, 1991, p. 491.
68. H. B. Scher, *ACS Symp. Ser.* **53**, 126 (1977).
69. S. Meghir, *Pestic. Sci.* **9**, 411 (1978).
70. A. Stoner, B. Ross, and W. T. Wilson, *Bee World* **63**, 72 (1982).
71. *Pest Control*, 24 (Mar. 1980).
72. H. B. Scher, *Proceed. Intern. Symp. Control. Rel. Bioact. Mater.* **12**, 110 (1985).
73. G. W. Bennett and R. D. Lund, *Pest Control*, 44 (Mar. 1980).
74. T. Ohtsubo, S. Tsuda, H. Takeda, and K. Tsuji, *J. Pesticide Sci.* **16**, 609 (1991).
75. B. B. Petersen and P. J. Shea, *Weed Science* **37**, 719 (1989).
76. B. E. Greenwold, F. Pereiro, T. J. Purnell, and H. B. Scher, *Proceedings of the 1980 British Crop Protection Conference on Weeds*, British Crop Protection Society, London, p. 185.
77. D. Trimnell, R. E. Wing, M. E. Carr, and W. D. Doane, *Starch/Starke* **43**, 146 (1991).

78. D. A. Jones, D. L. Holland, and S. Jabborie, *Applied Biochem. Biotech.* **10**, 275 (1984).
79. U.S. Pat. 4,104,910 (Aug. 8, 1978), Y. Ogata and N. Hfbina (to Fuji Photo Film Co.).
80. M. Magill, *Cosmet. Toiletries* **105**, 59 (1990).

CURT THIES
Washington University

MICROMACHINING.   See NANOTECHNOLOGY; SUPPLEMENT.

MICROPLANTS.   See PILOT PLANTS AND MICROPLANTS.

MICROSCOPY

Microscopy involves the production and study of magnified images of objects too small to resolve with the unaided eye. There is no single microscope; rather, there are dozens of types, having in common only the ability to present an enlarged image. Microscopes vary in the mechanism by which enlarged images are formed, and in the information presented for each enlarged object. Some microscopes use visible light; others use infrared, ultraviolet, or x-ray radiation; electrons are also used. There are a variety of scanning microscopes, some of which scan the object with an electron beam, a laser beam, or ultraviolet light. Most recently, an entire family of microscopes has been developed that scan a surface, maintaining a constant, tiny distance from that surface and thus plotting a surface map of objects as small as single atoms.

Microscopes are also classified by the type of information they present: size, shape, transparency, crystallinity, color, anisotropy, refractive indices and dispersion, elemental analyses, and fluorescence, as well as infrared, visible, or ultraviolet absorption frequencies, etc. One or more of these microscopes are used in every area of the physical sciences, ie, biology, chemistry, and physics, and also in their subsciences, mineralogy, histology, cytology, pathology, metallography, etc.

Microscopy is an unusual scientific discipline, involving as it does a wide variety of microscopes and techniques. All have in common the ability to image and enlarge tiny objects to macroscopic size for study, comparison, evaluation, and identification. Few industries or research laboratories can afford to ignore microscopy, although each may use only a small fraction of the various types. Microscopy review articles appear every two years in *Analytical Chemistry* (1,2). Whereas the style of the *Enclyclopedia* employs lower case abbreviations for analytical techniques and instruments, eg, sem for scanning electron microscope, in this article capital letters will be used, eg, SEM.

Microscopists in every technical field use the microscope to characterize, compare, and identify a wide variety of substances, eg, protozoa, bacteria, viruses, and plant and animal tissue, as well as minerals, building materials, ceramics, metals, abrasives, pigments, foods, drugs, explosives, fibers, hairs, and even single atoms. In addition, microscopists help to solve production and process problems, control quality, and handle trouble-shooting problems and customer complaints. Microscopists also do basic research in instrumentation, new techniques, specimen preparation, and applications of microscopy. The areas of application include forensic trace evidence, contamination analysis, art conservation and authentication, and asbestos control, among others.

History

Microscopy as a field follows only astronomy in the history of science and technology. Stelluti in 1630 published the oldest known microscopical observations which were on the microstructure of the bee and the weevil. The compound microscope (based on a two-lens, two-step magnification by objective and ocular) was proposed by Zacharias Jansen in 1590. However, because of spherical and chromatic aberrations and other faults of the compound microscope, most microscopists preferred the simple microscope, ie, a one-lens magnifying glass, for the next 200 years. Even so, Antoni van Leeuwenhoek (1632–1723) in the late seventeenth and early eighteenth centuries developed simple lenses capable of magnifications nearly as high as 300×. The resolution was adequate for the detection and identification of blood cells, spermatozoa, and even bacteria, all reported by Leeuwenhoek in a long succession of letters to the Royal Society in London. Robert Hooke (1635–1703) made similar studies in England, principally on biological materials.

The development of corrected (achromatic) lenses and the addition of polarizing Nicol prisms, both around 1830, marked the real beginning of light microscopy. The most important applications then followed rapidly. Toxicology, involving microchemical reactions for arsenic, lead, mercury, etc, was soon followed by forensic applications to blood, hair, fibers, etc. During the period 1850–1900, optical crystallography, mineralogy, and metallography developed into full-fledged fields of study. Many of these developments were the work of Henry Sorby (1826–1908), who also first fitted a spectroscope onto a microscope to better characterize and identify colored substances, eg, blood, dyes, and pigments. Others, including Otto Lehmann during the late 1800s, added a heating stage to the polarized light microscope and studied phase and composition diagrams, as well as, more particularly, liquid crystals.

By 1900, Émile Chamot (1866–1950) at Cornell University was teaching microscopy, with courses in water and food microscopy as well as microchemistry. After 1900, Chamot continued to develop microchemical procedures, his first volume on chemical microscopy appearing in 1916, with its most recent edition in 1983 (3). During the same period, Simon Gage, also at Cornell, was actively developing the biological applications of the light microscope (4). The combined works of these two teachers served as the foundation for the development and application of microscopy in the world of science and technology during the first half of the twentieth century. They were not alone, however, and Hartshorne, Allen, and others also contributed to spreading the use and furthering the development of microscopy (5,6).

During most of the nineteenth century, the microscope-type characteristic of the time of Lister and Nicol (ca 1835) met the needs of most microscopists. There were, however, stirrings of interest in improvements. The problems faced by biologists differed only in degree from those of the materials scientists, ie, chemists, mineralogists, metallographers, etc. Basically, the needs common to both groups could be summarized as follows: (*1*) improved specimen contrast; (*2*) increased resolving power; and (*3*) obtaining more characterization data. The microscopes used by biologists and materials scientists during the nineteenth and the early twentieth centuries were essentially identical, differing principally in the presence of two polarizing elements and a rotating stage on the materials science microscope. This duality and the attendant small differences have continued to the present day in the basic light microscope. The proliferation of microscopes and microscopy can be considered under the three listed needs.

Improving Specimen Contrast

It was learned very early that the angular aperture of the substage condenser controls specimen contrast. Decreasing that aperture, usually with a continuously adjustable iris diaphram, greatly increases contrast. It was not, however, appreciated fully until Ernst Abbe's classic contributions (7,8) in the period ca 1880–1889 that decreasing the aperture to increase contrast also decreases the resolving power of the microscope.

The choice of the mounting medium usually obviates the need for any new instrumental technique for enhancing contrast. A particle mounted in a liquid medium having the same refractive index (n_D) may disappear completely. The darkness of the borders (contrast), however, increases rapidly as the refractive index difference increases. Choosing the proper refractive index liquid is the first step in contrast enhancement. Unfortunately, biologists are often restricted by their specimens in their choice of mountant. Water ($n_D = 1.33$) is far superior to glycerol ($n_D = 1.47$) or canada balsam ($n_D = 1.54 \pm 0.01$) for most biological substances whose refractive indices lie in a narrow range near 1.54. More specialized contrast-enhancing microscopes have been developed during the twentieth century, especially after 1930. They are also useful to both biologists and materials scientists. An excellent survey of the standard methods of improving contrast up to the video age is available (9).

Darkfield Illumination. During most of the nineteenth century, biological microscopists were having trouble seeing the finer structures of the plant and animal tissue they wished to study. Although water is satisfactory optically as a mounting medium, permanent aqueous preparations are nearly impossible. The standard medium used by nineteenth-century biologists was canada balsam. Its refractive index, however, is close to that of nearly all biological materials; hence, with matching refractive indices all such materials tend to disappear, ie, show low contrast images. This problem had already resulted by 1900 in the use of darkfield illumination, whereby a strong hollow illuminating cone whose minimum angular aperture exceeds the maximum angular aperture of the objective yields a completely dark field when no object is present in the field of view. The hollow cone of light from the substage condenser is focused on the specimen plane, as is the objective. Light striking the specimen is then

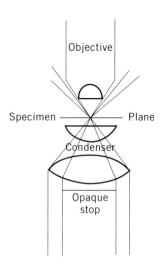

Fig. 1. Darkfield illumination; the image is formed by light scattered from the specimen detail against a dark field.

reflected, diffracted, or refracted by object detail into the otherwise dark field of the objective aperture (Fig. 1). With a bright light and fully dark field, all particle and tissue detail appear brightly outlined on a black background. In fact, detail much smaller than the resolving power limit of the microscope optics becomes visible.

The contrast for specimen detail in the field of view is greatly enhanced by darkfield illumination (10). The degree of contrast and sensitivity of detection of small-object details depend on the relative indices of the specimen and the mounting liquid and on the intensity of the illumination. Darkfield illumination is not, however, a satisfactory solution for biologists who need direct transmitted light in order to observe specimens, especially stained specimens. It is, however, very useful in detecting asbestos fibrils that often exist in floor tiles or water and air samples as 20-nm fibers (10 times finer than the resolution of an asbestos analyst's usual microscope) (11). Darkfield illumination yields an unnatural appearance and difficulties in interpretation; hence, a need for better contrast methods still exists.

The electron microscope is also used in the darkfield mode, in which it improves contrast between different chemical phases or different orientations of a given solid phase.

Phase Contrast. In 1934, Fritz Zernike resolved most of these problems with phase contrast (9). Like darkfield illumination, phase contrast also involves a hollow illuminating cone, but within the objective aperture rather than outside as with darkfield illumination. This hollow cone of light is a result of a transparent annulus in an otherwise opaque stop in the substage condenser. The light in the phase-contrast hollow cone interacts with each element of specimen detail to yield both zero-order and higher orders of diffracted beams of light. The zero-order beams in the hollow cone all pass into the objective where they then pass through a phase ring (a conjugate to the phase annulus in the substage condenser). The diffracted higher order rays (now out of phase by $1/4\lambda$ relative to the zero-order rays) deviate at some of the diffraction angles into the dark cone and through the phase plate, but not through the phase ring. The optical path difference between the direct rays passing through the phase ring and the diffracted rays passing through other areas of the phase plate introduces an additional $1/4\lambda$ path difference (and therefore a phase difference) between the direct zero-order beams and the diffracted first- and higher order beams from the same object detail in the specimen. On recombination in the microscope image, this results in destructive interference for those rays one-half wavelength out of phase and thus greatly improves the contrast.

Phase contrast is most useful to the biologist who often needs to differentiate between biological entities, most of which have closely similar refractive indices. Because the phase condenser annulus lies in the outer region of the condenser aperture, the resulting oblique rays result in full resolving power, corresponding to a fully open substage aperture diaphram. Chemical microscopists also profit from the use of phase contrast by increasing the sensitivity of refractive index determinations using the Becke line technique. Polymer microscopists find phase contrast helpful in the study of emulsions and multilayer polymer films.

Differential Interference Contrast.   Phase contrast is still routinely used by many biologists and by those who observe and count asbestos fibers, although other, better-contrast methods have been developed. In 1950, Georges Nomarski invented differential interference contrast (DIC) (12). Although phase contrast is also an interference method, the separation of a given beam by DIC into two beams continuing along light paths of different length before recombining in image formation is accomplished differently. Nomarski used the double refraction of a calcite crystal (actually a Nomarski-modified Wollaston prism) to separate light into the ordinary and extraordinary rays that travel through the calcite at different angles and at greatly different velocities (Fig. 2). Emerging from the calcite crystal, they have a separation distance proportional to the thickness of that crystal, but purposely less than the resolution of the objective to be used. The two rays traverse different optical pathlengths in the object space as, for example, one ray passing through the edge of any object detail and the paired ray passing through the mounting medium immediately adjacent to the object detail. This develops a phase difference and, on recombination in a second Wollaston prism, interference contrast results.

Unfortunately, in a sense, DIC images are sheared images. The two beams from the lower prism lie in a specific vertical plane. A maximum effect is observed for object detail having maximum refractive index differences in the same vertical plane. There is no contrast effect perpendicular to the plane of the two illuminating ray directions. Still, there is an advantage to the shearing method. In addition to a 3-D effect, the contrast effects observed in the shearing plane identify the higher and lower index areas. This furnishes an excellent way of comparing refractive indices, eg, of glass samples recovered in a crime lab hit-and-run case. When heated in a hotstage, a particle of glass, mounted in a properly chosen known refractive index liquid, will disappear at a particular temperature. With DIC, the contrast effect at the particle boundary will reverse at that specific temperature from light-to-dark to dark-to-light across the boundary of the particle.

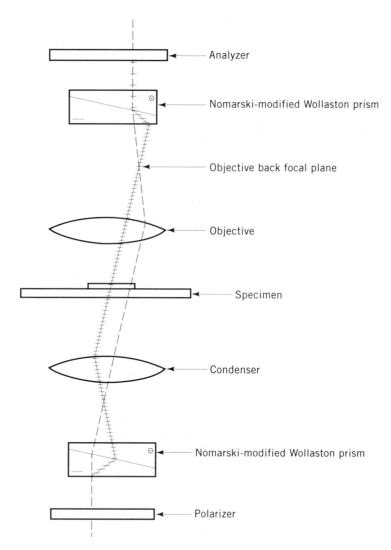

Fig. 2. Differential interference contrast (Nomarski) microscope, showing exaggerated light path (12).

Because DIC is an interference method, interference colors are produced. This means that one can estimate the retardation of light passing through a particle relative to light passing through the adjacent medium. The pattern of retardation reveals at a glance whether a given particle has an index higher or lower than the medium. This makes DIC a useful procedure for the mineralogist examining mineral grains. Most of the applications of DIC, however, involve biologists who wish to see structural details of plant and animal tissue or forensic microscopists who wish to characterize and compare glass samples.

Interference Microscope. There is an interference microscope based on the same general idea as DIC (13), ie, separation of a light beam into two beams that then traverse different paths through the object space. However, the separation of the two rays is much greater than with DIC, and some interference microscopes use other means than the Wollaston prism to separate the light beam into two parallel beams. Because the result then is not specifically increased contrast but increased physical characterization data, it will be discussed later.

Modulation Contrast. In 1975, modulation contrast, a light "modulator" that obscures most of the substage condenser aperture, was invented (14). A fully open aperture is usually considered essential for optimum resolution, yet the modulator forms a small rectangular aperture made up of three smaller areas differing in light absorption. Corresponding conjugate areas in the objective back focal plane and in an auxiliary polarizer control the degree of contrast. The resultant contrast and resolution are excellent and the equipment cost moderate.

Modulation contrast, like DIC, is a shearing method with the same advantages and disadvantages except that interference colors are absent with modulation contrast. The resolution, in spite of the restricted condenser aperture, is certainly not reduced. If anything, resolution seems to be improved and the images obtained have an excellent quality, especially in regard to contrast, one not seen with any other technique except the much more expensive video-enhanced imaging.

Video-Enhanced Contrast. This technique is more expensive but much more effective than any other contrast-enhancing technique (15). Since the 1970s, the development of video processing of microscopical images has resulted in electronic control of contrast. As Shinya Inoué, author of a classic text in the field, states: "We can now see objects that are far too thin to be resolved, and extract clear images from scenes that appeared too fuzzy, too pale, or too dim, or that appeared to be nothing but noise" (16). The depth of the in-focus field can now be expanded or confined, very thin but very sharp optical sections can be produced, and a vertical succession of these images can be accumulated to reconstruct thicker structures in three dimensions (16).

Inoué also describes the performance of image analysis such as counting particles and sorting them by size, color, and shape, automatically tracking moving objects, and mapping a given sample area for distribution and concentration of calcium ions or pH in living cells (16). This requires the use of video and computers but is almost without question the most important advance in microscopy since the invention of the transmission electron microscope in 1932. One obvious advantage of video-enhanced resolution is the ability to leave the microscope substage aperture diaphragm fully open. Without video imaging it must usually be partly to almost completely closed in order for the object to be seen. With electronic control of the contrast, the fully open substage aperture permits realization of the full resolution of the microscope.

Video goes far beyond that goal. Video microscopy can achieve full contrast even after a deliberate effort is made to make an object invisible by closely matching the refractive index. It can also render visible objects that are far smaller than the resolution limit of the light microscope. A retardation as small as 1 nm for birefringent objects or an ultrafaint fluorescent image is enhanced to full visibility. Hence, extremely fine microtubules or fine asbestos fibrils only a few nanometers in diameter can be detected and rendered visible. They are in no sense resolved, but they are visible although they appear to be the same diameter as the best resolution of the microscope used; this is 160–220 nm, depending on the microscope system, for particles or fibers at least 10 times thinner. Still, they are detected.

It is interesting that given very well-crossed polars or very good darkfield illumination (to ensure a very black background), long photomicrographic time exposures (100–200 times normal exposures) also make 20-nm asbestos fibrils visible. This technique does not replace video-enhanced contrast, but it is effective and several orders of magnitude less costly. A completely satisfactory dark-field microscope can be produced by sticky-taping an opaque stop (7–15 mm in diameter, depending on the objective used) to the bottom of the substage condenser of a standard compound light microscope. This creates a dark central cone and a completely black field of view, but the annular space around the opaque stop allows a strong beam of light to focus on the specimen detail and be refracted, reflected, or diffracted onto the dark field of view.

Confocal Microscopy.   Confocal imaging not only improves contrast but it also results in a distinct improvement in resolving power (17). The main feature of confocal microscopy is restriction of the illuminating beam to a hollow cone of illuminating rays that illuminate only the in-focus object space. Out-of-focus details are left dark and are therefore invisible. In normal microscopy, all levels of the specimen are illuminated; hence, the fuzzy above- and below-focus details make it impossible to obtain a sharp high contrast image of the in-focus plane of the specimen.

A confocal microscope accomplishes this task by creating a large-angle hollow cone of light whose apex coincides with the in-focus specimen plane. There are several basic ways to creating a confocal microscope. One, the tandem scanning microscope (TSM) involves a Nipkow disk that has thousands of tiny apertures arranged in an Archimedes spiral near the outer edge of the disk. These apertures beam a large number of very narrow light rays to the in-focus plane of the specimen. Each of these beams lies in the surface of the hollow cone that illuminates the object. A second Nipkow disk, synchronously rotating above the specimen, transmits each of these beams of light through tiny apertures conjugate with corresponding apertures in the substage disk. These beams transmit the in-focus image to the viewer. Again, all out-of-focus detail in the specimen is unilluminated and hence invisible. The absence of the fuzzy out-of-focus images leaves a sharp, high contrast, well-resolved, in-focus image.

The TSM can be modified for reflected light confocal microscopy (TSRLM). In that case, one Nipkow disk above the objective serves both for the illuminating beams and the reflected image-forming rays.

There are also laser-scanning confocal microscopes rapidly overtaking the TSM as a means of confocal microscopy. There is also a computer software program that produces a "confocal" image by recognizing the shapes of out-of-focus detail, ie, halos, and subtracting these from the in-focus image.

One of the advantages of the confocal microscopes is the ability to obtain excellent images deep within transparent specimens, eg, bone, teeth, tissue, ceramics, paint films, etc. This ability then makes possible accumulation of a set of in-depth images, micrometer by micrometer, combining them into a single image many micrometers thick. A final variation of this technique involves making two sets of in-depth images as stereo-pairs and then producing a single stereo-pair showing many-micrometer depth in the transparent matrix of choice.

Another significant advantage of confocal microscopy is the increase in resolution achieved by the absence of the out-of-focus images. A microscope yielding a resolution of 220 nm, if confocal, yields a resolution of 160 nm.

Improving Resolving Power

The resolution of the best of the early nineteenth century microscopes was not too different from the corresponding present-day microscope (about 225 nm), although few early microscopists were able to align their microscopes to achieve this resolution. Only late in the nineteenth century, with the invention of highly corrected (apochromatic) objectives by Ernst Abbe (8), and the sound reasoning of August Köhler (18), were microscopists generally able to optimize the illumination conditions to assure the best resolution possible. Ernst Abbe, in 1883, presented his diffraction theory of resolution for the light microscope (19). This made it possible for microscopists to understand and use the microscope as a high resolution instrument. His equation emphasizes that resolution (RP) improves with higher apertures (NA) and lower wavelengths,

$$RP = 0.61 \ \lambda/NA \tag{1}$$

where RP = the resolving power, the distance between two points just resolved; 0.61 = a Rayleigh constant related to the detectability of resolution; λ = wavelength in nm; NA = the numerical aperture, in turn = n sin AA/2; AA = the average of the angular apertures of objective and condenser; and n = the refractive index of the medium between the objective and coverslip. During the 1800s, attempts had been made to increase resolution by using higher refractive index objective front lenses (even diamond) to increase NA; however, the difficulties outweigh the advantages. Table 1 shows the effects of NA on resolution for a common set of objectives at wavelengths of 500 and 225 nm.

Ultraviolet Light Microscope. Table 1 shows the obvious improvement in resolution by using ultraviolet light, a factor of better than 2 down to 0.10 μm for the oil immersion objective and condenser, and $\lambda = 225$ nm (20). The use of ultraviolet light necessitates the use of ultraviolet–transparent quartz optics, or optics utilizing reflection rather than transmission optics based on refraction.

Table 1. Resolution as a Function of NA and Wavelength

| Objective | NA | Resolution, nm | |
		$\lambda = 500$	$\lambda = 225$
5×	0.05	6100	2740
10×	0.30	1000	460
20×	0.50	610	270
40×	0.65	470	210
40×[a]	0.85	390	160
100×	1.40	220	100

[a]The standard 40× objective is the 0.65-NA version, rather than the higher resolution but shorter working distance 0.85-NA objective.

The confocal microscope improves the resolution of the 40×, 0.65-NA objective to about the same level as the 40×, 0.85-NA objective, although not to the degree achieved by using an ultraviolet objective of the same NA.

A confocal microscope using ultraviolet light and a 1.30-NA objective is expected to produce a resolution of about 0.07 μm (70 nm), but no such instrument has been developed. There are confocal attachments that fit on almost any compound microscope. If one of the early twentieth century ultraviolet microscopes or a Burch reflected optics scope can be found, the shorter wavelength and improved contrast would make possible better resolution than any compound light microscope.

Transmission Electron Microscope.   The use of UV light and high NA condensers and objectives was state-of-the-art in 1929 when Ernst Ruska produced the first transmission electron microscope (TEM) and, with it, achieved resolution of finer detail than the best light microscope. The TEM, although diffraction-limited as is the light microscope, betters the light microscope in resolving power by a factor of about 200 by virtue of its very short wavelength, in spite of its low numerical aperture (21). Ernst Abbe is responsible for showing (1873) that fine detail is resolved by an objective that includes within its angular aperture a direct beam of light and at least one of the diffracted beams from that detail. Because electrons have the properties of waves, Abbe's equation (eq. 1) therefore holds and is used to calculate the theoretical resolution of the transmission electron microscope.

Transmission electron microscopy is very widely used by biologists as well as materials scientists. The advantage of being able to resolve 0.2 nm outweighs the disadvantages of TEM. The disadvantages include the inability of the common 100-kV electron beam to penetrate more than a few tenths of a micrometer (a 1000-kV beam, rarely used, penetrates specimens about 10 times thicker). Specimen preparation for the TEM is difficult because of the thickness limitation.

Ultramicrotomes have been developed to cut soft materials into <0.5 μm sections. The cost of a TEM (about $300,000) may also be considered a disadvantage. The advantages, however, are the ability to observe specimen detail down to the single-atom level and the ability to obtain electron diffraction patterns on subpicogram particles. It is also possible to add energy-dispersed x-ray detectors for an elemental analysis covering about 90% of the elements in the Periodic Table. A special detector of the electrons penetrating the sample shows measurable electron energy losses that also characterize the elements in the sample. This is the electron energy loss spectroscopy (EELS).

The TEM is one of the most generally useful microscopes; many thousands of them are in daily use throughout the world. They are applicable to the study of ultrafine particles (eg, pigments abrasives and carbon blacks) as well as microtomed thin sections of plant and animal tissue, paper, polymers, composites of all kinds, foods, industrial materials, etc. Even metals can be thinned to sections thin enough for detailed examination.

An excellent historical account of the beginning of transmission electron microscopy in North America is available (22).

Scanning Electron Microscope.   Manfred von Ardenne is credited with invention in 1938 of the scanning electron microscope (23). This instrument, although also diffraction-limited, achieves its resolution by scanning a very finely focused beam of very short-wavelength electrons across a surface and by the detection of either the back-scattered or secondary electrons in a raster pattern in order to build up an image on a television monitor. By 1993, the SEM was able to resolve 6-nm detail, at least 20× better than the best light microscope. Resolution of the transmission electron microscope is about 200× better than the light microscope.

Because of its very low numerical aperture, the SEM has an extreme depth of field so that thick objects can be visualized in focus from top to bottom. As mentioned above, the confocal light microscope has to accumulate dozens of successive single images about 1-μm thick and combine them by video imaging into a many micrometers thick image entirely in focus. The SEM, unlike light microscopes, is limited to a surface view only. Nothing is learned visually about the interior of the specimen. It does, however, permit the production of dramatic black-and-white images and, if fitted with an x-ray energy-dispersive detector, an elemental analysis based on the x-rays emitted by the sample during electron bombardment. The energies of those emitted x-rays are characteristic of the elements in the sample. Maps of the sample may also be produced by the x-rays characteristic of sodium and any desired higher atomic number elements.

The SEM is very useful in any microscopy laboratory, perhaps especially in police crime labs. The possibility of confirming the common origin of fibers and hairs from a crime scene and those subsequently taken from a suspect is convincingly demonstrated by the SEM. The use of the energy dispersive x-ray detector (EDS) is also very helpful for identification of many materials, especially unknown minerals. Trace-element composition is different for most substances of the same gross composition, eg, human hair, lead white pigments, and asbestos minerals. Each has a number of trace elements detectable by SEM or EDS and this approach can help to indicate or deny the possibility of common origin. Admittedly, the SEM is not sensitive enough to detect all of the 40 or more trace elements in most human hairs but it would indicate enough difference to suggest a closer look by mass spectroscopy or an electron or ion microprobe.

Photon Tunneling Microscopy. Although total reflection occurs when light strikes a higher index interface at greater than the critical angle, a short-range electrical field extends a fraction of a wavelength through that interface, where it can interact with very near surfaces. Such electrical fields are termed evanescent waves. These waves image any surface lying within their narrow range. Photon tunneling decreases exponentially with distance from the totally reflecting surface; hence, height variations in any surface within the evanescent wave range are imaged (24).

Commercially available photon tunneling microscopes have a lateral resolution of 160 nm but subnanometer vertical resolution. The nondestructive, instantaneous 3-D viewing of a surface (no scanning) yields real-time imaging as one traverses a given sample. The sample must be a dielectric, but transparent polymer replicas of opaque samples can be studied.

X-Ray Microscopy. Because of the short wavelength of x-rays, they have, for nearly 100 years, held out the hope of being utilized in order to significantly lower the diffraction limit of resolution when visible light is used. The difficulties of focusing x-rays and the relative weakness of x-ray sources have, until recently, frustrated efforts to reach that goal (25).

The development of more intense sources (eg, plasma sources, soft x-ray lasers, and synchrotron sources) has made possible highly effective instruments both for x-ray microscopy and x-ray diffraction on a few cubic nanometer sample. The optical problem of focusing x-rays is accomplished by the use of zone plates or by improved grazing incidence or multilayer reflectors.

There are four main approaches to x-ray imaging: contact radiography, scanning x-ray microscopy, holographic x-ray microscopy, and shadow projection x-ray microscopy. In the future, there will likely be phase-contrast imaging and photoelectron x-ray microscopy.

Contact x-ray microradiography using photographic film was at one time used to map the presence of particular elements in a thin section of ceramics, rocks, etc. The method is still used in the 1990s, but most often with a photoresist as detector. Two images are produced by x-rays from two different elemental targets having x-ray wavelengths on either side of a strong absorption edge of the element to be mapped. One of the two images shows particles or concentrated areas of the element; the other image shows no absorption of its characteristic x-rays and hence the same particles are barely if at all visible. Copper K-alpha and cobalt K-alpha x-rays differentiate concentrations of iron by differential absorption of the copper and cobalt x-rays as the imaging medium. It is most useful for detection of copper, nickel, cobalt, iron, manganese, chromium, and vanadium, although potassium, calcium, scandium, and titanium may also be detected. Submicrometer resolution is possible with high energy x-ray sources and poly(glycyl methacrylate–ethyl acrylate) (PBS) resists.

Contact radiography has an interesting application. The sample, usually thin biological tissue placed on a clean PBS film, is imaged on it utilizing a short, single-shot exposure of intense, soft x-rays. The x-ray beam, large enough to cover the specimen, hardens the film differentially, depending on the amount of absorption by the specimen. The resist is then "developed" using a solvent, eg, 1:1 solution of methyl isobutyl ketone and 2-propanol, leaving a surface relief image that can be examined by SEM, or TEM if thin enough. Although single-atom imaging may never be possible with x-rays, the 10-nm region has already been resolved.

Soft x-rays with wavelengths of 1–10 nm are used for scanning x-ray microscopy. A zone plate is used to focus the x-ray beam to a diameter of a few tens of nanometers. This parameter fixes and limits the resolution. Holographic x-ray microscopy also utilizes soft x-rays with photoresist as detector. With a strong source of x-rays, eg, synchrotron, resolution is in the 5–20-nm range. Shadow projection x-ray microscopy is a commercially established method. The sample, a thin film or thin section, is placed very close to a point source of x-rays. The "shadow" is projected onto a detector, usually photographic film. The spot size is usually about 1 μm in diameter, hence the resolution cannot be better than that.

All of these x-ray methods are being intensively studied and improved. A detailed survey of most of these techniques is available (26).

Video-Enhanced Imaging.   The idea of video-enhanced light microscopy as an approach to increased resolution (15,16) is a misconception. Video-enhanced imaging does, however, make otherwise invisible detail visible, thus enhancing the ability to achieve the best resolution that is possible with the light scope. It also greatly improves the visibility of very fine (down to 10–20-nm) detail. For these reasons it must be mentioned as a means of achieving improved resolution, although it is perhaps more appropriately treated as a contrast-enhancing tool. This topic has been discussed under the methods of increasing contrast, where it is without peer. It is used in particular by biologists who have serious contrast problems that make it impossible for them to see what their microscopes have resolved.

Nuclear Magnetic Resonance. In addition to human body scans by magnetic resonance imaging (MRI), there is a future prospect of NMR imaging of microscopic objects with resolution of object detail as small as 1 μm. This is not as good as the light microscope, but the advantages of no specimen preparation, and the ability to scan live specimens without damage and specimens too thick to be penetrated by the light microscope furnishes the incentive for the improvement of resolution by NMR microscopy (26).

Scanning Tunneling Microscopy. As of 1995, there are at least two dozen different microscopes based on a very tiny probe that is able to scan a surface at resolutions down to single-atom levels. All have been developed since 1980, and most after 1985. The proliferation, improvement, and commercialization continues unabated. Only the near-field scanning optical microscope (NSOM) among the probe instruments uses a light beam, but it delivers an enlarged image, hence qualifying as a microscope. Many of them are capable of imaging single atoms. The earliest of these, the scanning tunneling microscope (STM), was invented in 1981 by IBM (Zürich) (27). It involves a very sharp Pt–Ir or W tip scanned only a few tenths of a nm above a conducting surface with a small tunneling current. The distance is kept constant by piezoelectric scanning, with an electronic feedback loop to maintain a constant current as the tip scans the surface. The output of the feedback loop plots a relief map of the surface. The resolution is about 0.2 nm laterally but 0.01–0.001 nm vertically (28).

Atomic Force Microscopy. An important modification of the STM is the atomic force microscope (AFM) (29). Instead of monitoring the tunneling current, which requires a conducting sample, the AFM monitors the repulsive force between tip and surface. Hence it is used on insulating or conducting materials to attain a resolution comparable to that of the STM. The same scanning and feedback principles used for STM and AFM apply to a wide variety of probe-type microscopes based on different methods of sensing the surface and controlling the surface-to-probe tip distance. This list includes, besides STM and NSOM, a laser-driven STM, scanning probes based on capacitance, thermal effects, magnetic force, frictional force, electrostatic force, inelastic tunneling, ballistic electron emission, inverse photoemission, near-field acoustics, "noise," spin-precession, ion conductance, electrochemical forces, absorption (spectroscopy), phonon absorption, chemical potential, photovoltage, and Kelvin probe force. These have been discussed (30).

Most of these are finding specialized applications. The scanning electrochemical microscope uses a conducting wire tip in a solution containing ions or other electroactive species to scan a surface while monitoring a faraday current. It has been applied to electrochemically active surfaces such as living mitochondria, conducting polymers, semiconductors, and metal electrodes. The resolution is not yet in the STM or AFM range.

Scanning thermal microscopy uses the world's smallest thermometer, actually a tiny thermocouple tip, to measure temperature variations as small as 10 microdegrees on a scale of <100 nm.

The molecular dipstick microscope is related to the AFM. It measures lubricant film thickness. The probe is lowered into the oil film on a surface (like the automobile engine crankcase dipstick). The tip is attracted to the surface by the surface tension of the film but repelled by van der Waal's forces from the hard substrate. By noting the height of the probe from the two surfaces as it makes contact, the film thickness can be measured with a precision of about 0.5 nm.

Other Probe Microscopes. The probe microscope field is advancing very rapidly. One of the most useful and most sophisticated of the probe scopes is PoSAP, a Position-Sensitive Atom Probe. By combining the probe with a field-ion microscope and a time-of-flight mass spectrometer, PoSAP can yield chemical composition and position of single atoms in a sample. It plots the positions of 40,000 atoms on any surface. By pulsing a voltage between tip and specimen, a monolayer is expelled, revealing a second layer to characterize. Any particular atom may be ejected into the time-of-flight mass spectrometer for identification. It has already been used for grain boundary composition of metals, surface segregation studies, and the study of catalysts, semiconductors, and superconductors.

Near-Field Scanning Optical Microscope. The near-field scanning optical microscope (NSOM) should, strictly speaking, be NSLM for near-field scanning light microscopy because "optical" includes electron optical as well as light optical and NSOM is a light microscope.

Using a visible light probe NSOM is the earliest of the probe scopes, at least in conception, and is another apparent exception to the diffraction-limited resolution rule, in that NSOM illuminates an object with a beam of visible light smaller than the diffraction limit. The resolution then is limited only by the size of that beam. To achieve this, light issuing from a very tiny aperture at the end of a glass capillary scans a very near sample. The tip must be located on the order of $\lambda/2$ from that surface. Resolution in the range of 10–20 nm has been achieved (31).

Although the idea on which NSOM is based goes back more than 50 years (32), D. W. Pohl first believed it could be achieved with visible light and brought the concept to do it to fruition in 1984 (33). There is considerable interest in NSOM, and two commercial instruments have already been announced. A recent application involves using NSOM for localized absorption spectroscopy and fluorescence imaging of living cells (33).

Field-Emission Electron Microscope. The probe microscopes are certainly impressive in their ability to demonstrate single-atom resolution. However, the field-emission electron microscope (FEEM) (34) and the field ion-emission microscope (35), like the probe microscopes, utilize a fine probe-like tip of tungsten (or later, other conducting substances) at a high negative potential, in a high vacuum. The high charge at the tip causes electrons to be emitted in the field-emission microscope, normal to the surface of the tip. The expanding pattern of electrons travels to a fluorescent screen some centimeters distant, yielding an image of the tip. The image shows that surface at a magnification calculated from the tip curvature radius and the projection distance, usually corresponding to a resolution of 10–20 nm.

Field Ion Microscope. The field ion microscope (FIM) differs from the FEEM only in the presence of 2 μm (pressure) of helium in the otherwise high vacuum ion chamber (35). The helium atoms are ionized at the tip and ejected, again normal to the tip, and impinged onto a fluorescent screen to form an image of the tip. The resolution is now in the single-atom range. Different atoms, such as impurity atoms, on the tip surface are identified by observing where they are and then arranging a tiny aperture in the image plane to encompass an image of the impurity atom. By increasing the voltage, that atom is ionized and ejected to and through the aperture, which is the sample port for a time-of-flight mass spectrometer. The mass, so determined, identifies the element. It was the first microscope capable of identifying a single atom although PoSAP has duplicated this feat.

Unfortunately, both FEEM and FIM microscopes require a conducting sample, usually metallic, capable of being fashioned into a very fine point. The microscopes are used for study of crystal defects, purity, and, with FIM, the identification of single impurity atoms.

Obtaining Additional Characterization Data

The ability to enlarge tiny objects to macroscopic dimensions immediately suggests the need to make measurements and other observations helpful in documenting what is seen and thus enabling others to confirm that a specimen has been identified with certainty. Many physical and chemical properties of a microscopic substance can be measured, even on particles nearing atomic dimensions.

Polarized Light Microscopy. The most useful and characteristic properties presently employed by light microscopists are also among the earliest actual improvements, dating from the 1830s. These were made possible by adding polarizing elements and a rotating stage to a biological microscope to make a polarized light microscope (PLM), the most generally useful of all forms of microscope, whether light, electron, or probe (36).

Many things can be done with the polarized light microscope (PLM). In fact, no other analytical tool or technique yields the variety and extent of information about small objects. An emission spectrograph gives a spectrum identifying the chemical elements present; an absorption spectrograph gives only the functional groups present (eg, carbonyl, amino, hydroxyl, etc); a mass spectrometer gives only the nuclear masses of the constituent elements or their combinations; a scanning electron microscope shows an image only of the surface of a sample and, when properly modified, an elemental analysis; and a transmission electron microscope gives a silhouette and, with very thin substances, a transmission view and an electron diffraction pattern. It may also be fitted for elemental analysis (EDS) and electron energy loss spectra (EELS).

The PLM can be used in a reflection or a transmission mode. With either mode, light of various wavelengths from ultraviolet to infrared, polarized or unpolarized, is used to yield a wide variety of physical measurements. With just ordinary white light, a particle or any object detail down to about 0.5 μm

(500 nm) in diameter can be observed to detect shape, size, color, refractive index, melting point, and solubility in a group of solvents, all nondestructively. Somewhat larger particles yield UV, visible, or IR absorption spectra.

With the polarizing filter (a Polaroid film), refractive index or color variation (pleochroism) as a function of direction can be measured within <1-ng particles. Adding a second polar (crossed polars) permits determination of a variety of highly specific identifying characteristics: anisotropy, sign of elongation, polarization colors and retardation, birefringence, extinction angles, interference figures with optic axial angles, optic sign, dispersion of the optic axes, etc. No two substances could match each other in all of these parameters. These rapid and certain procedures are particularly valuable for the identification of settled or suspended dust particles, for the comparison of trace evidence in the crime laboratory (eg, glass, natural or synthetic fibers and hairs, minerals, drugs, and explosives); and for corrosion products and contaminant particles in drugs, foods, polymers, paint, etc. It is also valuable for product and process failure investigations for quality control and customer complaints.

A significant advantage of the PLM is in the differentiation and recognition of various forms of the same chemical substance; polymorphic forms, eg, brookite, rutile, and anatase, three forms of titanium dioxide; calcite, aragonite and vaterite, all forms of calcium carbonate; Forms I, II, III, and IV of HMX (a high explosive), etc. This is an important application because most elements and compounds possess different crystal forms with very different physical properties. PLM is the only instrument mandated by the U.S. Environmental Protection Agency (EPA) for the detection and identification of the six forms of asbestos (qv) and other fibers in bulk samples.

A cathodoluminescence stage for a polarized light microscope that will take advantage of the x-rays generated by the electrons to detect the elements excited from single small (10–20-μm) particles (EDS) is under development (see LUMINESCENT MATERIALS, CHEMILUMINESCENCE).

The two most useful supplementary techniques for the light microscope are EDS and FTIR microscopy. Energy dispersed x-ray systems (EDS) and Fourier-transform infrared absorption (FTIR) are used by chemical microscopists for elemental analyses (EDS) of inorganic compounds and for organic function group analyses (FTIR) of organic compounds. Insofar as they are able to characterize a tiny sample microscopically by PLM, EDS and FTIR ensure rapid and dependable identification when applied by a trained chemical microscopist.

Additions to the PLM include monochromatic filters or a monochromator to obtain dispersion data (eg, the variation in refractive index with wavelength). By the middle of the twentieth century, ultraviolet and infrared radiation were used to increase the identification parameters. In 1995 the FTIR microscope gives a view of the sample and an infrared absorption pattern on selected 100-μm^3 areas (about 2–5-ng samples) (37).

Thermal Microscopy.   Otto Lehman, by 1880, had added a hotstage to his microscope and used it to study crystals (38). He measured melting points and polymorph transition temperatures, studied phase and composition diagrams, and published the first detailed studies of liquid crystals. Later, additional techniques and applications were developed. The behavior of crystals on heating is quite distinctive, and a rich variety of thermal and optical properties are quickly measured by thermal microscopy, sometimes termed fusion methods (39). It is also an excellent way to recrystallize a compound to yield well-formed crystals of different polymorphs or solvates for use in observing crystal morphology and optical properties. These include refractive indices, birefringence, extinction angles, crystal system, axes, and forms as well as optic sign, optical axial angle, and dispersion thereof. Although there are snowflakes that look alike, contrary to common belief, no two compounds will ever be identical in all of the parameters measured by PLM.

Fluorescence Microscope. A useful light microscope utilizes UV light to induce fluorescence in microscopic samples (40). Because fluorescence is often the result of trace components in a given sample rather than intrinsic fluorescence of the principal component, it is useful in the crime laboratory for the comparison of particles and fibers from suspect and crime scene. Particles of the same substance from different sources almost certainly show a different group of trace elements. It is also very useful in biology where fluorescent compounds can be absorbed on (and therefore locate and identify) components of a tissue section.

Dispersion Staining. Dispersion staining (41), introduced by Germain Crossmon during the 1940s, involved the observation of colored particle boundaries for small particles immersed in mounting media having suitable refractive indices and dispersion. Specifically, the dispersion of refractive index must be different for the particle and the mounting liquid, and the two curves relating refractive index to wavelength must cross within or close to the visible range (400–700 nm). Light rays of different wavelength are refracted differently at particle boundaries, and by suitable masking, images of the particle are observed and show edge colors corresponding to the matching wavelength. A typical application is the identification of chrysotile asbestos by the dispersion staining colors of magenta (for the lengthwise vibration direction) and blue (for the crosswise direction) by observing the fibers in a high dispersion Cargille refractive index medium having $n_D = 1.550$. Dispersion staining is also used in forensic laboratories for the rapid identification of drugs, explosives, fibers, and minerals.

Dispersion staining is useful for rapid determination of refractive index and dispersion. It is applied most often, however, for needle-in-a-haystack detection of any particular substance in a mixture such as chrysotile in insulation, cocaine in dust samples, quartz in mine samples, or any particular mineral, eg, tourmaline, in a forensic soil sample.

Schlieren Microscope. Other important developments were proposed during the 1960s and 1970s, including a unique schlieren microscope, a simple interference microscope (42), and a novel holographic microscope (43). These novel developments were, however, lost in the general eclipse of light microscopy that occurred as the transmission and scanning electron microscopes and their accessory techniques of elemental analysis by energy or wavelength dispersive analysis took precedence over PLM in most laboratories.

The schlieren microscope is able to detect refractive index variations to six decimal places. Any small difference in optical path (index difference, film thickness, etc) is very precisely detected by the schlieren microscope, especially in the Dodd modification. It is, in effect, a darkfield method. The specimen is illuminated with light in a portion of the illuminating cone and that direct light is

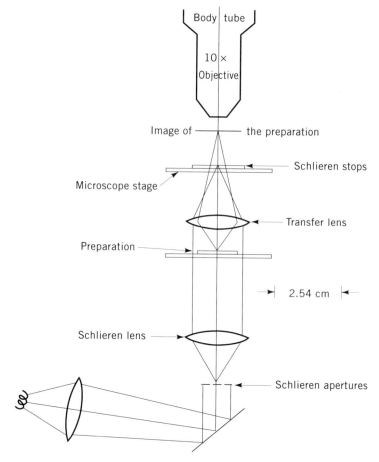

Fig. 3. Schematic diagram for a compact schlieren microscope (44).

masked in the conjugate back focal plane of the objective (Fig. 3). The only light
to pass through this plane is refracted, reflected, or diffracted by the specimen.

Dodd's contribution was to create hundreds, if not thousands, of tiny apertures (literally pinholes in aluminum foil) placed in the substage aperture plane.
A fine-grained photographic film is then placed in the objective back focal plane
to register the image of those pinholes. The developed film is replaced in the
objective back focal plane and the black spots (schlieren stops) are then registered with the schlieren apertures, eliminating all light unless the preparation
deflects the pinhole beams around the conjugate schlieren stops. The result is a
very sensitive schlieren microscope.

These effects were enhanced in a modification (44) in which a set of many
more and smaller schlieren apertures is prepared by placing a Nuclepore filter
with 8-μm pores on a coverslip and evaporating a metal onto the filter, perpendicular to it. The metal then coats and renders opaque the top of the filter and the
bottoms of the pores on the coverslip. This creates a set of schlieren apertures
(the metallized filter) and a conjugate set of schlieren stops on the coverslip.

These are placed, respectively, in the front focal plane of the condenser and the back focal plane of the objective to create a schlieren microscope.

The resulting images are excellent schlieren images, but surprisingly, this schlieren microscope also functions as an interference microscope. This results from diffraction at the edges of the schlieren apertures, creating zero- and first-order diffraction beams that traverse different areas of the specimen and recombine in the image plane to show interference. This microscope also acts as a very sensitive dispersion-staining microscope and shows colored particle borders even when the refractive indices of particle and liquid are almost precisely identical at all wavelengths.

Interference Microscope. There are many variations in the design of interference microscopes, eg, Baker, Jamin-Lebedeff, and Mach-Zender. In general, they are relatively sophisticated, complex instruments. The interference microscope designed by Dodd (42) is, like other instruments originated by Dodd, simple and low cost. The beam splitter is simply a glass plate arranged to reflect the image-forming rays from the ocular from either side of the plate to a focus at the film plane. The path difference for the two reflected beams yields the interference pattern at the film plane of the photomicrographic camera.

Holographic Microscope. Holography (qv) can be done microscopically and produces three-dimensional images with excellent detail. The hologram stores both amplitude and phase information in a 2-D image which is then restored to a 3-D image by viewing that image with a reconstruction beam of light. A great advantage of holography is that the hologram can be studied later at various magnifications to make measurements of size, etc, as one focuses through or scans laterally any portion of the holographic image.

An interesting application is the study of aerosols, such as sprayed paint. A flash hologram of the paint spray at a convenient magnification records positions and sizes of the particles in the hologram. This image can then be studied at leisure at higher magnification for size and shape of individual particles.

One holographic microscope (43) has the advantage of simplicity and low cost. It is, in fact, an attachment fitting any compound light microscope. The attachment consists simply of a modified eyepiece which uses half of the microscope image rays as a reference wave. Together with an inexpensive He–Ne laser, the attachment permits holography (qv) using any microscope equipped with a camera. There is no requirement of coherence for reconstruction, and incoherent white light is used to produce excellent reconstructions.

Laser Raman Microprobe. A more sophisticated microscope is the Laser Raman Microprobe, sometimes referred to as MOLE (the molecular orbital laser examiner). This instrument is designed around a light microscope to yield a Raman spectrum (45) on selected areas or particles, often <1 μm^3 in volume. The data are related, at least distantly, to infrared absorption, since the difference between the frequency of the exciting laser and the observed Raman frequency is the frequency of one of the IR absorption peaks. Both, however, result from rotational and vibrational states. Unfortunately, strong IR absorption bands are weak Raman scatterers and vice versa; hence there is no exact correspondence between the two.

MOLE, however, is more sensitive than FTIR (<1 μm^3 samples compared to about 100 μm^3). With surface-enhanced Raman spectroscopy the Raman signal is enhanced by several orders of magnitude. This requires that the sample be absorbed on a metal surface (eg, Ag, Cu, or Au). It also yields sophisticated characterization data for the polytypes of silicon carbide, graphite, etc.

Cathodoluminescence Microscope. There is also a cathodoluminescence microscope (CLM) (46). An electron source built into the stage of a PLM bombards microscopic samples and yields a spectrum of the resulting characteristic luminescence for many substances. A milliprobe designed to perform elemental analyses on small particle samples or areas was under development in 1994. This unit will be an addition to the CLM. Since the sample is already bombarded with electrons in order for the luminescence to be seen, x-rays characteristic of the elements in the sample are also being generated. With this unit on the PLM stage, any particle in the field of view can be centered in the electron beam to obtain an x-ray energy-dispersive spectrum (EDS). The goal is to do PLM characterization as well as CLM and EDS on samples as small as 500 μm^3 (ie, 5 $\times$ 10 $\times$ 10 μm).

Infrared Microscopy. Many substances, opaque or nearly so in the visible spectrum, are transparent in the near infrared (eg, silicon, iodine, and potassium permanganate). Silicon chips can be examined for internal defects or for strongly absorbing substances. Many normally opaque particles and thin films can be examined by infrared microscopy. One method of doing this is to image the subject with an enhanced-infrared CCD video camera. A microscope field of view can be shown on a monitor by replacing the observer's eye with such a CCD camera. A less costly alternative is to use a photocathode image intensifier. This device (ie, the Army's "snooperscope," used to enhance night vision) collects photons from the microscope eyepiece on a photocathode. This converts the photons to electrons, which are multiplied by a microchannel plate and converted back to photons by a fluorescent screen. The result can then be viewed directly, further magnified, or coupled to a CCD camera and monitor.

Acoustic Microscope. In acoustic microscopy, magnified images are formed using sound waves (47). This process replaces x-ray radiography as a way of seeing inside opaque objects such as metals, ceramics, polymers, etc. The acoustic microscope has the advantage of being able to detect and image voids, cracks, and phase and chemical regions in light-transmitting or opaque materials. It is nondestructive.

BIBLIOGRAPHY

1. P. M. Cooke, *Chem. Microscopy Rev., Anal. Chem.* **64**(12), 219R–243R (1992).
2. P. M. Cooke, *Chem. Microscopy Rev., Anal. Chem.* **66**(12), 558R–594R (1994).
3. C. W. Mason, *Handbook of Chemical Microscopy* 4th ed., John Wiley & Sons, Inc., New York, 1983.
4. S. H. Gage, *The Microscope*, Comstock, Ithaca, N.Y., 1908, 1917, 1920, 1925, 1932, 1936.
5. N. H. Hartshorne and A. Stuart, *Crystals and the Polarizing Microscope*, Edward Arnold, London, 1934.
6. R. M. Allen, *The Microscope*, Van Nostrand, New York, 1940.
7. E. Abbe, *J. Roy. Microscop. Soc.*, 790–812 (1883).
8. *Ibid.*, 20–34 (1887).
9. R. B. McLaughlin, *Special Methods in Light Microscopy*, McCrone Research Institute, Chicago, Ill., 1977.

10. G. H. Needham, *The Practical Use of the Microscopy*, Charles C. Thomas, Springfield, Ill., 1958.
11. W. C. McCrone and R. D. Moss, *Microscope* **38**(1), 1–8 (1990).
12. Ref. 9, pp. 114–121.
13. Ref. 9, pp. 110–114.
14. W. C. McCrone, *The Particle Atlas*, Vol. V, 1979, p. 1155 (out-of-print); available on CD-ROM, McCrone Research Institute, Chicago, Ill.
15. S. Inoué, *Video Microscopy*, Plenum Publishing Corp., New York, 1986.
16. S. Inoué, T. Inoué, and B. Gunning, *Amer. Lab.*, 52–81 (Apr. 1969).
17. T. Wilson, ed., *Confocal Microscopy*, Academic Press, London, 1990.
18. A. Köhler, *Zeitschr. f. Wiss. Mikr.*, 433–440 (1893); abstract, *J. Roy. Microscop. Soc.*, 261–262 (1894).
19. Ref. 7, pp. 790–812.
20. B. W. Rossiter and J. F. Hamilton, eds., *Physical Methods of Chemistry*, 2nd ed., Vol. IV, John Wiley & Sons, Inc., New York, 1991, p. 123.
21. J. W. Stephenson, *Trans. R.M.S.*, 82–88 (1877).
22. F. W. Doane, G. T. Simon, and J. H. L. Watson, *Canadian Contributions to Microscopy*, Microscopy Society of Canada, 1993.
23. Ref. 20, p. 285.
24. J. M. Guerra, *Appl. Opt.* **29**, 3741 (1993).
25. R. E. Burge, *Proc. R. Instr. C.B.* **64**, 137–168 (1992); *Opt. Sci.* **67**, 3–10 (1992).
26. P. J. Duke and A. G. Michette, *Modern Microscopies*, Plenum Publishing Corp., New York, 1990.
27. G. Binning, H. Rohrer, C. Gerber, and E. Weibel, *Phys. Rev. Lett.* **49**, 57 (1982).
28. J. H. Stroscio and W. J. Kaiser, eds., *Scanning Tunneling Microscopy*, Academic Press, Inc., San Diego, Calif., 1993.
29. A. Pidduck, *Proc. Roy. Microscop. Soc.* **28**(3), 133–128 (1993).
30. H. K. Wickramasingle, in Ref. 28.
31. E. Betzig and co-workers, *Science*, **251** 1468–1470 (Mar. 22, 1991).
32. E. H. Synge, *Phil. Mag.* **6**, 356 (1928).
33. D. W. Pohl, W. Denk, and M. Lenz, *Appl. Phys. Lett.* **44**, 651 (1984).
34. E. W. Müller, *Z. Physik* **106**, 541 (1937).
35. E. W. Müller, *J. Appl. Phys.* **27**, 473 (1956); **28**, 1 (1957).
36. W. C. McCrone, L. B. McCrone, and J. G. Delly, *Polarized Light Microscopy*, McCrone Research Institute, Chicago, Ill., 1987.
37. J. A. Reffner, *Microscopy Today*, **3**, 6–7 (1993).
38. O. Lehmann, *Die Krystallanalyse*, Englemann, Leipzig, Germany, 1890.
39. W. C. McCrone, *Fusion Methods in Chemical Microscopy*, Interscience Publishers, New York, 1957.
40. Ref. 10, pp. 101–133.
41. Ref. 36, pp. 169–196.
42. J. G. Dodd and J. J. Dodd, *Microscope* **27**, 31–40 (1979).
43. J. G. Dodd, *Microscope* **25**, 55–63 (1977).
44. W. C. McCrone, *Microscope* **20**, 309–318 (1972).
45. W. C. McCrone and co-workers in Ref. 14, pp. 1147–1154.
46. V. I. Petrov, *Phys. Status Solidi A* **133**(2), 189–230 (1992).
47. C. A. D. Briggs, *Rep. Prog. Phys.* **55**(7), 851–909 (1992).

General References

G. L. Clark, ed. *The Encyclopedia of Microscopy*, Reinhold Publishing Corp., New York, 1961.

G. H. Needham, *The Practical Use of The Microscope*, Charles C. Thomas, Springfield, Ill., 1958.

C. W. Mason, *Handbook of Chemical Microscopy*, 4th ed., John Wiley & Sons, Inc., New York, 1983.

W. C. McCrone, L. B. McCrone, and J. G. Delly, *Polarized Light Microscopy*, McCrone Research Institute, Chicago, Ill., Ninth printing, 1995.

R. D. McLaughlin, *Special Methods in Light Microscopy*, Microscope Publications, Chicago, Ill., 1977.

S. Inoué, *Video Microscopy*, Plenum Publishing Corp., New York, 1986.

WALTER C. MCCRONE
McCrone Research Institute

MICROWAVE TECHNOLOGY

The application of electrical or electromagnetic (EM) energy to materials as part of some chemical process is a broad subject that has a long history in chemical technology. Electrical energy can be delivered to materials through conductive, near-field coupling, or radiative techniques. The microwave portion of the electromagnetic spectrum in its role as a power source for chemical and materials processing is discussed herein. Microwave is the name applied to the central portion of the nonionizing radiation part of the electromagnetic spectrum. Nonionizing radiation has quantum energy too low to ionize an atom on a single event basis (1) and conventionally is defined as ranging from d-c power to visible light. This portion of the spectrum can be divided into five regions in order of increasing frequency: static, quasistatic, microwave, quasioptical (nanowave), and optical.

Microwaves may be used to ionize gases when sufficient power is applied, but only through the intermediate process of classical acceleration of plasma electrons. The electrons must have energy values exceeding the ionization potential of molecules in the gas (see PLASMA TECHNOLOGY). Ionizing radiation exhibits more biological-effect potential whatever the power flux levels (2).

The term radio waves has evolved to refer to coherent generation of energy in the nonionizing spectral region and its application for information processes such as communication, broadcasting, and target location, eg, radar. The development of radio-wave technology began at lower frequencies where electromechanical, then vacuum-tube, and later solid-state techniques are easiest to apply. The term microwaves was introduced just before World War II (3). Vacuum-tube development had proceeded far enough toward higher frequency to permit significant power levels to be transmitted in hollow pipes (waveguides) of practical laboratory size (ca 15-cm dia) and high (>20 dB) antenna gain to become feasible using practical rotating diameters, eg, <5 m. This led to definitions of the microwave spectrum as >1000 MHz. Over the years, these definitions have been variously specified to include frequencies as low as 100 MHz to as high as 3000 GHz. The more scientific meaning of microwaves refers to the principles and techniques applying to electromagnetic systems (4) where the principal dimensions are of the order of a wavelength, γ, or more broadly ca 0.1–10 γ.

For many materials, in particular biological tissue, maximum penetration of the electromagnetic energy irradiating objects of macroscopic size occurs in the microwave range (5). At low frequencies, the human body acts as a conductor and the electric field is shunted out, ie, the body has a shielding effect. At high frequencies, the penetration depth (decay by $1/e$ in field strength) rapidly becomes much smaller than body dimensions. Only in the microwave range, ca 100 MHz, is deep penetration significant. At low frequencies, penetration of magnetic fields into most materials is efficient. Almost all microwave interactions with lossy materials, ie, those absorbing microwave energy, involve the electric field which penetrates only in the microwave range, however. The property of resonance in lossy materials is related to resonant conducting antennas and resonant low loss dielectric modes. Most materials of potential interest for microwave processing have dielectric properties that permit interpretation in terms of geometric resonance. A consequence is the scaling of processing frequency having some characteristic size of the material or the container in which it is processed.

Another reason for interest in microwaves in chemical technology involves the fields of dielectric spectrometry, electron spin resonance (esr), or nuclear magnetic resonance (nmr) (see MAGNETIC SPIN RESONANCE). Applications in chemical technology relating to microwave quantum effects are of a diagnostic nature and are not reviewed herein.

Microwave power is an important factor in the commercial processing of materials. These processes almost exclusively utilize classical interactions such as dielectric heating in solids or plasma heating. Other interactions may also be possible. The field of microwave power applications as distinct from information processing is relatively new. Development and possibilities were recognized in 1965 (6). Interest followed (7–9). The International Microwave Power Institute (IMPI), founded in 1966 in Canada, has offices in Manassas, Virginia. Publications of IMPI include two newsletters, symposium digests and proceedings, and workshop proceedings as well as the *Journal of Microwave Power*. A number of reviews (10–13) and books (14,15) are available covering microwave power applications in a wide range of areas including food processing (qv), ceramics (qv) processing, biological tissue fixation, chemical analysis and processing, and plasma applications as well as microwave power transmission.

Microwave Power Applications

Frequency Allocations. Under ideal conditions, an optimum frequency or frequency band should be selected for each application of microwave power. Historically, however, development of the radio spectrum has been predominantly for communications and information processing purposes, eg, radar or radio location. Thus within each country and to some degree through international agreements, a complex list of frequency allocations and regulations on permitted radiated or conducted signals has been generated. Frequency allocations developed later on a much smaller scale for industrial, scientific, and medical (ISM) applications.

In 1979, the ISM frequency allocations were revised as a result of the World Administrative Radio Conference (WARC) (16). A considerable effort was made to increase the number and worldwide uniformity of ISM frequency allocations. Most of those proposals were rejected. The resulting allocations are listed in Table 1.

The three frequencies 13.56, 27.12, and 40.68 MHz are well-known allocations for radio-frequency (r-f) heating using quasistatic coupling to loads, such as a capacitor arrangement. The most popular of these assignments is 27.12 MHz for which the broadest bandwidth is authorized. This frequency also overlaps the frequency allocations for citizen's band (CB) radio in the United States. The techniques used at these frequencies are not generally considered microwave for small loads, but would be considered microwave if energy were applied to sufficiently large loads, as in the tentative attempts to heat oil shale (qv) *in situ*. The frequency 433.92 MHz, a harmonic of these three frequencies, is allocated only in some European countries (see Table 1). It is used mostly for medical diathermy and hyperthermia.

The remaining ISM allocations above 433.92 MHz are not harmonically related. This is unfortunate in terms of the problem of minimizing radio-frequency interference (RFI), except for the harmonic relation in the millimeter wave range.

Table 1. Frequency Allocations for ISM Applications[a,b]

Frequency, MHz	Region	Conditions
6.765–6.795	worldwide	special authorization with CCIR[c] limits; both in-band and out-of-band
13.553–13.567 26.957–27.283 40.66–40.70	worldwide	free radiation bands
433.05–434.79	selected countries in Region 1[d]	free radiation bands
433.05–434.79	rest of Region 1[d]	special authorization with CCIR[c] limits
902–928	Region 2[e]	free radiation band
2.40–2.50 × 10³	worldwide	free radiation band
5.725–5.875	worldwide	free radiation band
24.0–24.25	worldwide	free radiation band
61.0–61.5 122–123 244–246	worldwide	special authorization with CCIR[c] limits; both in-band and out-of-band

[a] Ref. 16.
[b] ISM = industrial, scientific, and medical.
[c] CCIR = International Radio Consultative Committee of the International Telecommunications Union (ITU).
[d] Region 1 comprises Europe and parts of Asia; the selected countries are Germany, Austria, Liechtenstein, Portugal, Switzerland, and Yugoslavia.
[e] Region 2 comprises the western hemisphere.

The frequency bands at 915 and 2450 MHz (2375 MHz in the past in eastern Europe) are the most developed bands for microwave power applications. Microwave ovens are almost all at 2450 MHz. Many industrial heating applications are at 915 MHz. After the 1979 WARC, eastern Europe adopted 2450 MHz in place of 2375.

The higher frequencies, 5800 and 24,125 MHz, have been allocated for ISM for some years but have not been further developed. The unavailability of inexpensive efficient power sources (tubes) at those frequencies is the limiting factor. At the 1979 WARC, additional ISM allocations were adopted at 61.25, 122.5, and 245 GHz in anticipation of future applications. Power sources, especially gyrotrons, are available >30 GHz, but are expensive and generally <50% in efficiency, hindering the exploitation of this higher frequency region called millimeter waves.

In general, ISM applications at allocated frequencies are permitted to freely radiate energy within the allocated band. Any other users of this band must tolerate possible interference from such emissions. Emissions outside of these bands, however, must be limited to values specified by regulatory bodies such as the Federal Communications Commission (FCC) in the United States and verified by certain certification procedures (17). Typically, field strengths for signals within a 5 MHz band must be below 10 μV/m at a distance of 1.6 km from industrial heating equipment. In other countries, the RFI regulations must conform to similar limitations. In many cases, the Comité International Spécial de Perturbations Radioélectriques (CISPR) limits, are applicable (18).

Experiments and even production operations can be conducted at any frequency providing the radiated and conducted signals meet the applicable rfi limits for ISM equipment. Tests to certify this stipulation must be carried out before inception of operations. This implies well-shielded enclosures at high power levels which is expensive but justified in certain applications.

There has been increasing co-use of ISM bands by communications interests, technically under the condition that any interference that may result from co-existing ISM applications be accepted. Proposals to reallocate part of the ISM bands for communication purposes have been made and the FCC had proposed auctioning a part (2402–2417 MHz) of the 2.45 GHz ISM band for primary use under co-use conditions. The number of microwave ovens has grown enormously (ca 200 million as of 1995). The CISPR has therefore been studying new limits in the 1· 18 GHz range in order to minimize interference with wireless communications systems expected to occupy allocations across the band from 1 to 3 GHz by the late 1990s. The resulting regulations are likely to have some impact on microwave oven and higher power equipment design because of the well-known excess noise produced by magnetrons in sideband regions, ie, 2.2–2.4 and 2.5–2.7 GHz for a 2.45 GHz source (19).

Principles in Processing Materials. In most practical applications of microwave power, the material to be processed is adequately specified in terms of its dielectric permittivity and conductivity. The permittivity is generally taken as complex to reflect loss mechanisms of the dielectric polarization process; the conductivity may be specified separately to designate free carriers. For simplicity, it is common to lump all loss or absorption processes under one constitutive parameter (20) which can be alternatively labeled a conductivity, σ, or an imaginary part of the complex dielectric constant, ϵ_i, as expressed in the following equations for complex permittivity:

$$\epsilon = \epsilon_0(\epsilon_r + j\epsilon_i) = \epsilon_0(\epsilon_r + j\sigma/\omega\epsilon_0) \tag{1}$$

where ϵ is the complex dielectric permittivity in F/m, $\epsilon_0 = 8.86 \times 10^{-12}$ F/m, the permittivity of free space, ϵ_r, is the real part of the relative dielectric constant, and σ is the conductivity in S/m (mhos/m) which is equivalent to the following:

$$\epsilon_i = \sigma/\omega\epsilon_0$$

where ω is the assumed radian frequency of the fields. It is convenient to define auxiliary terms like the loss tangent, tan δ:

$$\tan \delta = \epsilon_i/\epsilon_r = \sigma/\omega\epsilon_r\epsilon_0 \tag{2}$$

From Maxwell's equation (21), the current density J in A/m^2 is related to the internal electric field, E_i, by equation 3:

$$J = (\sigma - j\omega\epsilon_r\epsilon_0)E_i \tag{3}$$

thus the rate of internal density of absorbed energy, or power, P, is given by equation 4, where rp = real part of:

$$P = \text{rp}(J \cdot E^*) = \sigma|E_i|^2 \tag{4}$$

or simply,

$$P = \omega\epsilon_r\epsilon_0 \tan \delta|E_i|^2 \tag{5}$$

Equation 5 is the practical equation for computing power dissipation in materials and objects of uniform composition adequately described by the simple dielectric parameters.

The internal field is that microwave field which is generally the object for solution when Maxwell's equations are applied to an object of arbitrary geometry and placed in a certain electromagnetic environment. The E_i is to be distinguished from the local field seen by a single molecule which is not necessarily the same (22). The dielectric permittivity as a function of frequency can be described by theoretical models (23) and measured by well-developed techniques for uniform (homogeneous) materials (24).

The dielectric permittivity as a function of frequency may show resonance behavior in the case of gas molecules as studied in microwave spectroscopy (25) or more likely relaxation phenomena in solids associated with the dissipative processes of polarization of molecules, be they nonpolar, dipolar, etc. There are exceptional circumstances of ferromagnetic resonance, electron magnetic resonance, or nmr. In most microwave treatments, the power dissipation or absorption process is described phenomenologically by equation 5, whatever the detailed molecular processes.

The general engineering task in most applications of microwave power to materials or chemicals is to deduce from the geometry of samples and the electromagnetic (EM) environment (applicator), the internal field distribution, $E_i(r)$, and hence the distribution, $P(r)$, of absorbed power. From this, the temperature distribution can be calculated, and processing time and applied power required for the desired result deduced.

The electromagnetic problem is one of solving Maxwell's equations under various boundary conditions (21). If the object is small, the applied EM field may be little perturbed and perturbation theory is adequate (26). If the object is large in terms of penetration depth, quasioptical radiation calculations are valid. If the object is of the order of a wavelength in dimension, geometric resonance can apply using a moderately enhanced absorption cross section. Many calculations for simple models of biological tissue are available (27). There are also many computer programs, software, or codes for the detailed numerical solution for electromagnetic fields. The capabilities and pitfalls of modern computer solutions have been reviewed (28,29).

Penetration depth, D, at which fields are reduced by a factor of $1/e$, is given by the following formula:

$$D = 0.225\lambda/\epsilon_r^{1/2}\cdot\left((1 + \tan^2 \delta)^{1/2} - 1\right)^{1/2} \tag{6}$$

or for low loss materials, where $\tan \delta < 1$:

$$D \cong \frac{0.318\,\lambda}{\epsilon_r^{1/2}(\tan \delta)} \tag{7}$$

where λ is the free-space wavelength of the microwave radiation.

For low frequencies or small objects, the components of electrical fields normal to a material boundary are related by equation 8:

$$\left|\frac{E_i}{E_0}\right| \cong (\omega\epsilon_0/\sigma) \tag{8}$$

where E_i and E_0 are the internal and external fields, respectively. The material is characterized by σ, and the outside volume is free space. If the applied field E_0 is parallel to the surface, then the internal field is equal to E_0. Thus if the applied field is parallel to the long axis, the internal E_i is ca E_0. Otherwise, if E_0 is perpendicular to the long axis, the internal field E_i is given by equation 8 and $E_i < E_0$ for even moderate conductivity ca 1 S/m.

In the typical application of microwave power, the engineering task is to solve for the internal fields of an object for the given system, compute its heating distribution vs time and resulting change of state, and similar problems. The dielectric parameters are key data for such a calculation, particularly the dependence on temperature. Literature references to such dielectric data are available (30,31). These do not include extensive data on temperature dependence except for water. Because of the concentrated interest in microwave cooking at 2450 MHz, more data are available at this frequency. Temperature dependence of typical food dielectric properties are shown in Figure 1 (32). Table 2 contains information on dielectric properties for foods and other materials. Other references and techniques for measuring dielectric parameters are given in the literature (33).

A basic problem in microwave heating applications is evident in Figure 1b. If tan δ increases with temperature, runaway conditions may be created in hot spots, although there are possible mitigating factors. Because water shows the

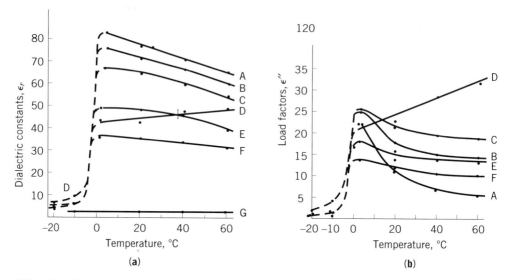

Fig. 1. Properties of foods near 2.45 GHz as a function of temperature, where A represents distilled water; B, cooked carrots; C, mashed potatoes; D, cooked ham; E, raw beef; F, cooked beef; and G, corn oil: (**a**) dielectric constants and (**b**) load factors, $\epsilon'' = \epsilon \tan \delta$ (32).

Table 2. Dielectric Properties of Foods and Other Materials at 2450 MHz and 20°C

Material	ϵ	tan δ
distilled water	78	0.16
raw beef	49	0.33
paper	2–3	0.05–0.1
wood	1.2–5.0	0.01–0.1
alumina	7.8	0.001
borosilicate glass	4.5	0.004–0.007
neoceram	6.2	0.003
plastics		
ABS	2.85	0.006
Ultem	3.0	0.001–0.004
polysulfone	2.1	0.006
polypropylene	2.2	0.0005
Teflon	2.0	0.0004
liquid crystal polymer	3.9	0.007

opposite behavior above freezing, it is presumably conducive to even cooking. Below freezing, the loss of water is much lower, hence there is a basic conceptual difficulty in designing a process for uniform thawing by microwave heating. Heating to a point short of the melting point, on the other hand, is aided by a negative value of the curve slope and underlies the commercial success of meat tempering. The behavior of ham (Fig. 1**b**) reflects the effect of salts and increased ion conductivity with temperature.

Most demonstrated and useful effects of microwave irradiation of materials are explainable as heating influences. The results may be unique to microwave heating because most conventional heating techniques are not associated with deep penetration of the radiant energy. Speculations on more esoteric, sometimes called nonthermal or field-force, influences have arisen. The only well-demonstrated effect in this category is that of thermoelastic conversion accompanying short pulses or rapid changes of microwave radiation levels (34,35). Other speculative mechanisms, generally considered to be very weak, are those of electrostriction (36), dielectrophoresis (37), and pearl-chain effects. In all these cases the effects are expected to be proportional to $|E|^2$. A review of these mechanisms in biological systems is available (38).

The electromagnetic interactions that make use of the microwave magnetic fields are generally of less interest commercially. There is, however, considerable scientific and possibly medical interest. Magnetic fields, H, play a key role in weak mechanical forces induced by electromagnetic waves (39). The interactions in esr and nmr are of scientific interest and have been successfully developed into the medical imaging technology (qv) called magnetic resonance imaging (mri) (40).

The most widespread applications of interactions with microwave magnetic fields are those of ferromagnetic materials. The dissipation of microwaves in systems of ferromagnets can be quite complicated especially in the presence of nonreciprocal transmission properties in biased ferrite systems (see FERRITES; MAGNETIC MATERIALS) (41). In the simplest cases, eg, linearly polarized or operations far from ferromagnetic resonance, the localized absorption, P_{abs}, in analogy with equation 5, is given by equation 9:

$$P_{abs} = \omega \mu'' |H|^2 \tag{9}$$

where μ'' is the imaginary part of the magnetic permeability and the contributions from the imaginary part of the cross-components of the permeability tensor are negligible. Dissipation is temperature dependent and disappears above the Curie temperature of the material. This is the basis of some browning dishes developed for microwave oven use, designed for browning at a Curie temperature of ca 230°C (42).

The interaction of microwaves with ferrites (qv) has many complicating features. Low field loss mechanism (41), nonlinear effects, and losses at high power levels (41,43) as well as dielectric losses are among these.

The application of microwave power to gaseous plasmas is also of interest (see PLASMA TECHNOLOGY). The basic microwave engineering procedure is first to calculate the microwave fields internal to the plasma and then calculate the internal power absorption given the externally applied fields. The constitutive dielectric parameters are useful in such calculations. In the absence of d-c magnetic fields, the dielectric permittivity, ϵ, of a plasma is given by equation 10:

$$\epsilon = \epsilon_0\left(1 - \frac{Ne^2}{m(\omega^2 + v_c^2)\epsilon_0}\right)$$ (10)

and the conductivity, σ, by

$$\sigma = \frac{Ne^2v_c^2}{m(\omega^2 + v_c^2)}$$ (11)

where N is the electron volume density, ω is the microwave radian frequency, v_c is the collision frequency, and e and m are the electron charge, 1.602×10^{-19} C, and mass, 9.11×10^{-31} kg, respectively.

Maximum power transfer to electrons for a given internal field occurs when $v_c = \omega$. The plasma frequency, ω_p, is the frequency at which $\epsilon = 0$:

$$\omega_p = \left(\frac{Ne^2}{m\epsilon_0}\right)^{1/2}$$ (12)

Because the permittivity is negative for $\omega < \omega_p$, transmission through the plasma is cut off and penetration is only by means of evanescent waves.

Various data sources (44) on plasma parameters can be used to calculate conditions for plasma excitation and resulting properties for microwave coupling. Interactions in a d-c magnetic field are more complicated and offer a rich array of means for microwave power transfer (45). The literature offers many data sources for dielectric or magnetic permittivities or permeability of materials (30,31,46). Because these properties vary considerably with frequency and temperature, available experimental data are insufficient to satisfy all proposed applications. In these cases, available theories can be applied or the dielectric parameters can be determined experimentally (47).

Instrumentation

Power Sources. The development of electron tubes, including those for the microwave range, is a mature field (6). It is feasible to generate almost any desired power for most microwave frequencies of practical interest, limited only by costs. Power limits for various devices are shown in Figure 2 (48). Below 500 MHz large power is typically generated by gridded tubes. Above 1 GHz, the feasible power drops roughly as f^{-5}, where f is frequency, because of fundamental limitations on electron beam current density and circuit losses. The boundary is similar for microwave tubes such as magnetrons, klystrons, traveling wave tubes, and backward wave oscillators, except it is at a frequency roughly 30 times higher.

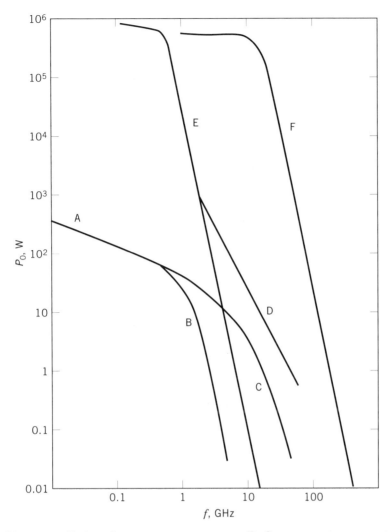

Fig. 2. Maximum limits of average output power P_0 from power sources at average wattages or continuous wave (CW) where A is the solid state; B, transistors; C, varactors and solid-state avalanche diodes; D, theoretical limit for solid-state devices; E, gridded tubes; and F, microwave tubes.

Power sources in the millimeter wave range are mostly in the category of extended interaction klystrons or narrow band backward wave oscillators (49). Power outputs of tens of watts or even more than 100 W are feasible but available tubes are quite expensive and suffer from low life and efficiency compared to what is available at low microwave frequencies. Very high powers of many kilowatts continuous wave (CW), and hundreds of kilowatts low duty cycle, have been generated by gyrotrons (50). These have been developed mostly for fusion energy (qv) at government laboratories and reflect large development costs. Because of the unavailability of inexpensive and efficient tubes for millimeter wave frequencies, ISM applications in that range are essentially nil.

Significant power is generated only below 300 GHz. Above this frequency, the expectation has always been that useful laser sources would be eventually developed (51). Thus the most difficult region for power generation appears to be that of submillimeter waves or the far infrared, ie, 300–3000 GHz (1000–100 μm).

The remaining class depicted in Figure 2 is that of solid-state devices, ie, transistors, various types of semiconductor diode amplifiers, etc. At frequencies below 1 GHz, generation of hundreds or even at the lower frequencies, kilowatts, is feasible by solid state. Above 1 GHz power capability of solid-state sources drops. Development of efficient ($\sim$50%) sources at about the 50 W level at S-band (2 GHz) has been demonstrated. It is reasonable to expect solid-state sources to replace tubes for low frequency and low ($<$100 W) power applications (52). For high power or high frequency, however, tube sources should continue to prevail.

The most dramatic evolution of a microwave power source is that of the cooker magnetron for microwave ovens (48). These magnetrons are air-cooled, weigh 1.2 kg, generate well over 700 W at 2.45 GHz into a matched load, and exhibit a tube efficiency on the order of 70%. Application is enhanced by the availability of comparatively inexpensive microwave power and microwave oven hardware (53). The cost of these tubes has consistently dropped (11) since their introduction in the early 1970s. As of this writing (ca 1995), cost is $<$\$15/tube for large quantities. For small quantities the price is $<$\$100/tube.

The availability of a low cost source of microwave power has led to an explosion of work at 2.45 GHz on newer microwave power applications. For many applications at 2.45 GHz, it is feasible to utilize a number of low cost tubes to generate large total powers, eg, 25 or 50 kW. In such cases, multiple feeds or antennas to a heating chamber are used. At 2.45 or 0.915 GHz, it is preferable to use larger power sources at 5, 25, or even 50 kW. Table 3 shows available tubes commonly considered for use at 0.915 or 2.45 GHz. Most of these tubes are magnetrons, although klystrons become competitive at the higher power levels. These tubes are designed to meet the requirements of government agencies on out-of-band spurious emissions. Hence, filter boxes are used around the high voltage terminals of the tubes.

Microwave tubes for other ISM bands are not commonly available as tubes designed specifically for ISM use. Available tubes, generally of military and communications types, are more expensive. Reasonably priced tubes exist at 0.915 GHz at high ($>$25 kW) power, but not at 5.8 GHz and higher, or at low ($<$1 kW) power at 0.915 GHz.

Use of traveling wave tube (TWT) amplifiers at power levels of hundreds of watts has been proposed (54) for power applications, at least when the heating chamber is well shielded. The potential advantage is an improved uniformity of heating when a broad band of frequency is used, ie, excitation of many modes. Disadvantages are high cost and lower ($<$50%) efficiency of the TWT.

Applicators and Instruments. The basic elements of a microwave power system for materials processing are indicated schematically in Figure 3. A power supply drives the microwave tubes with an applied d-c voltage or even raw rectified voltage (50 or 60 Hz). The former may be required of klystrons (55), whereas the latter has been perfected in the form of voltage doubler power supplies for microwave ovens (56). The microwave tube source is thus generally

Table 3. Magnetrons for ISM Applications[a]

Tube manufacturer and type		Frequency, MHz	Nominal power, kW	Voltage, kV
Hitachi	2M107A	2455	0.900	4.2
	2M214	2455	0.900	4.25
	2M120	2455	1.45	4.5
Toshiba	2M172A	2460	0.850	4.0
	2M229	2460	0.850	4.0
	2M240	2460	0.850	4.0
	2M248	2460	1.000	4.35
Matsushita	2M167B	2455	0.900	4.1
	2M137	2455	1.260	4.5
Samsung	2M204	2455	0.900	4.1
	OM75S	2460	0.870	4.1
Goldstar	2M613	2460	1.420	4.5
Richardson	NL10251	2450	1.50	3.6
	YJ1442	2450	3.0	6.0
	YJ1191A	2450	6.0	7.3
	NL10257	915	1–5	6.5
	NL10258-1	915	5–75	17.0
	C9460E	915	60	16.1
Burle	C9460E	915	60	16.0
California	CWM-60L	915	60	16.0
Tube Lab.	CWM-3L	915	2.5	3.7

[a]ISM = industrial, scientific, and medical.

operated as a continuous wave (CW), ie, unmodulated signal, except for the possible incidental 60-Hz amplitude modulation for power applications. Only for a few applications is a capability of amplitude or frequency modulation required and these modifications would entail a significant departure from the present low cost sources. The spectrum of typical microwave power sources is not tightly

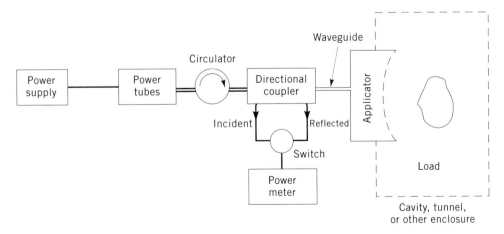

Fig. 3. Basic elements of a microwave power system for processing of materials.

controlled. Considerable frequency drift and signal broadening is tolerable within the most popular designated ISM bands.

Some power tubes can be operated without the need for a protective ferrite isolator. One example is the cooker magnetron (700 W) used in modern microwave ovens (57). At higher power levels, such as 25 kW, it is more common to employ a protective ferrite device, particularly in the form of a circulator (58), as shown in Figure 3. This results in a power loss equivalent to a few percentage points in system efficiency. The ferrite circulator prevents reflected power from returning to the power tube and instead directs it into an auxiliary dummy load. The pulling of tube frequency is thus minimized.

It is common to employ microwave power monitoring by means of a dual-directional coupler in the waveguide transmission system between the power tube and the useful load. Part of the coupled signals may be used for examination with spectrum analyzers, frequency meters, and other microwave instrumentation for special purposes. Generally, this is not necessary in a practical application. Many microwave measurement techniques have been described (59,60). Availability of components, plumbing, and instrumentation is well described in trade journals.

The useful load is the material to be heated by microwave or otherwise affected by exposure to microwave radiation. Some type of applicator, the feed (transmitter), or antenna arrangements that couple microwave energy into the load, is also needed. The object is to maximize the coupling. This is indicated by zero-reflected power in the waveguide monitoring system. Various waveguide components used to produce an effectively matched applicator, $P_{ref} = 0$, are not indicated in Figure 3 (61). The power transmitted by the applicator is not necessarily all captured by the load, ie, the circuit efficiency in the heating enclosure is not necessarily 100%. Generally it can be designed to be 90% or better, especially for large loads.

The heating system is calculated by taking into account the power supply efficiency (generally >90%), tube efficiency (60–90%), waveguide (including circulator) transmission efficiency (generally >90%), and circuit efficiency of the heating enclosure. The net efficiency of such systems, ie, useful power into load divided by line input power, varies significantly among systems, but a typical value is ca 50%. The U.S. Department of Energy (DOE) is proposing that manufacturers improve the efficiencies of consumer microwave ovens through improvements in high voltage transformer efficiency, circuit efficiency of oven cavity and waveguide, and the cooling fans (62).

The use of inverter-type power supplies (63) in place of doubler-type 60-Hz supplies results in significant reduction in the weight of microwave ovens. Disadvantages are primarily cost for the consumer market in addition to somewhat less efficiency and increased noise. Their use in commercial or industrial equipment is more attractive.

Different applicator types serve various forms, shapes, and sizes of the material being processed or heated. The latter may be gaseous plasma enclosed in a glass tube passing through a cavity or waveguide, thin films passing through a single cavity, or large-size boxes of frozen meat passing in a conveyor through a long metal enclosure. The aim to achieve efficient and uniform heating (or temperature) in the product is at the heart of the design of microwave power systems, and is reflected in much of the patent literature in this field (1). The number of U.S. patents related to microwave power applications has stayed at about 100/yr since the mid-1980s.

The large variety of possible applicators can be classified in various ways. Most fall into one of the following categories: single-mode cavity; folded guide; slow-wave structures; multimode cavity, batch, and conveyor fed; and radiating antenna. In the single-mode cavity, classical microwave theory and techniques about resonant cavities apply. Generally, the material being heated can be used to calculate the shift in cavity resonance frequency and change in cavity, Q, caused by the insertion of the material. This technique is better suited for measuring dielectric properties than most materials processing. It has been applied in special cases for heating (64) of thin fibers passing along the axis of a cavity of cylindrical symmetry, such that the fiber axis is aligned with the maximum E-field.

The folded guide applicator is formed by bending a rectangular guide in its E-plane to form a meander. Longitudinal slots in the broad walls in the center of the meander permit continuous passage of material in sheet form through the successive portions of guide with the sheets in the E-plane for good coupling. The meandering properties permit efficient heating without resonant cavity properties of high E-field and cavity losses. The folded guide is described in the literature (65).

Slow-wave structures (66,67) are related to the folded waveguide because the phase velocity of waves along an axis passing through all the guides, v, ie, through the slots where the sheet material is fed, is less than that in free space, c, $v < c$. A variety of planar structures, meander lines, interdigital lines, and vane lines produce a region of fringing field where E or H vary as $e^{-\gamma x}$ where γ ca β ca ω/v is the slow-wave propagation constant. In some cases, such structures are suitable for heating sheet material with more space flexibility than the folded guide. Similar slow-wave structures exist in cylindrical form with a fringing field.

It should be possible to achieve greater penetration into a load material using the fringing field of a slow-wave than can be achieved by plane-wave propagation (68). There are, however, no reports of practical application of these principles.

Perhaps the most generally used applicator is the multimode cavity, designed for both batch or conveyor fed processing. The theoretical properties of large loads have never been completely established. A few initial studies were reported (69,70). Earlier developments were marked by the use of waveguide stirrers (dump feeds) having rotating blades somewhere in the cavity acting as mode stirrers (71,72). It was later found that rotating antenna feeds achieve a more uniform heating of the load material (73,74).

A rotating turntable is a simple way to improve heating uniformity (75). A related technique is the use of a linearly moving load-bearing surface, ie, a conveyor belt. In carrying materials through a long cavity, each portion of the load undergoes, in principle, a similar history of field pattern and consequent heating. In Figure 4 a practical conveyor-fed cavity, typical of systems operating at 915 MHz, is depicted. In order to accept substantially sized boxes of material, the conveyor belt passes through long entrance and exit tunnels that are designed to minimize microwave leakage by various techniques, including liquid absorbers along the walls of the tunnels.

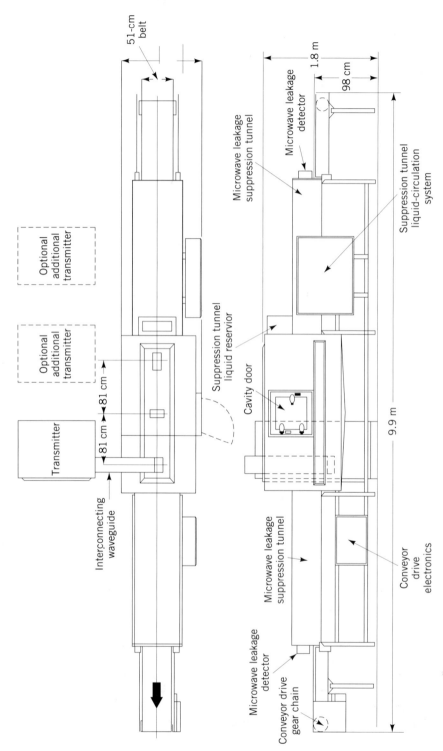

Fig. 4. Top and front views of typical high power conveyor-type industrial microwave equipment.

In some applications, such as the repair of road surfaces, the load cannot be enclosed in a cavity and radiating applicators must be used. These could be in the form of simple horns (76) or various modified waveguide apertures designed for microwave diathermy (77,78). They can be designed for efficient coupling for a closely spaced load surface as well as minimizing leakage or loss of energy to the sides.

In addition to the measurements of incident and reflected power, measurements related to field strengths, field patterns, and resulting heating may be desired. In simple cavity waveguide or slow-wave systems, perturbation techniques (59) can be used to determine the E-field strengths on the load. This is done by measuring cavity resonance or structure-phase shifts. For multimode cavities the use of conventional E- or H-probes is laborious and not generally valid because of the perturbation of the feed lines for the probes. The electric field can be measured using small-discharge lamps (79). Various techniques, none completely satisfactory, have been used to determine field patterns in a plane or volume of a multimode cavity in the absence of the load (80–82).

The most satisfactory measurement is the determination of fields or temperature distribution in the load. Historically conventional temperature measurements were made at the end of the heating (microwaves off) or during the microwave process by arranging probe leads to be perpendicular to the microwave E-field (82). Because this is in general impossible, nonperturbing probes of various degrees of effectiveness have been developed. These include fiber-optic leads with liquid crystal sensors (qv) (83), semiconductors (qv) (84), and fluorescent materials (85). In addition, small thermocouple and thermistor sensors have been used with high resistance leads (86). These techniques permit measurement of internal temperatures during microwave heating and, in conjunction with miniature nonperturbing E-field probes (87), are important tools for better design. Commercially available fiber-optic probes have become the preferred means for measuring temperature during microwave heating (see FIBER OPTICS). The most prevalent type (88) works on the basis of the temperature-dependent properties of fluorescence in a small sensor attached to the tip of a fiber. The other type (89) senses the infrared radiation at the tip of the fiber. The fluorescent sensor can also be used as an E-field probe in the presence of strong electric fields (90).

The determination of surface temperature and temperature patterns can be made noninvasively using infrared pyrometers (91) or infrared cameras (92) (see INFRARED TECHNOLOGY AND RAMAN SPECTROSCOPY). Such cameras have been bulky and expensive. A practical portable camera has become available for monitoring surface temperatures (93). An appropriately designed window, transparent to infrared radiation but reflecting microwaves, as well as appropriate optics, is needed for this measurement to be carried out during heating (see TEMPERATURE MEASUREMENT).

Economic Aspects

The cost of microwave power systems is generally considerably greater than that of conventional heating systems. For example, a 120-kW conveyor system may cost $150,000–200,000. Costs of installation, operating power, maintenance including replacement of power tubes, and financing must be weighed against potential savings in labor, space, yield, productivity, and energy. One additional benefit of increasing worth is the reduction in chemical pollution which accompanies some microwave applications.

A basic choice is that of operating frequency. In principle, operation can take place at any frequency at the cost of suppression of electromagnetic leakage to regulatory limits on RFI, eg, 25 μV/m at 304 m. This cost is avoided, however, by operating within assigned ISM bands. Minimum cost results in bands of considerable use where components are readily available. In the United States, these popular microwave bands are 915 and 2450 MHz.

The price of cooker magnetrons at ca 700 W was in the range of several hundred dollars in the 1960s (11). As of the mid-1990s, this price is <$15. Total sales of microwave ovens worldwide exceed 20×10^6/yr. The majority of homes in Japan and the United States have microwave ovens (94). European homes should follow before the year 2000 (95).

The alternative of lower cost r-f systems, ie, induction and r-f heating systems at 40 MHz and below, should be considered (96) (see also FURNACES, ELECTRIC). More extensive discussions of the economic aspects of microwave systems and payback calculations are available (97,98).

Health and Safety Factors

There are some unique safety considerations in microwave systems. Microwave voltage breakdown can occur in microwave systems and waveguides at power levels far below the theoretical values for ideal systems, ie, by a factor of at least 100 below theoretical breakdown (99,100). This is often the result of impurities or dirt particles that overheat and cause a breakdown or a spurious high quality factor, Q, resonance in the system which builds up high fields. In addition, the presence of sharp metal objects, accidental small gaps, and other situations often can induce localized arcing or corona which may or may not lead to a basic system breakdown. In this case, the plasma region of the breakdown travels down the feed waveguide toward the source and may cause failure of the tube through cracking of the output window. Therefore flammable materials should not be processed in microwave systems. Precautions can be taken, however.

In most cases, microwave, like conventional, heating produces the highest temperatures at or near the surface of load objects, except that microwave heating is more penetrating. In special situations, such as the heating of small objects of a few centimeters diameter, microwave frequencies produce rather unique heating patterns where maximum temperatures occur at the center of the object (101,102). This would be a genuine heating from the inside out, a characteristic commonly attributed to microwave heating, but not generally true. Superheating of small volumes of water can occur followed by mild explosions. In principle this phenomenon could occur for any size object if the material dielectric parameters and frequency are at certain values. No reports exist except for small water-like objects.

The most serious hazard that can occur from leakage of microwave energy is interference with other systems (103,104). This could be caused by out-of-band radiation, ie, a violation of RFI regulations, or by high power effects where the offending radiation is out of the band of the affected system, but still effectively interferes because of its intense level (105). An example is the incidental interference with cardiac pacemakers (106). This problem was partly caused by insufficient protection from RFI in the pacemaker unit and has been reduced (107) because of governmental supervision (108). Concern has arisen (109) over potential RFI from cellular phones disrupting medical electronics, both in the hospital and in portable units.

Radiated r-f energy is a hazard to systems containing flammable fuel or electroexplosive devices (EED) used for construction blasting or for military purposes (110). It is recommended that users of large amounts of microwave/r-f energy be aware of guidelines on safe distances of EEDs from sources of radiated power (111).

The hazard of exposure of personnel to microwave energy has been thoroughly reviewed (112,113). Exposure safety limits in Western countries are reasonably similar (Fig. 5), at least in the microwave frequency range. The most advanced standard includes detailed rules dependent on frequency, degree of partial body exposure, exposure time, contact conditions, etc (114). In the microwave range the recommended limits are in the range of 0.3 to 10 mW/cm^2 as averaged over any 6 or 30 minute period depending both on frequency and whether the environment is controlled or uncontrolled. Limits in eastern Europe for many years were much more stringent at microwave frequencies (115).

The nature of potential exposure hazards of low level microwave energy continues to be investigated (116–118). In the United States, leakage emission from microwave ovens is regulated to the stringent limit of 5 mW/cm^2 at 5 cm (119). There is no federal limit on emission from industrial systems but the IMPI has set a voluntary standard which specifies 10 mW/cm^2 at 5 cm (120). Emission values are equivalent to personnel exposures at several meters, well below limits that had previously prevailed in eastern Europe. This conclusion, derived for microwave ovens, should be valid for all microwave systems (121).

Leakage through door-seal areas is reduced through choke techniques (122) as well as absorbing materials (123) described in manufacturers literature on lossy gaskets. Leakage through holes in viewing screens is kept to acceptable limits by well-known limits on hole sizes (124). Generally the technology for minimizing microwave leakage is effectively utilized (125).

Uses

Food. The most successful application of microwave power is that of food processing (qv), cooking, and reheating. The consumer industry surpasses all other microwave power applications. Essentially all microwave ovens operate at 2450 MHz except for a few U.S. combination range models that operate at 915 MHz. The success of this appliance resulted from the development of low cost magnetrons producing over 700 W for oven powers of 500–800 W (Table 3).

The dielectric properties of most foods, at least near 2450 MHz, parallel those of water, the principal lossy constituent of food (Fig. 1). The dielectric properties of free water are well known (30), and presumably serve as the basis for absorption in most foods as the dipole of the water molecule interacts with the microwave electric field. By comparison, ice and water of crystallization absorb

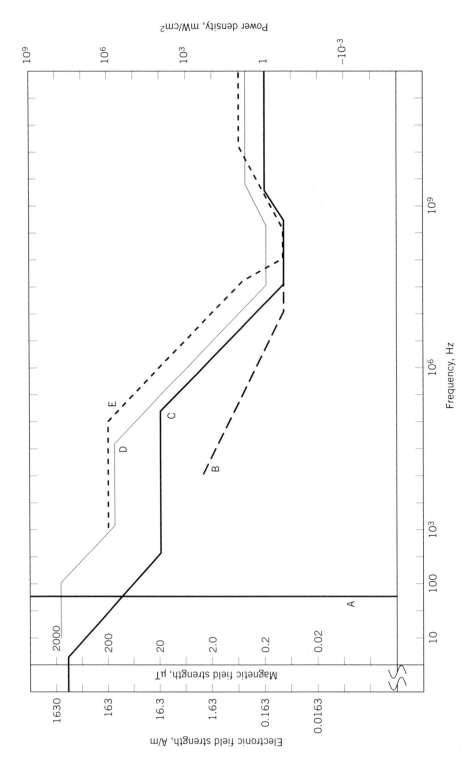

Fig. 5. Long-term magnetic field exposure limits for general public exposures in uncontrolled environment, where A is the 50/60 Hz window, according to B, IRPA; C, VDE (Germany); D, NRPB (U.K.); and E (IEEE C95.1, 1991). To convert T to G, multiply by 10^4.

very little microwave energy. Adsorbed water, however, can retain its liquid character below 0°C and absorb microwaves (126).

It had been suggested that microwave interaction with foods should be similar to interaction with ionic solutions such as sodium chloride (127,128). Thus the dielectric loss for a 0.5 N solution below ca 2450 MHz is mostly owing to ionic conductivity, whereas above 2450 MHz the loss results mostly from water dipole relaxation. At low frequencies the dielectric loss increases with temperature, whereas at higher frequencies, but still below the water-relaxation frequency, ca 20 GHz, the dielectric loss decreases with temperature. Increased ionic conductivity in salted products such as ham tend to show increasing dielectric loss with temperature (Fig. 1). Generally the other temperature dependence prevails, however, and is believed to stabilize food heating and avoid runaway heating. The suitability of the higher frequencies for stable heating was one argument put forth in an unsuccessful application for a microwave oven frequency allocation at ca 10.5 GHz (129).

Advances in the technology of microwave ovens include techniques for achieving uniform heating, eg, mode stirrers and turntables (130); temperature (131) and humidity monitoring (132); and the use of microprocessors (133) in programming cooking time, defrost, and variable power levels. In addition, a wide variety of ceramic, glass, and plastic, as well as paperboard products have been developed for utensils, shelves, and food packaging (qv) (134). A growing number of accessory products have been developed (135), eg, browning dishes or utensils, popcorn poppers, coffee-makers, etc. Combination ranges have been developed (136) for microwave electric, microwave gas, and microwave convection, this last in counter-top arrangement.

A unique and successful innovation in food packaging for microwave heating is the microwave susceptor. This is usually the incorporation of a thin vapor-deposited aluminum film in a multilayer construction on paperboard or some other packaging dielectric material (137,138). This susceptor heats up preferentially and contributes browning or enhanced heating. It is used for a variety of foods such as popcorn, pizza, etc. Marketing to consumers of such susceptor paper on rolls, much as the packaging of aluminum foil, plastic wrap, etc, has begun.

The more general food processing applications require data on dielectric and thermal properties (139). Considerable effort has been expended by food companies in the design of food for the microwave oven. These principles have been reviewed (140). The microwave oven at 2450 MHz, used for reheating, cooking, and thawing foods, may also be used for drying (qv), eg, flowers or food materials (141). Commercial microwave ovens are used extensively in restaurants and fast-food establishments.

Food Processing. Only a few industrial food-processing applications are commercialized, although many have been investigated (11,15,142). Some, such as the drying of potato chips, were too costly; others, such as freeze drying (143), encountered technical problems like vacuum breakdown. There have been reports (144) however of the successful resumption of a potato chip processing application. Microwave ovens have been used extensively for thawing. No effective solution has been found for the runaway heating caused by the great increase in dielectric loss at 0°C (144) (see Fig. 1**b**).

Reviews of food processing applications are available in general (10,11,15) as well as in specific reviews (145,146). The most successful food-processing application is that of meat tempering, particularly at 915 MHz (147), although systems exist at 2450 MHz (148) (see MEAT PRODUCTS). Microwave power in the range of 25–120 kW is applied to frozen food conveyed in tunnels so as to achieve a fairly uniform temperature in the range of -5 to $-2°C$ where the product can be cut and otherwise processed mechanically before it is returned to a freezer in its final form. It is estimated that about 200 such installations exist worldwide. Some process over 7000 metric tons of meat or fish annually. Another successful food processing application is that of bacon cooking (149). Precooked bacon is a desirable product for restaurants and institutional kitchens. Bacon fat provides a valuable by-product for the rendering industry.

Nutritional quality of food cooked or heated by microwaves generally does not differ greatly from food cooked or heated by other means (150). Scorching is minimized in microwave heating. Underheating or undercooking can be avoided only in ovens having superior mode-stirrer techniques or through occasional manual rotation.

Studies of food processing by microwave heating include blanching effects on food (151), the processing of stale bread (152), and studies of thawing (153). Most food processing applications involve heating. However, microwaves may be used for oyster shucking (154). This may involve thermal shock to the oyster, thus reducing the mechanical resistance to shucking.

Biological, Medical, and Agricultural Applications.   Diathermy at both 27.33 and 2450 MHz has been extensively used in physical therapy since the 1950s; its popularity has greatly declined in the 1990s (155). An extension of diathermy heating techniques has been investigated for application in hyperthermia as an adjunct to cancer therapy (156). The basis for preferential destruction of tumor cells was studied by hyperthermia alone, and in conjunction with ionizing radiation or chemotherapy. A variety of applicators have been designed, including multiple focused antennas, injected probe antennas, and contact applicators.

Microwaves have successfully been used for rewarming of blood for medical applications (157). Another successful application, not commercialized as of this writing, is the use of microwave heating for rapid tissue fixation (158,159). This procedure appears to reduce the time for tissue sample analysis from many hours to minutes.

Microwaves are also used for the rapid inactivation of brain enzymes in rodents (160). Microwave power at high levels of kilowatts is applied by means of a waveguide applicator to achieve a rapid sacrifice of the rodent.

Most, if not all, microwave biological effects and potential medical applications are believed to be the result of heating, ie, thermal effects. The phenomenon of microwave hearing, ie, the hearing of clicking sounds when exposed to an intense radar-like pulse, is generally believed to be a thermoelastic effect (161). Excellent reviews of the field of microwave bioeffects are available (162,163).

Reports of sterilization (qv) against bacteria by nonthermal effects have appeared, but it is generally believed that the effect is only that of heating (164). Because microwave heating often is not uniform, studies in this area can be seriously flawed by simplistic assumptions of uniform sample temperature.

The use of microwaves has been investigated to affect plant growth by irradiating seeds or to achieve insect control (165). Most useful effects result from heating, however. Studies of agricultural applications of microwaves, eg, the drying of maize (166), exist. There has also been some investigation of the use of microwaves in the processing of pharmaceuticals (qv) (167).

Chemical Applications. Chemical analysis can be conveniently accelerated by heating of samples in small pressure-tight plastic containers (168–171).

Catalysis can sometimes be improved through the use of microwaves, particularly pulsed microwaves (172). An important component of this process is believed to be an appropriate metallized combination catalyst–susceptor (173). Microwave catalysis is an active area of research (174).

Other chemical applications being studied include the use of microwaves in the petroleum (qv) industry (175), chemical synthesis (176,177), preparation of semiconductor materials (178), and the processing of polymers (179).

Plastics Fabrication and Processing. The application of microwave or r-f energy in various stages of fabrication of plastic products is widespread. Most applications operate at low frequency, ca 27 MHz. The established processes are well known (180).

Microwaves have also been studied for a variety of curing, bonding, and drying applications for plastics and composite materials (qv) (181).

Textiles. Microwave drying of textiles is under investigation, in addition to the possible uses for curing of impregnated and dyed fabrics (182). A microwave clothes dryer for consumer or commercial application is also under discussion (183). Considerable developmental work and media publicity have occurred. Problems remain, however, particularly relating to arcing and resonant heating of metal objects that may be present in a load of clothes. These problems may be alleviated by operation at 915 rather than 2450 MHz (184).

Drying of Castings and Other Products. The use of microwaves in the curing and drying of foundry cores is well established (185). The best example is the use of microwaves for drying water-based core washes at 2450 MHz with up to 150 kW. These applications have not, however, found application in manufacturing. Many similar drying applications have been examined (186,187).

Rubber Products. Microwaves are used in the preheating, curing, and drying of rubber products (188). The most successful applications are those for large products where heating time is greatly reduced because of the penetration properties of microwave energy. Batch oven units at microwave power over 30 kW at 915 MHz are used for preheating of giant tires. Developmental studies continue in this area (189).

Waste Treatment. Microwave energy has been studied for the desulfurization of coal (qv) and treatment of wastes (190). Developments in microwave incinerators for medical and radioactive wastes have occurred (191,192). Even a consumer unit for consumption of solid household waste has been proposed (193). Economic factors remain a key barrier in these developments.

Microwave Plasmas. Classical applications of microwave plasmas have been in the areas of torches (194), deposition of organic films (see THIN FILMS) (195), fusion energy (196), and plasma chemistry, eg, synthesis or decomposition in nitrogen discharges (197). These developments continue (198), but emphasis has shifted to other areas, in particular the growing of diamond films. Beginning with some early work in the former USSR (199), the investigation of chemical vapor deposition (CVD) of diamond under the assistance of a microwave excited plasma (199,200) has continued. Commercially available equipment at 2450 MHz (201) is used by a majority of the investigators. The results are impressive and likely to find application for hardening of machine tools, reducing abrasion on the surface of missile domes, and possibly the refinement of diamond as a semiconductor material. In addition diamond is a useful heat-sink material (see CARBON, DIAMOND).

Another development is microwave excited discharges for lamps. The most impressive result is the production of about 500,000 lumens of visible light by the application of 3.6 kW of 2450 MHz energy to a quartz lamp containing some gases and sulfur pellets (202). These lamps are undergoing testing in U.S. government buildings and show promise because of efficiency and pleasing spectral quality of light.

Ceramics Processing. One of the largest application areas for microwaves is in ceramics (qv) processing. Microwaves are used for sintering, drying, and enhancement of certain materials properties (203–207). Most of this work has been done at 2450 MHz, often in ordinary microwave ovens outfitted with an appropriate casket for the ceramic sample. There is some controversy over the amount of specific improvement offered by microwave heating over ordinary means of heating. Large amount of powers at many frequencies have been employed. Advantages of broad-band sources such as TWTs have been explored for superior uniformity of heating of large ceramic loads (208).

Other Applications. The use of lower frequency energy (rf) has been explored for *in situ* heating of oil shale (qv) (209). Other power applications address the potential application to electrified roadways (210) and microwave power transmission over large distances on land (211) as well as from space to earth (212). In this latter application the goal would be the generation in space of 10 GW of microwave power and its transmission to earth supplying enough electricity for a city (213).

The field of microwave technology is expected to increase as better and cheaper microwave systems are developed. In particular, uses for 5800 and 2450 MHz and the millimeter wave frequencies await the development of inexpensive efficient sources of power at those frequencies.

BIBLIOGRAPHY

"Microwave Technology" in *ECT* 3rd ed., Vol. 15, pp. 494–522, by J. M. Osepchuk, Raytheon Co.

1. G. M. Wilkening, in *The Industrial Environment—Its Evaluation and Control*, U.S. Dept. of HEW, National Institute for Occupational Safety and Health, Washington, D.C., 1973, Chapt. 28.
2. T. Lauriston, *Legislative History of Radiation Control for Health and Safety Act of 1968*, testimony before U.S. Congress, DHEW Publication (FDA) 75-8033, Washington, D.C., May 1975, pp. 157, 487, and 1025.
3. H. E. M. Barlow, *Microwaves and Waveguides*, Constable, London, 1948.
4. W. Cheung, *Microwaves Made Simple, Principles and Applications*, Adtech Books Co. Ltd., London, 1985.

5. J. M. Osepchuk, *Bull. N.Y. Acad. Med.* **55**, 976 (Dec. 1979).
6. E. W. Herold, *IEEE Spectrum* **2**, 50 (Jan. 1965).
7. E. Okress and co-workers, *IEEE Spectrum* **1**, 76 (Oct. 1964).
8. P. L. Kapitsa, *Uspekhi fiz. Nauk* **78**, 181 (1962).
9. E. C. Okress, *Microwave Power Engineering*, Vols. 1 and 2, Academic Press, New York, 1968.
10. M. Stuchly and S. Stuchly, *IEE Proc. A* **130**, 467 (Nov. 1983).
11. J. M. Osepchuk, *IEEE Trans. Mic. Theory Tech.* **MTT-32**, 1200 (Sept. 1984).
12. K. Oguro and Y. Kase, *J. Mic. Power* **13**, 115 (1978).
13. M. Kachmar, *Microwaves RF* **41** (Sept. 1992), **39** (Oct. 1992).
14. R. A. Metaxas and R. Meredith, *Industrial Microwave Heating*, Peter Peregrinus Ltd., London, 1983.
15. J. Thuéry, *Microwaves: Industrial, Scientific and Medical Applications*, Artech House, Boston, Mass., 1992.
16. *Final Acts of the World Administrative Radio Conference (1979)*, Vols. I and II, International Telecommunications Union, Geneva, Switzerland, Dec. 1979.
17. "Industrial, Scientific and Medical Equipment," *Rules and Regulations*, Federal Communications Commission, Vol. II, Part 18, subpart H, Washington, D.C., Oct. 1992, pp. 595–602.
18. *Limits and Methods of Measurement of Radio Interference Characteristics of Industrial, Scientific, and Medical (ISM) Radio Frequency Equipment*, International Electrotechnical Commission, Special Committee on Radio Interference (CISPR), Publication 11, rev. 1990, Geneva, Switzerland.
19. P. E. Gawthrop and co-workers, *Radio Spectrum Measurements of Individual Microwave Ovens*, Vols. 1 and 2, NTIA Reports 94-303-1 and 94-303-2, U.S. Dept. of Commerce, Washington, D.C., Mar. 1994.
20. R. W. P. King, *Electromagnetic Engineering*, Vol. 1, McGraw-Hill Book Co., Inc., New York, 1948.
21. J. A. Stratton, *Electromagnetic Theory*, McGraw-Hill Book Co., Inc., New York, 1941.
22. F. N. H. Robinson, *Macroscopic Electromagnetism*, Pergamon Press, New York, 1973.
23. H. Fröhlich, *Theory of Dielectrics*, Oxford Press, Oxford, U.K., 1958.
24. M. N. Afsar and co-workers, *Proc. IEEE* **74**, 183 (Jan. 1986).
25. C. H. Townes and A. L. Schawlow, *Microwave Spectroscopy*, McGraw-Hill Book Co., Inc., New York, 1955.
26. G. Birnbaum and J. Franeau, *J. Appl. Phys.* **20**, 817 (1949).
27. C. H. Durney and co-workers, *Radiofrequency Radiation Dosimetry Handbook*, 4th ed., Report SAM-TR-85-73, USAF, Brooks Air Force Base, Tex., Oct. 1986.
28. R. F. Harrington and co-workers, *Computational Methods in Electromagnetics*, SCEEE Press, St. Cloud, Fla., 1981.
29. E. K. Miller, *IEEE Trans.* **AP-36**(9), 1281 (Sept. 1988).
30. A. R. Von Hippel, ed., *Dielectric Materials and Applications*, MIT Press, Cambridge, Mass., 1954.
31. W. R. Tinga and S. O. Nelson, *J. Microwave Power.* **8**(1), 23 (1973).
32. N. E. Bengtsson and P. O. Risman, *J. Microwave Power* **6**(2), 107 (1971).
33. E. H. Grant, R. J. Sheppard, and G. P. South, *Dielectric Behavior of Biological Molecules in Solutions*, Clarendon Press, Oxford, U.K., 1978.
34. R. M. White, *J. Appl. Phys.* **34**, 3559 (1963).
35. J. Lin, *Proc. IEEE* **68**(1), 69 (1980).
36. A. W. Guy, C. K. Chou, J. C. Lin, and D. Christensen, *Ann. N. Y. Acad. Sci.* **247**, 194 (1975).
37. H. A. Pohl, "AC Field Effects of and by Living Cells," in A. Chiabrera, C. Nicolini, and H. P. Schwan, eds., *Interactions Between Electromagnetic Fields and Cells*, Plenum Press, New York, 1985.

38. H. P. Schwan, in Ref. 37.
39. G. Franschetti and C. H. Papas, *Appl. Phys.* **23**, 153 (1980).
40. P. C. Lauterbur, *IEEE Trans. Nuc. Sci.* **N5-26**, 2808 (Apr. 1979).
41. J. Helszajn, *Principles of Microwave Ferrite Engineering*, Wiley-Interscience, New York, 1969.
42. E. A. Maguire and D. W. Readey, *J. Am. Ceram. Soc.* **59**, 434 (1976).
43. A. J. Baden Fuller, *Ferrites at Microwave Frequencies*, Peter Peregrinus Ltd., U.K., 1987.
44. S. C. Brown, *Basic Data of Plasma Physics*, MIT Press, Cambridge, Mass., 1966.
45. A. F. Harvey, *Microwave Engineering*, Academic Press, Inc., New York, 1963, p. 897.
46. W. Von Aulock and J. H. Rowen, *Bell System Tech. J.* **36**, 427 (1957).
47. J. C. Bennett and co-workers, *IEE Proc.-Radar, Sonar Navig.* **141**(6), 337 (Dec. 1994).
48. J. M. Osepchuk, *Microwave J.* **21**, 51 (Nov. 1978).
49. J. T. Coleman, *Microwave Devices*, Reston Publishing Co., Reston, Va., 1982.
50. V. Granatstein and I. Alexeff, eds., *High Power Microwave Sources*, Artech House, Boston, Mass., 1991.
51. A. S. Gilmour, Jr., *Microwave Tubes*, Artech House, Boston, Mass., 1986.
52. E. L. Holzman, *Solid-State Microwave Power Oscillator Design*, Artech House, Boston, Mass., 1992.
53. C. Buffler, *Microwave Cooking and Processing: Engineering Fundamentals for the Food Scientist*, Van Nostrand Reinhold, New York, 1992.
54. R. J. Lauf and co-workers, *Microwave J.* **36**, 24 (Nov. 1993).
55. E. D. Moloney and G. Faillon, *J. Microwave Power* **9**, 231 (Sept. 1974).
56. R. E. Davis and T. Maeda, *Appl. Manu.* **13**, 87 (Feb. 1979).
57. I. Oguro, *J. Microwave Power* **13**(1), 27 (Mar. 1978).
58. C. Forterre and co-workers, *J. Microwave Power* **13**(1), 65 (Mar. 1978).
59. F. Gardiol, *Introduction to Microwaves*, Adtech Book Co., Ltd., London, 1984.
60. A. E. Bailey, *Microwave Measurements*, Peter Peregrinus, Ltd., London, 1985.
61. A. F. Harvey, *Microwave Engineering*, Academic Press, Inc., New York, 1963, Chapt. 3.
62. *Fed. Reg.* **59**(43), 10493 (Mar. 4, 1994).
63. R. D. Petkov and L. Hobson, *Int. J. Electronics* **77**(5), 793 (1994).
64. A. L. Van Koughnett, *J. Microwave Power* **7**(1), 17 (1972).
65. *Applications Industrielles des Microondes*, Thomson-CSF, Division Tubes Electroniques, Paris, June 1980.
66. D. Dunn, *J. Microwave Power* **2**(1), 7 (1967).
67. A. L. Van Koughnett and co-workers, *J. Microwave Power* **10**(4), 451 (1975).
68. L. S. Taylor, *IEEE Trans.* **AP-32**(10), 1138 (Oct. 1984).
69. T. G. Mihran, *IEEE Trans.* **MTT-26**, 381 (June 1978).
70. M. Watanabe and co-workers, *J. Microwave Power* **13**(2), 173 (June 1978).
71. D. A. Copson, *Microwave Heating*, Avi Publishing Co., Westport, Conn., 1975, Chapt. 11.
72. P. Bhartia and co-workers, *J. Microwave Power* **6**, 221 (1971).
73. W. W. Teich and K. Dudley, *Proceedings of the 1980 Symposium on Microwave Power*, IMPI, Manassas, Va., pp. 122–124.
74. J. Simpson, *Microwave J.* **23**, 47 (Jan. 1980).
75. D. A. Copson and R. V. Decareau, "Ovens" in E. C. Okress, ed., *Microwave Power Engineering*, Vol. 2, Academic Press, Inc., New York, 1968.
76. S. C. Kashyap and W. Wyslouzil, *J. Microwave Power* **12**(3), 223 (1977).
77. G. Kantor and T. C. Cetas, *Radio Science* **12**(6s), 111 (Nov.–Dec. 1977).
78. A. W. Guy and co-workers, *IEEE Trans.* **MTT-26**, 550 (Aug. 1978).
79. J. K. White, *J. Microwave Power* **5**(2), 145 (1970).

80. D. A. Copson, *Microwave Heating*, 2nd ed., Avi Publishing Co., Westport, Conn., 1975, Chapt. 18.
81. A. Hiratsuka and co-workers, *J. Microwave Power* **13**(2), 189 (June 1978).
82. X. Jia, *J. Microwave Power* **28**(1), 25 (1993).
83. T. C. Rozzell and co-workers, *J. Microwave Power* **9**, 241 (Sept. 1974).
84. D. A. Christensen, *J. Biomed. Eng.* **1**, 541 (1977).
85. K. A. Wickersheim and R. B. Alves, *Ind. Res./Dev.* **21**, 82 (Dec. 1979).
86. R. G. Olsen and R. R. Bowman, *Bioelectromagnetics* **10**, 209 (1989).
87. H. Bassen and co-workers, *Radio Science* **12**(6), 15 (Nov.–Dec. 1977).
88. K. A. Wickersheim and M. H. Sun, *J. Microwave Power* **22**(2), 85 (1987).
89. *Fiberoptic Sensor Probes*, Metricor Co., Eoodinville, Wash., 1994.
90. K. A. Wickersheim and co-workers, *J. Microwave Power* **25**(3), 141 (1990).
91. C. W. Price, *IEEE Trans.* **IA-15**(3), 319 (May–June 1979).
92. T. Ohlsson and P. O. Risman, *J. Microwave Power* **13**, 303 (Dec. 1978).
93. *Radiance 1*, Amber Co., Goleta, Calif., 1994.
94. "A Portrait of the U.S. Appliance Industry," *Appliance* (Sept. 1994).
95. G. Andrews, *Microwave World* **10**(1), 5 (1989).
96. P. J. Hulls, *J. Microwave Power* **17**(1), 29 (Mar. 1982).
97. J. A. Jolly, *J. Microwave Power* **11**, 233 (1976).
98. R. Edgar and R. Snider, in D. E. Clark and co-workers, eds., *Microwaves: Theory and Application in Materials Processing*, American Ceramics Society, Westerville, Ohio, 1991.
99. W. Beust and W. L. Ford, *Microwave J.*, 91 (Oct. 1961).
100. J. Benford and J. Swegle, *High Power Microwaves*, Artech House, Boston, Mass., 1987.
101. H. Kritikos and co-workers, *J. Microwave Power* **16**(3–4) (1981).
102. R. E. Apfel and R. L. Day, *Nature* **321**, 657 (June 12, 1986).
103. C. R. Paul, *Introduction to Electromagnetic Compatibility*, John Wiley & Sons, Inc., New York, 1992.
104. L. O. Hoeft and co-workers, *1994 IEEE Int. Sympos. Electr. Comp. Record.*, 264 (Aug. 1994).
105. W. E. Ours, *Proceedings of the 1971 IEEE EMC Symposium*, IEEE Press, New York, pp. 4–7.
106. J. C. Mitchell, in J. M. Osepchuk, ed., *Biological Effects of Electromagnetic Radiation*, IEEE Press, New York, 1983, pp. 557–584.
107. C. H. Swanson, *Statement for the Health Industry Manufacturers Association*, before the House Government Operations Subcommittee in Information, Justice, Transportation, and Agriculture, Washington, D.C., Oct. 5, 1994.
108. D. P. Sager, *Heart Lung* **16**(2), 211 (Mar. 1987).
109. T. Knudson and W. M. Bukeley, *Wall St. J.*, 1 (June 15, 1994).
110. S. S. J. Robertson and co-workers, *IEE Proc. Pt. A* **128**(9), 607 (Dec. 1981).
111. *Safety Guide for the Prevention of Radio-Frequency Hazards in the Use of Electric Blasting Caps*, ANSI C95.4 Guide, New York, 1971.
112. *IEEE Committee on Man and Radiation (COMAR)*, IEEE-USA, Washington, D.C., 1990.
113. WHO/IRPA Task Group on Electromagnetic Fields, *Environmental Health Criteria*, World Health Organization, Geneva, 1993, p. 137.
114. IEEE Standards Coordinating Committee 28, *Safety Levels with Respect to Human Exposure to Radio Frequency Electromagnetic Fields, 3 kHz to 300 GHz*, ANSI/IEEE C95.1-1992, IEEE Standards Dept., Piscataway, N.J., 1992.
115. J. M. Osepchuk, "Panel Discussion on Standards," in J. C. Lin, ed., *Electromagnetic Interaction with Biological Systems*, Plenum Publishing Co., New York, 1989.

116. J. M. Osepchuk, "Some Misconceptions about Electromagnetic Fields and Their Effects and Hazards," in O. P. Gandhi, ed., *Biological Effects and Medical Applications of Electromagnetic Energy*, Prentice-Hall, Englewood Cliffs, N.J., 1990.

117. Oak Ridge Associated Universities Panel for the Committee on Interagency Radiation Research and Policy Coordination, *Health Effects of Low Frequency Electric and Magnetic Fields*, Washington, D.C., June 1992.

118. *Electromagnetic Fields and the Risk of Cancer, Report of an Advisory Group on Non-Ionizing Radiation*, documents of the National Radiological Protection Board, U.K., Vol. 3, Mar. 1992.

119. *Regulations for Administration and Enforcement of the Radiation Control for Health and Safety Act of 1968*, Part 1030, U.S. Dept. of HHS/FDA, DHHS Publication No. FDA 76-8035, Washington, D.C., Jan. 1976.

120. *IMPI Performance Standard on Leakage from Industrial Microwave Systems*, IMPI Publication IS-1, IMPI, Manassas, Va., Aug. 1973.

121. J. M. Osepchuk and co-workers, "Computation of Personnel Exposure in Microwave Leakage Fields and Comparison with Personnel Exposure Standards," *Digest of the 1973 Microwave Power Symposium*, IMPI, Manassas, Va., 1973.

122. J. M. Osepchuk and co-workers, *J. Microwave Power* **8**, 295 (1973).

123. Technical data, Robar Industries, Conshohocken, Pa.; Penton Publishing Co., Hasbrouck Heights, N.J., Horizon House, Dedham, Mass.

124. T. Y. Otoshi, *IEEE Trans.* **MTT-20**, 235 (Mar. 1972).

125. J. M. Osepchuk, *J. Microwave Power* **13**(1), 13 (Mar. 1978).

126. H. P. Schwan, *Ann. N.Y. Acad, Sci.* **125**, 344 (Oct. 1965).

127. S. A. Goldblith, ed., *Freeze Drying Advanced Food Technology*, Academic Press, London, 1975.

128. B. D. Roebuck, S. A. Goldblith, and W. Westphal, *J. Food Sci.* **37**(2), 199 (1972).

129. *In the Matter of Allocation of the 10,500 to 10,700 MHz Band for the Use of Microwave Ovens*, Litton Microwave Cooking Products, petition to FCC for Rule Making, Washington, D.C., Nov. 1976.

130. S. M. Bakanowski and L. H. Belden, *Proceedings of the 1979 Microwave Power Symposium*, IMPI, Manassas, Va., p. 14.

131. K. Sato and co-workers, in Ref. 130, pp. 14–16.

132. S. Nagamato and co-workers, in Ref. 130, pp. 17–19.

133. D. Winstead, *J. Microwave Power* **13**(1), 7 (Mar. 1978).

134. J. A. McCormick, *Microwave World* **13**(1), 17 (Summer 1992).

135. R. F. Bowen and G. Freedman, in Ref. 73, pp. 37–39.

136. *Microwave World* **10**(1), 14–15 (1989).

137. C. Turpin, *Microwave World*, **10**(6) (1989).

138. J. A. McCormick, *1994 Microwave Power Symposium Digest*, IMPI, Manassas, Va.

139. R. W. Dickerson, Jr., *The Freezing of Foods*, 4th ed., Vol. 2, Avi Publishing Co., Westport, Conn., 1967, Chapt. 2.

140. G. Cooper, *Food Manuf.* **67**(2), 27 (1992).

141. D. W. Hansen, *Microwave World* **15**(1), 8 (1994).

142. N. E. Bengtsson and T. Ohlsson, *Proc. IEEE* **62**, 44 (Jan. 1974).

143. J. W. Gould and E. M. Kenyon, *J. Microwave Power* **6**(2), 151 (1971).

144. G. Collins, *New York Times*, D3 (Jan. 18, 1994).

145. R. V. Decareau, *Microwaves in the Food Processing Industry*, Academic Press, Inc., New York, 1985.

146. R. E. Mudgett, *Food Technol.* **40**, 84 (1986).

147. A. F. Bezanson, *Food Technol.* **30**, 34 (Dec. 1976).

148. M. N. Meisel, *Microwave Energy Appl. Newsletter* **5**(3), 3 (1972).

149. R. Edgar, in D. E. Clark and co-workers, eds., *Microwaves: Theory and Application in Materials Processing*, Vol. II, American Ceramics Society, Westerville, Ohio, 1993.

150. J. Augustin and co-workers, *J. Food Sci.* **45**, 814 (1980).
151. M. S. Brewer and co-workers, *J. Food Quality* **17**, 294 (1994).
152. H. Yamanchi, *J. Che. Eng. of Japan* **26**(6), 749 (Dec. 1993).
153. X. Zeng, *J. Heat Transfer* **116**(2), 446 (May 1994).
154. J. M. Mendelsohn and co-workers, *Fishery Ind. Res.* **4**, 241 (1969).
155. A. W. Guy and C. K. Chou, in E. R. Adair, ed., *Microwaves and Thermoregulation*, Academic Press, Inc., New York, 1983, pp. 57–93.
156. F. Sterzer and co-workers, *Microwave J.* **29**(7), 147 (1986).
157. K. Carr, *Microwave J.* **37**(7), 24 (July 1994).
158. G. R. Login and A. M. Dvorak, *Am. J. Pathol.* **120**, 230 (1985).
159. M. E. Boon, *Microwave Cookbook of Pathology: the Art of Microscopic Visualization*, Coulomb Press, Leyden, the Netherlands, 1988.
160. D. Scheinder and co-workers, *J. Neurochem.* **38**, 749 (1982).
161. J. C. Lin, *Proc. IEEE* **68**, 67 (Jan. 1980).
162. *Proc. IEEE* **68**(1) (Jan. 1980).
163. S. M. Michaelson and J. C. Lin, *Biological Effects and Health Implications of Radiofrequency Radiation*, Plenum Press, New York, 1987.
164. R. A. Heddleson and S. Doores, *J. Food Protection* **57**(11), 1025 (Nov. 1994).
165. K. J. Lessman and A. W. Kirleis, *Crop Sci.* **19**(2), 189 (1979).
166. V. S. Shirhan, *J. Agric. Eng. Res.* **57**(3), 199 (Mar. 1994).
167. N. Elander and co-workers, *J. Label Comp. Radiopharmaceuticals* **34**(10), 949 (Oct. 1994).
168. W. T. Westbrook and R. H. Jefferson, *J. Microwave Power* **21**, 25 (1986).
169. H. M. Kingston and L. B. Jassie, eds., *Introduction to Microwave Sample Preparation*, American Chemical Society, Washington, D.C. 1988.
170. J. Emsley, *New Scientist*, 55, (Nov. 12, 1988).
171. M. Feinberg, *Chemometrics Intell. Lab. Syst.* **22**(1), 37 (Jan. 1994).
172. J. Wan and co-workers, *J. Microwave Power* **25**, 32 (1990).
173. P. Marquardt and co-workers, *J. Physics Lett.* **114**, 39 (1986).
174. C. Bond and co-workers, *Catal. Today* **17**, 427 (1993).
175. M. S. Ioffe, *Petroleum Chemistry* (transl. *Neftekhimiya*) **33**(6), 594 (1993).
176. D. M. Minger, *Chem. Indust.* (15), 596 (Aug. 1994).
177. C. Landry and A. R. Barron, *Science* **260**, 1653 (June 11, 1993).
178. D. A. Lewis and co-workers, *Polymer Reprints* **29**(1) (1988).
179. T. Weidman and A. Joshi, *Appl. Phys. Lett.* **62**(4), 372 (Jan. 25, 1993).
180. A. F. Readdy, Jr., *Plastics Fabrication by Ultraviolet, Infrared, Induction, Dielectric and Microwave Radiation Methods*, Plastec Report R43, Picatinny Arsenal, Dover, N.J., Apr. 1976.
181. K. T. Ohnier and co-workers, *Polymer Compos.* **15**(3), 231 (June 1994).
182. A. Bugos and co-workers, *J. Microwave Power* **25**(4), 230 (1990).
183. M. Hamid, *J. Microwave Power* **26**(2), 107 (1991).
184. *The Magnetron*, Astex/Gerling Laboratories, Inc., Modesto, Calif., July 1994.
185. J. Crowley and J. Apelbaum, *Electronic Progress*, Vol. XVIII, Raytheon Co., Lexington, Mass., 1976, p. 13.
186. T. Nishio, *J. Mat. Sci.* **29**(13), 3408 (July 1, 1994).
187. B. Al-Duri and S. McIntyre, *J. Food Eng.* **15**, 139 (1992).
188. R. Edgar, *J. Microwave Power* **21**(3), 194 (1986).
189. P. Luypaert and P. Reusens, *J. Microwave Power* **21**(2), 75 (1986).
190. K. P. Singh, *Res. Indust.* **39**(3), 198 (Sept. 1994).
191. L. Dauerman and G. Windgasse, *TIZ Int.* **114**, 503 (1990).
192. C. Shibata and H. Tamai, *Proceedings of the International Conference on High Frequency/Microwave Processing and Heating*, Tokyo, Japan, 1989, pp. 5.2.1–5.2.5.

193. J. Suzuki and co-workers, *J. Microwave Power* **25**(3), 168 (1990).

194. M. Moisan and co-workers, *J. Microwave Power* **14**(1), 57 (Mar. 1979).

195. J. Wightman, *Proc. IEEE* **62**(4), 4 (Jan. 1974).

196. D. B. Batchelor, *INIS Atomindex* **10**(6), Abstr. No. 434525 (1979).

197. S. Susuki and co-workers, *Nippon Kagaku Kaishi* (7), 1037 (1978).

198. R. Claude and co-workers, *Appl. Phys. Lett.* **50**(25), 1797 (June 22, 1987).

199. M. Simpson, *New Scientist*, 50, (Mar. 10, 1988).

200. R. Koba, in R. F. Davis, ed., *Diamond Films and Coatings*, Noyes Publisher, Park Ridge, N.J., 1993.

201. M. Maeda, *Diamond Rel. Mater.* **3**(7), 1072 (May 1994).

202. J. T. Dolan, M. G. Urey, and C. H. Wood, in L. Bartha and F. J. Kedves, eds., *The Sixth International Symposium on the Science and Technology of Light Sources* **6**, 301 (1992).

203. W. H. Sutton and co-workers, eds., *Microwave Processing of Materials; Materials Research Society Proceedings*, Vol. 124, Pittsburgh, Pa., 1988.

204. J. McKittrick, *J. Mat. Sci.* **29**(8), 2119 (Apr. 15, 1994).

205. H. D. Kimrey and co-workers, *Mat. Res. Soc. Proc.* **189** (1990).

206. C. E. Holcombe and N. L. Dykes, *J. Mat. Sci.* **26**, 3730 (1991).

207. D. E. Clark and co-workers, eds., *Microwaves: Theory and Applications in Materials Processing* Vol. II, American Ceramics Society, Westerville, Ohio, 1993.

208. R. J. Lauf, in Ref. 205.

209. J. P. Casey and R. Bansai, *J. Appl. Phys.* **61**(12), 5455 (June 15, 1987).

210. *Highway Electrification and Automation Development Program*, TR-093-003, H. R. Ross Industries, Inc., Los Angeles, Feb. 10, 1993.

211. W. C. Brown, *IEEE Trans.* **MTT-32**(9), 1230 (Sept. 1984).

212. P. E. Glaser, *Microwave J.* **29**(12), 44 (Dec. 1986).

213. P. E. Glaser, *Solar-Power Satellites: The Emerging Energy Option*, Ellis Howard, New York, 1993.

JOHN OSEPCHUK
Raytheon Company

MILK AND MILK PRODUCTS

Milk has been a source for food for humans since the beginning of recorded history. Although the use of fresh milk has increased with economic development, the majority of consumption occurs after milk has been heated, processed, or made into butter. The milk industry became a commercial enterprise when methods for preservation of fluid milk were introduced. The successful evolution of the dairy industry from small to large units of production, ie, the farm to the dairy plant, depended on sanitation of animals, products, and equipment; cooling facilities; health standards for animals and workers; transportation systems; construction materials for process machinery and product containers; pasteurization and sterilization methods; containers for distribution; and refrigeration for products in stores and homes.

Composition and Properties

Milk consists of 85–89% water and 11–15% total solids (Table 1); the latter comprises solids-not-fat (SNF) and fat. Milk having a higher fat content also has higher SNF, with an increase of 0.4% SNF for each 1% fat increase. The principal components of SNF are protein, lactose, and minerals (ash). The fat content and other constituents of the milk vary with the animal species, and the composition of milk varies with feed, stage of lactation, health of the animal, location of withdrawal from the udder, and seasonal and environmental conditions. The nonfat solids, fat solids, and moisture relationships are well established and can be used as a basis for detecting adulteration with water (qv). Physical properties of milk are given in Table 2.

Nutritional Content. To assure that milk provides the necessary nutrients it may be fortified with vitamins. Vitamin D [1406-16-2] milk has been sold since the 1920s when it was fortified with vitamin D by irradiation or by feeding irradiated yeast to cows. Ergosterol [57-87-4] is converted to vitamin D by ultraviolet irradiation. Presently, vitamin D is added directly to the milk to provide 400 *U.S. Pharmacopoeia* (USP) units/L. Vitamin A [68-26-8] may be added to low fat skimmed milk to provide 1000 retinol equivalents (RE) per liter. Multivitamin, mineral fortified milk provides the recommended daily requirements. The vitamin content of milk from various mammals is given in Table 3 (see VITAMINS). The daily nutritional needs for an adult are given in Table 4.

Fat. Milk fat is a mixture of triglycerides and diglycerides (see FATS AND FATTY OILS). The triglycerides are short-chain, C_{24}–C_{46}; medium-chain, C_{34}–C_{54}; and long-chain, C_{40}–C_{60}. Milk fat contains more fatty acids than those in vegetables. In addition to being classified according to the number of carbon atoms, fatty acids in milk may be classified as saturated or unsaturated and soluble or insoluble. Fat carries numerous lipids (Table 5) and vitamins A, D, E,

Table 1. Constituents of Milk from Various Mammals, Average, wt %

Species	Water	Fat	Protein	Lactose	Ash	Nonfat solids	Total solids
human	87.4	3.75	1.63	6.98	0.21	8.82	12.57
cows							
Holstein	88.1	3.44	3.11	4.61	0.71	8.43	11.87
Ayrshire	87.4	3.93	3.47	4.48	0.73	8.68	12.61
Brown Swiss	87.3	3.97	3.37	4.63	0.72	8.72	12.69
Guernsey	86.4	4.5	3.6	4.79	0.75	9.14	13.64
Jersey	85.6	5.15	3.7	4.75	0.74	9.19	14.34
goat	87.0	4.25	3.52	4.27	0.86	8.65	12.90
buffalo (India)	82.76	7.38	3.6	5.48	0.78	9.86	17.24
camel	87.61	5.38	2.98	3.26	0.70	6.94	12.32
mare	89.04	1.59	2.69	6.14	0.51	9.34	10.93
ass	89.03	2.53	2.01	6.07	0.41	8.49	11.02
reindeer	63.3	22.46	10.3	2.50	1.44	14.24	36.70

Table 2. Physical Properties of Milk

Property	Value
density at 20°C with 3–5% fat, average, g/cm^3	1.032
weight at 20°C, kg/L[a]	1.03
milk serum at 20°C, 0.025% fat	
density, g/cm^3	1.035
weight, kg/L[a]	1.03
freezing point, °C	−0.540
boiling point, °C	100.17
maximum density at °C	−5.2
electrical conductivity, S($= \Omega^{-1}$)	$(45-48) \times 10^{-8}$
specific heat at 15°C, kJ/(kg·K)[b]	
skim	3.94
whole	3.92
40% cream	3.22
fat	1.95
relative volumes	
4% milk at 20°C = 1, volume at 25°C	1.002
40% cream 20°C = 1.0010, volume at 25°C	1.0065
viscosity at 20°C, mPa·s($=$cP)	
skim	1.5
whole	2.0
whey	1.2
surface tension of whole milk at 20°C, mN/m($=$dyn/cm)	50
acidity, pH	6.3–6.9
titratable acid, %	0.12–0.15
refractive index at 20°C	1.3440–1.3485

[a]To convert kg/L to lb/gal, multiply by 8.34.
[b]To convert kJ/(kg·K) to Btu/(lb·°F), divide by 4.183.

and K, which are fat soluble. Tables 6–8 give fatty, saturated, and unsaturated acid contents of milk fat.

Processing

The processing operations for fluid or manufactured milk products include cooling, centrifugal sediment removal and cream (a mixture of fat and milk serum)separation, standardization, homogenization, pasteurization or steriliza-tion, and packaging, handling, and storing.

    Cooling. After removal from the cow by a mechanical milking machine, (at $\sim$ 34°C), the milk is rapidly cooled to $\leq$ 4.4°C to maintain quality. At this low temperature, enzyme activity and microorganism growth are minimized. Commercial dairy production operations usually consist of a milking machine, a pipeline to convey the milk directly to the tank, and a refrigerated bulk milk tank in which the milk is cooled and stored for later pickup. Rancidity is avoided by preventing air from passing through the warm milk, via air leaks and long risers in the pipeline. The pipelines, made of glass or stainless steel, are usually cleaned by a cleaning-in-place (CIP) process. Bulk milk is pumped from the refrigerated bulk milk tank to a tanker and transported to a processing plant.

Table 3. Vitamin Content of Milk from Various Mammals, mg/L[a]

| | Vitamins | | | | | | | | | | |
Species	A, RE[b]	B_6	B_{12}	C	Thiamine	Riboflavin	Nicotinic acid	Pantothenic acid	Biotin	Folic acid
cow	312	0.48	0.0056	16	0.42	1.57	0.85	3.50	0.035	0.0023
goat	415	0.07	0.0006	15	0.40	1.84	1.87	3.44	0.039	0.0024
sheep	292		0.0064	43	0.69	3.82	4.27	3.64	0.093	0.0024
horse	160	0.21	0.0012	100	0.30	0.33	0.58	3.02	0.022	0.0012
human	380	0.10	0.0003	43	0.16	0.36	1.47	1.84	0.008	0.0020
pig	207	0.40	0.0016	140	0.70	2.21	8.35	5.28	0.014	0.0020
whale	1439	1.10	0.0085	70	1.16	0.96	20.40	13.10	0.050	0.0039

[a]Ref. 1.
[b]Ref. 2. Vitamin A is reported as retinol [68-26-8] equivalents/L. RE = 1 μg of all *trans*-retinol, 6 μg of all *trans*-β-carotene, and 12 μg of other provitamin A cartenoids, with older definitions giving 3.33 IU vitamin A from retinol and 10 IU vitamin A activity from β-carotene.

Table 4. Nutritional Content (for Adults) of Cow Milk[a]

Nutrient	Recommended daily allowance	Supplied by 1 L, %
energy, kJ[b]	11,720	96
protein, g	56	49
calcium, g	0.8	155
phosphorus, g	0.8	115
iron, mg	10	4.5
vitamin A, RE[c]	1,000	31
thiamine, mg	1.4	30
riboflavin, mg	1.7	92
niacin, mg	18.5	5
ascorbic acid, mg	60	27
vitamin D, IU	200	200[d]

[a]Ref. 2.
[b]To convert kJ to kcal, divide by 4.184; 1 food Calorie = 1 kcal.
[c]RE = retinol equivalent, the standard for vitamin A; 1 RE = 1 μm of all *trans*-retinol.
[d]Fortified milk.

Table 5. Composition of Lipids in Cow Milk[a]

Class of lipid	Range of occurrence
triglycerides of fatty acids, %	97.0–98.0
diglycerides, %	0.25–0.48
monoglycerides, %	0.016–0.038
keto acid glycerides, %	0.85–1.28
aldehydrogenic glycerides, %	0.011–0.015
glyceryl ethers, %	0.011–0.023
free fatty acids, %	0.10–0.44
phospholipids, %	0.2–1.0
cerebrosides, %	0.013–0.066
sterols, %	0.22–0.41
free neutral carbonyls, ppm	0.1–0.8
squalene, ppm	70
carotenoids, ppm	7–9
vitamin A, ppm	6–9
vitamin D, ppm	0.0085–0.021
vitamin E, ppm	24
vitamin K, ppm	1

[a]Ref. 3.

Centrifugation. Centrifugal devices include clarifiers for removal of sediment and extraneous particulates, and separators for removal of fat (cream) from milk (see SEPARATION, CENTRIFUGAL).

Clarification. A standardizing clarifier removes fat to provide a certain fat content while removing sediment, a clarifixator partially homogenizes while separating the fat, and a high speed clarifier removes bacteria cells in a bactofugation process. Clarifiers have replaced filters in the dairy plant for removing sediment, although the milk may have been previously strained or filtered while

Table 6. Fatty Acids in Samples of Milk Fat for Cows Fed Normal Rations

| Fatty acid | Acid content[a] | |
	Range	Average
butyric (4:0)[b]	2.4–4.23	2.93
hexanoic (6:0)	1.29–2.40	1.90
octanoic (8:0)	0.53–1.04	0.79
decanoic (10:0)	1.19–2.01	1.57
lauric (12:0)	4.53–7.69	5.84
myristic (14:0)	15.56–22.62	19.78
oleic (18:1)	25.27–40.31	31.90
palmitic (16:0)	5.78–29.0	15.17
stearic (18:0)	7.80–20.37	14.91

[a]Percent of total acids.
[b]A shorthand designation for fatty acids is used. For example, 18:0 = saturated C_{18}; 18:1 = C_{18} acid with one double bond; 18:2 = C_{18} acid with two double bonds; 18:0 br = branched-chain saturated C_{18} acid; etc (see CARBOXYLIC ACIDS).

Table 7. Saturated Acids as % of Total Acids of Milk Fat[a]

| Even | | Odd | |
Acid[b]	%	Acid[b]	%
4:0	2.79	5:0	0.01
6:0	2.34	7:0	0.02
8:0	1.06	9:0	0.03
10:0	3.04	11:0	0.03
12:0	2.87	13:0	0.06
14:0	8.94	13:0 br	0.04
14:0 br	0.10	15:0	0.79
16:0	23.80	15:0 br A[c]	0.24
16:0 br	0.17	15:0 br B[c]	0.38
18:0	13.20	17:0	0.70
18:0 br	trace	17:0 br A[c]	0.35
20:0	0.28	17:0 br B[c]	0.25
20:0 br	trace	19:0	0.27
22:0	0.11	21:0	0.04
24:0	0.07	23:0	0.03
26:0	0.07	25:0	0.01

[a]Ref. 4.
[b]See footnote b in Table 6.
[c]A and B designate isomers.

on the farm. A clarifier has a rotating bowl with conical disks between which the product is forced. The sediment is forced to the outside of the rotating bowl where the sludge or sediment remains. Some clarifiers have dislodging devices to flush out the accumulated material. The clarified milk leaves through a spout or outlet.

Clarification is usually performed at 4.4°C, although a wide range of temperatures is used. The clarifier may be used in numerous positions in the milk processing system, depending on the temperature, standardization procedure,

Table 8. Unsaturated Acids as % of Total Acids of Milk Fat[a]

Even				Odd	
Acid	%	Acid	%	Acid	%
10:1[b]	0.27	20:2	0.05	15:1	0.07
12:1[c]	0.14	20:3	0.11	17:1	0.27
14:1[c]	0.76	20:4	0.14	19:1	0.06
16:1[d]	1.79	20:5	0.04	21:1	0.02
18:1[d]	29.60	22:1	0.03	23:1	0.03
18:2	2.11	22:2	0.01		
18:2 c,t conj[e]	0.63	22:3	0.02		
18:2 t,t conj[e]	0.09	22:4	0.05		
18:3	0.50	22:5	0.06		
18:3 conj	0.01	24:1	0.01		
20:1	0.22				

[a] Ref. 4.
[b] Terminal double bond.
[c] Includes cis, trans, and terminal double-bond isomers.
[d] Includes cis and trans isomers.
[e] c,t = cis−trans isomer; t,t = trans−trans isomer; conj = conjugated.

flow rate, and use of the clarified product. The clarifier may be between the bulk milk tanker and raw milk storage tank, the raw milk receiving tanks and raw milk storage tank, the storage tank and standardizing tank, the standardizing tank and high temperature−short time (HTST) pasteurizer, the preheater or regenerative heater for raw milk and the heating sections of the HTST pasteurizer, or the regenerative cooler for the pasteurized milk side and the final cooling sections of the HTST pasteurizer, which is rarely used because of possible post-pasteurization contamination.

To avoid the accumulation of sediment following homogenization, the clarifier generally is used before homogenization to clarify the incoming raw milk. Clarification at this point provides milk ready for pasteurization, particularly if standardized; permits longer operation of the clarifier without stopping or cleaning, because sediment builds up more rapidly with a warm product; and when used as an operation independent from pasteurization, does not interfere with the pasteurization if maintenance is necessary.

Bactofugation. This process is not used for ordinary fluid milk, but for sterile milk or cheese. Although no longer used in the United States, bactofugation is a specialized process of clarification in which two high velocity centrifugal bactofuges operate at 20,000 rpm in series. The first device removes 90% of the bacteria, and the second removes 90% of the remaining bacteria, providing a 99% bacteria-free product. The milk is heated to 77°C to reduce viscosity. From the centrifugal bowl there is a continuous discharge of bacteria and a high density nonfat portion of the milk (1−1.5%).

Separation. In 1890, continuous-flow centrifugal cream separators using cone disks in a bowl were introduced. Originally, cream separators were basic plant equipment, and dairy plants were known as creameries. The original gravity-fed units incorporated air to produce foam and separators developed 5,000−10,000 times the force of gravity to separate the fat (cream) from the milk. Skimmed milk was discarded or returned to the farm as animal feed, and the cream was used for butter and other fat-based dairy products. Separators in the 1990s are pressure- or forced-fed sealed airtight units. The separator removes all or a portion of the fat, and the skimmed milk or reduced fat milk is sold as a beverage or ingredient in other formulated foods.

Separation is done between 32 and 38°C, although temperatures as high as 71°C are acceptable. Cold milk separators, which have less capacity at lower temperatures, may be used in processing systems in which the milk is not heated.

Separating fat globules from milk serum is proportional to difference in densities, the square of the radius of the fat globule, and centrifugal force; and inversely proportional to flow resistance of the fat globule in serum, viscosity of the product through which the fat globule must pass, and speed of flow through the separator.

The ease with which the separated products leave the bowl determines the richness of the fat. Fluid whole milk enters the separator under pressure from a positive displacement pump or centrifugal pump with flow control (Fig. 1). The fat (cream) is separated and moves toward the center of the bowl, while the skimmed milk passes to the outer space. There are two spouts or outlets, one for cream and one for skimmed milk. Cream leaves the center of the bowl with the percentage of fat (~ 30–40%) controlled by the adjustment of a valve, called a cream or skim milk screw, that controls the flow of the product leaving the field of centrifugal force and thus affects the separation.

Standardization.   Standardization is the process of adjusting the ratio of butterfat and solids-not-fat (SNF) to meet legal or industry standards. Adding cream of high butterfat milk into serum of low butterfat milk might result in a product with low SNF, thus careful control must be exercised.

A standardizing clarifier and separator are equipped with two discharge spouts. The higher fat product is removed at the center and the lower fat product at the outside of the centrifugal bowl. The standardizing clarifier removes sediment and a smaller portion of the fat than the conventional separator that leaves only 0.25% fat. Fat in the milk discharge of a standardizing clarifier is only slightly less than that of the entering milk; the reduction is ~ 10% from 4.0–3.6% fat. Accurate standardization is performed by sampling a storage tank of milk and adding appropriate fat or solids, or by putting the product through a standardizing clarifier and then into a tank for adjustment of fat and SNF.

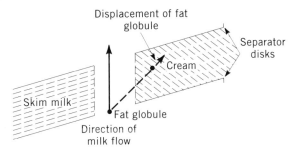

Fig. 1. Diagrammatic representation of fat globule separation in a centrifugal separator (5).

Homogenization. Homogenization is an integral part of continuous HTST pasteurization. It is the process by which a mixture of components is treated mechanically to give a uniform product that does not separate. In milk, the fat globules are broken up into small particles that form a more stable emulsion in the milk. In homogenized milk, the fat globules do not rise by gravity to form a creamline as with untreated whole milk. The fat globules in raw milk are 1–15 μm in diameter; they are reduced to 1–2 μm by homogenization. The U.S. Public Health Service defines homogenized milk as "milk that has been treated to insure the breakup of fat globule to such an extent that after 48 h of quiescent storage at 45°F (7°C) no visible cream separation occurs in the milk..." (6). Most fluid milk is homogenized.

Milk is homogenized in a homogenizer or viscolizer. It is forced at high pressure through the small openings of a homogenizing valve by a simple valve or a seat, or a disposable compressed stainless steel conical valve in the flow stream (Fig. 2). The globules are broken up as a result of shearing, impingement on the wall adjacent to the valve, and to some extent by the effects of cavitation and explosion after the product passes through the valve. In a two-stage homogenizer, the first valve is at a pressure of 10.3–17.2 MPa (1500–2500 psi) and the second valve at ~ 3.5 MPa (500 psi). The latter functions primarily to break

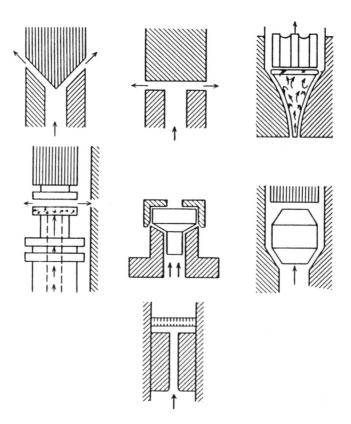

Fig. 2. Types of homogenizer valves based on velocity and impact (7).

up clumps of homogenized fat particles, and is particularly applicable for cream and products with more than 6–8% fat.

A homogenizer is a high pressure positive pump with three, five, or seven pistons, that is driven by a motor and equipped with an adjustable homogenizing valve. Smoother flow and greater capacity are obtained with more pistons, which force the product into a chamber that feeds the valve. In design and operation, it is desirable to minimize the power requirements for obtaining an acceptable level of homogenization. At 17.2 MPa (2500 psi) and a volume of 0.91 t/h (2000 lb/h), a 56-kW (75-hp) motor is required.

Several operating factors should be considered: (1) before homogenization, milk is heated to break up fat globules and prevent undesirable lipase activity; (2) as the temperature of the milk is increased, the size of the globules decreases; (3) viscosity of fluid milk is not greatly influenced by homogenization, whereas viscosity of cream is increased; (4) clarification before or after homogenization prevents the formation of sediment which otherwise adheres to the fat; and (5) it is difficult to separate cream from homogenized milk to make butter.

The homogenizer must be placed appropriately in the system to assure the proper temperature of the incoming product, provide for clarification, and avoid air incorporation that would cause excessive foaming. The homogenizer also may be used as a pump in the pasteurization circuit.

Pasteurization. Pasteurization is the process of heating milk to kill pathogenic bacteria, and most other bacteria, without greatly altering the flavor. It also inactivates certain enzymes, eg, phosphatase, thus the degree of pasteurization can be determined by measuring the phosphatase present. The principles were developed by and named after Pasteur and his work in 1860–1864. Since then, stringent codes have been developed to assure that pasteurization is done properly. The basic regulations are included in the U.S. Public Health Service Pasteurized Milk Ordinance (6) which has been adopted by most local and state jurisdictions. The quality of milk depends on the care of the animals that produce it, the environment on the farm, and the care of the product throughout.

Pasteurization may be carried out by batch- or continuous-flow processes. In the batch process, each particle of milk must be heated to at least 63°C and held continuously at this temperature for at least 30 min. In the continuous process, milk is heated to at least 72°C for at least 15 s in what is known as high temperature–short time (HTST) pasteurization, the primary method used for fluid milk. For milk products having a fat content above that of milk or that contain added sweeteners, 66°C is required for the batch process and 75°C for the HTST process. For either method, following pasteurization the product should be cooled quickly to ≤ 7.2°C. Time–temperature relationships have been established for other products including ice cream mix, which is heated to 78°C for 15 s, and eggnog, which must be pasteurized at 69°C for 30 min or 80°C for 25 s.

Another continuous pasteurization process, known as ultrahigh temperature (UHT), employs a shorter time (2 s) and a higher temperature (minimum 138°C). The UHT process approaches aseptic processing (Fig. 3).

Batch Holding. The milk in the batch holding tanks is heated in a flooded tank around which hot water or steam is circulated, or by coils surrounding the liner through which the heating medium is pumped at a high velocity. Two

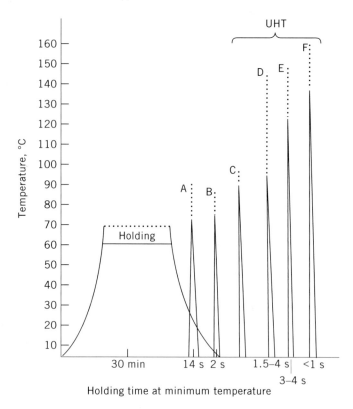

Fig. 3. Pasteurization by various methods (8): A, HTST; B, quick time; C, vacuum; D, modified tubular; E, small-diameter tube; and F, steam injection.

other methods include spraying hot water on the tank liner holding the milk, and pumping hot water through a large-diameter coil that circulates in the milk. A self-acting regulator closely controls the temperature of the water, usually heated with steam. Table 9 gives the overall heat-transfer coefficients (U-values) for these methods.

An airspace heater ejects steam into the airspace above the product and into the foam, maintaining a temperature at least 5°C above the minimum holding temperature of 63°C. The time–temperature exposure is recorded on a chart which must be kept for proof of treatment. If the lid is opened, and

Table 9. U-Values[a] for Holding Methods of Batch Pasteurization

Method	kW/(m²·K)[b]	Remarks
water spray[c]	0.350	heat from 10–63°C in 25 min, hot water at 71°C
coil vat[c]	0.350	coils turns at 130 rpm, water through coil at 100 rpm
flooded	0.350	gravity circulation, agitator
high velocity	0.525	requires more energy to pump heating fluid

[a]Overall heat-transfer value.
[b]To convert kW/(m²·K) to Btu/(h·ft²·°F), multiply by 571.2.
[c]No longer used in the United States.

the milk temperature falls below 63°C, the exposure is interrupted causing the pasteurization cycle to restart.

Valves are mounted so that the plug of the valve is flush with the tank to avoid a pocket of unpasteurized milk, and a leak detector valve permits drainage of the milk trapped in the plug of the valve. All covers, piping, and tubing must drain away from the pasteurizer.

Agitators provide adequate mixing without churning, assist in heat transfer by sweeping the milk over the heated surface, and assure that all particles are properly pasteurized.

High Temperature–Short Time Pasteurizers. The principal continuous-flow process is the high temperature–short time (HTST) method. The product is heated to at least 72°C and held at that temperature for not less than 15 s. Other features are similar to the batch holding method.

The equipment needed includes a balance tank, regenerative heating unit, positive pump, plates for heating to pasteurization temperature, tube or plates for holding the product for the specified time, a flow-diversion valve (FDV), and a cooling unit (Fig. 4). Often the homogenizer and booster pump also are incorporated into the HTST circuit.

The balance or float tank collects raw milk entering the unit, receives milk returned from the flow-diversion valve that has not been adequately heated, and maintains a uniform product elevation on the pasteurizer intake.

The heat-regeneration system partially heats the incoming cold product and partially cools the outgoing pasteurized product. The regenerator is a stainless steel plate heat exchanger, usually of the product-to-product type. The configuration is so arranged that the outgoing pasteurized product is at a higher pressure to avoid contamination. A pump in the circuit moves the milk from the raw milk side and the discharge to the final heater. Heat regenerators are usually 80–90% efficient. The regeneration efficiency may be improved by increasing the number of regenerator plates, and although this increases the energy for pumping, it also increases the cost for additional heat-exchanger plates.

The final heater increases the regeneration temperature ($\sim 60°C$) to pasteurization temperature (at least 72°C) with hot water. The hot water is ~ 1–2°C above the highest product temperature (73°C). Four to six times as much hot water is circulated compared to the amount of product circulated on the opposite side of the plates.

The holder or holding tube is at the discharge of the heater. Its length and diameter assure that fluid milk is exposed to the minimum time–temperature (72°C for 15 s). Glass or stainless steel tubing, or plate heat exchangers, may be used for holders. Holding tubes must be designed for continuous uphill flow (0.64 cm/m) from the start of the tube to the FDV.

On the outlet of the holder tube, the FDV directs the pasteurized product to the regenerator and then to the final cooling section (forward flow). Alternatively, if the product is below the temperature of pasteurization, it is diverted back to the balance tank (diverted flow). The FDV is controlled by the safety thermal-limit recorder.

The final cooling section is usually a plate heat exchanger cooled by water chilled through brine or compression refrigeration. Milk leaves the regenerator and enters the cooling section at ~ 18–24°C and is cooled to 4.4°C by glycol, or

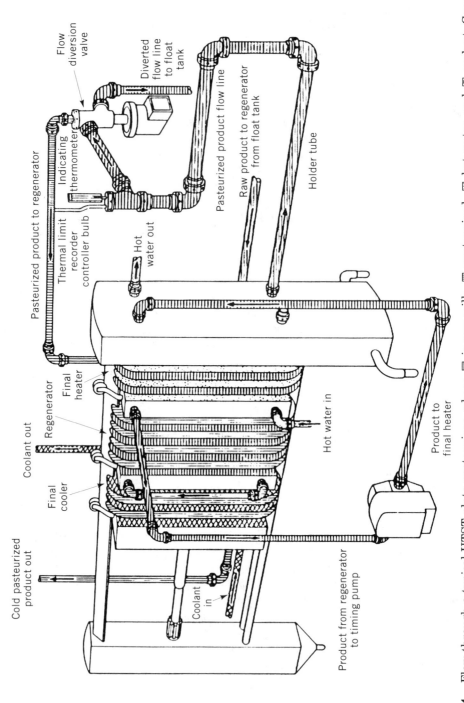

Fig. 4. Flow through a typical HTST plate pasteurizer, where ▦ is raw milk, ▦ pasteurized, ▦ hot water, and ▨ coolant. Courtesy of St. Regis Crepaco (now APV Crepaco).

Flow diversion valve

Diverted flow line to float tank

Pasteurized product flow line

Raw product to regenerator from float tank

Holder tube

Pasteurized product to regenerator

Indicating thermometer

Thermal limit recorder controller bulb

Hot water out

Coolant out

Regenerator

Final heater

Final cooler

Hot water in

Product to final heater

Cold pasteurized product out

Coolant in

Product from regenerator to timing pump

water circulating at 1°C. The relationships of regenerator, heater, and cooler for flow, number of plates, and pressure drop are given in Table 10.

The heat-transfer sections of the HTST pasteurizer, ie, regenerator, heater, and cooler, are usually stainless steel plates ~ 0.635–0.91 mm thick. Plates for different sections are separated by a terminal that includes piping connections to direct product into and out of the spaces between plates. The plates hang on a support from above and can be moved along with the terminals, for inspection or for closing the unit, and a screw assembly can be operated, manually or mechanically, to hold the plates together during operation. The plates are mounted and connected in such a manner that the product can flow through ports connecting alternate plates. The heat-transfer medium flows between every other set of plates.

The stainless steel plates are separated (ca 3 μm between) by nonabsorbent vulcanized gaskets. Various profiles and configurations, including raised knobs, crescents, channels, or diamonds, provide a rapid, uniform heat-transfer plate surface. During operation the plates must be pressed together to provide a seal, and mounted and connected in such a manner that air is eliminated and that the product drains from the plates without opening.

Various arrangements and configurations are available for the HTST pasteurizer. For regeneration, the milk-to-milk regenerator is most common. A

Table 10. Representative Capacities of HTST Plate Pasteurizers[a]

Capacity, L/h	3,800	7,600	11,360	15,140	18,930
regenerator, 84%[b]					
plates, number	31	51	71	91	111
pressure drop milk, kPa[c]	62	90	103	103	117
heater[d]					
plates, number	9	15	21	29	33
water, L/min	261	522	587	787	492
pressure drop milk, kPa[c]	55	76	76	69	96
pressure drop water, kPa[c]	83	117	76	69	165
cooler[e]					
plates, number	9	17	31	41	49
water, L/min	326	662	462	643	772
pressure drop milk, kPa[c]	55	55	117	117	145
pressure drop water, kPa[c]	131	131	165	165	179
total, 84% regeneration					
plates, number	49	83	123	161	193
pressure drop, milk, kPa[c]	172	221	296	289	358
size of frame, m	1.22	1.52	1.83	2.13	2.13
total, 90% regeneration					
plates, number	73	109	147	189	239
pressure drop, milk, kPa[c]	131	200	214	221	207
size of frame, m	1.52	1.83	1.83	2.13	2.44

[a]Courtesy of Crepaco, Inc. (now APV Crepaco).
[b]Up temperature = 4–65°C; down temperature, 77–16°C.
[c]To convert kPa to mm Hg, multiply by 7.5.
[d]Milk temperature = 65–77°C; water temperature, 79–77°C.
[e]Milk temperature = 16–3°C; water temperature, 1–4°C.

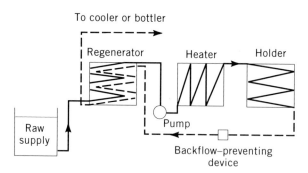

Fig. 5. Milk-to-milk regenerator with both sides closed to atmosphere (9). Courtesy of the U.S. Dept. of Health and Human Services.

heat-transfer medium, usually water, provides a milk–water–milk system. Both sides may be closed (Fig. 5) or the raw milk supply may be open.

A homogenizer or rotary positive pump may be used as a timing or metering pump to provide a positive, fixed flow through the pasteurization system (Fig. 6). The pump is placed ahead of the heater and the holding section. Various control drives assure that the pasteurized side of the heat exchanger is at a higher (7 kPa (1 psi)) pressure than the opposite side.

The homogenizer can be used as a timing pump as it is homogenizing the product (Fig. 7), or both the timing pump and homogenizer can be used in the

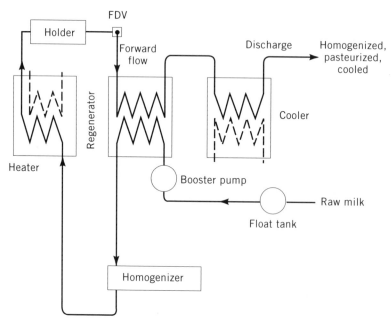

Fig. 6. Homogenizer used as a timing pump for HTST pasteurization. Details of bypass, relief lines, equalizer, and check valves are not included (10).

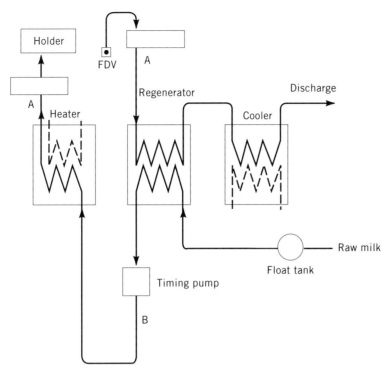

Fig. 7. Homogenization of regenerated milk, A, after HTST heat treatment, and B, before HTST pasteurization. Details of bypass, relief lines, equalizer, and check valves are not included (10).

same system; in the latter, appropriate connections and relief valves must be provided to permit the product to bypass any unit that is not operating.

Booster Pump. Use of a centrifugal booster pump avoids a low intake pressure, particularly for large, high volume units. A low pressure (> 26.6 kPa (200 mm Hg)) on the intake of a timing pump can cause vaporization of the product. The booster pump is in the circuit ahead of the timing pump and operates only when the FDV is in forward flow, the metering pump is in operation, and the pasteurized product is at least 7 kPa (1 psi) above the maximum pressure developed by the booster pump (Fig. 8).

Separator. Fat is normally separated from the milk before the HTST; however, in one system the airtight separator is placed after the FDV, following pasteurization. A restricting device and several control combinations are placed in the line after the FDV to ensure that constant flow is maintained, that vacuum does not develop in the line, that the timing pump stops if the separator stops, and that the legal holding time is met.

Control System. For quality control, a complete record of the control and operation of the HTST is kept with a safety thermal-limit recorder–controller (Fig. 9). The temperature of product leaving the holder tube, ahead of the FDV, is recorded and the forward or diverted flow of the FDV is determined. Various visual indicators, operator temperature calibration records, and thermometers also are provided.

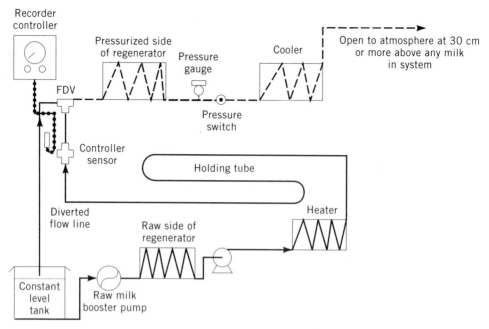

Fig. 8. Booster pump for milk-to-milk regeneration, where (---) is pasteurized milk, (—) is raw milk, and (-•-) is capillary tubing (11).

Utilities. Electricity, water, steam refrigeration, and compressed air must be provided to the pasteurizer for heating, cooling, and cleaning of water. The water is heated by steam injection or an enclosed heating and circulating unit. The controller, sensing the hot water temperature, permits heating until the pre-set temperature is reached, usually 1–2°C above the pasteurization temperature. A diaphragm valve, directed by the controller, maintains the maximum temperature of the hot water by control of the steam. Water is cooled with a direct expansion refrigeration system and may be cooled directly or over an ice bank formed by direct expansion refrigeration. The compressed air should be clean, relatively dry, and supplied at ~ 138 kPa (20 psi) to operate valves and controls.

Other Continuous Processes. Various pasteurization heat treatments are identified by names such as quick time, vacuum treatment (vacreator), modified tubular (Roswell), small-diameter tube (Mallorizer), and steam injection. The last three methods are ultrahigh temperature (UHT) processes (see Fig. 3). Higher treatment temperatures with shorter times, approaching two seconds, are preferred because the product has to be cooled quickly to prevent deleterious heat effects.

Vacuum Treatment. Milk can be exposed to a vacuum to remove low boiling substances, eg, onions, garlic, and some silage, which may impart off-flavors to the milk, particularly the fat portion. A three-stage vacuum unit, known as a vacreator, produces pressures of 17, 51–68, and 88–95 kPa (127, 381–508, and 660–711 mm Hg). A continuous vacuum unit in the HTST system may consist of one or two chambers and be heated by live steam, with an equivalent release of water by evaporation, or flash steam to carry off the

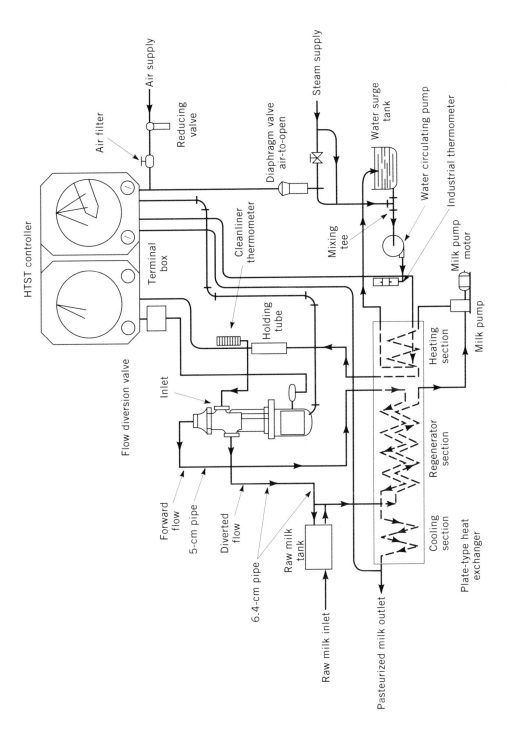

Fig. 9. HTST control system. Courtesy of Taylor Instrument Co. (now Taylor Instrument, Combustion Engineering, Inc.).

volatiles. If live steam is used, it must be culinary steam which is produced by heating potable water with an indirect heat exchanger. Dry saturated steam is desired for food processing operations.

Product Heat Treatment. Equivalent heat treatment for destruction of microorganisms or inactivation of enzymes can be represented by plotting the logarithm of time versus temperature. These relationships were originally developed for sterilization of food at 121.1°C, therefore the time to destroy the microorganism is the F_0 value at 121.1°C (250°F). The slope of the curve is z, and the temperature span is one log cycle. The heat treatment at 131°C for one minute is equivalent to 121.1°C for 10 minutes (Fig. 10).

Irradiation. Although no irradiation systems for pasteurization have been approved by the U.S. Food and Drug Administration, milk can be pasteurized or sterilized by β-rays produced by an electron accelerator or γ-rays produced by cobalt-60. Bacteria and enzymes in milk are more resistant to irradiation than higher life forms. For pasteurization, 5000–7500 Gy (500,000–750,000 rad) are required, and for inactivating enzymes at least 20,000 Gy (2,000,000 rad). Much lower radiation, about 70 Gy (7000 rad), causes an off-flavor. A combination of heat treatment and irradiation may prove to be the most acceptable approach.

Equipment. Equipment is designed according to 3A Sanitary Standards established by a committee of users, manufacturers, and sanitarians in the food industry. The objective of the committee is to provide interchangeable parts and equipment, establish standards for inspection, and provide knowledge of accept-

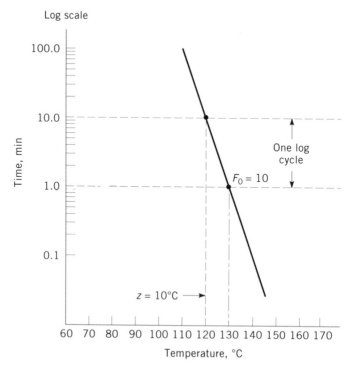

Fig. 10. Representation of z and F values (12). F_0 is the zero point for identifying the sterilization value at 121.1°C (250°F): $F_0/t = e^{2.3/z(T-121.1)} = 10^{(T-121.1)/z}$.

able design and materials, primarily to fulfill sanitary requirements. Sanitary equipment design requires that the material of construction is 18–8 stainless steel, with a carbon content of not more than 0.12%, although equally corrosion-resistant material is acceptable; the metal gauge for various applications is specified; surfaces fabricated from sheets have a No. 4 finish or equivalent; weld areas are substantially as corrosion resistant as the parent material; minimum radii are often specified, eg, for a storage tank, 0.62 cm for inside corners of permanent attachments; no threads are in contact with food; and threads are Acme threads (flat-headed instead of V-shaped).

Materials of Construction.   Stainless Steel.   The use of stainless steel (qv) for flat surfaces, tubing, coils, and castings in milk and dairy equipment has advanced since the 1950s. Previously metal-coated materials such as tinned copper were used for most applications, and copper alloys were used for castings and fittings (see COPPER AND COPPER ALLOYS). In the 1990s, the contact surfaces of milk and dairy equipment are primarily stainless steel, which permits cleaning-in-place (CIP), automation, continuous operations, and aseptic processing and packaging.

Many types of stainless steels are available. The type most widely used in the dairy industry is 18–8 (18% chromium, 8% nickel plus iron). Small amounts of silicon, molybdenum, manganese, carbon, sulfur, and phosphorus may be included to obtain characteristics desired for specific applications.

The most important stainless steel [*12597-68-1*] series are the 200-, 300-, and 400-series. The 300-series, primarily 302, 304, and 316, is used in the dairy industry, whereas the 400-series is used for special applications, such as pump impellers, plungers, cutting blades, scrapers, and bearings (Table 11). Surface finishes are specified from No. 1 to No. 8 (highly polished); the No. 4 finish is most commonly used.

Stainless steel develops a passive protective layer ($\leq$ 5-nm thick) of chromium oxide [*1118-57-3*] which must be maintained or permitted to rebuild after it is removed by product flow or cleaning. The passive layer may be removed by electric current flow across the surface as a result of dissimilar metals being in contact. The creation of an electrolytic cell with subsequent current flow and corrosion has to be avoided in construction. Corrosion may occur in welds, between dissimilar materials, at points under stress, and in places where the passive layer is removed; it may be caused by food material, residues, cleaning solutions, and brushes on material surfaces (see CORROSION AND CORROSION CONTROL).

Cleaning.   Equipment is cleaned to prevent contamination of subsequent dairy processing operations and damage to the surface. In cleaning stainless steel, surface contaminants are removed that would otherwise destroy the protective passive layer. The surface is dried and exposed to air to rebuild the protective passive chromium oxide layer. Metal adhering to the stainless steel surface should be removed with the least abrasive material, and after cleaning, the surface should be washed with hot water and left to dry. Equipment should be sanitized with 200-ppm chlorine solution within 30 minutes before use, not necessarily after cleaning, to avoid corrosion resulting from chlorine on the surface for an extended period of time. For cleaning-in-place (CIP), the velocity of the cleaning solution over the surfaces should be $\leq$ 1.5 m/s. Excessive velocities

Table 11. Stainless Steels Used in Food Processing Equipment

Identification	Alloy content, wt %		Characteristics	Uses
	Chromium[a]	Nickel[a]		
300 Series[b]				
301	16–18	6–8	ductile; lower resistance to corrosion, particularly as temperature increases	
302	17–19	8	good corrosion resistance; can be cold worked and drawn; anneal following welding to avoid intergranular corrosion in corrosive environment	general-purpose, used widely
304	18–20	8–12	better corrosion resistance than 302	most widely used for food
310	24–26	19–22	scale-resisting properties at elevated temperatures	high temperature applications
316	16–18, 2–3% Mb	10–14	superior corrosion resistance of all stainless steels	in contact with brine and various acids; gaining import-ance in food industry
400 Series[c]				
410	11.5–13.5, 0.15% C	0.15	basic martensitic alloy hardenable by heat treatment	roofing, siding, blades on freezers
416	12–14 C	0.15 C	easily machinable	valve stems, plugs, and gates
420	12–14		hardenable by heat treatment	cladding over steel; high spring temper
430	14–18		nonhardenable, good corrosion resistance	trim, structural, and decorative purposes
440	16–18 C	0.60 C	harder than others; generally not recommended for welding	pumps, plungers, gears, seal rings, cutlery, bearings

[a] Or as indicated.
[b] Nonmagnetic or slightly magnetic.
[c] Magnetic.

can cause erosion of the surface and reduction of the protective layer. Excessive time of contact of the cleaning solution may cause corrosion, depending on the strength of the cleaning solution.

Piping and Tubing.   Piping size is designated by a nominal rather than an exact inside diameter (see PIPING SYSTEMS), ie, a pipe of 2.5-cm diameter can have an inside diameter slightly more or less than 2.5 cm, depending on the wall thickness. Tubing size is designated by the outside diameter, ie, a tube of 2.5-cm diameter has an outside diameter of 2.5 cm, and as the thickness of the tubing increases the inside diameter decreases and is always less than 2.5 cm. Both piping and tubing have fixed but different outside diameters for a particular size, and standard fittings can be used with different wall thicknesses.

The food industry uses stainless steel tubing or piping extensively for moving food products; conventional steel, cast iron, copper, plastic, glass (qv), aluminum, and other alloys are used for utilities.

Most piping and tubing systems are designed for in-place cleaning. Classification is based on the type of connections for assembly: welded joints for permanent connections; ground joints with Acme threads and hexagonal nuts having gaskets for connections that are opened daily or periodically; and clamp-type joints.

Corrosion between the support device and the pipeline must be avoided. Drainage is provided by the pipeline slope, normally 0.48–0.96 cm/m of length, and gaskets must be nonabsorbent and of a type that does not affect the food product.

Fittings.   Fittings connect pipes and provide for the attachment of equipment to change flow direction. They must be easily cleaned inside and out, have no exposed pipe threads, and, if of the detachable type, have an appropriate gasket. The fittings are constructed of the same or similar materials as the pipeline and are installed on tubing. Standard shapes and sizes are specified by the 3A Standards Committee.

An air valve, sometimes called the air-activated valve, is widely used for automated food handling operations. Although electronic or electric control boxes may be a part of the system, the valve itself generally is air-activated, and is more reliable than other types. Air-operated valves are used for in-place cleaning systems, and for the transfer and flow control of various products.

Plastic.   Plastic tubing is used for farm-to-receiving operations rather than for permanent food handling installations. It is widely used to transport water for cleaning and sanitizing.

Pumps.   The flow of fluids through a dairy processing plant is maintained by a centrifugal (nonpositive) or a displacement (positive) pump. Positive displacement pumps are either of the piston or plunger type, which are usually equipped with multiple pistons, or of the rotary positive type. The pump is selected on the basis of the quantity of product to be moved against a specified head. Generally, a hardenable 400-series stainless steel is used for the moving parts which chip easily and must be handled carefully during disassembly, cleaning, and assembly.

Centrifugal Pump.   The centrifugal pump consists of a directly connected impeller which operates in a casing at high speed. Fluid enters the center and is discharged at the outer edge of the casing. The centrifugal pump is used with for moving products against low discharge heads or where it is necessary to regulate the flow of product through a throttling valve or restriction. Pumps for a CIP system include a self-cleaning diaphragm.

Positive Pumps. Positive pumps employed by the food industry have a rotating cavity between two lobes, two gears that rotate in opposite directions, or a crescent or stationary cavity and a rotor. Rotary positive pumps operate at relatively low speed. Fluid enters the cavity by gravity flow or from a centrifugal pump. The positive pump also may use a reciprocating cavity, and may be a plunger or piston pump. These pumps are not truly positive with respect to displacement, but are used for metering product flow.

Speed Devices. Many displacement pumps are connected by variable speed drives. When these pumps are used as a time device on a homogenizer, the setting is fixed, ie, the maximum speed is limited in order to meet the requirements of pasteurization.

Pump Suction. The net positive suction head required (NPSHR) affects the resistance on the suction side of the pump. If it drops to or near the vapor pressure of the fluid being handled, cavitation and loss of performance occurs (13). The NPSHR is affected by temperature and barometric pressure and is of most concern on evaporator CIP units where high cleaning temperatures might be used. A centrifugal booster pump may be installed on a homogenizer or on the intake of a timing pump to prevent low suction pressures.

Cleaning Systems. Both manual and automatic methods are used for cleaning food processing (qv) equipment.

Cleaning-In-Place. In dairy plants, the equipment surfaces and pipelines are cleaned in place at least once every 24 hours. Cleaning-in-place (CIP) systems evolved from recirculating cleaning solutions in pipelines and equipment to a highly automatic system with valves, controls, and timers. The results of cleaning in place are influenced by equipment surfaces, time of exposure, and the temperature and concentration of the solution being circulated. Cleaning is a mechanical–chemical operation.

In the CIP procedure, a cold or tempered aqueous prerinse is followed by circulation of a cleaning solution for 10 minutes to one hour at 54–82°C. The temperature of the cleaning solution should be as low as possible, because hot water rinses may harden the food product on the surface being cleaned, but high enough to avoid excess cleaning chemicals. A wide variety of cleaning solutions may be used, depending on the food product, hardness of water, and equipment.

A CIP system includes pipelines, interconnected with valves to direct fluid to appropriate locations, and the control circuit, which consists of interlines to control the valves that direct the cleaning solutions and water through the lines, and air lines which control and move the valves. A programmer controls the timing and the air flow to the valves on a set schedule. The 3A Standards for CIP components, equipment, and installation have been developed. A simple CIP system circuit is shown in Figure 11.

Economic Aspects

In 1990, U.S. milk production was 67.4×10^6 t from nearly 10.1×10^6 cows. In the United States there has been an increase in quantity of production

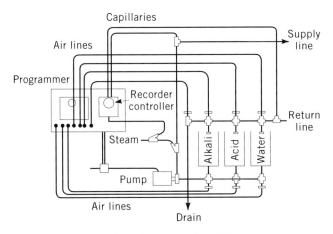

Fig. 11. Simple circuit for CIP system.

with a decrease in number of dairy cows. The world production in 1987 was 463.4×10^6 t from $\sim 50 \times 10^6$ cows. Table 12 gives the 1990 production and utilization of fluid and milk products (15). Table 13 gives the total U.S. per capita consumption.

The leading states in milk production in decreasing order are Wisconsin, Minnesota, New York, California, and Pennsylvania. These states produce $\sim 45\%$

Table 12. U.S. Milk Production and Utilization, 10^6 t[a]

Product	1982	1990
Production		
fed to calves	0.69	0.69
used for consumer milk, cream, and butter	0.38	0.24
Total produced on farms	*61.6*	*67.4*
Sold		
as whole milk	59.9	66.05
directly to consumers	0.58	0.43
Total production	*60.5*	*66.47*
Utilization[b]		
creamery butter	1.52	1.98
cheese[c]	18.2	36.3
canned milk[d]	0.67	0.61
bulk condensed whole milk	0.32	0.27
dry milk[e]	0.35	0.59
ice cream and frozen products	6.49	6.51
other manufactured products	0.30	0.15
Total manufactured products	*36.6*	*40.2*
Total fluid milk sales	*22.4*	*25.17*

[a]Ref. 14.
[b]Milk equivalent.
[c]American, cottage, other.
[d]Evaporated and sweetened condensed.
[e]Whole and nonfat.

Table 13. U.S. Per Capita, kg, Consumption of Dairy Products[a]

Products	1982	1990
fluid milk and cream	107	106
plain and flavored whole milk	60.6	41.1
plain low fat, less than 2%	33.4	44.6
plain skim milk	4.8	10.4
flavored low fat and skim milk	2.5	3.0
buttermilk	1.86	1.59
yogurt, excluding frozen	1.18	1.86
cream: heavy, light, half-and-half	1.59	2.1
sour cream and dip	0.86	1.14
cheese	9.1	11.23
American	5.13	5.05
Cheddar	3.95	4.14
Italian	2.18	4.14
condensed and evaporated milk		
whole	1.82	1.41
skim	1.36	2.18
ice cream	8.0	7.14
Total, milk equivalent	*252.1*	*259.4*

[a]Ref. 16.

of the U.S. milk supply. Less than 5% of the total production is used on farms and the remainder is sold for commercial purposes. Whereas milk and cream were formerly shipped in 19-, 30-, or 38-L cans from the farm to the plant, in the 1990s most commercial production, particularly for fluid milk, is moved in bulk from the cows to refrigerated farm tanks to insulated bulk truck tankers and to the manufacturing plant. The investment in equipment and the cost of hired labor are associated with large, capital-intensive production centers.

Storage, Cooling, Shipping, and Packaging

Bulk Milk Tanks.   Commercial dairy production enterprises generally employ tanks in which the milk is cooled and stored. In some operations, the warm milk is first cooled and then stored in a tank; 3A Standards have been established for their design and operation. Among other requirements, the milk must be cooled to 4.4°C within two hours after milking. The temperature must not be permitted to increase above 10°C when warm milk from the following milking is placed in the tank. Bulk milk tanks are classified according to method of refrigeration, ie, direct expansion (DX) or ice bank (IB); pressure in tank, ie, atmospheric or vacuum; regularity of pickup, ie, every day or every other day; capacity, in liters, when full or at amount which can be received per milking; shape, ie, cylindrical, half-cylindrical, or rectangular; position, ie, vertical or horizontal; and method of cooling refrigeration condenser, ie, by water, air, or both.

Cooling.   A compression refrigeration system, driven by an electric motor, supplies cooling for either direct expansion or ice bank systems (Fig. 12). In the former, the milk is cooled by the evaporator (cooling coils) on the bulk tank liner opposite the milk side of the liner. The compressor must have the capacity to cool the milk as rapidly as it enters the tank.

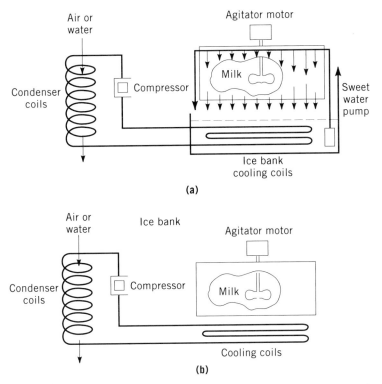

Fig. 12. Compression refrigeration supplying cooling for (**a**) the ice bank, where (↓) represents the flow of sweet water and (–––), the water level, or (**b**) the direct expansion systems (17).

In the ice bank system, ice is formed over the evaporator coils. Water is pumped over the ice bank and circulated over the inner liner of the tank to cool the milk. The water is returned to the ice bank compartment. This system provides a means of building refrigeration capacity for later cooling; therefore a smaller compressor and motor can be used, although the unit operates two to three times as long as a direct expansion system for the same cooling capacity. Off-peak electricity might be used for the ice bank system, thus reducing operating costs.

Important features of bulk milk tanks include a measuring device, generally a calibrated rod or meter; cleaning and sanitizing facilities; and stirring with an appropriate agitator to cool and maintain cool milk temperatures.

Surface Coolers. Milk coming from cows may be rapidly cooled over a stainless steel surface cooler before entering a bulk tank. The cooler may either use compression refrigeration or have two sections, one using cold water followed by a section using compression refrigeration.

Shipping. Bulk milk is hauled to the processing plant in insulated tanks using truck tanks or trailer tankers. The milk is transferred from the bulk tank to the tanker with a positive or centrifugal-type pump. For routes of some distance, pick-up every other day reduces handling costs.

Receiving Operations. Bulk milk-receiving operations consist primarily of transferring milk from the tanker to a storage tank in the plant. Practically all

Grade A milk is handled in bulk. The handling of milk in 38-L cans requires equipment and space for quality and quantity check of the product, washing of cans, and conveyors for moving and storing the cans.

Packaging. Aseptic packaging was developed in conjunction with high temperature processing and has contributed to make sterilized milk and milk products a commercial reality.

The objective in packaging cool sterilized products is to maintain the product under aseptic conditions, to sterilize the container and its lid, and to place the product into the container and seal it without contamination. Contamination of the head space between the product and closure is avoided by the use of superheated steam, maintaining a high internal pressure, spraying the container surface with a bactericide such as chlorine, irradiation with a bactericidal lamp, or filling the space with an inert sterile gas such as nitrogen.

Analysis and Testing

Milk and its products can be subjected to a variety of tests to determine composition, microbial quality, adequacy of pasteurization, contamination with antibiotics and pesticides (qv), and radioactivity (18).

Microbial Quality. The microbial quality of dairy products is related to the number of viable organisms present. A high number of microorganisms in raw milk suggests that it was produced under unsanitary conditions or that it was not adequately cooled after removal from the cow. If noncultured dairy products contain excessive numbers of bacteria, post-pasteurization contamination probably occurred or the product was held at a temperature permitting substantial microbial growth. Raw milk, as well as other milk products, is commonly examined for its concentration of microorganisms by a dye reaction test, ie, the methylene blue or resazurin [635-78-9] methods, the agar plate test, or the direct microscopic method.

The methylene blue and resazurin reduction methods indirectly measure bacterial densities in milk and cream in terms of the time interval required, after starting incubation, for a dye–milk mixture to change color (methylene blue, from blue to white; resazurin, from blue through purple and mauve to pink). In general, reduction time is inversely proportional to bacterial content of the sample when incubation starts.

The agar [9002-18-0] plate method consists of adding a known quantity of sample, usually 1.0 or 0.1 mL, depending on the concentration of bacteria, to a sterile petri plate and then mixing the sample with a sterile nutrient medium. After the agar medium solidifies, the petri plate is incubated at 32°C for 48 hours after which the bacterial colonies are counted and the number expressed in terms of a 1 mL or 1 g sample. This procedure measures the number of viable organisms present and able to grow under test conditions, ie, 32°C.

The direct microscopic count determines the number of viable and dead microorganisms in a milk sample. A small amount (0.01 mL) of milk is spread over a 1.0 cm^2 area on a microscope slide and allowed to dry. After staining with an appropriate dye, usually methylene blue, the slide is examined with the aid of a microscope (oil immersion lens). The number of bacterial cells and clumps of cells per microscopic field is determined and, by appropriate calculations, is expressed as the number of organisms per milliliter of sample.

Coliform Bacteria.   Pasteurized products are tested for numbers of coliform bacteria in order to detect significant bacterial recontamination resulting from improper processing, damaged or poorly sanitized equipment, condensate dripping into pasteurized milk, and direct or indirect contamination of equipment by insects or hands or garments of workers. Coliform bacteria are detected by using the agar plate method and a selective culture medium (violet-red bile agar). A liquid medium (brilliant green lactose bile broth) can also be used to detect this group of organisms. Coliform bacteria are not present in properly processed products that have not been recontaminated.

Thermoduric, Thermophilic, and Psychrophilic Bacteria.   Thermoduric bacteria survive but do not grow at pasteurization temperatures. They are largely nonspore-forming, heat-resistant types that develop on surfaces of unclean equipment. These bacteria are determined by subjecting a sample to laboratory pasteurization and examining it by the agar plate method.

Thermophilic bacteria are able to grow at 55°C. They are spore-forming bacilli that can enter milk from a variety of farm sources. Thermophiles grow in milk held at elevated temperatures. Their presence in milk is determined by means of the agar plate method and incubation at 55°C.

Psychrophilic bacteria can grow relatively rapidly at low temperatures, commonly within a range of 2–10°C. They are particularly important in the keeping quality of products held at refrigerator storage temperatures, and their growth is associated with the development of fruity, putrid, and rancid off-flavors. These bacteria can be detected and counted by the agar plate method with incubation at 7°C for 10 days.

Inhibitory Substances.   When antibiotics or other chemicals appear in milk, starter culture growth in such milk may be inhibited. To test for the presence of such chemicals, an agar medium is inoculated with spores of *Bacillus stearothermophilus*. A thin layer of the medium is poured into a petri dish and allowed to harden. Filter disks (1.25-cm in diameter) are dipped into milk samples and placed on the surface of the agar medium. After appropriate incubation, plates are examined for a zone of growth inhibition surrounding the disks; the presence of such a zone suggests that the milk contains an antibiotic or other inhibitory agent.

Sediment.   The sediment test consists of filtering a definite quantity of milk through a white cotton sediment test disk and observing the character and amount of residue. Efficient use of single-service strainers on dairy farms has reduced the use of sediment tests on milk as delivered to receiving plants. Although the presence of sediment in milk indicates unsanitary production or handling, its absence does not prove that sanitary conditions always existed.

Phosphatase Test.   The phosphatase [*9001-78-9*] test is a chemical method for measuring the efficiency of pasteurization. All raw milk contains phosphatase and the thermal resistance of this enzyme is greater than that of pathogens over the range of time and temperature of heat treatments recognized for proper pasteurization. Phosphatase tests are based on the principle that alkaline phosphatase is able, under proper conditions of temperature and pH, to liberate phenol [*108-95-2*] from a disodium phenyl phosphate substrate.

The amount of liberated phenol, which is proportional to the amount of enzyme present, is determined by the reaction of liberated phenol with 2,6-dichloroquinone chloroimide and colorimetric measurement of the indophenol blue formed. Under-pasteurization as well as contamination of a properly pasteurized product with raw milk can be detected by this test.

Pesticides. Chlorinated hydrocarbon pesticides (qv) are often found in feed or water consumed by cows (19,20); subsequently, they may appear in the milk, where they are not permitted. Tests for pesticides are seldom carried out in the dairy plant, but are most often done in regulatory or private specialized laboratories. Examining milk for insecticide residues involves extraction of fat, because the insecticide is contained in the fat, partitioning with acetonitrile, cleanup (Florisil [26686-77-1] column) and concentration, saponification if necessary, and determination by means of paper, thin-layer, microcoulometric gas, or electron capture gas chromatography (see TRACE AND RESIDUE ANALYSIS).

Fat Content of Milk. Raw milk as well as many dairy products are routinely analyzed for their fat content. The Babcock test, or one of its modifications, has been a standard direct measure for many years and is being replaced by indirect means, particularly for production operations. The Babcock test employs a bottle with an extended and calibrated neck, milk plus sulfuric acid [7664-93-9] to digest the protein, and a centrifuge to concentrate the fat into the calibrated neck. The percentage of fat in the milk is read directly from the neck of the bottle with a divider or caliper, reading to ±0.05% (21).

Other direct tests for measuring the fat in milk and dairy products include the Mojonnier method, which employs thermostatically controlled vacuum drying ovens and hot plates together with desiccators whose temperature is controlled by circulating water; the Gerber test, developed and used extensively in Europe (21,22), which employs sulfuric acid to dissolve solids other than fat, amyl alcohol [71-41-0] to prevent charring of fat, and centrifuging to separate the fat into the calibrated neck of the Gerber test bottle; and the DPS detergent test, based on the principle that the selected detergent(s) dissolves readily in both fat and water phases of milk and then leaves the solution upon application of heat and/or salt, thereby liberating the accumulated fat for measurement. The official AOAC Te Sa test, a rapid detergent method using alkaline buffering agents and test bottles fitted with a side arm and plunger, is essentially a chemical extraction method applicable to a variety of animal and vegetable fat products. These fat tests are described (23).

Indirect methods for determination of fat and solids-not-fat include infrared spectroscopy and turbidity or light scattering (see INFRARED TECHNOLOGY AND RAMAN SPECTROSCOPY). An infrared spectroscopy unit can measure fat (5.73 μm), protein (6.46 μm), and lactose (9.6 μm) and print out results at 180 samples per hour (22). Light scattering methods include extensive homogenization of milk before passing light through the material to minimize the effect of different sizes of fat globules.

Protein Content. The protein content of milk can be determined using a variety of methods including gasometric, Kjeldahl, titration, colorimetric, and optical procedures (see PROTEINS). Because most of the techniques are too cumbersome for routine use in a dairy plant, payment for milk has seldom been made on the basis of its protein content. Dye-binding tests have been applied to milk for determination of its protein content; these are relatively simple to perform and can be carried out in dairy plant laboratories. More emphasis will be given to assessing the nutritional value of milk, and the dependence on fat content as a basis for payment will most likely change.

In dye-binding tests, milk is mixed with excess acidic dye solution where the protein binds the dye in a constant ratio and forms a precipitate. After the dye–protein interaction takes place, the mixture is centrifuged and the optical density of the supernatant is determined. Utilization of the dye is thus measured and from it the protein content determined. Several methods for application of dye-binding techniques to milk are given (24,25).

Health and Safety Factors

Milk may be a carrier of diseases from animals or from other sources to humans. To avoid contamination before pasteurization, healthy animals should be separated from sick animals or those with infected udders. The animals should be clean, kept in clean housing with clean air, and handled by workers and equipment under strictly sanitary conditions. Post-pasteurization contamination can occur as a result of improper handling, due to exposure to contaminated air, improperly sanitized equipment, or an infected worker.

Proper refrigeration prevents the growth of some microorganisms, such as Salmonella, and the production of toxins, such as *Staphylococcus aureus*. The growth of bacteria *Escherichia coli* and *Bacillus cereus* is substantially checked by proper cooling and handling of milk. Table 14 lists diseases transmitted by cows to humans. Pasteurization is the best means of prevention.

Manufactured Products

In the United States, 62% of fluid milk production is used for manufactured products, mainly cheese, evaporated and sweetened condensed milk, nonfat dry milk, and ice cream. Evaporated and condensed milk and dry milk are made from milk only; other ingredients are added to make ice cream and sweetened condensed milk.

Evaporated and Condensed Milk. Evaporated milk is produced by removing moisture from milk, under a vacuum, followed by packaging and sterilizing in cans. The milk is condensed to half its volume in single- or multiple-effect evaporators. The final product has a fat to solids-not-fat ratio of 1:2.28, and is standardized before and after evaporation. It must have at least 7.9% fat and 25.9% total milk solids, including fat. The process for making evaporated skim milk is similar. A key operation is sterilization in the container at 116–118°C for 15–20 minutes; subsequent cooling with cold water should be completed in 15 minutes. The cans are continuously turned and moved through the sterilizing unit. Sterilization in the can imparts a distinct cooked flavor to the product. Higher temperatures and shorter treatment (ie, UHT) lessen this effect. Standards and definitions for evaporated and condensed milk have been set by the World Health Organization (WHO) and the Food and Agriculture Organization (FAO) of the United Nations (26).

Evaporated milk is a liquid product obtained by the partial removal of water only from milk. It has a minimum milk-fat content of 7.5 mol % and a

Table 14. Diseases Transmitted by Milk to Humans

Disease	Microorganism	Carrier
	Direct transmission	
tuberculosis (cow)	*Mycobacterium bovis*	udder and manure of infected cows
brucellosis	*Brucella abortus*	milk
foot-and-mouth	virus	blood to udder
milk sickness	white snakeroot	in forage
anthrax	*Bacillus anthracis*	udder by systemic disease; organisms live in soil
Q fever	*Rickettsiae burneti*, also called *Coxiella burneti*	spread by ticks and inhalation
mastitis	*Streptococcus agalactiae*, plus several other bacteria	manure, soil, forage, udder
gastroenteritis	*Escherichia coli, Bacillus subtilus*, and salmonella of many types	udder
	Indirect transmission	
tuberculosis, human	*Mycobacterium tuberculosis*	sputum, breath droplets
typhoid fever	*Salmonella typhi*	human excreta, flies, polluted water
paratyphoid fever	*Salmonella paratyphi*	feces and urine
scarlet fever	hemolytic streptococcus	udder infection
salmonellosis	salmonella of many types	water, milk, feces, other animals
staphylococcal infections	*Staphylococcus aureus*	udder, human infection
diphtheria	*Corynebacterium diphtheriae*	throat, nose, tonsils[a]
dysentery		
bacillary	*Shigella dysenteriae*	bowel discharge[a]
amoebic	*Entamoeba histolytica*	bowel discharge[a]

[a]Of humans.

minimum milk-solids content of 25.0 mol %. Evaporated skimmed milk is a liquid product obtained by the partial removal of water only from skimmed milk. It has a minimum milk-solids content of 20.0 mol %. Sweetened condensed milk is a product obtained by the partial removal of water only from milk with the addition of sugars. It has a minimum milk-fat content of 8.0 mol % and a minimum milk-solids content of 28.0 mol %. Skimmed sweetened condensed milk is a product obtained by the partial removal of water only from skimmed milk with the addition of sugars. It has a minimum milk-solids content of 24.0 mol %. All may contain food additives (qv) as stabilizers, in maximum amounts, including sodium, potassium, and calcium salts of hydrochloric acid at 2000 mg/kg singly; citric acid, carbonic acid, orthophosphoric acid, and polyphosphoric acid at 3000 mg/kg in combination, expressed as anhydrous substances; and in the evaporated milk carrageenin may be added at 150 mg/kg.

In addition to sections 1, 2, 4, and 6 of the General Standards for the Labeling of Prepackaged Foods (Ref. No. CAC/RS 1-1969), the following specific provisions apply. The name of the product shall be "Evaporated milk," "Evaporated whole milk," "Evaporated full cream milk," "Unsweetened condensed whole milk," "Unsweetened full cream condensed milk," "Evaporated skimmed milk," "Unsweetened condensed skimmed milk," "Sweetened condensed milk," "Sweetened condensed whole milk," "Sweetened full cream condensed milk," "Skimmed sweetened condensed milk," or "Sweetened condensed skimmed milk," as appropriate. Where milk other than cow's milk is used for the manufacture of the product or any part thereof, a word or words denoting the animal or animals from which the milk has been derived should be inserted immediately before or after the designation of the product, except that no such insertion need be made if the consumer would not be misled by its omission. In sweetened milks, when one or several sugars are used, the name of each sugar shall be declared on the label (26).

Vitamin A (845 RE/L) and vitamin D (913 RE/L) may be added to fortify evaporated milk. Other possible ingredients are sodium citrate, disodium phosphate, and salts of carrageenan. Phosphate ions maintain an appropriate salt balance to prevent coagulation of the protein (casein) during sterilization. The amount of phosphate added depends on the amount of calcium and magnesium present.

Large quantities of evaporated milk are used to manufacture ice cream, bakery products, and confectionery products (see BAKERY PROCESSES AND LEAVENING AGENTS). When used for manufacturing other foods, evaporated milk is not sterilized, but placed in bulk containers, refrigerated, and used fresh. This product is called condensed milk. Skimmed milk may be used as a feedstock to produce evaporated skimmed milk. The moisture content of other liquid milk products can be reduced by evaporation to produce condensed whey, condensed buttermilk, and concentrated sour milk.

Sweetened Condensed Milk. For sweetened condensed milk, unlike evaporated milk that is sterilized, sugar is added as a preservative and provides keeping quality. The equipment is similar to that used for evaporated milk, except that sugar is added in a hot well before condensing (evaporating) the liquid. Preheating pasteurizes the product and no sterilization is needed. According to standards, sweetened condensed milk must contain a minimum of 8.5% fat and 28% total milk solids, including fat (fat to solids-not-fat ratio = 1:2.3). The final product contains 43–45% sugar. Sweetened condensed skimmed milk has not less than 24% total milk solids, but up to 50% sugar may be added.

Age-thinning and age-thickening defects occur in sweetened condensed products because of the preheating temperature before evaporation of the water. A low temperature can result in thinning, a high temperature in thickening. The optimum preheating temperature is in the range of 60–81°C.

Dry Milk. Dry milk provides long-term storage capabilities, supplies a product that can be used for food manufacturing operations, and because of its reduced volume and weight, transportation and storage costs are reduced. Dry milk has been used for manufactured products, but is used to a much greater extent for beverage products. Its properties are listed in Table 15.

Dry milk is generally made using the spray process or the so-called roller drum process. These processes generally follow condensing of milk in an evaporator. The moisture content for nonfat dry milk, the principal dry product, is less than 5.0% for standard grade and less than 4.0% for extra grade. Dry whole

Table 15. Properties of Dry Milk[a,b]

Property	Value
moisture content, nonfat, wt %	4–5
apparent or bulk density, including voids, g/cm^3	
drum dried, nonfat	0.3–0.5
spray dried, nonfat	0.5–0.6
true density without voids	
dry milk	1.31–1.32
nonfat dry milk	1.44–1.46
coefficient of friction at 20°C, 5 wt % fat	0.64
porosity, spray dried, nonfat, wt %	0.482
solubility index, spray process	1.2
vapor pressure	
5% moisture, nonfat, 38°C, kPa[c]	1.17
5% moisture, 13% fat, 38°C, kPa[c]	0.75
threshold radiation level to produce off-flavor	
dry whole milk, Gy[d]	590
dry nonfat milk, Gy[d]	1280
titratable acidity, wt %	0.15
specific heat, kJ/(kg·K)[e]	1.04
thermal conductivity, k, W/(m·K)[f]	
4.2% at 40°C	0.05
at 65°C	0.06

[a]Approximate values.
[b]Atomization of one liter of condensed product to an average particle size of 50 μm dia equals 341,000 cm^2 surface.
[c]To convert kPa to mm Hg, multiply by 7.5.
[d]To convert Gy to rad, multiply by 100.
[e]To convert kJ/(kg·K) to Btu/(lb·°F), divide by 4.184.
[f]To convert W/(m·K) to Btu/(h·ft·°F), multiply by 1.874.

milk contains less than 3.0% moisture. Other drying methods include the use of foam sprays, jet sprays, freeze-drying, and tall towers.

Clarification and homogenization precede evaporating and drying. Homogenization of whole milk at 63–74°C with pressures of 17–24 MPa (2500–3500 psi) is particularly desirable for reconstitution and the preservation of quality.

Standards and definitions for whole milk powder, partly skimmed milk powder, and skimmed milk powder have been set by WHO. This standard applies exclusively to dried milk products as defined, having a fat content of not more than 40 mol %.

Dry milk was referred to as milk powder until the mid-1960s, when the designation was changed by the American Dry Milk Institute to dry milk in the United States. Milk powder, having a milk-fat content of 26–40 mol %, is a product obtained by the removal of water only from milk, partly skimmed milk (powder having a milk-fat content of 1.5–26 mol %), or skimmed milk (powder having a maximum milk-fat content of 1.5 mol %). All have a maximum water content of 5 mol %. All may contain food additives as stabilizers, in maximum amounts, including sodium, potassium, and calcium salts of hydrochloric acid, citric acid, carbonic acid, orthophosphoric acid, and polyphosphoric acid at 5000 mg/kg singly or in combination expressed as anhydrous substances.

Emulsifiers in instant milk powders include monoglycerides and diglycerides at 2500 mg/kg and lecithin (qv) at 5000 mg/kg. Anticaking agents in milk powders intended to be dispensed in vending machines include tricalcium phosphate, silicon dioxide (amorphous), calcium carbonate, magnesium oxide, magnesium carbonate, magnesium phosphate, and silicates of aluminum, calcium, magnesium, and sodium–aluminate, at 10 g/kg singly or in combination (26).

Dry whole milk should be vacuum or gas packed to maintain the quality while in storage. Products with milk fats deteriorate in the presence of oxygen, giving oxidation off-flavor. Several factors may be involved in oxidative deterioration, such as preheating of product, storage temperature, presence of metallic ions, particularly copper and iron, presence of oxygen (air) in product, and light. Antioxidants (qv) of many kinds have been used with various degrees of success, but a universally acceptable antioxidant which meets the requirements for food additives has not been found.

Drum Drying. The drum or roller dryers used for milk operate on the same principles as for other products. A thin layer or film of product is dried over an internally steam-heated drum with steam pressures up to 620 kPa (90 psi) and 149°C. Approximately 1.2–1.3 kg of steam are required per kilogram of water evaporated. The dry film produced on the roller is scraped from the surface, moved from the dryer by conveyor, and pulverized, sized, cooled, and put into a container.

The operating variables for a drum or roller dryer include condensation of incoming product in an evaporator, temperature of incoming product, steam pressure (temperature) in drum, speed of drum, and height of product over drum. The capacity of the dryer is increased by increasing the steam pressure, the temperature of the milk feed, the height of milk over the drums, the gap between drums (double), and the speed of rotation of the drums. Increasing the capacity is limited by the effect on the product quality.

Drum-dried products are more affected by heat than spray-dried products. Drying in a vacuum chamber decreases the temperature and thus the heat effect on the product, although the atmospheric dryers are used more widely.

Drum-dried products, mostly nonfat, make up only 5–10% of dried milk products. Because of the high temperature and longer contact time, considerable protein denaturation occurs. Drum-dried products are identified as high heat dry milk and as such have a lower solubility index, lower protein nitrogen content, and a darker color.

Spray Drying. The spray dryer provides a chamber in which the milk or milk product is atomized in a heated air stream that removes most of the moisture. The dry product is separated from the air stream and removed from the chamber. The process involves condensing the product from 3 to 2:1, preheating or reheating at 63–74°C, pumping at 17.2–20.7 MPa (2500–3000 psi), atomizing, spray drying with an outlet air temperature of 82–85°C, separating air and product, cooling product at 32–38°C, sifting, packaging (vacuum plus nitrogen for whole milk), and storing.

In spite of the higher energy requirements, the spray dryer has gained in popularity because of the reduced heat effect on the product as compared to the drum dryer. Modifications such as foam spraying are being developed to reduce the heat effect further.

In the manufacture of dry milk by the spray process, a condensed product is pumped to an atomizer in order to produce a large surface area to enhance drying. A high pressure nozzle or centrifugal device, such as a rotating disk or wheel, is used for atomization. The air is filtered, heated to 149–260°C and moved over the atomizing product, saturated with water, and exhausted from the dryer. The dry product is removed from the air in a mechanical centrifugal separator and filtered outside the drying chamber. In order to minimize heat effects, the dried product is removed as rapidly as possible from the chamber and cooled. Considerable variation exists in the operation of spray dryers, depending on the product and the dryer. A low heat, nonfat dry milk product is obtained by minimizing heating before and after drying.

Foam spray drying consists of forcing gas, usually air or nitrogen, into the product stream at 1.38 MPa (200 psi) ahead of the pump in the normal spray dryer circuit. This method improves some of the characteristics of dried milk, such as dispersibility, bulk density, and uniformity. The foam–spray dryer can accept a condensed product with 60% total solids, as compared to 50% without the foam process. The usual neutralization of acid whey is avoided with the foam–spray dryer (see DRYING; FOAMS; SPRAYS).

Agglomeration. The process of treating dried products, particularly nonfat products, in order to increase speed and ability to reconstitute those products, is known as instantizing or agglomeration. Particles are agglomerated into larger particles which dissolve more easily than small particles. In this process the dry particle surface is first wetted; this is followed by agglomeration and drying. Instantized products can also be obtained by foam–spray drying. Instantized products have a lower density, are more fragile than conventional products, and must be handled with extra care. They are of particular importance to the fast-food market. The process is also used for various beverage and milk products.

Packaging. Dry milk is packaged in large bulk or small retail containers. A suitable container keeps out moisture, light, and air (oxygen). For dry whole milk, oxygen is removed by vacuum, and an inert gas, such as nitrogen, is inserted in the heat space. An oxygen level of $\leq 2.0\%$ is required by U.S. standard for premium quality.

Cream. Cream is a high fat product which is secured by gravity or mechanical separation through differential density of the fat and the serum. Fat content may range from 10 to 40%, depending on use and federal and state laws. The U.S. Public Health Service (6) milk ordinance defines cream as a product that contains not less than 18% milk fat. Whipping cream has a fat content of 30–40%, and light cream has a fat content of 18–30%. Half-and-half, suggesting a mixture of cream and milk, has not less than 10.5% milk fat, and in some states up to 12%. Cream is standardized in the same manner as milk, following separation. The addition of whole milk rather than serum is preferred.

The sale of fresh cream as a table item for serving has decreased greatly since the 1970s, primarily as a result of changing customer demand based on diet. A variety of cream and fat substitutes are available for spreads, toppings, whiteners, and cooking (see DAIRY SUBSTITUTES).

Anhydrous Milk Fat. One high milk-fat material is butter oil (99.7% fat), also called anhydrous milk fat or anhydrous butter oil if less than 0.2% moisture is present. Although the terms are used interchangeably, anhydrous butter oil is made from butter and anhydrous milk fat is made from whole milk. For milk and cream there is an emulsion of fat-in-serum, for butter oil and anhydrous milk fat there is an emulsion of serum-in-fat, such as with butter. It is easier to remove moisture in the final stages to make anhydrous milk fat with the serum-in-fat emulsion.

Butter. In the United States about 10 wt % of edible fats used are butter. Butter is defined as a product that contains 80% milk fat with not more than 16% moisture. It is made of cream with 25–40% milk fat. The process is primarily a mechanical one in which the cream, an emulsion of fat-in-serum, is changed to butter, an emulsion of serum-in-fat. The process is accomplished by churning or by a continuous operation with automatic controls. Some physical properties are given in Table 16 (see EMULSIONS).

Butter, fresh and salted, was once a primary trade commodity, but is no longer in as high a demand. There has been a shift in emphasis from fat

Table 16. Physical Properties of Milk Fat and Butter[a]

Property	Value
fat content, wt %	80
size of fat globules, μm	1–20
melting point of milk fat, °C	31–36
solidification of milk fat, °C	19–24
apparent specific heat, kJ/(kg·K)[b]	
at 0°C	2.14
15°C	2.20
40°C	2.32
60°C	2.42
density of milk fat, g/cm^3	
at 34°C, >mp	0.91–0.95
60°C	0.896
viscosity of milk fat, mPa·s(= cP)	
at 30°C	25.8
50°C	12.4
70°C	7.1
viscosity of butter 21°C, mPa·s(= cP)[c]	3.1×10^5
iodine number, normal butter	30.5
melting point of butter, °C	33.3
spreadability	
at 21°C	good
7–16°C	desirable
4°C	difficult
ratio of firmness to butter:firmness of butterfat	
summer	1.97:1
winter	1.48:1
coefficient of expansion of liquid pure butterfat, at 30–60°C	0.00076
free acidity, fresh butterfat	0.05–0.10%

[a] Refs. 27 and 28.
[b] To convert kJ/(kg·K) to Btu/(lb·°F), divide by 4.184.
[c] Brookfield at 1 rpm.

content to the protein, mineral, and vitamin content of milk and milk products, particularly in developed countries.

Buttermilk. Buttermilk is drained from butter (churn) after butter granules are formed; as such, it is the fluid other than the fat which is removed by churning. Buttermilk may be used as a beverage or may be dried and used for baking. Buttermilk from churning is ~ 91% water and 9% total solids. Total solids include lactose [598-82-3], 4.5%; nitrogenous matter, 3.4%; ash, 0.7%; and fat, 0.4%. Table 17 gives the U.S. specifications for dry buttermilk (DBM) and whey.

Cultured buttermilk is that which is produced by the fermentation (qv) of skimmed milk, often with some cream added. The principal fermentation organisms used are *Lactococcus lactis* subsp. *cremoris*, *Lactococcus lactis* subsp. *lactis*, and *Leuconostoc citrovorum*. The effect of the high processing temperature and the lactic acid provide an easily digestible product.

Dried buttermilk is made by either the drum or spray process. Buttermilk is usually pasteurized before drying, even though the milk was previously pasteurized before churning. Dried buttermilk is used primarily for baking, confectionery, and dairy products.

Cheese. The making of cheese is based on the coagulation of casein from milk, and to a minor extent the proteins of whey. The casein is precipitated by acidification which can be accomplished by natural souring of milk. The procedures for making cheese vary greatly and cheese products are countless. The composition and handling of the original milk, bacterial flora, and starter culture are the basis variables, which along with heat treatments, flavoring, salting, and forming, affect the final product.

Membrane Separation. The separation of components of liquid milk products can be accomplished with semipermeable membranes by either ultrafiltration (qv) or hyperfiltration, also called reverse osmosis (qv) (30). With ultrafiltration (UF) the membrane selectively prevents the passage of large

Table 17. U.S. Specifications for Dry Buttermilk and Dry Whey[a]

Property	Spray process DBM		Roller process DBM		Dry whey,[b] extra
	Extra	Standard	Extra	Standard	
moisture, wt %	≤5.0	≤5.0	≤5.0	≤5.0	≤5.0
milk fat, wt %	≥4.5	≥4.5	≥4.5	≥4.5	≤1.25%
solubility index, mL	≤1.25	≤2.0	≤15.0	≤15.0	≤1.25
scorched particles, mg	≤15.0	≤22.5	≤22.5	≤32.5	≤15.0
titratable acidity, wt %	0.10–0.18	0.10–0.20	0.10–0.18	0.10–0.18	≤0.16
bacteria count, per g	≤50,000	≤200,000	≤50,000	≤200,000	≤50,000
ash alkalinity, mL of 0.1 N HCl/100 g	≤125	≤125	≤125	≤125	≤125

[a]Ref. 29.
[b]Not applicable to cottage cheese whey.

molecules such as protein. In reverse osmosis (RO) different small, low molecular weight molecules are separated. Both procedures require that pressure be maintained and that the energy needed is a cost item. The materials from which the membranes are made are similar for both processes and include cellulose acetate, poly(vinyl chloride), poly(vinylidene difluoride), nylon, and polyamide (see MEMBRANE TECHNOLOGY). Membranes are commonly used for the concentration of whey and milk for cheesemaking (31). For example, membranes with 100 and 200 μm are used to obtain a 4:1 reduction of skimmed milk.

Four configurations for membranes are plate, hollow fine fiber, spiral wound, and tubular (32). With a variety of shapes, sizes, and materials many options exist for meeting the various needs in the dairy industry.

Ultrafiltration. Membranes are used that are capable of selectively passing large molecules (> 500 daltons). Pressures of 0.1–1.4 MPa ($\leq$ 200 psi) are exerted over the solution to overcome the osmotic pressure, while providing an adequate flow through the membrane for use. Ultrafiltration (qv) has been particularly successful for the separation of whey from cheese. It separates protein from lactose and mineral salts, protein being the concentrate. Ultrafiltration is also used to obtain a protein-rich concentrate of skimmed milk from which cheese is made. The whey protein obtained by ultrafiltration is 50–80% protein which can be spray dried.

Reverse Osmosis. Membranes are used for the separation of smaller components (< 500 daltons). They have smaller pore space and are tighter than those used for ultrafiltration. High pressure pumps, usually of the positive piston or multistage centrifugal type, provide pressures up to 4.14 MPa (600 psi).

Following ultrafiltration of whey, the permeate passes over a reverse osmosis (qv) membrane to separate the lactose from other components of the permeate. Reverse osmosis can be used to remove water and concentrate solids in a dairy plant, giving a product with 18% solids and thus decreasing the difficulty of waste disposal. Concentration of rinse water gives a product with 4–5% total solids. Proper maintenance of the membrane allows for use up to two years. Membranes are available for use up to 100°C with pH ranges from 1 to 14; the usual temperature range is 0–50°C.

Cheddar Cheese. Milk is heated to 30°C and a lactic acid-producing starter is added. The milk is held for about one hour, during which time the acidity increases. Rennet extract is mixed with the milk that produces a curd in approximately 30 minutes. The curd is cut into cubes and the whey expressed. The curd solidifies and is stirred and heated slowly. The heating is continued until the curd becomes completely firm, and the whey is drained and separated by forming channels. With the development of lactic acid and the removal of whey, the curd becomes a solid mass and is cut, with the pieces moved to continue the removal and drainage of the whey. The whey increases from 0.1% acid at the time of cutting, to 0.5% acid at the end of drainage. Cheddared cheese is put through a curd mill to reduce the curd sizes.

Cottage Cheese. Cottage cheese is made from skimmed milk. As compared to most other cheeses, cottage cheese has a short shelf-life and must be refrigerated to maintain quality, usually $\leq$ 4.4°C to provide a shelf-life of three weeks or more. Cottage cheese is a soft uncured cheese which contains not more than 80% moisture.

Several procedures can be used for making cottage cheese. In general, pasteurized skim milk is inoculated with lactic acid culture and rennet starter to coagulate the protein. The coagulated material is divided or cut and the resulting curd cooked to expel the whey. The whey is drained and the curd washed with water. Mechanized operations are used for large-scale production. The conditions for manufacture are given in Table 18.

Horizontal vats are employed for manual and mechanized operations. The starter may be blended with the incoming product or added at the vat. The setting temperature of the treated whey is typically 30°C and is held for 4.5–5 hours. The curd is cut when the titratable acidity is 0.52% for lactic acid milk with 9.0% nonfat milk solids, or pH 4.6–4.7. The acidity controls the calcium level of the casein that determines many of the characteristics of the curd; low acidity causes a rubbery curd, and high acidity causes a tender curd that shatters easily. The curd is cut by moving a knife first horizontally, then vertically, and finally crosswise through the vat. The cut curd is cooked about 30 minutes after cutting is finished. The temperature is gradually increased in increments of 0.5–1.0°C every 3–5 minutes to avoid the formation of a hardened protein layer that would inhibit moisture removal. After cooking, the whey is drained off and the curd is washed successively with cooler water, pasteurized or treated with chlorine, and rinsed at 4.4°C for firmness. Curd pumps move the curd to the blender where salt, cream, and stabilizer may be added. Creamed cottage cheese that has a fat content of at least 4% is produced by mixing in 12–14% fat cream.

Yogurt. Yogurt is a fermented milk product that is rapidly increasing in consumption in the United States. Milk is fermented with *Lactobacillus bulgaricus* and *Streptococcus thermophilous* organisms that produce lactic acid. Usually some cream or nonfat dried milk is added to the milk in order to obtain a heavy-bodied product.

Yogurt is manufactured by procedures similar to buttermilk. Milk with a fat content of 1–5% and solids-not-fat (SNF) content of 11–14% is heated to ca 82°C and held for 30 minutes. After homogenization the milk is cooled to 43–46°C and inoculated with 2% culture. The product is incubated at 43°C for three hours in a vat or in the final container. The yogurt is cooled and held at < 4.4°C. The cooled product should have a titratable acidity of not less than 0.9% and a pH of 4.3–4.4. The titratable acidity is expressed in terms of percentage of lactic acid [598-82-3], which is determined by the amount of 0.1 N NaOH/100 mL required to neutralize the substance. Thus 10 mL of 0.1 N NaOH represents 0.10% acidity. Yogurts with less than 2% fat are popular. Fruit-flavored yogurts

Table 18. Manufacture of Cottage Cheese

Conditions	Value
amount of starter, wt %	0.5–5
setting temperature, °C	21–32
coagulating time, h	4–12
size of curd cubes, cm^3	0.25–2.00
cooking temperature, °C	49–60
rennet extract, g/500 kg milk	0.5–1.0

are also common in which 30–50 g of fruit are placed in the carton before or with the yogurt.

Frozen Desserts. Ice cream is the principal frozen dessert produced in the United States. It is known as the American dessert and was first sold in New York City in 1777. Frozen yogurt is also gaining in acceptance as a dessert. The composition of various frozen desserts is given in Table 19.

Ice Cream. Ice cream is a frozen food dessert prepared from a mixture of dairy ingredients (16–35%), sweeteners (13–20%), stabilizers, emulsifiers, flavoring, and fruits and nuts (qv). Ice cream has 10–20% milk fat and 8–15% nonfat solids with 38.3% (36–43%) total solids. These ingredients can be varied, but the dairy ingredient solids must total 20%. The dairy ingredients are milk or cream, and milk fat supplied by milk, cream butter, or butter oil, as well as SNF supplied by condensed whole or nonfat milk or dry milk. The quantities of these products are specified by standards. The milk fat provides the characteristic texture and body in ice cream. Sweeteners are a blend of cane or beet sugar and corn syrup solids. The quantity of these vary depending on the sweetness desired and the cost.

Stabilizers to improve the body of the ice cream include gelatin (qv), sodium alginate (alginic acid sodium), certain pectins, guar gum, locust bean gum, and carboxymethylcellulose. Emulsifiers such as lecithin (qv), monoglycerides, and diglycerides assist the incorporation of air and improve the whipping properties. The mixture of components for making ice cream is called ice cream mix and is often sold as a commercial product to those who make ice cream. Ice cream mix in dry powder form is also available. The properties of ice cream are given in Table 20.

Preceded by a blending operation and pasteurization, the ingredients are mixed in a freezer that whips the mix to incorporate air and freezes a portion of

Table 19. Composition of Frozen Desserts,[a] %

Component	Ice cream Premium[b]	Ice cream Average	Ice milk	Sherbet	Ice	Soft-serve
milk fat[c]	16.0	10.5	3.0	1.5		6.0
milk solids, nonfat	9.0	11.0	12.0	3.5		12.0
sucrose	16.0	12.5	12.0	19.0	23.0	9.0
corn syrup solids		5.5	7.0	9.0	7.0	6.0
stabilizer[d]	0.1	0.3	0.3	0.5	0.3	0.3
emulsifier[d]		0.1	0.15			0.2
total solids, kg/L	41.1	39.9	34.45	33.5	30.3	33.5
	1.09	1.12	1.13	1.14	1.13	1.11
overrun, %	65–70	95–100	90–95	50	10	40
~kg/L, from freezer	0.64	0.55	0.57	0.74	1.01	0.77

[a]Frozen desserts containing vegetable fat (mellorine-type) are permitted in some states. A wide variation of composition exists depending on individual state standards.
[b]To be classified as custard or French, product must contain ≥ 1.4% egg yolk solids.
[c]Milk-fat content regulated by individual state.
[d]Usage level as recommended by manufacturer.

Table 20. Properties of Ice Cream and Ice Cream Mix[a]

Property	Value
structural constituents,[a] particle diameter, μm	
ice crystals	45–56
air cells	110–185
unfrozen materials	6–8
average distance between air cells	100–150
lamellae thickness	30–300
lactose crystals (when apparent to tongue feel)	16–30
individual fat globules	0.5–2.0
small fat globules	≤ 20
agglomerated fat	≤ 25
coalesced fat	≥ 25
weight per 3.9 L, kg, 100% overrun	2.04
specific gravity, 100% overrun, g/cm^3	0.54
specific heat, kJ/(kg·K)[b]	
ice cream	1.88
ice cream mix	3.35
fuel values, kJ[b]	8.70
overrun, %	60–100
temperature at which freezing begins, °C	3.3
water in frozen ice cream	
at −5 to −6°C	50%
−30°C	90%
ice cream mix	
pH	6.3
acidity, %	0.19
specific gravity	1.054–1.123
surface tension, mN/m(=dyn/cm)	50×10^{-3}
composition of SNF of mix, %	
protein	36.7
lactose	55.5
minerals	7.8

[a]Ref. 33; values are approximate.
[b]To convert kJ to Btu, divide by 1.054.

the water. Freezers may be of a batch or continuous type. Commercial ice cream is produced mostly in continuous operation.

The incorporation of air decreases the density and improves the consistency. If one-half of the final volume is occupied by air, the ice cream is said to have 100% overrun, and 4 L will have a weight of 2.17 kg. Ice cream from the freezer is at ca −5.5°C with one-half of the water frozen, preferably in small crystals.

Containerized ice cream is hardened on a stationary or continuous refrigerated plate-contact hardener or by convection air blast as the product is carried on a conveyor or through a tunnel. Air temperatures for hardening are −40 to −50°C. The temperature at the center of the container as well as the storage temperature should be ≤ -26°C. Approximately one-half of the heat is removed at the freezer and the remainder in the hardening process.

Other Frozen Desserts. Although ice cream is by far the most important frozen dessert, other frozen desserts such as frozen yogurt, ice milk, sherbet, and mellorine-type products are also popular. The consumption of frozen yogurt has been increasing rapidly.

Ice milk is a frozen product which has less fat (2–7%) and slightly more nonfat milk solids than ice cream. Stabilizers and emulsifiers are added. About half of ice milk produced is made as a soft-serve dessert, produced in freezers with an overrun of 40–100%.

Sherbets have a low fat content (1–2%), low milk solids (2–5%), and a sweet but tart flavor. Ice cream mix and water ice can be mixed to obtain a sherbet. The overrun in making sherbets is about 40–60%.

Mellorine is similar to ice cream except that the milk fat is replaced with vegetable fat (6% min). The total solids in mellorine are 35–39%, of which there are 10–12% milk solids.

Other frozen desserts are parfait, souffle, ice cream pudding, punch, and mousse. These are often classified with the sherbets and ices.

By-Products From Milk. Milk is a source for numerous by-products resulting from the separation or alteration of the components. These components may be used in other so-called nondairy manufactured foods, dietary foods, pharmaceuticals (qv), and as a feedstock for numerous industries, such as casein for glue.

Lactose. Lactose [63-42-3] (milk sugar), $C_{12}H_{22}O_{11} \cdot H_2O$, makes up about 5% of cow's milk. Lactose is a disaccharide composed of D-glucose and D-galactose. Compared to sucrose, lactose has about one-sixth the sweetening strength (see SUGAR). Because of its low solubility, lactose is limited in its application; however, it is soluble in milk serums and can be removed from whey. Upon fermentation by bacteria lactose is converted to lactic acid [598-82-3], and is therefore of particular importance in producing fermented or cultured dairy products, such as buttermilk, cheeses, and yogurt.

The ratio of α-lactose [10039-26-6] and β-lactose in dry milk and whey varies according to the speed and temperature of drying. An aqueous solution at equilibrium at 25°C contains 35% α- and 63% β-lactose. The latter is more soluble and sweeter than DL-lactose and is obtained by heating an 80% DL-lactose [63-42-3] solution above 93.5°C, followed by drying on a drum or roller dryer. Lactose is used for foods and pharmaceutical products.

Casein. Milk contains proteins and essential amino acids lacking in many other foods. Casein is the principal protein in the skimmed milk (nonfat) portion of milk (3–4% of the weight). After it is removed from the liquid portion of milk, whey remains. Whey can be denatured by heat treatment of 85°C for 15 minutes. Various protein fractions are identified as α-, β-, and γ-casein, and δ-lactoglobulin; and blood–serum albumin, each having specific characteristics for various uses. Table 21 gives the concentration and composition of milk proteins.

Casein is used to fortify flour, bread, and cereals. Casein also is used for glues and microbiological media. Calcium caseinate is made from a pressed casein, by rinsing, treating with calcium hydroxide [1305-62-0], heating, and mixing followed by spray drying. A product of 2–4% moisture is obtained.

Casein hydrolyzates are produced from dried casein. With appropriate heat treatment and the addition of alkalies and enzymes, digestion proceeds. Follow-

Table 21. Composition and Concentrations of Milk Protein[a]

Component	Whole	Casein α-	Casein β-	Casein γ-	β-Lactoglobulin	α-Lactalbumin	Blood serum albumin	Euglobulin	Pseudoglobulin[b] A	Pseudoglobulin[b] B
					Concentration, g/100 mL					
	2.23–8.84	1.4–2.3	0.5–1.0	0.06–0.22	0.20–0.42	0.07–0.15	0.02–0.05	0.03–0.06	0.02–0.05	
					Composition, g/100 g					
N	15.63	15.53	15.33	15.40	15.60	15.86	16.07	16.05	15.29	15.9
amino N	0.93	0.99	0.72	0.67	1.24		0.78			
amide N	1.6	1.6	1.6	1.6	1.07					
P	0.86	0.99	0.61	0.11	0.00	0.02	0.00	0.00	0.00	
S	0.80	0.72	0.86	1.03	1.60	1.91	1.92	1.01	1.00	1.1
hexose								2.93	2.96	
hexosamine								1.58	1.45	
Gly[c]	2.7	2.8	2.4	1.5	1.4	3.2	1.8			
Ala	3.0	3.7	1.7	2.3	7.4	2.1	6.2			
Val	7.2	6.3	10.2	10.5	5.8	4.7	5.9	10.4	9.6	8.7
Leu	9.2	7.9	11.6	12.0	15.6	11.5	12.3	10.4	9.6	8.5
Ile	6.1	6.4	5.5	4.4	6.1	6.8	2.6	3.0	3.0	4.2
Pro	11.3	8.2	16.0	17.0	4.1	1.5	4.8			10.0
Phe	5.0	4.6	5.8	5.8	3.5	4.5	6.6	3.6	3.9	3.9
Cys$_2$	0.34	0.43	0.0–0.1	0.0	2.3	6.4	5.7	3.3	3.0	
Cys	0.0	0.0	0.0	0.0	1.1	0.0	0.3	0.0	0.0	
Met	2.8	2.5	3.4	4.1	3.2	1.0	0.8	0.9	0.9	1.3
Trp	1.7	2.2	0.83	1.2	1.9	7.0	0.7	2.4	2.7	3.2
Arg	4.1	4.3	3.4	1.9	2.9	1.2	5.9	5.1	3.3	5.6
His	3.1	2.9	3.1	3.7	1.6	2.9	4.0	2.0	2.1	2.3
Lys	8.2	8.9	6.5	6.2	11.4	11.5	12.8	6.3	7.1	6.1
Asp	7.1	8.4	4.9	4.0	11.4	18.7	10.9			9.4
Glu	22.4	22.5	23.2	22.9	19.5	12.9	16.5			12.3
Ser	6.3	6.3	6.8	5.5	5.0	4.8	4.2			
Thr	4.9	4.9	5.1	4.4	5.8	5.5	5.8	10.6	10.3	9.0
Tyr	6.3	8.1	3.2	3.7	3.8	5.4	5.1			6.7

[a]Ref. 34. [b]A, from milk; B, from colostrum. [c]See AMINO ACIDS, SURVEY for abbreviations.

ing pasteurization, evaporation (qv), and spray drying, a dried product of 2–4% is obtained. Many so-called nondairy products such as coffee cream, topping, and icings utilize caseinates (see DAIRY SUBSTITUTES). In addition to fulfilling a nutritional role, the caseinates impart creaminess, firmness, smoothness, and consistency of products. Imitation meats and soups use caseinates as an extender and to improve moistness and smoothness.

Nutritional Value of Milk Products. Milk is considered one of the principal sources of nutrition for humans. Some people are intolerant to one or more components of milk so must avoid the product or consume a treated product. One example is intolerance to lactose in milk. Fluid milk is available in which the lactose has been treated to make it more digestible. The consumption of milk fat, either in fluid milk or in products derived from milk, has decreased markedly in the 1990s. Whole milk sales decreased 12% between 1985 and 1988, whereas the sales of low fat milk increased 165%, and skimmed milk sales increased 48% (35). Nutritionists have recommended that fat consumed provide no more than 30 calories, and that consumption of calories be reduced. Generally, a daily diet of 2000–3000 cal/d is needed depending on many variables, such as gender, type of work, age, body responses, exercise, etc. Further, there is concern about cholesterol [57-88-5] and density of fat consumed. Complete information on the nutritive value of milk and milk products is provided on product labels (36) (see also Table 4).

The concern by consumers about cholesterol has stimulated the development of methods for its removal. Three principal approaches are in the pilot-plant stages: use of enzymes, supercritical fluid extraction, and steam distillation. Using known techniques, it is not possible to remove all cholesterol from milk. Therefore, FDA guidelines identify cholesterol-free foods as containing less than 2 mg cholesterol per serving, and low cholesterol foods as containing from 2 to 20 mg (37).

Biotechnology

Biotechnology is being applied in the dairy industry. A significant and controversial development is the technique of producing transgenic animals, ie, animals in which hereditary deoxyribonucleic acid (DNA) has been augmented by DNA from another source, using recombinant DNA (rDNA) techniques.

One technology uses bovine somatotropin (bST) produced by recombinant technology (38). Somatotropin [9002-72-6] is a growth hormone. The bST-supplemented cows provide an increase in milk output per cow or an increased feed efficiency. Recombinant bST, also known as recombinant bovine growth hormone (rBGH) is the synthetic analogue of a natural hormone that increases milk production in cows (39). The use of recombinant technology was approved by the FDA in 1993. The Commission of the European Community has recommended that the moratorium on commercial use of BGH be delayed until the year 2000.

The same principle has been applied to other mammals, particularly ewes and goats, as well as dairy cows (Table 22), for various purposes, although the primary objective has been to develop pharmaceuticals (qv). New biotechnology products are also being developed for food processing (qv). Genetically engineered enzymes have been approved by FDA for cheese manufacturing. Engineering

Table 22. Use of Milk-Producing Animals for Biotechnology

Animal	Product	Use
dairy cow[a]	rBGH[b] bovine growth hormone, also known as rbST, bovine somatotropin	growth hormone
dairy calf[c]	human lactoferrin (HLF)	human protein
ewe (sheep)[d]	α_1-antitrypsin[e]	replace human protein deficiency
goat[c]	tissue plasinogen activator (TPA)[f]	human blood clot-reducing compound

[a]Ref. 38.
[b]Ref. 39.
[c]Ref. 40.
[d]Ref. 41.
[e]35 g/L active agent produced.
[f]2–3 g/L active agent produced.

microorganisms will be available to produce enzymes to be added to curd for ripening cheese. Various applications of biotechnology include production of milk that can be ingested by lactose-intolerant people, improved fermented products, production of natural preservatives in milk, and methods for treating and processing waste products for further use or nondamaging disposal (38).

BIBLIOGRAPHY

"Dairy Products" in *ECT* 1st ed., Vol. 4, pp. 774–846, by A. H. Johnson, National Dairy Research Laboratories, Inc.; "Milk and Milk Products" in *ECT* 2nd ed., Vol. 13, pp. 506–576, by E. H. Marth, University of Wisconsin, and R. V. Husong, L. F. Cremers, J. H. Guth, L. D. Hilker, H. W. Jackson, O. J. Krett, E. G. Simpson, R. A. Sullivan, and L. Tumerman, National Dairy Products Corp.; in *ECT* 3rd ed., Vol. 15, pp. 522–570, by C. W. Hall.

1. B. H. Webb, A. H. Johnson, and J. A. Alford, *Fundamentals of Dairy Chemistry*, 2nd ed., Avi Publishing Co., Westport, Conn., 1974, p. 396.
2. *Recommended Dietary Allowances*, 10th ed., National Research Council, National Academy Press, Washington, D.C., 1989, p. 289.
3. Ref. 1, p. 125.
4. S. F. Herb, P. Magidman, F. E. Luddy, and R. W. Riemenschneidet, *J. Am. Oil Chem. Soc.* **39**, 142 (1962).
5. W. J. Harper and C. W. Hall, *Dairy Technology and Engineering*, Avi Publishing Co., Westport, Conn., 1976, p. 413.
6. *Grade A Pasteurized Milk Ordinance*, U.S. Department of Health and Human Services, Public Health Service Publication No. 229, Rev. ed., Washington, D.C., 1989.
7. Ref. 5, p. 426.
8. C. W. Hall and G. M. Trout, *Milk Pasteurization*, Avi Publishing Co., Westport, Conn., 1968, p. 51.
9. Ref. 8, p. 64.
10. C. W. Hall, G. M. Trout, and A. L. Rippen, *Michigan Agr. Exp. Stn. Q. Bull.* **43**, 634 (1961).
11. Ref. 8, p. 73.
12. Ref. 8, p. 125.

13. Ref. 5, pp. 411–412.
14. *Agricultural Statistics*, U.S. Department of Agriculture, Washington, D.C., 1991, pp. 305, 321.
15. *Statistical Abstracts of the United States*, U.S. Department of Commerce, Bureau of the Census, Washington, D.C., 1991, 986 pp.
16. Ref. 14, p. 479.
17. C. W. Hall and D. C. Davis, *Processing Equipment for Agricultural Products*, 2nd ed., Avi Publishing Co., Westport, Conn., 1979, p. 49.
18. R. T. Marshall, *Standard Methods for the Examination of Dairy Products*, American Public Health Association, Washington, D.C., 1993, 546 pp.
19. E. H. Marth, *J. Milk Food Technol.* **25**, 72 (1962).
20. E. H. Marth and B. E. Ellickson, *J. Milk Food Technol.* **22**, 112, 145 (1959).
21. R. S. Kirk and R. Sawyer, *Pearson's Composition and Analysis of Foods*, Longmans Scientific and Technical Books, Essex, U.K., 1991, p. 537.
22. Y. Pomeranz and C. E. Meloan, *Food Analysis*, Von Nostrand Reinhold, New York, 1987, p. 708.
23. *Laboratory Manual, Methods of Analysis of Milk and Milk Products*, The Milk Industry Foundation, Washington, D.C., 1959.
24. R. M. Dolby, *J. Dairy Res.* **28**, 43 (1961).
25. R. W. Weik, M. Goehle, H. A. Morris, and R. Jenness, *J. Dairy Sci.* **47**, 192 (1964).
26. *Code of Principles Concerning Milk and Milk Products*, FAO/WHO, Food and Agriculture Organization of the U.N., Rome, Italy, 1973, pp. 27–32.
27. C. W. Hall, A. W. Farrall, and A. L. Rippen, eds., *Encyclopedia of Food Engineering*, 2nd ed., Avi Publishing Co., Westport, Conn., 1986, pp. 84–85.
28. F. H. McDowell, *The Buttermakers Manual*, New Zealand University Press, Wellington, N.Z., 1953, pp. 51–58.
29. C. W. Hall and T. I. Hedrick, *Drying of Milk and Milk Products*, 2nd ed., Avi Publishing Co., Westport, Conn., 1971, pp. 212–213.
30. R. F. Madsen, in S. A. Goldblith, L. Rey, and W. W. Rothmayr, eds., *Freeze Drying and Advanced Food Technology*, Academic Press, Inc., New York, 1975, pp. 575–587.
31. J. H. Woychik, P. Cooke, and D. Lu, *J. Food Sci.* **57**, 46–58 (1992).
32. D. R. Heldman and D. B. Lund, eds., *Handbook of Food Engineering*, Marcel Dekker, Inc., New York, 1992, p. 423.
33. C. W. Hall, A. W. Farrall, and A. L. Rippen, *Michigan Agr. Exp. Stn. Q. Bull.* **43**, 433 (1961).
34. R. Jenness and S. Patton, *Principles of Dairy Chemistry*, R. E. Krieger Publishing Co., Huntington, N.Y., 1976, 446 pp.
35. B. Shroder and R. J. Baer, *Food Technol.* **44**, 145 (1990).
36. U.S. Department of Agriculture, *Nutritive Value of Foods*, Home and Garden Bulletin No. 72, U.S. Government Printing Office, Washington, D.C., 1991, pp. 10–14.
37. F. Kosilowski, *Food Technol.* **44**, 134 (1990).
38. U.S. Congress, Office of Technology Assessment, *U.S. Dairy Industry at a Crossroad: Biotechnology and Policy Choices, Special Report*, OTA-F-470, U.S. Government Printing Office, Washington, D.C., May 1991, pp. 51–52.
39. W. Rouse, *Tech. Rev.* **94**(5), 28–34 (1991).
40. R. Seltzer, *Chem. Eng. News* **69**(35), 7 (1991).
41. A. Coghlin, *New Sci.* **1321**(1791), 22 (1991).

General References

W. S. Arbuckle, *Ice Cream*, Avi Publishing Co., Westport, Conn., 1986, 483 pp.
J. G. Brennan and co-workers, *Food Engineering Operations*, 2nd ed., Applied Science Publishers, Ltd., London, 1990, 700 pp.

A. W. Farrall, *Engineering for Dairy and Food Products*, John Wiley & Sons, Inc., New York, 1963, 674 pp.

C. W. Hall, A. W. Farrall, and A. L. Rippen, eds., *Encyclopedia of Food Engineering*, 2nd ed., Avi Publishing Co., Westport, Conn., 1986, 882 pp.

C. W. Hall and T. I. Hedrick, *Drying of Milk and Milk Products*, 2nd ed., Avi Publishing Co., Westport, Conn., 1976, 631 pp.

D. R. Heldman and D. R. Lund, eds., *Handbook of Food Engineering*, Marcel Dekker, Inc., New York, 1992, 756 pp.

J. L. Henderson, *The Fluid Milk Industry*, 3rd ed., Avi Publishing Co., Westport, Conn., 1971, 677 pp.

M. Loncin and R. L. Merson, *Food Engineering—Principles and Selected Applications*, Academic Press, Inc., New York, 1979, 494 pp.

R. T. Marshall, ed., *Standard Methods for the Examination of Dairy Products*, 16th ed., American Public Health Association, Washington, D.C., 1993, 546 pp.

R. P. Singh and D. R. Heldman, *Introduction to Food Engineering*, 2nd ed., Academic Press, San Diego, Calif., 1993, 499 pp.

Dairy Science Abstracts

Food Science and Technology Abstracts (U.K.)

CARL W. HALL
Engineering Information Services

MINERAL NUTRIENTS

Minerals that are essential to life are the source of metals and other inorganic elements involved in the most fundamental processes. For example, oxygen, required by the cells of animals, is utilized with the aid of metal complexes. In humans both iron-containing hemoglobin and zinc-containing carbonic anhydrase play pivotal roles in binding oxygen and delivering it to the cells. Moreover, enzymes developed to protect cells from high levels of oxygen also contain metals. One such class of protective enzymes is known as the superoxide dismutases (SODs). These contain metals such as manganese, copper, zinc, and iron (1). Mutations in the copper- and zinc-containing superoxide dismutase gene have been linked to amyotrophic lateral sclerosis (2).

The human skeleton is composed of calcium and phosphorus and traces of other ions, eg, magnesium, sulfur, and sodium, embedded in an organic matrix. The regulation of body fluid volume and acid–base balance requires the cations sodium, potassium, magnesium, and calcium. The principal anion is chloride. Calcium also plays a role in neuromuscular excitability and blood coagulation. Metabolic energy, cellular homeostasis, and most enzyme activities are dependent on phosphorus. The electron-transport chain requires copper and iron. Several vitamins (qv) contain sulfur; one contains cobalt. Hormones (qv) contain iodine, sulfur, and zinc. Each cell contains a complex set of enzymes, many requiring metal ions, either as part of the basic structure, eg, copper, iron, molybdenum, selenium, and zinc, or as activators, eg, chromium, magnesium, and manganese.

As for other biological substances, states of dynamic equilibrium exist for the various mineral nutrients as well as mechanisms whereby a system can adjust to varying amounts of these minerals in the diet. In forms usually found in foods, and under circumstances of normal human metabolism, most nutrient minerals are not toxic when ingested orally. Amounts considerably greater than the recommended dietary allowances (RDAs) can generally be eaten without concern for safety (Table 1) (3).

Some elements found in body tissues have no apparent physiological role, but have not been shown to be toxic. Examples are rubidium, strontium, titanium, niobium, germanium, and lanthanum. Other elements are toxic when found in greater than trace amounts, and sometimes in trace amounts. These latter elements include arsenic, mercury, lead, cadmium, silver, zirconium, beryllium, and thallium. Numerous other elements are used in medicine in non-nutrient roles. These include lithium, bismuth, antimony, bromine, platinum, and gold (Fig. 1). The interactions of mineral nutrients with carbohydrates, fats, and proteins, minerals with vitamins (qv), and mineral nutrients with toxic elements are areas of active investigation (7–9).

The amount of each element required in daily dietary intake varies with the individual bioavailability of the mineral nutrient. Bioavailability depends both on body need as determined by absorption and excretion patterns of the element and by general solubility, and on the absence of substances that may cause formation of insoluble products, eg, calcium phosphate, $Ca_3(PO_4)_2$. In some cases, additional requirements exist either for transport of substances or for uptake or binding. For example, calcium-binding proteins are involved in calcium transport; an intrinsic factor is needed for vitamin B_{12}, ie, cobalt, uptake (see VITAMINS, VITAMIN B_{12}).

The essential mineral nutrients are classified either as principal elements or as trace and ultratrace elements. The distinction between these groups is the relative amounts in the dietary requirement (see Table 1).

Normal blood plasma or serum levels of the mineral nutrients and the usual form in circulating blood are given in Table 2. Modes of absorption and excretion are summarized in Table 3. Standard treatises on mineral nutrients (4–6, 10–21) and standard sources of nutrient composition (22,23) are available in the literature.

The Principal Elements

Calcium. Calcium, the most abundant mineral element in mammals, comprises 1.5–2.0 wt % of the adult human body, over 99 wt % of which is present in bones and teeth (24). About 48% of serum calcium is ionic, ca 46% is bound to blood proteins, the rest is present as diffusible complexes, eg, of citrate (24). The calcium ion level must be maintained within definite limits (see Table 2). Common food sources rich in calcium are listed in Table 4 (see also CALCIUM COMPOUNDS).

Table 1. Essential Mineral Nutrients[a]

Element[b]	Body content, mg/kg body wt	Daily requirement,[c] mg
Principal elements		
calcium	14,000–20,000	800–1,200[d,e]
phosphorus	11,000–12,000	800–1,200[d,e]
sulfur	1,600–2,500	[f]
potassium	2,000–3,500	2,000[g]
sodium	1,500–1,600	500[g]
chlorine	1,200–1,500	750[g]
magnesium	270–500	280[d,e,h];350[d,i]
Trace and ultratrace elements		
iron	60–66	10[d,i];15[d,h]
fluorine	37	1.5–4.0[d,j]
zinc	33–50	12[d,e,h];15[d,i]
silicon	15–16	5–20
copper	1.0–2.5	1.5–3.0[j]
boron	0.69	0.5–1.0
selenium	0.2–0.3	0.055[d,e,h];0.07[d,i]
iodine	0.2–0.4	0.15[d,e]
manganese	0.2–4.0	2.0–5.0[d,j]
molybdenum	0.1–0.5	0.075–0.25[d,j]
chromium	0.06–0.2	0.05–0.2[d,j]
cobalt	0.02	0.003[k]
tin	0.2	
vanadium	0.14	<0.01
nickel	0.07–0.14	<0.10

[a] Refs. 4–6.
[b] Generally not ingested in elemental form.
[c] RDA values from Ref. 3 unless otherwise noted.
[d] Values are for adults.
[e] Increased amounts are required during pregnancy and lactation.
[f] Adequate intake with adequate intake of protein.
[g] Estimated minimum requirement from Ref. 3.
[h] Value for females.
[i] Value for males.
[j] Estimated safe and adequate daily intake from Ref. 3.
[k] As vitamin B_{12}.

Metabolic Functions. Bones act as a reservoir of certain ions, in particular Ca^{2+} and PO_4^{3-}, which readily exchange between bones and blood. Bone structure comprises a strong organic matrix combined with an inorganic phase which is principally hydroxyapatite [1306-06-5], $3Ca_3(PO_4)_2 \cdot Ca(OH)_2$. Bones contain two forms of hydroxyapatite. The less soluble crystalline form contributes to the rigidity of the structure. The crystals are quite stable, but because of the small size present a very large surface area available for rapid exchange of ions and molecules with other tissues. There is also a more soluble intercrystalline frac-

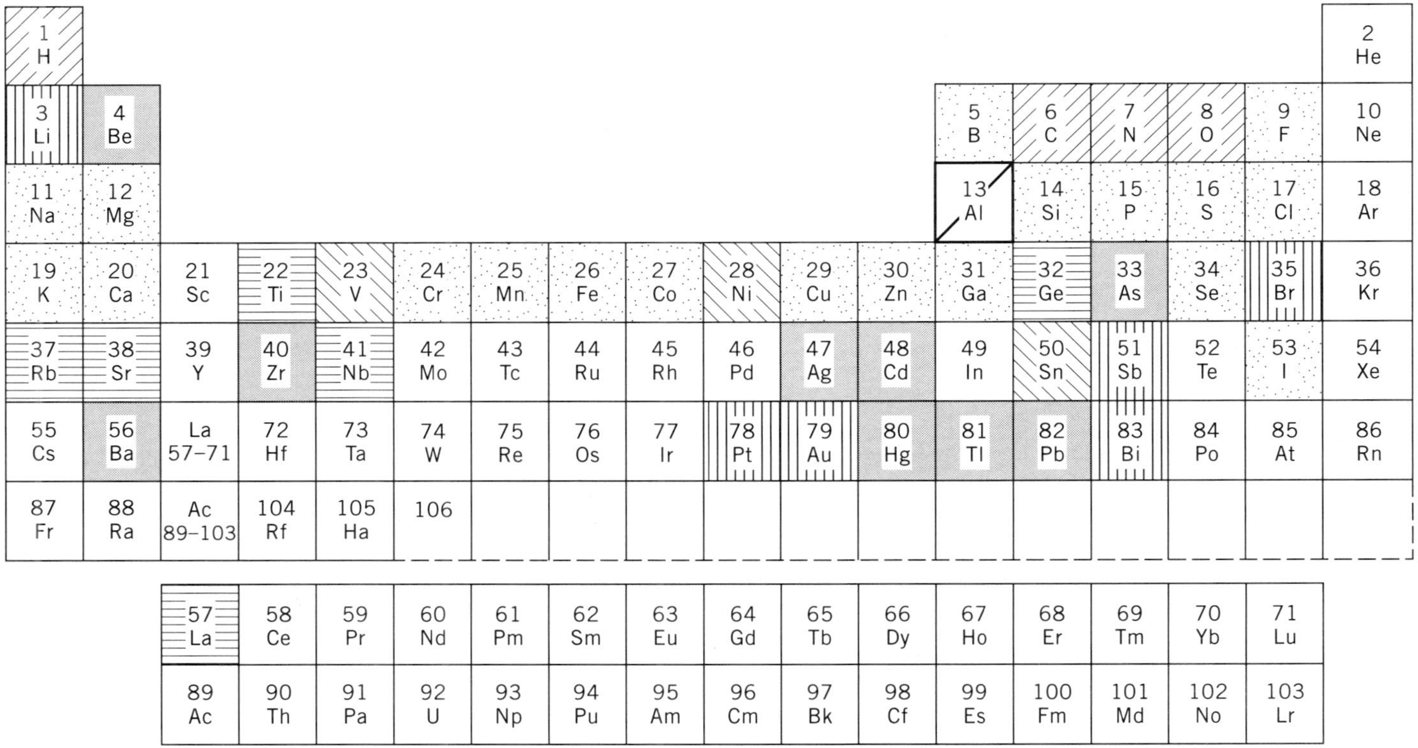

Fig. 1. Periodic Table showing elements of importance in biological systems: ▨ principal element of bioorganic compounds; ▨ essential mineral nutrients for humans and other animals; ▨ essential mineral nutrient for animals, probably for humans; ▦ present in body, not known to be a nutrient or toxic element; ▨ element used in medicine; ▨ element generally poisonous; and ▨ present in body, possibly toxic.

749

Table 2. Blood Values and Carriers for Mineral Nutrients[a]

Mineral nutrient	Concentration,[b] mg/100 mL	Form in circulating blood
	Principal elements	
calcium	9.0–10.6	free Ca^{2+}; chelated to organic acids; bound to prealbumin
phosphorus	3.0–4.5	70% in organic phospholipids; as orthophosphate: $H_2PO_4^-$ and $H_2PO_4^{2-}$
sulfur	2.9–3.5[c]	free SO_4^{2-}; bound in protein
potassium	14–20	free K^+
sodium	310–340	free Na^+
chlorine	360–375	free Cl^-
magnesium	1.3	free Mg^{2+}; chelated to organic acids; bound to albumin
	Trace and ultratrace elements	
iron	0.065–0.175	bound to transferrin
fluorine	0.280[d]	in albumin
zinc	0.072–0.120[c] 0.408–1.170[d]	in albumin; in α_1-, α_2-macroglobulins
silicon	0.00016–0.00131	monosilicic acid
copper	0.100–0.200	bound to ceruloplasmin; albumin; amino acids
boron		probably borate ion
selenium	0.00013–0.0034[d]	bound in protein
iodine	0.004–0.008	mainly as thyroid compounds T_3 and T_4
manganese	0.0024–0.0069[d]	in transferrin
molybdenum	0.00135–0.00159[d]	bound in protein; in α_2-macroglobulin
chromium	0.0005–0.0031	bound to transferrin
cobalt	0.00035–0.0063[d]	bound to albumin
vanadium	0.0005–0.0023[e]	bound to transferrin
nickel	0.002–0.004[e]	some free Ni^{2+}; bound to albumin

[a] Refs. 4–18.
[b] In serum unless otherwise noted.
[c] In serum or plasma.
[d] In whole blood.
[e] In plasma.

tion. Bone salts also contain small amounts of magnesium, sodium, carbonate, citrate, chloride, and fluoride (25). Osteoporosis is reported to result when bone resorption is relatively faster than bone formation (24).

The calcium ion, necessary for blood-clot formation, stimulates release of bloodclotting factors from platelets (see BLOOD, COAGULANTS AND ANTICOAGULANTS) (25). Neuromuscular excitability also depends on the relative concentrations of Na^+, K^+, Ca^{2+}, Mg^{2+}, and H^+ (26). Upon a decrease in Ca^{2+} concentration, termed hypocalcemia, excitability increases. If this condition is not corrected, the symptoms of tetany, ie, muscular spasm, tremor, and even convulsions, can appear. Too great an increase in Ca^{2+} concentration, hypercalcemia,

Table 3. Primary Sites of Absorption and Excretion of Mineral Nutrients

Nutrient	Absorption[a]	Excretion[b]
Principal elements		
calcium	duodenum and jejunum (a,f)	kidney; intestine as digestive juices
phosphorus	small intestine (a)	kidney
sulfur	small intestine (a,f transport of S-containing amino acids)	kidney; intestine as bile acids
potassium	small intestine (p)	kidney; skin
sodium	large intestine, ileum (a); jejunum (f); stomach, skin (p); some two-step absorption in parts of small intestine (p or a)	kidney; intestine; some from skin as perspiration
chlorine	absorbed with Na^+, K^+, and Ca^{2+}	kidney; intestine; and skin
magnesium	ileum (a)	kidney; skin; very small amount from intestine
Trace and ultratrace elements		
iron	duodenum and jejunum (a,f)	no significant excretion mechanism; skin as perspiration; exfoliation of cells, eg, intestinal, etc
fluorine	stomach (p); possibly intestine	kidney
zinc	duodenum (f)	intestine as bile and pancreatic juices; skin as perspiration; almost none from kidney
silicon	intestine	kidney
copper	stomach and upper intestine with low pH (p,f)	intestine as bile and pancreatic enzymes; kidney
boron	across gastrointestinal epithelia (p)[c]	kidney
selenium	duodenum (p)	kidney; intestine as bile and pancreatic juices; lungs (in expired air) if excess is ingested
iodine	small intestine, entire gastrointestinal tract as I (p)	kidney
manganese	small intestine in two-step mechanism (p or a)	intestine as bile and pancreatic juices
molybdenum	small intestine (f)[c]	kidney; skin as perspiration
chromium	small intestine (p)[c]	kidney
cobalt	as B_{12} in ileum (f); as inorganic Co (a,f)	kidney; skin as perspiration
tin	possibly small intestine (u)	kidney
vanadium	small intestine (u)	kidney
nickel	small intestine (u)	kidney

[a]Role is (a), active; (p), passive; (f), facilitated; or (u), mechanism not reported.
[b]Usually some fraction of the mineral nutrient ingested is not absorbed and passes into the feces. Modes of excretions listed pertain only to the fraction of mineral nutrient absorbed.
[c]Mechanism is possible.

Table 4. Common Food Sources Rich in Calcium[a]

Food[b]	Calcium in serving, mg
canned sockeye salmon with bones	543
yogurt, nonfat	452
sardines, Atlantic[c]	433
flour, self-rising	423
cheese[d]	
Parmesan	390
Romano	301
Swiss	272
Cheddar	204
Muenster	203
milk	
goat	325
whole cow	291
rhubarb, cooked	348
figs, dried[e]	269
turnip greens, cooked	198
broccoli, cooked	136

[a]Refs. 22 and 23.
[b]Serving corresponds to 236 mL (1 cup) unless otherwise noted.
[c]Serving corresponds to 113 g (4 oz).
[d]Serving corresponds to 28 g (1 oz).
[e]Serving corresponds to 187 g (10 whole figs).

may impair muscle function to such an extent that respiratory or cardiac failure may occur.

Contraction of muscle follows an increase of Ca^{2+} in the muscle cell as a result of nerve stimulation. This initiates processes which cause the proteins myosin and actin to be drawn together making the cell shorter and thicker. The return of the Ca^{2+} to its storage site, the sarcoplasmic reticulum, by an active pump mechanism allows the contracted muscle to relax (27). Calcium ion, also a factor in the release of acetylcholine on stimulation of nerve cells, influences the permeability of cell membranes; activates enzymes, such as adenosine triphosphatase (ATPase), lipase, and some proteolytic enzymes; and facilitates intestinal absorption of vitamin B_{12} [68-19-9] (28).

Blood Calcium Ion Level.   In normal adults, the blood Ca^{2+} level is established by an equilibrium between blood Ca^{2+} and the more soluble intercrystalline calcium salts of the bone. Additionally, a subtle and intricate feedback mechanism responsive to the Ca^{2+} concentration of the blood that involves the less soluble crystalline hydroxyapatite comes into play. The thyroid and parathyroid glands, the liver, kidney, and intestine also participate in Ca^{2+} control. The salient features of this mechanism are summarized in Figure 2 (29–31).

Factors controlling calcium homeostasis are calcitonin, parathyroid hormone (PTH), and a vitamin D metabolite. Calcitonin, a polypeptide of 32 amino acid residues, mol wt ~3600, is synthesized by the thyroid gland. Release is stimulated by small increases in blood Ca^{2+} concentration. The sites of action

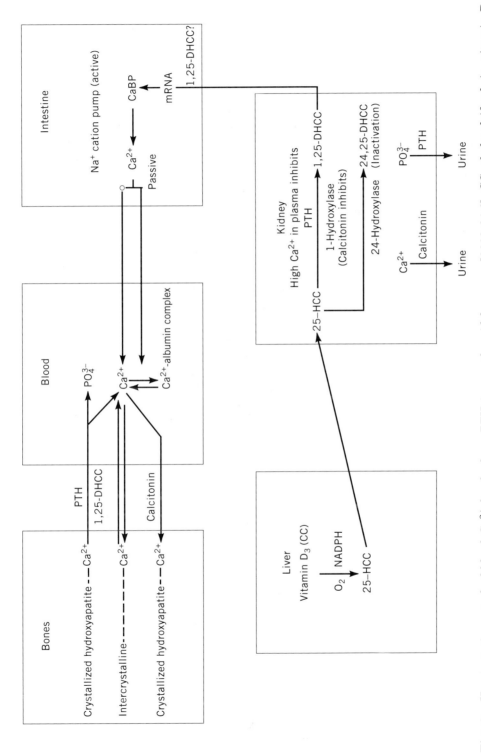

Fig. 2. Homeostatic control of blood Ca²⁺ level where PTH is parathyroid hormone [9002-64-6]; CC, cholecalciferol, ie, vitamin D₃; HCC, hydroxycholecalciferol; DHCC, dihydroxycholecalciferol; CaBP, calcium-binding protein; NADPH, protonated nicotinamide-adenine dinucleotide phosphate; and mRNA, messenger ribonucleic acid.

of calcitonin are the bones and kidneys. Calcitonin increases bone calcification, thereby inhibiting resorption. In the kidney, it inhibits Ca^{2+} reabsorption and increases Ca^{2+} excretion in urine. Calcitonin operates via a cyclic adenosine monophosphate (cAMP) mechanism.

Parathyroid hormone, a polypeptide of 83 amino acid residues, mol wt 9500, is produced by the parathyroid glands. Release of PTH is activated by a decrease of blood Ca^{2+} to below normal levels. PTH increases blood Ca^{2+} concentration by increasing resorption of bone, renal reabsorption of calcium, and absorption of calcium from the intestine. A cAMP mechanism is also involved in the action of PTH. Parathyroid hormone induces formation of 1-hydroxylase in the kidney, required in formation of the active metabolite of vitamin D (see VITAMINS, VITAMIN D).

Metabolites of vitamin D, eg, cholecalciferol (CC), are essential in maintaining the appropriate blood level of Ca^{2+}. The active metabolite, 1,25-dihydroxycholecalciferol (1,25-DHCC), is synthesized in two steps. In the liver, CC is hydroxylated to 25-hydroxycholecalciferol (25-HCC) which, in combination with a globulin carrier, is transported to the kidney where it is converted to 1,25-DHCC. This step, which requires 1-hydroxylase formation, induced by PTH, may be the controlling step in regulating Ca^{2+} concentration. The sites of action of 1,25-DHCC are the bones and the intestine. Formation of 1,25-DHCC is limited by an inactivation process, ie, conversion of 25-HCC to 24,25-DHCC, catalyzed by 24-hydroxylase.

Calcium is absorbed from the intestine by facilitated diffusion and active transport. In the former, Ca^{2+} moves from the mucosal to the serosal compartments along a concentration gradient. The active transport system requires a cation pump. In both processes, a calcium-binding protein (CaBP) is thought to be required for the transport. Synthesis of CaBP is activated by 1,25-DHCC. In the active transport, release of Ca^{2+} from the mucosal cell into the serosal fluid requires Na^{+}.

Paget's Disease of Bone. Paget's disease, *osteitis deformans*, occurs mainly in people over 40. About twice as many men as women are affected. The disease, caused by faulty utilization of Ca^{2+}, may be mild and asymptomatic requiring little or no treatment. Clinical signs are high alkaline phosphatase and high urine hydroxyproline as well as abnormal bone structure which usually goes unrecognized until discovered accidentally by routine x-ray examination (32).

About 10% of the cases are highly symptomatic, ie, considerable disability and even crippling may occur. In these cases, the disease affects many bones including the long bones of the legs, the pelvis, and the skull. The bones soften and buckle, the skull may become enlarged, and height may decrease if the spine is involved resulting in a bent-over stooped posture. In addition there may be severe bone pain and other neurologic complications such as deafness.

One method of treatment is to inject calcitonin, which decreases blood Ca^{2+} concentration and increases bone calcification (33). Another is to increase the release of calcitonin into the blood by increasing the blood level of Ca^{2+} (34). This latter treatment is accomplished by increasing Ca^{2+} absorption from the intestine requiring dietary calcium supplements and avoidance of high phosphate diets. The latter decrease Ca^{2+} absorption by precipitation of the insoluble calcium phosphate.

Other Calcium Disorders. In addition to hypocalcemia, tremors, osteoporosis, and muscle spasms (tetary), calcium deficiency can lead to rickets, osteomalacia, and possibly heart disease. These, as well as Paget's disease, can also result from faulty utilization of calcium. Calcium excess can lead to excess secretion of calcitonin, possible calcification of soft tissues, and kidney stones when combined with magnesium deficiency.

Phosphorus. Eighty-five percent of the phosphorus, the second most abundant element in the human body, is located in bones and teeth (24,35). Whereas there is constant exchange of calcium and phosphorus between bones and blood, there is very little turnover in teeth (25). The Ca:P ratio in bones is constant at about 2:1. Every tissue and cell contains phosphorus, generally as a salt or ester of mono-, di-, or tribasic phosphoric acid, as phospholipids, or as phosphorylated sugars (24). Phosphorus is involved in a large number and wide variety of metabolic functions. Examples are carbohydrate metabolism (36,37), adenosine triphosphate (ATP) from fatty acid metabolism (38), and oxidative phosphorylation (36,39). Common food sources rich in phosphorus are listed in Table 5 (see also PHOSPHORUS COMPOUNDS).

Table 5. Common Food Sources Rich in Phosphorus[a]

Food[b]	Phosphorus in serving, mg
pumpkin kernels, roasted	2658
sunflower seeds, roasted	1548
almonds, dry roasted	756
peanuts, shelled	744
wheat bran	608
black walnuts	580
sardines, Atlantic[c]	555
split peas, cooked	536
brains[c]	438
soybeans, cooked	422
chicken, liver[c]	352
whitefish[c]	323

[a]Refs. 22 and 23.
[b]Serving corresponds to 236 mL (1 cup) unless otherwise noted.
[c]Serving corresponds to 113 g (4 oz).

Energy-Rich Compounds. Reactions of energy-rich compounds are required to drive the many endergonic metabolic processes, such as active transport, muscle contraction, and biosynthesis of fats and macromolecules, eg, nucleic acids (qv) and proteins (qv). Energy-rich compounds contain high energy bonds where the negative free energy resulting from breaking these bonds is large. Most of the high energy compounds are phosphates, eg, adenosine triphosphate (ATP) (**1**), which can undergo hydrolysis of the P–O bond.

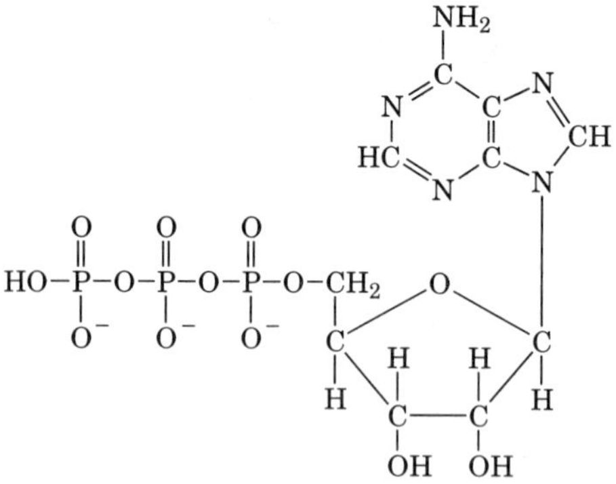

$$\text{(chemical structure of ATP)}$$

(1)

Two and twelve moles of ATP are produced, respectively, per mole of glucose consumed in the glycolytic pathway and each turn of the Krebs (citrate) cycle. In fat metabolism, many high energy bonds are produced per mole of fatty ester oxidized. For example, 129 high energy phosphate bonds are produced per mole of palmitate. Oxidative phosphorylation has a remarkable 75% efficiency. Three moles of ATP are utilized per transfer of two electrons, compared to the theoretical four. The process occurs via a series of reactions involving flavoproteins, quinones such as coenzyme Q, and cytochromes.

Metabolic Functions.   The formation of phosphate esters is the essential initial process in carbohydrate metabolism (see CARBOHYDRATES). The glycolytic, ie, anaerobic or Embden-Meyerhof pathway comprises a series of nine such esters. The phosphogluconate pathway, starting with glucose, comprises a succession of 12 phosphate esters.

Cyclic adenosine monophosphate (cAMP), produced from ATP, is involved in a large number of cellular reactions including glycogenolysis, lipolysis, active transport of amino acids, and synthesis of protein (40). Inorganic phosphate ions are involved in controlling the pH of blood (41). The principal anion of intercellular fluid is HPO_4^{2-} (Fig. 3) (41).

Phospholipids.   Phospholipids, components of every cell membrane, are active determinants of membrane permeability. They are sources of energy, components of certain enzyme systems, and involved in lipid transport in plasma. Because of their polar nature, phospholipids can act as emulsifying agents (42). The structure of most phospholipids resembles that of triglycerides except that one fatty acid radical has been replaced by a radical derived from phosphoric acid and a nitrogen base, eg, choline or serine.

Nucleic Acids.   Phosphorus is an essential component of nucleic acids, polymers consisting of chains of nucleosides, a sugar plus a nitrogenous base, and joined by phosphate groups (43,44). In ribonucleic acid (RNA), the sugar is D-ribose; in deoxyribonucleic acids (DNA), the sugar is 2-deoxy-D-ribose.

Phosphorus Disorders.   Phosphorus nutrient deficiency can lead to rickets, osteomalacia, and osteoporosis, whereas an excess can produce hypocalcemia. Faulty utilization of phosphorus results in rickets, osteomalacia, osteoporosis, and Paget's disease, and renal or vitamin D-resistant rickets.

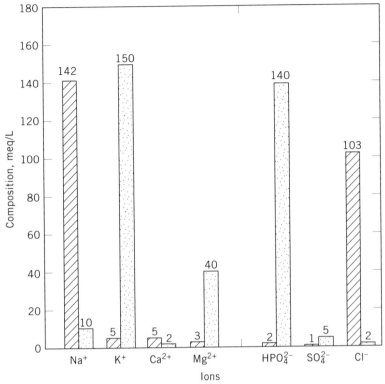

Fig. 3. Cation and anion composition of extracellular ▨ and intracellular ▯ fluids.

Sulfur. Sulfur is present in every cell in the body, primarily in proteins containing the amino acids methionine, cystine, and cysteine. Inorganic sulfates and sulfides occur in small amounts relative to total body sulfur, but the compounds that contain them are important to metabolism (45,46). Sulfur intake is thought to be adequate if protein intake is adequate and sulfur deficiency has not been reported. Common food sources rich in sulfur are listed in Table 6.

Sulfur is part of several vitamins and co-factors, eg, thiamin, pantothenic acid [79-83-4], biotin [58-85-5], and lipoic acid. Mucopolysaccharides, eg, heparin [9005-49-6] and chondroitin sulfate [9007-28-7], contain a monoester of sulfuric acid having an HSO_3^- group. Sulfur-containing lipids isolated from brain and other tissues usually are sulfate esters of glycolipids. The sulfur-containing amino acid taurine [107-35-7] is conjugated to bile acids (45). Labile sulfur is attached to nonheme iron in stoichiometric amounts in the respiratory chain where it is associated with the flavoproteins and cytochrome b (47).

Disulfides. As shown in Figure 4, the A- and B-chains of insulin are connected by two disulfide bridges and there is an intrachain cyclic disulfide link on the A-chain (see INSULIN AND OTHER ANTIDIABETIC DRUGS). Vasopressin [9034-50-8] and oxytocin [50-56-6] also contain disulfide links (48). Oxidation of thiols to disulfides and reduction of the latter back to thiols are quite common and important in biological systems, eg, cysteine to cystine or reduced

Table 6. Common Food Sources Rich in Sulfur[a,b]

Food[c]	Sulfur in serving, mg
peanuts, roasted	555
brazil nuts	406
sardines[d]	350
soybean flour	348
pork chops, lean[d]	339
turkey[d]	328
beef, lean[d]	305
chicken[d]	288
lamb[d]	271
whole grain flour	228
wheat germ[e]	136
navy beans	136
brewer's yeast[f]	76
molasses, blackstrap[f]	70
eggs, whole[g]	67
cheese, Cheddar[h]	64

[a] Refs. 22 and 23.
[b] Calculated as sulfur-containing amino acids, methionine plus cystine.
[c] Serving corresponds to 236 mL (1 cup) unless otherwise noted.
[d] Serving corresponds to 113 g (4 oz).
[e] Serving corresponds to 140 mL (1/2 cup).
[f] Serving corresponds to 20 g (2 tbsp).
[g] Serving corresponds to one medium 48-g egg.
[h] Serving corresponds to 28 g (1 oz).

lipoic acid to oxidized lipoic acid. Many enzymes depend on free SH groups for activation–deactivation reactions. The oxidation–reduction of glutathione (Glu-Cys-Gly) depends on the sulfhydryl group from cysteine.

Sulfur in Fat Metabolism.   Although sulfur is in the same group of the Periodic Table, Group 16(VIA), as oxygen, sulfur functions much more like phosphorus, Group 15(VA), in biological systems. In fat metabolism, sulfur plays a key role analogous to that of phosphorus in carbohydrate metabolism. Fatty acid synthesis and degradation begin and end with the same compound, acetyl-S coenzyme A (acetyl–SCoA) (49).

Detoxification.   Detoxification systems in the human body often involve reactions that utilize sulfur-containing compounds. For example, reactions in which sulfate esters of potentially toxic compounds are formed, rendering these less toxic or nontoxic, are common as are acetylation reactions involving acetyl–SCoA (45). Another important compound is *S*-adenosylmethionine [*29908-03-0*] (SAM), the active form of methionine. SAM acts as a methylating agent, eg, in detoxification reactions such as the methylation of pyridine derivatives, and in the formation of choline (qv), creatine [*60-27-5*], carnitine [*461-06-3*], and epinephrine [*329-65-7*] (50).

Sulfur Disorders.   Sulfur nutrient deficiency results in retarded growth, and faulty utilization in homocystinuria.

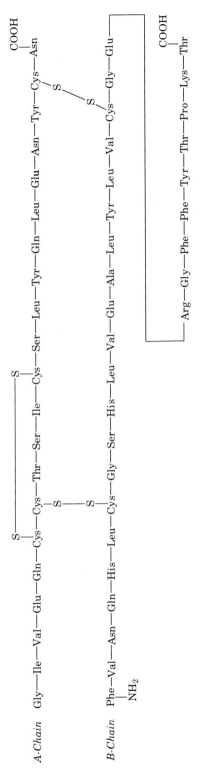

Fig. 4. Sulfur bridges in human insulin.

Sodium and Potassium.   Whereas sodium ion is the most abundant cation in the extracellular fluid, potassium ion is the most abundant in the intracellular fluid. Small amounts of K^+ are required in the extracellular fluid to maintain normal muscle activity. Some sodium ion is also present in intracellular fluid (see Fig. 5). Common food sources rich in potassium may be found in Table 7. Those rich in sodium are listed in Table 8.

Metabolic Functions.   Sodium ion acts in concert with other electrolytes, in particular K^+, to regulate the osmotic pressure and to maintain the appropriate water and pH balance of the body. Homeostatic control of these functions is accomplished by the lungs and kidneys interacting by way of the blood (51,52). Sodium is essential for glucose absorption and transport of other substances across cell membranes. It is also involved, as is K^+, in transmitting nerve impulses and in muscle relaxation. Potassium ion acts as a catalyst in the intracellular fluid, in energy metabolism, and is required for carbohydrate and protein metabolism.

Active Transport.   Maintenance of the appropriate concentrations of K^+ and Na^+ in the intra- and extracellular fluids involves active transport, ie, a process requiring energy (53). Sodium ion in the extracellular fluid ($0.136-0.145\ M\ Na^+$) diffuses passively and continuously into the intracellular fluid ($<0.01\ M\ Na^+$) and must be removed. This sodium ion is pumped from the intracellular to the extracellular fluid, while K^+ is pumped from the extracellular (ca $0.004\ M\ K^+$) to the intracellular fluid (ca $0.14\ M\ K^+$) (53–55). The energy for these processes is provided by hydrolysis of adenosine triphosphate (ATP) and requires the enzyme Na^+–K^+ ATPase, a membrane-bound enzyme which is widely distributed in the body. In some cells, eg, brain and kidney, 60–70 wt % of the ATP is used to maintain the required Na^+–K^+ distribution.

Sodium and potassium ions are actively absorbed from the intestine. As a consequence of the electrical potential caused by transport of these ions,

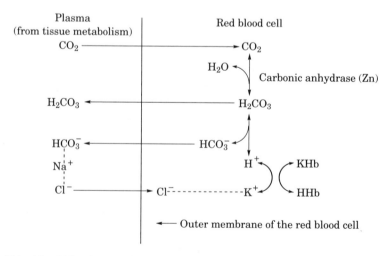

Fig. 5.   Chloride shift where KHb is potassium hemoglobin and HHb is acid hemoglobin (16).

Table 7. Common Food Sources Rich in Potassium[a]

Food[b]	Potassium in serving, mg
pumpkin kernels	1829
figs, dried[c]	1331
pistachio, dry roasted	1241
apricots, dried and cooked	1222
raisins, not packed[d]	1088
almonds, whole, dried	1039
peanuts, shelled	982
soybeans, cooked	972
Swiss chard, cooked	961
potato, baked[e]	903
buckwheat, whole grain	805
split peas, cooked	710
prunes, dried	710
kidney beans, canned	658
avocado[f]	622
Florida	742
California	548
banana[g]	594
orange juice[h]	496

[a] Refs. 22 and 23.
[b] Serving corresponds to 236 mL (1 cup) unless otherwise noted.
[c] Serving corresponds to 187 g (10 whole figs).
[d] Serving corresponds to 145 g (1 cup).
[e] Serving corresponds to one large potato (202 g).
[f] Serving corresponds to 1/2 an average (Florida, 152 g; California, 86.5 g) avocado.
[g] Serving corresponds to an average banana (150 g).
[h] Serving corresponds to 227 g (8 oz).

Table 8. Common Food Sources Rich in Sodium and Chloride[a]

Food[b]	Composition, mg	
	Na$^+$	Cl^{-c}
table salt[d]	2132	3324
Canadian bacon, cooked	1745	2690
pickle[e]	833	1284
ham, cured	1364	2103
corned beef	1139	1756
beans with frankfurters[f]	1105	1703
cheese, processed American[g]	406	626
tuna, canned	400	617
vegetables, canned (beans, carrots, and peas)[f]	350	540
hamburger	86	133

[a] Refs. 22 and 23.
[b] Serving corresponds to 113 g (4 oz) unless otherwise noted.
[c] Calculated from sodium.
[d] Serving corresponds to 5.5 g (1 tsp).
[e] Serving corresponds to one large pickle (65 g).
[f] Serving corresponds to 236 mL (1 cup).
[g] Serving corresponds to 28 g (1 oz).

an equivalent quantity of Cl^- is absorbed. The resulting osmotic effect causes absorption of water (56).

Excretion and Reabsorption of Na^+ and K^+. Selective excretion and reabsorption of Na^+ and K^+ are accomplished by means of the kidney tubular cell membranes (51,55). Water, Na^+, and Cl^- passively diffuse into the proximal tubular cells. Potassium ion is pumped into the cells and Na^+ is pumped out by the Na^+-K^+ pump. In the extracellular fluid, Na^+ and K^+ account for 90–92 wt % and 3 wt %, respectively, of the cations; in the intracellular fluid, the distribution is ca 70–80 wt % K^+ and 6 wt % Na^+. In the adult, ca 160 L of fluid is filtered daily. Urinary volume is 0.5–2.5 L/d so that ca 99 wt % of the filtered water is reabsorbed.

The volume of extracellular fluid is directly related to the Na^+ concentration which is closely controlled by the kidneys. Homeostatic control of Na^+ concentration depends on the hormone aldosterone. The kidney secretes a proteolytic enzyme, rennin, which is essential in the first of a series of reactions leading to aldosterone. In response to a decrease in plasma volume and Na^+ concentration, the secretion of rennin stimulates the production of aldosterone resulting in increased sodium retention and increased volume of extracellular fluid (51,55).

Sodium and Hypertension. Salt-free or low salt diets often are prescribed for hypertensive patients (57). However, sodium chloride increases the blood pressure in some individuals but not in others. Conversely, restriction of dietary NaCl lowers the blood pressure of some hypertensives, but not of others. Genetic factors and other nutrients, eg, Ca^{2+} and K^+, may be involved. The optimal intakes of Na^+ and K^+ remain to be established (58,59).

Other Potassium and Sodium Disorders. Potassium and/or sodium deficiency can lead to muscle weakness and sodium deficiency to nausea. Hyperkalemia resulting in cardiac arrest is possible from 18 g/d of potassium combined with inadequate kidney function. Faulty utilization of K^+ and/or Na^+ can lead to Addison's or Cushing's disease.

Chlorine. Foods rich in chloride are listed in Table 8.

Metabolic Functions. The chlorides are essential in the homeostatic processes maintaining fluid volume, osmotic pressure, and acid–base equilibria (11). Most chloride is present in body fluids; a little is in bone salts. Chloride is the principal anion accompanying Na^+ in the extracellular fluid. Less than 15 wt % of the Cl^- is associated with K^+ in the intracellular fluid. Chloride passively and freely diffuses between intra- and extracellular fluids through the cell membrane. If chloride diffuses freely, but most Cl^- remains in the extracellular fluid, it follows that there is some restriction on the diffusion of phosphate. As of this writing (ca 1994), the nature of this restriction has not been conclusively established. There may be a transport device (60), or cell membranes may not be very permeable to phosphate ions minimizing the loss of HPO_4^{2-} from intracellular fluid (61).

Some of the blood Cl^- is used for formation in the gastric glands of hydrochloric acid, HCl, required for digestion. Hydrochloric acid is secreted into the stomach where it acts with gastric enzymes in the digestive processes. The chloride is then reabsorbed with other nutrients into the blood stream. Chloride is actively transported in gastric and intestinal mucosa. In the kidney, chloride is passively reabsorbed in the thin ascending loop of Henle and actively reabsorbed in the thick segment of the ascending loop, ie, the distal tubule.

In the chloride shift, Cl^- plays an important role in the transport of carbon dioxide (qv). In the plasma, CO_2 is present as HCO_3^-, produced in the erythrocytes from CO_2. The diffusion of HCO_3^- requires the counterdiffusion of another anion to maintain electrical neutrality. This function is performed by Cl^- which readily diffuses into and out of the erythrocytes (Fig. 5). The carbonic anhydrase-mediated Cl^-–HCO_3^- exchange is also important for cellular *de novo* fatty acid synthesis and myelination in the brain (62).

Numerous neurotransmitter receptors, eg, glutamate, γ-aminobutyric acid (GABA), and benzodiazepine (called the valium receptor), have been identified as chloride channel proteins. The genetic defect in cystic fibrosis involves defective-functioning chloride channel proteins with excessive Cl^- loss. Deficient Cl^- during development adversely affects language skills in humans (63), as well as impaired growth in infants and metabolic alkalosis.

Fruit and vegetable juices high in potassium have been recommended to correct hypokalemic alkalosis in patients on diuretic therapy. Apparently the efficacy of this treatment is questionable. A possible reason for ineffectiveness is the low Cl^- content of most of these juices. Because Cl^- is high only in juices in which Na^+ is high, these have to be excluded (64).

Magnesium. In the adult human, 50–70% of the magnesium is in the bones associated with calcium and phosphorus. The rest is widely distributed in the soft tissues and body fluids. Most of the nonbone Mg^{2+}, like K^+, is located in the intracellular fluid where it is the most abundant divalent cation. Magnesium ion is efficiently retained by the kidney when the plasma concentration of Mg^{2-} falls; in this respect it resembles Na^+. The functions of Na^+, K^+, Mg^{2+}, and Ca^{2+} are interrelated so that a deficiency of Mg^{2+} affects the metabolism of the other three ions (26). Foods rich in magnesium are listed in Table 9.

Table 9. Common Food Sources Rich in Magnesium[a]

Food[b]	Magnesium in serving, mg
pumpkin kernels	1212
sunflower seeds	510
buckwheat, whole grain	404
almonds, whole, dried	400
wheat germ, toasted	362
wheat bran	366
spinach, cooked	157
Swiss chard, cooked	150
oysters, eastern	135
corn meal, whole grain	125
navy beans, cooked from dry	107
chocolate, baking[c]	82
molasses, blackstrap[d]	52

[a] Refs. 22 and 23.
[b] Servings correspond to 236 mL (1 cup) unless otherwise noted.
[c] Servings correspond to 28 g (1 oz).
[d] Servings correspond to 15 mL (1 tbsp).

Metabolic Functions. Magnesium is essential in numerous metabolic processes. It is the activator of many enzymes, eg, adenyl cyclase, alkaline phosphatases, and the phosphokinases, pyrophosphatases, and thiokinases (9, 65–67). Because the phosphokinases are required for the hydrolysis and transfer of phosphate groups, magnesium is essential in glycolysis and in oxidative phosphorylation. The thiokinases are required for the initiation of fatty acid degradation. Magnesium is also required in systems in which thiamine pyrophosphate is a coenzyme.

As an activator of the phosphokinases, magnesium is essential in energy-requiring biological processes, such as activation of amino acids, acetate, and succinate; synthesis of proteins, fats, coenzymes, and nucleic acids; generation and transmission of nerve impulses; and muscle contraction (67).

Regulation of Serum Mg^{2+} Concentration. Regulation of serum Mg^{2+} appears to result from a balance among intestinal absorption, renal reabsorption, and excretion (64,68). The controlling factor is probably the renal threshold (65). In the normal adult, intestinal absorption is, to a large extent, proportional to the Mg^{2+} supplied in the diet. The system responds to a wide range of dietary intake by increasing or decreasing urinary excretion; the plasma Mg^{2+} concentration varies only within the normal range. Although exchange of bone salt Mg^{2+} with blood Mg^{2+} is slow as compared to exchange of Ca^{2+}, the bone salts serve as a reservoir of Mg^{2+} to buffer depletions developing over long (weeks or months) periods (69). Parathyroid hormone may be involved in mobilizing bone Mg^{2+} and increasing tubular reabsorption of Mg^{2+}, but not to as great an extent as with Ca^{2+}. PTH may also increase absorption of Mg^{2+} from the intestine.

Magnesium Deficiency. A severe magnesium deficiency in humans is seldom encountered except as a secondary effect resulting from numerous disease states, eg, chronic alcoholism with malnutrition, acute or chronic renal disease, long-term Mg^{2+}-free parenteral feeding, protein–calorie malnutrition, and hyperthyroidism. In these situations, it is difficult to attribute specific clinical manifestations to magnesium deficiency (64). The specific role of magnesium in cardiovascular disease, eg, arrythmia, spasms, or ischemia, remains a subject of conflicting research findings (64).

Magnesium ion is essential for normal Ca^{2+} and K^+ metabolism. In acute experimental magnesium deficiency in humans, hypocalcemia occurs despite adequate calcium intake and absorption and despite normal renal and parathyroid functions. Negative K^+ balance is also observed. All biochemical and clinical abnormalities disappear upon restoration of adequate amounts of magnesium to the diet (64).

Magnesium supplements, such as MgO, have been used successfully in the treatment of patients with a history of calcium oxalate stone formation (70–73). Marginal magnesium deficiencies may occur in areas where food crops are grown on magnesium-deficient soil. One such area is a narrow strip of the Atlantic coast of the United States extending from Pennsylvania to Florida, sometimes called the stone belt because of the high incidence of calcium oxalate kidney stones (70). Stone formation may be the consequence of low magnesium intake. Evidently a high Mg^{2+} concentration in the kidney increases the solubility of calcium oxalate.

Other Magnesium Disorders. Neuromuscular irritability, convulsions, muscle tremors, mental changes such as confusion, disorientation, and hallucinations, heart disease, and kidney stones have all been attributed to magnesium deficiency. Excess Mg^{2+} can lead to intoxication exemplified by drowsiness, stupor, and eventually coma.

Trace Elements and Ultratrace Elements

Iron. The total body content of iron, ie, 3–5 g, is recycled more efficiently than other metals. There is no mechanism for excretion of iron and what little iron is lost daily, ie, ca 1 mg in the male and 1.5 mg in the menstruating female, is lost mainly through exfoliated mucosal, skin, or hair cells, and menstrual blood (74–76). Common food sources rich in iron and other trace elements are listed in Table 10.

Metabolic Functions. A large percentage of the iron in the human body is in hemoglobin: 85 wt % in the adult female, 60 wt % in the adult male (75). The remainder is present in other iron-containing compounds involved in basic metabolic functions, or in iron transport or storage compounds. Myoglobin, the cytochromes, catalase, sulfite oxidase, and peroxidase are all heme iron enzymes. NADH-dehydrogenase, succinate dehydrogenase, α-glycerophosphate dehydrogenase, monoamine oxidase, xanthine oxidase, and alcohol dehydrogenase are nonheme metalloflavoproteins that contain iron. Aconitase and microsomal lipid peroxidase do not contain iron but do require it as a co-factor. The transport and storage proteins for iron are transferrin, ferritin, and hemosiderin (Fig. 6) (74–76). The iron in transferrin in the blood is a combination of freshly absorbed iron and recycled iron.

The hemoglobin molecule (mol wt 65,000) is a tetramer containing four iron atoms (77). The iron reversibly binds oxygen, allowing the hemoglobin molecule to function as an oxygen carrier to the tissues (78). Myoglobin, a monomer containing one iron atom, functions in the muscle by accepting oxygen from the blood and storing it for use during muscle contraction. The iron in both hemoglobin and myoglobin is in the ferrous state, Fe^{2+}. The iron in methemoglobin and metmyoglobin is oxidized to the ferric state, Fe^{3+}. These latter species do not bind oxygen (77).

The ability of iron to exist in two stable oxidation states, ie, the ferrous, Fe^{2+}, and ferric, Fe^{3+}, states in aqueous solutions, is important to the role of iron as a biocatalyst (79) (see IRON COMPOUNDS). Although the cytochromes of the electron-transport chain contain porphyrins like hemoglobin and myoglobin, the iron ions therein are involved in oxidation–reduction reactions (78). Catalase is a tetramer containing four atoms of iron; peroxidase is a monomer having one atom of iron. The iron in these enzymes also undergoes oxidation and reduction (80).

Homeostatic Control of Iron Levels. Absorption of iron from food to maintain homeostasis, tightly controlled, increases in instances of increased demands, such as during pregnancy and lactation, and iron-deficiency states which are the result of blood loss or iron-deficiency anemia resulting from inadequate iron intake (74–76). Iron absorption is greatly reduced in the normal individual when iron stores are adequate or excessive. Absorption is enhanced by acid conditions and reducing agents. Heme iron from animal sources is absorbed more readily than nonheme iron from cereals and vegetables (74–76).

Table 10. Common Food Sources Rich in Trace Elements[a]

Food[b]	In serving, mg
Iron[c]	
fortified cereal	33.0
Grapenuts cereal	32.6
bran flakes fortified	24.8
liver[d]	
lamb	20.2
veal	16.1
chicken	9.6
wheat germ, toasted	10.3
prune juice	10.2
oysters, fried[d]	9.2
molasses, blackstrap[e]	6.4
wheat bran, raw	6.3
apricots, dried	6.1
Fluorine	
dried seaweed[f]	9.1
tea	7.7
fluoridated water	0.2
mackerel[d]	2.1
sardines[d]	1.2
salmon[d]	0.7
Zinc	
raw oysters[d]	167.9
wheat germ, toasted	14.8
beef[d]	5.0
crab, steamed[d]	4.9
pork[d]	4.3
almonds	4.0
walnuts and pecans	3.0
lobster[d]	2.1
chicken[d]	2.0
tuna, canned	1.8
Silicon[g]	
high fiber grains (eg, oats)	
Barbados brown sugar	
organ meats	
Copper	
oysters, raw[d]	15.5
beef or veal liver[d]	13.6
lamb liver[d]	11.3
cocoa powder, nonfat	4.8
brazil nuts	3.4
sunflower seeds	2.6
lobster[d]	2.0
wheat germ	1.0
brewers' yeast[f]	0.9

Table 10. (*Continued*)

Food[b]	In serving, mg
Boron	
apple sauce	0.690
grape juice	0.5111
apple juice	0.4665
peaches, canned	0.4643
cherries, dark	0.3592
broccoli flowerette, frozen	0.3408
pears, canned	0.3026
orange juice	0.1021
cinnamon, ground[h]	0.0518
parsley flakes[i]	0.0269
grape jelly[e]	0.0265
catsup[e]	0.0130
Selenium	
herring, kippered[d]	0.1600
brazil nuts	0.1483
lobster[d]	0.1775
lamb liver[d]	0.1140
grains grown on seleniferous soils[d]	0.0907
torula yeast[j]	0.0246
butter[e]	0.0219
brewer's yeast[j]	0.0182
coconut, fresh grated	0.0158
Chromium	
liver[d]	0.0565
beef[d]	0.0362
brown sugar	0.0261
eggs, whole[k]	0.0250
wheat bran	0.0240
brewer's yeast[j]	0.0236
molasses, blackstrap[e]	0.0230
oysters, raw[d]	0.0226
wheat germ[l]	0.0141
Manganese	
chestnuts	5.9
filberts	5.7
Brazil nuts, raw	3.9
barley, dry	3.4
brown rice	3.2
almonds, raw	2.7
sunflower seeds	2.6
sesame seeds	2.4
peanuts, roasted	2.2
carrot[m]	2.2

Table 10. (*Continued*)

Food[b]	In serving, mg
Cobalt[d,n]	
liver	
beef	127.0
veal	98.7
lamb	91.7
turkey	53.8
chicken	22.0
brains	17.2
liverwurst	15.2
beef, lean	2.8
lamb	2.2
chicken breast	0.4

[a]Refs. 22 and 23.
[b]Serving corresponds to 236 mL (1 cup) unless otherwise noted.
[c]Not all iron from different sources is equally absorbed; see text.
[d]Serving corresponds to 113 g (4 oz).
[e]Serving corresponds to 15–20 g (1 tbsp).
[f]Serving corresponds to 28 g (1 oz).
[g]Accurate quantitative data unavailable; best reported sources listed.
[h]Serving corresponds to 5 g (1 tsp).
[i]Serving corresponds to 59 mL (1/4 cup).
[j]Serving corresponds to 20 g (2 tbsp).
[k]Serving corresponds to one medium egg (48 g).
[l]Serving corresponds to 118 mL (1/2 cup).
[m]Serving corresponds to one large carrot.
[n]Cobalt as μg of vitamin B_{12}.

A system of internal iron exchange exists which is dominated by the iron required for hemoglobin synthesis. For formation of red blood cells, iron stores can furnish 10–40 mg/d of iron, as compared to 1–3 mg from dietary sources (74). Only ca 10 wt % of ingested iron actually is absorbed. Transferrin is essential for movement of iron and without it, as in genetic absence of transferrin, iron overload occurs in tissues. This hereditary atransferrinemia is coupled with iron-deficiency anemia. The iron overload in hereditary or acquired hemochromatosis results in fully saturated transferrin and is treated by phlebotomy (10).

Iron Deficiency and Toxicity. Iron deficiency is a significant worldwide nutritional problem and cause of anemia which can also lead to a decreased resistance to infection. Insufficient dietary iron intake; iron losses, eg, bleeding and parasite infestation; and malabsorption of iron are the principal causes. The groups at greatest risk for developing iron-deficiency anemia are menstruating females, pregnant or nursing females, and young children. Children can experience impaired psychomotor development and intellectual performance.

Iron toxicity resulting from excess absorbable iron ingestion is rare except in Africa where fermented beverages made in large iron pots have levels of iron approaching 80 mg/L in a brew where the pH is very low. This results in Bantu siderosis which can result in hemochromatosis, ie, damage to various organs from excessive storage of iron. This condition can cause numerous disease states, eg, hepatic fibrosis and diabetes in 80% of the cases of idiopathic hemochromato-

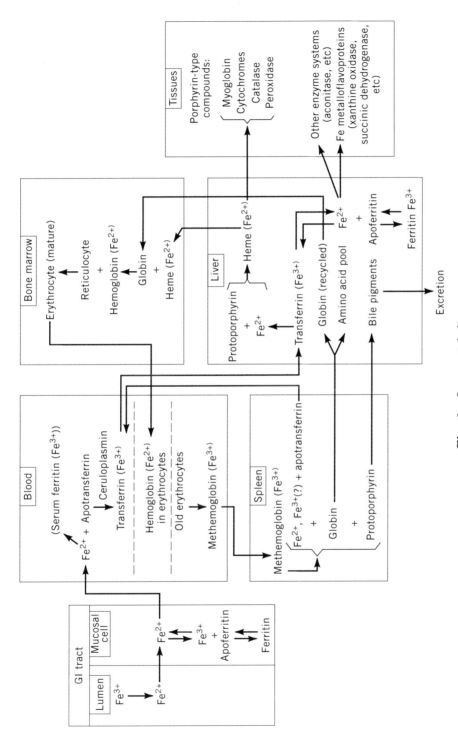

Fig. 6. Iron metabolism.

sis patients (74–76). Iron overload is frequently a complication of repeated blood transfusions in anemias, eg, thalassemia (74,76). The lethal dose of ferrous sulfate for a two-year old is 2 g; for an adult the lethal dose is from 200–300 g.

Fluorine. Fluoride is present in the bones and teeth in very small quantities. Human ingestion is from 0.7–3.4 mg/d from food and water. Evidence for the essentiality of fluorine was obtained by maintaining rats on a fluoride-free diet, resulting in decreased growth rate, decreased fertility, and anemia. These impairments were remedied by supplementing the diets with fluoride (81). Similar effects have been reported in goats (82).

Fluoridation and Dental Caries. Fluoridation of public water supplies, a common practice throughout much of the United States, may be an effective means of significantly reducing the incidence of dental caries (83–85) (see DENTAL MATERIALS; FLUORINE COMPOUNDS, INORGANIC). Concern regarding the narrow range of safety between effective and toxic fluoride concentrations has been expressed and poisoning from excessive fluoride, fluorosis, added to public water has been reported (86). Assertions that fluoridation of water supplies increases the incidence of cancer have not been substantiated (87).

Other Effects of Fluoride. Excess fluoride ingestion damages developing teeth, causing mottling, chalky-white coloration, and pitting (83,84). Adding fluoride to animal feed leads to fragile and brittle teeth and bones (88). Fluoride is an inhibitor of enzymes, especially enolase in the glycolytic pathway (88). Fluoride may have some effect on osteoporosis (83). Fluoride supplementation results in poorly mineralized bone unless very high pharmacologic levels, ie, 50,000 IU of vitamin D and 900 mg of calcium are included in the regimen (89).

Zinc. The 2–3 g of zinc in the human body are widely distributed in every tissue and tissue fluid (90–92). About 90 wt % is in muscle and bone; unusually high concentrations are in the choroid of the eye and in the prostate gland (93). Almost all of the zinc in the blood is associated with carbonic anhydrase in the erythrocytes (94). Zinc is concentrated in nucleic acids (90), and found in the nuclear, mitochondrial, and supernatant fractions of all cells.

Metabolic Functions. Zinc is essential for the function of many enzymes, either in the active site, ie, as a nondialyzable component, of numerous metalloenzymes or as a dialyzable activator in various other enzyme systems (91,92). Well-characterized zinc metalloenzymes are the carboxypeptidases A and B, thermolysin, neutral protease, leucine amino peptidase, carbonic anhydrase, alkaline phosphatase, aldolase (yeast), alcohol dehydrogenase, superoxide dismutases, and aspartate transcarbamylase. Other enzymes reported to contain zinc include the RNA and DNA polymerases and a number of dehydrogenases, eg, lactic, malic, and glutamic (91–93). Generally, enzymes that contain zinc in one species often contain zinc in other species. This is true across species for carbonic anhydrase, but it is not true for aldolase which is a zinc enzyme in yeast but not in mammalian muscle (93). In addition to its role in the various enzyme activities, zinc is a membrane stabilizer and a participant in electron-transfer processes (93).

Zinc–hormone interactions include hormonal influence on absorption, distribution, transport, and excretion of zinc and zinc influence on synthesis, secretion, receptor binding, and function of numerous hormones (qv). Zinc enhances pituitary activity by increasing circulating levels of growth hormone, thyroid-stimulating hormone, luteinizing hormone, follicle-stimulating hormone, and adrenocorticotropin (93). The role of zinc in insulin action is recognized but not well understood (93). Zinc is required for maintenance of normal plasma concentrations of vitamin A and for normal mobilization of vitamin A from the liver (93,95).

Zinc Deficiency. Zinc was confirmed as essential for humans in 1956 (92, 94) and deficiency symptoms were reported in 1961 (96). The size of the human fetus is correlated with zinc concentration in the amniotic fluid and habitual low zinc intake in the pregnant female is thought to be related to several congenital anomalies in humans (92,95). Low zinc intakes result in hypogonadism, dwarfism, mental retardation, low serum and red blood cell zinc in humans and animals, and retarded growth and teratogenic effects on the nervous system in rats.

In children suffering from marginal zinc deficiency, impaired taste acuity, poor appetite, and suboptimal growth can be reversed upon zinc supplementation (93,95). Accelerated wound healing occurs in humans upon zinc supplementation (95), suggesting that marginal zinc deficiency in humans may be more widespread than has been thought. Zinc supplementation has also been effective in alleviating symptoms of active rheumatoid arthritis in clinical trials (97). Acrodermatitis enteropathica, a hereditary disease that involves aberrant zinc metabolism, responds to oral zinc supplementation (93,95) (see Table 10). Excessive zinc intake may interfere with copper metabolism.

Silicon. Silicon comes mainly from ingestion of silicates, primarily from vegetables. It is found in the serum as silicic acid [*1343-98-2*], $Si(OH)_4$, and normal blood serum levels are ca 1 mg/100 mL regardless of intake because of efficient kidney excretion of excess (98). Silicon is necessary for calcification, growth, and as cross-linking material in mucopolysaccharide formation (98). Essentiality of silicon in rats and chicks has been established (98–101). Silicon, recognized as essential for humans in 1991, is especially helpful in situations where the diet is low in calcium or high in aluminum, or thyroid function is inadequate (98). The human requirement may be 5–20 mg/d (98). Silicon deficiency may lead to altered metabolism of connective tissue and bone and/or aluminum accumulation in the brain.

Copper. All human tissues contain copper. The highest amounts are found in the liver, brain, heart, and kidney (102). In blood, plasma and erythrocytes contain almost equal amounts of copper, ie, ca 110 and 115 mg/100 mL, respectively.

Metabolic Functions. In plasma, ca 90 wt % of copper is in the metalloprotein ceruloplasmin, also known as a_2-globulin, mol wt 151,000, which contains 8 atoms of copper per molecule (103). Ceruloplasmin has been identified as a ferroxidase(I) which catalyses the oxidation of aromatic amines and of Fe^{2+} to Fe^{3+} (79,104). The ferric ion is then incorporated into transferrin which is necessary for the transport of iron to tissues involved in the synthesis of iron-containing compounds, eg, hemoglobin. Lowered levels of ceruloplasmin interfere with hemoglobin synthesis.

Erythrocuprein, which contains about 60 wt % of the erythrocyte copper, hepatocuprein, and cerebrocuprein act as superoxide dismutases. Each contains two atoms of copper per molecule, having mol wt ca 34,000. The superoxide ion,

O_2^+, and peroxide, O_2^{2-}, are the two main toxic by-products of oxygen reduction in the body (1,105). Superoxide dismutase catalyzes the dismutation, ie, the simultaneous oxidation, reduction, and decarboxylation, of superoxide, a free-radical anion, thus protecting the cell from oxidative damage.

The oxidation of the ϵ-amino groups of lysine is required for the cross-linking of polypeptide chains of collagen and elastin. The catalyst for this reaction is the copper metalloenzyme lysyl oxidase (103). Copper deficiency is characterized by poorly formed collagen which leads to bone fragility and spontaneous bone fractures in animals, and also results in cardiac hypertrophy (103). Abnormal electrocardiographs have been noted when low copper diets were fed to humans (103). Anemia, neutropenia, and bone disease have been reported in children having protein calorie malnutrition (PCM) and accompanying hypocupremia (102). Some other copper metalloproteins are cytochrome c oxidase, dopamine B hydroxylase, urate oxidase, tyrosinase, and ascorbic acid oxidase. Most copper enzymes are involved in redox reactions (102).

Genetic Disease.   At least two genetic diseases involving copper are known. Wilson's disease, an autosomal recessive disease, is usually detected in adulthood. There is a toxic increase in copper storage upon neurological and liver damage, but a decrease in the amount of circulating copper because of decreased ceruloplasmin (103,104,106). Menkes' kinky-hair syndrome is an x-linked defect of copper transport out of the intestinal cell that results in lowered activity of several copper-dependent enzymes, lowered copper levels in the serum, progressive mental deterioration, defective keratinization of hair, and degenerative changes in the aorta (103,104,107,108).

Dietary Copper.   Analytical data indicate that many diets contain less than the RDA for copper (109). Excessive copper has been reported to be fatal for oral dose levels of copper sulfate of 200 mg/kg body weight for a child and 50 mg/kg for adults.

Boron.   The essentiality of boron, first accepted for higher plants in 1923, then for animals, was recognized in 1981 for human metabolism (110,111). Boron is reported to help maintain function or stability of cell membranes and is thought to be involved with hormone reception and transmembrane signaling (101). Boron forms complexes with organic compounds having hydroxyl groups, especially those having more than two such groups, such as sugars and polysaccharides, adenosine-5-phosphate, pyridoxine, riboflavin, dehydroascorbic acid, and pyridine nucleotides (112). Enhanced need for boron may develop with nutritional or metabolic stress involving other nutrients, eg, magnesium deprivation or physiological changes in calcium metabolism (101). Boron depletion impairs cognitive function (112). Organs known to contain the highest levels of boron are bone, spleen, and thyroid (112). An excess of boron can, however, cause seizures in infants, riboflavinuria, and gastrointestinal upset.

Selenium.   Selenium, thought to be widely distributed throughout body tissues, is present mostly as selenocysteine in selenoproteins or as selenomethionine (113,114). Animal experiments suggest that greater concentrations are in the kidney, liver, and pancreas and lesser amounts are in the lungs, heart, spleen, skin, brain, and carcass (115).

Metabolic Functions.   The most clearly documented role for selenium is as a necessary component of glutathione peroxidase (Fig. 7) (113,114). Glutathione

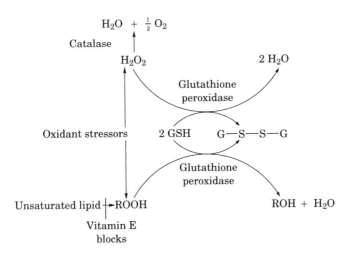

Fig. 7. The glutathione peroxidase (a selenium enzyme) system where GSH = *N*-(*N*-ʟ-γ-glutamyl-ʟ-cysteinyl)glycine and G–S–S–G, the disulfide.

peroxidase reduces hydrogen peroxide, which is formed by free-radical oxi-dant–stresser reactions, to H_2O and reduces organic peroxides, eg, those formed by the peroxidation of unsaturated fatty acids, to alcohols and H_2O (79,113,116). Phospholipid hydroperoxide glutathione peroxidase inhibits lipid peroxidation (114). Selenium is also involved in the functions of additional enzymes, eg, type 1 iodothyronine deiodinase (114), leukocyte acid phosphatase, and glucuronidases (117). A role for selenium in electron transfer has been suggested (115,118) as has involvement in nonheme iron proteins (114,115). Selenium and vitamin E appear to be necessary for proper functioning of lysosomal membranes (119). A role for selenium in metabolism of thyroid hormone has been confirmed (114).

Toxicity, Deficiency, and Medicinal Aspects. Interest in biological effects of selenium developed upon recognition that alkali disease and blind staggers of grazing livestock in the western United States were the result of selenium poisoning. There are unusually high concentrations of selenium in certain plants, because of selenium-accumulating properties, or in ordinary plants growing on highly seleniferous soils. No treatment for this type of poisoning is known (114), thus excess selenium in the animal diet must be avoided. Prolonged ingestion of up to 600 mg/d of selenium did not produce toxic effects in humans (120). Toxic effects in humans have been reported, however, from chronic ingestion of food in China supplying 5 mg/d of Se, and from supplements in the United States of 27–2387 mg/d of Se (113,114). Effects include hair loss, changes in the nails, gastrointestinal upset, and peripheral neuropathy.

Pure selenium deficiency, without concurrent vitamin E deficiency, is not generally seen except in animals on experimental diets (113). In China, selenium deficiency in humans has been associated with Keshan disease, a cardiomyopa-thy seen in children and in women of child-bearing ages, and Kashin-Beck dis-ease, an endemic osteoarthritis in adolescents (113).

Selenium may have anticarcinogenic effects possibly because of the anti-oxidant properties of selenium compounds (114,115). Animals involved in early carcinogenicity studies lived significantly longer when fed supplementary selenium. The role of selenium in the prevention of cancer and other chronic diseases, eg, heart conditions, and as an antiaging and antimutagenic agent are under investigation (113).

Iodine. Of the 10–20 mg of iodine in the adult body, 70–80 wt % is in the thyroid gland (see THYROID AND ANTITHYROID PREPARATIONS). The essentiality of iodine, present in all tissues, depends solely on utilization by the thyroid gland to produce thyroxine [51-48-9] and related compounds. Well-known consequences of faulty thyroid function are hypothyroidism, hyperthyroidism, and goiter. Dietary iodine is obtained from eating seafoods and kelp and from using iodized salt.

Metabolic Functions. The functions of the thyroid hormones and thus of iodine are control of energy transductions (121). These hormones increase oxygen consumption and basal metabolic rate by accelerating reactions in nearly all cells of the body. A part of this effect is attributed to increase in activity of many enzymes. Additionally, protein synthesis is affected by the thyroid hormones (121,122).

Thyroid Hormones. Iodine, absorbed as I^-, is oxidized in the thyroid and bound to a thyroglobulin. The resultant glycoprotein, mol wt 670,000, contains 120 tyrosine residues of which ca two-thirds are available for binding iodine in several ways. Proteolysis introduces the active hormones 3,5,3'-triiodothyronine (T_3) and 3,5,3',5'-tetraiodothyronine (T_4), (thyroxine) in the ratio $T_4:T_3$ of 4:1 (121,122).

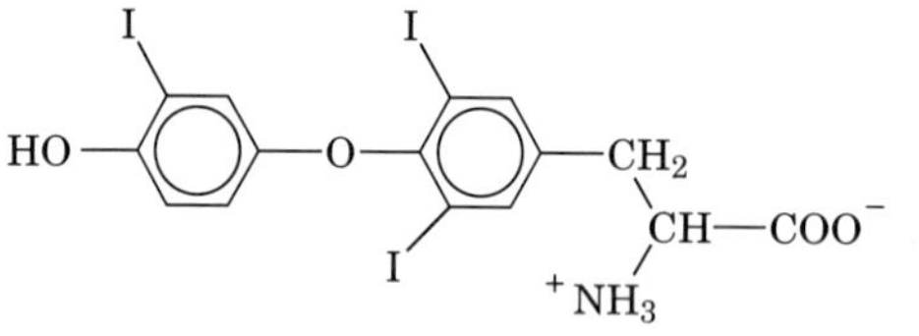

T_3 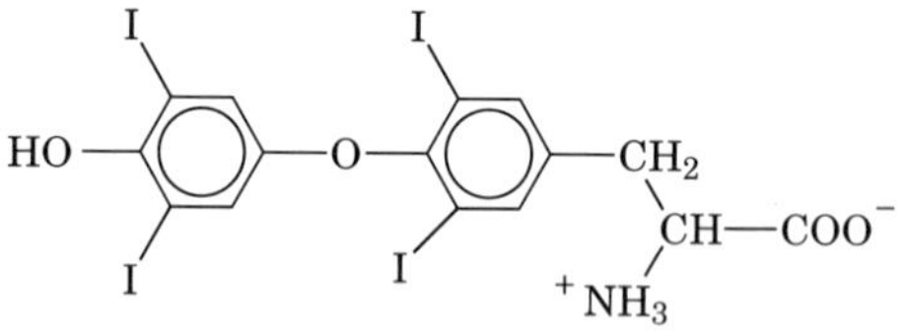 T_4

Only small amounts of free T_4 are present in plasma. Most T_4 is bound to the specific carrier, ie, thyroxine-binding protein. T_3, which is very loosely bound to protein, passes rapidly from blood to cells, and accounts for 30–40% of total thyroid hormone activity (121). Most of the T_3 may be produced by conversion of T_4 at the site of action of the hormone by the selenoenzyme deiodinase (114). That is, T_4 may be a prehormone requiring conversion to T_3 to exert its metabolic effect (123).

Thyroid-stimulating hormone (TSH), also called thyrotropin, influences thyroid activity by a feedback mechanism (121–123). TSH, secreted by the pituitary gland, stimulates the thyroid to increase production of thyroid hormones when the blood hormone level is low. This results in an increase in size of thyroid cells. If this continues, the increase in size of the thyroid becomes noticeable as simple goiter. In many parts of the world, simple goiter is endemic and usually results from dietary iodine deficiency (121) or goiterogens in foods which bind iodine. Iodine deficiency disorders (IDD) include cretinism, myxedema, hypothyroidism, and goiter (121,122). In technologically advanced countries, the problem of iodine deficiency has been minimized by the use of iodized salt (121,122). Faulty utilization of iodine can lead to Grave's disease. A sodium iodide excess can also produce goiter and an excess of 500 mg/kg body weight can be fatal.

Manganese. The adult human body contains ca 10–20 mg of manganese (124,125), widely distributed throughout the body. The largest Mg^{2+} concentration is in the mitochondria of the soft tissues, especially in the liver, pancreas, and kidneys (124,126). Manganese concentration in bone varies widely with dietary intake (126) (see Table 10).

Metabolic Functions. Manganese is essential for normal body structure, reproduction, normal functioning of the central nervous system, and activation of numerous enzymes (126). Synthesis of the mucopolysaccharide chondroitin sulfate involves a series of reactions where manganese is required in at least five steps (127). These reactions are responsible for formation of polysaccharides and linkage between the polysaccharide and proteins that form the mucopolysaccharide of cartilage (124). In addition to the glycosyl transferases of mucopolysaccharide synthesis that require manganese, a number of metalloenzymes contain Mn^{2+} (124,126,127). Superoxide dismutase and pyruvic carboxylase contain two and four atoms of manganese per molecule, respectively. Most enzymes that require magnesium, eg, kinases, can use manganese *in vitro*, and Mg^{2+} can also substitute for Mn^{2+} in some enzyme activations (124,126). An excess of manganese can lead to neural damage and possible impaired insulin production.

Manganese Deficiency. In animals, manganese deficiency results in wide-ranging disorders, eg, impaired growth, abnormal skeletal structure, disturbances of reproduction, and defective lipid and carbohydrate metabolism (124). The common denominator appears to be the impairment of activity of enzymes that have a specific requirement for manganese. Although overt manganese deficiency has not been induced in humans, some forms of epilepsy in humans and animals and a decrease in glucose tolerance in animals have been linked to low levels of manganese in the tissues (128,129).

Molybdenum. Molybdenum is a component of the metalloenzymes xanthine oxidase, aldehyde oxidase, and sulfite oxidase in mammals (130). Two other molybdenum metalloenzymes present in nitrifying bacteria have been characterized: nitrogenase and nitrate reductase (131). The molybdenum in the oxidases, is involved in redox reactions. The heme iron in sulfite oxidase also is involved in electron transfer (132).

Foods rich in molybdenum include legumes, dark green vegetables, liver, whole-grain cereals, and milk.

Xanthine oxidase, mol wt ca 275,000, present in milk, liver, and intestinal mucosa (131), is required in the catabolism of nucleotides. The free bases guanine and hypoxanthine from the nucleotides are converted to uric acid and xanthine in the intermediate. Xanthine oxidase catalyzes oxidation of hypoxanthine to xanthine and xanthine to uric acid. In these processes and in the oxidations catalyzed by aldehyde oxidase, molecular oxygen is reduced to H_2O_2 (133). Xanthine oxidase is also involved in iron metabolism. Release of iron from ferritin requires reduction of Fe^{3+} to Fe^{2+} and reduced xanthine oxidase participates in this conversion (133).

Copper–Molybdenum Antagonism. A copper–molybdenum antagonism involving sulfate occurs in animals, ie, large amounts of molybdenum and sulfate can depress copper absorption (133). Cattle grazing on pasturage of high Mo content succumb to teart or peat scours, characterized by diarrhea and general wasting. Control involves increasing copper intake. The Cu–Mo antagonism has been observed in humans (133,134). Significant increases in urinary copper excretion have been observed with increasing Mo intake (135).

Deficiency or Toxicity in Humans. Molybdenum deficiency in humans results in deranged metabolism of sulfur and purines and symptoms of mental disturbances (130). Toxic levels produce elevated uric acid in blood, gout, anemia, and growth depression. Faulty utilization results in sulfite oxidase deficiency, a lethal inborn error.

Chromium. The history of the investigations establishing the essentiality of chromium has been reviewed (136). An effect of brewer's yeast in preventing or curing impaired glucose tolerance in rats was revealed, and the active factor was identified as a Cr(III) organic complex, glucose tolerance factor (GTF) (137,138).

Metabolic Functions. Chromium(III) potentiates the action of insulin and may be considered a cofactor for insulin (137,138). In *in vitro* tests of epididymal fat tissue of chromium-deficient rats, Cr(III) increases the uptake of glucose only in the presence of insulin (137). The interaction of Cr(III) and insulin also is demonstrated by experimental results indicating an effect of Cr(III) in translocation of sugars into cells at the first step of sugar metabolism. Chromium is thought to form a complex with insulin and insulin receptors (136).

There appears to be a chromium pool in individuals who are not chromium deficient (136). When there is an increase in level of circulating insulin in response to a glucose load, an increase in circulating chromium occurs over a period of 0.5–2 h. This is followed by a decline and excretion of chromium in urine increases. Chromium deficiency is indicated when no increase or a small increase in blood chromium level or urine chromium occurs.

Test Results with Humans. Studies of elderly people and mildly diabetic patients showed significant improvement in the glucose tolerance test (GTT) when chromium supplementation of 150–200 μg/d was given (136,139–141). In other tests, these positive results were not obtained (142). It is possible that not all subjects are capable of utilizing inorganic chromium to the same extent. Some may require a preformed GTF. Chromium chloride supplementation has been effective in normalizing impaired glucose tolerance in malnourished children and in patients receiving total parenteral nutrition for a long time (143,144). The most available form of chromium is GTF obtained from brewer's yeast. In human studies in which GTF was administered, one of the most significant results was normalization of the exaggerated insulin responses to glucose loads (145).

Attempts to isolate GTF from brewer's yeast have resulted in production of very active concentrates, but the substance is too labile to be obtained in the solid state (136). However, it has been shown that GTF is a Cr(III) complex containing two coordinated nicotinate radicals and other amino acid anions (146). Active preparations containing similar complexes have been synthesized (147). Chromium deficiency may also lead to atherosclerosis and peripheral neuropathy.

Chromium(III) Chemistry. The most characteristic reactions of Cr(III) in aqueous solution at >4 pH, eg, in the intestine and blood, and hydrolysis and olation (147). As a consequence, inorganic polymeric molecules form that probably are not able to diffuse through membranes. This may be prevented by ligands capable of competing for coordination sites on Cr(III) (see COORDINATION COMPOUNDS) (147). Thus any large fraction of ingested Cr(III) should be absorbed. Chromium(III) in the form of GTF may be more efficiently absorbed.

Cobalt. Cobalt is nutritionally available only as vitamin B_{12} (148). Although Co^{2+} can function as a replacement *in vitro* for other divalent cations, in particular Zn^{2+}, no *in vivo* function for inorganic cobalt is known for humans (149). In ruminant animals, B_{12} is synthesized by bacteria in the rumen.

B_{12} Vitamins. In foods, vitamin B_{12} (partial structure, **2**) occurs only in animal products (148). A particularly rich source is liver from which it was originally isolated following the successful use of liver in treatment of pernicious anemia, a form of megalo-blastic anemia not responsive to iron supplementation. Vitamin B_{12} is produced commercially by bacterial fermentation in the form of cyanocobalamin for use as a dietary supplement and for injection in the treatment of pernicious anemia. In vitamin B_{12}, the cobalt atom is at the center of the corrin ring coordinated to nitrogen atoms of four five-membered heterocyclic rings. In cyanocobalamin, the source of the cyanide is apparently contamination from the reagents used. Naturally occurring forms of B_{12} contain little or no cyanide (150). In coenzyme B_{12} the 5'-C of the deoxyadenosyl group is bonded to the cobalt atom. The chemical and biological activities of coenzyme B_{12} are dependent on this carbon–metal bond (151).

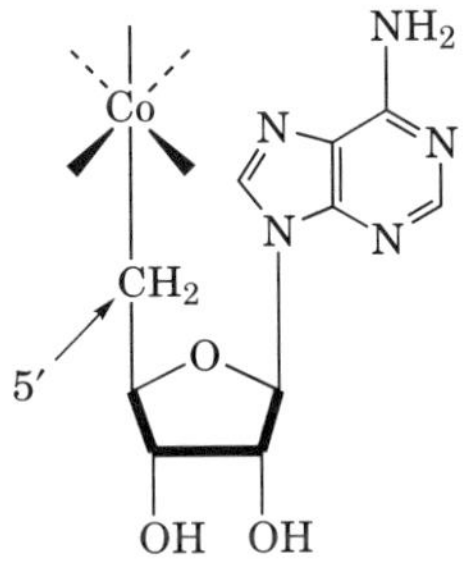

(**2**)

Metabolic Functions. In pernicious anemia, the bone marrow fails to produce mature erythrocytes as a result of defective cell division, a consequence of impaired DNA synthesis which requires vitamin B_{12}. If the disease goes untreated, extensive neurological damage, eg, irreversible degeneration of the spinal cord by demyelinization, may occur because of faulty fatty acid metabolism (148,151). Coenzyme B_{12} is required in mammalian enzyme systems involving methylmalonyl CoA and methyl transferase (transmethylase). Methylmalonyl CoA is an intermediate in the metabolism of propionate which is converted to succinyl CoA. Transmethylase is required for methylation of homocysteine to methionine (150,152).

Intrinsic Factor. Vitamin B_{12} deficiency commonly is caused by inadequate absorption resulting from a lack or insufficient intrinsic factor (IF) (153).

Intrinsic factor is a glycoprotein, mol wt ca 50,000, which binds vitamin B_{12} in a 1:1 molar ratio. The B_{12}–IF complex, formed in the stomach, is absorbed in the ileum. Absorption in this part of the intestine occurs because of the specific characteristics of the cells of the microvilli (brush border) of the ileum (153). The IF remains in the intestine attached to the epithelial cells. Transport of B_{12} into the blood stream requires Ca^{2+}. In the blood, B_{12} is bound to transcobalamin II (transport protein). Whatever bound B_{12} is not utilized immediately is stored in the liver. With increasing quantities of dietary B_{12}, the fraction that is absorbed decreases. Generally, vitamin B_{12} is excreted in the urine, but with large intake, some is excreted in the bile. A nutritional excess of cobalt can lead to polycythemia.

Tin.   The widespread use of canned foods results in a daily intake of tin that is ca 1–17 mg for an adult male (154). At this level it has not been shown to be toxic. Some grains also contain tin. Too much tin can adversely affect zinc balance and iron metabolism. Essentiality has not been confirmed for humans. It has been shown for the rat. An enhanced growth rate results from tin supplementation of low tin diets (85). Animals on deficient diets exhibit poor growth and decreased feed efficiency (155).

Vanadium.   Vanadium is essential in rats and chicks (85,156). Estimated human intake is less than 4 mg/d. In animals, deficiency results in impaired growth, reproduction, and lipid metabolism (157), and altered thyroid peroxidase activities (112). The levels of coenzyme A and coenzyme Q_{10} in rats are reduced and monoamine oxidase activity is increased when rats are given excess vanadium (157). Vanadium may play a role in the regulation of (NaK)–ATPase, phosphoryl transferases, adenylate cyclase, and protein kinases (112).

Nickel.   There is considerable evidence for the essentiality of nickel in animals. Various pathological manifestations of nickel deficiencies have been observed in chicks, cows, goats, pigs, rats, and sheep (112,158,159). Average intake is reported to be about 60–260 μg/d, and a dietary requirement for humans of less than 100 μg/d has been suggested (101,112). In vitro studies have shown nickel to be an activator of several enzymes. Nickel stabilizes RNA and DNA against thermal denaturation and may have a role in membrane structure or metabolism (158). Nickel may be required for metabolism of odd-chain fatty acids (112). A nickel metalloprotein has been isolated from human serum (158).

Arsenic.   Arsenic is under consideration for inclusion as an essential element. No clear role has been established, but aresenic, long thought to be a poison, may be involved in methylation of macromolecules and as an effector of methionine metabolism (158,160). Most research has focused on the toxicity or pharmaceutical properties of arsenic (158).

Health and Safety Factors

Under unusual circumstances, toxicity may arise from ingestion of excess amounts of minerals. This is uncommon except in the cases of fluorine, molybdenum, selenium, copper, iron, vanadium, and arsenic. Toxicosis may also result from exposure to industrial compounds containing various chemical forms of some of the minerals. Aspects of toxicity of essential elements have been published (161).

Efficient homeostatic controls of mammalians generally prevent serious toxicity from ingestion of the mineral nutrients. Toxicity may occur under conditions far removed from those of nutritional significance or for individuals suffering from some pathological conditions. Because of very low concentrations in foods, the trace elements are not toxic under normal nutritional conditions. Exceptions are selenium and iron (162).

BIBLIOGRAPHY

"Mineral Nutrients" in *ECT* 3rd ed., Vol. 15, pp. 570–603, by C. L. Rollinson and M. G. Enig, University of Maryland.

1. I. Fridovich. *Am. Sci.* **63**, 54 (1975).
2. *Nutr. Rev.* **51**, 243 (1993).
3. Recommended *Dietary Allowances*, Food and Nutrition Board, National Academy of Sciences National Research Council, Washington, D.C., 1989.
4. M. Brown, ed., *Present Knowledge in Nutrition*, 6th ed., International Life Sciences Institute/Nutrition Foundation, Washington, D.C., 1990.
5. M. C. Linder, ed., *Nutritional Biochemistry and Metabolism with Clinical Applications*, Elsevier, New York, 1985.
6. M. E. Shils, J. A. Olson, and M. Shike, eds., *Modern Nutrition in Health and Disease*, 8th ed., Vols. 1 and 2, Lea & Febiger, Philadelphia, Pa., 1993.
7. *Conference on Micronutrient Interactions: Vitamins, Minerals and Hazardous Elements*, The New York Academy of Sciences, Feb. 20–22, 1980.
8. W. Mertz, ed., *Trace Elements in Human and Animal Nutrition*. Vols. 1 and 2, Academic Press, Inc., San Diego, Calif., 1987.
9. E. J. Calabrese. *Nutrition and Environmental Health: The Influence of Nutritional Status on Pollutant Toxicity and Carcinogenicity*, Vol. II, *Minerals and Macronutrients*, Wiley-Interscience, New York, 1980.
10. T. M. Devlin, ed., *Textbook of Biochemistry with Clinical Correlations*, John Wiley & Sons, Inc., New York, 1986.
11. A. L. Lehninger, D. L. Nelson, and M. M. Cox, *Principles of Biochemistry*, 2nd ed., Worth Publishers, Inc., New York, 1993.
12. R. Montgomery, R. L. Dryer, T. W. Conway, and A. A. Spector, *Biochemistry, A Case-Oriented Approach*, 2nd ed., The C. V. Mosby Co., St. Louis, Mo., 1977.
13. J. M. Orten and O. W. Neuhaus, *Human Biochemistry*, 9th ed., The C. V. Mosby Co., St. Louis, Mo., 1975.
14. A. White, P. Handler, and E. L. Smith, *Principles of Biochemistry*, 5th ed., McGraw-Hill Book Co., Inc., New York, 1973.
15. A. C. Guyton, *Textbook of Medical Physiology*, 5th ed., W. B. Saunders Co., Philadelphia, Pa., 1976.
16. H. A. Harper, *Review of Physiological Chemistry*, Lange Medical Publications, Los Altos, Calif., 1975.
17. R. L. Pike and M. L. Brown, *Nutrition: An Integrated Approach*, 2nd ed., John Wiley & Sons, Inc., New York, 1975.
18. A. H. Ensminger, M. E. Ensminger, J. Konlande, and J. R. K. Robson, *Food and Nutrition Encyclopedia*, 2nd ed., CRC Press, Boca Raton, Fla., 1994.
19. H. S. Mitchell, H. J. Rynbergen, L. Anderson, and M. V. Dibble, *Nutrition in Health and Disease*, J. B. Lippincott Co., Philadelphia, Pa., 1976.
20. H. J. M. Bowen, *Trace Elements in Biochemistry*, Academic Press, Inc., New York, 1966.

21. H. A. Schroeder, *The Trace Elements and Man*, The Devin-Adair Co., Old Greenwich, Conn., 1973.

22. *Food Processor II® Nutrient Analysis System*, Version 3.04, ESHA Research, Salem, Oreg., 1990; C. D. Hunt, T. R. Shuler, and L. M. Mullen, *J. Am. Diet. Assoc.* **91**, 558 (1991).

23. *Composition of Foods: Raw, Processed, Prepared, Agricultural Handbook* No. 8-1 to 8-21, Washington, D.C., Consumer and Food Economics Institute, United States Department of Agriculture, 1989 to 1992.

24. L. H. Allen and R. J. Wood, in Ref. 10, Chapt. 7, pp. 144–163.

25. Ref. 19, p. 52.

26. J. B. Peterson, *Limestone* (Fall 1980).

27. Ref. 16, p. 660.

28. Ref. 16, p. 182.

29. Ref. 12, p. 182.

30. C. R. Martin, *Textbook of Endocrine Physiology*, Oxford University Press, Inc., New York, 1976, p. 155.

31. H. F. DeLuca, *J. Steroid Biochem.* **11**, 35 (1979).

32. S. Wallach, ed., *Paget's Disease of Bone*, Armour Pharmaceutical Co., Phoenix, Ariz., 1979.

33. A. Avramides, *Clin. Orthop.* **127**, 78 (1977); W. C. Sturtridge, J. E. Harrison, and D. R. Wilson, *Can. Med. Assoc.* **117**, 1031 (1977).

34. R. A. Evans, *Aust. N.Z. J. Med.* **7**, 259 (1977).

35. C. D. Arnaud and S. D. Sanchez, in Ref. 4, Chapt. 24, p. 212.

36. Ref. 13, p. 173.

37. T. P. Bennett and E. Frieden, *Modern Topics in Biochemistry*, The MacMillan Co., New York, 1966, p. 81.

38. Ref. 37, p. 117.

39. Ref. 37, p. 70.

40. Ref. 17, p. 74.

41. Ref. 11, p. 157.

42. Ref. 17, p. 42.

43. Ref. 37, p. 120.

44. Ref. 13, p. 29.

45. Ref. 17, p. 191.

46. O. H. Muth and J. E. Oldfield, eds., *Symposium: Sulfur in Nutrition*, The Avi Publishing Co., Inc., Westport, Conn., 1970.

47. Ref. 15, p. 181.

48. Ref. 13, p. 390.

49. Ref. 17, p. 113.

50. Ref. 13, p. 340; Ref. 15, p. 370; Ref. 11, p. 428.

51. Ref. 11, p. 157.

52. Ref. 19, pp. 180, 190.

53. Ref. 11, p. 198.

54. Ref. 17, p. 198.

55. C. R. Martin, *Textbook of Endocrine Physiology*, Oxford University Press, Inc., New York, 1976, p. 118.

56. Ref. 14, p. 881.

57. H. R. Knapp, in Ref. 4, p. 355.

58. "Research Needs for Establishing Dietary Guidelines for Sodium" in *Research Needs for Establishing Dietary Guidelines for the U.S. Population*, The National Research Council, National Academy of Sciences, Washington, D.C., 1979.

59. T. A. Ketchen and J. M. Ketchen, in Ref. 10, p. 1293.

60. Ref. 14, p. 789.

61. H. Netter, *Theoretical Biochemistry*, John Wiley & Sons, Inc., New York, 1969, p. 784.

62. V. S. Sapirstein, P. Strocchi, and J. M. Gilbert, in R. E. Tashian and D. Hewett-Emmett, eds,. *Biology and Chemistry of the Carbonic Anhydrases*, Vol. 429, The New York Academy of Sciences, New York, 1984, p. 481.

63. *Maximizing Human Potential: Decade of the Brain 1990–2000*, Report of the Subcommittee on Brain and Behavioral Sciences, Office of Science and Technology Policy, Washington, D.C., 1991, p. 76; A. M. Chutorian, C. P. LaScala, C. N. Ores, and R. Nass, *Pediat Neurol.* **1**, 335 (1985); C. S. Wing, *Lang. Speech, Hear. Serv. Schools* **21**, 22 (1990).

64. S. A. Miller, P. A. Roche, P. Srinavasan, and V. Vertes, *Am. J. Clin. Nutr.* **32**, 1757 (1979).

65. M. E. Shils, in Ref. 4, p. 224.

66. Ref. 17, p. 185.

67. W. E. C. Wacker, *Ann. N. Y. Acad. Sci.* **162**, 717 (1969).

68. W. E. C. Wacker and A. F. Parisi, *New Eng. J. Med.* **278**, 658, 712, 772 (1968).

69. B. A. Barnes, *Ann. N. Y. Acad. Sci.* **162**, 786 (1969).

70. I. Melnick, R. R. Landes, A. A. Hoffman, and J. F. Burch, *J. Urol.* **105**, 119 (1971).

71. P. F. De Albuquerque and M. Tuma, *J. Urol.* **87**, 504 (1962).

72. E. L. Prien and S. F. Gershoff, *J. Urol.* **112**, 509 (1974).

73. S. N. Gershoff and E. L. Prien, *Am. J. Clin. Nutr.* **20**, 393 (1967).

74. P. R. Dallman, in Ref. 4, p. 241.

75. *Iron*, Subcommittee on Iron, Committee on Medical and Biologic Effects of Environmental Pollutants, National Research Council, National Academy of Sciences, University Park Press, Baltimore, Md., 1979, p. 79.

76. V. F. Fairbanks, in Ref. 6, p. 185.

77. R. F. Dickerson and I. Geis, *Hemoglobin: Structure, Function, Evolution and Pathology*, Benjamin/Cummings, Menlo Park, Calif., 1983.

78. H. M. Goff, in B. King, ed., *Encyclopedia of Inorganic Chemistry*, Vol. 4, John Wiley & Sons, Inc., New York, 1994, p. 1635.

79. W. G. Hoekstra, J. W. Suttie, H. E. Ganther, and W. Mertz, eds., *Trace Element Metabolism in Man 2 (Tema-2)*, University Park Press, Baltimore, Md., 1974.

80. A. M. English, in Ref. 78, p. 1683.

81. K. Schwarz, in Ref. 79, p. 355; H. H. Messer, W. D. Armstrong, and L. Singer, in Ref. 79, p. 425.

82. M. Anke, B. Groppel, U. Krause, in B. Momcilovic, ed., *Trace Elements in Man and Animals*, Zagreb, IMI, 1991, p. 26.28.

83. R. H. Ophaug, in Ref. 4, p. 274; F. H. Nielsen, in Ref. 6, p. 284.

84. E. J. Underwood and W. Mertz, in W. Mertz, ed., *Trace Elements in Human and Animal Nutrition*, 5th ed., Academic Press, Inc., San Diego, Calif., 1987, p. 1.

85. H. J. Sanders, *Chem. Eng. News*, 30 (Feb. 25, 1980); B. A. Bart, *Chem. Eng. News*, 56 (Oct. 22, 1979).

86. G. I. Waldbott, P. A. Coleman, and M. B. Schacter, *Chem. Eng. News* **2**, 3 (Dec. 17, 1979); J. R. Lee, *Chem. Eng. News*, 4 (Jan. 28, 1980); B. D. Gessner and co-workers, *New Eng. J. Med.* **330**, 95 (1994).

87. D. R. Taves, in H. H. Hiatt, J. D. Watson, and J. A. Winsten, eds., *Origins of Human Cancer*, Vol. 4, Cold Spring Harbor Conferences on Cell Proliferation, Cold Spring Harbor Laboratory, Maine, 1977, p. 357.

88. Ref. 13, p. 549.

89. J. Jowsey, B. L. Riggs, P. J. Kelly, and D. L. Hoffman, *Am. J. Med.* **53**, 43 (1972).

90. Ref. 19, p. 65; Ref. 17, p. 206; Ref. 11, p. 951.

91. R. J. Cousins and J. M. Hempe, in Ref. 4, p. 251.

92. J. C. King and C. L. Keen, in Ref. 6, p. 214.

93. *Zinc*, Subcommittee on Zinc, Committee on Medical and Biologic Effects of Environmental Pollutants, National Research Council, National Academy of Sciences, University Park Press, Baltimore, Md., 1979, p. 123.

94. B. L. Vallee, F. L. Hoch, S. J. Adelstein, and W. E. C. Wacker, *J. Am. Chem. Soc.* **78**, 5879 (1956).

95. Ref. 93, pp. 173, 225; R. E. Burch and J. F. Sullivan, in R. E. Burch and J. F. Sullivan, eds., *The Medical Clinics of North America*, Vol. 60, No. 4 (Symposium on Trace Elements), W. B. Saunders Co., Philadelphia, Pa., 1976, p. 675.

96. A. S. Prasad, J. A. Halsted, and M. Nadimi, *Am. J. Med.* **31**, 532 (1961).

97. P. A. Simkin, *Lancet ii*, **539** (1976); *Prog. Clin. Biol. Res.* **14**, 343 (1977).

98. F. H. Nielsen, *FASEB J.* **5**, 2661 (1991); F. H. Nielsen, in Ref. 10, p. 281.

99. E. M. Carlisle, in Ref. 4, p. 337.

100. E. M. Carlisle, in Ref. 79, p. 407.

101. K. Schwarz and C. M. Foltz, *J. Am. Chem. Soc.* **79**, 3292 (1957); K. Schwarz, *Proc. Natl. Acad. Sci.* **70**, 1608 (1973).

102. Ref. 17, p. 196.

103. B. L. O'Dell, in Ref. 4, p. 261.

104. J. R. Turnlund, in Ref. 6, p. 231.

105. E. M. Gregory and I. Fridovich, in Ref. 73, p. 486.

106. Ref. 19, p. 466.

107. Ref. 19, p. 66.

108. N. A. Holtzman, *Fed. Proc.* **35**, 2276 (1976).

109. L. M. Klevay and co-workers, *Am. J. Clin. Nutr.* **33**, 45 (1980).

110. F. H. Nielsen, in Ref. 4, p. 296.

111. F. H. Nielsen, *Nutr. Today* **23**, 4 (1988).

112. F. H. Nielsen, in Ref. 6, p. 272.

113. O. A. Levander and R. F. Burk, in Ref. 4, p. 268.

114. O. A. Levander and R. F. Burk, in Ref. 6, p. 242.

115. *Selenium*, Subcommittee on Selenium, Committee on Medical and Biological Effects of Environmental Pollutants, National Research Council, National Academy of Sciences, Washington, D.C., 1976, p. 51.

116. G. N. Schrauzer, D. A. White, and C. J. Schneider, *Bioinorg. Chem.* **8**, 387 (1978).

117. J. R. Chen and J. M. Anderson, *Science* **206**, 1426 (1979).

118. T. C. Stadtman, *Science* **183**, 915 (1974).

119. Ref. 17, p. 542.

120. G. N. Schrauzer and D. A. White, *Bioinorg. Chem.* **8**, 303 (1978).

121. G. A. Clugston and B. S. Hetzel, in Ref. 6, Chapt. 13, p. 252; Ref. 17, p. 198.

122. Ref. 55, p. 189.

123. Ref. 12, p. 615.

124. C. L. Keen and S. Zidenberg-Sherr, in Ref. 4.

125. Ref. 17, p. 201.

126. F. H. Nielsen, in Ref. 6, p. 275.

127. M. F. Utter, in Ref. 95, p. 713.

128. Y. Tanaka, C. Dupont, and E. R. Harpur, *Abstracts of 174th American Chemical Society Meeting*, Abstract No. 130, Chicago, Ill., Aug. 28–Sept. 2, 1977.

129. G. J. Everson and R. E. Schrader, *J. Nutr.* **94**, 89 (1968).

130. F. H. Nielsen, in Ref. 4, p. 297; Ref. 6, p. 277.

131. Ref. 84, p. 109.

132. P. D. Boyer, ed., *The Enzymes*, Academic Press, Inc., New York, 1970.

133. Ref. 11, p. 546; Ref. 17, p. 202.

134. A. Galli, *Ann. Biol. Clin.* **26**, 976 (1968).

135. Y. G. Doestahale and C. Gopalan, *Br. J. Nutr.* **31**, 351 (1974).

136. W. Mertz, in D. Shapcott and J. Hubert, eds., *Chromium in Nutrition and Metabolism, Developments in Nutrition and Metabolism*, Vol. 2, Elsevier-North Holland Biomedical Press, Amsterdam, the Netherlands, 1979, p. 1; W. Mertz, *J. Nutr.* **123**, 626 (1993); R. A. Anderson, *Sci. Total Environ.* **86**, 75 (1989).

137. B. J. Stoecker, in Ref. 4, p. 287.
138. F. H. Nielsen, in Ref. 6, p. 264.
139. R. A. Levine, D. H. P. Streeten, and R. J. Doisy, *Metabolism* **17**, 114 (1968).
140. L. L. Hopkins, Jr. and M. G. Price, *Proceedings Western Hemisphere Nutrition Congress*, Vol. 11, Puerto Rico, 1968, p. 40.
141. H. Schroeder, *Am. J. Clin. Nutr.* **21**, 230 (1968).
142. L. Sherman, J. A. Glennon, W. J. Brech, G. H. Klomberg, and E. S. Gordon, *Metabolism* **17**, 439 (1968).
143. L. L. Hopkins, Jr., O. Ransome-Kuti, and A. S. Majaj, *Am. J. Clin. Nutr.* **21**, 203 (1968); C. T. Gurseil and G. Saner, *Am. J. Clin. Nutr.* **24**, 1313 (1971); *Am. J. Clin. Nutr.* **26**, 988 (1973).
144. K. N. Jeejeebhoy, R. C. Chu, E. B. Marliss, G. R. Greenberg, and A. Bruce-Robertson, *Am. J. Clin. Nutr.* **30**, 531 (1977); H. Freund, S. Atamian, and J. E. Fisher, *J. Am. Med. Assoc.* **241**, 496 (1979).
145. V. J. K. Liu and J. S. Morris, *Am. J. Clin. Nutr.* **31**, 972 (1978); R. A. Anderson, M. M. Polansky, N. A. Bryden, S. J. Bhathena, J. J. Canary, *Metabolism* **36** 351 (1987); R. A. Anderson, N. A. Bryden, M. M. Polansky, S. Reiser, *Am. J. Clin. Nutr.* **51**, 864 (1990).
146. E. W. Toepfer, W. Mertz, M. M. Polansky, E. E. Roginski, and W. R. Wolf, *J. Agri. Food Chem.* **25**, 162 (1977).
147. C. L. Rollinson, in *Comprehensive Inorganic Chemistry*, Vol. 3, Pergamon Press, Oxford, U.K., 1973, p. 676; A. J. Gould, ed., *Radioactive Pharmaceuticals*, U.S. Atomic Energy Commission, Division of Technical Information, Oak Ridge, Tenn., 1966, p. 429; C. L. Rollinson and E. W. Rosenbloom, in S. Kirschner, ed., *Coordination Chemistry: Papers Presented in Honor of Prof. John C. Bailar, Jr.*, Plenum Press, New York, 1969, p. 108.
148. V. Herbert, in Ref. 4, p. 170.
149. B. L. Vallee, in S. K. Dhar, ed., *Advances in Experimental Medicine and Biology*, Vol. 40, Plenum Press, New York, 1973, p. 1.
150. V. Herbert and K. C. Das, in Ref. 6, p. 402.
151. Ref. 11, p. 22.
152. Ref. 11, p. 262.
153. Ref. 17, p. 256.
154. Ref. 84, p. 449.
155. K. Yokoi, M. Kimura, and Y. Itokawa, *Biol. Trace Elem. Res.* **24**, 223 (1990).
156. L. L. Hopkins, Jr., in Ref. 79, p. 397.
157. Ref. 84, p. 388.
158. Ref. 84, p. 159.
159. F. H. Nielsen, in Ref. 4, p. 299; in Ref. 6, p. 279.
160. F. H. Nielsen, in Ref. 4, p. 294; Ref. 6, p. 270.
161. M. Abdulla, B. M. Nair, and R. K. Chandra, eds., *Proceedings of An International Symposium*, Health Effects and Interactions of Essential and Toxic Elements, *Nutrition Research*, Suppl. 1, Pergamon Press, New York, 1985.
162. R. J. Lewis, Sr. and R. L. Tatkin, eds., *Registry of Toxic Effects of Chemical Substances*, 8th ed., National Institute for Occupational Safety and Health, Public Health Service Center for Disease Control, U.S. Government Printing Office, Washington, D.C., 1979.

MARY G. ENIG
Enig Associates, Inc.

MINERALS RECOVERY AND PROCESSING

Minerals, critically important to the development of civilization, are derived from the earth's crust, a relatively thin shell of siliceous material about 13 km deep. The art of minerals processing can be traced back to the beginnings of civilization as evidenced by distinct historic periods known as the Stone Age, Copper Age, Bronze Age, and Iron Age. In the sixteenth century Agricola in his treatise *De Re Metallica* noted "The art of metal making is one of the most ancient, the most necessary and the most profitable to mankind. Without doubt, none of the arts is older than agriculture, but that of the metals is not less ancient; in fact they are at least equal and coeval, for no mortal man ever tilled a field without implements" (1). Mining can be considered the second largest industry in the world, next to agriculture. In terms of importance, mining is considered equal to agriculture.

The word mineral comes from *minera*, which once meant a mine or an ore specimen. It commonly refers to inorganic substances occurring naturally in the earth. It also applies to some organic material such as coal (qv). In its broadest sense, mineral means any natural substance that is neither plant nor animal. Minerals make up rocks and ores. Biotite granite, for instance, is composed of three principal minerals of different composition: light colored feldspar, grey quartz, and black mica (biotite). Magnetite, on the other hand, is virtually a monomineral, ie, composed of crystalline magnetite grains. Genetically, minerals are natural chemical compounds of elements and in a few cases, native elements. Minerals are natural products of various physico-chemical processes occurring in the earth's crust including the life activity of various organisms. Although with some degree of approximation, natural minerals may be regarded as homogeneous crystalline masses, no mineral is perfectly homogeneous, either physically or chemically.

The occurrence of minerals or elements in the earth's crust is not uniform. Rather, minerals concentrate in particular areas (called deposits) as a result of geological conditions and activity. Such a deposit is referred to as an ore deposit, ore body, or when it is large enough, ie, the percentage of the useful component is high enough to make mining worthwhile for economic recovery of the mineral, simply as an ore. Most minerals or elements are present in the earth's crust in very small amounts. The concentrations of selected metals are given in Table 1. Eight elements, O, Si, Al, Fe, Ca, Na, Mg, and K, account for more than 99% of the crust. The abundance of a given element or a mineral often has no bearing on its industrial importance or its price (see also METALLURGY, EXTRACTIVE METALLURGY).

Ores which comprise a variety of minerals are, as a rule, heterogeneous. An ore body is usually named for the most important mineral(s) in the rock, referred to as value minerals, mineral values, or simply values. Some minerals contain metals, which are extracted by concentration and smelting. Other minerals, such as diamond, asbestos (qv), quartz (see SILICON COMPOUNDS), feldspars, micas (see MICA), gypsum, soda, mirabillite, clays (qv), etc, may be used either as found, with some or no pretreatment, or as stock materials for industrial compounds or building materials (qv) (3).

Table 1. Abundance of Metals in the Earth's Crust[a,b]

Element	Abundance, %	Amount in 3.5 km of crust, t	Element	Abundance, %	Amount in 3.5 km of crust, t
silicon	28.2	$10^{16}-10^{18}$	chromium	0.010	$10^{14}-10^{15}$
aluminium	8.2		nickel	0.007 5	
iron	5.6		zinc	0.007 0	
calcium	4.1		copper	0.005 5	$10^{13}-10^{14}$
sodium	2.4		cobalt	0.002 5	
magnesium	2.3	$10^{16}-10^{18}$	lead	0.001 3	
potassium	2.1		uranium	0.000 27	
titanium	0.57		tin	0.000 20	
manganese	0.095	$10^{15}-10^{16}$	tungsten	0.000 15	$10^{11}-10^{13}$
barium	0.043		mercury	8×10^{-6}	
strontium	0.038		silver	7×10^{-6}	
rare earths	0.023		gold	$< 5 \times 10^{-6}$	
zirconium	0.017	$10^{14}-10^{16}$	platinum	$< 5 \times 10^{-6}$	$< 10^{11}$
vanadium	0.014		metals		

[a]Ref. 2. Courtesy of Pergamon Press.
[b]By comparison, oxygen has an abundance of 46.4%.

Most ores as mined require some processing before they can be converted into usable, final mineral products. Processing of ores by physical or chemical methods is described as mineral processing. For convenience, it is often the practice to restrict minerals processing to physical methods, and to treat chemical methods under the realm of extractive metallurgy. This distinction is a fine one, however, and hydrometallurgy (see SUPPLEMENT), which comprises chemical methods of minerals treatment in aqueous medium, has much in common with minerals processing. Therefore, hydrometallurgy can often be considered as minerals processing. Until the early part of the twentieth century, minerals processing was much more of an art than a science. Rapid developments in understanding the fundamentals of processing have given minerals processing a sound scientific basis.

The size of the minerals industry is demonstrated by world production figures of the principal mineral commodities (Table 2). The total tonnage of commodities listed amounted to approximately 8.5 billion metric tons for 1992. A conservative estimate of the total tonnage of earth moved in order to produce these 8.5×10^9 t of mineral products is ca $85-850 \times 10^9$ in one year. A large copper (qv) flotation (qv) plant alone treats from 100,000 to 150,000 t/d of ore. This ore contains less than 1% copper and may involve removing an equivalent tonnage in the form of overburden or waste rock. Vast amounts of energy, water, and chemicals are also used. Materials for U.S. flotation plants are given in Table 3.

The minerals processing industry has made contributions to all areas of technology, both in terms of products and processing. Technologies developed in the mineral industry are used extensively in the chemicals industry as well as in municipal and industrial waste treatment and recycling industry, eg, scrap

Table 2. 1992 World Production of Principal Mineral Commodities

Commodity[a,b]	Production, t
aluminum, ingot metal	19,219,000
antimony, mine output, Sb content	75,659
arsenic trioxide	47,600
beryl concentrate	7,002
bismuth, mine output, Bi content	2,998
cadmium, smelter output	18,750
chromite	10,896,000
cobalt, mine output, Co content	21,924
columbium–tantalum concentrate	35,193
copper, mine output, Cu content	9,290,000
gold, mine output, Au content	2,247
iron and steel	1,256,796,000
lead, mine output, Pb content	3,242,000
magnesium, smelter output	382,542
manganese ore	19,929,000
mercury, mine output, Hg content	3,014
molybdenum, mine output, Mo content	111,667
monazite concentrate, rare earths and Th	16,400
nickel, mine output, Ni content	921,929
platinum-group metals, mine output, metals content	280
selenium, refinery	1,724
silver, mine output, Ag content	15,345
tellurium, refinery	91
tin, mine output, Sn content	179,446
titanium concentrate	5,660,000
tungsten, mine output, W content	31,555
uranium, mine output, U_3O_8 content	32,792
vanadium, mine output, V content	21,917
zinc, mine output, Zn content	7,137,000
zirconium concentrate	807,000
asbestos	3,121,000
barite	5,436,000
boron materials	2,608,000
bromine	379,000
cement, hydraulic	1,252,501,000
clays	33,650,000
corundum, natural	6,000
diatomite	1,581,000
feldspar	5,771,000
fluorspar	3,846,000
graphite	567,390
gypsum	97,791,000
iodine	16,930
lime	128,730,000
magnesite	11,129,000
mica	186,000
nitrogen, N content of ammonia	95,532,000
perlite	1,397,000
phosphate rock	143,753,000
potash, marketable, K_2O equivalent	24,327,000

Table 2. (*Continued*)

Commodity[a,b]	Production, t
pumice	11,142,000
salt	184,854,000
sand and gravel, industrial	106,308,000
sodium compounds	35,364,000
strontium materials	217,100
sulfur, elemental basis	52,409,000
talc	8,864,000
vermiculite	474,649
carbon black	4,800,000
coal	4,568,000,000
coke, metallurgical	325,118,000

[a] Ref. 4.
[b] Diamond production is 107,771,000 carats (21.6 t). Fuel products such as natural gas, peat, petroleum, etc, are not included.

Table 3. U.S. Flotation Plants[a]

Plant	1970	1991
flotation plants	240	133[b]
ore treated, t	405,300,000	470,250,000
different ore types	27	17
concentrate produced, t	54,100,000	81,907,000
energy consumed, kWh	6,000,000,000	7,708,000,000
water used, L	1,843,295,000,000	2,918,844,385,000
reagents consumed, kg	952,553,750	599,553,660

[a] Ref. 4.
[b] Plants that responded to the survey of U.S. Bureau of Mines.

recycling, processing of domestic refuse, automobiles, electronic scrap, battery scrap, and decontamination of soils.

Ores

Deposits.   Ore deposits (5) are classified as igneous, metamorphic, or sedimentary. A further classification is into sulfides, oxides, metallic, nonmetallics, etc, depending on the nature of occurrence of the value mineral and use of the mineral product. Oxides can encompass oxides, hydrated oxides, sulfates, carbonates, and silicates. Metallic ores are those that are processed to produce a metal as a final product. Nonmetallic ores are not used for metal production, although they may contain a metal, eg, bauxite which is nonmetallic when used for production of alum or alumina (see ALUMINUM COMPOUNDS). Considerable effort goes into estimating the size and geometry of ore deposits. These are important factors in economic assessment. The size of ore deposits is described in terms of proven or measured reserve which is well-outlined by extensive drilling; indicated reserve, implied by a limited amount of drilling; and potential reserve, based on geological data with no drilling information. Potential reserves could also include well-outlined deposits that are not economical to treat at present.

Ores are mined (6) by open-pit, underground, alluvial, and solution mining methods. The first two are forms of hard rock mining. Open-pit mining is preferred because it is more economical and practical, especially when the ore grade is low and large tonnages have to be handled. It can, however, be practiced only when the ore body is close to the surface. Underground mining, the most expensive method of mining, has to be used when the ore body is at a considerable depth from the surface. Often a combination of open-pit and underground mining is used on the same ore body. Alluvial and solution mining are inexpensive, but can be used only on sedimentary, alluvial, or placer deposits such as river beds, dried-up oceans, and lakes. Dredges or water pressure are used to extract the ore. Unlike subsequent processing, the mining method has comparatively little dependence on the mineralogical nature of the valuable minerals in the ore although the nature of the gangue minerals may affect the actual method of hard rock mining used.

Mineralogy. Ores are inherently heterogeneous and unique. Any given element can exist in a variety of compounds or minerals of distinctly different composition. There is considerable variation in minerals composition within an ore body and from one ore body to another. For the purposes of minerals processing, any ore can be considered to be made up of valuable minerals and gangue (waste) minerals, and these are associated with one another intimately in the deposit. The valuable minerals may contain more than one valuable metal. Ores may be designated simple or complex, depending on the ease and extent of mineral liberation and subsequent processing. The choice of a processing method is entirely dependent on ore mineralogy (7–9), ie, the specific minerals present, their relative proportions, composition, and the mode of physical occurrence of the various mineral constituents such as size and extent of liberation. A thorough knowledge of mineralogy is also important in monitoring the efficiency of day-to-day operation of the mill because mineralogical factors can change frequently. Often the chemical analysis of the ore may remain essentially the same, but the mineralogy differs vastly. A simple example is that of a sulfide copper ore. At the same chemical analysis for copper, a significant amount of copper may be present in the oxide mineral form and this may not be recovered in the sulfide flotation circuit. Mineralogical information dictates the optimization of processing steps such as grinding and concentration. Without mineralogical examination, it may be impossible to determine whether a poor separation is the result of inadequate grinding, mineralogical characteristics, or inefficient plant operation.

Mineralogical information is routinely gathered by collecting representative samples of the ore, or of the various flow streams in an operating plant, and studying polished or thin sections of these samples using optical microscopy (qv) in either the reflected light mode or the transmitted light mode. The electron microprobe and x-ray diffraction are also used extensively in conjunction with the optical microscope (see X-RAY TECHNOLOGY). A more recent development, qem*sem, is a fully automated image analyzer combining a scanning electron microscope and electron microprobe (9). The type of information sought in mineralogical examination includes the grain size of the value minerals in the host matrix rock; associations or locking between the value minerals and gangue minerals and between value minerals; existence of trace elements in the lattice of the value minerals; presence of oxidation or alteration of the mineral surfaces;

and the occurrence of minor amounts of potentially valuable metals or minerals. In the reflected light mode, mineral identification is made using many optical properties of the minerals, most notably color. Polarized light is used for obtaining additional information and confirmation because this reveals anisotropy of minerals. Chemical etching or staining techniques exploit differences in chemical properties of the minerals, and these can also reveal texture, cleavage, or the existence of trace elements. Dark-field illumination is used to obtain information on translucent minerals. Transmitted light microscopy is useful for the study of gangue minerals which are difficult to identify by reflected light. Heavy liquid fractionation using organic and inorganic liquids of various specific gravities may be invaluable for physically separating the minerals in the laboratory to assess the degree of particle locking as well as the potential for separation by gravity concentration methods. Ferromagnetic fluids are also used for separation based on specific gravity (10).

Liberation. In most ores, the value minerals are intimately locked or associated with gangue minerals. One of the most important prerequisites to the success of any physical separation process is liberation of value minerals from the ore matrix. Liberation, achieved by size reduction (qv) operations, can occur either because of intergranular or transgranular fracture (6,11,12). The degree of liberation of a particular mineral species, ie, the percentage of that mineral occurring as free particles (in a macroscopic sense) in relation to the total quantity of the mineral, is dictated by the type, performance, and economics of the processing operation, ie, throughput rate, recovery, product grade, processing costs, and profits. In many plants liberation and processing are performed in stages to optimize efficiency and economics. For example, coarse grinding separation is followed by fine grinding separation, thus many concentration units deal with a significant proportion of middlings, ie, locked, particles which contain varying amounts of one or more value minerals still associated with one or more gangue minerals. As can be expected, middlings decrease the sharpness of any separation. Both grade and recovery could be affected depending on the type of separation. Recovery of middlings constitutes one of the primary challenges in concentration separation. Middlings can be characterized by many types, but some of the common types are the occurrence of one mineral as veins, blebs, or inclusions in another mineral, or as two large areas of minerals making contact with each other. Mineral liberation has been studied extensively, and numerous liberation models have been developed. Microscopy of polished sections of sized fractions of crushed or ground ore is valuable in assessing the degree of liberation.

Processing

Flow Sheets. All minerals processing operations function on the basis of a flow sheet depicting the flow of solids and liquids in the entire plant (6,13,14). The complexity of a flow sheet depends on the nature of the ore treated and the specifications for the final product. The basic operations in a flow sheet are size reduction (qv) (comminution) and/or size separation (see SEPARATION, SIZE), minerals separation, solid–liquid separation, and materials handling. The overall flow sheet depends on whether the specification for the final mineral product is size, chemical composition, ie, grade, or both. Products from a quarry, for example, may have a size specification only, whereas metal concentrates have a grade specification.

Each basic operation can be divided into one or more unit operations. Size reduction involves crushing and grinding depending on the size of material handled, and these may be carried out in stages. Separations can be either solids from solids, based on size or mineral composition, or solids from liquids, ie, dewatering (qv). Size separation or classification is an integral part of any flow sheet, not only to meet product size specifications, but also to ensure a narrow size distribution for subsequent minerals separation circuits and to decrease the load and improve the efficiency of size reduction units which are energy intensive.

Minerals separation, also known as concentration separation, operations, such as gravity concentration, flotation (qv), and magnetic separation (see SEPARATIONS, MAGNETIC SEPARATION) are carried out most frequently in stages called roughing, scavenging, and cleaning. Roughing, the first stage in a separation, produces a concentrate (value product) and rejects (tails). Rougher rejects are treated in scavengers to recover additional mineral values. Scavenger rejects constitute the final plant tails or feed to another separation circuit. The rougher and scavenger concentrates can be combined or treated separately. For many ores, the scavenger concentrate represents the middlings or unliberated or partially liberated particles and, therefore, may require further grinding before treatment. Even the rougher concentrate may require further grinding before it enters the cleaning circuit. There may be additional stages such as cleaning of a scavenger product, scavenging of a cleaner product, or recleaning of a cleaner product, depending on the metallurgical objectives of the plant and the type of mineral product produced.

In addition to encompassing all of the unit operations in the plant, the plant flow sheets may also include materials handling operations associated with the transport and storage of materials in and around the mill. Typically, flow sheets provide quantitative information regarding water and slurry flows, tonnages, and assays.

Metallurgical Performance.   Most minerals separations, whether by size or composition, are not perfect. There are invariably misplaced values and gangue minerals. The performance of a separation unit is, therefore, described by both recovery and grade (2,10,15). These are similar to yield and purity used in the chemical industry. In addition, other parameters such as ratio of concentration, ie, ratio of weight of feed to weight of concentrate; enrichment ratio, ratio of grade of concentrate to grade of feed; economic efficiency; and separation efficiency, recovery of value minus recovery of gangue, are also used sometimes. Recovery is a measure of the efficiency, and the grade the measure of selectivity of a unit operation. Grade is expressed in a variety of ways including physical, eg, brightness of clays (qv) or of the calcium carbonate used in the paper industry, or chemical properties, eg, metal assays of concentrates or P_2O_5 content of a phosphate concentrate. Recovery and grade are not independent, and there is a tradeoff at every stage of the separation operation. Recovery is the primary objective in both roughers and scavengers; grade is the primary objective of cleaners.

Material Balances. Material balances in the plant flow sheet provide an assessment of the efficiency of operation of individual unit operations and the plant on the whole (2,10,16). Information necessary for trouble-shooting is also provided. Basically, materials balance involves calculation of input, output, and accumulation around any given circuit of a flow sheet. Many unit operations are characterized by high accumulations (circulating loads), sometimes by design. Most units are described as separation or junction points (nodes). Materials balance equations are presented as two-, three-, or n-product formulas depending on the number of products a flow stream consists of or is split into. The bases for materials balance are usually mass, assay (grade), volume, mineral composition, or energy. Balancing a circuit is usually quite a tedious task. There is either inadequate information about flow streams (leading to errors) or too much data (leading to conflicting results). Computer programs are available to make the task easier and to handle a complex set of nodes that typically exist in a plant. These can also provide for reconciliation of excess data.

There are many sources of errors in the plant. The principal ones are related to sampling (qv), mass flow rates, assaying, and deviations from steady state. Collecting representative samples at every stage of the flow sheet constitutes a significant task. Numerous methods and equipment are available (10,16,17).

Economic Aspects

The estimated value of world crude mineral production in 1992 was $1.6 trillion in terms of 1992 dollars, a tenfold increase since 1950 in terms of constant dollars (4). This represents the value of mineral materials as mined or otherwise extracted from the earth, and does not reflect the value added through processing. The annual world consumption of selected mineral commodities for 1992 is given in Table 4.

A conservative estimate of the total value of the products from the mineral industry is ca $3.9 trillion in terms of 1992 dollars (4). This estimate does not include the value of products derived from secondary sources such as recycling (qv) or reclamation. Secondary recovery is significant for certain commodities.

Table 4. Worldwide Consumption of Mineral Commodities[a]

Commodity	1992 Consumption, t $\times$ 10^3
ferrous metals	1,175,000
aluminum, refined	19,219
cadmium	20
copper, refined	10,805
lead, refined	5,150
magnesium, primary	318
nickel	750
tin, refined	200
zinc, slab	6,472
cement	1,250,000
fertilizers	145,598
sulfur, elemental S equivalent	56,503

[a] Ref. 4.

For example, in 1992 ca 30% of the world steel (qv) production, 46% of the world refined lead output, 15% of the world refined copper (qv) production, and ca 30% of the aluminum (see ALUMINUM AND ALUMINUM ALLOYS) output from the Western world were clearly identified as being derived from scrap. The value of the world mineral commodity export trade in 1992 was ca $616,698 million in 1992 dollars. This accounted for ca 18% of all commodities exported (4).

The economics of minerals processing and production (6,18) are determined not only by supply and demand, but also by socio-political factors, including those tied into energy supply and environmental aspects, and by the need for a minimum product quality. The use of a mineral product has to justify all costs involved in producing it. Those costs include mining, processing, transportation, and marketing, as well as those costs associated with meeting environmental regulations.

Mining of the ore deposit constitutes a significant cost, especially in hard rock mining. Mining costs vary considerably from ore to ore and from a few cents to well over $100/t mined. Underground mining is the most expensive; hydraulic mining of sedimentary deposits is the least expensive.

Processing costs include those for size reduction, size classification, minerals concentration and separations, solid–liquid separation (dewatering), materials handling and transportation, and tailings disposal. Size reduction, one of the most expensive unit operations in minerals processing, could account for as much as 50% of the total energy consumed. This cost varies considerably from deposit to deposit and quite often from one area of a deposit to another. Ore bodies are extremely heterogeneous and the associated minerals liberation, complex.

Minerals concentration operations also constitute a primary cost item. The grade of the ore deposit has a strong influence on the size of the processing operation. When using a lower grade ore, a large-scale operation is required to produce a given amount of product economically. Operating cost per ton of ore treated is smaller for large-scale operation. Physical methods consume less energy than do chemical methods. The location of the mineral processing plant or mill affects the costs of fuel, power, water, transportation, and labor, as well as the cost of the mill itself. Mills and tailings pond are usually located close to the mine site, although environmental concerns may require pumping of the plant tails to considerable distances from the plant.

The valuation of mineral products is complex because it is influenced by composition, which inherently lacks uniformity. Credits are given for desirable features in the mineral product and penalties for undesirable features. For example, in a metal sulfide concentrate, credits are paid for contained gold and silver even if the amounts are very small, and penalties are paid for contained detrimental or toxic elements such as arsenic and antimony, or other species that incur problems or metal losses in smelting. Government legislation and environmental regulations are also becoming increasingly important in product valuation.

The economic importance of an ore deposit itself is largely affected by mineral or metal prices. Mine closures and reopenings are a common event in the mineral industry for this reason. Economics can also be affected by the ore composition, for example, by unacceptable levels of penalty elements in the ore. The assessment of overall economics of exploiting a given ore deposit is similar to that for any large-scale industry. The various cost components are those associated with equipment, labor, utilities, contingencies, operation and production, transportation, working capital, supplies, maintenance, depreciation, and increasingly important mine and tailings decommissioning and closure. Environmental awareness, concerns, and regulations have had a significant impact on mineral industry economies in the latter 1900s and are expected to continue to play an increasingly important role into the twenty-first century (see also ECONOMIC EVALUATIONS; ENERGY MANAGEMENT).

Environmental Aspects

Environmental issues, concerns, and management are a primary preoccupation of the mining and minerals industry. The goal is to minimize the impact of the industry to the world ecosystem. The mining and mineral industry continue to face and deal with increasingly demanding challenges and liabilities from regulations pertaining to environmental matters which periodically change and increase in complexity.

Principal areas of environmental management (19) in minerals processing are tailings and other waste treatment and disposal, water discharge and protection of ground and surface waters, acid mine drainage, land reclamation, restoration and abatement, mine subsidence, mine closure, dust control, air emissions, and the elements or minerals that cause environmental pollution in subsequent operations such as smelting or burning of coal. Such regulations as the Clean Air Act have high impact on the industry. Laws in the 1990s require that permits for new mine openings be covered by a huge performance bond. The release of the bond is contingent on the completion of all required reclamation, restoration, and abatement work on the permit area. A significant amount of research and development is being conducted to address the environmental issues in mineral processing. Many newer technologies have already become available.

Size Reduction

Size reduction (qv) or comminution is the first and very important step in the processing of most minerals (2,6,10,20–24). It also involves large expenditures for heavy equipment, energy, operation, and maintenance. Size reduction is necessary because the value minerals are intimately associated with gangue and need to be liberated, and/or because most minerals processing/separation methods require the ore mass to be of certain size and/or shape. Size reduction is also required in the case of quarry products to produce material of controlled particle size (see SIZE MEASUREMENT OF PARTICLES). In some instances, liberation of valuables or impurities from the ore matrix is achieved without any apparent size reduction. Scrubbers and attritors used in the industrial minerals plants, eg, phosphate, rutile, glass sands, or clay, are examples.

Size reduction is conducted in stages, the first occurring during mining of the ore body by using either explosives, eg, for hard rock mining or mechanical means, eg, for hydraulic mining of sedimentary deposits such as clays (qv) and phosphates (see EXPLOSIVES AND PROPELLANTS). Subsequent size reduction stages involving crushing and grinding are grouped according to the particle size. General types of size reduction equipment include crushers, tumbling mills, impact mills, fixed-path mills, and fluid energy mills. A wide variety of crushers and grinding mills has been developed as necessitated by the ore type and rock hardness. The extent of size reduction achieved by any of these units is described by the reduction ratio, the ratio of the feed size to the product size. Particle size is defined as the size of the separator through which a certain percentage by weight, typically 80%, of the particles pass.

The three basic types of size reduction circuits used to produce a fine product are shown in Figure 1. The final stages of the grinding circuit are typically operated in closed circuit, at comparatively high circulating loads, so that the material has little chance of being broken a second time before it is removed from the circuit by a classifier. Rod mills are operated normally in an open circuit.

Size reduction is the most energy intensive unit operation in minerals processing. In the late 1970s, approximately 29 billion kWh/yr of electrical input was used for size reduction and an additional 3.7 billion kWh/yr in contained energy consumables such as grinding media and liners (23). This is because most of the input energy is absorbed by the machine itself in vibration, generation of heat, etc, and only a small fraction (<1% to a few per cent) of the total energy is available for breaking the rock. The probability of breakage in comminution

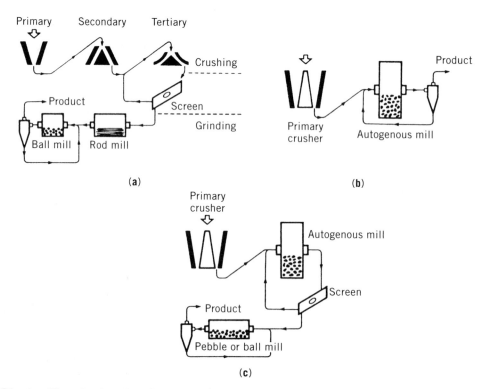

Fig. 1.   Three basic types of size reduction circuits: (**a**) conventional, (**b**) autogenous, and (**c**) autogenous plus separate fine grinding (6).

decreases, and the energy required per unit mass increases rapidly as the particle size decreases.

Crushing.   Crushing handles the mine product (run-of-mine ore) and reduces the rock size from over 1 m to approximately 10–25 mm, often in stages termed primary, secondary, or tertiary. Crushing, carried out by compression or impact methods, is usually conducted dry, but water-flush crushers are emerging as a newer technology (22). There are many types of crushers: jaw, gyratory, roll, rotary, and impact. All are typically massive. The jaw and gyratory crushers dominate the primary crushing stage which is operated in open circuit, often in conjunction with scalping screens or grizzlies which remove undersize. In all crushers the energy is applied directly to the particles.

Jaw crushers are characterized by a fixed and moving jaw (Fig. 2**a**). The latter allows for changes in the feed and discharge openings. Common types are the Blake crusher, which has the swing jaw pivoted at the top and a variable discharge opening; the Dodge crusher, used only in the laboratory, which pivots at the bottom and has variable feed area; and the Universal crusher, which pivots at an intermediate position and has variable feed and discharge areas. Jaw crushers are rated according to receiving areas, ie, the width (150–2100 mm) of the plates, and the gape, the distance (125–1600 mm) between the jaws at the feed opening. A large crusher can weigh up to 200 tons. A machine of size 1680 mm (gape) by 2130 mm (width) can crush ore (1.2 m) at a rate of 725 t/h at a 203-mm set (discharge opening) (2).

Gyratory crushers (Fig. 2**b**) consist of a long spindle seated in an eccentric sleeve, housed in a fixed conical shell. Material is crushed as it gets nipped

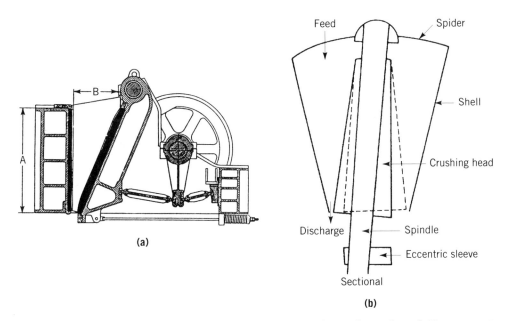

Fig. 2.   (**a**) Cross section of a double-toggle jaw crusher where A and B represent dimensions (10). (**b**) Functional diagram of a gyratory crusher (2).

between the rotating spindle and the fixed crushing shell. The gyratory crusher has a much higher capacity than a jaw crusher and can be considered a continuum of jaw crushers. Common types are the suspended spindle (short or long shaft), the supported spindle (most used), and the fixed spindle. Large crushers having 0.2–0.7 MW (250–1000 hp) motors, weighing between 150–350 tons, have a gape of 1000–2130 mm, and can crush ore at a rate of 500–7000 t/h depending on the gape and discharge setting (10,13).

Cone crushers, which are modified gyratory crushers, perform the bulk of secondary crushing. These have a flat crushing chamber giving higher capacity and reduction ratio, a short spindle operated at much higher speeds, and a higher capacity. They are classified into standard (secondary) cone crushers, shorthead (Symons design, tertiary crushing), and fine crushers (such as Gyradisc, Hydrofine, etc) (2,10). Crusher sizes are reported as the crushing-head (mantle) diameter; typical range is 610–3050 mm. Weights vary in the range of 5–200 t and connected power ranges from 7–500 kW (10–700 hp). A 3050-mm Symons Standard cone crusher can crush ore at a rate of up to 3000 t/h at a discharge setting of 64 mm (13).

Other types of crushers such as rolls and impact crushers (10), ie, hammer or stamp mills, have limited application in metal mining, but are more common in the quarrying industry, especially for relatively soft, friable, and sticky rocks such as phosphate, limestone (see LIME AND LIMESTONE), clay, asbestos (qv), coal, etc. These are characterized by larger wear and smaller (250–1000 t/h) capacity than the gyratory crushers. Some of the more recent crusher designs include the Tidco Barmac crusher, high pressure or high compression rolls, and the rotary coal breaker. High pressure rolls, for example, were introduced in 1985 and are used extensively in the cement (qv) industry. The largest units are capable of using 100 kN of pressing force per cm of roll width and handling over 1000 t/h. Water flash cone crushers are an example of a wet crusher (product is 30–50% solids) (2).

Chemical and electrical, such as spark and microwave, methods of size reduction have also been suggested. Considerable effort has been made in crushing circuit control and automation to reduce capital and operating costs and to improve crusher efficiency (21,23). Larger crushing units are also used increasingly because these are considered to be more energy efficient and to provide savings in capital costs. Large gyratory crushers have replaced jaw crushers in many applications. Another trend in crushers is to provide mobile (in-pit) crushers to enhance flexibility (22).

Grinding. Grinding, which refers to size reduction of the crushed material, is carried out in stages, if necessary, until the required liberation is achieved (2,6,10,20–24). This liberation is dictated by the mineralogy of the ore deposit, the processing method, and prevailing economic conditions. Grinding is achieved by abrasion and impact in tumbling mills (Fig. 3) using steel rods or balls, ceramic pebbles, or large pieces of the ore itself. The latter process is called autogenous grinding. Crushing circuits often provide media for autogenous milling. A variation used more commonly is the semiautogenous grinding wherein a small number (usually 10% of the total loading by weight) of large steel balls are used to break down certain intermediate sizes of particles that otherwise would build up in the mill.

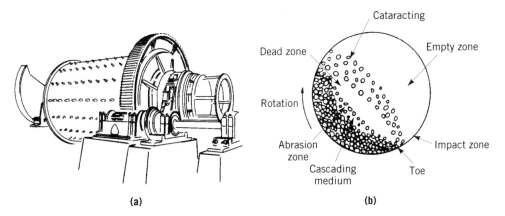

(a)   **(b)**

Fig. 3. (**a**) Tumbling mill; (**b**) motion of charge within it (2).

Tumbling mills are classified into ball mills or rod mills, based on the type of tumbling media used. Tumbling mills vary considerably in size, autogenous mills being the largest, ie, 1.5–11 m in diameter and 3–10 m in length. Nominally ball mills have a length-to-diameter ratio of 1.5:1 or less, and rod mills have a ratio 1.5:1 or greater. Grinding balls range from 19 to 127 mm in diameter. Grinding rods range from 38 to 100 mm in diameter and 3 to 6 m long. Grinding media occupy less than half the volume of the mill and could weigh 5–500 tons. Typical drive motors have horsepowers in the range of 75 to 8000 or greater. Some of the large autogenous mills have 15,000 hp motors and handle over 35,000 t/d of ore (22). Tumbling mills are lined with suitable abrasion-resistant material to minimize wear of the mill shell and to reduce slip between the shell and the grinding media. Liners are replaced periodically. The most commonly used materials are cast or rolled steel and rubber.

Grinding is a continuous process. It may be either wet or dry, but wet grinding (at 50–80% solids) is most common and requires less power per ton of material ground. Media consumption is significantly higher, however. Wet grinding is also the choice when subsequent processing is wet. In certain instances, depending on use, dry grinding is necessary, especially in the case of certain industrial minerals or cement (qv). The feed to a tumbling mill enters through a hollow trunnion at the center of the feed end, and the product is discharged at the other end in a variety of ways, ie, hollow trunnion, grate discharge, peripheral discharge, etc. Particles are broken as a result of a combination of cascading and cataracting (free fall) motions of the media. At high tumbling speeds, cataracting dominates, and a large part of grinding occurs by impact; at lower speeds, cascading dominates, and grinding is by attrition. Typically, particles between 5 and 250 mm are reduced to between 10 and 300 μm. In rod mills the tendency is for large particles to be broken selectively and the fine particles to be passed selectively, whereas ball mills tend to grind both large and small particles. In tumbling mills most of the energy is consumed in keeping the mill shell, the media, and the mineral mass in motion; fracture occurs as a by-product of passage through the mill and is a statistical process. The average energy consumption for grinding can be about 11.6 kWh/t compared to about 2.2 for crushing and 2.6 for flotation (2). As a result of wear, all tumbling mills have a range of media sizes. Makeup media are added regularly. Media and liner wear, which could be as high as 0.5–1 kg of media and 0.05 kg of liner per ton of ore ground, constitute a significant cost in the mill operation (10). Wear is significantly greater in wet grinding. Autogenous and semiautogenous mills have a distinct advantage over conventional rod and ball mills in terms of savings in media consumption.

As for crushers, the trend is to increasing the size of grinding mills. Very large autogenous and semiautogenous grinding mills having up to 11 m dia driven by 11 MW motor, have already been in use (2). Much effort has been made in grinding circuit control and automation (21,23).

Other types of mills include fluid energy mills, vibratory mills, centrifugal mills, tower mills, roller mills, and attrition mills (2,10,22,25). These mills are specifically designed to produce a very fine product upon high efficiency. In the fluid energy mills, the solids are entrained in high velocity gaseous or vaporous streams. Reducing particle size occurs by impact and attrition against other particles or a target, eg, a Micronizer. Vibratory mills consist of a nonrigidly supported chamber filled with grinding media (up to 80% of volume) and material to be ground, vibrated at frequencies up to 1800 min^{-1} by an eccentric mechanism. Attrition mills are of the rotating-disk (colloid mills), patterned-fluid (fluid energy mills), or abrading-sand (sand grinders) type. Size reduction is achieved by particles breaking each other after the particles have acquired the necessary energy from a solid or fluid impeller. In tower mills, grinding occurs by attrition and abrasion in a vertical chamber containing steel balls or pebbles. An internal screw flight provides medium agitation. These mills are characterized by efficient energy usage, small installation area, and low operating costs. A feed of 6 mm top size can be ground to a product in the size range of 74 to 2 μm or finer at 100 t/h.

Size Separation

Sizing of the crushed and ground product is a necessary step prior to any mineral processing operation, and in the production of a product having a specific size. Controlling the size of material fed to other equipment is important. All equipment has an optimum size range of material that it can handle most efficiently. Size separation can be achieved either by screening (for coarser particles) or by classification (for fines) (see also SEPARATION, SIZE).

Screening or Sieving. Screening or sieving (2,6,10,26) is accomplished by passing the crushed or ground ore through a mesh of perforated plates or woven metal wires/rods or profile bars providing a uniform distribution of fixed-size apertures. Screening is a continuous process, whereas sieving is a batch process used to determine the performance of crushing and grinding processes. Material retained on the screen is the oversize or the plus fraction, and the material passing is the undersize or the minus fraction. The product specifications are usually in terms of percentage of material passing a certain screen or sieve size. Although the principle and the process of screening and sieving are simple, separation is seldom perfect for any sizing operation. The most obvious contributing factor is the irregular three-dimensional shape of the particles. Each particle at any given time presents only two dimensions to the screen surface.

Laboratory sieves are made from woven wire of standard square apertures that follow a geometric progression from one screen to the next. The sieve size is designated in micrometers or mesh number. The latter denotes the number of openings per linear inch. In the laboratory determination of size distribution of a crushed or ground product, standard sieves in the desired size range are stacked. Successively smaller apertures are arranged from top to bottom, the crushed or ground material is placed on the coarsest sieve at the top, and the entire stack is shaken on vibratory equipment providing both vertical and circular motion to the particles. Material retained on each screen is then weighed, and the weight percent passing each screen is plotted against the sieve size. This is typically a log–log plot, but many other techniques have been developed (2). Wet screening is usually conducted before final dry sieving to remove very fine particles, which tend to adhere to coarse particles or to each other during dry sieving and lead to errors.

Industrial screening is used essentially for separations over 0.2 mm and in conjunction with crushers because the efficiency decreases rapidly as particle size decreases. The main objective is to remove undersize material that should not be circulated back to the crushers, or to remove (scalp) oversize material or trash that should not report to the subsequent processing step. Other applications of screening include production of a specification size material (as in quarrys), dewatering, and trash removal from processed material.

Industrial screens can be classified as stationary or moving, which in turn can be either conventional (apertures) or probability (based on statistical principles). The applicable size ranges for screens are shown in Figure 4. The op-

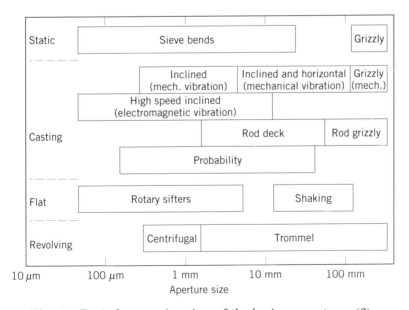

Fig. 4.   Typical separation sizes of the basic screen types (6).

erating mechanism and the screen surfaces of probability screens are basically the same as that of conventional screens, only the design is different. Principal screen types include stationary and vibrating grizzly (Fig. 5a), Hukki screen, roll grizzly, revolving screens, vibrating screens (Fig. 5b), shaking screens, reciprocating screens, sieve bends (Fig. 5d), and rotary sifters (2,6,10,26). Vibrating screens, the most widely used, have numerous designs, ie, inclined, horizontal, etc, and vibration can be generated either mechanically or electromagnetically. They can handle material from 250 μm to 250 mm. The probability type has a series of relatively small inclined screen surfaces. Separation is based on probability rather than physical constraint. This type of screen has higher capacity and efficiency than the other vibrating screens. The selection of a particular screen design depends on the application and the size cut-off. Grizzlies are used generally for very coarse material. Sieve bends have extended the applicability of industrial screening into fine sizes down to 50 μm. The upper size can be as large as 12,000 μm and capacities as high as 2000–5000 t/d. These bends consist of a curved or inclined screen surface having horizontal wedge bars. The size of the unit varies in height from 1.5 to 2.5 m and width from 0.5 to 2.5 m. They have relatively high capacity and efficiency. Among the revolving screens, the trommel is a slightly inclined, rotating cylindrical screen and it can be used wet or dry. High wear is a big disadvantage. Other types of revolving screens are the centrifugal and probability screens (Fig. 5c). The former is a vertically mounted cylindrical screen; high wear is again a disadvantage. The probability screen consists of rotating radiating bars; the speed of rotation determines cut size. This high capacity screen can also produce fine separations. The rotary sifters can be either reciprocating or gyrating. In the former, a rectangular screen surface with a slight incline is used with a reciprocating motion at the discharge end, whereas in the gyrating type, a circular screen is used with a circular motion.

There are three basic types of screen surfaces: perforated or punched plate, woven cloth, and profile bars. Woven cloth surfaces are the most common. Perforated plates are made up of hardened steels, stainless steels, Monel, rubber, or plastic. Woven cloth can be made from high carbon steels, tempered steels, manganese steels, galvanized steel, Monel, copper, bronze, or reinforced synthetic cloths of polyurethane rubber. Screens can handle material ranging from the fineness of talc through boulders as large as 2 × 2 m weighing as much as 10–12 tons. Screen openings range from 0.1 mm to as large as 500 mm.

Capacity and efficiency are the two criteria used to assess screen performance. Whereas many empirical formulas for measuring these criteria are available, none is entirely satisfactory. The efficiency is usually expressed in the form of a partition or performance curve which is a plot of percent oversize vs the geometric mean size on a log scale (2). For a perfect separation, such a curve would be a vertical line at the cut-off size, but in practice it is an S-shaped curve. On such a plot, the separation or cut-off size is that corresponding to 50% oversize. A number of factors affect screen efficiency. These include feed rate, vibration rate or intensity, particle shape, screen design, the amount of moisture in the feed, and the amount of near-mesh particles which tend to blind or plug the apertures. Wet screening is more efficient than dry screening, but the decision to wet or dry screen depends largely on the application.

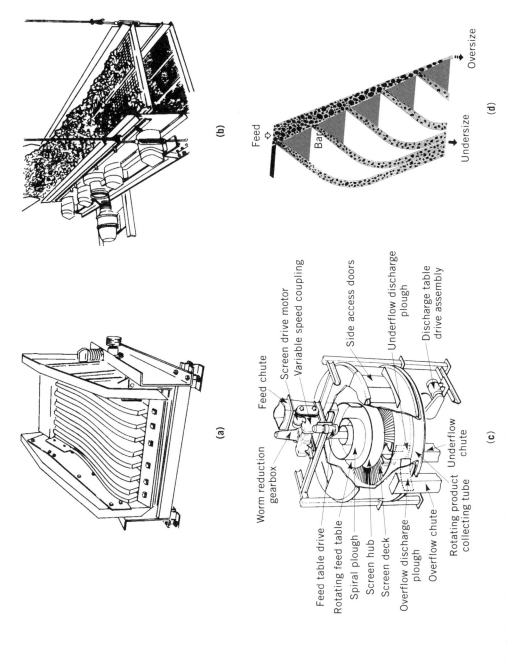

(a)

(b)

Worm reduction
gearbox

Feed chute
Screen drive motor
Variable speed coupling

Side access doors

Underflow discharge
plough

Discharge table
drive assembly

Feed table drive
Rotating feed table
Spiral plough
Screen hub
Screen deck
Overflow discharge
plough
Overflow chute
Rotating product Underflow
collecting tube chute

(c)

Feed

Ba

Undersize

Oversize

(d)

Fig. 5. Schematic diagrams of screens: (**a**) vibrating grizzly, (**b**) double-deck vibrating screen, and (**c**) a rotating probability screen (2). A schematic illustration of the operating principle of a sieve bend is shown in (**d**) (6).

Classification. Classification covers a broad range of size separation methods that rely on the differences in settling or sedimentation velocities of particles in a fluid (air or water). Because these velocities are affected by size, shape, and density of the particles, classification is in reality a sorting rather than sizing method (2,3,6,10,27). Moreover, there is incomplete liberation of minerals in the ground product and the pulp invariably contains composite (locked) particles of varying specific gravities. Classification is more suitable for particles in the fine size range where the performance of screens is poor. The products are an oversize (underflow, heavies, sands) and an undersize (overflow, lights, slimes). An intermediate size can also be produced by varying the effective separating force. Separation size may be defined either as a specific size in the overflow screen analysis, eg, 5% retained on 65 mesh screen or 45% passing 200 mesh screen, or as a d_{50}, defined as a cut-off or separation size at which 50% of the particles report to the oversize or undersize. The efficiency of a classifier is represented by a performance or partition curve (2,6), similar to that used for screens, which relates the particle size to the percentage of each size in the feed that reports to the underflow.

In a simplistic view, a classifier may consist of a column with rising fluid and falling particles. Particles either sink (sedimentation or settling) or are carried upward with the fluid depending on the terminal velocity of the particles as dictated by gravitational and frictional forces. Particles experience either free or hindered settling depending on the solid:fluid ratio. Density differences have a more pronounced effect on classification at coarser size ranges and size differences dominate at fine size ranges. Free-settling conditions accentuate the effect of size, and the hindered settling conditions accentuate the effect of density. Depending on the conditions, classifiers can, therefore, separate relatively coarse from relatively fine particles, or separate smaller heavier particles from larger lighter ones. They can also split a wide size distribution into manageable fractions.

Classifiers can be grouped into horizontal current and vertical current types, or into mechanical, nonmechanical, sedimentation, and hydraulic or fluidized-bed types depending on the design of the equipment. The available equipment, their sizes, capacities, and their uses are given in Table 5.

Horizontal current classifiers such as the mechanical classifiers are essentially of the free-settling type, whereas the vertical current classifiers are of the hindered-settling type. Sands discharge is by mechanical means against gravity in mechanical-type classifiers. Gravity or centrifugal forces are used for sands removal in the nonmechanical classifiers. In hydraulic or fluidized-bed classifiers, particles in the pulp settle under hindered settling conditions against a countercurrent flow of additional water in a series of columns. Products can be produced from each column: coarser and denser particles from the first column, finer and lighter particles from subsequent columns, and the slimes reporting to the final overflow. Hydraulic classifiers are used extensively in conjunction with gravity separators. In general, the fluidized-bed classifier is capable of giving a sharper separation and a more precise d_{50} than a sedimentation classifier, but this is achieved at the expense of capacity. Some of the fluidized-bed classifiers are essentially sedimentation classifiers using only small amounts of hydraulic water. In sedimentation classifiers, particles settle through a pool of water formed

Table 5. Types of Classifiers[a]

Equipment	Type[b]	Size, m			Feed rate, t/h	Uses
		Width	Dia	Length		
sloping tank classifier (spiral, rake, drag)	M,S	0.3–7	2.4	14	5–850	closed-circuit grinding, washing, dewatering, desliming
log washer	M,S	0.8–2.6	0.6–1.1	4.6–11	40–450	removing trash, clay from sand; break down agglomerates
bowl classifier	M,S	0.5–6	1.2–15	12	5–225	closed-circuit grinding, washing
hydraulic bowl classifier	M,F	1.2–3.7	1.2–4.3	12	5–225	closed-circuit grinding, washing
cylindrical tank classifier	M,S		3–45		5–625	primary dewatering
hydraulic cylindrical tank classifier	M,F		1.0–40		1–150	washing, desliming, closed-circuit grinding
cone classifier	N,S		0.6–3.7		2–100	desliming, primary dewatering
hydraulic cone classifier	M,F		0.6–1.6		10–120	closed-circuit grinding
hydrocyclone	N,S		0.01–12		≤20[c]	closed-circuit grinding, desliming, degritting, dewatering, washing
air separator	N,S		0.5–7.5		≤2100	dry classification[d]
solid bowl centrifuge	M,S		0.3–1.4	1.8	0.04–2.5[c]	fine size separations
countercurrent classifier	M,F		0.5–3.3	12	3–600	clean sands, washing, desliming
eutriator, pocket classifier	N,F				4–120	clean sands, washing, desliming

[a] Ref. 6.
[b] M = mechanical transport of sands to discharge; N = nonmechanical (gravity or pressure); S = sedimentation classifier; and F = fluidized-bed classifier.
[c] Units are m^3/min.
[d] For cement.

803

from the feed stream in an inclined trough (Fig. 6). The settled mass is dragged upward by mechanical means, typically through a spiral, rake, or drag, and the fines go with the overflow. Widely used in closed-circuit grinding, these classifiers are also used to produce a clean, ie, free of slimes and fines, sized final product. The capacity and separation efficiency are influenced by feed rate, the speed of the rake or the spiral, the height of the overflow weir, and dilution of the pulp. One of the principal disadvantages is that the overflow tends to be very dilute for many subsequent operations. Mechanical sedimentation types of classifiers, such as the spiral classifiers, are large units and occupy substantial floor space, in contrast to the nonmechanical units.

The hydrocyclone, commonly referred to simply as cyclone, is a nonmechanical sedimentation-type classifier (2,6,10,27) (Fig. 7). It has no moving parts or power attachments directly connected to it. The hydrocyclone has become the workhorse of most mineral processing operations because of its simplicity, short residence time, compactness, and low cost of operation. It is, however, characterized by lack of sharpness of separation. Equipment consists of a cone having an open apex attached to a cylindrical section which has a tangential feed inlet at the top (Fig. 7a). The pressure feed generates centrifugal action to give high separation forces and discharge. Overflow is through a pipe mounted axially on the plate that covers the cylindrical section. A short extension of this pipe into the cylindrical section acts as a vortex finder which prevents short circuiting of the feed directly into the overflow. The flow pattern in a hydrocyclone can be described as a spiral within a spiral generated by the tangential feed. An air core exists in the center along the vertical axis extending through the apex. The centrifugal force developed accelerates the settling rate of particles, and particles are separated according to size and specific gravity. Faster settling particles exit through the apex and slower settling particles exit through the overflow (see also SEPARATION, CENTRIFUGAL).

The hydrocyclone is used in closed-circuit grinding and is efficient for separation at fine sizes. It can be used not only as a classifier, but also as a thickener, a desliming unit, or a concentrator (6). The disadvantages are the lack of

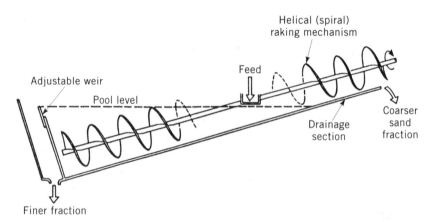

Fig. 6. Schematic of a mechanical classifier with submerged spiral rake where (– – –) is the pool level and slope = 1:4 to 1:3 (27).

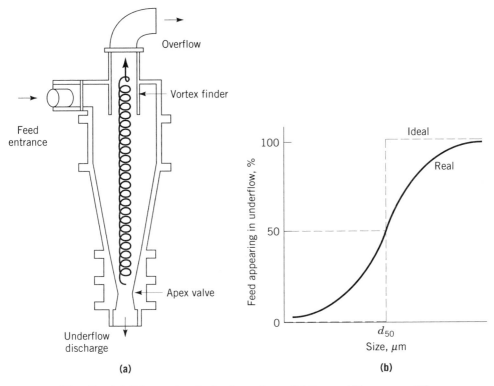

Fig. 7. (a) Schematic of a hydrocyclone; (b) its partition curve (2).

sharpness of separation and the amount of material that passes through without classification. When the required cut-off size is coarse, the plant throughput may not be large enough to satisfy the cyclone feed requirements. It is also generally impractical to operate when cut points are <2 μm. In spite of simple design and operation, the actual processes occurring in a hydrocyclone are far from simple. A complete rigorous analysis is unavailable, but numerous empirical formulations have been developed to describe its operation (2,6). The separation is not perfect and the performance curve is typically S-shaped (Fig. 7b). The slope of the central portion of this performance curve represents the efficiency or the sharpness of separation. The closer to vertical the slope, the greater the efficiency.

Empirical improvements have been made to the performance curve (2,6). Corrected performance curves, ie, corrected for part of the inlet stream passing out of the classifier without being classified, especially fines reporting in the underflow, and reduced performance curves, wherein the corrected performance curve is normalized by dividing the size scale by d_{50}, have resulted. The latter is largely independent of the nature of particles, and is a characteristic of the type of classifier within a reasonable range of design. In a practical mineral slurry, which is invariably nonhomogeneous, there is more than one mineral, either as liberated or as middling particles, and a number of different sized particles have the same settling rate. Therefore, in a given classifier, each mineral has its own performance curve resulting from differing densities or shape or both. This has important implications, either beneficial or adverse depending on the type of ore being treated, in closed-circuit grinding and the subsequent concentration separation.

Although performance curves are valuable in assessing classifier performance, frequently the cyclone overflow size analysis is used more than the d_{50} of the cyclone. In practice, clusters of cyclones (in parallel) are used to handle large capacities. Cyclones are manufactured in sizes ranging from 0.01 to 1.2 m in cyclone diameter, ie, the cylindrical section at the top (2,10). Capacities run from 75 to 23,000 L/min. Materials of construction vary widely. Rubber-lined or all-polyurethane cyclones are used when abrasion is a problem.

Minerals Concentration

Although the size separation/classification methods are adequate in some cases to produce a final saleable mineral product, in a vast majority of cases these produce little separation of valuable minerals from gangue. Minerals can be separated from one another based on both physical and chemical properties (Fig. 8). Physical properties utilized in concentration include specific gravity, magnetic susceptibility, electrical conductivity, color, surface reflectance, and radioactivity level. Among the chemical properties, those of particle surfaces have been exploited in physico-chemical concentration methods such as flotation and flocculation. The main objective of concentration is to separate the valuable minerals into a small, concentrated mass which can be treated further to produce final mineral products. In some cases, these methods also produce a saleable product, especially in the case of industrial minerals.

Selection of a particular concentration method depends entirely on the mineral and metal in question, the nature and mineralogy of the ore deposit, particle size at economic liberation, and the prevailing socio-economic factors. For many centuries, sorting by hand and gravity concentration were the only methods available. By the end of the nineteenth century magnetic and electrostatic methods had been introduced. Minerals processing underwent a revolution when the flotation method was developed in the early twentieth century because large tonnages of a wide variety of ores at various grades could be processed and complex mineral separations that were not possible by any other method could be performed.

Ore Sorting. Sorting methods rely on differences in the physical properties of the various mineral components in an ore (2,6,10,28,29). Such properties include optical characteristics, magnetic susceptibilities, x-ray fluorescence, electrical and thermal conductivity and charging, and radioactivity. Sorting techniques became successful once the sorter was able to analyze individual particles. Sorters can function efficiently only at coarse (typically > ca 10 mm) sizes, and in a narrow size range. Particles up to 160 mm can be sorted at capacities of 180 t/h (6). The extent of liberation of the valuable mineral from the gangue at such coarse sizes determines whether a sorter can be used. The nature of distribution of the value mineral in the ore is also important. For some sorters, the desired mineral must be exposed on the surface of the particle and not buried inside the particle. In many cases, limitations are imposed by the type and capacity of the mining equipment. The particle surfaces presented to the sorters must be clean of dust and slime free not only to ensure that the surface physical

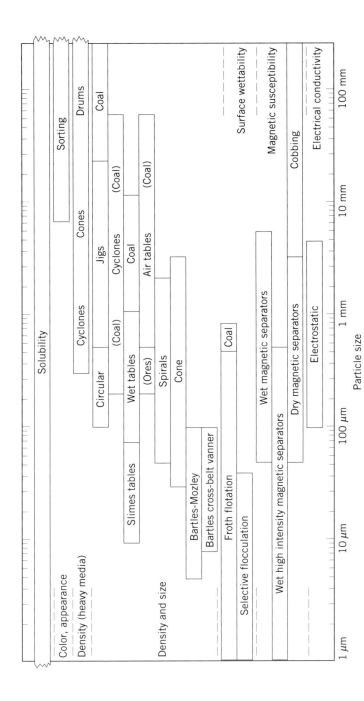

Fig. 8. Particle size ranges for concentrating equipment based on mineral properties (6).

properties of the minerals are not obscured, but also to protect the optics of the detection system. Such requirements are less critical for sorters that rely on the bulk properties of the mineral such as electrical and thermal conductivity, magnetic properties, and radioactivity. In these cases, a random distribution of the valuable mineral in the particle is preferred. Attempts to quantify ore sortability are limited.

The primary features of a typical sorting machine are shown in Figure 9. The detection system in a sorter is the most important unit and sorters are, therefore, grouped according to the detection system. Advances in the electronic industry have greatly facilitated the development of sophisticated sorting machines. The modern sorter has three distinct functions: singulation, detection, and ejection. Singulation ensures that individual particles are presented to the detection system. The detection system evaluates the selected physical property for the particle presented and sends a signal to a combination of electrical and mechanical ejection systems to classify the particle as valuable or gangue. The singulation and ejection systems determine the capacity of the sorter and the detection system determines the efficiency of the separation. A perfect synchronization between the detector and the ejector is necessary in the decision making process which incorporates the time delay between the detector and the ejector, and the size and position of the particle in the stream. After the ejection of one particle is complete, the ejector must return to its original position as rapidly as possible. Most sorters use solenoid controlled air valves to eject selected particles. Particles are presented to the detector either single file or single layer using a vibrating hopper and a conveyor belt. The single layer mode provides for increased capacity. Scanning speeds can be as high as $2000 \ s^{-1}$ for the ore stream, equivalent to one scan for every 2 mm of rock length (2).

Optical and photometric sorters (10) are the most widely used. Optical properties in the detection system include reflectivity, color, transparency, and fluorescence. Photometric sorters have all of the features shown in Figure 9 and utilize a laser light source in a scanning mode to detect reflected light using a

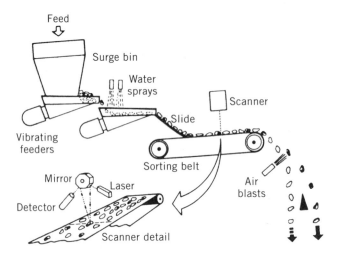

Fig. 9. Primary features of a sorting machine (6).

sensitive photomultiplier (see LASERS; PHOTODETECTORS). A typical application includes upgrading rock salt, magnesite, barite, gypsum, marble, diamond, aggregates, asbestos, talc, and limestone. Color forms the basis for sorting white mineral-bearing quartz from quartzite of various colors. Other applications include separation of gold, cassiterite, and wolframite in quartz from quartzite, and white quartz containing tetrahedrite and other minerals, ie, Ag and Cu values, from quartzite. Scheelite sorting is achieved by sensing fluorescence under ultraviolet light. For minerals that do not exhibit fluorescence naturally, selective chemicals that adsorb on the value minerals and provide fluorescence have been developed. Asbestos (qv) is sorted using sequential heating and infrared scanning.

Radiometric ore sorting has been used successfully for some uranium ores because uranium minerals emit gamma rays which may be detected by a scintillation counter (2). In this application, the distribution of uranium is such that a large fraction of the ore containing less than some specified cut-off grade can be discarded with little loss of uranium values. Radioactivity can also be induced in certain minerals, eg, boron and beryllium ores, by bombarding with neutrons or gamma rays.

Other sorters such as x-ray, conductivity, and magnetic sorters have found only limited application (10). Sorting based on mineral conductivity has been used for iron ores and native copper. Diamonds emit visible light when bombarded with x-rays. This property has been used for the sorting of diamond ores in conjunction with the traditional grease tables. X-ray fluorescence has also been used in sorting. One commercial sorter measures electrical conductivity and magnetic susceptibility of individual particles and the detectors respond to only slight variations in such properties. It also uses an optical system to track the size and shape of the particle. A microprocessor estimates the grade of the particle from its size and measured bulk properties for the decision making process. Such a sorter would be suitable for a preconcentration step and would be applicable for a wide variety of ores because of bulk property measurement. Gamma radiation scattering analyses form the basis of another sorter (2). The most suitable metals are chromium, iron, cobalt, nickel, copper, and zinc or any combination of these.

Gravity Methods. Gravity methods date back to antiquity, and by sixteenth century these were well advanced (2,6,10,30–32). A considerable amount of gravity concentration occurs in nature in the form of concentration of heavy minerals in placer or beach sand deposits as a result of moving water. Gravity methods waned in popularity with the advent of flotation (qv) in the early part of the twentieth century. A revival, either as a primary technique or a preconcentration step, has occurred because of the simplicity, low cost, and advances made in the design of equipment (2,6). Gravity methods have been extended into the 50–10 μm range although capacity and throughput can still be a problem. These methods are also being evaluated for the treatment of large dumps of flotation tailings that contain heavy minerals. In 1975, approximately 490×10^6 t of coal and ore were treated by gravity methods compared to the 423×10^6 t treated by flotation (10).

Gravity methods, ie, gravity concentration and dense medium separation, rely on the differences in specific gravities between minerals in fluids. In gravity concentration, the fluid is water or air; in dense medium separation, the medium of separation is a liquid or fluid of specific gravity greater than that of water. Gravity methods are used to treat a variety of ores, including industrial minerals, coal, iron ore, phosphate ore, precious metals, tungsten ores, diamonds, and heavy-metal sulfides.

Gravity Concentration.   Gravity concentration devices or gravity separators have much in common with size separators and can be used interchangeably under certain conditions. The principles are similar except that in gravity separators care is taken to emphasize specific gravity differences of the materials being processed. The performance of gravity separators is best when a narrow range of feed size is used, just as the performance of size classifiers is best for a narrow range of densities. It is, therefore, important that a properly size-classified and consistent pulp sample be fed to the gravity circuit. Care is taken to eliminate or minimize slimes in the feed as these tend to decrease the sharpness of separation. The efficiency of gravity separators increases as particle size increases, but other factors, such as degree of liberation of values, is also important in determining the optimum size range of the feed to the gravity circuit. A gravity concentration criterion (Dh−Df)/(Dl−Df), where Dh, Dl, and Df are specific gravities of the heavy mineral, the light mineral, and fluid medium, respectively, is often used to assess the type of separation possible (2). If this ratio is 2.5 or greater, gravity concentration is very efficient. The efficiency decreases as the ratio decreases. Gravity concentration can occasionally be used to produce a final saleable product. More typically, it is used as a preconcentration method; the final gravity concentrate requires cleaning by other methods, such as flotation, magnetic and electrostatic separation, and leaching, to remove impurity minerals. A typical gravity circuit may include roughing, and several stages of cleaning of the rougher concentrate with regrinding in between, although regrinding of middlings in the rougher stage may also be required.

The performance of gravity concentrators can be measured empirically by performance or partition curves similar to those for classifiers (Fig. 7b) except that a cut-off gravity is used instead of a cut-off size (6). Much of the available information, however, has been obtained only for coal cleaning (see COAL CONVERSION PROCESSES), because of the difficulty in carrying out measurements on high density minerals. The reduced performance curve is primarily a characteristic of the particle size being treated and of the device. Available empirical correlations are rather limited, however, unlike the case of size classifiers.

Gravity separators can be grouped according to either the operational feed size range or the manner in which the particles and fluid move relative to one another in the separator. Gravity separators fall into three groups: jigs, shaking concentrators, and gravity flow concentrators (6,10). Many of these are rather simple devices. The size, capacity, and uses of selected gravity separators are given in Table 6.

Jigs, generally very effective for relatively coarse (typically 0.5–200 mm) material, have relatively high unit capacities. The sharpness of separation is a function of size distribution. Fine specific gravity separation is possible for a closely sized material. The principal usage of jigs is in coal beneficiation. Other applications include concentration of cassiterite, tungsten, gold, barites, and iron

Table 6. Types of Gravity Separators[a]

Equipment	Size,[b] m²	Capacity,[c] t/h	Uses
		Jigging	
diaphragm or plunger mineral jig	≤1.2 × 1.1	4[d]	roughing, cleaning, scavenging relatively coarse cassiterite, gold, scheelite
Baum jig	≤17.6[e,f]	20[d]	mainly for coal washing
Batac jig	30[g]	12–24[d]	mainly for coal washing
Wemco-Remer circular jig	1.5 × 4.9	7[d]	primarily for aggregate production
	41.7 (7.5 dia[e])	10[d]	extensively on tin dredges
pneumatic jig	1.8 × 3.8	2–3[d]	for coals, when dry product is an advantage
		Shaking	
shaking table	2.0 × 4.6	0.05–0.25[d]	coal, cassiterite, scheelite, and other heavy minerals
Holmans slimes table	2.0 × 4.6	0.01–0.06[d]	for particles too fine for conventional table; also for cleaning concentrate from Bartles-Mozley
Bartles-Mozley concentrator	1.2 × 1.5	2.5	rougher concentrator for very fine heavy minerals
Bartles cross-belt vanner	2.75 × 2.4	0.5	similar applications to slimes table
		Flowing	
Humphrey's spiral	0.6[e] dia, 2.9[h]	1–5	beach sands, iron ore, and other heavy minerals
pinched sluice	≤1.8 × 0.4	2–4	beach sands, phosphate ore
Reichert cone	2 dia[e]	60–90	beach sands, coal, iron ore, trace heavy minerals from tailings

[a]Ref. 6. [b]Dimensions are width by length unless otherwise noted. [c]Capacities depend mainly on type of minerals treated and their particle size range. [d]Value given is per square meter of jig or table. [e]Value is in meters. [f]Arrangement of 2 × 6 cells in parallel. [g]Arrangement of six 5 × 1 cells. [h]Height of spiral in meters.

ores. The basic jig (Fig. 10) has a large tank or hutch divided in the upper portion into two main sections. One section contains the stationary screen with the mineral bed on it, the bed depth being many times the thickness of the largest particle, and the other section contains the pulsating device, most commonly using air pressure rather than mechanical means. Mineral separation is achieved by applying a vertical oscillatory (pulsating) motion to the solids–fluid bed. This pulsating motion produces dilation of the bed and subsequent stratification. The denser and larger particles form a lower layer whereas the finer lighter particles are on top. The processes occurring in a full cycle of operation may be considered differential initial acceleration, hindered settling, and consolidation trickling. Several other theories have been developed, however, notably the center-of-gravity theory (6,10). The pulsing action is supplemented by using additional water in the hutch during the settling period. This extends the open state of the bed for a longer time. The dense minerals are collected either on the screen or under the screen depending on the screen aperture size. In the latter case, a layer of dense (ragging) particles larger than the aperture size are placed on the screen to regulate the collection of dense fraction. Examples are feldspar in coal cleaning, and hematite in cassiterite and scheelite separation. Several stages of jigging are used to achieve efficient separation. The commercial jigs have a variety of designs for the pulsating device and the removal of products. Some examples are shown in Figure 11. The Batac jig, which uses multiple air chambers under the screen, is the industry standard in coal cleaning. A more recent development is the Kelsey centrifugal jig (33,34). The circular or radial jig is a variation of the conventional rectangular design of hutches in series. The pulp is fed at the center and flows radially over the jig bed and exits at the circumference. A raking mechanism ensures an even bed depth throughout. It is mechanically simple and has very high capacity, up to 300 m³/h for a maximum particle size of 25 mm. It achieves a fast compression/slow suction stroke with virtually no hutch water. The slow suction stroke allows more time for the fines to settle to the bed.

In the second group of gravity concentrators, eg, the shaking table, the vanner, the Bartles-Mozley concentrator, and the miner's pan, a horizontal shear is applied to the solids–fluid stream by vibrating the surface under the stream. Despite early popularity, conventional tables have limited usage in the minerals industry mainly because of low capacities and the large floor area they occupy,

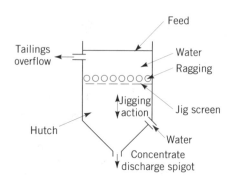

Fig. 10. Basic jig construction (2).

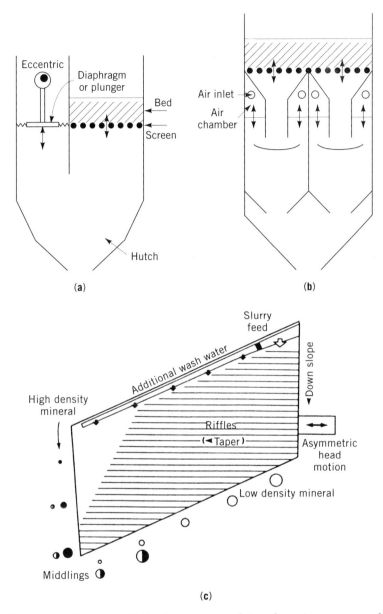

Fig. 11. Schematic diagrams of the basic types of jig where ↕ represents the jigging action: (**a**) Denver/Harz and (**b**) Batac. (**c**) Schmatic of a shaking table, showing the distribution of products (6).

and because of the availability of many alternative concentrating methods. Tables are employed for processing coal and, to a lesser extent, many industrial minerals. In general, tables are efficient separators used for difficult flow streams and for producing finished concentrates from products of other separators. The shaking table is effective in separating small dense particles from coarse light particles, and is capable of treating finer particles than jigs, but at the expense of capacity.

In a typical shaking table (Fig. 11**c**) the feed enters through a distribution box along part of the upper edge and spreads out over the table as a result of the differential shaking action and the wash water. Product discharge occurs along the opposite edge and end. The surface of the table is a suitably smooth material, eg, rubber or fiber glass, and has an appropriate arrangement of riffles on it, which decrease in height along their length toward the discharge end. Various riffle arrangements can be used to emphasize grade or recovery. The table also has an adjustable slope or tilt of about 0–6 degrees from the feed edge down to the discharge edge to regulate distribution of material. During tabling, minerals are subjected to lateral (table motion) and longitudinal (flowing film of water) forces. The net effect is a diagonal movement from the feed end. Separation of minerals is effected as a result of flowing film concentration, hindered settling, consolidation trickling, and asymmetrical acceleration. Small dense particles move more slowly than coarse light ones and depending on the size and density, the particles fan out on the tables. Several products can, therefore, be collected.

Double- and triple-deck shaking tables (2,6) have much higher area/capacity ratio at the expense of some flexibility and control. The Bartles-Mozley concentrator is an example of a multiple-deck device developed to recover fine cassiterite that cannot be recovered by other devices. It consists of a suspended assembly of 40 fiber glass decks arranged in two sandwiches of 20, each deck separated by a 13-mm space that also defines the pulp channel. Feed is distributed to all 40 decks for a period of up to 35 minutes, after which the flow is interrupted briefly while the table is tilted for concentrate removal and the feed resumes. The concentrator has high capacity and is capable of recovering a majority of fine (ca 10-μm particles) cassiterite in low grade slurries. Other devices include pneumatic tables and the duplex concentrator (2).

In gravity flow concentrators, eg, sluices, troughs, spirals, and the Reichert cone, a layer of slurry flows under gravity down an inclined surface (2,10,31). The separation is one of both size and density and has been used for centuries, including the natural processes in placer deposits such as those of tin, gold, and beach sand minerals. The simplest device is the pinched sluice, an inclined, tapered launder on which pulp flows gently and stratifies as it descends. Although relatively simple and inexpensive to operate, usage is limited because of the development of more efficient, higher capacity concentrators. The Reichert cone (Fig. 12) is a high capacity device developed initially to treat titanium-bearing beach sands. The efficiency is rather low, but this cone is effective as a roughing device for large tonnages. The principle of operation is similar to that of a pinched sluice. It comprises several cone sections stacked on each other which permit separation in stages. Separation takes place as the feed descends from the periphery of the cone to the center. It is essential to maintain a high feed density (55–70% solids). These concentrators are most efficient in the 100–600-μm size range (2).

Another gravity flow concentrator is the spiral (Fig. 12**b**), initially known as the Humphrey's spiral, which has also been used largely for mineral sands containing rutile, ilmenite, and zircon, and for iron ore. As the slurry flows down a spiral surface, particles are stratified owing to the combined effect of centrifugal force, differential settling rates, and trickling through the flowing particle bed. The primary process is hindered settling. The addition of secondary

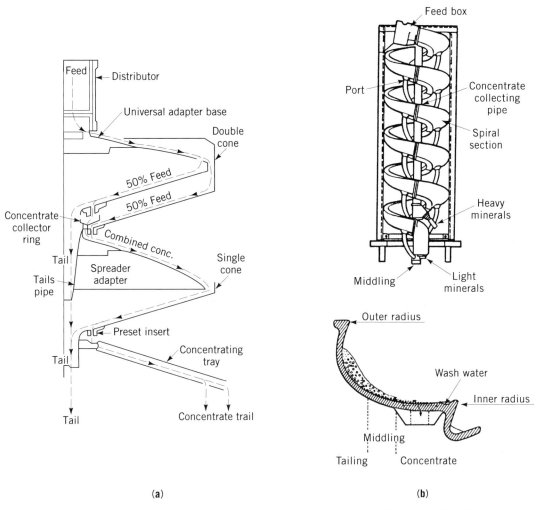

Fig. 12. Gravity flow concentrators: (**a**) Reichert cone (2), and (**b**) spiral, where (∘) represent particles of low density and (•), particles of high density (27).

wash water at the center of the spiral aids the already present river bend action resulting from the circular path of the slurry. This results in the larger lighter particles reporting to the periphery of the spiral. The denser material goes near the center, where it is removed through the discharge ports. The spiral requires lower (15–45%) pulp densities than the Reichert cone and can handle a size range of 3 mm to 75 μm (2). Modern spirals are made of fiber glass and may comprise two separate spirals intertwined to save floor space. High capacity spirals are a more recent development and operate at high pulp densities using no wash water. These have a flatter profile and have no splitters. These newer spirals have increased spiral usage.

More recent advances in the centrifugal gravity separators are the Mozley multigravity separators (MGS), the Knelson concentrator, the Falcon concentrator, and the Knudsen bowl centrifugal concentrator (33,34). These separators are particularly suitable for fine particles and applications include precious metals, coal, many industrial minerals, coal desulfurization, soil remediation, and harbor silt clean-up. They are all characterized by a high capacity/area ratio. The MGS is essentially a conventional shaking table that has been wrapped into a drum. By rotating the drum, forces many times the gravity can be exerted on the particles in the film flowing across the surface. The large-scale unit comprises two drums mounted back to back, rotating at speeds 90–150 rpm with shaking oscillations of 4–6 cps, and enabling forces of 5 to 15 G. The Knelson concentrator, a compact unit with an active fluidized bed to capture heavy minerals, has gained popularity for precious metal ores. It consists of a tapered bowl in which a ribbed inner cone is rotated at high speeds. Capacities up to 40 t/h (for 760-mm dia unit) and forces up to 60 G are possible. The Falcon concentrator, which can generate up to 300 G centrifugal field, can have capacities of 0.5 to 100 t/h.

Dense Medium Separation. Dense medium separation (DMS) (2,6,10) is a gravity concentration method in which a medium of density higher than that of water, but between the densities of the minerals to be separated, is used. It is also referred to as heavy media separation or the sink–float process. DMS is used in the concentration of a variety of minerals, essentially as a preconcentration step. The principal use is in the cleaning of coal to produce a final product. The conventional dense medium separators, not highly efficient for the separation of fine particles, are restricted in use to ores in which either the valuable or gangue minerals liberate at a relatively large size. The advantages of this method include sharp separations, high efficiency even in the presence of large amounts of near-density material, ease of changing the separation density, and ease of control. This method, however, is relatively expensive.

In the simplest form of the method, heavy organic or inorganic liquids, the latter usually dissolved salts in water, of appropriate (1.5–5) specific gravities are used (2,10). Higher density minerals are collected in a sink product; lower density minerals are collected in a float product. A third middlings product is also sometimes collected. Examples of heavy liquids are tri- and tetrachloroethane, tri- and tetrabromoethane, di- and triiodomethane, and aqueous solutions of sodium polytungstate, and thallium formate–malonate. The specific gravities of the medium can be varied by mixing in liquids of lower specific gravities such as carbon tetrachloride or triethyl orthophosphate. In view of the toxicity and high cost of most of these liquids, use is restricted to laboratory testing of ore to assess suitability of ores for gravity concentration and to determine the economic separation density and minerals liberation.

The medium used in industrial separations is a suspension in water of finely divided high density particles, most commonly fine magnetite (sp gr 5.1), ferrosilicon (sp gr 6.7–6.9), or a mixture of the two, although quartz (sp gr 2.65, Chance process for coal) and other solids have been used (10). Medium specific gravities in the range of 1.5–3.4 can be obtained readily. Both magnetite and ferrosilicon are physically hard and chemically stable, form low viscosity fluids, and can be recovered readily using magnetic separation. The size of the medium particles is generally in the range of 95% −150 to −40 μm. Magnetite is used mainly for coal, and ferrosilicon for metalliferous ores.

Industrial separations are conducted in gravity or bath separators for a coarse feed, and in centrifugal separators for a fine feed (2,6,10). In gravity-type separators the feed and medium are introduced to the surface of a large quiescent pool of the medium. The float material overflows or is scraped from the pool surface. The heavy particles sink to the bottom of the separator and are removed using a pump or compressed air. The drum separator (Fig. 13), up to 4.6 m dia and 7 m long, processes approximately 800 t/h, and treats feed of size up to 30 cm dia, operates in the gravity or the static bath mode. Feed enters at one end of the drum and the floats exit from the other end. The sink product is removed continuously from the rotating drum through the use of lifters attached to the drum which empty into a launder as they move to the top. A modification of the simple drum separator is the two-compartment drum separator which allows a two-stage separation. In the cone-type separator (up to 6.1 m in dia and 450 t/h) feed is introduced at the top. The medium in the cone is kept in suspension by gentle agitation. The sink product is removed from the bottom of the cone either directly or by airlift in the center of the cone. The maximum particle size that can be separated is limited to 10 cm. Other separators include the Drewboy bath and the Norwalt bath (2).

The settling rate for smaller particles would be too low for efficient separation in gravity separators. This is overcome by the use of centrifugal forces. Acceleration up to 20 times that of gravity can be obtained. Hydrocyclones, similar to those used for fine feed classification, are used as the centrifugal separators after certain design modification. Separations can be made with feed sizes in the range of 0.5–30 mm at capacities approximately 75 t/h. Hydrocyclones are used for the separation of a wide variety of feeds, most notably coal, where many advantages are clearly realized because the separation is extended into finer size ranges. Pyrite removal from coal and upgrading of oxidized coal are also achieved more effectively than using flotation. Separation can be extended down to 0.1 mm. The Dutch State Mines cyclone is able to treat ores and coal in the size range 40–0.5 mm (2). The ore suspension in fine ferrosilicon or magnetite is fed tangentially to the cyclone under pressure. The sink product is removed through the central vortex finder. In the Vorsyl separator the feed makes an involute entry under pressure (10). The lighter, clean coal exits through the vortex finder, and the heavier particles move in a spiral path against the wall and are removed at the base of the vessel. The large coal dense medium separator (LARCODEMS) (Fig. 14) comprises a cylindrical vessel (1.2-m dia and 3-m long)

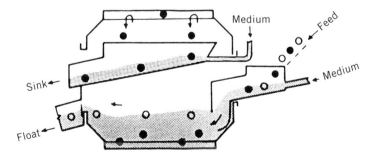

Fig. 13. Gravity dense medium drum separator where (○) is the float and (●), the sink (2).

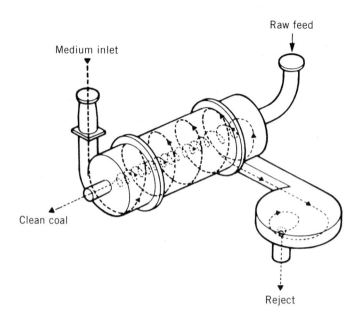

Fig. 14. LARCODEMS separator (2).

inclined at 30° to the horizontal (2). Feed medium makes an involute entry. It can handle coal in the size range 0.5–100 mm, and the capacity can be as high as 250 t/h. The Dyna whirlpool (10), the precursor to the LARCODEMS, is used to treat fine coal, diamonds, fluorspar, manganese, tin, and lead–zinc ores. It has many advantages over the DSM cyclone and has a high sinks capacity. The Tri-Flo separator (2) is similar to the Dyna whirlpool but has two separators in series. It can be operated with two media of differing densities in order to produce sink products of individual controllable densities, or with a single medium in which case a two-stage separation is possible resulting in increased recovery. There are also autogenous dense medium separators wherein fine particles of high and intermediate density from the ore to be treated form the autogenous dense medium in the cone. These can separate coal particles in the size range of 600–150 μm. They have lower capital and operating costs because of the elimination of medium separation and recovery costs.

Desliming of the feed to the dense medium separators is important because slimes interfere with the separation. Low amounts of clay contaminations have been known to stabilize the suspension, however. The particle size and rheology of the medium are two important factors. Significant advances have been made in automatic control of the medium consistency. Density of the medium can be controlled to within ±0.005 specific gravity units (6). Because DMS does not depend on surface properties of particles, this could often be a significant advantage over flotation and other methods. Also the energy requirement for grinding, flotation, and dewatering is sufficiently high to make dense medium separator an attractive concentration method, at least as a preconcentration step.

The efficient recovery and recycle of the medium is a factor in the economics of the DMS process. Typical losses of medium per ton of feed treated are about 0.5 kg for magnetite and 0.25 kg for ferrosilicon (10). Losses occur because of inadequate washing of the separated products and the need for periodic replacement owing to buildup of fine mineral particles resulting in lower medium density. The recovery involves screening of feed to remove fine particles, screening of both sink and float products in two stages, and concentration of diluted medium by wet, low intensity magnetic separation of the drum type using permanent magnets. The feed to the drum separator contains 10–15% solids which are 75–90% magnetics. A 760-mm diameter concurrent single-drum separator can give magnetic recovery of 99% or better (10). Double-drum units are used when feed volumes are large.

Dry dense medium (pneumatic fluidized-bed) separation has been used, but has not received wide attention by the industry. An area of promise for future development is the use of magnetically stabilized dense medium beds by using ferro or magnetic fluids (2,10). Laboratory and pilot-scale units such as Magstream are available. In this unit, material is fed into a rotating column of water-based magnetic fluid. Particles experience centrifugal forces and opposing buoyant forces which are magnetically derived. These forces are balanced so that lighter fraction is biased toward the center and the heavies are biased outwardly, and these fractions are collected separately. The unit can treat particles in the size range 1 mm to 53 μm and the specific gravity split point can be in the range of 1.3–21.

Magnetic Separation. Magnetic separation, based on the differences in magnetic susceptibilities between minerals, has been used since the early 1800s (2,6,10,25,30,35). Usage has increased steadily in the latter twentieth century because of significant advances made in applications technology and design of equipment. One application is in the concentration of iron ores by the removal of nonmagnetic, nonvalue minerals. Another is the removal of iron and iron-bearing nonvalue minerals from valuable nonmagnetics and for many nonferrous minerals (see MAGNETIC MATERIALS).

Magnetic susceptibility, a bulk property, is the ratio of the intensity of magnetization (M) produced in the mineral to the magnetic field (H) which produces the magnetization. In addition to field intensity, field gradient, ie, the rate at which field intensity increases toward the magnet surface, is also important (2). Minerals may be divided into ferromagnetic, paramagnetic, and diamagnetic depending on how strongly they interact with an applied magnetic field. Iron and magnetite are ferromagnetic; hematite, ilmenite, pyrrhotite, wolframite, and chromite are paramagnetic; quartz and feldspar are diamagnetic. Magnetic susceptibility is ca -0.001 for quartz and ca 0.01 for hematite (2). It is much higher for ferromagnetic materials and is a function of the magnetic field.

Magnetic separation is carried out either wet or dry in low or high intensity separators depending on the type of separation. A steep field strength gradient is maintained in both types of separators by using different pole designs. The field intensity is regulated by changing either the space between poles or the current (in electromagnets). Belts or drums are used to transport the feed through the field. Some examples of the equipment used are shown in Table 7 and Figures 15 and 16.

High magnetic susceptibility minerals, ie, ferromagnetics and paramagnetics of high susceptibility, are primarily separated in wet low intensity separators (<0.2 T (2 kG)) using permanent magnets of the Alnico or barium–strontium

Table 7. Types of Magnetic Separators[a]

Equipment	Feed rate, m³/(min·m)	Field strength at 5 cm, T[b]	Size, mm Dia	Size, mm Width	Application
					Tramp removal and cobbing
suspended magnets				≤1010	removal of tramp iron in crushers and other process equipment, and iron from foundry sands; recovery of iron from slags
plate and grate magnets					tramp iron removal
magnetic pulley	≤1500[d]		910[e] [c]		similar to suspended magnets
cobbing drum			300–916		coarse magnetic cobbing, tramp iron removal, iron recovery from slags, cleaning of scrap
					Wet low intensity
concurrent	50–350	0.06–0.7	760–1200	1525–3050	high grade magnetite from coarse ore; taconites; dense medium recovery
counter-rotation	50–250	0.5–0.6 m/s	760–1200	1525–3050	retreating tailings from a concurrent drum such as dense medium recovery; taconites
countercurrent	20–250	0.05–0.06	760–1200	1525–3050	finishing separator; good recovery and clean concentrate
					Dry low intensity
high speed drum	0.05–0.45	0.04–0.05	400–916	300–3000	magnetites; dry grinding circuits
Ball-Norton type	0.13–0.5		≤760	1525	coarse magnetite in dry separation
					Wet high intensity
carousal type	0.01–1[f]	<2			paramagnetic minerals; hematite and chromite; iron from China clay and other minerals; pyrite from coal
canister type	0.07–0.3[f]	2	2130		primarily iron from China clay; other uses similar to carousal type
					Dry high intensity
induced roll	0.01–0.1	<2.1	250–1000		dry paramagnetic materials, beach sands, wolframite, monazite, cassiterite, cleaning silica sand and feldspar
cross-belt	0.02	2		450	similar use to induced roll, but limited to high value minerals; extremely selective separation; simultaneously separate many minerals with a range of susceptibilities

[a]Ref. 6. [b]To convert tesla to gauss, multiply by 1×10^{-4}. [c]Size of grate = 1 m². [d]Value is in units of m³/h. [e]Maximum belt speed is 150 m/min. [f]Value is in units of t/min.

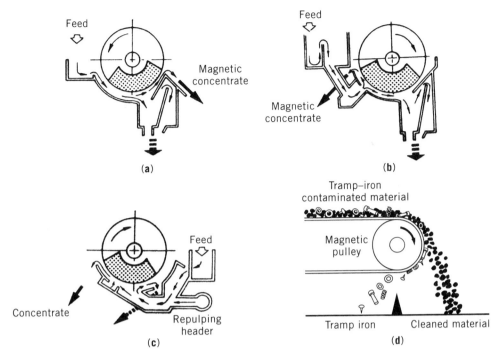

Fig. 15. Wet drum low intensity magnetic separator tank designs: (**a**) concurrent, (**b**) counter-rotation, and (**c**) countercurrent (6); (**d**) shows the operating principle of a magnetic pulley. Courtesy of Eriez Magnetics.

ferrite (6,10). Electromagnets are sometimes used when high field strengths are required. Wet separators are more common.

Dry low intensity separators, such as the magnetic pulleys, drum-type, suspended magnets, and plate and grate magnets, are used for coarse tramp removal from many feed systems to protect downstream equipment such as crushers, screens, and conveyors (10). The magnetic pulley (permanent or electromagnets) is one of the simplest devices (Fig. 15**d**). Installed at the head end of a conveyor belt, it provides continuous removal and automatic discharge of tramp iron. Electromagnetic pulleys having d-c power supply up to 5000 W are used for handling large volumes and removal of larger pieces of tramp iron. Tramp iron drums from either permanent or electromagnets handle material in a reasonably thick layer. These are built with radial poles capable of projecting a deep magnetic field and holding tramp on the drum face until discharged. Suspended or over-the-belt rectangular magnets are used when belt speeds and tramp burden are much greater. These can also be located over chutes or at the end of shaking screen discharges. They can be made self-cleaning when a moving belt is incorporated over the rectangular magnet. Plate magnets are placed in the bottom of chutes and hoppers. Grate magnets consist of parallel tubes or grids of permanent magnets, and are mounted over chutes or troughs handling dry or wet streams. These are gaining popularity because they can handle any volume passing through chutes and are easy to install.

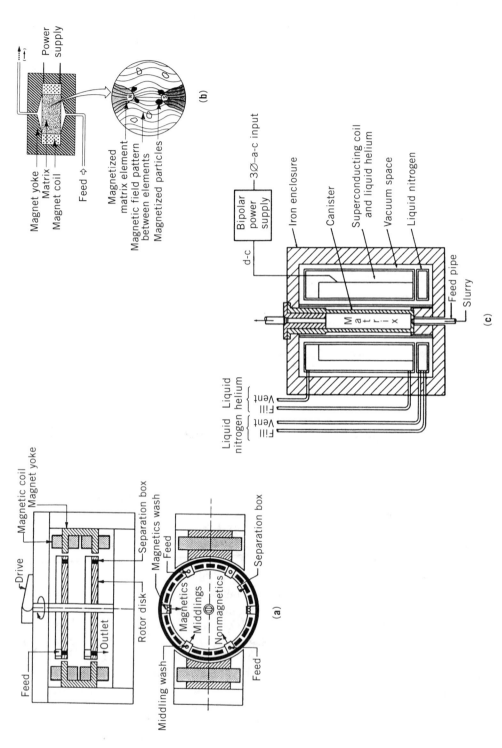

Fig. 16. Operating components of various separators: (**a**) Jones continuous high intensity wet magnetic separator (10), (**b**) canister-type high intensity magnetic (6), and (**c**) superconducting magnetic separator (2).

Dry low intensity separators are extensively used in concentration (or cobbing) of coarse, strongly magnetic material (2,10). For example, magnetite, nickel ore, recovery of dense medium from DMS circuits, recovery of iron values from blast furnace and steelmaking slags, and for processing low grade taconites. Wet separation using drums is more common, especially in the finer (<5 mm) size range. The drum separator (Fig. 15) consists essentially of a rotating non-magnetic drum equipped with several stationary, permanent or electromagnets having field intensities up to 0.7 T (7 kG), of alternating poles (2). Variations of this design are available. Feed enters on one side of the drum; magnetic particles travel through the field close to the drum and are discharged at the end, and the nonmagnetics or tailing discharge at the bottom of the drum. The feed flow can be either in the same direction as, or the opposite of, drum rotation. In countercurrent design, the tailings are made to travel in the direction opposite that of drum rotation. Magnetic flocculation often occurs when ferromagnetic particles are present because the particles can behave as tiny magnets and attract each other (6). Magnetic flocculation is used in the concentration of magnetite and the cleaning of steel plant waste water. Such flocs can, however, entrain nonmagnetic particles and thus lower separation efficiency. The flocs can also cause problems in subsequent mineral processing operation (see FLOCCULATING AGENTS). Demagnetizing coils are used for the depolarization of the particles.

In order to process paramagnetic minerals of low magnetic susceptibility, both a high magnetic field intensity (1–2 T (10–20 kG)) and a high field gradient (1 T/mm (10 kG/mm)) are required (2,10). Dry separation is common, although wet high intensity and high gradient separation are gaining importance rapidly and have widened the application of magnetic separation. Modern units have capacities as high as 120–180 t/h. The common separators are the induced roll, carousel (Fig. 16a), the cross-belt, and the canister (Fig. 16b). The induced roll separator consists of a series of revolving laminated rolls of alternating magnetic and nonmagnetic disks, of slightly different sizes, compressed together on a nonmagnetic stainless steel shaft. These rolls are magnetized by induction using a high intensity stationary electromagnet. Very high field strengths are generated in the gap between the pole and the roll, and the gap size is adjustable. They are typically used for coarse feed (>75 μm) and find application in the concentration of beach sands and other weakly magnetic minerals, and removal of iron contamination from many industrial minerals.

In the well-known Jones high intensity separator (Fig. 16a) a carousel of grooved plates revolves in a magnetic field generated by electromagnets enclosed in air-cooled cases (2,10). The grooved plates concentrate the magnetic field at the tip of the ridges and increase the collection area for magnetic particles significantly compared with the grooved rotor of the induced rolls. Feed enters at two points on the carousel. Paramagnetic particles remain on the plates and the nonmagnetic particles flow through the plate grooves. The paramagnetics are collected at points where the magnetic field is essentially zero. Variations of this design include the use of steel balls, steel wool, or sheets of expanded metal as the ferromagnetic matrix. These separators are used extensively in processing low grade hematite ores. Other uses include processing of ores containing siderite, ilmenite, chromium, manganese, tungsten, etc, and removal of magnetite impurities from cassiterite concentrates, glass sands, asbestos, scheelite, kaolin, talc, removal of pyrite from coal, and processing wolframite, some sulfide ores, and beach sands.

The cross-belt separator is one of the oldest types of separators (10). The feed flows over a conveyor belt, and the magnetics are picked up by another belt perpendicular to the feed belt and moving over the sharp edged upper poles of an electromagnet. The lower poles are situated below the belts and are flat. The disk separator is a modification of the cross-belt and consists of a series of disks containing concentrating grooves and revolving above the feed belt. Electro-magnets are used to magnetize the concentrating grooves by induction. This design lends itself to excellent control, sharpness of separation, and selectivity. It is capable of producing a separate middlings product.

Very high (1 T/mm (10 kg/mm)) field strength gradients are achieved in the canister-type separators (Fig. 16**b**) and particles of very low susceptibility can be concentrated (2). Canisters are used typically for very fine particles. A solenoid is used instead of the conventional magnetic circuit design. The core is filled with a matrix of secondary poles such as steel wool in a canister, and these poles produce high field gradients. The weakly magnetic particles are captured in the matrix and are removed periodically. They are used to remove very fine iron-containing impurities from kaolin clays.

Many improvements have occurred in magnetic separators, especially in the high intensity and high gradient units. Many powerful magnet materials have also been developed, eg, neodymium–iron–boron. Magnetic separators using a superconducting magnet have been used (Fig. 16**c**), eg, processing of kaolin clays (2,51). The industrial units operate at temperatures near absolute zero, achieve magnetic fields of 5 T, and are characterized by greatly reduced power costs (up to 90% over conventional units), high capacities (20–40 t/h), and superior performance. These types of separators are expected to gain in importance as advances in high temperature superconductors are made (see SUPERCONDUCTING MATERIALS). Progress has also been made in the use of magnetohydrodynamic and magnetohydrostatic methods in minerals processing (2,6,25). These methods can separate minerals by density, magnetic susceptibility, and electrical conduc-tivity simultaneously (see MAGNETOHYDRODYNAMICS).

Many other devices are available for laboratory use. These include the Davis tube, Frantz isodynamic separator, laboratory drum-type separators, low intensity rotating field separator, and superconducting high gradient separator (2).

Electrostatic Separation. Electrostatic separators exploit the differences in electrical conductivities, triboelectric effects, and polarizabilities between min-erals (2,6,10,13,37). These have a limited number of applications in minerals pro-cessing. They are used extensively in dust removal from gas streams, and are quite successful where applied. They are frequently used in combination with gravity and magnetic separation. Whereas early usage was in the separation of high conductivity gold and metallic sulfides from low conductivity siliceous gangue and separation of sphalerite from galena, principal use in the 1990s is in processing beach sands and alluvial deposits containing titanium minerals (rutile and ilmenite are separated from zircon and monazite) (2). Other concen-tration methods have been less successful in these latter systems because of the similarities in surface properties and specific gravities of the minerals to be separated. Other uses include beneficiation of cassiterite, columbite, and ilmenite; iron ores; separating halite and sylvite; shape separation of vermiculite and gangue minerals; and industrial waste recovery, eg, plastics from scrap. Minerals pinned on a rotor in an electric field are apatite, barite, calcite, corundum, garnet, gypsum, kyanite, monazite, quartz, scheelite, sillimonite, spinel, tourmaline, and zircon. Those thrown from the rotor are cassiterite, chromite, diamond, fluorspar, galena, gold, hematite, ilmenite, limonite, magnetite, pyrite, rutile, sphalerite, stibnite, tantalite, and wolframite (2).

Electrostatic separations can be either the electrophoresis type, involving charge-transfer to or from a particle, depending on the differences in conductivities and triboelectric properties; or the dielectrophoresis type, involving induced polarization owing to differences in dielectric constant, shape, and structure (37). Electrophoresis is the basis of all of the applications in minerals processing. Dielectrophoresis is an emerging technology as of the mid-1990s limited to laboratory and pilot plant (see also ELECTROSEPARATIONS).

The basic processes in electrophoresis electrostatic separation are partial charging and subsequent separation of charged particles either using a grounded surface or an attracting electrode. Particles can be charged by contacting dissimilar particles, which occurs during bulk movement of particles; ion bombardment or charging in an ionizing field; or induction in a nonionizing field. Usually a combination of charging mechanisms exists in separators, but the last two are the most important. The separation that occurs at the grounded surface results from the combination of electrical, centrifugal, and gravity forces.

Electrostatic separators are either of the electrodynamic or the electrostatic type (6). Early machines were of the electrostatic type and based on electrostatic processes in charged fields. Charged particles were attracted to an electrode of opposite charge and were lifted from the particle stream toward the electrode (lifting effect). An example is the separation of the negatively charged quartz from other nonconductors using a positively charged electrode. Such separations are inefficient and are sensitive to humidity and temperature.

Most of the separators used as of this writing (ca 1995) are the high tension or electrodynamic type (Fig. 17**a**), based on the principle of corona discharge in an ionizing field utilizing the focusing or the beam-type electrode design (2,6). The particle stream is fed on to a rotating metal drum or rotor which is grounded. Particles enter a field of charged ionizing electrode assembly which spans the entire length of the drum and supplied with a d-c voltage of up to 50 kV and a negative polarity (current flow is usually 5–15 mA/m rotor length). The electrode assembly itself comprises a fine wire electrode to produce the ionizing field and a large-diameter electrode to produce a dense nondischarging field. Particles are charged by ion bombardment. Particles that have high conductivity lose their charge to the grounded drum and therefore are thrown from the drum surface by centrifugal force and aided by a nondischarging static electrode that is placed after the ionizing electrode. Particles having low conductivity and nonconductors retain their charge and are pinned to the drum surface by their own image charges. These particles slowly lose their charge and fall off the drum, the middling particles losing their charge faster than the nonconductors. Any nonconductors still pinned to the drum are removed by a brush, in some cases aided by an additional electrode. The voltage supplied to the electrode

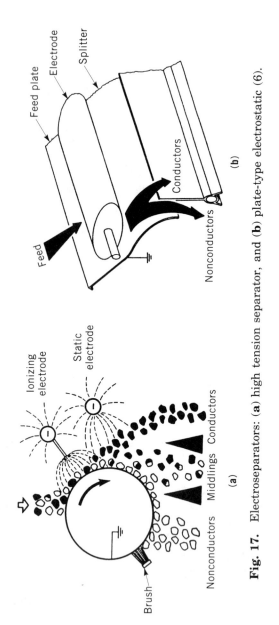

Fig. 17. Electroseparators: (**a**) high tension separator, and (**b**) plate-type electrostatic (6).

assembly is sufficiently high to promote ionization but no arcing. The voltage range of stable corona discharge is narrow. The separation is influenced not only by the conductivities of the particles, but also by the degree of liberation and the particle size distribution because of differences in the charge per unit mass on fine and coarse particles. There is a tendency for the fine particles to be pinned to the drum for a longer time and report to the middlings or the nonconductors fraction. Several stages of cleaning are therefore required.

Plant capacity is a function of feed size distribution and liberation. Separators can accept a size range as wide as 50–1000 μm. Capacities are typically 1000–2500 kg/(h·m) based on rotor length which could be up to 3 m and have dia 150–250 mm. The feed should be as dry as possible because moisture interferes seriously with separation. Heaters are usually provided before the feed enters the charged field. Final cleaning is often conducted in electrostatic-type separators. Electrostatic shape separation, a newer form of ion bombardment separation, involves separation of particles based on shape and density without consideration to conductivities (37).

Electrostatic separators are either rotor or plate (6). The former is similar in appearance to the high tension separator. However, there is no ionizing electrode in the electrode assembly, instead there is a large single electrode producing an electric field. Particle charging is by induction in this case. The paths followed by conductors and nonconductors are similar to that in high tension separators. Modern plate-type separators are either plate (Fig. 17**b**) or screen electrostatic types. The particle stream passes over a sloping, curved, grounded plate into an electrostatic field induced by a large curved electrode. The conductor particles acquire a charge opposite to the electrode and are lifted toward it. The particles go over a splitter or a screen. The nonconductor particles continue down the plate or through the screen. Because the action of the field is mainly on the conductors, a sharp separation can be made. Fine particles are more affected by the field than the coarse particles; therefore the latter are readily rejected, unlike in high tension separators. The plate-type separators are used for removing small amounts of nonconductors from large amounts of conductors and vice versa for the screen-type separators. The main usage of electrostatic-type separators is in final cleaning of concentrates produced by other methods.

Electrostatic separation is sensitive to humidity and moisture, temperature, and any organic coatings. The feed is usually cleaned by washing, attrition scrubbing, or caustic scrubbing and then dried. Desliming is also practiced where necessary because slimes interfere with the charging of particles. Improved selectivity can often be achieved at elevated temperatures. Normal practice in rutile separation from zircon is to operate at 90°C or higher. Electrostatic separators are equipped with heating coils and lamps.

Froth Flotation. Flotation (qv) is the most extensively used primary mineral concentration technique (38–42). Its introduction in the early 1990s revolutionized the industry. The attributes of flotation are its applicability to a wide range of minerals systems, ore types, and size ranges, and its high versatility and selectivity. Flotation is used to treat all of the sulfide ores, precious metals ores, nonsulfide metallic minerals, many industrial minerals, and coal (qv). In flotation, one or more types of particles are separated from others by actually floating the particles against gravitational forces with the aid of air bubbles introduced into the separator. The chemistry of flotation is complex though the process itself is seemingly simple. Most minerals are naturally hydrophilic. The surface chemical properties of the minerals to be floated are changed selectively to make the surfaces hydrophobic (water repelling) by the use of organic reagents called collectors.

A simple example is the flotation of a small (<5%) amount of copper sulfide minerals, eg, chalcopyrite, chalcocite, etc, and molybdenite from a copper ore containing siliceous gangue minerals constituting >95% of the ore. Xanthates are used to float metal sulfides so that these particles can attach to air bubbles, and together the bubble–particle aggregates can float to the surface of the pulp where they form a stable froth (Fig. 18). This froth is removed continuously by either natural overflow or mechanical means. A stable froth is ensured by the addition of suitable chemicals, called frothers, which also facilitate the production of a fine dispersion of bubbles in the pulp. The froth must be stable only long enough to complete the removal of the floated particles and break readily in the launders where it is collected. In order for the collector to adsorb selectively on the desired mineral, the surfaces of this mineral must often be modified. This is usually achieved by addition of modifiers such as activators, depressants, pH modifiers, and dispersants. Activators modify the surfaces of the desired mineral directly to promote collector adsorption which otherwise would not occur. Depressants act on other minerals in order to prevent collector adsorption on these. Dispersants have many roles to play. One of them is to ensure that very fine gangue minerals (or slimes), such as clays and other silicates, do not interfere with collector adsorption on desired minerals and their subsequent transfer to the froth phase. pH modifiers are common acids and bases such as lime, caustic, soda ash, and sulfuric acid (see HYDROGEN-ION ACTIVITY).

It is the buoyancy of the bubble–particle aggregate that determines flotation, not the specific gravity of the particles or the particle size. Thus even the heaviest, eg, native gold, sp gr 19.3, and coarsest, eg, ≤3 mm for sylvite flotation, of the particles can be made to float if they can attach to a large enough bubble or if sufficient number of bubbles can be attached to each particle. In a loose sense, flotation can be considered a variation of a gravity separation tech-

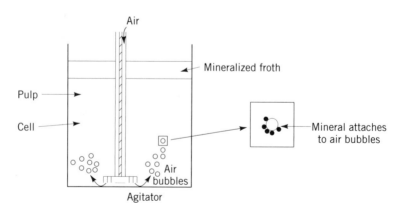

Fig. 18.   Principles of froth flotation (2).

nique. Large differences in gravitational forces acting on mineral-laden bubbles and other (unattached) particles are exploited. It is generally preferable to float a small mass of particles, either value minerals or impurities, away from the rest of the ore. As for any other concentration method, flotation efficiency is affected by particle size, falling off at both very coarse and very fine sizes. The size range of optimum flotation is different for different mineral systems, but successful separations can be made down to 1 μm (2,6). The 10–150 μm range is considered the best. Flotation efficiency and kinetics are a function of both the chemical and hydrodynamic (physical–mechanical) factors.

Collectors and Frothers. Collectors play a critical role in flotation (41). These are heteropolar organic molecules characterized by a polar functional group that has a high affinity for the desired mineral, and a hydrocarbon group, usually a simple 2–18 carbon atom hydrocarbon chain, that imparts hydrophobicity to the minerals surface after the molecule has adsorbed. Most collectors are weak acids or bases or their salts, and are either ionic or neutral. The mode of interaction between the functional group and the mineral surface may involve a chemical reaction, for example, chemisorption, or a physical interaction such as electrostatic attraction.

In general, collectors for the flotation separation of sulfides and precious metals contain at least one sulfur atom in an appropriate bonding state. The functional group in collectors for nonsulfide minerals is characterized by the presence of either a N (amines) or an O (carboxylic acids, sulfonates, etc) as the donor atoms. In addition to these, straight hydrocarbons, such as fuel oil, diesel, kerosene, etc, are also used extensively either as auxiliary or secondary collectors, or as primary collectors for coal and molybdenite flotation. The chain length of the hydrocarbon group is generally short (2–8 C) for the sulfide collectors, and long (10–20 C) for nonsulfide collectors, because sulfides are generally more hydrophobic than most nonsulfide minerals (10).

Xanthates and dithiophosphates dominate sulfide flotation usage, though several other collectors including more recently developed ones are gaining acceptance rapidly (43). As of this writing, this is an active area of research. Many of the sulfide collectors were first used in the rubber industry as vulcanizers (16). Fatty acids, amines, and sulfonates dominate the nonsulfide flotation usage. The fatty acids are by-products from natural plant or animal fat sources (see FATS AND FATTY OILS). Similarly petroleum sulfonates are by-products of the wood (qv) pulp (qv) industry, and amines are generally fatty amines derived from fatty acids.

The amount of collector used is necessarily very small because surface coverages of a monomolecular layer or less are required to impart sufficient hydrophobicity to the mineral. The usages typically range from 1–100 g of collector per ton of ore treated for sulfide flotation (typically 0.2–10% value metal content in the ore) and 100–1000 g/t for nonsulfide flotation (1–20% value mineral content) (10).

Frothers are generally alcohols (C_5–C_8), glycols (qv), or polyethylene or polypropylene glycol ethers (41). They are, therefore, heteropolar molecules having surface-active properties. Some collectors such as fatty acids and amines exhibit frothing properties that are sufficient in some applications so as not to warrant the use of a separate frother. The role of a frother in flotation is rather complex despite its seemingly simple function. Although frothers and collectors appear to have exclusive roles, this is seldom the case. There is much evidence for their interaction in the presence of hydrophobic particles (41). This aspect is not well understood, but has been recognized to be important, especially in the context of rate of flotation of fine and coarse particles, and selectivity of separation. Frother requirements are strongly dependent on the one being treated and the collector used. A typical range for sulfide flotation is 5–100 g/t.

The adsorption of sulfide collectors on sulfide minerals can best be described by electrochemical reactions wherein the mineral, the collector, or both are known to undergo redox reactions (30). This process is unique to sulfide mineral systems. Most sulfides exhibit metallic properties and undergo electrochemical reactions that are much like corrosion reactions exhibited by metals. Redox reactions are not relevant in nonsulfide mineral flotation systems. In these latter systems, the adsorption is generally a chemisorption, surface chemical reaction, or a physisorption, ie, electrostatic attraction between oppositely charged mineral surface and collector, and is often a combination of these processes (30). Surface charge on the minerals, as approximated by zeta potentials, is therefore important in nonsulfide systems. Because surface charge on oxides and silicates is strongly dependent on pH, the adsorption mechanism for a given collector, the choice of a collector, and the flotation selectivity are all influenced strongly by pH.

Sulfide collectors in general show little affinity for nonsulfide minerals, thus separation of one sulfide from another becomes the main issue. The nonsulfide collectors are in general less selective and this is accentuated by the large similarities in surface properties between the various nonsulfide minerals (42). Some examples of sulfide flotation are copper sulfides flotation from siliceous gangue; sequential flotation of sulfides of copper, lead, and zinc from complex and massive sulfide ores; and flotation recovery of extremely small (a few ppm) amounts of precious metals. Examples of nonsulfide flotation include separation of sylvite, KCl, from halite, NaCl, which are two soluble minerals having similar properties; selective flocculation–flotation separation of iron oxides from silica; separation of feldspar from silica, silicates, and oxides; phosphate rock separation from silica and carbonates; and coal flotation.

Modifiers. Modifiers assume a critical role in many separations (41). Most modifiers used in flotation are inorganic compounds, although many organic molecules and polymers such as polysaccharides are also used. In addition to the use of modifiers, numerous other techniques such as high intensity and high solids conditioning (attrition, scrubbing), desliming, selective flocculation of slimes, etc, have been developed, especially for nonsulfide mineral systems.

Flotation Equipment. Numerous designs of flotation machines are available (2,6,30,34,44). Mechanical flotation machines are the most widely used. Flotation columns have seen a rapid acceptance in the industry, however. Pneumatic machines, popular in the early years of flotation, are used only to a limited extent as of the mid-1990s. Most flotation is carried out in banks of flotation cells. The primary functions of the flotation machine are to maintain all particles in suspension, to disperse fine air bubbles throughout the pulp and promote particle–bubble collision, to provide a quiescent pulp region close to the froth phase to minimize mechanical entrainment of unwanted particles into the froth phase, or to prevent turbulent disruption of the froth layer. The froth phase should be of sufficient depth to permit drainage of entrained particles. The selection of a flotation machine is dictated by metallurgical performance (grade and recovery of valuable minerals), capacity, operating costs, and ease of operation.

The main feature of a mechanical flotation machine is the impeller surrounded by baffles, both designed to provide adequate suspension of pulp, in a rectangular tank. Aeration (qv) is via either self-induced or more commonly forced air, in both cases through the hollow impeller shaft or through a standpipe surrounding the impeller shaft, to provide good dispersion and particle–bubble collisions. There are numerous designs of impeller/baffle assembly, each claimed to be better than others (30,42,44). All provide the same basic requirements: a turbulent zone for particle–bubble collision and a quiescent zone below the froth phase which allows for efficient transfer of the mineral-laden bubbles to the froth. A dramatic increase in the size of the flotation cells has occurred such that $14-100$ m^3 cells are typically used (34). The main advantages of these large cells are better efficiency, smaller space requirement, higher throughput, and lower power consumption. Froth collection from the mechanical cells is via either direct overflow or using paddles. Pulp flow from cell to cell down a bank is most commonly accomplished by gravity. Individual cells may or may not be separated from the others by baffles and overflow weirs. Flotation cells can handle pulp densities in the range of 10–50% solids, and are dictated by the type of ore treated and the metallurgical objectives.

In pneumatic machines (Fig. 19**a**) air is either mixed with pulp by turbulent pulp addition or blown or induced in cells. In any case air is used not only to produce froth but also to keep the pulp in suspension. One of the more recent designs is the Ekoflot pneumatic cell, very similar to the Jameson cell in which aerated pulp is introduced near the bottom of a cylindro-conical vessel at a controlled speed. Froth overflows at the top and the tailings are collected at the bottom. The largest unit has a 6-m diameter vessel and can handle up to 1500 m^3/h and 100–700 t/h depending on application. Dissolved air flotation (6), which can be grouped under pneumatic flotation, utilizes air dissolved under pressure in the pulp to form bubbles on fine solids, and it is used primarily in the treatment of industrial effluents. In froth separators (2) developed in Russia, aerated pulp is introduced on top of a froth bed in an inverted pyramidal tank. Hydrophobic particles are retained in the froth and the remaining pulp descends and is discharged at the bottom.

In a flotation column (Fig. 19**b**), feed slurry is introduced in the top half of the column, and air, sometimes premixed with pulp, is introduced in the bottom half of the column. Numerous variations of feed and air inlet designs have been used. Flotation columns can be up to 18 m high (2,34), having diameters of up to 3.5 m. Columns having much smaller height/diameter ratios have also been developed, eg, the Jameson cell. In this cell, air and feed are mixed at the head of a long vertical pipe and the aerated pulp is introduced near the bottom of a cylindrical vessel with a conical bottom (2). Flotation columns generally have no agitators or moving parts inside. Water sprays are used invariably above the froth phase in order to provide efficient drainage of the mechanically entrained particles. A vast amount of research and development has been conducted on column flotation. Columns have become an integral part of a flotation circuit in

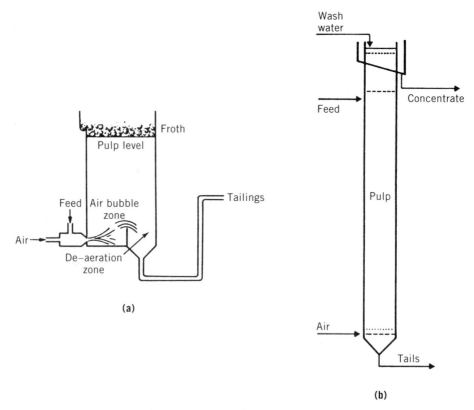

Fig. 19. (**a**) Davcra pneumatic flotation cell, and (**b**) schematic diagram of a flotation column cell (2).

most operations. One of the attractive features of a column cell is its ability to make excellent separations even in the very fine (<20 μm) particle size range.

A considerable amount of research on novel flotation systems has been carried out. Several have been developed, eg, the air-sparged or air-injected hydrocyclones, and the rapid flotation device (34). The air-sparged hydrocyclones have a high throughput per unit volume. Research and development into modeling, automation, and control of flotation circuits has also been done (15,42) Numerous on-line monitoring, measurement, and control systems are part of a flotation circuit. The basic flotation circuit comprises roughers, scavengers, and cleaners. Recovery of the values is the primary emphasis in the roughers and scavengers; grade is the primary emphasis in the cleaners. The most common method of assessing flotation circuit performance is via the mass and metallurgical balance which provides the recovery and grade of value minerals recovered at each stage of the circuit. Several excellent techniques have been developed to perform the metallurgical balances. Much effort has been made to ensure a proper estimation of recirculating loads and the mass balance of the various flow streams. Grade-recovery relationships are most commonly used to assess the efficiency and selectivity of separation.

Flocculation and Agglomeration.   Selective flocculation and selective agglomeration are two other processes based on surface chemistry changes of mineral particles (2,6,25). Both are used commercially to a limited extent, the former for hematite, clay, and potash ores, and the latter for coal and fine metallic oxide minerals. In selective flocculation, water-soluble polymers (qv) such as polysaccharides or synthetic polymers are used to selectively flocculate fine particles of one mineral. The nonflocculated minerals are removed subsequently by desliming or flotation. In selective agglomeration, also referred to as hydrophobic coagulation, one type of mineral particle is rendered hydrophobic, or is naturally hydrophobic, as is coal, and is forced to agglomerate by using a combination of hydrocarbon oils and high shear conditions. The agglomerate size is increased until a subsequent size separation is possible either by flotation or sedimentation technique.

Solid–Liquid Separation

Most minerals processing operations are conducted in large quantities of water. A typical copper ore flotation plant uses about 3800 L/t of ore treated (2,6,10,13,25,45,46). Water usage can be as high as 23,000 L/t in glass sands flotation. Dewatering (qv) of mineral slurries to varying degrees becomes necessary for a variety of reasons. For example, subsequent treatment by pyrometallurgical operations, such as pelletizing and smelting, transportation, disposal, water recovery for recycle, etc, require dewatering.

Dewatering is performed by using one or more of the following methods: sedimentation, also known as settling or thickening; filtration (qv); and thermal drying (qv). Frequently all three are used, in that order, on the same slurry to ensure that the final product has a low moisture content. Thickening is generally the most economical method for tailings dewatering before disposal in the tailings pond.

Sedimentation.   Sedimentation (2,6) is used to remove the bulk of water from streams (thickening). The thickened product is usually 55–65% solids. Sedimentation is also used to remove suspended solids from a relatively dilute stream to produce a liquid phase that is as clear as possible (clarification). Although the same equipment can often be used as a thickener or a clarifier, the distinction is based on operation. Thickeners operate with a clear solid–liquid interface and their capacity is dictated by the underflow conditions. In clarifiers there is no well-defined interface, and the capacity is dictated by overflow clarity. Polymeric flocculants are used in most of the thickeners to enhance the settling rates of particles by flocculating the solids, especially the fines. Synthetic flocculants, such as polyacrylamides and polyacrylates, are used extensively (see FLOCCULATING AGENTS). Polysaccharides are also used to a significant extent in certain systems. The synthetic flocculants are available in a wide range (several thousands to over 20 million) of molecular weights and charge (0–100 mol %) (2). The type of flocculant, ie, anionic, cationic, or nonionic, depends on the type of minerals involved. Often a combination is used. Anionic polymers having very high molecular weights can be produced; whereas cationics tend to be of much lower molecular weight. Electrostatic interactions between the charged minerals and the polymers and particle bridging are two important mechanisms leading to flocculation of solids. The physical characteristics of the flocs formed, ie, floc density, size, etc, have a strong influence on the settling and consolidation of the sludge. Flocs are generally quite fragile. Although flocculation leads to significant increase in settling rate or improves filtration rate, it is often detrimental to final consolidation of the sludge or to producing low moisture filter cakes.

There are three types of thickener designs: cylindrical, lamella, and deep cone. The cylindrical design is the most common (Fig. 20). It is also continuous. It comprises a large (up to 200 m dia, 1–7 m deep) cylindrical tank, a shallow conical base (80–140 mm/m), and a central structure carrying sludge-raking arms. Feed enters the thickener through a central feed well and clarified liquor overflows into a launder around the periphery of the cylindrical tank. Thickened sludge collects in the shallow conical base and is raked by the slowly revolving mechanism to a central discharge point. There are many designs of the feed well: conventional, bottom-fed, counterflow, perforated, and deep feed well. The rakes help to move the sludge toward the central discharge and help improve settling and consolidation of the sludge. Typical tip speeds are 5–8 m/min and torque ratings up to 13,000 kNm (890 lbf/ft) (46). Several designs are available for the rake support and drive mechanism.

The other types of thickeners include the high capacity/rate/compression tray, lamella, and the deep cone type (13,46). These are designed to reduce floor area requirements and take up only 5–20% of the floor area required for a conventional thickener and produce denser sludges and clearer overflow. In the high capacity thickener, feed, and the flocculant enter the hollow drive shaft and are mixed by staged mechanical mixing. The flocculated feed is injected into the sludge blanket where further flocculation occurs. Settling occurs along inclined plates. The lamella thickener has packs of sloping parallel plates in the settling area which reduce the settling distance and increase the effective settling area.

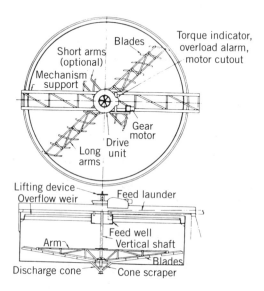

Fig. 20.   Cylindrical thickener with mechanism supported by superstructure (2).

The unit can also be vibrated to enhance dewatering. The same unit can function as both a thickener and a clarifier. The tray thickener is a series of independent conventional thickener units stacked vertically having a common central drive shaft (2,10).

Filtration. In many mineral processing operations, filtration follows thickening and it is used primarily to produce a solid product that is very low in moisture. Filtration equipment can be either continuous or batch type and either constant pressure (vacuum) or constant rate. In the constant pressure type, filtration rate decreases gradually as the cake builds up, whereas in the constant rate type the pressure is increased gradually to maintain a certain filtration rate as the cake resistance builds. The size of the device is specified by the required filter surface area.

Most continuous vacuum filters are the constant pressure type. Their main use is in dewatering concentrated slurries such as concentrates. They belong to three classes: the disk, drum, and horizontal filters. Disk, and to a lesser extent, drum filters, are the mainstay for most final dewatering. These filters remove most fine particles from a process stream.

The drum filter consists of a horizontal cylindrical drum, having from 1 to 5 m diameter, that rotates while partially submerged in an open feed slurry tank. The filter medium is wrapped tightly around the drum surface. The drum shell is divided into compartments and drain lines are connected to the central valve system allowing either vacuum or pressure dictated by the cycle. A normal cycle comprises filtration, draining, and discharge by a blast of air or by mechanical means. Cake washing and filter cloth cleaning can also be part of a cycle. Several variations of the standard drum filters exist, including hyperbaric filters, ie, pressure filtration, up to 600 kPa (87 psi), which are continuous and give high filtration rates and drier cakes (34).

The disk filter is similar to the drum in operation, but filtration is conducted using a series of large diameter filter disks that carry the filter medium on both sides of the disk. They are connected to the main horizontal shaft and partly immersed in the feed slurry. The central shaft is connected by a set of valves which serve to provide vacuum and air as in drum filters. As the disk sections submerge during rotation, vacuum is applied to form a cake on both sides of the disk. The cycle of operation is similar to that in a drum filter. One unit can have as many as 12 disks of up to 5-m diameter. Disk filters, both compact and cost effective, are used extensively in the iron ore industry to dewater magnetite concentrates.

In the horizontal continuous vacuum filters, also called belt filters (Fig. 21), filtration, washing, and drying occur on a traveling belt filter cloth which is provided with suction boxes underneath (2,10). Many variations are available. The belt can be linear or circular. Slurry is fed on the belt at the beginning. Filtration is by both gravity and suction. Filter cake is discharged from the belt using scrapers before the belt reverses. Some advantages of belt filters are relatively low capital and operating costs and excellent washability, but they require high flocculant dosage and greater floor area for a given filter area.

Pressure filters or filter presses are commonly of the batch type (2,47). These are characterized by high filtration rates, smaller floor area, and lower capital cost. Dryer cakes are produced. These filters are more widely used in the

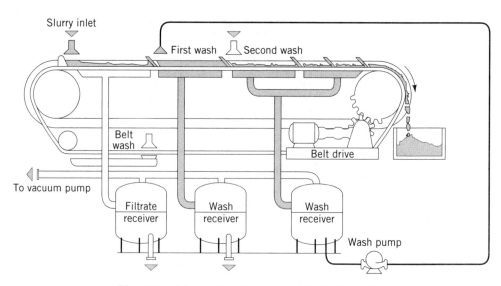

Fig. 21. Schematic of horizontal belt filter (2).

chemical industry than in mineral processing because this is a batch operation. The plate and frame presses (Fig. 22) and the chamber presses are the most common types of pressure filters used. These consist of a series of vertical, alternating parallel frames and plates. The filter cloth is held against the plate. Cake formation occurs in the hollow frame. The fully automatic Larox chamber filter, reported to reduce dewatering energy requirements, is a more recent

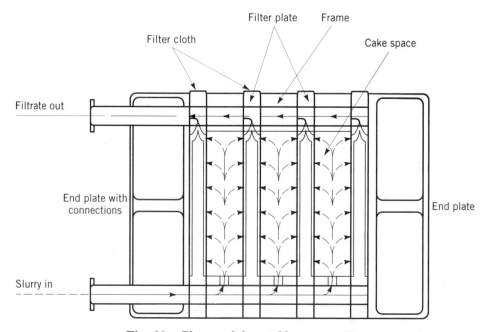

Fig. 22. Plate and frame filter press (2).

development (47). Other developments are the tube press or the pinch press which can operate at 10,000 kPa of dewatering pressure. These units produce very low moisture filter cakes even in the absence of dewatering acids (51).

Flocculants and surfactants (qv) are used frequently as filter aids, particularly when slimes are present or when the particles to be filtered are very fine and difficult to filter. Low molecular weight polymers are more commonly used. These form small, dense flocs which provide higher cake porosity. Blinding of the filter cloth by fine particles or slimes is reduced. Surfactants are also used to enhance flow through the filter cake pores.

Centrifugal Separations. Centrifuges are used when conventional dewatering methods are not applicable because the settling rates are too slow, as in the case of clays, or when low moisture levels are required before the next unit operation (2,6). The centrifugal force is used in these devices to enhance solid–liquid separation. These are high capital cost and high maintenance units, but can perform many functions, eg, as classifiers, thickeners, clarifiers, and filters. The hydrocyclones are an example of centrifugal devices used for classification or thickening that are simple and inexpensive. These do not, however, produce very high solids concentration in the underflow and their efficiency falls off rapidly at fine size ranges, ie, <10 μm particles invariably appear in the overflow. Hydrocyclones are still valuable for initial dewatering before thickening. Many other classifiers can also be used for primary dewatering.

Centrifuges are either the solid bowl or the perforated basket type. The solid bowl centrifuge consists of a horizontal bowl rotating at a high (1000–6000 rpm) speed. Slurry enters in the center of one end of the bowl, the liquid discharges at the bottom of the same end, and the solids are discharged at the other end by means of a rotating helical scroll. Further dewatering occurs as the solids move along the scroll. Washing can be incorporated at this stage. The size of the bowl is dictated by the extent of dewatering required and the application. These are particularly suited for clarification.

In the perforated basket centrifuge, material transport is through vertical vibrations in the basket. These vibrations also loosen the bed of particles, aiding drainage, thus allowing lower (550–750 rpm) speeds than are necessary for solid bowl centrifuges. These are commonly used in dewatering coal. Because of the perforations in the basket, these centrifuges are not suitable for feeds having a significant proportion of fines.

Thermal Dewatering. Thermal drying is typically the last stage in dewatering (2,6,10). It is an expensive operation because energy is wasted in heating solids, but is necessary in many cases because mechanical dewatering cannot reduce the moisture level below a certain limit. The extent of reduction in moisture by thermal drying is dictated by economics and specifications required for the mineral product in terms of flow properties of particles, dust prevention, etc. A moisture level of 5% is often acceptable and complete drying is not necessary.

The common types of dryers are rotary, hearth, flash (spray), and fluidized beds (10). Hot gases are used invariably to remove moisture. The gas flow can be either cocurrent or countercurrent to the flow of solids, the former tends to be more efficient. In the hearths, the gas flow is countercurrent as the solids are raked down from one hearth to the next below. Flash dryers are very rapid because the solids are exposed only briefly to the hot gases. Fluidized-bed dryers, which use hot gases to suspend the solids, are rapid and efficient, but require elaborate dust collection systems. These are preferred when fine solids are involved, and are used commonly for drying fine coal. Indirect-fired dryers are used when the solids are heat sensitive or combustible.

Materials Handling

The management of transportation, storage, feeding, washing, and packing of processing streams to final products in a mineral processing operation constitutes a significant effort and cost factor. Materials handling comprises dry solids and slurry handling, and tailings disposal.

Dry solids, such as as-mined ore, crushed ore, and dried concentrates, are transported using trucks, rail cars, ore passes, conveyor belts (see CONVEYING), or slurry pipelines (qv) as dictated by the logistics, distances involved, and capacity. Within the mill, conveyor belts are more common, but for fine particles, tailings, and coal, slurry transportation is more typical.

Storage of dry solids provides for surge capacity between unit operations and for blending ore types. Stockpiles are preferred for coarse material and large capacity such as crushed ore, sized products, and dried concentrates ready for shipment. Bins are preferred for smaller capacities. Stockpiles are formed on a concrete or earthen pad using a variety of equipment such as fixed stackers, tripper conveyors, reversing shuttle conveyors, traveling winged stackers, and radial stackers. Material from a stockpile is reclaimed using bottom tunnels, bucket-wheel reclaimers, scraper trucks, or front-end loaders. Bins and hoppers provide a greater flexibility in terms of transportation, storage, blending, and controlled feed to subsequent stage and their design is crucial. Improper design for a specific application may result in total system failure or extensive operation problems associated with intermittent feed. Feeders are usually part of the bin and hopper unit. Feeder design types are belt or apron, screw, rotary table, vibratory, star, and rotary plow. Belt and apron and the vibratory types are the most common. Material from a stockpile or a bin is transported using chutes; mechanical conveyors, ie, belt, screw, chain, and vibratory type, that are horizontal or inclined; pneumatic conveyors; skip hoists, used widely to haul ore from underground; and bucket elevators (see CONVEYING).

Much effort is made to obtain a representative sample from bulk dry solids. This can be a difficult task for very coarse material and from large stockpiles, bins, and hoppers as segregation is inevitable. It is, therefore, preferable to collect samples or subsamples when the material enters the stockpile. Cross-chute and the rotary samplers are the most commonly used (see SAMPLING).

Slurry handling in minerals processing includes transportation and suspension in tanks during processing. Transportation can be from mine site to the processing plant, as for hydraulically mined ores such as clays, phosphates, beach sands, and cassiterite; within the processing plant, eg, from grinding to dewatering; and from the processing plant to another location, eg, transport of tails to a tailings pond, or transport of concentrate slurries through a pipeline to a dewatering plant or a shipping port. Powerful pumps (qv) are required at every stage, and the selection of appropriate pumps and pipeline sizing constitutes a large task in the design of a plant. Agitated tanks are used within the operating mill essentially as sumps for collection, holding, scrubbing of mineral surfaces, conditioning of reagents, and distribution of slurries as dictated by the logistics and the circuit design. These are typically large cylindrical tanks provided with an impeller or a propeller and baffles. Efficient suspension of solids is an important criterion in the design of a tank (see TANKS AND PRESSURE VESSELS).

Tailings Disposal. In many operations the bulk of the ore becomes tailings (2,6,13,48). In porphyry copper ores, for example, as much as 90–98% of the ore is discarded as tailings. In other operations tailings could be 20–50% of the ore. Tailings must be treated in an economical and environmentally safe manner. In a majority of operations, tailings are collected in a tailings pond (Fig. 23), the design of which constitutes a significant task requiring long-term planning.

Tailings are usually thickened before discharging into a pond. This provides rapid water recycle and reduces the volume of tailings transported to the pond. Coarse tailings are used for the construction of a dam which is built up continuously (Fig. 23c). Hydrocyclones are used at the pond site to effect a rough size separation so that coarse stream discharges onto the dam and the fines report to the pond. Several measures are taken to prevent a dam break. Waterproof pads, ie, plastic sheeting and clay, are used in some ponds to prevent seepage of

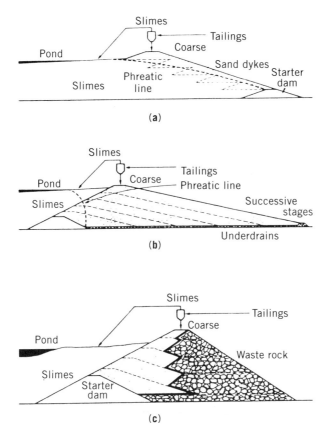

Fig. 23. Methods of tailings dam construction: (**a**) upstream method, (**b**) downstream method, and (**c**) mine waste rock dam construction (6).

contaminated water and potentially hazardous chemicals such as cyanide from gold leaching plants or flotation chemicals, into the underground water system or to rivers. In some plants around the world tailings are still discharged into rivers, although such practice is rapidly becoming extinct. Coarse solids are also used for enhancing the settling of fine solids in the pond.

In many plants, the main purpose of the tailings pond is for water management and recovery. Most plants in the United States recycle all of the process water, ie, achieve zero discharge. Another purpose is to collect the solids and use them for backfill of an excavated site or an abandoned underground mine as part of the overall land reclamation plan. Mine overburden is also used for this purpose. In many mines the overburden could be a much larger volume than the ore deposit itself. Tailings ponds also provide adequate time for biodegradation of potentially hazardous chemicals and for precipitation of heavy metals which, if present in high concentrations in the recycled water, may be detrimental to minerals separations. Settling of solids in a tailings pond is not a problem generally, unless clays and other silicate slimes are involved. Extreme cases of slow settling ponds are those of clay tailings of Florida phosphate operations and the Kimberly diamond operations in Africa.

Process Control

The most significant developments in minerals processing in the latter 1900s have been in the area of automation and computer control of minerals processing plants (2,6,49–51). Rapid advances in the electronics area are largely responsible for the introduction of on-stream analyzers and sensors (qv). Process control (qv) is an extremely difficult task because of the heterogeneous and complex nature of ores, the extreme variability in any given circuit, and the severity of conditions in the operation. Frequently data needed for process control are available only after the fact. Obtaining representatitive samples from various flow streams is a prerequisite to obtaining reliable data.

The primary goals in process control are to improve the efficiency and/or selectivity and to reduce the operating cost of each unit operation. Overall goals are also strongly influenced by economic factors prevailing in the market. Flexibility is therefore a key factor in process control. As for any process control system, the key elements are measurement or sensing, comparison to a target value, manipulation of the variable value, and feedback to the controller. Separate algorithms are developed for each control unit based on empirical factors and experimentation. Continuous improvements and corrections are often made as data are accumulated and as ore characteristics change. Control strategies become more effective when predictions can be made for any unit operation at a high degree of confidence. The more modern control systems are based on multivariable control and model-based concept and digital instrumentation (50). Present trends are toward knowledge-based and artificial intelligence (see EXPERT SYSTEMS), controlled systems which optimize overall performance rather than performance of individual unit operations. These are rule-based systems that attempt to implement human expert knowledge or a rule-of-thumb approach and the uncertainty inherent in human decision making involving linguistic variables (fuzzy terms) and subjective interpretation. Expert system programs or shells have already been marketed and are used in many plants.

A development in the 1960s was that of on-line elemental analysis of slurries using x-ray fluorescence. These have become the industry standard. Both in-stream probes and centralized analyzers are available. The latter is used in large-scale operations. The success of the analyzer depends on how representative the sample is and how accurate the calibration standards are. Neutron activation analyzers are also available (45,51). These are especially suitable for light element analysis. On-stream analyzers are used extensively in base metal flotation plants as well as in coal plants for ash analysis. Although elemental analysis provides important data, it does not provide information on mineral composition which is most crucial for all separation processes. Devices that can give mineral composition are under development.

Dry ore tonnage is measured by belt scales mounted along the conveyor belt line. These are continuous weighing devices and take into account the belt load and the belt speed. The output is a flow rate. An integrator in the scale calculates the total tonnage. Slurry flow rates are measured using magnetic or ultrasonic flow meters (Fig. 24). Slurry densities are routinely measured by batch operation by collecting a representative slurry sample in a liter vessel and weighing it on a density scale. Continuous pulp density measurement is made using a nuclear density meter (Fig. 25) or a gamma-gauge which measures the transmission of gamma rays from a radioactive source through the slurry using an ionization chamber-type detector. Transmittance is inversely proportional to the slurry density. Particle size is measured routinely, in a batch operation, by collecting a representative sample and using laboratory standard sieves. Numerous devices are also available for continuous particle size measurement. One of the more common devices uses the attenuation of ultrasonic energy that occurs on transmission through a slurry, eg, the PSM systems for on-line measurement (2) (see ULTRASONICS). For fine particles, laser-based size analyzers, which are light-scattering devices, can be used, eg, the PAR-TEC system which operates as a scanning laser microscope (50) and the Microtrac system (34). Various other

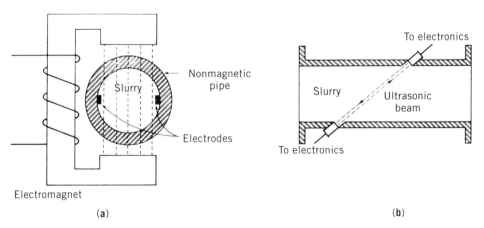

Fig. 24. Flow meters for on-line measurement of slurry flow rates: (**a**) magnetic and (**b**) ultrasonic (6).

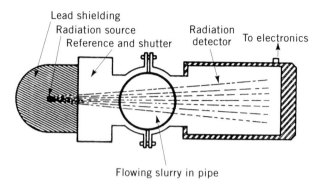

Fig. 25. Nuclear density meter for on-line measurement of slurry density (6).

components of plant control and automation include the elaborate alarm systems and shutdown mechanisms for crushers based on bearing pressure and temperature; crusher power and ore level; grinding and classifier controls comprising ore feed rate, water addition rate, classifier feed rate and pulp density, particle size distribution, mill power, and load; flotation controls comprising aeration rate, pulp and froth level; reagents addition rates; and pH and lime addition. Considerable effort is going into improving the performance of existing measurement devices and sensors, and developing new ones. One area of active work is in the development and implementation of reliable redox control systems for sulfide flotation circuits (51). Color (or vision) sensors for on-line analysis of flotation froths and slurries is also under development (50).

BIBLIOGRAPHY

1. G. Agricola, *De Re Metallica*, 1556, transl. H. C. Hoover and L. H. Hoover, Dover Publications, New York, 1950.
2. B. A. Wills, *Mineral Processing Technology*, 5th ed., Butterworth-Heinemann Ltd., Oxford, U.K., 1992.
3. S. J. Lefond, ed., *Industrial Minerals and Rocks*, 5th ed., AIMME, New York, 1983.
4. *Minerals Yearbook*, U.S. Bureau of Mines, Dept. of the Interior, Washington, D.C., published annually.
5. R. Edwards and K. Atkinson, *Ore Deposit Geology*, Chapman & Hall, London, 1986.
6. E. G. Kelly and D. J. Spottiswood, *Introduction to Minerals Processing*, John Wiley & Sons, Inc., New York, 1982.
7. A. Betekhtin, *A Course of Mineralogy*, transl. V. Agol, Peace Publishers, Moscow, Russia, 1964.
8. D. M. Hausen and W. C. Park, eds., *Process Mineralogy*, Metallurgical Society of AIME, AIMME, New York, 1981.
9. W. C. Park, D. M. Hausen, and R. D. Hagni, eds., *Applied Mineralogy*, The Metallurgical Society of AIME, AIMME, New York, 1985.
10. N. L. Weiss, ed., *SME Mineral Processing Handbook*, AIMME, New York, 1985.
11. A. M. Gaudin, *Principles of Mineral Dressing*, McGraw-Hill Book Co., Inc., New York, 1939.
12. R. P. King, *Mineral. Eng.* **7**(2–3), 129–140 (1994).
13. A. L. Mular and R. B. Bhappu, eds., *Mineral Processing Plant Design*, 2nd ed., AIMME, New York, 1980.

14. R. Thomas, ed., *E/MJ Operating Handbook of Mineral Processing*, McGraw-Hill Book Publishing Co., Inc., New York, 1977.
15. D. Malhotra, R. Klimpel, and A. L. Mular, eds., *Evaluation and Optimization of Metallurgical Performance*, SME, Littleton, Colo., 1991.
16. A. F. Taggart, *Handbook of Mineral Dressing*, John Wiley & Sons, Inc., New York, 1945.
17. P. M. Gy, *Sampling of Particulate Materials: Theory and Practice*, 2nd ed., Elsevier, Amsterdam, the Netherlands, 1982.
18. W. A. Vogely, ed., *Economics of the Mineral Industry*, 4th ed., AIMME, New York, 1985.
19. D. J. Lootens, W. M. Greenslade, and J. M. Barker, eds., *Environmental Management for the 1990s*, SME, Littleton, Colo., 1991.
20. G. C. Lowrison, *Crushing and Grinding*, Butterworths, London, 1974.
21. A. J. Lynch, ed., *Mineral Crushing and Grinding Circuits*, Elsevier, Amsterdam, the Netherlands, 1977.
22. K. R. Suttill, *Eng. Mining J.*, 28–38 (Apr. 1991).
23. S. K. Kawatra, ed., *Comminution—Theory and Practice*, SME, Littleton, Colo., 1992.
24. C. L. Prasher, *Crushing and Grinding Process Handbook*, John Wiley & Sons, Ltd., Chichester, U.K., 1987.
25. P. Somasundaran, ed., *Fine Particles Processing*, Vols. 1 and 2, AIMME, New York, 1980.
26. K. R. Suttill, *Eng. Mining J.*, 18–22 (Feb. 1990).
27. M. R. Smith and R. Gochin, *Mining Mag.*, 27–44 (July 1984).
28. J. D. Salter and N. P. G. Wyatt, *Mineral. Eng.* **4**(7–11), 779 (1991).
29. R. Sivamohan and E. Forssberg, in Ref. 28, p. 797.
30. P. Somasundaran, ed., *Advances in Mineral Processing*, SME, Littleton, Colo., 1986.
31. R. O. Burt, *Gravity Concentration Technology*, Elsevier, Amsterdam, the Netherlands, 1984.
32. B. A. Wills, *Mining Mag.*, 325–341 (Oct. 1984).
33. B. A. Wills, ed., "Proceedings of the International Symposium on Gravity Separation Technology," *Mineral. Eng.* **4**(3–4) (1991).
34. *Mining Annual Review*, published annually by the Mining Journal Ltd., London; useful reference for obtaining recent information, also contains annual *Buyer's Guide* for list of manufacturers and products for mining and mineral processing.
35. J. Svoboda, *Magnetic Methods for the Treatment of Minerals*, Elsevier, Amsterdam, the Netherlands, 1987.
36. U. Andres, *Magnetohydrodynamic and Magnetohydrostatic Methods of Mineral Separation*, John Wiley & Sons, Inc., New York, 1976.
37. F. S. Knoll and J. B. Taylor, *Mineral. Metallurg. Proc.*, 106–114 (May 1985).
38. K. L. Sutherland and I. W. Wark, *Principles of Flotation*, Australian Institute of Mining and Metallurgy, Melbourne, 1955.
39. A. M. Gaudin, *Flotation*, McGraw-Hill Book Co., Inc., New York, 1957.
40. D. W. Fuerstenau, ed., *Froth Flotation*, 50th Anniversary Volume, AIMME, New York, 1962.
41. J. Leja, *Surface Chemistry of Froth Flotation*, Plenum Press, New York, 1982.
42. M. C. Fuerstenau, ed., *Flotation*, Vols. 1 and 2, AIMME, New York, 1986.
43. P. S. Mulkutla, ed., *Reagents for Better Metallurgy*, SME Littleton, Colo., 1994.
44. V. A. Glembotskii, V. I. Klassen, and I. N. Plaksin, *Flotation*, Primary Sources, New York, 1963.
45. A. L. Mular and M. A. Anderson, eds., *Design and Installation of Concentration and Dewatering Circuits*, SME, Littleton, Colo., 1986; B. M. Moudgil and B. J. Scheiner, eds., "Flocculation and Dewatering," *Proceedings of the Engineering Foundation Conference, Jan. 1988*, Engineering Foundation, New York, 1989.

46. K. R. Suttill, *Eng. Mining J.*, 20–26 (Feb. 1991).

47. R. J. M. Wyllie, *Eng. Mining J.*, 22–27 (Oct. 1988).

48. M. E. Chalkley and co-workers, eds., "Tailings and Effluent Management," *Proceedings of the International Symposium*, Pergamon Press, New York, 1989.

49. J. A. Herbst, ed., *Control '84—Minerals/Metallurgical Processing*, AIMME, New York, 1984.

50. R. K. Rajamani and J. A. Herbst, eds., *Control '90—Mineral and Metallurgical Processing*, SME, Littleton, Colo., 1990.

51. B. J. Scheiner, D. A. Stanley, and C. L. Karr, eds., *Emerging Computer Techniques for the Minerals Industry*, SME, Littleton, Colo., 1993.

D. R. NAGARAJ
Cytec Industries

MINERAL WOOL.   See REFRACTORY FIBERS.

MINIUM.   See LEAD COMPOUNDS.

MISCHMETAL.   See CERIUM AND CERIUM COMPOUNDS.

MIXING AND BLENDING

Fluid mixing is a unit operation carried out to homogenize fluids in terms of concentration of components, physical properties, and temperature, and create dispersions of mutually insoluble phases. It is frequently encountered in the process industry using various physical operations and mass-transfer/reaction systems (Table 1). These industries include petroleum (qv), chemical, food, pharmaceutical, paper (qv), and mining. The fundamental mechanism of this most common industrial operation involves physical movement of material between various parts of the whole mass (see SUPPLEMENT). This is achieved by transmitting mechanical energy to force the fluid motion.

Mixing systems are broadly divided into single-phase systems involving miscible liquids, and multiphase systems such as solid–liquid, mutually insoluble liquids, and gas–liquid. Viscous liquids and heat-transfer systems are treated differently because of the need for unique equipment designs. This article discusses fundamental mixing concepts and suitable mixer types in addition to design and scale-up issues for various mixing classes as shown in Figure 1.

These mixing systems offer high flexibility because they can be operated in batch, semibatch, or continuous modes. Adequate mixing is a prerequisite for the success of chemical processes in terms of minimizing investment and operating costs. In addition, chemical reactions with mass-transfer limitation can

Table 1. Classes of Mixing Applications

Mixing class	Physical	Mass-transfer/reaction
miscible liquids	blending: lube oils, gasoline additives, for pH control, dilution	slow batch chemical reactions, fast reactions in in-line mixers
liquid–solid	preparing homogeneous slurries of light and heavy solids such as polymers, catalyst, etc	dissolving, crystallization, liquid–solid reactions, solvent extraction
immiscible liquids	washing liquids with immiscible solvents, cosmetics, salad dressing	hydrolysis/neutralization reactions, extraction, suspension polymerization
gas–liquid	gas scrubbing, steam heating of liquids	absorption, stripping, oxidizing liquids, hydrogenation, oxonation of olefins, chlorination, fermentation
viscous liquids Newtonian non-Newtonian	blending polymer solutions, paints and pigments, food products	solution polymerization, de-ashing of catalyst from polymers
heat transfer	heating and cooling through jackets and internal coils	

be enhanced to provide high yields. Good mixing, therefore, plays a significant role in the profitability of the process industry.

The desired mixing in a commercial process is achieved with different types of equipment, eg, agitators, jets, static mixers, air lifts, etc. The design approach requires defining process mixing requirements; specifying mixer type and size, and other internals such as baffles; and designing mechanical components such as impeller blades, shaft, drive assembly, bearings, and supports.

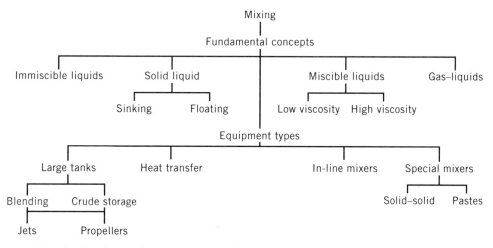

Fig. 1.   Design and scale-up issues for various mixing classes (see Table 1).

Fundamental Concepts

Pumping, Velocity Head, and Power. Mechanical mixers can be compared with pumps (1) because they produce circulating capacity Q and velocity head H. The analogy between a pump and mixer can be appreciated by comparing a pumping loop with a mixing tank (Fig. 2). Power input P to a pump is represented by

$$P = Q(P_1 - P_2)$$

The parameter Q in a mixer represents internal recirculation which is not confined and directed as in a pipe. The pressure drop in a mixer is analogous to velocity head H which can also be considered degradation of kinetic energy. H is proportional to shear in mixing because head from kinetic energy generates shear through the jet or pulsating motion of the fluid. Q in a mixer is related to speed N and the impeller diameter D by

$$Q = N_q ND^3$$

The pumping number N_q is a function of impeller type, the impeller/tank diameter ratio (D/T), and mixing Reynolds number $Re = \rho ND^2/\mu$. Figure 3 shows the relationship (2) for a 45° pitched blade turbine (PBT). The total flow in a mixing tank is the sum of the impeller flow and flow entrained by the liquid jet. The entrainment depends on the mixer geometry and impeller diameter. For large-size impellers, enhancement of total flow by entrainment is lower (Fig. 4) compared with small impellers.

The velocity head H in a pipe flow is related to liquid velocity by $H = v^2/2g_c$. The liquid velocity in a mixing tank is proportional to impeller tip speed πND. Therefore, H in a mixing tank is proportional to $N^2 D^2$. The power consumed by a mixer can be obtained by multiplying Q and H and is given by

$$P = N_p \rho N^3 D^5 / g_c$$

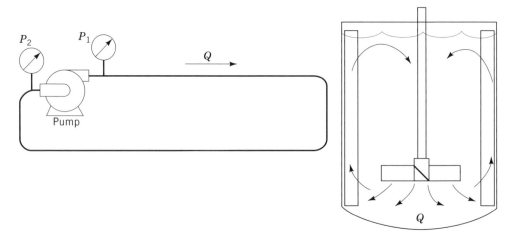

Fig. 2. Pump/mixer analogy.

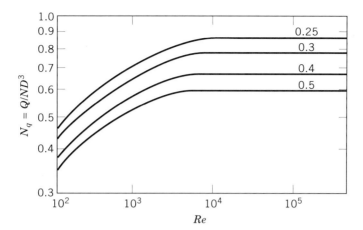

Fig. 3. Pumping numbers, N_q, vs mixing Re for a pitched blade turbine. The numbers on the curves represent the ratio D/T.

The power number N_p depends on impeller type and mixing Reynolds number. Figure 5 shows this relationship for six commonly used impellers. Similar plots for other impellers can be found in the literature. The functionality between N_p and Re can be described as $N_p \propto Re^{-1}$ in laminar regime and depends on μ; N_p in turbulent regime is constant and independent of μ.

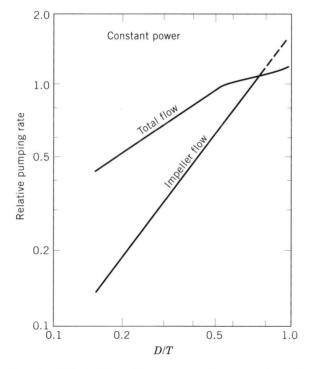

Fig. 4. Effect of impeller size on mixer pumping rate.

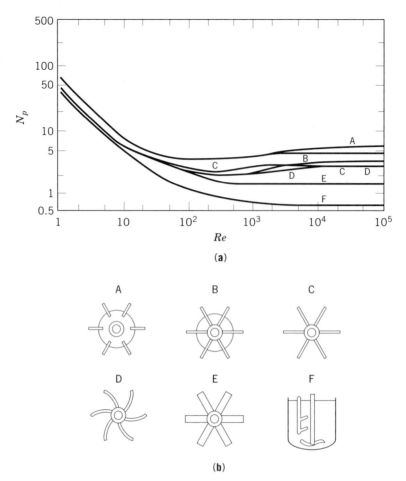

Fig. 5. (**a**) Power number vs Reynolds number; (**b**) W/D for impeller A = 1/5, and for impellers B–E = 1/8. A and B are Rushton impellers, C is FBT, D is curve blade, E is PBT, and F is three-blade retreat curve.

Other parameters, such as blade width W, number of blades n, and blade angle β, also affect the power as

$$P \alpha W n^{0.2} (\sin \beta)^{2.5}$$

The number of wall baffles N_b and their width B also have a significant effect on N_p. As $N_b B$ increases, N_p increases (Fig. **6a**) up to the N_p of the conventional configuration with four baffles having width equal to $T/10$. At higher $N_b B$ the power number is constant at a level which depends on D/T. For multiple impellers, N_p increases as spacing S_i between the impellers is increased (Fig. **6b**). For S_i/D greater than 1, the power consumption reaches a plateau. This effect of S_i on N_p is different for PBT and a flat blade turbine (FBT), the latter showing a maximum at $S_i/D = 0.3$. Also, in the range of S_i/D between 0.1 and 1.0, the N_p of two FBTs is more than twice that of a single FBT.

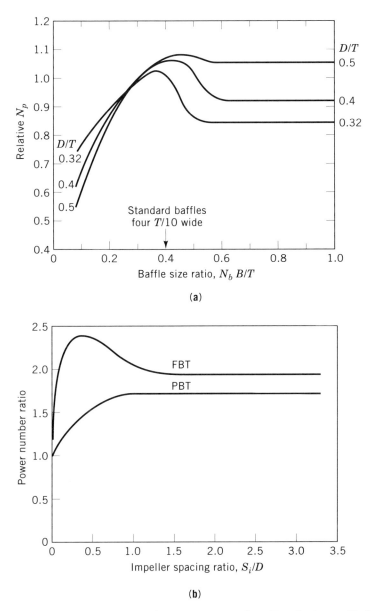

Fig. 6. (a) Effect of baffling and D/T on power number N_p relative to N_p for standard configuration; and (b) effect of dual impeller spacing on power number ratio of two impellers to one impeller.

Depending on the process needs, the combination of pumping and shear can be changed by merely changing the impeller diameter and the mixer speed. A larger diameter impeller at slow speed would provide high pumping action necessary for systems such as blending and solids suspension. On the other hand, a small diameter impeller and high mixer speed would be more suited for high shear systems where mass transfer is important. At constant power input the ratio of pumping capacity to head is related to impeller diameter as

$$Q/H \propto D^{8/3}$$

Shear Rate/Shear Stress.   Whenever there is relative motion of liquid layers, shearing forces exist that are related to flow velocities. These forces, represented by shear stress, carry out the mixing process and are responsible for producing fluid intermixing, dispersing gas bubbles, and stretching or breaking liquid drops. The shear stress is a complex function of shear rate defined by the velocity gradients $\Delta V/\Delta Y$. These velocity gradients can be caused by either entraining liquid with a propeller jet (Fig. 7) or by creating velocity profiles with rotating impellers. Velocity profiles in the propeller jet are proportional to the jet velocity at the blades. The total flow increases by entrainment of the fluid surrounding the jet, whereas the velocity decreases as the jet expands. The velocity profile near the impeller blade (Fig. 8) can be obtained by measuring time averaged velocities. Shear rate can then be calculated by taking the slopes.

Obviously shear rate in different parts of a mixing tank are different, and therefore there are several definitions of shear rate: (*1*) for average shear rate in the impeller region, $\propto N$, the proportionality constant varies between 8 and 14 for all impeller types; (*2*) maximum shear rate, $\propto$ tip speed (πND), occurs near

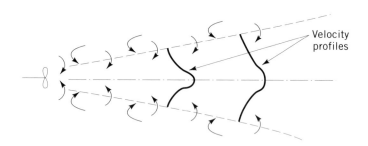

Fig. 7.   Liquid entrainment and velocity profiles in propeller jet.

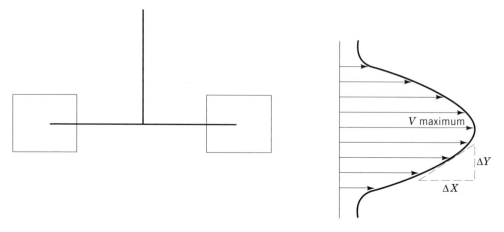

Fig. 8.   Vertical velocity profile near impeller blade tip where the shear rate = $\Delta V/\Delta Y$.

the blade tip; (3) average shear rate in the entire tank is an order of magnitude less than case 1; and (4) minimum shear rate is about 25% of case 3.

In turbulent mixing, the liquid velocity at any point u can be considered the sum of an average velocity $\bar{u}$ and a fluctuating (with time) velocity u':

$$u = \bar{u} + u'$$

Because u' can be positive or negative, its impact on mixing is best quantified by calculating the root mean square (RMS) fluctuation. The RMS velocity is obtained by squaring the fluctuation, time averaging, and then taking the square root. Turbulent intensity is defined as ratio of RMS velocity to average velocity. Typically, turbulent intensity ranges from 0.5–0.7 near the impeller tip, to 0.05–0.15 in other parts of the tank.

The shear rate between the average velocities is called macroscale shear present in eddies of 500 μm or larger in size. The shear rate between the fluctuating velocities, present in smaller than 100 μm eddies, is called microscale shear. A mixing tank, therefore, has several types of shear, ie, macroscale and microscale, maximum and minimum, average in the impeller zone and in the entire tank.

Macromixing vs Micromixing. Mixing in an agitated tank is considered to occur at two levels, macromixing and micromixing. Macromixing is established by the mean convective flow pattern. The flow is divided into different circulation loops or zones created by the mean flow field. The material is exchanged between zones, increasing homogeneity. Micromixing, on the other hand, occurs by turbulent diffusion. Each circulation zone is further divided into a series of back-mixed or plug flow cells between which complete intermingling of molecules takes place.

Macromixing is sufficient for certain mixing duties such as blending oils for concentration homogeneity, whereas micromixing is necessary for fast reacting systems or where enhancement of mass transfer is desired. In a turbulent field, eddies are generated in varying sizes having a maximum scale of the size of the equipment to a minimum length scale defined as Kolmogoroff microscale η:

$$\eta = [\nu^3/\epsilon]^{1/4}$$

where ν is kinematic viscosity of solution and ϵ is energy dissipation rate per unit mass of fluid. Mixing energy is transferred from the largest eddies to the smallest eddies until it is eventually dissipated through friction in viscous and turbulent shear stresses and finally appears as heat in the system.

Dimensionless Numbers. With impeller diameter D as length scale and mixer speed N as time scale, common dimensionless numbers encountered in mixing depend on several controlling phenomena (Table 2). These quantities are useful in characterizing hydrodynamics in mixing tanks and when scaling up mixing systems.

Flow Patterns. There are two main classes of turbine impellers (Fig. 9) based on the flow patterns they generate: axial flow and radial flow (3). Axial flow impellers produce a flow pattern (Fig. 9c) involving full tank volume as a

Table 2. Dimensionless Numbers Used in Mixing

Number	Formula	Important for
flow, Fl	Q_g/ND^3	gas dispersion
Froude, Fr	N^2D/g	free surface or vortexing
Nusselt, Nu	hT/k	heat transfer
power, N_p	$P/\rho N^3 D^5$	power consumption
Prandtl, Pr	$C_p\mu/k$	heat transfer
pumping, N_q	Q/ND^3	blending
Reynolds, Re	$\rho ND^2/\mu$	laminar or turbulent flow
Richardson, Ri	$g\Delta\rho l/\rho\mu^2$	blending
viscosity ratio, V_i	μ_1/μ_2	viscosity differences
Weber, We	$\rho N^2 D^3/\sigma$	emulsification
Weissenberg, Wi	$\sigma_1/N\mu$	viscoelastic effects

single stage. Radial flow impellers, on the other hand, produce two circulating loops, one below and one above the impeller (Fig. 9a). Mixing occurs between the two loops, but less intensely than within each loop. True axial flow is usually created with hydrofoil impellers (Fig. 9b), which provide a confined flow similar to that created in a draft tube. These differences in flow patterns cause

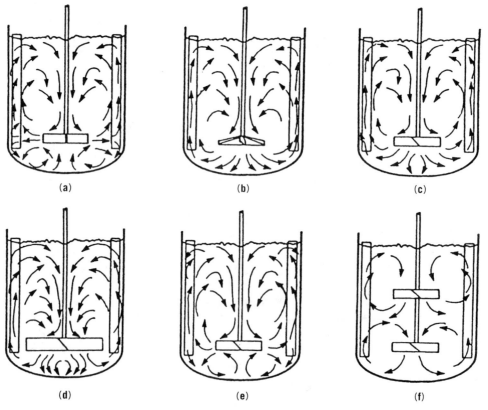

(a)   (b)   (c)

(d)   (e)   (f)

Fig. 9. Flow patterns with different impeller types, sizes, and liquid viscosity: (**a**) FBT; (**b**) hydrofoil; (**c**) PBT; (**d**) PBT, large diameter; (**e**) PBT, high velocity; and (**f**) two PBTs.

variations in shear rate distributions in the tank so that the process result is highly impacted by the impeller flow patterns. The flow patterns within a given impeller are altered by parameters such as impeller diameter, liquid viscosity, and use of multiple impellers. For example, the flow pattern with a PBT becomes closer to radial as the impeller diameter is increased (Fig. 9**d**) or liquid viscosity is increased (Fig. 9**e**). Multiple impellers are used when liquid depth-to-tank diameter ratio is higher than 1. In that case more circulation loops are formed, eg, two loops with PBT (Fig. 9**f**). Radial flow impellers give two circulation loops with each impeller.

Wall Baffles. Because rotating impellers mainly create tangential flows, wall baffles are needed to transform them to vertical flows and hence top-to-bottom liquid recirculation. Vertical baffles located along the tank wall are necessary for providing top-to-bottom mixing without a swirl and for eliminating cavitation. The standard baffle configuration consists of four vertical plates having width equal to 8–10% of tank diameter. Narrower baffles are sometimes used for high viscosity systems, buoyant particle entrainment (width = 2% of T), or when a small vortex is desired. A small spacing between baffles and the tank wall (1.5% of T) is allowed to minimize dead zones particularly in solid–liquid systems. The presence of wall baffles causes an increase in power consumption (Fig. 6**a**), but generally enhances the process result. In square and rectangular shaped tanks, the corners break up the tangential flow pattern thus providing the baffling effect, and wall baffles may not be needed.

Scale-Up Principles. The main objective of scale-up is to achieve the same quality of mixing in a commercial size mixing tank as in a laboratory test tank, or at least have an understanding of the differences expected in the commercial process result. Unfortunately it is not possible to maintain the same combination of flow and shear distributions in commercial mixers as in small-scale tanks. Several scale-up methods have been developed depending on the process type and mixing requirements, but all methods emphasize geometric similarity. Otherwise correction factors are required which lead to high uncertainties in the prediction of the process result. Geometric similarity refers to maintaining the same impeller type, and relative dimensions of impeller, liquid height, and baffles.

The most commonly used scale-up criterion uses constant P/V, which is usually conservative. Depending on the process requirement, P/V should be increased or decreased with increase in volume according to $P/V = $ constant V^y. The scale-up exponent y can be negative or positive and should be determined either through pilot-plant testing or from recommended values in Figure 10. With constant P/V scale-up, the average shear rate decreases while the maximum shear rate increases. The effect of these shear rate changes on the process result must be established experimentally. Scale-up based on constant torque per unit volume T_q/V is considered by some vendors to be more desirable because it is used as a direct index of the mixer size. Torque is related to the mixer power by $T_q = P/2\pi N$. Similar to P/V scale-up, T_q/V is changed proportional to the volume to an exponent which can be obtained either experimentally or from standard values.

When choosing the scale-up method, changes in other flow/power parameters and their impact on the process result must be considered. Figure 11 shows changes in important parameters for different scale-up bases. For example,

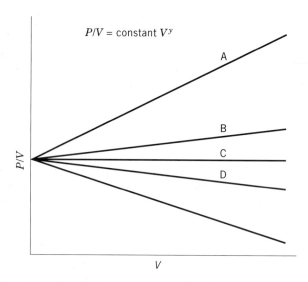

Fig. 10. Log–log plot scale-up by power per unit volume where for A, constant blend time, $y = 2/3$; B, same vortex, $y = 1/6$; C, dispersion, $y = 0$; D, solids suspension, $y = -1/12$; and E, constant flow velocity, $y = -1/3$.

scale-up based on same tip speed maintains the T_q/V but decreases P/V by 80%. T_q/V is almost always increased on scale-up. Scale-up based on the same P/V means a reduction in mixer speed by 66%, which also translates to an increase in blend time by a factor of 3. This is acceptable in most applications except in very fast reacting systems which are scaled-up based on constant mixer speed.

When the process controlling mixing parameter is unknown, laboratory and pilot-plant experiments should be carried out at different mixer speeds. Variability of the desired process result with P can be a good indicator of the controlling regime. For example, no effect with increasing P indicates kinetic regime in a homogeneous reaction. Increase in the process result proportional to $P^{0.5-1}$ means mass-transfer control. For suspension of solids with high settling velocity, constant P/V is required in scale-up. However, constant tip speed suffices for slow settling solids. The criterion valid for dispersion processes involving

N, Q/V	Tip speed	Re	Q	T_q/V	We	P/V	P
①1	5	25	125	25	125	25	3125
0.2	①1	5	25	①1	5	0.2	25
0.34	1.7	8.5	42.7	2.9	14.6	①1	125

Fig. 11. Changes in mixing parameters on scale-up to 125 times the volume of the pilot plant.

liquid–liquid and gas–liquid systems is constant P/V. This method maintains the same interfacial area per unit volume, hence the same mass-transfer rate. However, considerable deviation from constant P/V is needed if other factors such as coalescence, velocities, static pressure, etc, affect the process result.

Impeller Types. There are literally hundreds of impeller types in commercial use. Only the most common and general types are shown in Figure 12. Turbine impellers are characterized based on flow patterns as axial flow (Fig. 12**a**) and radial flow (Fig. 12**b**). Developments in impeller technology have been focused on increasing axial flow at reduced shear. These impellers use the hydrofoil profile of blades for efficient, more streamlined pumping (Fig. 12**c**). For high viscosity applications, anchor and helical ribbon impellers (Fig. 12**d**) are more suitable. They are typically large-diameter, nearly the same as tank diameter, close-clearance impellers. In addition to the differences in flow patterns, these impellers provide different combinations of shear and pumping. Turbine impellers have the widest use in low and medium viscosity liquid applications, solids suspension, liquid dispersion, and gas–liquid contacting. The oldest axial flow impeller design is the marine propeller which is used as a side entering mixer in large tanks and top entering in small tanks. Top entering mixers for large tanks also use pitched blade turbines (PBT) for axial flow and flat blade turbines (FBT) for radial flow. Some hydrofoil impellers, such as Lightnin A315 and Prochem Maxflo, are designed with high solidity ratio for gas dispersion and blending of non-Newtonian liquids. The solidity ratio is the ratio of projected blade area to the area swept by the impeller. Although the number of turbine blades can vary from 2 to 12, two blades are normally not mechanically stable and there is no advantage to more than four blades in hydrofoil impellers.

Blending of Miscible Liquids

Mutually soluble liquids are blended to provide a desired degree of uniformity in an acceptable mixing time. Agitator effects are critical particularly in commercial operations when both product quality and production rates are important. Although mixing time and pumping rate for a given impeller are related, this relationship can be different with different impeller types. For designing axial flow turbines there are two useful methods using number of tank turnovers and bulk fluid velocity, both based on impeller pumping rate. Pumping rate of an impeller can be estimated from impeller diameter and mixer speed, and the pumping number N_q for a PBT can be obtained from Figure 3. Turnover rate in a mixing tank is obtained by dividing pumping rate Q by the tank volume V. Blend time can then be estimated by defining the number of turnovers needed for the process result. It is recommended that the turnover requirements be defined through laboratory testing on the basis of a specific process result. The approximate numbers of turnovers required for different liquid viscosity ranges are as follows.

liquid viscosity, mPa·s	up to 100	100–1000	1000–5000	>5000
number of turnovers	3	10	50	>100

The bulk fluid velocity method relates a blending quality Chemscale number C_n to a qualitative description of mixing (Table 3). The value of C_n is equal

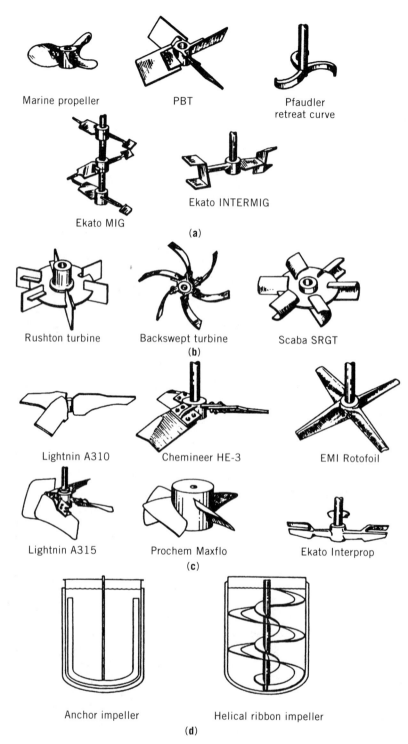

Marine propeller

PBT

Pfaudler retreat curve

Ekato MIG

Ekato INTERMIG

(a)

Rushton turbine

Backswept turbine

Scaba SRGT

(b)

Lightnin A310

Chemineer HE-3

EMI Rotofoil

Lightnin A315

Prochem Maxflo

Ekato Interprop

(c)

Anchor impeller

Helical ribbon impeller

(d)

Fig. 12. Common impeller types: (**a**) axial, (**b**) radial, (**c**) hydrofoil, and (**d**) close-clearance.

Table 3. Mixing Quality Descriptions for Chemscale Numbers, C_n

C_n	Specific gravity difference, $\Delta\rho$	Viscosity ratio, μ_1/μ_2	Solids suspension for $<2\%$ slurries	Surface motion
2	<0.1	<100	no	fluid surface flat but moving
6	<0.6	$<10,000$	settling rates 0.6–1.2 m/min	rippling surface at low μ
10	<1.0	$<100,000$	settling rates 1.2–1.8 m/min	surging surface at low μ

to one-sixth of the bulk fluid velocity defined by pumping rate divided by cross-sectional area of the tank (4).

When a liquid-filled tank is stratified with denser liquid at the bottom, higher mixing energy is needed to cause homogeneity. The time to break up stratification depends on the rate of erosion of the interface. This interface rises to the liquid surface at a rate based on horizontal velocities at the interface. The entrainment velocity is correlated with the Richardson number, Ri, which is the ratio of buoyant to inertial forces. When $Ri < 7$ the rise of the interface is fast enough to ignore the time taken for breakup of the layers. Therefore, the blend time t_b mainly depends on the time taken for a desired liquid homogeneity, eg, 95% homogeneity, and can be estimated from the following:

$$t_b = N^{-1}(D/T)^{-2.3}f(Re)(\Delta\rho/\rho)^{0.9}$$

If mixing quality better than 95% is desired, the blend time can be estimated using a variance decay model for amplitude of concentration variation:

$$A_1 = 2e^{-k_1 t}$$

The decay constant k_1 is the key to the mixing rate number defined for $Re > 10^4$ by $N/k_1(D/T)^{2.3} = 0.5$ for a disk flat blade turbine (DFBT), and $N/k_1(D/T)^2 = 0.9$ for the propeller. Therefore, power required to achieve any degree of uniformity in a fixed time is proportional to $D^{-1.9}$ and D^{-1} for DFBT and propeller, respectively.

Feedpipe Backmixing. Backmixing of fluid into the feedpipe can result in lower yields of the desired product in a system with fast competitive/consecutive reactions. This is because some reactions can occur in unmixed conditions in the feedpipe where the reacting species may be in nonstoichiometric ratios. Therefore, it is important to design the feed injector nozzles to provide minimum jet velocity v_f to prevent backmixing into the feedpipe; v_f depends on impeller type, mixer speed, and the feedpipe location. For a given impeller type and feedpipe location, the ratio of minimum v_f to tip speed v_t is nearly constant. The minimum values of this ratio v_f/v_t for preventing feedpipe backmixing are 1.9 for DFBT radial feed, 0.25 for DFBT feed above impeller, 0.1 for HE-3 radial feed, and 0.15 for HE-3 feed above impeller.

Viscous Liquid Blending.   As liquid viscosity is increased, the discharge stream velocity from an impeller is more rapidly dissipated by liquid friction. As a result, the flow pattern flattens out and axial flow impellers generate radial flows. This change significantly affects blending quality. For example, if an axial flow impeller circulates low viscosity liquid to a height of four impeller diameters above and below, it circulates a 50,000 mPa·s liquid to only $D/2$ height above and below. This indicates good mixing in a small volume around the impeller while little interchange of material occurs outside of this volume. Under these conditions mixing can best be achieved with proximity or close-clearance impellers (see Fig. 12).

Viscous liquids are classified based on their rheological behavior character- ized by the relationship of shear stress with shear rate. For Newtonian liquids, the viscosity represented by the ratio of shear stress to shear rate is inde- pendent of shear rate, whereas non-Newtonian liquid viscosity changes with shear rate. Non-Newtonian liquids are further divided into three categories: time-independent, time-dependent, and viscoelastic. A detailed discussion of these rheologically complex liquids is given elsewhere (see RHEOLOGICAL MEASUREMENTS).

The selection of mixer type and mixing system design method changes with ranges of viscosity. Figure 13 shows relative blending capacities of turbine impellers and close-clearance impellers such as anchors and helical ribbons. For low viscosity liquids, axial or radial flow impellers provide excellent mixing, whereas anchors are also good but consume high energy. Anchors (Fig. 12**d**) are suitable for liquid viscosity between 5,000 and 50,000 mPa·s(= cP) because at low viscosity there is not enough viscous drag at the wall to provide pumping. Above 50,000 mPa·s, pumping capacity of the anchor declines and the impeller simply "slips" in the liquid. In jacketed tanks anchor blades are designed with wipers for scraping the tank wall and enhancing heat transfer.

Helical impellers (Fig. 12**d**) provide better top-to-bottom turnover compared to anchors, particularly at viscosities higher than 50,000 mPa·s. They can be

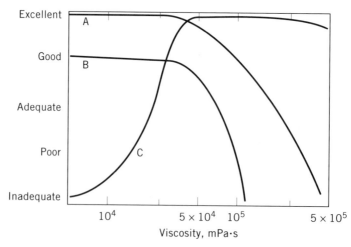

Fig. 13.   Blending qualities with different impellers (A, turbines; B, anchors; C, helicals) for different viscosity ranges.

made with or without an inner helix pumping in the opposite direction, and sometimes with more than one helix. Use of the inner helix significantly enhances blending, particularly for non-Newtonian liquids. The most commonly used pitch for the helix impeller is 0.5. Higher pitch reduces top-to-bottom mixing whereas lower pitch causes excess friction and energy consumption. Impeller blade width normally varies between $T/12$ and $T/6$. The product of mixer speed and mixing time is constant for high viscosity liquids under laminar conditions. Therefore, mixing time θ can simply be estimated from this constant for the impeller used. For example, $N\theta = 33$ for a helical ribbon impeller of standard dimensions.

Suspension of Solids

Solids are suspended in liquid-filled agitated tanks for the purpose of dissolution, accelerating chemical reactions or preparing a homogeneous slurry. Typical industrial applications include catalyst suspension, rubber particle slurrying, TiO_2 slurry for paper coatings (qv), activated carbon suspension in water treatment, or slurrying in leaching of metals. In most applications it is important to select a suitable mixer type that can be designed to provide the desired process result at minimum energy consumption. This requires understanding of the mixing mechanism of solids suspension and possibly laboratory testing to develop design guidelines. The critical mixing parameters depend on whether the particles sink or float under unmixed conditions, and therefore are discussed separately for the two types of solids.

    Sinking Solids. Lifting and distribution of solids heavier than the liquid is accomplished by inducing necessary flow patterns by expending mechanical energy. This energy, supplied by a rotating impeller, is dissipated in turbulent eddies having a variety of sizes. The largest velocity scale of turbulent eddies is on the order of maximum tip speed, whereas the largest length scale of eddies is on the order of the impeller size. According to the Kolmogoroff principle of isotropic turbulence, small eddies are 30–100 μm in size. Because most solids of industrial importance are 1 mm or larger, small eddies are ineffective in suspending them; only large-scale liquid motion is responsible for their suspension. When the velocity scale of large eddies is higher than the free settling velocity of particles, entrainment occurs. For particle suspension from the rest position, this free settling velocity must be overcome. The energy requirement for maintaining a suspension is, however, much less.

    Measurement of single particle settling velocity in a turbulent field is not easy. However, it is known to be a function of free settling velocity u_t, which for spherical particles can be estimated from the following:

$$u_t = \sqrt{\frac{4gD_p\Delta\rho}{3\rho C_D}}$$

The drag coefficient C_D has different functionalities with particle Reynolds number Re_p in three different regimes (Fig. 14), which results in the following expressions (1).

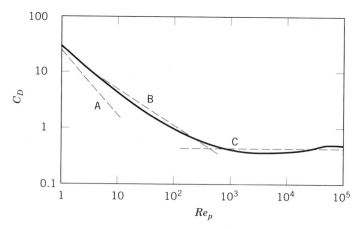

Fig. 14. Drag coefficient C_D for terminal settling velocity correlation (single particle) where A represents Stokes law; B, intermediate law; and C, Newton's law (1).

Region	Stokes	Intermediate	Newton
Re_p	< 0.3	$0.3\text{–}10^3$	$10^3\text{–}10^5$
u_t	$\dfrac{gD_p^2\Delta\rho}{18\mu}$	$\dfrac{0.15g^{0.7}D_p^{1.14}\Delta\rho^{0.7}}{\rho^{0.3}\mu^{0.4}}$	$1.74\sqrt{\dfrac{gD_p\Delta\rho}{\rho}}$

Quite often the settling velocity is modified by the presence of a large number of particles. This hindered settling velocity is a function of solids concentration and should be measured experimentally or estimated from literature correlations (1).

In solids suspension applications it is desirable to have at least movement of particles on the tank floor, although this type of mixing is rarely sufficient. For most situations off-bottom suspension is necessary, ie, all particles are suspended even though they may temporarily settle before being lifted again. Some processes require complete slurry uniformity which for practical purposes is achievable only up to 95–98% of liquid height. Between off-bottom suspension and complete uniformity there are several mixing quality levels defined by the ratio of solids concentration near the top to bulk concentration. In mass-transfer operations, as mixer power is increased, the mass-transfer coefficient, k_L, increases rapidly up to the point of off-bottom suspension. With further increase in mixer power, k_L still increases but at a slower rate providing reduced incremental benefit. Impeller speed for off-bottom suspension, N_c, is the most important parameter in designing mixers.

In order to measure N_c for a given system, several laboratory techniques are used which are classified into three categories: visual, light transmission, and rate measurement (Fig. 15). The visual method is the most common and inexpensive, and consists of observing solids movement in a transparent tank and recording the results on still or motion pictures for quantification. The light transmission method is more sophisticated and requires transmitting light through the tank and measuring transmitted light on the opposite side. This measurement is inversely proportional to solids concentration in the plane of

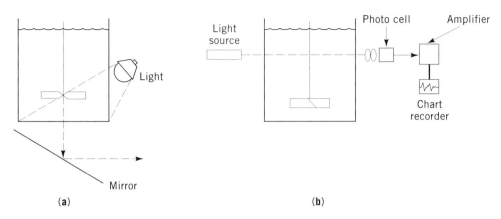

Fig. 15. Laboratory methods for measuring solids suspension quality: (**a**) visual, and (**b**) light transmission.

the light passage. A typical light source is a helium–neon laser (see LASERS) and microscope lamp; the receiver consists of a silicon or selenium photocell. For measurement of N_c, a fiber optic probe can be pointed at the tank floor and the impeller speed recorded at maximum light transmission (see FIBER OPTICS). The rate measurement method is used when mass transfer can be easily measured, eg, dissolution of solids. Analyzing the liquid composition at different batch times provides evaluation of the mixer performance. This technique is a direct measure of the process result, and therefore is recommended if such tests can be conveniently carried out.

The correlation for N_c based on experimental data has been well researched. The pioneering work was carried out in 1958 and a correlation with different impeller types and tank sizes was developed (5):

$$N_c = s\nu^{0.1}D_p^{0.2}D^{-0.85}\left(\frac{g\Delta\rho}{\rho}\right)^{0.45} C_s^{0.13}$$

The constant s depends on impeller configuration and C_s is solids concentration. Several other researchers have since developed similar correlations which include effects of a few additional parameters such as D/T, C/D, and H_l/T. The corresponding exponents of these parameters vary in range: -1 to -1.7 for D/T, 0.15 to 0.2 for C/D, and ~ 0.33 for H_l/T.

Some studies (6) have been carried out to measure distribution of solids in mixing tanks. Local solids concentrations at various heights are measured at different impeller speeds. Typical data (Fig. 16) demonstrate that very high mixer speeds are needed to raise the solids to high levels. At low levels, solids concentration can exceed the average concentration at low mixer speeds. These solids distributions depend on the impeller diameter, particle size, and physical properties.

Floating Solids. There are several applications where solids are lighter than the carrying liquid and are also sticky at process conditions. These solids either have density less than that of the liquid or are highly porous. In order to

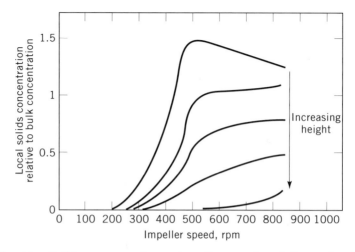

Fig. 16. Solids concentration vs impeller speed at different heights.

keep these particles from rising to the surface and agglomerating, pulling down action must be provided by the mixer. Such floating solids are encountered in polymeric processes and in slurrying of difficult-to-wet porous solids.

The mechanism for entrainment of floating solids is quite different from that for suspension of heavy solids, and requires much higher mixing energy. A force balance on a floating particle (7) about to be incorporated consists of an upward buoyant force opposed by downward drag and lift forces. For entrainment to occur a net downward force is required. Observations of the motion of floating solids have revealed that liquid swirl created in unbaffled tanks provides these necessary forces (Fig. 17), whereas tanks with full baffles do not have such swirl. The swirl produces centrifugal forces that move the light particles along the liquid surface radially into the cone of the vortex where velocities are high enough to cause entrainment. Unbaffled agitated tanks require high energy to provide the necessary axial velocities below the vortex. In addition, large unbaffled tanks are subject to lateral swaying because of large surface waves.

Mixing tanks equipped with modified baffles called partial baffles have been used successfully for entrainment of floating solids at reasonably low energy

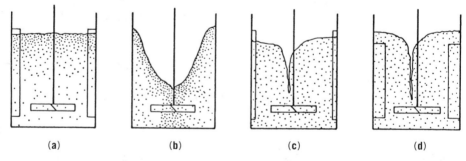

Fig. 17. Mixing of floating solids in agitated tanks: (**a**) no surface movement, full baffles width = $T/12$; (**b**) deep vortex, no baffles need high energy which causes tank to sway; (**c**) precessing vortex, partial baffles width = $T/50$; and (**d**) submerged partial baffles.

with few mechanical problems. These partial baffles can be full length having width equal to 1/50 of the tank diameter (Fig. 17**c**) or submerged below the liquid level and having conventional width (Fig. 17**d**). There are other equally effective partial baffles which vary in configuration, eg, the single baffle or finger baffles of different shapes positioned at the surface. Scale-up of this mixing system is based on maintaining a constant Froude number, N^2D/g, which results in increasing the mixer tip speed in proportion to $T^{0.5}$.

Immiscible Liquid–Liquid Mixing

Mixing of immiscible liquids is frequently encountered in chemical, petroleum, pharmaceutical, cosmetics, food, and mining industries. Several reacting and nonreacting systems include extraction (qv), alkylation (qv), interfacial and suspension polymerization, emulsifications, and phase-transfer catalysis (see CATALYSIS, PHASE-TRANSFER). When mutually insoluble liquids are mixed in an agitated tank, a dispersion of one phase is produced, thereby increasing the interfacial area manyfold. For example, in a 1320-L (350-gal) tank an interfacial area of 4047 m^2 (1 acre) is achieved with equal volumes of the two phases and 1-mm diameter dispersed drops. Mixing energy spent in maximizing this interfacial area is generally cheaper than the resultant enhancement in the mass-transfer rate. However, optimization of mixing energy is necessary because too much energy can result in formation of undesirable emulsions, increasing temperature and reducing product quality. In order to design a mixer for a given process, it is important to define the process needs, eg, homogenization of phases, fine dispersions for fast reacting systems, dispersions with narrow drop sizes, minimize diffusional resistance in the continuous phase, induce convection within the drops, generate very fine stable emulsions, etc.

When an impeller is rotated in an agitated tank containing two immiscible liquids, two processes take place. One consists of breakup of dispersed drops due to shearing near the impeller, and the other is coalescence of drops as they move to low shear zones. The drop size distribution (DSD) is decided when the two competing processes are in balance. During the transition, the DSD curve shifts to the left with time, as shown in Figure 18. Time required to reach the equilibrium DSD depends on system properties and can sometimes be longer than the process time.

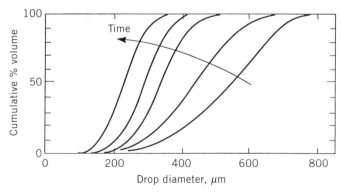

Fig. 18.　Variation in drop size distribution with time.

Drop breakage occurs when surrounding fluid stresses exceed the surface resistance of drops. Drops are first elongated as a result of pressure fluctuations and then split into small drops with a possibility of additional smaller fragments (Fig. 19). Two types of fluid stresses cause dispersions, viscous shear and turbulence. In considering viscous shear effects, it is assumed that the drop size is smaller than the Kolmogoroff microscale, η.

The largest drop that can exist in a shear field is related to the maximum value of the shear rate by

$$d_{max} = \frac{2\sigma(p + 1)f(p)}{\dot{\gamma}_{max}\mu_c(1.19p + 1)}$$

where p is the viscosity ratio of dispersed to continuous phase (8). The function $f(p)$ has been experimentally determined (9). Drop breakup by turbulence occurs when fluid turbulent energy exceeds the stabilizing surface energy of drops. Thus the maximum stable drop diameter is determined by equating the two energies and can be expressed as follows:

$$d_{max} \, \alpha \, (\sigma/\rho_c)^{0.6}\epsilon^{-0.4}$$

Drops coalesce because of collisions and drainage of liquid trapped between colliding drops. Therefore, coalescence frequency can be defined as the product of collision frequency and efficiency per collision. The collision frequency depends on number of drops and flow parameters such as shear rate and fluid forces. The collision efficiency is a function of liquid drainage rate, surface forces, and attractive forces such as van der Waal's. Because dispersed phase drop size depends on physical properties which are sometimes difficult to measure, it becomes necessary to carry out laboratory experiments to define the process mixing requirements. A suitable mixing system can then be designed based on satisfying these requirements.

Dispersion Measurements.   Several laboratory techniques have been successfully used to measure dispersion qualities which include phase homogenization, freezing of drops, direct photography, light transmittance, reaction/mass transfer, and conductivity. Homogenization is the simplest technique and is applicable when mass-transfer rate is less important. The experiments are carried out by observing the location of the interface as mixer speed is increased in a transparent tank. This measurement is aimed at determining the minimum agitator speed for homogeneous dispersion. Dispersed drops in a mixing tank can

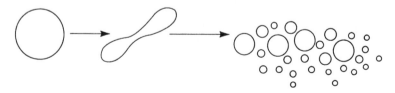

Fig. 19.   Breakup of dispersed drops.

be frozen by forming a protective film around them, thereby preventing coalescence. A sample of stabilized emulsion can then be drawn and optically analyzed for drop size distribution. The protective film around the drops can be formed by a reaction of poly(acrylic acid) in the aqueous phase and polyisobutylene succinic acid in the oil phase. There are other reactions which are equally effective.

Direct photography of drops in done with the use of fiber optic probes using either direct or reflected light. Still or video pictures can be obtained for detailed analysis. The light transmittance method uses three components: a light source to provide a uniform collimated beam, a sensitive light detector, and an electronic circuit to measure the amplified output of the detector. The ratio of incident light intensity to transmitted intensity is related to interfacial area per unit volume.

The reaction/mass-transfer technique is based on Danckwerts theory of mass transfer accompanied by a fast pseudo first-order reaction (10):

$$Ra = aS\sqrt{\mathcal{D}k + k_L^2}$$

when

$$\sqrt{\mathcal{D}k}/k_L << C_R/ZS$$

R is rate of reaction per unit area, a is interfacial area per unit volume, S is solubility of solute in continuous phase, $\mathcal{D}$ is diffusivity of solute, k is rate constant, k_L is mass-transfer coefficient, C_R is concentration of reactive species, and Z is stoichiometric coefficient. When $\mathcal{D}k$ is considerably greater (10 times) than k_L^2, $Ra = aS\sqrt{\mathcal{D}k}$.

The integral values of effective interfacial area can thus be obtained by measuring the reaction (extraction) rate and using physico-chemical properties of the reactants. A reaction satisfying the above conditions consists of hydrolysis of hexyl formate (11):

$$HCOOC_6H_{13} \text{ (org)} + NaOH \text{ (aq)} \longrightarrow HCOONa \text{ (aq)} + C_6H_{13}OH \text{ (org)}$$

The value of $S\sqrt{\mathcal{D}k}$ for this reaction at 30°C is 1.72×10^{-4} mol/m^2 per second.

Turbines. Turbine agitators provide the desired mixing conditions for immiscible liquids. Even in high viscosity liquid emulsifications, turbines are more effective than agitators conventionally used for blending of viscous liquids. There are two types of correlations reported in the literature (18) for turbine impellers, one for minimum speed for homogeneous dispersion and the other for Sauter mean diameter d_{32}. The minimum agitator speed N_m for suspension can be estimated from the following:

$$N_m = C_1 D^{C_2}(\mu_c/\mu_d)^{1/9}\sigma^{0.3}\Delta\rho^{0.25}$$

The value of C_2 is around $-2/3$ and C_1 depends on impeller type and location. This correlation does not impact on drop sizes and is applicable when the system is not mass-transfer limited.

Theoretically, d_{32} can be correlated to interfacial tension, continuous-phase density, and power per unit mass P_M swept by the impeller:

$$d_{32} \, \alpha \, (\sigma/\rho)^{0.6} P_m^{-0.4}$$

This correlation is valid when turbulent conditions exist in an agitated vessel, drop diameter is significantly bigger than the Kolmogoroff eddy length, and at low dispersed phase holdup. The most commonly reported correlation is based on the Weber number:

$$d_{32}/D = C_3(1 + C_4\phi)We^{-0.6}$$

C_4 is termed the turbulence damping or coalescence factor because of high holdup, and its reported values vary in the 2.5–5.4 range. C_3 depends on impeller type and diameter.

The criterion of maintaining equal power per unit volume has been commonly used for duplicating dispersion qualities on the two scales of mixing. However, this criterion would be conservative if only dispersion homogeneity is desired. The scale-up criterion based on laminar shear mechanism (9) consists of constant $T^{1/2}(ND)^{-3/2}$, typical for suspension polymerization. The turbulence model gives constant tip speed πND for scale-up.

Phase Inversion.   This is a phenomenon caused by changing mixing conditions so that the dispersed and continuous phases interchange. The phase to become dispersed depends on the volume concentration of the two liquids, their physical properties, and the dynamic characteristics of the mixing process. There is always a range of volume fractions throughout which either component remains stably dispersed, and this is called the range of ambivalence. The limits of this range are influenced by the size and shape of the vessel, mixer speed, and physical properties of the liquids.

Gas–Liquid Mixing

Mechanically agitated gas–liquid contactors are widely used in industrial processes. Typical nonreacting processes include absorption and stripping, whereas reacting systems include oxidation, hydrogenation, chlorination, etc. They are also used for carrying out biochemical processes such as aerobic fermentation (qv), manufacture of protein, and wastewater treatment. The fractional hold-up of gas ϕ in these contactors is a basic measure of their efficiency. This, in conjunction with Sauter mean bubble diameter d_{32}, determines the interfacial area, $a = 6\phi/d_{32}$, and hence the mass-transfer rate. Knowledge of ϕ also gives the residence time for each phase. One of the most commonly used devices is the disk flat blade turbine (DFBT), also called the Rushton turbine, which can create large interfacial areas by means of the turbulent radial flow pattern. Axial flow impellers and hydrofoil impellers have also been used when high liquid recirculation is desired.

Process performance of gas–liquid contactors depends on the hydrodynamic conditions inside the vessel. The various hydrodynamic conditions observed during the process of gas dispersion are flooding, gas dispersion, and bubble recirculation. These conditions depend on superficial gas velocity, impeller type and size, mixer speed, and sparger design and location. When an impeller is rotated in a gas-sparged liquid-filled tank, it creates rapidly spinning trailing vortices behind the impeller blades. These vortices contain centrifugal force fields and shear rates, and lead to pressures low enough to form vortex cavities. As the aeration number ($Fl = Q_g/ND^3$ where Q_g is gas rate) is increased (Fig. 20), the vortex cavities change size and shape from clinging cavities (Fig. 20**a**) to large cavities (Fig. 20**b**), alternating 3–3 cavities (Fig. 20**c**), until they bridge from one blade to the next (Fig. 20**e**).

Mixer power under gassed conditions, P_g, varies as Fl is increased as shown in Figure 21. The initial small drop in power consumption is associated with a steadily increasing but limited reduction in the pumping capacity of the turbine as the gas bubbles in the vortex cavities increase in size. If gas throughput is further increased, at an $Fl = 0.025$, there is a clear point of inflection in the curve which is associated with formation of alternating large and clinging cavities. Fl alone is not sufficient to correlate P_g because the curves as shown in Figure 21 are different at different mixer speeds (12).

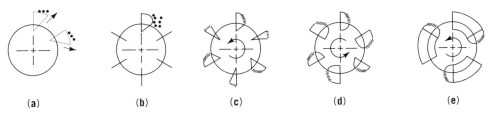

| (a) | (b) | (c) | (d) | (e) |

Fig. 20. Cavity formations behind impeller blades where (**a**) illustrates clinging cavities, (**b**) a large cavity, (**c**) 3–3 cavities, (**d**) alternating large and larger cavities, and (**e**) bridging cavities.

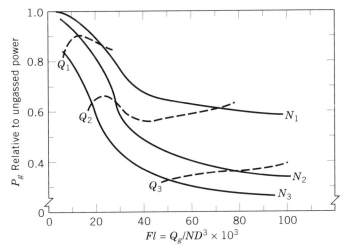

Fig. 21. Gassed power curves for constant Q_g and N where mixer speeds $= N_1 < N_2 < N_3$ and gas rates $= Q_1 < Q_2 < Q_3$.

When flow regime dominated by impeller pumping is changed to one controlled by buoyancy effects of gas leaving the sparger, the hydrodynamic condition is called flooding. The results of increasing mixer speed with constant gas rate have been associated with changes from flooding to gas recirculation (13) (Fig. 22). At flooding (Fig. 22**a**) gas bubbles are not recirculated by the impeller, whereas at high mixer speeds (Fig. 22**e**) gas bubbles are recirculated both above and below the impeller. The conditions in Figure 22**b–d** represent increasing degrees of dispersion.

A general flow map of different hydrodynamic conditions (Fig. 23) consists of regions of flooding, dispersion, and recirculation on a plot of N vs Q_g for a Rushton turbine. For a low viscosity aqueous/air system, the gas flow numbers for the three conditions are given by $Fl_f = 30\, Fr(D/T)^{3.5}$ for flooding, $Fl_{cd} = 0.2\, Fr^{0.5}(D/T)^{0.5}$ for complete dispersion, and $Fl_r = 13\, Fr^2(D/T)^5$ for recirculation.

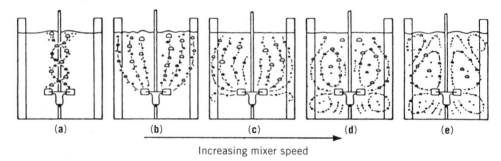

(a)   (b)   (c)   (d)   (e)

Increasing mixer speed

Fig. 22. Bulk flow patterns with increasing N at constant Q_g, where (**a**) shows flooding; (**b**) to (**d**), increasing degrees of dispersion, and (**e**) complete recirculation.

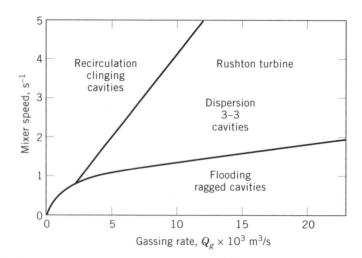

Fig. 23. General flow map of a gas-sparged agitated tank.

Gas holdup with Rushton turbine can be estimated from the following correlation:

$$\phi = 0.1(D/T)^{1.25}(ReFrFl)^{0.35}$$

At a given gas sparging rate, interfacial area a is constant at low mixer speeds. When mixer speed is increased above a critical speed N_o, a starts increasing and varies linearly with N. For Rushton turbines this critical speed, as determined in an O_2/sodium sulfate system, is given by the following:

$$\frac{N_o D}{(\sigma g/\rho)^{0.25}} = 1.22 + 1.25(T/D)$$

The interfacial area in the region with agitation effect $N > N_o$ and liquid height, H_l, is given by

$$\frac{aH_l}{(1 - \phi)} = 0.79\mu(N - N_o)D\sqrt{\frac{\rho T}{\sigma}}$$

Radial flow impellers such as the Rushton turbine and FBT are most suitable for dispersing gas; axial flow impellers are inferior. Gas dispersion with a 45° PBT provides an unstable hydrodynamic regime because large cavities develop leading to severe fluctuations in the power demand. For optimum mixer design, ie, maximum a at lowest power input, small-diameter impellers at high mixer speeds should be used. Scaling up for equal a must be done on the basis of equal tip speed and geometric similarity. For maximizing gas holdup, it is recommended that a ring sparger of diameter equal to 0.8 D located below the impeller be used.

When the gas used in the process is hazardous and/or expensive, it is recycled from the vapor space using gas-inducing impellers. Typical applications include hydrogenation, chlorination, and phosgenation processes. A gas-inducing impeller uses the acceleration of the liquid phase over the blades to locally reduce the pressure at an orifice and induce a flow of gas through a hollow shaft. The critical speed at which gas flow is first induced is given by the following:

$$N_g = \sqrt{\frac{2gh_s}{k_h(\pi D)^2}}$$

where h_s is submergence of the orifice and k_h is head loss constant for different impeller types having values in the range of 0.9–1.3 (14). At speeds greater than N_g, the flow rate of induced gas Q_i increases with increasing mixer speed. Also, Q_i is higher with larger impellers. Turbine impellers require lower N_g and induce more gas compared to hydrofoil impellers.

Blending in Large Tanks

Blending of miscible liquids in 6–90-m (20–300-ft) diameter tanks is carried out with jet mixers and side-entering propellers (SEP). Jet mixers are used in conjunction with a pump that serves as the source of the required mixing energy. With suitable pumps, jet mixers are attractive for use in liquefied natural gas (LNG), liquefied petroleum gas (LPG), gasoline, jet fuel, and distillate fuel storage and blending tanks; otherwise SEP mixers must be used. The basic requirements for blending in large storage tanks are that the entire contents of the tank be mixed and that the mixing be completed in the desired time. In LNG and LPG tanks it is possible to add a dense layer to the bottom of a tank which, upon warm-up during storage, can become less dense than the upper layer resulting in a roll-over of the tank contents. This can result in dangerously high vapor release rates due to spontaneous flashing. Therefore, such tanks should be adequately mixed to eliminate stratification.

Jet Mixers.   These mixers can be used for continuously blending miscible fluids as they enter a tank, or batchwise by recirculating a portion of the tank contents through the jet. The jet mixer is a streamlined nozzle installed in the side or center of a tank near the bottom or top. It can be pointed horizontally or at an angle diametrically across the tank (Fig. 24**a**). Jet eductors (Fig. 24**b**) provide higher entrainment of the surrounding liquid compared to nozzles. Typical jet nozzle configurations (Fig. 25) include side-entry and axial. Either configuration can be effective if the jet nozzles are adequately designed. However, underdesigned systems can have poorly mixed zones as indicated in Figure 25.

Jets mix by entrainment of the surrounding fluid into the jet. The induced flow within the tank is therefore greater than the jet flow itself and can lead to rapid mixing. The amount of fluid entrained by a jet is a function of jet Reynolds number $Re_j = dU_o\rho/\mu$, jet expansion angle, and jet length. For most effective

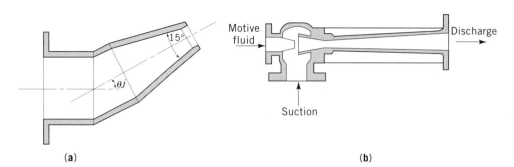

Fig. 24.   Types of jet nozzles: (**a**) inclined and (**b**) eductor.

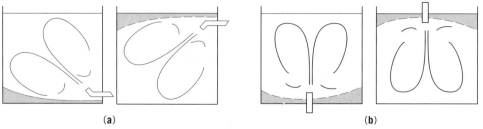

Fig. 25.   Flow patterns in jet mixed tanks where ▨ represents zones that are poorly mixed: (**a**) side entry and (**b**) axial.

mixing, commercial jet mixers (15,16) are designed for turbulent jet operation at $Re_j > 3000$. The minimum required nozzle discharge velocity, U_o, for heavy into light liquid is given by

$$U_o = \left(\frac{2gFh_s\Delta\rho}{\sin^2(\theta_j + 5)\rho} \right)^{1/2}$$

where factor F depends on the initial condition of tank homogeneity and is calculated by the following:

For initially homogeneous tank

$$F = 0.25(\Delta\rho/\rho)^{-0.34}(H_s/d)^{0.65} \qquad \text{for } H_s/d \leq 100$$
$$F = 7.5(\Delta\rho/\rho)^{-0.24} \qquad\qquad\qquad \text{for } H_s/d > 100$$

For initially stratified tank

$$F = 0.59(H_s/d) - 4\ln(\Delta\rho/\rho) - 32.7 \qquad \text{for} \quad H_s/d > 50$$

For a known flow rate the nozzle diameter is set by the largest size that satisfies the turbulent jet requirement for both heavy and light jets, ie,

$$\frac{dU_o\rho}{\mu} \geq 3000$$

Additionally, for adequate penetration of the jet across the tank, the nozzle diameter should be at least as large as that given by

$$d \geq \frac{H_s}{400\sin\theta_j}$$

The pressure drop through a jet mixing nozzle is given by $\Delta P = 0.9\rho Q^2/d^4$.

For blending of low viscosity liquids to practical homogeneity, turnover of three tank volumes is normally adequate. For viscous liquids or for a higher degree of liquid homogeneity, a greater number of turnovers is needed.

SEP Mixers. These mixers induce a spiral jet flow across the floor of the tank continually entraining liquid from other areas of the tank. This jet stream initially only agitates the denser liquid at the bottom, but gradually penetrates the upper layers of the tank with sufficient velocity to generate both full top-to-bottom flow and break the interface between various density strata to achieve a full homogeneous mix. This gradual breakdown of the interfaces is illustrated in Figure 26 by flow patterns and changes in liquid specific gravity as a function of blending time at different tank heights. The design of the mixer can be based on the discharge capacity of the propeller and liquid entrainment in relation to tank volume and desired blending time.

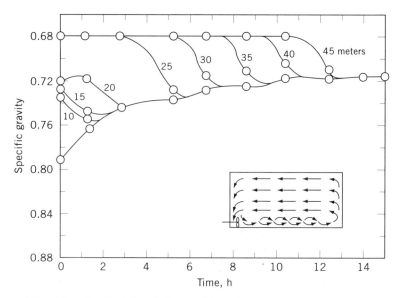

Fig. 26. Gradual breakdown of interface in a stratified tank.

Crude Tank Sludge Control

Depending on the type and concentration of sludge in petroleum crude receipt, settling of sludge can occur in crude storage tanks at a rate of 25–150 mm height per month. Without its suspension, this sludge can harden due to packing of top layers, and at high levels mounds are formed because of sludge shifting during crude receipts and pumpouts. Mixing energy, therefore, must be supplied on a continuous and frequent basis to maintain tank cleanliness. For achieving good on-stream sludge control, two competing technologies are available: SEP mixers and Butterworth P-43 rotating jet machines. Adequately designed and operated SEP mixers can prevent sludge settling by establishing movement of crude throughout the tank. They cannot, however, be expected to resuspend sludge accumulated for an extended period. The P-43 submerged jet nozzle system can prevent sludge accumulation and resuspend large sludge volumes. Selection of the mixing system depends on the nature of existing facilities.

SEP mixers are horizontally installed on the side of the tank and near the bottom. By rotating the propeller, a spiralling jet is produced (Fig. 27) near the tank floor which provides the desired thrust to dislodge and entrain the sediments. SEP mixers should be designed to provide the suspension velocity at the farthest distance on the tank floor.

For effective utilization of mixing energy the SEP shaft must be off-center so that the angle between the shaft and the diameter is 7–12 degrees. With clockwise mixer rotation (looking from motor to propeller), this angle shift should be counterclockwise, resulting in tank circulation without a swirling vortex. The swirling action of liquid consumes less energy but causes sludge to concentrate at the center of the tank. In addition, tank operation is impaired at low liquid levels.

Commercial SEP mixer sizes are limited to 56 kW and therefore more than one mixer is needed for large tanks in order to provide the desired mixing energy.

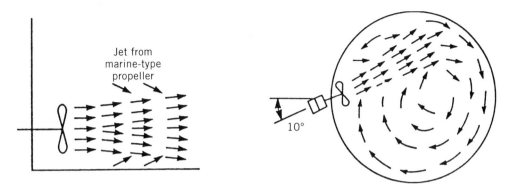

Fig. 27. Flow patterns with side-entry mixers.

The arrangement of these mixers determines the nature of the liquid circulation path. For example, clustered mixers provide a combined jet which spreads in different directions toward the opposite wall (Fig. 28). In the distributed configuration, each mixer creates nearly independent liquid circulation in an arc. With both arrangements, it is recommended that mixers of the same size be used, and the number of mixers should be chosen based on the tank diameter using the following guidelines:

Tank diameter, m	<30	30–45	45–60	>60
Number of mixers	1	2	3	4 or 5

When using the clustered layout of multiple mixers, a mixer spacing of 22.5° is commonly used. It is important to position the clustered mixers opposite the crude outlet in order to benefit from liquid velocities during pump out.

With fixed-angle mixers, the sludge may be shifted to low velocity areas particularly when mixers are underpowered. The mixer performance can then

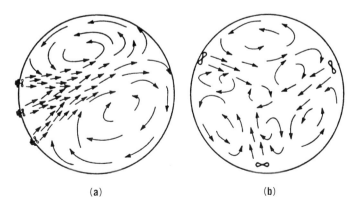

(a)   (b)

Fig. 28. Flow patterns with (**a**) clustered and (**b**) distributed mixers.

be improved by using swivel-angle mixers and operating them sequentially at different angles moving from left to right. For the same mixing quality, the power required by swivel-angle mixers is less than that required by fixed-angle mixers. For common crudes the mixer power requirement is 1.88 W/m^3. Higher mixing energy is required for heavier and waxy crudes.

The Butterworth P-43 system uses the rotating submerged jet concept to remove deposited sludge from crude oil storage tanks. The P-43 machine is a twin-nozzle jet device that directs single or opposed jets horizontally across the tank bottom. The nozzles slowly rotate to provide 360° coverage of the tank floor. The jets are designed to provide a concentrated force on the tank floor needed for suspension of settled sludge. Once suspended, the sludge is homogeneously mixed with oil and processed downstream, thus recovering its hydrocarbon value. The P-43 operation consists of pumping crude oil through the jet nozzles to provide the minimum suspension velocity at the farthest point. A portion of crude flow is directed to an impeller, which through a series of gears rotates the machine. Slow rotation of the jets, 1.5–3.5°/min, provides the necessary residence time for jet penetration through the sludge.

The P-43 installations can be center-mounted (CM) or shell-mounted (SM). The CM configuration uses a single P-43, but there can be different numbers of machines with a SM configuration. A CM machine is placed at the tank center and each jet is designed to suspend sludge at the tank wall with the jet cleaning radius equal to the tank radius. SM machines can be installed near the tank wall and can be single or multiple depending on tank size. A three-machine SM configuration has a cleaning radius equal to the tank radius and can be operated sequentially with one jet at a time. The P-43 machines should be located as close to the tank floor as possible for maximum effectiveness. Tank internals such as heating coils, water drawoff lines, and roof drains should be secured to prevent dislodgement from supports by direct jet impingement.

Inline Motionless Mixers

Inline motionless mixers derive the fluid motion or energy dissipation needed for mixing from the flowing fluid itself. These mixers include orifice mixing columns, mixing valves, and static mixers.

Orifice mixing columns consist of a series of orifice plates contained in a pipe. The pipe normally is fabricated of two vertical legs connected by a U-bend at the bottom with orifice plates installed between flanges in the legs. Typical use is for cocurrent contacting in caustic and water-washing operations. It is generally designed at mixing energy of 50–200 kJ/m^3 (12–48 kcal/m^3) and contact times in the range of 10–50 seconds. Typical pressure drop per orifice is in the range of 5–15 kPa (37.5–112.5 mm Hg).

A mixing valve in the form of a conventional globe valve is simple and economical. A typical service involves caustic washing of gas oil and water–oil mixing upstream of a desalter. The valve is normally specified to handle a pressure drop in the range of 20–350 kPa (0.2–3.5 atm).

Static mixers are used in the chemical industries for plastics and synthetic fibers, eg, continuous polymerization, homogenization of melts, and blending of additives in extruders; food manufacture, eg, oils, juices, beverages, milk, sauces, emulsifications, and heat transfer; cosmetics, eg, shampoos, liquid soaps, cleaning liquids, and creams; petrochemicals, eg, fuels and greases; environmental control, eg, effluent aeration, flue gas/air mixing, and pH control; and paints, etc.

These motionless mixers provide complete transverse uniformity and minimize longitudinal mixing, therefore their performance approaches perfect plug flow conditions. They consist of repeated structures called mixing elements attached inside a pipe. These mixing elements alternately divide and recombine fluids passing through. As a result they create shearing action at the cost of pressure drop which causes mixing of single- and multi-phase systems.

Static mixers are usually classified as operating either in laminar or turbulent flow regimes. There are many proprietary designs marketed, a few are shown in Figure 29. These mixers generate a process of division, rotation, and reversal of fluid which create shear and mixing. Homogeneity is accomplished by producing striations (Fig. 30). The number of striations is equal to 2^n, where n is number of elements. There are other designs which produce a greater number of striations.

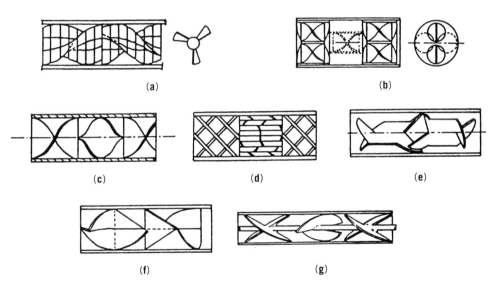

(a)   (b)

(c)   (d)   (e)

(f)   (g)

Fig. 29. Various proprietary static mixer designs: (**a**) Lightnin, (**b**) Toray Hi-Mixer, (**c**) Kenics, (**d**) Koch/Sulzer SMX, (**e**) Komax, (**f**) Etoflo, and (**g**) Ross LLPD.

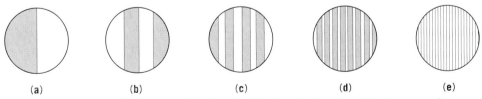

(a)   (b)   (c)   (d)   (e)

Fig. 30. Mechanism for laminar blending in Kenics static mixer (**a**) after one element, (**b**) two elements, (**c**) three elements, (**d**) four elements, and (**e**) five elements, where ▨ represents liquid A and □, liquid B.

The energy consumed by a static mixer is given by $P = Q \Delta P_{sm}$. This energy is supplied by the pump used to create flow of the fluid through the mixer. For homogenization of two or more liquids, static mixers reduce standard deviation or variance. The reduction of variance is a product of shear rate and time, and therefore equal to L/D_s. The pressure drop depends on flow rate and liquid viscosity. For a Kenics mixer, ΔP_{sm} is about six times that of an empty pipe in laminar flow; for a Koch/Sulzer SMX mixer it is 64 times higher.

The degree of mixing is described as variation coefficient $\sigma_s/\overline{X}$ where σ_s is standard deviation, X is fraction of additive, and $\overline{X}$ is average fraction of additive.

$$\sigma_s \text{ of unmixed liquid} = [X(1 - X)]^{0.5}$$

$$\text{initial } \sigma_s/\overline{X} = [(1 - X)/X]^{0.5}$$

If the material starts with an additive composition of 10% ($X = 0.1$) corresponding to an initial variation coefficient of 3, the variance is reduced logarithmically as L/D_s increases. Figure 31 shows a comparison of the degree of mixing produced by various static mixers as a function of L/D_s in laminar flow. These mixers appear to give markedly different performance, but when compared on a ΔP_{sm} basis are almost equivalent. Many manufacturers suggest that satisfactory homogeneity is when $\sigma_s/\overline{X}$ is reduced to 0.05. In many applications lower $\sigma_s/\overline{X}$

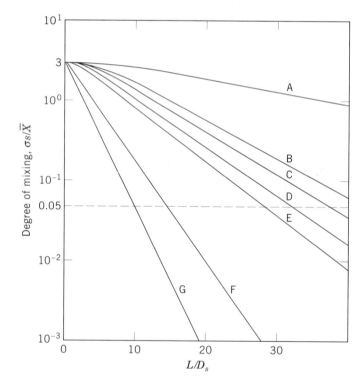

Fig. 31. Comparison of mixing rates with different static mixers: A, Lightnin; B, Komax; C, Etoflo HV; D, Kenics; E, SMXL; F, Hi-Mixer; and G, SMX. Satisfactory homogeneity occurs at $\sigma_s/\overline{X} = 0.05$.

may be required. The exact degree of mixing depends on the product and must be determined by experience and experimentation. The smaller the amount added the larger the initial variation coefficient and the more reduction required to achieve a certain degree of uniformity. When the ratio of viscosity of additive to the main liquid is greater than 10, mixing becomes much harder as the additive slips past the main stream. Mixers with high ΔP_{sm} handle this better than those with low ΔP_{sm}. The Koch/Sulzer SMX has a correlation factor for design based on log(viscosity ratio of the two liquids).

Pressure drop in static mixers depends very strongly on geometric arrangement of the inserts. It is simply defined in relation to the pressure drop ΔP in an empty tube given by Darcy's equation:

$$\Delta P = \frac{5 f \rho L v^2}{g_c D_s} \qquad\qquad \Delta P_{\mathrm{sm}} = K \Delta P$$

K depends on the mixer type and its value can be obtained from the vendor literature.

Static mixing of immiscible liquids can provide excellent enhancement of the interphase area for increasing mass-transfer rate. The drop size distribution is relatively narrow compared to agitated tanks. Three forces are known to influence the formation of drops in a static mixer: shear stress, surface tension, and viscous stress in the dispersed phase. Dimensional analysis shows that the drop size of the dispersed phase is controlled by the Weber number. The average drop size, d_{avg}, in a Kenics mixer is a function of Weber number $We = \rho_c v^2 d_h / \sigma$, and the ratio of dispersed to continuous-phase viscosities (Fig. 32).

Under turbulent flow conditions, the Sauter mean diameter from two static mixers can be obtained from the following:

Koch/Sulzer SMV $\qquad\qquad d_{32}/d_h = 0.21 We^{-0.5} Re^{0.15}$

Kenics $\qquad\qquad\qquad\quad d_{32}/d_h = C We^{-0.6} Re^{0.1} \ \text{ for } Re > 3000$

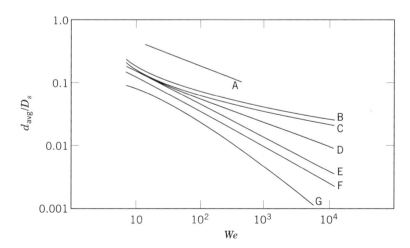

Fig. 32. Dimensionless drop size vs Weber number: A, empty pipe at $\mu_d/\mu_c = 1$; B through G, Kenics mixer at $\mu_d/\mu_c = 25$, 10, 2, 1, 0.75, and 0.5, respectively.

where d_h is hydraulic diameter of the static mixer, Re is Reynolds number $Re = \rho_c v d_h/\mu_c$, and C is a constant that depends on mixer type.

Static mixing of gas–liquid systems can provide good interphase contacting for mass transfer and heat transfer. Specific interfacial area for the SMV (Koch/Sulzer) mixer is related to gas velocity v_g and gas holdup ϕ by the following:

$$a = C_5 v_g^{0.85} \phi^{0.15}$$

The constant C_5 depends on the hydraulic diameter of the static mixer. The mass-transfer coefficient expressed as a Sherwood number $Sh = k_L d_h/\mathcal{D}$ is related to the pipe Reynolds number $Re = D_s v \rho/\mu$ and Schmidt number $Sc = \mu/\rho D_s$ by $Sh = 0.0062 Re^{1.22} Sc^{1/3}$.

The pressure drop for gas–liquid flow is determined by the Lockhart-Martinelli method. It is assumed that the ΔP for two-phase flow is proportional to that of the single phase times a function of the single-phase pressure drop ratio P_r.

$$\Delta P_{sm} = \phi_L^2 \Delta P_L = \phi_G^2 \Delta P_G$$

The functions ϕ_L and ϕ_G are plotted against the pressure drop ratio P_r in Figure 33.

Heat transfer in static mixers is intensified by turbulence causing inserts. For the Kenics mixer, the heat-transfer coefficient h is two to three times greater, whereas for Sulzer mixers it is five times greater, and for polymer applications it is 15 times greater than the coefficient for low viscosity flow in an open pipe. The heat-transfer coefficient is expressed in the form of Nusselt number $Nu = hD_s/k$ as a function of system properties and flow conditions.

$$Nu = Nu_o + C_6 Re^A Pr^B \left(\frac{D_s}{L}\right)^C$$

Values of Nu_o, C_6, A, B, and C can be obtained from respective vendors.

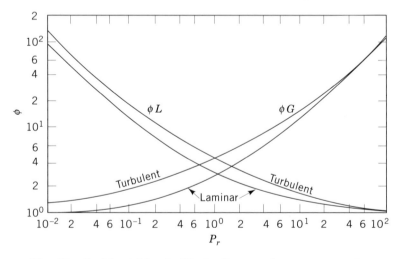

Fig. 33. Lockhart-Martinelli plot for two-phase pressure drop.

Heat Transfer in Agitated Tanks

Frequently the contents of an agitated vessel must be heated or cooled to a given operating temperature by heat transfer to or from the fluid in the vessel. Agitators which force large amounts of fluid to circulate near the heat-transfer surfaces provide the most efficient heat transfer. In the case of close-clearance impellers used for high viscosity liquids, heat transfer is promoted by the thinning of the stagnant fluid layer near the wall. Rubber scrapers can be attached to these impellers to keep the wall surface free from deposits.

Commonly used heat-transfer surfaces are internal coils and external jackets. Coils are particularly suitable for low viscosity liquids in combination with turbine impellers, but are unsuitable with process liquids that foul. Jackets are more effective when using close-clearance impellers for high viscosity fluids. For jacketed vessels, wall baffles should be used with turbines if the fluid viscosity is less than 5 Pa·s (50 P). For vessels equipped with coils, wall baffles should be used if the clear space between turns is at least twice the outside diameter of the coil tubing and the fluid viscosity is less than 1 Pa·s (10 P). Otherwise the baffles should be located inside the coil helix. A conventional jacket consists of a vessel outside the main vessel with a gap for the flow of heat-transfer fluid. Half-pipe jackets are useful for high pressures up to 4 MPa (600 psi). They are better for liquid than for vapor service fluids and can be easily zoned. Dimple jackets are suitable for larger vessels and process conditions up to 2 MPa (300 psi) and 370°C. Internal coils can be either helical or baffle coils (Fig. 34).

The rate of heat-transfer q through the jacket or coil heat-transfer area A is estimated from log mean temperature difference ΔT_m by $q = UA\Delta T_m$. The overall heat-transfer coefficient U depends on thermal conductivity of metal, fouling factors, and heat-transfer coefficients on service and process sides. The

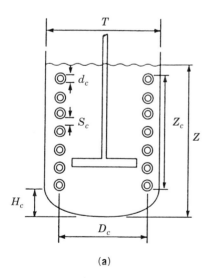
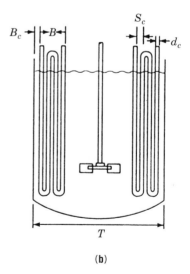

(a)   (b)

Fig. 34. Internal coil configurations for heat-transfer surfaces: (**a**) helical coil where $D_c = 0.02T$, $S_c \geq d_c$, $H_c = 0.15T$, and $Z_c = 0.65Z$; (**b**) baffle coil where $d_c = 0.02T$, $S_c = 0.03T$, $B_c = 0.02T$, and $B = 0.2T$.

process side heat-transfer coefficient depends on the mixing system design (17) and can be calculated from the correlations for turbines in Figure 35a.

For multiple turbines (i in number) the sum of impeller blade widths $\sum W_i$ should be used for W, and the average impeller height $\sum C_i/i$ should be used for C in the equations which include these terms. With turbines having different diameters on the same shaft, a weighted average diameter based on the exponents of D in the appropriate equations should be used.

When concentric banks of helical coils are used, the process side heat-transfer coefficient h for the coil bank closest to the impeller is as given by the foregoing equation. For the second and third banks, the heat-transfer coefficient is 70 and 40% of the calculated value, respectively.

For close-clearance impellers the correlations in Figure 35b apply to installations without scrapers. If scrapers are used, the values given by the equations should be multiplied by a factor of 1.3.

For non-Newtonian fluids the correlations in Figure 35 can be used with generally acceptable accuracy when the process fluid viscosity is replaced by the

External jackets

$$Nu = 1.4 \, Re^{2/3} Pr^{1/3} \left(\frac{\mu}{\mu_w}\right)^{0.14} \left(\frac{D}{T}\right)^{-0.3} \left(\frac{W}{T}\right)^{0.45} n^{0.2} \left(\frac{C}{Z}\right)^{0.2} \left(\frac{H_l}{T}\right)^{-0.6} (\sin \alpha)^{0.5}$$

Internal coils

$$Nu = 2.68 \, Re^{0.56} Pr^{1/3} \left(\frac{\mu}{\mu_w}\right)^{0.14} \left(\frac{D}{T}\right)^{-0.3} \left(\frac{W}{T}\right)^{0.3} n^{0.2} \left(\frac{C}{Z}\right)^{0.15} \left(\frac{H_l}{T}\right)^{-0.5} (\sin \alpha)^{0.5}$$

Baffle coils $$Nu = 0.1 \, Re^{0.65} Pr^{1/3} \left(\frac{\mu}{\mu_w}\right)^{0.14} \left(\frac{D}{T}\right)^{1/3} n^{-0.2}$$

(**a**)

Anchors $$Nu = 1.5 \, Re^{0.5} Pr^{1/3} \left(\frac{\mu}{\mu_w}\right)^{0.14}$$ for $Re < 200$

$$Nu = 0.36 \, Re^{2/3} Pr^{1/3} \left(\frac{\mu}{\mu_w}\right)^{0.14}$$ for $Re > 200$

Helical ribbons $$Nu = 4.2 \, Re^{1/3} Pr^{1/3} \left(\frac{\mu}{\mu_w}\right)^{0.2}$$ for $1 < Re < 1000$

$$Nu = 0.42 \, Re^{2/3} Pr^{1/3} \left(\frac{\mu}{\mu_w}\right)^{0.14}$$ for $Re > 1000$

(**b**)

Fig. 35. Correlations for calculating heat-transfer coefficients for (**a**) turbine external jackets, internal coils, and baffle coils, and (**b**) for close-clearance impellers without scrapers.

apparent viscosity. For non-Newtonian fluids having power law behavior, the apparent viscosity can be obtained from shear rate estimated by $\dot{\gamma} = 10\,N$.

Specially Designed Mixers

There are many specially designed mixers for unique nonconventional systems, eg, those in Figure 36. The air-lift mixer consists of a tank containing a draft tube. Air is injected from the bottom into the draft tube. Because of buoyancy, air bubbles rise inducing a liquid upflow. The liquid then flows down through the annulus. The air bubbles escape through the surface. The draft tube circulator consists of a flat-bottom tank containing a draft tube which is sized in the range of 20–40% of the tank diameter. In the draft tube an axial flow impeller is placed that pumps down the tube producing return flow in the annulus. The mixer emulsifier consists of a high speed rotor and a stator, and generates shearing action and pumping. This device is used in pharmaceutical andcosmetics processing. The vortex mixer uses centrifugal flow to cause homogenization of two or more liquids.

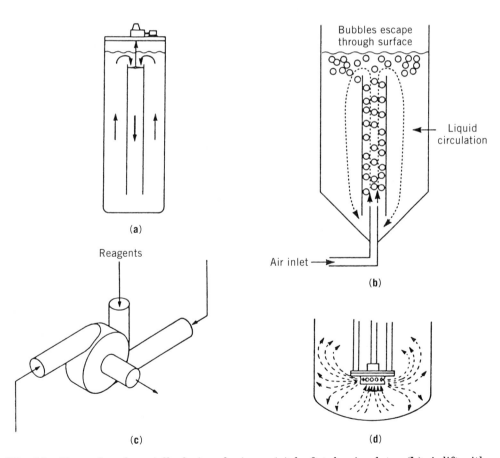

Fig. 36. Examples of specially designed mixers: (**a**) draft tube circulator, (**b**) air lift with draft tube, (**c**) Fluidics vortex mixer, and (**d**) mixer emulsifier.

Mixing of Dry Solids and Pastes

Dry solids are mixed in processes associated with food, pharmaceuticals, fertilizer, tobacco, cement (qv), rubber products, ceramics (qv), soap, and many other industries. Viscous pastes are frequently handled in polymer and petroleum (qv) processes. Mixing equipment used in such systems are uniquely designed to divide and recombine the materials to attain uniformity. Moving agitator components often scrape the walls because the material does not flow easily. The energy requirements may be very high because of the work involved in dividing and shearing the material. Mixing machinery is selected according to its capacity to shear material at low speed and to wipe, smear, fold, stretch, or knead the mass to be handled. Mixers with intermeshing blades are sometimes required to keep the material from clinging unmixed to the lee side of the blade. Wiping of heat-transfer surfaces promotes addition or removal of heat. Some mixing devices break down solids in pastes and thus have the character of mills.

Solids. For mixing of solids the mixers can be categorized according to the mixing mechanism used.

Tumbling. Gentle mixing by a tumbling action causes materials to cascade from the top of the rotating vessel. Common types offer various vessel configurations including drum, container, V (Fig. 37**a**), and cone. Intake and discharge of materials in these mixers take place through an opening in the vessel end. Dry and partly dry powders, granules, and crystalline substances are readily mixed in such equipment. Intensive blending is possible with the addition of a high speed mix bar.

Convection. In these mixers an impeller operates within a static shell and particles are moved from one location to another within the bulk.

Ribbon Type. Spiral or other blade styles transfer materials from one end to the other or from both ends to the center for discharge (Fig. 37**b**). This mixer can be used for dry materials or pastes of heavy consistency. It can be jacketed for heating or cooling. Blades can be smoothly contoured and highly polished when cleanliness is an important process requirement.

Spiral Elevator. Materials are moved upward by the centrally located spiral-type conveyor in a cylindrical or cone-shaped Nautamix vessel (Fig. 37**c** and **d**). Blending occurs by the downward movement at the outer walls of the vessel. The vessel serves the dual purposes of blending and storage. In these mixers the screw impeller actively agitates only a small portion of the mixture

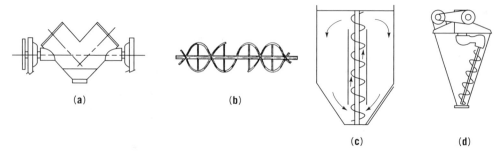

| (a) | (b) | (c) | (d) |

Fig. 37. Various mixer types for dry solids: (**a**) the V-mixer, (**b**) ribbon blender, (**c**) vertical screw mixer, and (**d**) Nautamix.

and natural circulation is used to ensure all the mixture passes through the impeller zone. In the case of Nautamix, an Archimedian screw lifts powder from the base of a conical hopper while progressing around the hopper wall.

Paddle Type. This type is similar to the ribbon type except that interrupted flight blades or paddles transfer materials from one end to the other, or from both ends to the center for discharge. The paddle-type mixer can be used for dry materials or pastes of heavy consistency. It can be jacketed for heating or cooling.

Planetary Type. Paddles or whips of various configurations are mounted in an off-center head that moves around the central axis of a bowl or vessel. Material is mixed locally and moved inward from the bowl side, causing intermixing. This mixer handles dry materials or pastes.

Pan Type. The mulling action of this mixer is similar to the action of a mortar and pestle. Scrapers move the materials from the center and side of a pan into the path of rotating wheels where mixing takes place. The pan may be of the fixed or rotating type. Discharge is through an opening in the pan. The flow type uses rotating plows in a rotating pan to locally mix and intermix by the rotation of plows and pan, respectively.

Fluidization. Particles suspended in a gas stream behave like a liquid. They can be mixed by turbulent motion in a fluidized bed. This mixer is used for mixing and drying, or mixing and reaction.

Pastes. For blending of viscous pastes, mixers are classified as batch or continuous. Most convection-type mixers for dry solids are also used for thick pastes.

Batch Mixers. These mixers are preferred when batch identity must be maintained, eg, in pharmaceutical products; when frequent product changes occur and off-spec products must be minimized, eg, in dye and pigment manufacture; and when a multitude of ingredients is required with accurate additions, eg, in adhesives (qv) and caulking compounds. Also, the batch mode is best when various changes of state are involved, eg, a reaction followed by the pulling of a vacuum to drive off volatiles or when very long mixing and reacting times are required.

Change-Can Mixers. In change-can mixers one or more blades cover all regions of the can either by a planetary motion of the blades or a rotation of the can (Fig. 38**a**). The blades may be lowered into the can or the can may be raised to the mixing head. Separate cans allow the ingredients to be measured carefully before the mixing operation begins, and can be used to transport the finished batch to the next operation while the next batch is being mixed.

Stationary Tank Mixers. These are recommended when the particular advantages of the change-can mixer are not required. The agitator may be particular to a specific industry, eg, the soap crutcher, or for general use as single- or double-rake mixers (Fig. 38**b** and **c**). In the latter, some part of the agitator moves in close proximity to the vessel walls or stationary bar baffles. The impeller may also consist of a single- or double-helical blade to promote top-to-bottom turnover, minimizing the amount of hardware that must be moved through the viscous mass.

Double-Arm Kneading Mixers. The material in these mixers (Fig. 38**d**) is carried by two counter-rotating blades over the saddle section of a W-shaped

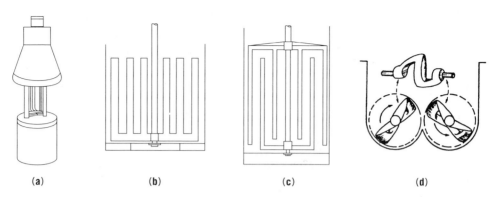

Fig. 38. Various mixer types for viscous pastes: (**a**) the change-can mixer, (**b**) rake mixer, (**c**) double-rake mixer, and (**d**) sigma-blade kneader.

trough. Randomness is introduced by the differences in blade speed and end-to-end mixing by differences in the length of the arms on the sigma-shaped blades. Other blade shapes are used for specific end purposes, such as smearing or cutting edges on the blade faces. Discharge is usually by tilting the trough or by a door in the bottom of the trough. Double-arm kneading mixers are also available with a centrally located screw to discharge the contents.

Intensive Mixers. Intensive mixers such as the Banbury (Fig. 39**a**) are similar in principle to the double-arm kneading mixers, but are capable of much higher torques. Used extensively in the rubber and plastics industries, the Banbury mixer is operated with a ram cover so that the charge can be forced into the relatively small-volume mixing zone. The largest of these mixers holds only 500 kg but is equipped with a 2000 kW motor.

Roll Mills. When dispersion is required in exceedingly viscous materials, the large surface area and small mixing volume of roll mills allow maximum shear to be maintained as the thin layer of material passing through the nip is continuously cooled. The rolls rotate at different speeds and temperatures to generate the shear force with preferential adhesion to the warmer roll.

Continuous Mixers. In most continuous mixers one or more screw or paddle rotors operate in an open or closed trough. Discharge may be restricted at the end of the trough to control holdup and degree of mixing. Some ingredients may be added stagewise along the trough or barrel. The rotors may be cored to provide additional heat-transfer area. The rotors may have interrupted flights to permit interaction with pins or baffles protruding inward from the barrel wall.

Single-Screw Extruders. These incorporate ingredients such as antioxidants (qv), stabilizers, pigments, and other fillers into plastics and elastomers (Fig. 39**b**). In order to provide a uniform distribution of these additives, the polymer is melted primarily by the work energy imparted in the extruder, rather than by heat transfer through the barrel wall. In addition to melting the polymer, the extruder is used as a melt pump to generate pressure for extrusion through a die, shaping the molten product to a specific profile, or into strands or pellets. The extruder screw drags the polymer through the barrel, generating shear between screw and barrel. In addition, some axial mixing occurs in response to a back flow along the screw channel caused by the pressure required

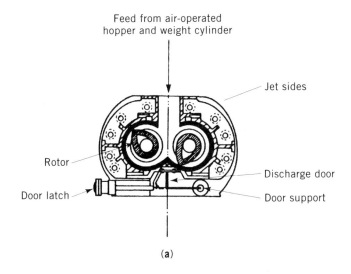

(a)

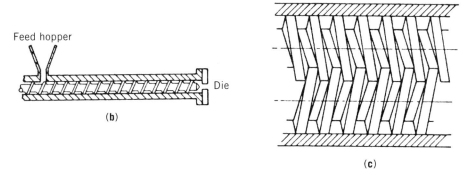

Fig. 39. Various mixer types for viscous pastes: (**a**) Banbury mixer, (**b**) single-screw extruder, and (**c**) co-rotating twin-screw extruder.

to get through the die. Mixing in a single-screw extruder can be enhanced by interrupting the flow pattern within the screw flight channel. This can be done by variations in the screw flight channel width or depth, by causing an interchange with spiral grooves in the barrel, or by interengaging teeth.

   Twin-Screw Mixers.   These continuous mixers provide more radial mixing by interchange of material between the screws, rather than just acting as screw conveyors. Intermeshing co-rotating twin-screw mixers (Fig. 39**c**) have the additional advantage that the two rotors wipe each other as well as the barrel wall. This action eliminates any possibility of dead zones or unmixed regions. In addition to variations in screw helix angles, these mixers can be fitted with kneading paddles that interactively cause a series of compressions and expansions to increase the intensity of mixing. Such mixers are used for a variety of pastes and doughs, as well as in plastics compounding.

BIBLIOGRAPHY

"Mixing and Agitating" in *ECT* 1st ed., Vol. 9, pp. 133–166, by J. H. Rushton and C. W. Selheimer, Illinois Institute of Technology, and R. D. Boutros, Mixing Equipment Co., Inc.; "Mixing and Blending" in *ECT* 2nd ed., Vol. 13, pp. 577–613, by J. H. Rushton, Purdue University, and R. D. Boutros, Mixing Equipment Co., Inc.; in *ECT* 3rd ed., Vol. 15, pp. 604–637, by J. Y. Oldshue, Mixing Equipment Co., Inc., and D. B. Todd, Baker Perkins, Inc.

1. R. R. Hemrajani, *Chem. Proc.* **50**, 22 (July 1987).
2. D. S. Dickey and J. G. Fenic, *Chem. Eng.* **83**(1), 139 (1976).
3. D. S. Dickey and R. R. Hemrajani, *Chem. Eng.* **99**(3), 82 (1992).
4. R. W. Hicks, J. R. Morton, and J. G. Fenic, *Chem. Eng.* **83**(9), 102 (1976).
5. Th. N. Zwietering, *CES* **8**, 244 (1958).
6. P. A. Shamlou and E. Koutsakos, *CES* **44**(3), 529 (1989).
7. R. R. Hemrajani, D. L. Smith, R. M. Koros, and B. L. Tarmy, *6th European Conference on Mixing*, Pavia, Italy, May 24, 1988.
8. D. E. Leng and G. J. Quarderer, *Chem. Eng. Commun.* **14**, 177 (1982).
9. H. J. Karam and J. C. Bellinger, *I&EC Fund.* **1**, 576 (1968).
10. P. V. Danckwerts, *Trans. Faraday Soc.* **46**, 300 (1950).
11. J. B. Fernandes and M. M. Sharma, *CES* **22**, 1267 (1967).
12. M. M. C. G. Warmoeskerken and J. M. Smith, *CES* **40**(11), 2063 (1985).
13. A. W. Nienow, D. J. Wisdom, and J. C. Middleton, *2nd European Conference on Mixing*, Cambridge, U.K., Mar. 30–Apr. 1, 1977.
14. G. Q. Martin, *Ind. Eng. Chem. Proc. Des. Dev.* **11**(3), 397 (1972).
15. H. Fossett, *Trans. Inst. Chem. Engrs.* **29**, 322 (1951).
16. E. A. Fox and V. E. Gex, *AIChE J.* **2**(4), 539 (1956).
17. M. F. Edwards and W. L. Wilkinson, *Heat Transfer to Newtonian and Non-Newtonian Fluids in Agitated Vessels*, HTFS-DR 27, Harwell, Berkshire, U.K., 1972.
18. A. H. P. Skelland and G. G. Ramsay, *Ind. Eng. Chem.* **26**(1), 77 (1987).

General References

Handbook of Mixing Technology, Ekato Ruhr-und Mischtechnik GmbH, Schopfheim, Germany, 1991.
N. Harnby, M. F. Edwards, and A. W. Nienow, *Mixing in the Process Industries*, Butterworths, London, 1985.
N. Harnby, *Fluid Mixing III*, The Institution of Chemical Engineers, Symposium, Series No. 108, Hemisphere Publishers, New York, 1988.
F. A. Holland and F. S. Chapman, *Liquid Mixing and Processing in Stirred Tanks*, Van Nostrand Reinhold, New York, 1966.
S. Nagata, *Mixing Principles and Applications*, John Wiley & Sons, Inc., New York, 1975.
J. Y. Oldshue, *Fluid Mixing Technology*, McGraw-Hill, Inc., New York, 1983.
Z. Sterbacek and P. Tausk, trans. K. Mayer, *Mixing in the Chemical Industry*, Pergamon Press, New York, 1965.
G. B. Tatterson, *Fluid Mixing and Gas Dispersion in Agitated Tanks*, McGraw-Hill, Inc., New York, 1991.
V. W. Uhl and J. B. Gray, *Mixing-Theory and Practice*, Vols. I & II, Academic Press, Inc., 1966–1967; V. W. Uhl and J. A. Von Essen, Vol. III, 1986.
J. J. Ulbrecht and G. K. Patterson, *Mixing of Liquids by Mechanical Agitation*, Gordon & Breach Science Publishers, New York, 1985.

General concepts

R. R. Corpstein, R. A. Dove, and D. S. Dickey, *CEP* **75**(2), 66 (1979).
D. S. Dickey, *CEP* **87**(12), 22 (1991).
J. B. Joshi, A. B. Pandit, and M. M. Sharma, *CES* **37**(6), 813 (1982).
Y. Kawase and M. Moo-Young, *Chem. Eng. J.* **43**, B19 (1990).
D. E. Leng, *CEP* **87**(6), 23 (1991).
K. W. Norwood and A. B. Metzner, *AIChE J.* **6**(3), 432 (1960).
J. H. Rushton, E. W. Costich, and H. J. Everett, *CEP* **46**(8), 395 (1950).
G. B. Tatterson, R. S. Brodkey, and R. V. Calabrese, *CEP* **87**(6), 45 (1991).
V. W. Uhl, *Chem. Proc.* **47**, 26 (1984).

Liquid blending

J. B. Fasano and W. R. Penney, *CEP* **87**(12), 46 (1991).
J. B. Fasano and W. R. Penney, *CEP* **87**(10), 56 (1991).
M. C. Jo, W. R. Penney, and J. B. Fasano, *1993 AIChE Annual Meeting*, St. Louis, Mo.
S. J. Khang and O. Levenspiel, *Chem. Eng.* **83**(10), 141 (1976).
H. Kramers, G. M. Baars, and W. H. Knoll, *CES* **2**, 35 (1953).
J. Y. Oldshue, H. E. Hirschland, and A. T. Gretton, *CEP* **52**(11), 481 (1956).

Solid–liquid

M. Bohnet and G. Niesmak, *Ger. Chem. Eng.* **3**, 57 (1980).
L. E. Gates, J. R. Morton, and P. L. Fondy, *Chem. Eng.* **83**(11), 144 (1976).
T. Hobler and J. Zablocki, *Chem. Tech. (Leipzig)* **18**, 650 (1966).
A. W. Nienow, *CES* **23**, 1453 (1968).

Liquid–liquid

M. M. Clark, *CES* **43**(3), 671 (1988).
J. C. Godfrey, F. I. N. Obi, and R. N. Reeve, *CEP* **85**(12), 61 (1989).
R. R. Hemrajani, *5th European Conference on Mixing*, Wurzburg, June 10–12, 1985.
W. J. McManamey, *CES* **34**, 432 (1979).
R. Shinnar and J. M. Church, *Ind. Eng. Chem.* **53**, 479 (1961).
V. G. Trice and W. A. Rodger, *AIChE J.* **2**(2), 205 (1956).
J. W. Van Heuven and J. C. Hoevenaar, *Proceedings of 4th European Symposium on Chemical Reaction Engineering*, Brussels, 1968.
T. Vermeulen, G. M. Williams, and G. E. Langlois, *CEP* **51**(2), 85-F (1955).

Gas–liquid

R. W. Hicks and L. E. Gates, *Chem. Eng.* **83**(7), 141 (1976).
V. B. Rewatkar and J. B. Joshi, *Can. J. Chem. Eng.* **71**, 278 (1993).
C. D. Rielly, G. M. Evans, J. F. Davidson, and K. J. Carpenter, *CES* **47**, 3395 (1992).

Jet mixing

J. H. Rushton and J. Y. Oldshue, *CEP* **49**(4), 161 (1953).

RAMESH R. HEMRAJANI
Exxon Research and Engineering Company

MOLASSES. See SYRUPS.

MOLD-RELEASE AGENTS. See RELEASE AGENTS.

MOLECULAR BEAM EPITAXY. See THIN FILMS, FILM FORMATION
TECHNIQUES.

MOLECULAR MODELING. See SUPPLEMENT.

MOLECULAR SIEVES

In the broadest sense, any material that can exclude pmolecular species by size
can be considered a molecular sieve. However, in this article the term molecular
sieve is restricted to inorganic materials that possess uniform pores with diameters in either the micro- (< 2 nm) or meso- (2–20 nm) size range. The most
technologically important molecular sieves are zeolites, ie, crystalline silicate or
aluminosilicate framework structures with channels of diameters < 1.2 nm (1).
Several of these topologies, with boron, gallium, or iron replacing aluminum, or
germanium replacing silicon, have also been prepared. The chemical composition of microporous framework structures has been expanded considerably with
the substitution of phosphorus for silicon, and new families of aluminophosphate
(2) and silicoaluminophosphate (3) structures have been synthesized in the laboratory. Some of these new frameworks have zeolite analogues, whereas others
are unique. The addition of elements such as Mg, Ti, Mn, Co, Fe, or Zn into
these structures has made it possible to generate metalloaluminophosphates (4),
metallosilicoaluminophosphates (5), etc. Microporous sulfide-based framework
structures are also possible (6).

Considerable synthesis effort has been devoted to developing frameworks
with pore diameters within the mesoporous range; the largest synthesized are
the phosphate-based AlPO-8 (14-membered ring) (7), VPI-5 (18-MR) (8), and
cloverite (20-MR) (9), which have pore diameters within the 0.8–1.3 nm range.
Cacoxenite, a natural ferroaluminophosphate, has been structurally characterized as having ~1.4-nm channels that approach the mesoporous size range (10).
A new family of mesoporous molecular sieves designated M41S has been discovered (11). Although not framework structures like zeolites, silicate and aluminosilicate M41S materials possess very uniform mesopores. Other chemical
compositions are possible (12).

The technological applications of molecular sieves are as varied as their
chemical makeup. Heterogeneous catalysis and adsorption processes make extensive use of molecular sieves. The utility of the latter materials lies in their
microstructures, which allow access to large internal surfaces, and cavities that
enhance catalytic activity and adsorptive capacity.

Zeolites

Molecular-sieve zeolites of the most important aluminosilicate variety can be represented by the chemical formula $M_{2/n}O \cdot Al_2O_3 \cdot ySiO_2 \cdot wH_2O$, where y is 2 or greater, M is the charge balancing cation, such as sodium, potassium, magnesium, and calcium, n is the cation valence, and w represents the moles of water contained in the zeolitic voids. The zeolite framework is made up of SiO_4 tetrahedra linked together by sharing of oxygen ions. Substitution of Al for Si generates a charge imbalance, necessitating the inclusion of a cation. The structures contain channels or interconnected voids that are occupied by the cations and water molecules. The water may be removed reversibly, generally by the application of heat, which leaves intact the crystalline host structure permeated with micropores that may account for $> 50\%$ of the microcrystal's volume. In some zeolites, dehydration may produce some perturbation of the structure, such as cation movement, and some degree of framework distortion.

Zeolites were first recognized as a new type of mineral in 1756 by the Swedish mineralogist A. F. Cronstedt (13). The word "zeolite" was derived from two Greek words, $\zeta\epsilon\iota\nu$ and $\lambda\iota\theta o\sigma$, meaning to boil and a stone. More than 200 synthetic zeolite types and 50 natural zeolites are known. The nomenclature of zeolite minerals follows established procedures. Both synthetic and natural zeolites of different topologies are given three-letter codes by the International Zeolite Association.

Zeolite minerals are formed over much of the earth's surface, including the sea bottom (14). However, with a few notable exceptions, they are exceedingly rare. Until the early 1960s, zeolite minerals were thought to occur mainly in cavities of basaltic and volcanic rocks. Since then, several zeolite minerals formed by the natural alteration of volcanic ash in alkaline environments over long periods of time, eg, in Cenozoic lakes, such as the alkaline lake beds of the western United States, have been identified through the use of x-ray diffraction techniques. The more common topological types that have been identified there include analcime (ANA), clinoptilolite (HEU), mordenite (MOR), chabazite (CHA), erionite (ERI), and phillipsite (PHI) (Table 1). Owing to the hydrolysis of the alkaline constituents of the volcanic ash, the water in these lakes became salty and alkaline (pH up to 9.5), which resulted in the crystallization of zeolites from this deposit. The resulting zeolites were produced as readily accessible flat-lying beds. Of the ~ 50 known zeolite minerals, chabazite, erionite, mordenite, and clinoptilolite occur in sufficient quantity and purity to allow their use as commercial products.

A less ancient zeolite formation process occurred through the percolation of surface water through appropriate sediments. The city of Naples, Italy, is underlain by a zeolitic deposit some 200 km^2 (77 mi^2) in area which is only 5–10 thousand years old. Two large-pore zeolites recently discovered in Oregon, boggsite (BOG) and tschernichite, the latter a natural version of zeolite beta, were formed by percolation of rainwater through basaltic rock (15).

In general, high grade zeolite ore is mined, then processed by crushing, drying, powdering, and screening. Depending on its intended use, it may then be chemically modified, eg, by ion exchange (qv), acid extraction, etc, and calcined at 350–550°C.

Table 1. Zeolite Compositions

Zeolite	CAS Registry Number	Typical formula
	Natural	
chabazite	[12251-32-0]	$Ca_2[(AlO_2)_4(SiO_2)_8]\cdot 13H_2O$
mordenite	[12173-98-7]	$Na_8[(AlO_2)_8(SiO_2)_{40}]\cdot 24H_2O$
erionite	[12150-42-8]	$(Ca, Mg, Na_2, K_2)_{4.5}[(AlO_2)_9(SiO_2)_{27}]\cdot 27H_2O$
faujasite	[12173-28-3]	$(Ca, Mg, Na_2, K_2)_{29.5}[(AlO_2)_{59}(SiO_2)_{133}]\cdot 235H_2O$
clinoptilolite	[12321-85-6]	$Na_6[(AlO_2)_6(SiO_2)_{30}]\cdot 24H_2O$
phillipsite		$(0.5Ca, Na, K)_3[(AlO_2)_3(SiO_2)_5]\cdot 6H_2O$
	Synthetic	
zeolite A		$Na_{12}[(AlO_2)_{12}(SiO_2)_{12}]\cdot 27H_2O$
zeolite X	[68989-23-1]	$Na_{86}[(AlO_2)_{86}(SiO_2)_{106}]\cdot 264H_2O$
zeolite Y		$Na_{56}[(AlO_2)_{56}(SiO_2)_{136}]\cdot 250H_2O$
zeolite L		$K_9[(AlO_2)_9(SiO_2)_{27}]\cdot 22H_2O$
zeolite omega		$Na_{6.8}TMA_{1.6}[(AlO_2)_8(SiO_2)_{28}]\cdot 21H_2O^a$
ZSM-5	[58339-99-4]	$(Na, TPA)_3[(AlO_2)_3(SiO_2)_{93}]\cdot 16H_2O^b$

[a]TMA = tetramethylammonium.
[b]TPA = tetrapropylammonium.

Structure

Of the approximately 120 known framework aluminosilicates, ~ 50 occur naturally, the rest being synthetic. There are 56 structural types known. Understanding the complexities of zeolite structures is made easier by recognizing three important structural keys: the basic arrangement of the individual structural units in space, which defines the framework topology; the location of the charge-balancing cations; and the channel-filling material, such as water or an organic template, which is incorporated as the zeolite is formed. After the channel-filling material is removed, the void space can be used for the adsorption of gases, liquids, salts, elements, metal complexes, etc. In turn, this void-filling property makes zeolites commercially useful in ion exchange, catalysis, etc. The current concepts of zeolite structure were developed by Pauling in 1930 (16). Modern characterization tools such as x-ray, electron and neutron diffraction, infrared and Raman spectroscopies, nuclear magnetic resonance, and electron microscopy have provided a very detailed description of many structures. Computer-based structure determination techniques have been developed, and computer-aided tools, such as ftir, are being used for characterization of zeolites.

There are two types of structures: one provides an internal pore system comprising interconnected cage-like voids; the second provides a system of uniform channels which, in some instances, are one-dimensional and in others intersect with similar channels to produce two- or three-dimensional channel systems. The preferred type has two- or three-dimensional channel systems to provide rapid intracrystalline diffusion in adsorption and catalytic applications.

In most zeolite structures, the primary structural units, tetrahedra, are assembled into secondary building units, which may be simple polyhedra such as cubes, hexagonal prisms, or truncated octahedra. The final framework structure consists of assemblages of the secondary units. These assemblages of secondary units are a convenient means of grouping frameworks by similar units, but should not necessarily be used as a model for zeolite nucleation processes. Models of the structures are often constructed of skeletal tetrahedra. Space-filling models demonstrating the preponderance of oxygen ions are more realistic, but more difficult to construct (Fig. 1).

Zeolite Minerals. Crystal structures of zeolite minerals are illustrated by the zeolites chabazite and mordenite. The structure of chabazite is hexagonal and the framework consists of double six-membered rings of $(Si,Al)O_4$ tetrahedra arranged in parallel layers in an AABBCC sequence. These tetrahedra are cross-linked by four-membered rings, as shown in Figure 2, resulting in cavities, 0.67×1.0 nm, each of which is entered by elliptical apertures

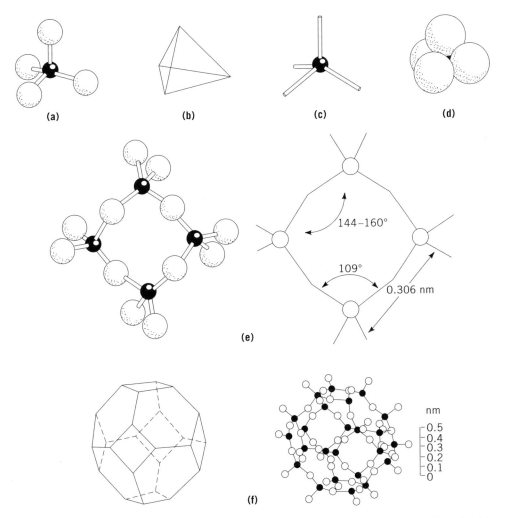

Fig. 1. Methods for representing SiO_4 and AlO_4 tetrahedra by means of (**a**) ball-and-stick model, (**b**) solid tetrahedron, (**c**) skeletal tetrahedron, and (**d**) space-filling of packed spheres (1). (**e**) Linking of four tetrahedra in a four-membered ring. (**f**) Secondary building unit called truncated octahedron as represented by a solid model, left, and a ball-and-stick model, right.

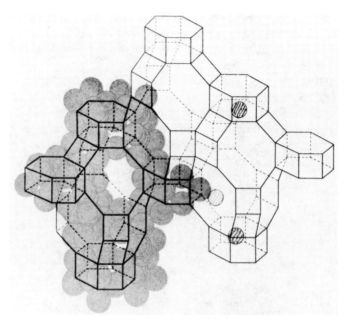

Fig. 2.   Structure of the mineral zeolite chabazite is depicted by packing model, left, and skeletal model, right. The silicon and aluminum atoms lie at the corners of the framework depicted by solid lines. In this figure, and Figure 1, the solid lines do not depict chemical bonds. Oxygen atoms lie near the midpoint of the lines connecting framework corners. Cation sites are shown in three different locations referred to as sites I, II, and III. Courtesy of *Scientific American*.

0.44×0.31 nm. Exchangeable metal ions, such as calcium, occupy positions within or near the double six-membered rings. The mineral mordenite is more complex, as illustrated in Figure 3, and provides for a one-dimensional channel of about 0.6×0.7 nm. The framework itself is built from chains of five-membered rings which are cross-linked. The mineral does not exhibit adsorption properties commensurate with the channel size, apparently because of occluded material.

Several properties of zeolite minerals were studied, including adsorption and ion exchange. This led to the preparation of amorphous aluminosilicate ion exchangers for use in water softening. Studies of the gas-adsorption properties of dehydrated natural zeolite crystals in the 1920s led to the discovery of their molecular-sieve behavior (17). As microporous solids with uniform pore sizes that range from 0.3 to 0.8 nm, these materials can selectively adsorb or reject molecules based on their molecular size. This effect, with obvious commercial potential leading to novel processes for separation of materials, inspired attempts to duplicate the natural materials by synthesis.

Synthetic Zeolites.   Many new crystalline zeolites have been synthesized and several fulfill important functions in the chemical and petroleum industries and in consumer products such as detergents. The structural formula of a zeolite is based on the crystal unit cell, the smallest unit of structure, represented by $M_{x/n}[(AlO_2)_x(SiO_2)_y] \cdot w H_2O$, where n is the valence of cation M, w is the number of water molecules per unit cell, x and y are, respectively, the number of AlO_4 and SiO_4 tetrahedra per unit cell, and y/x usually has values of 1–5. Examples

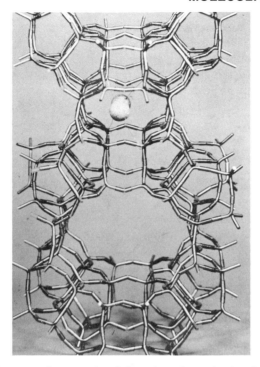

Fig. 3. Model of the crystal structure of the mineral mordenite showing the main channel formed by 12-membered ring and small channels which contain some of the sodium cations. Synthetic types of mordenite exhibit the adsorption behavior of a 12-membered ring, whereas the mineral does not, probably owing to channel blocking.

of important synthetic zeolites are shown in Table 1. These include zeolites A, X, Y, and Zeolon, a synthetic form of mordenite. Subsequently synthesized high silica zeolites include ZSM-5 (18) and ZSM-11, in which y/x is $10->5000$ and molecular sieves consisting essentially of silica have been prepared; the first one of this type, a high silica form of ZSM-5, was named silicalite (19).

The secondary structure unit in zeolites A, X, and Y is the truncated octahedron shown in Figure 1f. These polyhedral units are linked in three-dimensional space through the four- or six-membered rings. The former linkage produces the zeolite A structure, and the latter the topology of zeolites X and Y and of the mineral faujasite. In zeolite A, the internal cavity is 1.1 nm in diameter and is entered by six circular apertures 0.42 nm in size. The interlinked cavities form three-dimensional, unduloid channels with a 0.42-nm minimum diameter. The sodium ions are located in the six-membered rings and near the eight-membered rings (20). The faujasite-type structure consists of a tetrahedral arrangement of the truncated octahedra by joining of the six-membered rings. The resulting cages are 1.3 nm in size, and each is entered by a 12-membered ring, 0.74 nm in diameter. This is the largest known pore size in zeolites. Figure 4 illustrates several types of cation positions; some lie within the smaller polyhedra and some are exposed on the internal surface (1). A slightly different stacking of the truncated cubooctahedrahexagonal prisms combination results in the hexagonal structure of EMC-3 (EMT) having one straight channel formed

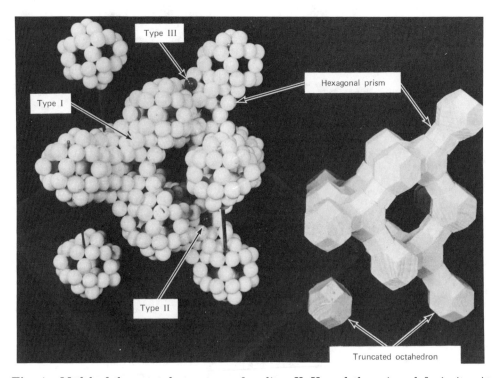

Fig. 4. Model of the crystal structure of zeolites X, Y, and the mineral faujasite. At the right is shown the tetrahedral arrangement of truncated octahedra surrounding one large cavity. On the left the packing model of zeolite X is shown, containing three types of Na^+ cations.

by 12-membered rings of 0.71-nm crystallographic diameter. These channels are connected in the a-direction by 12-membered rings of 0.74×0.65 nm in size. The similarity of these two structures permits intergrowth of the different unit cells within the same crystal, eg, ZSM-20. The structure of ZSM-5 (Fig. 5) contains a high concentration of five-membered rings and provides two sets of intersecting channels with pore sizes of 0.56×0.53 and 0.55×0.51 nm (19).

Structure Modification. Several types of structural defects or variants can occur which figure in adsorption and catalysis: (1) surface defects due to termination of the crystal surface and hydrolysis of surface cations; (2) structural defects due to imperfect stacking of the secondary units, which may result in blocked channels; (3) ionic species, eg, OH^-, AlO_2^-, Na^+, SiO_4^{4-}, may be left stranded in the structure during synthesis; (4) the cation form, acting as the salt of a weak acid, hydrolyzes in aqueous suspension to produce free hydroxide and cations in solution; and (5) hydroxyl groups in place of metal cations may be introduced by ammonium ion exchange, followed by thermal deammoniation.

Properties

Adsorption. Although several types of microporous solids are used as adsorbents for the separation of vapor or liquid mixtures, the distribution of

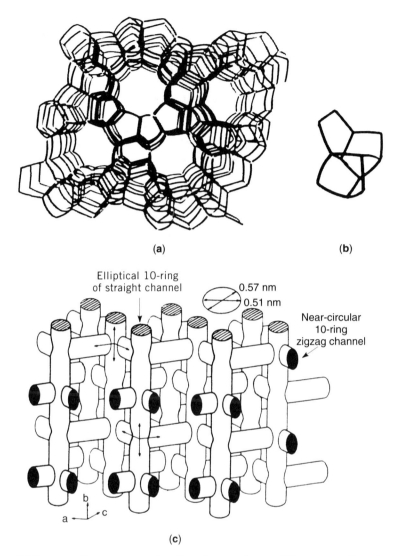

(a)   (b)

Elliptical 10-ring
of straight channel

0.57 nm
0.51 nm

Near-circular
10-ring
zigzag channel

b
a   c

(c)

Fig. 5. (a) Framework structure showing the topology of the molecular sieve Z5M-5 (silicalite) viewed in the direction of the main channel. (b) The 12-tetrahedra secondary building unit. (c) Idealized channel system in ZSM-5 (19). Courtesy of *Nature*.

pore diameters does not enable separations based on the molecular-sieve effect. The most important molecular-sieve effects are shown by crystalline zeolites, which selectively adsorb or reject molecules based on differences in molecular size, shape, and other properties such as polarity. The sieve effect may be total or partial.

Activated diffusion of the adsorbate is of interest in many cases. As the size of the diffusing molecule approaches that of the zeolite channels, the interaction energy becomes increasingly important. If the aperture is small relative to the molecular size, then the repulsive interaction is dominant and the diffusing species needs a specific activation energy to pass through the aperture. Similar shape-selective effects are shown in both catalysis and ion exchange, two important applications of these materials (21).

During the adsorption or occlusion of various molecules, the micropores fill and empty reversibly. Adsorption in zeolites is a matter of pore filling, and the usual surface area concepts are not applicable. The pore volume of a dehydrated zeolite and other microporous solids which have type I isotherms may be related by the Gurvitch rule, ie, the quantity of material adsorbed is assumed to fill the micropores as a liquid having its normal density. The total pore volume V_p is given by

$$V_p = x_s/d_a$$

where d_a = the density of the liquid adsorbate in g/cm^3, x_s = amount adsorbed at saturation in g/g, and V_p in cm^3/g.

The channels in zeolites are only a few molecular diameters in size, and overlapping potential fields from opposite walls result in a flat adsorption isotherm, which is characterized by a long horizontal section as the relative pressure approaches unity (Fig. 6). The adsorption isotherms do not exhibit hysteresis as do those in many other microporous adsorbents. Adsorption and desorption are reversible, and the contour of the desorption isotherm follows that of adsorption.

In order to utilize the absorption properties of the synthetic zeolite crystals in processes, the commercial materials are prepared as pelleted aggregates combining a high percentage of the crystalline zeolite with an inert binder.

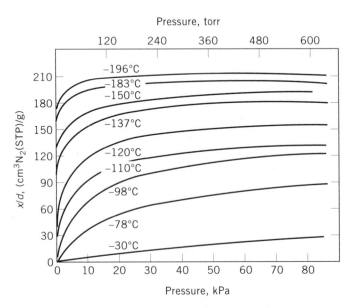

Fig. 6. Family of adsorption isotherms for adsorption of nitrogen on zeolite X at temperatures of −30 to 196°C (1).

The formation of these aggregates introduces macropores in the pellet which may result in some capillary condensation at high adsorbate concentrations. In commercial materials, the macropores contribute diffusion paths. However, the main part of the adsorption capacity is contained in the voids within the crystals.

Zeolites are high capacity, selective adsorbents capable of separating molecules based on the size and shape of the molecule relative to the size and geometry of the main apertures of the structure. They adsorb molecules, in particular those with a permanent dipole moment which show other interaction effects, with a selectivity that is not found in other solid adsorbents. Separation may be based on the molecular-sieve effect or may involve the preferential or selective adsorption of one molecular species over another. These separations are governed by several factors. The basic framework structure, or topology, of the zeolite determines the pore size and the void volume. The exchange cations, in terms of their specific location in the structure, their population density, their charge and size, affect the molecular-sieve behavior and adsorption selectivity of the zeolite. By changing the cation types and number, the selectivity of the zeolite in a given separation can be tailored or modified, within certain limits.

The cations, depending on their locations, contribute electric field effects that interact with the adsorbate molecules. The effect of the temperature of the adsorbent is pronounced in cases involving activated diffusion. Sieving by dehydrated zeolite crystals is based on the size and shape differences between the crystal apertures and the adsorbate molecule. The aperture size and shape in a zeolite may change during dehydration and adsorption because of framework distortion or cation movement. In some instances, the aperture is circular, such as in zeolite A. In others, it may take the form of an ellipse such as in dehydrated chabazite. In this case, subtle differences in the adsorption of various molecules result from a shape factor. Some typical molecular dimensions are shown in Figure 7, based on the Lennard-Jones (20–26) potential function (1).

As shown in Figure 7, the calcium-exchanged form of zeolite A has a pore diameter of 0.42 nm, which compares well with the value of 0.43 nm for the kinetic diameter of normal paraffin hydrocarbons and 0.44 nm for dichloromethane, both of which are absorbed. The apparent pore size, therefore, varies between 0.42–0.44 nm. This molecular sieve is referred to as 5A. For the sodium A zeolite (NaA), because of the higher cation population, the apparent pore diameter is 0.36–0.40 nm, depending on temperature (4A). The potassium form of zeolite A (KA), when highly exchanged, adsorbs some carbon dioxide and, at lower degrees of exchange, ethylene. A pore diameter of 0.33 nm is appropriate to these results (3A).

The zeolite sodium X (type 13X) has a crystallographic aperture of 0.74 nm. This compares well with the adsorbate value of 0.81 nm. Zeolite calcium X exhibits a smaller apparent pore size of 0.78 nm (10X). This difference is probably due to some distortion of the aluminosilicate framework upon dehydration and calcium ion migration.

When two or more molecular species involved in a separation are both adsorbed, selectivity effects become important because of interaction between the zeolite and the adsorbate molecule. These interaction energies include dispersion and short-range repulsion energies (ϕ_D and ϕ_R), polarization energy (ϕ_P), and components attributed to electrostatic interactions.

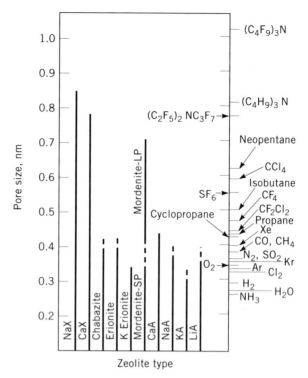

Fig. 7. Molecular dimension and zeolite pore size. Chart showing a correlation between effective pore size of various zeolites over temperatures of 77–420 K (---) with the kinetic diameters of various molecules (1). M–A is a cation–zeolite A system. M–X is a cation–zeolite X system.

Types of Separations. The first type of adsorption separation is based on differences in the size and shape of molecules (1,23). The molecular sieve separation of hydrocarbons by zeolite calcium A (5A) is used in commercial processes for the recovery of normal paraffins from hydrocarbon feedstocks (27) (see ADSORPTION, LIQUID SEPARATION). Paraffin isomers and cyclic hydrocarbons are too large to be adsorbed and are excluded. The recovered *n*-paraffins are utilized, for example, in the manufacture of biodegradable detergents. The affinity of zeolites for water and other polar molecules, decreases with rising SiO_2/Al_2O_3. Therefore, low silica zeolites can be used for drying gases and liquids. However, in many instances, secondary reactions such as polymerization of a coadsorbed olefin may take place. This is avoided by using the potassium-exchanged form of zeolite A (3A) for the removal of water from unsaturated hydrocarbon streams. The effective pore size of 3A excludes all hydrocarbons, including ethene. The molecular-sieve effect for water removal is also utilized in the drying of refrigerants. In this instance, zeolite A (4A) is employed because the size of the refrigerant molecules, such as refrigerant-12 with a kinetic diameter of 0.44 nm, is too large to be adsorbed. The second type of adsorption separation is based on differences in the relative selectivity of two or more coadsorbed gases or vapors. An outstanding example of this type is the production of oxygen-enriched air by the selective adsorption of nitrogen at ambient temperature on various molecular-sieve zeolites, including calcium A, calcium X, and various types of mordenite. Many separations based on relative selectivity range from simple drying processes to the separation of sulfur compounds from natural gas, and of aromatics from saturated hydrocarbons.

Tailoring of Zeolite Adsorption Selectivities. It is possible to tailor the zeolite adsorption characteristics in terms of size selectivity or the selectivity caused by other interactions, including cation exchange; cation removal or decationization; the presorption of a very strongly held polar molecule, such as water; pore-closure effects, that is, effects which alter the size of the openings to the internal pore volume; and the introduction of various defects such as removal of framework aluminum and an increase in the silicon/aluminum ratio (24). When synthetic mordenite is dealuminated by acid treatment, the SiO_2/Al_2O_3 ratio can increase to about 100 with the result that the water-adsorption capacity is essentially eliminated, and the zeolite becomes hydrophobic (24). The limit is attained in high silica ZSM-5 also known as silicalite. This material is capable of removing organic compounds from water (19).

Adsorption Kinetics. In zeolite adsorption processes the adsorbates migrate into the zeolite crystals. First, transport must occur between crystals contained in a compact or pellet, and second, diffusion must occur within the crystals. Diffusion coefficients are measured by various methods, including the measurement of adsorption rates and the determination of jump times as derived from nmr results. Factors affecting kinetics and diffusion include channel geometry and dimensions; molecular size, shape, and polarity; zeolite cation distribution and charge; temperature; adsorbate concentration; impurity molecules; and crystal-surface defects.

Catalytic Properties. In zeolites, catalysis takes place preferentially within the intracrystalline voids. Catalytic reactions are affected by aperture size and type of channel system, through which reactants and products must diffuse. Modification techniques include ion exchange, variation of Si/Al ratio, hydrothermal dealumination or stabilization, which produces Lewis acidity, introduction of acidic groups such as bridging Si(OH)Al, which impart Brønsted acidity, and introducing dispersed metal phases such as noble metals. In addition, the zeolite framework structure determines shape-selective effects. Several types have been demonstrated including reactant selectivity, product selectivity, and restricted transition-state selectivity (28). Nonshape-selective surface activity is observed on very small crystals, and it may be desirable to poison these sites selectively, eg, with bulky heterocyclic compounds unable to penetrate the channel apertures, or by surface silation.

Some current and possible future zeolite catalyst applications are as follows: alkylation, cracking, hydrocracking, dewaxing, isomerization, hydrogenation and dehydrogenation, hydrodealkylation, methanation, shape-selective reforming, dehydration, methanol to gasoline, methanol to olefins, organic catalysis, inorganic reactions, H_2S oxidation, NH_3 reduction of NO, $H_2O \rightarrow 1/2\ O_2 + H_2$, and CO oxidation.

Acid Sites. Acidic zeolites have outstanding catalytic activity. The hydrogen form may be produced by ammonium ion exchange, followed by thermal deammoniation. The unsolvated proton forms an OH group with a bridging O:

$$NH_4^+ \quad - NH_3 \qquad H^+$$

$$\equiv Si-O-Al^- \equiv \; \rightleftharpoons [\equiv Si-O-Al^- \equiv] \rightleftharpoons \; \equiv Si-OH \; Al \equiv$$

$$+ \; NH_3$$

The presence of an electron donor causes the equilibrium to shift to the left. The acidity represented by this mechanism is important in hydrocarbon conversion reactions. Acidity may also be introduced in certain high silica zeolites, eg, mordenite, by hydrogen-ion exchange, or by hydrolysis of a zeolite containing multivalent cations during dehydration, eg,

$$Ce^{3+}Z + H_2O \longrightarrow Ce(OH)^{2+}HZ$$

where Z = zeolite. The number and acid strength of these Brønsted acid sites are both important. The hydroxyl groups thus generated have been well characterized by several techniques, eg, ir spectroscopy. It has been demonstrated, for example, that the acid hydroxyl groups have a characteristic frequency that is determined by the charge density of the framework (25). Lewis sites may be generated by the formation of hexacoordinated aluminum atoms at cationic positions in extra-lattice positions (25).

Stabilized Zeolites. It is desirable that a zeolite to be used as a catalyst be stable during reaction and regeneration. For many catalytic applications, adequate thermal and hydrothermal stability of the zeolite is required. The structural stability of a zeolite, eg, zeolite Y, increases with Si/Al ratio. Further stabilization, particularly of the acid function, can be achieved by exchange with polyvalent cations such as rare-earth ions. Mixed rare-earth-exchanged zeolite Y is used in cracking catalysts. Increased stability is also obtained by hydrothermal treatment of the ammonium or rare-earth-exchanged form. When heated at high temperatures in the presence of water vapor, tetrahedral aluminum atoms are removed from the framework by hydrolysis with the formation of extra-framework cationic Al species such as $Al(OH)^{2+}$ or AlO^+, which impart additional structural and acid stability (29). The stabilizing effect of aluminum cations is similar to that of rare-earth cations. Such cationic aluminum species can be removed by ion exchange. Careful chemical treatment with acids or chelating agents may also be used to dealuminate the framework (30).

In shape-selective catalysis, the pore size of the zeolite is important. For example, the ZSM-5 framework contains 10-membered rings with ~ 0.6-nm pore size. This material is used in xylene isomerization, ethylbenzene synthesis, dewaxing of lubricatius oils and light fuel oil, ie, diesel and jet fuel, and the conversion of methanol to liquid hydrocarbon fuels (21).

The zeolites used for catalysis are principally modified forms of zeolite Y, acid forms of synthetic mordenite, and ZSM-5.

Dispersed Metals. Bifunctional zeolite catalysts, principally zeolite Y, are used in commercial processes such as hydrocracking. These are acidic zeolites containing dispersed metals such as platinum or palladium. The metals are introduced by cation exchange of the ammine complexes, followed by a reductive decomposition (21):

$$Na_2Y + Pt(NH_3)_4^{2+} \longrightarrow Pt(NH_3)_4^{2+} Y^{2-} + 2\ Na^+$$

$$Pt(NH_3)_4^{2+} Y^{2-} + H_2 \longrightarrow Pt^0(H^+)_2 Y^{2-} + 4\ NH_3$$

A bidisperse system involving metal agglomerates in the supercages and some crystallites at the external surface is commonly the result of this reaction. This migration during reduction with hydrogen may be attributable to formation of mobile platinum hydride, $Pt(NH_3)_2H_2$. Migration can be prevented by adding an olefin, which appears to act as a hydride trap, to the hydrogen gas (31). Other transition-metal ions such as Cd, Zn, Ni, and Ag are introduced by ion exchange followed by reduction with hydrogen. Agglomeration and migration to the external surface can also occur with these metals. Dehydrated zeolites can be loaded with metals by adsorption of neutral compounds such as carbonyls, followed by thermal decomposition. Molybdenum, ruthenium, and nickel have been loaded by this method into large-pore zeolites (26).

Ion Exchange. The exchange behavior of nonframework cations in zeolites, eg, selectivity and degree of exchange, depends on the nature of the cation, eg, the size and charge of the hydrated cation, on the temperature, the concentration, and, to some degree, on the anion species. Cation exchange may produce considerable change in various other properties, such as thermal stability, adsorption behavior, and catalytic activity.

The ion-exchange process is represented by

$$Z_A B^{Z_B}(z) + Z_B A^{Z_A}(s) \rightleftharpoons Z_A B^{Z_B}(s) + Z_B A^{Z_A}(z)$$

where Z_A and Z_B are the ionic charge of cations A and B and (z) and (s) represent the zeolite and solution. The ion-exchange isotherm (Fig. 8) is constructed by plotting A_z versus A_s, where A_z and A_s represent the mole fractions of cation A in the zeolite and solution, respectively. Similarly, with B_z and B_s representing the mole fraction of cation B in the zeolite and solution, the preference of the zeolite for ion A is given by the separation factor:

$$\alpha_B^A \equiv \frac{A_z B_s}{B_z A_s}$$

Ion-exchange isotherms assume different shapes depending on the selectivity factor and the variations in A_s with the level of exchange A_z. The rational selectivity coefficient K_B^A includes the ionic charge and is given by

$$K_B^A \equiv \frac{A_z^{Z_B} B_s^{Z_A}}{B_z^{Z_A} A_s^{Z_B}}$$

Typical exchange isotherms are given in Figure 9 for zeolite X. The exchange capacity of various zeolites is given in Table 2. In many cases, complete

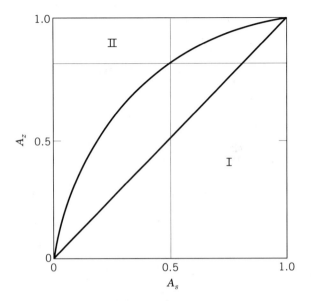

Fig. 8. Ion-exchange isotherm. The separation factor α_B^A is given by the ratio of area I/area II (1). See text.

exchange does not take place, such as for dipositive and tripositive ions in zeolite Y because of nonoccupancy of cation sites type I and I', located in the small cages. This corresponds to a maximum level of 0.68. Similarly, the level of exchange diminishes with the size and volume of the cation since the intracrystalline volume available does not permit full cation-site occupancy.

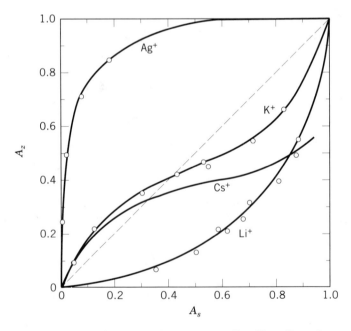

Fig. 9. Ion-exchange isotherms on zeolite X, sodium form.

Table 2. Ion-Exchange Capacity of Various Zeolites[a]

Zeolite	Si/Al ratio	meq/g
chabazite	2	5
mordenite	5	2.6
erionite	3	3.8
clinoptilolite	4.5	2.6
zeolite A	1	7.0
zeolite X	1.25	6.4
zeolite Y	2.0	5.0
ZSM-5	2.0	0.78
ZSM-5	3.5	0.46

[a] Anhydrous powder basis.

Kinetics. Ion-exchange rates in zeolites are controlled by ion diffusion within the zeolite structure, and are affected by particle radii, ionic diffusion coefficients, and temperature. For example, in zeolite NaX, diffusion of sodium ions occurs by migration from type II sites (in the six-membered rings) into the large cages, followed by diffusion to the surface through the large channels. As a result, complete replacement of sodium by calcium can be accomplished only at elevated temperatures.

Selectivity. Cation sieving in zeolites is attributed to cation size, distribution of charge in the zeolite structures, and size of the hydrated ion in aqueous solution. Solvation influences exchange since, for the ion to diffuse into the crystal, exchange of solvent molecules such as H_2O must always occur.

The selectivity coefficient K_B^A varies with the Si/Al ratio of the zeolite. In zeolites A (Si/Al = 1) and X (Si/Al = 1.2), the selectivity series for unipositive ions is Na > K > Rb > Cs. In zeolite Y (Si/Al = 2.8), the selectivity series is Cs > Rb > K > Na > Li. Even in unipositive–dipositive ion exchange (qv), the normal preference for the dipositive ion shown by zeolites A and X is reversed in zeolites with Si/Al ratio of about 3.

Framework Modification

The zeolite framework can be stabilized by hydrothermal treatment, which removes aluminum from the framework and forms aluminum cations. During this steaming process, the tetrahedral vacancies left behind in the framework are gradually refilled with silicon, which appears to migrate as a form of silicic acid from other parts of the framework and contributes to the stabilization by repairing the damaged framework. Simultaneously, such other parts of the framework disappear under formation of mesopores (32). Cationic aluminum can be extracted with an acid, and a subsequent steaming causes further dealumination of the framework and migration of silicon into the vacancies. Carefully controlled conditions can produce high silica forms of zeolites, eg, zeolite Y (33). Since the Si–O bond is shorter than the Al–O bond, hydrothermal dealumination causes the unit cell parameter to decrease.

Mesopores can be avoided by replacing the aluminum directly with external silicon, eg, by treatment with silicon tetrachloride (34):

$$Na[AlO_2(SiO_2)_x] + SiCl_4 \rightarrow [(SiO_2)_{x+1}] + NaAlCl_4$$

The NaAlCl$_4$ formed in this process needs to be removed by washing with water, in order to achieve the desired improved stability. A more convenient method for replacing framework-Al with Si is the reaction with ammonium hexafluorosilicate (35).

In a reversal of the reaction with SiCl$_4$, aluminum can be introduced into the framework by reaction of the hydrogen or ammonium form with gaseous AlCl$_3$ (36). Similarly, reaction with aqueous ammonium fluoroaluminates replaces framework-Si with Al (37). When alumina-bound high silica zeolites are hydrothermally treated, aluminum migrates into framework positions and generates catalytically active acid sites (38). The reaction can be accelerated by raising the pH of the aqueous phase.

The presence of internal silanol groups (39) and hydroxyl nests (40) in high silica zeolites facilitates the insertion of aluminum (hydroxyl nests can be generated, eg, by facile removal of boron from the framework of [B]ZSM-5). In analogy to the reaction with AlCl$_3$, boron, gallium, and indium have been inserted into the zeolite framework by treatment with the respective chlorides (41). Beryllium has been inserted by treatment with (NH$_4$)$_2$BeF$_4$ (42). Although some insertion of Fe may have been achieved, Fe, Cr, and Ti are incompletely inserted, or grafted to silanol groups by treatment with the respective complex fluorides.

Manufacture

Zeolites are formed under hydrothermal conditions, defined here in a broad sense to include zeolite crystallization from aqueous systems containing various types of reactants. Most synthetic zeolites are produced under nonequilibrium conditions, and must be considered as metastable phases in a thermodynamic sense.

More than 150 synthetic zeolites have been reported, and many important types have no natural mineral counterpart. Conversely, synthetic counterparts of many zeolite minerals are not yet known. The conditions generally used in synthesis are reactive starting materials such as freshly co-precipitated gels, or amorphous solids; relatively high pH introduced in the form of an alkali metal hydroxide or other strong base, including tetraalkylammonium hydroxides; low temperature hydrothermal conditions with concurrent low autogenous pressure at saturated water vapor pressure; and a high degree of supersaturation of the gel components, leading to nucleation of a large number of crystals.

A gel is defined as a hydrous metal aluminosilicate prepared from either aqueous solutions, reactive solids, colloidal sols, or reactive aluminosilicates such as the residue structure of metakaolin and glasses.

The gels are crystallized in a closed hydrothermal system at temperature varying from room temperature to about 200°C. The time required for crystallization varies from a few hours to several days. When prepared, the aluminosilicate gels differ in appearance, from stiff and translucent to opaque gelatinous precipitates and heterogeneous mixtures of an amorphous solid dispersed in an aqueous solution. The alkali metals form soluble hydroxides, aluminates, and silicates. These materials are well suited for the preparation of homogeneous mixtures.

Gel preparation and crystallization is represented schematically using the Na$_2$O—Al$_2$O$_3$—SiO$_2$—H$_2$O system as an example (1).

$$\text{NaOH (aq)} + \text{NaAl(OH)}_4 \text{ (aq)} + \text{Na}_2\text{SiO}_3 \text{ (aq)} \xrightarrow{25^\circ\text{C}} [\text{Na}_a(\text{AlO}_2)_b(\text{SiO}_2)_c \cdot \text{NaOH} \cdot \text{H}_2\text{O}] \text{ gel}$$

$$\xrightarrow{25-175^\circ\text{C}} \text{Na}_x [(\text{AlO}_2)_x(\text{SiO}_2)_y] \cdot m\text{H}_2\text{O}$$

Typical gels are prepared from aqueous solutions of reactants such as sodium aluminate, NaOH, and sodium silicate; other reactants include alumina trihydrate ($\text{Al}_2\text{O}_3 \cdot 3\text{H}_2\text{O}$), colloidal silica, and silicic acid. Some synthetic zeolites prepared from sodium aluminosilicate gels are given in Table 3.

When the reaction mixtures are prepared from colloidal silica sol or amorphous silica, additional zeolites may form which do not readily crystallize from the homogeneous sodium silicate–aluminosilicate gels. The temperature strongly influences the crystallization time of even the most reactive gels; for example, zeolite X crystallizes in 800 h at 25°C and in 6 h at 100°C.

Synthesis mechanisms of the typical low silica zeolites, such as A, X, and Y, are apparently different from the high silica zeolites such as ZSM-5. In the low silica zeolites, nuclei are formed consisting of alkali metal-ion complexes

Table 3. Some Synthetic Zeolites Prepared from Sodium Aluminosilicate Gels

Zeolite type	Typical composition, mol/mol Al_2O_3			Reactants	Reactant temp, °C	Zeolite product composition, mol/mol Al_2O_3		
	Na_2O	SiO_2	H_2O			Na_2O	SiO_2	H_2O
A	2	2	35	NaAlO_2 NaOH sodium silicate	20–175	1	2	4.5
X	3.6	3	144	NaAlO_2 NaOH sodium silicate	20–120	1	2.0–3.0	6
Y	8	20	320	NaAlO_2 colloidal SiO_2 NaOH	20–175	1	3.0–6.0	9
mordenite, Zeolon	6.3	27	61	NaAlO_2 diatomite sodium silicate	100	1	9–10	6.7
omega	5.60[a]	20	280	colloidal SiO_2 Al(OH)_3 TMAOH[b] NaOH	100	0.71 0.36 TMA	7.3	6.3
ZSM-5	10[c]	7.7	453	NaAlO_2 SiO_2 TPAOH[d]	150	0.89	31.1[e]	2.0

[a]Also 1.4 TMA_2O.
[b]TMA = tetramethylammonium.
[c]Also 8.6 TPA_2O.
[d]TPA = tetrapropylammonium.
[e]After calcination at 1000°C.

of the aluminosilicate species. Structural units consisting of four-membered rings, six-membered rings, and cages coordinated with cations are thought to be involved in the nucleation and crystallization. In the high silica zeolites, the mechanism appears to be a templating type (43) where an alkylammonium cation complexes with silica by hydrogen bonding. These complexes cause the structures to replicate by hydrogen bonding of the organic cation with framework oxygen atoms (44).

In general, zeolite crystallization consists of three stages: (1) formation of precursors, ie, building blocks that can generate nuclei; (2) nucleation; and (3) crystal growth.

X-ray diffraction can monitor only the progress of crystal growth, and stages 1 and 2 are frequently combined as the "induction period" (45). Nucleation without appreciable crystal growth can be accomplished by aging at low temperatures, eg, room temperature for zeolite Y, while crystal growth is accelerated at high temperatures. In some cases, crystallization of a zeolitic impurity can be prevented by aging. Another method for inducing crystal growth is seeding, ie, the introduction of seed crystals or precrystal nuclei. Using this procedure, the induction period is eliminated, and crystal growth starts immediately, so that the crystallization of the desired zeolite may be complete before nucleation of a zeolitic impurity reaches a stage at which growth commences (Fig. 10) (46).

Processes.   Manufacturing processes for commercial molecular sieve products may be classified into three groups, as shown in Table 4 (1): the preparation of molecular sieve zeolites as high purity crystalline powders or as preformed pellets from reactive aluminosilicate gels or hydrogels; the conversion of clay minerals into zeolites, either in the form of high purity powders or as binderless high purity preformed pellets; and processes based on the use of other naturally occurring raw materials. The hydrogel and clay conversion processes may also

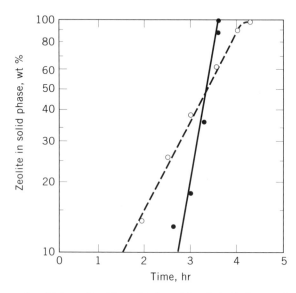

Fig. 10.   The growth of (∘) zeolite X and (•) zeolite B (46). Prenucleation or seeding of zeolite X moves the growth curve to the left.

Table 4. Processes for Molecular Sieve Zeolites

Process	Reactants	Products
hydrogel	reactive oxides	high purity powders
	soluble silicates	gel preform
	soluble aluminates	zeolite in gel matrix
	caustic	
clay conversion	raw kaolin	
	meta-kaolin	
	calcined kaolin	low to high purity powder
	acid-treated clay	binderless, high purity preform
	soluble silicate	zeolite in clay-derived matrix
	caustic	
	sodium chloride	
other	natural SiO_2	
	amorphous minerals	low to high purity powder
	volcanic glass	zeolite on ceramic support
	caustic	binderless preforms

be used to manufacture products that contain the zeolite as a major or a minor component in a gel matrix, a clay matrix, or a clay-derived matrix. Powdered products are often bonded with inorganic oxides or minerals into agglomerated particles for ease in handling and use.

Hydrogel Processes. The first commercial process for preparing synthetic zeolites on a large scale was based on a laboratory synthesis using amorphous hydrogels. The typical starting materials included an aqueous solution of sodium silicate, sodium aluminate, and sodium hydroxide. Hydrogel processes are based either on homogeneous gels, that is, hydrogels prepared from solutions of soluble reactants, or on heterogeneous hydrogels which are prepared from reactive alumina or silica in a solid form, eg, solid amorphous silica powder.

In gel-forming processes, the reactive aluminosilicate gel is first formed into a pellet which reacts with sodium aluminate solution and caustic solution. The zeolite crystallizes *in situ* within an essentially self-bonded pellet, or as a component in an unconverted amorphous matrix.

A process flow sheet for the manufacture of types 4A, 13X, and Y as high purity crystalline zeolite powders is shown in Figure 11.

A typical material balance as well as chemical compositions are given in Table 5. The raw materials are metered into the makeup tanks in the proper ratios. Crystallization takes place in a separate crystallizer. An intermediate aging step at ambient temperature may be required for the synthesis of certain high purity zeolites.

The process appears to be simple in terms of equipment and experimental conditions; however, because of the metastability of zeolite species formed from typical reactant systems, problems may arise when large-scale synthesis is attempted. For low silica zeolites, the crystallization temperature is usually near the boiling point of water; higher temperatures are commonly applied in the synthesis of high silica zeolites. After the digestion period, the slurry of crystals in the mother liquor is filtered in a rotary filter.

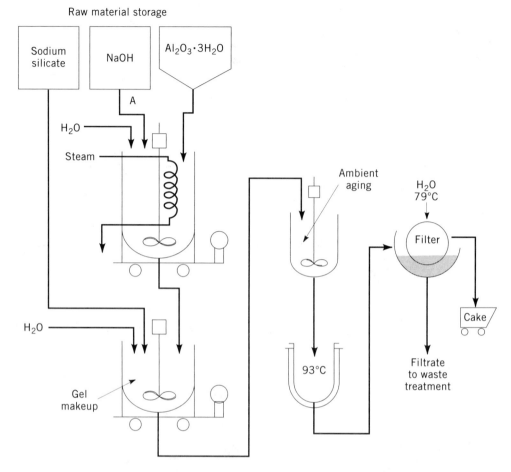

Fig. 11. Hydrogel process. Process flow sheet for the manufacture of zeolites 4A, 13X, and Y from reactant hydrogels (1).

Clay Conversion. The starting material for this process is kaolin, which usually must be dehydroxylated to *meta*-kaolin by air calcination. At 500–600°C, *meta*-kaolin forms, followed by a mullitized kaolin at 1000–1050°C.

$$Al_2Si_2O_5(OH)_4 \xrightarrow{550°C} Al_2Si_2O_7 + 2 H_2O$$
$$\text{kaolin} \qquad\qquad \textit{meta}\text{-kaolin}$$

$$3 Al_2Si_2O_7 \xrightarrow{1050°C} Si_2Al_6O_{13} + 4 SiO_2$$
$$\textit{meta}\text{-kaolin} \qquad \text{mullite} \qquad \text{cristobalite}$$

The zeolites are prepared as essentially binderless preformed particles. The kaolin is shaped in the desired form of the finished product and is converted

Table 5. Typical Material Balance for Hydrogel Process, kg^a

	Zeolite		
Raw materials	A	X	Y
sodium silicate^b	1350	2000	
SiO$_2$ powder^c			1450
alumina trihydrate^d	575	500	340
caustic, 50% NaOH	870	1600	1400
water	3135	7687	5300
	Gel composition, mole ratio		
Na$_2$O	2.04	4.09	4.0
Al$_2$O$_3$	1	1	1
SiO$_2$	1.75	3.0	10.6
H$_2$O	70	176	161

^{a}To produce 1000 kg, dry basis.
^{b}9.4% Na$_2$O, 28.4% SiO$_2$.
^{c}95% SiO$_2$.
^{d}65% Al$_2$O$_3$, 35% H$_2$O.

in situ in the pellet by treatment with suitable alkali hydroxide solutions. Preformed pellets of zeolite A are prepared by this method. These pellets may be converted by ion exchange to other forms such as molecular sieve Type 5A (1). Zeolites of higher SiO$_2$/Al$_2$O$_3$ ratios, eg, zeolite Y, can be obtained by the same method, when sodium metasilicate is incorporated in the preshaped pellets, or when acid-leached metakaolin is used.

Agglomeration of Synthetic Zeolite Powders. High purity zeolite crystals used in adsorption processing must be formed into agglomerates having high physical strength and attrition resistance. The crystalline powders are blended with an inorganic binder, generally a clay for adsorbents or, eg, alumina for catalysts, and wetted. The kneaded binder–zeolite–water mixture is extruded into cylindrical pellets that are subsequently calcined to form a strong composite. Or an aqueous binder–zeolite slurry is spray-dried to form a fluid catalyst in the shape of microspheres.

The preparation of zeolite–binder agglomerates as spheres or cylindrical pellets having high mechanical attrition resistance is not difficult. However, in order to utilize the zeolite in a process of adsorption or catalysis, the diffusion characteristics must not be unduly affected. Consequently, the binder system must permit a macroporosity which does not unduly increase diffusion resistance. The problem, therefore, is to optimize the binder–zeolite combination to achieve a particle of maximum density (to produce a high volumetric adsorption capacity) with maximum mechanical attrition resistance and minimum diffusion resistance.

Economic Aspects

Zeolites have high potential for protecting ecosystems, from facilitating wastewater and gas treatment to providing water softeners in detergents to replace the undesirable polyphosphate.

Catalytic cracking is an example of a use in which zeolites have had great impact on both crude oil and cracking unit economy. The application of zeolites made it possible to greatly increase catalytic activity, which allowed further improvement in fluid catalytic cracking (fcc) riser technology. Simultaneously, the cracking reactions became much more selective for desired products such as gasoline, while coke and gas productions were reduced. The gasoline yield improvements as induced by zeolite technology since the 1960s translate into a worldwide added product value of $15–20 \times 10^9$ annually (1988). Application of zeolitic fcc catalysts permitted more economical use of crude oil resources to reach the desired gasoline output. Moreover, new zeolitic cracking catalysts were developed to produce fcc gasoline with higher octane number (47) (see CATALYSIS).

The consumption of zeolites as detergent builders in 1990 in the United States, Canada, western Europe, and Japan was about 63.8×10^4 t, an increase of more than 20% over 1986. Zeolites accounted for all powder detergent builder use in Japan in 1990. The second largest use of zeolites is in catalysis, mainly for cracking in the petroleum industry; this amounted to $71–74 \times 10^3$ t in 1990 (48). Various modified Y-type zeolite products as well as ZSM-5 used for catalytic purposes are considered specialty chemicals with relatively low production volumes, and consequently they demand much higher prices than commodity zeolites, such as zeolite A. About 32×10^3 t of synthetic zeolites was used in 1990 for adsorbents and desiccants. In 1988, considerable amounts of synthetic zeolites were used in Japan for soil improvement and other agricultural applications to an estimated total of 50×10^3 t; the U.S. consumption was about 10×10^3 t, for a variety of applications (47). Both synthetic and natural zeolites are being used at approximately 1% as supplement in animal feeds, mainly in Japan, but to some extent also in the United States. In other agricultural applications, natural zeolites are used for odor and pollution control.

The U.S. capacity for synthetic zeolites at the end of 1991 was 182×10^3 t for detergent use (expansion of $>145 \times 10^3$ t by mid-1992); 86×10^3 t for catalysts; and about 30×10^3 t for use as adsorbents and desiccants. As of mid-1991, the price of builder-grade zeolite A to the largest customers was about $0.60/kg on an anhydrous basis (list price: $0.73), whereas zeolite 4A adsorbent sold for $2.29/kg as of January 1991. The price of zeolite-containing fcc catalyst was about $1650/t in early 1991.

Analytical Procedures

Identification. Each zeolite has a characteristic x-ray powder diffraction pattern, which is used for identification and determination of the purity or quality of zeolite present in a composite such as a catalyst. Generally, powder patterns are determined over a 2θ range of 4 to 56° since these materials have large unit cells and, correspondingly, exhibit the strongest lines at low diffraction angles. However, peak intensities and, to some extent, positions, vary with dehydration or cation exchange. Suitable procedures for x-ray analysis have been developed by The American Society for Testing and Materials (ASTM). Other procedures are based on vibrational spectroscopy (infrared and Raman), thermal analysis, and standard chemical analyses (49). Magic angle spinning nuclear magnetic resonance spectroscopy (mas nmr) permits tetrahedrally and octahedrally coordinated aluminum to be determined separately, a procedure particularly important for evaluation of modified zeolites in catalysts (50). The location of adsorbed xenon in the zeolite cavities can be determined by Xe nmr, and the obtained data can give information about the available intracrystalline space and the presence of extraneous matter in the cavities (50).

Adsorption. The BET (Brunauer-Emmett-Teller) method for surface-area measurement commonly employed to characterize adsorbents and catalysts is not relevant for zeolites, since adsorption on zeolites occurs by a pore-filling mechanism. Oxygen adsorption at low temperature ($-183°C$) is employed as a method for determining zeolite content, utilizing an appropriate reference. Since the structure of the zeolite is known, the void volume and oxygen capacity can be calculated as a reference value. Nitrogen could be used but, because cation–nitrogen interactions would contribute to additional adsorption capacity, oxygen is preferred. A gravimetric microbalance of the McBain-Bakr type is used. Before adsorption, the zeolite sample is outgassed at 350–450°C under a reduced pressure of 1.3 mPa (10^{-5} mm Hg) for 9–16 h. Complete isotherms should be measured in order to determine deviation from the type I contour. In the petroleum industry, hydrocarbons are frequently employed as adsorbates (51). These have the advantage that different-size molecules can distinguish between zeolites in a mixture, eg, high silica zeolites of type A topology and zeolite Y; n-hexane is sorbed by both zeolites, whereas only the latter sorbs cyclohexane.

Health and Safety Factors and Toxicology

Zeolites have applications in food, drugs, cosmetic products, and detergents. Thus, extensive toxicological and environmental studies have been carried out (52). In single oral-intubation studies, rats have survived a single massive dose equivalent to 32 g/kg of body weight (powder form of type 4A, 5A, 13X, and Y). Feeding of 5.0 g/kg of body weight for seven days produced no ill effect (53). There is no contraindication to the use of zeolite A (Sasil) in detergents. No negative effect on biological wastewater treatment was found. In toxicity studies using algae, macroinvertebrates, and fish, zeolite A showed no evidence of acute toxicity to four species of freshwater fish. No mortality was found for either cold- or warm-water fish exposed to suspensions of 680 mg/L.

Uses

Some commercially available molecular-sieve products and related materials are shown in Table 6, classified according to the basic zeolite structure types. In most cases, the water content of the commercial product is below 1.5–2.5 wt %; certain products, however, are sold as fully hydrated crystalline powders.

Adsorption. Since the 1960s, molecular-sieve adsorbents have become firmly established as a means of performing difficult separations, including gases from gases, liquids from liquids, and solutes from solutions (27). They are supplied as pellets, granules, or beads, and occasionally as powders. The adsorbents may be used once and discarded or, more commonly, may be regenerated and used for many cycles. They are generally stored in cylindrical vessels through

Table 6. Commercial Molecular Sieve Products[a,b]

Zeolite type	Designation	Cation	Effective pore diameter, Å	Unit cell parameter,[c] Å
A, KA	3A[d,e]	K	3	UOP
	Zeolum A-3[d,e]			TOSOH
	Sylosiv A3[e]			Grace Davison
A, NaA	4A[d,e]	Na	4	UOP
	Zeolum A-4[d,e]			TOSOH
	Valfor G100			PQ
	Sylosiv A4[e]			Grace Davison
A, CaA	5A[d,e]	Ca	5	UOP
	Zeolum A-5[d,e]			TOSOH
X, NaX	13X[d]	Na	10	UOP
	Zeolum F-9[d,e]			TOSOH
	Sylosiv A10[e]			Grace Davison
Y, NaY	LZY-54[d]	Na	10	UOP (5.0); $a_0 = 24.68$
	HSZ-320NAA[d,e]			TOSOH (5.5); $a_0 = 24.64$
	CBV 100			PQ (5.2); $a_0 = 24.64$
Y, NH$_4$Y	LZY-64, -84[d]	NH$_4$	10	UOP (5.1, 5.9); $a_0 = 24.70$, 24.57
	CBV 300			PQ (5.2); $a_0 = 24.68$
Y, HY	LZY-74	H	10	UOP (5.2); $a_0 = 24.52$
	HZS-320HOA[d]			TOSOH (5.5); $a_0 = 24.50$
	HZS-330HSA[d]			TOSOH (6); $a_0 = 24.50$
	CBV 400, 500			PQ (5.1, 5.2); $a_0 = 24.50$, 24.53
Y, USY	LZ-10, -20	H, Al	10	UOP (5.5, 5.6); $a_0 = 24.30$, 24.35
	HSZ-330HUA[d]			TOSOH (5.6); $a_0 = 24.40$
	CBV 600			PQ (5.2); $a_0 = 24.33$
Y, dealuminate	LZ-210	H	10	UOP (6.5-18); a_0 varies
	HSZ-360,390HUA[d]			TOSOH(14,600); $a_0 = 24.30$, 24.27
	CBV 712-780			PQ (11.5-80); $a_0 = 24.33 - 24.24$
L	LZ-KL	K	8	UOP (6.3)
	HSZ-500KOA			TOSOH (6.2)
Mordenite, small-pore	AW-300	Na, mixed	4	UOP
Mordenite, large-pore	HSZ-610-640NAA	Na	7	TOSOH (12,15,20)
	CBV 10A			PQ (13)
	CBV 20, 30A	NH$_4$		PQ (20,35)
	LZM-5 -8	H		UOP (10.7, 18)
	HSZ620, 640HOA			TOSOH (15, 16)

Table 6. (*Continued*)

Zeolite type	Designation	Cation	Effective pore diameter, Å	Unit cell parameter,[c] Å
Chabazite	AW-500	mixed	5	UOP
Ferrierite	HSZ-720KOA	K, Na	4	TOSOH (16.8)
ZSM-5	MFI, S-115	H	6	UOP (30–45, 180–400)
	HSZ-690HUA			TOSOH (>200)
	CBV 3020, 5020,			PQ (30, 50, 80, 150)
	8020, 1502			
Beta	CP 806B-25	Na, T[f]	7	PQ (25)
	CP 806BL-25	NH$_4$, T[f]		PQ (25)
	CP 811BL-25	H		PQ (25)
Zeolite F	Ionsiv F80	K, Na	4	UOP (2)
Zeolite W	Ionsiv W85	K, Na	4	UOP (3.6)

[a]Because of commercial usage, Angstrom units are shown here. To convert Å to nm, divide by 10.
[b]All zeolite types are available as powders unless otherwise indicated. Chabazite and Mordenite, small-pore are available as extrudates only.
[c]For (SiO_2/Al_2O_3) by manufacturer indicated.
[d]Also available as extrudate.
[e]Also available as bead.
[f]T = template.

which the stream to be treated is passed. For regeneration, two or more beds are usually employed with suitable valving, in order to obtain a continuous process. As a unit operation, adsorption (qv) is unique in several respects. In some cases, one separation is equivalent to hundreds of mass-transfer units. In others, the adsorbent allows the selective removal of one component from a mixture, based on molecular size differences, which would be nearly impossible to perform by any other means. In addition, contaminants can be removed from fluid streams to attain virtually undetectable impurity concentrations. Adsorbents are used in applications requiring a few grams to several tons.

Regenerative adsorption units can be operated by a thermal-swing cycle, pressure-swing cycle, displacement-purge cycle, or inert-purge cycle. Combinations of these are frequently employed.

Purification. Purification refers to separations wherein the feed stream is upgraded by the removal of a few percent or even traces of a contaminant (Fig. 12). A heated purge gas is usually employed for this purpose.

Water. The dehydration of natural gas and air was the first of the gas-purification applications of molecular sieves. Because of their high adsorptive selectivity for water and high capacity at low water partial pressures, molecular sieves were an obvious choice for water removal from natural gas and air before cryogenic extraction of helium and cryogenic separation of oxygen, nitrogen, and the rare gases, respectively. Molecular-sieve ehydration is used in the cryogenic production of liquified natural gas (LNG), for small-peak demand-type storage facilities and giant base-load facilities. In addition, molecular sieves have proved to be the most effective dehydration technique for the cryogenic recovery of ethane and heavier liquids from natural gas.

Molecular sieves have had increasing use in the dehydration of cracked gases in ethylene plants before low temperature fractionation for olefin produc-

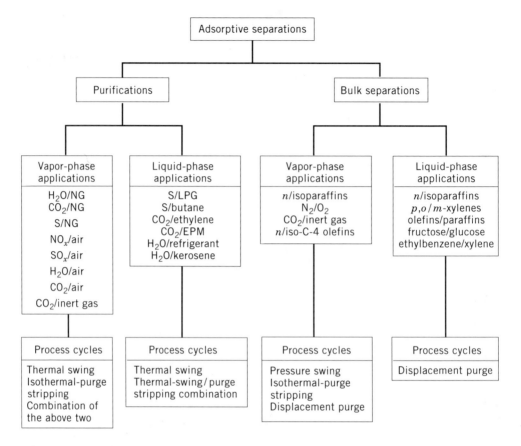

Fig. 12.   Classification of adsorptive separations where NG = natural gas and S = sulfur.

tion. The Type 3A molecular sieve is size-selective for water molecules and does not co-adsorb the olefin molecules.

The unique features of molecular sieves are demonstrated in the removal of water from natural gas streams containing high percentages of acid gases, eg, H_2S and CO_2. Other dry-bed adsorbents degrade in highly acidic environments. However, acid-resistant molecular sieves have been developed which maintain dehydration capacities over long periods of on-stream use. They also are used to dehydrate gas streams containing corrosive components such as chlorine, sulfur dioxide, and hydrogen chloride.

Molecular sieves are also used widely in the dehydration of liquid streams. Both batch-type and continuous processes have been developed for drying a variety of hydrocarbon and chemical liquids.

For nonregenerative drying, the molecular sieve is designed for the lifetime of the unit. A typical example is refrigerant drying and purification. A suitable-size cartridge of the proper molecular sieve, installed in the circulating refrigerant stream, adequately protects the refrigeration system from freeze-ups and corrosion for the life of the unit by adsorbing water and the acidic decomposition products of the refrigerant.

Another nonregenerative drying application for molecular sieves is their use as an adsorbent for water and solvent in dual-pane insulated glass windows. The molecular sieve is loaded into the spacer frame used to separate the panes. Once the window has been sealed, low hydrocarbon and water dew points are maintained within the enclosed space for the lifetime of the unit. Consequently, no condensation or fogging occurs within this space to cloud the window.

Gas and liquid dehydrators employing molecular sieves provide product gas streams of <0.1 ppmv water and product liquid streams routinely to <10 ppmv water. Applicable pressures range from less than one to several hundred times atmospheric pressure. Temperatures range from subzero to several hundred °C. Processing units range in capacity from as little as 10 m^3/h to as much as 10^8 m^3/d in multiple-train units.

Carbon Dioxide.   Molecular sieves are used to purify gas streams containing carbon dioxide in cryogenic applications where freeze-out of CO_2 would cause fouling of low temperature equipment. They are also employed for air purification in cryogenic air-separation plants where one front-end purifier unit can be used for the simultaneous removal of both water and CO_2. Peakshaving natural gas liquefaction is employed by utilities to store LNG during the summer. Molecular-sieve adsorbents remove water and CO_2 before liquefaction. Commercial processes are available for the removal of CO_2 from air, natural gas, ethylene, ethane–propane mix, and synthesis gases. Operations cover wide ranges of pressure and temperature.

Sulfur Compounds.   Various gas streams are treated by molecular sieves to remove sulfur contaminants. In the desulfurization of wellhead natural gas, the unit is designed to remove sulfur compounds selectively, but not carbon dioxide, which would occur in liquid scrubbing processes. Molecular sieve treatment offers advantages over liquid scrubbing processes in reduced equipment size because the acid gas load is smaller; in production economics because there is no gas shrinkage (leaving CO_2 in the residue gas); and in the fact that the gas is also fully dehydrated, alleviating the need for downstream dehydration.

Molecular sieves are being used to treat refinery hydrogen streams containing trace amounts of H_2S. A single molecular sieve unit may be designed to remove trace water and H_2S in the recycle hydrogen loop of a catalytic reformer to protect the catalyst from poisoning. Then, during catalyst regeneration, the same unit acts as a dryer for treating the inert gas used in regenerating the reforming catalyst.

A large use of molecular sieves in the natural gas industry is LPG sweetening, in which H_2S and other sulfur compounds are removed. Sweetening and dehydration are combined in one unit and the problem associated with the disposal of caustic wastes from liquid treating systems is eliminated. The regeneration medium is typically natural gas. Commercial plants are processing from as little as ca 30 m^3/d (200 bbl/d) to over 8000 m^3/d (50,000 bbl/d).

Bulk Separation.   The adsorptive separation of process streams into two or more main components is termed bulk separation (see Fig. 12). The development of processes and products is complex. Consequently, these processes are proprietary and are purchased as a complete package under licensing agreements. High purities and yields can be achieved.

Separation of Normal and Isoparaffins.   The recovery of normal paraffins from mixed refinery streams was one of the first commercial applications of molecular sieves. Using Type 5A molecular sieve, the *n*-paraffins can be adsorbed and the branched and cyclic hydrocarbons rejected. During the adsorption step, the effluent contains isoparaffins. During the desorption step, the *n*-paraffins are recovered. Isothermal operation is typical.

Regeneration is carried out by a pressure-swing process for a process separating light hydrocarbons (≤C-7). For heavier streams, a displacement-purge cycle employing lighter *n*-paraffins is used. The *n*-paraffins are separated by distillation from the regeneration effluent and recycled.

There are seven commercial processes in operation; six operate in the vapor phase. The Universal Oil Products process operates in the liquid phase and is unique in the simulation of a moving bed. The adsorption unit consists of one vessel segmented into sections with multiple inlet and outlet ports. Flow to the various segments is accomplished by means of a rotary valve which allows each bed segment to proceed sequentially through all the adsorption/desorption steps.

The normal paraffins produced are raw materials for the manufacture of biodegradable detergents, plasticizers, alcohols, and synthetic proteins. Removal of the *n*-paraffins upgrades gasoline by improving the octane rating.

Xylene Separation.   *p*-Xylene is separated from mixed xylenes and ethylbenzene by means of the Parex process (Universal Oil Products Co.). A proprietary adsorbent and process cycle are employed in a simulated moving-bed system. High purity *p*-xylene is produced.

Olefin Separation.   Olefin-containing streams are separated either by the OlefinSiv process (Union Carbide Corp.) separating *n*-butenes from isobutenes in the vapor phase, or the Olex process (Universal Oil Product) a liquid-phase process.

Oxygen from Air.   A demand has developed for oxygen for processes needing from a few to about 50 t/d. In this size-range, a pressure-swing adsorption process is often competitive with the conventional cryogenic separation route. Oxygen of 95% purity can be obtained; the main impurities are the inert gases found in air.

Catalysis.   As of mid-1995, zeolite-based catalysts are employed in catalytic cracking, hydrocracking, isomerization of paraffins and substituted aromatics, disproportionation and alkylation of aromatics, dewaxing of distillate fuels and lube basestocks, and in a process for converting methanol to hydrocarbons (54).

Catalytic Cracking.   The addition of relatively small amounts of hydrothermally stable acidic zeolites to conventional cracking catalyst formulations significantly increases both the yield and the quality of the products from fluidized-bed and moving-bed cracking reactors. At present, catalytic cracking is the largest scale industrial process employing zeolite catalysts. Catalysts containing the rare-earth-exchanged form or steam-stabilized forms of zeolite Y with or without rare-earth-exchange are marketed. The commercial catalysts comprise 5–40% zeolite dispersed in a matrix of synthetic silica–alumina, semisynthetic clay-derived gel, or natural clay. Such composites can be prepared either by blending a synthetic zeolite with a binder, or by chemical treatment of suitable clays to produce the zeolite component *in situ*.

Zeolite-promoted cracking catalysts offer the advantage of high rates of intermolecular hydrogen transfer coupled with extremely high intrinsic cracking activity, and the high thermal stability of the zeolite-cracking catalyst. Zeolite catalysts increase the yield of light cycle oil as well as the yield and octane of the gasoline product fraction, and decrease the production of coke and gas. The high cracking rates obtained with zeolite catalysts result in greatly reduced contact times for given conversion levels, thus further increasing liquid product yields. Additional advantages are increased tolerance to poisons and greater operating flexibility. The relationships between catalyst properties, feedstock composition, and reactor operating conditions are very complex (55). Additional incorporation of the shape-selective zeolite ZSM-5 in the cracking catalyst yields gasoline of improved octane rating (56).

Hydrocracking for Fuels Production.   Hydrocracking is catalytic cracking in the presence of hydrogen using a dual-function catalyst possessing both cracking and hydrogenation–dehydrogenation activity. At present, it is the second largest use for zeolite-containing catalysts. In general, such catalysts consist of an acidic, hydrothermally stable, large-pore zeolite loaded with a small amount of a noble metal, or admixed with a relatively large amount of an active hydrogenation system such as $NiO + MoO_3$ or WO_3.

Although several proprietary hydrocracking technologies are in use, the Union Oil Co. Unicracking process exemplifies the value of zeolite catalysts in broadening the range of feedstocks that can be handled and in simplifying hydrocracker design and operation. The most elaborate and versatile Unicracking process scheme is the two-stage configuration shown in Figure 13. Feedstock is mixed with hydrogen and admitted to reactor R-1, a conventional hydrotreater in which it is substantially freed of nitrogen and sulfur compounds. The product then enters the first-stage Unicracker, R-2, in which it is hydrocracked, typically at a per-pass conversion of 40–70%. The zeolite catalysts developed for the Unicracking process can operate stably and efficiently in the presence of hydrogen sulfide and ammonia. Thus, separation of these substances from the R-2 feed is not necessary. The fractionator bottoms are recycled through a second-stage reactor, R-3. The recycled gas which is mixed with the R-3 feed is essentially ammonia-free; thus the reactor can be operated efficiently at comparatively low temperatures and pressures while maintaining conversion at 50–80% per pass. In single-stage Unicracking, the R-2 product is treated in liquid–gas separators and fractionated.

In hydrocracking, as with all multipurpose refinery processes, the relationships between feedstock properties, catalyst type, operating conditions, and product yield and character are complex. With zeolite catalysts, a variety of feedstocks is converted into a range of fuels, including LPG, medium-octane unleaded gasoline, and jet, diesel, and heating oils, or into feedstocks suitable for catalytic reforming or petrochemical manufacture. Zeolite hydrocracking catalysts are noted for permitting long (2–6 yr) periods of highly efficient reactor operation at moderate conditions. Following such service, some of these catalysts can be completely restored to their original levels of activity.

Dewaxing of Distillate Fuels and Lube Basestocks. Removal of waxy molecules improves the cold-flow properties (pour point, freezing point, cloud point, and cold filter plugging point) of distillate fuels, which include jet and

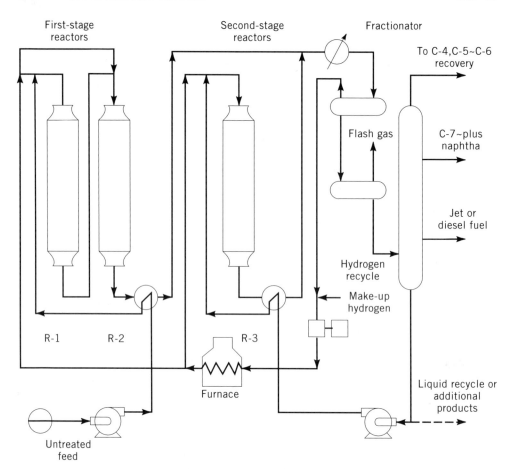

Fig. 13. Two-stage Unicracking unit (21). Courtesy of American Chemical Society.

diesel fuels. The Mobil Distillate Dewaxing process, employing ZSM-5 catalyst, was commercialized in 1974 and is in use worldwide. Typical operating conditions are 260–430°C, 2–5.5 MPa (20–55 atm), 250–440 m^3 H$_2$/m^3 distillate (21). The waxy molecules, mainly consisting of straight-chain and methyl-substituted straight-chain hydrocarbons, are selectively cracked by this process. Two-thirds of the cracked by-product is gasoline of sufficiently high octane rating, to be blended directly into the gasoline pool.

Paraffin Isomerization.   Another well-established commercial process which employs zeolite catalysts is the isomerization of normal paraffins into higher octane, branched isomers. The catalyst for the Hysomer process of the Shell Oil Co. is dual-functional, and consists of a highly acidic, large-pore zeolite loaded with a small amount of a noble-metal hydrogenation component. This catalyst possesses the same hydrogenation–dehydrogenation and acid functions as hydrocracking catalysts. However, since hydrocracking of the Hysomer feedstock, which is usually a light, straight-run naphtha containing principally pentanes and hexanes, is not desired because it represents a direct reduction of gasoline yield, the process operating conditions are more stringently defined than those for hydrocracking. In general, reaction conditions are adjusted to allow the lowest possible reactor temperature consistent with economical production rates, since lower temperatures favor higher equilibrium concentrations of branched isomers, as well as the lowest hydrogen pressure consistent with economy of operation. In this way, yields of the highest octane highly branched isomers are maximized, and hydrocracking is minimized.

The Hysomer process produces an increase of about 12 octane numbers in suitable naphtha feedstocks. The process can be operated in conjunction with the Isosiv process (Union Carbide Corp.) for the separation of normal and isoparaffins, achieving complete isomerization of a C-5–C-6 stream. The combined process is trade named TIP (total isomerization process), and results in increases in octane numbers of about 20, rather than the 12 obtained with a once-through Hysomer treatment.

Catalysis of Aromatic Reactions. In the 1960s, zeolite syntheses that included organic cations such as quaternary ammonium species in addition to the traditional alkali metal ions were developed, leading to a number of new molecular sieves. Mobil Corp. laboratories developed, among other zeolites, the zeolite ZSM-5. This zeolite is characterized by an unusual pore structure, the openings of which are 10-membered rings, and highly siliceous frameworks, with $SiO_2:Al_2O_3$ molar ratios of 20 to $>10,000$. The exceptional shape-selectivity of ZSM-5 has been utilized to prepare some unique catalysts for a variety of processes (21).

Selective Toluene Disproportionation. Toluene disproportionates over ZSM-5 to benzene and a mixture of xylenes. Unlike this reaction over amorphous silica–alumina catalyst, ZSM-5 produces a xylene mixture with increased *p*-isomer content compared with the thermodynamic equilibrium. Chemical modification of the zeolite causing the pore diameter to be reduced produces catalysts that achieve almost 100% selectivity to *p*-xylene. This favorable result is explained by the greatly reduced diffusivity of *o*- and *m*-xylene compared with that of the less bulky *p*-isomer. For the same reason, large crystals (3 μm) of ZSM-5 produce a higher ratio of *p*-xylene:total xylenes than smaller crystallites (28,57).

Xylene Isomerization. The objective of C-8-aromatics processing is the conversion of the usual four-component feedstream (ethylbenzene and the three xylenes) into an isomerically pure xylene. Although the bulk of current demand is for *p*-xylene isomer for polyester fiber manufacture, significant markets for the other isomers exist. The primary problem is separation of the 8–40% ethylbenzene that is present in the usual feedstocks, a task that is complicated by the closeness of the boiling points of ethylbenzene and *p*-xylene. In addition, the equilibrium concentrations of the xylenes present in the isomer separation train raffinate have to be reestablished to maximize the yield of the desired isomer.

In the most readily adopted C-8-aromatics process based on ZSM-5 and related zeolites, the catalyst is an acid form of the sieve added optionally with a Group VIII metal to avoid hydrogenation of aromatic rings, among other goals (28). Such catalysts can be used in existing plants in place of the Pt/silica–alumina materials originally developed for the process. They have the advantages of higher throughput, owing to the ability to operate at low H_2/hydrocarbon ratios, and longer run life between regenerations. The shape-selectivity of the catalyst greatly favoring diffusion of the *p*-isomer permits higher than equilibrium concentrations of p-xylene to be produced. Ethylbenzene present in the feedstock is converted by transalkylation and dealkylation, yielding benzene of nitration-grade purity (28).

Ethylbenzene Synthesis.   The synthesis of ethylbenzene for styrene production is another process in which ZSM-5 catalysts are employed. Although some ethylbenzene is obtained directly from petroleum, about 90% is synthetic. In earlier processes, benzene was alkylated with high purity ethylene in liquid-phase slurry reactors with promoted $AlCl_3$ catalysts or the vapor-phase reaction of benzene with a dilute ethylene-containing feedstock with a BF_3 catalyst supported on alumina. Both of these catalysts are corrosive and their handling presents problems.

These operations have been gradually replaced by the Mobil-Badger process (28), which employs an acidic ZSM-5 catalyst and produces ethylbenzene using both pure and dilute ethylene sources. In both cases, the alkylation is accomplished under vapor-phase conditions of about 425°C, 1.5–2 MPa (15–20 atm), 300–400 kg benzene per kg catalyst per h, and a benzene:ethylene feed ratio of about 30. ZSM-5 inhibits formation of polyalkylated benzenes produced with nonshape-selective catalysts. With both ethylene sources, raw material efficiency exceeds 99%, and heat recovery efficiency is high (see XYLENES AND ETHYLBENZENE).

Synthesis of p-Ethyltoluene.   *para*-Ethyltoluene, the feedstock for p-methylstyrene, is difficult to separate from the products of toluene alkylation with ethane using conventional acidic catalysts. The unique configurational diffusion effect of ZSM-5 permits p-dialkylbenzenes to be produced in one step. In the alkylation of toluene with ethene over a chemically modified ZSM-5, p-ethyltoluene is obtained at 97% purity (58).

The Methanol-to-Gasoline Process.   Mobil Corp. laboratories have reported the conversion of methanol and some other oxygenated organic compounds to hydrocarbons using an acidic ZSM-5 zeolite catalyst (21). In the case of methanol, the highly exothermic reaction sequence includes the formation of dimethyl ether. The process is quite selective in producing an aromatics-rich gasoline-range mixture of hydrocarbons free of oxygen compounds. Several mechanisms have been proposed for this reaction, but none has been shown unequivocally to be correct (21). The hydrocarbon product typically contains no compounds with more than 10 carbon atoms, and this, in conjunction with the lower than predicted concentrations of certain polyalkylated benzenes, is construed as evidence of shape selectivity owing to the moderate pore size of the catalyst. The process has been in commercial operation for many years in New Zealand, which has no crude oil deposits, and where methanol is produced from abundant natural gas. It is not competitive with the refining of crude oil.

A variation of this process is Mobil's methanol-to-olefins (MTO) process, in which up to 80% C_2–C_5 olefins are produced over ZSM-5 of reduced acidity and at much higher temperatures.

Ion Exchange.   Crystalline molecular sieve ion exchangers do not follow the typical rules and patterns exhibited by organic and other inorganic ion exchangers. Many provide combinations of selectivity, capacity, and stability superior to the more common cation exchangers. Their commercial utilization has been based on these unique properties (59).

Cesium and Strontium Radioisotopes. Because of their stability in the presence of ionizing radiation and in aqueous solutions at high temperatures, molecular sieve ion exchangers offer significant advantages in the separation and purification of radioisotopes. Their low solubility over wide pH ranges, together with their rigid frameworks and dimensional stability and attrition resistance, have endowed zeolites with properties which generally surpass those of the other inorganic ion exchangers. The high selectivities and capacities of several zeolites for cesium and strontium radioisotopes resulted in the development of processes currently used by nuclear processing plants.

Ammonium Ion Removal. A fixed-bed molecular-sieve ion-exchange process has been commercialized for the removal of ammonium ions from secondary wastewater treatment effluents. This application takes advantage of the superior selectivity of molecular-sieve ion exchangers for ammonium ions. The first plants employed clinoptilolite as a potentially low cost material because of its availability in natural deposits. The bed is regenerated with a lime-salt solution that can be reused after the ammonia is removed by pH adjustment and air stripping. The ammonia is subsequently removed from the air stream by acid scrubbing.

Detergent Builders. The prime function of phosphates in detergents is to reduce the activity of the hardness ions, Ca^{2+} and Mg^{2+}, in the wash water by complexing. Zeolite ion exchangers in powder form replace Ca^{2+} and Mg^{2+} in the solution with ions such as Na^+. Heavy-duty detergents employ the sodium form of Type A zeolite for this purpose in low or zero-phosphate formulations. The zeolite powder is incorporated into the detergent powder during formulation. Large amounts of zeolite are used in this application.

New Trends

Aluminosilicate of faujasite topology, but higher Si/Al ratio (≤ 5), has been synthesized with the use of a crown ether, 15-crown-5, as the directing agent. The same Si/Al ratio was obtained when 18-crown-6 was applied, but the topology, although related to faujasite, had a different stacking order of the sodalite cages, so that the structure has hexagonal instead of cubic symmetry (60). This product, sometimes called hexagonal faujasite, has the designation EMT.

The foregoing discussion has focused on the most important commercial molecular sieves, zeolites. New directions in the preparation of framework structures of different chemical composition and of large-pore molecular sieves have also appeared.

Phosphate-Containing Molecular Sieves. The discovery of novel aluminophosphate ($AlPO_4$) (2), silicoaluminophosphate (SAPO) (3), metal-containing aluminophosphate (MeAPO) (4), and metal-containing silicoaluminophosphate (MeAPSO) (5) molecular sieves announced by the Union Carbide Corp. provided an extraordinary expansion of new chemical compositions with known topologies as well as unique topologies. Table 7 lists some of these interesting materials. These materials are synthesized via techniques utilized for aluminosilicate zeolites (autoclaves, gels, etc) including the use of organic directing agents such as tri-*n*-propylamine, tetraethyl- or tetrapropylammonium, quinuclidine, etc.

Table 7. List of Phosphate-Containing Molecular Sieves[a]

Framework designation	Topology	Ring size	Pore diameter, nm
AlPO4-5	novel	12	0.73
AlPO4-8	novel	14	0.79 × 0.87
AlPO4-11	novel	10 or puckered 12	~0.6
AlPO4-17	Erionite		0.43
AlPO4-20	Sodalite		0.26
AlPO4-31	novel	12	0.54
AlPO4-41	novel	10	0.43 × 0.70
SAPO-5	novel	12	0.73
SAPO-11	novel	10 or puckered 12	~0.6
SAPO-20	Sodalite		0.3
SAPO-34	Chabazite		0.43
SAPO-37	Faujasite		0.74
SAPO-35	Levynite		0.43
SAPO-40	novel	12	0.67 × 0.69
SAPO-42	Zeolite A		0.43
CoAPO-50	novel	12, 8	0.61–0.40 × 0.43

[a]Thirteen structures of various compositions, as $AlPO_4$, SAPO, MeAPO, and MeAPSO, are available from UOP.

The introduction of phosphate into the framework has led to an extraordinary expansion of the possible pore diameters in framework molecular sieves. The discovery of VPI-5, the first 16-membered-ring aluminophosphate framework, was a landmark (8). Although it is a neutral framework, it pointed to the

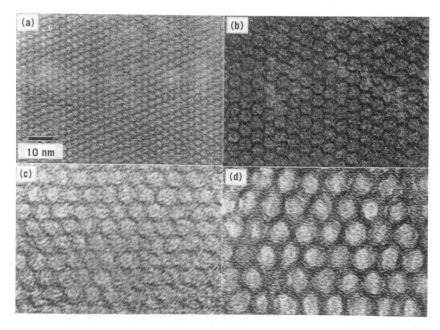

Fig. 14. Transmission electron micrographs of MCM-41 having pore diameters of (**a**) 2 nm, (**b**) 4 nm, (**c**) 6 nm, and (**d**) 8 nm (63).

possibility of larger pore-diameter materials. Since the VPI-5 discovery, at least two new phosphate structures have been announced: Cloverite, an 18-membered-ring gallophosphate (9), and JDF-20 (61), a 20-membered-ring aluminophosphate. Beryllo- and zinco-phosphate versions of faujasite have been reported (62).

Mesoporous Molecular Sieves. A new family of molecular sieves with pore diameters in the mesoporous range (~200–2000 nm) have been reported by the Mobil Corp. (11,63). The designation given to this extensive family of molecular sieves is M41S. Although the materials described in these reports are silicates and aluminosilicates, the chemical composition of these materials is quite broad. Of particular interest are two members of this family: MCM-41 (for Mobil Composition of Matter number 41), which possesses a hexagonal array of uniform mesopore, and MCM-48, which is a cubic phase (Ia3d space group). The x-ray diffraction patterns of these materials consist of a strong low angle line with a d-spacing that correlates, but is not identical, with the pore diameter, and several higher order lines of lower intensity. Transmission electron microscopic images of MCM-41 are shown in Figure 14. The discovery of these materials and the liquid crystal templating mechanism invoked for their formation open a new chapter in molecular sieve technology. Other compositions are possible (12).

BIBLIOGRAPHY

"Molecular Sieves" in *ECT* 3rd ed., Vol. 15, pp. 638–669, by D. W. Breck and R. A. Anderson, Union Carbide Corp.

1. D. W. Breck, *Zeolite Molecular Sieves, Structure, Chemistry, and Use*, John Wiley & Sons, Inc., New York, 1974.
2. S. T. Wilson and co-workers, *J. Amer. Chem. Soc.* **104**, 1146 (1982); S. T. Wilson, B. M. Lok, C. A. Messina, and E. M. Flanigen, *ACS Symp. Ser.* **218**, 79 (1983).
3. B. M. Lok and co-workers, *J. Amer. Chem. Soc.* **106**, 6092 (1984).
4. U.S. Pat. 4,554,143 (Nov. 19, 1985), C. A. Messina, B. M. Lok, and E. M. Flanigen (to Union Carbide Corporation); U.S. Pat. 4,567,029 (Jan. 28, 1986), S. T. Wilson and E. M. Flanigen (to Union Carbide Corporation); S. T. Wilson and E. M. Flanigen, *ACS Symp. Ser.* **398**, 329 (1989).
5. E. M. Flanigen, B. M. Lok, R. L. Patton, and S. T. Wilson, *Pure & Appl. Chem.* **58**, 1351 (1986).
6. R. L. Bedard and co-workers, *Stud. Surf. Sci. Catal.* **49A**, 375 (1989).
7. R. M. Dessau, J. L. Schlenker, and J. B. Higgins, *Zeolites* **10**, 522 (1990).
8. M. E. Davis, C. Montes, and J. M. Garces, *ACS Symp. Ser.* **398**, 291 (1989).
9. J. Patarin and co-workers, *Proc. 9th Intern. Zeolite Conf.* **I**, 263 (1993).
10. P. B. Moore and J. Shen, *Nature* **306**, 356 (1983).
11. C. T. Kresge and co-workers, *Nature* **359**, 710 (1992).
12. Q. Huo and co-workers, *Nature* **368**, 317 (1994).
13. J. L. Schlenker and G. H. Kühl, *Proc. 9th Inter. Zeolite Conf.* **I**, 3 (1993).
14. L. B. Sand and F. A. Mumpton, eds., *Natural Zeolites, Occurrence, Properties, Use*, Pergamon Press, Oxford, UK, 1978, pp. 135–350.
15. J. J. Pluth and J. V. Smith, *Amer. Mineral.* **75**, 501 (1990); J. V. Smith, J. J. Pluth, R. C. Boggs, and D. G. Howard, *J. Chem. Soc., Chem. Commun.*, 363 (1991).
16. L. Pauling, *Proc. Nat. Acad. Sci.* **16**, 453 (1930).
17. F. A. Mumpton, in L. B. Sand and F. A. Mumpton, eds., *Natural Zeolites, Occurrence, Properties, Use*, Pergamon Press, New York, 1978, p. 3.

18. G. T. Kokotailo, S. L. Lawton, D. H. Olson, and W. M. Meier, *Nature* **272**, 437 (1978).
19. E. M. Flanigen and co-workers, *Nature* **271**, 512 (1978).
20. J. V. Smith, *ACS Monogr.* **171**, 3 (1976).
21. N. Y. Chen, W. E. Garwood, and F. G. Dwyer, *Shape–Selective Catalysis in Industrial Applications*, Marcel Dekker, Inc., New York and Basel, 1989.
22. W. M. Meier and D. H. Olson, *Atlas of Zeolite Structure Types*, 3rd rev. ed., Structure Commission of the International Zeolite Association, Butterworth-Heinemann, London, 1992.
23. R. M. Barrer, *Zeolites and Clay Minerals as Sorbents and Molecular Sieves*, Academic Press, London, 1978.
24. N. Y. Chen, *J. Phys. Chem.* **80**, 60 (1976).
25. P. A. Jacobs, *Carboniogenic Activity of Zeolites*, Elsevier Science Publishing Co., Inc., New York, 1977.
26. J. B. Uytterhoeven, *Acta Phys. Chem.* **24**, 53 (1978).
27. R. A. Anderson, *ACS Symp. Ser.* **40**, 637 (1977).
28. D. H. Olson and W. O. Haag, *ACS Symp. Ser.* **248**, 275 (1984).
29. G. T. Kerr, *J. Catal.* **15**, 200 (1969); G. H. Kühl, *J. Phys. Chem. Solids* **38**, 1259 (1977).
30. G. T. Kerr, A. W. Chester, and D. H. Olson, *Acta Phys. Chem.* **24**, 169 (1978).
31. R. M. Dessau, *J. Catal.* **89**, 520 (1984).
32. U. Lohse and co-workers, *Z. Anorg. Allg. Chem.* **460**, 179 (1980).
33. J. Scherzer, *J. Catal.* **54**, 285 (1978).
34. H. K. Beyer and co-workers, *J. Chem. Soc., Faraday Trans. I* **81**, 2889.
35. G. W. Skeels, and D. W. Breck, in D. H. Olson and A. Bisio, eds., *Proceeding of the 6th International Zeolite Conference, Reno, Nev.*, Butterworths, Guildford, U.K., 1984, p. 87.
36. R. M. Dessau and G. T. Kerr, *Zeolites* **4**, 315 (1984).
37. C. D. Chang and co-workers, *J. Am. Chem. Soc.* **106**, 8143 (1984).
38. D. S. Shihabi and co-workers, *J. Catal.* **93**, 471 (1985).
39. A. W. Chester and co-workers, *J. Chem. Soc., Chem. Commun.* **289** (1985).
40. R. M. Barrer and B. Coughlin, *Molecular Sieves*, Society of Chemical Industry, London, 1968, p. 141.
41. K. Yamagishi, S. Namba, and T. Yahima, *Bull. Chem. Soc. Jpn.* **64**, 949 (1991).
42. S. Han, K. D. Schmitt, D. S. Shihabi, and C. D. Chang, *J. Chem. Soc., Chem. Commun.*, 1287 (1993).
43. L. D. Rollmann, *Adv. Chem. Ser.* **173**, 387 (1979);
44. E. M. Flanigen, *Adv. Chem. Ser.* **121**, 119 (1973); S. P. Zhdanov, *Adv. Chem. Ser.* **101**, 20 (1971); F. G. Dwyer, P. H. Schipper, and F. Gorra, *Nat. Petr. Ref. Assoc.*, AM-87-63 (Mar. 1987); C. L. Angell and W. H. Flank, *ACS Symp. Ser.* **40**, 194 (1977).
45. G. T. Kerr, *J. Phys. Chem.* **70**, 1047 (1966).
46. G. T. Kerr, *J. Phys. Chem.* **72**, 1385 (1968).
47. L. Moscou, *Stud. Surf. Sci. Catal.* **58**, 1 (1991).
48. M. Smart, H. Janshekar and Y. Yoshida, *Zeolites*, Marketing Research Report, *Chemical Economics Handbook*, SRI International, 1992.
49. A. P. Bolton, in R. B. Anderson and P. T. Dawson, eds., *Experimental Methods in Catalysis Research*, Vol. 2, Academic Press, Inc., New York, 1976, p. 1.
50. G. Engelhardt and D. Michel, *High-Resolution Solid-State NMR of Silicates and Zeolites*, John Wiley & Sons, Inc., New York, 1987.
51. G. R. Landolt, *Anal. Chem.* **43**, 613 (1973).
52. P. Berth, *J. Amer. Oil Chem. Soc.* **55**, 52 (1978).
53. *Adsorbent Data Sheets, Nos. 3797, 4172, 4174, and 4175*, Union Carbide Corp., New York.
54. J. A. Rabo, R. D. Bezman, and M. L. Poutsma, *Acta Phys. Chem.* **24**(1–2), 39 (1978).

55. P. B. Venuto and E. T. Habib, Jr., *Fluid Catalytic Cracking with Zeolite Catalysts*, Marcel Dekker, Inc., New York, 1979.
56. C. D. Anderson, F. G. Dwyer, G. Koch, and P. Niitranen, *Proc. 9th Iberoamerican Conf. on Catalysis*, Lisbon (1984).
57. N. Y. Chen, W. W. Kaeding, and F. G. Dwyer, *J. Am. Chem. Soc.* **101**, 6783 (1979).
58. W. W. Kaeding, L. B. Young, and A. G. Prapas, *Chemtech.* **12**, 556 (1982).
59. J. D. Sherman, *A.I.Ch.E. Symp. Ser.* **179**, 74, 98 (1978).
60. F. Delprato and co-workers, in J. C. Jansen, L. Moscou, and M. F. M. Post, eds., Recent Research Reports of the 8th International Zeolite Conference, 1989, p. 127.
61. Q. Huo and co-workers, *J. Chem. Soc., Chem. Commun.* 875 (1992).
62. T. E. Gier and G. D. Stucky, *Nature* **349**, 508 (1991).
63. J. S. Beck and co-workers, *J. Amer. Chem. Soc.* **114**, 10834 (1992).

GÜNTER H. KÜHL
CHARLES T. KRESGE
Mobil Research and Development Corp.

MOLLUSCICIDES. See PESTICIDES.

MOLYBDENUM AND MOLYBDENUM ALLOYS

Molybdenum [7439-98-7], identified as a discrete element in 1778 by Scheele, remained a laboratory curiosity until the late 1880s when French metallurgists produced a molybdenum-containing armor-plate steel. Molybdenum was also found to be useful as an additive to tool steels and in chemical dyes (see STEEL; TOOL MATERIALS). The first significant industrial applications of molybdenum were developed during World War I when it was employed in armor-plate steels, tool steels, and high strength steels for aircraft engines.

After World War I, the demand for molybdenum dropped to very low levels. In the 1920s, the Society of Automotive Engineers adopted the 4100 series of chromium–molybdenum steels. From that beginning, the importance of molybdenum both as an alloying element and as a refractory metal grew steadily (see REFRACTORIES). Along with the many metallurgical uses of the metal, molybdenum compounds (qv) are used in such chemical applications as catalysis (qv), corrosion protection (see CORROSION AND CORROSION CONTROL), and lubrication (see LUBRICATION AND LUBRICANTS). A series of conferences (1) on the chemistry and uses of molybdenum provides excellent background information, including discussion of the role of molybdenum in life processes.

Sources and Supply

Most of the world's supply of molybdenum comes as a by-product or co-product from copper (qv) mining. Only about one quarter of the supply comes from primary mines. A small but significant supply of molybdenum is also obtained from the processing of spent petroleum catalysts (see CATALYSTS, REGENERATION; RECYCLING, NONFERROUS METALS). The most abundant mineral, and the only one of commercial significance, is molybdenite [1309-56-4], MoS_2. The minerals powellite [14020-57-0], $Ca(MoW)O_4$, and wulfenite [14913-82-7], $PbMoO_4$, also are known but are not sources of the metal. Primary ore bodies in the western hemisphere contain ca 0.2–0.4% molybdenum and give a recovery of 2–4 kg/t of ore. Copper ores typically contain less than 0.1% Mo.

The largest share of molybdenum supply comes from North America, once the only significant source. Sizeable amounts also come from Latin America (mostly Chile), China, and countries of the former Soviet Union.

In the early 1980s mine capacity approached a level of 136,000 t/yr Mo, far more than demand. Several mines have since closed. As of the early 1990s total molybdenum capacity remained well above total demand of about 90,000 t/yr. Estimated 1993 production in units of metric ton of molybdenum was in Armenia, 450 t; in Canada, 10,000 t; in Chile, 15,000 t; in China, 16,000 t; in Iran, 1,400 t; in Kazakhstan, 1,400 t; in Mexico, 1,800 t; in Mongolia, 1,400 t; in Peru, 2,700 t; in Russia, 5,000 t; in Uzbekistan, 900 t; and in the United States, 37,000 t.

Market Demand

The Western World demand for molybdenum as of the mid-1990s was slightly below 90,000 t/yr. Demand rises and falls with the fluctuating economic climate. World political changes in the early 1990s, including the dissolution of the former USSR and the large growth in the Chinese economy, have had a significant effect on worldwide demand patterns for molybdenum. Whereas it is difficult to determine molybdenum consumption in those two particular geographic sectors, corresponding demand patterns for Mo in the Western World from 1988 to 1992 are shown in Table 1.

Table 1. Western World Molybdenum Demand, t × 10³ Mo[a]

	Year		
Geographic area	1988	1990	1992[b]
United States	23	24	23
Western Europe	37	38	33
Japan	15	17	14
other Western	8.2	8.6	9
Total	*83.2*	*87.6*	*79*

[a]Ref. 2.
[b]Values are estimated.

Metallurgical applications remain the largest use for molybdenum, as shown in Table 2.

Table 2. Molybdenum Demand in 1992[a]

Usage	Quantity of Mo, t $\times$ 10^3	Demand, %
cast iron	4.5	6
chemicals	11.8	15
constructional alloy steels	26.3	33
Mo-base and superalloys	6.4	8
stainless steel	24.5	31
tool steel	5.4	7

[a]Usage in former Eastern-bloc countries having centralized economies is excluded.

Processing Ores into Commercial Products

Molybdenite is concentrated by first crushing and grinding the ore, and passing the finely ground material through a series of flotation (qv) cells (see MINERALS RECOVERY AND PROCESSING). Operations which recover molybdenum as a by-product of copper mining produce a concentrate containing both metals. Molybdenite is separated from the copper minerals by differential flotation.

Molybdenite concentrate contains about 90% MoS_2. The remainder is primarily silica, with lesser amounts of Fe, Al, and Cu. The concentrate is roasted to convert the sulfide to technical molybdic oxide. Molybdenum is added to steel in the form of this oxide. In modern molybdenum conversion plants, the oxidized sulfur formed by roasting MoS_2 is converted to sulfuric acid.

Technical molybdic oxide can be reduced by reaction of ferrosilicon in a thermite-type reaction. The resulting product contains about 60% molybdenum and 40% iron. Foundries generally use ferromolybdenum for adding molybdenum to cast iron and steel, and steel mills may prefer ferromolybdenum to technical molybdic oxide for some types of steels.

A small portion of molybdenite concentrate production is purified to yield lubricant-grade molybdenum disulfide, a widely used solid-state lubricant.

Chemical products are produced from technical-grade oxide in two very different ways. Molybdenum trioxide can be purified by a sublimation process because molybdenum trioxide has an appreciable vapor pressure above 650°C, a temperature at which most impurities have very low volatility. The alternative process uses wet chemical methods in which the molybdenum oxide is dissolved in ammonium hydroxide, leaving the gangue impurities behind. An ammonium molybdate is crystallized from the resulting solution. The ammonium molybdate can be used either directly or thermally decomposed to produce the pure oxide, MoO_3.

Analytical Methods. Molybdenum contents in ore concentrates and technical oxide are most accurately determined gravimetrically by precipitating lead molybdate. Molybdenum content is usually not determined on pure compounds or metal. Instead, spectrographic methods are used to measure impurity elements that must be controlled. Carbon and oxygen in metal products are measured by standard gas analysis methods.

Environmental and Safety Considerations

Because of its position in the Periodic Table, molybdenum has sometimes been linked to chromium (see CHROMIUM AND CHROMIUM ALLOYS) or to other heavy metals. However, unlike those elements, molybdenum and its compounds have relatively low toxicity, as shown in Table 3. On the other hand, molybdenum has been identified as a micronutrient essential to plant life (11,12) (see FERTILIZERS), and plays a principal biochemical role in animal health as a constituent of several important enzyme systems (see MINERAL NUTRIENTS).

Information on the toxic effects of molybdenum in humans is scarce. A high incidence of gout was reported in a locale in Armenia where the soil contained exceptionally high levels of both molybdenum and copper (15). However, the significance of the suggested correlation is questionable because of the lack of information on the study population and the absence of a control group.

Industrial exposure of humans to molybdenum has been reported in both the former Soviet Union and the United States. In the former, workers having the highest exposure had the highest blood level of uric acid (16). In the latter,

Table 3. Toxicity of Molybdenum Compounds

Compound	Species	Time, h	LC_{50}, mg/L	Reference
	Freshwater fish			
ammonium dimolybdate	bluegill	48	157	3
	rainbow trout	48	135	3
molybdenum trioxide	bluegill	48	>87 <120	3
	rainbow trout	48	>65 <87	
	Daphnia			
sodium molybdate dihydrate	all species	96	3940	4
	Saltwater species			
sodium molybdate dihydrate	mysid shrimp	96	3997	5
	sheepshead minnow	96	6590	
	pink shrimp	96	3997	
	American oyster		3526	
ammonium molybdate	marine shore crab	48	1018[a]	6
	Mammals			
molybdenum trioxide	rat[b]		2.73[c]	7
molybdenum trioxide (technical grade)	rat[b]		6.66[c]	8
sodium molybdate	[d]		344[e]	9
molybdenum disulfide	rat[b]		[f]	10

[a]Value is mg of Mo.
[b]Compound given orally.
[c]Units are g/kg.
[d]Compound given intraperitoneally or intramuscularly.
[e]Units are mg/kg.
[f]No effect level >15 g/kg given orally.

urinary and plasma molybdenum levels were higher for the worker group than for a control group (17). However, the urea level of the exposed group was still within the range of the control group, and no gout-like symptoms were reported.

Molybdenum Metal

Physical Properties.   Molybdenum has many unique properties, leading to its importance as a refractory metal (see REFRACTORIES). Molybdenum, atomic no. 42, is in Group 6 (VIB) of the Periodic Table between chromium and tungsten vertically and niobium and technetium horizontally. It has a silvery gray appearance. The most stable valence states are +6, +4, and 0; lower, less stable valence states are +5, +3, and +2.

Molybdenum, a typical transition element, has the maximum number, five, of unpaired $4d$ electrons, which account for its high melting point, strength, and high modulus of elasticity. There are many similarities between molybdenum and its horizontal and vertical neighbors in the periodic system.

The melting point of molybdenum is about 2626°C, 1100°C above that of iron. Only two other commercially significant elements, tungsten and tantalum, have higher melting points than molybdenum. As a result of its high melting temperature, molybdenum metal has strength characteristics at temperatures where most metals are in the molten state, and some applications, such as for furnace parts, rocket nozzles, welding tips, thermocouples, glass melting electrodes, dies, and molds, are based on this property. Atomic properties are given in Table 4, thermal properties in Table 5, and electrical, magnetic, and optical properties are given in Table 6.

Chemical Properties.   Molybdenum has good resistance to chemical attack by mineral acids, provided that oxidizing agents are not present. The metal also offers excellent resistance to attack by several liquid metals. The approximate temperature limits for molybdenum to be considered for long-time service while in contact with various metals in the liquid state are as follows:

Metal	Limit, °C	Metal	Limit, °C
bismuth	1425	mercury	600
gallium	400	potassium	1100
lead	1200	sodium (liquid and vapor)	1500
lithium	925		
magnesium	700		

In addition, molybdenum has high resistance to a number of alloys of these metals and also to copper, gold, and silver. Among the molten metals that severely attack molybdenum are tin (at 1000°C), aluminum, nickel, iron, and cobalt. Molybdenum has moderately good resistance to molten zinc, but a molybdenum–30% tungsten alloy is practically completely resistant to molten zinc at temperatures up to 800°C. Molybdenum metal is substantially resistant to many types of molten glass and to most nonferrous slags. It is also resistant to liquid sulfur up to 440°C.

Table 4. Physical Properties of Molybdenum[a]

Property	Value
isotopes	
natural	92,94,95,96,97,98,100
artificial	90,91,93,99,101,102,105
atomic weight	95.94
atomic radius, coordination number 8, pm	136
ionic radius, pm	
Mo^{3+}	92
Mo^{6+}	62
atomic volume, cm^3/mol	9.41
lattice	
type	body-centered cubic
constant at 25°C, pm	314.05
fast-neutron absorption cross section, 10/250 keV, m^2[b]	9×10^{-28}
ionization potential, eV	7.2
work function, eV	
apparent positive-ion	8.6
apparent electron	4.2
density at 20°C, g/cm^3	10.22
coefficient of friction vs steel at HRC[c] 44	
dry	
static	0.271
dynamic	0.370
humid	
static	0.405
dynamic	0.465
compressibility at 293°C, cm^2/kg	3.6×10^{-7}
velocity of sound at 2.25 MHz, cm/s	
longitudinal wave, V_L	$6.37 \pm 0.02 \times 10^5$
shear wave, V_S	$3.41 \pm 0.06 \times 10^5$
thin rod, V_O	5.50×10^5
surface tension at melting point, mN/m(=dyn/cm)	2240

[a]Refs. 18–22.
[b]To convert m^2 to barn, divide by 10^{-28}.
[c]HRC = Rockwell C hardness.

    Above ca 600°C, unprotected molybdenum oxidizes so rapidly in air or oxidizing atmospheres with formation of volatile MoO_3, that extended use under these conditions is impractical. Because no molybdenum-base alloy that combines high oxidation resistance and good high temperature properties has been discovered, protective coatings are required where oxidation is a problem. Various coatings, differing in maximum time–temperature capabilities and in physical and mechanical characteristics, are available. The most widely used coatings depend on the formation of a thin surface layer of $MoSi_2$ on the molybdenum metal part. This compound has outstanding oxidation resistance at temperatures up to about 1650°C (see REFRACTORY COATINGS).

    In a vacuum, uncoated molybdenum metal has an unlimited life at high temperatures. This is also true under the vacuum-like conditions of outer space. Pure hydrogen, argon, and helium atmospheres are completely inert to molyb-

Table 5. Thermal Properties of Molybdenum[a]

Property	Value
melting point, °C	2626 ± 9
heat of fusion,[b] kJ/mol[c]	28
boiling point, °C	5560
heat of vaporization, kJ/mol[c]	491
entropy of crystals, $S^{\circ}_{298.16}$, J/(mol·K)[c]	28.6
vapor pressure, Pa[d]	
at 1725°C	3.95×10^{-9}
2225°C	4.35×10^{-6}
2610°C	1.72×10^{-4}
2725°C	4.05×10^{-4}
3225°C	8.71×10^{-3}
3725°C	8.41×10^{-2}
4225°C	4.76×10^{-1}
4725°C	1.82
5225°C	5.57
rate of evaporation, r_{ev}, g/(cm²·s)	$\log r_{ev} =$
diffusivity, cm²/s	$17.11 - 38,600\,T^e - 1.76 \log T^e$
at 200°C	0.43
540°C	0.40
870°C	0.38
specific heat at 100°C, kJ/(kg·K)[c]	0.27
coefficient of linear expansion, %	
at 0–400°C	0.23
0–800°C	0.46
0–1200°C	0.72
thermal conductivity, W/(m·K)	
at 500°C	122
1000°C	101
1500°C	82

[a]Refs. 18–22.
[b]Value is estimated.
[c]To convert J to cal, divide by 4.186.
[d]To convert Pa to mm Hg, multiply by 0.0075.
[e]T in Kelvin.

denum at all temperatures, whereas water vapor, sulfur dioxide, and nitrous and nitric oxides have an oxidizing action at elevated temperatures. Molybdenum is relatively inert to carbon dioxide, ammonia, and nitrogen atmospheres up to about 1100°C; a superficial nitride film may be formed at higher temperatures in the latter two gases. Hydrocarbons and carbon monoxide may carburize molybdenum at temperatures above 1100°C.

In a reducing atmosphere, molybdenum is resistant at elevated temperatures to hydrogen sulfide, which forms a thin adherent sulfide coating. In an oxidizing atmosphere, however, molybdenum is rapidly corroded by sulfur-containing gases. Molybdenum has excellent resistance to iodine vapor up to about 800°C. It is also resistant to bromine up to about 450°C and to chlorine up to about 200°C. Fluorine, the most reactive of the halogens, attacks molybdenum at room temperature.

Table 6. Electrical, Magnetic, and Optical Properties[a]

Property	Value
electrical resistivity, $n\Omega \cdot m$	
at 0°C	50
1000°C	320
2000°C	610
Franz-Wiedemann or Lorenz constant	2.72
hydrogen overpotential, 1×10^{-2} A/cm^2, V	0.44
electrochemical equivalent, Mo^{6+}, mg/C	0.1658
minimum arcing amperage, A	
at 24 V	10
110 V	1.5
220 V	1.0
magnetic susceptibility,[b] κ_p, m^3/kg	
at 25°C	0.93×10^{-6}
1825°C	1.11×10^{-6}
optical reflectivity, %	
at 500 nm	46
10,000 nm	93
total hemispherical emissivity	
at 1200 K	0.104
1600 K	0.163
2000 K	0.209
2400 K	0.236
thermionic emission of high vacuum, mA/cm^2	
at 1600°C	ca 0.7
2000°C	85
spectral emissivity	
at 390 nm	ca 0.43
6700 nm	0.40
radiation for 550 nm at 20°C, % of black-body radiation	54
total radiation, W/cm^2	
at 527°C	ca 0.2
1127°C	3.0
1727°C	19
2327°C	68

[a]Refs. 18–22.
[b]Mo is paramagnetic.

Manufacturing Processes. Ammonium molybdate or molybdenum trioxide is reduced to molybdenum metal powder by hydrogen in a two-stage process. In the first stage, MoO$_3$ or ammonium molybdate is reduced to molybdenum dioxide, MoO$_2$, at temperatures around 600°C; in the second stage, the dioxide is reduced to metallic powder at temperatures near 1100°C. Both rotary and boat-and-tube types of furnaces are used for first-stage reduction. Boat-and-tube furnaces are used for the second stage.

Wire. Molybdenum wire is produced by a long-established powder metallurgy process (see METALLURGY, POWDER). Molybdenum powder is compacted in dies in a hydraulic press at pressures in the 210–280 MPa (30,000–40,000 psi) range to produce bars approximately 30 mm square. These bars are sintered by electrical resistance heating in hydrogen atmosphere chambers or bells. Currents are adjusted to about 90% of that required to cause melting; the bars attain temperatures in the range 2200–2300°C. The sintered bars are hot-rolled or swaged to small-diameter rods which are subsequently drawn to fine wire. Tungsten carbide dies are used for drawing heavy-gauge wire and diamond dies for fine-gauge wire. Drawing of the coarse sizes is accomplished at elevated temperatures; final drawing of fine wire is performed at ambient temperatures. Lubrication is provided by suspensions of graphite or molybdenum disulfide. At various stages in the reduction process, the wire may be annealed at 800–850°C to facilitate further working.

Mill Products. Two consolidation processes are used to produce molybdenum mill products such as forging billets, bars, rods, plate, sheet, and foil: powder metallurgy and arc-casting or vacuum-arc melting. In the powder metallurgy process, molybdenum powder is compacted isostatically in hydraulic pressure chambers to cylindrical bars or billets and to rectangular sheet bars. The green compacts are subsequently sintered in hydrogen atmosphere muffle furnaces for several hours at 1600–1700°C. The longer time at lower temperature results in about the same 95% density produced by bell sintering. Sintered bars up to 75 mm in diameter are rolled directly to smaller bars; larger consolidations of 155–300 mm diameter are extruded to bars for further processing. Pressed and sintered sheet bars are rolled directly to plate, sheet, and foil. As mechanical work proceeds, the density of powder metallurgy molybdenum improves to the full theoretical density.

In the arc-casting process developed in the 1940s, a consumable electrode of compacted molybdenum powder is melted by an alternating current arc inside a water-cooled copper tube, or mold, to form an ingot. A blend of molybdenum powder, carbon for deoxidation, recycled machine chips of pure molybdenum, and any required alloying elements is charged into a hopper. The hopper feeds a device that continuously compacts the loose bulk charge into a column of hexagonal wafers about 67 mm across flats and 15–20 mm thick. Following compaction the column is sintered by resistance heating to temperatures near 1200°C to strengthen the compacted metal and to weld adjacent wafers together. The pressed and sintered electrode extends downward along the axis of the water-cooled copper mold to the position of the arc between it and the molten pool of metal at the top of the ingot. Compacting, sintering, and melting are performed continuously and consecutively in connected chambers evacuated to pressures of 1.3–6.5 Pa (10–50 μm Hg). Ingots weighing 820 kg have been produced consistently in the pressing, sintering, and melting (PSM) machine. The cast ingots have full theoretical density. The distribution of alloying elements is microscopically uniform throughout the entire ingot.

Metalworking. Molybdenum metal can be mechanically worked by almost any process: forging, extrusion, rolling, bending, punching, stamping, deep drawing, spinning, conventional forming, and power roll forming. Except for fine wire and thin sheet, it is recommended that mill products be heated moderately for most shaping operations. Infrared lamps or hot plates often provide adequate heating for thinner gauges of molybdenum sheet; for heavier gauges, it is advantageous to heat the metal in a furnace or to use an oxyacetylene torch.

Using proper tool angles, molybdenum can be machined without difficulty. Although high speed steel tools are recommended for heavy, intermittent cuts on lathes, sintered carbide tools are preferred for most turning operations, shaping, or milling. Speeds up to 182 surface m/min can be used on lathes having carbide tools.

For many applications requiring joints of molybdenum, mechanical methods and brazing are satisfactory. Mechanical joints can be produced without deterioration of the properties developed by cold-working the metal. Brazed joints suffer from loss of strength at high temperatures and lowering of the melting range at the juncture. Nickel foil has been used successfully as the brazing metal to join flat surfaces of molybdenum.

Molybdenum can be welded to itself by a number of fusion welding (qv) processes, including electric arc (gas–tungsten arc and gas–metal arc), electric resistance (spot welding and flash welding), and electron beam. In these welding processes, all the physical properties of molybdenum are retained but the tensile properties developed by cold-working are reduced. In fusion welding there is always the cast weld metal and a heat-affected zone (haz) comprising recrystallized, coarse grains. Each of these structures has lower strength and lower ductility than the parent metal. The high melting point of molybdenum and its high thermal conductivity necessitate high energy input and rapid travel to minimize the heat-affected zone for optimally welded joints. To improve the strength and ductility of a fusion weld, it is beneficial to warm-work the joint whenever feasible.

Because the ductility of molybdenum is adversely affected by even minute amounts of oxygen, stringent precautions must be taken to prevent it from being present during fusion welding. Inert gas atmospheres must be as free of oxygen as possible. Commercial argon, helium, and hydrogen may contain oxygen levels which can cause cracking and porosity unless specific purification measures are provided. Thorough cleaning of the faying surfaces is essential for optimum ductility. Such surfaces have usually become contaminated during prior processing and handling. Wrought arc-cast molybdenum can be welded with less difficulty than powder metallurgy products because the arc-cast metal has been out-gassed and deoxidized during melting.

Excellent high strength welds have been produced by inertia-welding, or friction-welding, which develops essentially no heat-affected zone. The actual interface is wrought molybdenum because the molten metal and the adjacent solid metal that has been raised to very high temperatures have been expelled from the joint.

Uses. Molybdenum metal is the most widely used electrical resistance element in furnaces where temperatures beyond the limits of ordinary resistance alloys are required. Such furnaces are generally used for temperatures up to about 1650°C, but some are in successful operation at 2200°C. The elements, which may be wire, rod, ribbon, or expanded sheet, must be protected from oxidation by a reducing or inert atmosphere, or a vacuum. Hydrogen is commonly employed. Under these conditions, molybdenum has a long life and seldom is the limiting factor in the durability of the furnace. The furnace industry also consumes sizable amounts of molybdenum sheet for susceptors in high frequency units, as well as radiation shields, baffles, structural supports, muffle liners, skids, hearths, boats, and firing trays in all types of high temperature vacuum and controlled atmosphere furnaces.

The same properties that make molybdenum metal effective in high temperature furnace applications make it useful as support wires for tungsten filaments in incandescent light bulbs and as targets in x-ray tubes.

The glass (qv) industry has become an important user of large molybdenum parts. The advantages of molybdenum include its high melting point, high strength at elevated temperature, good electrical conductivity, resistance to attack by most molten glasses, as well as the fact that any oxide formed is colorless under normal operation conditions. The main application is in electrically heated glass furnaces, where it serves as electrodes which may be either of the plate or rod type. Stirrers, pumps, bowl liners, wear parts, and molds are also made from molybdenum metal.

Because of its high modulus of elasticity, molybdenum is used in machine-tool accessories such as boring bars and grinding quills. Molybdenum metal also has good thermal-shock resistance because of its low coefficient of thermal expansion combined with high thermal conductivity. This combination accounts for its use in casting dies and in some electrical and electronic applications.

Molybdenum Base Alloys

The strength of molybdenum depends on work-hardening. The greater the amount of cold-working, ie, percent reduction of area below the recrystallization temperature, the higher the yield and tensile strengths at all temperatures and the lower the creep rates at elevated temperatures. Exposure of the work-hardened structure to temperatures high enough to cause recrystallization produces a drastic reduction in tensile properties, and a loss of ductility as measured by bending tests or notched-bar impact tests. Unalloyed molybdenum cold-worked by 90% reduction of area fully recrystallizes at 1175°C in 1 h.

Additions of selected alloying elements raise the recrystallization temperature, extending to higher temperature regimes the tensile properties of the cold-worked molybdenum metal. The simultaneous additions of 0.5% titanium and 0.1% zirconium produce the TZM alloy, which has a corresponding recrystallization temperature of 1500°C and which cold-works to higher hardness and strength than unalloyed molybdenum. The alloy is recognized as a standard by the American Society for Testing and Materials for critical, high temperature structural applications (ASTM B386, B387); its superior properties justify the higher cost of processing. Tensile strength data for molybdenum and TZM are given in Table 7.

An alloy of molybdenum containing 1.2% hafnium with carbon at the level of 0.08–0.10% has a slight advantage over TZM. This alloy has been produced in small quantities for special extrusion dies and ejector pins in the isothermal forging of superalloys.

Tungsten has little effect on recrystallization temperature or the high temperature properties of molybdenum. However, the Mo–30% W alloy is recognized as a standard commercial alloy for stirrers, pipes, and other equipment that is required to be in contact with molten zinc during processing of the metal and in galvanizing and die casting operations.

Table 7. Tensile Strength of Rolled Bars[a]

Temperature, °C	Tensile strength, MPa[b]	
	Unalloyed molybdenum	TZM alloy
ambient	790	830
200	750	790
400	660	730
600	570	670
800	480	600
1000	370	530
1200	180	450

[a]Ref. 23.
[b]To convert MPa to psi, multiply by 145.

Molybdenum–rhenium alloys received increased attention during the 1980s for special low volume applications. The addition of 40–50% rhenium increases the tolerance for oxygen, improves ductility at ambient temperatures, and enhances weldability. The alloy containing 41% Re is used for critical high temperature structures in aerospace applications, especially where welding is required (see HIGH TEMPERATURE ALLOYS). The Mo–47.5% Re alloy is produced for high temperature thermocouple wire, for radar equipment, and for microwave communication applications. These alloys are produced by powder metallurgy operations. The high and unstable price of rhenium powder has limited the market for Mo–Re alloys.

Molybdenum as an Alloying Element

Molybdenum, an unusually versatile alloying element, imparts numerous beneficial properties to irons and steels and to some alloy systems based on cobalt, nickel, or titanium. Comprehensive summaries of uses through 1948 (24) and 1980 (25) are available.

Steels develop excellent combinations of strength and toughness if heat-treated by quenching to martensite, followed by tempering (see STEEL). Steels having adequate hardenability develop martensitic structures in practical section sizes. Molybdenum is a potent contributor to hardenability, and has been shown to be even more effective in the presence of carefully selected amounts of other alloying elements (26). The end-quench test has become the accepted method for measuring hardenability, and the data can be correlated with section size. Technical societies worldwide have standardized hardenability limits (bands) for a large number of carbon and alloy steels; standards of the Society of Automotive Engineers are examples (27).

Many steels used for gears and bearings are surface-hardened by carburizing, quenching, and tempering. Molybdenum is frequently used in carburized steels, and carburized Ni–Mo steels have been shown to provide optimum resistance to fatigue and impact effects (28).

The microstructures which can be developed using a wide range of heat treating conditions can be predicted from data presented in continuous-cooling transformation (CCT) diagrams and from isothermal transformation (IT or TTT) diagrams. These diagrams represent an extension of the hardenability concept.

Many such diagrams exist (29). Reviews of specific applications show the ability of molybdenum, in combination with other alloying elements, to develop the desired microstructure (30,31).

The concentration of molybdenum in constructional alloy steels is usually in the range of 0.15–0.30%. A few specialty steels contain 0.30–0.80% Mo. The best-known steels are the Cr–Mo and Ni–Cr–Mo grades. Molybdenum is particularly effective in reducing the susceptibility of Cr–Ni steels to embrittlement following tempering at >600°C or exposure to temperatures ca 500°C for extended times. Embrittlement is attributed to the presence of undesirable trace elements in the grain boundaries. Molybdenum tends to concentrate at the grain boundaries, rendering the trace elements less harmful.

Molybdenum-alloyed steels have been found to perform better than other steels in service in the oil industry (see PETROLEUM). Drill pipe and casing for deep wells are exposed to high concentrations of hydrogen sulfide, which causes sulfide stress cracking (SSC). Steels containing molybdenum in amounts as high as 0.80% can be heat-treated to high strength levels and still resist SSC (32).

An early application of molybdenum was its substitution for tungsten in the high speed steels popular before World War I. The red hardness of a steel containing 9% molybdenum is as high as that of a steel containing 18% tungsten. Subsequently, a series of high speed steels containing 3.75–9.5% molybdenum was developed. Substantially all hacksaws and twist drills made in the United States are manufactured from molybdenum-containing high speed steels.

High strength, low alloy (HSLA) steels often contain 0.10–0.30% molybdenum. These steels exhibit toughness at low temperatures and good weldability. They are used extensively for undersea pipelines (qv) transporting gas and oil from offshore wells to pumping stations on shore, and are also used extensively in remote Arctic environments.

Molybdenum improves the corrosion resistance of stainless steels that are alloyed with 17–29% chromium. The addition of 1–4% molybdenum results in high resistance to pitting in corrosive environments, such as those found in pulp (qv) and paper (qv) processing (33), as well as in food preparation, petrochemical, and pollution control systems.

A simple, low cost steel for high temperature service in electric power generation (qv) is the C–0.5% Mo steel known as carbon–half moly, which was widely used for many years. The power industry and oil refineries have turned to 1.25% Cr–0.5% Mo and 2.25% Cr–1% Mo steels for high stress and high temperature service, because these steels have improved resistance to graphitization and oxidation, as well as higher creep and rupture strength.

In cast irons, alloying with molybdenum increases hardness and strength. Such irons have been used for crankshafts in large diesel engines for power generation, in ship propulsion and linepipe pumps (qv), and in some special railroad castings. There is increasing interest in specially heat-treated nodular graphite (ductile) irons as alternatives to steel in automotive parts such as gears, crankshafts, and pistons (34,35). Additions of molybdenum are important to the development of the unique acicular microstructure which characterizes these irons.

Abrasion-resistant white cast irons are used in ore processing (see MINERALS RECOVERY AND PROCESSING), in the crushing and pulverizing of coal (qv), and in the grinding of cement (qv) clinker. These white irons are alloyed with 15–28% chromium and 2–3% molybdenum to provide effective wear resistance. Although the highly alloyed irons are sometimes heat-treated to develop the desired martensite–austenite structure, castings often develop a suitable matrix structure in the mold and can be put into service without heat treatment, providing substantial savings in fuel and handling costs. The relevant effect of molybdenum in these castings is its ability to stabilize the structure so effectively that undesirable pearlite does not form during relatively slow cooling to room temperature.

BIBLIOGRAPHY

"Molybdenum and Molybdenum Alloys" in *ECT* 1st ed., Vol. 9, pp. 191–199, by R. I. Jaffee, Battelle Memorial Institute; in *ECT* 2nd ed., Vol. 13, pp. 634–644, by J. Z. Briggs, Climax Molybdenum Co.; in *ECT* 3rd ed., Vol. 15, pp. 670–682, by R. Q. Barr, Climax Molybdenum Company.

1. *First International Conference on the Chemistry and Uses of Molybdenum*, Reading University, Reading, U.K., 1973; *Second International Conference on the Chemistry and Uses of Molybdenum*, New College, Oxford, U.K., 1976; *Third International Conference on the Chemistry and Uses of Molybdenum*, University of Michigan, 1979; *Fourth International Conference on the Chemistry and Uses of Molybdenum*, Colorado School of Mines, 1982; *Proceedings of the Fifth International Conference on the Chemistry and Uses of Molybdenum*, University of Newcastle-upon-Tyne, U.K., 1985, *Polyhedron* **5**(1/2), (1986).
2. J. B. Hartland, *Eng. Mining J.* (April 1992).
3. *Acute Toxicity of Ammonium Molybdate and Molybdic Trioxide to Bluegill (Lepomis macrochirus) and Rainbow Trout (Salmo gairdneri)*, Bioassay Report submitted to AMAX Inc., Bionomics, Inc., Wareham, Mass., Jan. 1975.
4. *Acute Toxicity of Sodium Molybdate to Bluegill (Lepomis macrochirus), Rainbow Trout (Salmo gairdneri), Fathead Minnow (Pimephales promeals), Channel Catfish (Ictalurus punctatus), Water Flea (Daphnia magna) and Scud (Gammarus fasciatus)*, Bioassay Report submitted to Climax Molybdenum Co. of Michigan, Bionomics, Inc., Wareham, Mass., Dec. 1973.
5. D. W. Knothe, G. G. Van Riper, *Bull. Environ. Contamin. Toxicol.* **40**, 785 (1988).
6. O. J. Abbott: *Marine Poll. Bull.* **8**(9), 204–205 (1977).
7. *Acute Oral LD$_{50}$ Assay in Rats*, FDRL ID:81-0393 for AMAX Inc., Greenwich, Conn., Aug. 1981.
8. *Acute Oral LD$_{50}$ Assay in Rats*, FDRL ID:81-0394 for AMAX Inc., Greenwich, Conn., Aug. 1981.
9. N. I. Sax, *Dangerous Properties of Industrial Materials*, 6th ed., Van Nostrand Reinhold Co., New York, 1984, p. 1953.
10. *Acute Oral Toxicity in Rats of MoS$_2$*, FDRL ID:9589A for AMAX Inc., Greenwich, Conn., Nov. 29, 1987.
11. A. J. Anderson, *J. Aust. Inst. Agric. Sci.*, 873 (1942).
12. M. Neenan, *Proc. Soil Sci. Soc. Fla.* **13**, 178 (1953).
13. E. C. DeRenzo and co-workers, *Arch. Biochem. Biophys.* **45**, 247 (1953).
14. D. A. Richert and W. W. Westerfield, *J. Biol. Chem.* **203**, 915 (1953).
15. V. Kovalskii, G. Yarovaya, and D. Schmavonyan, *Z. Obsc. Biol.* **22**, 179 (1961).
16. O. Akopajan, *Some Biological Shifts in the Bodies of Workers in Contact with Molybdenum Dust*, Second Scientific Conference of the Institute of Labor of Hygiene and

Occupational Diseases on Problems of Labor Hygiene and Occupational Pathology, Erevan, 1963, pp. 103–106.

17. P. S. Walravens and co-workers, *Arch. Environ. Health* **34**, 302–308 (1979).
18. *Molybdenum Metal*, Climax Molybdenum Co., Ann Arbor, Mich., 1960.
19. *Metals Handbook*, 9th ed., Vol. 2, American Society for Metals, Metals Park, Ohio, 1979, p. 771.
20. *Aerospace Structural Metals Handbook*, Mechanical Properties Data Center, Department of Defense, Belfour Stulen, Inc., Code 5301, Mar. 1963, p. 5.
21. V. E. Peletskii and V. P. Druzhinin, *High Temp. High Press.* **2**, 69 (1970).
22. M. M. Kenisarin, B. Ya. Berezin, and V. Ya. Chekhovskoi, *High Temp. High Press.* **4**, 707 (1972).
23. *Molybdenum Mill Products*, AMAX Specialty Metals Division, Greenwich, Conn., 1971.
24. R. S. Archer, J. Z. Briggs, and C. M. Loeb, *Molybdenum Steels, Irons, Alloys*, Climax Molybdenum Co., New York, 1948.
25. A. Sutulov, *International Molybdenum Encyclopedia*, Vol. III, Intermet Publications, Santiago, Chile, 1980.
26. C. A. Siebert, D. V. Doane, and D. H. Breen, *The Hardenability of Steels—Concepts, Metallurgical Influences and Industrial Applications*, American Society for Metals, Metals Park, Ohio, 1977.
27. *SAE J406—Methods of Determining Hardenability of Steels*; *SAE J1268— Hardenability Bands for Carbon and Alloy Steels*; *SAE J1868—Restricted Hardenability Bands for Selected Alloy Steels*, Society of Automotive Engineers Standards for Hardenability, SAE, Warrendale, Pa.
28. D. V. Doane, in G. Krauss, ed., "Carburizing, Processing and Performance," *Proceedings of International Conference*, ASM International, 1989, p. 169.
29. G. F. Vander Voort, ed., *Atlas of Time-Temperature Diagrams for Irons and Steels*, ASM International, 1991.
30. J. A. Straatmann, D. V. Doane, and Y. J. Park, in A. J. DeArdo, ed., *Processing Microstructure and Properties of HSLA Steels*, The Metallurgical Society, Warrendale, Pa., 1988.
31. D. V. Doane, in M. E. Finn, ed., *Factors Influencing Machining and Their Controls*, Proceedings of an International Conference and Workshop, ASM International, 1990.
32. P. J. Grobner, D. L. Sponseller, and D. E. Diesburg, Paper No. 40, in *Corrosion/78*, International Corrosion Forum, sponsored by the National Association of Corrosion Engineers, Mar. 1978.
33. A. Poznansky and E. A. Lizlovs, Paper No. 24, in *Corrosion/84*, NACE, New Orleans, La., Apr. 2–6, 1984.
34. *Proceedings of 1st International Conference on Austempered Ductile Iron*, American Society for Metals, 1984.
35. *Proceedings of 2nd International Conference on Austempered Ductile Iron*, Gear Research Institute, Lisle, Il., 1986.

D. V. DOANE
Consultant

G. A. TIMMONS
Consultant

C. J. HALLADA
Climax Molybdenum Company

MOLYBDENUM COMPOUNDS

The chemistry of molybdenum, Mo, is among the most diverse of the transition elements. In its compounds, molybdenum exhibits coordination numbers from four to eight, oxidation numbers from $-II$ to VI, and numerous states of aggregation (nuclearity). Molybdenum forms binary compounds with many nonmetallic elements, and a number of these, namely the halides, oxides, sulfides, carbides, nitrides, and silicides, are of technological interest. In contrast to its congeners, chromium and tungsten, molybdenum is found naturally in the form of its sulfide molybdenite [1309-56-4], MoS_2. Similarly, in the enzymes in which molybdenum is found, the active site Mo is generally in a high sulfur environment. This thiophilicity of Mo also plays a role in a number of its technological uses.

In biology molybdenum is a component of fertilizer and nutrient formulations (see FERTILIZERS; MINERAL NUTRIENTS). Over 20 enzymes have been found to have molybdenum as a component of their active sites. The roles of molybdenum in nitrogen fixation (qv) and nitrate reduction establish this metal as a key element of a biological nitrogen cycle. In technology various solid and soluble molybdenum compounds have found use in lubrication (see LUBRICATION AND LUBRICANTS); hydrodesulfurization, hydrogenation, and oxidation catalysis; anticorrosion and coatings (qv); flame and smoke retardancy (see FLAME RETARDANTS); and various forms of pigmentation.

The most important molybdenum oxidation states are VI, V, IV, III, II, and 0. The higher oxidation states are usually characterized by molybdenum binding to electronegative atoms, such as oxygen and the halogens. The lowest oxidation states are largely in the realm of organometallic chemistry, wherein the Mo is bound directly to the carbon atom of carbon monoxide (qv), to organic phosphines, and/or to a variety of unsaturated carbonaceous ligands.

Molybdenum(VI)

The chemistry of hexavalent molybdenum is very prominent in both biological and industrial systems. Oxygen coordination of molybdenum is most common in this oxidation state (1–3). Molybdenum trioxide [1313-27-5], MoO_3, is a key intermediate in the technological utilization of molybdenum (Fig. 1). In the refining of Mo, molybdenite ore, MoS_2, which contains tetravalent Mo, is first roasted in air to form impure MoO_3. The MoO_3 is then reduced to the metal with hydrogen from 500–1150°C. The trioxide melts at 795°C but sublimes significantly below that temperature. The structure of MoO_3 is a complex, layered arrangement in which each of the six-coordinate Mo(VI) atoms shares the face of an octahedron with another Mo(VI) atom. The MoO_3 reacts with base to produce a variety of molybdate salts, the simplest of which are of the form M_2MoO_4. Sodium molybdate [7631-95-0] is an example. These water-soluble salts serve as the starting materials for the synthesis of a wide variety of compounds.

The molybdate ion, MoO_4^{2-}, is a d^0, four-coordinate, tetrahedral anion. The structure (Fig. 2a) resembles that of other Group 6 (VIB) and Group 16 (VIA) ions, such as CrO_4^{2-}, WO_4^{2-}, SO_4^{2-}, and SeO_4^{2-}. The discrete dimolybdate ion [19282-23-6], $Mo_2O_7^{2-}$, exists in $N(C_4H_9)_4^+$ salts (see QUATERNARY AMMONIUM

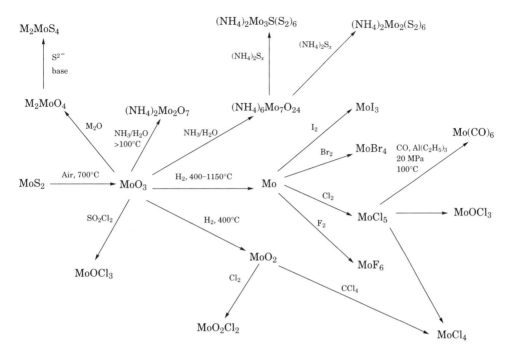

Fig. 1. Scheme for the preparation of technologically important compounds of molybdenum, where M = Li, Na, K, Rb, Cs, and NH$_4$. To convert MPa to psi, multiply by 145.

COMPOUNDS). Diammonium dimolybdate [*27546-07-2*], (NH$_4$)$_2$Mo$_2$O$_7$, available commercially as the tetrahydrate and prepared from MoO$_3$ and excess NH$_3$ in aqueous solution at 100°C, has an infinite chain structure based on MoO$_6$ octahedra. In aqueous solution the behavior of Mo(VI) is extremely pH-dependent (4). Above pH 7 molybdenum(VI) occurs as the tetrahedral oxyanion MoO$_4^{2-}$, but below pH 7 a complex series of concentration-, temperature-, and pH-dependent equilibria exist. The best known of these equilibria lead to the formation of the heptamolybdate, Mo$_7$O$_{24}^{6-}$ (Fig. 2**h**), and octamolybdate, Mo$_8$O$_{26}^{4-}$, ions. Even larger aggregates may be present in solution and in salts. Both Mo$_{12}$O$_{37}^{2-}$ and Mo$_{36}$O$_{112}$(H$_2$O)$_{16}^{8-}$ have been isolated and crystallographically characterized (4). At sufficiently low pH in very dilute solutions, cationic forms such as MoO$_2^{2+}$ and MoO^{4+} are present.

The polymolybdate and heteropolymolybdate ions constitute a broad and commercially significant class. In these ions molybdenum is six-coordinate with octahedral geometry (4–8). Oxo (O^{2-}) groups bridge the Mo atoms and serve as terminal ligands on some of the Mo ions. When other atoms are present during the acidification of molybdate solutions, a series of heteropolymolybdates is formed. For example, cations such as Cr^{3+} or Co^{2+}, or anions such as PO$_4^{3-}$ or AsO$_4^{3-}$, form the heteropoly anions H$_6$CrMo$_6$O$_{24}^{3-}$, H$_6$CoMo$_6$O$_{24}^{4-}$, PMo$_{12}$O$_{40}^{3-}$, and AsMo$_{12}$O$_{40}^{3-}$, respectively. The yellow ion, PMo$_{12}$O$_{40}^{3-}$, is analytically useful, being formed in the molybdenum test for phosphate ion. Poly- and heteropolymolybdate ions are used in the precipitation of dyes. The protonated forms of the ions are strongly acidic and many poly- and heteropolymolybdate

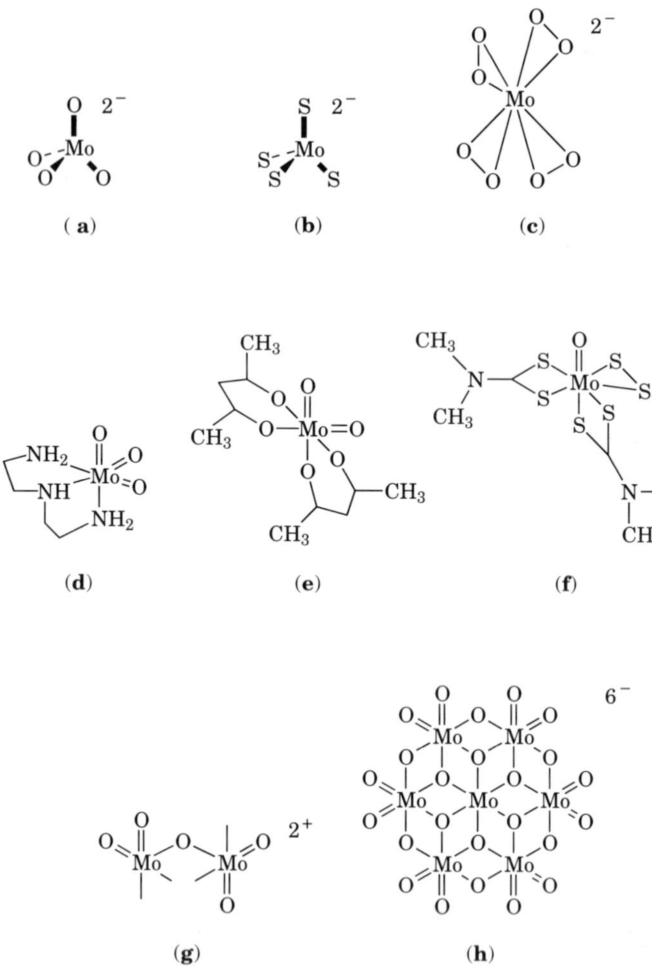

Fig. 2. Representative structures for compounds of molybdenum(VI): (**a**) molybdate(VI), MoO_4^{2-}; (**b**) tetrathiomolybdate(VI), MoS_4^{2-}; (**c**) tetrakis(peroxo)molybdate(VI), $Mo(O_2)_4^{2-}$; (**d**) *cis*-trioxodiethylenetriaminemolybdenum(VI), (MoO₃(dien)), $C_4H_{13}N_3MoO_3$; (**e**) *cis*-bis(acetylacetonato)dioxomolybdenum(VI), $MoO_2(C_5H_7O_2)_2$; (**f**) bis(dialkyldithiocarbamato)disulfidooxomolybdenum(VI), $MoO(S_2)(S_2CNR_2)_2$ (R = alkyl); (**g**) the dinuclear core structure for $Mo_2O_5^{2+}$ complexes; (**h**) heptamolybdate(VI), $Mo_7O_{24}^{6-}$.

compounds have catalytic activity that is attributable to their acid–base or redox properties.

The reduction of molybdate salts in acidic solutions leads to the formation of the molybdenum blues (9). Reductants include dithionite, stannous ion, hydrazine, and ascorbate. The molybdenum blues are mixed-valence compounds where the blue color presumably arises from the intervalence $Mo(V) \rightarrow Mo(VI)$ electronic transition. These can be viewed as intermediate members of the class of mixed oxy hydroxides the end members of which are $Mo(VI)O_3$ and $Mo(V)O(OH)_3$ [27845-91-6]. MoO_3 and Mo(VI) solutions have been used as effective detectors of reductants because formation of the blue color can be monitored spectrophotometrically. The nonprotonic oxides of average oxidation state between V and VI are the molybdenum bronzes, known for their metallic luster and used in the formulation of bronze paints (see PAINT).

Reaction of the molybdate ion with organic ligands leads to a wide variety of (mostly) mononuclear Mo(VI) complexes (10,11). The largest number of these compounds contain the cis-MoO$_2^{2+}$ core and four additional ligands, or one or two chelates, fill the remaining octahedral coordination sites. Examples, in addition to MoO$_2$(acac)$_2$ (Fig. 2**e**), are (N(CH$_3$)$_4$)$_2$MoO$_2$(NCS)$_4$, MoO$_2$(dtc)$_2$, and MoO$_2$(CH$_3$)$_2$(bipy), where dtc = dialkyl dithiocarbamate and bipy = 2,2'-bipyridine. Complexes containing the MoO core, eg, MoOS$_2$(dtc)$_2$ (Fig. 2**f**), MoOCl$_2$(dtc)$_2$, [MoO(dtc)$_3$]BF$_4$, and MoOCl$_4$; the cis-MoO$_3$ core, eg, MoO$_3$(dien) (Fig. 2**d**); or the dinuclear Mo$_2$O$_5^{2+}$ core (Fig. 2**g**), eg, (NH$_4$)$_2$[Mo$_2$O$_5$(cat)$_2$], where cat = catecholate, are also known. The compounds MoO$_2$Cl$_2$ [13637-68-8], formed by the reaction of MoO$_2$ and Cl$_2$, and MoO$_2$(acac)$_2$ serve as useful starting materials for the synthesis of new Mo(VI) compounds.

Peroxo, O$_2^{2-}$, molybdate complexes are also well established in the chemistry of Mo(VI) (12). The prototypical complex is Mo(O$_2$)$_4^{2-}$ (Fig. 2**c**). The peroxo ligands are side-on coordinated to Mo to form a formally eight-coordinate structure, although the centroids of the O–O bonds form a tetrahedral array about Mo. Numerous complexes are known in which MoO(O$_2$)$^{2+}$ or MoO(O$_2$)$_2$ cores are present and additional ligands complete six- or seven-coordinate structures. One example is K$_2$[MoO(O$_2$)$_2$C$_2$O$_4$]. These complexes are of great interest because of their ability to transfer oxygen atoms to olefins, ie, act as peroxidation reagents, and to other organic molecules.

The tetrathiomolybdate ion [16330-92-0] (Fig. 2**b**), which has received great attention in the late 20th century, was first reported by Berzelius in 1838. The simple preparation from molybdate in basic aqueous solution occurs in high yield according to:

$$MoO_4^{2-} + 4\,S^{2-} \longrightarrow MoS_4^{2-} + 4\,O^{2-}$$

The red tetrathiomolybdate ion appears to be a principal participant in the biological Cu–Mo antagonism and is reactive toward other transition-metal ions to produce a wide variety of heteronuclear transition-metal sulfide complexes and clusters (13,14). For example, tetrathiomolybdate serves as a bidentate ligand for Co, forming Co(MoS$_4$)$_2^{3-}$. Tetrathiomolybdates and their mixed metal complexes are of interest as catalyst precursors for the hydrotreating of petroleum (qv) (15) and the hydroliquefaction of coal (see COAL CONVERSION PROCESSES) (16). The intermediate forms MoOS$_3^{2-}$, MoO$_2$S$_2^{2-}$, and MoO$_3$S^{2-} have also been prepared (17).

The tris(dithiolene) complexes of Mo can be formed by reaction of the corresponding dithiol and molybdate in acid solution. The intense green compound, Mo(tdt)$_3$, where tdt = toluene-3,4-dithiolate, possesses a trigonal prismatic six-coordination about Mo and has found great use in the analytical determination of molybdenum.

Molybdenum(V)

Molybdenum(V) compounds generally occur as mononuclear or dinuclear species. Molybdenum pentachloride [*10241-05-1*], $MoCl_5$, formed by combination of the elements, serves as a useful and reactive starting material (Fig. 1). $MoCl_5$ has a dinuclear structure (Fig. 3) in the solid state but is mononuclear in the gas phase. In solution or in the solid state the compound, actually Mo_2Cl_{10} (Fig. 3**a**), is readily hydrolyzed in air to form $MoOCl_3$ [*13814-74-9*]. The compound $MoOCl_3(thf)$, where thf is tetrahydrofuran, is also known.

Mononuclear Mo(V) compounds have a $4d^1$ electronic configuration, and the single unpaired electron gives rise to distinct magnetic and electron paramagnetic resonance (epr) signatures (see MAGNETIC SPIN RESONANCE) (10,18). The g and A values of the epr spectra are characteristic of the ligand donor set and the coordination geometry. This technique has been valuable in detecting

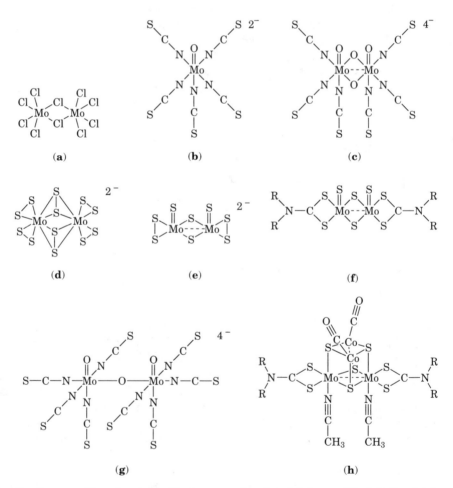

Fig. 3. Representative structures for compounds of molybdenum(V): (**a**) dimolybdenum dodecachloride [*26814-39-1*], Mo_2Cl_{10}, the dimer of molybdenum pentachloride; (**b**) pentakis(thiocyanato)oxomolybdenum(V), $MoO(NCS)_5^{2-}$; (**c**) $Mo_2O_4(NCS)_6^{4-}$; (**d**) $Mo_2(S_2)_6^{2-}$; (**e**) $Mo_2S_8^{2-}$; (**f**) $Mo_2S_4(S_2CNR_2)_2$; (**g**) $Mo_2O_3(NCS)_8^{4-}$; (**h**) $Mo_2Co_2S_4(S_2CNR_2)_2(CO)_2(CH_3CN)_2$.

the presence of the Mo(V) oxidation state in catalysts and enzymes. Mono oxo molybdenum(V) complexes such as $(NH_4)_2MoOCl_5$ are quite common (see also Fig. 3**b**).

Virtually all Mo(V) dinuclear complexes are diamagnetic and contain bridges between the two Mo atoms. The bridge can be a linear Mo–O–Mo linkage as in the $Mo_2O_3(NCS)_8^{4-}$ complex (Fig. 3**g**), which illustrates the $Mo_2O_3^{4+}$ core structure. Compounds such as $Mo_2O_3(acac)_4$ and $Mo_2O_3(dtc)_4$ also exist. More commonly, a double bridge is found in which the bridging atoms are oxo or sulfido ligands and a single molybdenum–molybdenum bond is also present. Here the Mo coordination sphere can be completed by a variety of nonoxo and nonsulfido ligands. The complexes $Mo_2O_4(NCS)_6^{4-}$ (Fig. 3**c**) and $Mo_2S_4(dtc)_2$ (Fig. 3**f**) present examples of the $Mo_2O_4^{2+}$ and $Mo_2S_4^{2+}$ core structures, respectively. Mixtures of $Mo_2S_4(dtc)_2$, $Mo_2O_4(dtc)_2$, and intermediate species with $Mo_2O_xS_{4-x}^{2+}$ cores and related dithiophosphate complexes are useful as additives in lubricant formulations.

In many of the dinuclear complexes the Mo atom is five-coordinate, not counting the metal–metal bond. In a few examples such as in $Mo_2O_4(L\text{-}cysteine)_2^{2-}$, the molybdenum is six-coordinate, not including the metal–metal bond. Also known are $Mo_2O_4(dtc)_2$ and $Mo_2O_2S_2(dtc)_2$. In general, the position trans to the terminal Mo oxo (or Mo sulfido) ligand is either empty or the ligand binds only weakly. If present, the trans ligand is kinetically labile, leading to facile substitution reactions.

Interestingly, the $Mo_2S_4^{2+}$ (Fig. 3**f**) core structure can be viewed as occupying six of the eight vertices of a distorted cube. Reaction of the dinuclear complexes having the $Mo_2S_4^{2+}$ core with appropriate metal ions leads to the planned assembly of $M_2Mo_2S_4$ thiocubane structures (19,20). When M = Co (Fig. 3**h**) the compounds are potential precursors for hydrodesulfurization catalysts (15).

The dinuclear ion $Mo_2(S_2)_6^{2-}$ (Fig. 3**d**) prepared from the reaction of molybdate and polysulfide solution (13) is a useful starting material for the preparation of dinuclear sulfur complexes. These disulfide ligands are reactive toward replacement or reduction to give complexes containing the $Mo_2S_4^{2+}$ core (Fig. 3**f**).

Molybdenum(IV)

Representative compounds for the +4 oxidation state are shown in Figure 4. The violet tetravalent molybdenum dioxide [*18868-43-4*], MoO_2, is formed by the reduction of MoO_3 with H_2 at temperatures below which Mo metal is formed or MoO_3 is volatile (ca 450°C). $MoCl_4$ [*13320-71-3*] is formed upon treatment of MoO_2 at 250°C with CCl_4 (see Fig. 1).

The most important compound of Mo(IV) is molybdenum disulfide [*1317-33-5*], MoS_2 (21). The layered structure of MoS_2 is reflected in the flat plate-like hexagonal gray-black crystallites found in natural and synthetic samples. The structure consists of pairs of close-packed layers of sulfur which are eclipsed with respect to each other. The close-packed sulfur surfaces are naturally hydrophobic, which facilitates the extraction of MoS_2 ore by flotation.

In the structure of MoS_2 molybdenum atoms occupy trigonal prismatic holes between the eclipsed sulfur layers with every other hole occupied. The molyb-

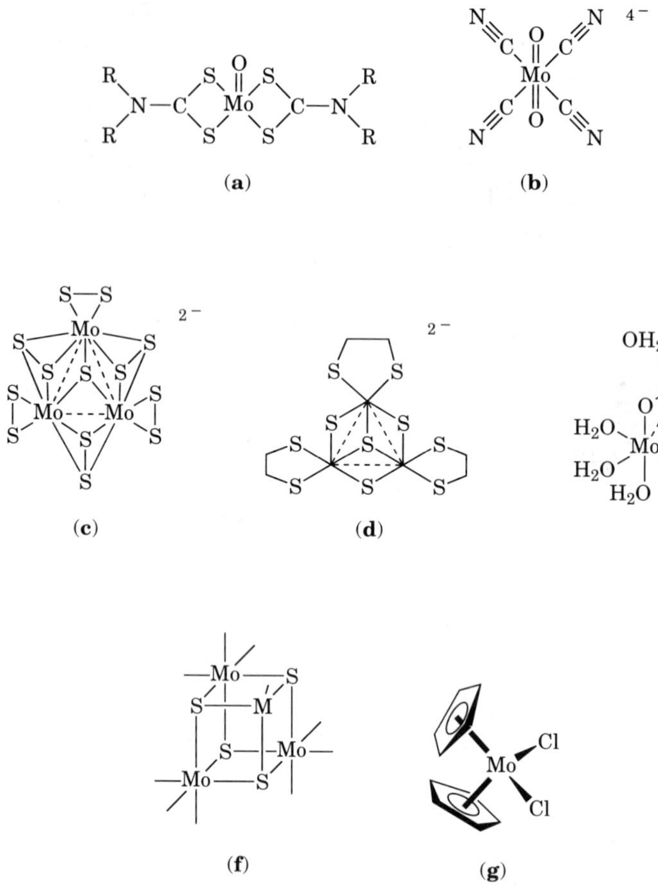

Fig. 4. Representative structures for compounds of molybdenum(IV): (**a**) bis(dialkyl-dithiocarbamato)oxomolybdenum(IV), $MoO(S_2CNR_2)_2$, where R = alkyl; (**b**) *trans*-tetracyanodioxomolybdenum(IV), $MoO_2(CN)_4^{4-}$; (**c**) $Mo_3S_{13}^{2-}$; (**d**) $Mo_3S_4(SCH_2CH_2)_3^{2-}$; (**e**) $Mo_3O_4(H_2O)_9^{4+}$; (**f**) the $Mo_3M'S_4$ thiocubane core structure; (**g**) bis(cyclopenta-dienyl)dichloromolybdenum(IV), Cp_2MoCl_2, where Cp = cyclopentadienyl.

denum is thereby coordinated to six sulfur atoms from a trigonal prism the sides of which are nearly square. The commercial importance of MoS_2 in lubrication and catalysis is a direct reflection of its solid-state structure. Molybdenum diselenide [*12078-18-3*], $MoSe_2$, and molybdenum ditelluride, $MoTe_2$, share the same basic structure but are not as important technologically. All the molybdenum dichalcogenides are semiconductors and have interesting photochemical, electrochemical, and photovoltaic properties (22).

The similarity of MoS_2 to graphite has been noted. Like elemental carbon, which has been found to form nanotubular structures, MoS_2 has also been found to form nested structures upon exposure to the electron beam in an electron microscope (23). Moreover, MoS_2 displays a variety of intercalation reactions typical of layered materials. Single-layer MoS_2 has been successfully prepared and manipulated (22).

The well-studied eight-coordinate octacyanide ion, $Mo(CN)_8^{4-}$, prepared by treatment of the oxo thiocyanate complexes of Mo(V) with KCN, has a reversible one-electron redox relationship with the Mo(V) species $Mo(CN)_8^{3-}$. Photolysis of $Mo(CN)_8^{4-}$ yields $MoO_2(CN)_4^{4-}$ (Fig. **4b**) which has a *trans*-dioxo configuration. The principal type of oxo complex of molybdenum(IV) contains a single oxo ligand and has a distorted octahedral or tetragonal pyramidal, five-coordinate structure. One example is $MoO(S_2CNR_2)_2$ (Fig. **4a**). The strong *trans*-activating effect of the oxo ligand causes the position trans to Mo—O (terminal bond) to be weakly binding and kinetically labile or empty.

For molybdenum(IV), trinuclear species are common and often contain $Mo_3O_4^{4+}$ (Fig. **4e**), $Mo_3S_4^{4+}$ (Fig. **4d**), or $Mo_3S_7^{4+}$ (Fig. **4c**) core structures (24). The ion $Mo_3S_{13}^{2-}$ prepared from the reaction of molybdate with polysulfide solutions (13) serves as a useful starting material for the formation of trinuclear sulfide bridged clusters which have threefold symmetric structures. The core structures for the $Mo_3O_4^{4+}$ (Fig. **4e**) and $Mo_3S_4^{4+}$ (Fig. **4d**) compounds (or mixed oxo–sulfido analogues) can be viewed as containing seven of the eight vertices of a distorted cube. Indeed, the sulfido structures readily react with metal ions to form complete "thiocubane" structures such as that shown in Figure **4f** (25). The trinuclear clusters serve as precursors to and molecular analogues of MoS_2 (26).

Other molybdenum(IV) compounds include $Mo(NR_2)_4$, $Mo(OR)_4$, $MoCl_4$-$(NCCH_3)_2$, $(NH_4)_2[MoCl_6]$, $Mo(S$-t-$C_4H_9)_4$, $Mo(dtc)_4$, $(NH_4)_2[MoS(S_2)_6]$, $[(C_6H_5)_4]_2[MoS(S_4)_2]$, $MoO(tpp)$, and $MoO(pcn)$, where tpp = tetraphenylporphyrin and pcn = phthalocyanin.

Molybdenum(III)

Molybdenum(III) complexes include the molybdenum trihalides. Molybdenum trichoride [*13478-18-7*], trifluoride [*20193-58-2*], tribromide [*13446-57-6*], and triiodide [*14055-75-5*] are all known. The oxide dimolybdenum trioxide [*1313-29-7*], Mo_2O_3, and the seldom-studied sulfide analogue [*12033-33-9*], Mo_2S_3, are formally trivalent.

Molecular examples of trivalent molybdenum are known in mononuclear, dinuclear, and tetranuclear complexes, as illustrated in Figure 5. The hexachloride ion, $MoCl_6^{3-}$ (Fig. **5a**) is generated by the electrolysis of Mo(VI) in concentrated HCl. Hydrolysis of $MoCl_6^{3-}$ in acid gives the hexaaquamolybdenum(III) ion, $Mo(H_2O)_6^{3+}$, which is obtainable in solution of poorly coordinating acids, such as triflic acid (17). Several molybdenum(III) organometallic compounds are known. These contain a single cyclopentadienyl ligand (Cp) attached to Mo (Fig. **5d**) (27).

Dinuclear structures are known for molybdenum(III) in a series of air and moisture sensitive compound containing multiple Mo—Mo bonds. Examples include $Mo_2(N(CH_3)_2)_6$ (Fig. **5b**) and $Mo_2(CH_2Si(CH_3)_3)_6$ in which there is a strong Mo—Mo bond, presumably of triple-bond character (28).

The tetranuclear Mo_4S_4 (Fig. **5c**) core structures are known with a variety of ligands including aqua, ammonia, cyanido, and a variety of cyclopentadienyl and chelating 1,1-dithiolate ligands (25). The core structures are of the thiocubane form and readily undergo redox reactions to form a variety of mixed

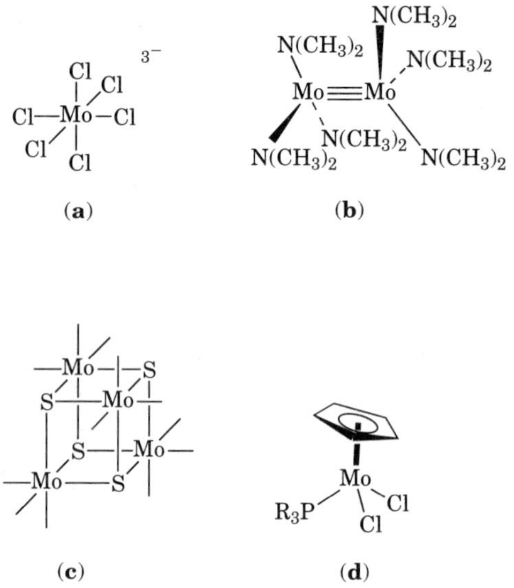

Fig. 5. Representative structures for compounds of molybdenum(III): (**a**) hexacholoromolybdenum(III) ion, $MoCl_6^{3-}$; (**b**) hexakis(dimethylamido)dimolybdenum(III), $Mo_2(N(CH_3)_2)_6$; (**c**) the Mo_4S_4 thiocubane core structure; (**d**) dichlorocyclopentadienyl trialkylphosphinedichloromolybdenum(III), $CpMo(PR_3)Cl_2$, where Cp = cyclopentadienyl and R = alkyl.

valence species, eg, $Mo_4S_4(dtc)_6$, a Mo(III)/Mo(V) complex. Analogues containing the Mo_4O_4 core have been formed in solid-state structures of molybdenum oxide phosphate phases (29).

Molybdenum(II)

Divalent molybdenum compounds occur in mononuclear, dinuclear, and hexanuclear forms. Selected examples are shown in Figure 6. The mononuclear compounds are mostly in the realm of organometallic chemistry (see ORGANOMETALLICS) (30–32). Seven-coordinate complexes are common and include $MoX_2(CO)_3(PR_3)_2$, where X = Cl, Br, and I, and R = alkyl; $MoCl_2(P(CH_3)_3)_4$, heptakis(isonitrile) complexes of the form $Mo(CNR)_7^{2+}$ (Fig. 6d), and their chlorosubstituted derivatives, eg, $Mo(CNR)_6Cl^+$. The latter undergo reductive coupling to form C–C bonds in the molybdenum coordination sphere (33).

The dinuclear compounds of molybdenum(II) have strong, quadruple, molybdenum–molybdenum bonds and eclipsed structures, such as those of $Mo_2Cl_8^{4-}$ (Fig. 6a) and $Mo_2(O_2CCH_3)_4$ (Fig. 6b). These quadruply bonded complexes have been intensely studied from the point of view of their structures, electronic structures, and reactivity (34,35).

The dichloride of molybdenum(II) [13478-17-6], $MoCl_2$, contains $Mo_6Cl_8^{4+}$ core units (Fig. 6c) having chloride bridges in its solid-state structure. Similar or identical hexanuclear units are known in soluble species such as $Mo_6Cl_{14}^{2-}$ and other derivatives containing the $Mo_6Cl_8^{4+}$ core. These compounds have been under investigation because of their photochemical and photoluminescent activity

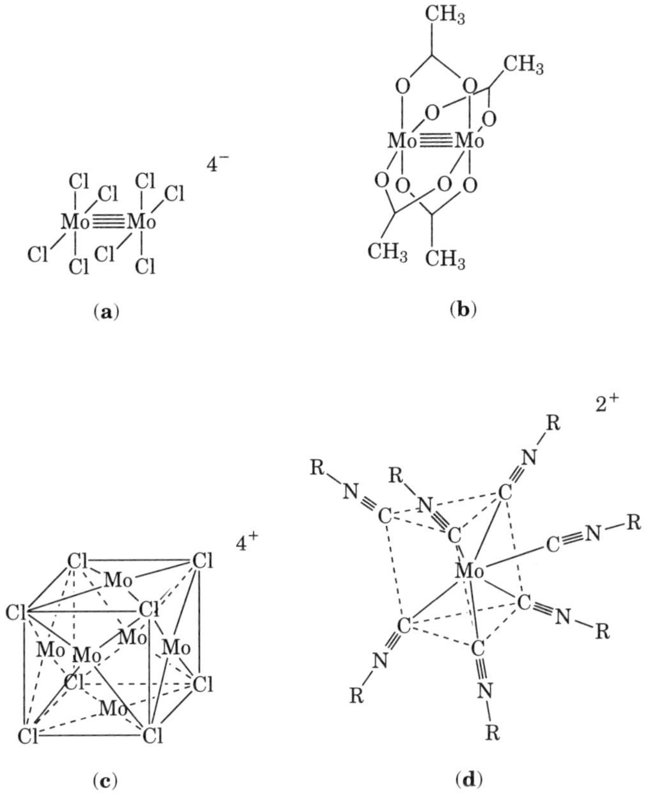

Fig. 6. Representative structures for compounds of molybdenum(II): (**a**) octachlorodimolybdenum(II) ion, $Mo_2Cl_8^{4-}$; (**b**) tetrakis(acetato)dimolybdenum(II), $Mo_2(O_2CCH_3)_4$; (**c**) the $Mo_6Cl_8^{4+}$ core; (**d**) heptakis(isocyanide)molybdenum(II), $Mo(CNR)_7^{2+}$, where R = alkyl.

(see PHOTOCHEMICAL TECHNOLOGY) (36,37). The hexanuclear clusters consist of octahedra of Mo embedded in cubes of chlorine. This structure is similar to that of solid-state Chevrel phase materials, which have cubes of sulfur in place of chlorine. The Chevrel phases have elicited great interest in light of their superconductivity and interesting magnetic properties (see MAGNETIC MATERIALS; SUPERCONDUCTING MATERIALS) (38). Related molecular clusters, such as $Mo_6S_8L_6$ where L is pyridine or $P(C_2H_5)_3$, have been reported (39).

Other molybdenum(II) compounds of interest include the nitric oxide complexes $Mo(NO)Cl_3$ and $Mo(NO)(dtc)_3$, trans-$MoH_2(dppe)_2$ where dppe = 1,2-bis(diphenylphosphino)ethane and $K_4[Mo_2(SO_4)_4]\cdot 2H_2O$.

Molybdenum(0)

Molybdenum hexacarbonyl [13939-06-5] (Fig. 7a) is the starting material for the synthesis of most organometallic compounds of molybdenum. White crystalline $Mo(CO)_6$ is prepared by the reduction of $MoCl_5$ and Zn in the presence of a high pressure of carbon monoxide (see CARBONYLS). The hexacarbonyl melts at 150–151°C, but readily sublimes at lower temperatures and is soluble in nonpolar organic solvents. $Mo(CO)_6$ reacts readily in polar solvents such as acetonitrile

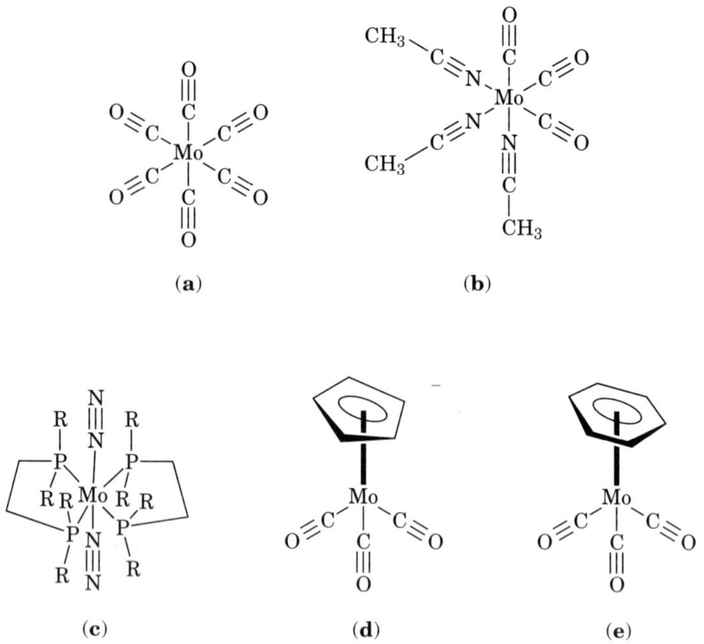

(a)

(b)

(c)

(d)

(e)

Fig. 7. Representive structures for compounds of molybdenum(0): (**a**) $Mo(CO)_6$; (**b**) tris(acetonitrile)tris(carbonyl)molybdenum(0); (**c**) bis(1,2-diphenylphosphinoethane) bis(dinitrogen) molybdenum(0), $[R_2PCH_2CH_2PR_2]_2Mo(N_2)_2$, where R = C_6H_5, also known as $Mo(dppe)_2(N_2)_2$, where dppe = 1,2-diphenylphosphinoethane; (**d**) cyclopentadienyl tricarbonyl molybdenum(0) anion, $CpMo(CO)_3^-$, where Cp = cyclopentadienyl; (**e**) benzene-tricarbonyl molybdenum(0), $(C_6H_6)Mo(CO)_3$.

to produce trisubstituted products such as $Mo(CO)_3(CH_3CN)_3$ (Fig. 7**b**), which serve as synthetically useful starting materials. Numerous other organometallic complexes have been prepared and structurally characterized. A few are shown in Figure 7 (31,32). A notable example is $Mo(CO)_3(PR_3)_2(H_2)$, one of the first examples of a coordinated dihydrogen complex (40). Many of the organometallic compounds of Mo display interesting organic reactivity, and some have been used as precursors for a variety of homogeneous and heterogeneous catalysts.

Molybdenum(0) also forms a variety of dinitrogen complexes (41), especially when there are phosphine ligands in the molybdenum coordination sphere (see Fig. 7**c**). This type of complex has been extensively studied because the coordinated dinitrogen is reduced to ammonia upon acidification.

Other molybdenum(0) compounds of interest include $MoCl_2(NO)_2$ and $MoCl_2(NO)_2(bipy)$ where bipy = 2,2'-bipyridine. The compound $(C_5H_5)Mo(CO)_2NO$, similar in structure to Figure 7**d**, is also known.

Chemistry of Molybdenum Compounds

Metal–Metal Bonding. The degree of nuclearity exhibited as a function of the oxidation state of molybdenum is shown in Table 1. In the highest oxidation state, Mo(VI), the tendency is to form mononuclear or a wide variety of polynuclear complexes in which there are no molybdenum–molybdenum bonds. The

Table 1. Metal–Metal Bonding in Molybdenum Complexes and Clusters

Metal	Oxidation state				
	Mo(VI)	Mo(V)	Mo(IV)	Mo(III)	Mo(II)
d-electron configuration	d^0	d^1	d^2	d^3	d^4
number of metal–metal bonds	0	1	2	3	4
formulation mononuclear or dinuclear	Mo	Mo —Mo	Mo $=$ Mo	Mo $\equiv$ Mo	Mo $\equiv\!\!\equiv$ Mo
polynuclear		Mo —Mo	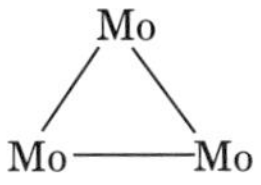	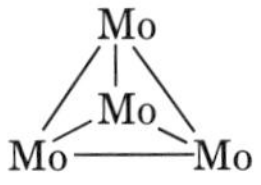	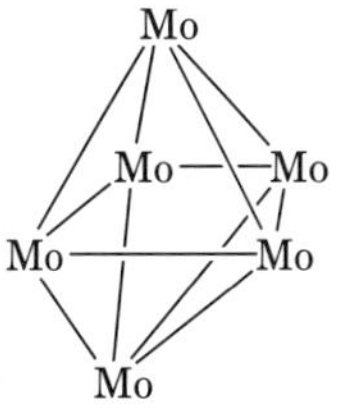
polynuclear geometry		linear	triangle	tetrahedron	octahedron

absence of metal–metal bonding is attributable to the $4d^0$ electronic configuration of molybdenum, ie, there are no d electrons available for bonding. The Mo(V) complexes (Fig. 3), on the other hand, have a strong tendency to form dimers which have a single metal–metal bond. This is the result of the $4d^1$–$4d^1$ configurations of the two metal centers. The complexes in this case are usually also doubly or triply bridged by ligands such as oxide, sulfide, or halide. For Mo(IV) the $4d^2$ electronic configuration allows molybdenum to enter into two metal–metal bonds. Whereas dinuclear complexes containing a Mo–Mo double bond could be formed, a trinuclear structure in which each Mo atom is bound to two others in a triangular array is more often the observed case.

For Mo in the trivalent state, the $4d^3$ configuration leads to the possibility of three metal–metal bonds being formed per molybdenum. Some dinuclear structures in which the Mo atoms are triply bonded to each other in a relatively strong bond, generally in an unbridged complex, are known. Alternatively, each metal can achieve full binding capacity by binding to three other metals. This leads to the stabilization of a tetranuclear cluster in which Mo atoms are found at the corners of a tetrahedron. Because the faces of the tetrahedron are usually occupied by sulfur, or sometimes selenium or oxygen, the resulting overall structure is called a thiocubane. The four molybdenum and four sulfur atoms form a distorted cube (see Fig. 5c).

In the case of Mo(II), the $4d^4$ electronic configuration allows the formation of four metal–metal bonds. Such bonding can be accomplished in a dinuclear complex by the formation of a quadruple bond between the Mo atoms. Alternatively, the Mo atoms in Mo(II) compounds can form four metal–metal bonds by constructing an octahedral Mo_6 cluster in which each Mo atom is bonded to four molybdenum atoms. The resultant clusters are well known in certain Chevrel phases, such as $PbMo_6S_8$ [39432-49-0] (38), and in molecular clusters such as $Mo_6Cl_8^{4+}$ (37).

Halides of Molybdenum. The halides of molybdenum are solids that are quite reactive and useful starting points for further syntheses. Only fluoride supports the highest (hexavalent) oxidation state. Molybdenum hexafluoride [7783-77-9], MoF_6, and derivative salts containing MoF_7^- exist. The highest Mo oxidation states for chloride, bromide, and iodide are V, IV, and III, respectively. The finding that MoF_6, $MoCl_5$, $MoBr_4$, and MoI_3 are, respectively, the highest halides formed for F, Cl, Br, and I is consistent with the ease of oxidation of the respective halides. The higher halides in the presence of higher oxidation states of molybdenum are susceptible to internal oxidation, eg, MoI_4 or MoI_5 would be unstable with respect to the formation of the MoI_3 and elemental iodine (I_2). The properties of the molybdenum halides have been described in some detail (42,43).

Aqueous Chemistry. Molybdenum has well-characterized aqueous chemistry in the five oxidation states, VI, V, IV, III, and II. A listing of aqua ions is given in Table 2. Except for the Mo(VI) species all of the aqua ions are only soluble or stable in acidic media (17). The range of aqueous ions known for molybdenum is far broader than that of other elements.

Table 2. Aqueous Ions of Molybdenum in Acid Solutions

Molybdenum oxidation state	Ion	Color	Mo−Mo bonding
VI	MoO_2^{2+} and others	colorless	none
V	$Mo_2O_4(H_2O)_6^{2+}$	yellow	one single bond
IV	$Mo_3O_4(H_2O)_9^{4+}$	green	two double bonds
III	$Mo_4O_4(H_2O)_{12}^{4+}$		six single bonds
	$Mo(H_2O)_6^{3+}$	yellow	none
II	$Mo_2(H_2O)_8^{4+}$	red	quadruple

Biological Aspects

Molybdenum, recognized as an essential trace element for plants, animals, and most bacteria, is present in a variety of metallo enzymes (44–46). Indeed, the absence of Mo, and in particular its co-factor, in humans leads to severe debility or early death (47,48). Molybdenum in the diet has been implicated as having a role in lowering the incidence of dental caries and in the prevention of certain cancers (49,50). To aid the growth of plants, Mo has been used as a fertilizer and as a coating for legume seeds (51,52) (see FERTILIZERS; MINERAL NUTRIENTS).

Environmentally, the presence of molybdenum has been of concern only in isolated instances (53). Reports of molybdenum toxicity have been rare (54). Molybdenum is involved in copper–molybdenum antagonism wherein excess molybdenum in the soil elicits a copper (qv) deficiency in animals (especially ruminants) that graze on the vegetation (55). Conversely, excess copper in the soil induces a molybdenum deficiency in ruminant animals that graze on the vegetation (see FEEDS AND FEED ADDITIVES, RUMINANT FEEDS). The problem is exacerbated in soils that have high levels of sulfate. The mechanism of the toxicity is understood. The anaerobic rumen of the affected animals contains a population of sulfate-reducing bacteria that reduces the soil sulfate to hydrogen sulfide (sulfide ions in solution). The sulfide ions react with molybdates to form thiomolybdates, which are known to be excellent ligands for copper. The resultant copper molybdenum compounds are insoluble and therefore the elements in them are not available to satisfy the nutritional needs of the animals. Fortunately, supplementation with the deficient element, eg, copper, alleviates the toxic effects.

Molybdate is also known as an inhibitor of the important enzyme ATP sulfurylase where ATP is adenosine triphosphate, which activates sulfate for participation in biosynthetic pathways (56). The tetrahedral molybdate dianion, MoO_4^{2-}, substitutes for the tetrahedral sulfate dianion, SO_4^{2-}, and leads to futile cycling of the enzyme and total inhibition of sulfate activation. Molybdate is also a co-effector in the receptor for steroids (qv) in mammalian systems, a biochemical finding that may also have physiological implications (57).

The clearest manifestation of molybdenum in biology is its presence in over 20 enzymes which participate in a wide variety of redox processes (44–46). Some of the Mo enzymes and their occurrence are as follows:

Enzyme	Occurrence
Nitrogen metabolism	
nitrogenase	bacteria (including symbionts)
nitrate reductase	plants, fungi, algae, bacteria
trimethylamine *N*-oxide reductase	bacteria
xanthine oxidase	cow's milk, mammalian liver, kidney
xanthine dehydrogenase	chicken liver, bacteria
quinoline oxidoreductase	bacteria
picolinic acid dehydrogenase	bacteria
Carbon metabolism	
aldehyde oxidase	mammalian liver
formate dehydrogenase	fungi, yeast, bacteria, plants
carbon monoxide oxidoreductase	bacteria
formylmethanofuran dehydrogenase	bacteria
Sulfur metabolism	
sulfite oxidase	mammalian liver, bacteria
dimethyl sulfoxide (DMSO) rectuctase	bacteria
biotin sulfoxide reductase	bacteria
tetrathionite reductase	bacteria
Others	
arsenite oxidase	bacteria
chlorate reductase	bacteria

The enzyme nitrogenase (58) catalyzes the reduction of dinitrogen to ammonia. The process known as nitrogen fixation (qv) is of critical importance to crops such as soybeans and other legumes (see SOYBEANS AND OTHER OIL SEEDS). Nitrogenase is found in the symbiotic bacteria that live in root nodules of legumes. These bacteria reduce atmospheric dinitrogen to ammonia, which is incorporated into the metabolic pathways of both the bacterium and the plant to synthesize proteins (qv), nucleic acids (qv), and other biomolecules that contain nitrogen. The crystal structure (59) of the iron–molybdenum protein of nitrogenase reveals an iron–molybdenum cluster (FeMoco) of unusual structure, which is the active site responsible for reduction of N_2 in the enzyme.

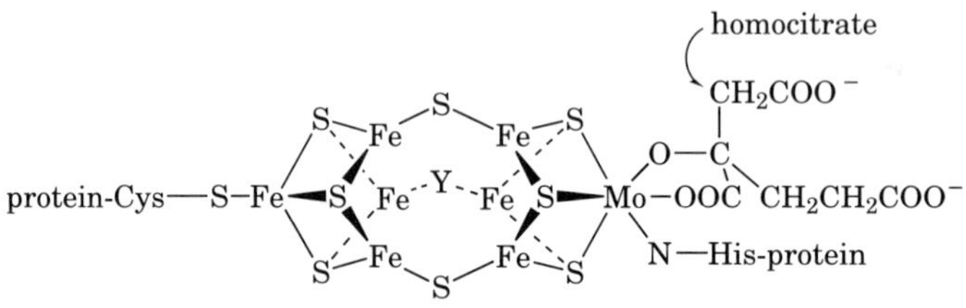

FeMoco 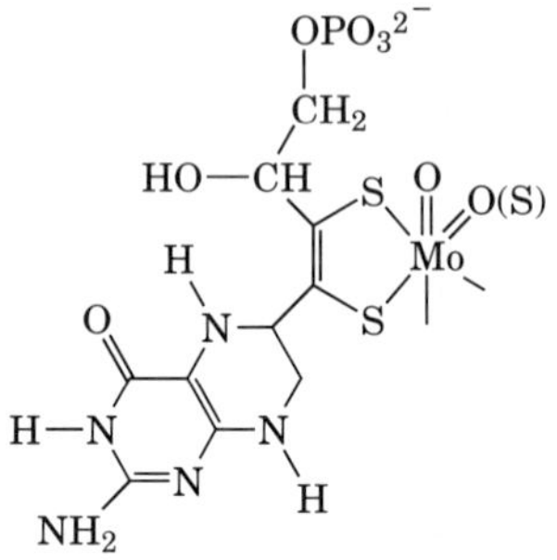   Moco

The remaining Mo enzymes all contain the molybdenum co-factor (Moco) which is chemically, biochemically, and genetically distinct from the nitrogenase co-factor, FeMoco. In Moco the molybdenum is bound by a pterin-enedithiolate ligand (60). The Moco enzymes (61,62) catalyze such important reactions as the oxidation of sulfite to sulfate; the reduction of nitrate to nitrite; the oxidation of xanthine to uric acid, ie, the last step in purine metabolism in humans; and quinoline hydroxylation which is crucial for the biodegradation of petroleum residues. The sulfite oxidation reaction is critical in human metabolism. Sulfite is toxic and must be removed from the body as it is produced metabolically or ingested. Children born without the sulfite oxidase activity are severely compromised and generally do not survive (47). The nitrate reduction reaction is critical to plant health, as all plants that do not harbor nitrogen-fixing symbionts require this enzyme as the first step in their assimilation of inorganic nitrogen. Molybdenum clearly plays a principal role in the biological nitrogen cycle.

Most of the Moco enzymes catalyze oxygen atom addition or removal from their substrates. Molybdenum usually alternates between oxidation states VI and IV. The Mo(V) state forms as an intermediate as the active site is reconstituted by coupled proton–electron transfer processes (62). The working of the Moco enzymes depends on the oxo chemistry of Mo(VI), Mo(V), and Mo(IV).

Although molybdenum is an essential element, excess levels can have deleterious effects. The LD_{50} and TLV values of the most common Mo compounds are listed in Table 3 (63,64). In general the toxicity of Mo compounds is considered to be low. For example, MoS_2 has been found to be virtually nontoxic even at high levels. Certain Mo compounds such as $MoCl_5$ and $Mo(CO)_6$, have higher toxicity because of the chemical nature and reactivity of these compounds rather than the Mo content. Supplementary dietary Cu^{2+}, thiosulfate, methionine, and cysteine have been shown to be effective in alleviating Mo toxicity in animals.

Table 3. Toxicity of Molybdenum Compounds[a]

Compound	Molecular formula	LD_{50} (rat, oral), mg/kg	TLV(TWA),[b] mg/m^3
ammonium heptamolyb-date	$(NH_4)_6Mo_7O_{24}\cdot 4H_2O$	333	
molybdenum trioxide	MoO_3	2689	5
sodium molybdate	Na_2MoO_4	4000	5
molybdenum disulfide, molybdenite	MoS_2	nontoxic[c]	

[a]Refs. 63 and 64.
[b]On the basis of weight of Mo.
[c]Rats ingesting 500 mg/d for 44 days showed no toxic signs.

Economic Aspects

The total world mine production of molybdenum was approximately 110,000 t in 1992 and 1993. The newly merged Cyprus Climax is by far the largest producer at ca 65,000 t. Kennecott, Thompson Creek Metals, and Molycorp are also significant producers (65). Roughly 30% of the molybdenum processed goes into compounds used in nonmetallurgical applications. The diversity of molybdenum compounds, coupled with potential environmental advantages and reduced costs of molybdenum relative to the noble metals, leads to projections for the increased use of molybdenum, especially in catalysis (66). Most of the bulk molybdenum chemicals, eg, MoO_3, Na_2MoO_4, $(NH_4)_2Mo_2O_7$, $(NH_4)_6Mo_7O_{24}$, and MoS_2, sell for \$6–9/kg (65). The annual value of these bulk chemicals is between \$200 and \$300 × 10^6.

Uses

In most of the nonmetallurgical uses of molybdenum compounds the metal is coordinated by oxygen or sulfur ligands. Molybdenum nitrides, carbides, and silicides are, however, coming under increasing study for various applications. Roughly 75% of all molybdenum compounds are used as catalysts in the petroleum and chemicals industries.

Lubrication. Molybdenum is found naturally mostly as the disulfide, MoS_2. Once purified, the graphite-like layered structure and the thermal and oxidative stability of MoS_2 make it extremely useful as a lubricant or lubricant component. MoS_2 is used directly as a solid or in coatings (qv) that are bonded onto the metal surface by burnishing; by vapor deposition, eg, sputtering; or by bonding processes that use binders, solvents, and mechanochemical procedures. Sputtered MoS_2 films have been found to have remarkably low friction coefficients (67) (see THIN FILMS). As a solid lubricant, MoS_2 is preferable to graphite in applications that involve high pressure, high vacuum, and radiation exposure. MoS_2 has been found effective in the lubrication of ceramic and polymer, as well as metal surfaces. Extensive use has been seen in high vacuum applications, including space vehicles where MoS_2 is the most widely used lubricant (68). Solid MoS_2 is a component of self-lubricating polymers and rubbers, and is often suspended in greases, pastes, and oils (69). MoS_2 is considered an intrinsic lubricant.

Unlike graphite to which MoS_2 is often compared, no additives or adsorbates are required for efficacy.

More recently, molecular molybdenum-sulfur complexes and clusters have been used as soluble precursors for MoS_2 in the formulation of lubricating oils for a variety of applications (70). Presumably, the oil-soluble molybdenum–sulfur-containing precursors decompose under shear, pressure, or temperature stress at the wear surface to give beneficial coatings. In several cases it has been shown that the soluble precursors are trifunctional in that they not only display antifriction properties, but have antiwear and antioxidant characteristics as well. In most cases, the ligands for the Mo are of the 1,1-dithiolate type, including dithiocarbamates, dithiophosphates, and xanthates (55,71).

Hydrodesulfurization and Hydrotreating Catalysis. Hydrotreating is used to remove sulfur, nitrogen, oxygen, and metals, mostly vanadium and nickel, from petroleum fractions. In hydrodesulfurization, the hydrogenolysis of C—S bonds in organosulfur compounds, such as thiophene, benzo-, or dibenzothiophene, yields H_2S and a corresponding hydrocarbon. In hydrodenitrogenation, the hydrogenolysis of C—N bonds in organic molecules, eg, pyridines, and carbazoles, yields ammonia (qv) and the hydrocarbon. Related hydrogenating processes lead to the saturation of aromatics and coal liquefaction (see COAL CONVERSION PROCESSES) (16). An important attribute of the catalysts for these processes is sulfur tolerance. The metal most closely associated with all the processes in use is molybdenum. Molybdenum sulfide catalysts are not only sulfur-tolerant but actually require sulfur for their activity.

The industry mainstay for the hydrodesulfurization of petroleum is the cobalt–molybdenum (co–moly) on alumina catalyst (72,73). Estimates of world molybdenum consumption for hydrotreating catalysis was over 2000 t in 1993 (66). The classic preparative technique involves wet impregnation of molybdates on alumina to produce a molybdenum oxide on alumina. Subsequent treatment using (usually divalent) cobalt leads to a cobalt oxide coating. This material is then sulfided using $H_2S–H_2$ or organosulfur compounds. Increasingly, presulfided (passivated) catalysts, preferred for their convenience and for environmental considerations, are being used (74). Environmental considerations leading to tighter sulfur specifications on petroleum products should lead to an increase in the use of molybdenum catalysts. Moreover, the recycling of spent catalyst should grow in importance as environmental trends disallow classical methods of disposal (see CATALYSTS, REGENERATION).

The active hydrotreating catalyst appears to have MoS_2-like aggregates as the active species, with the edges of the MoS_2 responsible for the catalytic activity (72). The cobalt, which is said to be a promoter of MoS_2 activity, is associated with the edges of the MoS_2 in groupings that are sometimes referred to as the CoMoS phase (75). Alternative approaches to making the CoMo catalysts are being developed including the use of specific precursors that already contain Co, Mo, and S in a single compound (15). Low valent Co compounds such as $Co_2(CO)_8$ have been found to promote MoS_2 activity by binding directly at the edge surfaces (76).

Oxidation Catalysis. The multiple oxidation states available in molybdenum oxide species make these excellent catalysts in oxidation reactions. The oxidation of methanol (qv) to formaldehyde (qv) is generally carried out commercially on mixed ferric molybdate–molybdenum trioxide catalysts. The oxidation of propylene (qv) to acrolein (77) and the ammoxidation of propylene to acrylonitrile (qv) (78) are each carried out over bismuth–molybdenum oxide catalyst systems. The latter (Sohio) process produces in excess of 3.6×10^6 t/yr of acrylonitrile, which finds use in the production of fibers (qv), elastomers (qv), and water-soluble polymers.

The addition of an oxygen atom to an olefin to generate an epoxide is often catalyzed by soluble molybdenum complexes. The use of alkyl hydroperoxides such as *tert*-butyl hydroperoxide leads to the efficient production of propylene oxide (qv) from propylene in the so-called Oxirane (Halcon or ARCO) process (79).

In addition to these principal commercial uses of molybdenum catalysts, there is great research interest in molybdenum oxides, often supported on silica, ie, MoO_3–SiO_2, as partial oxidation catalysts for such processes as methane-to-methanol or methane-to-formaldehyde (80). Both O_2 and N_2O have been used as oxidants, and photochemical activation of the MoO_3 catalyst has been reported (81). The research is driven by the increased use of natural gas as a feedstock for liquid fuels and chemicals (82). Various heteropolymolybdates (83), MoO_3-containing ultrastable Y-zeolites (84), and certain mixed metal molybdates, eg, $MnMoO_4$, $Fe_2(MoO)_3$, photoactivated $CuMoO_4$, and $ZnMoO_4$, have also been studied as partial oxidation catalysts for methane conversion to methanol or formaldehyde (80) and for the oxidation of C-4-hydrocarbons to maleic anhydride (85). Heteropolymolybdates have also been shown to effect ethylene (qv) conversion to acetaldehyde (qv) in a possible replacement for the Wacker process.

Other Catalytic Applications. Molybdenum disulfide is used as a catalyst for the decomposition of NaN_3 and other alkali and alkaline-earth azides. This leads to rapid N_2 production, useful in the rapid inflation of airbags in passive automotive restraint systems. Molybdenum compounds are also under continued study for the catalysis of a wide variety of reactions (86), including hydrogenation; hydrogenolysis; dehydrogenation; olefin metathesis, including ring-opening metathesis polymerization (ROMP); olefin, and especially, alkyne polymerization; dehydration; hydration, eg, isobutene to *tert*-butanol; hydrolysis; isomerization of alkanes (*n*-pentane to isopentane, cyclohexane to methylcyclopentane); water gas shift; Fischer-Tropsch synthesis of alcohols; naphtha reforming; and methanation. The varied chemistry of molybdenum is clearly reflected in the diverse array of reactions catalyzed by the compounds of this element.

Advanced Structural and Heating Materials. Molybdenum silicide [*12136-78-6*] and composites of $MoSi_2$ and silicon carbide, SiC, have properties that allow use as high temperature structural materials that are stable in oxidizing environments (see COMPOSITE MATERIALS; METAL-MATRIX COMPOSITES). Molybdenum disilicide also finds use in resistance heating elements (87,88).

Anticorrosion Agents. Sodium molybdate is finding increasing use in corrosion inhibition in cooling systems (see CORROSION AND CORROSION CONTROL), automotive antifreeze (see ANTIFREEZES AND DEICING FLUIDS), and cutting fluids. A big incentive in air conditioning cooling towers involves the replacement of the relatively toxic chromate ion with the far less toxic molybdate ion. Molybdate, a component of vitamin preparations and fertilizers, is generally considered to be a nutrient rather than a toxin (89). The mechanism of corrosion protection presumably involves adsorption of the molybdate on the internal surface of the tower to form mixed metal molybdate phases that resist corrosion (90).

Coatings, Paints, and Pigments. Various slightly soluble molybdates, such as those of zinc, calcium, and strontium, provide long-term corrosion control as undercoatings on ferrous metals (90–92). The mechanism of action presumably involves the slow release of molybdate ion, which forms an insoluble ferric molybdate protective layer. This layer is insoluble in neutral or basic solution. A primary impetus for the use of molybdenum, generally in place of chromium, is the lower toxicity of the molybdenum compound.

Molybdate orange and red are pigments (qv) that contain lead(II) molybdate [10190-55-3], $PbMoO_4$, formulated in mixed phases with $PbCrO_4$ and $PbSO_4$. The mixed phase is more intensely colored than any of the component phases. Concerns about lead content are lessening the use of these materials (see also PAINT). Various organic dyes are precipitated with heteropolymolybdates. This process allows the fixation of the dye in various fabrics. The molybdenum anion generally imparts light stability to the colorant as well (91).

Other Industrial Uses. *Flame and Smoke Retardants.* Molybdenum compounds are used extensively as flame retardants (qv) (93,94) in the formulation of halogenated polymers such as PVC, polyolefins, and other plastics; elastomers; and fabrics. An incentive for the use of molybdenum oxide and other molybdenum smoke and flame retardants is the elimination of the use of arsenic trioxide. Although hydrated inorganics are often used as flame retardants, and thought to work by releasing water of crystallization, anhydrous molybdenum oxides are effective. Presumably the molybdenum oxides rapidly form surface char, providing a barrier that prevents additional thermal or oxidative damage. The molybdenum is apparently not volatilized in the process.

Pyrotechnics. Molybdenum-containing formulations are used in delay elements of pyrotechnic display devices (see PYROTECHNICS) (95).

Battery Electrodes. Molybdenum disulfide in amorphous form and molybdenum trisulfide [12033-29-3] are useful as electrodes in Li batteries (qv). The lithium presumably intercalates between the molybdenum sulfide layers during charging, and then deintercalates upon discharging (96).

Biological Usage. *Soil Nutrient.* Molybdenum has been widely used to increase crop productivity in many soils worldwide (see FERTILIZERS). It is the heaviest element needed for plant productivity and stimulates both nitrogen fixation and nitrate reduction (51,52). The effects are particularly significant in leguminous crops, where symbiotic bacteria responsible for nitrogen fixation provide the principal nitrogen input to the plant. Molybdenum deficiency is usually more prominent in acidic soils, where Mo(VI) is less soluble and more easily reduced to insoluble, and hence unavailable, forms. Above pH 7, the soluble anionic, and hence available, molybdate ion is the principal species.

Biomedical Uses. The molybdate ion is added to total parenteral nutrition protocols and appears to alleviate toxicity of some of the amino acid components in these preparations (see MINERAL NUTRIENTS) (97). Molybdenum supplements have been shown to reduce nitrosamine-induced mammary carcinomas in rats (50). A number of studies have shown that certain heteropolymolybdates (98) and organometallic molybdenum compounds (99) have antiviral, including anti-

AIDS, and antitumor activity (see ANTIVIRAL AGENTS; CHEMOTHERAPEUTICS, ANTICANCER).

The isotope molybdenum-99 is produced in large quantity as the precursor to technetium-99m, a radionucleide used in numerous medical imaging procedures such as those of bone and the heart (see MEDICAL IMAGING TECHNOLOGY). The molybdenum-99 is either recovered from the fission of uranium or made from lighter Mo isotopes by neutron capture. Typically, a Mo-99 cow consists of MoO_4^{2-} adsorbed on a lead-shielded alumina column. The TcO_4^- formed upon the decay of Mo-99 by β-decay, $t_{1/2} = 66$ h, has less affinity for the column and is eluted or milked and either used directly or appropriately chemically derivatized for the particular diagnostic test (100).

BIBLIOGRAPHY

"Molybdenum Compounds" in *ECT* 1st ed., Vol. 9, pp. 199–214, by A. Linz, Climax Molybdenum Co.; in *ECT* 2nd ed., Vol. 13, pp. 645–659, by J. C. Bacon, Climax Molybdenum Co.; in *ECT* 3rd ed., Vol. 15, pp. 683–697, by H. F. Barry, Climax Molybdenum Co. of Michigan.

1. K.-H. Tytko, W.-D. Fleischmann, D. Gras, and E. Warkentin, in H. Katscher and F. Schröder, eds., *Gmelin Handbook of Inorganic Chemistry*, Vol. B4, Springer-Verlag, Berlin, 1985.
2. K.-H. Tytko and U. Trobish, in H. Katscher and F. Schröder, eds., *Gmelin Handbook of Inorganic Chemistry*, Suppl. Vol. B3a, Springer-Verlag, Berlin, 1987.
3. K.-H. Tytko and D. Gras, in H. Katscher and F. Schröder, eds., *Gmelin Handbook of Inorganic Chemistry*, Vol. B4, Springer-Verlag, Berlin, 1989.
4. M. T. Pope, in G. Wilkinson, R. D. Gillard, and J. A. McCleverty, eds., *Comprehensive Coordination Chemistry*, Vol. 3, Pergamon Press, Oxford, UK, 1987, pp. 1023–1058.
5. Q. Chen and J. Zubieta, *Coord. Chem. Rev.* **114**, 107–162 (1992).
6. M. T. Pope and A. Müller, *Angew. Chem. Int. Ed. Engl.* **30**, 34–48 (1991).
7. M. T. Pope and A. Müller, eds., *Polyoxometallates: From Platonic Solids to Anti-Retroviral Activity*, Kluwer Academic Publishers, Dordrecht, the Netherlands, 1994.
8. Y. Jeannin, G. Hervé, and A. Proust, *Inorg. Chim. Acta*, **198–200**, 319–336 (1992).
9. R. I. Buckley and R. J. H. Clark, *Coord. Chem. Rev.* **65**, 167–218 (1985).
10. E. I. Stiefel, *Progr. Inorg. Chem.* **22**, 1 (1977).
11. E. I. Stiefel, in Ref. 4, pp. 1375–1420.
12. M. H. Dickman and M. T. Pope, *Chem. Rev.* **94**, 589–594 (1994).
13. A. Müller, E. Diemann, R. Jostes, and H. Bögge, *Angew. Chem. Int. Ed. Eng.* **20**, 934–955 (1981).
14. C. D. Garner, in Ref. 4, pp. 1421–1444.
15. T. C. Ho, in M. C. Oballa and S. S. Shih, eds., *Catalytic Hydroprocessing of Petroleum and Distillates*, Marcel Dekker, Inc., New York, 1994.
16. S. W. Weller, *Energy Fuels* **8**, 415–420 (1994).
17. A. G. Sykes, in Ref. 4, pp. 1229–1264.
18. C. D. Garner and J. M. Charnock, in Ref. 4, pp. 1329–1374.
19. H. Brunner and J. Wachter, *J. Organomet. Chem.* **240**, C41–C44 (1982).
20. T. R. Halbert, S. A. Cohen, and E. I. Stiefel, *Organometallics* **4**, 1689–1690 (1985).
21. T. J. Risdon, *Properties of Molybdenum Disulfide, MoS₂ (Molybdenite)* Bulletin C-5c, Climax Molybdenum Co., Ann Arbor, Mich., Aug. 1989.
22. S. K. Srivastava and B. N. Avasthi, *J. Sci. Ind. Res.* **41**, 656–664 (1982).
23. L. Margulis, G. Salitra, R. Tenne, and M. Talianker, *Nature* **365**, 114 (1993).

24. A. Müller, R. Jostes, and F. A. Cotton, *Angew. Chem. Int. Ed. Eng.* **19**, 875–882 (1980).

25. T. Shibahara, *Coord. Chem. Rev.* **123**, 73–147 (1993).

26. A. Müller and E. Diemann, *Chimia* **39**, 312–313 (1985).

27. R. Poli, *J. Coord. Chem.* 29, 121–173 (1993).

28. M. H. Chisholm, *Acc. Chem. Res.* **23**, 419–425 (1990).

29. R. C. Haushalter and L. A. Mundi, *Chem. Mater.* **4**, 31–48 (1992).

30. G. J. Leigh and R. L. Richards, in Ref. 4, pp. 1265–1299.

31. G. Wilkinson, F. G. A. Stone, and E. W. Abel, eds., *Comprehensive Organometallic Chemistry*, Vol. 6, Pergamon Press, Oxford, U.K., 1982, pp. 1079–1253.

32. H. Schumann, in M. Winter, ed., *Gmelin Handbook of Inorganic Chemistry*, Pt. 6, Springer-Verglag, Berlin, 1990.

33. E. M. Carnahan, J. D. Prostasiewicz, and S. J. Lippard, *Acc. Chem. Res.* **26**, 90–97 (1993).

34. F. A. Cotton and R. A. Walton, *Multiple Bonds Between Metal Atoms*, 2nd ed., Oxford University Press, Oxford, U.K., 1993.

35. C. D. Garner, in Ref. 4, pp. 1301–1328.

36. G. Ferraudi, *Bol. Soc. Chil. Quim.* **32**, 23–44 (1987).

37. T. C. Zeitlow, M. D. Hopkins, and H. B. Gray, *J. Solid State Chem.* **57**, 112–119 (1985).

38. R. Chevrel, M. Hirrien, and M. Sergent, *Polyhedron* **5**, 87–94 (1986).

39. S. J. Hilsenbeck, V. G. Young, Jr., and R. E. McCarley, *Inorg. Chem.* **33**, 1822–1832 (1994).

40. G. J. Kubas, *Acc. Chem. Res.* **21**, 120–127 (1988).

41. G. J. Leigh, *Acc. Chem. Res.* **25**, 177–181 (1991).

42. H. Tenn, W. Kurtz, and J. Wagner, in H. Katscher, W. Kurtz, and F. Schröder, eds., *Gmelin Handbook of Inorganic Chemistry*, Suppl. Vol. B5, 1990.

43. I. Haas, S. Jäger, H. Katscher, D. Schneider, and J. Wagner, in U. W. Gewarth and H. Katscher, eds., *Gmelin Handbook of Inorganic Chemistry*, Suppl. Vol. B6, 1990.

44. E. I. Stiefel, in E. I. Stiefel, D. Coucouvanis, and W. E. Newton, eds., *Molybdenum Enzymes, Cofactors, and Model Systems*, ACS Symposium Series, Vol. 535, Washington, D.C., 1993, pp. 1–19.

45. E. I. Stiefel, D. Coucouvanis, and W. E. Newton, eds., *Molybdenum Enzymes, Cofactors, and Model Systems*, ACS Symposium Series, Vol. 535, Washington, D.C., 1993.

46. S. J. N. Burgmayer and E. I. Stiefel, *J. Chem. Educ.* **62**, 943–953 (1985).

47. J. L. Johnson and S. K. Wadman, in C. R. Scriver and co-workers, eds., *Metabolic Basis of Inherited Disease*, McGraw-Hill, New York, 1989, Chapt. 56, pp. 1463–1475.

48. C. F. Mills and G. F. Davis, in W. Mertz, ed., *Trace Elements in Human and Animal Nutrition*, 5th ed., Academic Press, New York, 1987, pp. 429–463.

49. J. Lener and B. Bíbr, *J. Hyg. Epidem. Microbiol. Immunol.* **28**, 405–419 (1984).

50. C. C. Seaborn and S. P. Yang, *Biol. Trace. Elem.* **39**, 245–256 (1993).

51. E. R. Purvis, *J. Agric. Food Chem.* **3**, 666–669 (1955).

52. U. C. Gupta and J. Lippsett, *Adv. Agron.* **34**, 73–115 (1981).

53. G. K. Davis, in E. Merion, ed., *Metals and Their Compounds in the Environment*, VCH Publishers, Weinheim, Germany, 1991, pp. 1089–1100.

54. R. Eisler, *Molybdenum Hazards to Fish, Wildlife, and Invertebrates: A Synoptic Review*, U.S. Fish and Wildlife Service Biology Report No. 85 (1.19), 1989, 61 pp.

55. J. Mason, *Toxicology* **42**, 99–109 (1986).

56. L. G. Wilson and R. S. Bandurski, *J. Biol. Chem.* **233**, 975–981 (1958).

57. P. V. Bodine and G. Litwack, *Mol. Cell. Endocrin.* **74**, C77–C81 (1990).

58. E. I. Stiefel and G. N. George, in I. Bertini, H. B. Gray, S. J. Lippard, and J. S. Valentine, eds., *Bioinorganic Chemistry*, University Science Books, Mill Valley, Calif., 1994, pp. 365–453.

59. J. Kim and D. C. Rees, *Biochemistry* **33**, 387–397 (1994).

60. K. V. Rajagopalan, in Ref. 44, pp. 38–49.

61. J. C. Wooton, R. E. Nicolson, J. M. Cock, D. E. Walters, J. F. Burke, W. A. Doyle, and R. C. Bray, *Biochim. Biophys. Acta* **1057**, 157–185 (1991).

62. R. S. Pilato and E. I. Stiefel, in J. Reedijk, ed., *Bioinorganic Catalysis*, Marcel Dekker, New York, 1993, pp. 131–188.

63. *Hazardous Substances Data Base*, National Library of Medicine (NTIS), Springfield, Va., 1994.

64. R. L. Lewis and D. V. Sweet, eds., *Registry of Toxic Effects of Chemical Substances*, National Institute for Occupational Safety and Health, Cincinnati, Ohio, 1994.

65. *Chemical Economics Handbook*, SRI International, Menlo Park, Calif., Sept. 1994.

66. L. M. Cohn, *Amer. Metal. Mkt.*, 6, (Apr. 13, 1993).

67. T. Spalvins, *J. Mater. Eng. Perf.* **I**, 347–352 (1992).

68. M. R. Hinton and P. D. Fleischauer, *Surface Coatings Technol.* **54/55**, 435–441 (1992).

69. R. F. Sebenek, *NLGI Spokesman* **57**, 96–106 (1993).

70. P. C. H. Mitchell, *Wear* **100**, 281–300 (1984).

71. R. Sarin, A. K. Gupta, A. V. Sureshbabu, V. Martin, A. K. Misra, and A. K. Bhatnagar, *Lubr. Sci.* **5**, 213–239 (1993).

72. R. R. Chianelli, M. Daage, and M. J. Ledoux, *Adv. Catal.* **40**, 177–232 (1994).

73. R. Prins, V. H. J. De Beer, and G. A. Somorjai, *Catal. Rev. Sci. Eng.* **31**, 1–41 (1989).

74. E. M. de Wind, J. J. Heinerman, S. L. Lee, F. J. Plantenga, C. C. Johnson, and D. C. Woodward, *Oil Gas J.*, 49–53 (Feb. 24, 1992).

75. H. Topsøe and B. S. Clausen, *Catal. Rev.-Sci Eng.* **26**, 395–420 (1984).

76. T. R. Halbert, T. C. Ho, E. I. Stiefel, R. R. Chianelli, and M. Daage, *J. Catal.* **130**, 116–129 (1991).

77. T. P. Snyder and C. G. Hill, Jr., *Catal. Rev.-Sci. Eng.* **31**, 43–95 (1989).

78. R. K. Grasselli, *J. Chem Educ.* **63**, 216–221 (1986).

79. H. Mimoun, in G. Wilkinson, R. D. Gillard, and J. A. McCleverty, eds., *Comprehensive Coordination Chemistry*, Vol. 6, Pergamon Press, Oxford, UK, 1987, pp. 317–410.

80. V. Krylov, *Catal. Today* **18**, 209–302 (1993).

81. M. Ward, J. D. Bradzil, A. P. Mehandru, and A. A. Anderson, *J. Phys. Chem.* **91**, 6515–6521 (1987).

82. N. D. Parkyns, C. I. Warburton, and J. D. Wilson, *Catal. Today* **18**, 385–442 (1993).

83. S. Kasztelan and J. B. Moffat, *J. Catal.* **116**, 83–94 (1989).

84. M. A. Bañares, B. Pawelec, and J. L. G. Fierro, *Zeolites* **12**, 882–888 (1992).

85. U. Ozkan and G. L. Schrader, *J. Catal.* **95**, 120–154 (1985).

86. J. Haber, *The Role of Molybdenum in Catalysis*, Climax Molybdenum Co. Ltd., London, 1981.

87. V. Bizzarri, B. Lindner, and N. Lindskog, *Sprechsaal Ceram.* **122**, 568–571 (1989).

88. J. J. Petrovic and R. E. Honnell, *Ceram. Trans.* **19**, 817–830 (1991).

89. C. Zavodni, *Amer. Metal Mkt.* **71**(101), 204–205 (1993).

90. C. H. Simpson, *Amer. Paint. Coating J.*, 66 (1992).

91. E. R. Braithwaite, in R. Thompson, ed., *Specialty Inorganic Chemicals*, Royal Society of Chemistry, Burlington House, London, 1980, pp. 346–374.

92. E. R. Braithwaite, *Chem. Ind.*, 405–412 (1978).

93. W. J. Kroenke, *J. Appl. Poly. Sci.* **32**, 4255–4168 (1986).

94. G. A. Skinner and P. J. Haines, *Fire Mater.* **10**, 63–69 (1986).

95. L. M. Tsai, S. J. Wang, and K. Lin, *J. Energetic Mat.* **10**, 17–41 (1992).

96. C. Julien, S. I. Saikh, and M. Balkanski, *Mater. Sci. Eng.*, **B14**, 121–126 (1992).

97. N. N. Abumrad, A. J. Schneider, D. Steel, and L. S. Rogers, *Amer. J. Clin. Nutr.* **34**, 2551–2559 (1981).

98. T. Yamase, K. Tomita, Y. Seto, and H. Fujita, *Polym. Med: Biomed. Pharm. Appl.*, 185–212 (1992).
99. P. Köpf-Maier and T. Klopötke, *J. Cancer Res. Clin. Oncol.* **118**, 216–221 (1992).
100. E. Shikata and A. Iguchi, *J. Radioanal. Nucl. Chem.* **102**, 533–550 (1986).

EDWARD I. STIEFEL
Exxon Research and Engineering Company

MONAZITE.   See CERIUM AND CERIUM COMPOUNDS; THORIUM.

MONOCLONAL ANTIBODIES.   See ENZYME APPLICATIONS; VACCINE TECHNOLOGY.

MONOSODIUM GLUTAMATE.   See AMINO ACIDS (MSG).

MORPHOLINE.   See AMINES.

MOUTHWASHES.   See DENTIFRICES.

MUCILAGES.   See GUMS.

MUNTZ METAL.   See COPPER ALLOYS, WROUGHT COPPER AND ALLOYS.

MUSCLE RELAXANTS.   See NEUROREGULATORS; PSYCHOPHARMACO-LOGICAL AGENTS.

NANOTECHNOLOGY. See SUPPLEMENT.

NAPHTHALENE

This article deals mainly with naphthalene [*91-20-3*]. The hydrogenated naph-thalenes, the alkylnaphthalenes (particularly methyl- and isopropylnaph-thalenes), and acenaphthene also are discussed (see also NAPHTHALENE DERIVA-TIVES).

Properties

The accepted configuration of naphthalene, ie, two fused benzene rings sharing two common carbon atoms in the ortho position, was established in 1869 and was based on its oxidation product, phthalic acid (1). Based on its fused-ring configuration, naphthalene is the first member in a class of aromatic compounds with condensed nuclei. Naphthalene is a resonance hybrid:

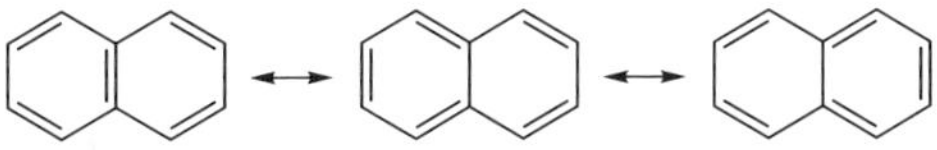

In chemical reactions, naphthalene usually acts as though the bonds were fixed in the positions as shown in the first structure above at the left. For most purposes, the conventional formula (**1**) is adequate; the numbers represent the carbon atoms with attached hydrogen atoms.

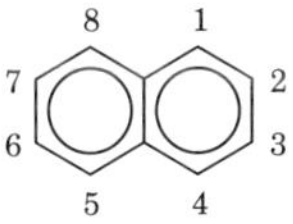

(**1**) 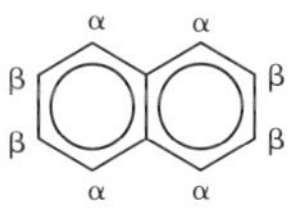   (**2**)

The two carbons that bear no numbers are common to both rings and carry no hydrogen atoms. From the symmetrical configuration of the naphthalene molecule, it should be possible for only two isomers to exist when one hydrogen atom is replaced by another atom or group. Therefore, positions 1, 4, 5, and 8 are identical and often are designated as "α" positions; likewise, positions 2, 3, 6, and 7 are identical and are designated as "β" positions, as shown in (**2**).

Some selected chemical and physical properties of naphthalene are given in Table 1. Selected values from the vapor pressure–temperature relationship for naphthalene are listed in Table 2, as are selected viscosity–temperature relationships for liquid naphthalene. Naphthalene forms azeotropes with several compounds; some of these mixtures are listed in Table 3.

Naphthalene is very slightly soluble in water but is appreciably soluble in many organic solvents, eg, 1,2,3,4-tetrahydronaphthalene, phenols, ethers, carbon disulfide, chloroform, benzene, coal-tar naphtha, carbon tetrachloride, acetone, and decahydronaphthalene. Selected solubility data are presented in Table 4.

The ir, uv, mass, nmr, and ^{13}C-nmr spectral data for naphthalene and other related hydrocarbons have been reported (7–11). Additional information regarding the properties of naphthalene has been published (3,6,12,13).

Table 1. Properties of Naphthalene

Property	Value	Ref.
molecular wt	120.1732	
mp, °C	80.290	2
normal bp at 101.3 kPa[a], °C	217.993	3
triple point (t_{tp}), °C	80.28	3
critical temperature (t_c), °C	475.2	3
critical pressure (p_c), kPa[a]	4051.0	3
flash point (closed cup), °C	79	4
ignition temperature, °C	526	4
flammable limits, vol %		
upper	5.9	5
lower	0.9	5
heat of vaporization, kJ/mol[b]	43.5	3
heat of fusion at triple point, kJ/mol[b]	18.979	3
heat of combustion, at 15.5°C and 101.3 kPa[a], kJ/mol[b]	−5158.41	3
heat capacity, at 15.5°C and 101.3 kPa[a], J/(mol·K)[b]	159.28	3
heat of formation at 25°C, kJ/mol[b]		
solid	78.53	3
gas	150.58	3
density, g/mL		
at 25°C	1.175	3
at 90°C	0.97021	3

[a]To convert kPa to atm, divide by 101.3.
[b]To convert J to cal, divide by 4.184.

Table 2. Selected Values of Vapor Pressure–Temperature and Viscosity–Temperature Relationships for Naphthalene[a]

Vapor pressure data		Viscosity data	
Temperature, °C	Pressure, kPa[b]	Temperature, °C	Viscosity, mPa·s(=cP)
0	0.0008	80.3	0.96
10	0.003	90	0.846
20	0.007	100	0.754
40	0.043	110	0.678
87.6	1.33	120	0.616
119.1	5.33	150	0.482
166.3	26.66	180	0.394
191.3	53.33	220	0.320
214.3	93.33		
218.0	101.33		
230.5	133.32		
250.6	199.98		

[a] Ref. 6.
[b] To convert kPa to mm Hg, multiply by 7.5.

Table 3. Azeotropes of Naphthalene[a]

Compound	Bp[b] of azeotrope, °C	Naphthalene, wt %	Bp[b] of azeotrope, °C
water	100.0	16.0	98.8
dodecane	216.3	60.5	140.2
dipropylene glycol	231.8	87.6	142.9
ethylene glycol	197.4	49.0	183.9
benzyl alcohol	205.2	40.0	204.1
p-ethylphenol	218.8	55.0	215.0
p-chlorophenol	219.8	63.5	216.3
diethylene glycol	245.5	78.0	216.6
3,4-dimethylphenol	226.8	84.0	217.6
benzoic acid	249.2	95.0	217.7

[a] Ref. 3.
[b] At 101.3 kPa = 1 atm.

Table 4. Naphthalene Solubility Data[a]

Solvent	Solubility at 25°C, mol fraction
ethylbenzene	0.2926
benzene	0.2946
toluene	0.2920
cyclohexane	0.1487
carbon tetrachloride	0.2591
n-hexane	0.1168
water	0.18×10^{-5}

[a] Ref. 3.

Reactions

Substitution.   Substitution products retain the same nuclear configuration as naphthalene. They are formed by the substitution of one or more hydrogen atoms with other functional groups. Substituted naphthalenes of commercial importance have been obtained by sulfonation, sulfonation and alkali fusion, alkylation, nitration and reduction, and chlorination.

The hydrogen atoms in the 1 position of naphthalene can be substituted somewhat more easily than hydrogen atoms in benzene, and they tend to do so under mild conditions. For example, 1-chloronaphthalene [90-13-1] can be formed by direct substitution with little or no catalyst, and 1-nitronaphthalene [86-57-7] can be prepared using dilute nitric acid. Sulfonation of naphthalene also occurs readily in the 1 position but can be influenced by temperature. When a second group substitutes, its position also is influenced by the nature and position of the first group. In the case of the substitution of the second identical group during nitration or sulfonation, it predominately attaches to the unsubstituted ring. For identical groups, there are seven possible disubstituted isomers. If the two substituting groups are different, 14 disubstitution isomers are possible. The number of possibilities is still larger when the substituents are different and three or more hydrogen atoms are replaced. The number of possible isomeric substitution products from naphthalene has been calculated and reported (14). In naming substituted naphthalenes, positions are designated numerically according to the lowest position number for the first substituent, eg (**3**), 1,3-dinitronaphthalene [606-37-1]. The sum of the position numbers should be the lowest possible, eg (**4**), 1,4,5-trinitronaphthalene (not 1,4,8-trinitronaphthalene).

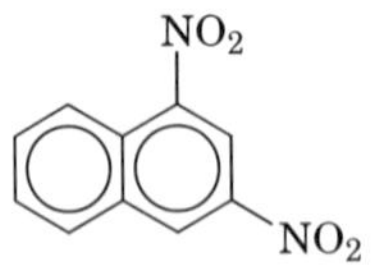

(**3**) 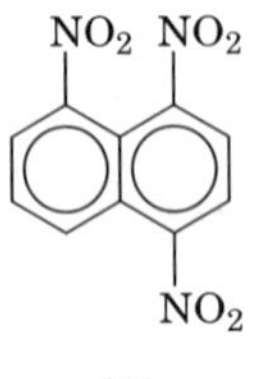   (**4**)

Sulfonation.   Sulfonation of naphthalene with sulfuric acid produces mono-, di-, tri-, and tetranaphthalenesulfonic acids (see NAPHTHALENE DERIVATIVES). All of the naphthalenesulfonic acids form salts with most bases. Naphthalenesulfonic acids are important starting materials in the manufacture of organic dyes (15) (see AZO DYES). They also are intermediates used in reactions, eg, caustic fusion to yield naphthols, nitration to yield nitronaphthalenesulfonic acids, etc.

Nitration.   Naphthalene is easily nitrated with mixed acids, eg, nitric and sulfuric, at moderate temperatures to give mostly 1-nitronaphthalene and small quantities, 3–5%, of 2-nitronaphthalene. 2-Nitronaphthalene [581-89-5] is not made in substantial amounts by direct nitration and must be produced by indirect methods, eg, the Bucherer reaction starting with 2-naphthalenol (2-naphthol [135-19-3]). However, the 2-naphthylamine [91-59-8] made using this route is a carcinogen; thus the Bucherer method is seldom used in the United States.

Halogenation. Under mild catalytic conditions, halogen substitution occurs, and all of the hydrogen atoms of the naphthalene molecule can be replaced. The only commercially significant halogenated naphthalene products are the mixed chlorinated naphthalenes. Naphthalene is chlorinated readily by introducing gaseous chlorine into molten naphthalene at ambient pressure and temperatures of 100–220°C in the presence of small amounts of a catalyst, eg, ferric chloride. The chlorination of molten naphthalene gives a mixture of mono-, di-, and polychloronaphthalenes; the degree of chlorination is controlled by monitoring the specific gravity or melting point of the crude reaction product.

The commercial products are mixtures ranging from liquids, eg, mono- and mixed mono- and dichloronaphthalenes, to various wax-like solids, which contain di-, tri-, and polychloronaphthalenes, with high melting points, ie, 90–185°C. Chloronaphthalenes are flame resistant and have high dielectric constants. Their high degree of chemical stability is indicated by their resistance to most acids and alkalies as well as to dehydrochlorination. They also are resistant to attack by fungi and insects. The U.S. demand for chloronaphthalenes has declined steadily. Koppers Co., Inc., the sole U.S. producer, ceased manufacturing their chloronaphthalene products (Halowax) in 1977.

The toxicity of chloronaphthalenes requires that special attention and caution be used during their manufacture and use; acne is the most common result of excessive skin exposure to them and the most frequently affected areas are the face and neck (16). Liver damage has occurred in workers who have been exposed repeatedly to vapors, particularly to those of penta- and hexachloronaphthalene [1335-87-1] (17,18). Uses for the chlorinated naphthalenes include solvents, gauge and instrument fluids, capacitor impregnants, components in electric insulating compounds, and electroplating stop-off compounds.

Alkylation. Naphthalene can be easily alkylated using various alkylating agents, eg, alkyl halides, olefins, or alcohols, in the presence of a suitable catalyst (19,20). In vapor-phase alkylations, phosphoric acid on kieselguhr and silica–alumina catalysts are useful. In liquid-phase reactions, acid catalysts, eg, sulfuric acid, hydrofluoric acid, and phosphoric acid, are used widely when olefins are the alkylating agent. Aluminum chloride and other metal halides, eg, iron and zinc chlorides, are also active alkylating catalysts; however, their use often involves reactions that yield undesirable resinous by-products. Sulfuric acid is the preferred catalyst for alkylations with alcohols, since the active aluminum chloride forms a complex with alcohols and must be used in prohibitively large quantities.

Isopropylnaphthalenes produced by alkylation of naphthalene with propylene have gained commercial importance as chemical intermediates, eg, 2-isopropylnaphthalene [2027-17-0], and as multipurpose solvents, eg, mixed isopropylnaphthalenes. Alkylation of naphthalene with alkyl halides (except methyl halides), acid chlorides, and acid anhydrides proceeds in the presence of anhydrous aluminum chloride by Friedel-Crafts reactions (qv). The products are alkylnaphthalenes or alkyl naphthyl ketones, respectively (see ALKYLATION).

Chloromethylation. The reactive intermediate, 1-chloromethylnaphthalene [86-52-2], has been produced by the reaction of naphthalene in glacial acetic acid and phosphoric acid with formaldehyde and hydrochloric acid. Heating of these ingredients at 80–85°C at 101.3 kPa (1 atm) with stirring for ca 6 h is required. The potential hazard of such chloromethylation reactions, which results from the possible production of small amounts of the powerful carcinogen methyl chloromethyl ether [107-30-2], has been reported (21).

Addition.   The most important addition products of naphthalene are the hydrogenated compounds. Of less commercial significance are those made by the addition of chlorine.

Hydrogenation.   Hydrogen is added to the naphthalene nucleus by reagents that do not affect benzene. Two, four, six, eight, or ten hydrogen atoms may add. Of these, only the tetra- and decahydronaphthalenes are commercially significant. In addition to the commercially important 1,2,3,4-tetrahydronaphthalene, the 1,4,5,8-isomer has been reported. A review of the other hydronaphthalenes is available (22). Some chemical and physical properties of 1,2,3,4-tetrahydronaphthalene and the decahydronaphthalenes are given in Table 5.

1,2,3,4-Tetrahydronaphthalene [119-64-2] (Tetralin) is a water-white liquid that is insoluble in water, slightly soluble in methyl alcohol, and completely soluble in other monohydric alcohols, ethyl ether, and most other organic solvents. It is a powerful solvent for oils, resins, waxes, rubber, asphalt, and aromatic hydrocarbons, eg, naphthalene and anthracene. Its high flash point and low vapor pressure make it useful in the manufacture of paints, lacquers, and varnishes; for cleaning printing ink from rollers and type; in the manufacture of shoe creams and floor waxes; as a solvent in the textile industry; and for the removal of naphthalene deposits in gas-distribution systems (25). The commercial product typically has a tetrahydronaphthalene content of ≥97 wt %, with some decahydronaphthalene and naphthalene as the principal impurities.

1,2,3,4-Tetrahydronaphthalene reacts similarly to an alkylbenzene because its structure contains only one aromatic nucleus. It can be sulfonated readily,

Table 5. Properties of Tetra- and Decahydronaphthalenes[a]

Property	Tetralin	*cis*	Decalin Mixed isomers	*trans*
mol wt	132.2048	138.2522		
mp, °C	−35.749[b]	−42.98		−30.38
normal bp at 101.3 kPa[c], °C	207.62	195.815		187.310
density at 25°C, g/mL	0.9659[b]	0.8929		0.8660
viscosity at 25°C, mPa·s(=cP)	2.012[b]	2.99		1.936
flash point, closed cup, °C	71[d]		57.8	
ignition temperature, °C	385[d]		250	
flammable limits, vol %				
upper	5.0[d,e]		4.9[d,f]	
lower	0.8[d,f]		0.7[d,f]	

[a]Tetralin and decalin, respectively; data from Ref. 24 unless otherwise noted.
[b]Ref. 23.
[c]To convert kPa to mm Hg, multiply by 7.5.
[d]Ref. 4.
[e]At 150°C.
[f]At 100°C.

nitrated, oxidized, and hydrogenated. Sulfonation occurs first in the 6 and, to some extent, in the 5 position. Nitration with mixed acid yields the 5- and 6-nitro compounds in the cold; 4,6- and 5,7-dinitro compounds are formed at 35–40°C (22). 1,2,3,4-Tetrahydronaphthalene is oxidized to a hydroperoxide by passing air or oxygen through the warm liquid (26,27).

Further hydrogenation under pressure in the presence of a nickel catalyst gives a mixture of *cis*- and *trans*-decahydronaphthalene [*493-02-7*] (Decalin). 1,2,3,4-Tetrahydronaphthalene dehydrogenates to naphthalene at 200–300°C in the presence of a catalyst; thermal dehydrogenation takes place at ca 450°C and is accompanied by cracking to compounds, such as toluene and xylene.

Tetrahydronaphthalene is produced by the catalytic treatment of naphthalene with hydrogen. Various processes have been used, eg, vapor-phase reactions at 101.3 kPa (1 atm) as well as higher pressure liquid-phase hydrogenation where the conditions are dependent upon the particular catalyst used. Nickel or modified nickel catalysts generally are used commercially; however, they are sensitive to sulfur, and only naphthalene that has very low sulfur levels can be used. Thus many naphthalene producers purify their product to remove the thionaphthene, which is the principal sulfur compound present. Sodium treatment and catalytic hydrodesulfurization processes have been used for the removal of sulfur from naphthalene; the latter treatment is preferred because of the hazardous nature of sodium treatment.

1,2,3,4-Tetrahydronaphthalene is not a highly toxic compound. A threshold limit value of 25 ppm or 135 mg/m^3 has been suggested for Tetralin. Tetralin vapor is an irritant to the eyes, nose, and throat, and dermatitis has been reported in painters working with it (28). The single-dose oral toxicity LD$_{50}$ for rats is 2.9 g/kg (29).

Decahydronaphthalene [*91-17-8*] (Decalin) is the product of complete hydrogenation of naphthalene. Like Tetralin, it is a clear, colorless liquid with excellent solvent properties. It is produced commercially by the catalytic hydrogenation of naphthalene or 1,2,3,4-tetrahydronaphthalene and consists of a mixture of cis and trans isomers (22,30–32). The commercial product typically has a decahydronaphthalene content of ≥97 wt %, with the principal impurity being 1,2,3,4-tetrahydronaphthalene. Decahydronaphthalene can be converted to naphthalene by heating with platinum, palladium, or nickel catalyst at 300°C (33).

Decahydronaphthalene is slightly to moderately toxic (29). The vapors are irritating to the eyes, nose, and throat. Excessive exposure to high concentrations causes numbness, nausea, headache, and vomiting. Dermatitis has been noted among painters handling decahydronaphthalene. No serious cases of industrial poisoning have been reported (34). A threshold limit value has not been established, although a value of 25 ppm has been suggested (28).

The uses of decahydronaphthalene are similar to those of 1,2,3,4-tetrahydronaphthalene. Mixtures of the two are used for certain applications where a synergistic solvency effect is noted.

Some selected chemical and physical properties of 1,2,3,4-tetrahydronaphthalene and the decahydronaphthalenes are listed in Table 5.

Chlorine Addition. Chlorine addition and some chlorine substitution occurs at normal or slightly elevated temperatures in the absence of catalysts. The chlorination of molten naphthalene under such conditions yields a mixture of naphthalene tetrachlorides, a monochloronaphthalene tetrachloride, and a dichloronaphthalene tetrachloride, as well as mono- and dichloronaphthalenes (35). Sunlight or uv radiation initiates the addition reaction of chlorine and naphthalene resulting in the production of the di- and tetrachlorides (36). These addition products are relatively unstable and, at ca 40–50°C, they decompose to form the mono- and dichloronaphthalenes.

Oxidation. Naphthalene may be oxidized directly to 1-naphthalenol (1-naphthol [90-15-3]) and 1,4-naphthoquinone, but yields are not good. Further oxidation beyond 1,4-naphthoquinone [130-15-4] results in the formation of *ortho*-phthalic acid [88-99-3], which can be dehydrated to form phthalic anhydride [85-44-9]. The vapor-phase reaction of naphthalene over a catalyst based on vanadium pentoxide is the commercial route used throughout the world. In the United States, the one phthalic anhydride plant currently operating on naphthalene feedstock utilizes a fixed catalyst bed. The fluid-bed process plants have all been shut down, and the preferred route used in the world is the fixed-bed process.

The naphthalene is vaporized, mixed with air, and fed to the top of the reactor. This process also allows for mixtures of *ortho*-xylene [95-47-6] to be mixed with the naphthalene and air, which permits the use of dual feedstocks. Both feedstocks are oxidized to phthalic anhydride. The typical range of reactor temperature is 340–380°C. The reactor temperatures are controlled by an external molten salt.

The quality of naphthalene required for phthalic anhydride manufacture is generally 95% minimum purity. The fixed plants do not require the high (>98%) purity naphthalene product and low (<50 ppm) sulfur. The typical commercial coal-tar naphthalene having a purity ca 95% (freezing point, 77.5°C), a sulfur content of ca 0.5%, and other miscellaneous impurities, is acceptable feedstock for the fixed-bed catalyst process based on naphthalene.

Manufacture

Two sources of naphthalene exist in the United States; coal tar and petroleum (qv). Coal tar was the traditional source until the late 1950s, when it was in short supply. In 1960, the first petroleum-naphthalene plant was brought on stream, and by the late 1960s, petroleum naphthalene accounted for over 40% of total naphthalene production. The availability of large quantities of o-xylene at competitive prices during the 1970s affected the position of naphthalene as the prime raw material for phthalic anhydride. In 1971, 45% of U.S. phthalic anhydride capacity was based on naphthalene, as compared to only 29% in 1979 and 17% in 1990. Production for 1992 was less than 50% of the levels in the early 1980s. The last dehydroalkylation plant for petroleum naphthalene was shut down late in 1991. Coal tar has stabilized at around 85×10^3 t/yr, and petroleum-naphthalene production is around $6–8 \times 10^3$ t/yr. The reduction of petroleum production has opened the door for import naphthalene, mainly from Canada. The 1993 United States naphthalene capacities are given in Table 6.

Coal-Tar Process. Coal tar is condensed and separated from the coke-oven gases formed during the high temperature carbonization of bituminous coal in coke plants (see COAL; COAL CONVERSION PROCESSES). Although some

Table 6. U.S. Naphthalene Capacities, 1993

Producer	Location	Coal tar or petroleum	Capacity, t
AlliedSignal, Inc.	Ironton, Ohio	coal tar	34,000
Crosscreek Industries	Baytown, Tex.	petroleum	10,000
Koppers Industries, Inc.	Follansbee, W. Va.	coal tar	80,000
Total			*124,000*

naphthalene is present in the oven gases after tar separation and is removed in subsequent water-cooling and scrubbing steps, the amounts are of minor importance. The largest quantities of naphthalene are obtained from the coal tar that is separated from the coke-oven gases. A typical dry coal tar obtained in the United States contains ca 10 wt % naphthalene. The naphthalene content of the tar varies somewhat depending on the coal source, the coke-oven operating temperature, and the coking cycle times: the higher the coking temperature and rate, the higher the naphthalene content of the tar.

The coal tar first is processed through a tar-distillation step where ca the first 20 wt % of distillate, ie, chemical oil, is removed. The chemical oil, which contains practically all the naphthalene present in the tar, is reserved for further processing, and the remainder of the tar is distilled further to remove additional creosote oil fractions until a coal-tar pitch of desirable consistency and properties is obtained.

The chemical oil contains ca 50 wt % naphthalene, 6 wt % tar acids, 3 wt % tar bases, and numerous other aromatic compounds. The chemical oil is processed to remove the tar acids by contacting with dilute sodium hydroxide and, in a few cases, is next treated to remove tar bases by washing with sulfuric acid.

Principal U.S. producers obtain their crude naphthalene product by fractional distillation of the tar acid-free chemical oil. This distillation may be accomplished in either a batch or continuous fashion. One such method for the continuous recovery of naphthalene by distillation is shown in Figure 1. The tar acid-free chemical oil is charged to the system where most of the low boiling components, eg, benzene, xylene, and toluene, are removed in the light-solvent column. The chemical oil next is fed to the solvent column, which is operated under vacuum, where a product containing the prenaphthalene components is taken overhead. This product, which is called coal-tar naphtha or crude heavy solvent, typically has a boiling range of ca 130–200 °C and is used as a general solvent and as a feedstock for hydrocarbon-resin manufacture because of its high content of resinifiables, eg, indene and coumarone (see HYDROCARBON RESINS). The naphthalene-rich bottoms from the solvent column then are fed to the naphthalene column where a naphthalene product (95% naphthalene) is produced. The naphthalene column is operated at near atmospheric pressure to avoid difficulties which are inherent to vacuum distillation of this product, eg, naphthalene-filled vacuum jets and lines. A side stream which is rich in methylnaphthalenes may be taken near the bottom of the naphthalene column.

However, since the naphthalene produced from petroleum is of high purity and quality, the production of refined naphthalene by such chemical treatments essentially has ceased in the United States. Not only are such treatments expen-

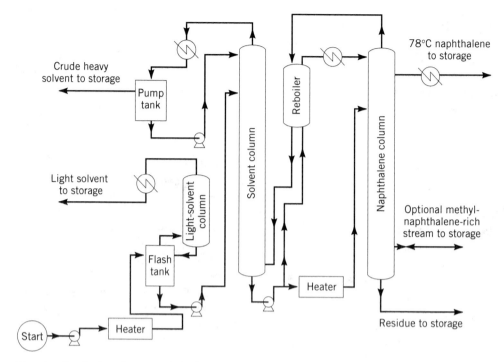

Fig. 1. Typical coal-tar naphthalene distillation, starting with naphthalene-rich chemical oil.

sive, but they also generate a significant amount of waste sludge, which creates additional costs for appropriate waste-disposal facilities.

The main impurity in crude 78°C coal-tar naphthalene is sulfur which is present in the form of thionaphthene (1–3%). Methyl- and dimethylnaphthalenes also are present (1–2%) with lesser amounts of indene, methylindenes, tar acids, and tar bases.

Economic Aspects

Total nameplace capacity for all U.S. naphthalene producers in 1993 was 124 × 10^3 t, with 114 × 10^3 t produced from coal tar and 10 × 10^3 t from petroleum. Naphthalene production from 1968 to 1993 is listed in Table 7.

The decline in naphthalene production in 1973 primarily resulted from competition with o-xylene as the feedstock for phthalic anhydride. Periods of feedstock shortages and the loss of one principal producer also affected petroleum naphthalene output.

Naphthalene imports provided about 10–20% of the material consumed in the United States until ca 1963, when that percentage dropped to and leveled off at less than 5%. Imports increased again in 1992 to 10–15%, owing to the closure of Texaco's plant.

The economics of naphthalene recovery from coal tar can vary significantly, depending on the particular processing operation used. A significant factor is the

Table 7. U.S. Naphthalene Production and Price History[a]

Year	Coal-tar naphthalene, 10^3 t	Petroleum naphthalene, 10^3 t	Total, 10^3 t	Crude coal-tar naphthalene, ¢/kg	Petroleum naphthalene, ¢/kg
1968	206	171	377		
1969	208	162	370		
1970	188	132	320		
1971	170	117	287		
1972	181	105	286		
1973	163	109	272		
1974	170	91	261		
1975	159	50	209	19.8	28.7–33.1
1976	159	49	208	26.5	30.1–35.3
1977	150	68	218	26.5	30.1–35.3
1978	141	78	219	26.5	30.1–35.3
1979	152	74	226	26.5	30.1–35.3
1980	142	59	201	26.5	30.1–35.3
1981	159	64	223	45.2	44.1–50.7
1982	104	57	161	48.5	44.1–61.7
1983	105	42	147	48.5	44.1–61.7
1984	85	38	123	48.5	44.1–63.9
1985	84	23	107	37.5	44.1–63.9
1986	86	24	110	35.3	44.1–63.9
1987	85	24	109	33.1	44.1–63.9
1988	84	24	108	35.3	44.1–68.3
1989	86	24	110	39.7	44.1–70.5
1990	82	23	105	39.7	44.1–70.5
1991	80	23	103	44.1	44.1–70.5
1992	78	14	92	44.1	44.1–77.2
1993	77[b]	10[b]	87[b]	44.1	44.1–77.2
1994	90	11	101	44.1	45–55

[a]The prices shown are ca mid-year list prices; whenever a range of prices is given in the source, the lowest of those given is listed in the table.
[b]Estimated.

cost of the coal tar. As the price of fuel oil increases, the value of coal tar also increases. The price history of naphthalene from 1975 to 1993 is given in Table 7.

The higher price of the petroleum product results from its higher quality, ie, higher purity, lower sulfur content, etc. The price of crude coal-tar naphthalene is primarily associated with that of o-xylene, its chief competitor as phthalic anhydride feedstock.

The preferred route to higher purity naphthalene, either coal-tar or petroleum, is crystallization. This process has demonstrated significant energy cost savings and yield improvements. There are several commercial processes available: Sulzer-MWB, Brodie type, Betz, and Recochem (37).

Specifications and Test Methods

Naphthalene usually is sold commercially according to its freezing or solidification point, because there is a correlation between the freezing point and the naphthalene content of the product; the correlation depends on the type and relative amount of impurities which are present (38). Because the freezing point can be changed appreciably by the presence of water, values and specifications are listed on a dry, wet, or as-received basis, using an appropriate method agreed upon between buyer and seller, eg, ASTM D1493.

Gas–liquid chromatography is used extensively to determine the naphthalene content of mixtures. Naphthalene can be separated easily from thionaphthene, the methyl- and dimethylnaphthalenes, and other aromatics. Analysis of the various other impurities may require the use of high resolution capillary columns.

Other tests that are routinely performed on commercial grades of naphthalene include evaporation residues (ASTM D2232), APHA color (ASTM D1686), water (ASTM D95), and acid-wash color (ASTM D2279). Three methods used to measure sulfur content are the oxygen-bomb combustion method (ASTM D129), the lamp-combustion method (ASTM D1266), and the Raney nickel reduction technique (39). Some typical specifications of commercially available grades of coal-tar (40) and petroleum naphthalene and the ASTM specifications for refined naphthalene are listed in Table 8.

Table 8. Naphthalene Specifications

Analysis	Crude coal tar[a]	Petroleum[b]	ASTM refined[c]
solidification point (min), °C	78.0	79.9	79.8
assay (min), wt %	95.9	99.6	
sulfur (max), wt %	0.9		
color (max), APHA			100
acid wash color (max)			10
nonvolatiles (max), wt %	0.25		0.10
water (max), wt %	0.25		

[a]Ref. 40.
[b]80° Naphthalene, ASTM D2386 and D1491. Courtesy of Advanced Aromatics, Inc.
[c]ASTM D2431–2468.

Health and Safety Factors

Handling. Naphthalene is generally transported in molten form in tank trucks or tank cars that are equipped with steam coils. Depending on the transportation distance and the insulation on the car or truck, the naphthalene may solidify and require reheating before unloading. Without inert-gas blanketing and at the temperature normally used for the storage of molten naphthalene, ie, 90°C, the vapors above the liquid are within the flammability limits. Thus, storage tanks containing molten naphthalene have a combustible mixture in the vapor space and care must be taken to eliminate all sources of ignition around such systems. Naphthalene dust also can form explosive mixtures with air, which necessitates care in the design and operation of solid handling systems. Perhaps the greatest hazard to the worker is the potential for operating or maintenance personnel to be accidentally splashed with hot molten naphthalene while taking samples or disassembling process lines (ASTM D3438). Molten naphthalene tank vents must be adequately heated and insulated to prevent the accumulation of sublimed and solidified naphthalene. A collapsed tank can result easily from pumping from a tank with a plugged vent.

Toxicology. The acute oral and dermal toxicity of naphthalene is low with LD_{50} values for rats from 1780–2500 mg/kg orally (41) and greater than 2000 mg/kg dermally. The inhalation of naphthalene vapors may cause headache, nausea, confusion, and profuse perspiration, and if exposure is severe, vomiting, optic neuritis, and hematuria may occur (28). Chronic exposure studies conducted by the NTP in mice for two years showed that naphthalene caused irritation to the nasal passages, but no other overt toxicity was noted. Rabbits that received 1–2 g/d of naphthalene either orally or hypodermically developed changes in the lens of the eye after a few days, followed by definite opacity of the lens after several days (41). Rare cases of such corneal epithelium damage in humans have been reported (28). Naphthalene can be irritating to the skin, and hypersensitivity does occur.

In reports submitted to the U.S. Environmental Protection Agency (EPA), there was no mutagenicity in the Ames test, no mutagenicity in the mouse micronucleus test, and no mutagenicity in the rat hepatocyte–DNA repair test. It was confirmed in other trials that naphthalene is nonmutagenic in the Salmonella microsome mutagenicity (Ames) test (42,43). In tests submitted to the EPA, there were no developmental or teratogenic effects noted in rabbits treated while pregnant.

The octanol–water partition coefficient, which is used as an indicator of the tendency of an organic chemical to accumulate in living tissue, was low. This indicates that naphthalene is unlikely to accumulate in the body.

In additional EPA studies, subchronic inhalation was evaluated in the rat for 4 and 13 weeks, respectively, and no adverse effects other than nasal irritation were noted. In the above-mentioned NTP chronic toxicity study in mice, no chronic toxic effects other than those resulting from bronchial irritation were noted. There was no treatment-related increase in tumors in male mice, but female mice had a slight increase in bronchial tumors. Neither species had an increase in cancer. Naphthalene showed no biological activity in other chemical carcinogen tests, indicating little cancer risk (44). No incidents of chronic effects have been reported as a result of industrial exposure to naphthalene (28,41).

Because naphthalene vapors can cause eye irritation at concentrations of 15 ppm in air and because continued exposure may result in adverse effects to the eye, a threshold limit value of 10 ppm (50 mg/m^3) has been set by the ACGIH (45). This amount is about 30% of the air-saturation value at 27°C.

Uses

The U.S. naphthalene consumption by markets for 1992 is listed in Table 9. The production of phthalic anhydride by vapor-phase catalytic oxidation has been the main use for naphthalene. Although its use has declined in favor of o-xylene, naphthalene is expected to maintain its present share of this market, ie, ca 18%. Both petroleum naphthalene and coal-tar naphthalene can be used for phthalic anhydride manufacture. U.S. phthalic anhydride capacity was 465×10^3 t in 1992 (38).

Table 9. U.S. Naphthalene Consumption, 1992

Use	Consumption, 10^3t	% of Total
phthalic anhydride	69	63
surfactants	19	17
insecticides	13	12
moth repellants	6	6
miscellaneous	2	2
Total	*109*	*100*

Naphthalenesulfonates represent a large and growing outlet for naphthalene, ie, ca 17% of supply in 1994, and growing by 7–9% annually. The products are used as wetting agents and dispersants in paints and coatings and in a variety of pesticides and cleaner formulations. Their application as surfactants is expected to continue as a growth item in uses such as concrete and gypsum board additives (see SURFACTANTS).

Another large use of naphthalene is as a raw material for the manufacture of 1-naphthyl-*N*-methylcarbamate [63-25-2] (carbaryl, Sevin). Crude or semirefined coal-tar or petroleum naphthalene can be used for carbaryl manufacture. Carbaryl is used extensively as a replacement for DDT and other products that have become environmentally unacceptable (see INSECT CONTROL TECHNOLOGY).

Miscellaneous uses include several organic compounds and intermediates, eg, 1-naphthalenol, 1-naphthylamine [134-32-7], 1,2,3,4-tetrahydronaphthalene, decahydronaphthalene, and chlorinated naphthalenes.

Alkylnaphthalenes

Methyl- and dimethylnaphthalenes are contained in coke-oven tar and in certain petroleum fractions in significant amounts. A typical high temperature coke-oven coal tar, for example, contains ca 3 wt % of combined methyl- and dimethylnaphthalenes (6). In the United States, separation of individual isomers is seldom attempted; instead a methylnaphthalene-rich fraction is produced for commercial purposes. Such mixtures are used for solvents for pesticides, sulfur, and various aromatic compounds. They also can be used as low freezing, stable heat-transfer fluids. Mixtures that are rich in monomethylnaphthalene content have been used as dye carriers (qv) for color intensification in the dyeing of synthetic fibers, eg, polyester. They also are used as the feedstock to make naphthalene in dealkylation processes. Phthalic anhydride also can be made from methylnaphthalene mixtures by an oxidation process that is similar to that used for naphthalene.

A mixed monomethylnaphthalene-rich material can be produced by distillation and can be used as feedstock for further processing. By cooling this material to about 0°C, an appreciable amount of 2-methylnaphthalene crystallizes, leaving a mother liquor consisting of approximately equal quantities of 1- and 2-methylnaphthalene. Pure 2-methylnaphthalene [91-57-6] (bp = 341.1°C; mp = 34.58°C) is used primarily as a raw material for the production of vitamin K preparations. Oxidation produces 2-methyl-1,4-naphthoquinone [58-27-5] (menadione, vitamin K_3), which itself and in the form of water-soluble sodium hydrogen sulfite adducts shows similar antihemorrhagic effects similar to the natural vitamin K_1 (see BLOOD, COAGULANTS AND ANTICOAGULANTS). Other compounds of the vitamin K series can be prepared from menadione (46) (see VITAMINS).

1-Methylnaphthalene [90-12-0] (bp = 244.6°C; mp = −30.6°C) can be used as a general solvent because of its low melting point. It also is used as a test substance for the determination of the cetane number of diesel fuels (see GASOLINE AND OTHER MOTOR FUELS). By side-chain chlorination of 1-methylnaphthalene to 1-chloromethylnaphthalene and formation of naphthaleneacetonitrile, it is possible to produce 1-naphthylacetic acid which is a growth regulator for plants, a germination suppressor for potatoes, and an intermediate for drug manufacture (46) (see GROWTH REGULATORS, PLANT).

Of the individual dimethylnaphthalenes, 2,6-dimethylnaphthalene [28804-88-8] has been of particular interest as a precursor to 2,6-naphthalenedicarboxylic acid [1141-38-4], a potentially valuable monomer for polyesters.

Isopropylnaphthalenes can be prepared readily by the catalytic alkylation of naphthalene with propylene. 2-Isopropylnaphthalene [2027-17-0] is an important intermediate used in the manufacture of 2-naphthol (see NAPHTHALENE DERIVATIVES). The alkylation of naphthalene with propylene, preferably in an inert solvent at 40–100°C with an aluminum chloride, hydrogen fluoride, or boron trifluoride–phosphoric acid catalyst, gives 90–95% wt % 2-isopropylnaphthalene; however, a considerable amount of polyalkylate also is produced. Preferably, the propylation of naphthalene is carried out in the vapor phase in a continuous manner, over a phosphoric acid on kieselguhr catalyst under pressure at ca 220–250°C. The alkylate, which is low in di- and polyisopropylnaphthalenes, then is isomerized by recycling over the same catalyst at 240°C or by using aluminum chloride catalyst at 80°C. After distillation, a product containing >90 wt % 2-isopropylnaphthalene is obtained (47).

Mixtures containing various concentrations of mono-, di-, and polyisopropylnaphthalenes have been prepared by treating molten naphthalene with concentrated sulfuric acid and propylene at 150–200°C followed by distillation (39). Products comprised of such isomeric mixtures have extremely low pour points, ie, ca −50°C, are excellent multipurpose solvents, and have been evaluated as possible liquid-phase heat-transfer oils.

Of the higher alkylnaphthalenes, those of importance are the amyl-, diamyl-, polyamyl-, nonyl-, and dinonylnaphthalenes. These alkylnaphthalenes are used in sulfonated form as surfactants and detergent products.

Acenaphthene. Acenaphthene [83-32-9] is a hydrocarbon, $C_{12}H_{10}$, present in high temperature coal tar (6). Acenaphthene may be halogenated, sulfonated, and nitrated in a manner similar to naphthalene (41). Oxidation first yields acenaphthenequinone, followed by 1,8-naphthalenedicarboxylic acid anhydride [81-84-5], an important intermediate for dyes, pigments, fluorescent whiteners, and pesticides. Acenaphthylene [208-96-8] is formed upon catalytic dehydrogenation of acenaphthene (42). Acenaphthene can be isolated and recovered from a tar-distillation fraction by concentrating it by fractional distillation followed by crystallization, to give ca 40% recovery of 98–99% pure acenaphthene. This material can be further purified by recrystallization from a suitable solvent, eg, ethanol (43).

BIBLIOGRAPHY

"Naphthalene" in *ECT* 1st ed., Vol. 9, pp. 216–231, by G. Riethof and A. Pozefsky, Gulf Research & Development Co., in *ECT* 2nd ed., Vol. 13, pp. 670–690, by G. Thiessen, Koppers Co., Inc.; in *ECT* 3rd ed., Vol. 15, pp. 698–719, by R. M. Gaydos, Koppers Co., Inc.

1. J. R. Partington, *A Short History of Chemistry*, Macmillan & Company, Ltd., London, 1957, pp. 293, 316.
2. J. P. McCullough and co-workers, *J. Phys. Chem.* **61**, 1105 (1957).
3. *Naphthalene*, American Petroleum Institute Monograph Series, Publication 707, API, Washington, D.C., Oct. 1978.
4. *Fire Protection Guide on Hazardous Materials*, 7th ed., National Fire Protection Association, Boston, Mass., 1978, pp. 49–212, 213, 225M-61, 173.
5. G. W. Jones and G. S. Scott, *U.S. Bur. Mines Rep. Invest.*, 3881 (1946).
6. *The Coal Tar Data Book*, 2nd ed., The Coal Tar Research Association, Gomersal, Leeds, U.K., 1965, B-2, p. 62; A-1, pp. 3, 4; B-2, p. 74.
7. B. J. Zwolinski and co-workers, *Selected Infrared Spectral Data*, American Petroleum Institute Research Project 44, Thermodynamics Research Center, Texas A&M University, College Station, Tex., 1954.
8. B. J. Zwolinski and co-workers, *Selected Ultraviolet Spectral Data*, American Petroleum Institute Research Project 44, Thermodynamics Research Center, Texas A&M University, College Station, Tex., 1961.
9. B. J. Zwolinski and co-workers, *Selected Mass Spectral Data*, American Petroleum Institute Research Project 44, Thermodynamics Research Center, Texas A&M University, College Center, Tex., 1949.
10. B. J. Zwolinski and co-workers, *Selected Nuclear Magnetic Resonance Spectral Data*, American Petroleum Institute Research Project 44, Thermodynamics Research Center, Texas A&M University, College Station, Tex., 1962.
11. B. J. Zwolinski and co-workers, *Selected ^{13}C Nuclear Magnetic Resonance Spectral Data*, American Petroleum Institute Research Project 44, Thermodynamics Research Center, Texas A&M University, College Station, Tex., 1974.
12. H. C. Anderson and W. R. K. Wu, *U.S. Bur. Mines Bull.* **606**, 304 (1963).
13. C. L. Yaws and A. C. Turnbough, *Chem. Eng.* **82**, 107 (Sept. 1).
14. R. F. Evans and W. J. LeQuesne, *J. Org. Chem.* **15**, 19 (1950).
15. H. E. Fierz-David and L. Blangley, Fundamental Processes of Dye Chemistry, Interscience Publishers, Inc., New York, 1949, pp. 125, 440–453.
16. *Chloronaphthalenes*, Hygienic Guide Series, American Industrial Hygiene Association, Jan.–Feb. 1966.
17. F. B. Flinn and N. E. Jarvik, *Proc. Soc. Exp. Biol. Med.* **35**, 118 (1936).
18. M. Mayers and A. Smith, *N. Y. Ind. Bull.*, 21, 30 (Jan. 1942).
19. A. N. Sachanen, *Conversion of Petroleum*, 2nd ed., Reinhold Publishing Corp., New York, 1948, pp. 550–565.
20. G. A. Olah, *Friedel-Crafts and Related Reactions*, Vols. 1–4, Wiley-Interscience, New York, 1963–1965.
21. *Current Report*, The Bureau of National Affairs, Nov. 22, 1979, 589; C. C. Yao and G. Miller, NIOSH Contract No. 210-75-0056 (Research Study on Bis(Chloromethyl) Ether), Jan. 1979.
22. N. Donaldson, *The Chemistry and Technology of Naphthalene Compounds*, Edward Arnold Publishers, London, 1958, pp. 455–473.
23. *Tetralin*, American Petroleum Institute Monograph Series, Publication 705, API, Washington, D.C., Oct. 1978.
24. *Cis and Trans-Decaline*, American Petroleum Institute. Monograph Series, Publication 706, API, Washington, D.C., Oct. 1978.

25. *Tetralin and Decalin Solvents*, Bulletin, E. I. du Pont de Nemours & Co. Inc., Organic and Chemicals Division, Wilmington, Del., 1976.

26. J. S. Bogen and G. C. Wilson, *Petrol. Refiner* **23**, 118 (1944).

27. U.S. Pat. 2,462,103 (Feb. 22, 1949), R. Johnson (to Koppers Co., Inc.).

28. E. E. Sandmeyer, in G. D. Clayton and F. E. Clayton, eds., *Patty's Industrial Hygiene and Toxicology*, 3rd rev. ed., Vol. II, Wiley-Interscience, New York, 1981, Chapt. 46.

29. N. I. Sax, *Dangerous Properties of Industrial Materials*, 4th ed., Van Nostrand Reinhold Co., New York, 1975, pp. 348, 599, 948–949, 1152.

30. Millard, *Ann. l'Off. Nat. Combust. Liq.* **9**, 1013 (1934); **10**, 95 (1935).

31. R. Baker and R. Schultz, *J. Amer. Chem. Soc.* **69**, 1250 (1947).

32. W. Seyer and R. Walker, *J. Amer. Chem. Soc.* **60**, 2125 (1938).

33. Wessely and Grill, *Montatsh. Chem.* **77**, 282 (1947).

34. C. Marsden and S. Mann, *Solvents Guide*, 2nd ed., Wiley-Interscience, New York, 1963, p. 161.

35. Faust and Saame, *Ann. Chem.* **160**, 67 (1871).

36. A. Leeds and E. Everhart, *J. Amer. Chem. Soc.* **2**, 205 (1880).

37. H. G. Franck, J. W. Stadelhofer, *Industrial Aromatic Chemistry*, Springer-Verlag, Berlin, 1987, pp. 302–305.

38. W. Kirby, *J. Soc. Chem. Ind.* **59**, 168 (1940).

39. U.S. Pat. 3,962,365 (May 28, 1975), R. M. Gaydos and co-workers (to Koppers Co., Inc.).

40. Technical data, Koppers Industries, Inc., Pittsburgh, Pa.

41. A. E. Everest, *The Higher Coal Tar Hydrocarbons*, Longmans Green & Co., 1927, pp. 1–57.

42. M. Kaufman and A. Williams, *J. Appl. Chem.* **I**, 489 (1951).

43. G. Markus and T. Kraatsover, *Coke Chem. USSR* **5**, 37 (1971).

44. I. Purchase and co-workers, *Br. J. Cancer* **37**, 873 (1978).

45. *Documentation of the Threshold Limit Values*, 3rd ed., ACGIH, Cincinnati, Ohio, 1971.

46. H. Franck and G. Collin, *Erzeuginisse aus Steinkohlenteer*, Springer-Verlag, Berlin, 1968, p. 173.

47. K. Handrick, *Erdoel and Kohle, Compendium* 76/77, p. 308.

ROBERT T. MASON
Koppers Industries, Inc.

NAPHTHALENE CARBOXYLIC ACIDS.   See NAPHTHALENE DERIVATIVES.

NAPHTHALENE DERIVATIVES

Naphthalene derivatives are of diverse importance as intermediates for agricultural, construction, pharmaceutical, photographic, rubber, tanning, and textile chemicals. In this article production figures, economics, and processes are discussed for most commercially important compounds. Sources for a more comprehensive study of naphthalene derivatives are available (1–8).

Several systems of nomenclature have been used for naphthalene, and many trivial and trade names are well established. The *Chemical Abstracts Index Guide* is employed in this article. The numbering of the naphthalene nucleus is shown in (**1**); older practices are given in (**2**) and (**3**).

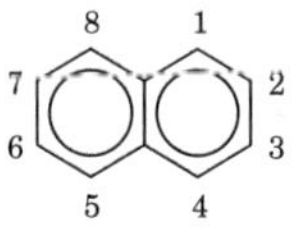

(**1**) 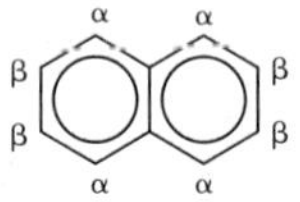   (**2**) 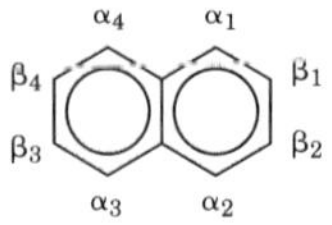   (**3**)

Substituents in the 1,8 or 2,6 positions of naphthalene are in the peri or amphi positions, respectively.

The number of naphthalene derivatives is very large, since the number of positional isomers is large: 2 for monosubstitution, 10 for disubstitution–same substituent, 14 for disubstitition–different substituents, 14 for trisubstitution–same substituent, 42 for trisubstitution–two different substituents, 84 for trisubstitution–three different substituents, and so on with multiplying complexity.

Naphthalenesulfonic Acids

Naphthalenesulfonic acids are important chemical precursors for dye intermediates, wetting agents and dispersants, naphthols, and air-entrainment agents for concrete. The production of many intermediates used for making azo, azoic, and triphenylmethane dyes (qv) involves naphthalene sulfonation and one or more unit operations, eg, caustic fusion, nitration, reduction, or amination.

Commercially, sulfonation is carried out by the classic method with sulfuric acid. Modern reactors are glass-lined; older equipment was made from cast iron or coated with enamel. Processes often use chlorosulfonic acid or sulfur trioxide to minimize the need of excess sulfuric acid. Improved analytical methods have contributed to the success of process optimization (9–12).

Generally, the sulfonation of naphthalene leads to a mixture of products. Naphthalene sulfonation at less than ca 100°C is kinetically controlled and produces predominantly 1-naphthalenesulfonic acid (**4**). Sulfonation of naphthalene at above ca 150°C provides thermodynamic control of the reaction and 2-naphthalenesulfonic acid as the main product. Reaction conditions for the sulfonation of naphthalene to yield desired products are given in Figure 1; alternative paths are possible. A list of naphthalenesulfonic acids and some of their properties is given in Table 1.

1-Naphthalenesulfonic Acid.   The sulfonation of naphthalene with excess 96 wt % sulfuric acid at <80°C gives >85 wt % 1-naphthalenesulfonic acid

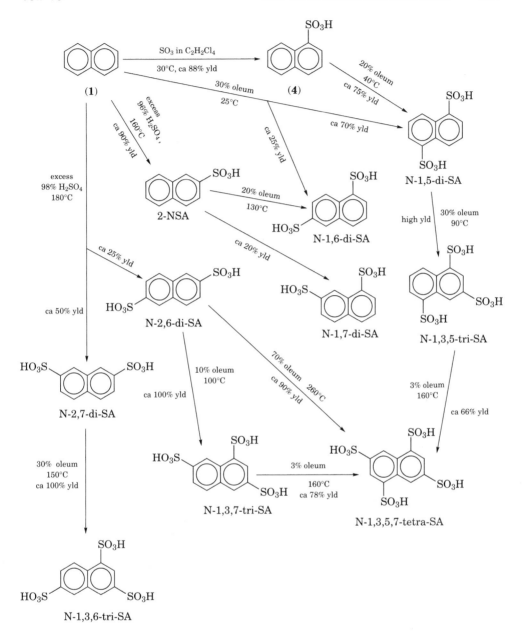

Fig. 1. Selected paths to naphthalenesulfonic acids where N = naphthalene, SA = sulfonic acid, and yld = yield.

(α-acid); the balance is mainly the 2-isomer (β-acid). An older German commercial process is based on the reaction of naphthalene with 96 wt % sulfuric acid at 20–50°C (13). The product can be used unpurified to make dyestuff intermediates by nitration or can be sulfonated further. The sodium salt of 1-naphthalenesulfonic acid is required, for example, for the conversion of 1-naphthalenol (1-naphthol) by caustic fusion. In this case, the excess sulfuric

Table 1. Melting Points of Naphthalenesulfonic Acids

Compound	CAS Registry Number	Mp, °C	Mp of corresponding sulfonyl chloride, °C
1-naphthalenesulfonic acid	[85-47-2]	139–140	68
1-naphthalenesulfonic acid dihydrate	[6036-48-2]	90	
2-naphthalenesulfonic acid	[120-18-3]	139–140	76
2-naphthalenesulfonic acid hydrate	[76530-12-6]	124–125	
2-naphthalenesulfonic acid trihydrate	[17558-84-8]	83	
1,2-naphthalenedisulfonic acid	[25167-78-6]		160
1,3-naphthalenedisulfonic acid	[6094-26-4]		137.5
1,4-naphthalenedisulfonic acid	[46859-22-7]	240–245 dec	162
1,5-naphthalenedisulfonic acid	[81-04-9]	125 dec	183
1,6-naphthalenedisulfonic acid	[525-37-1]		129
1,7-naphthalenedisulfonic acid	[5724-16-3]		123
2,6-naphthalenedisulfonic acid	[581-75-9]	199 dec	228–229
2,7-naphthalenedisulfonic acid	[92-41-1]		159.5
1,3,5-naphthalenetrisulfonic acid	[6654-64-4]		146
1,3,6-naphthalenetrisulfonic acid	[86-66-8]		194–197
1,3,7-naphthalenetrisulfonic acid	[85-49-4]		165–166
1,4,5-naphthalenetrisulfonic acid	[60913-37-3]		156–157
1,3,5,7-naphthalenetetra-sulfonic acid	[6654-67-7]		261–262

acid first is separated by the addition of lime and is filtered to remove the insoluble calcium sulfate; the filtrate is treated with sodium carbonate to precipitate calcium carbonate and leave the sodium 1-naphthalenesulfonate [130-14-3] in solution. The dry salt then is recovered, typically, by spray-drying the solution.

The older methods have been replaced by methods which require less, if any, excess sulfuric acid. For example, sulfonation of naphthalene can be carried out in tetrachloroethane solution with the stoichiometric amount of sulfur trioxide at no greater than 30°C, followed by separation of the precipitated 1-naphthalenesulfonic acid; the filtrate can be reused as the solvent for the next batch (14). The purification of 1-naphthalenesulfonic acid by extraction or washing the cake with 2,6-dimethyl-4-heptanone (diisobutyl ketone) or a C-1–4 alcohol has been described (15,16). The selective insoluble salt formation of 1-naphthalenesulfonic acid in the sulfonation mixture with 2,3-dimethylaniline has been patented (17).

1-Naphthalenesulfonic acid can be converted to 1-naphthalenethiol [529-36-2] by reduction of the related sulfonyl chloride; this product has some utility as a dye intermediate, and is converted by reaction with alkyl isocyanates to S-naphthyl-N-alkylthiocarbamates, which have pesticidal and herbicidal activity (see HERBICIDES; PESTICIDES) (18). Either 1- or 2-naphthalenethiol reacts with acetic anhydride in the presence of sulfuric acid to produce naphthalene thioesters, used as substrates for a continuous, nondestructive kinetic assay of esterases associated with insecticide resistance (19).

2-Naphthalenesulfonic Acid. The standard manufacture of 2-naphthalenesulfonic acid involves the batch reaction of naphthalene with 96 wt % sulfuric acid at ca 160°C for ca 2 h (13). The product contains the 1- and 2-isomers in a ratio of ca 15:85. Because of its faster rate of desulfonation, 1-naphthalenesulfonic acid can be hydrolyzed selectively by dilution of the charge with water and agitating for 1 h at 150°C; the naphthalene that is formed can be removed by steam distillation.

Sulfonation can be conducted with naphthalene–92 wt % H_2SO_4 in a 1:1.1 mole ratio with staged acid addition at 160°C over 2.5 h to give a 93% yield of the desired product (20). Continuous monosulfonation of naphthalene with 96 wt % sulfuric acid in a cascade reactor at ca 160°C gives 2-naphthalenesulfonic acid and small amounts of by-product naphthalenedisulfonic acids (21). The purification of 2-naphthalenesulfonic acid by hydrolysis of the 1-isomer can be done in a continuous manner (22,23).

In the manufacture of 2-naphthalenol, 2-naphthalenesulfonic acid must be converted to its sodium salt; this can be done by adding sodium chloride to the acid, and by neutralizing with aqueous sodium hydroxide or neutralizing with the sodium sulfite by-product obtained in the caustic fusion of the sulfonate. The crude sulfonation product, without isolation or purification of 2-naphthalenesulfonic acid, is used to make 1,6-, 2,6-, and 2,7-naphthalenedisulfonic acids and 1,3,6-naphthalenetrisulfonic acid by further sulfonation. By nitration, 5- and 8-nitro-2-naphthalenesulfonic acids, [89-69-1] and [117-41-9], respectively, are obtained, which are intermediates for Cleve's acid. All are dye intermediates. The crude sulfonation product can be condensed with formaldehyde or alcohols or olefins to make valuable wetting, dispersing, and tanning agents.

1,5- and 1,6-Naphthalenedisulfonic Acid. 1,5- and 1,6-Naphthalenedisulfonic acids are co-products in the low temperature disulfonation of naphthalene. They are known by the trivial names Armstrong acid (1,5) and Eiver-Pick acid (1,6). A typical process involves the sulfonation of naphthalene with an excess of 20–30 wt % oleum at not over 25°C (24). The sulfonation mass is diluted with water, and the sodium salt of the disulfonic acid is formed by reaction with sodium sulfate. On cooling, the dihydrate of the disodium 1,5-naphthalenedisulfonate [76758-30-0] precipitates (ca 70% yield) and is recovered by filtration or centrifugation. The filtrate contains sodium 1,6-naphthalenedisulfonate [1655-43-2] which can be recovered by lime addition to precipitate sulfate as calcium sulfate and by filtration and evaporation of the filtrate. The sulfonation of naphthalene at −5 to 40°C in a chlorinated alkane solvent with SO_3 or first with chlorosulfonic acid, followed by sulfur trioxide, gives 1,5-naphthalenedisulfonic acid in excellent yields and purity (25).

2,6- and 2,7-Naphthalenedisulfonic Acids. 2,6-Naphthalenedisulfonic acid (Ebert-Merz β-acid) is an important isomer and is useful as an intermediate for 2,6-naphthalenediol and as an additive for cleansing cosmetics (qv). The sulfonation of naphthalene with excess 98 wt % sulfuric acid at 135–180°C gives a mixture of 2,6- and 2,7-naphthalenedisulfonic acids; the mixture is diluted with water and converted to the sodium salt by the addition of NaCl or Na_2SO_4 (13). At first, sodium 2,6-naphthalenedisulfonate [1655-45-4] precipitates and is recovered by filtration in ca 25% yield; further salt addition and cooling precipitates sodium 2,7-naphthalenedisulfonate [1655-35-2] in ca 50% yield.

A naphthalene sulfonation product that is rich in the 2,6-isomer and low in sulfuric acid is formed by the reaction of naphthalene with excess sulfuric acid at 125°C and by passing the resultant solution through a continuous wiped-film evaporator at 245°C at 400 Pa (3 mm Hg) (26). The separation in high yield of 99% pure 2,6-naphthalenedisulfonate, as its anilinium salt from a crude sulfonation product, has been claimed (27). A process has been patented for the separation of 2,6-naphthalenedisulfonic acid from its isomers by treatment with phenylenediamine (28).

2,7-Naphthalenedisulfonic acid (Ebert-Merz α-acid) is partially isomerized in sulfuric acid at 160°C to 2,6-naphthalenedisulfonic acid. The reaction takes place by a desulfonation–resulfonation mechanism.

1,3,5- and 1,3,6-Naphthalenetrisulfonic Acids. The sulfonation of 1,5-naphthalenedisulfonic acid with oleum at 90°C gives 1,3,5-naphthalenetrisulfonic acid in good yield (29). 1,3,6-Naphthalenetrisulfonic acid can be made by sulfonation of 1,6- or 2,7-naphthalenedisulfonic acid but this is a fairly costly process (30). A more acceptable manufacturing method involves the time–temperature–acid concentration-programmed sulfonation of naphthalene with sulfuric acid and oleum (31).

Alkylnaphthalenesulfonic Acids. The alkylnaphthalenesulfonic acids can be made by sulfonation of alkylnaphthalenes, eg, with sulfuric acid at 160°C, or by alkylation of naphthalenesulfonic acids with alcohols or olefins. These products, as the acids or their sodium salts, are commercially important as textile auxiliaries, surfactants (qv), wetting agents, dispersants (qv), and emulsifying aids, eg, for dyes (qv), wettable powder pesticides, tars, clays (qv), and hydrotropes.

Naphthalenesulfonic Acid–Formaldehyde Condensates. The sodium salts of the condensation products of naphthalenesulfonic acid with formaldehyde constitute an important class of compounds which are mainly used in the area of concrete additives (32,33), agricultural formulations, rubber formulations, and synthetic tanning agents. They are also used in photographic materials (34). Hampshire Chemical Co. and Henkel of America, Inc., are the largest suppliers of naphthalene sulfonate in concrete additives (superplasticizer) and reportedly hold 75–80% of this market. It was estimated that naphthalene sulfonate demand from U.S. producers would reach approximately 15×10^3 t/yr for all surfactant and dispersant applications by 1994, representing an average annual growth rate of approximately 5%; most of the demand is from the growth of naphthalene sulfonate formaldehydes (35). In addition, the domestic consumption of naphthalene sulfonate syntans for leather tanning was estimated to be more than 7×10^3 t/yr.

In 1994 estimated naphthalene consumption in western Europe and Japan for the production of alkylnaphthalene sulfonates, naphthalene sulfonate formaldehyde condensates, and synthetic tanning agents was 34×10^3 and 17×10^3 t/yr, respectively. G. Bozzetto (Italy), part of the Ruetgers Werke group, produces about 15×10^3 t/yr of naphthalene sulfonate condensates (35).

Hydronaphthalenesulfonic Acid. Sodium tetralinsulfonate [37837-69-7] (sodium 1,2,3,4-tetrahydronaphthalenesulfonate) is marketed by Du Pont as a dispersing and solubilizing agent under the trade name Alkanol S. Poly(dihydronaphthalene) sulfonates have been proposed for use as ion-exchange resins and antistatic agents for thermoplastics (36).

Nitronaphthalenes and Nitronaphthalenesulfonic Acids

The nitro group does not undergo migration of the naphthalene ring during the usual nitration procedures. Therefore, mono- and polynitration of naphthalene is similar to low temperature sulfonation. The nitronaphthalenes and some of their physical properties are listed in Table 2. Many of these compounds are not accessible by direct nitration of naphthalene but are made by indirect methods, eg, nitrite displacement of diazonium halide groups in the presence of a copper catalyst, decarboxylation of nitronaphthalenecarboxylic acids, or deamination of nitronaphthalene amines.

1-Nitronaphthalene. 1-Nitronaphthalene is manufactured by nitrating naphthalene with nitric and sulfuric acids at ca 40–50°C (37). The product is

Table 2. Melting Point of Nitronaphthalenes

Compound	CAS Registry Number	Mp, °C
1-nitronaphthalene	[86-57-7]	52[a]; 57.8[b]
2-nitronaphthalene[c]	[581-89-5]	78.7[d]
1,2-dinitronaphthalene[c]	[24934-47-2]	161–162
1,3-dinitronaphthalene[c]	[606-37-1]	148
1,4-dinitronaphthalene[c]	[6921-26-2]	134
1,5-dinitronaphthalene	[605-71-0]	219
1,6-dinitronaphthalene[c]	[607-46-5]	166.5[e]
1,7-dinitronaphthalene[c]	[24824-25-7]	156
1,8-dinitronaphthalene	[602-38-0]	172
2,3-dinitronaphthalene[c]	[1875-63-4]	174.5–175
2,6-dinitronaphthalene[c]	[24824-26-8]	279
2,7-dinitronaphthalene[c]	[24824-27-9]	234
1,2,3-trinitronaphthalene[c]	[76530-13-7]	190
1,2,4-trinitronaphthalene[c]	[76530-14-8]	258
1,3,5-trinitronaphthalene	[2243-94-9]	122
1,3,6-trinitronaphthalene[c]	[59054-75-0]	186
1,3,8-trinitronaphthalene	[2364-46-7]	218
1,4,5-trinitronaphthalene	[2243-95-0]	149
1,3,5,7-tetranitronaphthalene[c]	[60619-96-7]	260
1,3,5,8-tetranitronaphthalene	[2217-58-5]	194–195
1,3,6,8-tetranitronaphthalene	[28995-89-3]	203
1,4,5,8-tetranitronaphthalene	[4793-98-0]	340–345 dec

[a]Metastable form.
[b]Bp 304°C (169°C at 1.6 kPa (12 mm Hg)).
[c]Made by indirect methods, not by the direct nitration of naphthalene or naphthalene-nitration products.
[d]Bp 312.5°C at 97.8 kPa (733 mm Hg) and 165°C at 2.0 kPa (15 mm Hg).
[e]Bp 370°C (235°C at 1.3 kPa (9.75 mm Hg)).

obtained in very high yield and contains ca 3–10 wt % 2-nitronaphthalene and traces of dinitronaphthalene; the product can be purified by distillation or by recrystallization from alcohol. 1-Nitronaphthalene is important for the manufacture of 1-naphthaleneamine. Photochemical nitration of naphthalene by tetranitromethane in dichloromethane and acetonitrile to give 1-nitronaphthalene has been described (38).

2-Nitronaphthalene is metabolized to the carcinogenic 2-naphthylamine in the human body (39). Respirators, protective clothing, proper engineering controls, and medical monitoring programs for workers involved in making byproduct 2-nitronaphthalene should be used.

1,5- and 1,8-Dinitronaphthalenes. 1,5- and 1,8-dinitronaphthalenes, (**5**) and (**6**), respectively, can be made by nitration of 1-nitronaphthalene in a ~40:60 ratio. Similar results are obtained by the direct dinitration of naphthalene with mixed acid at 40–80°C and separation of isomers by fractional crystallization from ethylene dichloride (13). Process studies involve improvements in the separation of 1,5- and 1,8-dinitronaphthalenes by solvent extraction (40). The analysis of the mono- and dinitronaphthalenes can be done by gas–liquid chromatography (glc).

The reaction of the corresponding diamine of 1,5-dinitronaphthalene with phosgene produces 1,5-naphthalenediisocyanate [3173-72-6] (41,42). 1,8-Dinitronaphthalene is reduced to 1,8-diaminonaphthalene, which is an intermediate for making phthaloperinone [6925-69-5] (**7**), an orange colorant for plastics.

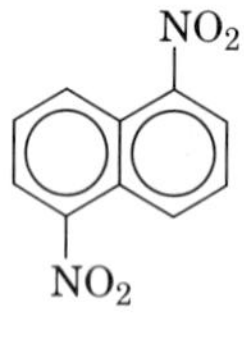

(**5**) 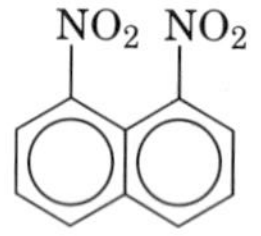   (**6**) 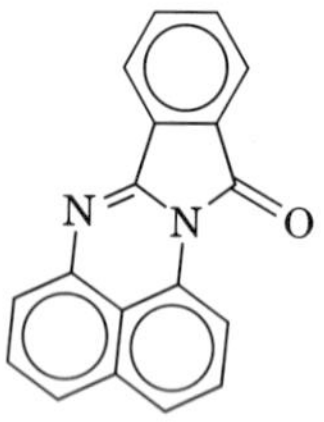   (**7**)

Nitronaphthalenesulfonic Acids. Nitronaphthalenesulfonic acids can be obtained either by the sulfonation of 1-nitronaphthalene or by the nitration of 1- or 2-naphthalenesulfonic acid. Thus the sulfonation of 1-nitronaphthalene with oleum at ca 25°C gives mainly 5-nitro-1-naphthalenesulfonic acid [17521-00-5]. The mononitration of 1-naphthalenesulfonic acid gives mainly 5- and 8-nitro-1-naphthalenesulfonic acid [117-41-9] and mononitration of 2-naphthalenesulfonic acid gives mainly 5-nitro-2-naphthalenesulfonic acid [86-69-1] and 8-nitro-2-naphthalenesulfonic acid [18425-74-6]. These compounds seldom are isolated; usually, the nitro group is reduced to the amino group to obtain dye intermediates.

Naphthaleneamines and Naphthalenediamines

Selected physical properties of naphthaleneamines and naphthalenediamines are listed in Table 3.

Table 3. Physical Properties of Naphthaleneamines and Naphthalenediamines

Compound	CAS Registry Number	Mp, °C	Density	Other
1-naphthaleneamine	[134-32-7]	50	1.13_4^{14}	flash pt, 157°C; sol 0.496 g/L H_2O; vol with steam; bp 301°C (160°C at 1.6 kPa^a)
2-naphthaleneamine	[91-59-8]	111–113	1.061_4^{98}	sol hot water; vol with steam; bp 306°C (175.8°C at 2.7 kPa^a)
1,2-naphthalenediamine	[938-25-0]	96–98		sol hot water, alc, ether; bp at 0.01 kPa^a 150–151°C
1,4-naphthalenediamine	[2243-61-0]	120		sl sol hot water
1,5-naphthalenediamine	[2243-62-1]	189.5		sol hot water, alc
1,6-naphthalenediamine	[2243-63-2]	78	$1.147_4^{99.4}$	sol hot water, alc
1,7-naphthalenediamine	[2243-64-3]	117.5		sol alc
1,8-naphthalenediamine	[479-27-6]	66.5	$1.127_4^{99.4}$	sol alc, ether; bp at 1.6 kPa^a 205°C
2,3-naphthalenediamine	[771-97-1]	191		sol alc, ether
2,6-naphthalenediamine	[2243-67-6]	216–218		sparingly sol alc, ether
2,7-naphthalenediamine	[613-76-3]	159		

^{a}To convert kPa to mm Hg, multiply by 7.5.

1-Naphthaleneamine. 1-Naphthylamine or α-naphthylamine [139-32-7] can be made from 1-nitronaphthalene by reduction with iron–dilute HCl, or by catalytic hydrogenation; it is purified by distillation and the content of 2-naphthylamine can be reduced as low as 8–10 ppm. Electroreduction of 1-nitronaphthalene to 1-naphthylamine using titania–titanium composite electrode has been described (43). Photoinduced reduction of 1-nitronaphthalene on semiconductor (eg, anatase) particles produces 1-naphthylamine in 77% yield (44). 1-Naphthylamine [134-32-7] can also be prepared by treating 1-naphthol with NH_3 in the presence of a catalyst at elevated temperature. The sanitary working conditions are improved by gas-phase reaction at 200–450°C using a dehydration catalyst consisting of aluminosilicate, Al_2O_3, or silica gel (45). 1-Naphthaleneamine is also toxic (LD_{50} (dogs) = 400 mg/kg) and a suspected human carcinogen, which conditions mandate that appropriate precautions be followed in manufacture and use.

1-Naphthaleneamine is a dye intermediate and is used as the starting material in the manufacture of the rodenticide, Antu (**8**), 1-naphthalenethiourea [86-88-4], which is prepared by heating a mixture of 1-naphthylamine hydrochloride, NH_4SCN, and a large amount water for 14–16 h while keeping its volume constant by adding an additional amount of water, to give a 97% yield. Its LD_{50} is 600 mg in squirrels (46).

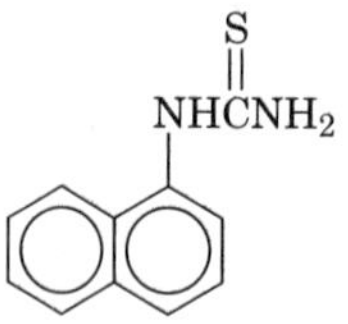

(8) 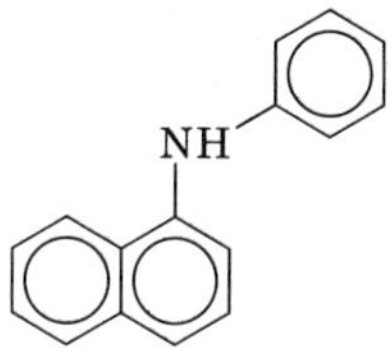   (9) 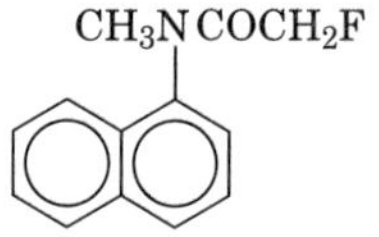   (10)

1-Naphthaleneamine is also the starting material for making the rubber antioxidant, N-phenyl-1-naphthaleneamine [90-30-2] (9), made by the condensation of 1-naphthaleneamine or 1-naphthalenol with aniline. Fluoroacetamidonaphthalene [5903-13-9] (Nissol) (10), an insecticide and miticide, is made from 1-naphthaleneamine by adding FCH_2COCl dropwise to a mixture of 1-N-methylnaphthylamine and benzene and refluxing for 2 h (47), or in an improved method, PCl_3 and $ClCH_2COOH$ are heated in xylene at 50–60°C to give $ClCH_2COCl$, which is treated with a solution of 1-naphthylamine in xylene 70–78°C and then at 100–120°C to give 92% N-1-naphthylchloroacetamide of 93% purity. N-1-Naphthylchloroacetamide is then methylated with $(CH_3)_2SO_4$ and fluorinated with KF to give 78–80% Nissol of 93–95% purity (48). The herbicide, Naptalam (11) Alanap, or Dyanap, or N-1-naphthylphthalamic acid [132-66-1] is usually prepared by adding 1-naphthylamine to an aqueous slurry of phthalic anhydride. The yield of the product is about 96% (49).

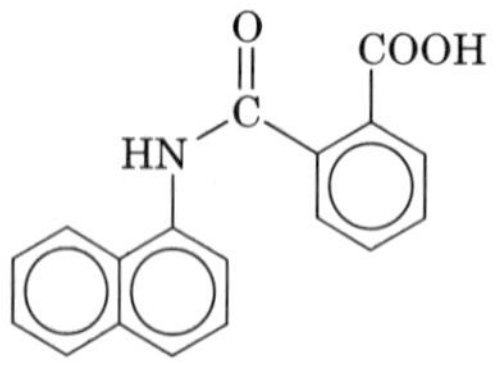

(11) 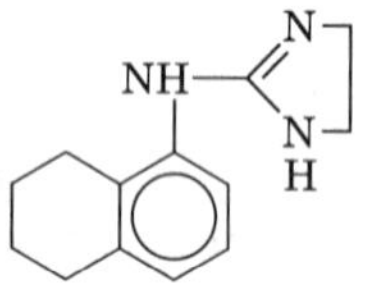   (12)

A tetrahydronaphthaleneamine derivative, 2-(5,6,7,8-tetrahydro-1-naphthaleneamine)-2-imidazoline (12), is used in the hydrochloride form as an adrenergic agent, ie, tramazoline hydrocholoride, also called KB 227 and Rhinaspray.

2-Naphthaleneamine.   2-Naphthylamine or β-naphthylamine [91-59-8], has been recognized as a human carcinogen, producing bladder cancer on prolonged exposure. Thus 2-naphthaleneamine as such is no longer commercially produced or used in the United States. Before its commercial demise, 2-naphthaleneamine was used in the production of dyes. An important derivative is the rubber antioxidant, N-phenyl-β-naphthaleneamine [135-88-6] (PBNA) (13), which is made by the condensation of 2-naphthol and with aniline in the presence of an acid catalyst. N-Phenyl-2-naphthaleneamine is metabolized in the human body to 2-naphthylamine [91-59-8]; the National Institute of Occupational Safety and Health (NIOSH) has published recommendations for working with this product (36).

1,5-Naphthalenediamine. 1,5-Naphthylenediamine (**14**) is manufactured by metal-acid, eg, iron–acetic acid, reduction, or by catalytic hydrogenation of 1,5-dinitronaphthalene (50). Aside from its possible use as an intermediate for azo dyes, 1,5-naphthalenediamine is used for the manufacture of 1,5-naphthalene diisocyanate [3173-72-6] (**15**) by the phosgenation route. This diisocyanate is used for making high grade urethane cast elastomers.

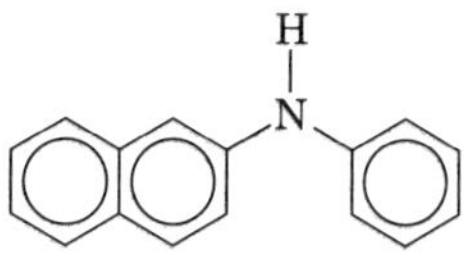

(13) 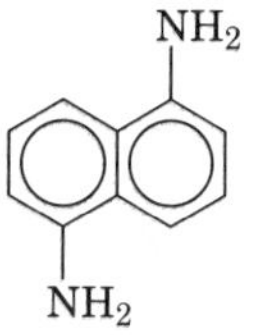   (14) 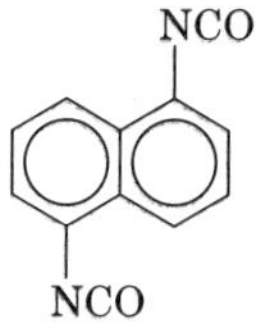   (15) 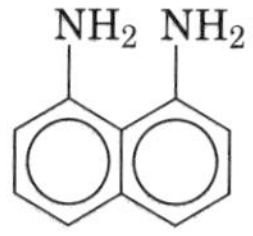   (16)

1,8-Naphthylenediamine. 1,8-Naphthylenediamine (**16**) is produced by metal-acid reduction or by catalytic hydrogenation of 1,8-dinitronaphthalene (50). The most important use of 1,8-naphthylenediamine is for the manufacture, by condensation with phthalic anhydride, of phthaloperinone (**7**), an orange colorant for plastics used in automobile turn-signal and warning-light lenses.

Aminonaphthalenesulfonic Acids

Many aminonaphthalenesulfonic acids are important in the manufacture of azo dyes (qv) or are used to make intermediates for azo acid dyes, direct, and fiber-reactive dyes (see DYES, REACTIVE). Usually, the aminonaphthalenesulfonic acids are made by either the sulfonation of naphthalenamines, the nitration–reduction of naphthalenesulfonic acids, the Bucherer-type amination of naphtholsulfonic acids, or the desulfonation of an aminonaphthalenedi- or trisulfonic acid. Most of these processes produce by-products or mixtures which often are separated in subsequent purification steps. A summary of commercially important aminonaphthalenesulfonic acids is given in Table 4.

H-acid, 1-hydroxy-3,6,8-trisulfonic acid, which is one of the most important letter acids, is prepared as naphthalene is sulfonated with sulfuric acid to trisulfonic acid. The product is then nitrated and neutralized with lime to produce the calcium salt of 1-nitronaphthalene-3,6,8-trisulfonic acid, which is then reduced to T-acid (Koch acid) with Fe and HCl; modern processes use continuous catalytical hydrogenation with Ni catalyst. Hydrogenation has been performed in aqueous medium in the presence of Raney nickel or Raney Ni–Fe catalyst with a low catalyst consumption and better yield (51). Fusion of the T-acid with sodium hydroxide and neutralization with sulfuric acid yields H-acid. Azo dyes such as Direct Blue 15 [2429-74-5] (**17**) and Acid Black [1064-48-8] (**18**) are the coupling products of H-acid (52).

Table 4. Manufacture, Production, and Application Data for Selected Aminonaphthalenesulfonic Acids

Acid	Trivial name	CAS Registry Number	Manufacturing method[a-e]	Intermediate for
1-amino-2-naphthalenesulfonic		[81-06-1]	a or e using naphthionic acid	azo dyes, eg, CI Direct Violet 11
4-amino-2-naphthalenesulfonic		[134-54-3]	d	4-hydroxy-2-naphthalenesulfonic acid
4-amino-1-naphthalenesulfonic	Piria's acid; naphthionic acid	[84-86-6]	a	azo dyes, eg, CI Acid Red 88 and Acid Brown 14; 4-hydroxy-1-naphthalenesulfonic acid
5-amino-1-naphthalenesulfonic	Laurent's acid	[84-89-9]	a or b with Peri acid as the co-product	azo dyes, eg, CI Acid Black 24 and Mordant Brown 1; 5-hydroxy-1-naphthalenesulfonic acid; 5-amino-1,3-naphthalenedisulfonic acid; M-acid; 5-amino-1-naphthalenol
5-amino-2-naphthalenesulfonic	1,6-Cleve's acid	[119-79-9]	b with 1,7-Cleve's acid as the co-product	azo dyes, eg, CI Direct Blue 120; 5-amino-8-acetamino-2-naphthalenesulfonic acid
8-amino-2-naphthalenesulfonic	1,7-Cleve's acid	[119-28-8]	b with 1,6-Cleve's acid	azo dyes, eg, CI Direct Green 51; 8-amino-2-naphthalenol; 4-amino-1,6-naphthalenedisulfonic acid
5- and 8-amino-2-naphthalenesulfonic	Cleve's acid (mixed)		b	azo dyes, eg, CI Direct Brown 62
8-amino-1-naphthalenesulfonic	Peri acid	[82-75-7]	b with Laurent's acid as the primary by-product	azo dyes, eg, CI Acid Black 35; 8-hydroxy-1-naphthalenesulfonic acid; 4-amino-1,5-naphthalenedisulfonic acid; 1,8-naphthosultam; 4-amino-1,3,5-naphthalenetrisulfonic acid
8-phenylamino-1-naphthalenesulfonic	Phenyl Peri acid	[82-76-8]	by condensation of aniline with Peri acid	azo dyes, eg, CI Acid Blue 113
2-amino-1-naphthalenesulfonic	Tobias acid	[81-16-3]	c	pigments, eg, CI Pigment Red 49; 6-amino-1-naphthalenesulfonic acid; 6-amino-1,3-naphthalenedisulfonic acid

Compound	Common name	CAS number	Note	Uses
6-amino-1-naphthalenesulfonic	Dahl's acid	[81-05-0]	d using Tobias acid	azo dyes, eg, CI Acid Green 12, one of six listed; 6-hydroxy-1-naphthalenesulfonic acid
6-amino-2-naphthalenesulfonic	Broenner's acid	[93-00-5]	c	azo dyes, eg, CI Direct Red 4
7-amino-2-naphthalenesulfonic	F-acid	[92-40-0]	c	azo dyes, eg, CI Direct Red 22
7-amino-1-naphthalenesulfonic	Badische acid	[86-60-2]	c	azo dyes, eg, CI Direct Green 33
1-amino-2,7-naphthalenedisulfonic	Kalle's acid	[486-54-4]	d	triphenylmethane dye
4-amino-2,7-naphthalenesulfonic	1,3,6-Freund's acid	[6521-07-6]	b with 1,3,7-Freund's acid	azo dyes, eg, CI Acid Black 7; 5-hydroxy-2,7-naphthalenedisulfonic acid
4-amino-2,6-naphthalenedisulfonic	1,3,7-Freund's acid	[6362-05-6]	b 1,3,6-Freund's acid is co-product	azo dyes, eg, CI Direct Orange 49
8-amino-1,6-naphthalenedisulfonic	amino-ε-acid	[129-91-9]	b	4-amino-2-naphthalenesulfonic acid; 8-hydroxy-1,6-naphthalenedisulfonic acid; 4-hydroxy-2-naphthalenesulfonic acid
4-amino-1,7-naphthalenedisulfonic	Dahl's acid II	[85-74-5]	a using naphthionic acid or 1,6-Cleve's acid	azo dyes, eg, CI Direct Orange 69
4-amino-1,6-naphthalenedisulfonic	Dahl's acid III	[85-75-6]	a as by-product of Dahl's acid II	azo dyes, eg, CI Direct Orange 49; 4-hydroxy-1,6-naphthalenedisulfonic acid
8-amino-1,5-naphthalenedisulfonic		[117-55-5]	a or b	8-hydroxy-1,5-naphthalenedisulfonic acid; 4-amino-5-hydroxy-1-naphthalenesulfonic acid
5-amino-1,3-naphthalenedisulfonic		[13306-42-8]	d	4-hydroxy-8-amino-2-naphthalenesulfonic acid
3-amino-2,7-naphthalenedisulfonic	amino-R-acid	[92-28-4]	c	azo dyes, eg, CI Direct Orange 13; 6-hydroxy-7-amino-2-naphthalenesulfonic acid
3-amino-1,5-naphthalenedisulfonic	Cassella acid	[131-27-1]	b	azo dyes, eg, CI Direct Red 15; 7-hydroxy-1,5-naphthalenedisulfonic acid
6-amino-1,3-naphthalenedisulfonic	amino J-acid	[118-33-2]	d	J-acid (3-hydroxy-6-amino-2-naphthalenesulfonic acid)

Table 4. (*Continued*)

Acid	Trivial name	CAS Registry Number	Manufacturing method[a–e]	Intermediate for
7-amino-1,3-naphthalenedisulfonic	amino G-acid	[86-65-7]	c	azo dyes, eg, CI Direct Orange 74; γ-acid (4-hydroxy-6-amino-2-naphthalenesulfonic acid)
4-amino-1,3,5-naphthalenetrisulfonic (as the sultam)		[76530-15-9]	a	Chicago acid (4-amino-5-hydroxy-1,3-naphthalene-disulfonic acid)
8-amino-1,3,6-naphthalenetrisulfonic	Koch's acid	[117-42-0]	b	H-acid; chromotropic acid; 8-hydroxy-1,3,6-naphthalenetrisulfonic acid
8-amino-1,3,5-naphthalenetrisulfonic	B-acid	[17894-99-4]	b	K-acid (4-amino-5-hydroxy-1,7-naphthalenedi-sulfonic acid)
6-amino-1,3,5-naphthalenetrisulfonic		[55524-84-0]	a	6-amino-1,3-naphthalenedisulfonic acid; -amino-5-2hydroxy-1,7-naphthalenedisulfonic acid
7-amino-1,3,6-naphthalenetrisulfonic	2R amino acid	[118-03-6]	a	3-amino-5-hydroxy-2,7-naphthalenedisulfonic acid
6,8-di(phenylamino)-1-naphthalene-sulfonic	diphenyl-ε-acid	[129-93-1]	from aniline and 8-amino-1,6-naphthalenedi-sulfonic acid	safranine dyes, eg, CI Acid Blue 61

[a]By sulfonation of the appropriate naphthaleneamine or aminonaphthalenesulfonic acid.
[b]By nitration/reduction of the appropriate naphthalene(poly)sulfonic acid.
[c]By amination of the appropriate hydroxynaphthalenesulfonic acid.
[d]By the desulfonation of an aminonaphthalenedi- or trisulfonic acid.
[e]By rearrangement of another aminonaphthalenesulfonic acid.

992

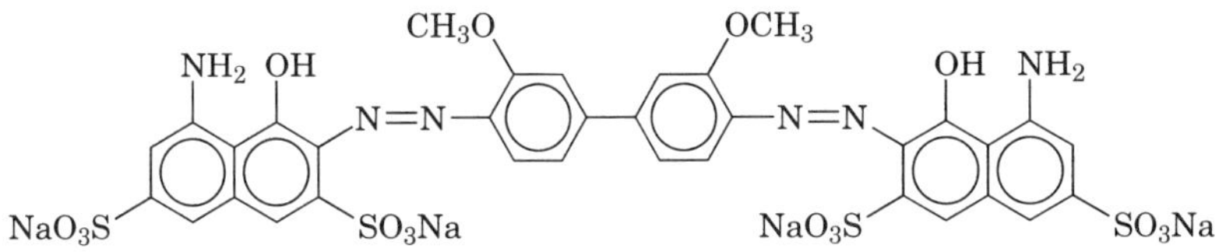

(**17**)

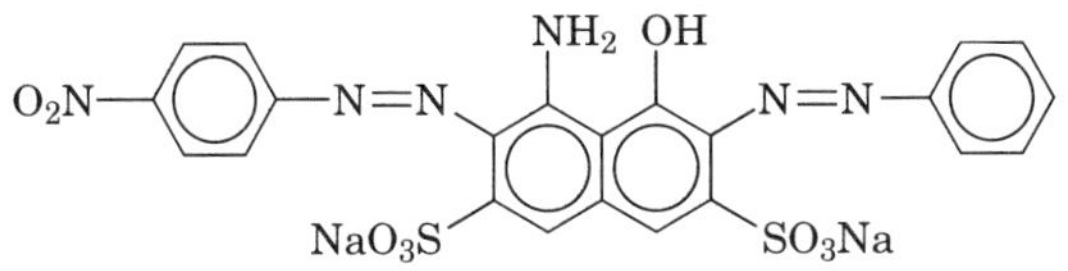

(**18**) 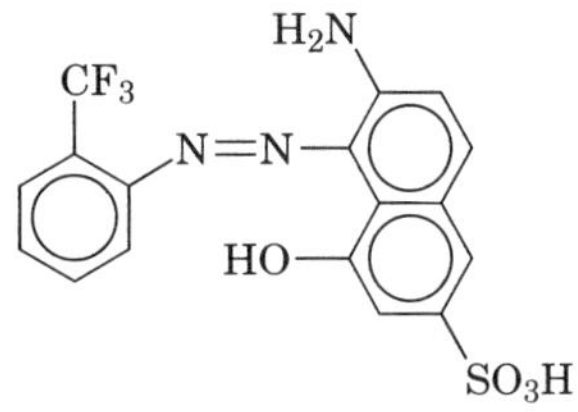   (**19**)

An example of an azo dyestuff in which γ-acid is the coupling component is Acid Red 337 (**19**), which is obtained through the reaction of diazotized o-trifluoromethylaniline and γ-acid (52).

7-Amino-1,3-naphthalenedisulfonic acid (**20**) is made by the Bucherer amination route (53). A mixture of dipotassium 7-hydroxy-1,3-naphthalene-disulfonate [842-18-2], excess aqueous ammonia, and ammonium sulfite is heated slowly in an autoclave to 185°C and is maintained at this temperature for ca 16 h. Aqueous (50 wt %) sodium hydroxide is added to the charge to liberate ammonia, which is recovered for recycling. The charge is neutralized with a mineral acid. The yield of amino G-salt is 97% of theoretical yield. Ammoniation of the di-NH$_4$ salt of 7-hydroxy-1,3-naphthalenedisulfonic acid in the presence of sodium sulfite gives the 7-amino-1,3-naphthalenedisulfonic acid in increased purity and requires only one-fourth of the reaction time compared with the process starting from the di-K salt of 7-hydroxy-1,3-naphthalenedisulfonic acid (54).

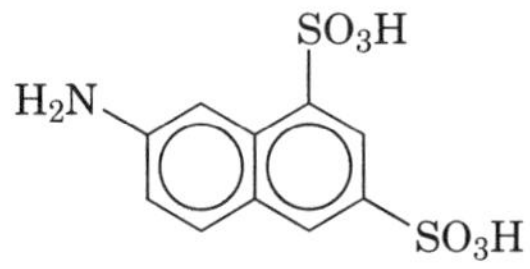

(**20**) 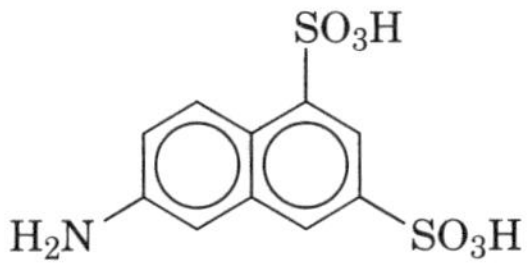   (**21**)

Another example of manufacture in this series is the sulfonation of an aminonaphthalenesulfonic acid, followed by selected desulfonation, to make 6-amino-1,3-naphthalenedisulfonic acid (**21**). Thus, 2-amino-1-naphthalenesulfonic acid made by amination of 2-hydroxy-1-naphthalenesulfonic acid is added to 20 wt % oleum at ca 35°C. At this temperature, 65 wt % oleum is added and the charge is stirred for 2 h, is then slowly heated to 100°C and is maintained for 12 h to produce 6-amino-1,3,5-naphthalenetrisulfonic acid. The mass is diluted with water and maintained for 3 h at 105°C to remove the sulfo group adjacent to the amino group. After cooling to ca 20°C and filtration, 6-amino-1,3-naphthalenedisulfonic acid is obtained in 80% yield (55).

Naphthalenols and Naphthalenediols

Naphthalenols, naphthalenediols, and their sulfonated and amino derivatives are important intermediates for dyes, agricultural chemicals, drugs, perfumes, and surfactants. The methods of manufacture include caustic fusion of naphthalene-1-sulfonic acid, hydrolysis of 1-chloro- or bromonaphthalene, pressure hydrolysis of 1-naphthaleneamine, oxidation–aromatization of tetralin, and hydroperoxidation of 2-isopropylnaphthalene [2027-17-0]. As the toxic hazard of the 1-naphthaleneamine was recognized, its commercial use was minimized. The sulfonation–caustic fusion process is more difficult to operate than in the past because of increasing difficulties posed by product purity requirements, high investment and replacement cost, and by-product effluent handling problems. In the United States, the naphthalenols are made by hydrocarbon oxidation routes.

The chemical properties of the naphthaleneols are similar to those of phenol and resorcinol, with added reactivity and complexity of substitution because of the condensed ring system. Some of the naphthols and naphthalenediols are listed with some of their physical properties in Table 5.

1-Naphthalenol. 1-Naphthol, α-naphthol, or 1-hydroxynaphthalene [90-15-3] forms colorless needles, mp 96°C, bp 288°C, which tend to become colored on exposure to air or light. It is almost insoluble in water, but readily soluble in alcohol, ether, and benzene. 1-Naphthol and 2-naphthol are found in coal tar (56).

Acid-catalyzed hydroxylation of naphthalene with 90% hydrogen peroxide gives either 1-naphthol or 2-naphthiol at a 98% yield, depending on the acidity of the system and the solvent used. In anhydrous hydrogen fluoride or 70% HF–30% pyridine solution at −10 to +20°C, 1-naphthol is the product formed in >98% selectivity. In contrast, 2-naphthol is obtained in hydroxylation in super acid (HF–BF$_3$, HF–SbF$_5$, HF–TaF$_5$, FSO$_3$H–SbF$_5$) solution at −60 to −78°C in >98% selectivity (57). Of the three commercial methods of manufacture, the pressure hydrolysis of 1-naphthaleneamine with aqueous sulfuric acid at 180°C has been abandoned, at least in the United States. The caustic fusion of sodium 1-naphthalenesulfonate with 50 wt % aqueous sodium hydroxide at ca 290°C followed by the neutralization gives 1-naphthalenol in a ca 90% yield.

The most important process to produce 1-naphthalenol was developed by Union Carbide and subsequently sold to Rhône-Poulenc. It is the oxidation of tetralin, 1,2,3,4-tetrahydronaphthalene [119-64-2], in the presence of a transition-metal catalyst, presumably to 1-tetralol–1-tetralone by way of the 1-hydroperoxide, and dehydrogenation of the intermediate ie, 1-tetralol to 1-tetralone and aromatization of 1-tetralone to 1-naphthalenol, using a noble-metal catalyst (58). 1-Naphthol production in the Western world is around 15×10^3 t/yr, with the United States as the largest producer (52).

1-Naphthol is mainly used in the manufacture of the insecticide carbaryl (59), 1-naphthyl N-methylcarbamate [63-25-2] (Sevin) (**22**), which is produced by the reaction of 1-naphthol with methyl isocyanate. Methyl isocyanate is usually

Table 5. Properties of Naphthalenols and Naphthalenediols

Compound	CAS Registry Number	Mp, °C	Density	Other
1-naphthalenol	[90-15-3]	95.8–96.0	1.224_4^4 1.099_4^{99}	sublimes; sol 0.03 g/100 mL H_2O at 25°C; readily sol alc, ether, benzene; bp 280°C (158°C at 2.6 kPa^a)
2-naphthalenol	[135-19-3]	122	1.078_4^{130} 1.22_4^{25}	sublimes; sol 0.075 g/100 mL H_2O at 25°C; readily sol alc, ether, benzene; flash pt 161°C; bp 295°C (161.8°C at 2.6 kPa^a)
1,2-naphthalenediol	[574-00-5]	103–104		
1,3-naphthalenediol	⌐132-86-5]	124		
1,4-naphthalenediol	[571-60-8]	195		heat of combusion 4.77 MJ^b
1,5-naphthalenediol	[83-56-7]	258		sublimes; sparingly sol water; readily sol ether, acetone
1,6-naphthalenediol	[575-44-0]	137–138		
1,7-naphthalenediol	[575-38-2]	181		
1,8-naphthalenediol	[569-42-6]	144		
2,3-naphthalenediol	[92-44-4]	159		
2,6-naphthalenediol	[581-43-1]	222		
2,7-naphthalenediol	[582-17-2]	194		sol boiling water

^{a}To convert kPa to mm Hg, multiply by 7.5.
^{b}To convert MJ to kcal, divide by 4.184×10^{-3}.

prepared by treating methylamine with phosgene. Methyl isocyanate is a very toxic liquid, boiling at 38°C, and should not be stored for long periods of time (Bhopal accident, India). India has developed a process for the preparation of aryl esters of *N*-alkyl carbamic acids. Thus 1-naphthyl methylcarbamate is prepared by refluxing 1-naphthol with ethyl methylcarbamate and $POCl_3$ in toluene (60). In 1992, carbaryl production totaled $>11.4 \times 10^3$ t (35). Rhône-Poulenc, at its Institute, W. Va., facility is the only carbaryl producer in United States.

Devrinol, 2-(1-naphthoxy)-*N,N*-diethylpropionamide [16299-99-7] (naprop-amide) (**23**), which is prepared from 1-naphthol, is used as a herbicide (61). Another agricultural chemical, 1-naphthoxyacetic acid [2976-75-2] (**24**), is pre-pared by stirring 1-naphthol with monochloroacetic acid and sodium hydroxide in water at 100–110°C for several minutes. After treatment with concentrated HCl about 94% of the product is obtained (62).

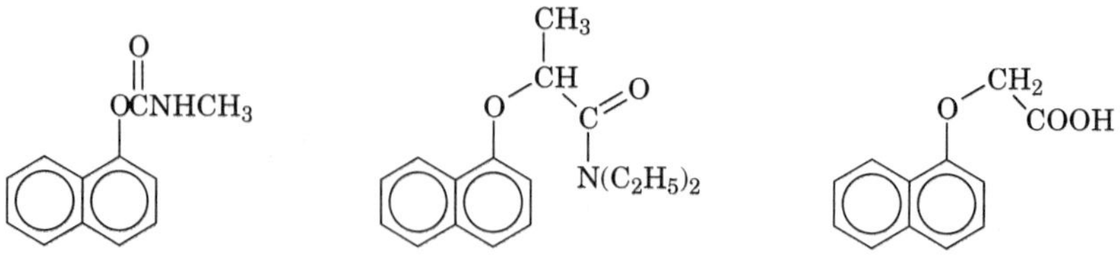

(22)   (23)   (24)

Several biologically and pharmacologically active compounds have been prepared from the condensation of the acid chloride of 1-naphthoxyacetic acid with carbazole, indole, or pyrrole in 2N NaOH solution in ethanol (63). Also, naphthyloxy derivatives of imidazole, benzimidazole, and benzotriazoles have been synthesized and screened for their antimicrobial, analgesic, and antiinflammatory activities. 2-Naphthyloxy derivatives are comparatively more active than 1-naphthyloxy derivatives (64).

Several drugs are derived from 1-naphthalenol: the magnesium salt of 3-(4-methoxy-1-naphthoyl)propionic acid [6643-66-6] (Hepalande) (25) is used as a choleretic (65). Propranolol (Inderal), 1-isopropylamino-3-(1-naphthoxy)-2-propanol [3506-09-0] (26), is an important adrenergic blocking agent used in the treatment of angina and cardiac arrhythmias, with a worldwide production of approximately 500 t/yr (52). It is prepared by the reaction of 1-naphthol with epichlorohydrin, followed by substitution of the chlorine in 1-chloro-3-(1-naphthoxy)-2-propanol with isopropylamine (66). 1-Naphthyl salicylate [550-97-0] (Alphol) (27) has been used as an antiseptic and antirheumatic and is prepared by the acylation of phenols with salicylic acid using polyphosphoric acid (67).

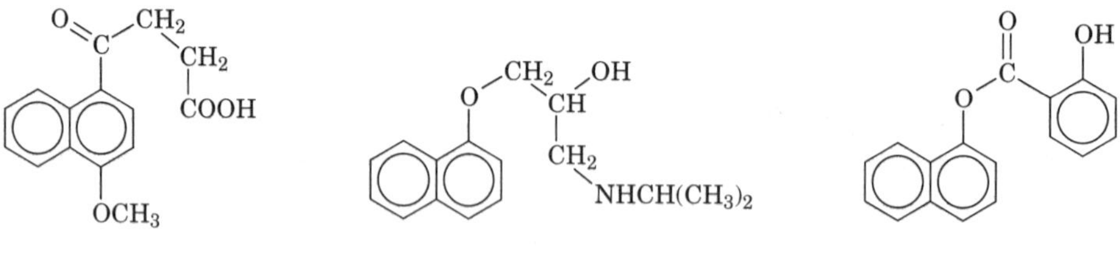

(25)   (26)   (27)

1-Naphthalenol also is used in the preparation of azo, indigoid, and nitro, eg, 2,4-dinitro-1-naphthol, dyes, and in making dye intermediates, eg, naphtholsulfonic acids, 4-chloro-1-naphthalenol, and 1-hydroxy-2-naphthoic acid. 1-Naphthalenol is an antioxidant for gasoline, and some of its alkylated derivatives are stabilizers for plastics and rubber (68).

2-Naphthalenol. 2-Naphthol or β-naphthol or 2-hydroxynaphthalene [135-19-3] melts at 122°C and boils at 295°C, and forms colorless crystals of characteristic, phenolic odor which darken on exposure to air or light. 2-Naphthol [135-19-3] is manufactured by fusion of sodium 2-naphthalenesulfonate with sodium hydroxide at ca 325°C, acidification of the drowned fusion mass which is quenched in water, isolation and water-washing of the 2-naphthalenol, and vacuum distillation and flaking of the product. A continuous process of this type has been patented (69). The high sulfate content in the primary effluent from 2-naphthol production is greatly reduced in modern production plants by the recovery of sodium sulfate.

Another method of manufacture involves the oxidation of 2-isopropylnaphthalene in the presence of a few percent of 2-isopropylnaphthalene hydroperoxide [6682-22-0] as the initiator, some alkali, and perhaps a transition-metal catalyst, with oxygen or air at ca 90–100°C, to ca 20–40% conversion to the hydroperoxide; the oxidation product is cleaved, using a small amount of ca 50 wt % sulfuric acid as the catalyst at ca 60°C to give 2-naphthalenol and acetone in high yield (70). The yields of both 2-naphthalenol and acetone from the hydroperoxide are 90% or better.

A process variation of the extraction of 2-isopropylnaphthalene hydroperoxide from the crude oxidation product with an alkylene glycol has been patented (71). The 2-naphthalenol plant of American Cyanamid, which was using the hydroperoxidation process and had a 14×10^3 t/yr capacity (72), ceased production in 1982, leaving the United States without a domestic producer of 2-naphthol. The 2-naphthol capacity in the Western world is approximately 50×10^3 t/yr, with ACNA, Italy and Hoechst AG, Germany operating the largest plants. China produces about 7×10^3 t/yr. Other important producing countries are Poland, Romania, the former Czechoslovakia, and India (35,52).

The principal uses for 2-naphthalenol are in the dyes and pigments industries, eg, as a coupling component for azo dyes, and to make important intermediates, such as 3-hydroxy-2-naphthalenecarboxylic acid (BON) (**28**) and its anilide (naphthol AS), 2-naphtholsulfonic acids, aminonaphtholsulfonic acids, and 1-nitroso-2-naphthol [131-91-9] (**29**).

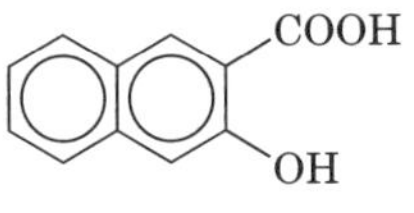

(**28**) 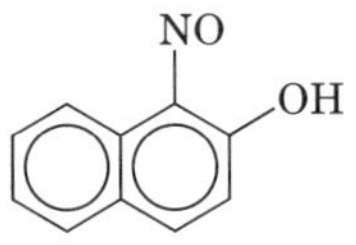   (**29**) 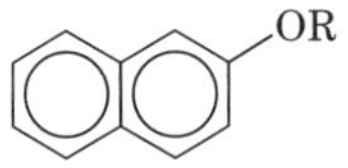   (**30**)

Naphthalenolsulfonic acid formaldehyde condensates are used in tanning agents. Other uses of 2-naphthalenol are in the manufacture of perfuming agents, eg, 2-naphthyl methyl ether [93-04-9] (Yara Yara), R = CH_3 (**30**), and 2-naphthyl ethyl ether [93-18-5] (nerolin, Bromelia) R = C_2H_5 (**30**); an antioxidant for polyolefins, ie, thio-1,1-bis(2-naphthol) [17096-15-0] (**31**); the intestinal antiseptic, 2-naphthyl lactate [93-43-6] (Lactol, Lactonaphthol) (**32**); a gastrointestinal and genitourinary antiinfective, ie, 2-naphthyl salicylate [*613-78-5*] (Betol, Salinaphthol), R = OH (**33**); a semisynthetic penicillin, sodium 6-(2-ethoxy-1-

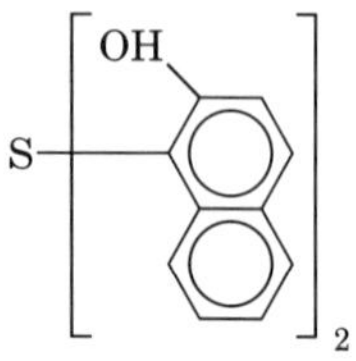

(**31**) 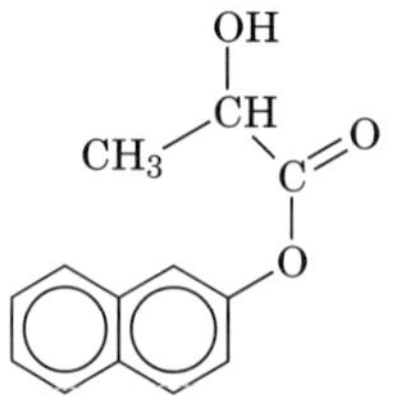   (**32**) 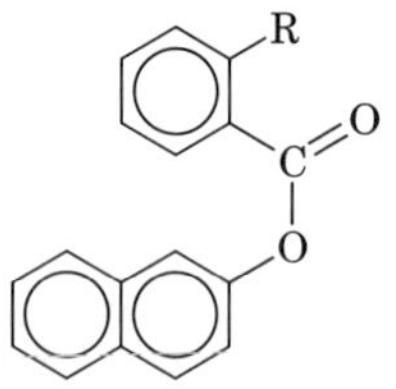   (**33**)

naphthamido) penicillanate [*985-16-0*] (nafcillin sodium, Naptopen, Unipen) (**34**); an intestinal antiseptic, 2-naphthyl benzoate [*93-44-7*] (Lintrin, Haertolan) R = H (**33**); and a topical antifungal agent, tolnaftate [*2398-96-1*] (*m,N*-dimethylthiocarbanilic acid *O*-naphthyl ester) (**35**), which is prepared from 2-naphthol, thiophosgene, and *N*-methyl-*m*-toluidine. Naproxen [*22204-53-1*] (**36**), an antirheumatic, is also prepared from 2-naphthol by the Friedel-Crafts acylation of 2-methoxynaphthalene and subsequent Willgerodt-Kindler reaction. The S-configuration which is obtained from its racemic mixture with the alkaloid cinchonidine is the effective isomer.

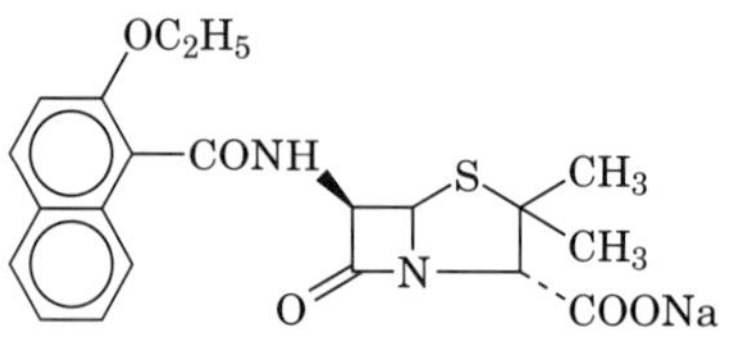

(**34**) 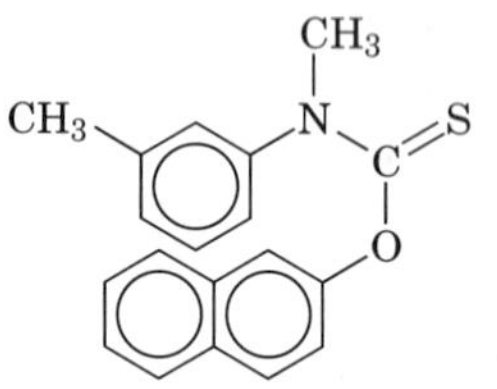   (**35**) 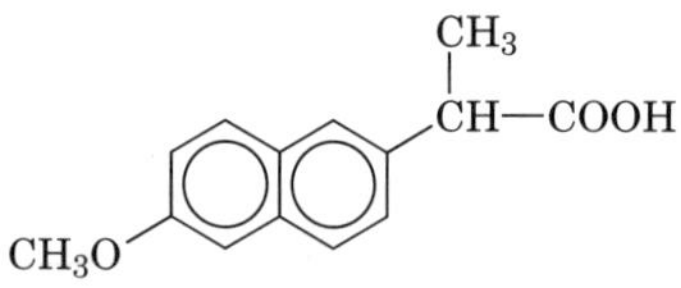   (**36**)

1,4-Naphthalenediol. This diol can be prepared by the chemical or catalytic reduction of 1,4-naphthoquinone. Both the diol and quinone are of interest because of their relation to the vitamin K family. Carboxylation of 1,4-naphthalenediol with $CO_2-K_2CO_3$ followed by neutralization gives 1,4-dihydroxy-2-naphthoic acid (DHNA). DHNA and its aryl esters are useful as intermediates for photochemicals, dyes, and pigments. Phenyl 1,4-dihydroxy-2-naphthoate (PDNA) (**37**) has been prepared by heating DHNA with triphenyl phosphites at 110°C for 10 h (73). The yield is approximately 70%. The principal

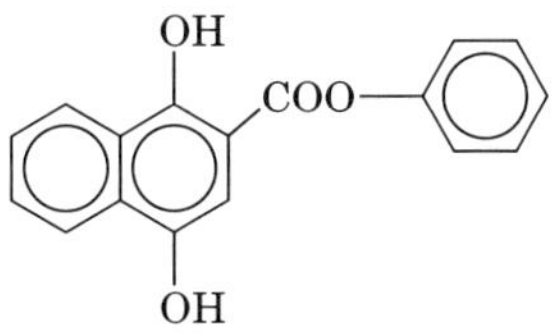

(37)

by-product, which is about 30%, is the ester formed by reaction of PDNA with the starting acid DHNA. An aqueous NaOH solution is added dropwise to an aqueous suspension of this ester at 40–70°C over 1 h and the reaction mixture kept for 2 h to give 86.6% DHNA of 98.7% purity (74), which is then esterified with $(C_6H_5O)_3P$ to obtain PDNA. The esterification process is dramatically improved by adding a small amount of inorganic or organic acid, preferably methanesulfonic acid, benzene sulfonic acid, or naphthalene sulfonic acid; subsequent isolation and crystallization gives a pure product (75).

1,5-Naphthalenediol. 1,5-Dihydroxynaphthalene or Azurol is a colorless material which darkens on exposure to air. It is manufactured by the fusion of disodium 1,5-naphthalenedisulfonate with sodium hydroxide at ca 320°C in high yield. 1,5-Naphthalenediol is an important coupling component, giving ortho-azo dyes which form complexes with chromium. The metallized dyes produce fast black shades on wool. 1,5-Naphthalenediol can be aminated with ammonia under pressure to 1,5-naphthalenediamine.

1,8-Naphthalenediol. This compound darkens rapidly in air. It can be made by fusion of the sultone of 8-hydroxy-1-naphthalenesulfonic acid with 50 wt % sodium hydroxide at 200–230°C, or by the hydrolytic desulfonation of 1,8-dihydroxy-4-naphthalenesulfonic acid. The diol also reacts with ammonia to give 1,8-naphthalenediamine.

2,3-Naphthalenediol. This diol is made by the hydrolytic desulfonation of 2,3-naphthalenediol-6-sulfonic acid at ca 180°C. It is used as a coupler forming azo dyes which are applied in reprographic processes.

2,6-Naphthalenediol. This diol is prepared by the alkali fusion of 2-hydroxynaphthalene-6-sulfonic acid (Schaffer acid) at 290–295°C. Schaffer acid is usually produced by sulfonation of 2-naphthol with the addition of sodium sulfate at 85–105°C. This acid is also used as a coupling component in the production of azo dyes such as Acid Black 26. 2,6-Naphthalenediol is used as a component in the manufacture of aromatic polyesters which, as is also true of the corresponding amides, display liquid crystal characteristics (52).

2,7-Naphthalenediol. This diol is made by the fusion of sodium 2,7-naphthalenedisulfonate with molten sodium hydroxide at 280–300°C in ca 80% yield. A formaldehyde resin prepared from this diol has excellent erosion resistance, strength, and chemical inertness; it is used as an ablative material in rocket-exhaust environments (76).

Hydroxynaphthalenesulfonic Acids

Hydroxynaphthalenesulfonic acids are important as intermediates either for coupling components for azo dyes or azo components, as well as for synthetic tanning agents. Hydroxynaphthalenesulfonic acids can be manufactured either by sulfonation of naphthols or hydroxynaphthalenesulfonic acids, by acid hydrolysis of aminonaphthalenesulfonic acids, by fusion of sodium naphthalenepolysulfonates with sodium hydroxide, or by desulfonation or rearrangement of hydroxynaphthalenesulfonic acids (Table 6).

In the production of sodium 3-hydroxy-2,7-naphthalenedisulfonate [*135-59-3*] (R-salt) (**38**), 2-naphthol is stirred with excess 98 wt % sulfuric acid at 60°C, sodium sulfate is added, and the mixture is stirred and heated for 36 h at 105–122°C (77). The charge is diluted with water and salted out with ca 15 wt % sodium chloride at 60°C to give R-salt in 68% yield.

Another example of manufacture of a hydroxynaphthalenesulfonic acid is a caustic fusion process to make 7-hydroxy-2-naphthalenesulfonic acid (**39**) (78). Sodium 2,7-naphthalenedisulfonate, which is made by mixing the 40 wt % disulfonic acid paste with ca 70 wt % caustic, is fused with excess sodium hydroxide in an agitated autoclave at 230–265°C for 10 h. The charge is drowned in water, brought to pH 8 with hydrochloric acid, diluted with water, boiled and treated with carbon, and filtered hot. The product is isolated by filtration after cooling at 30°C. Additional product can be obtained by adding sodium chloride to the filtrate to give a combined yield of 90% of sodium 7-hydroxy-2-naphthalenesulfonate [*135-55-7*].

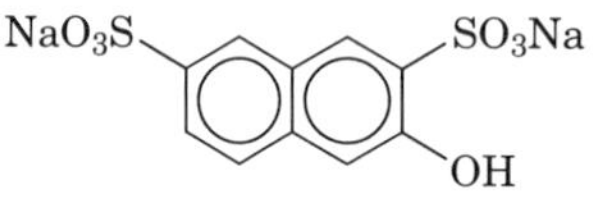

(**38**) 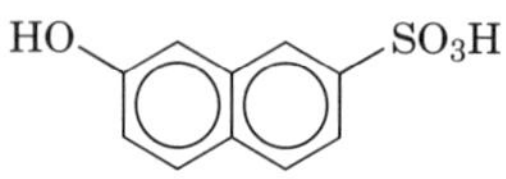   (**39**)

Aminonaphthols and Aminonaphtholsulfonic Acids

The aminonaphthols are of minor use but the aminohydroxynaphthalenesulfonic acids are intermediates for dyes, eg, fiber-reactive azo dyes and plain and metallized azo dyes (Table 7).

The manufacture of 4-hydroxy-6-amino-2-naphthalenesulfonic acid in ca 85% of the theoretical yield has been described (76).

Manufacture of 3-hydroxy-4-amino-1-naphthalenesulfonic acid involves the nitrosation of 2-naphthalenol, bisulfite addition, and reduction of the nitroso to the amino group by sulfur dioxide generated *in situ* (47). 3-Hydroxy-4-amino-1-naphthalenesulfonic acid is obtained in 80% yield.

A number of *N*-acyl-, *N*-alkyl-, and *N*-arylaminonaphthalenolsulfonic acids are used as couplers for azo dyes. Examples of such intermediates are shown in Table 8.

Naphthalenecarboxylic Acids

Physical properties for naphthalene mono-, di-, tri-, and tetracarboxylic acids are summarized in Table 9. Most of the naphthalene di- or polycarboxylic acids have been made by simple routes such as the oxidation of the appropriate di-

Table 6. Manufacture, Production, and Application Data for Selected Hydroxynaphthalenesulfonic Acids

Compound	Trivial name	CAS Registry Number	Manufacturing method[a–f]	Intermediate for
4-hydroxy-2-naphthalene-sulfonic acid	Armstrong & Wynne's acid; 1,3-oxy-acid	[3771-14-0]	a,b	azo dyes, eg, CI Direct Blue 127
4-hydroxy-1-naphthalene-sulfonic acid	Nevile-Winther acid; 1,4-oxy-acid	[84-87-7]	c,d	azo dyes, eg, CI Acid Red 14; tanning agents
5-hydroxy-1-naphthalene-sulfonic acid	L-acid	[117-59-9]	d,e	azo dyes and pigments, eg, CI Pigment Red 54, toner; 1,5-naph-thal enediol
8-hydroxy-1-naphthalene-sulfonic acid		[117-22-6]	f	metallized o,o'-dihydroxyazo dyes, eg, CI Acid Blue 58
2-hydroxy-1-naphthalene-sulfonic acid	oxy-Tobias acid	[567-47-5]	c	Tobias acid; J-acid
6-hydroxy-2-naphthalene-sulfonic acid	Schaeffer's acid	[93-01-6]	c	azo dyes, eg, CI Acid Orange 12; synthetic tanning agents
7-hydroxy-2-naphthalene-sulfonic acid	F-acid	[92-40-0]	e	azo dyes, eg, CI Direct Blue 128
7-hydroxy-1-naphthalene-sulfonic acid	Crocein acid; Baeyer's acid	[132-57-0]	c	azo dyes, eg, CI Acid Red 70
4,5-dihydroxy-1-naphthalene-sulfonic acid	dioxy S-acid	[83-65-8]	e	azo dyes, eg, CI Direct Blue 26
6,7-dihydroxy-2-naphthalene-sulfonic acid	dioxy R-acid	[92-27-3]	e	2,3-dihydroxy-naphthalene
5-hydroxy-2,7-naphthalene-disulfonic acid	RG-acid; violet acid	[578-85-8]	e	azo dyes, eg, CI Acid Red 99
8-hydroxy-1,6-naphthalene-disulfonic acid	ε-acid; Andresen's acid	[117-43-1]	f	azo dyes, eg, CI Direct Blue 98
4-hydroxy-1,6-naphthalene-disulfonic acid	Dahl's acid; D-acid	[6361-37-1]	a,d	nitro coloring matter, eg, CI Acid Yellow 1

Table 6. (Continued)

Compound	Trivial name	CAS Registry Number	Manu-facturing method[a–f]	Intermediate for
4-hydroxy-1,5-naphthalene-disulfonic acid	Schoellkopf's acid; CS-acid; δ-acid	[82-75-7]	f	azo dyes, eg, CI Acid Blue 169
3-hydroxy-2,7-naphthalene-disulfonic acid	R-acid	[148-75-4]	c	azo dyes, eg, CI Acid Red 115, Acid Red 26
7-hydroxy-1,3-naphthalene-disulfonic acid	G-acid	[118-32-1]	c	azo dyes, eg, CI Acid Red 73; triphenyl-methane dyes
4,5-dihydroxy-2,7-naphthal-enedisulfonic acid	chromotropic acid	[148-25-4]	a,e	azo dyes, eg, CI Acid Violet 3
8-hydroxy-1,3,6-naphthalene-trisulfonic acid	oxy-Koch's acid	[3316-02-7]	a	azo dyes, eg, CI Direct Blue 27; chromotropic acid
7-hydroxy-1,3,6-naphthalene-trisulfonic acid		[6259-66-1]	c	azo dyes, eg, CI Acid Red 41

[a]By the hydrolysis of the corresponding aminonaphthalenesulfonic acid.
[b]By the desulfonation of 8-hydroxy-1,6-naphthalenedisulfonic acid.
[c]By the sulfonation of the appropriate (1- or 2-) naphthalenol.
[d]By the Bucherer reaction (with sulfite) of the appropriate aminonaphthalenesulfonic acid.
[e]By the alkali fusion or alkaline hydrolysis under pressure of the appropriate naphthalenedisulfonic or naphthalenetrisulfonic acid or hydroxynaphthalenedisulfonic acid.
[f]By the alkaline hydrolysis of the sultone formed on boiling an aqueous solution of the diazonium salt of 8-amino-1-naphthalenesulfonic acid or its appropriate derivatives.

or polymethylnaphthalenes, or by complex routes, eg, the Sandmeyer reaction of the selected aminonaphthalenesulfonic acid, to give a cyanonaphthalenesulfonic acid followed by fusion of the latter with an alkali cyanide, with simultaneous or subsequent hydrolysis of the nitrile groups.

1- and 2-Naphthalenecarboxylic Acids. Naphthalenecarboxylic acids are useful intermediates for dyes and photographic materials. These acids are also used in the preparation of antitumor agents (79) and also in the preparation of cholecystokinin-agonist tetrapeptide (80). The acids are prepared readily by the oxidation of 1- or 2-alkylnaphthalenes with dilute nitric acid, chromic acid, or permanganate. The oxygen or air oxidation of alkylnaphthalenes in an alkanoic acid solvent in the presence of a Ce-, Co-, or Mn-containing catalyst and a Br-containing catalyst gives good results (81–83). The direct carboxylation of naphthalene with CO and oxygen in the presence of a Pd–carboxylate catalyst also has been patented (84). The photo carboxylation of naphthalene in the presence of carbon dioxide and an electron donor has been described. About 67% naphthoic acids were obtained by this method, upon visible light irradiation

Table 7. Selected Aminonaphthalenols and Aminohydroxynaphthalenesulfonic Acids

Compound	Trivial name	CAS Registry Number	Manu-facturing method[a–c]	Intermediate for
5-amino-1-naphthalenol	Purpurol	[83-55-6]	a	azo dyes, eg, CI Acid Blue 70; sulfur dyes
7-amino-2-naphthalenol	Cyanol	[118-46-7]	a	azo dyes, eg, CI Mordant Brown 65
3-hydroxy-4-amino-1-naphthalenesulfonic acid	1,2,4-acid; Boeniger acid	[116-63-2]	b	azo dyes, eg, CI Acid Red 186, Mordant Red 7; chrome complex dyes
5-amino-6-hydroxy-2-naphthalenesulfonic acid	Amino-Schaeffer acid	[5639-34-9]	c	photographic developer; rarely used for dyes
4-hydroxy-8-amino-2-naphthalenesulfonic acid	M-acid	[489-78-1]	a	azo dyes, eg, CI Direct Green 42
4-hydroxy-7-amino-2-naphthalenesulfonic acid	J-acid	[87-02-5]	a	azo dyes, eg, CI Direct Blue 71, Direct Red 16; direct dyes using N-phenyl J-acid and J-acid imide
4-hydroxy-6-amino-2-naphthalenesulfonic acid	γ-acid	[90-51-7]	a	azo dyes, eg, CI Direct Black 22
4-amino-5-hydroxy-2,7-naphthalenedisulfonic acid	H-acid	[90-20-0]	a	azo dyes, eg, CI Direct Black 19, Direct Blue 15
4-amino-5-hydroxy-1,3-naphthalenedisulfonic acid	Chicago acid; SS-acid; 2S-acid	[82-47-3]	a	azo dyes, eg, CI Acid Blue 42
4-amino-5-hydroxy-1,7-naphthalenedisulfonic acid	K-acid	[130-23-4]	a	azo dyes, eg, Sulfon Acid Blue G, CI 13400
3-amino-5-hydroxy-2,7-naphthalenedisulfonic acid	RR-acid; 2R-acid	[90-40-4]	a	azo dyes, eg, CI Direct Brown 31

[a]By the alkali fusion or hydrolysis of the appropriate aminonaphthalenesulfonic acid.
[b]By the nitrosation of 2-naphthalenol and the reaction of the nitroso compound with sodium bisulfite.
[c]By nitrosation/reduction of 6-hydroxy-2-naphthalenesulfonic acid.

Table 8. Selected *N*-Substituted Aminohydroxynaphthalenesulfonic Acids

Compound	Structure	Trivial name	CAS Registry Number	Intermediate for
7,7'-ureylene-bis-4-hydroxy-2-naphthalenesulfonic acid	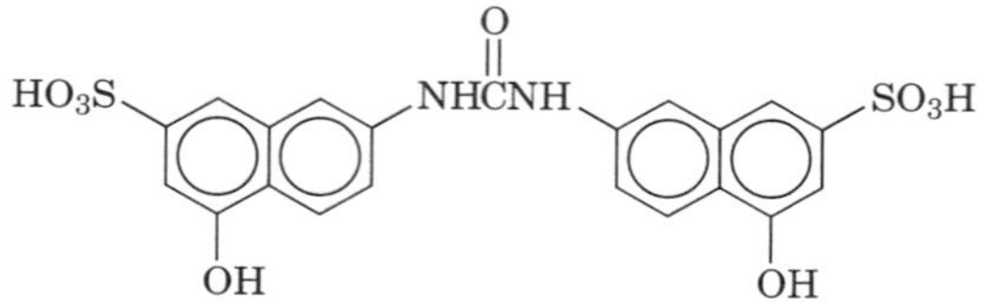 *(structure)*	J-acid urea	[137-47-4]	azo dyes, eg, CI Direct Orange 26
7-benzamido-4-hydroxy-2-naphthalenesulfonic acid	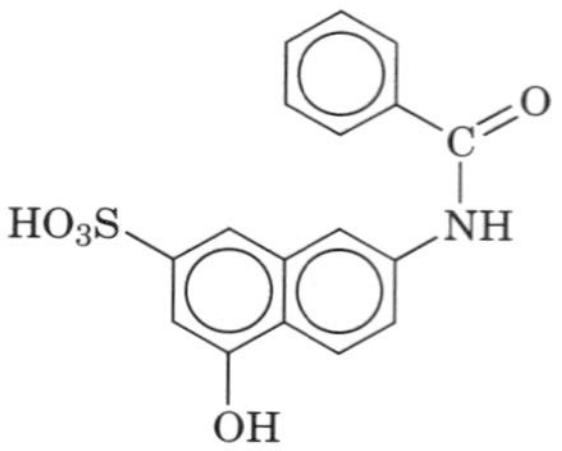 *(structure)*	*N*-benzoyl J-acid	[132-87-6]	azo dyes, eg, CI Direct Red 81
7-phenylamino-4-hydroxy-2-naphthalenesulfonic acid	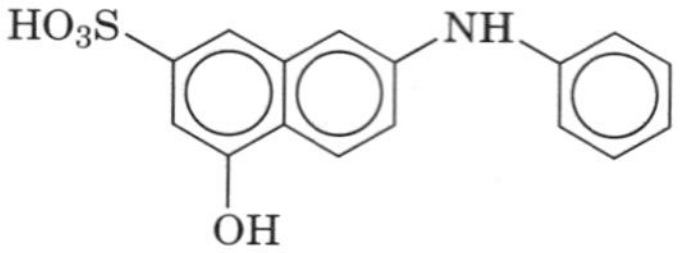 *(structure)*	*N*-phenyl J-acid	[119-40-4]	azo dyes, eg, CI Direct Violet 7
7,7'-imino-bis-4-hydroxy-2-naphthalenesulfonic acid	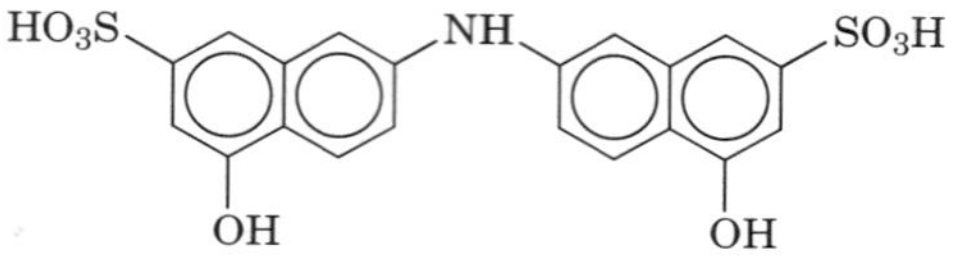 *(structure)*	di-J-acid; J-acid imide	[87-03-6]	azo dyes, eg, CI Direct Red 149

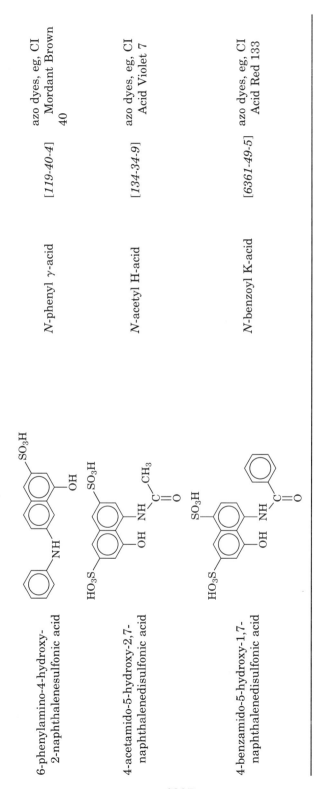

6-phenylamino-4-hydroxy-2-naphthalenesulfonic acid   *N*-phenyl γ-acid   [*119-40-4*]   azo dyes, eg, CI Mordant Brown 40

4-acetamido-5-hydroxy-2,7-naphthalenedisulfonic acid   *N*-acetyl H-acid   [*134-34-9*]   azo dyes, eg, CI Acid Violet 7

4-benzamido-5-hydroxy-1,7-naphthalenedisulfonic acid   *N*-benzoyl K-acid   [*6361-49-5*]   azo dyes, eg, CI Acid Red 133

Table 9. Selected Properties of Naphthalenecarboxylic Acids

Compound	CAS Registry Number	Mp, °C	Other
1-naphthalene-carboxylic acid	[86-55-5]	162	sol ethanol; sparingly sol water; $K_a = 2.04 \times 10^{-4}$ at 25°C; bp at 6.7 kPa[a] 231°C
2-naphthalene-carboxylic acid	[93-09-4]	184–185	sol ethanol, ether, chloroform; $K_a = 6.78 \times 10^{-5}$ at 25°C; bp >300°C
1,2-naphthalenedi-carboxylic acid	[2088-87-1]	175 dec	sol ethanol, ether, acetic acid; mp anhydride 168–169°C
1,3-naphthalenedi-carboxylic acid	[2089-93-2]	267–268	
1,4-naphthalenedi-carboxylic acid	[605-70-9]	309	sol ethanol; insol boiling water
1,5-naphthalenedi-carboxylic acid	[7315-96-0]	315–320 dec	insol common solvents
1,6-naphthalenedi-carboxylic acid	[2089-87-4]	310	sol hot ethanol, acetic acid
1,7-naphthalenedi-carboxylic acid	[2089-91-0]	308	sol common organic solvents
1,8-naphthalenedi-carboxylic acid	[518-05-8]	on heating converts to anhydride (mp 274°C)	sol warm ethanol; bp anhydride at 440 Pa[b] 215°C
2,3-naphthalenedi-carboxylic acid	[2169-87-1]	239–241 dec	sol hot ethanol; mp anhydride 246°C
2,6-naphthalenedi-carboxylic acid	[1141-38-4]	310–313 dec	sol aq alc
2,7-naphthalenedi-carboxylic acid	[2089-89-6]	>300	sol ethanol
1,2,5-naphthalenetri-carboxylic acid	[36439-99-3]	270–272	sol methanol
1,3,8-naphthalenetri-carboxylic acid	[36440-24-1]		mp 1,8-anhydride 289–290°C
1,4,5-naphthalenetri-carboxylic acid	[28445-09-2]	on heating, forms anhydride (mp undefined)	mp 4,5-anhydride 274°C
1,2,4,5-naphthalene-tetracarboxylic acid	[22246-61-3]	263	mp dianhydride 263°C
1,4,5,8-naphthalene-tetracarboxylic acid	[128-97-2]	on heating, forms anhydride	sol acetone; dianhydride sublimes >300°C

[a]To convert kPa to mm Hg, multiply by 7.5.
[b]To convert Pa to mm Hg, divide by 133.3.

with phenazine as a sensitizer. Over 90% of the naphthoic acids was 1-naphthoic acid (85).

4-Alkyl-N,N-dialkyl-1-naphthalenecarboxamides are useful herbicides (86) and the 2,2-dimethylhydrazide of 1-naphthalenecarboxylic acid has been patented as a plant growth regulator (87). 2-Propynyl-2-naphthalenecarboxylate [53548-27-9] and similar esters are insecticides (88). 1-Naphthaleneacetic acid, the plant growth regulator, has been prepared from naphthalene, concentrated HCl, and paraformaldehyde without isolation of intermediate 1-chloromethylnaphthalene or 1-naphthaleneacetonitrile (89).

1-Naphthaleneacetic acid has also been prepared by the carbonyl-insertion reaction of 1-chloromethylnaphthalene catalyzed by carbonyl cobalt cation (90,91). Carboxylation of 1-chloromethylnaphthalene in the presence of the catalyst Pd[P(C$_6$H$_5$)$_3$]$_2$Cl$_2$ under phase-transfer conditions gave 1-naphthaleneacetic acid in 78% yield (92).

Tetrahydrozoline [84-22-0], 2-(1,2,3,4-tetrahydro-1-naphthyl)2-imidazolin, (Tysine, Visine) (**40**), a sympathomimetic and nasal decongestant, is made by the condensation of 1,2,3,4-tetrahydro-1-naphthoic acid or its methyl ester with 1,2-ethylenediamine.

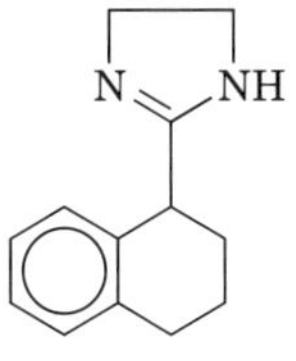

(**40**) 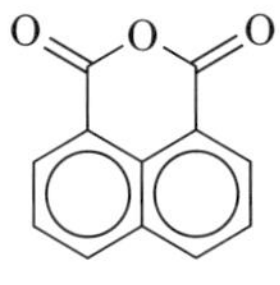 (**41**) 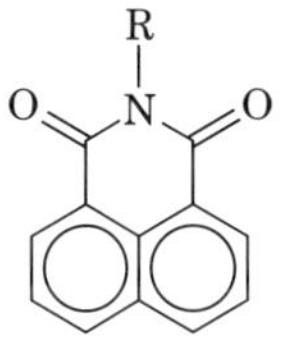 (**42**)

1,8-Naphthalenedicarboxylic Acid. Naphthalic acid readily dehydrates on heating to 1,8-naphthalenedicarboxylic acid anhydride [81-84-5] (naphthalic anhydride) (**41**). The anhydride and its imide naphthalimide [81-83-4] (R = H) (**42**) are intermediates for important dyes, pigments, optical bleaches, and biologically active compounds.

The anhydride can be made by the liquid-phase oxidation of acenaphthene [83-32-9] with chromic acid in aqueous sulfuric acid or acetic acid (93). A post-oxidation of the crude oxidation product with hydrogen peroxide or an alkali hypochlorite is advantageous (94). An alternative liquid-phase oxidation process involves the reaction of acenaphthene, molten or in alkanoic acid solvent, with oxygen or acid at ca 70–200°C in the presence of Mn resinate or stearate or Co or Mn salts and a bromide. Addition of an aliphatic anhydride accelerates the oxidation (95).

The anhydride of 1,8-naphthalenedicarboxylic acid is obtained in ca 95–116 wt % yield by the vapor-phase air-oxidation of acenaphthene at ca 330–450°C, using unsupported or supported vanadium oxide catalysts, with or without modifiers (96).

The anhydride of 1,8-naphthalenedicarboxylic acid has fungicidal properties (97). This anhydride has been commercially introduced, under the trade name

Protect, as a seed treatment (eg, for corn) to prevent injury to the seed by thio-carbamate herbicides. The effectiveness of the antidote 1,8-naphthalic anhydride has also been successfully studied with several plants and herbicides (98,99).

4-Halogenated and 4,5-halogenated derivatives of 1,8-naphthalene-dicarboxylic acid anhydride are useful intermediates for dyes, pigments, and fluorescent whiteners for polymers.

Imides of 1,8-naphthalenedicarboxylic acid are used as drugs; an an-thelmintic for animals, eg, Naphthalophos; and as rodenticides (100,101). Other imides are useful fluorescent whiteners for polyesters and acrylonitrile polymers (102). The imide of 1,8-naphthalenedicarboxylic acid gives, by oxidative alkali fusion, the diimide of 3,4,9,10-perylenetetracarboxylic acid, the parent compound of an important class of red dyes and pigments for plastics and coatings with high color fastness. The diimides are useful electrical conductors and semiconductors, eg, in solar-cell systems (103). Some benzimidazole derivatives of 1,8-naphthalenedicarboxylic acid are excellent thickening agents for high temperature greases (104).

2,6-Naphthalenedicarboxylic Acid. This dicarboxylic acid, a potential monomer in the production of polyester fibers and plastics with superior properties (105), and of thermotropic liquid crystal polymers (106), is manufactured by the oxidation of 2,6-dialkylnaphthalenes (107,108).

The interest in 2,6-dialkylnaphthalenes such as dimethyl, diethyl, diiso-propyl, dihexyl, etc, is shown by the increasing number of patents relevant to their preparation, isomerization, and separation (109–111). No method for selectively preparing 2,6-dialkylnaphthalene has to date been discovered. The efforts that have been made have each failed, not only with conventional Fridel-Crafts catalysts (106,112,113), but also with other catalysts such as silica–alumina (114–116) or zeolites in the gas phase (117–120) as well as in the liquid phase (121–123). The Amoco Corp. has patented a process for the selective gas-phase isomerization of the 2,6-dimethylnaphthalene over a lower acidity, supported, molecular sieve-based catalyst composition (124). Amoco has also developed a purification process which uses a monocarboxylic acid anhydride as solvent for adsorption, oxidation, and reduction processes, followed by recrystallization or hydrolysis (125). Purification of dimethyl 2,6-naphthalenedicarboxylate [840-65-3] has been patented by Mitsubishi Gas Chemical Co., Inc. (126). Crude 2,6-naphthalenedicarboxylic acid (95% pure), prepared by oxidation of 2,6-diisopropylnaphthalene, was dissolved in an aqueous KOH solution at 70°C and filtered. The filtrate was treated with dimethylacetamide at ordinary temperature and the crystal obtained was dissolved in H_2O, then treated with an aqueous HCl solution to give 87% 2,6-naphthalenedicarboxylic acid of 99.9% purity (127). A Henkel-type process involves either the isomerization under pressure of 1,8-naphthalenedicarboxylic acid at ca 400–500°C in the presence of a CdO catalyst and K_2CO_3 or the carboxylation–isomerization of a naphthalenecar-boxylic acid in a similar system. Teijin Ltd. has produced polyester film with good elevated temperature resistance based on 2,6-naphthalenedicarboxylic acid. Teijin's process for the diacid involves the carboxylation–isomerization of the naphthalenecarboxylic acid (128). Cadmium iodide-catalyzed transcarboxylation of naphthalene using potassium salt of benzene carboxylic acid has been investigated at 400°C under pressure of CO_2 (8.1 MPa = 80 atm) to give a high selectivity (90%) of dipotassium 2,6-naphthalenedicarboxylate (129). Also, 2,6-naphthalenedicarboxylic acid can be prepared in high selectivity by the reaction of sodium 2-naphthalenecarboxylate with carbon monoxide, sodium carbonate, and sodium formate at ca 300°C and 2.1–4.8 MPa (300–700 psi) under CO_2 (130). Manufacturing methods for the diacid have been reviewed (131).

2,3,6-Naphthalenetricarboxylic Acid. Among the tricarboxylic acids, 2,3,6-naphthalenetricarboxylic acid was found to have an excellent use in materials for functional resins. Thus, Friedel-Crafts acylation of 2,6-dimethyltetralin with acetyl chloride in CH_2Cl_2 in the presence of $AlCl_3$ at 20°C gave 95% 2,6-dimethyl-7-acetyltetralin, which was air-oxidized in acetic acid in the presence of $Co(OOCCH_3)_2 \cdot 4H_2O$, $Mn(OOCCH_3)_2 \cdot 4H_2O$, and KBr at 200°C and 2.9 MPa to give 82% 2,3,6-naphthalenetricarboxylic acid (132).

1,4,5,8-Naphthalenetetracarboxylic Acid. Traditionally, the tetracarboxylic acid (**43**) has been manufactured by oxidation of the coal-tar component, pyrene, eg, with chromic acid, or by a chlorination–oleum hydrolysis–oxidation sequence. Also, oxidizing pyrene with aqueous sodium dichromate–sulfuric acid in a sand mill gave 1,6- and 1,8-pyrenequinone in 96% combined yield, further oxidation of which with aqueous KOCl at 80°C gave 51% 1,4,5,8-naphthalene tetracarboxylic acid dianhydride (133). Alternative processes start with acylation of acenaphthene in the 5,6-position followed by an oxidation of naphthalene ring substituents.

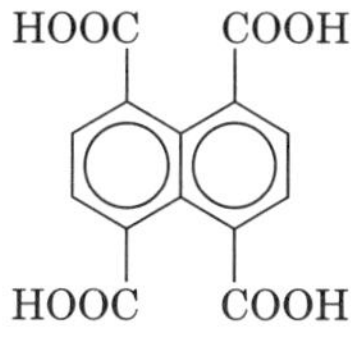

HOOC   COOH

HOOC   COOH

(**43**)

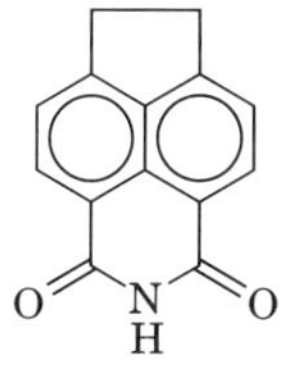

(**44**)

For example, 5,6-acenaphthenedicarboximide (**44**) can be prepared in 84% yield by the reaction of acenaphthene with excess sodium cyanate in anhydrous HF (78). The intermediate can be oxidized to the tetracarboxylic acid.

The dianhydride of 1,4,5,8-naphthalene tetracarboxylic acid [*81-30-1*] has been of research interest for the preparation of high temperature polymers, ie, polyimides. The condensation of the dianhydride with *o*-phenylenediamines gives vat dyes and pigments of the benzimidazole type.

2,3,6,7-Naphthalenetetracarboxylic Acid. The dianhydride of 2,3,6,7-naphthalenetetracarboxylic acid has also been of interest for the preparation of high temperature polymers, eg, polyimides. It has been prepared by thermal reaction of naphthalene with sodium aromatic carboxylates at 400–500°C in the presence of transcarboxylic catalysts under CO_2 pressure. Thus, autoclaving a mixture of naphthalene, disodium *o*-phthalate, CsI, and CdI_2 at 440°C under 2.45–7.36 MPa (356–1067 psi) CO_2 for 18 h gave 21% naphthalenecarboxylic acids containing 28% 2,3,6,7-naphthalenetetracarboxylic acid (134). 2,3,6,7-Naphthalene carboxylic acid was purified by partial esterification, followed by selective crystallization and separation (135).

Hydroxynaphthalenecarboxylic and Aminonaphthalenecarboxylic Acids

Some properties of selected hydroxynaphthalenecarboxylic acids are presented in Table 10.

2-Hydroxy-1-Naphthalenecarboxylic Acid. 2-Hydroxy-1-naphthoic acid is manufactured by a Kolbe-type process, ie, by reaction of the thoroughly dried potassium or sodium 2-naphthalenolate with CO_2 at ca 115–130°C in an autoclave at ca 300–460 kPa (3.0–4.5 atm) for 10–16 h. It decarboxylates readily, eg, in water starting at ca 50°C.

1-Hydroxy-2-Naphthalenecarboxylic Acid. 1-Hydroxy-2-naphthoic acid is made similarly to the isomer (2-hydroxy-1-naphthoic acid) by reaction of dry sodium 1-naphthalenolate with CO_2 in an autoclave at ca 125°C. It has been used in making triphenylmethane dyes and metallizable azo dyes. Alkylamides and arylamides of 1-hydroxy-2-naphthalenecarboxylic acid are cyan couplers, ie, components used in indoaniline dye formation in color films (see COLOR PHOTOGRAPHY).

2-Hydroxy-6-Naphthalenecarboxylic Acid. 2-Hydroxy-6-naphthoic acid, useful as an intermediate for dyes and a starting material for polyesters, is pre-

Table 10. Selected Properties of Hydroxynaphthalenecarboxylic Acids

Carboxylic acid	CAS Registry Number	Mp, °C	Other
2-hydroxy-1-naphthalene-	[2283-08-1]	157–159	sparing sol H_2O; sol alcohol, benzene
3-hydroxy-1-naphthalene-	[19700-42-6]	248–249	
4-hydroxy-1-naphthalene-	[7474-97-7]	188–188	
5-hydroxy-1-naphthalene-	[2437-16-3]	236	
6-hydroxy-1-naphthalene-	[2437-17-4]	213	
7-hydroxy-1-naphthalene-	[2623-37-2]	256–257	
8-hydroxy-1-naphthalene-	[1769-88-6]	169l	acetone, mp 108°C
1-hydroxy-2-naphthalene-	[86-48-6]	2000.55 wt %	sol in boiling water, alcohol, ether, benzene
3-hydroxy-2-naphthalene-	[92-70-6]	222–2230.1 wt %	sol in water at 25°C, ether, benzene, chloroform
4-hydroxy-2-naphthalene-	[1573-91-7]	225–226	
5-hydroxy-2-naphthalene-	[2437-18-5]	215–216	
6-hydroxy-2-naphthalene-	[16712-64-4]	245–248	
7-hydroxy-2-naphthalene-	[613-17-2]	274–275	
8-hydroxy-2-naphthalene-	[5776-28-3]	229	

pared by stirring the potassium salt of 2-naphthol with the potassium salt of *m*-cresol and high bp distillates of light oil in the presence of 294 kPa (43 psig) CO_2 at 260°C for 1 h (136). A U.S. patent describes the preparation of metal naphthoxide by treating 2-naphthol with cesium or rubidium hydroxide. Metal naphthoxide was dried and then treated with CO_2 ($\sim$140–700 kPa = 20–100 psig) in the presence of cesium or rubidium carbonate and a high boiling solvent to give 36% 2-hydroxy-6-naphthoic acid (137). In another method, sodium salt of 2-naphthol was treated with CO_2 in organophosphine oxide solvents, eg, $(C_4H_9)PO$ at 75–180°C for 3 h to give 2-hydroxy-6-naphthoic acid and 2-hydroxy-3-naphthoic acid in 50% and 40% selectivity (138).

3-Hydroxy-2-Naphthalenecarboxylic Acid. 3-Hydroxy-2-naphthoic acid or β-oxynaphthoic acid (BON; BONA; Developer 8) (**28**) is the principal commercial product among the hydroxynaphthalenecarboxylic acids. To produce BON, 2-naphthol is first transformed into sodium 2-naphthoate with a 50% sodium hydroxide solution, which is followed by Kolbe-Schmitt carboxylation with CO_2 at temperatures of 235 to 255°C and a pressure of 15 bar; the yield is 90–95% (52,65). Western European production of 3-hydroxy-2-naphthalene carboxylic acid was over 8×10^3 t in 1985 (52).

By reaction of BON with aniline at 80°C in toluene in the presence of PCl_3, the corresponding amide (**45**) of 3-hydroxy-2-naphthalenecarboxylic acid is produced. The product anilide (Azoic Coupling Component 2) is the base compound of the naphthol AS dyestuffs. Naphthol AS dyestuff components are distinguished from the unsubstituted 2-naphthol derivatives by their increased affinity for the dyed substrate and by higher chemical stability in the atmosphere. Usually the coupling reaction with the diazo salt takes place in the 1-position. Griesheim Red and Indra Red are two earlier used naphthol AS dyestuffs. Also, pigment Red 7 and pigment Red 112 are important naphthol AS pigments. The worldwide production of 2-naphthol AS dyestuffs is about 25×10^3 t/yr (52).

These amides (**45**), which are of the Naphthol AS type, are important coupling components that are applied to fiber, eg, cotton (qv). They then react with a diazo component on the fiber to produce insoluble azo dyes of high washfastness and lightfastness. A wide range of arylamides of 3-hydroxy-2-naphthalenecarboxylic acids and diazo components is available; azo pigments of similar structure are made from Naphthol AS by coupling a diazo component with a 3-hydroxy-2-naphthalenecarboxamide.

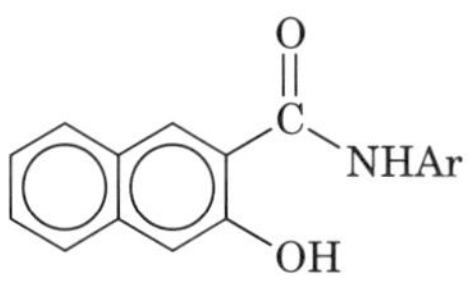

(**45**) 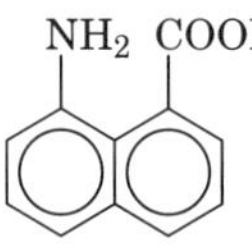   (**46**) 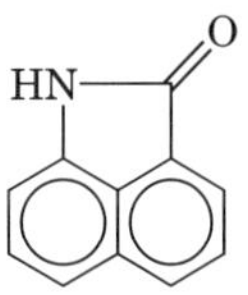   (**47**)

8-Amino-1-Naphthalenecarboxylic Acid. In most methods of 8-amino-1-naphthalenecarboxylic acid (**46**) manufacture, the lactam naphthostyril [130-00-7] (**47**) is obtained. For example, 8-amino-1-naphthalenesulfonic acid is converted to 8-cyano-1-naphthalenesulfonic acid in the Sandmeyer reaction, and the nitrile is treated with concentrated alkali at 185°C to form the lactam. The lactam can be hydrolyzed to the amino acid by treatment with dilute alkali at 100°C (139). A potentially useful process is the reported treatment of the imide (**42**) of 1,8-naphthalenedicarboxylic acid with sodium hypochlorite to give the lactam (**47**) in 57% yield (140). A procedure has been developed in Germany in which 1,8-naphthalimide is dissolved in an aqueous solution of LiOH and KOH with warming to 40–80°C, the solution is cooled and then chlorine bleaching liquor is added while maintaining the temperature between 10 and 20°C. After the reaction, excess active chlorine is removed reductively, and the pH is adjusted by adding acid to pH 2.0 to give 8-amino-1-naphthalenecarboxylic acid (141). A practical method of manufacture is the preparation of the imide by the reaction of 1-naphthaleneisocyanate with anhydrous aluminum chloride followed by hydrolysis (142).

8-Amino-1-naphthalenecarboxylic acid can be converted, by diazotization and treatment with ammoniacal cuprous oxide, to 1,1'-binaphthalene-8,8'-dicarboxylic acid [29878-91-9] (**48**). Treatment of (**48**) with concentrated sulfuric acid yields anthranthrone. The dihalogenated anthranthrones are valuable vat dyes.

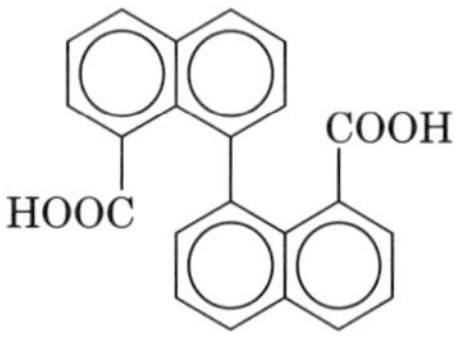

(**48**)

The (*N*-alkylated) lactam of 8-aminonaphthalenecarboxylic acid (**47**) also is a valuable dye intermediate, eg, for cyclomethine-type dyes used for dyeing polyacrylonitrile fibers and other synthetics. 1,8-Naphtholactams are prepared in high yield and purity by the reaction of naphtholactones with RNH_2 (R = H, C1–4 alkyl, cycloalkyl, or optionally substituted aryl) in aqueous medium, usually in the presence of bisulfite at 150°C over a period of 15 h (143).

BIBLIOGRAPHY

"Amino Naphthols and Amino Naphtholsulfonic Acids" in *ECT* 1st ed., Vol. 1, pp. 730–737; by J. Werner and A. W. Dawes, General Aniline Works Div., General Aniline & Film Corp.; "Naphthalenesulfonic Acids," "Naphthols and Naphthosulfonic Acids," and "Naphthylamines and Naphthylaminesulfonic Acids" in *ECT* 1st ed., Vol. 9, pp. 232–240, 248–258, and 258–270, by J. Werner, General Aniline Works Div., General Aniline, "Naphthalene Derivatives" in *ECT* 3rd ed., Vol. 15, pp. 719–749, by H. Dressler, Koppers Co., Inc.

  1. F. Radt, ed., *Elsevier's Encyclopedia of Organic Chemistry*, Sec. III, Vols. 12B, Elsevier, Amsterdam, the Netherlands, 1949–1955.

2. N. Donaldson, *The Chemistry and Technology of Naphthalene Compounds*, E. Arnold Ltd., London, 1958.

3. E. Muller, ed., *Methoden Der Organischen Chemie (Houben-Weyl)*, 4th ed., George Thieme Verlag, Stuttgart, Germany, 1958–present.

4. *Ullmann's Encyklopädie Der Techischen Chemie*, 4th ed., Verlag Chemie, Weinheim/Bergstr., Germany, 1972–present.

5. *Color Index*, 3rd ed., The Society of Dyers & Colorists Publishers, Bradford, U.K., and Research Triangle Park, N.C., 1971; Rev. 3rd ed., 1975.

6. H. A. Lubs, *The Chemistry of Synthetic Dyes and Pigments*, ACS Monograph Series No. 127, Reinhold Publishing Corp., New York, 1955.

7. K. Venkataraman, ed., *The Chemistry of Synthetic Dyes*, 8 vols., Academic Press, Inc., New York, 1952–1978.

8. H. Cerfontain, *Mechanistic Aspects in Aromatic Sulfonation and Desulfonation*, Wiley-Interscience, New York, 1968.

9. N. A. Korneeva, O. M. Prokhorova, and V. V. Kozlov, *Izv. Vyssh. Ucheb. Zaved. Khim. Khim. Tekhnol.* **19**, 171 (1976).

10. K. Kaufmann and F. Wolf, *Z. Chem.* **11**, 352 (1971).

11. H. Bretscher, G. Eigemann, and E. Plattner, *Chimia* **32**(5), 180 (1978).

12. R. H. Schreuder, A. Martijn, and C. J. Van de Kraats, *Chromatographia* **467**(1), 177–184 (1989).

13. *B.I.O.S. Final Report No. 1152*, Item No. 22, London.

14. Neth. Pat. 138,100 (Feb. 15, 1973), (to Sandoz, AG).

15. Ger. Offen. 2,337,395 (Feb. 15, 1973), (to Farbwerke Hoechst).

16. Jpn. Pat. 75 13,792 (Mar. 23, 1970), (to A. Ito and H. Hiyama).

17. Jpn. Pat. 53 071,046 (June 24, 1978), (to Sugai Kagaku Kogyo).

18. U.S. Pat. 4,059,609 (Nov. 22, 1977), J. K. Rinehart (to PPG Industries).

19. Y. A. I. Abdel-Aal, E. P. Lampert, M. A. Wolff and R. M. Roe, *Experientia* **49**, 571 (1993).

20. B. V. Passet, V. A. Kholodnov, and A. V. Matveev, *Zh. Prikl. Khim.* **51**, 1606 (1978).

21. U.S. Pat. 4,110,365 (Aug. 29, 1978), S. Bildstein, R. Lademann, S. Peitzsch, and G. Schaeffer (to Hoechst AG).

22. U.S. Pat. 3,655,739 (Nov. 4, 1972), H. Clasen (to Farbwerke Hoechst AG).

23. Ger. Offen. DE 3,937,748 (May 16, 1991), S. Bildstein, J. Heck, D. Schmid, K. Schmiedel, (to Hoechst AG).

24. *FIAT Final Rep. 1016*, 45 (1947).

25. Jpn. Pat. 73 38,699 (Nov. 19, 1973), (to Sumitomo Chemical Co.); USSR Pat. 596,565 (Sept. 27, 1977), (to Leningrad Chem. Pharm.).

26. U.S. Pat. 3,546,280 (Dec. 8, 1970), H. Dressler and K. G. Reabe (to Koppers Co.).

27. Jpn. Kokai 78 50,149 (May 8, 1978) H. Fujii, T. Nagashima, and H. Oguri (to Sugai Chemical Industries).

28. Jpn. Kokai Tokkyo Koho JP 63,280,055 (Nov. 17, 1988), A. Yamuchi and co-workers (to Kawasaki Steel Corp.).

29. F. Allison, G. Brunner, and H. E. Feirz-David, *Helv. Chim. Acta* **35**, 2139 (1952).

30. Jpn. Pat. 54 467,559 (Apr. 12, 1979), (to Sugai K. K.).

31. Belg. Pat. 866,223 (Oct. 23, 1978), (to Bayer AG).

32. *Chem. Eng. News* **54**, 11 (June 23, 1975); *Chem. Age*, 3 (Feb. 13, 1976).

33. USSR SU 1,742,254 (June 23, 1992), S. T. Babaev and co-workers (to Scientific Research Institute of Concrete and Reinforced Concrete).

34. Eur. Pat. Appl. EP 488,217 (June 3, 1992), H. Fuzimoto, T. Ishikawa, and K. Yoshida.

35. *Chemical Economics Handbook*, SRI International, Menlo Park, Calif., 1994.

36. USSR Pat. 267,889 (July 24, 1970), (to Phys. Org. Chem. Inst. Acad. Sci. Beloruss, SSR); USSR Pat 443,885 (June 5, 1975), T. I. Vasilenok, and co-workers.

37. *B.I.O.S. Final Report 1143*, London,   .

38. L. Eberson, M. P. Hartshorn, Radner, *Finn. J. Chem. Sec., Perkin Trans.* **2**(10) 1793–1798 (1992).

39. *Chem. Eng. News* **55**, 7 (Jan. 3, 1977).

40. U.S. Pat. 4,053,526 (Apr. 19, 1975), H. U. Blank, F. Duerholz and G. Skipka (to Bayer AG); Jpn. Pat. 54,016,460 (Feb. 2, 1979), (Nippon Synthetic Chemicals Industries).

41. Brit. Pat. 1142628 690212 (1969), S. Suzuki, M. Kurata, A. Akiyoshi, S. Aoshima, D. Hirohiko, and N. Matsuoka.

42. Brit. Pat. 1173890 691210 (1969), A. A. Artem'ev, Y. A. Strepikheev, Y. A. Shmidt, B. M. Babkin.

43. V. Vijayakumaran, S. Muralidharan, C. Ravichandran, S. Chellammal, and P. N. Anantharan, *Bull. Electrochem.* **6**(5), 522–523 (1990).

44. F. Mahdavi, T. C. Bruton, and Y. Li, *J. Org. Chem.* **58**, 744–746 (1993).

45. USSR SU 292,474 (Apr. 23, 1984), S. V. Dobrovolskii and co-workers.

46. X. Yang, *Huaxue Shijie* **25**(1), 10–12, (1984).

47. Jpn. Pat. 43008808 680408 (1968), S. Kano and co-workers (to Japan Soda Co., Ltd.).

48. Chn. Pat. 10-59905 A 920401 (1992), Y. Jingguo (to Donghe Chemical Co., Ltd.).

49. Hung. Pat. 18824 800927, (1980), I. Besan and co-workers (to Nehezvegyipari Kutato Intezet).

50. Ger. Offen. 2,523,351 (Dec. 9, 1976), H. U. Blank, F. Duerholz and G. Skipka (to Bayer AG); Jpn. Pat 50 0970,954 (July 26, 1975), (to Mitsui Toastu Chemical Co.).

51. Ger. Offen. DE 4,025,131; (Feb. 1992), G. Steffan, (to Bayer AG).

52. H.-G. Franck and J. W. Stadelhofer, *Industrial Aromatic Chemistry*, 1988, Chapt. 9.

53. *B.I.O.S. Document FD 4637/47*, frames 58–60, London.

54. Czech. CS 256,840 (Feb. 1, 1989), J. Kroupa and co-workers.

55. PB Report 74197, Washington, D.C.

56. Jpn. Kokai Tokkyo Koho JP 03,106,843 (May 7, 1991), S. Yamauchi and Y. Isuda (to Osaka Gas Co., Ltd.).

57. G. A. Olah and co-workers, *J. Org. Chem.* **56**, 6148–6151 (1991).

58. V. I. Trofimov, Y. Levkov, and A. M. Yakubson, *Sov. Chem. Ind.* **3**, 168 (1973).

59. *Chem. Eng. News* **53**, 25 (July 28, 1975).

60. U.S. Pat. 5066819A 911119, (1991), G. H. Kulkarni, R. H. Naik, and S. Rajappa (to Council of Scientific and Industrial Research (India)).

61. U.S. Pat. 3,480,671 (Nov. 25, 1969), H. Files and co-workers (to Stauffer).

62. S. Huang and P. Liu, *Huaxue Shiji* **14**(5), 313, (1992).

63. P. K. Jain and S. K. Srivastava, *J. Indian Chem. Soc.* **69**(7), 402–403 (1992).

64. M. Purohit and S. K. Srivastava, *Proc. Nat. Acad. Sci., India, Sect. A.* **61**(4), 461–464 (1991).

65. K. Moersdorf and G. Wolf, *Dtsch. Med. J.* **17**(10), 303–306, (1967).

66. U.S. Pats. 3,337,628 (Aug. 22, 1967) and 3,520,919 (July 21, 1970), S. Crowther (to I. C. I.).

67. S. P. Kamat, S. K. Paknikar, *Indian J. Chem., Sect. B*, **27B**(8), 773–774, (1988).

68. S. P. Starkov and Y. I. Mostyaev, *Sov. Chem. Ind.* **5**, 302 (1973).

69. USSR Pat. 340,270 (Jan. 31, 1973), (to Novosibirsk Org. Chem. Inst. Siberian Academy of Sciences USSR).

70. Brit. Pat. 654,035 (May 30, 1951), W. Webster and D. C. Quin (to the Distillers Co. Ltd.); U.S. Pat. 3,804,723 (Apr. 16, 1974), J. P. Dundon, H. R. Kemme, and E. J. Scharf (to American Cyanamid Co.); U.S. Pat. 3,848,001 (Nov. 12, 1974), J. P. Dunden (to American Cyanamid Co.); U.S. Pat. 3,939,211 (Feb. 17, 1976), R. H. Spector and R. K. Madison (to American Cyanamid Co.); U.S. Pat. 4,021,495 (May 3, 1977), H. Hosaka, K. Tamaka, and Y. Ueda (to Sumitomo Chemical Co.); U.S. Pat. 4,049,720 (Sept. 20, 1977), H. Hosaka, K. Tamimoto, and H. Yamachika (to Sumitomo Chemical

Co.); Ger. Offen. 2,207,915 (Aug. 23, 1973), M. Elstner, H. Hoever, and M. Fremery (to Union Rheinishe Braunishe Braunkohlen Kraftsoff AG): Ger. Offen. 2,517,591 (Oct. 23, 1975), M. Takahashi, T. Yamanchi, and T. Imura (to Kureha Chemical Industries); USSR Pat. 498,292 (Jan. 5, 1976), M. I. Ferberov, B. N. Bychkov, G. N. Shustovskaya, and G. D. Mantyukov (to Yaroslavl Polytechnic Institute); USSR Pat. 593,729 (Mar. 9, 1976); M. I. Ferberov, B. N. Bychkov, G. N. Shustovskaya, and G. D. Mantyukov (to Yaroslavl Polytechnic Institute).

71. Jpn. Kokai 75 0076,054 (June 21, 1975), K. Kobayaschi, I. Dogane, and Y. Nagao (to Sumitomo Chemical Industries).

72. *J. Commerce*, 325 (June 26, 1972).

73. Jpn. Kokai Tokkyo Koho JP 01 45,341 (Feb. 17, 1989), T. Sonoda and co-workers (to Nippon Shokubai Kagaku Kogyo Co., Ltd.).

74. Jpn. Kokai Tokkyo Koho JP 01 45,342 (Feb. 17, 1989), T. Sonoda and co-workers (to Nippon Shokubai Kagaku Kogyo Co., Ltd.).

75. Eur. Pat. Appl. EP 405,928 (Jan. 2, 1991) Y. Maegawa and Y. Nishida (to Sumitomo Chemical Co., Ltd., and Deiei Chemical Co., Ltd.).

76. U.S. Pat. 3,391,117 (July 2, 1968), N. Bilow and L. J. Miller (to U.S. Department of the Air Force); *Chem. Week*, 84 (Aug. 21, 1965).

77. *FIAT Final Report No. 1016*, 22–23.

78. Ger. Offen. 2,304,873 (Aug. 8, 1974), K. Eiglmeier and H. Luebbers (to Farbwerke Hoechst AG).

79. Eur. Pat. Appl. EP 390,181 (Oct. 3, 1990), T. Kishi and co-workers (to Kyowa Hakko Kogyo Co. Ltd.).

80. PCT Int. Appl. WO 90 06, 937 (June 28, 1990), U.S. Pat. Appl. 287,955 (Dec. 21, 1988), K. Shiosaki and co-workers (to Abbott Laboratories).

81. Fr. Pat. 1,543,144 (Oct. 18, 1868), G. J. Rolman (to Ashland Oil); USSR Pat. 225,249 (Oct. 29, 1970), N. D. Rusyanova and N. S. Mulyaeva (to Eastern Scientific Research Institute of Coal Chemicals); Jpn. Pats. 73 43,891 and 73 43,893 (Dec. 21, 1973), G. Yamashita and K. Yamamoto (to Teijin Ltd.).

82. Jpn. Kokai Tokkyo Koho JP 02,240,046 (Sept. 25, 1990), T. Sato and co-workers (to Sumikin Coke Co., Ltd.).

83. Jpn. Kokai Tokkyo Koho JP 02,250,850 (Oct. 8, 1990), Y. Doko and Y. Teruaki (to Sumikin Coke Co., Ltd.).

84. U.S. Pat. 3,920,734 (Nov. 18, 1975), Y. Ichikawa and Y. Yamaji (to Teijin Ltd.).

85. H. Tagaya and co-workers, *Bull. Chem. Soc. Jpn.* **63**(11), 3233–3237 (1990).

86. Ger. Offen. 2,028,555 (Dec. 17, 1970), E. Arsuro and co-workers (to Montecatini Edison SpA).

87. U.S. Pat. 3,855,289 (Dec. 17, 1974), G. H. Alt (to Monsanto Co.).

88. Ger. Offen. 2,824,988 (Dec. 21, 1978), M. Suchy (to F. Hoffmann-La Roche and Co.).

89. Hung. Pat. 50095 A2 891228 (1989), L. Balogh and co-workers (to Alkaloida Veg-yeszeti Gyar, Hung. Teljes).

90. Jpn. Kokai Tokkyo Koho, JP 62158243 A2 870714, (1987), T. Ikeda and co-workers (to Denki Kogaku Kogyo K.K.).

91. X. Shen and co-workers, *Ziran Kexueban* **1**, 6–13 (1981).

92. X. Huang and co-workers, *Ziran Kexueban* **17**(4), 499–500 (1990).

93. USSR Pat. 517,579 (Sept. 6, 1976), L. A. Kozorez and co-workers (to Voroshilovgrad Mechanical Institute).

94. Brit. Pat. 1,224,418 (Mar. 10, 1971), (to Mitsubishi Chem. Ind.); U.S. Pat. 3,646,069 (Feb. 29, 1972), H. Okada and co-workers (to Sandoz Ltd.); Jpn. Kokai 76 11,749 (Jan. 30, 1976), Y. Wakisada and M. Sumitami (to Nippon Kayaku Co.).

95. Ger. Pat. 1,262,268 (Mar. 7, 1968); A. Marx and co-workers (to Rutgerswerke AG); USSR Pat. 291,910 (Jan. 6, 1971), V. L. Plakin and co-workers; Ger. Offen. 2,520,094

(Nov. 20, 1975), H. Okushima, I. Nitta, and R. Furuya (to Mitsui Petrochem, K.K.); Jpn. Pat. 53 141,253 (Dec. 12, 1978), (to Nippon Kayaku K.K.).

96. Belg. Pat. 525,660 (June 8, 1956), (to Rutgerswerke AG); Jpn. Pat. 71 13,737 (Apr. 13, 1971), Y. Shimada, H. Yoshizumi, and Namikiri (to Japan Catalytic Chemical Industrial Co.); U.S. Pat. 3,708,504 (Jan. 2, 1973), O. Kratzar, H. Suter, and F. Wirth (to Badische Anilin and Soda Fabrik AG); Jpn. Kokai 75 39,292 (Apr. 11, 1975), Y. Nanba and co-workers (to Japan Catalytic Chemical Industrial Co.); USSR Pat. 535,306 (Dec. 20, 1976), D. Kh. Sembaev and co-workers (to AS Kazakistan Chemical Sciences).

97. U.S. Pat. 3,860,720 (Jan. 14, 1975), M. F. Covey (to Gulf Research & Development Co.).

98. M. Jain and co-workers, *Indian J. Plant Physiol.* **32**(2), 185–187 (1989).

99. D. I. Chkanikov and co-workers, *Fiziol Rast.* (*Moscow*) **38**(2), 290–296 (1991).

100. U.S. Pats. 3,940,397 and 3,940,398 (Feb. 24, 1976); 3,947,452 (Mar. 30, 1976); 3,959,286 (May 25, 1976); 3,996,363 (Dec. 7, 1976); 4,007,191 (Feb. 8, 1977); 4,051,246 (Sept. 27, 1977); 4,062,953 (Dec. 13, 1977); 4,070,465 (Jan. 24, 1977), P. C. Wade and B. R. Vogt (to E. R. Squibb & Sons, Inc.).

101. Can. Pat. 1,030,147 (Apr. 25, 1978), R. Martinez and co-workers (to Laboratories Made SA).

102. Ger. Offen. 2,507,459 (Sept. 2, 1976), (to Badische Aniline and Soda Fabrik AG).

103. Ger. Offen. 2,636,421 (Feb. 16, 1978), (to Badische Aniline and Soda Fabrik AG).

104. U.S. Pat. 4,040,968 (Aug. 9, 1977), H. A. Harris (to Shell Oil Co.).

105. Jpn. Kokai Tokkyo Koyo 75 0076 054 (1975), K. Kobayaschi and co-workers.

106. U.S. Pat. 4288646, (1979), G. A. Olah.

107. Jpn. Kokai Tokkyo Koho 57203032 (1981), (to Toray Industries Inc.).

108. Jpn. Kokai Tokkyo Koho JP 04 99,749 (Mar. 31, 1992) Y. Koko and R. Minami (to Sumikin Kako K.K).

109. PCT Int. Appl. Wo 90 03961, (1990), J. D. Fellman and co-workers.

110. Jpn. Kokai Tokkyo Koyo 02 264 733, (1990) K. Tate, Y. Sasaki, and T. Sasaki.

111. Jpn. Kokai Tokkyo Koyo 03 167 139, (1991), T. Fujita and co-workers.

112. F. Radt, in F. Radt, ed; *Elsevier's Encyclopedia of Organic Chemistry*, Vol. 12B, Elsevier Publishing Co., New York, 1948, pp. 132–161.

113. G. A. Olah and J. A. Olah, *J. Amer. Chem. Soc.* **98**, 1839 (1976).

114. W. M. Kutz and B. B. Corson, *J. Amer. Chem. Soc.* **67**, 1312 (1945).

115. Jpn. Pat. 74 48949 (1974), (to Teijin Co. Ltd.).

116. A. V. Topchiev and co-workers, *Dokl. Akad. Nauk SSSR* **139**, 124 (1961).

117. B. Chiche and co-workers, *J. Org. Chem.* **51**, 2128 (1986).

118. W. N. Kaeding and co-workers, *J. Catal.* **67**, 159 (1981).

119. D. Fraenkel and co-workers, *J. Catal.* **101**, 273 (1986).

120. Ger. Offen. DE 3334084 (1985), K. Eichler and E. I. Leupold.

121. P. Moreau and co-workers, *J. Catal*, in press.

122. A. Katayama and co-workers, *J. Chem. Soc., Chem. Commun.*, 39 (1991)

123. P. Moreau, *J. Org. Chem.* **57**, 5040 (1992).

124. U.S. Pat. 4,783,569 (Nov. 8, 1988), G. P. Hussmann and P. E. McMahon (to Amoco Corp.).

125. U.S. Pat. 5,097,066 (Mar. 17, 1992) J. K. Holzhauer and G. E. Kuhlmann (to Amoco Corp.).

126. Jpn. Kokai Tokkyo Koho JP 03,258,753 (Nov. 19, 1991), T. Tanaka and H. Fujita (to MGC Co. Inc.).

127. Jpn. Kokai Tokkyo Koho Jp 02,243,652 (Sept. 27, 1990), M. Morita and co-workers (to Nippon Steel Chemical Co., Ltd.).

128. *Chem. Week* **112**, 43 (Apr. 25, 1973).

129. K. Kudo, S. Mori, and N. Sugita, *Bull. Inst. Chem. Res., Kyoto Univ.* **70**(3), 284 (1992).

130. U.S. Pat. 3,718,690 (Feb. 278, 1973), R. D. Bushick, O. L. Norman, and H. J. Spinnelli.

131. I. I. Kiiko, *Khim. Prom.* **1**, 13 (1975).

132. Jpn. Kokai Tokkyo Koho JP 01,228,939 (Sept. 12, 1989), Y. Sato and Y. Tanaka (to Sumikin Kako Co., Ltd.).

133. V. A. Shigalevskii and co-workers, *Zh. Org. Khim.* **21**(7), 1506–1513 (1985).

134. Jpn. Kokai Tokkyo Koho JP 04,224,544 (Aug. 13, 1992), Y. Tachibana and co-workers (to Nippon Kokan K.K.).

135. Jpn. Kokai Tokkyo Koho Jp 04 89,453 (Mar. 23, 1992), Y. Tachibana and co-workers (to Nippon Kokan K.K.).

136. Jpn. Kokai Tokkyo Koho JP 03,223,230 (Oct. 2, 1991), T. Suzuki and co-workers (to Kawasaki Steel Corp.).

137. U.S. Pat. 5,075,496 (Dec. 24, 1991), J. Pugach and D. T. Derussy (to Aristech Chemical Corp.).

138. Jpn. Kokai Tokkyo Koho JP 03,200,741 (Sept. 2, 1991), T. Nakanishi and T. Miura.

139. *Product Bulletin IC-4*, American Cyanamid Co., July 1978.

140. *B.I.O.S. Final Rep. No. 1152*, London, pp. 118–123.

141. Ger. Pat. 3535482 AI 870409 (1987), O. Arndt and T. Papenfuhs (to Hoechst AG).

142. T. Maki, S. Hashimoto, and K. Kamada, *J. Chem. Soc. Jpn. Ind. Chem. Sect.* **55**, 4835 (1952).

143. Eur. Pat. 84853 AI 830803 (1983), R. Begrich (to Ciba-Geigy AG).

MANNAN TALUKDER
CURTIS R. KATES
Advanced Aromatics, Inc.

NAPHTHENIC ACIDS

The term naphthenic acid, as commonly used in the petroleum industry, refers collectively to all of the carboxylic acids present in crude oil. Naphthenic acids [1338-24-5] are classified as monobasic carboxylic acids of the general formula RCOOH, where R represents the naphthene moiety consisting of cyclopentane and cyclohexane derivatives. Naphthenic acids are composed predominantly of alkyl-substituted cycloaliphatic carboxylic acids, with smaller amounts of acyclic aliphatic (paraffinic or fatty) acids. Aromatic, olefinic, hydroxy, and dibasic acids are considered to be minor components. Commercial naphthenic acids also contain varying amounts of unsaponifiable hydrocarbons, phenolic compounds, sulfur compounds, and water. The complex mixture of acids is derived from straight-run distillates of petroleum, mostly from kerosene and diesel fractions (see PETROLEUM).

Naphthenic acids have been the topic of numerous studies extending over many years. Originally recovered from the petroleum distillates to minimize corrosion of refinery equipment, they have found wide use as articles of commerce in metal naphthenates and other derivatives. A comprehensive overview of the uses of naphthenic acid and its derivatives can be found in References 1 and 2. A review of the extensive research on carboxylic acids in petroleum conducted up to 1955 is available (3), as is a more recent review of purification, identification, and uses of naphthenic acid (4).

Chemical Structure

The name naphthenic acid is derived from the early discovery of monobasic carboxylic acids in petroleum, with these acids being based on a saturated single-ring structure. The low molecular weight naphthenic acids contain alkylated cyclopentane carboxylic acids, with smaller amounts of cyclohexane derivatives occurring. The carboxyl group is usually attached to a side chain rather than directly attached to the cycloalkane. The simplest naphthenic acid is cyclopentane acetic acid [*1123-00-8*], (**1**, $n = 1$).

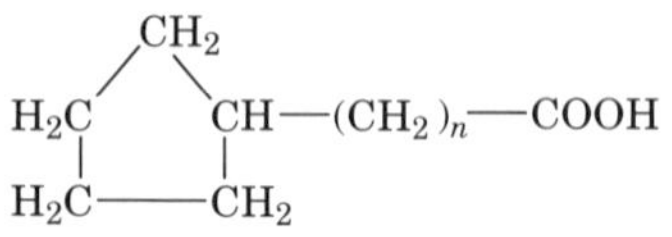

$$H_2C \underset{H_2C \underline{\quad\quad} CH_2}{\overset{\overset{\displaystyle CH_2}{\diagup \quad \diagdown}}{\quad}} CH - (CH_2)_n - COOH$$

(**1**)

Naphthenic acids are represented by a general formula $C_n H_{2n-z} O_2$, where n indicates the carbon number and z specifies a homologous series. The z is equal to 0 for saturated, acyclic acids and increases to 2 in monocyclic naphthenic acids, to 4 in bicyclic naphthenic acids, to 6 in tricyclic acids, and to 8 in tetracyclic acids. Typical structures for the homologues of naphthenic acids are shown in Figure 1. Naphthenic acids in the range of C-7 to C-12 consist mainly of monocyclic acids. The more complex acids contain larger proportions of multicyclic condensed compounds.

The fundamental nature of naphthenic acids was determined by the 1960s (3,5,6). As recently as 1955, only two naphthenic acids with as many as 10 carbon atoms had been positively identified. Later, more extensive laboratory studies revealed an astonishing variety of organic acids present in crude oil, including fatty acids as low in molecular weight as acetic acid [*64-19-7*], as well as saturated and unsaturated acids based on single and multiple five- and six-membered rings. One study identified ~1500 different organic acids in a single California crude oil, ranging in molecular weight between 200 and 700. The peak of the molecular weight distribution is between 300 and 400 (7). Early studies concentrated on identification of individual low molecular weight acids; more recent work has focused largely on geochemical correlations and biodegradation mechanisms of petroleum (8). C-14 to C-20 acyclic isoprenoid acids and C-10 to C-11 monocyclic naphthenic acids have been isolated from California crude petroleum (9–12). California crudes have also yielded C-22 to C-24 steroid acids, affirming the biological origin of petroleum (13).

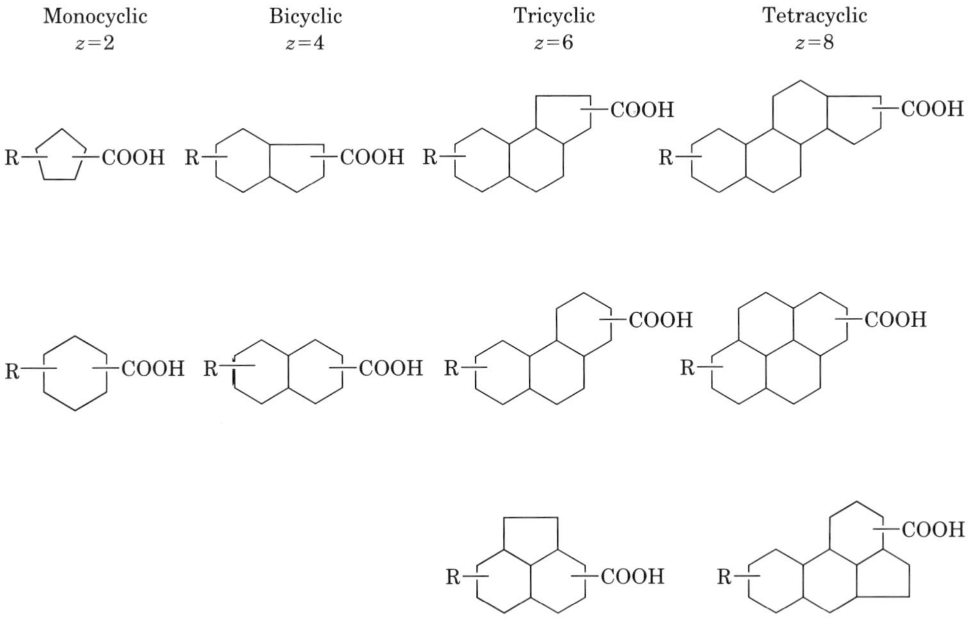

Fig. 1. Typical naphthenic acid structures, where R = alkyl. For the acyclic case $z = 0$ and the structure is simply R—COOH.

Physical and Chemical Properties

Naphthenic acids are viscous liquids, with phenolic and sulfur impurities present that are largely responsible for their characteristic odor. Their colors range from pale yellow to dark amber. An odor develops upon storage of the refined acids. Naphthenic acids have wide boiling point ranges at high temperatures (250–350°C). They are completely soluble in organic solvents and oils but are insoluble (<50 mg/L) in water. Commercial naphthenic acids are available in various grades and are marketed by acid number, impurity level, and color. Typical compositions and properties of three grades of naphthenic acids are given in Table 1. The product of refractive index and specific gravity, approximately 1.40–1.45, can be used to distinguish naphthenic acids from unsaponifiables (ca 1.2) and from other acidic compounds found in petroleum such as aliphatic acids (ca 1.3) and phenols (>1.5) (14).

Chemically, naphthenic acids behave like typical carboxylic acids with similar acid strength as the higher fatty acids. Dissociation constants are on the order of 10^{-5} to 10^{-6} (15). They are slightly weaker acids than low molecular weight carboxylic acids such as acetic, but are stronger acids than phenol [108-95-2] and cresylic acid [1319-77-3]. Whereas the principal use of naphthenic acids has been in the production of metal salts, they also can react to form esters, amine salts, amides, imidazolines, and many other derivatives (see ESTERS, ORGANIC; AMIDES). Derivatives of naphthenic acid often have advantageous physical properties over fatty acid derivatives, particularly their high stability to oxidation and solubility in hydrocarbons.

Table 1. Properties of Commercial Naphthenic Acid

Property	Grade		
	Crude	Refined	Highly refined
acid number, mg KOH/g	150–200	220–260	225–310
acid number (oil-free)	170–230	225–270	230–315
unsaponifiables, wt %	10–20	4–10	1–3
phenolic compounds, wt %	2–15	0.1–0.4	0.05–0.4
water, wt %	0.3–1.0	0.01–0.1	0.01–0.08
specific gravity at 20°C	0.95–0.98	0.95–0.98	0.95–0.98
viscosity at 40°C, mPa·s(=cP)	40–80	40–100	50–100
color, Gardner	black	6–8	5–6
refractive index, n_D^{20}	1.482	1.478	1.475
avg. mol wt (oil-free)	240–330	210–250	180–250

Naphthenic acid corrosion has been a problem in petroleum-refining operations since the early 1900s. Naphthenic acid corrosion data have been reported for various materials of construction (16), and correlations have been found relating corrosion rates to temperature and total acid number (17). Refineries processing highly naphthenic crudes must use steel alloys; 316 stainless steel [11107-04-3] is the material of choice. Conversely, naphthenic acid derivatives find use as corrosion inhibitors in oil-well and petroleum refinery applications.

Occurrence

Naphthenic acids are normal constituents of nearly all crude oils, but not all crudes contain sufficient quantities of usable acids to make recovery an economic process. Heavy crudes from geologically young formations have the highest acid content, and paraffinic crudes usually have low acid content. The acid content of crude petroleum varies from 0–3%, with crudes from California, Venezuela, Russia, and Romania having the highest content. Smaller amounts are found in U.S. Gulf Coast crudes, whereas little or no naphthenic acids are found in Pennsylvania, Iraq, or Saudi Arabia crudes. Typical concentrations are shown in Table 2. Minor amounts of naphthenic acids are also found in bituminous oil sands, but these are not economically recoverable. Identification of naphthenic

Table 2. Acid Content of Various Crudes[a]

Crude oil source	Petroleum acids, wt %
Pennsylvania	0.03
West Texas	0.4
Gulf Coast	0.6
California	1.5
Russia, Balakhany light	1.0
Russia, Balakhany heavy	1.6
Romania, waxy	0.2
Romania, asphaltic	1.6
Venezuela, Lagunillas	1.2

[a]Ref. 3.

acids in water from oil-bearing strata is being examined as a potential method of petroleum exploration (18).

Naphthenic acids occur in a wide boiling range of crude oil fractions, with acid content increasing with boiling point to a maximum in the gas oil fraction (ca 325°C). Jet fuel, kerosene, and diesel fractions are the source of most commercial naphthenic acid. The acid number of the naphthenic acids decreases as heavier petroleum fractions are isolated, ranging from 255 mg KOH/g for acids recovered from kerosene and 170 from diesel, to 108 from heavy fuel oil (19). The amount of unsaturation as indicated by iodine number also increases in the high molecular weight acids recovered from heavier distillation cuts.

Manufacture

The commercial production of naphthenic acid from petroleum in based on the formation of sodium naphthenate [61790-13-4]. Although the separation of naphthenic acids from hydrocarbon fractions is a relatively simple process, extraction from crude petroleum is not feasible because of the low percentage of acids in the oil, the large volumes of fluids to be handled, and the abundance of other caustic-extractable compounds found in most crudes. The low water solubility of high molecular weight naphthenate soaps and the tendency of sodium naphthenate to emulsify also contribute to making direct recovery of acids from crude oil difficult. Although naphthenic acids can be removed from crude oil and heavy fractions by reaction with caustic soda [1310-73-2] prior to and during their distillation, the acids are probably unusable since they are mixed (as salts) with large quantities of asphalt or residual pitch. The numerous patents on this subject are basically modifications to an early patent for distilling oil in the presence of sodium hydroxide (20). The alkali residues are extracted with alcohol to remove unsaponifiables, followed by acidification to recover the naphthenic acids (21).

Naphthenic acids are generally obtained by caustic extraction of petroleum distillates boiling between 200 and 370°C. A continuous process has been developed for removing naphthenic acids from refinery streams by caustic washing (22). Caustic extraction also removes other acidic components of the petroleum fraction, including phenol and cresols (cresylic acid), mercaptans, and thiophenols. In addition to reducing corrosion in the refinery, the caustic wash is necessary to improve the burning qualities, storage stability, and odor of the finished kerosene and diesel fuels. The petroleum fractions are extracted with dilute (2–10%) sodium hydroxide since the sodium naphthenate salts are emulsifying agents. Stronger caustic strengths increase the solubility of hydrocarbon oils (unsaponifiables) in the sodium naphthenate. The Fiber-Film contacting process patented by Merichem reduces emulsification and caustic carryover during removal of naphthenic acid from petroleum fractions and achieves low acid number specifications in a single stage (23,24). Noncaustic processes for recovery of naphthenic acids from petroleum distillates, including ammonia [7664-41-7] (qv) (25), triethylene glycol [112-27-6] (26), ion-exchange resins (27), and aluminosilicate zeolites (28), have also been reported but not commercialized (see ION EXCHANGE).

The aqueous sodium naphthenate phase is decanted from the hydrocarbon phase and treated with acid to regenerate the crude naphthenic acids. Sulfuric acid [7664-93-9] is used almost exclusively, for economic reasons. The wet crude naphthenic acid phase separates and is decanted from the sodium sulfate [7757-82-6] brine. The volume of sodium sulfate brine produced from dilute sodium naphthenate solutions is significant, on the order of 10 L per L of crude naphthenic acid. The brine contains some phenolic compounds and must be treated or disposed of in an environmentally sound manner. Sodium phenolates can be selectively neutralized using carbon dioxide [124-38-9] and recovered before the sodium naphthenate is finally acidified with mineral acid (29). Recovery of naphthenic acid from aqueous sodium naphthenate solutions using ion-exchange resins has also been reported (30).

The other acidic compounds extracted by caustic remain in the crude naphthenic acid after acidification of the sodium naphthenate has occurred. These phenolic and sulfur compounds are objectionable because of their odor and color-forming properties. Although numerous methods for naphthenic acid purification have been suggested, only those which combine low cost and relatively simple operation are used commercially. Crude naphthenic acids are dried and distilled under reduced pressure to produce the refined products. Distillation removes some of the phenolic and unsaponifiable impurities, but the refined products still contain significant levels of impurities boiling at the same temperature. Because it is difficult and costly to remove all of the hydrocarbon impurities, commercial refined naphthenic acids typically contain 5–10% unsaponifiables. Unsaponifiables can be reduced to <5% by distilling the hydrocarbons from the alkali salts of the naphthenic acids (31). Naphthenic acids can also be purified by treatment with concentrated sulfuric acid to precipitate sludge-forming materials, which reduces the content of phenolics (32). The naphthenic acid is then distilled for final purification. Liquid–liquid extraction processes reduce unsaponifiable content by extracting naphthenic acids into water–alcohol mixtures (33,34).

Interest in synthetic naphthenic acid has grown as the supply of natural product has fluctuated. Oxidation of naphthene-based hydrocarbons has been studied extensively (35–37), but no commercially viable processes are known. Extensive purification schemes must be employed to maximize naphthene content in the feedstock and remove hydroxy acids and nonacidic by-products from the oxidation product. Free-radical addition of carboxylic acids to olefins (38,39) and addition of unsaturated fatty acids to cycloparaffins (40) have also been studied but have not been commercialized.

Production

Nameplate capacities of naphthenic acid producers in North America are 9000 metric tons of crude and refined acid at Merichem (Tuscaloosa, Ala.), and 3600 t of crude acid at Hewchem (Gulfport, Miss.) (41). However, actual production capacity may vary widely as a result of the mix of feedstocks being processed. Some feedstocks require significantly greater processing time to achieve high grade finished product. Naphthenic acid products are shipped

in tank cars, tank trucks, and drums under DOT 9137/UN 3082 identification numbers. Principal producers in Japan are Sanko Yuka Kogyo, Yamato,
Taniguchi Oil, and Union Oil (a subsidiary of Idimitsu); in Europe, Nord Import, Oleochimie, Imperial Oil, and Corn Van Loocke.

Economic Aspects

Roughly 5500–6000 t of naphthenic acid were consumed in North America in
1992. Less than 500 t of finished product demand was met by imports in spite of
the large quantities of naphthenic acid produced overseas. The average unit value
was $1.56/kg. After much price volatility in the late 1970s and early 1980s, prices
have risen moderately in the 1990s. For fob bulk shipments, current refined acid
prices range from $1.94–2.09/kg for acid numbers 220 to 300. Crude acid prices
range from $0.99–1.17/kg for acid numbers 150 to 180 (41).

   Naphthenic acid availability exceeds demand, although some minor market
disruptions occurred in North America during the early 1990s. In 1990, the U.S.
Environmental Protection Agency (EPA) put into place low sulfur requirements
for highway diesel fuel to be met by October 1993. To meet the specification,
several feedstock producers were prompted to hydrotreat their diesel fractions,
since caustic extraction does not reduce sulfur to the required level. Hydrotreating destroys naphthenic acids, so the result was a decrease in feedstock availability. Raw materials from previously unused feedstock sources in Asia, Europe,
and South America were then imported into North America to achieve market
balance. Long-term yearly feedstock supply is expected to meet market growth
projections with large discoveries of high naphthenic-content oil.

Analysis

Naphthenic acid concentration in crude oil and distillates is typically measured
by titrating with potassium hydroxide [1310-58-3] for neutralization number
using potentiometric (ASTM D664) or colorimetric (ASTM D974) methods. The
same procedures are used to measure the acid number of crude or refined
naphthenic acids after recovery from the petroleum fraction. The neutralization
or acid numbers are expressed numerically as mg KOH (formula wt = 56.1)
required to neutralize the acidity in 1 g of sample. This value for the acid on an
impurity-free basis is readily converted to equivalent weight by dividing 100 $\times$
the formula weight of KOH (56,100) by the acid number. Naphthenic acid having
a high acid number indicates a low molecular or equivalent weight product. Crude
oil or a petroleum fraction would have a very low acid number, since the acid
content is low.

   Naphthenic acids comprise a highly complex mixture containing hundreds
of compounds that are impossible to separate into individual components, even
by high resolution gas chromatography. Progress in obtaining information on the
structure of naphthenic acids has been retarded because of this complexity. Anal-

ysis by gas chromatography of methyl esters is useful in detecting naphthenic acids adulterated with synthetic or vegetable oil-based fatty acids. Infrared spectroscopy is unable to distinguish clearly between naphthenic and fatty acids.

Two mass spectrometric methods have been developed which improve the characterization of naphthenic acids. Chemical ionization using fluoride ion as a reactant gas (42) and fast-atom bombardment (43) both use the negative ion detection mode. The two methods detect the $(M - 1)^-$ ion in carboxylic acids, including high molecular weight species with no fragmentation. These methods allow the identification of naphthenic acids based on carbon number and z-series distributions, providing a fingerprint useful in correlating the sources of crude oils.

Health and Safety Factors

Naphthenic acids are only slightly toxic to mammals but are toxic to fish, bacteria, and wood-destroying insects. The lethal oral dose for humans is approximately 1 L (44), and the oral LD_{50} is 3.0–5.2 g/kg in rats (45). The deaths appeared to result from gastrointestinal disturbances. Naphthenic acid is non-mutagenic by the Ames mutagenicity test (46), and is not listed as a carcinogen by the International Agency for Research on Cancer, the National Toxicology Program, or the U.S. Occupational Safety and Health Act.

Commercial Uses

More than two-thirds of the naphthenic acid produced is used to make metal salts, with the largest volume being used for copper naphthenate [*1338-02-9*], consumed in the wood preservative industry (see WOOD). Metal salts used as paint driers accounted for only 16% of the naphthenic acid market in 1993 (see PAINT). This is a dramatic contrast with 1977 usage, when 75% of the naphthenates went into the paint drier market. An overall view of the 1993 naphthenic acid market in North America shows the following uses:

Use	Percentage
wood	40.1
oil field	27.7
paint	15.8
tires	8.5
miscellaneous	7.9

Oil field uses are primarily imidazolines for surfactant and corrosion inhibition (see PETROLEUM). Besides the lubrication market for metal salts, the miscellaneous market is comprised of free acids used in concrete additives, motor oil lubricants, and asphalt-paving applications (47) (see ASPHALT; LUBRICATION AND

Table 3. Metal Naphthenate Uses

Name	CAS Registry Number	Applications
copper naphthenate	[1338-02-9]	wood and textile preservative, catalyst
zinc naphthenate	[12001-85-3]	wood and textile preservative, lubricant, wetting agent
cobalt naphthenate	[61789-51-3]	paint drier, tires, ink drier, catalyst
manganese naphthenate	[1336-93-2]	paint drier, catalyst, fuel additive
lead naphthenate	[61790-14-5]	paint drier, wetting agent, lubricant additive
calcium naphthenate	[61789-36-4]	paint drier, catalyst, lube additive
iron naphthenate	[1338-14-3]	paint drier, fuel additive, catalyst
zirconium naphthenate	[72854-21-8]	paint drier, electrophotographic developer
cerium naphthenate	[68514-63-6]	paint drier, catalyst, fuel additive
vanadium naphthenate	[68815-09-8]	paint drier, catalyst, corrosion inhibitor
sodium naphthenate	[61790-13-4]	emulsifier, ore flotation, leather
potassium naphthenate	[66072-08-0]	emulsifier, plant growth modifier
aluminum naphthenate	[61789-64-8]	gelling agent, pigment wetting

LUBRICANTS). Naphthenic acid has also been studied in ore flotation for recovery of rare-earth metals (48) (see FLOTATION; LANTHANIDES).

Naphthenic acid is ideal for synthesizing metal carboxylates that require a ligand with some oxidative stability, solubility in hydrocarbons and oils, and insolubility in water. The general chemical structure of the naphthenate salts is $M^{x+}(^-OOCR)_x$ where x = metal charge and molar stoichiometry of naphthenic acid to the metal M. Table 3 lists commercially available metal naphthenates and their uses. Metal carboxylates are used in the oil-based (alkyd) paint industry as catalysts to accelerate the drying of coatings; hence the name driers was given to them (see ALKYD RESINS; DRIERS AND METALLIC SOAPS). The metal naphthenates were originally used to replace the corresponding linoleates, resinates, and tallates. These naphthenate salts have the advantages of high metal content, low viscosity, better solubility in hydrocarbons and oils, and greater stability to oxidation. Each metal naphthenate has a specific function and is typically used in combination with other driers. This market is declining as a result of the 1990 Clean Air Act that reduces volatile organic compound (VOC) emissions. The paint industry is responding by replacing their oil-based paints with waterborne formulations.

An expanding market for naphthenic acid is in wood and textile preservatives (see TEXTILES; WOOD). Copper and zinc naphthenate are used to prevent dry rot, as well as fungi and insect attack in wood. Zinc naphthenate is typically used in the log home industry because it provides a colorless coating on the wood. The naphthenates are environmentally preferred replacements for creosote [8021-39-4], pentachlorophenol [87-86-5], and chromated copper–arsenic.

Copper naphthenate prevents textile rotting at lower concentrations than copper oleates or tallates, in part because the naphthene moiety itself has fungicidal properties (49). In North America, the 1992 production of 8% copper naphthenate solution was estimated to be 2600–2900 t at a unit value of $2.20–2.45/kg. The estimated market growth for this application is 12–14%/yr.

Another market application for naphthenic acid is the tire industry, where cobalt naphthenate is used as an adhesion promoter (see ADHESIVES; TIRE CORDS). Cobalt naphthenate improves the bonding of brass-plated steel cords to rubber, presumably by suppressing the de-zincification of brass (50). Its first reported use was in 1970 and the first patent for its use was issued in 1975 (51). About 900 t of cobalt naphthenate is used worldwide as an adhesion promoter, half of it in North America. The unit value fluctuates between $8.75–13.25/kg because of the volatility of cobalt prices. Although it is the industry standard, the use of cobalt naphthenate is declining with the advent of more economical high metal-containing substitutes.

In the other market areas, lead naphthenates are used on a limited basis in extreme pressure additives for lubricating oils and greases. Sodium and potassium naphthenates are used in emulsifiable oils, where they have the advantage over fatty acid soaps of having improved disinfectant properties. Catalyst uses include cobalt naphthenate as a cross-linking catalyst in adhesives (52) and manganese naphthenate as an oxidation catalyst (35). Metal naphthenates are also being used in the hydroconversion of heavy petroleum fractions (53,54) and bitumens (55).

The surface-active properties of many naphthenic acid derivatives allow them to function in a variety of applications, including corrosion inhibitors, emulsifiers, and defoamers (see CORROSION AND CORROSION CONTROL; EMULSIONS; FOAMS). These properties can be exploited in performance-oriented applications such as petroleum recovery, asphalt emulsification, and concrete aeration where chemical composition is not critical (see ASPHALT; PETROLEUM). Surfactants may be prepared from the acids themselves or from derivatives such as amides or esters. The general surfactant applications are dominated by the ethoxylated naphthenic acids and naphthenyl amides and alcohols. Ethoxylated and sulfated naphthenic acid derivatives are highly surface-active and have substantial detergency power. The petroleum industry utilizes naphthenic acid amine derivatives both as surfactants for enhanced (tertiary) oil recovery (56) and as corrosion inhibitors (57). Amides and imidazolines of naphthenic acid used as corrosion inhibitors are noted for oil solubility and flow characteristics that are superior to those of similar fatty acid-based derivatives. Naphthenyl amides and imidazolines based on polyamines such as diethylenetriamine [111-40-0] are also reported in such diverse applications as bitumen emulsifiers, bactericides (58), and lubricant additives (see LUBRICATION AND LUBRICANTS). The surfactant properties of naphthenyl amides improve adhesion of aggregate rock to the asphalt binder (59) and improve mechanical strength and plasticity in concrete (60). Alkaline flooding of oil sands improves bitumen extraction by producing anionic surfactants *in situ* from the naturally occurring naphthenic acids (61).

Simple esters are readily formed by the reaction of naphthenic acids with monohydric alcohols, olefins, or ethylene or propylene oxides. More complex esters are prepared by partial or complete esterification of polyhydric alcohols.

Naphthenic acid esters have been repeatedly cited as replacements for phthalates as plasticizers for PVC resins. Alkyl and glycol naphthenate esters improve the flexibility and workability of resins and are valued for their low volatility, compatibility, stability, and resistance to kerosene extraction (62). Naphthenic acid esters of multifunctional alcohols such as pentaerythritol [115-77-5] (63) and poly(ethylene glycol)s have been cited as lubricants or additives for fuels, and for improved oil recovery (64). Glycol esters and other esters of naphthenic acid are used as hydraulic fluids (65), plasticizers, and surfactants, and are finding increasing use in the fat-liquoring process in leather tanning (66) (see HYDRAULIC FLUIDS; LEATHER; PLASTICIZERS; SURFACTANTS). Triethylene glycol naphthenate [132885-83-7] improves dispersion of carbon black [1333-86-4] in water-borne conductive primer coatings (67).

Naphthenyl alcohols are formed by reduction of the acids or their simple esters. They are valuable as surfactants, solvents, and components of lubricants. The acid halides are of value mainly as chemical intermediates (1).

BIBLIOGRAPHY

"Naphthenic Acids" in *ECT* 1st ed., Vol. 9, pp. 241–246, by V. L. Shipp, Socony-Vacuum Oil Co; in *ECT* 2nd ed., Vol. 13, pp. 727–734, by S. E. Jolly, Sun Oil Co.; in *ECT* 3rd ed., Vol. 15, pp. 749–753, by W. E. Sisco, W. E. Bastian, and E. G. Weierich, CPS Chemical Co.

1. E. S. Lower, *Specialty Chem.* **7**, 76 (1987); **7**, 282 (1987); **8**, 174 (1988); **9**, 135 (1989); **9**, 267 (1989).
2. E. S. Lower, *Pigment Resin Technol.* **15**(5), 4 (1986); **15**(6), 4 (1986); **15**(7), 11 (1986); **15**(8), 6 (1986); **15**(9), 8 (1986); **15**(10), 9 (1986); **15**(11), 7 (1986); **15**(12), 6 (1986).
3. H. L. Lochte and E. R. Littmann, *The Petroleum Acids and Bases*, Chemical Publishing Co., Inc., New York, 1955.
4. G. Narmetova, B. Khamidov, N. Ryabova, and E. Aripov, *Purification, Identification, and Use of Naphthenic Acid*, Fan, Tashkent, former USSR, 1983.
5. N. Zelinsky and A. Chukasanov, *Ber.* **57**, 42 (1924).
6. J. Von Braun, *The Science of Petroleum*, Vol. II, Oxford University Press, London, 1938, pp. 1007–1015.
7. W. Seifert, in W. Herz, H. Grisebach, and G. Kirby, eds., *Progress in the Chemistry of Organic Natural Products*, Vol. 32, Springer-Verlag, New York, 1975, pp. 1–49.
8. A. Mackenzie, G. Wolff, and J. Maxwell, in M. Bjoroy and co-workers, eds., *Advances in Organic Geochemistry*, John Wiley & Sons, Ltd., Chichester, U.K., 1981, p. 637.
9. J. Cason and A. Khodair, *J. Org. Chem.* **32**, 3430 (1967).
10. J. Cason, *Tetrahedron* **21**, 471 (1965).
11. J. Cason and K. Liauw, *J. Org. Chem.* **30**, 1763 (1965).
12. J. Cason and A. Khodair, *J. Org. Chem.* **31**, 3618 (1966).
13. W. Seifert, A. Gallegos, and R. Teeter, *J. Am. Chem. Soc.* **94**, 5880 (1972).
14. H. Schutze, B. Shive, and H. Lochte, *Ind. Eng. Chem. Anal. Ed.* **12**, 262 (1940).
15. J. Petkovic and M. Ristic, *J. Serb. Chem. Soc.* **52**, 641 (1987).
16. J. Heler, *Mater. Protection* **2**, 90 (1963).
17. R. Piehl, *Mater. Performance* **27**, 37 (1988).
18. V. Shvets, *Org. Geokhim. Vod Poisk. Geokhim., Mater. Mezhdunar Kongr. Org. Geokhim.,* 8tn 1977, 47 (1982).

19. E. Pyhälä, *Z. Angew. Chem.* **27**, 407 (1914).

20. U.S. Pat. 623,006 (April 11, 1899), F. Berg.

21. U.S. Pat. 1,804,451 (May 12, 1931), T. Andrews and C. Lauer (to the Texas Co.).

22. D. Todd and F. Rac, *Hydrocarbon Process.* **46**(8), 115 (1967).

23. B. Norris and J. Files, *An. Congr. Lat.-Am. Eng. Equip. Ind. Pet. Petroquim.* 1st 1979, II/91-102 (1980).

24. R. Vazquez, *Pet. Int. (Tulsa, Okla.)* **48**(5), 18 (1990).

25. U.S. Pat. 4,634,519 (Jan. 6, 1987), M. Danzik (to Chevron Research Co.).

26. M. Mardanov and co-workers, *Azerb. Neft. Khoz.*, 54 (1980).

27. U.S. Pat. 2,936,424 (Dec. 6, 1960), G. Ayres and M. Krewer (to Pure Oil Co.).

28. R. Kuliev and co-workers, *Sb. Tr.-Akad. Nauk. Az. SSR* **16**, 76 (1987).

29. U.S. Pat. 2,357,252 (Aug. 29, 1944), H. Berger, E. Nygaard, and H. Angel (to Socony-Vacuum Oil Co.).

30. G. Defer, A. Abadie, and H. Roques, *Trib. CEBEDEAU* **28**, 379 (1975).

31. U.S. Pat. 5,011,579 (Apr. 30, 1991), G. Davis (to Merichem Co.).

32. U.S. Pat. 2.081,475 (May 25, 1937), D. Carr (to Union Oil Co. of California).

33. U.S. Pat. 2,391,729 (Dec. 25, 1945), W. McCorquodale (to Sun Oil Co.).

34. U.S. Pat. 2,610,209 (Sept. 9, 1952), E. Honeycutt (to Sun Oil Co.).

35. U.S. Pat. 4,227,020 (Oct. 7, 1980), B. Zeinalov and co-workers.

36. B. Zeinalov and co-workers, *Azerb. Neft. Khoz.* **9**, 34 (1991).

37. U.S. Pat. 2,519,309 (June 29, 1948), W. Denton (to Socony-Vacuum Oil Co.).

38. USSR Pat. 1,154,266 (May 7, 1985), B. Zeinalov and co-workers (to Institute of Petrochemical Processes).

39. S. Sadykh-Zade and R. Dzhalilov, *Dokl. Akad. Nauk. Azerb. SSR* **22**, 17 (1966).

40. U.S. Pat. 2,534,074 (May 2, 1947), A. Shmidl (to Standard Oil Development Co.).

41. *Chemical Mktg. Rep.* **244**, 45 (Sept. 27, 1993).

42. I. Dzidic, A. Somerville, J. Raia, and H. Hart, *Anal. Chem.* **60**, 1318 (1988).

43. T. Fan, *Energy & Fuels* **5**, 371 (1991).

44. W. Rockhold, *A.M.A. Arch. Ind. Health* **12**(5), 477 (1955).

45. *Dangerous Prop. Ind. Mater. Rep.* **7**, 105 (1987).

46. R. Crebelli and co-workers, *IARC Sci. Publ.* **59**, 289 (1984).

47. Brit. Pat. 1,086,390 (Oct. 11, 1967), O. Pordes and N. Bangert (to Shell International Research).

48. Y. Ge and Q. Wang, *Kuangye Gongcheng* **9**, 56 (1989).

49. P. Marsh, G. Greathouse, K. Bollenbacher, and M. Butler, *Ind. Eng. Chem.* **36**, 176 (1944).

50. W. van Ooij, W. Weening, and P. Murray, *Rubber Chem. Technol.* **54**, 227 (1981).

51. U.S. Pat. 3,897,583 (July 29, 1975), C. Bellamy (to Uniroyal SA).

52. J. Donalson, S. Clark, and S. Grimes, eds., *Cobalt in Catalysts*, Cobalt Development Institute, London, 1990, Chapt. 8.

53. Eur. Pat. 512,778 (Nov. 11, 1992), J. Harrison, A. Bhattacharya, M. Patel, and D. Rao (to Texaco Development Corp.).

54. P. LePerchec and co-workers, *Prepr.-Amer. Chem. Soc., Div. Pet. Chem.* **38**, 401 (1993).

55. H. Chen and co-workers, *AOSTRA J. Res.* **5**, 33 (1989).

56. U.S. Pat. 3,861,467 (Jan. 21, 1975), B. Harnsberger (to Texaco Inc.).

57. U.S. Pat. 3,997,469 (Dec. 14, 1976), C. Howle (to Nalco Chemical Co.).

58. U.S. Pat. 3,788,989 (Jan. 29, 1974), P. Adams and A. Petrocci (to Millmaster Onyx Corp.).

59. Ger. Pat. 1,103,223 (Feb. 15, 1955), M. Lhorty (to Bataasfe Petroleum).

60. USSR Pat. 1,054,322 (Nov. 15, 1983), V. Shestakov and co-workers.

61. L. Schramm and co-workers, *AOSTRA J. Res.* **1**, 5 (1984).

62. B. Zeinalov and co-workers, *J. Polymer Sci. Part C* **33**, 353 (1971).

63. A. Ismailov and B. Salimova, *Khim. Tekhnol.* **9**, 79 (1967).
64. Rom. Pat. 51,410 (Nov. 7, 1968), F. Popescu and co-workers (to Romania Institute for Drilling and Extraction Research).
65. U.S. Pat. 2,512,255 (June 20, 1950), W. Gilbert, R. Kline, and C. Montgomery (to Gulf Research and Development).
66. B. Kochetygov and P. Levenko, *Kozh. Obuv. Prom.* **10**, 42 (1983).
67. J. Carmine and R. Ryntz, *J. Coatings Technol.* **66**(836), 93–98 (1994).

General References

M. Naphthali, *Chemie, Technologie, und Analyse der Naphthensäuren*, Wissenschaftliche Verlagsgesellschaft, Stuttgart, Germany, 1927.
M. Naphthali, *Naphthensäuren und Naphthensulfosäuren*, Wissenschaftliche Verlagsgesellschaft, Stuttgart, Germany, 1934.
C. Ellis, *The Chemistry of Petroleum Derivatives*, The Chemical Catalog Co., Inc., New York, 1934, pp. 1062–1091.
A. N. Sachanen, *The Chemical Constituents of Petroleum*, Reinhold Publishing Corp., New York, 1945, pp. 315–330.
W. A. Gruse and D. R. Stevens, *The Chemical Technology of Petroleum*, 2nd ed., McGraw-Hill Book Co., Inc., New York, 1942, pp. 97–105.
E. R. Littman and J. R. M. Klotz, *Chem. Rev.* **30**(1), 97–111 (1942).
W. Maass, E. Buchspiess-Paulentz, and F. Stinsky, *Naphthensäuren und Naphthenate*, Verlag für Chemische Industrie H. Ziolkowsky, Augsburg, Germany, 1961.

JAMES A. BRIENT
PETER J. WESSNER
Merichem Company

MARY NOON DOYLE
Shepherd Chemical Company

NAVAL STORES. See TERPENOIDS.

NEODYMIUM. See LANTHANIDES.

NEPTUNIUM. See ACTINIDES AND TRANSACTINIDES.

NEUROPEPTIDES. See MEMORY-ENHANCING DRUGS; NEUROREGULATORS.

NEUROREGULATORS

Molecular regulation of tissue function and cellular homeostasis provides an organism, from mollusk to human, with the ability to adapt and react to a changing environment. A critical element in such adaptation and reaction is the nervous system. This not only acts to control muscles, organs, and behavior, but is itself regulated by sections of the cardiovascular, endocrine, and immune systems. Signaling within the nervous system may be acute, eg, cell depolarization, or may involve longer term regulatory functions, eg, gene expression. Thus the term neuroregulator encompasses a diversity of chemical substances involved in signaling: (1) between neurons; (2) between neurons and glia; (3) from neuron to effector target such as skeletal muscle, cardiac and smooth muscle, endocrine cell, or immunocompetent cell; and (4) from these diverse cell types to the neuron.

Two specialties of the nervous system are speed and localization, accomplished using highly developed electrical signaling and close cellular apposition. At specialized points of communication, such as the synapse and the neuromuscular junction, the cells are separated by a nanometer or less. For the purposes herein, synapse also is used to describe the neuromuscular junction. The neuron uses electrical signals, such as the action potential, to transmit from one end of the cell to the other, from the cell body to the synapse, whether the distance is relatively short as for interneurons, or relatively long as for the motor neuron which carries action potentials from the spinal cord to skeletal muscle. Transmission between cells, however, is mediated chemically in the majority of instances. Classically, these signaling molecules are known as neurotransmitters or neuromodulators, the latter term reflecting agents having longer term effects than those of the former. The neuron synthesizes and stores neurotransmitters in specialized areas of the cell, known as nerve terminals when they occur at the end of an axon, or varicosities when they occur as swellings arranged serially along the axon. Electrical impulses arriving in these areas depolarize the plasma membrane, causing voltage-dependent Ca^{2+} channels to open. This leads to a rapid influx of Ca^{2+} that triggers neurotransmitter release into the synapase. The neurotransmitter diffuses across the synapse, acting upon discrete recognition sites known as receptors in the plasma membrane of the post-synaptic cell. The neurotransmitter–receptor interaction then results in the depolarization or hyperpolarization of the post-synaptic membrane. The transmitter is removed from the synaptic milieu by metabolism, re-uptake, and/or diffusion, thus terminating its action as an intercellular messenger (Fig. 1). The integration of many such signals at the post-synaptic cell determines the response, for example, elicitation of an action potential or muscle contraction. The entire process occurs within milliseconds.

The concept of discrete neurotransmitter recognition sites or receptors on nerve cells was based on work on systems physiology and drug action (1). It was not until 1921 however, that it was shown that information could be transferred between neurons via a chemical, in this instance acetylcholine [51-84-3] (ACh), $C_7H_{16}NO_2$ (1).

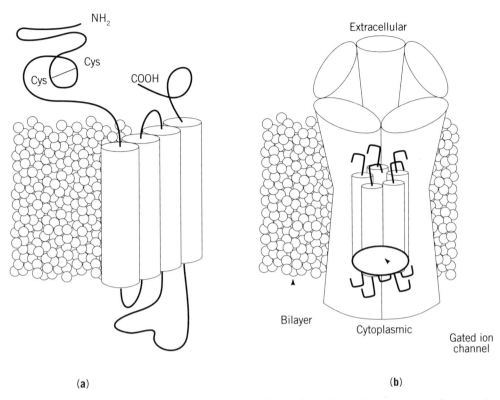

(a)　　　　　　　　　　　　　　　　　　　　　(b)

Fig. 1.　Model of a ligand gated ion channel (LGIC) where (**a**) is the structure of a generic LGIC subunit showing the two cysteine (Cys) residues common to all LGIC subunits, and (**b**) shows the arrangement of five such subunits as a pentamer having psuedo-cyclic symmetry delineating a gated, fluid-filled pathway for ions.

$$H_3C \overset{+}{\underset{H_3C}{\overset{H_3C}{N}}} CH_2 \, CH_2 \, O \, \overset{O}{\underset{||}{C}} CH_3$$

(1)

The fundamental concept of synaptic transmission has, to a large extent, been based on cholinergic transmission in skeletal muscle, heart, and the autonomic nervous system. This classical form of transmission clearly is important in the central as well the peripheral nervous system. However, numerous substances can act as important modulators and regulators, filtering, adjusting the gain, and even triggering agents that may lead to modification of the neuronal circuiting.

Classical rapidly acting neurotransmitter receptors are represented by ligand–gated ion channels (LGICs), which are multimeric protein complexes surrounding an ion channel (Fig. 1) that typically mediate fast responses. The neurotransmitter, eg, ACh, γ-aminobutyric acid (GABA), glutamate, and adenosine triphosphate (ATP), activate the receptor, probably through a conformational change, such that the ion channel is opened to specific ion permeation. LGICs can have a multiplicity of ligand recognition sites allowing modulation of the action of the primary neurotransmitter. For example, benzodiazepines like diazepam are positive modulators of the effects of GABA on the chloride channel associated with the $GABA_A$ receptor (2).

Another receptor class is the G-protein coupled receptors (GPCRs), which constitute the majority of receptors characterized to date. These have seven transmembrane helical structures connected by extracellular and intracellular loops (Fig. 2). The G-protein subunits act as transducer systems to couple receptor activation to membranous and intracellular actions. The latter usually involves a hierarchy of enzymes including adenylate cyclase, the various enzymes involved in phospholipid metabolism, and various protein kinases and phosphatases. Activation of GPCRs also can lead to early response gene induction. Thus ligands that activate GPCRs have the potential to mediate long-term modulatory and regulatory functions as well as synaptic neurotransmission.

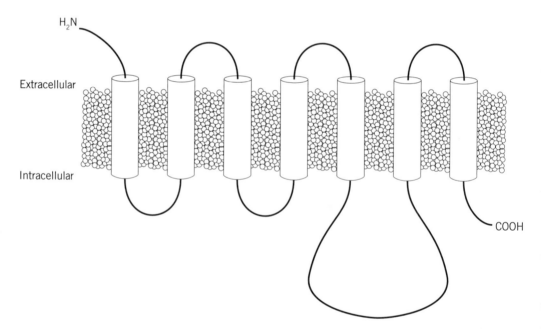

Fig. 2. Schematic of the G-protein coupled receptor (GPCR). The seven α-helical hydrophobic regions spanning the membrane are joined by extracellular and intracellular loops. The amino terminal is located extracellularly and the carboxy terminal intracellularly.

Neurotransmitter Criteria

A number of general characteristics can be defined that permit the definition of criteria for classifying an endogenous agent as a neurotransmitter (3,4).

The neurotransmitter must be present in presynaptic nerve terminals; and the precursors and enzymes necessary for its synthesis must be present in the neuron. For example, ACh is stored in vesicles specifically in cholinergic nerve terminals. It is synthesized from choline and acetyl-coenzyme A (acetyl-CoA) by the enzyme, choline acetyltransferase. Choline is taken up by a high affinity transporter specific to cholinergic nerve terminals. Choline uptake appears to be the rate-limiting step in ACh synthesis, and is regulated to keep pace with demands for the neurotransmitter. Dopamine [51-61-6] (2) is synthesized from tyrosine by tyrosine hydroxylase, which converts tyrosine to L-dopa (3,4-dihydroxy-L-phenylalanine) (3), and dopa decarboxylase, which converts L-dopa to dopamine.

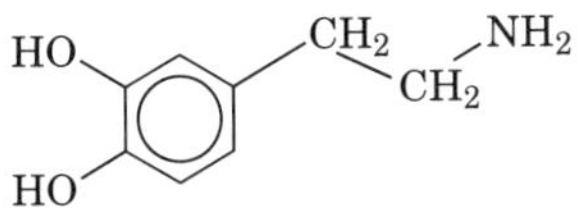

(2) 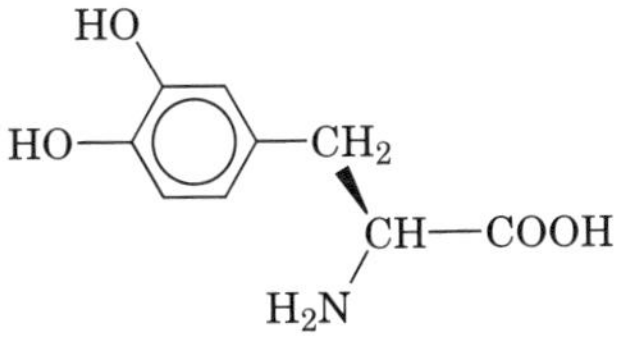   (3)

The dopamine is then concentrated in storage vesicles via an ATP-dependent process. Here the rate-limiting step appears not to be precursor uptake, under normal conditions, but tyrosine hydroxylase activity. This is regulated by protein phosphorylation and by *de novo* enzyme synthesis. The enzyme requires oxygen, ferrous iron, and tetrahydrobiopterin (BH_4). The enzymatic conversion of the precursor to the active agent and its subsequent storage in a vesicle are energy-dependent processes.

Stimulation of the neuron leading to electrical activation of the nerve terminal in a physiologically relevant manner should elicit a calcium-dependent release of the neurotransmitter. Although release is dependent on extracellular calcium, intracellular calcium homeostasis may also modulate the process. Neurotransmitter release that is independent of extracellular calcium is usually artifactual, or in some cases may represent release from a non-neuronal sources such as glia (3).

Neurotransmitter activation of a discrete receptor on the post-synaptic membrane should alter post-synaptic events in a concentration- or dose-dependent manner. Many neurotransmitter receptors are known to exist in a number of subtypes. For example, there are at least five subtypes of dopamine receptor, defined by molecular biological, pharmacological, and functional criteria. Pharmacological distinction depends on the discovery of potent ligands, synthetic or from natural sources, that are selective for specific subtypes. The processes of receptor characterization and ligand identification and characterization are frequently parallel events. The observation that new chemical entities elicit unusual responses in tissue and whole animal preparations can lead to the identification of new receptors or receptor subtypes. Identification of multiple receptor subtypes using recombinant DNA techniques provides a basis for interpretation of pharmacological results (see BIOTECHNOLOGY; GENETIC ENGINEERING).

Agonists are receptor ligands that act like the endogenous neurotransmitter and include the neurotransmitter itself as well as analogues and mimics that may have greater or lesser potency, efficacy, and/or receptor selectivity. A full agonist is one that produces a functional response identical to that seen with the endogenous transmitter or neuroeffector. Partial agonists are substances that mimic receptor activation but do not have full efficacy, eg, are not capable of producing the maximal response observed with a full agonist. Antagonists are receptor ligands that block the actions of agonists acting on the same receptor. By definition antagonists have zero efficacy and elicit no effect on their own (1). Although some antagonists are competitive in nature, interacting at the same site to which the agonist ligand binds to produce its effects, other antagonists are noncompetitive or uncompetitive in nature reflecting their ability to antagonize agonist actions by binding to sites distinct from the agonist binding site. Because full agonists have full efficacy, usually referred to as intrinsic activity of unity, and antagonists have zero efficacy, the range for efficacy for a partial agonist can extend from 0 to 1. Thus a partial agonist having an efficacy of 0.5 not only has half the efficacy of a full agonist, but also has intrinsic antagonist activity. Some receptors are subject to more complex ligand-mediated regulatory processes, such as the GABA–benzodiazepine LGIC receptor complex, which has numerous allosteric modulator sites (2). Studies of these receptors have resulted in identification of a more diversified group of ligands. These include the inverse agonists which are ligands that elicit an effect opposite to that elicited by agonists and consequently have negative efficacy, eg, -1.0. Inverse agonist effects at serotonin receptors have been described (5).

The neurotransmitter receptor should respond to pharmacological manipulation in a predictable manner. Direct application of the neurotransmitter should produce electrophysiological and/or biochemical responses identical to responses observed following presynaptic neuron stimulation. Neurotransmitter receptor antagonists should block the action of the neurotransmitter and related agonists. Agonists known to be selective for other receptors should not act as neurotransmitter mimics when applied at relevant concentrations. Antagonists known to be selective for other receptors likewise should not block the neurotransmitter or related agonists when present at relevant concentrations.

There should be specific, saturable binding to the receptor, accompanied by pharmacological characteristics appropriate to the functional effects, demonstrable using a radioactive, eg, tritium or iodine-125, ligand to label the receptor. Radioligand binding assays (1,6) have become a significant means by which to identify and characterize receptors and enzymes (see IMMUNOASSAYS; RADIOACTIVE TRACERS). Isolation of the receptor or expression of the receptor in another cell, eg, an oocyte can be used to confirm the existence of a discrete entity.

Specific systems, such as enzymatic inactivation or neurotransmitter reuptake, should exist to terminate the physiological actions of the neurotransmitter. Synaptic neurotransmission is a rapid and highly localized process, requiring that the effect of released neurotransmitter be terminated rapidly and prevented from affecting a widespread area. Thus diffusion of a typical chemical neurotransmitter is insufficient to terminate its actions. In the case of ACh, the neurotransmitter is rapidly hydrolyzed by acetylcholinesterase to the relatively inactive products, choline and acetate. About 50% of this choline is recaptured by the nerve terminal for resynthesis of acetylcholine. The effects of many neurotransmitters are terminated principally by active high affinity transporter systems that help to recapture a portion of the transmitter for subsequent release. Uptake processes in glia and catabolism of the neurotransmitter in neurons and glia also help to terminate and localize the neurotransmitter action. In Parkinson's disease, a neurodegenerative disorder resulting from a loss of certain dopaminergic neurons and presynaptic dopamine stores, the precursor, L-dopa can be used to increase brain levels of dopamine as long as presynaptic mechanisms are available to convert L-dopa to dopamine. Alternatively, inhibitors of catabolism or transport can augment dopamine availability or its synaptic concentration, provided some degree of dopaminergic neuronal function remains intact.

Numerous processes modulate neurotransmitter synthesis, presynaptic excitability, neurotransmitter release, post-synaptic receptors, and post-synaptic excitability.

The distinction between neuromodulator and neurotransmitter is sometimes difficult to define (7,8). Generally, however, the primary function of a neurotransmitter is thought to be the rapid transmission of impulses eliciting a brief depolarization or hyperpolarization of the post-synaptic cell. Summation of neurotransmitter signals lead either to electrical excitation or electrical blockade of excitation. Neuromodulators regulate these processes, and may or may not produce an electrical signal themselves. The same molecule may function as neurotransmitter or neuromodulator, depending on the receptors present. For example, ACh acts as a neurotransmitter via nicotinic receptors, although it can act as a neuromodulator via muscarinic receptors that regulate its release and post-synaptic excitability. Some substances such as steroids (qv) may act as neuromodulators, or may produce longer term effects through hormonal regulation.

Some of the neurotransmitter criteria may or may not apply to a particular neuromodulator. The source of the neuromodulator is not restricted to the nerve terminal. The neuromodulator may be stored in a form other than its released form; it may be released extrasynaptically, and it may or may not be metabolized or taken up particularly rapidly to terminate its action. However, neuromodulators are subject to agonist/antagonist pharmacological criteria. GPCRs are involved in many neuromodulatory functions. Other neuromodulators may act through receptive sites that are not typically considered receptors. For example, nitric oxide [10102-43-9], NO, a neuromodulator implicated in learning and memory (long-term potentiation) and cytotoxic phenomena, acts directly on the enzyme guanylate cyclase to increase activity (9,10).

Drug receptors represent another type of receptor family. The central nervous system (CNS) effects of the anxiolytic, diazepam, and the psychotropic actions of the cannabinoids and phencyclidine have resulted in the identification of specific receptors for these molecules. This has resulted in the search for an endogenous ligand for these receptors. Thus, in these situations, the pharmacological action has preceded the discovery of the receptor which, in turn, has provided clues in several instances to the endogenous ligand.

Another category of neuroregulator includes substances that do not directly regulate depolarization or hyperpolarization and may or may not be released from neurons, but do affect cellular function in neurons. These include trophic factors, growth factors, cytokines, and hormones (qv) that regulate neuronal morphology, synapse formation, or function through longer term effects on neurotransmitter synthesis, receptor expression, or ion channel expression.

Receptors for the thyroid hormone–retinoic acid (RA) superfamily, including steroid receptors, act through cell surface recognition sites (10). Ligands for this class of receptor are internalized and then interact with promoter regions on DNA, known as hormone responsive elements (HREs), to affect transcription processes in the target cell. This class of receptor is characterized by the presence of a zinc finger region. A class of intracellular drug targets that has yet to be exploited is represented by the G-protein transduction elements involved in coupling receptor events to activation of adenylate and guanylate cyclase as well as other enzyme systems and ion channels within the cell (11).

Another class of receptor that has been described is typified by receptors for cytokines such as tumor necrosis factor-α (TNF-α). Activation of this class of receptor leads to changes in intracellular target proteins including the so called receptor-activated factors of transcription (RAFTs), also known as signal transducers and activators of transcription (STATs), that include AP-1 and NFκB. These STATs can modulate the DNA transcription processes thought to be involved in cellular inflammatory responses and mechanisms related to apoptosis in the peripheral and central nervous systems. Cytokines have profound effects on target cell function including neurons and glia and may be involved in many disease states involving an inflammatory mediator component, eg, AIDS and Alzheimer's disease.

A newer receptor class, still being characterized, is that known as orphan receptors. These are receptors identified by cloning homology to GPCR and other receptors using a process known as receptor trolling. Whereas these receptors have the characteristic seven transmembrane (7TM) helical regions associated with the GPCR superfamily, they do not show sequence homology with other known receptors of this class. These are receptors in search of a ligand. The intracellular chicken ovalbumin upstream promotor transcription factor (COUP-TF) receptor is an orphan receptor (12), activation of which requires dopamine-dependent phosphorylation. An endogenous ligand has yet to be identified. Another orphan receptor for which the ligand has been isolated is thrombopoetin (13).

A subset of ion channels not gated by traditional neurotransmitters represents another receptor class. These include potassium, calcium, sodium, and cyclic adenosine monophosphate (cAMP)-gated channels (14–16) for which a large number of synthetic molecules exist that alter cellular function.

Despite quantal advances in technology in the twentieth century, the receptor–neurotransmitter concept remains the foundation of neuroregulator action and disease pathophysiology understanding. The drug discovery process typically has focused on neurotransmitter receptors, catabolic enzymes for neurotransmitters, and neurotransmitter transporters. Alterations in the process of chemical neurotransmission, either an over- or under-stimulation of the target receptor, occurring as a result of changes in the availability of the neurotransmitter or an alteration in the responsiveness of the signal transduction processes in the target cell, are thought to represent the molecular defects involved in many disease states. The physiology and pharmacology of such systems can be defined according to clear criteria, and often there is a pharmacological specificity that may be missing from more ubiquitous regulatory systems. Neurotransmission can be attenuated by post-synaptic receptor antagonists, including desensitizing agonists, or by activation of feedback inhibition processes. Augmentation of neurotransmitter–neuroregulator function can be effected by limiting feedback inhibition or the inactivation processes, catabolism or re-uptake. However, exogenous neurotransmitter agonists are unlikely to mimic the millisecond and micrometer resolution of the physiological system; this can be explained by considerations related to chaos theory (17). There have been many advances made in drug therapeutics and progress continues to be made. As of this writing (ca 1995), however, most approaches are more or less palliative. Transgenic animals that overexpress the β_2-adrenoceptor have been developed that reflect the potential for receptor replacement therapy (18). These animals show enhanced atrial contractility and increased left ventricular function even in the presence of β_2-adrenoceptor antagonists.

An increased focus on those processes that regulate neurotransmission and cellular functions both of neurons and of the other half of the brain, the glial cell family, is expected.

Neurotransmitters and Receptors

In general the receptor nomenclature used is consistent with the recommendations of the various International Union of Pharmacology (IUPHAR) Committee on receptor nomenclature (19,20). In some cases the human receptor has been cloned. By convention, pharmacologically defined receptors are shown in capital letters; cloned receptors in lower case letters.

The grouping of receptors on the basis of sequence homology has been easiest in the case of those receptor classes where pharmacological tools have enabled the elucidation of function consistent with structure.

Acetylcholine. Acetylcholine (ACh) (**1**) is a crystalline material that is very soluble in water and alcohol. ACh, synthesized by the enzyme choline acetyltransferase (3), interacts with two main classes of receptor in mammals: muscarinic (mAChR), defined on the basis of the agonist activity of the alkaloid muscarine (**4**), and nicotinic (nAChR), based on the agonist activity of nicotine (**5**) (Table 1). mAChRs are GPCRs (21); nAChRs are LGICs (22).

(**4**) (**5**)

mAChRs are divided into five subclasses, $M_1–M_5$. Carbachol (**6**) and oxotremorine-M (**7**) are prototypic agonists for all five classes. McN-A343 (**8**)

Table 1. Agonists and Antagonists of Acetylcholine and Receptors

Agonist/antagonist	CAS Registry Number	Molecular formula	Structure number
Muscarinic receptors			
muscarine	[300-54-9]	$C_9H_{20}NO_2$	(4)
carbachol	[51-83-2]	$C_6H_{15}ClN_2O_2$	(6)
oxotremorine-M	[63939-65-1]	$C_{11}H_{19}N_2O$	(7)
McN-A343		$C_{14}H_{18}ClN_2O_2$	(8)
pirenzepine	[28797-61-7]	$C_{19}H_{21}N_5O_2$	(9)
methoctramine	[104807-46-7]	$C_{36}M_{62}N_4O_2$	(10)
himbacine	[6879-74-9]	$C_{22}H_{35}NO_2$	(11)
hexahydrosiladifendiol	[98299-40-2]	$C_{20}H_{33}NOSi$	(12)
tropicamide	[1508-75-4]	$C_{17}H_{20}N_2O_2$	(13)
4-diphenylacetoxy-*N*-methylpiperidium	[106458-69-9]	$C_{21}H_{26}NO_2$	(14)
Nicotinic receptors			
nicotine	[54-11-5]	$C_{10}H_{14}N_2$	(5)
phenyltrimethyl-ammonium	[3426-74-2]	$C_9H_{14}N$	(15)
1,1-dimethyl-4-phenyl-piperazinium	[54-77-3]	$C_{12}H_{19}N_2$	(18)
cytisine	[485-35-8]	$C_{11}H_{14}N_2O$	(21)
ABT-418	[147402-53-7]	$C_9H_{14}N_2O$	(22)
methylcarbamylcholine	[14721-69-8]	$C_7H_{17}N_2O_2$	(23)
GTS 21	[148372-04-7]	$C_{19}H_{20}N_2O_2$	(24)
(±)-epibatidine	[140111-52-0]	$C_{11}H_{13}ClN_2$	(25)
d-tubocurarine	[57-95-4]	$C_{37}H_{41}N_2O_6$	(16)
α-bungarotoxin	[11032-79-4]	$C_{338}H_{529}N_{97}O_{105}S_{11}$	(17)
trimethaphan	[7187-66-8]	$C_{22}H_{25}N_2OS$	(19)
hexamethonium	[60-26-4]	$C_{12}H_{30}N_2$	(20)
dihydro-β-erythroidine	[23255-54-1]	$C_{16}H_2NO_3$	(26)
mecamylamine	[60-40-2]	$C_{11}H_{21}N$	(27)
methyllycaconitine	[21019-30-7]	$C_{37}H_{50}N_2O_{10}$	(28)

is a selective agonist for the M_1 receptor; pirenzepine (**9**) is a selective antagonist for the M_1 receptor. Methoctramine (**10**) and himbacine (**11**) are selective M_2 receptor antagonists. Hexahydrosiladifendiol (**12**) and tropicamide (**13**) are selective antagonists for the M_3 and M_4 mAChR, respectively. 4-Diphenylacetoxy-*N*-methylpiperidium (4-DAMP) (**14**) is an antagonist for the M_5 receptor (Table 1).

(**6**)   (**7**)

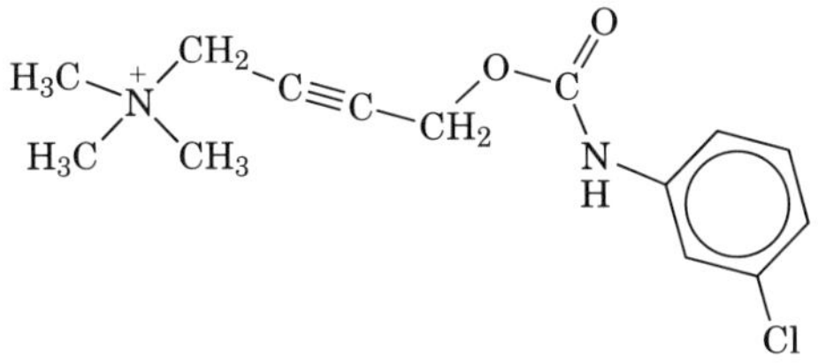

(8)

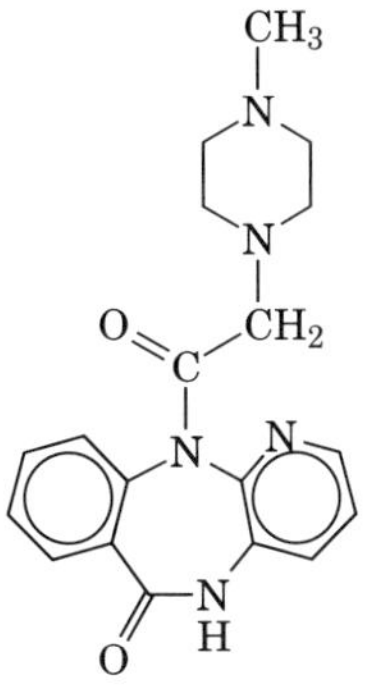

(9)

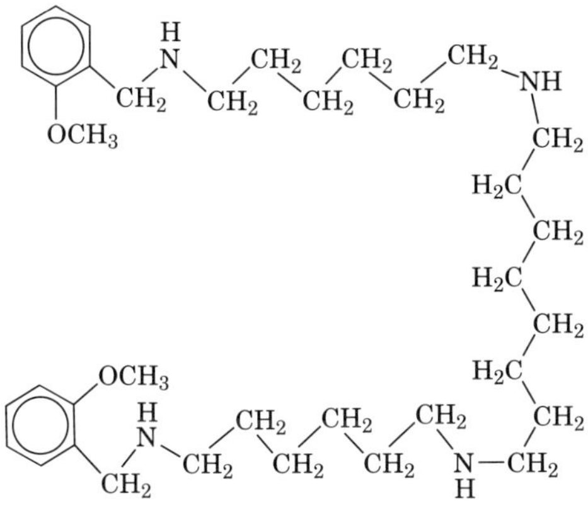

(10)

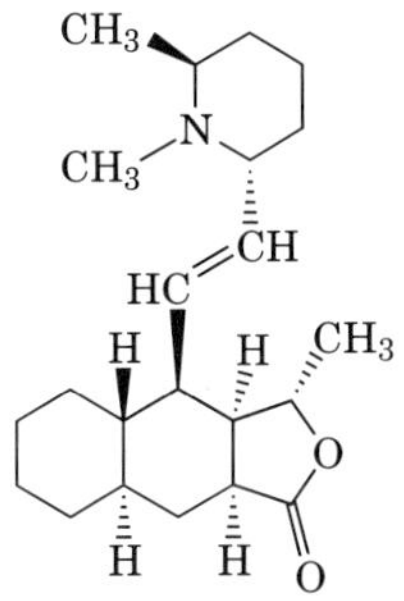

(11)

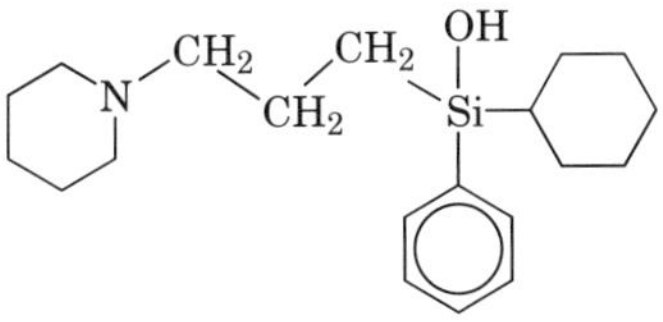

(12)

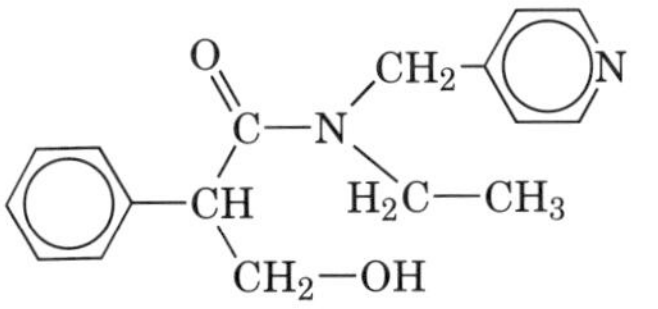

(13)

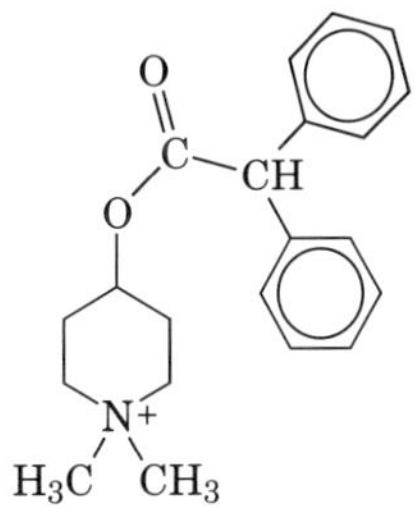

(14)

Selective M_1 receptor ligands have been targeted for therapeutic use in Alzheimer's disease and related dementias. There has been little success because of side effect liabilities (see MEMORY-ENHANCING DRUGS). M_2 antagonists may act as autoreceptor antagonists also promoting the release of ACh. Muscarinic receptor ligands have also been targeted for use in cardiovascular, gastrointestinal, and pulmonary disorders.

nAChRs are pentameric LGICs composed of α, β, γ, δ, and ϵ subunits, the composition of which is dependent on the tissue of origin. Eleven neuronal gene products ($\alpha 2-\alpha 9$; $\beta 2-\beta 4$) have been identified from the cloning of rat, chick, and human brain and sensory tissue cDNAs. At least eight combinations form putative neuronal nAChR subtypes (23).

A systematic nomenclature for nAChRs has yet to evolve. An N nomenclature describes receptors present in muscle as N_1. These are activated by phenyltrimethylammonium (PTMA) (**15**) and blocked by d-tubocurarine (**16**) and α-bungarotoxin (α-BgT) (**17**). N_2 receptors are present in ganglia and are activated by 1,1-dimethyl-4-phenylpiperazinium (DMPP) (**18**) and blocked by trimethaphan (**19**) and bis-quaternary agents, with hexamethonium (**20**) being the most potent.

In the brain, three nAChR subclasses have been identified: those having high affinity ($K_d = 0.5-5$ nM) for ($-$)-nicotine; those having high affinity ($K_d \sim$

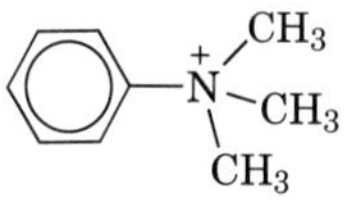

(15) 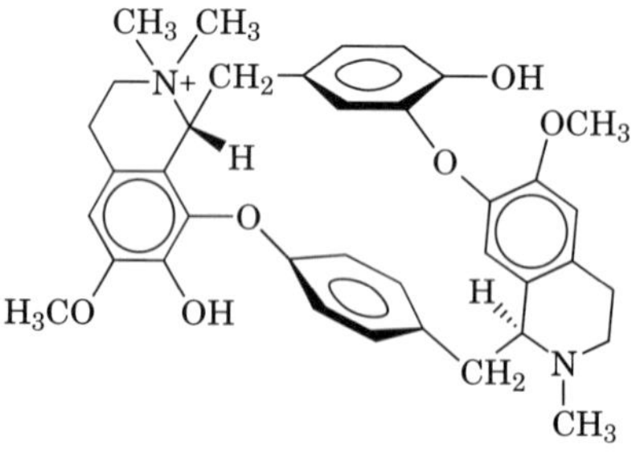   **(16)**

H-Ile-Val-Cys-His-Thr-Thr-Ala-Thr-Ile-Pro-Ser-Ser-Ala-Val-Thr-Cys-Pro-Pro-Gly-Glu-Asn-Leu-Cys-

Tyr-Arg-Lys-Met-Trp-Cys-Asp-Ala-Phe-Cys-Ser-Ser-Arg-Gly-Lys-Val-Val-Glu-Leu-Gly-Cys-Ala-Ala-

Thr-Cys-Pro-Ser-Lys-Lys-Pro-Tyr-Glu-Glu-Val-Thr-Cys-Cys-Ser-Thr-Asp-Lys-Cys-Asn-His-Pro-Pro-

Lys-Arg-Gln-Pro-Gly-OH

(17)

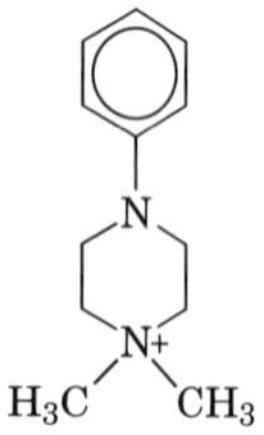

(18) 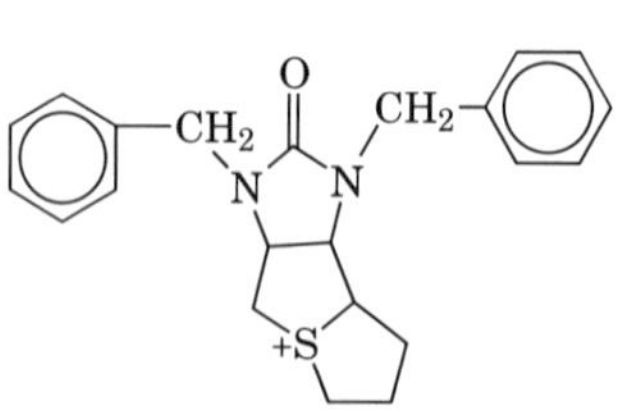   **(19)** 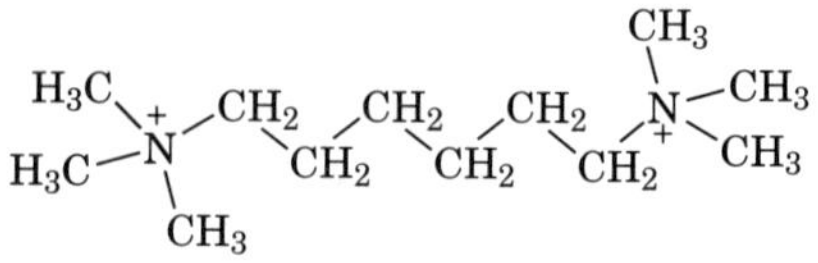   **(20)**

0.5 nM) for α-BgT (αBTnAChRs); and a population of receptors that display marked selectivity for neuronal bungarotoxin (n-BgT) (23). The distribution of the $\alpha 4 \beta 2$ subunit combination coincides somewhat with the distribution of high affinity [^{3}H]nicotine binding sites in rat brain. Agonists for the α-BTX insensitive receptors include nicotine (5), cytisine (21), ABT-418 (22), and methylcarbamylcholine (MCC) (23). The ability of the classical agonist, ($-$)-nicotine, to act as an agonist or an antagonist is dependent on the β-subunits present. GTS 21 (24) is a potent partial agonist at the α-BgT-sensitive α_7 subtype. The novel analgesic, ($+$)-epibatidine (25) is a potent ligand at high affinity [^{3}H]nicotine binding sites and the receptors labeled by n-BgT. Antagonists include dihydro-β-erythroidine (DHβE) (26), mecamylamine (27), hexamethonium (20), and methyllycaconitine (MLA) (28). MLA differentiates between αBgT sensitive sites on neuronal and muscle nAChRs.

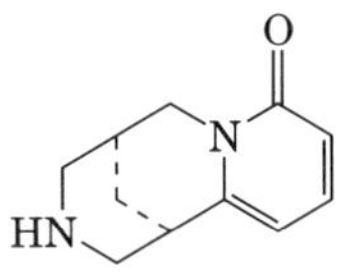

(21) 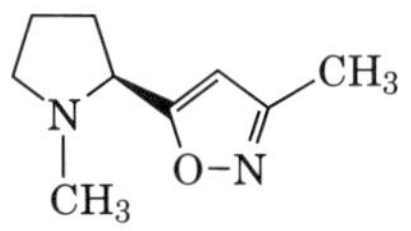   (22) 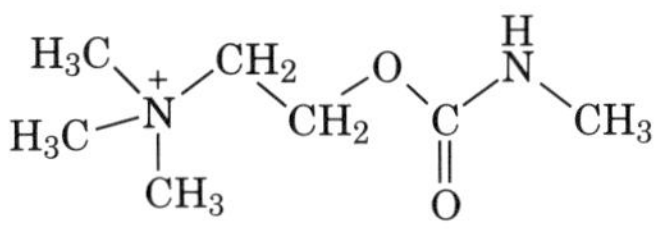   (23)

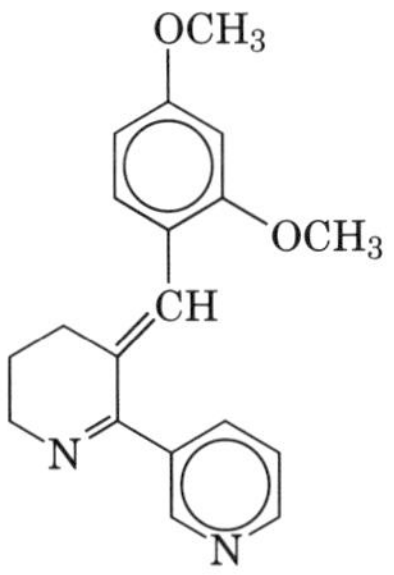

(24) 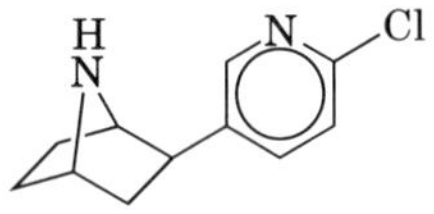   (25) 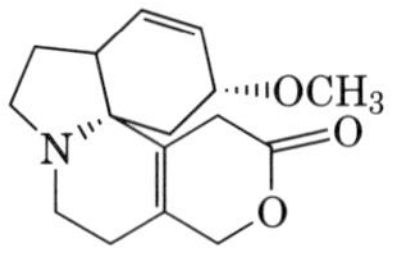   (26)

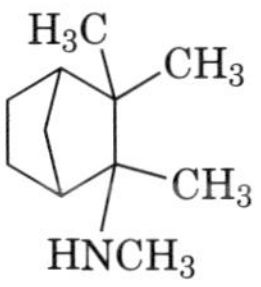

(27) 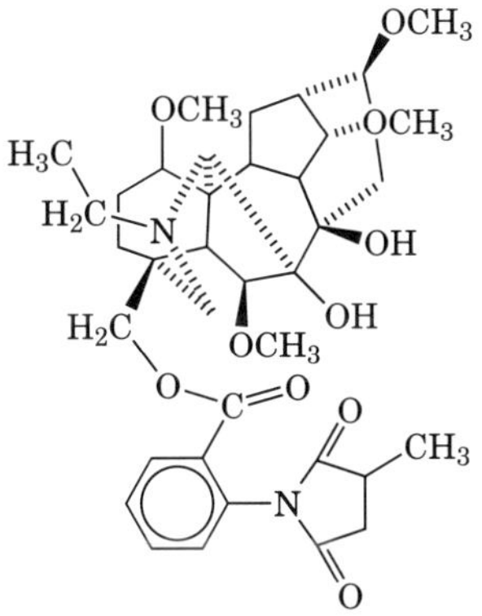   (28)

Like mAChRs, nAChRs have been implicated in the etiology of Alzheimer's disease and related dementias as well as in gastrointestinal and cardiovascular disorders. Nicotine has high abuse potential and is the primary component in reinforcing smoking behaviors. Nicotine patches have been developed for use as aids in smoking cessation, however, the usefulness of nicotine is hampered by dose-limiting side effects. Medicinal chemistry efforts in this emerging area are focused on the development of ligands, eg, ABT-418 and GTS-21, that are selective for nAChR subtypes and thus may have a reduced side effect profile as compared to nicotine. Such agents, collectively termed cholinergic channel modulators (ChCM), may be useful in the treatment of a number of neurological disorders, such as anxiety, schizophrenia, and analgesia.

Adenosine. Adenosine [58-61-7] (Ado), $C_{10}H_{13}N_5O_4$ (**29**), a purine nucleoside, is an intracellular constituent acting as both an enzyme cofactor and substrate and also as part of the basic energy cycle in the form of its phosphorylated derivative, ATP (**30**). Evidence for a role for adenosine as a neuromodulatory agent has accumulated for the better part of the twentieth century (24).

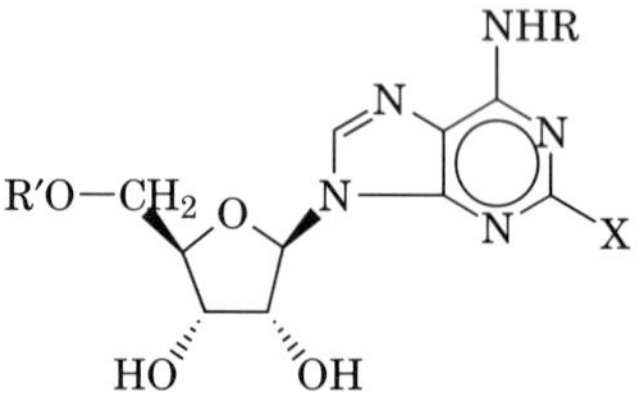

	R	X	R'
(**29**)	H	H	H
(**30**)	H	H	$(PO_3)_3^{4-}$
(**31**)	H	H	⬡
(**32**)	H	Cl	⬠

Adenosine is formed from ATP via a phosphatase cascade that sequentially involves the diphosphate, ADP, and the monophosphate, AMP. The actions of adenosine are terminated by uptake and rephosphorylation via adenosine kinase to AMP or by catabolism via adenosine deaminase to inosine and hypoxanthine.

Adenosine receptors are members of the P_1 purinoceptor GPCR family and can be classified into four subtypes: A_1, A_{2a}, A_{2b}, and A_3. At the A_1 receptor, N^6-substituted analogues of adenosine including N^6-cyclohexyladenosine (CHA) (**31**), and 2-chloro-N^6-cyclopentyladenosine (CCPA) (**32**) are potent and selective agonists. 5'-N-Ethylcarboxamidoadenosine (NECA) (**33**) is a potent agonist (K_i = 15 nM) at A_1, A_{2a}, and A_{2b} receptors. CGS 21680 (**34**) and DBXRM (**35**) are selective agonists for A_{2a} and A_3 receptors, respectively. Antagonists for the A_1 receptor include a large series of eight-substituted xanthines including cyclopentylxanthine (CPX) (**36**) and KFM 19 (**37**). The triazoloquinazoline, CGS 15943 (**38**) is a potent (K_i = 20 nM) nonselective, nonxanthine adenosine antagonist. The 8-styrylxanthine, KF 17837 (**39**) and the benzopyranopyrazolone, AMBP (**40**) are selective A_{2a} antagonists. The rat A_3 receptor, which is approximately 75% homologous to sheep and human A_3 receptors, is insensitive to xanthine blockade, whereas the sheep and human receptor can be blocked by acidic xanthines such as I-ABPOX (**41**) (Table 2).

Table 2. Agonists and Antagonists of Adenosine and Receptors

Agonist/antagonist	CAS Registry Number	Molecular formula	Structure number
N^6-cyclohexyladenosine	[36396-99-3]	$C_{16}H_{23}N_5O_4$	(31)
2-chloro-N^6-cyclopentyl-adenosine	[37739-05-2]	$C_{15}H_{20}ClN_5O_4$	(32)
5′-N-ethylcarboxamido-adenosine	[35920-39-9]	$C_{12}H_{16}N_6O_4$	(33)
CGS 21680	[120225-54-9]	$C_{23}H_{29}N_7O_6$	(34)
DBXRM		$C_{19}H_{29}N_5O_6$	(35)
cyclopentylxanthine	[102146-07-6]	$C_{16}H_{24}N_4O_2$	(36)
KFM 19	[133058-72-7]	$C_{16}H_{22}N_4O_3$	(37)
CGS 15943	[104615-18-1]	$C_{13}H_8ClN_5O$	(38)
KF 17837	[141807-96-7]	$C_{22}H_{28}N_4O_4$	(39)
AMBP		$C_{17}H_{13}N_3O_2$	(40)
I-ABPOX		$C_{23}H_{22}IN_5O_5$	(41)

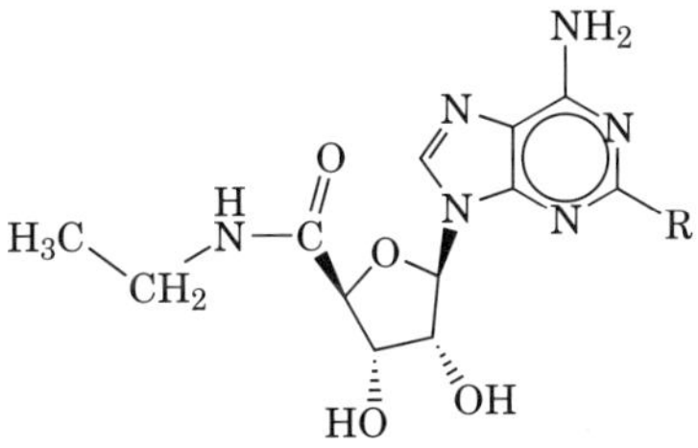

(33) R = H

(34) R = NH(CH$_2$)$_2$—⟨⟩—(CH$_2$)$_2$COOH

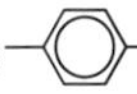

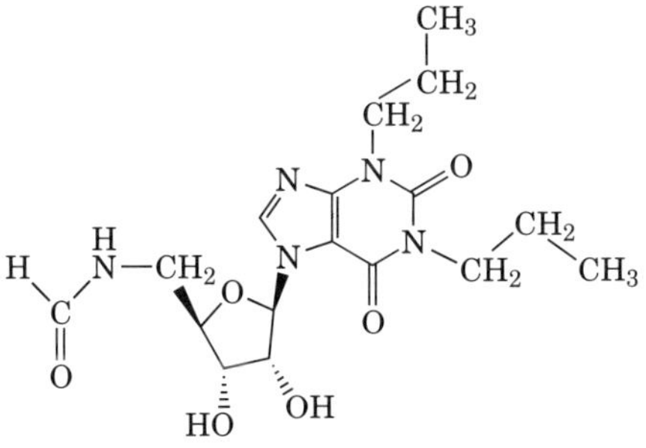

(35)

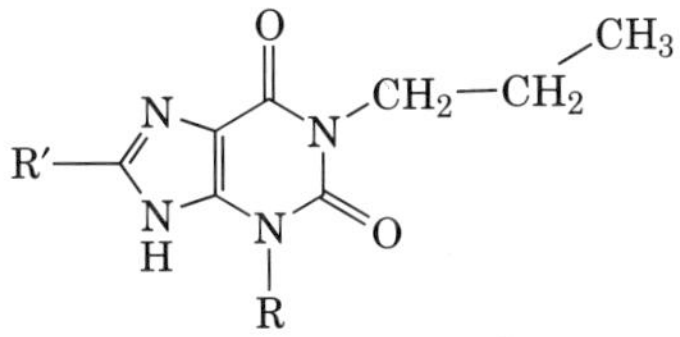

(36) R = C$_3$H$_7$, R′ = cyclopentyl

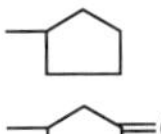

(37) R = C$_3$H$_7$, R′ = cyclopentanone

(41) R = CH$_2$—⟨⟩—NH$_2$, R = ⟨⟩—O—CH$_2$—COOH

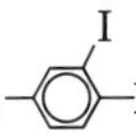

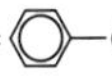

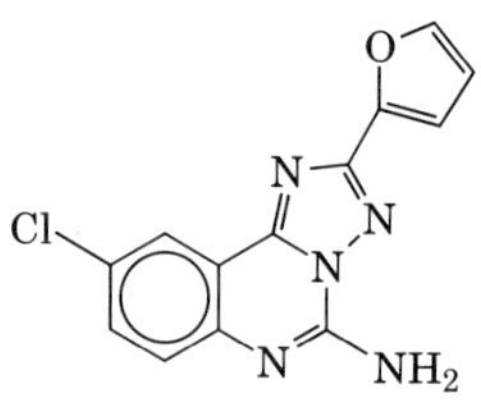

(38)

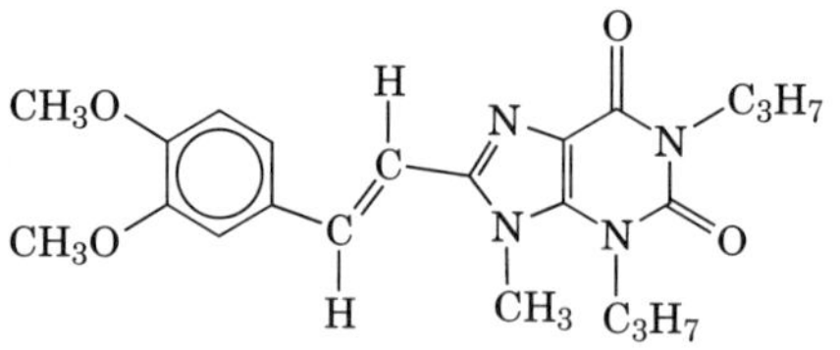

(39)

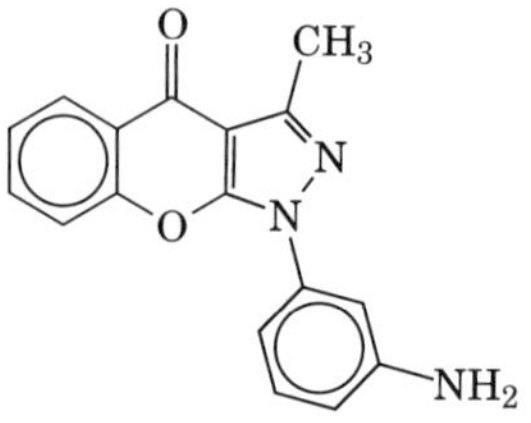

(40)

Adenosine is a ubiquitous neuromodulatory agent that is normally present in the extracellular milieu and functions to maintain tissue homeostasis. This is most evident in the daily intake of caffeine [58-08-2], a weak yet effective adenosine antagonist, that is the most widely used drug. Consumed in various beverages, caffeine acts to counteract the sedative actions of adenosine. A side effect of caffeine intake is the diuretic actions produced via the blockade of adenosine receptors in the kidney. More potent analogues of caffeine such as (37) are being evaluated for use as cognition enhancers and A_1 receptor agonists have potential use as antiischemic agents, both centrally and peripherally. Adenosine receptor ligands may have therapeutic potential in cardiovascular, pulmonary, renal, and immune system-associated disease states. The human A_3 receptor is implicated in the pathophysiology of asthma (see ANTIASTHMATIC AGENTS).

Adenosine Triphosphate. Adenosine triphosphate [56-65-5] (ATP), $C_{10}H_{16}N_5O_{13}P_3$ (30), like adenosine, is an important intracellular constituent. Its role as a neuromodulator has been firmly established (24). ATP is formed within the cell by a variety of energy-dependent phosphate exchanges. Co-released with ACh and catecholamines, its actions are terminated by the action of cell surface ectonucleotidases.

ATP receptors, initially classified as P_{2X} and P_{2Y} subtypes, have evolved into a number of subclasses including P_{2T}, P_{2Z}, P_{2U}/P_{2N}, and P_{2D}. Classification of P_2 purinoceptors has been limited by a lack of potent, selective, and bioavailable antagonists. Nonetheless a rational scheme for P_2 purinoceptor nomenclature divides P_2 receptors into two superfamilies: P_{2X}, an LGIC family having four subclasses; and P_{2Y}, a GPCR family having seven subclasses. A third receptor type, designated the P_{2Z}, is a nonselective ion pore.

α,β-Methylene–ATP (42) is an agonist at the P_{2X1}, P_{2X2}, P_{2X3}, and P_{2X4} receptors. The P_{2X3} receptor is located in vascular smooth muscle, and the P_{2X4} receptor is differentiated by its tissue location (25). 2-(4-Nitrophenylethylthio)ATP (**43**) is a selective P_{2X1} agonist. 5-Fluorouridine triphosphate (UTP) (**44**) is a selective P_{2X2} agonist. There are no selective agonists for P_{2X3} and P_{2X4} receptors. P_{2X} superfamily antagonists include suramin (**45**) and PPADS (**46**) (see Table 3).

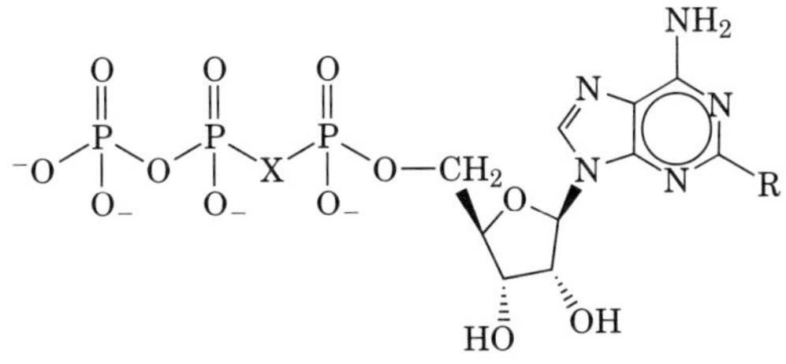

(**42**) R = H, X = CH_2

(**43**) R = CH_2—S—CH_2—⟨NO_2⟩ , X = O

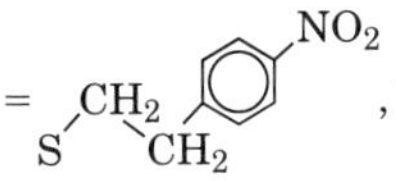

(**47**) R = SCH_3 , X = O

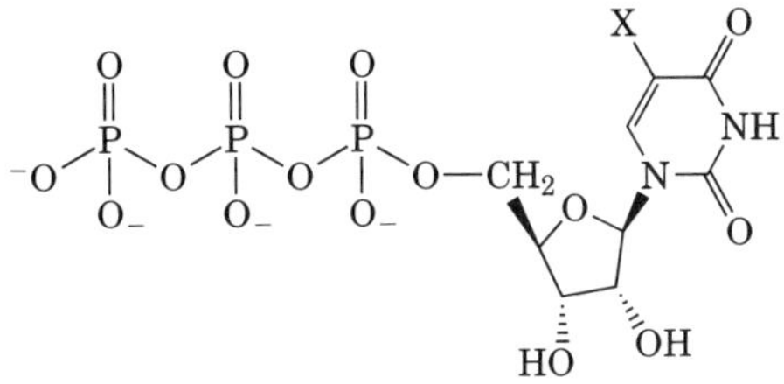

(**44**) X = F

(**48**) X = H

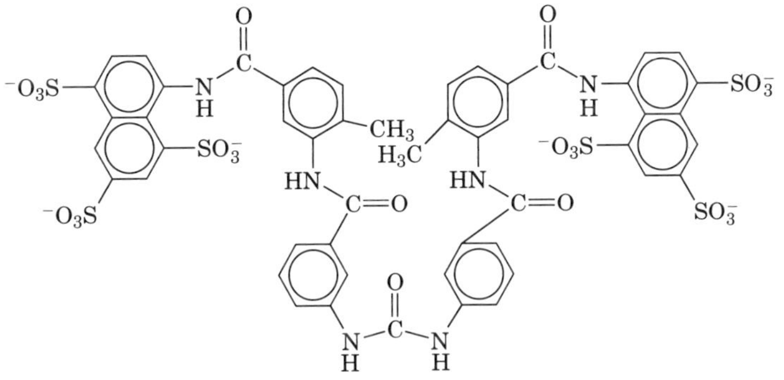

(**45**)

Table 3. Agonists and Antagonists of Adenosine Triphosphate and Receptors

Agonist/antagonist	CAS Registry Number	Molecular formula	Structure number
α,β-methylene–ATP	[7292-42-4]	$C_{11}H_{18}N_5O_{12}P_3$	(**42**)
2-(4-nitrophenylethylthio)–ATP		$C_{18}H_{19}N_6O_{15}P_3S$	(**43**)
5-fluorouridine triphosphate	[3828-96-4]	$C_9H_{14}FN_2O_{15}P_3$	(**44**)
2-methylthio–ATP	[43170-89-4]	$C_{11}H_{14}N_5O_{13}P_3S$	(**47**)
uridine triphosphate (UTP)	[63-39-8]	$C_9H_{15}N_2O_{15}P_3$	(**48**)
2-methylthio–ADP	[34983-48-7]	$C_{11}H_{14}N_5Na_3O_{10}P_2S$	(**49**)
Ap4A		$C_{20}H_{28}N_{10}O_{19}P_4$	(**50**)
suramin	[145-63-1]	$C_{51}H_{40}N_6O_{23}S_6$	(**45**)
PPADS		$C_{15}H_{11}N_2O_{12}PS_2$	(**46**)
FPL 67085		$C_{14}H_{18}Cl_2N_5Na_4O_{12}P_3S$	(**51**)

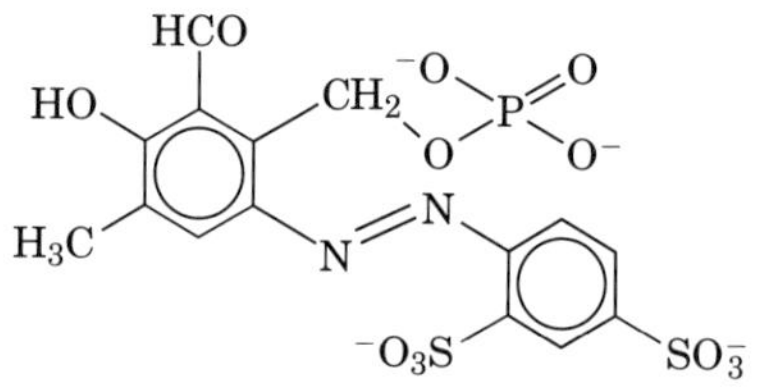

(**46**)

2-Methylthio-ATP (**47**) is an agonist at P_{2Y1}, P_{2Y4}, P_{2Y5}, and P_{2Y6} receptors. UTP (**48**) is equipotent with ATP at the P_{2Y2} receptor. 2-Methylthio-ADP (**49**) is a selective agonist for the platelet P_{2Y3} receptor. The differentiation between the P_{2Y4}, P_{2Y5}, and P_{2Y6} receptors is by agonist efficacy. (**47**) is much more effective than ATP at P_{2Y4}, and slightly more effective than ATP at P_{2Y5}. 2-(6-Cyanohexylthio)-ATP is an agonist at P_{2Y4}, but inactive at P_{2Y6}; 8-(6-aminohexyl)amino-ATP is an agonist at P_{2Y5} and P_{2Y6}. The P_{2Y7} receptor is sensitive to adenine dinucleotides, specifically Ap4A (**50**) and is involved in transmitter release regulation. ATP is a selective ligand for the P_{2Z} receptor. A UTP-sensitive receptor that does not respond to ATP and is thus distinct from the P_{2Y2} receptor has been described (26). Antagonists for the P_{2Y} family include (**45**) and (**46**). FPL 67085 (**51**) is a selective antagonist for the P_{2Y3} receptor.

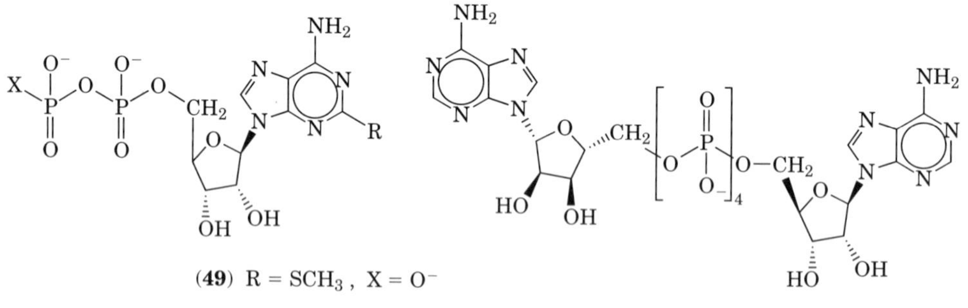

(**49**) R = SCH_3 , X = O^-

(**51**) R = SC_3H_7 , X = $CCl_2PO_3^{2-}$ (**50**)

Angiotensin. The octapeptide, angiotensin II [*4474-91-3, 11128-99-7*] (AT-II), H-Asp-Arg-Val-Tyr-Ile-His-Pro-Phe-OH, is the principal mediator of the renin−angiotensin system (RAS) that regulates hemodynamics and water and electrolyte balance (27). The aspartyl protease renin acts on the glycoprotein angiotensinogen to produce the decapeptide angiotensin I that is then cleaved by angiotensin-converting enzyme (ACE) to yield the active pressor agent, AT-II.

Two AT-II receptors, AT_1 and AT_2 are known and show wide distribution (27). The AT_1 receptor has been cloned and predominates in regions involved in the regulation of blood pressure and water and sodium retention, eg, the aorta, liver, adrenal cortex, and in the CNS in the paraventricular nucleus, area postrema, and nucleus of the solitary tract. AT_2 receptors are found primarily in the adrenal medulla, uterus, and in the brain in the locus coeruleus and the medial geniculate nucleus. AT_1 receptors are GCPRs inhibiting adenylate cyclase activity and stimulating phospholipases C, A_2, and D. AT_2 receptors use phosphotyrosine phosphatase as a transduction system.

Nonpeptide biphenyltetrazole AT_1 antagonists include losartan [*124750-99-8, 114798-26-4*] (DuP 753), $C_{22}H_{23}ClN_6O$ (**52**) and SKF 108566 [*133040-01-4*], $C_{23}H_{24}N_2O_4S$ (**53**). PD 123177 [*114785-12-5*], $C_{29}H_{28}N_4O_3$ (**54**) is a potent and selective nonpeptide AT_2 antagonist (Fig. 3).

AT-II exerts a wide range of physiological effects on the cardiovascular, renal, and endocrine systems, and the peripheral and central nervous systems. The main physiological effects of angiotensin are to increase blood pressure and heart rate and to cause retention of salt and water. The effects derive from a direct vasoconstrictor and myocardial effect of angiotensin, its ability to release other vasoconstrictors from endocrine and neuroendocrine tissues, and its ability to release sodium and water-retaining hormones, primarily aldosterone and vasopressin. In the brain, angiotensin II elicits a diversity of responses including an increase in blood pressure and water intake, stimulation of natriuresis and salt appetite, and secretion of pituitary hormones.

DuP 753 is an orally active AT_1 receptor antagonist and as of this writing is in clinical trials as an antihypertensive (see CARDIOVASCULAR AGENTS). AT-II

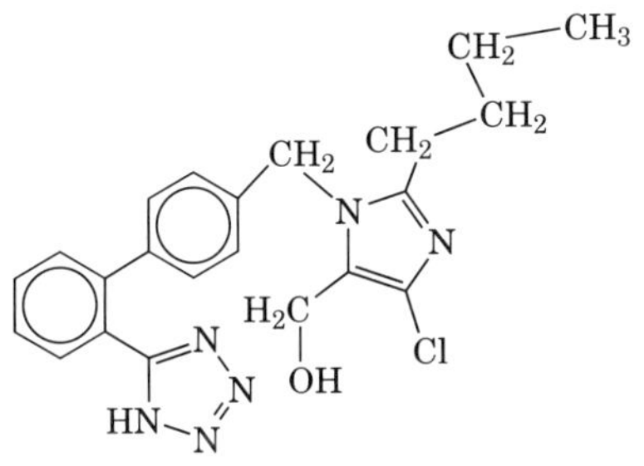

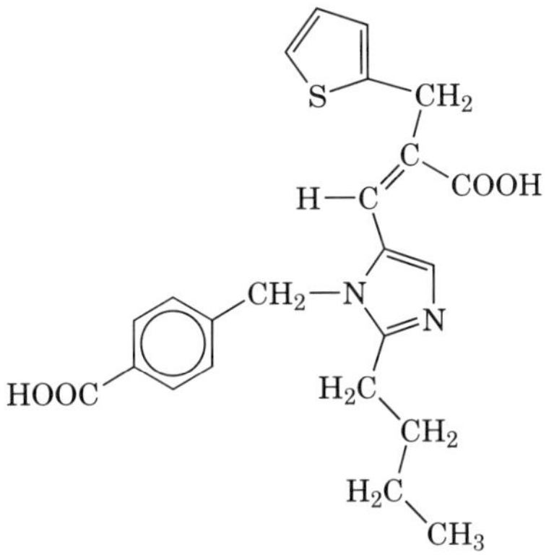

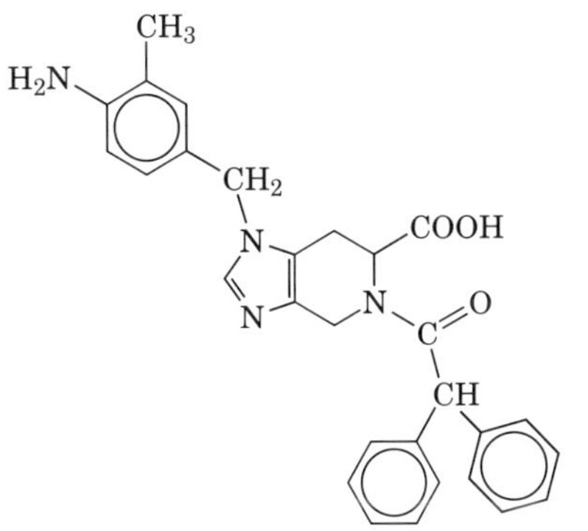

Fig. 3. Structures of angiotensin receptor antagonists.

antagonists affect the brain RAS system to enhance ACh release offering the possibility that these agents may function as cognition enhancers.

Atrial Natriuretic Peptide. α-Atrial natriuretic peptide [*85637-73-6*] (ANP) (**55**), also known as atrial natriuretic factor (ANF), brain natriuretic peptide (BNP) (**56**), and type C natriuretic peptide (CNP) (**57**) are members of the ANP family (28). These atrial peptides arise from a common 128 amino acid precursor where the active form of ANP is the 28 amino acid peptide at the C terminus.

H-Ser-Leu-Arg-Arg-Ser-Ser-Cys-Phe-Gly-Gly-Arg-Ile-Asp-Arg-Ile ⌐

HO-Tyr-Arg-Phe-Ser-Asn-Cys-Gly-Leu-Gly-Ser-Gln-Ala-Gly ⌐

(**55**)

H-Ser-Pro-Lys-Met-Val-Gln-Gly-Ser-Gly-Cys-Phe-Gly-Arg-Lys-Met-Asp-Arg-Ile ⌐

HO-His-Arg-Arg-Leu-Val-Lys-Cys-Gly-Leu-Gly-Ser-Ser-Ser-Ser ⌐

(**56**)

H-Asp-Leu-Arg-Val-Asp-Thr-Lys-Ser-Arg-Ala-Ala-Trp-Ala-Arg-Leu-Leu-Gln-Glu-
His-Pro-Asn-Ala-Arg-Lys-Tyr-Lys-Ile-Gly-Ala-Asn-Lys-Lys-Gly-Leu-Ser-Lys-Gly-
Cys-Phe-Gly-Leu-Lys-Leu-Asp-Arg-Gly-Ser-Met-Ser-Gly-Leu-Gly-Cys-OH

(**57**)

The ANP receptor exists in two forms, ANP_A and ANP_B, both of which have been cloned. These membrane-bound guanylate cyclases have a single transmembrane domain, an intracellular protein kinase-like domain, and a catalytic cyclase domain, activation of which results in the accumulation of cyclic guanosine monophosphate (cGMP). A third receptor subtype (ANP_C) has been identified that does not have intrinsic guanylate cyclase activity and may play a role in the clearance of ANP.

At low (1–10 n*M*) concentrations ANP activates ANP_A whereas ANP_B appears to be the physiological receptor for CNP. ANP and BNP are inactive at the latter subtype except at high micromolar concentrations. AP 811 [*124833-45-0*], $C_{46}H_{66}N_{12}O_8$ (**58**) is a selective ANP_C ligand. (L-α-Aminosuberic acid[7,23'])-b-ANP_{7-28} is an ANP_A antagonist.

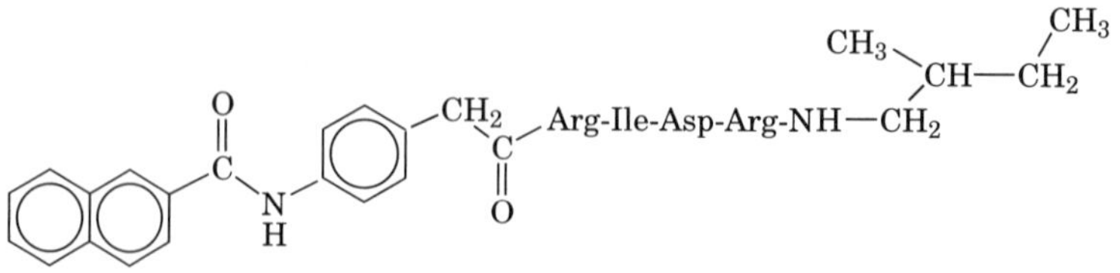

(**58**)

ANPs play an important role in the maintenance of cardiovascular homeostasis by counterbalancing the renin–angiotensin (RAS) system. ANP, the main circulating form of the natriuretic peptides, effectively relaxes vascular smooth muscle, promotes the excretion of sodium and water, and in the CNS inhibits vasopressin release and antagonizes AT-II induced thirst.

Benzodiazepines. The benzodiazepines (BZs) are a class of synthetic drugs widely used for the treatment of anxiety and epilepsy and also as muscle relaxants and sedative–hypnotics (see HYPNOTICS, SEDATIVES, ANTICONVULSANTS, AND ANXIOLYTICS). Diazepam (**59**) and chlordiazepoxide (**60**) (Table 4) were identified as CNS active agents in the early 1960s using empirical animal screening techniques. A unique BZ receptor was identified using classical radioligand binding techniques. A peripheral BZ receptor (PBR) has also been identified (29).

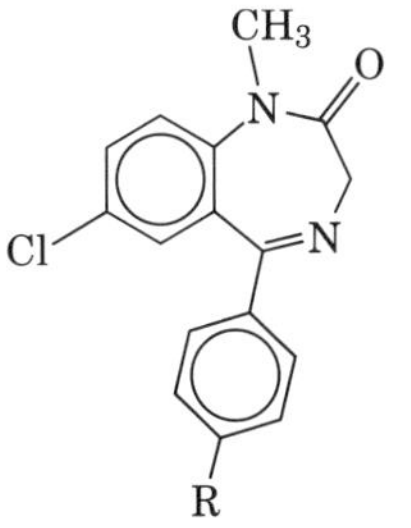

(**59**) R = H

(**62**) R = Cl

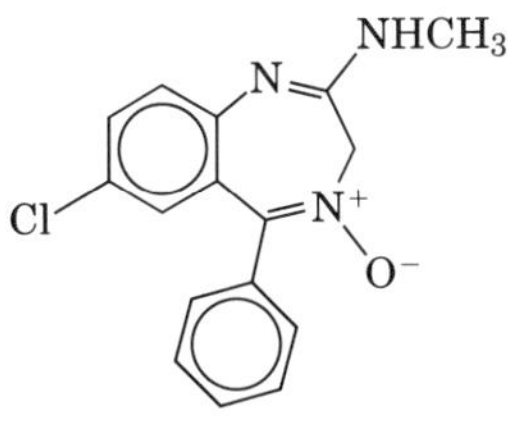

(**60**)

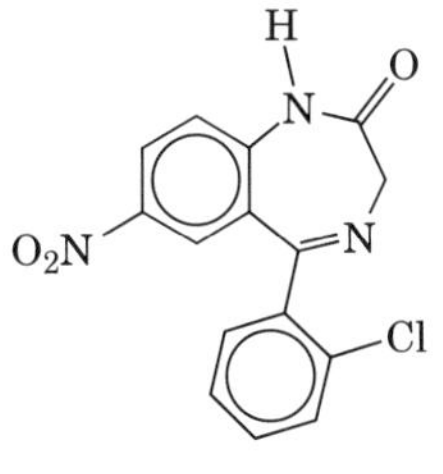

(**61**)

The endogenous ligand for the central BZ receptor has yet to be identified. Diazepam and clonazepam (**61**) are ligands for the central BZ receptor and Ro 5-4864 (**62**) and PK 11195 (**63**) are selective for the peripheral BZ receptor. Ro 15-1788 (flumazenil) (**64**) is a BZ inverse agonist used as an analeptic. The central

Table 4. Benzodiazepines and Related Compounds

Compound	CAS Registry Number	Molecular formula	Structure number
BZ receptor ligands			
diazepam	[*439-14-5*]	$C_{16}H_{13}ClN_2O$	(**59**)
chlordiazepoxide	[*58-25-3*]	$C_{16}H_{14}ClN_3O$	(**60**)
clonazepam	[*1622-61-3*]	$C_{15}H_{10}ClN_3O_3$	(**61**)
RO 5-4864	[*14439-61-3*]	$C_{16}H_{12}Cl_2N_2O$	(**62**)
PK 11195	[*85532-75-8*]	$C_{21}H_{21}ClN_2O$	(**63**)
flumazenil	[*78755-81-4*]	$C_{15}H_{14}FN_3O_3$	(**64**)
Peptide receptor ligands			
devazepide	[*103420-77-5*]	$C_{25}H_{20}N_4O_2$	(**65**)
midazolam	[*59467-70-8*]	$C_{18}H_{13}ClFN_3$	(**66**)
tifluadom	[*83386-35-0*]	$C_{22}H_{20}FN_3OS$	(**67**)
BZA-2B		$C_{24}H_{27}N_5O_5S_2$	(**68**)

BZ receptor is a pentameric LGIC associated with a chloride channel and a GABA$_A$ receptor, BZs acting to modulate the effects of GABA at the latter site. The pentameric structure can be drawn from six α-subunits ($\alpha 1-\alpha 6$), three β-subunits ($\beta 1-\beta 3$), three γ-subunits ($\gamma 1-\gamma 3$), and a δ-subunit. This provides the possibility of over 1000 permutations of the central BZ receptor having potentially different pharmacological and functional properties. However, the subunit stochiometry of endogenous BZ receptors is unknown. The central BZ receptor has several additional binding sites including those for barbiturates, picrotoxin, the avermectins, neurosteroids, etc. Compounds active at these sites modulate the effects of the BZs and GABA on chloride channel function.

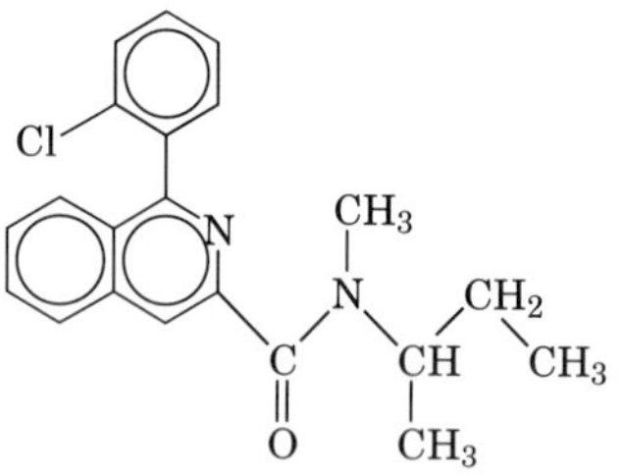

(63) 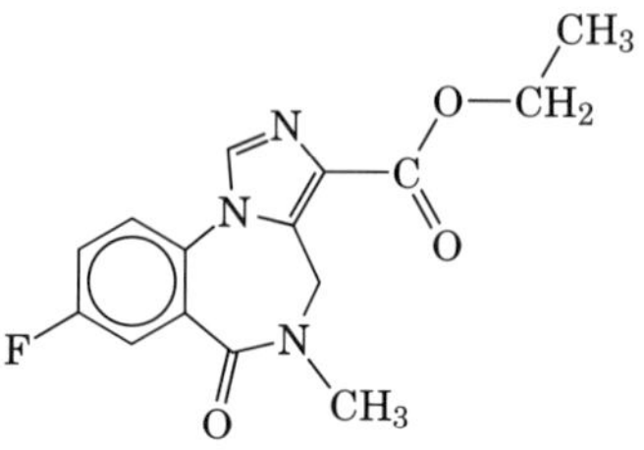   (64)

The PBR is distinct from the central BZ receptor although both can be present in the same tissues in differing ratios. PBRs are predominately localized on the outer mitochondrial membrane and are thus intracellular BZ recognition sites. The PBR is composed of three subunits: an 18,000 mol wt subunit that binds isoquinoline carboxamide derivatives; a 30,000 mol wt subunit that binds BZs; and a 32,000 mol wt voltage-dependent anion channel subunit. The porphyrins may be endogenous ligands for the PBR. PBRs are involved in the control of cell proliferation and differentiation and steroidogenesis.

The BZ structure also has provided a molecular scaffold for a number of peptide receptor ligands (26). Antagonists for the cholecystokinin (CCK-A) receptor, eg, devazepide (**65**), the thyrotropin-releasing hormone (TRH) receptor, eg, midazolam (**66**), and the κ-opiate receptor, eg, tifluadom (**67**), as well as a series of ras farnyl transferase inhibitors, eg, BZA-2B (**68**) (30) have been identified (Table 4).

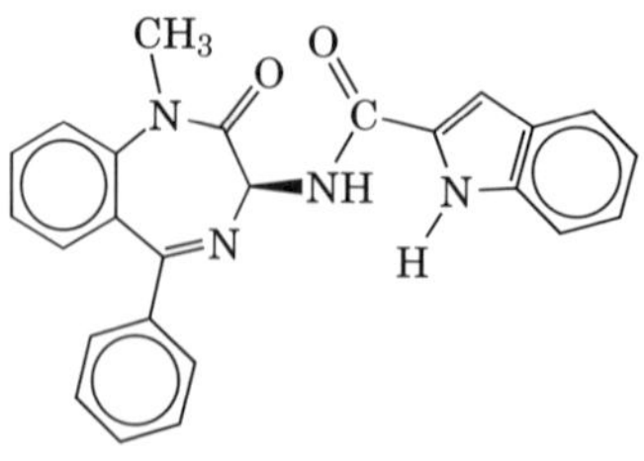

(65) 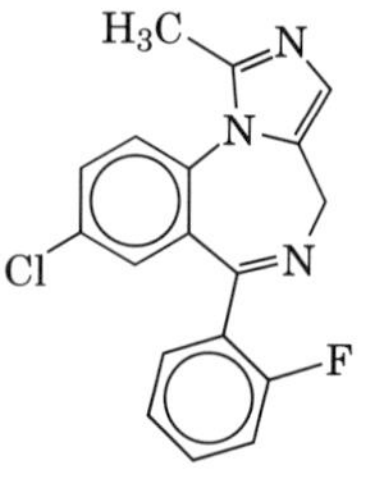   (66)

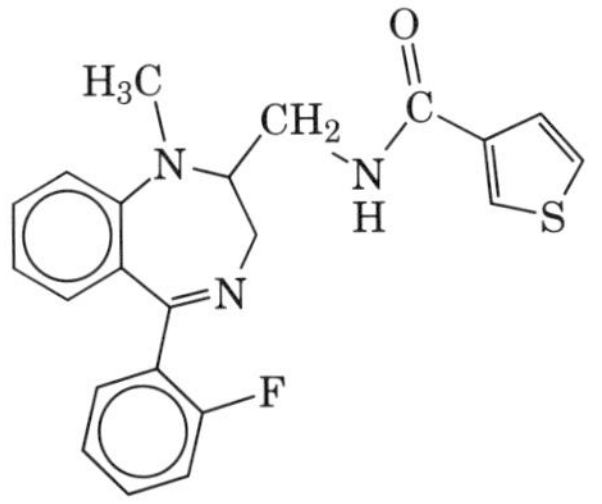

(67)

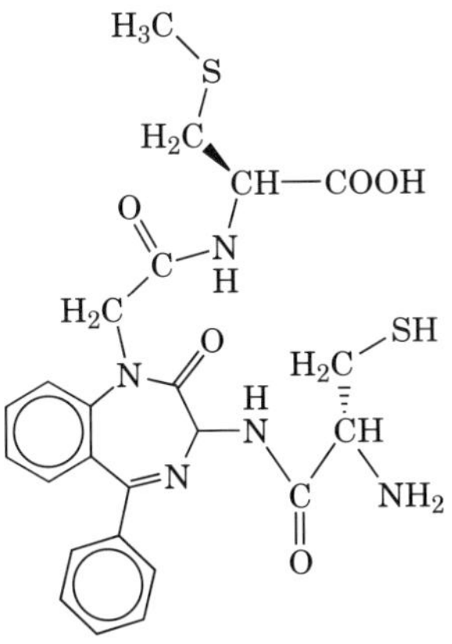

(68)

Bombesin.   Bombesin [*31362-50-2*], *p*-Glu-Gln-Arg-Leu-Gly-Asn-Gln-Trp-Ala-Val-Gly-His-Leu-Met-NH$_2$, is a tetradecapeptide isolated from the skin of the frog *Bombina bombina* (31). There are two mammalian bombesin-like peptides, the 27-amino acid gastrin-releasing peptide [*80043-53-4*], (GRP) (**69**), and the decapeptide, neuromedin B [*87096-84-2*] (NMB), H-Gly-Asn-Leu-Trp-Ala-Thr-Gly-His-Phe-Met-NH$_2$. These elicit a wide range of pharmacological activities including thermoregulation, smooth muscle contraction, stimulation of the release of numerous gastrointestinal (GI) peptides, and the regulation and maintenance of circadian rhythms in the suprachiasmatic nucleus. Two bombesin receptor subtypes have been identified, a receptor with high affinity ($K_i = 3$ nM) and 30-fold selectivity for NMB (BB$_1$) and one with high affinity ($K_i = 2$ nM) and 20-fold selectivity for GRP (BB$_2$). Both are GPCRs, utilizing IP$_3$/DAG for their transduction mechanisms, and are widely distributed in the central and peripheral nervous systems. A third bombesin receptor has been cloned from guinea pig and human. ICI 216,140 [*124001-41-8*], C$_{45}$H$_{65}$N$_{13}$O$_8$ (**70**) is a nonselective bombesin antagonist, whereas [D-Phe^6]bombesin$_{6-13}$ ethyl ester and AcGRP$_{20-26}$ ethyl ester are the most potent and selective BB$_2$ antagonists yet identified. CP 75998 [*149142-70-1*], C$_{16}$H$_9$BrCl$_2$N$_4$O$_3$ (**71**) is a nonpeptide bombesin antagonist. Potential therapeutic targets for bombesin agonists include satiety and analgesia.

H-Ala-Pro-Val-Ser-Val-Gly-Gly-Gly-Thr-Val-Leu-Ala-Lys-
Met-Tyr-Pro-Arg-Gly-Asn-His-Trp-Ala-Val-Gly-His-Leu-Met-NH$_2$

(69)

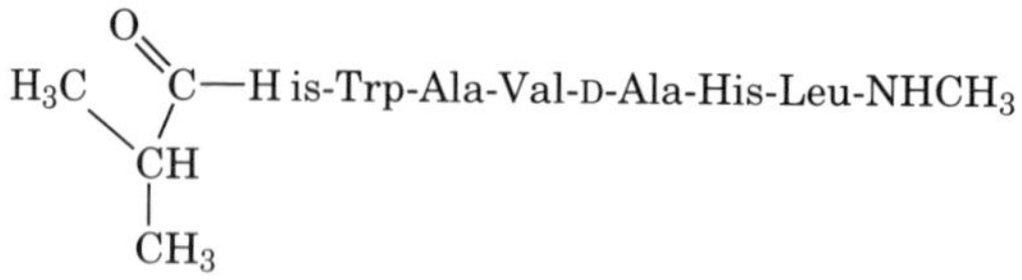

H₃C\C—H is-Trp-Ala-Val-D-Ala-His-Leu-NHCH₃

(70)

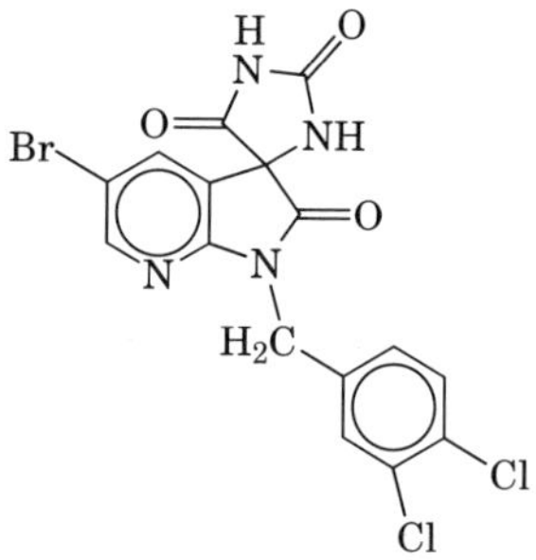

(71)

Bradykinin. Bradykinin [58-82-2] (BK), $C_{50}H_{73}N_{15}O_{11}$, H-Arg-Pro-Pro-Gly-Phe-Ser-Pro-Phe-Arg-OH, is a nonapeptide (25) released from plasma and exocrine glands by the action of kinin-generating enzymes, the kallikreins, on the α_2-globulins termed kininogens. Activation of plasma kallikrein leads to the formation of BK from high molecular weight but not from low molecular weight kininogen. Tissue kallikreins in contrast can liberate kinins from both forms of kininogen. LysylBK, also known as kallidin [342-10-9], $C_{56}H_{85}N_{17}O_{12}$, is an analogue of BK found in human tissues. T-kinin [86030-63-9] (Ile-Ser-BK), $C_{59}H_{89}N_{17}O_{14}$, is a BK analogue found only in rats. BK, kallidin, and T-kinin have similar pharmacological properties.

BK actions are mediated through at least two types of GPCR: B_1 and B_2. At the B_1 receptor, des-Arg⁹BK is more potent than BK. The converse is true at the B_2 receptor. The effects of BK are primarily mediated by activation of the B_2 receptor because the B_1 receptor has limited tissue distribution and is induced by noxious stimuli such as apamin or an inflammatory mediator-type response. The existence of a B_3 receptor was suggested on the basis of limited efficacy of known antagonists in some systems. A B_4 receptor may also exist. The human B_2 receptor has been cloned.

Hoe 140 [130308-48-4], $C_{59}H_{89}N_{19}O_{13}S$ (72) and NPC 567 [109333-26-8], $C_{60}H_{87}N_{19}O_{13}$, H-D-Arg-Arg-Pro-(trans-4-hydroxy)Pro-Gly-Phe-Ser-D-Phe-Phe-Arg-OH, are modified peptides which are selective antagonists for the B_2 receptor. WIN 64338, $C_{44}H_{68}N_4OP$ (73) is the first competitive, nonpeptide B_2 antagonist. [Leu⁸] des-Arg⁹BK and [des-Arg¹⁰] Hoe 140 are peptide analogues of BK that act as selective B_1 receptor antagonists.

BK and its congeners are involved in allergic reactions, asthma, viral rhinitis, hypertension, and septic shock. BK also is involved with the pathophysiological processes that accompany tissue damage, inflammation, and the production of pain. BK and kallidin are 100-fold more potent than histamine in increasing vascular permeability and thus are powerful mediators of edema by the release of release of platelet activating factor (PAF) and prostaglandins (qv). B_2 receptor antagonists thus may have analgesic and antiinflammatory actions in acute inflammatory pain.

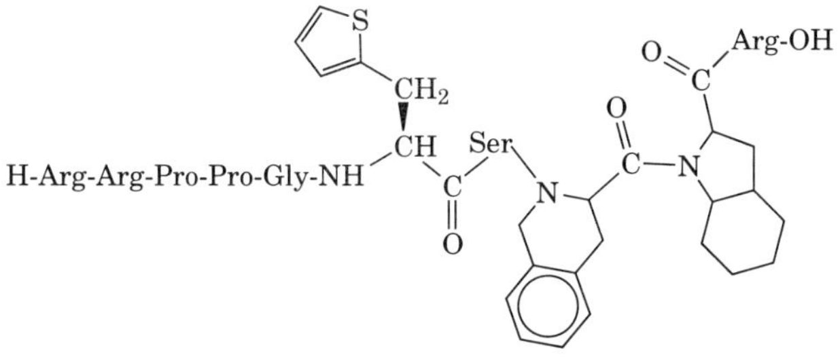

(**72**)

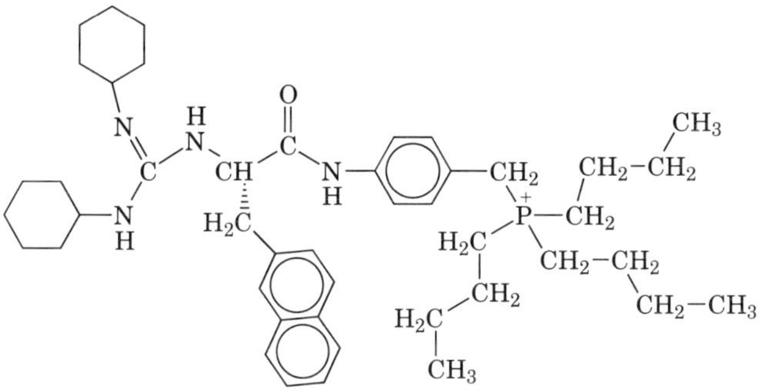

(**73**)

Calcitonin Gene-Related Peptide.   Calcitonin gene-related peptide (CGRP) [*83652-28-2*] (**74**) is a 37-amino acid peptide (Fig. 4) (32).

CGRP is widely distributed throughout the peripheral and central nervous systems and is found in sensory neurons and in the autonomic and enteric nervous systems. In many instances CGRP is co-localized with other neuroregulators, eg, ACh in motor neurons, substance P, somatostatin, vasoactive intestinal polypeptide (VIP), and galanin in sensory neurons. It is also present in the CNS, with ACh in the parabigeminal nucleus and with cholecystokinin (CCK) in the dorsal parabrachial area. CGRP functions as a neuromodulator or co-transmitter.

Amylin [*106602-62-4*] (**75**) (Fig. 4) is a 37-amino acid peptide having approximately 46% sequence similarity to CGRP (33). Amylin is present in pancreatic β-cells along with insulin. It may function as a hormone in glucoregulation and has been proposed as an etiologic factor in certain forms of diabetes. Amylin is also present in dorsal root ganglia (see INSULIN AND OTHER ANTIDIABETIC DRUGS).

CNS effects of CGRP include increased sympathetic outflow, increased temperature, reduced feeding and gastric secretion, reduced growth hormone release, reduced motor activity, and effects on nociception. In individual cells CGRP may be excitatory, potentiating transmitter release and/or increasing

H-Ala-Cys-Asp-Thr-Ala-Thr-Cys-Val-Thr-His-Arg-Leu-Ala-Gly-Leu-Leu-Ser-Arg-Ser-
Gly-Gly-Val-Val-Lys-Asn-Asn-Phe-Val-Pro-Thr-Asn-Val-Gly-Ser-Lys-Ala-Phe-NH$_2$

(**74**)

H-Lys-Cys-Asn-Thr-Ala-Thr-Cys-Ala-Thr-Gln-Arg-Leu-Ala-Asn-Phe-Leu-Val-Arg-Ser-
Ser-Asn-Asn-Leu-Gly-Pro-Val-Leu-Pro-Pro-Thr-Asn-Val-Gly-Ser-Asn-Thr-Tyr-NH$_2$

(**75**)

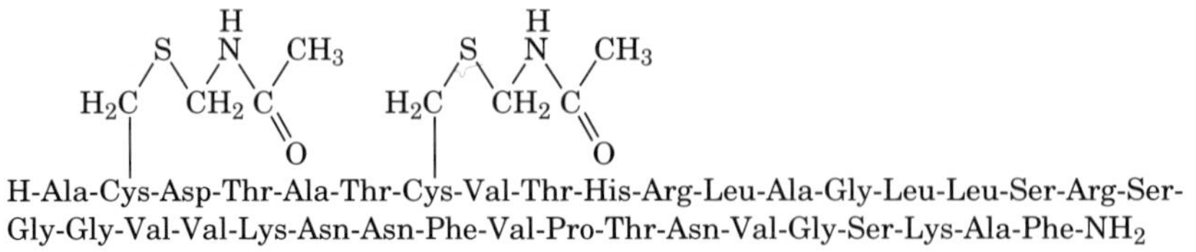

H-Ala-Cys-Asp-Thr-Ala-Thr-Cys-Val-Thr-His-Arg-Leu-Ala-Gly-Leu-Leu-Ser-Arg-Ser-
Gly-Gly-Val-Val-Lys-Asn-Asn-Phe-Val-Pro-Thr-Asn-Val-Gly-Ser-Lys-Ala-Phe-NH$_2$

(**76**)

H-Cys-Ser-Asn-Leu-Ser-Thr-Cys-Val-Leu-Gly-Lys-Leu-Ser-Gln-Glu-Leu-His-Lys-Leu-Gln-
Thr-Tyr-Pro-Arg-Thr-Asn-Thr-Gly-Ser-Gly-Thr-Pro-NH$_2$

(**77**)

Fig. 4.   Structures of human CGRP and related peptides.

specific calcium channel activity, or inhibitory, increasing specific potassium channel activity.

CGRP produces its effects (32) by two GPCR receptor subtypes, CGRP$_1$ and CGRP$_2$. These have been classified according to the selectivities of the fragment CGRP$_{8-37}$ which is a CGRP$_1$ antagonist and of [acetoamidomethylcysteine[2,7]] CGRP (**76**) which is a CGRP$_2$ agonist. In the nucleus accumbens there may be a third receptor sensitive to amylin and CGRP. This receptor is also sensitive to salmon calcitonin [*47931-85-1*], C$_{145}$H$_{240}$N$_{44}$O$_{48}$S$_2$ (**77**). Adenylate cyclase and Ca^{2+} and K$^+$ channel activation are involved in the transduction mechanism for CGRP$_1$. CGRP is vasodilatory and, in conjunction with other mediators, pro-inflammatory. In the heart, CGRP increases atrial contractile force and rate. In striated muscle, CGRP increases nicotinic receptor expression and the rate of desensitization, and augments ACh release from motor neurons (see Fig. 4).

Cannabinoids.   Like the BZ receptor, the cannabinoid receptor was initially identified using psychotropic alkaloids such as Δ^9-tetrahydrocannabinol (Δ-THC) (**78**) that were known to affect mammalian CNS function (see PSYCHOPHARMACOLOGICAL AGENTS). The CNS receptor, CB$_1$, was identified by radioligand binding techniques and subsequently cloned. A second receptor subclass, CB$_2$, has been identified in human spleen and also has been cloned (34). Table 5 lists cannabinoid receptor ligands.

Table 5. Cannabinoid Receptor Ligands

Compound	CAS Registry Number	Molecular formula	Structure number
Δ^9-tetrahydrocannabinol	[1972-08-3]	$C_{21}H_{30}O_2$	(78)
CP 55940	[83002-04-4]	$C_{24}H_{40}O_3$	(79)
nabilone	[51022-71-0]	$C_{24}H_{36}O_3$	(80)
anandamide	[94421-68-8]	$C_{22}H_{37}NO_2$	(81)
HU 243	[140835-14-9]	$C_{25}H_{40}O_3$	(82)
WIN 55212-2	[131543-22-1]	$C_{27}H_{26}N_2O_3$	(83)
3-(1,1-dimethylheptyl), Δ^6-tetrahydrocannabinol-1-carboxylic acid		$C_{23}H_{32}O_4$	(84)

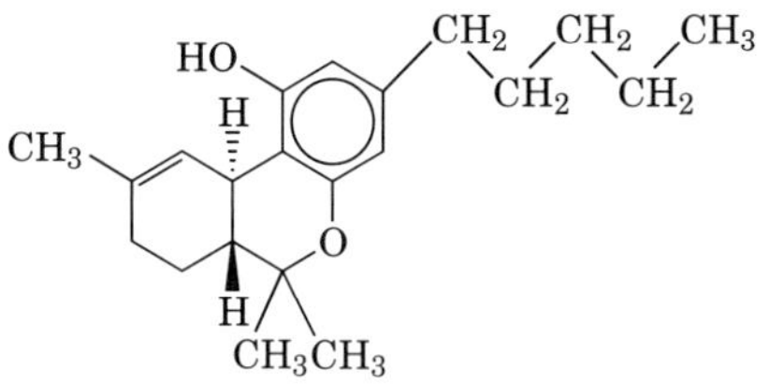

(78)

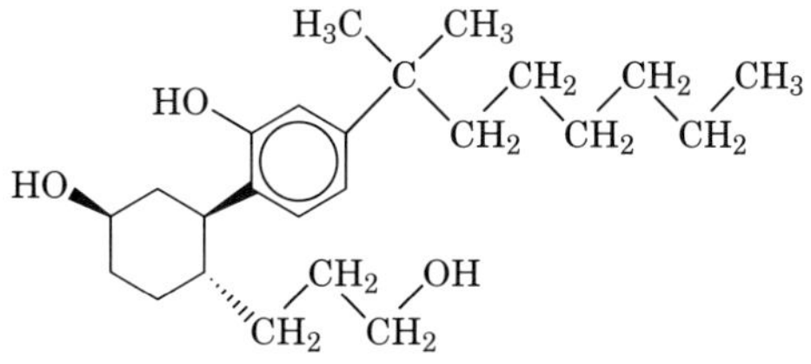

(79)

CP 55940 (**79**) and nabilone (**80**) are synthetic ligands for the cannabinoid receptor. However, the identification of the eicosanoid, anandamide (**81**), as an endogenous cannabimimetic has provided an important tool to study cannabinoid receptor function.

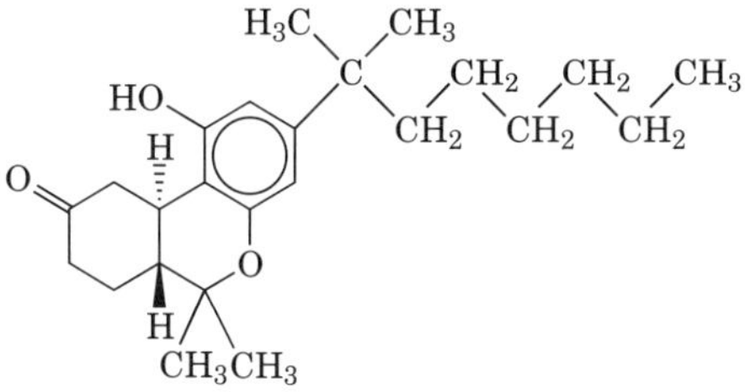

(80)

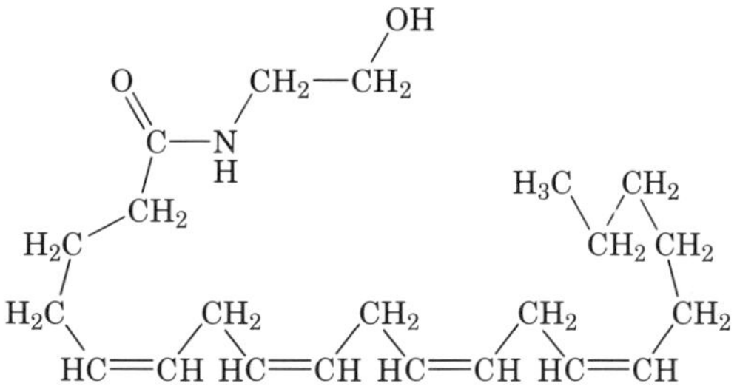

(81)

HU 243 (**82**) and WIN 55212-2 (**83**), newer cannabimimetics, may be involved in the regulation of neurotransmitter release. Like anandamide, these are potent blockers of N-type calcium channels. Cannabimimetics are psychotropics, effecting on time perception, euphoria, sedation, and causing hallucinations and a decrease in aggressive behavior. Short-term memory is impaired as is the ability to carry out tasks requiring mental processing. Δ-THC and nabilone are also antinociceptive agents and are used as antiemetics in cancer patients and as antiglaucoma agents. Nonpsychotropic cannabinoids, eg, 3-(1,1-dimethylheptyl),Δ^6-tetrahydrocannabinol-1-carboxylic acid (**84**) have antiinflammatory and leukocyte antiadhesion properties (34). Antagonists for cannabinoid receptors have yet to be identified.

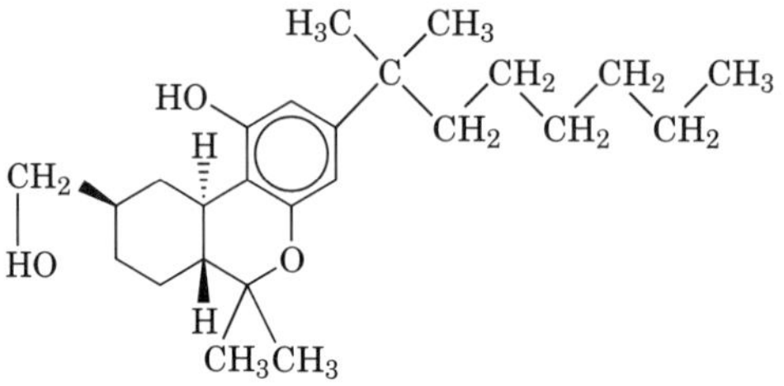

(**82**) 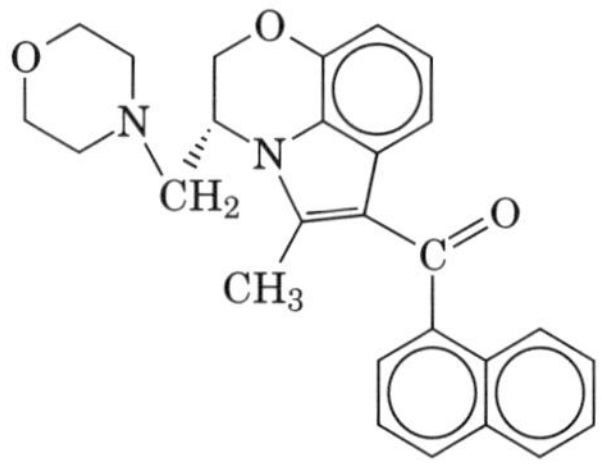   (**83**)

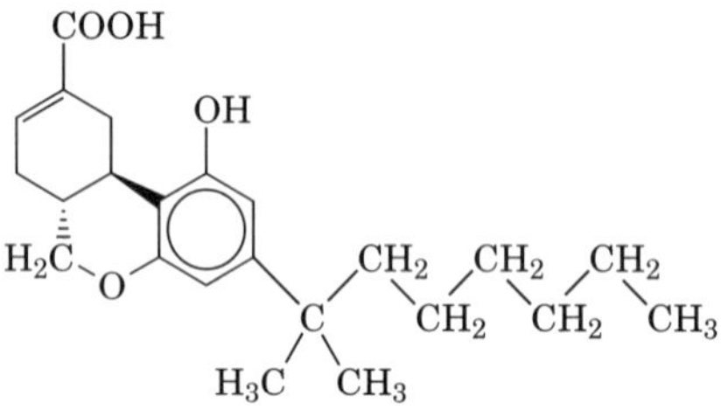

(**84**)

Catecholamines. The catecholamines, epinephrine (EPI; adrenaline) (**85**), norepinephrine (NE; noradrenaline) (**86**) (see EPINEPHRINE AND NOR-EPINEPHRINE), and dopamine (DA) (**2**), are produced from tyrosine by the sequential formation of L-dopa, DA, NE, and finally EPI. EPI and NE produce their physiological effects via α- and β-adrenoceptors. α-Adrenoceptors can be further divided into α_1- and α_2-subtypes which in turn are divided into α_{1A}, α_{1B}, and α_{1D} (clones α_{1a}, α_{1b}, and α_{1d}) and α_{2A}, α_{2B}, and α_{2C}. There is no $\alpha_{1C/c}$ receptor. β-Adrenoceptors are divided into β_1-, β_2-, and β_3-subtypes (35–37).

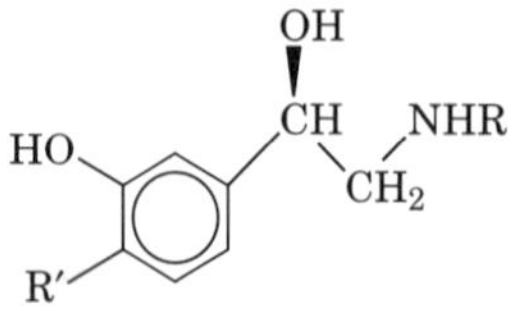

(**85**) R = CH_3 , R′ = OH

(**86**) R = H , R′ = OH

(**90**) R = CH_3 , R′ = H

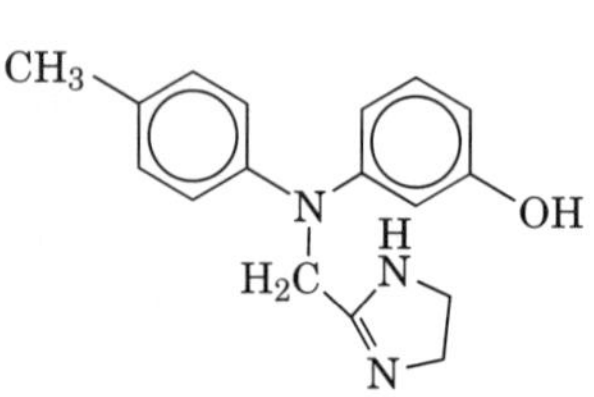

(**87**)

Phentolamine (**87**), WB 4101 (**88**), and the site directed alkylating agent, chloroethylclonidine (CEC) (**89**) have been traditionally used to define α_1-receptors. Table 6 lists the various catecholamines and adrenoreceptor agonists and antagonists.

(**88**)　　　　　　　　(**89**)

Phenylephrine (**90**) is a selective α_{1A} receptor agonist; (+)-niguldipine (**91**) is a selective antagonist for the α_{1A} receptor. Prazosin (**92**) and 5-methylurapidil

Table 6. Catecholamines and Adrenoceptor Agonists and Antagonists

Agonist/antagonist	CAS Registry Number	Molecular formula	Structure number
epinephrine	[51-43-4]	$C_9H_{13}NO_3$	(**85**)
norepinephrine	[51-41-2]	$C_8H_{11}NO_3$	(**86**)
phentolamine	[50-60-2]	$C_{17}H_{19}N_3O$	(**87**)
WB 4101	[613-67-2]	$C_{19}H_{23}NO_5$	(**88**)
chloroethylclonidine	[77472-95-8]	$C_{13}H_{17}Cl_3N_4$	(**89**)
phenylephrine	[59-42-7]	$C_9H_{13}NO_2$	(**90**)
(+)-niguldipine	[120054-86-6]	$C_{36}H_{39}N_3O_6$	(**91**)
prazosin	[19216-56-9]	$C_{19}H_{21}N_5O_4$	(**92**)
5-methylurapidil	[34661-85-3]	$C_{21}H_{31}N_5O_3$	(**93**)
UK 14,304	[59803-98-4]	$C_{11}H_{10}BrN_5$	(**94**)
clonidine	[4205-90-7]	$C_9H_9Cl_2N_3$	(**95**)
yohimbine	[146-48-5]	$C_{21}H_{26}N_2O_3$	(**96**)
rauwolscine	[131-03-3]	$C_{21}H_{26}N_2O_3$	(**97**)
sprioxatrine	[1054-88-2]	$C_{22}H_{25}N_3O_3$	(**98**)
imiloxan	[81167-16-0]	$C_{14}H_{16}N_2O_2$	(**99**)
BRL 44408	[118343-19-4]	$C_{13}H_{17}N_3$	(**100**)
idazoxan	[79944-58-4]	$C_{11}H_{12}N_2O_2$	(**101**)
L-agmatine	[306-60-5]	$C_5H_{14}N_4$	(**102**)
xamoterol	[81801-12-9]	$C_{16}H_{25}N_3O_5$	(**103**)
procaterol	[72332-33-3]	$C_{16}H_{22}N_2O_3$	(**104**)
BRL 37344	[90730-96-4]	$C_{19}H_{22}ClNO_4$	(**105**)
CL 316243	[138908-40-4]	$C_{20}H_{18}ClNO_7Na_2$	(**106**)
CGP 20712A	[81015-67-0, 105737-62-0]	$C_{23}H_{25}F_3N_4O_5$	(**107**)
betaxolol	[63659-18-7]	$C_{18}H_{29}NO_3$	(**108**)
atenolol	[29122-68-7]	$C_{14}H_{22}N_2O_3$	(**109**)
ICI 118551	[72795-19-8]	$C_{17}H_{27}NO_2$	(**110**)
butaxamine	[1937-89-9]	$C_{15}H_{25}NO_3$	(**111**)
(−)-pindolol	[26328-11-0]	$C_{14}H_{20}N_2O_2$	(**112**)

(**93**) are nonselective α_1-receptor antagonists. CEC can differentiate α_{1b} receptors from the other α_1 receptors. Prazosin has low and high affinity for α_{2A} and α_{2B} receptors, respectively.

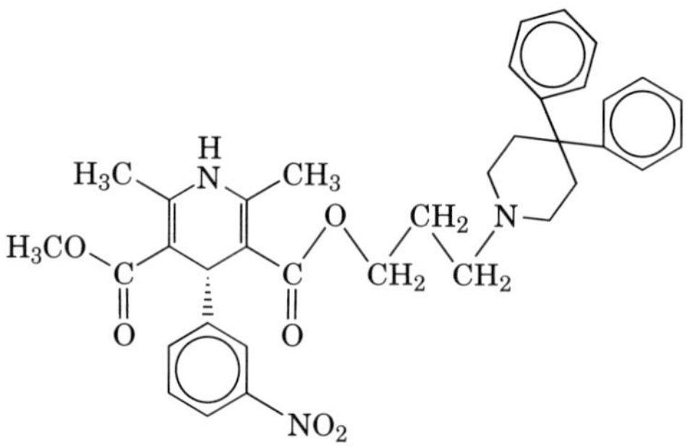

(**91**)

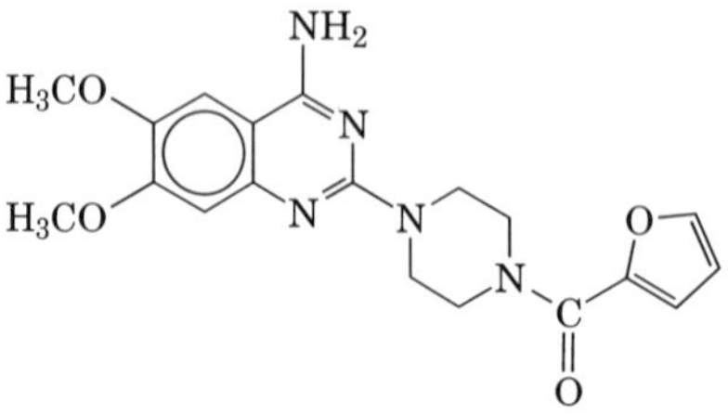

(**92**)

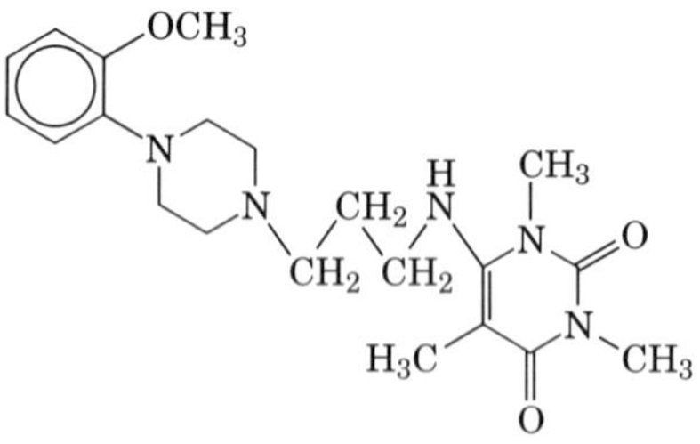

(**93**)

UK 14,304 (**94**), clonidine (**95**), yohimbine (**96**), and rauwolscine (**97**) interact with all three α_2 receptors. However, the α_{2C} receptor, unlike the α_{2A} and α_{2B} receptors, has high affinity for (**97**). Spiroxatrine (**98**) and imiloxan (**99**) bind to α_{2B} and α_{2C} receptors. BRL 44408 (**100**) is a selective α_{2A} antagonist.

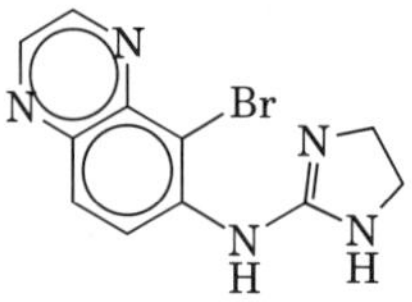

(**94**)

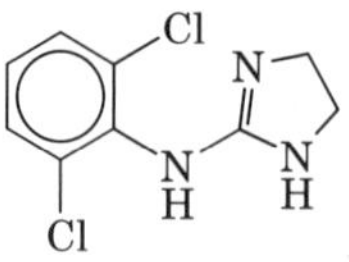

(**95**)

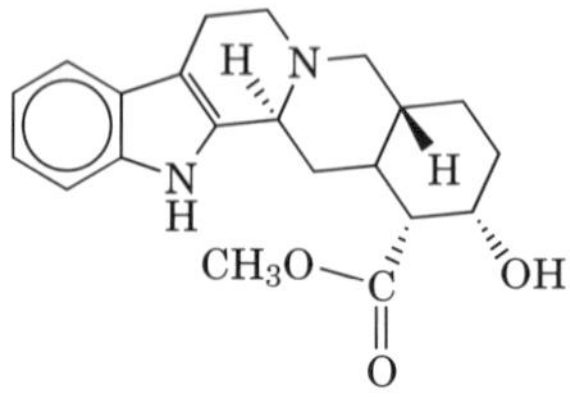

(**96**)

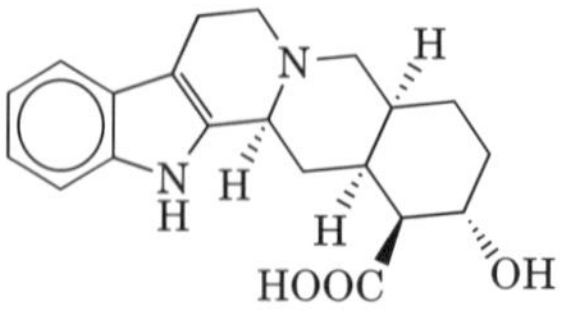

(**97**)

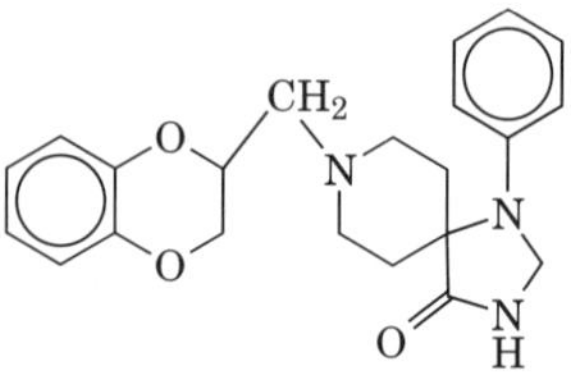

(**98**)

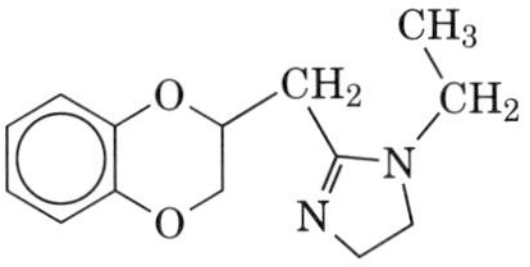

(99) 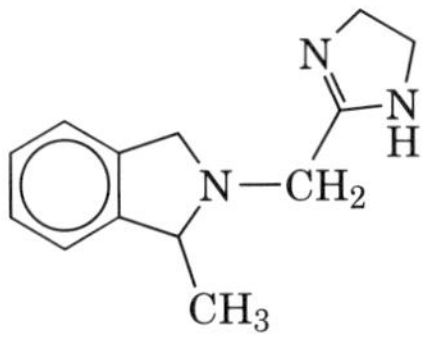   (100)

Imidazolines and imidazolidines, eg, clonidine, UK 14304, and idazoxan (**101**), represent a class of compounds that interact with α_2-adrenoceptors but the pharmacology is not fully explained by interactions with this receptor. There might be yet another adrenoceptor-like imidazoline receptor or clonidine-like receptor through which these agents mediate their antihypertensive actions. L-Agrnatine (**102**) has been identified as a potential endogenous clonidine-displacing substance (CDS) (38). Two nonadrenergic imidazoline binding sites or receptors identified as I_1 and I_2 have been identified.

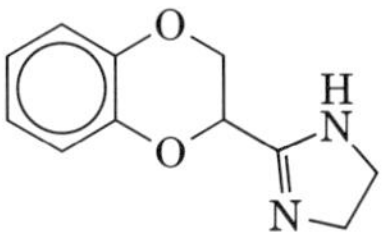

(101) 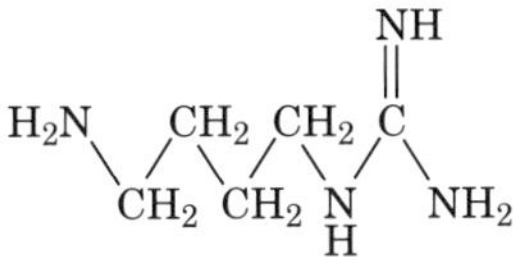   (102)

EPI and NE are also agonists at the β_1-adrenoceptor as is xamoterol (**103**). Procaterol (**104**) is a β_2-adrenoceptor agonist. BRL 37344 (**105**) and CL 316243 (**106**) are β_3-adrenoceptor agonists. CGP 20712A (**107**), betaxolol (**108**), and atenolol (**109**) are β_1-adrenoceptor antagonists. β_2-Adrenoceptor antagonists include ICI 118551 (**110**) and butaxamine (**111**). ($-$)-Pindolol (**112**) is a β_3-adrenoceptor partial agonist (Table 6).

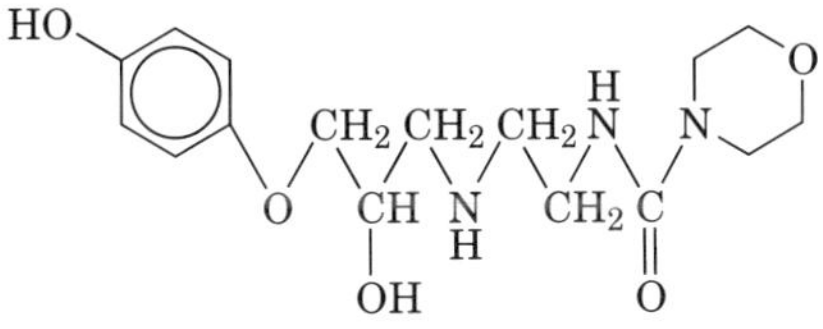

(103) 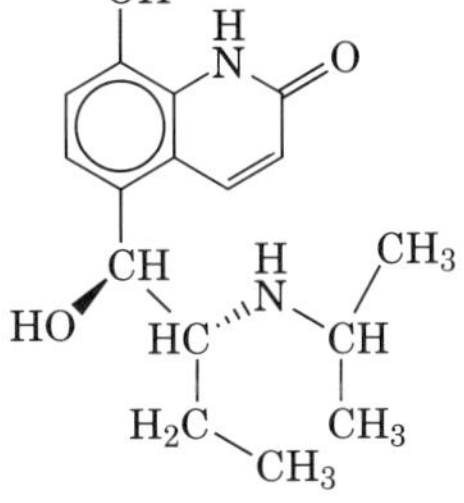   (104)

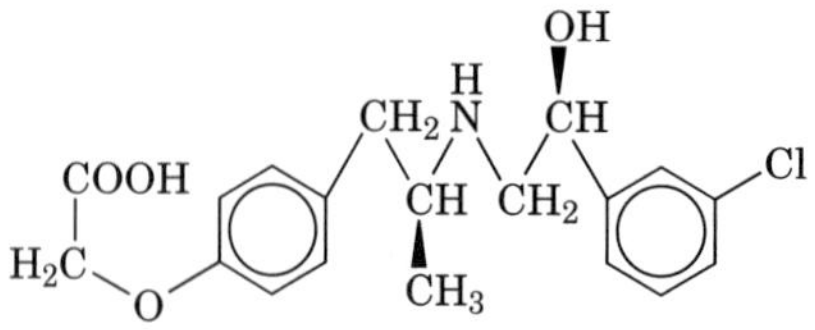

(105)

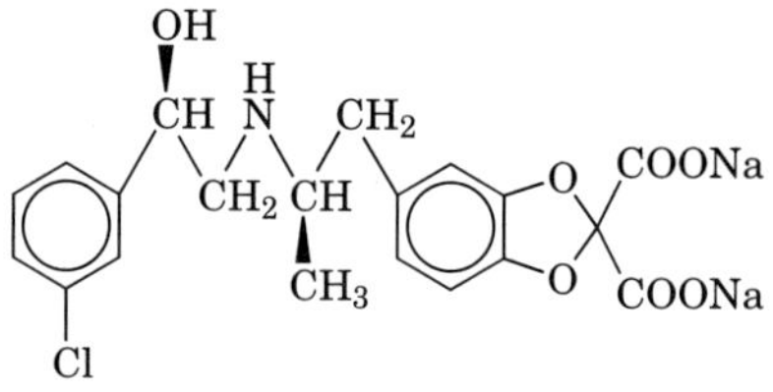

(106)

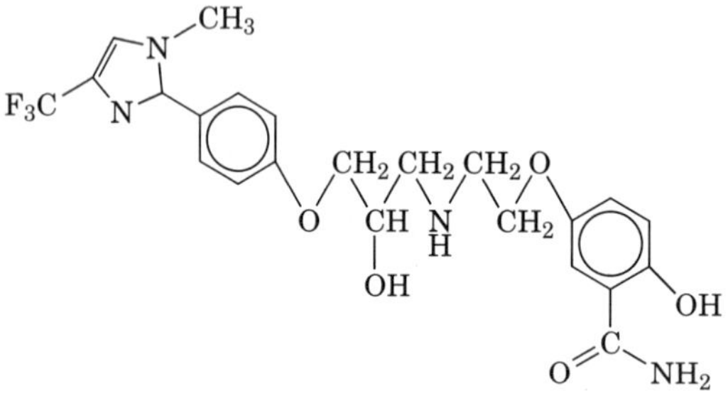

(107)

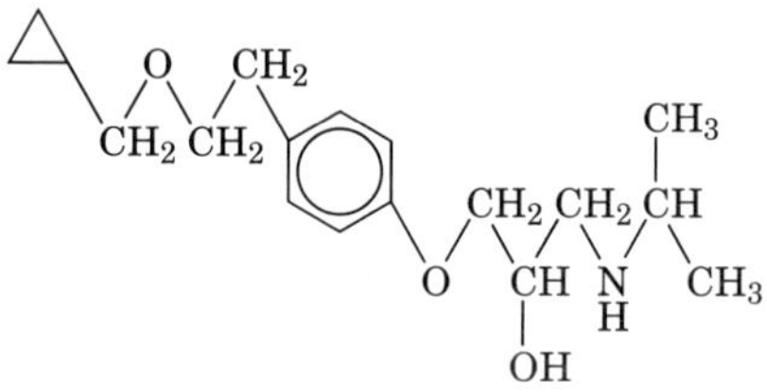

(108)

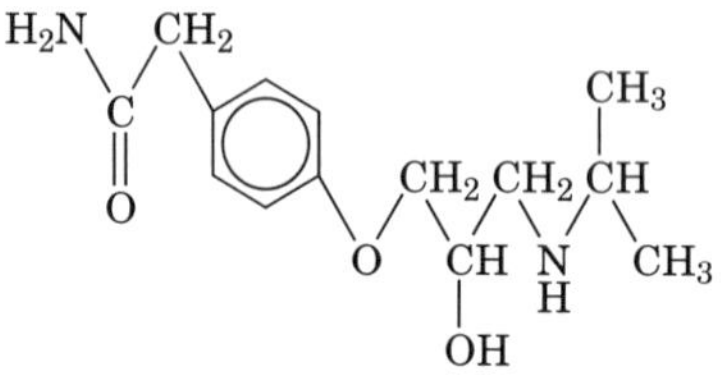

(109)

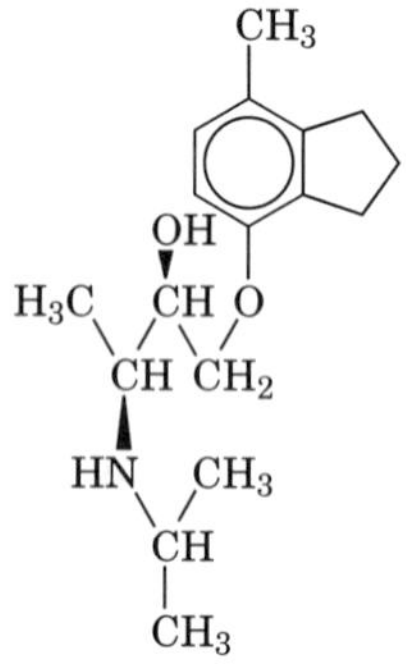

(110)

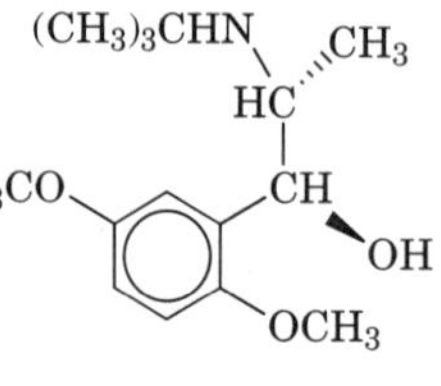

(111)

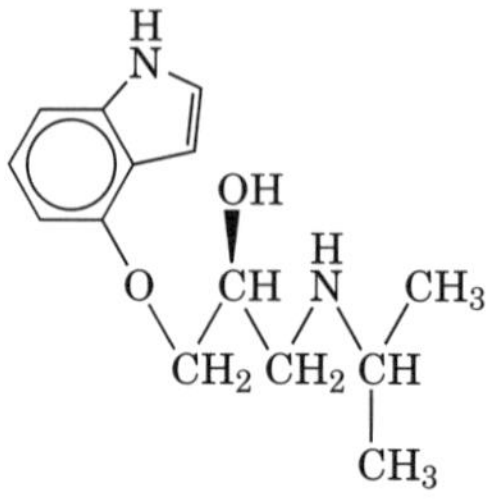

(112)

Adrenoceptors are involved in the regulation of blood pressure, myocardial contractile function, airway reactivity, smooth muscle tone, and a variety of metabolic functions. Presynaptic adrenoceptor antagonists can potentiate the effects of reuptake inhibitors by blocking feedback inhibition. Idazoxan has neuroprotective properties. β-Adrenoceptor antagonists are well known as antihypertensive agents. Agonists are used as antiasthmatics and may have potential as antiobesity agents or in the treatment of diabetes via β_3-adrenoceptor activation (see ANTIASTHMATIC AGENTS; ANTIOBESITY DRUGS).

Cholecystokinin. Cholecystokinin (CCK) is a peptide of gut/brain origin that is involved in digestive and homostatic functions (30). CCK originates from a 115-amino acid precursor, prepro-CCK, that is cleaved at a single arginine residue to yield active CCK moieties. The primary naturally occurring forms of CCK are CCK-8 [25126-32-3], $C_{49}H_{62}N_{10}O_{16}S_3$, H-Asp-Tyr(SO$_3$H)-Met-Gly-Trp-Met-Asp-Phe-NH$_2$; CCK-33 [96827-04-2], $C_{167}H_{263}N_{51}O_{52}S_4$ (**113**), and CCK-4, also known as tetragastrin [1947-37-1], $C_{29}H_{36}N_6O_6S$, H-Trp-Met-Asp-Phe-NH$_2$. CCK-8 predominates.

H-Lys-Ala-Pro-Ser-Gly-Arg-Met-Ser-Ile-Val-Lys-Asn-Leu-Gln-Asn-Leu-Asp-Pro-Ser-
His-Arg-Ile-Ser-Asp-Arg-Asp-Tyr(SO$_3$H)-Met-Gly-Trp-Met-Asp-Phe-NH$_2$

(**113**)

Two CCK receptor subtypes, CCK$_A$ and CCK$_B$ are known. A related receptor, the gastrin receptor, has also been described. CCK$_A$ receptors predominate in the gastrointestinal tract and pancreas and are also localized in discrete brain regions. CCK$_B$ receptors predominate in the brain. A 71623 [130408-77-4], $C_{44}H_{56}N_8O_9$ (**114**) is a selective CCK$_A$ agonist. Devazepide (**65**) is a selective antagonist (40). Desulfated CCK$_8$ [25679-24-7], $C_{49}H_{62}N_{10}O_{13}S_2$, H-Asp-Tyr-Met-Gly-Trp-Met-Asp-Phe-NH$_2$, and pentagastrin [5534-95-2], $C_{37}H_{49}N_7O_9S$, (CH$_3$)$_3$COCO-β-Ala-Trp-Met-Asp-Phe-NH$_2$, are selective CCK$_B$ receptor agonists. L 365260 [118101-09-0], $C_{24}H_{22}N_4O_2$ (**115**) and LY 262691, $C_{22}H_{18}BrN_3O_2$ (**116**) are species selective CCK$_B$ antagonists.

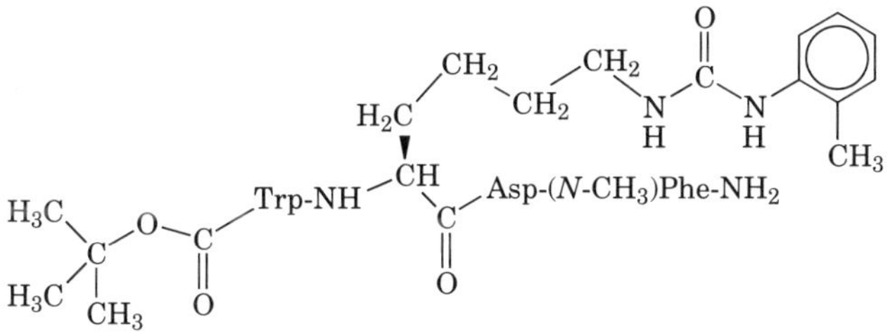

(**114**)

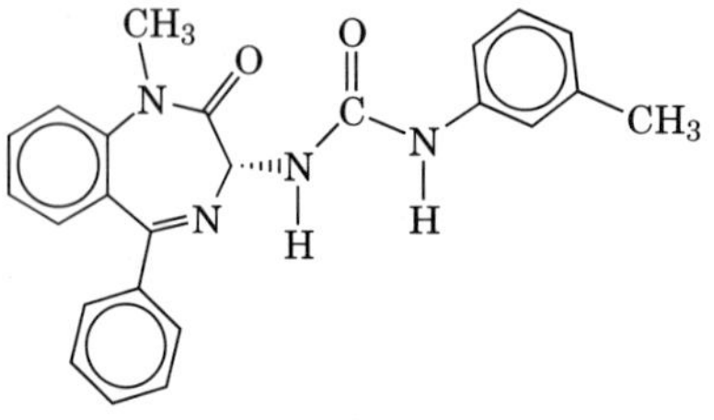

(115) 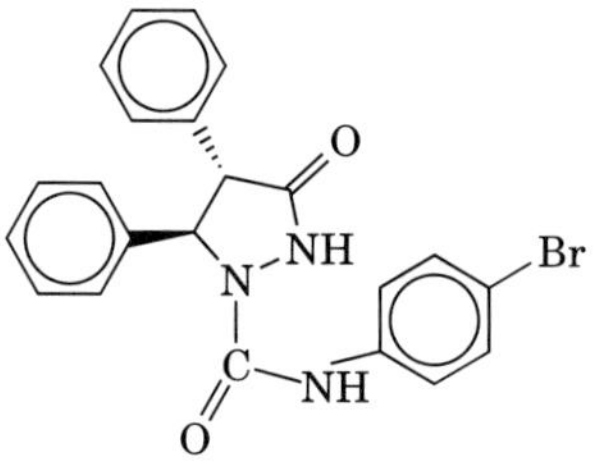   (116)

CCK is found in the digestive tract and the central and peripheral nervous systems. In the brain, CCK coexists with DA. In the peripheral nervous system, the two principal physiological actions of CCK are stimulation of gall bladder contraction and pancreatic enzyme secretion. CCK also stimulates glucose and amino acid transport, protein and DNA synthesis, and pancreatic hormone secretion. In the CNS, CCK induces hypothermia, analgesia, hyperglycemia, stimulation of pituitary hormone release, and a decrease in exploratory behavior. The CCK family of neuropeptides has been implicated in anxiety and panic disorders, psychoses, satiety, and gastric acid and pancreatic enzyme secretions.

Cytokines and Immunophilins. A large number of inflammatory mediators and related proteins including the cytokines, colony stimulating factors (CSFs), interferons (IFNs), tumor necrosis factors (TNFs), growth factors (see GROWTH REGULATORS), neurotrophic factors, and immunophilins are found in the mammalian CNS and appear to play a significant role in CNS function both in development and in aspects of brain homeostasis (40–43).

The cytokines are involved in the regulation of the growth, differentiation, and activation of the hematopoietic cells involved in the host immune response. Within the CNS, cytokines have been implicated in a number of hormonal, trophic, toxic, and immune functions and are classified into the CXC family where the first two conserved cysteine residues are separated by an amino acid, or into the CC family where the cysteines are adjacent (40). Members of the CXC family include interleukin (IL)-8, melanoma growth-stimulating activity (MGSA), and neutrophil-activating peptide-2 (NAP-2). Macrophage inflammatory proteins 1α and 1β (MIP-1α and -1β), monocyte chemotactic protein-1 (MCP-1), and regulated on activation normal T-cell and secreted (RANTES) are included in the CC family. The immunophilins interact with intracellular recognition sites for the immunosupressants, FK 506 [*104987-11-3*], $C_{44}H_{69}NO_{12}$, and cyclosporin-A [*59865-13-3*], $C_{62}H_{111}N_{11}O_{12}$, and are involved in cell signaling processes in a variety of tissues including the CNS (44) (see also IMMUNOTHERAPEUTIC AGENTS).

In the absence of selective antagonists, cytokines have been classified either on the basis of common functional properties or on the structural characteristics of their receptors.

The class 1 cytokine receptor family includes receptors for interleukins IL-2, IL-3, IL-4, IL-5, IL-6, IL-7, and IL-9, granulocyte macrophage colony stimulating factor (GM-CSF), granulocyte colony stimulating factor (G-CSF), erythropoietin

[*11096-26-7*] (EPO), leukemia inhibitory factor (LIF), and ciliary neurotrophic factor (CNTF). As of this writing, data suggest that CNS cytokine receptors are distinct from those seen in the periphery. CNTF is highly localized to the myelin forming Schwann cells of the peripheral nervous system as well as activated astrocytes of the central nervous system, but levels in brain are moderate to low. CNTF is thought to act as an injury factor, released by Schwann cells under pathological conditions.

Interleukin-1 (IL-1) exists in two forms, α and β, that are products of different genes. Both polypeptides are synthesized as 33,000 mol wt precursors that are proteolytically cleaved by the action of interleukin converting enzyme (ICE) to generate the mature biologically active 17,000 mol wt proteins. ICE has been implicated in cellular apoptosis (45). The primary peripheral source of IL-1 is the activated mononuclear phagocyte. Within the CNS, IL-1 is synthesized by astrocytes and microglia, and IL-1β has been localized in neurons. IL-1 induces fever and slow-wave sleep, reduces feeding, stimulates immune or glial reactivity in the CNS as well as in the periphery, and modulates the release of adrenocorticotrophic hormone (ACTH), luteinizing hormone (LH), and gonadotropin-releasing hormone (GRH). IL-1 may also function as a trophic factor and as a mediator of ischemic neurotoxicity and may stimulate the production of β-amyloid protein which is a principal component in Alzheimer's disease plaques. The biological effects of IL-1 are mediated via two subtypes of IL-1 receptor (types I and II). Activation of these receptors causes rapid translocation of a pre-existing complex, NF$-\kappa$B, from the cell cytoplasm to the nucleus where it binds to specific regulatory DNA sequences in the promoters of several cytokine-inducible genes. High levels of IL-1 receptor have been found in choroid plexus, hippocampus, dentate gyrus, and anterior pituitary. IL-1ra [*143090-92-0*] is a naturally occurring IL-1 antagonist (46).

Interleukin-2 [*85898-30-2*] (IL-2) ($\sim$15,000 mol wt) and its receptor occur in high levels in the hippocampus and striatum. Hippocampal IL-2 binding is increased following an excitotoxic lesion. IL-2 can inhibit ACh release and the formation of long-term potentiation in the hippocampus. IL-6 ($\sim$26,000) is present in astrocytes, microglia, and anterior pituitary cells and high levels have been found in the hypothalamus. IL-1, tumor necrosis factor-α (TNF-α) and interferon-γ (IFN-γ) stimulate the synthesis and secretion of IL-6 which acts as a trophic factor in cultured neurons, to induce nerve growth factor (NGF) secretion in astrocytes, to stimulate the release of hormones from anterior pituitary cells, and to reduce feeding.

The class II cytokine receptor family includes receptors for interferon α/β (IFNα/β) and γ (IFNγ) and IL-10. IFN-γ immunoreactivity has been found in neurons in the hypothalamus, cerebral cortex, mammilary nuclei, and dorsal tegmentum. Astrocytes and microglia *in vitro* can be stimulated to express class II histocompatibility complex (MHC-II) antigens by IFN-γ, which may be involved in the presentation of antigen to T-cells by astrocytes. Thus IFN-γ may be critical in CNS-immune function and dysfunction especially in regard to neuronal and glial apoptotic processes.

The class III cytokine receptor family includes two TNF receptors, the low affinity NGF receptor and 7-cell surface recognition sites that appear to play a role in proliferation, apoptosis, and immunodeficiency. TNF-α ($\sim$17,000 protein) is produced by astrocytes and microglia and can induce fever, induce slow-wave sleep, reduce feeding, stimulate prostaglandin synthesis, stimulate corticotrophin-releasing factor and prolactin secretion, and reduce thyroid hormone secretion. TNF-α stimulates IL-1 release, is cytotoxic to oligodendrocytes, and reduces myelination; this has been implicated in multiple sclerosis and encephalomyelitis. Astrocyte TNF-α receptors mediate effects on IL-6 expression and augment astrocytic expression of MHC in response to other stimulants such as IFN-γ.

Activation of immunophilin receptors in the CNS is thought to be involved in certain aspects of AIDS dementia and the CNS side effects seen with the immunosupressants, FK 506 and cyclophilin. FK 506 and related compounds also have antiischemic effects.

Dopamine. Dopamine (DA) (**2**) is an intermediate in the synthesis of NE and Epi from tyrosine. DA is localized to the basal ganglia of the brain and is involved in the regulation of motor activity and pituitary hormone release. The actions of DA are terminated by conversion to dihydroxyphenylacetic acid (DOPAC) by monoamine oxidase-A and -B (MAO-A and -B) in the neuron following reuptake, or conversion to homovanillic acid (HVA) through the sequential actions of catechol-O-methyl transferase (COMT) and MAO-A and -B in the synaptic cleft.

DA produces its effects through two GPCR families (47). The D_1 family which includes D_1 and D_5 receptors and the D_2 family which includes D_2, D_3, and D_4 receptors. All five receptors have been cloned. D_1 receptor agonists include SKF 82958 (**117**), dihydrexidine (**118**), and ABT-431 (**119**). D_2 receptor agonists include MK 458 (**120**) and pergolide (**121**). Quinpirole (**122**) and (R)-(+)-7-OH-DPAT (**123**) are D_3 agonists (Table 7).

Table 7. Dopamine Receptor Agonists and Antagonists

Agonist/antagonist	CAS Registry Number	Molecular formula	Structure number
SKF 82958	[80751-65-1]	$C_{19}H_{20}ClNO_2$	(**117**)
dihydrexidine	[123039-93-0]	$C_{17}H_{17}NO_2$	(**118**)
ABT 431		$C_{22}H_{25}NO_4S$	(**119**)
MK 458	[99705-65-4]	$C_{15}H_{21}NO_2$	(**120**)
pergolide	[66104-22-1]	$C_{19}H_{26}N_2S$	(**121**)
quinpirole	[80373-22-4]	$C_{13}H_{21}N_3$	(**122**)
(R)-(+)-7-OH-DPAT		$C_{16}H_{25}NO$	(**123**)
thioridazine	[50-52-2]	$C_{21}H_{26}N_2S_2$	(**124**)
haloperidol	[52-86-8]	$C_{21}H_{23}ClFNO_2$	(**125**)
chlorpromazine	[50-53-3]	$C_{17}H_{19}ClN_2S$	(**126**)
SCH 23390	[87075-17-0]	$C_{17}H_{18}ClNO$	(**127**)
NNC 687	[128022-68-4]	$C_{19}H_{20}N_2O_4$	(**128**)
risperidone	[106266-06-2]	$C_{23}H_{27}FN_4O_2$	(**129**)
sertindole	[106516-24-9]	$C_{24}H_{26}ClFN_4O$	(**130**)
olanzapine	[132539-06-1]	$C_{17}H_{20}N_4S$	(**131**)
clozapine	[5786-21-0]	$C_{18}H_{19}ClN_4$	(**132**)

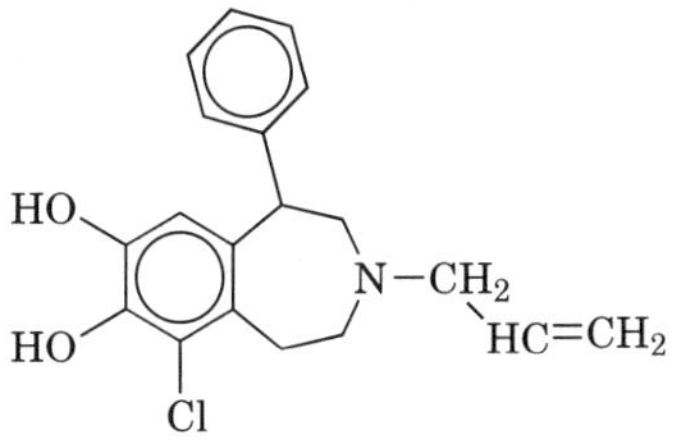

(117)

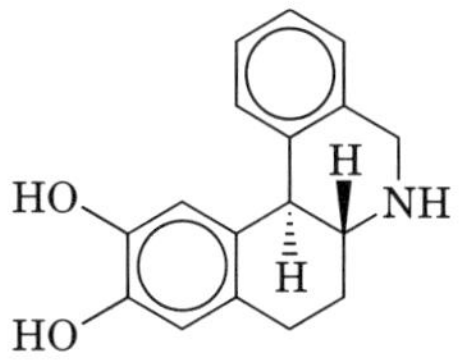

(118)

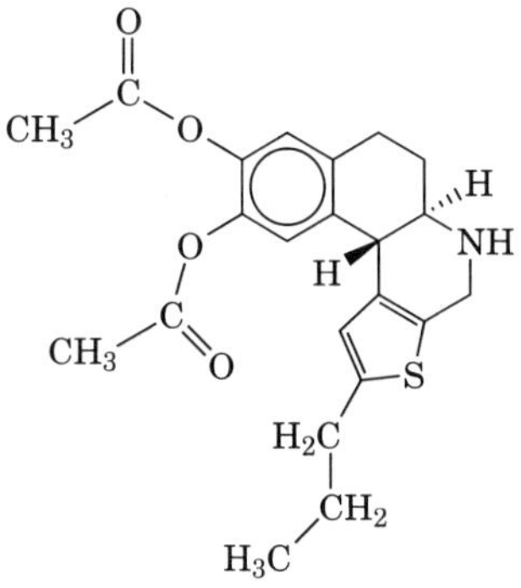

(119)

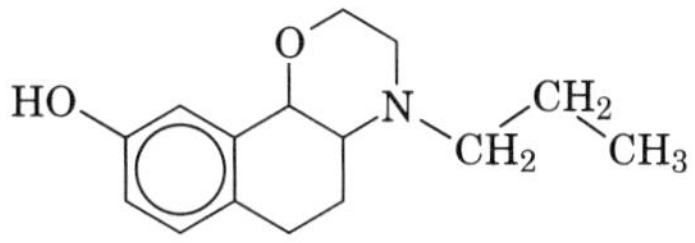

(120)

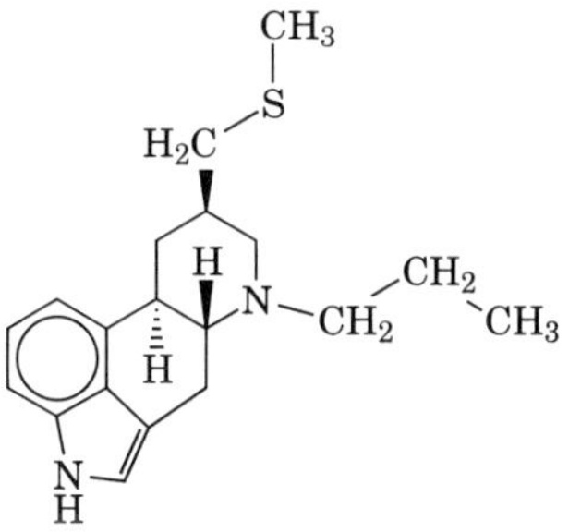

(121)

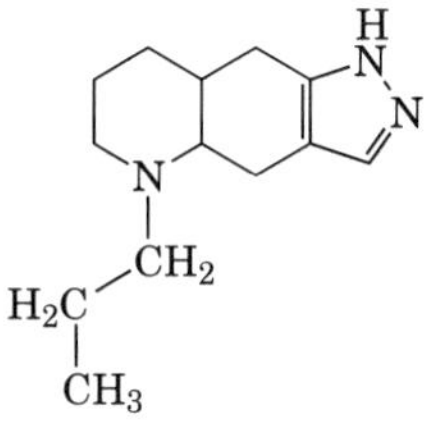

(122)

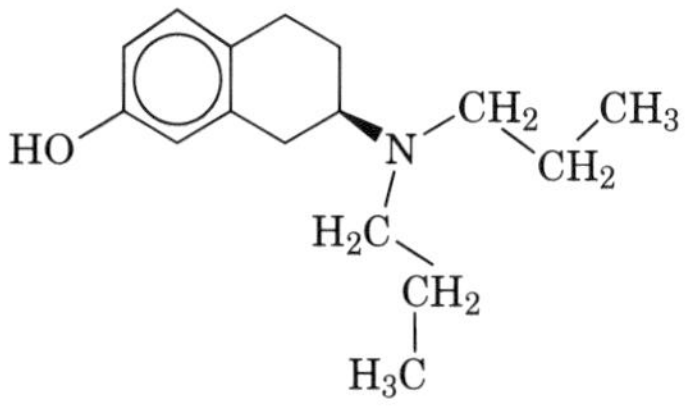

(123)

A loss of dopaminergic neurons in the basal ganglia underlies the etiology of Parkinson's disease (PD), a progressive neurodegenerative disorder, which is associated with motor deficits including tremors, muscular rigidity, a loss of postural reflexes, and freezing, ie, difficulty in initiating motor movement. Hyperactivity of brain dopaminergic systems is thought to be involved in the pathology of schizophrenia and many clinically useful antipsychotics produce their effects by blockade of post-synaptic dopamine receptor-mediated responses. DA is also produced in peripheral tissues and can modulate cardiovascular and renal function. Central dopaminergic systems also regulate pituitary hormone release.

DA replacement therapy is the primary palliative treatment regimen for PD (48) though it has limiting side effects including dyskinesias, hallucinations, sleep disturbances, and response fluctuations. Approaches to direct replacement to date have been limited to D_2 selective agents which are used as adjunct therapy with L-dopa (**3**). Dihydrexidine and ABT 431 are D_1 agonists for potential monotherapy in PD. DA antagonists include a broad range of neuroleptic agents used in the clinical treatment of schizophrenia. These include thioridazine (**124**), haloperidol (**125**), and chlorpromazine (**126**) which are D_2 selective antagonists. SCH 23390 (**127**) and NNC 687 (**128**) are D_1 selective antagonists. Because of their effects on DA-mediated motor function, the classical antipsychotics can produce extrapyramidal side effects (EPS), symptoms similar to those seen in PD. For this reason, there is a concerted effort to develop neuroleptics that have reduced side-effect liability. Agents selective for the D_1 receptor may represent compounds of this type although none have advanced very far in the clinic. Alternative approaches include compounds with combined $D_2/5\text{-}HT_2$ (serotonin) receptor blocking activity, that include risperidone (**129**), sertindole (**130**), and olanzapine (**131**). Clozapine (**132**) is an atypical antipsychotic that has a reduced EPS liability and has been reported to show preferential interactions with D_3 and D_4 receptors, although this is controversial (49). Clinically, clozapine elicits blood dyscrasias that can be fatal. There is an ongoing search for clozapine-like agents that lack this side effect. NGD 94-1 is a selective D_4 antagonist of undisclosed structure.

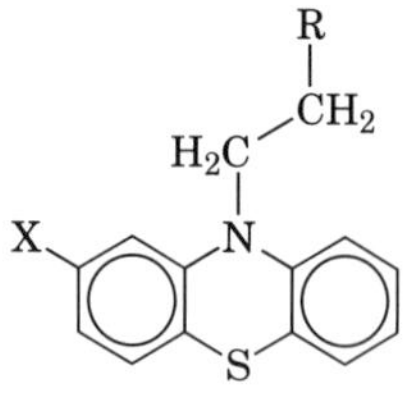

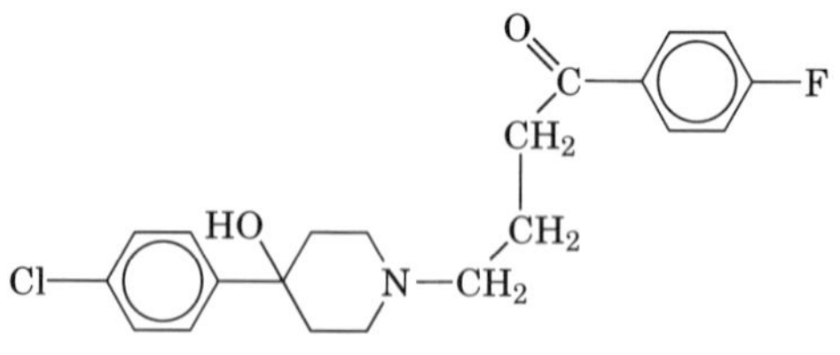

(**124**) X = SCH_3 , R = CH_3

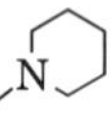

(**126**) X = Cl , R = $CH_2N(CH_3)_2$

(**125**)

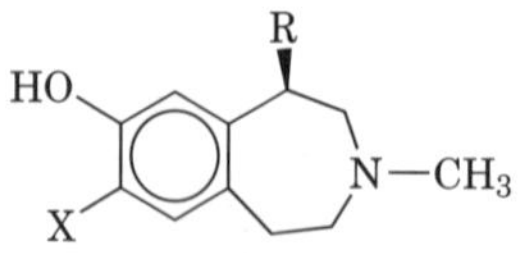

(**127**) R = ◯ , X = Cl

(**128**) R = ◯–O , X = NO_2

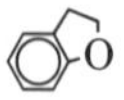

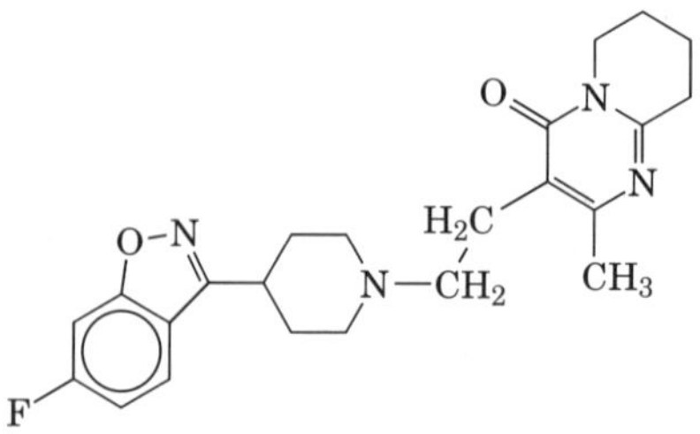

(**129**)

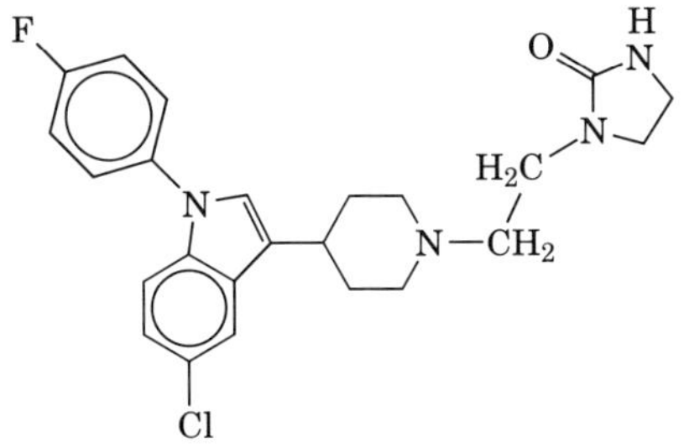

(130)

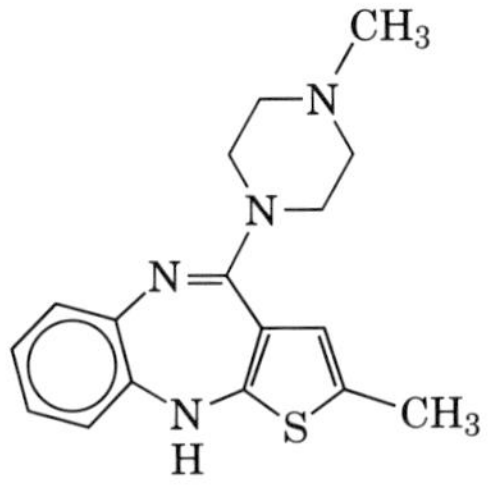

(131)

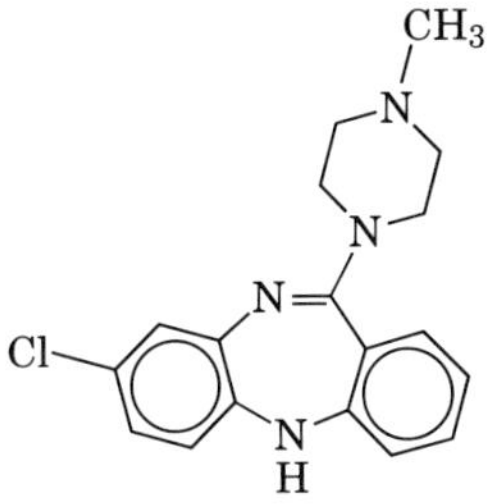

(132)

Endothelin. The endothelin (ET) peptide family (50) comprises three peptides: ET-1 (**133**), ET-2 (**134**), and ET-3 (**135**). ET-1, the most abundant, is a 21-amino acid peptide. A 203-amino acid peptide precursor, preproET, is cleaved after translation by endopeptidases to form a 38-amino acid proET which is converted to active ET by a putative endothelin-converting enzyme (ECE). ET-3 differs from ET-1 and ET-2 by six amino acids.

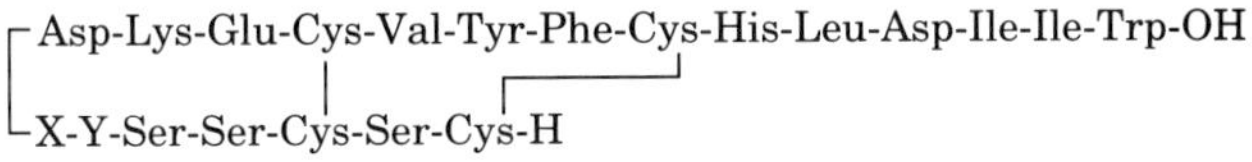

Asp-Lys-Glu-Cys-Val-Tyr-Phe-Cys-His-Leu-Asp-Ile-Ile-Trp-OH
X-Y-Ser-Ser-Cys-Ser-Cys-H

(**133**) X = Met , Y = Leu

(**134**) X = Leu , Y = Trp

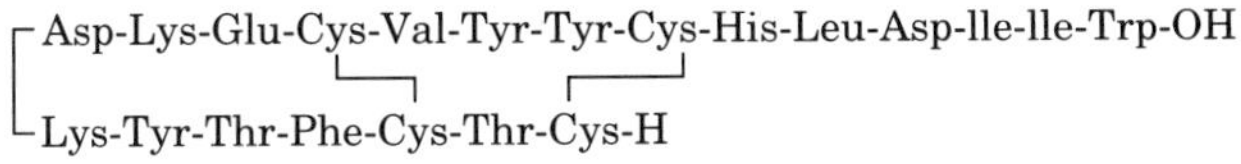

Asp-Lys-Glu-Cys-Val-Tyr-Tyr-Cys-His-Leu-Asp-Ile-Ile-Trp-OH
Lys-Tyr-Thr-Phe-Cys-Thr-Cys-H

(**135**)

Two ET GPCR subtypes, ET_A and ET_B, have been cloned from human tissues. Both receptors utilize IP_3/DAG for transduction. ET-1 and ET-2 have similar affinities for the ET_A subtype, whereas the affinity of ET-3 is much lower. All three peptides have similar affinities for the ET_B subtype. Both receptor subtypes are widely distributed, but ET_A receptors are more abundant in human heart, whereas ET_B receptors constitute 70% of the ET receptors found in kidney. BQ 123 [*136553-81-6*], $C_{31}H_{42}N_6O_7$, cyclo-[D-Asp-Pro-D-Val-Leu-D-Trp], and FR 139317 (**136**) are selective ET_A antagonists. [$Ala^{1,3,11,15}$]ET-1 and BQ 3020 (**137**) are selective ET_B agonists. [Cys^{11-15}]endothelin 1_{11-21}, IRL 1038 (**138**), and BQ 788 (**139**) are selective ET_B antagonists. Ro 46-2005 (**140**) and SB 209670 (**141**) are the first synthetic orally active endothelin receptor antagonists. The ET_C receptor is a third ET receptor. Peptides and receptors are listed in Table 8.

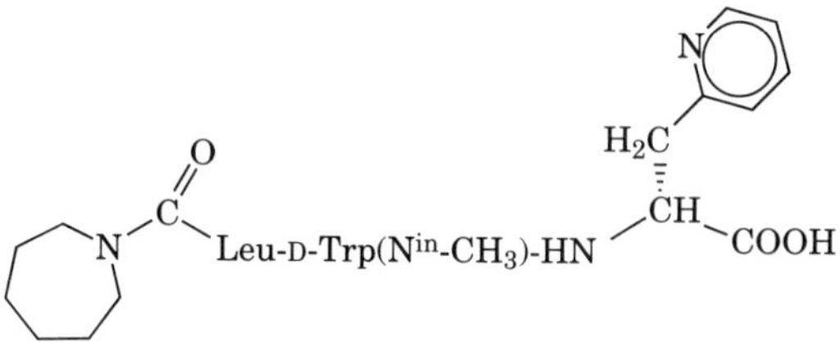

(**136**)

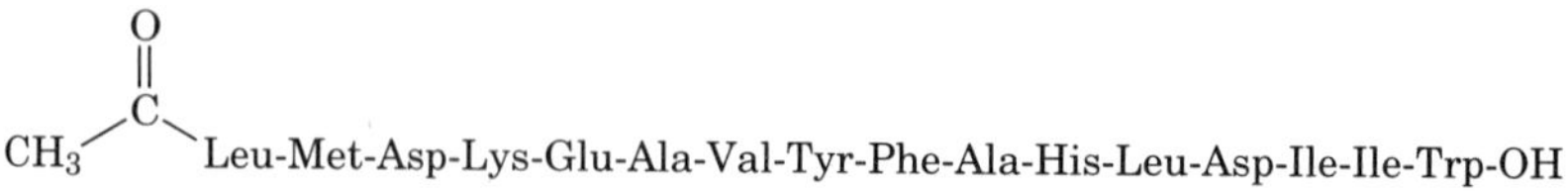

(**137**)

H-Cys-Val-Tyr-Phe-Cys-His-Leu-Asp-Ile-Ile-Trp-OH

(**138**)

Table 8. Endothelin Peptides and Receptors

Agonist/antagonist	CAS Registry Number	Molecular formula	Structure number
endothelin-1	[*117399-94-7*]	$C_{109}H_{159}N_{25}O_{32}S_5$	(**133**)
endothelin-2	[*122879-69-0*]		(**134**)
endothelin-3	[*125692-40-2*]		(**135**)
FR 139317	[*142375-60-8*]	$C_{33}H_{44}N_6O_5$	(**136**)[a]
BQ 3020	[*143113-45-5*]	$C_{96}H_{140}N_{20}O_{25}S$	(**137**)
IRL 1038	[*144602-02-8*]	$C_{68}H_{92}N_{14}O_{15}S_2$	(**138**)
BQ 788			(**139**)[a]
Ro 46-2005	[*150725-87-4*]	$C_{23}H_{27}N_3O_6S$	(**140**)
SB 209670	[*157659-79-5*]	$C_{29}H_{28}O_9$	(**141**)

[a] Trp(N^{in}-CH^3) = *N*-methyltryptophan where the methyl group is on the indole nitrogen.

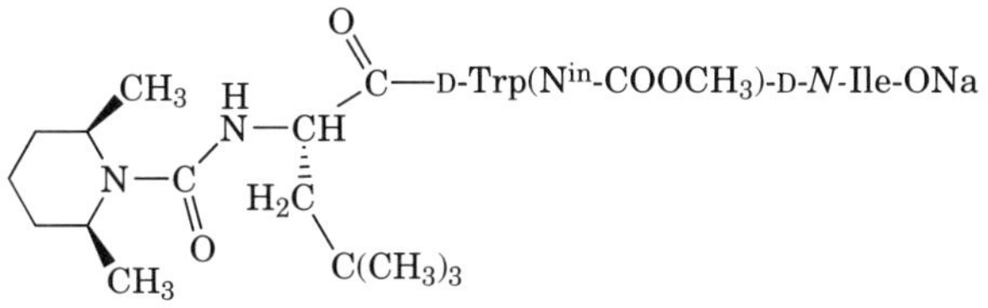

(139)

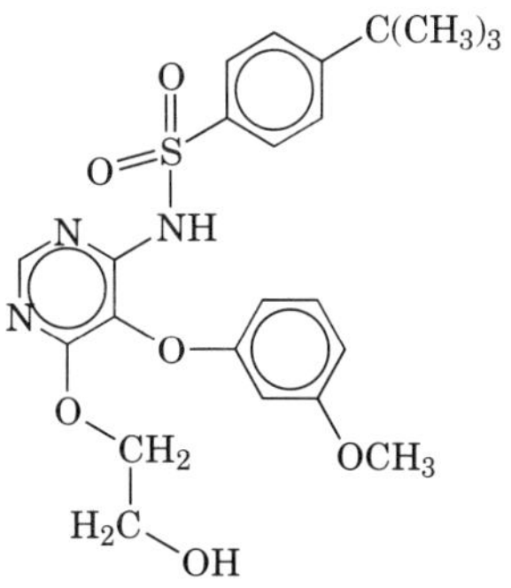

(140)

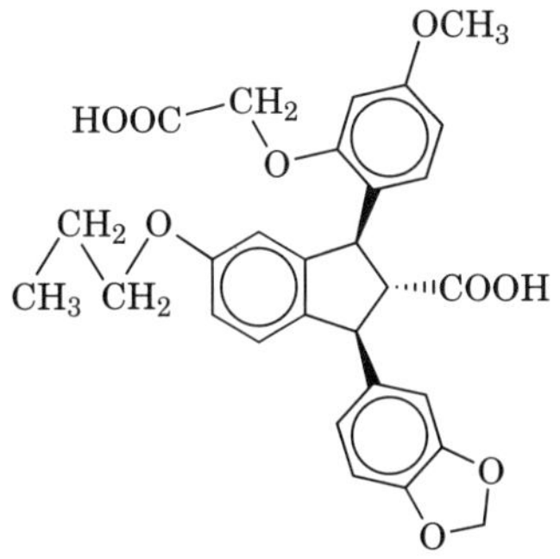

(141)

Long-lasting vasoconstriction is produced by the ETs in almost all arteries and veins and several studies have shown that ET-1 causes a reduction in renal blood flow and urinary sodium excretion. ET-1 has been reported to be a potent mitogen in fibroblasts and aortic smooth muscle cells and to cause contraction of rat stomach strips, rat colon and guinea pig ileum. In the central nervous system, ETs have been shown to modulate neurotransmitter release.

Enkephalins and Endorphins. Morphine (**142**), an alkaloid found in opium, was first isolated in the early nineteenth century and widely used in patent medicines of that era. It is pharmacologically potent and includes analgesic and mood altering effects. Endogenous opiates, the enkephalins, endorphins, and dynorphins were identified in the mid-1970s (3,51) (see OPIOIDS, ENDOGENOUS). Enkephalins and endorphins are listed in Table 9.

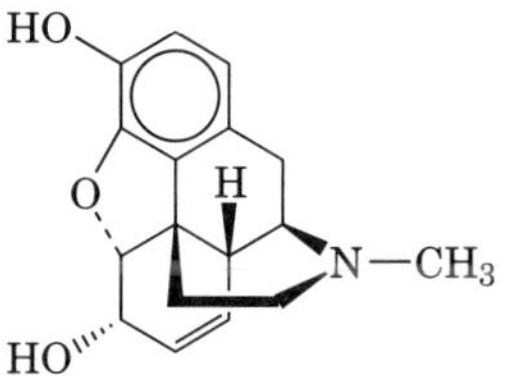

(142)

Table 9. Enkephalins and Endorphins

Agonist/antagonist	CAS Registry Number	Molecular formula	Structure number
morphine	[57-27-2]	$C_{17}H_{19}NO_3$	(**142**)
β-endorphin	[60118-07-2]		(**143**)
naloxone	[465-65-6]	$C_{19}H_{21}NO_4$	(**144**)
naltrexone	[16590-41-3]	$C_{20}H_{23}NO_4$	(**145**)
DAMGO	[78123-71-4]	$C_{26}H_{35}N_5O_6$	(**146**)
sufentanil	[56030-54-7]	$C_{22}H_{30}N_2O_2S$	(**147**)
DPDPE		$C_{30}H_{41}N_5O_7S_2$	(**148**)
U 69593	[96744-75-1]	$C_{22}H_{32}N_2O_2$	(**149**)
CI 977	[124439-07-2]	$C_{24}H_{32}N_2O_3$	(**150**)
β-FNA	[72782-05-9]	$C_{25}H_{30}N_2O_6$	(**151**)
ICI 174864	[89352-67-0]	$C_{34}H_{46}N_4O_6$	(**152**)
naltrindole	[111555-53-4]	$C_{26}H_{26}N_2O_3$	(**153**)
naltriben	[111555-58-9]	$C_{26}H_{25}NO_4$	(**154**)
norbinaltorphimine	[105618-26-6]	$C_{40}H_{43}N_3O_6$	(**155**)

β-Endorphin (**143**) is produced from pro-opiomelanocortin, a precursor for α-, β-, and γ-melanocyte stimulating hormones (MSH), adrenocorticotropin (ACTH), which includes the α-MSH sequence, and β-lipotropin (β-LPH), which includes β-MSH and β-endorphin sequences. (Met)-enkephalin [58569-55-4], $C_{27}H_{35}N_5O_7S$, H-Tyr-Gly-Gly-Phe-Met-OH, and (Leu)-enkephalin [61090-95-7], $C_{28}H_{37}N_5O_7$, H-Tyr-Gly-Gly-Phe-Leu-OH, are derived from proenkephalin, each molecule of which contains six (Met)-enkephalin sequences and one (Leu)-enkephalin sequence. Dynorphin-A, dynorphin-B, neoendorphin, and β-neoendorphin are all derived from prodynorphin and contain the Leu-enkephalin sequence at their N-terminal region. The opioid precursors and the opioids themselves may be processed to other peptidergic neuroregulators, depending on peptidases that may be regulated in a tissue- or activity-dependent manner. Enkephalin and dynorphin systems are widespread in the brain and spinal cord and are found in adrenal medulla and the enteric nervous system. Enkephalins are associated with areas thought to be involved in pain sensation, affective behavior, autonomic regulation, and endocrine regulation. The activity of the enkephalins, endorphins, and dynorphins is terminated by proteases, most notably the neutral endopeptidase 23.11, also known as enkephalinase.

H-Tyr-Gly-Gly-Phe-Met-Thr-Ser-Glu-Lys-Ser-Gln-Thr-Pro-Leu-Val-Thr-
Leu-Phe-Lys-Asn-Ala-Ile-Ile-Lys-Asn-Ala-Tyr-Lys-Lys-Gly-Glu-OH

(**143**)

Opiates interact with three principal classes of opioid GPCRs: μ-selective for the endorphins, δ-selective for enkephalins, and κ-selective for dynorphins (51). All three receptors have been cloned. Each inhibits adenylate cyclase, can activate potassium channels, and inhibit N-type calcium channels. The classical opiates, morphine and its antagonists naloxone (**144**) and naltrexone (**145**), have moderate selectivity for the μ-receptor. Pharmacological evidence suggests that there are two subtypes of the μ-receptor and three subtypes each of the δ- and κ-receptor. An ϵ-opiate receptor may also exist.

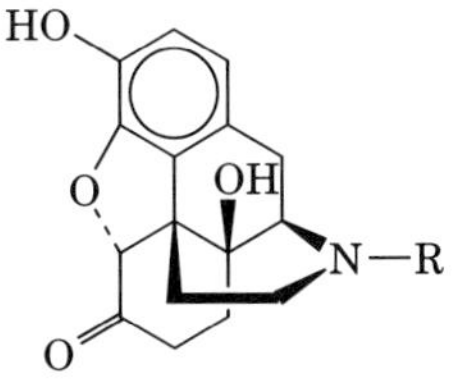

(**144**) R = CH$_2$CH=CH$_2$

(**145**) R = CH$_2$—◁

The search for nonpeptidic enkephalin-like analogues to replace morphine has been unsuccessful as of this writing. The majority of known enkephalin mimics are modified peptides or morphine congeners and include the selective μ-receptor agonists, DAMGO (**146**), sufentanil (**147**), and PL017 [*83397-56-2*], C$_{29}$H$_{37}$N$_5$O$_5$, H-Tyr-Pro-(N-Me)Phe-D-Pro-NH$_2$; the δ_1 agonists DPDPE (**148**), DSBULET, H-Tyr-D-Ser(Ot-Bu)-Gly-Phe-Leu-Thr-OH, and [D-Ala^2]-deltorphin [*122752-15-2*], H-Tyr-D-Ala-Phe-Asp-Val-Val-Gly-NH$_2$; the δ_2 receptor agonist, DSLET, H-Tyr-D-Ser-Gly-Phe-Leu-Thr-OH; and the κ-receptor agonists, U 69593 (**149**) and CI 977 (**150**). β-FNA (**151**) is a μ-receptor antagonist, ICI 174864 (**152**) and naltrindole (**153**) are δ_1 receptor antagonists, and naltriben (**154**) and norbinaltorphimine (**155**) are δ_2 and κ-antagonists, respectively.

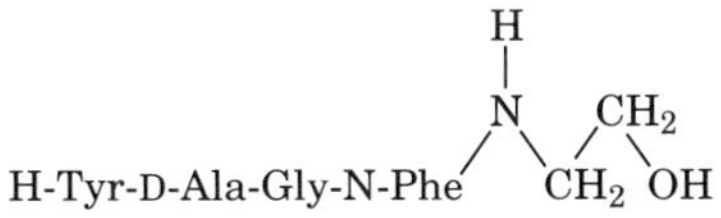

(**146**)

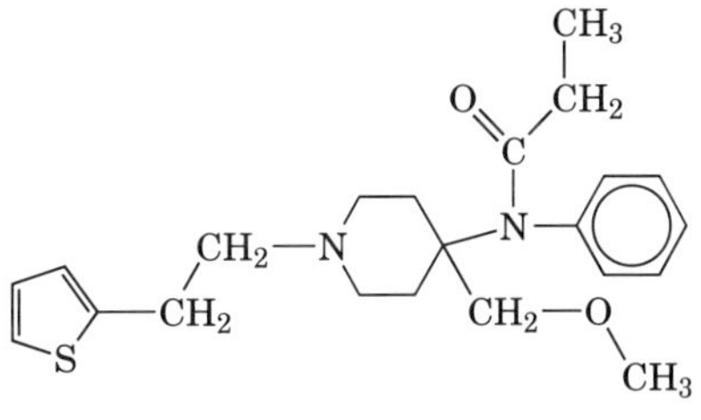

(**147**)

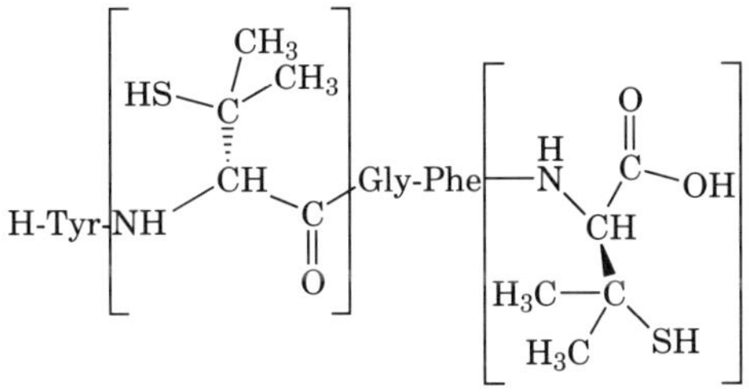

(**148**)

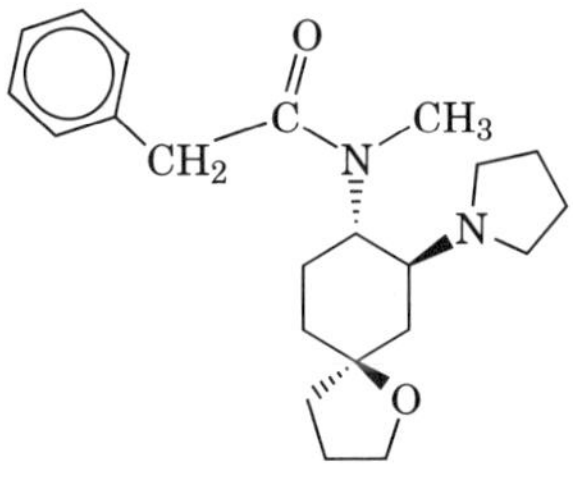

(**149**)

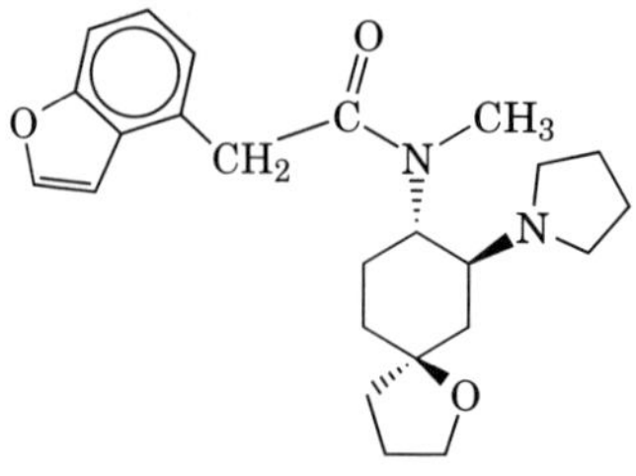

(**150**)

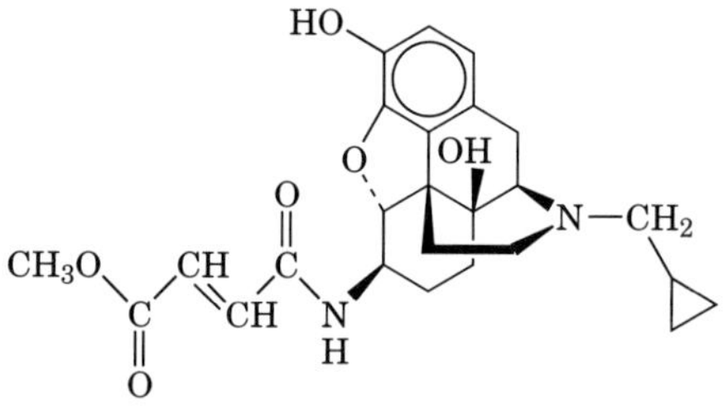

(**151**)

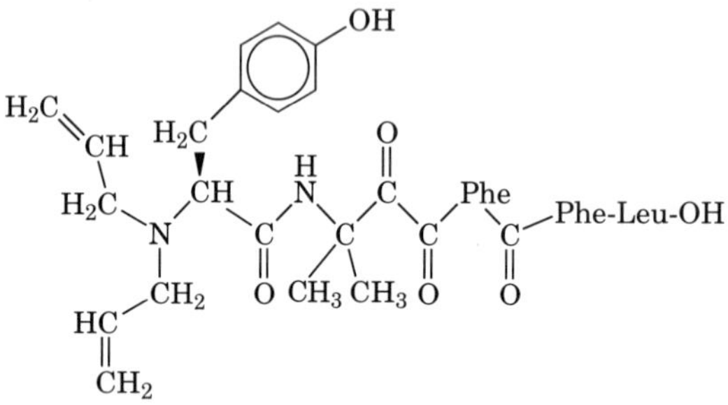

(**152**)

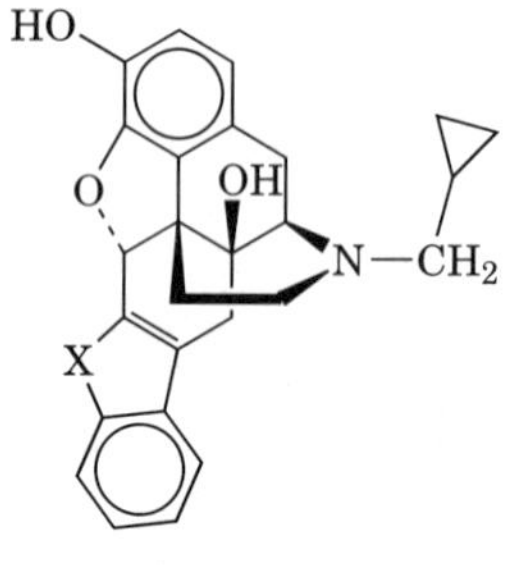

(**153**) X = NH

(**154**) X = O

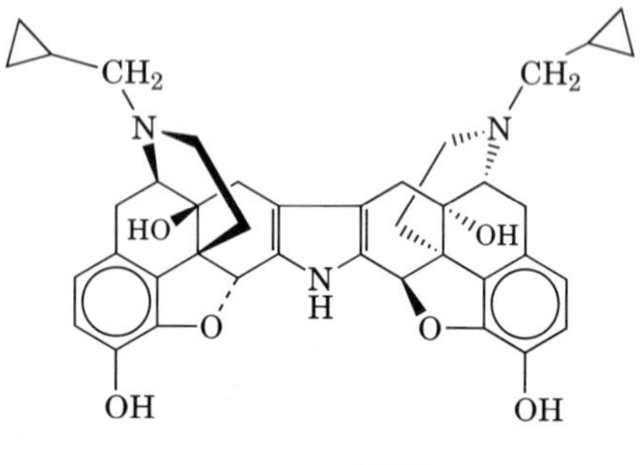

(**155**)

Opiates are useful analgesics because they reduce pain sensation without blocking feeling or other sensations. However, they also affect mood, induce euphoria, reduce mental acuity, and induce physical dependence. They can be immunosuppressive and disrupt other homeostatic processes through inhibition of autonomic and enteric nervous systems. The molecular basis of opiate drug dependence and these side-effect liabilities remain unclear except for specific roles for the various receptor subtypes. An analgesic opioid without significant liabilities has yet to be identified.

GABA. γ-Aminobutyric acid (GABA) (**156**) is the primary inhibitory neurotransmitter in the mammalian brain (52). It is formed by the α-decarboxylation of L-glutamate catalyzed by the enzyme glutamic acid decarboxylase (GAD). GABA also plays a role in the oxidative metabolism of carbohydrates in the Krebs cycle via GABA transaminase (GABA-T). The actions of GABA are terminated by reuptake and metabolism.

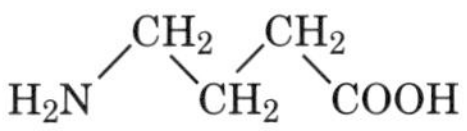

$$H_2N \overset{CH_2}{\diagdown} \overset{CH_2}{\underset{CH_2}{\diagup}} \overset{CH_2}{\underset{COOH}{\diagdown}}$$

(**156**)

GABA interacts with three subclasses of receptor: $GABA_A$ with an associated central BZ receptor; $GABA_B$, a GPCR that may exist in as many as four subclasses; and a newly described $GABA_C$ receptor also referred to as a non-$GABA_A$, non-$GABA_B$ receptor. The transduction mechanism for the $GABA_A$ receptor involves a Cl^- channel, and the transduction mechanism for the $GABA_{B1\alpha}$ receptor involves adenylate cyclase and K^+ and Ca^{2+} channels.

Muscimol (**157**), isoguvacine (**158**), and THIP (**159**) are selective $GABA_A$ agonists. (R)-(+)-Baclofen (**160**) and 3-APPA (**161**) are $GABA_B$ agonists. 3-APPA has selectivity for the $GABA_{B2}$ receptor. Baclofen is also active at the $GABA_C$ receptor, which is characterized as being GABA agonist-sensitive, but insensitive to isoguvacine and bicuculline. Bicuculline (**162**), SR 95531 (**163**), and Ro 5-3663 (**164**) are $GABA_A$ receptor antagonists. Phaclofen (**165**) and CGP 36742 (**166**) are $GABA_B$ receptor antagonists. CGP 35348 (**167**) is a $GABA_{B1\beta}$ antagonist. CACA (**168**) is an antagonist for the $GABA_C$ receptor (Table 10).

Table 10. GABA and GABA Receptor Agonists and Antagonists

Agonist/antagonist	CAS Registry Number	Molecular formula	Structure number
γ-aminobutyric acid	[56-12-2]	$C_4H_9NO_2$	(**156**)
muscimol	[2763-96-4]	$C_4H_6N_2O_2$	(**157**)
isoguvacine	[64603-90-3]	$C_6H_9NO_2$	(**158**)
THIP	[64603-91-4]	$C_6H_8N_2O_2$	(**159**)
(R)-(+)-baclofen	[69308-37-8]	$C_{10}H_{12}ClNO_2$	(**160**)
3-APPA	[103680-47-3]	$C_3H_{10}NO_2$	(**161**)
bicuculline	[485-49-4]	$C_{20}H_{17}NO_6$	(**162**)
SR 95531	[104104-50-9]	$C_{15}H_{17}N_3O_3$	(**163**)
Ro 5-3663	[70656-87-0]	$C_{10}H_{10}N_2O$	(**164**)
phaclofen	[114012-12-3]	$C_9H_{13}ClNO_3P$	(**165**)
CGP 36742	[123690-78-8]	$C_7H_{18}NO_2P$	(**166**)
CGP 35348	[123690-79-9]	$C_8H_{20}NO_4P$	(**167**)
CACA	[55199-25-2]	$C_4H_7NO_2$	(**168**)

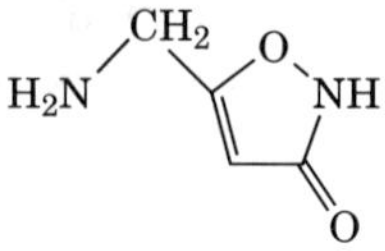

(**157**) 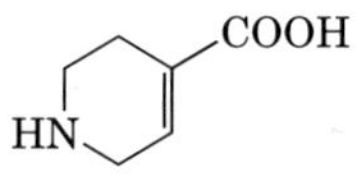   (**158**) 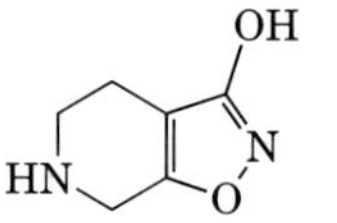   (**159**)

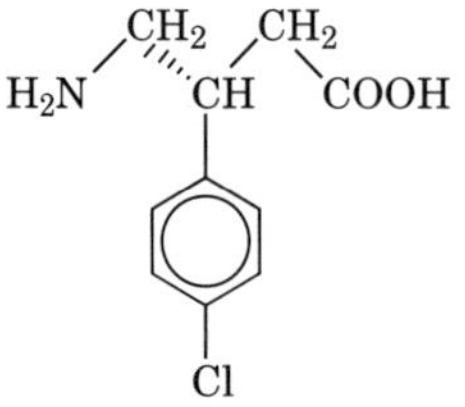

(**160**) 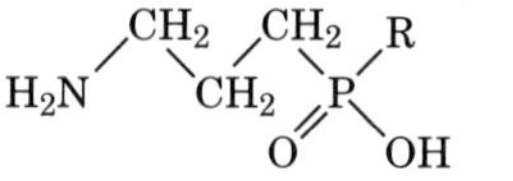   (**161**) R = H

                  (**166**) R = CH$_2$(CH$_2$)$_2$CH$_3$

                  (**167**) R = CH(OC$_2$H$_5$)$_2$ (**162**)

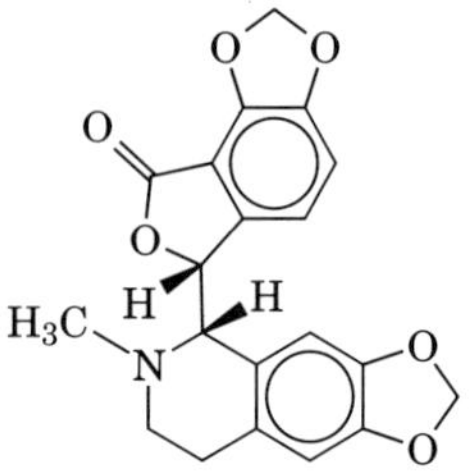

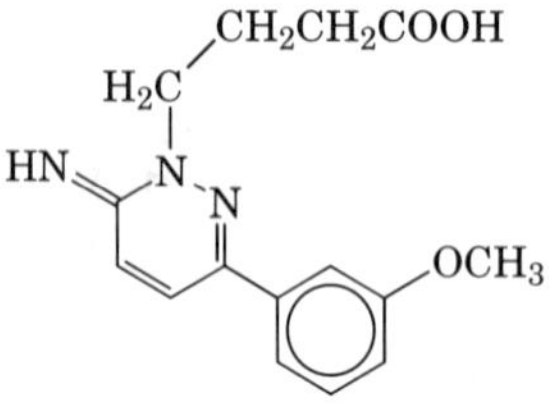

(**163**) 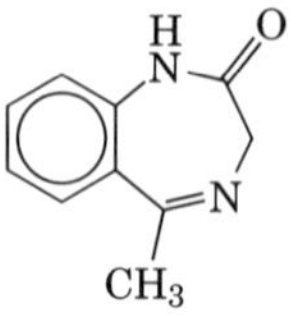   (**164**) 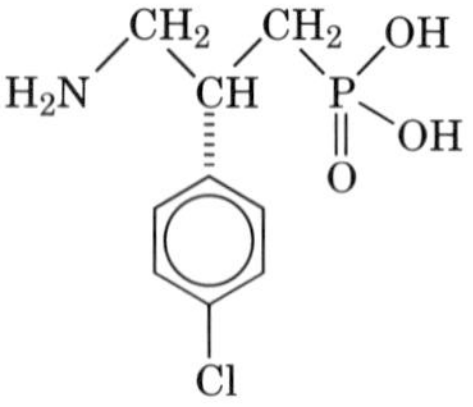   (**165**)

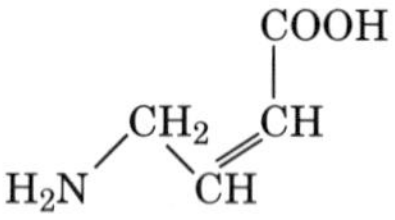

(**168**)

Agents that modulate GABA-ergic neurotransmission have been implicated in processes related to anxiety, hearing, pain sensation, and epilepsy. GABA$_A$ antagonists may also have cognition enhancing activity. GABA$_B$ receptor antagonists, eg, CGP 36742, may have cognition enhancing activity and utility in the treatment of absence seizures and depression.

Galanin.   Galanin [*119418-04-1*] (**169**) is a 29-amino acid neuropeptide derived from the precursor protein prepro-galanin (53). It is widely distributed throughout the peripheral and central nervous systems often co-existing with other neurotransmitters such as ACh, NE, and 5-HT. Prepro-galanin also contains a 59-amino acid C-terminal flanking peptide, a galanin message associated peptide (GMAP), whose function is as yet unknown.

Although numerous biological effects have been attributed to galanin, precise knowledge regarding galanin receptor function has yet to appear. No definitive evidence exists to demonstrate the existence of multiple galanin receptor subtypes. The galanin receptor is a member of the GPCR family; cAMP and Ca^{2+} and K$^+$ channels are involved in transduction. Potent nonpeptide agonists or antagonists of the galanin receptor have not yet been reported. The peptides [D-Thr^6, D-Trp^{8,9}]galanin(1–15)-ol (**170**) and [D-Trp^{8,9}]galanin(1–15)-ol (**171**) are potent galanin antagonists *in vitro* (54). The antagonistic properties of another galanin-receptor antagonist, galantide [*138579-66-5*], C$_{104}$H$_{151}$N$_{25}$O$_{26}$S (**172**), are controversial.

H-Gly-Trp-Thr-Leu-Asn-Ser-Ala-Gly-Tyr-Leu-Leu-Gly-Pro-His-Ala-Ile-Asp-Asn-His-Arg-Ser-Phe-His-Asp-Lys-Tyr-Gly-Leu-Ala-NH$_2$

(**169**)

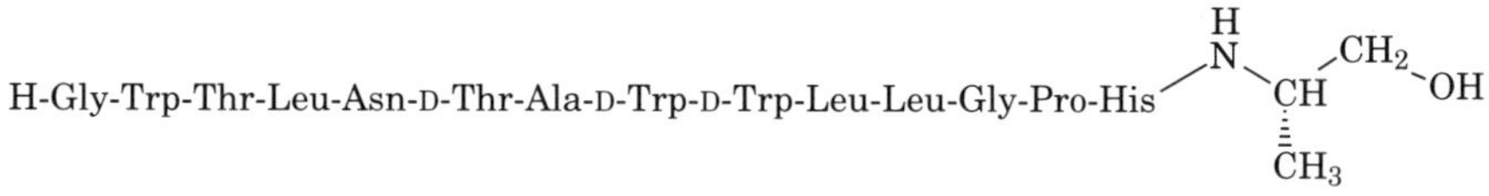

H-Gly-Trp-Thr-Leu-Asn-D-Thr-Ala-D-Trp-D-Trp-Leu-Leu-Gly-Pro-His

(**170**)

H-Gly-Trp-Thr-Leu-Asn-Ser-Ala-D-Trp-D-Trp-Leu-Leu-Gly-Pro-His

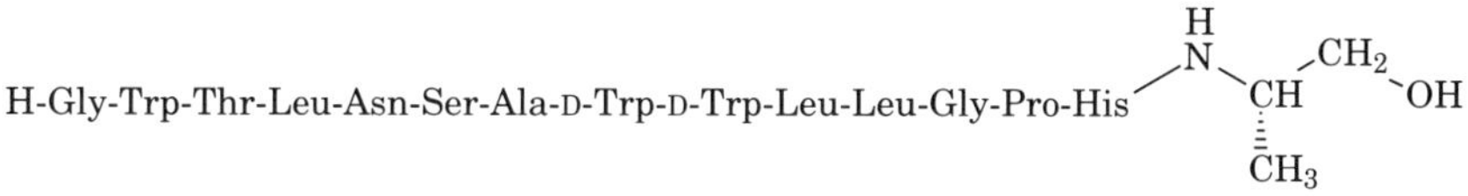

(**171**)

H-Gly-Trp-Thr-Leu-Asn-Ser-Ala-Gly-Tyr-Leu-Leu-Gly-Pro-Gln-Gln-Phe-Phe-Gly-Leu-Met-NH$_2$

(**172**)

Galanin receptor activation reduces intracellular free calcium in most cell types. Galanin is a potent inhibitor of glucose-induced insulin release and has been proposed to be the sympathetic transmitter inhibiting insulin release during stress. In the CNS, galanin is a potent inhibitor of locus coeruleus noradrenergic neuron firing. It also acts tonically as a presynaptic inhibitor of ACh release in the hippocampus which may lead to a decline in cognitive performance. There is abnormally high galanin innervation of the basal forebrain in Alzheimer's disease (AD). Galanin antagonists have therefore been considered for the treatment of AD. Galanin acts synergistically with opiates to suppress the nociceptive flexor reflex suggesting that related agents may prove useful in chronic pain.

Glutamate. Glutamate [*56-86-0*], C$_5$H$_9$NO$_4$ (**173**), the primary excitatory neurotransmitter in the CNS (3), is a direct precursor of GABA via the enzyme glutamic acid decarboxylase (GAD), and is also coupled to cellular energy processes by the tricarboxylic acid cycle (55,56). Glutamate is taken up by glia and neurons. Astrocytic uptake of glutamate appears to predominate over neuronal uptake and may play a predominant role in maintaining extracellular glutamate levels in the brain. Glutamate and glutamine can be interconverted by the enzymes glutamine synthase, which is found predominantly in astrocytes, and phosphate-activated glutaminase. Glial glutamine may serve as a storage form and precursor for neuronal glutamate.

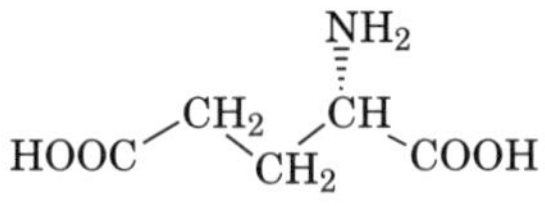

$$\text{HOOC}\overset{\displaystyle \overset{NH_2}{\underset{|}{CH}}}{\underset{\displaystyle CH_2}{\overset{\displaystyle CH_2}{\diagup}}}\text{COOH}$$

(173)

Glutamate interacts with two principal classes of neuronal receptor, an ionotropic LGIC class and the metabotropic GPCR class, each of which exists in multiple subtypes (57). Three distinct types of glutamate LGIC exist, designated AMPA, kainate, and NMDA receptors, each with additional subtypes based on pharmacological, electrophysiological, and structural criteria. Fourteen glutamate LGIC subunits and seven glutamate GPCRs have been cloned. The subunits comprising the LGIC superfamily, unlike those for nicotinic acetylcholine, GABA, and glycine receptors, do not appear to be derived from a common ancestral gene.

AMPA receptors have a high affinity for the agonist, D,L-α-amino-3-hydroxy-5-methyl-4-isoxazole-4-propionic acid (AMPA) (174) and relatively fast activation kinetics. There are four known AMPA receptor subunits. GluR-1, -3, and -4 subunits can form homomeric LGIC receptors in *in vitro* expression systems, but inclusion of GluR-2 with any one of the other three subunits is required for receptor and channel properties similar to most native ionotropic receptors. AMPA receptors lacking GluR-2 subunits are present in cerebellar glia. Numerous additional AMPA receptors can be formed through physiological editing of the RNA to alter an amino acid in transmembrane segment II, ie, the Q/R site that influences divalent ion permeability of the channel, through formation of a C-terminal splice variant of GluR-4, and through alternative splicing of an exon between transmembrane segments III and IV in all four AMPA receptor subunits. AMPA, quisqualate (175), and 5-fluorowillardiine (176) are AMPA agonists. AMPA antagonists include GYKI 52446 (177), LY 215490 (178), and NBQX (179) (Table 11).

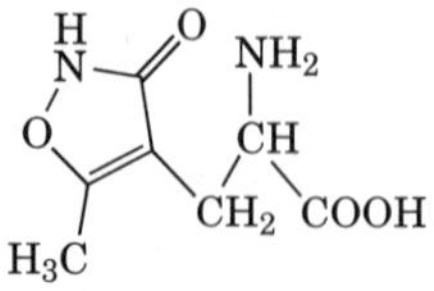

(174) 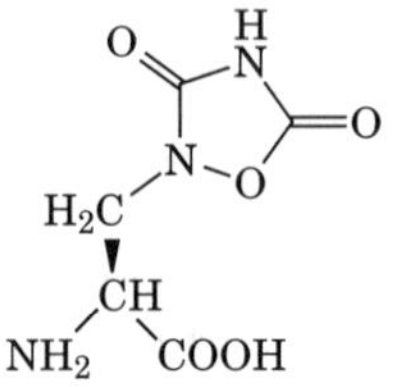   (175) 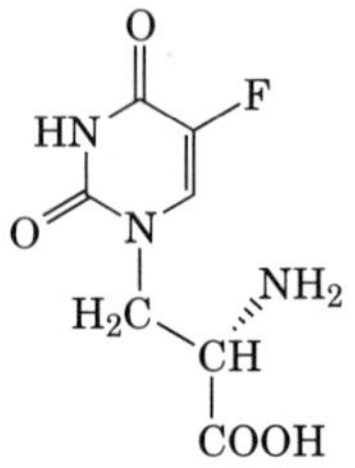   (176)

Table 11. Glutamate Receptor Agonists and Antagonists

Agonist/antagonist	CAS Registry Number	Molecular formula	Structure number
Ionotropic glutamate receptors			
D,L-α-amino-3-hydroxy-5-methyl-4-isoxazole-4-propionic acid	[77521-29-0]	$C_7H_{10}N_2O_4$	(**174**)
quisqualate	[52809-07-1]	$C_5H_7N_3O_5$	(**175**)
5-fluorowillardiine	[140187-23-1]	$C_7H_8FN_3O_4$	(**176**)
GYKI 52446	[102771-26-6]	$C_{17}H_{15}N_3O_2$	(**177**)
LY 215490	[150010-68-7]	$C_{13}H_{21}N_5O_2$	(**178**)
NBQX	[118876-58-7]	$C_{12}H_8N_4O_6S$	(**179**)
Kainate receptors			
kainic acid	[487-79-6]	$C_{10}H_{15}NO_4$	(**180**)
domoate	[14277-97-5]	$C_{15}H_{21}NO_6$	(**181**)
NMDA receptors			
N-methyl-D-aspartate	[6384-92-5]	$C_5H_9NO_4$	(**182**)
CGS 19755	[110347-85-8]	$C_7H_{14}NO_5P$	(**183**)
LY 233053	[125546-04-5]	$C_9H_{18}NO_5P$	(**184**)
phencylidine	[77-10-1]	$C_{17}H_{25}N$	(**185**)
MK 801	[77086-21-6]	$C_{16}H_{15}N$	(**186**)
glycine	[56-40-6]	$C_2H_5NO_2$	
serine	[56-45-1]	$C_3H_7NO_3$	
HA 966	[1003-51-6]	$C_4H_8N_2O_2$	(**187**)
ACBC	[22264-50-2]	$C_5H_9NO_2$	(**188**)
quinolinic acid	[89-00-9]	$C_7H_5NO_4$	(**189**)
kynurenic acid	[492-27-3]	$C_{10}H_7NO_3$	(**190**)
5,7-dichlorokynurenate	[131123-76-7]	$C_{10}H_5Cl_2NO_3$	(**191**)
ACEA 1021	[153506-21-5]	$C_8H_3Cl_2N_3O_4$	(**192**)
L 705,022		$C_9H_{12}ClNO_3S$	(**193**)
MNQX	[136529-54-9]	$C_8H_4N_4O_6$	(**194**)[a]
spermine	[71-44-3]	$C_{10}H_{26}N_4$	(**195**)
spermidine	[124-20-9]	$C_7H_{19}N_3$	(**196**)
eliprodil	[119431-25-3]	$C_{20}H_{23}ClFNO$	(**197**)
Metabotropic glutamate receptors			
1*S*,3(*R*)-ACPD	[111900-32-4]	$C_7H_{11}NO_4$	(**198**)
MCPG	[146665-29-6]	$C_{10}H_{11}NO_4$	(**199**)

[a]MNQX is also a metabotropic glutamate receptor ligand.

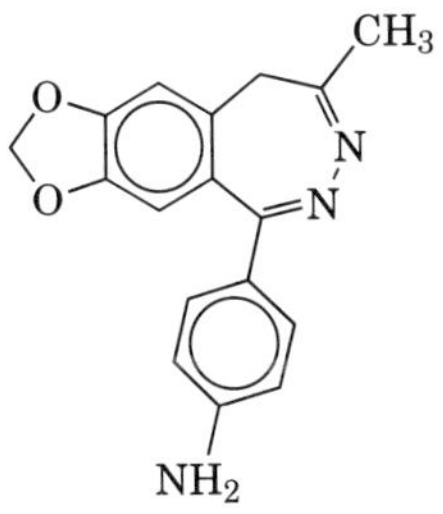

(**177**)

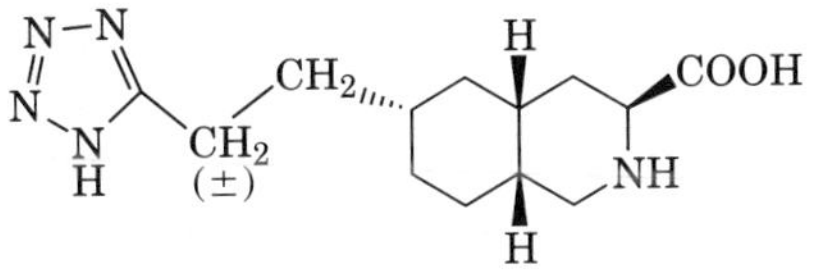

(**178**)

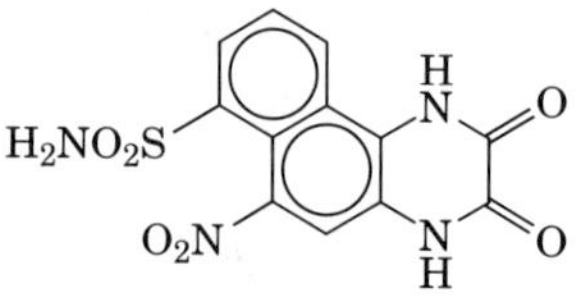

(**179**)

Kainate receptors are LGICs selectively activated by kainic acid (**180**) and domoate (**181**). Kainate receptors are formed by combination of the subunits GluR-5, -6, or -7 plus KA-1 or -2. GluR-5 and GluR-6, but not the other subunits, can be expressed as a homomeric LGIC. Coexpression of either KA subunit with GluR-5 or -6 affects the functional and pharmacologic properties of the complex. Seven different forms of GluR-6 derived from one DNA sequence arise from editing. Five splice variants of GluR-5 have also been identified. Selective antagonists have not yet been identified.

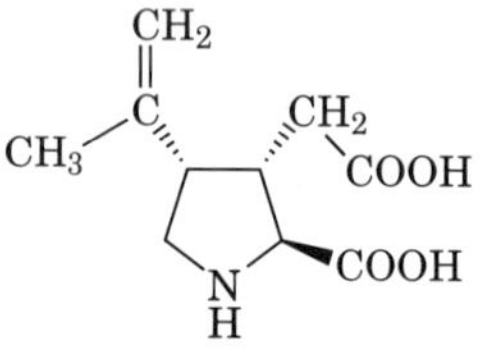

(**180**) 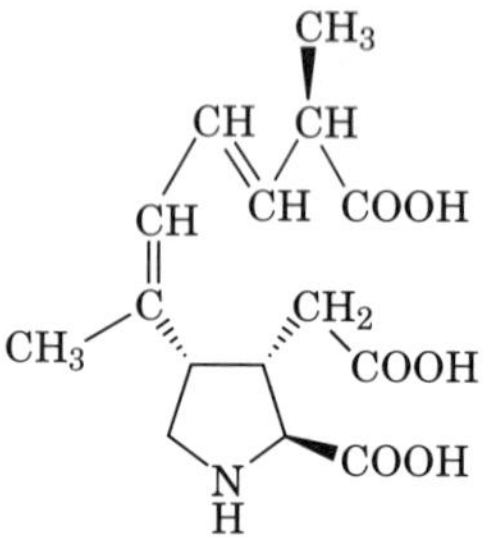   (**181**) 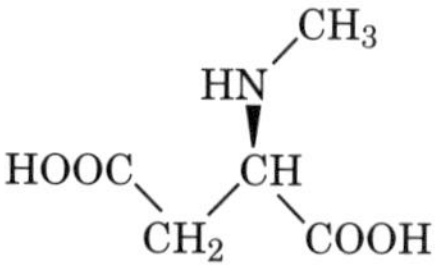   (**182**)

NMDA receptors are selectively activated by *N*-methyl-D-aspartate (NMDA) (**182**). NMDA receptor activation also requires glycine or other co-agonist occupation of an allosteric site. NMDAR-1, -2A, -2B, -2C, and -2D are the five NMDA receptor subunits known. Two forms of NMDAR-1 are generated by alternative splicing. NMDAR-1 proteins form homomeric ionotropic receptors in expression systems and may do so *in situ* in the CNS. Functional responses, however, are markedly augmented by co-expression of a NMDAR-2 and NMDAR-1 subunits. The kinetic and pharmacological properties of the NMDA receptor are influenced by the particular subunit composition.

The phosphonic acid derivatives, CGS 19755 (**183**) and LY 233053 (**184**) are NMDA antagonists, binding at the glutamate site. The channel associated with the NMDA receptor has a binding site for the psychotomimetic, phencyclidine (PCP) (**185**), and the noncompetitive antagonist, MK 801 (**186**). Because directly acting and channel blocking NMDA antagonists induce psychosis, there has been considerable interest in agents that modulate the NMDA receptor via NMDA receptor-associated allosteric modulatory sites, namely the glycine and polyamine sites. Glycine, serine, HA 966 (**187**), and the cyclobutane ACBC (**188**) are glycine-site agonists. Quinolinic acid (**189**) and kynurenic acid (**190**), endogenous tryptophan metabolites, are glycine-site antagonists suggesting that these or related ligands may function physiologically as antagonist neuromodulators of NMDA transmission. 5,7-Dichlorokynurenate (**191**), ACEA 1021 (**192**), L 705,022 (**193**), and MNQX (**194**) are also glycine-site antagonists. At the polyamine site, spermine (**195**), spermidine (**196**), and eliprodil (**197**) modulate NMDA receptor function.

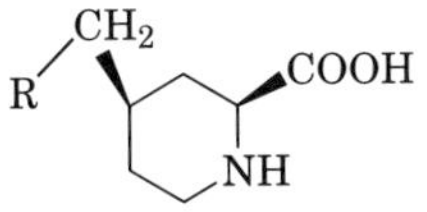

(**183**) R = PO$_3$H$_2$

(**184**) R = (±)

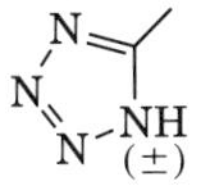

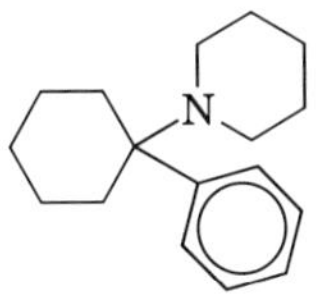

(**185**)

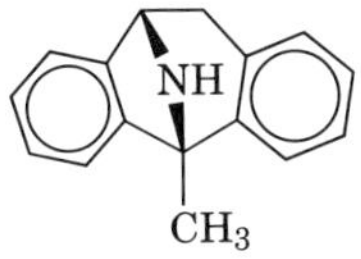

(**186**)

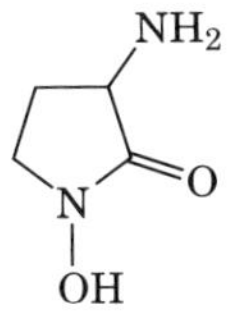

(**187**)

(**188**)

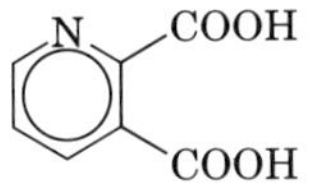

(**189**)

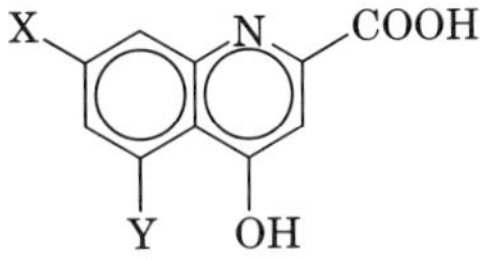

(**190**) X = Y = H

(**191**) X = Y = Cl

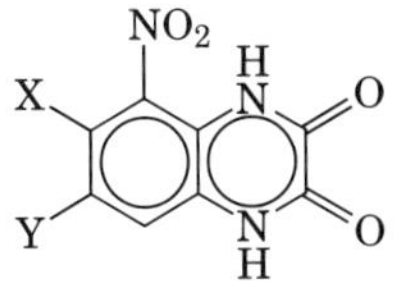

(**192**) X = Y = Cl

(**194**) X = H , Y = NO$_2$

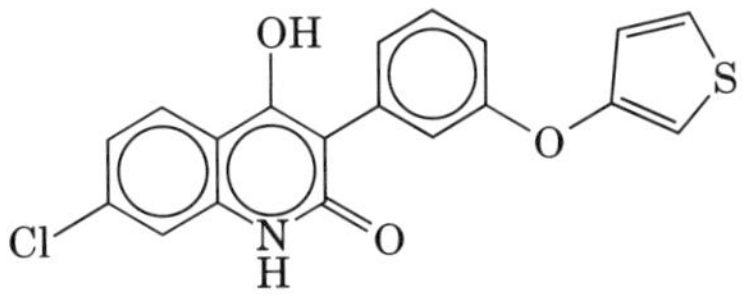

(**193**)

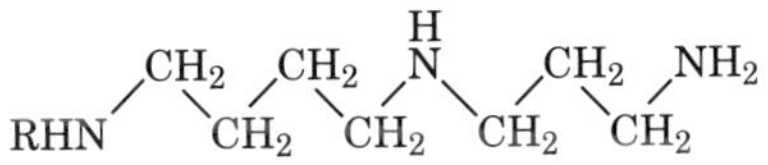

(**195**) R = CH$_2$CH$_2$CH$_2$NH$_2$

(**196**) R = H

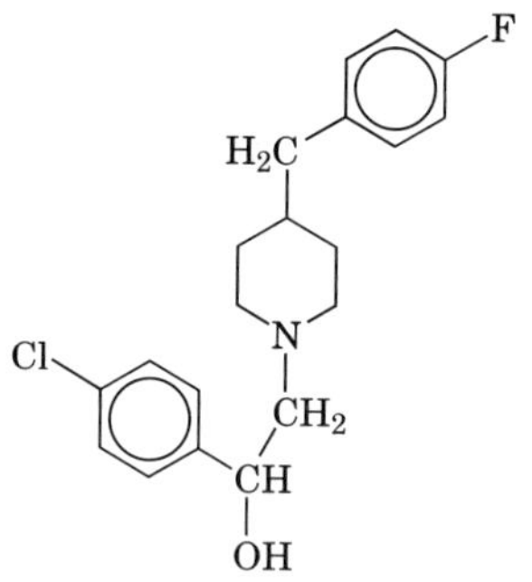

(197)

NMDA LGICs exhibit slow activation kinetics but are highly permeable to calcium relative to other glutamate receptors. They are known for their involvement in calcium-dependent phenomena such as the formation of long-term potentiation (LTP) and long-term depression (LTD), that are thought to underlie learning and memory. However, excess activation of NMDA receptors may contribute to pathologic processes such as excess excitability and calcium-dependent cell death following the hypoxia or ischemia associated with stroke. While MK 801 encountered problems in clinical trials for this indication, CGS 19755 may represent the first NMDA receptor ligand for the treatment of stroke. NMDA receptor modulators may also have potential as anxiolytics, antipsychotics, cognition enhancers, and in the treatment of certain types of pain.

Metabotropic glutamate receptors are GPCRs, although there is little sequence homology between these and other GPCRs (57,58). There are seven mammalian metabotropic glutamate receptor genes, mGluR-1 to -7, with three variants of mGluR-1 formed by alternative splicing. mGluR-1 may stimulate adenylate cyclase whereas the other metabotropic glutamate receptors inhibit cyclic AMP synthesis. mGluR-4 and mGluR-7 are candidate L-AP4 (L-amino-4-phosphonobutanoic acid) receptors, thought to be presynaptic inhibitory feedback autoreceptors.

Enantiomers of 1-aminocyclopentane dicarboxylic acid (ACPD), such as *trans*- or 1(*S*),3(*R*)-ACPD (**198**), activate metabotropic glutamate receptors selectively, but in some cases the activation may be weak and/or less potent than other glutamate agonists such as quisqualate. MNQX (**194**), an agonist at metabotropic glutamate receptors, is also an NMDA glycine-site antagonist. MCPG (**199**) is a metabotropic receptor antagonist.

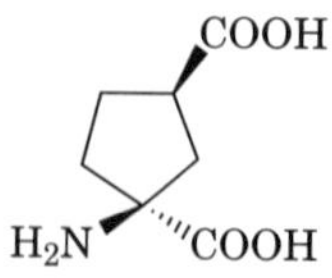

(198)

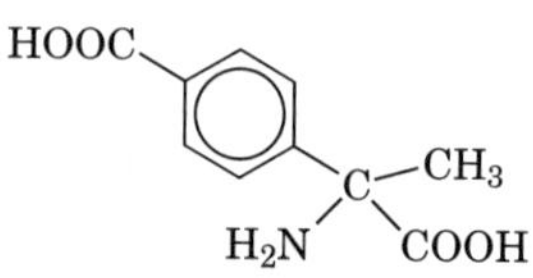

(199)

Metabotropic receptors may mediate increased excitability through inhibition of certain potassium channels, or decreased excitability, through activation of calcium-dependent potassium channels or inhibition of evoked glutamate release. Metabotropic receptors also may regulate the function of ionotropic glutamate receptors, eg, NMDA receptors, or other receptors. The roles of metabotropic glutamate receptors in the formation of long-term potentiation and the regulation of other synaptic processes are under active study (Table 11).

Histamine. Histamine [51-45-6], $C_5H_9N_3$ (**200**) is an inflammatory autacoid involved in allergic and anaphylactic reactions (3,39,59) (see HISTAMINE AND HISTAMINE ANTAGONISTS). It is formed from histidine by the enzyme L-histidine decarboxylase. In the periphery, histamine is stored in mast cells, basophils, cells of the gastric mucosa, and epidermal cells. In the CNS, histamine is released from nerve cells and acts as a neurotransmitter. The actions of histamine are terminated by methylation and subsequent oxidation via the enzymes histamine-N-methyltransferase and monoamine oxidase.

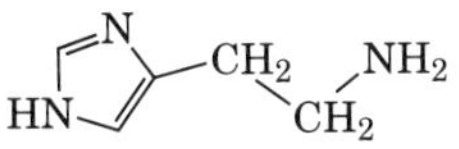

(**200**)

Histamine interacts with three distinct receptor subtypes, H_1, H_2, which have been cloned, and H_3. 2-Methylhistamine (**201**) and 2-(n-fluorophenyl) histamine are selective H_1 agonists (Table 12). Activation of the H_1 receptor results in smooth muscle contraction, bronchoconstriction, and gut contraction. Classical antihistamines like pyrilamine (**202**) are histamine H_1 receptor antagonists that block smooth muscle contraction and allergen release, limiting the scope of the

Table 12. Histamine Receptor Agonists and Antagonists

Agonist/antagonist	CAS Registry Number	Molecular formula	Structure number
	H_1 receptors		
2-methylhistamine	[34392-54-6]	$C_6H_{11}N_3$	(**201**)
pyrilamine	[91-84-9]	$C_{17}H_{23}N_3O$	(**202**)
astemizole	[68844-77-9]	$C_{28}H_{31}FN_4O$	(**203**)
terfenadine	[50679-08-8]	$C_{32}H_{41}NO_2$	(**204**)
	H_2 receptors		
dimaprit	[65119-89-3]	$C_6H_{15}N_3S$	(**205**)
impromidine	[55273-05-7]	$C_{14}H_{23}N_7S$	(**206**)
cimetidine	[51481-61-9]	$C_{10}H_{16}N_6S$	(**207**)
ranitidine	[66357-35-5]	$C_{13}H_{22}N_4O_3S$	(**208**)
	H_3 receptors		
(R)-α-methylhistamine	[75614-87-8]	$C_6H_{11}N_3$	(**209**)
imetit	[102203-18-9]	$C_6H_{10}N_4S$	(**210**)
thioperamide	[106243-16-7]	$C_{15}H_{24}N_4S$	(**211**)
clobenpropit	[145231-45-4]	$C_{14}H_{17}ClN_4S$	(**212**)

reaction. Such first-generation antagonists generally produce sedation and other CNS effects in addition to their beneficial effects. Second-generation peripherally selective H_1 antagonists, such as astemizole (**203**) and terfenadine (**204**), that do not cross the blood brain barrier have reduced sedative properties.

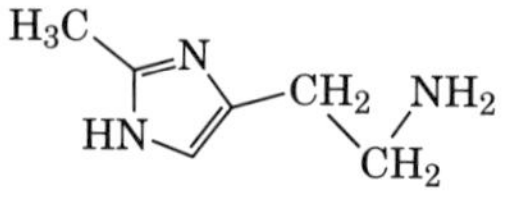

(**201**) 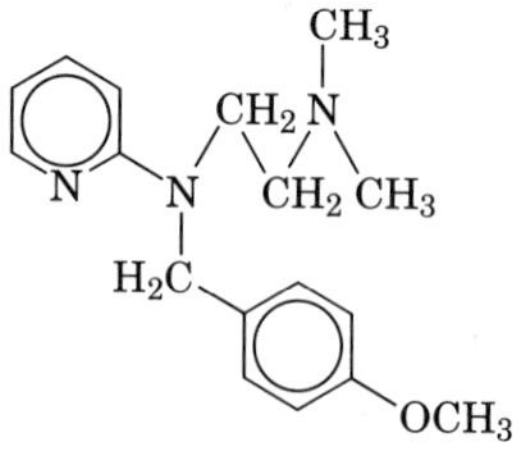   (**202**)

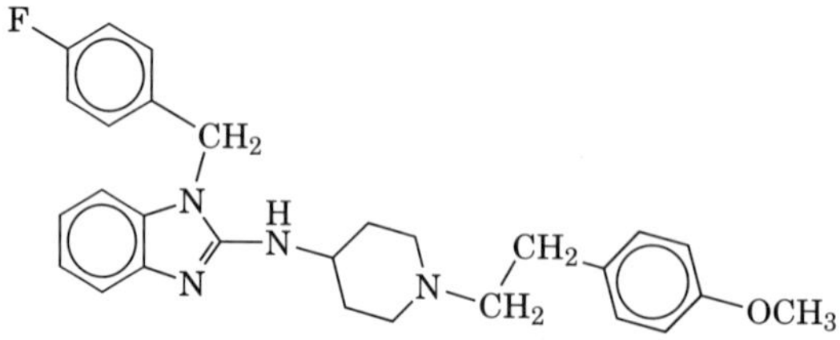

(**203**) 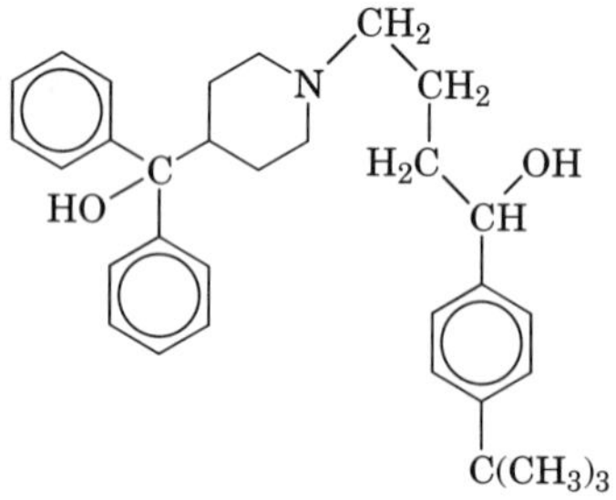   (**204**)

Activation of histamine H_2 receptors with histamine and selective agonists such as dimaprit (**205**) and impromidine (**206**) results in gastric acid secretion and ulcer formation. The selective H_2 blockers, cimetidine (**207**) and ranitidine (**208**) have revolutionized gastric ulcer treatment replacing expensive, debilitating surgery with cost-effective drug therapy (see GASTROINTESTINAL AGENTS).

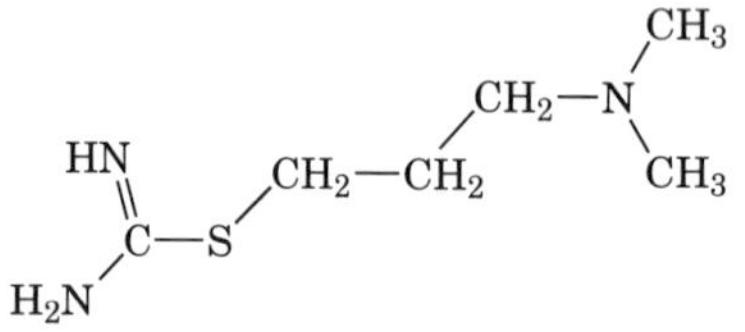

(**205**)

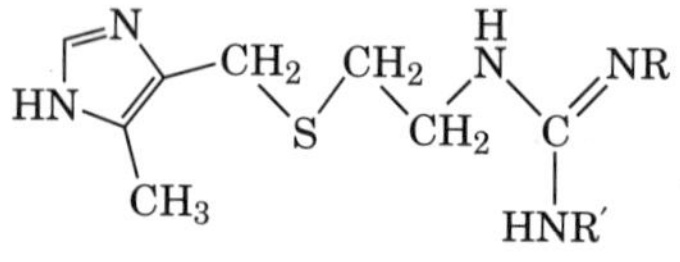

(**206**) R = H , R′ = CH$_2$CH$_2$SCH$_2$

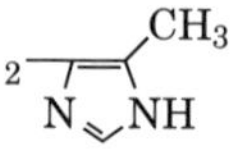

(**207**) R = CN , R′ = CH$_3$

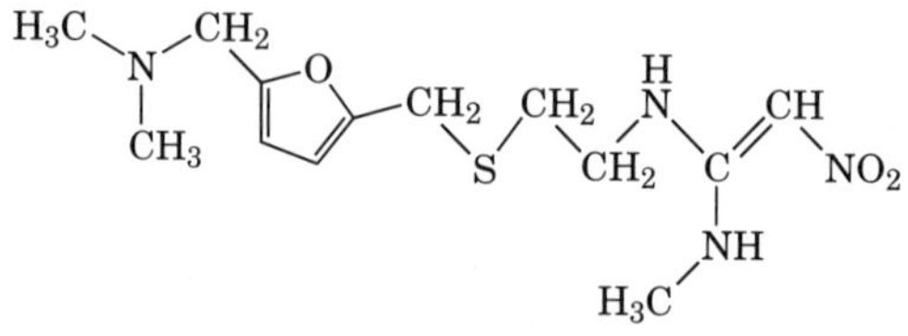

(**208**)

H_1 and H_2 receptors also evoke depressor and vasodilator responses. In the heart, ionotropic effects are H_2-mediated; the negative dromotropic effects of histamine appear to be H_1-mediated. All three types of histamine receptor are present in the CNS, and H_1 receptors in cortex, hippocampus, amygdala, caudate, and putamen are involved in sedative responses. CNS H_3 receptors are localized to cortex, striatum, hippocampus, and olfactory nuclei and appear to be involved in attention and cognition. Activation of presynaptic H_3 receptors by histamine leads to inhibition of both histamine release as well as inhibition of histamine synthesis. (R)-α-Methylhistamine (**209**) and imetit (**210**) are H_3 agonists. Thioperamide (**211**) and clobenpropit (**212**) are H_3 antagonists. All four ligands contain an imidazole moiety and consequently do not penetrate the blood brain barrier well.

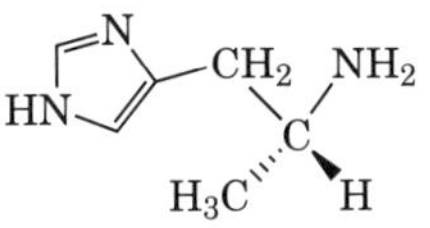

(**209**)

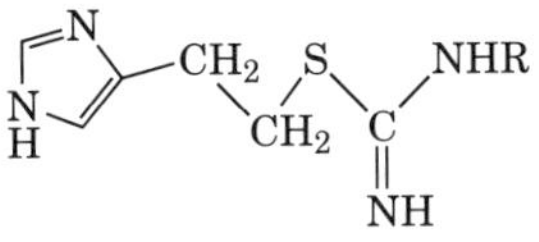

(**210**) R = H

(**212**) R = CH$_2$—⟨○⟩—Cl

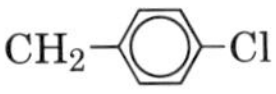

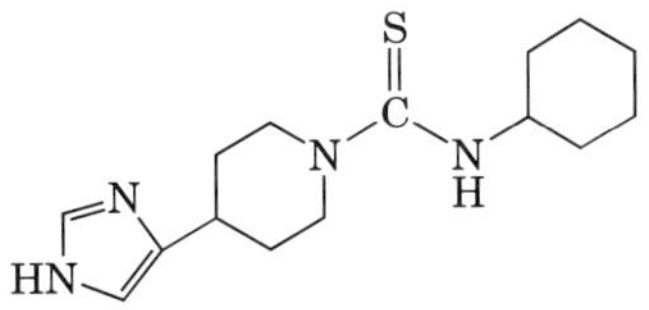

(**211**)

Insulin and Amylin. Insulin is a member of a family of related peptides, the insulin-like growth factors (IGFs), including IGF-I and IGF-II (60) and amylin (**75**), a 37-amino acid peptide that mimics the secretory pattern of insulin. Amylin is deficient in type 1 diabetes mellitus but is elevated in hyperinsulinemic states such as insulin resistance, mild glucose intolerance, and hypertension (33). Insulin is synthesized in pancreatic β cells from proinsulin, giving rise to the two peptide chains, A and B, of the insulin molecule. IGF-I and IGF-II have structures that are homologous to that of proinsulin (see INSULIN AND OTHER ANTIDIABETIC DRUGS).

Insulin elicits a remarkable array of biological responses in a number of tissues including liver, gut, and brain (61). Insulin and IGF-1 receptors (IGF1R) are ligand-activated tyrosine protein kinases. The insulin receptor does not directly bind effector molecules but rather phosphorylates its primary substrate, insulin receptor substrate-1 (IRS1), and IRS1 in turn binds effector molecules. Insulin and IGF-1 receptors are widely distributed throughout the brain and undergo discrete alterations in expression levels during development and post-natal differentiation. Both IGF-1 and -2 are present in the brain and participate in the growth and differentiation of neurons and astrocytes in developing organisms, in synapse formation, in repair processes, in the modulation of satiety, and in feedback regulation of growth hormone secretion. High affinity ($K_d = 28$ pM) binding sites for amylin have been identified in rat brain, particularly in nucleus accumbens. Amylin may be involved in aspects of amyloid formation in both the pancreas and brain (33).

Leukotrienes and Prostanoids. Arachidonic acid (AA) (**213**) and its metabolites are involved in cellular regulatory processes in all three principal chemical signaling systems: endocrine (see HORMONES), immune, and neuronal (62). Following receptor activation or increased intracellular calcium, AA is liberated from membrane phospholipids through the action of phospholipase A_2, or the sequential actions of phospholipase C and diacylglycerol lipase. AA may act within the cell or diffuse extracellularly. It is metabolized to a number of other secondary messenger substances through three main pathways. Cyclooxygenase (COX) is the key enzyme in the formation of the prostanoids, the prostaglandins, and thromboxanes. Two forms of COX are known: COX-1, a constitutive enzyme, and COX-2, an inducible form. The prostaglandins, PGD_2 (**214**), PGE_2 (**215**), $PGF_{2\alpha}$ (**216**), PGH_2 (**217**), and PGI_2 (**218**), and the prostanoid, thromboxane A_2 (**219**), are products of the COX pathway. Cytochrome P_{450} is the key enzyme in the generation of epoxyeicosatrienoic acids (EETs), eg, 14,15-EET (**220**) and hydroxyeicosatetraenoic acids, eg, 5-HETE (**221**). Various lipoxygenases (LOs) generate 5-, 12-, and 15-hydroperoxyeicosatetraenoic acids (5-, 12-, and 15-HPETE). 5-HPETE is metabolized to leukotriene A_4 (LTA$_4$) (**222**) which is a precursor to other leukotrienes. LTB$_4$ (**223**) is generated by LTA$_4$ hydrolase. The peptidoleukotrienes are generated sequentially: LTA$_4$ is converted to LTC$_4$ (**224**) by glutathione-S-transferase; LTC$_4$ to LTD$_4$ (**225**) by glutamyl transferase; and LTD$_4$ to LTE$_4$ (**226**) by a dipeptidase (Table 13).

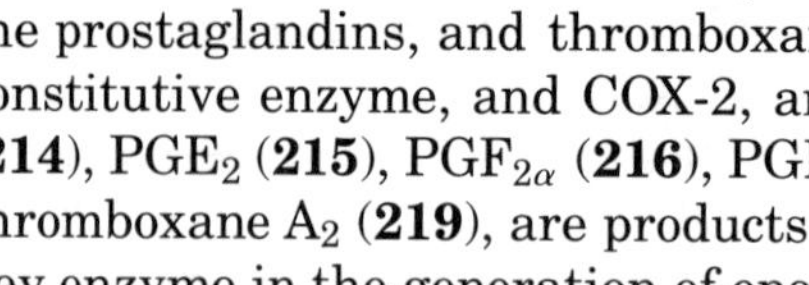

(**213**)

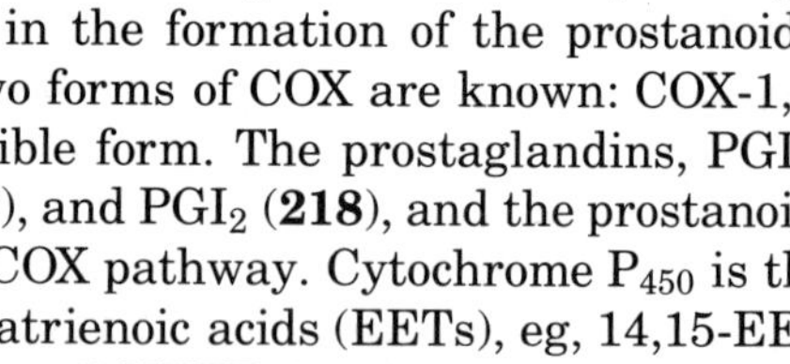

(**214**)

Table 13. Metabolites of Arachidonic Acid

Metabolite	CAS Registry Number	Molecular formula	Structure number
PGD$_2$	[41598-07-6]	$C_{20}H_{32}O_5$	(**214**)
PGE$_2$	[353-24-6]	$C_{20}H_{32}O_5$	(**215**)
PGF$_{2\alpha}$	[551-11-1]	$C_{20}H_{34}O_5$	(**216**)
PGH$_2$	[42935-17-1]	$C_{20}H_{32}O_5$	(**217**)
PGI$_2$	[35121-78-9]	$C_{20}H_{32}O_4$	(**218**)
thromboxane A$_2$	[57576-52-0]	$C_{20}H_{32}O_5$	(**219**)
14,15-EET	[155073-43-1]	$C_{20}H_{30}O_3$	(**220**)
5-HETE	[70608-72-9]	$C_{20}H_{32}O_3$	(**221**)
leukotriene A$_4$	[74807-57-1]	$C_{20}H_{30}O_3$	(**222**)
LTB$_4$	[71160-24-2]	$C_{20}H_{32}O_4$	(**223**)
LTC$_4$	[72025-60-6]	$C_{30}H_{37}N_3O_9S$	(**224**)
LTD$_4$	[73836-78-9]	$C_{25}H_{40}N_2O_6S$	(**225**)
LTE$_4$	[75715-89-8]	$C_{23}H_{37}NO_5S$	(**226**)

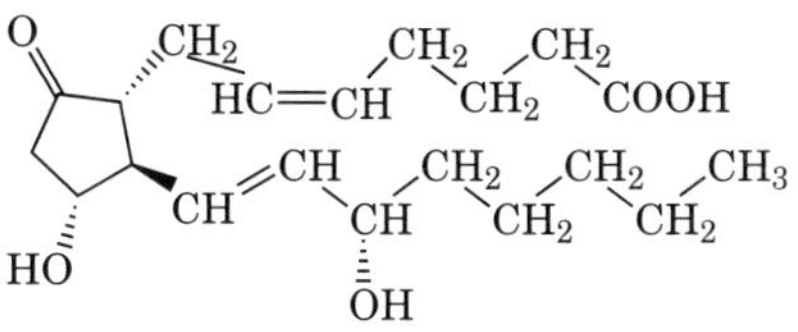

(215)

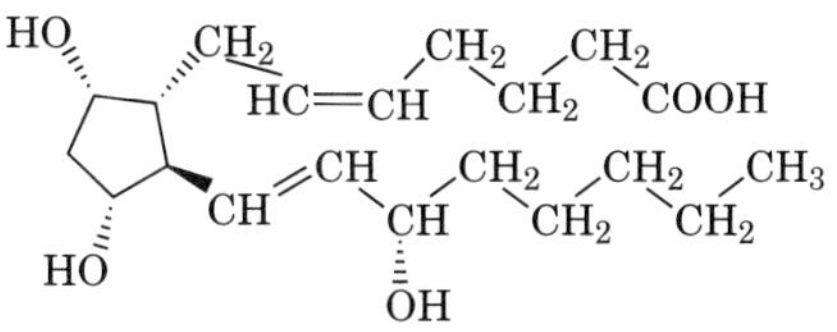

O

CH$_2$ CH$_2$ CH$_2$
HC=CH   CH$_2$ COOH

HO

CH   CH=CH   CH$_2$ CH$_2$ CH$_3$
CH   CH$_2$ CH$_2$
OH

(216)

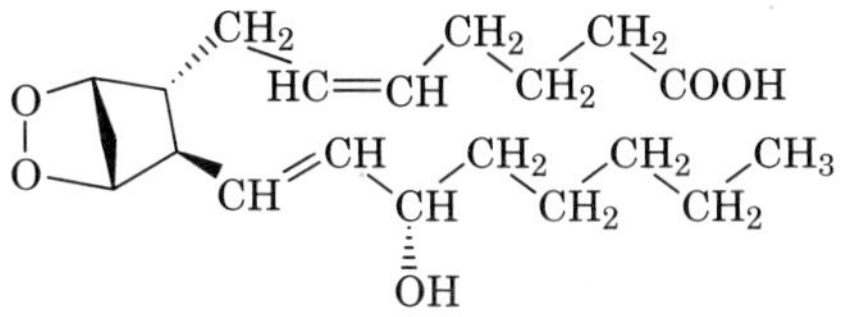

HO   CH$_2$ CH$_2$ CH$_2$
HC=CH   CH$_2$ COOH

HO

CH   CH=CH   CH$_2$ CH$_2$ CH$_3$
CH   CH$_2$ CH$_2$
OH

(217)

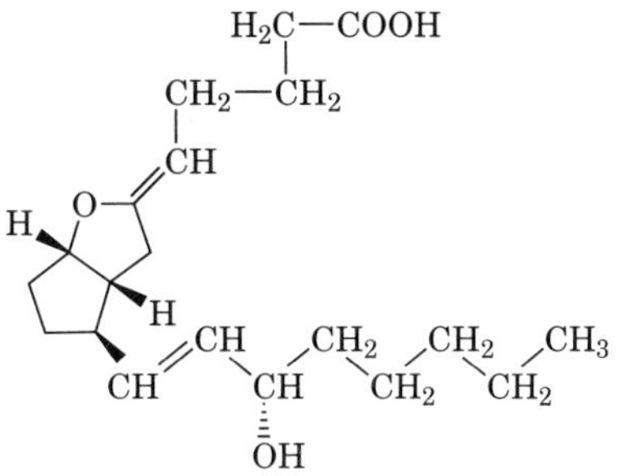

CH$_2$ CH$_2$ CH$_2$
O   HC=CH   CH$_2$ COOH
O

CH   CH=CH   CH$_2$ CH$_2$ CH$_3$
CH   CH$_2$ CH$_2$
OH

(218)

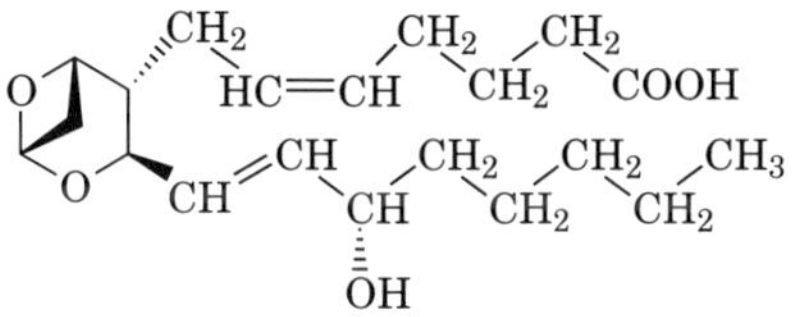

H$_2$C—COOH
CH$_2$—CH$_2$
CH
H   O
H
CH   CH=CH   CH$_2$ CH$_2$ CH$_3$
CH$_2$ CH$_2$
OH

(219)

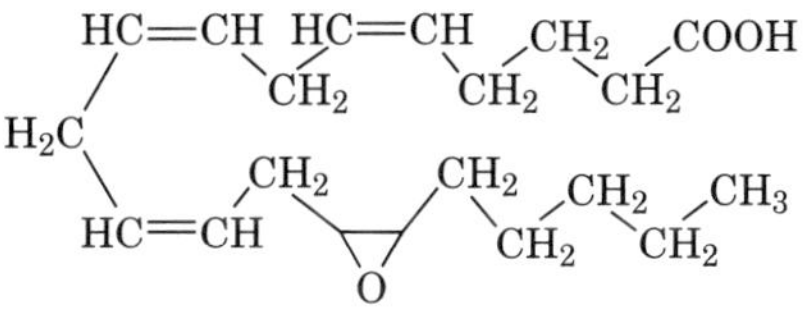

CH$_2$ CH$_2$ CH$_2$
O   HC=CH   CH$_2$ COOH
O
CH   CH=CH   CH$_2$ CH$_2$ CH$_3$
CH   CH$_2$ CH$_2$
OH

(220)

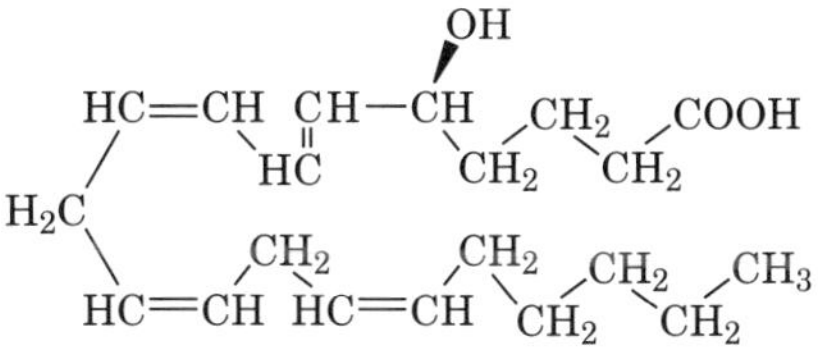

HC=CH HC=CH   CH$_2$ COOH
H$_2$C   CH$_2$ CH$_2$
CH$_2$ CH$_2$ CH$_2$ CH$_3$
HC=CH   O   CH$_2$

(221)

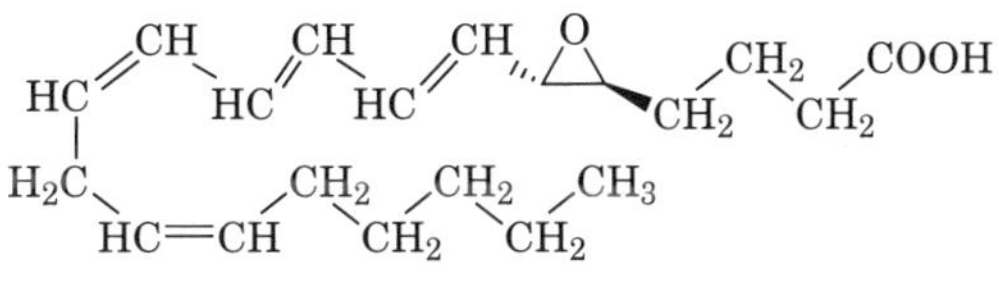

OH
HC=CH CH—CH   CH$_2$ COOH
H$_2$C   HC   CH$_2$ CH$_2$
CH$_2$ CH$_2$ CH$_2$ CH$_3$
HC=CH HC=CH   CH$_2$ CH$_2$

(222)

CH   CH   CH   O   CH$_2$ COOH
HC   HC   HC=CH   CH$_2$ CH$_2$
H$_2$C   CH$_2$ CH$_2$ CH$_3$
HC=CH   CH$_2$ CH$_2$

1085

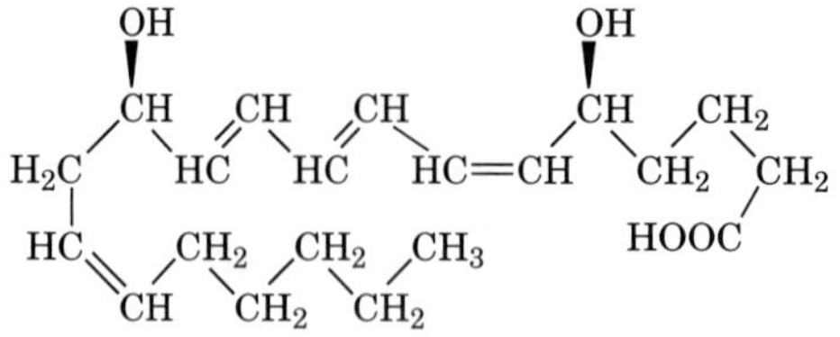

(223)

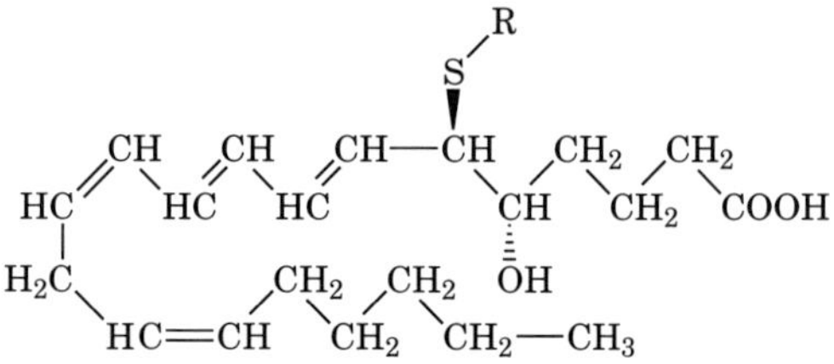

(224) R =

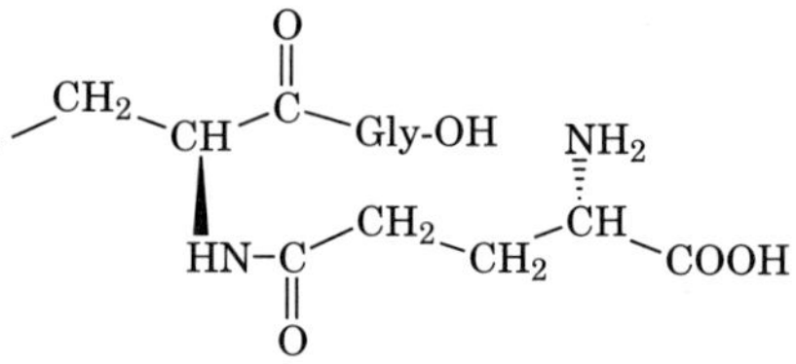

(225) R =

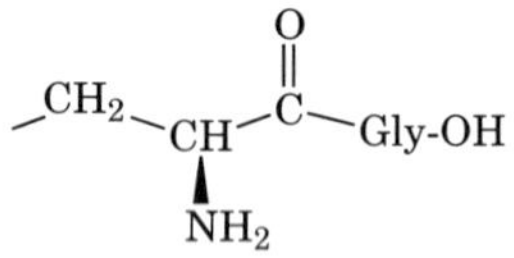

(226) R =

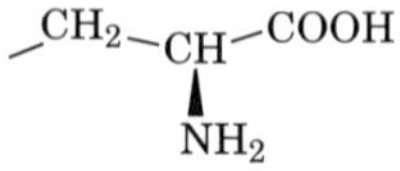

The prostanoids produce effects via five main subclasses of GPCR: DP, EP, FP, IP, and TP (63). The EP receptor exists in four subtypes, EP_1–EP_4. BW 245C (**227**), RS 93520 (**228**), and ZK 110841 (**229**) are DP receptor agonists. Iloprost and enprostil are EP receptor agonists. Fluprostenol (**230**) and cicaprost (**231**) are FP ($PGF_{2\alpha}$) and IP (PGI_2) agonists, respectively. U 46619 (**232**) and STA_2 (**233**) are TP (TXA_2) receptor agonists. GR 32191 (**234**), SQ 29548 (**235**), and ONO 3708 (**236**) are TP (TXA_2) antagonists. AY23626 [37786-01-9] is an EP_2 selective agonist; SC19220 [19395-87-0] is an EP_1 selective antagonist. No selective FP or IP antagonists are known.

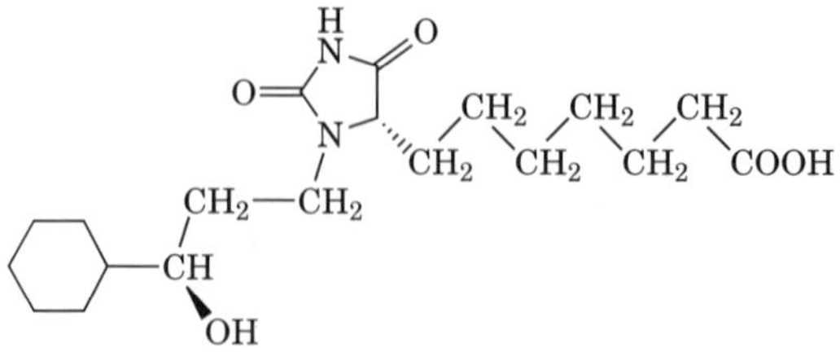

(227)

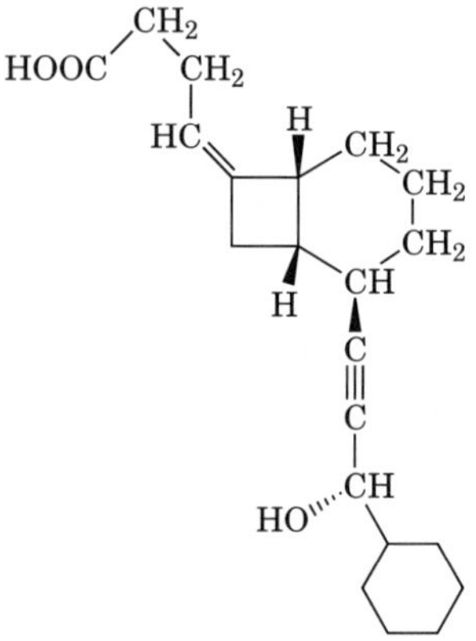

(228)

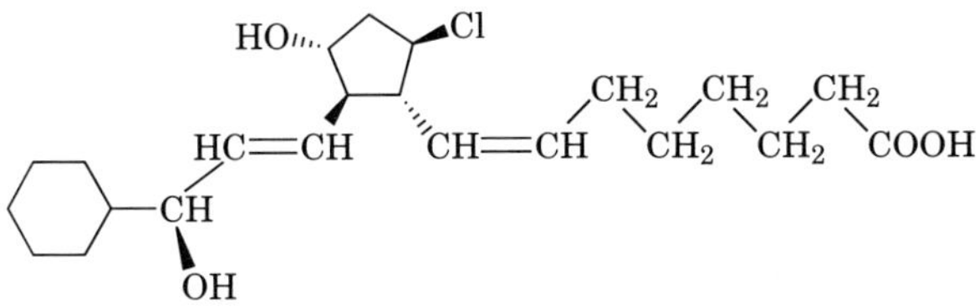

(**229**)

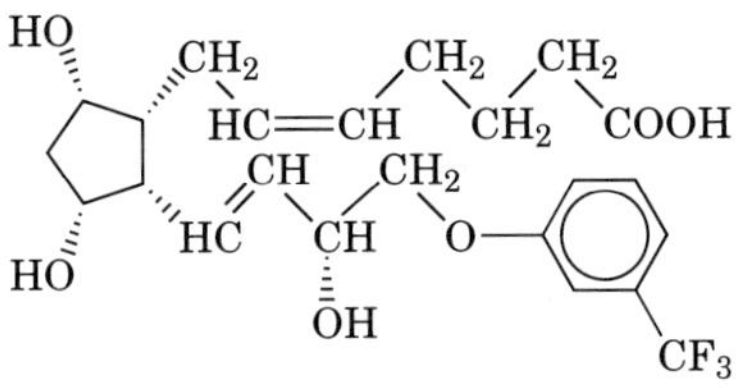

(**230**)

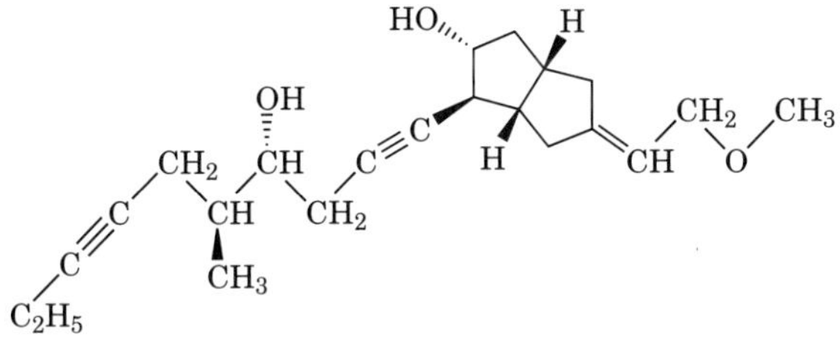

(**231**)

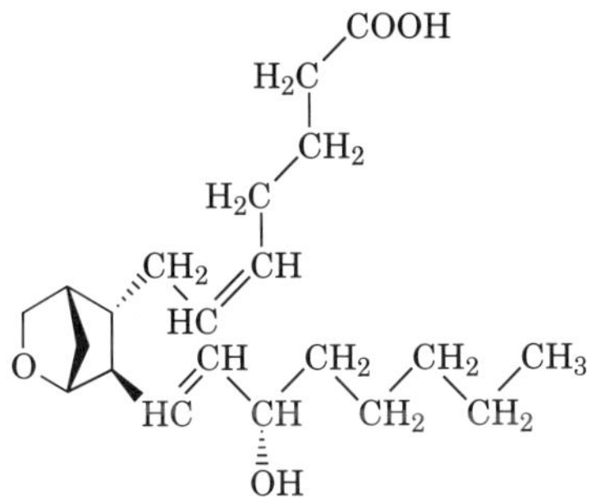

(**232**)

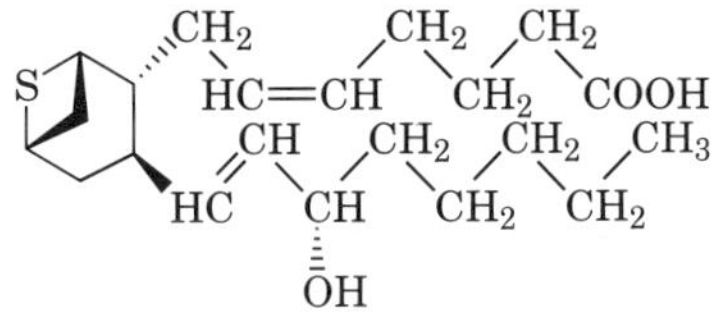

(**233**)

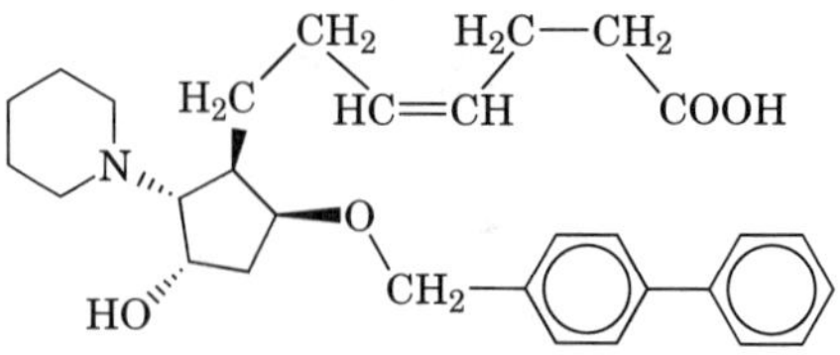

(234)

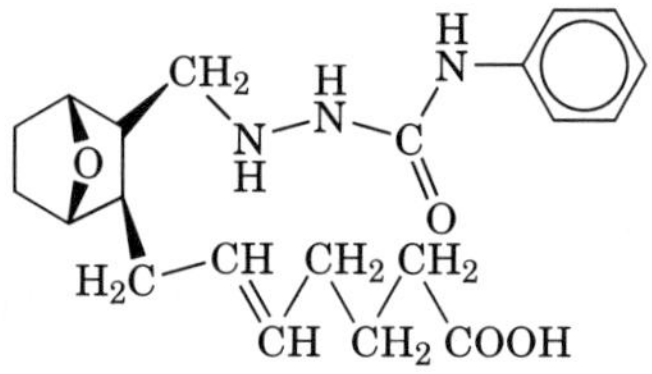

(235)

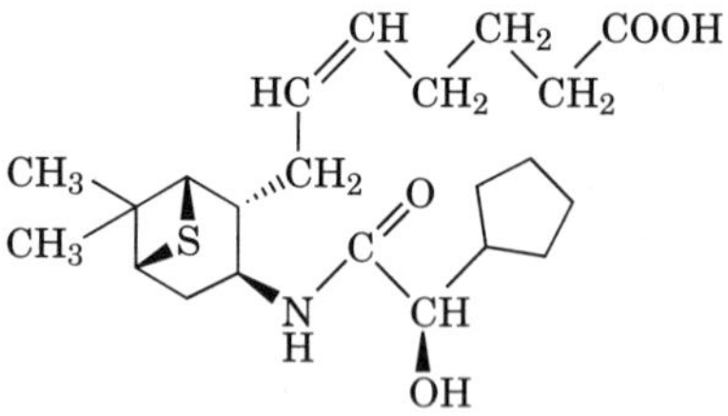

(236)

The leukotrienes also produce effects on tissue function via discrete GPCR subtypes. The leukotriene receptors (LTRs) comprise two main groups: OH LTRs that bind noncysteinyl, dihydroxy LTs such as LTB_4; and Cys-LTRs that bind the cysteinyl LTs, LTC_4, LTD_4, and LTE_4, and exist in two subtypes, Cys-LTR_1 and Cys-LTR_2. The transduction mechanisms for OH-LTR and Cys-LTR_2 involve IP_3/DAG. Leukotrienes are the only known agonists for LTRs. LY 255283 (**237**) and RP 69698 (**238**) are selective OH-LTR antagonists. ICI 198615 (**239**), ONO 1078 (**240**), MK 571 (**241**), and SKF 104353 (**242**) are selective Cys-LTR_1 antagonists. Bay u9773 (**243**) is the antagonist used to define the Cys-LTR_2 receptor (Table 14).

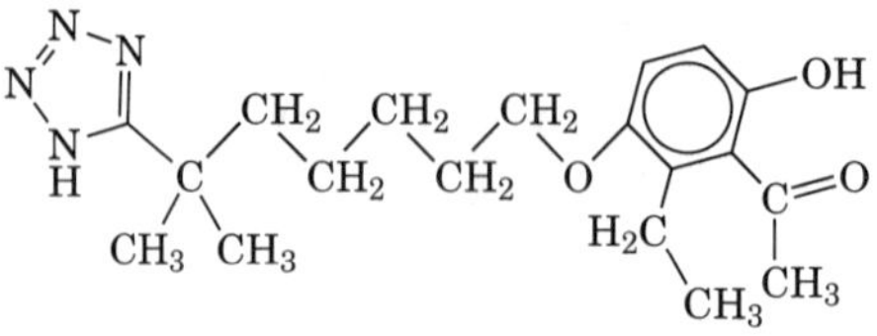

(237)

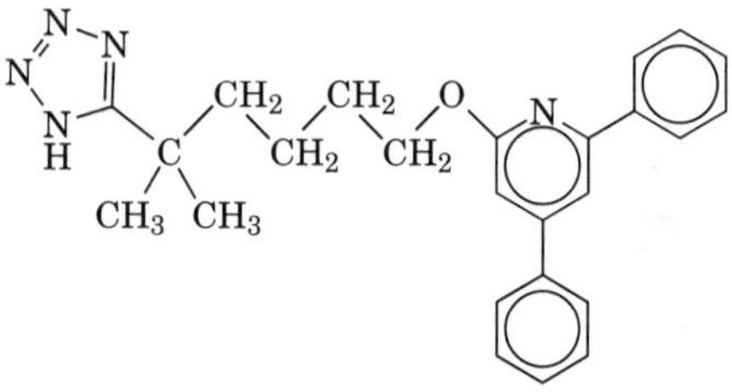

(238)

Table 14. Prostanoid and Leukotriene Receptor Agonists and Antagonists

Agonist/antagonist	CAS Registry Number	Molecular formula	Structure number
		Prostanoid receptors	
BW 245C	[72814-32-5]	$C_{19}H_{32}N_2O_5$	(227)
RS 93520	[105880-66-8]	$C_{21}H_{30}O_3$	(228)
ZK 110841	[105595-17-3]	$C_{22}H_{35}ClO_4$	(229)
fluprostenol	[40666-16-8]	$C_{23}H_{29}F_3O_6$	(230)
cicaprost	[94079-80-8]	$C_{22}H_{32}O_3$	(231)
U 46619	[56985-40-1]	$C_{21}H_{34}O_4$	(232)
STA$_2$	[89617-02-7]	$C_{21}H_{34}O_3S$	(233)
GR 32191	[85505-64-2]	$C_{30}H_{37}NO_4$	(234)
SQ 29548	[98672-91-4]	$C_{21}H_{29}N_3O_4$	(235)
ONO 3708	[102191-05-9]	$C_{22}H_{37}NO_4S$	(236)
		Leukotriene receptor antagonists	
LY 255283	[117690-79-6]	$C_{19}H_{28}N_4O_3$	(237)
RP 69698	[141748-00-7]	$C_{25}H_{27}N_5O$	(238)
ICI 198615	[104448-53-5]	$C_{28}H_{28}N_4O_6S$	(239)
ONO 1078	[103177-37-3]	$C_{27}H_{23}N_5O_4$	(240)
MK 571	[115104-28-4]	$C_{26}H_{27}ClN_2O_3S_2$	(241)
SKF 104353	[107023-41-6]	$C_{26}H_{34}O_5S$	(242)
Bay u9773		$C_{27}H_{36}O_5S$	(243)

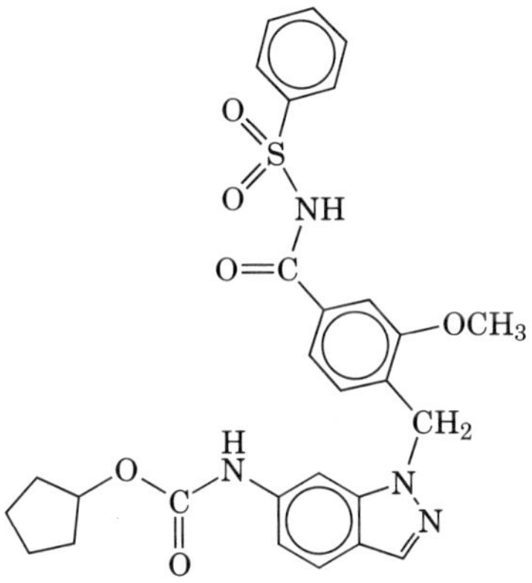

(239)

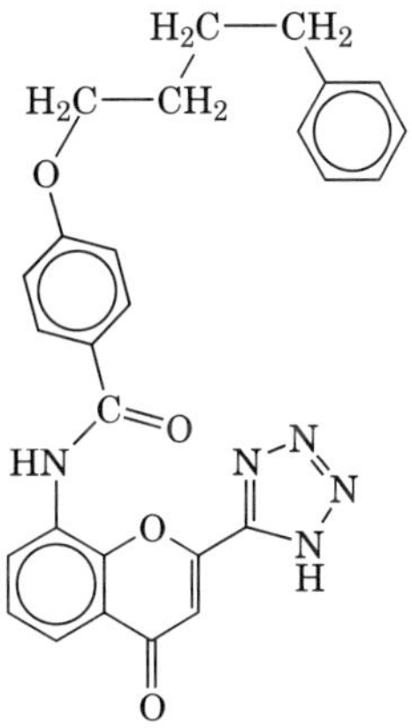

(240)

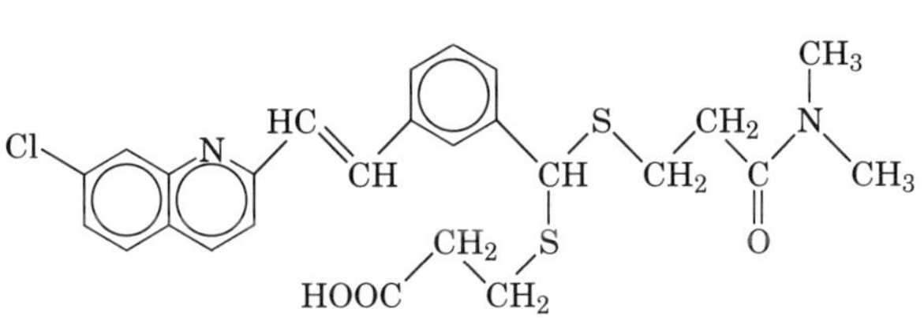

(241)

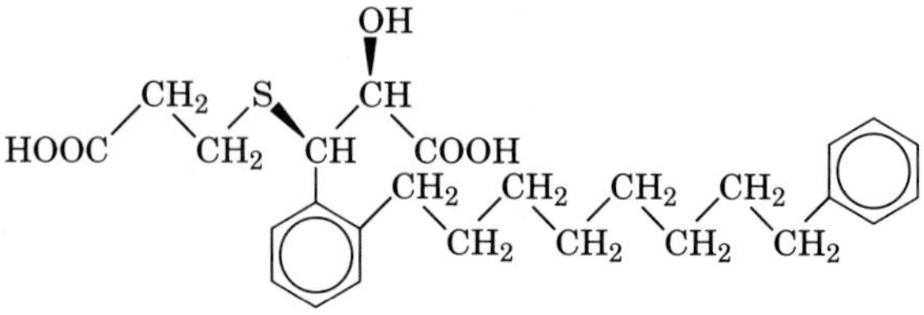

(242)

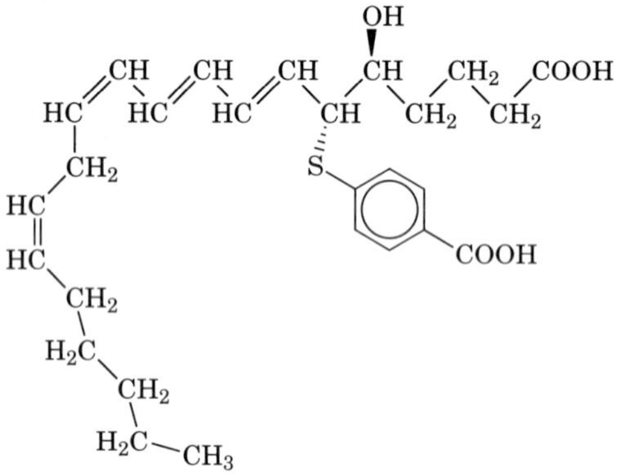

(243)

Leukotrienes are potent endogenous regulators of bronchial and vascular smooth muscle, vascular permeability, and allergic and inflammatory responses including leukocyte activation (64). As such, leukotrienes are of considerable interest physiologically and pharmaceutically in asthma, rheumatoid arthritis, psoriasis, allergic rhinitis, and ulcerative colitis. Most work has centered on peripheral airway, vascular, and isolated cell systems. Cerebral vasculature, mast cells, and other immune competent cells in the brain also may be relevant targets as AA is a second messenger in the CNS and may be intimately involved in neurodegenerative processes.

Direct receptor antagonists for the various products of either the COX or 5-LO pathways have not proven especially efficacious in the clinic. Thus therapeutic efforts related to modulation of the AA pathway have focused on enzyme inhibitors (see ENZYME INHIBITORS). COX and its inducible form, COX-2 are inhibited by nonsteroidal antiinflammatory agents like aspirin [50-78-2], $C_9H_8O_4$ (**244**) and ibuprofen [15687-27-1], $C_{13}H_{18}O_2$ (**245**). L-745,337, $C_{16}H_{13}F_2NO_3S$ (**246**) is a selective COX-2 inhibitor. 5-LO inhibitors include zileuton [111406-87-2], $C_{11}H_{12}N_2O_2S$ (**247**) and D 2138 [140841-32-3], $C_{22}H_{24}FNO_4$ (**248**). MK 886 [118414-82-7], $C_{26}H_{32}ClNO_2S$ (**249**) inhibits 5-LO by inhibiting its association with 5-LO activating protein (FLAP).

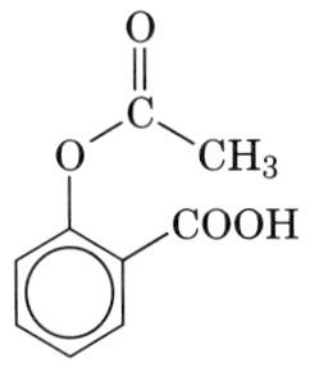

(244) 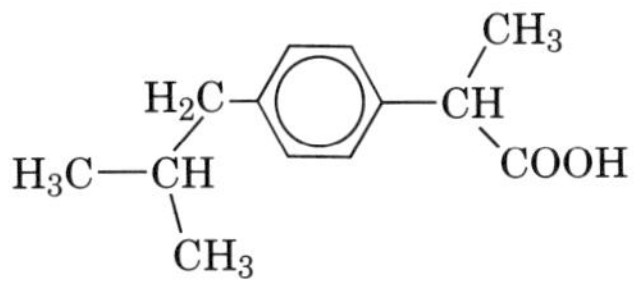   (245) 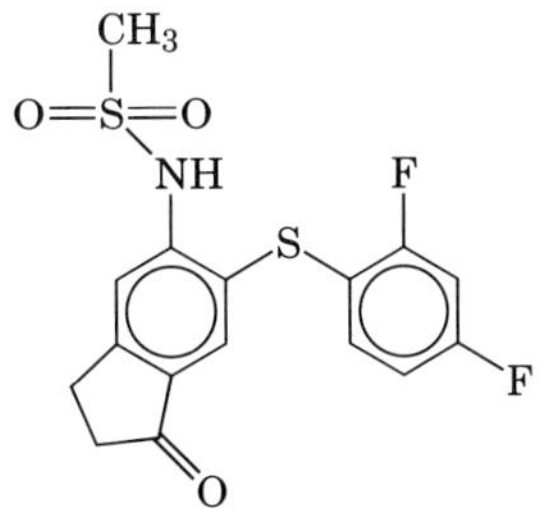   (246)

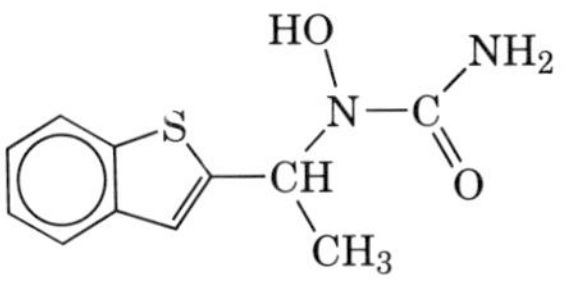

(247) 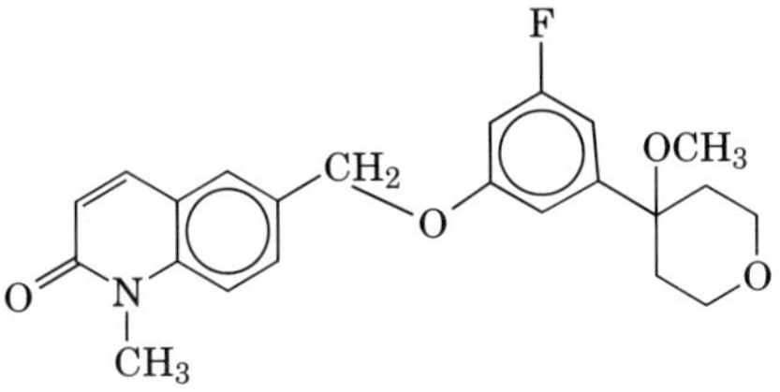   (248)

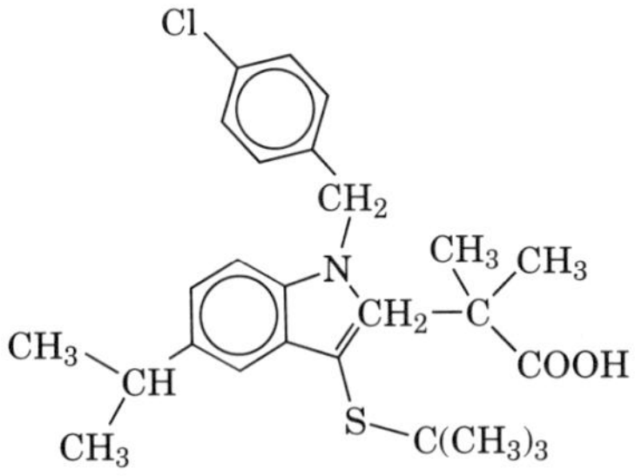

(249)

Melatonin.   Melatonin (*N*-acetyl-5-methoxytryptamine) [*73-31-4*], $C_{13}H_{16}$-N_2O_2 (**250**) is secreted from the pineal gland and retina during dark periods of the vertebrate circadian rhythm (65). Melatonin regulates biological rhythms and neuroendocrine function and is formed from serotonin (5-HT).

Melatonin produces its effects via the GPCR, ML-1. A second lower affinity form, ML-2, has been described on the basis of binding data. Activation of melatonin receptors can inhibit DA release in the retina.

Melatonin, 2-iodomelatonin [*93515-00-5*], $C_{13}H_{15}IN_2O_2$ (**251**) and S 20098 [*138112-76-2*], $C_{15}H_{17}NO_2$ (**252**) are ML-1 agonists. Luzindole [*117946-91-5*], $C_{19}H_{20}N_2O$ (**253**) is a melatonin antagonist. GR 135,531, $C_{14}H_{17}N_2O_3$ (**254**) is a selective ligand for the ML-2 receptor. In addition to a role in controlling circadian rhythms that may provide an approach to the treatment of the jet lag associated with air travel, melatonin may also be involved in the processes underlying migraine and cluster headaches.

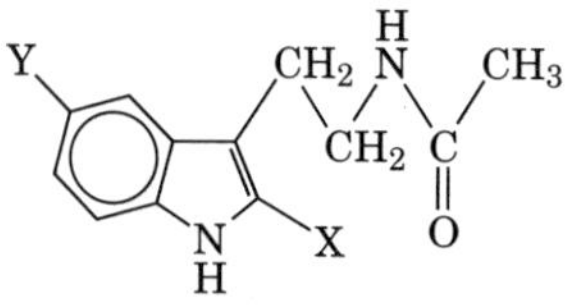

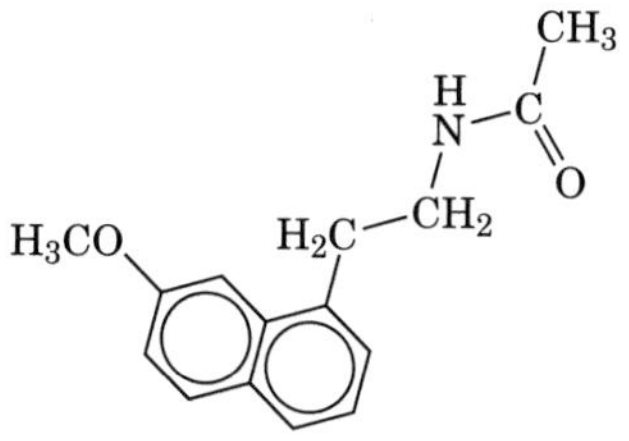

(**250**) X = H , Y = OCH$_3$

(**251**) X = I , Y = OCH$_3$

(**253**) X = CH$_2$C$_6$H$_5$, Y = H

(**254**) X = H , Y = NHCOOCH$_3$

(**252**)

Neuropeptide Y. Neuropeptide Y [*82785-45-3*] (NPY) (**255**) is a 36-amino acid peptide that is a member of a peptide family including peptide YY (PYY) [*81858-94-8, 106338-42-5*] (**256**) and pancreatic polypeptide (PPY) [*59763-91-6*] (**257**). In the periphery, NPY is present in most sympathetic nerve fibers, particularly around blood vessels and also in noradrenergic perivascular and selected parasympathetic nerves (66). Neurons containing NPY-like immunoreactivity are abundant in the central nervous system, particularly in limbic structures. Coexistence with somatostatin and NADPH-diaphorase, an enzyme associated with NO synthesis, is common in the cortex and striatum.

H-Tyr-Pro-Ser-Lys-Pro-Asp-Asn-Pro-Gly-Glu-Asp-Ala-Pro-Ala-Glu-Asp-Met-Ala-
Arg-Tyr-Tyr-Ser-Ala-Leu-Arg-His-Tyr-Ile-Asn-Leu-Ile-Thr-Arg-Gln-Arg-Tyr-NH$_2$

(**255**)

H-Tyr-Pro-Ile-Lys-Pro-Glu-Ala-Pro-Gly-Glu-Asp-Ala-Ser-Pro-Glu-Glu-Leu-Asn-
Arg-Tyr-Tyr-Ala-Ser-Leu-Arg-His-Tyr-Leu-Asn-LeuVal-Thr-Arg-Gln-Arg-Tyr-NH$_2$

(**256**)

H-Ala-Pro-Leu-Lys-Pro-Asp-Asn-Pro-Gly-Glu-Asp-Ala-Glu-Asp-Met-Ala-Arg-
Tyr-Tyr-Ser-Ala-Leu-Arg-His-Tyr-Ile-Asn-Leu-Ile-Thr-Arg-Gln-Arg-Tyr-NH$_2$

(**257**)

NPY exerts its physiological effects by three receptor subtypes, Y_1, Y_2, and Y_3. The Y_3 receptor recognizes NPY in preference to PPY and shows a rank order potency for agonists that is different from both Y_1 and Y_2 receptors. All three NPY receptor subtypes are GPCRs. The substituted analogues [Pro^{34}]NPY and [Leu^{31}]NPY are selective Y_1 agonists. NPY$_{18-36}$ and NPY$_{13-36}$ are Y_2 selective. No selective or potent NPY antagonists have been identified to date. In many cell types NPY raises intracellular calcium concentrations. Y_1 receptors are abundant on vasular smooth muscle cells where they mediate the vasoconstrictor effects of NPY. In rodent brain Y_1 receptors are localized primarily to discrete layers of the cerebral cortex, olfactory nucleus, and thalamic and hypothalamic nuclei where they are linked to NPY-induced stimulation of feeding behavior. Y_1 receptors appear to mediate the anxiolytic and sedative actions of NPY, although NPY is elevated in stress.

Neurotensin. Neurotensin [39379-15-2] (NT), p-Glu-Leu-Tyr-Glu-Asn-Lys-Pro-Arg-Arg-Pro-Try-Ile-Leu-OH, is a tridecapeptide that is cleaved from the ribosomally synthesized precursor, proneurotensin. NT is distributed through the peripheral and central nervous systems as well as in certain other cell types (3,67). NT is colocalized with catecholamines in some neurons.

Although high and low affinity NT binding sites have been described, only one high affinity NT receptor has been clearly demonstrated. It has been cloned from rat and human brain and is a member of the GPCR family. There are few pharmacologic tools for NT systems, but the nonpeptide NT antagonist, SR 48692 [14632-70-1], $C_{32}H_{31}ClN_4O_5$ (**258**) may help to delineate the physiological functions of NT.

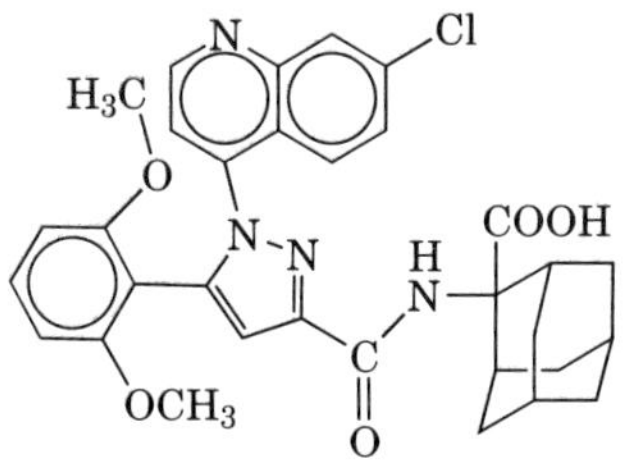

(**258**)

NT has been implicated in neuroendocrine function, thermal and circadian regulation, cardiovascular and digestive system function, nociception, and in psychoses as a DA modulator.

Nerve Growth Factor and Neurotrophins. Nerve growth factor [9061-61-4] (NGF) is a member of a family of neurotrophic peptides that interact with cell surface recognition sites on neurons to affect growth and maintain viability (68). This class of receptor agonist can supress apoptosis and can act either acutely, being liberated as the result of tissue trauma, or chronically in terms of differentiation and development. Other neurotrophins include brain-derived neurotrophic factor (BDNF), and neurotrophin-3 (NT-3), NT-4, NT-5, and NT-6. Cilliary neurotrophic factor (CNTF) is a member of the cytokine family.

NGF interacts with two distinct receptors. A high affinity receptor (HNGFR) (K_d = 25 pM) also known as trkA or p140$^{c\text{-}trk}$, is the proto-oncogene product of the *trk* gene and contains a tyrosine kinase in its internal domain. p75 is a low affinity NGF receptor (LNGFR) (K_d = 1 nM) that is linked to G-protein activation. p140$^{c\text{-}trk}$ is the main receptor for NGF (69). Other members of the NGF neurotrophin family also use *trk*A or *trk* homologues as their receptors indicating that tyrosine phosphorylation is the common transduction factor for neurotrophin receptor activation. The LNGFR is related to the TNF receptor, the lymphokine receptor, CD 40, and APO-1 (Fas antigen), a lymphocyte antigen involved in apoptosis.

The relationship between the two receptors for NGF is complex and not yet completely understood. It has been suggested that the functional form of the NGF receptor is a heterodimer of p75 and p140$^{c\text{-}trk}$ proteins. BDNF and NT-3 bind to p75, but the functional receptors for these neurotrophins are the proto-oncogene products of *trk*B and *trk*C.

In addition to cell surface recognition sites, NGF can also interact with nuclear chromatin receptors. Receptors for CNTF include CNTF-α, leukemia inhibitor factor receptor β (LIFRβ), and gp130. Little is known regarding pharmacological aspects of neurotrophin receptor activation or blockade. Levels of NGF can be increased by gene activation by a large number of conventional neurotransmitters including ACh. The involvement of neurotrophins in cell viability and apoptosis suggests that either the trophic factors themselves or agents that elicit production or mimic effects may be used in treating neurodegenerative diseases like Parkinson's or Alzheimer's disease, or be of use in reversing the effects of nerve trauma. Because NGF, BDNF, CNTF, and NT-3, NT-4, and NT-5 are peptidic in nature, usefulness as therapeutic agents is somewhat limited.

Nitric Oxide.   Nitric oxide [*10102-43-9*], NO, is a ubiquitous intracellular and intercellular messenger serving a variety of functions including vasodilation, cytotoxicity, neurotransmission, and neuromodulation (9). NO is a paramagnetic diatomic molecule that readily diffuses through aqueous and lipid compartments. Its locus of action is dictated by its chemical reactivity and the local environment. NO represents the first identified member of a series of gaseous second messengers that also includes CO.

The half life for NO in cellular systems ranges from 5–30 seconds. Superoxide, hemoglobin, and other radical trapping agents remove NO after it has been formed.

Nitric oxide synthases (NOS) (EC 1.14.13.39) are both constitutive and inducible and produce NO from L-arginine ($K_m = 1.5$–2.8 μM) (9). The endothelial isoform eNOS (type III NOS) and the brain or neuronal isoform nNOS (type I NOS) are constitutive. A third isoform is the inducible NOS (iNOS or type II) found in macrophages, astrocytes, and microglia. All isoforms require calcium and calmodulin for activity. NADPH, FAD, FMN, and tetrahydrobiopterin (BH$_4$) are required co-factors. Agents that increase intracellular calcium activate constitutive NOSs. Calmodulin antagonists such as calmidazolium (**259**) and diphenylene iodonium (**260**), an inhibitor of NADPH-dependent oxidase, inhibit NOS. NOS also contains a heme in the form of protoporphyrin. Both iNOS and nNOS are cytosolic and exist as dimers, whereas eNOS is monomeric and membrane bound by myristolylation.

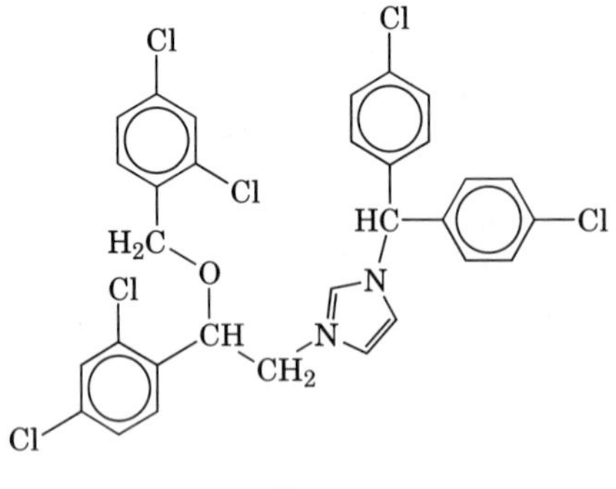

(**259**) 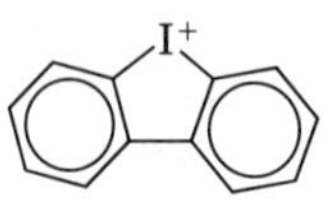   (**260**)

In the vascular system endothelial cells produce NO which diffuses into smooth muscle cells activating soluble guanylate cyclase (sGC) thus initiating vasodilation. Diffusion of NO into platelets inhibits aggregation. In neurons, NO has multiple transduction pathways including activation of sGC and stimulation of ADP ribosylation. NO is a retrograde messenger which has been implicated both in long-term depression (LTD) and long-term potentiation (LTP).

iNOS is induced in several cell types by cytokines and lipopolysaccharides. Macrophages utilize iNOS-produced NO as a cytotoxic agent. Reaction with iron-containing metabolic enzymes and oxygen and superoxide produces peroxynitrite (ONO_2^-), a potentially more cytotoxic agent. In the CNS microglia and astrocytes produce iNOS. This NO source has been implicated in a number of CNS pathologies. Only a limited number of CNS neurons contain nNOS.

NO synthons, including the vasodilators sodium nitroprusside (SNP) (**261**) and nitroglycerin (**262**), have been in clinical use since the 1970s. Newer synthons include molsidomine (**263**) and the NONOates, prodrug dimers of NO.

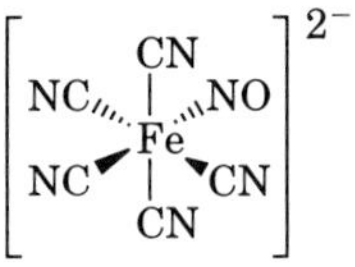

(**261**) 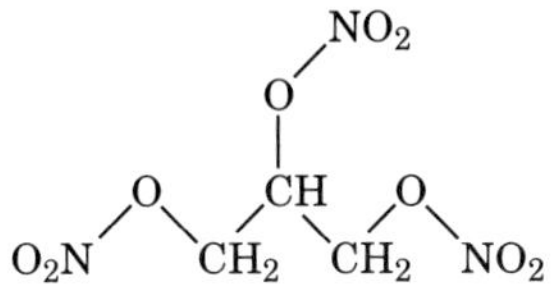   (**262**) 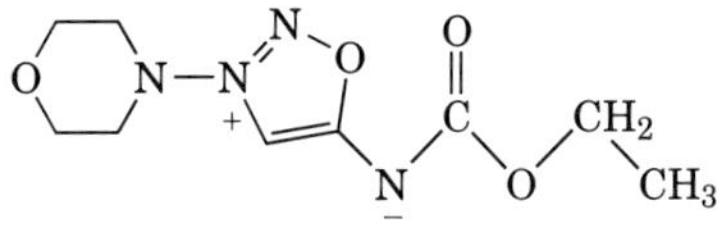   (**263**)

Most NOS inhibitors are structurally related to L-arginine and do not differentiate between the isoforms (Table 15). L-N^γ-Methylarginine (L-NMA) (**264**) is a competitive inhibitor and also irreversibly inhibits NOS. L-N^γ-Nitroarginine (L-NNA) (**265**), L-N^γ-nitroarginine methyl ester (L-NAME) (**266**), L-N^γ-cyclopropylarginine (**267**), and L-N^γ-aminoarginine (**268**) are also arginine-like inhibitors. L-N^δ-Iminoethylornithine (L-NIO) (**269**) has been reported to be an irreversible inhibitor (Table 15).

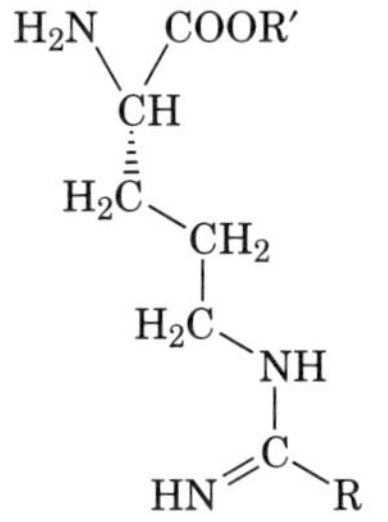

	R	R'
(**264**)	$NHCH_3$	H
(**265**)	$NHNO_2$	H
(**266**)	$NHNO_2$	CH_3
(**267**)	NH-cyclopropyl	H
(**268**)	$NHNH_2$	H
(**269**)	CH_3	H

Therapeutic opportunities for NO synthons include angina, for which nitroglycerin is effectively used, as well as penile erectile dysfunction. NOS inhibitors have demonstrated some protection in cerebral ischemia models and may be potentially beneficial in alleviating cell death associated with cerebral ischemia. L-NMA is under clinical study for treatment of sepsis.

Table 15. Nitric Oxide Synthons and Inhibitors

Material	CAS Registry Number	Molecular formula	Structure number
Calmodulin antagonists			
calmidazolium	[95013-41-5]	$C_{31}H_{24}C_{16}N_2O$	(259)
diphenylene iodonium	[244-54-2]	$C_{12}H_8I$	(260)
NO Synthons			
sodium nitroprusside	[14402-89-2]	$C_5FeN_6Na_2O$	(261)
nitroglycerin	[55-63-0]	$C_3H_5N_3O_9$	(262)
molsidomine	[25717-80-0]	$C_9H_{14}N_4O_4$	(263)
NOS inhibitors			
L-N^γ-methylarginine	[17035-90-4]	$C_7H_{16}N_4O_2$	(264)
L-N^γ-nitroarginine	[2149-70-4]	$C_7H_{13}N_5O_4$	(265)
L-N^γ-nitroarginine methyl ester	[50903-99-6]	$C_8H_{15}N_5O_4$	(266)
L-N^γ-cyclopropylarginine		$C_9H_{18}N_4O_2$	(267)
L-N^γ-aminoarginine	[57444-72-1]	$C_6H_{15}N_5O_2$	(268)
L-N^γ-iminoethylornithine	[36889-13-1]	$C_7H_{15}N_3O_2$	(269)

Octopamine.   Octopamine [104-14-3] (OA), $C_8H_{11}NO_2$ (270) is a mono-amine found in the insect CNS (70). It is involved in feeding behavior and in stimulating light production from the firefly light organ. The presence of octopamine in mammalian nervous tissue has yet to be determined.

Three classes of OA receptor, OA-1–OA-3, have been described on the basis of antagonist sensitivities and location (71). The OA-1 receptor is antagonized by the adrenoceptor antagonist, phentolamine (87) and the OA-2 receptor is blocked by mianserin [24219-97-4], $C_{18}H_{20}N_2$ (271). The OA-3 receptor is similar to the OA-2 receptor but is found in nerve cord and insect brain. TMP, $C_{12}H_{16}N_2$ (272) and NC5Z, $C_{13}H_{17}N_5$ (273) are more potent than OA at the OA-1 receptor. Tyramine [51-67-2], $C_8H_{11}NO$ (274) is an agonist at all three receptor subtypes. The OA-2 receptor is linked to activation of adenylate cyclase.

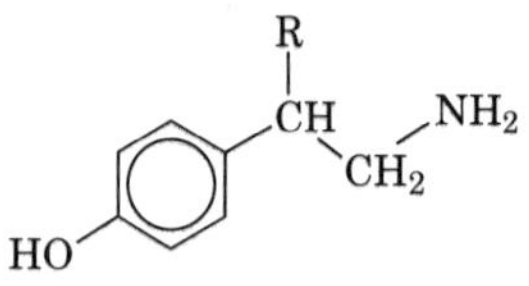

(270) R = OH

(274) R = H

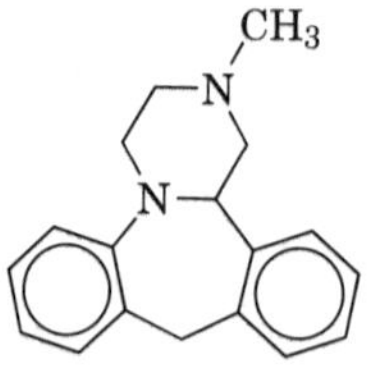

(271)

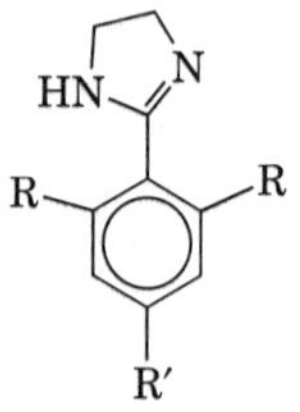

(272) R = CH₃ , R′ = CH₃

(273) R = C₂H₅ , R′ = N₃

Retinoic Acid and Thyroid Hormone.   The steroid hormones, vitamin D_3, the retinoic acids (RAs), *trans*-RA (275), 13-*cis*-RA (276), and 9-*cis*-RA (277), and thyroid hormone (TH) alter cell function by acting as ligand-controlled transactivating factors or signal transducers and activators of transcription (STATs).

They selectively interact with intracellular transcription factors to alter gene expression by interaction with hormone-responsive elements (HREs) on promotor regions of DNA (49,72). RA plays a pivotal role in development and embryogenesis; however, excessive doses are teratogenic. RA induces differentiation in neuronal cells *in vitro* and is likely to play a role in neuronal differentiation *in vivo*.

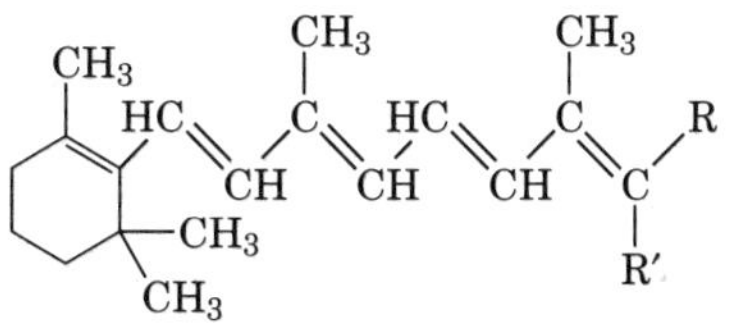

(**275**) R = COOH , R′ = H

(**276**) R = H , R′ = COOH

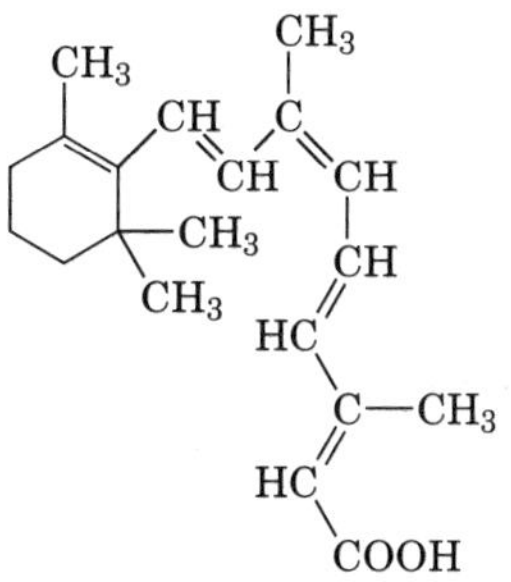

(**277**)

The steroid hormone receptor family includes Type I receptors that include estrogen, progesterone, and glucocorticoid receptor families. These bind to DNA at pallidromically arranged half-sites separated by three nucleotides and require ligand for initiation of DNA binding. The Type II (RAR) receptor group includes the RA, TH, vitamin D_3, and peroxisome proliferator activated receptor (PPAR) families (Table 16). RA receptors are subdivided into retinoic acid receptors (RARs) and retinoid X receptors (RXRs) based on differing affinities for 9-*cis* RA (**277**). Type II receptors activate transcription through DNA binding of closely related sequences arranged as direct repeats. They bind to such sites in the absence of ligand and require heterodimer formation with RXR for high affinity binding.

Table 16. Retinoic Acids, Vitamin D_3, and Type II Receptor Agonists and Antagonists

Agonist/antagonist	CAS Registry Number	Molecular formula	Structure number
trans-retinoic acid	[302-79-4]	$C_{20}H_{28}O_2$	(**275**)
13-*cis*-retinoic acid	[4759-48-2]	$C_{20}H_{28}O_2$	(**276**)
9-*cis*-retinoic acid	[5300-03-8]	$C_{20}H_{28}O_2$	(**277**)
Ro 13-7410	[71441-28-6]	$C_{24}H_{28}O_2$	(**278**)
Ro 41-5253	[144092-31-9]	$C_{28}H_{36}O_5S$	(**279**)
LGD 1069			(**280**)
TTAB	[107430-51-3]	$C_{25}H_{26}O_2$	(**281**)
Ro 10-9359	[54350-48-0]		(**282**)
1,25-dihydroxyvitamin D_3	[35211-63-0, 32222-06-3]	$C_{28}H_{46}O$	(**283**)
thyroxine	[51-48-9]	$C_{15}H_{11}I_4NO_4$	(**284**)
triiodothyronine	[6893-02-3]	$C_{15}H_{12}I_3NO_4$	(**285**)
TRIAC	[51-24-1]	$C_{14}H_9I_3O_4$	(**286**)

The RAR receptor family includes a variety of separate receptors, each having specific distribution and ligand binding specificities. Many of these receptors have been cloned. α, β, and γ forms of each subtype exist that may have several different isoforms differing in the 5'-untranslated region of the mRNA and/or the sequence encoding the A domain of the protein. Receptors have discrete domains for ligand binding, DNA binding, and transactivation. The presence of eight highly conserved Cys residues in the RAR receptor sequence has been correlated with the so-called zinc fingers necessary for DNA binding at HREs.

The primary endogenous ligand for RARs is *trans*-RA (**275**). 9-*cis* RA is the endogenous ligand for RXRs. Ro13-7410 (**278**) and Ro 41-5253 (**279**) are selective for RAR subtypes over RXR. Ro 41-5253 is characterized as RAR_α selective. LGD 1069 (**280**) is RXR receptor selective. TTAB (**281**) is RAR_γ selective. The characterization of RA ligands as agonist or antagonist is not always clear because ligands can interact with RAR or RXR subtypes to form inactive heterodimers that compete with transcription activation pathways. RAR_α is ubiquitous, RAR_γ is predominant in the skin and lungs, and RAR_β is expressed in the heart, lungs, and spleen (73). The antimalignant effects of retinoids are well documented. In addition, retinoids such as *trans*-RA, 13-*cis*-RA, and Ro 10-9359 (**282**) are used in the treatment of acne and psoriasis.

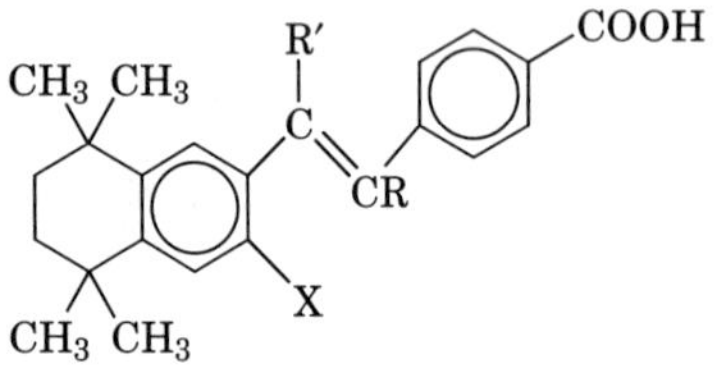

(**278**) R = X = H , R' = CH$_3$

(**279**) R = CH$_3$, R' = H , X = O(CH$_2$)$_6$CH$_3$

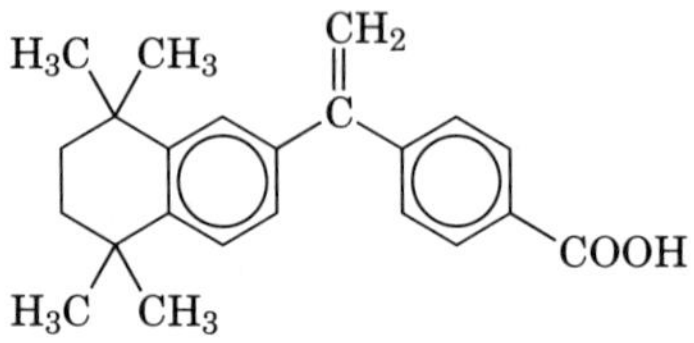

(**280**)

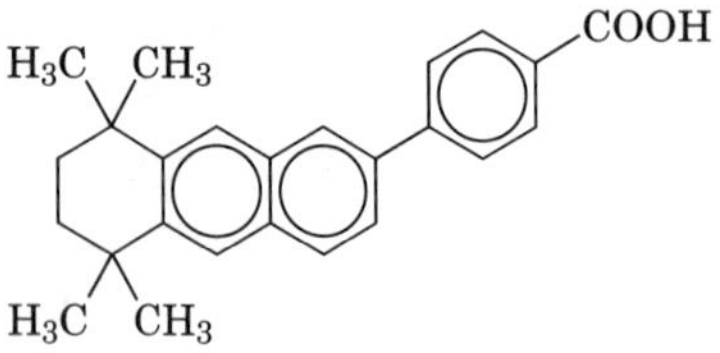

(**281**)

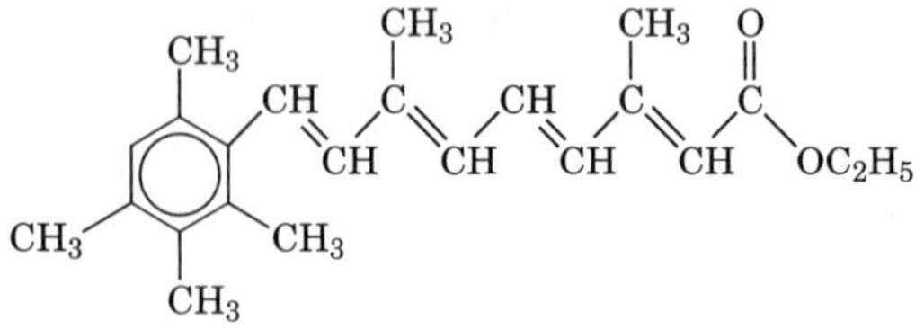

(**282**)

1,25-Dihydroxyvitamin D$_3$ (**283**) is the endogenous ligand for the vitamin D$_3$ receptor (VDR). It modulates genomic function in a tissue and developmentally specific manner and affects cell proliferation, differentiation, and mineral homeostasis (74). Vitamin D$_3$ mobilizes calcium from the bone to maintain plasma Ca^{2+} levels. Vitamin D$_3$ and VDR are present in the CNS where they may play a role in regulating Ca^{2+} homeostasis. Vitamin D$_3$ has potent immunomodulatory activity *in vivo*.

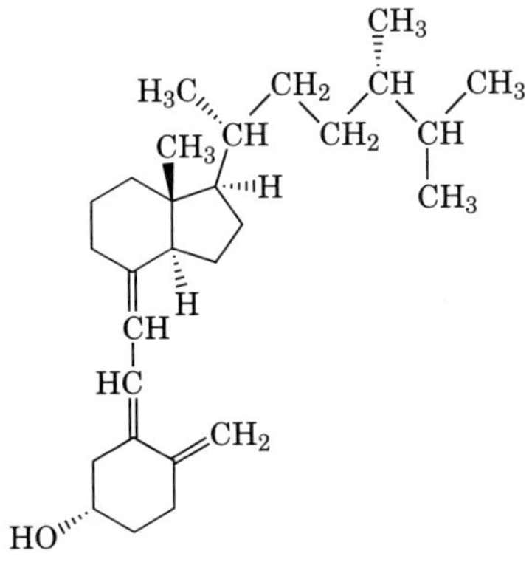

(**283**)

Thyroid hormone receptors (THRs) are subdivided into α and β types, each having two isoforms. In rat brain, THR$_\alpha$ mRNA is present in hippocampus, hypothalmus, cortex, cerebellum, and amygdala. Thyroxine (L-T$_4$) (**284**) and triiodothyronine (L-T$_3$) (**285**) are endogenous ligands for the THRs. TRIAC (**286**) is a THR antagonist. Selective ligands for PPARs have yet to be identified (Table 16).

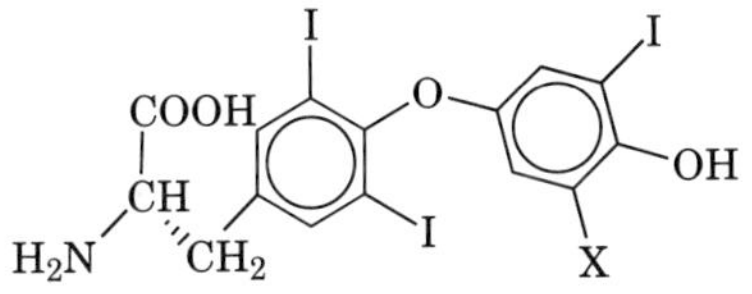

(**284**) X = I

(**285**) X = H

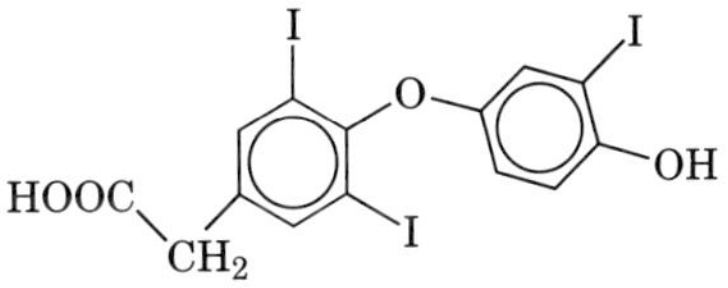

(**286**)

Serotonin. Serotonin [50-67-9] (5-HT), $C_{10}H_{12}N_2O$ (**287**) is a hydroxy-ethylaminoindole with widespread distribution. 5-HT is synthesized from L-tryptophan by hydroxylation to 5-hydroxy-L-tryptophan by the enzyme, tryptophan-5-hydroxylase. 5-Hydroxy-L-tryptophan is then rapidly decarboxylated by aromatic-L-amino acid deacarboxylase to 5-HT. The actions of 5-HT as a neurotransmitter are terminated by neuronal reuptake and metabolism.

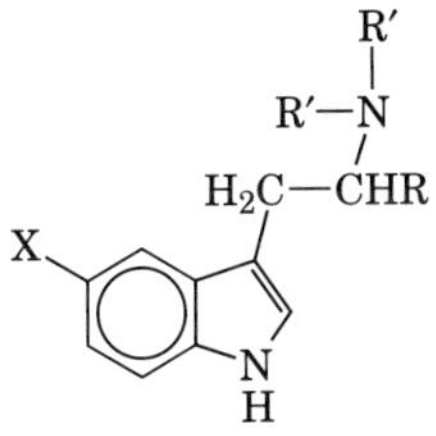

(**287**) R = R' = H , X = OH

(**292**) R = H , R' = CH$_3$, X = CH$_2$SO$_2$NHCH$_3$

(**295**) R = CH$_3$, R' = H , X = OH

5-HT produces its effects by a diversity of receptor subtypes (75) that are divided into seven pharmacologically distinct classes designated 5-HT_1–5-HT_7. The 5-HT_1, 5-HT_2, and 5-HT_5 subclasses can be further subdivided into five, three, and two subtypes, respectively, based on pharmacological and cloning criteria. With the exception of the 5-HT_3 receptor, which is an LGIC, all members of the 5-HT receptor superfamily are GPCRs. The 5-HT_{1A} receptor was identified using the selective agonist 8-OH-DPAT (**288**). WAY 100135 (**289**) is a 5-HT_{1A} receptor antagonist, CP 93129 (**290**) is a selective 5-HT_{1B} agonist, CGS 12066B (**291**) is a selective 5-HT_{1B} receptor antagonist, sumatriptan (**292**) is a 5-HT_{1D} agonist, and GR127935 (**293**) is a selective 5-HT_{1D} antagonist (Table 17).

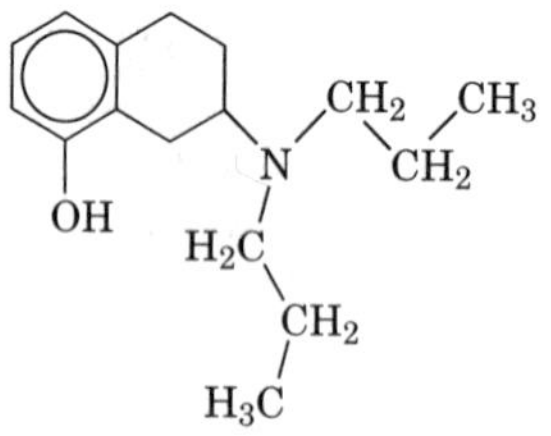

(**288**) 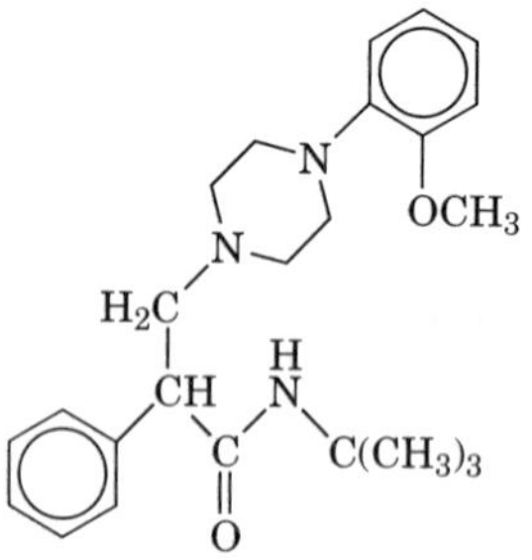   (**289**) 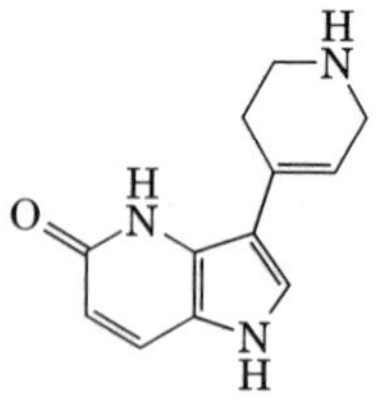   (**290**)

Table 17. Serotonin and Serotonin Receptor Agonists and Antagonists

Agonist/antagonist	CAS Registry Number	Molecular formula	Structure number
8-OH-DPAT	[78950-78-4]	$C_{16}H_{25}NO$	(**288**)
WAY 100135	[133025-23-7]	$C_{24}H_{33}N_3O_2$	(**289**)
CP 93129	[127792-75-0]	$C_{12}H_{13}N_3O$	(**290**)
CGS 12066B	[109028-10-0]	$C_{17}H_{17}F_3N_4$	(**291**)
sumatriptan	[103628-46-2]	$C_{14}H_{21}N_3O_2S$	(**292**)
GR 127935		$C_{29}H_{31}N_5O_3$	(**293**)
DOI	[82830-53-3]	$C_{11}H_{16}INO_2$	(**294**)
α-methyl-5-HT	[304-52-9]	$C_{11}H_{14}N_2O$	(**295**)
ketanserin	[74050-98-9]	$C_{22}H_{23}N_3O_3$	(**296**)
ritanserin	[87051-43-2]	$C_{27}H_{25}F_2N_3OS$	(**297**)
SB 200646	[143797-62-0]	$C_{15}H_{14}N_4O$	(**298**)
2-methyl-5-hydroxytryptamine	[78263-90-8]	$C_{11}H_{14}N_2O$	(**299**)
m-chlorophenylbiguanide	[48144-44-1]	$C_8H_{10}ClN_5$	(**300**)
ondansetron	[99614-02-5]	$C_{18}H_{19}N_3O$	(**301**)
BIMU 8	[134296-40-5]	$C_{19}H_{26}N_4O_2$	(**302**)
SB 204070A	[148688-01-1]	$C_{19}H_{27}ClN_2O_4$	(**303**)
GR 113808	[144625-51-4]	$C_{19}H_{27}N_3O_4S$	(**304**)
LSD	[50-37-3]	$C_{20}H_{25}N_3O$	(**305**)

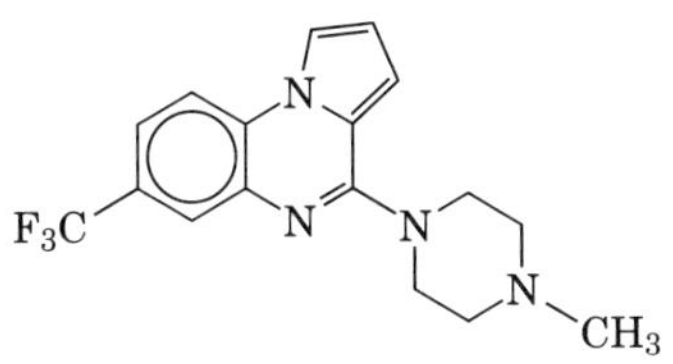

(291)

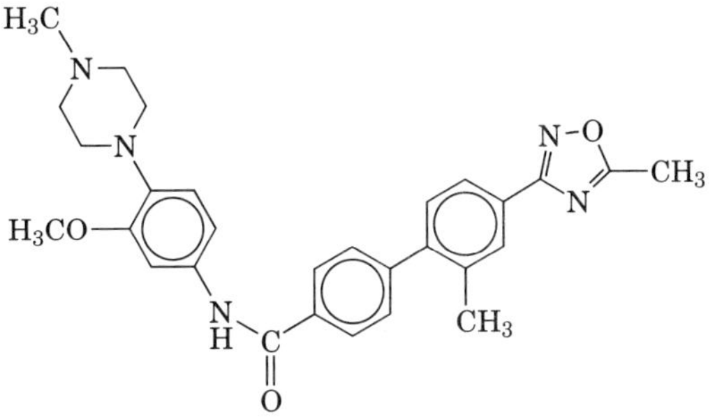

(293)

DOI (**294**) and α-methyl-5-HT (**295**) are selective 5-HT$_2$ receptor agonists. Ketanserin (**296**) and ritanserin (**297**) are potent and selective 5-HT$_{2A}$ antagonists. SB 200646 (**298**) is an antagonist which has greater selectivity toward 5-HT$_{2B}$ and 5-HT$_{2C}$ receptors compared to the 5-HT$_{2A}$ subtype. 2-Methyl-5- hydroxytryptamine (**299**) and m-chlorophenylbiguanide (**300**) are 5-HT$_3$ agonists. Ondansetron (**301**) is a selective 5-HT$_3$ antagonist. BIMU 8 (**302**) is a potent 5-HT$_4$ agonist, although the most selective antagonists at this subtype are SB 204070 (**303**) and GR 113808 (**304**). Selective pharmacological probes for the 5-HT$_5$, 5-HT$_6$, and 5-HT$_7$ subtypes have yet to be identified. LSD (**305**) is active at the 5-HT$_6$ receptor.

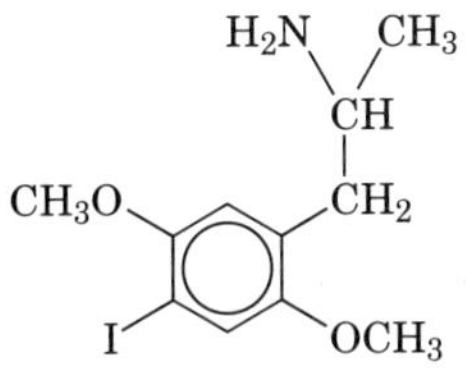

(294)

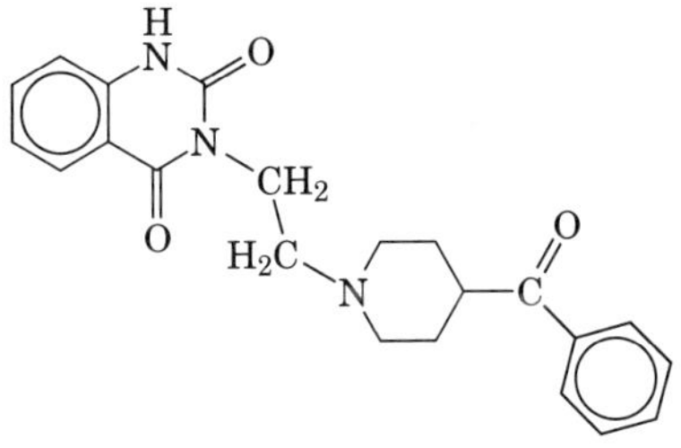

(296)

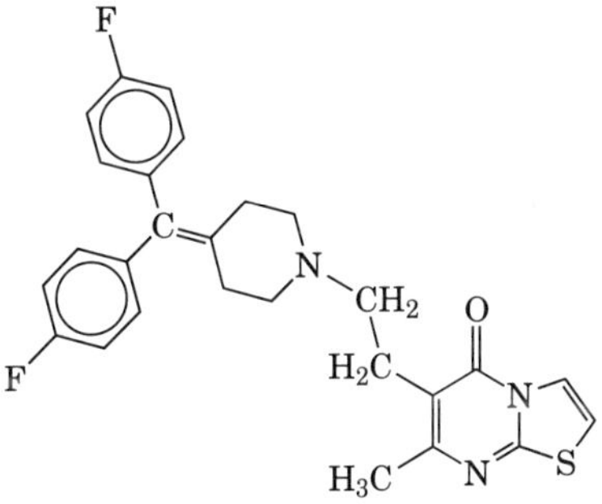

(297)

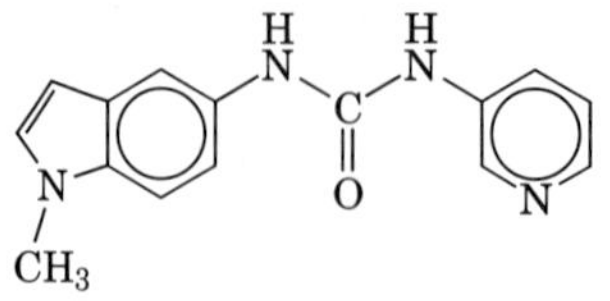

(298) 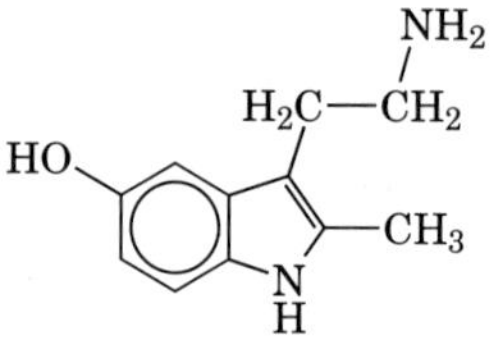   (299) 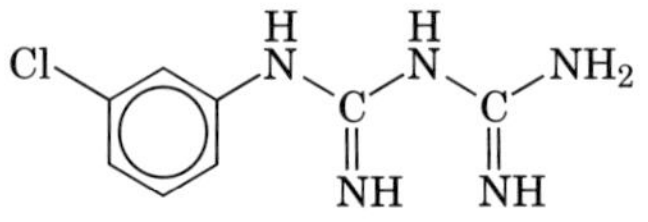   (300)

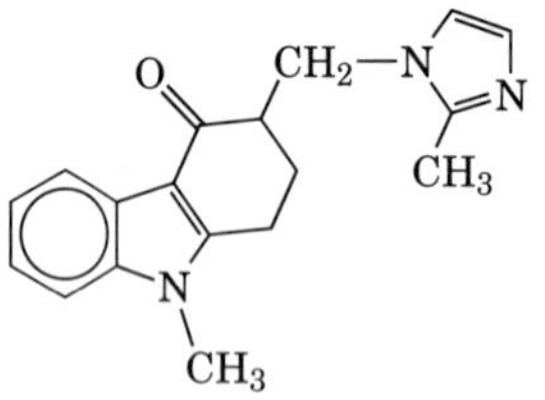

(301) 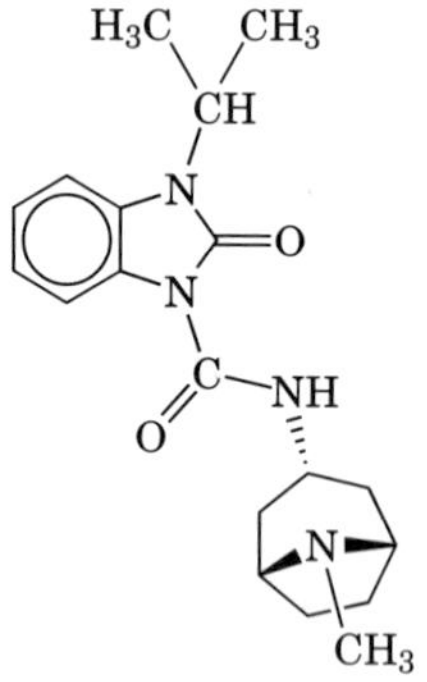   (302) 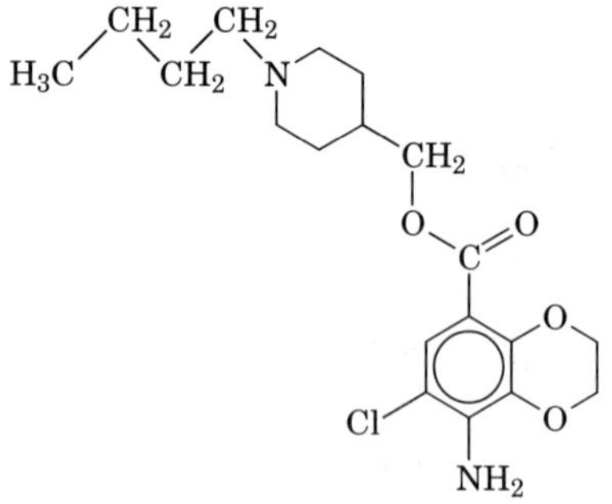   (303)

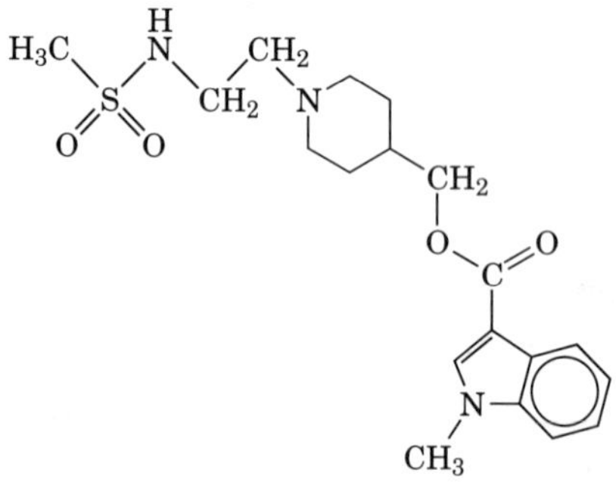

(304) 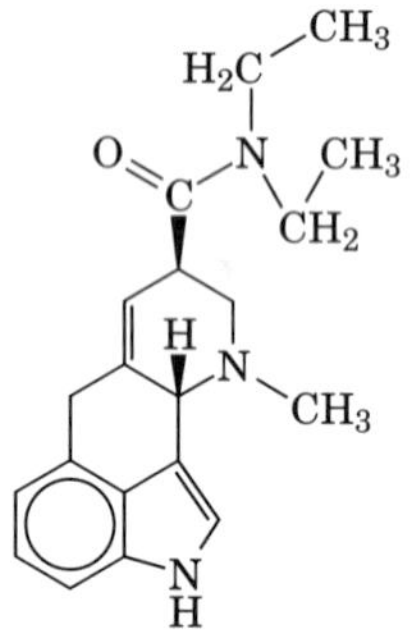   (305)

Serotonin is a key transmitter in CNS function. Altered serotonergic function has been implicated in many CNS disorders including depression, feeding behavior, sleep disorders, schizophrenia, and Alzheimer's disease.

Sigma Receptor Ligands. Sigma (σ-) receptors (76) were originally defined on the basis of the psychotomimetic effects of the benzomorphan opioid, SKF 10047 (**306**). Although σ-receptors were initially designated as opiate receptors, the binding of SKF 10047 to phencyclidine (PCP) (**185**) receptors suggested that σ and PCP receptors might be the same. The neuroleptic, haloperidol (**126**) also binds to σ-receptors. (+)-Pentazocine (**307**) distinguishes between

σ_1 and σ_2 sites. (+)-Pentazocine and BD 737 (**310**) have higher affinity for σ_1 sites (Table 18). The dopamine autoreceptor agonist, (+)-3-PPP (1-propyl-3-(3'-hydroxyphenyl)-piperidine) (**308**) labels two σ-sites with K_d values of 25 and 900 nM. A σ_3 receptor has also been described (77). The antitussive dextromethorphan (**309**), a ligand for the NMDA receptor complex, labels two sites termed DM$_1$ and DM$_2$, with the higher affinity DM$_1$ site corresponding to a σ-site.

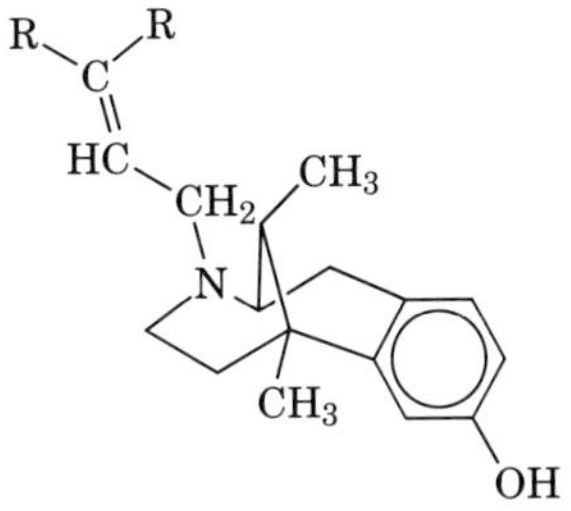

(**306**) R = H

(**307**) R = CH$_3$

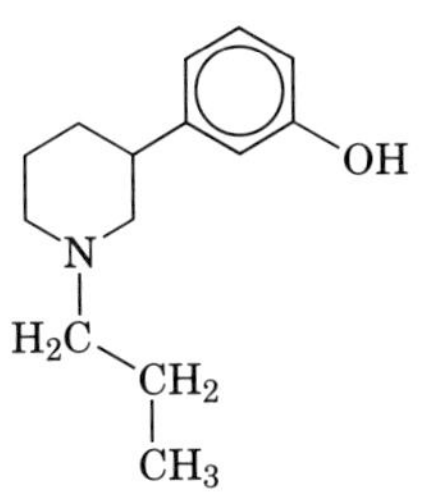

(**308**)

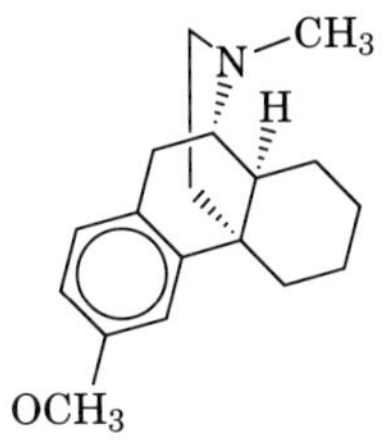

(**309**)

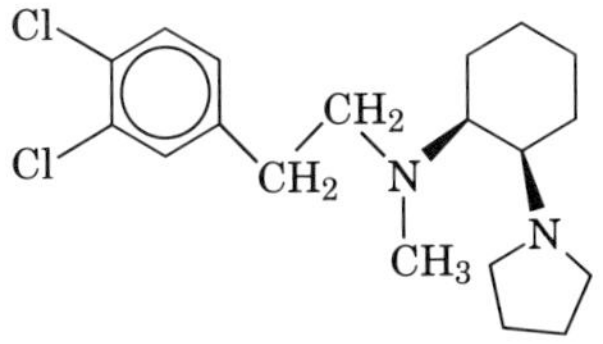

(**310**)

Table 18. Sigma-Receptor Ligands

Ligand	CAS Registry Number	Molecular formula	Structure number
SKF 10047	[14198-28-8]	$C_{17}H_{23}NO$	(**306**)
(+)-pentazocine	[7361-76-4]	$C_{19}H_{27}NO$	(**307**)
(+)-1-propyl-3-(3'-hydroxy-phenyl) piperidine	[75240-91-4]	$C_{14}H_{21}NO$	(**308**)
dextromethorphan	[125-71-3]	$C_{18}H_{25}NO$	(**309**)
BD 737	[130609-93-7]	$C_{19}H_{28}C_{12}N_2$	(**310**)
rimcazole	[75859-04-0]	$C_{21}H_{27}N_3$	(**311**)
remoxipride	[80125-14-0]	$C_{16}H_{23}BrN_2O_3$	(**312**)
gevotroline	[107266-06-8]	$C_{19}H_{20}FN_3$	(**313**)
GBR 12909	[67469-78-7]	$C_{28}H_{32}F_2N_2O$	(**314**)
ifenprodil	[23210-56-2]	$C_{21}H_{27}NO$	(**315**)
CNS 1102	[137160-11-3]	$C_{20}H_{21}N_3$	(**316**)
NPC 16377		$C_{27}H_{33}NO_4$	(**317**)

In addition to haloperidol, the putative neuroleptics, rimcazole (**311**), re-moxipride (**312**), and gevotroline (**313**) bind to σ-receptors as does the dopamine uptake blocker, GBR 12909 (**314**) and two ligands active at the NMDA receptor, ifenprodil (**315**) and CNS 1102 (**316**). NPC 16377, (**317**) is a selective σ-receptor ligand. MAO inhibitors and antidepressants also bind to σ-receptors. Some evidence indicates that σ-receptors in the brain are in fact a form of cytochrome P_{450} which may account for the diversity of ligands interacting with σ-sites.

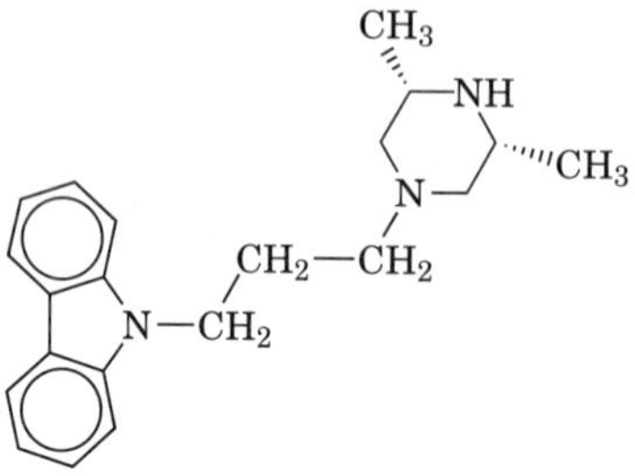

(**311**)

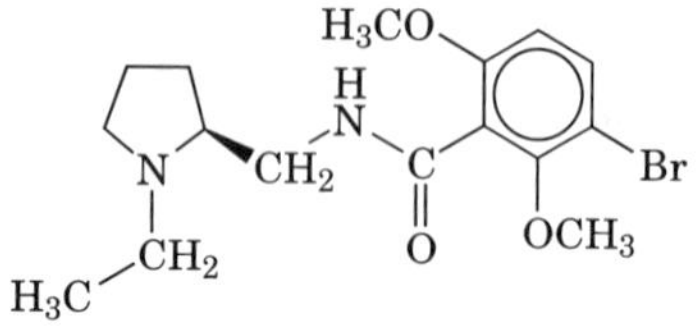

(**312**)

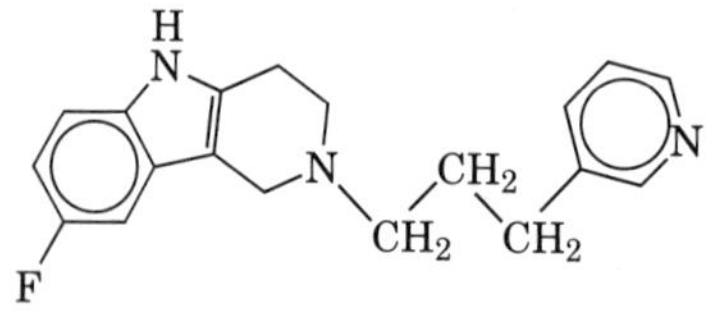

(**313**)

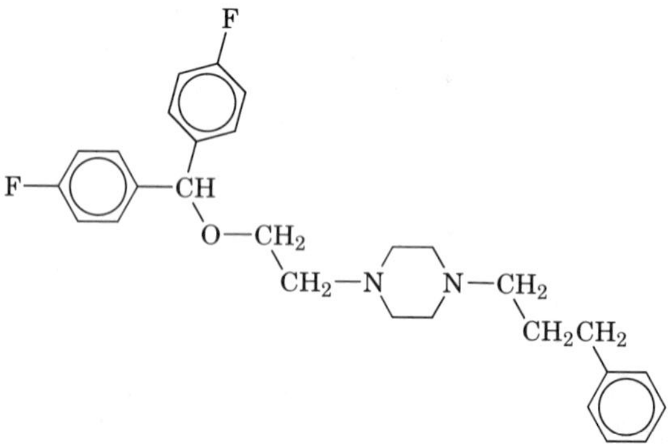

(**314**)

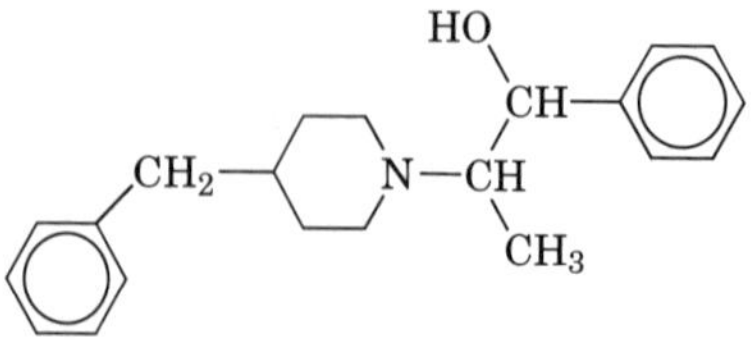

(**315**)

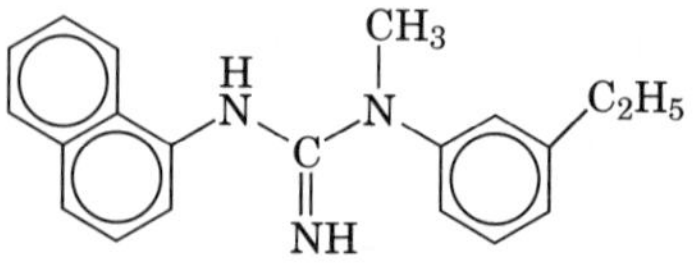

(**316**)

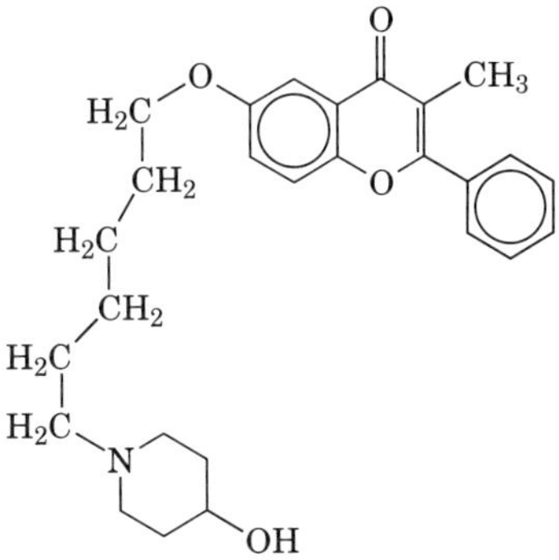

(**317**)

σ-Receptors are localized in the brain stem and limbic structure, regions associated with endocrine function (76). In the periphery, σ-receptors are found in the liver, heart, ileum, vas deferens, and on lymphocytes and thymocytes. Although there is insufficient evidence to clearly define the functional role of CNS σ-sites, based on the effects of PCP and the interaction of haloperidol with σ-sites, σ-receptor ligands may be antipsychotics or used for the treatment of substance abuse. Several σ-receptor ligands have shown neuroprotective effects *in vivo*. Ifenprodil (**315**) and CNS 1102 (**316**) are being developed for treatment of stroke (Table 18).

Steroid Hormones and Neurosteroids. Steroids (qv) can affect neuroendocrine function, stress responses, and behavioral sexual dimorphism (78,79) (see STEROIDS). Mineralocorticoid, glucocorticoid, androgen, estrogen, and progesterone receptors are localized in the brain and spinal cord. In addition to genomic actions, the neurosteroid can act more acutely to modulate the actions of other receptors or ion channels (80). Pregnenolone [145-13-1], $C_{21}H_{32}O_2$ (**318**) and dehydroepiandosterone [53-43-0], $C_{19}H_{26}O_2$ (**319**) are excitatory neurosteroids found in rat brain, independent of adrenal and gonadal sources, and show circadian fluctations in their CNS levels. Glia are a primary source of neurosteroids that inhibit the function of GABA$_A$ and glycine receptors at micromolar concentrations. CNS active steroids are also known to have anesthetic and sedative actions. Pregnenolone and dehydroepiandosterone have been reported to have memory-enhancing actions in male mice (see MEMORY-ENHANCING DRUGS).

Allopregnanolone [516-55-2], $C_{21}H_{34}O_2$ (**320**) and allotetrahydro-DOC, $C_{21}H_{34}O_3$ (**321**), metabolites of the steroids progesterone and deoxycorticosterone, augment GABA$_A$ receptor activation at low (10–50 nM) nanomolar concentrations and have anxiolytic- and antiepileptic-like activities as well as producing barbiturate-like effects. RU 5135 [78774–26–2], $C_{18}H_{28}N_2O_2$ (**322**), which lacks steroid activity, is a potent inhibitor of GABA and glycine receptor function. The epalons, a novel series of neurosteroids based on epiallopregnanolone [516-54-1], $C_{21}H_{34}O_2$ (**323**), are being developed as novel anticonvulsants, anxiolytics, and hypnotics (see HYPNOTICS, SEDATIVES, ANTICONVULSANTS, AND ANXIOLYTICS) (Fig. 5).

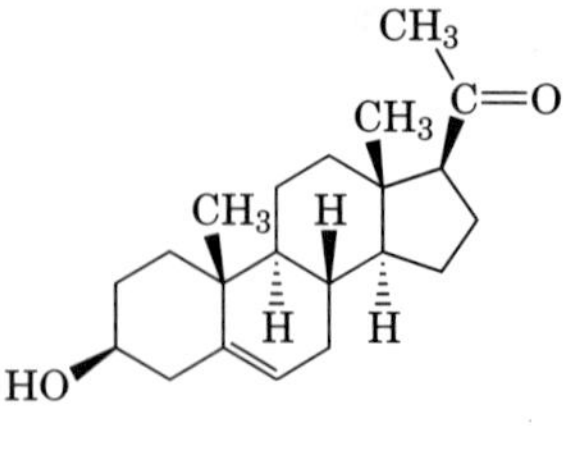

Fig. 5.   Structures of neurosteroids.

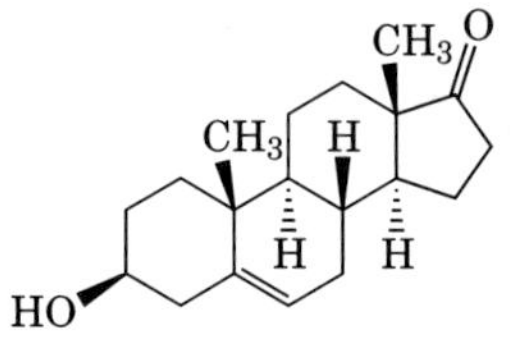

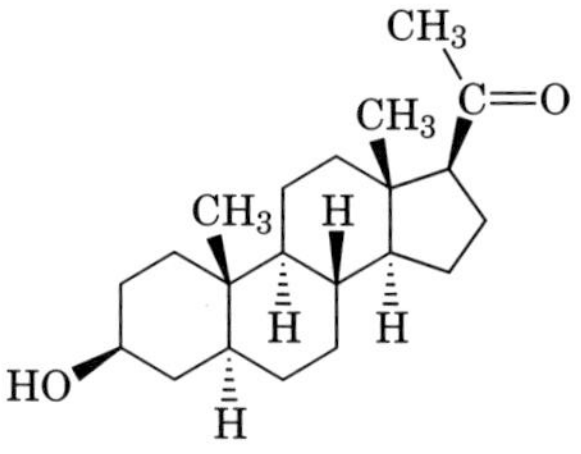

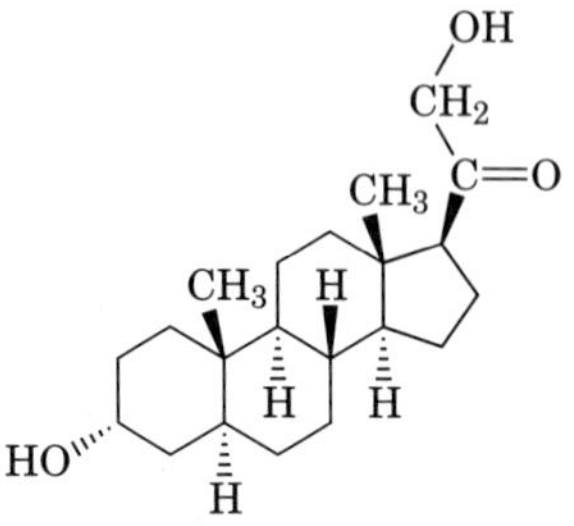

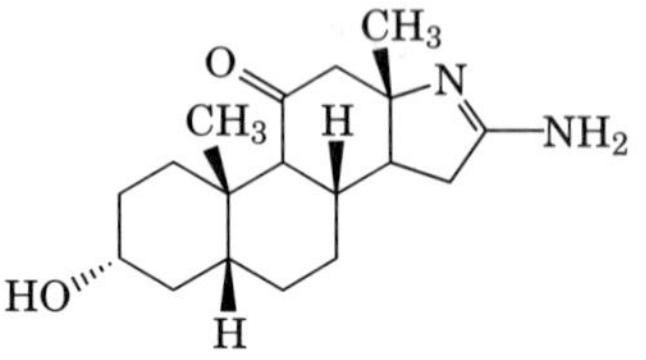

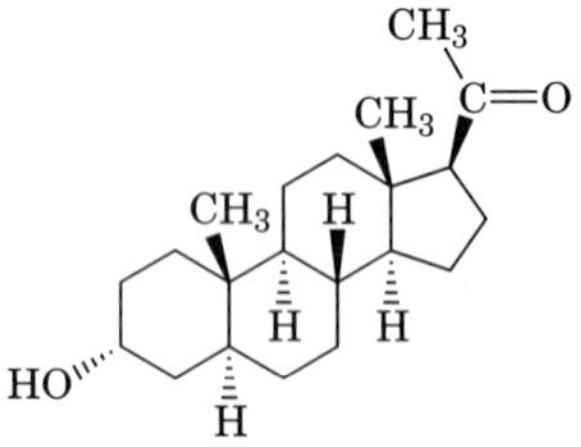

Somatostatin.   Somatostatin (SRIF or SS) is a cyclic peptide existing primarily in 14 and 28 amino acid forms SRIF$_{1-14}$ [38916-34-6], C$_{76}$H$_{104}$N$_{18}$O$_{19}$S (**324**) and SRIF$_{1-28}$ [75037-27-3] (**325**), respectively. SRIF was originally isolated from the hypothalmus and shown to regulate growth hormone (GH) secretion from the anterior pituitary (3). SRIF is present throughout the CNS where it modulates neuronal firing (81). It is also present in pancreas and gut where it regulates endocrine and exocrine secretions, particularly insulin and glucagon release.

SRIF produces its effects through two classes of GPCR, SRIF-1 and SRIF-2 that are structurally related to cloned opiate receptors. The agonists, seglitide [81377-02-8] (MK 678), cyclo((N-CH$_3$)Ala-Tyr-D-Trp-Lys-Val-Phe), and ocreotide [83150-76-9] (SMS 201-995), D-Phe-cyclo[Cys-Phe-D-Trp-Lys-Thr-Cys]-Thr-ol, are used to distinguish between the two SRIF receptor classes. These can be subdivided into three and two subtypes, respectively, based on pharmacology and cloning data. Originally termed SRIF$_{1A}$, SRIF$_{1B}$, SRIF$_{1C}$, SRIF$_{2A}$, and SRIF$_{2B}$, they have been renamed SSTR$_{1-5}$. Activation of SSTR receptors results in adenylate cyclase inhibition, modulation of K$^+$ and Ca^{2+} channel conductance, and regulation of tyrosine phosphatases and the Na$^+$–H$^+$ antiporter through pertussis toxin-insensitive mechanisms. High levels of mRNA for SSTR$_1$, SSTR$_3$, and SSTR$_5$ are found in the CNS; these levels tie SSTR$_1$ to the inhibition of GH secretion. SSTR$_1$, SSTR$_3$, and SSTR$_4$ are present in the pancreas and gut. The endogenous ligands for SSTR receptors are SRIF$_{1-14}$ (SS$_{1-14}$) and SRIF$_{1-28}$ (SS$_{1-28}$). SS$_{1-28}$ generally has greater affinity than SS$_{1-14}$ for SSTR receptors. There are no known nonpeptide ligands for SSTR receptors.

Seglitide readily distinguishes SSTR$_1$ with picomolar affinity. This compound has nanomolar affinity for SSTR$_2$ and is much weaker at the other subtypes. Ocreotide binds to SSTR$_1$, SSTR$_2$ and SSTR$_5$ receptor types. BIM 23052 [*133073-82-2*] D-Phe-Phe-Phe-D-Trp-Lys-Thr-Phe-Thr-NH$_2$, and BIM 23056 [*150155-61-6*], (**326**) differentiate the SSTR$_1$ and SSTR$_2$ subtypes. NC4 28B [*150155-58-1*] (**327**) and CGP 23996 [*86170-12-9*] *c*(Lys-Asn-Phe-Phe-Trp-Lys-Thr-Tyr-Thr-Ser-Asn), C$_{73}$H$_{99}$N$_{15}$O$_{18}$, are SSTR$_1$ agonists. L 362855 [*81710-71-6*] *c*(Ala-Phe-Trp-D-Trp-Lys-Thr-Phe), is a selective agonist for the SSTR$_4$ receptor. There are no known antagonists for any of the SSTR receptor classes.

SRIF acts as an excitatory neuromodulator in the CNS inhibiting the release of TRH, corticotropin-releasing hormone (CRH), growth hormone releasing factor (GHRH), and NE. It produces general arousal and hypotension. It inhibits the release of a number of peptides and modulators in the GI tract.

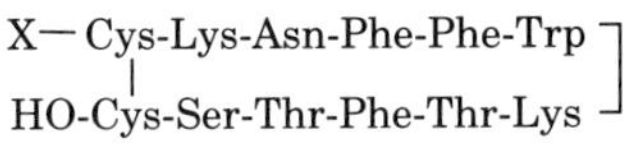

X— Cys-Lys-Asn-Phe-Phe-Trp ⌉
HO-Cys-Ser-Thr-Phe-Thr-Lys ⌋

(**324**)   X = H-Ala-Gly-

(**325**)   X = H-Ser-Ala-Asn-Ser-Asn-Pro-Ala-Met-Ala-Pro-Arg-Glu-Arg-Lys-Ala-Gly-

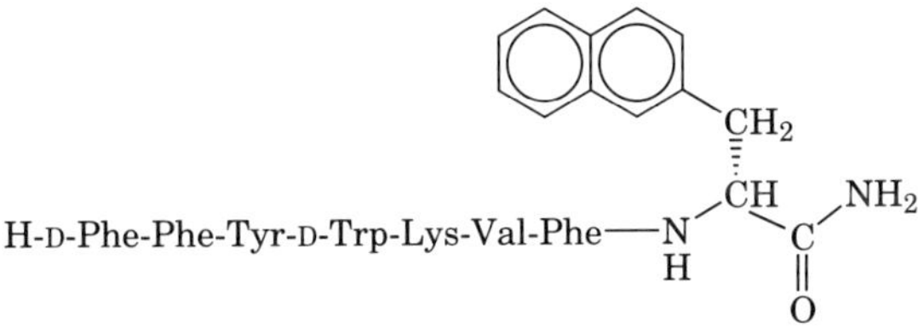

H-D-Phe-Phe-Tyr-D-Trp-Lys-Val-Phe—N(H)—CH(CH$_2$-naphthyl)—C(=O)—NH$_2$

(**326**)

H-D-Phe-Cys-Tyr-D-Trp-Lys-Thr-Cys—N(H)—CH(CH$_2$-naphthyl)—C(=O)—NH$_2$

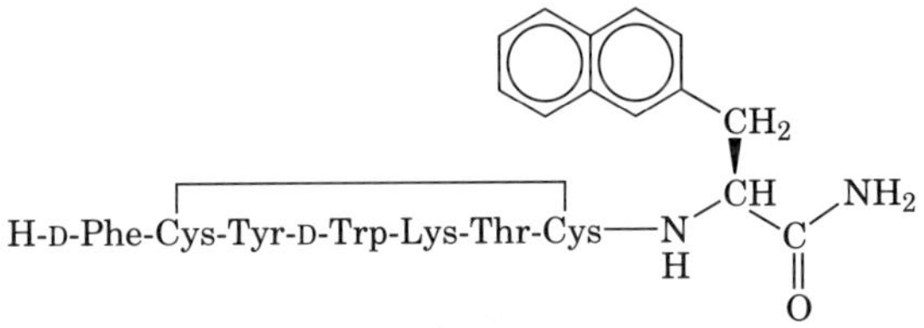

(**327**)

Tachykinins and Substance P. The tachykinins (82) include the undecapeptide, substance P, H-Arg-Pro-Lys-Pro-Gln-Gln-Phe-Phe-Gly-Leu-Met-NH$_2$, and the decapeptides, neurokinin A (NKA; also known as substance K, neuromedin L), H-His-Lys-Thr-Asp-Ser-Phe-Val-Gly-Leu-Met-NH$_2$, and neurokinin B (NKB, also known as neuromedin K), H-Asp-Met-His-Asp-Phe-Phe-Val-Gly-Leu-Met-NH$_2$. Physalaemin, eledoisin, kassinin, SCYI, and SCYII are nonmammalian tachykinins. Two larger peptides have been identified, neuropeptide K (**328**) and neuropeptide γ (**329**), both of which interact with tachykinin receptors (Table 19). The NKA sequence is contained within the carboxy-terminal

Table 19. Tachykinins and Receptor Agonists and Antagonists

Agonist/antagonist	CAS Registry Number	Molecular formula	Structure number
Tachykinins			
substance P	[33507-63-0]		
neurokinin A	[86933-74-6]		
neurokinin B	[102577-23-1]		
neuropeptide K	[106441-70-7]		(328)
neuropeptide γ	[123515-59-3]		(329)
NK_1 receptor			
SP methyl ester	[76260-78-1]	$C_{64}H_{99}N_{17}O_{14}S$	
[Sar^9, Met(O$_2$)^{11}] SP^a			
[Pro^9] SP			
L 668,169	[137012-28-3]	$C_{82}H_{108}N_{16}O_{14}S_2$	(330)
CP 99994	[136982-36-0]	$C_{19}H_{24}N_2O$	(331)
WIN 51708	[138091-24-4]	$C_{29}H_{33}N_3O$	(332)
SR 140333	[153050-21-6]	$C_{37}H_{45}Cl_2N_2O_2$	(333)
RP 67580	[135911-02-3]	$C_{29}H_{30}N_2O_2$	(334)
GR 82334	[129623-01-4]	$C_{69}H_{91}N_{15}O_{16}$	(335)
NK_2 receptor			
[β-Ala^8] NKA$_{4-10}$			
GR 64349	[137593-52-3]	$C_{42}H_{68}N_{10}O_{11}S$	(336)
[Lys^5, (N-CH$_3$)Leu^9, Nle^{10}] NKA$^b_{4-10}$			
SR 48968	[142001-63-6]	$C_{31}H_{35}Cl_2N_3O_2$	(337)
MEN 10376	[135306-85-3]	$C_{57}H_{68}N_{12}O_{10}$	
NK_3 receptor			
senktide	[106128-89-6]	$C_{40}H_{55}N_7O_{11}S$	(338)
[CH$_3$Phe^7] NKB			
[Pro^7] NKB			

^{a}Sar = sarcasine, Met(O$_2$) = $NH_2\overset{\displaystyle CHCH_2CH_2SO_2CH_3}{\underset{\displaystyle COOH}{|}}$

^{b}Nle = norleucine.

sequences of both neuropeptide K and neuropeptide γ. Like other neuroactive peptides, tachykinin peptide precursors are synthesized ribosomally and transported to nerve terminals where further processing occurs.

Asp-Ala-Asp-Ser-Ser-Glu-Lys-Gln-Val-Ala-Leu-Leu-Lys-Ala-Leu-Tyr-Gly-His-Gly-Gln-Ile-Ser-His-Lys-Arg-His-Lys-Thr-Asp-Ser-Phe-Val-Gly-Leu-Met-NH$_2$

(**328**)

Asp-Ala-Gly-His-Gly-Gln-Ile-Ser-His-Lys-Arg-His-Lys-Thr-Asp-Ser-Phe-Val-Gly-Leu-Met-NH$_2$

(**329**)

As a neurotransmitter in the sensory nervous system, high levels of substance P are found in the dorsal horn of the spinal cord as well as in peripheral sensory nerve terminals. However, substance P also plays a significant role as a neuromodulator in the central, sympathetic, and enteric nervous system. NKA and NKB are also localized selectively in the CNS.

Neurokinin effects are terminated by proteolysis. *In vitro*, acetylcholinesterase (ACE) and enkephalinase can hydrolyze substance P. However, there appears to be no clear evidence that either acetylcholinesterase or ACE limit the actions of released substance P. Enkephalinase inhibitors, eg, thiorphan, can augment substance P release or action in some systems but the distribution of enkephalinase in the brain does not precisely mirror that of substance P. There appears to be a substance P-selective enzyme in brain and spinal cord.

Capsaicin, an active ingredient in red pepper, is well known for its ability to release and deplete substance P in sensory C fibers. However, this action is not specific for substance P, as neurokinin A, calcitonin gene-related peptide (CGRP), and somatostatin also are released.

Three tachykinin GPCRs, NK_1, NK_2, and NK_3, have been identified and cloned. All are coupled to phosphatidylinositol hydrolysis. The NK_1 receptor is selective for substance P (SP) and is relatively abundant in the brain, spinal cord, and peripheral tissues. The NK_2 receptor is selective for NKA and is present in the gastrointestinal tract, urinary bladder, and adrenal gland but is low or absent in the CNS. The NK_3 receptor is selective for NKB and is present in low amounts in the gastrointestinal tract and urinary bladder, but is abundant in some areas of the CNS, ie, the spinal dorsal horn, solitary nucleus, and laminae IV and V of the cortex with moderate amounts in the interpeduncular nucleus. Mismatches in the distribution of the tachykinins and tachykinin receptors suggest the possibility of additional tachykinin receptor subtypes.

At the NK_1 receptor, SP, SP methyl ester, Sar^9, $Met(O_2)^{11}$ SP, and $[Pro^9]SP$ are agonists. L 668,169 (**330**) is a peptide NK_1 antagonist. CP 99994 (**331**), WIN 51708 (**332**), SR 140333 (**333**), RP 67580 (**334**), and GR 82334 (**335**) are nonpeptide antagonists. At the NK_2 receptor, NKA, $[\beta\text{-}Ala^8]$ NKA_{4-10}, GR 64349 (**336**), and $[Lys^5, (NMe)Leu^9, Nle^{10}]NKA_{4-10}$ are selective agonists. SR 48968 (**337**) and the peptide MEN 10376, H-Asp-Tyr-D-Trp-Val-D-Trp-D-Trp-Lys-NH_2, are selective antagonists. NKB, senktide (**338**), $[MePhe^7]NKB$ and $[Pro^7]NKB$ are agonists at the NK_3 receptor. There are no selective antagonists for this receptor subtype. The neurotransmitter actions of tachykinins are generally excitatory. The vasodilation caused by substance P results from the stimulation NO synthesis in the endothelium (Table 19).

(**330**)

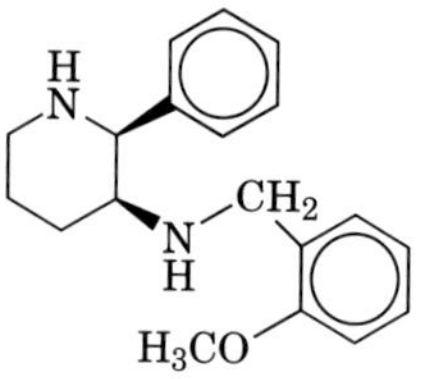

(331)

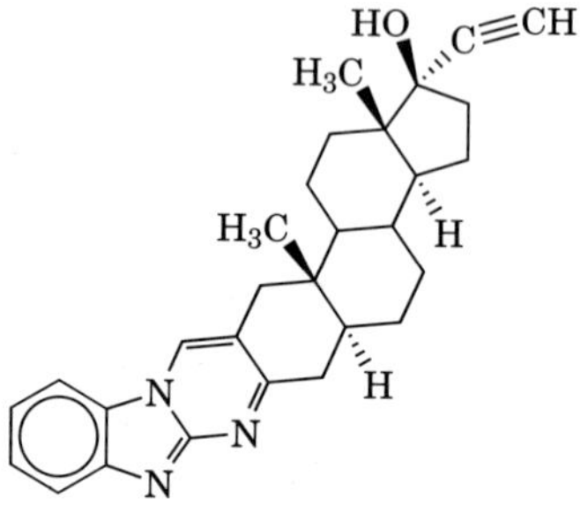

H
N

CH₂

N
H

H₃CO

(332)

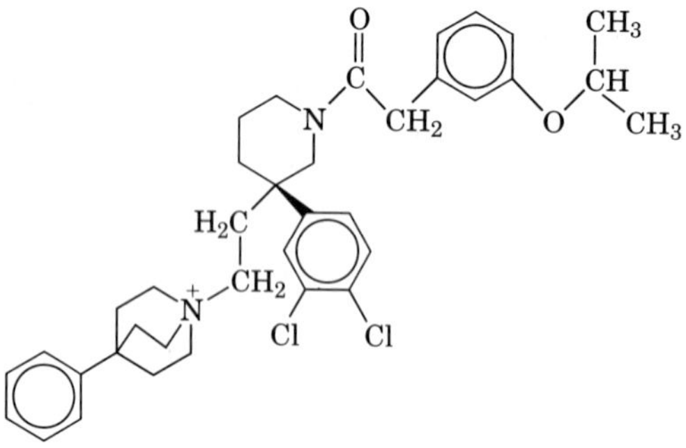

HO C≡CH

H₃C

H₃C

H

N

N

N

H

(333)

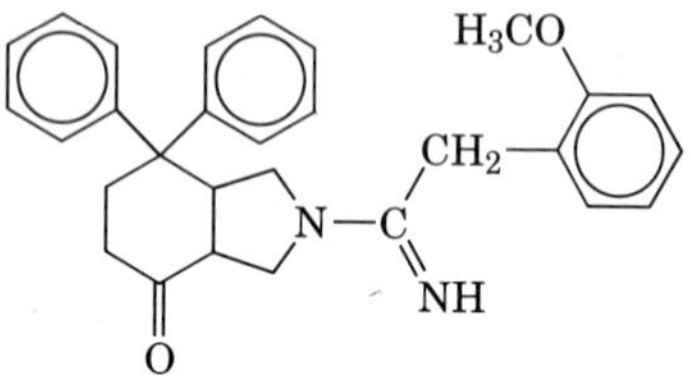

O

CH₃

CH

O

CH₃

CH₂

N

H₂C

CH₂

CH₂

N⁺

Cl Cl

(334)

H₃CO

CH₂

N

C

NH

O

(335)

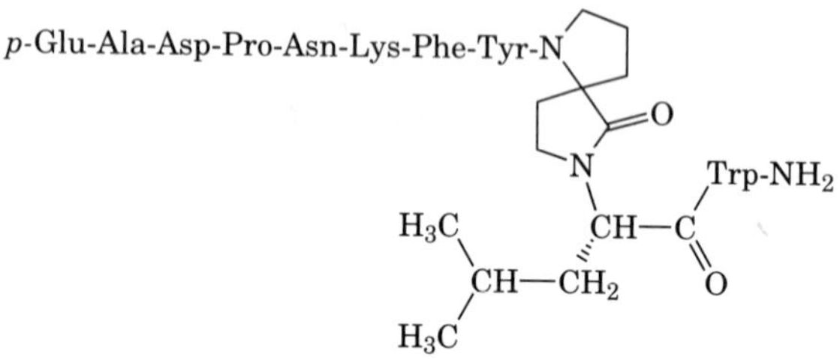

p-Glu-Ala-Asp-Pro-Asn-Lys-Phe-Tyr-N

O

N

Trp-NH₂

H₃C

CH—CH₂

CH—C

O

H₃C

(336)

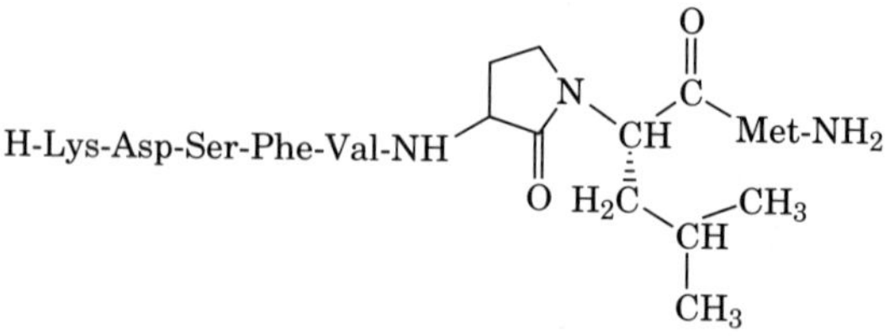

O

H-Lys-Asp-Ser-Phe-Val-NH

N

CH C

Met-NH₂

O H₂C CH₃

CH

CH₃

CH₃

1110

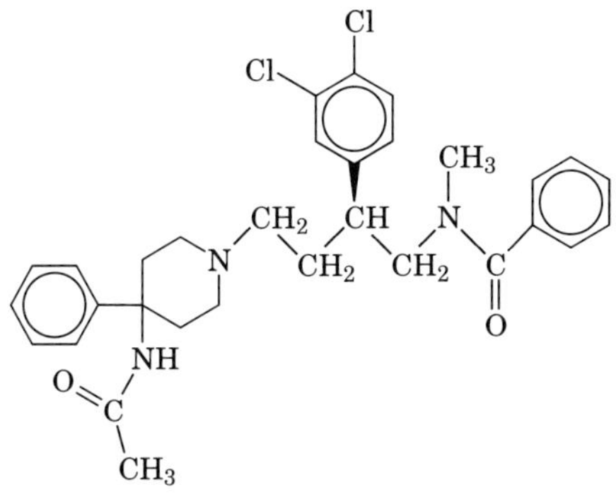

(**337**)

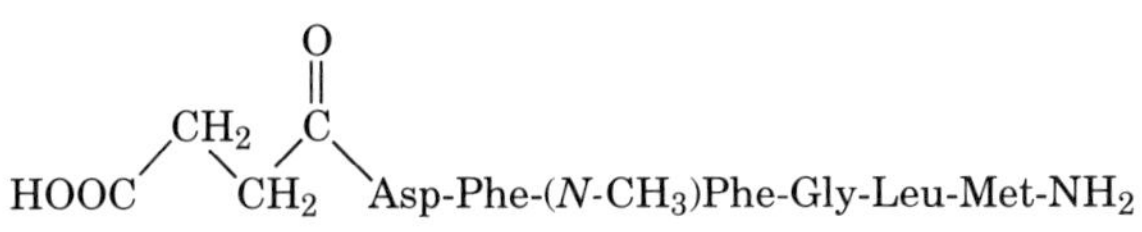

(**338**)

Vasoactive Intestinal Peptide and Pituitary Adenylate Cyclase Activating
Peptide. Vasoactive intestinal peptide (VIP) [*37221-79-7*] (**339**), a 28-amino acid peptide, is a member of a family of structurally related peptides that includes secretin [*1393-25-5*], (**340**), growth hormone releasing factor (GRF), and pituitary adenylate cyclase-activating peptide (PACAP) [*137061-48-4*] (**341**) (83).

H-His-Ser-Asp-Ala-Val-Phe-Thr-Asp-Asn-Tyr-Thr-Arg-Leu-Arg-Lys-
Gln-Met-Ala-Val-Lys-Lys-Tyr-Leu-Asn-Ser-Ile-Leu-Asn-NH$_2$

(**339**)

H-His-Ser-Asp-Gly-Thr-Phe-Thr-Phe-Thr-Ser-Glu-Leu-Ser-Arg-Leu-
Arg-Asp-Ser-Ala-Arg-Leu-Gln-Arg-Leu-Leu-Gln-Gly-Leu-Val-NH$_2$

(**340**)

H-His-Ser-Asp-Gly-Ile-Phe-Thr-Asp-Ser-Tyr-Ser-Arg-Tyr-Arg-Lys-Gln-Met-Ala-Val-
Lys-Lys-Tyr-Leu-Ala-Ala-Val-Leu-Gly-Lys-Arg-Tyr-Lys-Gln-Arg-Val-Lys-Asn-Lys-NH$_2$

(**341**)

The effects of VIP and PACAP are mediated by three GPCR subtypes, VIP$_1$, VIP$_2$, and PACAP receptor, coupled to the activation of adenylate cyclase (54). The VIP$_1$ subtype is localized in the lung, liver, and intestine, and the cortex, hippocampus, and olfactory bulb in the CNS. The VIP$_2$ receptor is most abundant in the CNS, in particular in the thalamus, hippocampus, hypothalamus, and suprachiasmatic nucleus. PACAP receptors have a wide distribution in the CNS with highest levels in the olfactory bulb, the dentate gyrus, and the cerebellum (84). The receptor is also present in the pituitary. The VIP$_1$ and PACAP receptors have been cloned.

There are few pharmacological tools to distinguish between the three receptors. Secretin has significant biological activity at VIP_1 receptors but is inactive at VIP_2 receptors, whereas VIP and PACAP have equivalent affinities for both the VIP_1 and VIP_2 receptors. PACAP displays marked selectivity for the PACAP receptor. $PACAP_{6-27}$ is the most potent and selective antagonist of the PACAP receptor identified as of this writing.

VIP and PACAP are found throughout the CNS and in the gastrointestinal, genitourinary, respiratory, and cardiovasular systems. VIP is a potent vasodilator and bronchodilator and plays a role in the control of prolactin secretion from the pituitary gland. PACAP may regulate the synthesis and secretion of catecholamines from adrenal medulla and can modulate pancreatic exocrine activity. The presence of PACAP receptors in the reproductive tract suggests that PACAP may be involved in the regulation of spermatogenesis.

Vasopressin and Oxytocin. Arginine[8]-vasopressin (AVP, vasopressin; also known as antidiuretic hormone, ADH) (**342**) is a nonapeptide amide that functions both as a neuroregulator and a hormone (84,85). Oxytocin (OT) (**343**) is a nonapeptide amide related to AVP.

H-Cys-Tyr-Phe-Gln-Asn-Cys-Pro-Arg-Gly-NH$_2$ H-Cys-Tyr-Ile-Gln-Asn-Cys-Pro-Leu-Gly-NH$_2$

(**342**)   (**343**)

AVP and OT protein precursors are separate gene products. The AVP precursor is processed post-translationally to generate AVP, neurophysin II which binds AVP, and a glycopeptide. A sexual dimorphism has been observed with testosterone increasing AVP expression. AVP is co-localized with a number of peptides in the hypothalamus, with galanin in the bed nucleus of the stria terminalis and the medial amygdala, and with NE in the locus coeruleus. AVP and OT are produced by magnocellular neurons of the hypothalamic paraventricular nucleus, supraoptic nucleus, and accessory nuclei that project to the posterior pituitary and release the hormones into the circulation.

AVP produces its effects via V_{1A}, V_{1B}, and V_2 receptors. There is one receptor for OT. V_{1A} and V_{1B} receptors are linked to phosphatidylinositol, hydrolysis, while the V_2 receptor is linked to cAMP formation. The brain contains predominantly V_{1A} receptors with little evidence for the presence of V_{1B} or V_2 receptors. V_{1A} receptors are distinguished from V_{1B} receptors on the basis of antagonist potency with $d(CH_2)_5[Tyr(CH_3)^2AVP]$ (**344**) being three orders of magnitude more potent at V_{1A} than at V_{1B} receptors, whereas [D-Pen[1],Tyr$(CH_3)^2$]AVP (**345**) has similar low nanomolar potency at both receptors. SR 49059 (**346**) and OPC 21268, (**347**) are nonpeptide V_1 antagonists (Table 20).

—Tyr(OCH$_3$)-Phe-Gln-Asn-Cys-Pro-Arg-Tyr-NH$_2$

(**344**)

Table 20. Vasopressin and Oxytocin and Receptor Agonists and Antagonists

Agonist/antagonist	CAS Registry Number	Molecular formula	Structure number
arginine[8] vasopressin	[9034-50-8]		(342)
oxytocin	[50-56-6]		(343)
d(CH$_2$)$_5$[Tyr(Me)^{2}AVP			(344)
[D-Pen^1, Tyr(Me)^2]AVP			(345)
SR 49059	[150375-75-0]	C$_{28}$H$_{27}$Cl$_2$N$_3$O$_7$S	(346)
OPC 21268		C$_{26}$H$_{31}$N$_3$O$_4$	(347)
d(D-Arg^8]VP			
d[Val^4]VP			
d(CH$_2$)$_5$[D-Ile^2,Ile^4]AVP			
[Thr^4, Gly^7] oxytocin			
d(CH$_2$)$_5$[Tyr(CH$_3$)^2, Thr^4, Orn^8] oxytocin (1−8)^a			
cyclo(D1-Nal, Ile, D-piperazyl, piperazyl, D-His, Pro)^b			
L-368,899	[148927-60-0]	C$_{27}$H$_{42}$N$_4$O$_5$S$_2$	(348)

^{a}Orn = ornithine.
^{b}Nal = naphthylalanine.

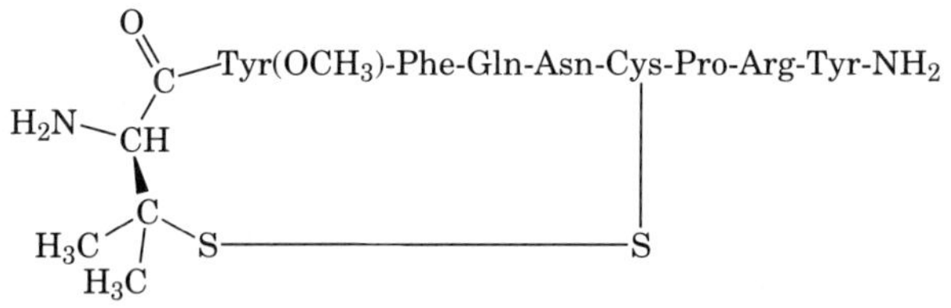

(345)

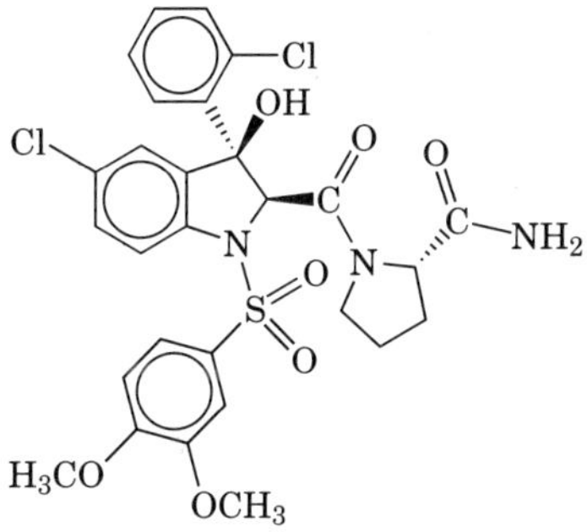

(346)

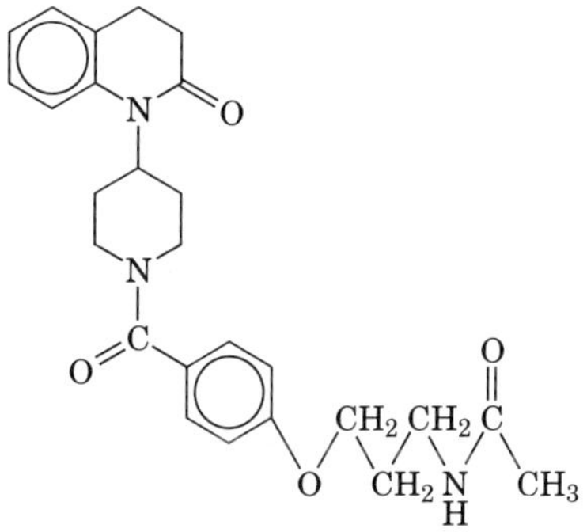

(347)

1113

OT receptors are localized in the brain hypothalamus, limbic system, cortex, striatum, olfactory system, and brain stem. In the periphery, OT is best known for its stimulation of uterine smooth muscle and the milk ejection reflex. $d(CH_2)_5[Tyr(CH_3)^2 \cdot Thr^4, Orn^8]$oxytocin(1–8), cyclo(D-1–Nal-Ile-D-pipecolic acid D-His-Pro) and L-368,899 (**348**) are OT antagonists, the latter being nonpeptidic and under evaluation for use in preterm labor (Table 20).

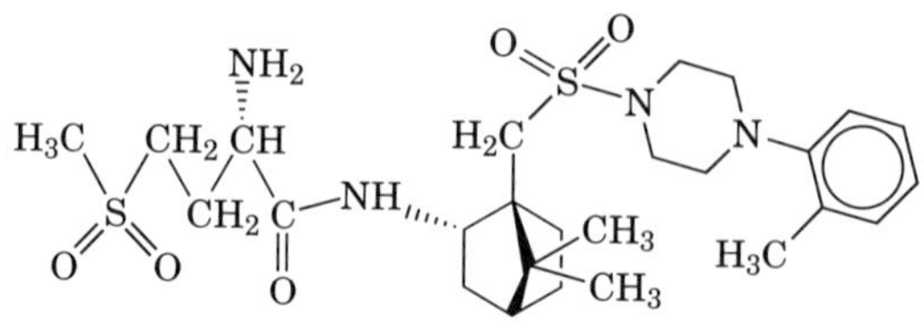

(**348**)

AVP is excitatory in the ventral hippocampus, either directly or by potentiation of glutamatergic responses. An inhibitory effect has been observed in C_{A1}. AVP may be involved in the formation of long-term potentiation and thus learning and memory. However, AVP is proconvulsive, may augment the formation of drug tolerance and dependence, and affects cardiovascular regulatory processes.

OT receptors in the hypothalamus are regulated by steroids. OT systems in the CNS are involved in homeostasis, reproduction, and related behavior. OT is also excitatory to neurons in the CNS at nanomolar concentrations (86), but relatively little is known about neuronal mechanisms and pharmacology.

Economic Aspects

Neuroregulators comprise a large portion of clinically effective human therapeutic agents in use as of this writing. Thus as an aggregate group, neuroregulators have an estimated global market in excess of $50–100 billion per year. The histamine H_2 blockers, cimetidine, and ranitidine have aggregate sales in their ethical and generic forms in excess of $6 billion/yr. The antidepressant drug market is approaching $5.0 billion; sales of drugs for Parkinson's disease are $1.0 billion; anxiolytics are $2 billion; drugs for pain, $5.0 billion; manic depression, $330 million; Alzheimer's, $1.0 billion, with sales for an effective agent in the $3–4 billion range; and other dementias $750 million. Sales of antipsychotic drugs are $1.5 billion; drugs for eating disorders and obesity, $310 million; migraine, $1 billion; attention hyperactivity deficit disorder, $460 million; epilepsy, $850 million; hearing loss, $1.1 billion; insomnia, $250 million; drug and alcohol abuse, $36 million, with sales for an effective agent in excess of $2 billion; neurotrauma, $180 million; and cardiovascular diseases, eg, hypertension, $5 billion.

BIBLIOGRAPHY

"Neuroregulators" in *ECT* 3rd ed., Vol. 15, pp. 754–786, by A. S. Horn, University of Groningen.

1. M. Williams and M. A. Sills, *Comp. Med. Chem.* **3**, 45–80 (1991).
2. F. A. Stephenson, in F. Hucho, ed., *Neurotransmitter Receptors*, Elsevier Science Publishers, Amsterdam, the Netherlands, 1993, pp. 183–207.
3. J. R. Cooper, F. E. Bloom, and R. H. Roth, *The Biochemical Basis of Neuropharmacology*, 5th ed., Oxford University Press, New York, 1991.
4. G. J. Siegel and co-workers, *Basic Neurochemistry*, 5th ed., Raven Press, New York, 1994.
5. E. L. Barker and co-workers, *J. Biol. Chem.* **269**, 11687–11690 (1994); R. A. Bond and co-workers, *Nature* **374**, 272–276 (1995); J. M. Stables and co-workers, *Pharmacol. Res.* **31** (Suppl.), 201 (1995).
6. P. M. Sweetnam and co-workers, *J. Nat. Prod.* **56**, 441–455 (1993).
7. F. O. Schmitt, in F. O. Schmitt, S. J. Bird, and F. E. Bloom, eds., *Molecular Genetic Neuroscience*, Raven Press, New York, 1982, pp. 1–9.
8. F. E. Bloom, *FASEB J.* **2**, 32–41 (1988).
9. J. F. Kerwin, Jr. and M. Heller, *Med. Res. Rev.* **14**, 23–74 (1994).
10. R. M. Evans, *Science* **240**, 889–895 (1988).
11. J. R. Hepler and A. G. Gilman, *Trends Biochem. Sci.* **17**, 383–387 (1992).
12. R. F. Power, O. M. Conneely, and B. W. O'Malley, *Trends Pharmacol. Sci.* **13**, 318–323 (1992).
13. D. Metcalf, *Nature* **369**, 519–520 (1994).
14. P. G. Strange, *Trends Pharmacol Sci.* **12**, 48–49 (1991).
15. R. H. Scott, H. A. Pearson, and A. C. Dolphin, *Prog. Neurobiol.* **36**, 485–520 (1991).
16. M. Inoue and M. Yoshii, *Prog. Neurobiol.* **38**, 203–230 (1992).
17. R. J. Tallarida, *Drug Dev. Res.* **19**, 257–274 (1990).
18. C. A. Milano and co-workers, *Science* **264**, 582–586 (1994).
19. T. P. Kenakin, R. A. Bond, and T. I. Bonner, *Pharmacol. Rev.* **44**, 351–362 (1992).
20. P. Vanhoutte and co-workers, *Pharmacol. Rev.* **46**, 111–116 (1994).
21. F. J. Ehlert, W. R. Roeske, and H. I. Yamamura, in F. E. Bloom and D. J. Kupfer, eds., *Psychopharmacology: 4th Generation of Progress*, Raven Press, New York, 1995, pp. 111–124.
22. S. P. Arneric, J. P. Sullivan, and M. Williams, in Ref. 21, pp. 94–110.
23. P. B. Sargent, *Ann. Rev. Neurosci.* **16**, 403–443 (1993); A. B. Elgoyhen and co-workers, *Cell* **79**, 705–715 (1994).
24. K. A. Jacobson, P. J. van Galen, and M. Williams, *J. Med. Chem.* **35**, 407–422 (1992); M. P. Abbrachio and G. Burnstock, *Pharmacol. Ther.* **64**, 445–475 (1994).
25. J. M. Bathon and D. Proud, *Ann. Rev. Pharmacol. Toxicol.* **31**, 129–162 (1991).
26. B. E. Evans and co-workers, *J. Med. Chem.* **30**, 1229–1239 (1987).
27. R. D. Smith and co-workers, *Ann. Rev. Pharmacol. Toxicol.* **32**, 135–165 (1992).
28. D. L. Garbers, *Cell* **71**, 1–4 (1992).
29. R. Weizman and M. Gavish, *Clin. Neuropharmacol.* **16**, 401–417 (1993).
30. N. Lindefors and co-workers, *Prog. Neurobiol.* **40**, 671–690 (1993).
31. N. E. Kohl and co-workers, *Science* **260**, 1934–1937 (1993).
32. D. R. Poyner, *Pharmacol. Ther.* **56**, 23–51 (1992).
33. T. J. Rink and co-workers, *Trends Pharmacol. Sci.* **14**, 113–118 (1993).
34. W. A. Devane, *Trends Pharmacol. Sci.* **15**, 40–41 (1994).
35. K. P. Minneman and T. A. Esbenshade, *Ann. Rev. Pharmacol. Toxicol.* **34**, 117–133 (1994).
36. A. D. Strosberg, *Protein Sci.* **2**, 1198–1209 (1993).

37. J. P. Hieble and R. R. Ruffolo, Jr., *Fundam. Clin. Pharmacol.* **6** (Suppl. 1), 7S–13S (1992).
38. G. Li and co-workers, *Science* **263**, 966–969 (1994).
39. R. C. Young and co-workers, *Eur. J. Med. Chem.* **28**, 201–211 (1993).
40. R. Horuk, *Trends Pharmacol. Sci.* **15**, 159–165 (1994).
41. E. N. Benveniste, *Am. J. Physiol. Cell Physiol.* **263**, C1–C16 (1992).
42. T. Bartfai and M. Schultzberg, *Neurochem. Int.* **22**, 435–444 (1993).
43. C. R. Plata-Salamán, *Neurosci. Biobehav. Rev.* **15**, 185–215 (1991).
44. T. M. Dawson and co-workers, *Proc. Natl. Acad. Sci. USA* **90**, 9808–9812 (1993).
45. V. Gagliardini and co-workers, *Science* **263**, 826–828 (1994).
46. N. J. Rothwell and J. K. Relton, *Neurosci. Biobehav. Rev.* **17**, 217–227 (1993).
47. O. Civelli, J. R. Bunzow, and D. K. Grandy, *Ann. Rev. Pharmacol. Toxicol.* **33**, 281–307 (1993).
48. I. J. Kopin, *Ann. Rev. Pharmacol. Toxicol.* **33**, 467–495 (1993).
49. Y. J. Wan, *Am. J. Surg.* **166**, 50–53 (1993).
50. K. Shiosaki and T. J. Opgenorth, *Drug News Perspect.* **7**, 593–602 (1994).
51. T. Reisine and G. I. Bell, *Trends Neurosci.* **16**, 506–510 (1993).
52. P. Krogsgaard-Larson and co-workers, *J. Med. Chem.* **37**, 2489–2505 (1994).
53. I. Merchenthaler, F. J. Lopez, and A. Negro-Vilar, *Prog. Neurobiol.* **40**, 711–769 (1993).
54. T. Harmar and E. Lutz, *Trends Pharmacol. Sci.* **15**, 97–99 (1994).
55. P. Kugler, *Int. Rev. Cytol.* **147**, 285–336 (1993).
56. F. Fonnum, *Prog. Biophys. Mol. Biol.* **60**, 47–57 (1993).
57. D. D. Schoepp and P. J. Conn, *Trends Pharmacol. Sci.* **14**, 13–20 (1993); P. H. Seeburg, *Trends Neurosci.* **16**, 359–365 (1993).
58. M. Hollmann and S. Heinemann, *Ann. Rev. Neurosci.* **17**, 31–108 (1994).
59. S. J. Hill, *Prog. Med. Chem.* **24**, 29–84 (1987).
60. J. Lee and P. F. Pilch, *Am. J. Physiol.* **266**, C319–C334 (1994).
61. J. Lee and P. F. Pilch, *Am. J. Physiol.* **266**, C319–C334 (1994).
62. D. Piomelli, *Crit. Rev. Neurobiol.* **8**, 65–83 (1994).
63. R. A. Coleman, W. L. Smith, and S. Narumiya, *Pharmacol. Rev.* **46**, 205–229 (1994).
64. S. Yamamoto, *Biochem. Biophys. Acta* **1128**, 117–131 (1992); D. Piomelli, *Crit. Rev. Neurobiol.* **8**, 65–83 (1994).
65. D. N. Krause and M. L. Dubocovich, *Ann. Rev. Pharmacol. Toxicol.* **31**, 549–568 (1991).
66. L. Grundemar and R. Hakanson, *Trends Pharmacol. Sci.* **15**, 153–159 (1994).
67. W.-X. Shi and B. S. Bunney, *Ann. N.Y. Acad. Sci.* **668**, 129–145 (1992).
68. S. Raffioni, R. A. Bradshaw, and S. E. Buxser, *Ann. Rev. Biochem.* **62**, 823–850 (1993).
69. S. Raffioni, R. A. Bradshaw, and S. E. Buxser, *Ann. Rev. Biochem.* **62**, 823–850 (1993).
70. T. Roeder and J. A. Nathanson, *Neurochem. Res.* **18**, 921–925 (1993).
71. J. A. Nathanson, *J. Pharmacol. Exp. Ther.* **265**, 509–515 (1993); T. Roeder and J. A. Nathanson, *Neurochem. Res.* **18**, 921–925 (1993).
72. D. J. Mangelsdorf and co-workers, *Recent Prog. Horm. Res.* **48**, 99–121 (1993).
73. J. Rees, *Br. J. Dermatol.* **126**, 97–104 (1992).
74. G. P. Studzinski, J. A. McLane, and M. R. Uskokovic, *Crit. Rev. Eukanyot. Gene Expr.* **3**, 279–312 (1993).
75. G. R. Martin and P. P. A. Humphrey, *Neuropharmacol.* **33**, 261–273 (1994).
76. M. Abou-Gharbia, S. Y. Ablordeppey, and R. A. Glennon, *Ann. Rep. Med. Chem.* **28**, 1–10 (1993).
77. R. Paul and co-workers, *J. Neuroimmunol.* **52**, 183–192 (1994).
78. B. S. McEwen and co-workers, *Cell. Mol. Neurobiol.* **13**, 457–482 (1993).

79. E. R. De Kloet, M. S. Oitzl, and M. Joëls, *Cell. Mol. Neurobiol.* **13**, 433–455 (1993).

80. P. Robel and E.-E. Baulieu, *Trends Endocrinol. Metab.* **5**, 1–8 (1994).

81. G. I. Bell and T. Reisine, *Trends Neurosci.* **16**, 34–38 (1993).

82. M. Otsuka and K. Yoshioka, *Physiol. Rev.* **73**, 229–308 (1993).

83. A. Arimura, *Regul. Pept.* **37**, 287–303 (1992).

84. D. de Wied and co-workers, *Front. Neuroendocrinol. Metab. Brain Dis.* **8**, 151–179 (1993).

85. M. Manning and W. H. Sawyer, *J. Recept. Res.* **13**, 195–214 (1993).

86. J. J. Dreifuss and M. Raggenbass, *Regul. Pept.* **45**, 109–114 (1993).

CLARK A. BRIGGS
MARK W. HOLLADAY
JAMES F. KERWIN, JR.
JAMES P. SULLIVAN
MICHAEL WILLIAMS
Abbott Laboratories